USEFUL FORMULAS

Lines (3.3)
1. Slope $m = \dfrac{y_2 - y_1}{x_2 - x_1}$
2. Point-slope formula
   $(y - y_0) = m(x - x_0)$
3. Slope-intercept formula
   $y = mx + b$
4. Vertical lines
   $x = a$
5. General formula
   $Ax + By + C = 0$
6. Parallel lines $m_1 = m_2$.
7. Perpendicular lines $m_1 m_2 = -1$

Parabola with vertex at (x_0, y_0) (3.4, 3.6)
1. $y - y_0 = a(x - x_0)^2$ (opens vertically)
2. $x - x_0 = A(y - y_0)^2$ (opens horizontally)

Ellipses and Hyperbolas centered at (h, k) (3.5, 3.6)
1. Ellipse: $\dfrac{(x-h)^2}{a^2} + \dfrac{(y-k)^2}{b^2} = 1$
2. Hyperbola: $\dfrac{(x-h)^2}{a^2} - \dfrac{(y-k)^2}{b^2} = 1$ (horizontal)
   or $\dfrac{(y-k)^2}{b^2} - \dfrac{(x-h)^2}{a^2} = 1$ (vertical)

Logarithms (6.2, 6.3, 6.5)
1. $u = \log_b v$ if and only if $v = b^u$
   ($v > 0$ and $0 < b \ne 1$)
2. $\log_b uv = \log_b u + \log_b v$
3. $\log_b\left(\dfrac{u}{v}\right) = \log_b u - \log_b v$
4. $\log_b u^r = r \log_b u$
5. $\log_b 1 = 0$
6. $\log_b\left(\dfrac{1}{u}\right) = -\log_b u$
7. $\log_b u = \dfrac{\log_a u}{\log_a b}$

Arithmetic Progressions (11.2, 11.3)
1. $a_n = a_{n-1} + d = a_1 + (n-1)d$
2. $a_1 + \cdots + a_n = \dfrac{n}{2}[a_1 + a_n]$

Geometric Progressions (11.2, 11.3)
1. $a_n = r \cdot a_{n-1} = a_1 \cdot r^{n-1}$
2. $\sum_{j=0}^{n-1} ar^j = a\left(\dfrac{1-r^n}{1-r}\right)$, $r \ne 1$

Summation Symbol (11.3, 11.4)
1. $\sum_{j=k}^{l} (a_j + b_j) = \sum_{j=k}^{l} a_j + \sum_{j=k}^{l} b_j$
2. $\sum_{j=k}^{l} (Ca_j) = C \sum_{j=k}^{l} a_j$

Factorial (11.4, 11.5)
$n! = 1 \cdot 2 \cdots n$

Bionomial Theorem (11.4)
$(a + b)^n = \sum_{l=0}^{n} \dfrac{n!}{(n-l)!\, l!} a^{n-l} b^l$

Permutations and Combinations (11.5)
1. ${}_nP_r = \dfrac{n!}{(n-r)!}$
2. ${}_nC_r = \dfrac{n!}{(n-r)!\, r!}$

Probability (11.6)
1. $P(E) = \dfrac{\text{number of ways in which } E \text{ could occur}}{\text{total number of possible outcomes}}$
2. $P(E_1 \text{ or } E_2) = P(E_1) + P(E_2) - P(E_1 \text{ and } E_2)$
3. $P(\sim E) = 1 - P(E)$
4. Independent events
   $P(E_1 \text{ and } \ldots \text{ and } E_n) = P(E_1) \cdot P(E_2) \cdots P(E_n)$
5. Mutually exclusive events
   $P(E_1 \text{ or } \ldots \text{ or } E_n) = P(E_1) + P(E_2) \cdots P(E_n)$

Angle Measurement (7.1)
$1° = \dfrac{\pi}{180}$ radians ≈ 0.017 radians

1 radian $= \left(\dfrac{180}{\pi}\right)° \approx 57.3°$

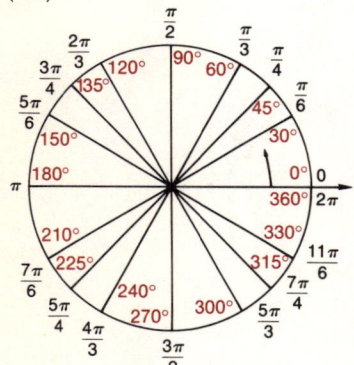

Arc Length, Area, and Radian Measure (7.1)

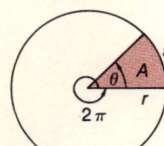

1. $\theta = \left(\dfrac{s}{r}\right)$
2. $s = r\theta$
3. $A = \left(\dfrac{1}{2}\right) r^2 \theta$

College Algebra
and Trigonometry

Second Edition

College Algebra *and* Trigonometry

Second Edition

Ralph C. Steinlage   PROFESSOR OF MATHEMATICS   UNIVERSITY OF DAYTON

WEST PUBLISHING COMPANY

St. Paul   New York   Los Angeles   San Francisco

Copyediting: Linda Thompson
Text Design: Christy Butterfield
Composition: Syntax International
Artwork: Signature Design
Cover Photograph: Timothy D. Schaedler
Cover Design: Christy Butterfield

Library of Congress Cataloging-in-Publication Data

Steinlage, Ralph C.
    College algebra and trigonometry.
    Includes index.
    1. Algebra.   2. Trigonometry.   I. Title.
QA154.2.S74   1987   512'.13   86-19099
ISBN 0-314-30080-5

COPYRIGHT © 1984 by
WEST PUBLISHING COMPANY
COPYRIGHT © 1988 by
WEST PUBLISHING COMPANY
50 W. Kellogg Boulevard
P.O. Box 64526
St. Paul, MN 55164-1003

All rights reserved

Printed in the United States of America

Photo Credits

Chapter 1 The Granger Collection;   **Chapter 2** The Granger Collection;   **Chapter 3** The Bettmann Archive, Inc.;   **Chapter 5** The Granger Collection;   **Chapter 6** (page 318) The Granger Collection;   **Chapter 6** (page 336) The Granger Collection;   **Chapter 10** The Bettmann Archive;   **Chapter 11** The Granger Collection.

Contents

Preface   x

1   Basic Algebra   1

1.1	The Real Number System	1
1.2	Order and Absolute Value	14
1.3	Integral Exponents	22
1.4	Radicals	31
1.5	Rational Exponents	38
1.6	Polynomials	43
1.7	Factoring	49
1.8	Rational Expressions	55
1.9	Complex Numbers	65
1.10	Chapter Review	72
1.11	Supplementary Exercises	75

2   Equations and Algebraic Inequalities   79

2.1	First Degree Equations in One Variable	80
2.2	Literal Equations and Word Problems	85
2.3	Quadratic Equations	96
2.4	Equations Solvable with Quadratic Techniques	104
2.5	Applications of Quadratic Equations	112
2.6	Proportionality; Variation	120
2.7	Algebraic Inequalities with One Variable	125
2.8	Equations and Inequalities Involving Absolute Values	138
2.9	Chapter Review	144
2.10	Supplementary Exercises	146

3 Graphs of Equations — 151

3.1	Cartesian Coordinates; Distance and Circles	151
3.2	Graphs	160
3.3	Lines and Linear Equations	168
3.4	Parabolas and Quadratic Equations	181
3.5	Central Ellipses and Hyperbolas (Optional)	189
3.6	The Conic Sections and Translation (Optional)	200
3.7	Chapter Review	210
3.8	Supplementary Exercises	213

4 Functions and Their Graphs — 217

4.1	Functions	217
4.2	Graphing Techniques	227
4.3	Operations on Functions	238
4.4	Inverse Functions	247
4.5	Chapter Review	262
4.6	Supplementary Exercises	264

5 Polynomial and Rational Functions — 269

5.1	Roots, Factors, and Synthetic Division	269
5.2	Real Roots of Polynomials	274
5.3	Complex Roots of Polynomials	283
5.4	Graphing Polynomial Functions	288
5.5	Rational Functions	297
5.6	Chapter Review	305
5.7	Supplementary Exercises	307

6 Exponential and Logarithmic Functions — 309

6.1	Exponential Functions	310
6.2	Logarithmic Functions and Their Graphs	320
6.3	Properties of Logarithms; Logarithmic Equations	328
6.4	Using Tables (Optional)	339
6.5	Applications of Exponential and Logarithmic Functions	345
6.6	Chapter Review	355
6.7	Supplementary Exercises	357

7 The Trigonometric Functions — 361

7.1	Angles	361
7.2	Right Triangle Trigonometry	371
7.3	Solving Right Triangles	381
7.4	Trigonometric Functions of Arbitrary Angles and Real Numbers	389
7.5	Elementary Properties of the Trigonometric Functions	399
7.6	Graphing the Sine and Cosine Functions	408
7.7	Other Trigonometric Graphs	420
7.8	Chapter Review	432
7.9	Supplementary Exercises	436

8 Trigonometric Identities and Equations; Inverse Trigonometric Functions — 439

8.1	Elementary Identities	439
8.2	Verifying Identities	444
8.3	Sum and Difference Formulas	448

8.4	Multiple Angles; Half-angles	457
8.5	Product-Sum Formulas	465
8.6	Trigonometric Equations	472
8.7	The Inverse Circular Functions	479
8.8	More Inverse Trigonometric Functions	488
8.9	Chapter Review	493
8.10	Supplementary Exercises	495

9 Further Applications of Trigonometry 499

9.1	The Law of Sines	499
9.2	The Law of Cosines	508
9.3	Vectors	513
9.4	The Complex Plane and Polar Representation	523
9.5	Powers and Roots of Complex Numbers	534
9.6	Polar Coordinates	541
9.7	Chapter Review	548
9.8	Supplementary Exercises	550

10 Systems of Equations and Inequalities 553

10.1	Two Equations in Two Unknowns	553
10.2	The Elimination Method of Solution	562
10.3	Nonlinear Systems	571
10.4	Matrix Methods	577
10.5	Determinants	589
10.6	Cramer's Rule	601
10.7	Partial Fractions (Optional)	605
10.8	The Algebra of Matrices	611
10.9	Systems of Inequalities	621
10.10	Linear Programming	629
10.11	Chapter Review	638
10.12	Supplementary Exercises	641

11 Mathematical Induction, Progressions, and Probability — 645

11.1	Mathematical Induction	645
11.2	Sequences and Progressions	652
11.3	The Summation Notation	659
11.4	The Binomial Theorem	668
11.5	Permutations and Combinations	673
11.6	Probability	685
11.7	Chapter Review	696
11.8	Supplementary Exercises	698

APPENDIX A — Tables: Exponents, Logarithms, and Trigonometric Functions — A.1

Table 1	Natural Logarithms (ln x)	A.2
Table 2	Common Logarithms ($\log_{10} x$)	A.4
Table 3	Powers of e	A.6
Table 4	The Circular Functions (θ in Radians)	A.7
Table 5	The Trigonometric Functions (θ in Degrees)	A.11

Solutions (Odd-Numbered Exercises) — A.17

Index — A.91

Preface

Overview

This text covers the traditional topics in a college Algebra and Trigonometry course; the primary emphasis is on development of skills in algebraic manipulation and on a geometric understanding of the trigonometric functions. Topics are presented in an intuitive manner, yet care is taken to insure that mathematical statements are precise. Word problems appear throughout the text to help students develop skills in translating applied problems into mathematics.

Revision Notes

In preparing the second edition of *College Algebra and Trigonometry*, I have been influenced heavily by comments and suggestions received from users and reviewers of the first edition. I have especially tried to make the text more inviting to students. Much of the material is now presented less formally.

Some examples have been rewritten at a lower level of difficulty and more elementary examples and exercises have been included. Comments have been added to many of the solutions to examples and special attention has been paid to the coordination between examples and exercises. The exercise sets have been restructured to improve the grading in level of difficulty, and exercise instructions have been clarified. Some of the more intimidating examples and exercises have been removed. The more challenging exercises that remain are now numbered in red.

Caution boxes describing typical errors are a new feature in this edition. The number of Calculator Comments boxes and Historical Perspectives has also been increased. Discussions of rounding and significant digits, and their effects on computations have been added. Checking of solutions is also given increased emphasis.

Organization

Chapter 1 begins with a review of the real number system followed by the basic development of algebra. Complex numbers are introduced in preparation for the discussion of complex roots obtained by the quadratic formula (Chapter 2). Depending on the background of the class, Chapter 1, either partially or in its entirety, may be treated as an optional review. Changes in this edition include the separation of radicals and fractional exponents into separate sections, significant expansion and reorganization of factoring into a separate section, combining of product expansions with polynomials, and placing of rational expressions in a separate section.

Chapter 2 is a thorough study of equations and inequalities; the techniques introduced here are used throughout the text. The first edition's section on word problems has been split. The first part (Section 2.2) describes a systematic approach to the solution of word problems, but is restricted essentially to linear word problems and expanded material on literal equations. Then a later word-problem section (Section 2.5) treats applications of quadratic equations. Additional non-science word problems have been included here and throughout the text. Scientific sophistication has been deemphasized throughout the text in both examples and exercises. Intervals are now emphasized as solutions to inequalities. Test points are now used to provide an expedient solution for higher-order inequalities and fractional inequalities.

Chapter 3 introduces the student to graphing in a Cartesian coordinate system. Circles, lines, and parabolas are studied extensively in the first part of the chapter. The remainder of the chapter treats the other conic sections: ellipses and hyperbolas.

This latter material can be postponed or even omitted if the instructor chooses to do so. No later topics are dependent on this material. In this edition, the discussion of symmetry has been moved to Chapter 3 from Chapter 4.

Chapter 4 introduces the concept of function. The techniques of modifying a known graph by translation, reflection, and scaling are still emphasized in the general approach to graphing functions but there has been some rearrangement of topics for this edition. One-to-one functions are now included with the discussion of inverse functions.

Polynomial and rational functions and their graphs are discussed in detail in Chapter 5. The chapter begins with an analysis of the roots of a polynomial and methods for finding these roots. The chapter concludes with the graphing of polynomial and rational functions. In graphing these functions, emphasis is placed on the importance of roots and on the graphing techniques developed in Chapter 4. In this edition, the summary boxes for graphing polynomial and rational functions have been redone and the range limitation test has been changed to standard form. This entire chapter is independent of the remainder of the text and may be postponed or omitted with no loss of continuity.

Chapter 6 emphasizes the uses of exponential and logarithmic functions, such as in the description of exponential growth, compound interest, and carbon dating. Many questions in these areas require finding a logarithm. Logarithms continue to be important even though their use as a computational tool has been usurped by inexpensive calculators. In this edition, more emphasis has been given to the inverse nature of the logarithmic and exponential functions.

Chapter 7 introduces the trigonometric functions by exploring the relationships among the sides and angles of a right triangle. This approach is easier and more natural for the student to grasp than the more analytical introduction via the unit circle. Immediately after the study of right triangles and the trigonometric properties of acute angles, these concepts are extended to arbitrary angles and to real numbers. The unit circle is presented at this stage. Thus even though the trigonometric functions are introduced through right triangles, the unit circle and its properties are also available quite early.

In the trigonometry chapters (7, 8, and 9), degrees have been deemphasized in favor of radians and real numbers as arguments for the trigonometric functions. However, degrees are still used as the preferred angle measure in the sections devoted to solving triangles. The 45° and 30°–60° right triangles are given increased emphasis in this edition.

The space devoted to graphing the trigonometric functions and their inverses has been expanded and this material is presented at a slower pace. The summary box regarding amplitude, period, and phase-shift has been redone. The concepts of damping and modulation have been added to the sections on graphing. The material on inverse functions and their graphs is now presented more visually.

A more elementary section on trigonometric identities has been added to Chapter 8. The format for displaying the verification of identities has been improved.

Chapter 9 includes an optional section on vectors and a discussion of the trigonometric properties of the complex plane and DeMoivre's theorem. An introduction to polar coordinates in the real plane and sketching of polar equations finish the treatment of trigonometry.

The material on systems of equations (Chapter 10) has been reordered and restructured. Echelon form now receives more emphasis. More elementary examples have been added. Matrices are introduced here as a means to simplify the solution of a system of equations. A section on matrix algebra is included for those who desire it but this section (10.7) could be omitted with no loss of continuity. Systems of inequalities are also studied in Chapter 10, and these lead naturally into linear programming. No topics later in the text depend on either of these sections, however.

Finally, Chapter 11 introduces the topics of discrete mathematics: mathematical induction, progressions, the binomial theorem, permutations, combinations, and probability. The topic of odds has been added to this edition. The emphasis in this chapter is on a structured approach to problems rather than memorization of formulas.

Features

Instructional Aids

1. **Chapter Introductions** An informal discussion of the concepts to be studied paves the way into the chapter and motivates the material in the chapter.

2. **Historical Perspectives** These are included to give the student an appreciation of the development of mathematics.

3. **Emphasis on Graphing** An intuitive approach is taken to the development of most topics. Whenever possible, graphs are used to illustrate the solution of a problem. Many of the exercises also emphasize graphs and graphing techniques.

4. **Calculators** The text recognizes and encourages the use of calculators without becoming a "calculator text." The use and operation of a calculator are explained in general terms (without reference to a specific model). Calculator Comments boxes describe calculator usage and warn students of erroneous answers that could be obtained by haphazardly using a calculator. Students are encouraged to use calculators rather than tables for computations involving exponents, logarithms, and trigonometric functions. For those who desire it, however, an optional section (6.4) detailing the use of tables is included; tables are again discussed lightly in conjunction with the trigonometric functions. Separate groups of calculator exercises have been added to this edition.

Study Aids

1. **Exercises** The exercise sets are extensive and are graded from easy to challenging. Students need not be expected to work all the exercises in the text. Supplementary exercise sets included at the end of each chapter may be used by the students as review before an exam. These extensive exercise sets should also serve as an excellent source of test problems for instructors. Answers to all odd-numbered exercises are provided at the end of the book.

2. **Rules Boxes** Throughout the text, rules and formulas are displayed in boxes and examples of their use are provided. In many cases, specific examples appear alongside the rules in these boxes.

3. **Listing of Steps** Whenever several steps are involved in a given procedure, each step is identified and listed so that the student can readily reference steps when needed.

4. **Emphasis on Orderly Procedure to Solve Long Problems** Solving large systems of equations can sometimes be frustrating. This is especially true when a minor arithmetic mistake has been made in the middle of a problem and a maze of computations must be checked. A record-keeping procedure is used to simplify the task of retracing one's steps in order to find an error in solving a system of equations or evaluating a large-order determinant.

5. **Chapter Reviews** Reviews summarize important terms and concepts, rules and formulas, and techniques introduced in the chapter. In many cases, a simple illustration or example appears alongside the concept, rule, or technique.

Accuracy

In order to guarantee the accuracy of the answers provided, all of the exercises and examples have been checked by an independent problem solver.

Applications

Numerous real-life applications are included throughout the text as examples and as exercises. These applications introduce and reinforce the concepts being studied, and serve to motivate the student whose primary interest is in a discipline other than mathematics. Many exercise sets also include applications problems.

Supplements

1. The Study Guide to accompany *College Algebra and Trigonometry* (by John Piccirillo and Paul Filegar) includes a checklist of key concepts, additional examples, and hints for problem solving. Complete worked solutions are provided for selected odd-numbered exercises.
2. The Instructor's Manual to accompany this text contains solutions to all exercises.
3. The Test Bank provides four sample tests (one a multiple choice test) for each chapter and four sample final examinations. Answer keys with solutions are also provided for these tests and examinations.
4. Transparency Masters for important illustrations and graphs are available for use with an overhead projector.

Acknowledgments

I wish to thank the people who reviewed this text in its various stages of development. Their advice and comments regarding style, inclusion of topics, order of presentation, and level of difficulty were especially helpful. Many pertinent suggestions relating to both the details and the overall organization of the text were obtained from these evaluations. The list of reviewers for the second edition includes the following people:

Karen Baker	Indiana University
Lynn Brown	Illinois State University
Ed Chandler	Phoenix College
Jeff Cole	Anoka-Ramsey Community College
Sumner Cotzin	Assumption College
Jim Delany	California Polytechnic State University
Alberto Delgado	Kansas State University
Barbara Duch	University of Delaware
Don I. Dwyer	Radford University
Bob Froyd	Eastern Oregon State College
Doug Hall	Michigan State University
Robbie Harris	Tulsa Junior College
Louise Hasty	Austin Community College
George Hilton	Loma Linda University
Don Johnson	Scottsdale Community College
Marilyn McCollum	North Carolina State University
Eldon Miller	University of Mississippi
Don Poulson	Mesa College
Fred Schuurman	Miami University
Judith Willoughby	Minneapolis Community College

A special note of appreciation is also due Mr. Peter Marshall, mathematics editor at West Publishing Company, who coordinated and guided the project from its inception through the first edition, and now into the second edition.

Finally, I wish to thank Jeff Cole of Anoka-Ramsey Community College, who checked every exercise and example in the book.

Ralph C. Steinlage
October, 1987

Basic Algebra

Preceding even the most primitive attempts to record history, numbers and the counting process developed in a natural way to communicate essential ideas such as the size of a herd of animals. As number systems continued to be developed, they became an absolute necessity for commerce and communication. For example, our monetary system is based on numbers and the arithmetic operations of addition, subtraction, multiplication, and division; space flights would not be possible without algebra, trigonometry, calculus, and so on.

Scientists express the laws of nature in terms of letters that can represent different numbers in different situations. An example of this is the mass-energy equation $E = mc^2$ developed by Albert Einstein (1879–1955). In **algebra,** letters and symbols are used to represent numbers and to state rules by indicating that a property holds for all numbers. Since letters are simply names for numbers, the rules of arithmetic apply. We will review the elementary operations of arithmetic as they apply to the system of real numbers.

Section 1.1

The Real Number System

Natural Numbers

The counting numbers 1, 2, 3, ... are called **natural numbers.** The sum and the product of two natural numbers is again a natural number. For example, $2 + 3 = 5$ and $3 \cdot 7 = 21$. Whenever one number is written as a product of other numbers as in $c = a \cdot b$, a and b are called **factors** or **divisors** of c; c is called a **multiple** of a and of b. Thus, 3 and 7 are factors of 21; we say that 21 can be **factored** as 3 times 7.

A natural number other than 1 whose only factors are 1 and itself is called a **prime number.** For example, 2, 3, 5, 7, 11, 13, 17, 19 are prime numbers, whereas $4 = 2 \cdot 2$, $9 = 3 \cdot 3$, and $12 = 3 \cdot 4$ are not. Every natural number that is not itself a prime number (these are called **composite numbers**) can be factored as a product of primes. For instance, $12 = 2 \cdot 2 \cdot 3$.

Factoring can facilitate finding the **greatest common divisor (GCD)** and **least common multiple (LCM)** of several numbers. For instance, we write

$$60 = 2 \cdot 2 \cdot 3 \cdot 5$$

and

$$84 = 2 \cdot 2 \cdot 3 \cdot 7$$

The factors 2, 2, and 3 divide both 60 and 84; thus, the GCD of 60 and 84 is $2 \cdot 2 \cdot 3 = 12$ since 12 is the largest natural number that divides evenly into both 60 and 84: $60 = 5 \cdot 12$ and $84 = 7 \cdot 12$. On the other hand, any multiple of 60 and 84 must have factors of 2, 2, 3, 5, and 7. The LCM of 60 and 84 is $2 \cdot 2 \cdot 3 \cdot 5 \cdot 7 = 420$ since 420 is the smallest natural number that is a multiple of both 60 and 84: $420 = 7 \cdot 60$ and $420 = 5 \cdot 84$. These techniques will be adapted to algebraic expressions later.

Integers

The **integers** are the natural numbers together with their negatives and 0:

$$\ldots, -3, -2, -1, 0, 1, 2, 3, \ldots$$

The minus sign (−) is used to denote *below zero* or *negative*. For instance, $-10°$ represents $10°$ below zero. Death Valley has an elevation of -282 feet; it is 282 feet below sea level.

Rational Numbers

A number that can be expressed as a ratio of integers m/n with $n \neq 0$ is called a **rational number.** Every integer m is also a rational number since $m = m/1$. Rational numbers are used to denote *fractional parts* or *fractions*. For instance, $\frac{2}{7}$ is used to indicate two pieces of something that has been divided into seven equal parts. In a fraction a/b, a is called the **numerator** and b is called the **denominator**.

Historical Perspective

One of the best-known theorems in mathematics is the **Pythagorean theorem,** which states that *the sum of the squares of the legs of a right triangle equals the square of the hypotenuse.* [The product of a number a with itself is denoted a^2 (read "a squared"). Thus, 7 squared is $7^2 = 7 \cdot 7 = 49$.]

The so-called Pythagorean school of mathematics, science, and philosophy was so well disciplined and organized that it survived for over 100 years after the death of Pythagoras (ca. 575–500 B.C.). All the works of the school were attributed to Pythagoras. However, there is no direct evidence that Pythagoras, or even the Pythagorean school, proved the Pythagorean theorem. In fact, the Babylonians knew of the "Pythagorean property" more than 1000 years before the time of Pythagoras, but they were unable to prove it.

The Pythagorean school evolved into a cult. Philosophy and mathematics became so intertwined that its members believed everything depended on the whole numbers. This is perhaps the only example in history of a religion based in mathematics. Their belief in the omnipotence of the whole numbers encompassed the fractional or rational numbers also; $\frac{1}{3}$ could be

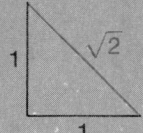

Figure 1.1
THE LENGTH $\sqrt{2}$

considered as being obtained by dividing a whole into three equal parts.

The Pythagorean theorem indicates the existence of a line segment whose length is $\sqrt{2}$. (The symbol $\sqrt{a}$ is used to denote a number that when multiplied by itself yields a; it is called the **square root** of a. Thus, $\sqrt{2} \cdot \sqrt{2} = 2$.) Here the $\sqrt{2}$ is the hypotenuse of a right triangle with legs of length 1 as shown in Figure 1.1. One of the greatest contributions of the Pythagoreans to the development of mathematics was their discovery that $\sqrt{2}$ could not be a rational number; it cannot be written as a fraction or ratio of integers m/n. Such numbers are said to be **irrational.** However, this discovery shook the Pythagorean school to its very core. It rendered invalid their basic belief that "everything depends on the whole numbers." Legend has it that Hippasus (ca. 500 B.C.) was set adrift at sea for revealing to outsiders the irrationality of $\sqrt{2}$.

Real Numbers

The real number line is used to describe numbers like $\sqrt{2}$, which arise from elementary geometry as lengths of line segments. (See the Historical Perspective.) Choosing a unit of length and marking an origin O on some line l, we assign to each point P on l a number equal to the length of the segment $\overline{OP}$. Points to the right of the origin are assigned positive numbers, and points to the left are assigned negative numbers. The number zero is assigned to the origin O. Each such number is called a **real number**. The line is called the **real number line**. The locations of several numbers on the real line are illustrated in Figure 1.2.

Figure 1.2
THE REAL NUMBER LINE

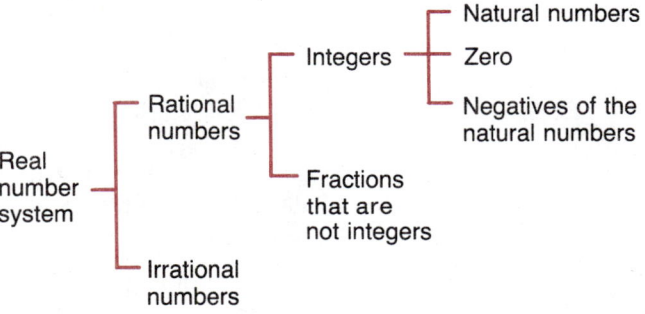

Figure 1.3 shows a breakdown of the number systems that we have discussed up to this point.

Figure 1.3
NUMBER SYSTEMS

```
                                              ┌─ Natural numbers
                          ┌─ Integers ────────┼─ Zero
          ┌─ Rational     │                   └─ Negatives of the
          │   numbers ────┤                      natural numbers
Real      │               └─ Fractions
number ───┤                  that are
system    │                  not integers
          │
          └─ Irrational
              numbers
```

One of the fundamental properties of the real number line is that it contains within itself the square roots, cube roots, and so on of each of its positive numbers. Later we shall describe the extension of the real numbers to a larger system, the complex numbers, which will include the square roots of all negative numbers as well.

The two basic arithmetic operations in the real number system are **addition** ($+$) and **multiplication** ($\cdot$). These operations are commutative and associative, and the distributive property holds.

SOME PROPERTIES OF REAL NUMBERS

	PROPERTY	EXAMPLE
Commutative	$a + b = b + a$ $ab = ba$	$5 + 4 = 4 + 5 = 9$ $5 \cdot 4 = 4 \cdot 5 = 20$
Associative	$a + (b + c) = (a + b) + c$ $a \cdot (b \cdot c) = (a \cdot b) \cdot c$	$5 + (4 + 3) = (5 + 4) + 3 = 12$ $5 \cdot (4 \cdot 3) = (5 \cdot 4) \cdot 3 = 60$
Distributive	$a \cdot (b + c) = ab + ac$ $(a + b) \cdot c = ac + bc$	$5 \cdot (4 + 3) = 5 \cdot 4 + 5 \cdot 3 = 35$ $(5 + 4) \cdot 3 = 5 \cdot 3 + 4 \cdot 3 = 27$
Identities	$a + 0 = a$ $1 \cdot a = a$	$5 + 0 = 5$ $1 \cdot 3 = 3$
Inverses	$a + (-a) = 0$ $a \cdot \left(\dfrac{1}{a}\right) = 1, \quad a \neq 0$	$3 + (-3) = 0$ $3 \cdot \left(\dfrac{1}{3}\right) = 1$

The number 0 is called the **additive identity** and 1 is called the **multiplicative identity**; $-a$ is called the **additive inverse** or **negative** of a, and $1/a$ is called the **multiplicative inverse** or **reciprocal** of a (when $a \neq 0$).

Multiplication by a natural number can be interpreted as repeated addition. For instance,

$$2a = (1 + 1) \cdot a$$
$$= 1 \cdot a + 1 \cdot a \qquad \text{(Distributive law)}$$
$$= a + a \qquad \text{(Multiplicative identity)}$$
$$3a = (2 + 1)a$$
$$= 2a + 1 \cdot a \qquad \text{(Distributive law)}$$
$$= (a + a) + a$$
$$= a + a + a \qquad \text{(Associative law)}$$

Since $(a + a) + a = a + (a + a)$, we simply write this as $a + a + a$; the parentheses are unnecessary. Similarly, $(a + b) + (c + d)$ and $a + (b + c + d)$ can be written as $a + b + c + d$; $(ab)(cd)$ and $a(bc)d$ can be written as $abcd$ by virtue of the associative properties.

We note also that $0 \cdot a = 0$ for every real number a. For a, $1 \cdot a = (1 + 0) \cdot a = 1 \cdot a + 0 \cdot a = a + 0 \cdot a$. Thus $0 \cdot a$ must be the additive identity 0.

Subtraction $(-)$ undoes the operation of addition. If b is added to something to obtain a, then subtracting b from a yields this other number.

DEFINITION

For any two numbers a and b, the **difference** $a - b$ is that number c which when added to b yields a:

$$a - b = c \quad \text{means} \quad a = b + c$$

For example,

$$5 - 3 = 2 \quad \text{since} \quad 5 = 3 + 2$$

This property of subtraction is sometimes used when making change for a purchase. The excess amount can be returned by counting from the purchase price up to the tendered amount.

The rules governing subtraction and negative numbers are summarized as follows.

RULES FOR NEGATIVE NUMBERS

RULE	EXAMPLE
1. $a + (-b) = a - b$	1. A loss is a negative profit. Thus, a profit of $10,000 plus a loss of $2000 yields a net profit of $8000: $\$10{,}000 + (-\$2000) = \$8000$ $= \$10{,}000 - \$2000$

RULE	EXAMPLE
2. $a - (-b) = a + b$	2. 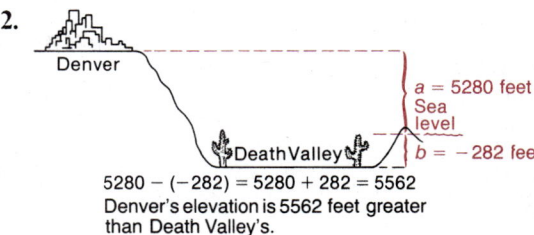 $5280 - (-282) = 5280 + 282 = 5562$ Denver's elevation is 5562 feet greater than Death Valley's.
3. $-(-a) = a$	3. $-(-2) = 2$; the opposite of a \$2 loss is a \$2 gain.
4. $a(-b) = -(ab) = (-a)b$	4. $\begin{cases} 2(-3) = (-3) + (-3) = -6 \\ (-2)(3) = (-2) + (-2) + (-2) = -6 \end{cases}$ $= -(2 \cdot 3)$
In particular $(-b) = (-1)b$	$(-1)b = -(1 \cdot b) = -b$
5. $(-a)(-b) = ab$	5. $(-2)(-7) = -[2(-7)] = -[-14]$ $= 14 = 2 \cdot 7$
6. $-(a + b) = -a - b$	6. $-(2 + 3) = -5 = -2 - 3$
7. a. $a(b - c) = ab - ac$	7. a. $3(6 - 2) = 3 \cdot 4 = 12$ and $3 \cdot 6 - 3 \cdot 2 = 18 - 6 = 12$
b. $(a - b)c = ac - bc$	b. $(7 - 4)5 = 3 \cdot 5 = 15$ and $7 \cdot 5 - 4 \cdot 5 = 35 - 20 = 15$

Division is the operation that undoes multiplication. If b is multiplied by another number c to yield a ($a = bc$), then dividing a by b yields this other number c. But $0 = 0 \cdot c$ for every real number c. Thus we cannot expect to recover the original number uniquely when $b = 0$. Hence, division by 0 is not defined.

DEFINITION

Division of a real number a by b, $b \neq 0$, $\left(\text{denoted by } a \div b, a/b, \text{ or } \dfrac{a}{b} \right)$ is that number c which when multiplied by b yields a:

$$\frac{a}{b} = c \quad \text{means} \quad a = bc, \quad (b \neq 0)$$

c is called the **quotient** of the **dividend** a and the **divisor** b.

For example, $8 \div 4 = 2$, since $8 = 4 \cdot 2$; 8 divided by 4 yields the quotient 2.

The rules governing division and rational numbers are summarized in the following table.

RULES FOR FRACTIONS

RULE	EXAMPLE
1. $\dfrac{k \cdot m}{k \cdot n} = \dfrac{m}{n},\ k \neq 0,\ n \neq 0$	1. $\dfrac{3}{6} = \dfrac{3 \cdot 1}{3 \cdot 2} = \dfrac{1}{2}$ 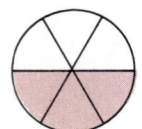
2. a. $\dfrac{a}{b} + \dfrac{c}{b} = \dfrac{a + c}{b}$	2. a. $\dfrac{2}{5} + \dfrac{1}{5} = \dfrac{2 + 1}{5} = \dfrac{3}{5}$ 
b. $\dfrac{a}{b} - \dfrac{c}{b} = \dfrac{a - c}{b}$	b. $\dfrac{2}{5} - \dfrac{1}{5} = \dfrac{2 - 1}{5} = \dfrac{1}{5}$ 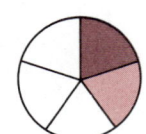
3. a. $\dfrac{a}{b} + \dfrac{c}{d} = \dfrac{ad + bc}{bd}$	3. a. $\dfrac{1}{2} + \dfrac{1}{3} = \dfrac{1 \cdot 3}{2 \cdot 3} + \dfrac{2 \cdot 1}{2 \cdot 3} = \dfrac{1 \cdot 3 + 2 \cdot 1}{2 \cdot 3} = \dfrac{5}{6}$ 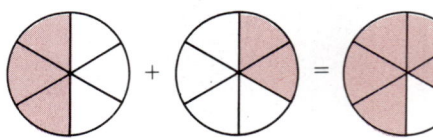
b. $\dfrac{a}{b} - \dfrac{c}{d} = \dfrac{ad - bc}{bd}$	b. $\dfrac{1}{2} - \dfrac{1}{3} = \dfrac{1 \cdot 3}{2 \cdot 3} - \dfrac{2 \cdot 1}{2 \cdot 3} = \dfrac{1 \cdot 3 - 2 \cdot 1}{2 \cdot 3} = \dfrac{1}{6}$ 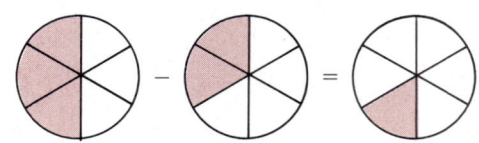
4. $-\dfrac{a}{b} = \dfrac{-a}{b} = \dfrac{a}{-b}$	4. $-\left(\dfrac{2}{3}\right) = \dfrac{-2}{3} = \dfrac{2}{-3}$ since when any of these is added to $\dfrac{2}{3}$, the result is 0
5. $\dfrac{a}{b} \cdot \dfrac{c}{d} = \dfrac{a \cdot c}{b \cdot d}$	5. $\dfrac{1}{5} \cdot \dfrac{2}{3} = \dfrac{1}{5} \cdot \dfrac{2 \cdot 5}{3 \cdot 5} = \dfrac{1}{5} \cdot \dfrac{10}{15} = \dfrac{2}{15}$ 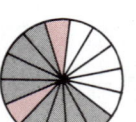
6. $\dfrac{\frac{a}{b}}{\frac{c}{d}} = \dfrac{a}{b} \cdot \dfrac{d}{c}$	6. $\dfrac{\frac{2}{3}}{\frac{5}{7}} = \dfrac{\frac{2}{3} \cdot \frac{7}{5}}{\frac{5}{7} \cdot \frac{7}{5}} = \dfrac{\frac{2}{3} \cdot \frac{7}{5}}{1} = \dfrac{2}{3} \cdot \dfrac{7}{5}$

Rule 1 is called the **cancellation law.** Factoring both numerator and denominator of a fraction and eliminating or "canceling" common factors yield a fraction equal to the original. The numerator and denominator of the resulting fraction will share no common factors other than ± 1. Such a fraction is said to be written in **lowest terms.** For instance, $\frac{30}{105}$ is reduced to lowest terms as follows:

$$\frac{30}{105} = \frac{2 \cdot \cancel{3} \cdot \cancel{5}}{\cancel{3} \cdot \cancel{5} \cdot 7} = \frac{2}{7}$$

Addition and subtraction of fractions are complicated by the fact that we must compare only like quantities. Although Rules 3a and b bypass this requirement, they can result in some very large numbers, especially in the denominator. But only a *common denominator* is really needed.

> **DEFINITION**
>
> The **least common denominator,** or **LCD,** of a collection of fractions is the LCM (least common multiple) of the denominators of these fractions.

EXAMPLE 1

Find $\dfrac{11}{36} - \dfrac{7}{30}$ using the LCD.

SOLUTION

$$\frac{11}{36} - \frac{7}{30} = \frac{11}{6 \cdot 6} - \frac{7}{5 \cdot 6} = \frac{5 \cdot 11}{5 \cdot 6 \cdot 6} - \frac{7 \cdot 6}{5 \cdot 6 \cdot 6} \quad \text{(Rule 1)}$$

$$= \frac{55 - 42}{5 \cdot 6 \cdot 6} = \frac{13}{180} \quad \text{(Rule 2b)}$$

■

Decimals

Each real number has a decimal representation. The long-division process can be used to find the decimal representation for any rational number, such as $\frac{13}{11}$, as follows.

$$\begin{array}{r} 1.1818\ldots \\ 11\overline{\smash{)}13.0000\ldots} \\ \underline{11} \\ 20 \\ \underline{11} \\ 90 \\ \underline{88} \\ 20 \\ \underline{11} \\ 90 \\ \underline{88} \\ 2\ldots \end{array}$$

$$\frac{13}{11} = 1.1818\ldots$$
$$= 1.\overline{18}$$

The expansion for $\frac{13}{11}$ is a repeating decimal. The bar over the 18 is used to indicate that the block 18 repeats itself without stopping. Every rational number has either a repeating decimal representation or a decimal representation that stops (terminates) after a finite number of places (for example, $\frac{1}{4} = 0.25$).

On the other hand, every repeating or terminating decimal represents a rational number. Thus the irrational numbers are precisely those that have non-terminating, nonrepeating decimal expansions.

Rounding Numbers

In practice, decimal numbers are oftentimes *rounded* to the degree of accuracy desired in a given calculation. For instance, $\frac{13}{11} = 1.1818\ldots$ can be rounded to the nearest thousandth as 1.182 (since it is closer to 1.182 than to 1.181). We write $\frac{13}{11} \approx 1.182$; the symbol $\approx$ denotes *approximately equal to*.

A decimal expansion that terminates with a 5 is midway between the two choices for rounding off the number. We could round up or round down in such cases. One current accepted practice is to round such a number so that the last digit of the rounded-off number is even. This prevents prejudicing large collections of data either upward or downward.

CALCULATOR COMMENTS

Calculators can accommodate only a finite number of digits or decimal places. Thus irrational numbers and repeating decimals must be approximated in order to be represented in the calculator. Some calculators round off the number while others simply *truncate* the number (drop the digits beyond the limitations of the calculator).

For instance, $\frac{2}{3}$, or $2 \div 3$, is represented on two different calculators as

$2 \div 3 \approx \boxed{0.666666667}$ (Rounded up)

$2 \div 3 \approx \boxed{0.6666666}$ (Truncated)

These characteristics of a calculator can affect the accuracy of further computations. Using the same two calculators as above, the computation $(\frac{2}{3}) \cdot 3$, or $(2 \div 3) \times 3$, appears on the display as

$(2 \div 3) \cdot 3 = \boxed{2.0}$

$(2 \div 3) \cdot 3 \approx \boxed{1.9999998}$

One way to reduce the effect of such round-off errors is to perform multiplications before division. In this example,

$$\left(\frac{2}{3}\right) \cdot 3 = \frac{2 \cdot 3}{3} = (2 \times 3) \div 3$$

Any calculator in good working order will yield the correct answer of $\boxed{2.0}$ for this computation.

Hierarchy of Operations

In an expression involving more than one operation, such as $2 \cdot 3 + 4$, parentheses are sometimes used to indicate that certain operations are to be performed before others. The order of computation can make a difference. For example,

$(2 \cdot 3) + 4 = 10$

$2 \cdot (3 + 4) = 14$

To prevent misunderstandings, mathematicians have adopted the following conventions.

Section 1.1 The Real Number System 9

> **ORDER OF OPERATIONS**
> 1. Expressions in parentheses are to be evaluated before being combined with numbers outside the parentheses.
> 2. Powers and roots (e.g., a^2 and $\sqrt{a}$) are to be calculated before multiplications and divisions.
> 3. Multiplications and divisions are performed before additions and subtractions and in the order in which they appear from left to right.
> 4. Then additions and subtractions are done in the order in which they appear from left to right.

EXAMPLE 2 Evaluate the following expressions.

a. $4 \div 2 - 4 \cdot 2 + 7$

b. $4 \div (2 - 4) \cdot 2 + 7$

SOLUTION

a. According to the hierarchy of operations, we have
$$4 \div 2 - 4 \cdot 2 + 7 = (4 \div 2) - (4 \cdot 2) + 7$$
$$= 2 - 8 + 7 = 1$$

b. Note that the division operation precedes multiplication when reading from left to right. Thus,
$$4 \div (2 - 4) \cdot 2 + 7 = 4 \div (-2) \cdot 2 + 7$$
$$= (-2) \cdot 2 + 7$$
$$= -4 + 7 = 3$$

CALCULATOR COMMENTS One must be careful when solving a problem that requires more than a single operation on a calculator. Some of the simpler calculators perform the operations in the order in which they are entered. On the other hand, scientific calculators are programmed to respect the hierarchy of operations as described earlier and will perform multiplication before addition. Some calculators also provide parentheses to enable the user to impose a desired operational order. But without parentheses one must be careful to enter the problem in such a way that the correct answer results. Read the owner's manual carefully, and *know your calculator. Be careful when using someone else's calculator.*

EXAMPLE	RESULT FROM A SCIENTIFIC CALCULATOR	POSSIBLE RESULT FROM A SIMPLE CALCULATOR	OBTAIN CORRECT RESULT ON A SIMPLE CALCULATOR BY ENTERING AS
1. $2 + 3 \cdot 4$	$2 + 12 = 14$	$(2 + 3) \cdot 4 = 20$	$3 \cdot 4 + 2 = 12 + 2 = 14$
2. $4\frac{1}{9} = 4 + 1 \div 9$	$4.111111111\ldots$	$(4 + 1) \div 9 \approx 0.555555\ldots$	$1 \div 9 + 4 \approx 4.1111\ldots$

Sets

There will be times when it will be convenient to use set notation. An intuitive understanding of a set as a collection of objects is adequate. The objects in a given set are said to be **members,** or **elements,** of the set. To say "x is a member of A (or an element of A)," we write $x \in A$; $x \notin A$ means x is not a member of A. Some sets are described by listing their members between braces, as in $A = \{a, b, c\}$, or by indicating the listing, as in the **set of natural numbers,** N:

$$N = \{1, 2, 3, \ldots\}$$

or the **set of integers,** Z:

$$Z = \{\ldots, -3, -2, -1, 0, 1, 2, 3, \ldots\}$$

If a set does not lend itself to a listing of its members, we can describe the set by stating properties that must be satisfied by its members. For example, the **set of rational numbers** Q is denoted by

$$Q = \left\{\frac{m}{n} : m \text{ and } n \text{ are integers, } n \neq 0\right\}$$

This is read "Q is the set of those numbers of the form m/n, where m and n are integers and n does not equal 0."

The **set of real numbers** is denoted by R. We can also describe the set of rational numbers as

$$Q = \{x : x \in R \text{ and } x \text{ has a terminating or repeating decimal representation}\}$$

The **set of irrational numbers** is

$$I = \{x : x \in R \text{ but } x \notin Q\}$$

which can also be described as

$$I = \{x : x \in R \text{ and } x \text{ has a nonterminating, nonrepeating decimal representation}\}$$

If every member of a set A is also a member of set B, we say A is a **subset** of B and write $A \subset B$. Thus if $A = \{1, 2, 3\}$ and $B = \{-1, 0, 3, 1, 4, 2\}$, we have $A \subset B$. Note that the order in which the members of a set are listed is immaterial. For the sets of numbers described above, we have

$$N \subset Z \subset Q \subset R \quad \text{and} \quad I \subset R$$

Historical Perspective

GEORG CANTOR
(1845–1918)

The theory of sets was initiated by Georg Cantor (1845–1918). With his theory of sets, Cantor was able to develop an arithmetic of "infinities," in which infinite quantities of various sizes could be added and multiplied. Cantor's ideas were rejected by his contemporaries. Many of his peers could not understand his concepts because they were so foreign to anything that had been seen before.

Bitter controversies ensued, especially with Leopold Kronecker (1823–1891). Cantor died in a mental institution, a broken man. He was a man ahead of his time. Only after his confinement to a mental institution did people begin to understand his ideas and to appreciate the power of his results. Today, all mathematics can be based on the theory of sets.

This relationship of the number systems is shown in Figure 1.4, a *Venn diagram*.

Figure 1.4
THE NUMBER SYSTEMS

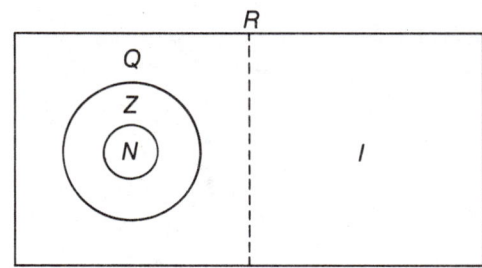

Three methods of obtaining new sets from two given sets are important. The **union** of A and B consists of all the members of A together with all the members of B:

$$A \cup B = \{x : x \in A \quad \text{or} \quad x \in B\}$$

The **intersection** of two sets consists of the common part of the sets in question:

$$A \cap B = \{x : x \in A \quad \text{and} \quad x \in B\}$$

Of course, if A and B have no members in common, then $A \cap B$ is empty; this is denoted by $A \cap B = \emptyset$. The **difference** $A \setminus B$ is obtained by removing the elements of B from the set A:

$$A \setminus B = \{x : x \in A \quad \text{but} \quad x \notin B\}$$

For our sets of numbers, we have

$$R = Q \cup I, \quad Q = R \setminus I, \quad I = R \setminus Q, \quad Q \cap I = \emptyset$$

EXAMPLE 3

Let $A = \{1, 2, 3, 4, 5\}$ and $B = \{-1, 0, 1, 3, 5, 7, 9\}$. Find each of the following.

a. $A \cup B$ **b.** $A \cap B$ **c.** $A \setminus B$ **d.** $B \setminus A$

SOLUTION

a. $A \cup B = \{-1, 0, 1, 2, 3, 4, 5, 7, 9\}$
b. $A \cap B = \{1, 3, 5\}$
c. $A \setminus B = \{2, 4\}$
d. $B \setminus A = \{-1, 0, 7, 9\}$

Section 1.1 Exercises

Factor the following into primes.

1. 24
2. 187
3. 396
4. 391
5. 819
6. 3850

Find the LCM and GCD of the following sets of numbers.

7. 45, 75
8. 68, 738
9. 129, 215
10. 76, 1425
11. 6, 15, 21
12. 36, 48, 60

Locate each of the following points on the number line.

13. 7.3
14. 9.45
15. -3.2
16. -4.65
17. $0.666\ldots$
18. $0.4999\ldots$

State which of the properties of the real numbers (commutative, associative, distributive, identity, inverse) are used in the following equalities.

19. $6 \cdot 1 = 6$
20. $(-6) + 6 = 0$
21. $3 + 5 = 5 + 3$
22. $3 + (4 + 0) = 3 + 4$

23. $\dfrac{1}{7} \cdot 7 = 1$

24. $2 + (3 + 5) = (2 + 3) + 5$

25. $(6 + 2) \cdot 3 = 6 \cdot 3 + 2 \cdot 3$

26. $7 \cdot 3 = 3 \cdot 7$

27. $[2 \cdot (-3)] \cdot 4 = 2 \cdot [(-3) \cdot 4]$

28. $6 \cdot (3 + 7) = 6 \cdot 3 + 6 \cdot 7$

Reduce each of the following to lowest terms.

29. $\dfrac{15}{27}$ 30. $\dfrac{-21}{28}$ 31. $\dfrac{57}{95}$ 32. $\dfrac{15}{-12}$

33. $\dfrac{63}{72}$ 34. $\dfrac{-30}{231}$ 35. $\dfrac{-660}{450}$ 36. $\dfrac{2310}{5005}$

Express each of the following as a single integer.

37. $-(-[-6])$

38. $-(2 - 7)$

39. $(-5) - (-7)$

40. $(-5) - (-3)$

41. $(-8)(-6)$

42. $(-2)(-4)(-3)$

43. $3(-11) + 4(-2)(-1) - 13(-2)$

44. $-(-2)[4 + 6 \cdot (-3)]$

45. $2 - 3 \cdot 5 - 6 \cdot 5 - 2 - 7$

46. $[2 - 3(5 - 6)][5 - (2 - 7)]$

47. $-5 \cdot 6 - (-10) + 3(-10) - 8$

48. $-5[6 - (-10) + 3(-10 - 8)]$

49. $10 + 6 \div 2 + 6 \div 2 + 2 \cdot 3$

50. $(10 + 6) \div (2 + 6) \div 2 + 2 \cdot 3$

51. $(10 + 6) \div [(2 + 6) \div (2 + 2)] \cdot 3$

52. $10 + [6 \div 2 + 6 \div 2 + 2] \cdot 3$

Perform the following operations and express the answers in lowest terms.

53. $\dfrac{1}{4} + \dfrac{3}{8}$

54. $\dfrac{1}{3} + \dfrac{1}{6}$

55. $\dfrac{1}{2} - \dfrac{1}{3}$

56. $\dfrac{5}{7} - \dfrac{5}{9}$

57. $\dfrac{3}{4} - \left(\dfrac{1}{2} + \dfrac{1}{6}\right)$

58. $\left(\dfrac{3}{4} - \dfrac{1}{2}\right) + \dfrac{1}{6}$

59. $\dfrac{3}{7} + \dfrac{4}{9} + \dfrac{5}{21}$

60. $\dfrac{3}{7} + \dfrac{-4}{9} - \dfrac{5}{21}$

61. $\dfrac{10}{14} \cdot \dfrac{3}{-8} \cdot \dfrac{35}{50}$

62. $2\dfrac{1}{16} \cdot 3\dfrac{1}{3} \cdot \left(-1\dfrac{1}{5}\right)$

$\left(\text{Note: } 2\dfrac{1}{16} = 2 + \dfrac{1}{16} = \dfrac{2 \cdot 16}{16} + \dfrac{1}{16} = \dfrac{33}{16}\right)$

63. $3\dfrac{1}{3} \cdot 4\dfrac{1}{2} \cdot 5\dfrac{1}{5}$

64. $\dfrac{\frac{9}{16}}{\frac{21}{32}}$

65. $\dfrac{2}{3} \div \dfrac{7}{3} \cdot \dfrac{7}{18}$

66. $\dfrac{5}{6} \div 4\dfrac{4}{9}$

67. $\left(\dfrac{3}{4} - \dfrac{1}{3}\right) \div \left(\dfrac{2}{3} - \dfrac{1}{2}\right)$

68. $\dfrac{\frac{1}{2} + \frac{1}{3}}{\frac{2}{3} + \frac{1}{6}}$

69. $\left(2\dfrac{1}{3} + 1\dfrac{1}{6}\right) \div \left(4\dfrac{1}{2} - 3\dfrac{1}{4}\right)$

70. $\left(1\dfrac{1}{4} - 2\dfrac{1}{3}\right) \div \left[\left(\dfrac{3}{4} \cdot \dfrac{1}{2}\right) - \dfrac{1}{8}\right]$

71. $\dfrac{3}{\frac{3}{5} + \frac{2}{4 - \frac{2}{3}}}$

72. $\dfrac{\frac{1}{5}}{7 - \dfrac{\frac{1}{6}}{\frac{1}{2} + \frac{2}{3}}}$

Express fractions as decimals when needed, then round each number to the nearest thousandth.

73. $\dfrac{6}{7}$ 74. $\dfrac{4}{9}$ 75. $\dfrac{23}{11}$

76. $\dfrac{9}{13}$ 77. $\dfrac{17}{24}$ 78. $\dfrac{27}{41}$

79. $0.4999\ldots$

80. $1.010101\ldots$

81. $5.1272727\ldots$

82. $3.27327\ldots$

83. $672.3456456\ldots$

84. $414.201420\ldots$

85. $29.768342\ldots$

"Mental magicians" are adept at exploiting the commutative, associative, and distributive laws. For example, a quick mental calculation of $6 \cdot 97$ can be accomplished as follows:

$$6 \cdot 97 = 6(100 - 3) = 600 - 18 = 582$$

Another use of these properties is illustrated in computing 18 · 25:

$$18 \cdot 25 = 9 \cdot 2 \cdot 25 = 9 \cdot 50 = 450$$

Use the commutative, associative, and distributive laws to evaluate the following mentally.

86. 5 · 297   87. 19 · 40   88. 7 · 69
89. 24 · 7   90. 99 · 98   91. 8 · 19

For

$A = \{a, b, c, d, e, f, g\}$
$B = \{c, d, e, f, g, h, i, j, k\}$
$C = \{a, c, e, g, i, k\}$

find the following sets.

92. $A \cup B$ 93. $A \cap B$
94. $A \setminus B$ 95. $B \setminus A$
96. $A \cup (B \cap C)$ 97. $A \cap (B \cup C)$
98. $B \setminus (A \cap C)$ 99. $C \setminus (A \setminus B)$
100. $B \cup (C \setminus A)$ 101. $C \cap (B \setminus C)$

102. When adding a column of numbers, we generally add from top to bottom. Then we check our work by adding from bottom to top. Which property or properties of the real number system are we using?

103. Is subtraction commutative? Associative?

104. Is division commutative? Associative?

105. Does division distribute over addition? That is to say, does $a \div (b + c) = (a \div b) + (a \div c)$?

106. Basic Building Blocks are sold by Creative Toys in rectangular packages of 24 square blocks. What different shapes could the company use for the packages?

107. Two hardware companies produce bolts of the same quality. Acme Fasteners packages 32 bolts per box; Better Bolts puts 40 in each box. If neither company will sell partial boxes, what is the smallest order that both companies could fill?

108. The highest and lowest points in the 48 continental United States lie within 85 miles of one another. Both are in California. The highest point is Mount Whitney at 14,494 feet; the lowest is, of course, Death Valley at −282 feet. What is the change in altitude in these 85 miles?

109. At the previous fill-up, the odometer on Bernice's car registered 26,739.2. At the time of this fill-up, requiring 18.7 gallons, it registers 26,986.8. How many miles per gallon did the car get?

110. Fixed expenses at Cinema X are $289 per day. If 62 patrons pay $1.50 each for the early show and 120 patrons pay $3.75 each for the evening show, how much profit is made?

111. Some supermarkets have introduced a "unit-pricing" concept. If a 16-ounce loaf of bread costs 83¢, what is its unit price, or price per ounce? If a 20-ounce loaf costs $1.05, which is the better buy?

112. The intelligence quotient or IQ is defined as 100 times mental age divided by chronological age:

$$IQ = 100 \, \frac{\text{mental age}}{\text{chronological age}}$$

a. Johnny is 10 years old and has the mental capabilities of a 12-year-old. What is his IQ?

b. Joan is 15 years old and has an IQ of 115. What is her mental age?

113. Paving a certain portion of highway requires 15,000 cubic yards of concrete. If a concrete truck can haul $6\frac{2}{3}$ cubic yards, how many truckloads are needed?

114. If a hen and a half can lay an egg and a half in a day and a half, how many eggs will 12 hens lay in 12 days?

Calculator Exercises

Use a calculator to perform the following operations. Be sure to enter data in such a way that the correct result appears.

115. a. $(-2)(4 + 6)(-3)$
     b. $(-2) \cdot 4 + 6 \cdot (-3)$

116. a. $1 \div (7 \cdot 13)$
     b. $1 \div 7 \cdot 13$

14 Chapter 1 Basic Algebra

117. a. $6 \div [3 - (1 + 4)] \cdot 5$
   b. $6 \div 3 - 1 + 4 \cdot 5$
118. a. $(5 + 3) \div (2 + 3) \div (2 + 2) \cdot 4$
   b. $5 + [3 \div 2 + 3 \div 2 + 2] \cdot 4$
119. a. $2 \div (11 \cdot 5)$
   b. $2 \div 11 \cdot 5$
120. a. $7 \cdot (3 + 2) - 5 \div (4 + 3) \cdot 9$
   b. $7 \cdot 3 + 2 - 5 \div 4 + 3 \cdot 9$

121. a. $12 \div (17 \cdot 5)$
   b. $12 \div 17 \cdot 5$
122. a. $(5 + 6)(3 - 8) \div 2$
   b. $5 + 6 \cdot 3 - 8 \div 2$
123. a. $(12 + 3) \div (10 + 5) \cdot (6 - 2)$
   b. $12 + 3 \div 10 + 5 \cdot 6 - 2$
124. a. $5 \div (9 \cdot 7)$
   b. $5 \div 9 \cdot 7$

Section 1.2

Order and Absolute Value

Order

The inequality symbols ($<, \leq, >, \geq$) are used to facilitate the comparison of numbers. Say, for instance, we want to express the fact that a particular business may go bankrupt if its debts exceed the value of its assets. We use the symbols $<$ and $>$ (which indicate **less than** and **greater than**, respectively) to describe the relationship between debts and assets. We might write $3 < 7$, $-3 < -2$, and $-1 < 2$, as illustrated in Figure 1.5.

Figure 1.5

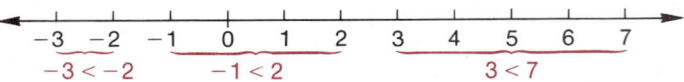

It should be clear that $a < b$ exactly when a lies to the left of b on the real number line. Thus **negative** numbers are less than zero and **positive** numbers are greater than zero. **Zero** is neither positive nor negative; zero is the only number for which $+0 = -0 = 0$.

The notations $a < b$ and $b > a$ are used interchangeably. If $a < b$, then a positive number must be added to a in order to obtain b; that is $b - a > 0$.

The following statements are equivalent:
1. a is less than b.
2. b is greater than a.
3. $a < b$.
4. $b > a$.
5. $b - a > 0$.
6. a lies to the left of b on the real number line.

An expression involving $<$ or $>$ is called an **inequality.** The following rules govern operations on inequalities.

RULES FOR INEQUALITIES

RULE	EXAMPLE
1. $a < b$ if and only if $b - a > 0$.	1. $2 < 5$ and $5 - 2 = 3 > 0$
2. If $a < b$ and $b < c$, then $a < c$.	2. $2 < 7$ and $7 < 9$, so $2 < 9$
3. If $a < b$, then $(a + c) < (b + c)$.	3. $2 < 7$ so $2 + 3 = 5 < 7 + 3 = 10$
4. If $a < b$, then $-a > -b$.	4. $2 < 5$ and $-2 > -5$

5. If $a < b$ and
  a. $x = 0$, then $ax = bx$.
  b. $x > 0$, then $ax < bx$.
  c. $x < 0$, then $ax > bx$.
	5. $3 < 5$ and
	  a. $3 \cdot 0 = 5 \cdot 0 = 0$
	  b. $3 \cdot 2 < 5 \cdot 2$ since $6 < 10$
	  c. $3(-2) > 5(-2)$ since $-6 > -10$

6. If $0 < a < b$, then $0 < \dfrac{1}{b} < \dfrac{1}{a}$.	6. Larger denominators make smaller fractions. $0 < 2 < 5$ and $0 < \dfrac{1}{5} < \dfrac{1}{2}$

Rule 5c should be emphasized. It says that *if we multiply or divide both sides of an inequality by a negative number, the direction of the inequality symbol is reversed.*

The symbols $\leq$ and $\geq$ are read *"less than or equal to"* and *"greater than or equal to,"* respectively. Thus,

$$a \leq b \quad \text{means} \quad (a < b \quad \text{or} \quad a = b)$$
$$a \geq b \quad \text{means} \quad (a > b \quad \text{or} \quad a = b)$$

For example, $3 \leq 3$ is a true statement since $3 = 3$ is true, even though $3 < 3$ is false.

Rules 1 through 6 also apply to $\leq$ and $\geq$. For example, rule 5c becomes: if $a \leq b$ and $x \leq 0$, then $ax \geq bx$. For if either $a = b$ or $x = 0$, it follows that $ax = bx$.

Intervals

An **interval** is the set of real numbers that lies between two given numbers a and b as follows:

$$(a, b) = \{x : a < x < b\}$$

The symbols $\{x : a < x < b\}$ are read "the set of all x such that x is greater than a and x is also less than b." The numbers a and b are called **endpoints** of this interval. The parentheses in the notation (a, b) indicate that the endpoints are not included in the interval. A square bracket, [or], is used to indicate that the endpoint is in the interval. See Figure 1.6.

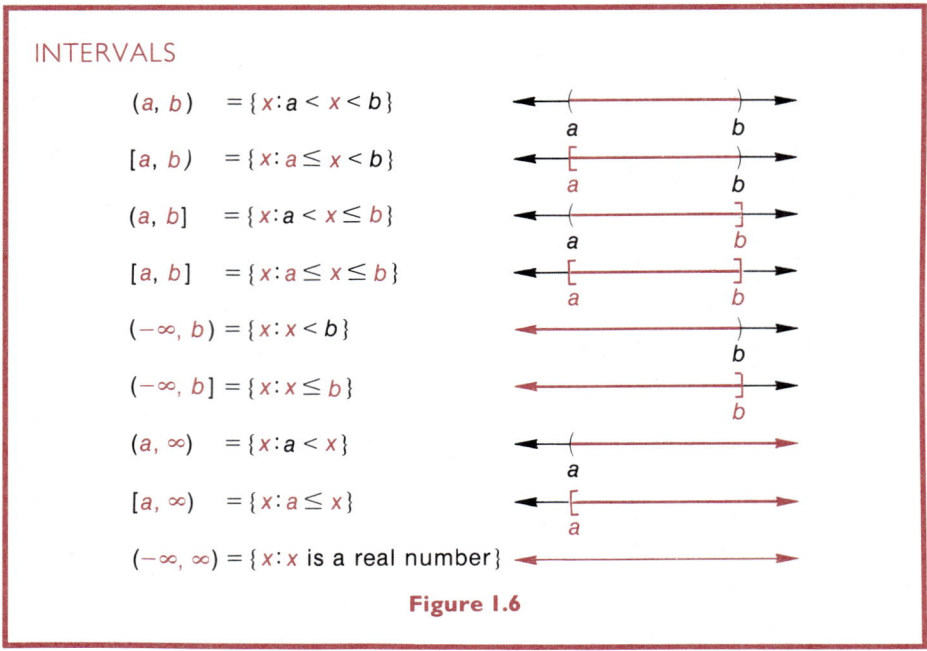

Figure 1.6

The symbols $-\infty$ and ∞ (read "negative infinity" and "infinity," respectively) are not regarded as numbers. They are used here merely to describe intervals that continue forever to the left or to the right, respectively. We never use a square bracket next to $-\infty$ or ∞.

Intervals that include their endpoints are called **closed intervals;** those that include neither endpoint are called **open intervals.** Intervals of the form $(a, b]$ or $[a, b)$ are called **half-open intervals.**

EXAMPLE 1

Sketch the following intervals.

a. $(2, 4)$ **b.** $(-2, 3]$ **c.** $[1, 5]$

SOLUTION

Figure 1.7 shows the intervals.

Figure 1.7

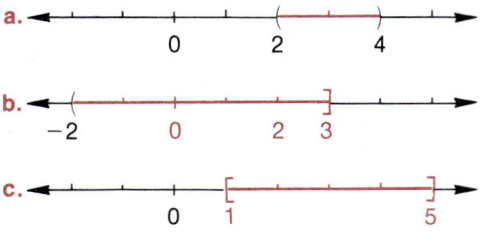

EXAMPLE 2 Let $A = (-2, 4)$ and $B = (1, 6)$. Describe each of the following as an interval.

a. $A \cap B$ **b.** $A \setminus B$ **c.** $A \cup B$

SOLUTION Figure 1.8 shows the intervals.

Figure 1.8

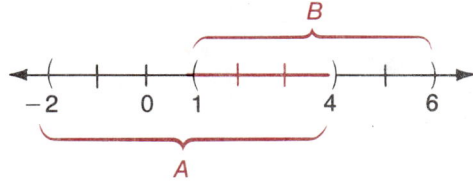

a. $A \cap B = (1, 4)$

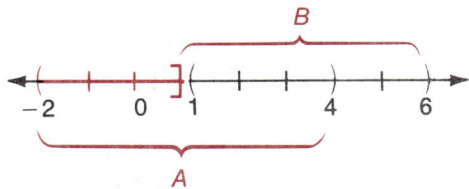

b. $A - B = (-2, 1]$

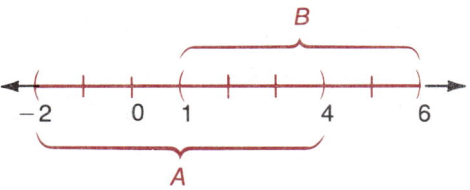

c. $A \cup B = (-2, 6)$

Absolute Value

The **absolute value** of a number is its distance (greater than or equal to 0) from 0 on the number line. For example, the absolute value of 7 is 7; the absolute value of -7 is also 7. Formally we define the absolute value of a number x, denoted by $|x|$, as follows:

DEFINITION

$$|x| = \begin{cases} x & \text{if } x \geq 0 \\ -x & \text{if } x < 0 \end{cases}$$

(*Note:* $-x > 0$ when $x < 0$.)

Many students are confused by the statement that $|x| = -x$ when $x < 0$. However, note that

$$|-7| = 7 = -(-7)$$

The absolute value of -7 is the negative of -7.

18 Chapter 1 Basic Algebra

The following rules govern operations on absolute values.

RULES FOR OPERATIONS ON ABSOLUTE VALUES

RULE	EXAMPLE
1. $\|a\| \geq 0$	1. $\|-3\| = 3 \geq 0$
2. $\|-a\| = \|a\|$	2. $\|-7\| = \|7\| = 7$
3. $\|x - y\| = \|y - x\|$	3. $\begin{cases} \|5 - 3\| = \|2\| = 2 \\ \|3 - 5\| = \|-2\| = 2 \end{cases}$
4. $\|ab\| = \|a\|\|b\|$	4. $\begin{cases} \|(-2) \cdot 7\| = \|-14\| = 14 \\ \|(-2)\|\|7\| = 2 \cdot 7 = 14 \end{cases}$
5. $\left\|\dfrac{a}{b}\right\| = \dfrac{\|a\|}{\|b\|}$	5. $\left\|\dfrac{8}{-4}\right\| = \dfrac{\|8\|}{\|-4\|}$ since $\|-2\| = \dfrac{8}{4}$

The **distance** between two numbers a and b on the real number line is defined to be the number (greater than or equal to 0) that must be added to the smaller to arrive at the larger. Thus, the distance is "larger − smaller." But since $(b - a) = -(a - b)$, the numbers $(b - a)$ and $(a - b)$ have the same magnitude (are of the same size) but have opposite signs $(+, -)$. Thus we can define distance as follows.

DISTANCE

The distance between two numbers a and b on the real number line is

$$d(a, b) = |b - a| = |a - b|$$

See Figure 1.9.

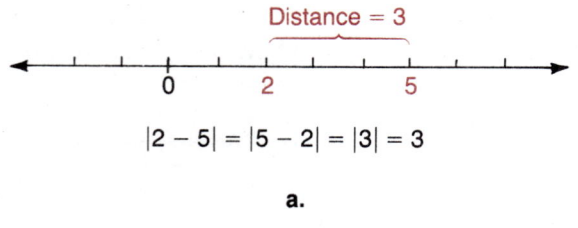

a.

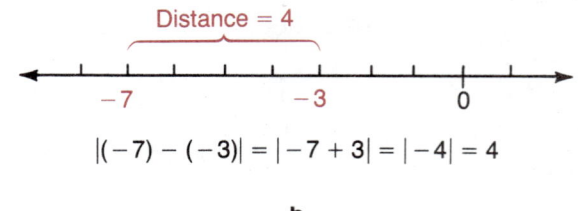

b.

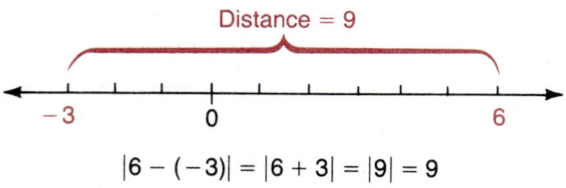

c.

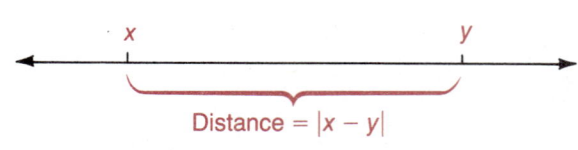

d.

Figure 1.9

Section 1.2 Order and Absolute Value

The distance from x to y should be no greater than the distance found by detouring through some third point z. This property is illustrated in Figure 1.10.

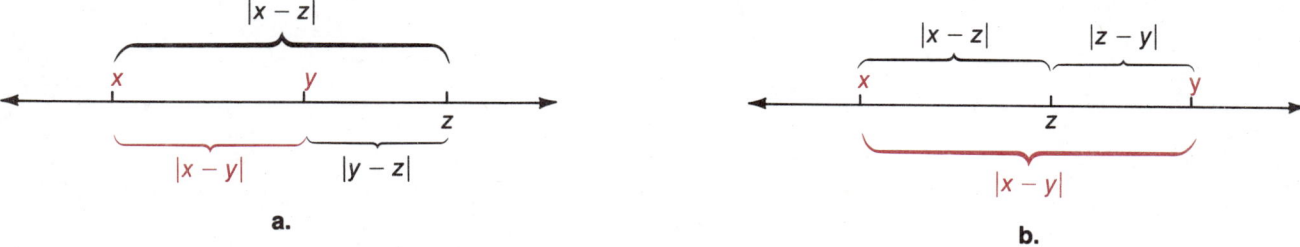

Figure 1.10

We see that

$$|x - y| < |x - z| + |z - y| \quad \text{(In Figure 1.10a)}$$
$$|x - y| = |y - z| + |z - y| \quad \text{(In Figure 1.10b)}$$

In any case,

$$|x - y| \leq |x - z| + |z - y| \quad (1)$$

Equality (=) holds in (1) if and only if z is between x and y.

Letting $a = x - z$ and $b = z - y$, we have $a + b = (x - z) + (z - y) = x - y$. Substituting these into (1), we obtain

$$|a + b| \leq |a| + |b| \quad (2)$$

Each of the inequalities (1) and (2) is a version of the **triangle inequality.**

TRIANGLE INEQUALITIES
1. $|x - y| \leq |x - z| + |z - y|$
2. $|a + b| \leq |a| + |b|$

Equality (=) prevails in (2) if and only if a and b have the same sign.

EXAMPLE 3 Verify the triangle inequalities in each case.

a. $x = 7, y = 4, z = 2$
b. $x = 7, y = 4, z = 5$
c. $a = 5, b = 3$
d. $a = 5, b = -3$

SOLUTION

Figure 1.11

a.

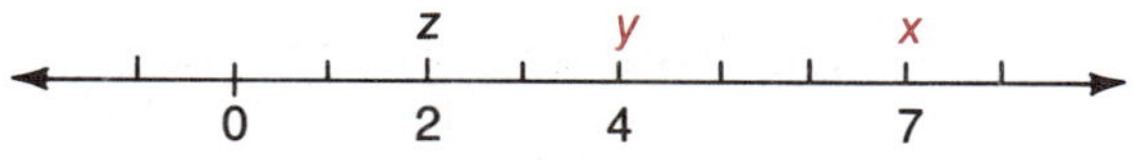

$$|x - y| = |7 - 4| = |3| = 3$$
$$|x - z| + |z - y| = |7 - 2| + |2 - 4|$$
$$= |5| + |-2| = 5 + 2 = 7$$

Note that $z = 2$ is not between $x = 7$ and $y = 4$ and that

$$|7 - 4| = 3 < 7 = |7 - 2| + |2 - 4|$$

b.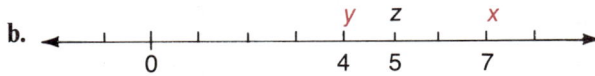

$$|x - y| = |7 - 4| = |3| = 3$$
$$|x - z| + |z - y| = |7 - 5| + |5 - 4|$$
$$= |2| + |1| = 2 + 1 = 3$$

Note that $z = 5$ is between $x = 7$ and $y = 4$ and that

$$|7 - 4| = 3 = |7 - 5| + |5 - 4|$$

c.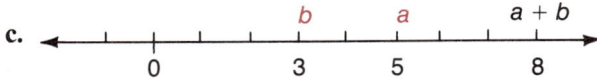

$$|a + b| = |5 + 3| = |8| = 8$$
$$|a| + |b| = |5| + |3| = 5 + 3 = 8$$

Note that $a = 5$ and $b = 3$ are both positive and that

$$|5 + 3| = 8 = |5| + |3|$$

d.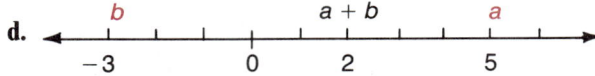

$$|a + b| = |5 + (-3)| = |5 - 3| = |2| = 2$$
$$|a| + |b| = |5| + |-3| = 5 + 3 = 8$$

Note that $a = 5$ and $b = -3$ have opposite signs and that

$$|5 + (-3)| = 2 < 8 = |5| + |-3|$$ ■

Section 1.2 Exercises

Express an order relation ($<$, $=$, $>$) *between the following numbers.*

1. 2, 7
2. 16, 5
3. -2, -6
4. 6, -2
5. 2, -6
6. -5, -4
7. -100, 17
8. 5, 5

Express an order relation between ax *and* bx *when a and b are as in Exercises 1–8 and* x *is given here for each pair respectively. For instance, if* $x = -3$, $a = 2$, $b = 4$, *then* $a < b$ *but* $ax = -6 > bx = -12$.

9. -1
10. 0
11. 2
12. 6
13. -2
14. -3
15. 3
16. -12

17. Write the absolute value of each of the numbers in Exercises 9–16.

18. For each of the pairs a, b in Exercises 1–8, verify that $|a + b| \leq |a| + |b|$ and that equality holds if and only if a and b have the same sign.

Use the inequality symbols ($<$, $>$, $\leq$, $\geq$) to express the following.

19. a is greater than or equal to 4.
20. b is less than or equal to -3.
21. a is nonnegative.
22. b is nonpositive.
23. x is strictly between 2 and 4.
24. y is strictly between 5 and 10.
25. a is strictly between 2 and -5.
26. b is strictly between 3 and -6.
27. r is strictly between -2 and -4.
28. s is strictly between -10 and -15.

Describe each of the following sets as an interval.

29. The set of all s such that $-3 < s \leq 5$.
30. The set of all t such that $-2 \leq t < 4$.
31. The set of all u such that $u > 10$.
32. The set of all v such that $v \leq 9$.
33. The set of all r such that $|r| < 2$.
34. The set of all x such that $|x| \leq 3$.
35. The set of all y such that $-3 \leq y \leq 5$.
36. The set of all z such that $-2 < z < -1$.
37. The set of all r such that $6 \geq r \geq -2$.
38. The set of all w such that $-3 \geq w \geq -10$.

Sketch the following intervals on the real number line.

39. $(-\infty, 2)$
40. $[3, 4]$
41. $(-2, 0]$
42. $(-\infty, 4]$
43. $(4, 6)$
44. $[-1, 5)$
45. $(3, \infty)$
46. $[1, \infty)$

Describe each of the following sets as an interval.

47. $(0, 2) \cap (1, 3)$
48. $(-1, 5) \cap (1, 8)$
49. $(-4, 2) \cup (-3, 5)$
50. $(4, 12) \cup (2, 8)$
51. $[3, 7) \cap (2, 6]$
52. $[-2, 8] \cap (-5, 0)$
53. $[4, 6] \cup (5, 7)$
54. $(-2, 3] \cup [2, 4)$
55. $[4, 6] \setminus (5, 7)$
56. $(-2, 3] \setminus [2, 4]$
57. $(1, 8) \setminus (-1, 5)$
58. $(-4, 2) \setminus [-3, 5]$

59. State the rules for inequalities in terms of the symbol $\leq$.

Arrange the following sets of numbers in order from lowest to highest. A bar indicates a repeating block of digits.

60. $-3, 5, -2, -6, 4$
61. $16, -12, -9, 42$
62. $\dfrac{1}{2}, -\dfrac{3}{4}, \dfrac{5}{6}, \dfrac{2}{3}$
63. $\dfrac{1}{4}, \dfrac{3}{8}, -\dfrac{1}{2}, -\dfrac{3}{5}$
64. $|-3|, -|5|, -(-2), |-6|, -4$
65. $|-16|, -|-12|, -9, |-15|$
66. $\sqrt{2}, -\sqrt{3}, -1.5, 1.414, 1.\overline{414}$
67. $4.8\overline{7}, 4.\overline{87}, 4.8, 4.87, 4.\overline{8}$
68. $2.\overline{367}, 2.3\overline{67}, 2.36\overline{7}, 2.367\overline{7}, 2.367$

Evaluate the following expressions for the given values of a and b.

69. $|2a - 3b|$, $a = -1$, $b = 2$
70. $|4b - 5a|$, $a = 2$, $b = -3$
71. $|2a| - 3|b|$, $a = -1$, $b = -2$
72. $|-3a| + |4b|$, $a = -3$, $b = 2$
73. $|a - b| - |b + 4a|$, $a = 2$, $b = 6$
74. $|2b - a| - |3a - 2b|$, $a = -6$, $b = 3$
75. $||2a + b| - |3b - a||$, $a = -3$, $b = 5$
76. $||b - 3a| - |4a + 6b||$, $a = 2$, $b = -1$

Write each of the following expressions without using the absolute value symbols.

77. $|x - 2|$, $x < 2$
78. $|x + 3|$, $x > -3$
79. $|3a - 9|$, $a > 3$
80. $|12 - 4y|$, $y > 4$
81. $|c^2 + 5|$
82. $|x^4 + 2|$
83. $|25 - x^2|$, $|x| \geq 5$
84. $|y^2 - 16|$, $|y| \leq 4$

22 Chapter 1 Basic Algebra

Find the distance between a *and* b *as given.*

85. $a = 6, b = 2$
86. $a = 8, b = 1$
87. $a = -3, b = -12$
88. $a = -4, b = -1$
89. $a = 5, b = -2$
90. $a = -6, b = 4$
91. $a = -15, b = 12$
92. $a = 4, b = -8$

93. If the base of a triangle is 10 feet and its area is between 60 and 70 square feet, describe its set of possible altitudes as an interval.

94. If Bill owns $\frac{2}{5}$ of his company's stock and Janice owns $\frac{3}{7}$ of her company's stock, who owns the larger share?

Calculator Exercises

95. Potato Ridgies come in several different sized packages. A 12-ounce package sells for $1.59, 10 ounces for $1.29, and $7\frac{1}{2}$ ounces for $1.09. Which is the best buy?

96. Generic label corn sells in 370-gram packages for 59¢, and Yellow Valley corn comes in 454-gram packages for 75¢. Which is more expensive?

97. Super Suds detergent in the economy size 5-pound package is priced at $5.23. The giant size $3\frac{1}{2}$-pound box sells for $3.59. The large size 24-ounce package is $1.59. Which package is the best buy?

98. Gasoline purchases are usually expressed to the nearest tenth of a gallon. Odometer readings are given to the nearest tenth of a mile. The previous odometer reading on Eileen's car was 23,468.7. She now purchases 17.3 gallons at an odometer reading of 23,684.3. Indicate her possible miles-per-gallon ratings in terms of inequalities.

Section 1.3

Integral Exponents

For each real number x and each *positive* integer n, the nth power of x is defined as the product of x with itself n times:

> **DEFINITION**
>
> $$x^n = \underbrace{x \cdot \cdots \cdot x}_{n \text{ times}}$$
>
> x is called the **base** and n is called the **exponent.** We call x^n the nth **power** of x.

In particular,

$$x^1 = x$$
$$x^2 = x \cdot x$$
$$2^3 = 2 \cdot 2 \cdot 2 = 8$$

Rules for working with positive integer exponents follow. Note that no rules are given regarding sums and differences. Such relationships do not follow any pattern. For example,

$$1^2 + 2^2 \neq 3^2 \quad \text{since} \quad 1 + 4 \neq 9$$

and

$$2^2 + 2^3 \neq 2^5 \quad \text{since} \quad 4 + 8 \neq 32$$

RULES FOR (POSITIVE) INTEGER EXPONENTS

RULE	NUMERICAL EXAMPLE	ALGEBRAIC EXAMPLE
1. $x^m x^n = x^{m+n}$	1. $2^3 \cdot 2^2 = (2 \cdot 2 \cdot 2)(2 \cdot 2)$ $= 2^5 = 2^{3+2}$	1. $x^2 x^3 = (x \cdot x)(x \cdot x \cdot x)$ $= x^5$
2. $(x^m)^n = (x^n)^m = x^{mn}$	2. $\left.\begin{array}{l}(2^2)^3 = 2^2 \cdot 2^2 \cdot 2^2 = 2^{2+2+2} \\ (2^3)^2 = 2^3 \cdot 2^3 = 2^{3+3}\end{array}\right\} = 2^6 = 2^{2 \cdot 3}$	2. $(x^3)^2 = x^3 x^3 = x^6$
3. a. $(xy)^n = x^n y^n$	3. a. $(2 \cdot 3)^2 = (2 \cdot 3)(2 \cdot 3)$ $= (2 \cdot 2)(3 \cdot 3) = 2^2 \cdot 3^2$	3. a. $(xy)^2 = (xy)(xy)$ $= (xx)(yy) = x^2 y^2$
b. $\left(\dfrac{x}{y}\right)^n = \dfrac{x^n}{y^n}$, $(y \neq 0)$	b. $\left(\dfrac{4}{2}\right)^3 = \dfrac{4}{2} \cdot \dfrac{4}{2} \cdot \dfrac{4}{2} = \dfrac{4 \cdot 4 \cdot 4}{2 \cdot 2 \cdot 2} = \dfrac{4^3}{2^3}$	b. $\left(\dfrac{x}{y}\right)^2 = \dfrac{x}{y} \cdot \dfrac{x}{y} = \dfrac{x \cdot x}{y \cdot y} = \dfrac{x^2}{y^2}$
4. $\dfrac{x^m}{x^n} = \begin{cases} x^{m-n} & \text{if } m > n \\ 1 & \text{if } m = n \\ \dfrac{1}{x^{n-m}} & \text{if } n > m \end{cases}$, $(x \neq 0)$	4. a. $\dfrac{5^3}{5^2} = \dfrac{5 \cdot 5 \cdot 5}{5 \cdot 5} = 5 = 5^1 = 5^{3-2}$ b. $\dfrac{3^2}{3^2} = \dfrac{3 \cdot 3}{3 \cdot 3} = 1$ c. $\dfrac{6^3}{6^5} = \dfrac{6 \cdot 6 \cdot 6}{6 \cdot 6 \cdot 6 \cdot 6 \cdot 6} = \dfrac{1}{6 \cdot 6} = \dfrac{1}{6^2} = \dfrac{1}{6^{5-3}}$	4. a. $\dfrac{x^3}{x^2} = \dfrac{x \cdot x \cdot x}{x \cdot x} = x = x^1 = x^{3-2}$ b. $\dfrac{x^2}{x^2} = 1$ c. $\dfrac{x^2}{x^4} = \dfrac{x \cdot x}{x \cdot x \cdot x \cdot x} = \dfrac{1}{x^2} = \dfrac{1}{x^{4-2}}$

EXAMPLE 1

Write each of the following using exponential notation.

a. $x \cdot x \cdot x \cdot x$
b. $(a+b) \cdot (a+b) \cdot (a+b)$

SOLUTION

a. $x \cdot x \cdot x \cdot x = x^4$
b. $(a+b) \cdot (a+b) \cdot (a+b) = (a+b)^3$ ∎

Hierarchy of Operations

Exponentiation, or *raising to a power*, takes precedence over ordinary multiplications and divisions. As usual, parentheses are used to impose another order of operations. For example

$$3x^2 = 3 \cdot x^2 = 3(x^2) = 3 \cdot x \cdot x$$

whereas

$$(3x)^2 = 3x \cdot 3x = 9 \cdot x \cdot x = 9x^2$$

Also note that

$$(-3)^2 = (-3)(-3) = 9$$

but
$$-3^2 = -3 \cdot 3 = -9$$

The hierarchy of operations is summarized briefly here; also see Section 1.1.

ORDER OF OPERATIONS

1. Expressions in parentheses.
2. Exponentiation, or raising to a power.
3. Multiplications and divisions from left to right.
4. Additions and subtractions from left to right.

EXAMPLE 2

Use the rules for (positive) integer exponents to simplify each of the following.

a. $\dfrac{(-2)^5}{(-2)^3}$ **b.** $\dfrac{(-2)^5}{2^3}$ **c.** $\dfrac{3x^2y^3z}{2x^2yz^2}$ **d.** $\dfrac{(k^2m^2)^3}{(k^3m)^2}$ **e.** $[(p^2q)^3(qp^3)^2]^4$

SOLUTION

a. $\dfrac{(-2)^5}{(-2)^3} = (-2)^{5-3} = (-2)^2 = 4$

b. $\dfrac{(-2)^5}{2^3} = \dfrac{(-2)(-2)(-2)(-2)(-2)}{2 \cdot 2 \cdot 2} = \dfrac{-32}{8} = -4$

Note that $\dfrac{(-2)^5}{2^3} \neq (-2)^{5-3} = (-2)^2 = 4$

c. $\dfrac{3x^2y^3z}{2x^2yz^2} = \dfrac{3}{2} \cdot \dfrac{x^2}{x^2} \cdot \dfrac{y^3}{y} \cdot \dfrac{z}{z^2} = \dfrac{3}{2} \cdot 1 \cdot y^2 \cdot \dfrac{1}{z} = \dfrac{3y^2}{2z}$

d. $\dfrac{(k^2m^2)^3}{(k^3m)^2} = \dfrac{(k^2)^3(m^2)^3}{(k^3)^2m^2} = \dfrac{k^6m^6}{k^6m^2} = 1 \cdot m^{6-2} = m^4$

e. $[(p^2q)^3(qp^3)^2]^4 = [(p^6q^3)(q^2p^6)]^4 = [p^{12}q^5]^4 = p^{48}q^{20}$ ∎

To define x^0 in a manner consistent with the properties of exponents, we must have

$$x^m \cdot x^0 = x^{m+0} = x^m \qquad \text{(Rule 1)}$$

and

$$\dfrac{x^m}{x^0} = x^{m-0} = x^m \qquad \text{(Rule 4)}$$

That is, multiplying or dividing by x^0 must be equivalent to multiplying or dividing by 1. Therefore, we have the following definition.

DEFINITION

$$x^0 = 1 \qquad \text{for any real number } x \neq 0$$

However, 0^0 is undefined.

The laws of exponents can also be extended to negative exponents. Then we must have, for example,

$$x^5 x^{-2} = x^{5+(-2)} = x^{5-2} = x^3 = \frac{x^5}{x^2}$$

Thus, multiplication by x^{-2} should be equivalent to dividing by x^2. Consequently, x^{-n} is defined as follows for any integer $n > 0$.

DEFINITION

$$x^{-n} = \frac{1}{x^n} \quad \text{for any real number } x \neq 0$$

Using these definitions of x^0 and x^{-n}, we absorb Rule 4 for exponents into Rule 1: $x^m x^n = x^{m+n}$. Specifically,

$$\frac{x^m}{x^n} = x^m x^{-n} = x^{m+(-n)} = x^{m-n}$$

$$= \begin{cases} x^{m-n} & \text{if } m > n \\ x^0 = 1 & \text{if } n = m \\ x^{-(n-m)} = \dfrac{1}{x^{n-m}} & \text{if } n > m \end{cases}$$

Furthermore, *all the rules for positive integer exponents mentioned earlier hold for negative integers as well.* For instance, $(x^m)^{-n} = x^{m(-n)}$ since

$$(x^m)^{-n} = \frac{1}{(x^m)^n} = \frac{1}{x^{mn}} = x^{-mn} = x^{m(-n)}$$

We have defined $x^{-2} = 1/x^2$; that is, *a negative exponent on a factor in the numerator is equivalent to a positive exponent on that factor in the denominator.* The opposite is also true: *A negative exponent in the denominator is equivalent to a positive exponent in the numerator.* For example,

$$\frac{1}{x^{-10}} = \frac{1}{\frac{1}{x^{10}}} = 1 \cdot \frac{x^{10}}{1} = x^{10}$$

Thus, $x^{-n} = 1/x^n$ (when $x \neq 0$) regardless of whether n is positive, negative, or zero.

CAUTION

$$\frac{x^{-3} + 2}{4} \neq \frac{2}{4x^3}$$

since x^{-3} is *not a factor* of the numerator. $\left(\text{Substitute } x = 1 \text{ to obtain } \dfrac{3}{4} \neq \dfrac{2}{4}.\right)$

EXAMPLE 3   Simplify the following and express the answer with nonnegative exponents.

a. $\dfrac{x^4}{x^{10}}$ **b.** $(x^{-2})^{-3}$ **c.** $\dfrac{12a^{-3}b^4}{4ab^{-2}}$ **d.** $x^{-1} + y^{-2}$

SOLUTION

a. $\dfrac{x^4}{x^{10}} = x^{4-10} = x^{-6} = \dfrac{1}{x^6}$

b. $(x^{-2})^{-3} = x^{(-2)(-3)} = x^6$

An alternate solution is

$$(x^{-2})^{-3} = \dfrac{1}{(x^{-2})^3} = \dfrac{1}{\left(\dfrac{1}{x^2}\right)^3} = \dfrac{1}{\dfrac{1}{x^6}} = 1 \cdot \dfrac{x^6}{1} = x^6$$

c. $\dfrac{12a^{-3}b^4}{4ab^{-2}} = \dfrac{12b^4b^2}{4aa^3} = \dfrac{3b^6}{a^4}$

d. $x^{-1} + y^{-2} = \dfrac{1}{x} + \dfrac{1}{y^2} = \dfrac{y^2}{xy^2} + \dfrac{x}{xy^2} = \dfrac{y^2 + x}{xy^2}$ ■

Scientific Notation

Scientists often use very small or very large numbers. For example, the frequency of some X rays is 300,000,000,000,000,000,000 cycles per second; the charge on an electron is 0.00000000000000000001602 coulombs. To facilitate working with such numbers, scientists express them as some number times a power of 10. *Multiplication or division by 10 moves the decimal point one place to the right or left, respectively.* We can write $300{,}000{,}000{,}000{,}000{,}000{,}000 = 3(10^{20})$ and we can write $0.00000000000000000001602 = 1.602(10^{-19})$. Other examples of large or small numbers follow.

Frequency of TV waves	$3(10^8)$ cycles per second
Threshold of human hearing	10^{-2} watts per square meter or 0 decibels
Height of Mt. Everest	$2.52(10^6)$ centimeters

Any number expressed in the form

$$b \cdot 10^k, \quad 1 \leq b < 10$$

is said to be written in **scientific notation.** Scientific notation enables us to compare numbers very quickly by looking at the respective powers of 10. Calculations involving such numbers are also simplified by using the laws of exponents on the powers of 10. Several examples of scientific notation are given in the following table.

DECIMAL NUMBER	SCIENTIFIC NOTATION
23,600	$2.36(10^4)$
0.00314	$3.14(10^{-3})$
451.63	$4.5163(10^2)$

Section 1.3 Integral Exponents

CALCULATOR COMMENTS

Many of the better calculators are programmed to work in scientific notation. In fact, some will even convert automatically to scientific notation when a certain number of digits are introduced into a calculation. Calculators usually suppress the base 10 and display scientific notation in one of the following forms.

NUMBER	SCIENTIFIC NOTATION	CALCULATOR DISPLAY (ONE FORM)	CALCULATOR DISPLAY (ANOTHER FORM)
236100	$2.361(10^5)$	2.361 E 5	2.361 5

See your calculator manual for instructions on entering numbers in scientific notation into the calculator.

Even the simplest calculators can be used to work with scientific notation. The powers of 10 can be manipulated by hand using the laws of exponents; the calculator need be used only to do the "hard arithmetic." For instance,

$$\frac{(0.000064)(637,000,000)}{(0.00000562)} = \frac{(6.4)(10^{-5})(6.37)(10^8)}{(5.62)(10^{-6})}$$
$$= \frac{(6.4)(6.37)}{5.62} 10^{-5} 10^8 10^6$$
$$= \frac{(6.4)(6.37)}{5.62} 10^9 \quad \text{Use a calculator.}$$
$$\approx (7.254)10^9$$
$$= 7,254,000,000$$

Significant Digits

Most physical measurements are approximations. For example, your height might be indicated on your driver's license as 5 feet 9 inches, when in reality you may be closer to 5 feet $9\frac{1}{8}$ inches tall. Decimal notation can be used to convey the accuracy of a given measurement. If we write $x \approx 13.76$ we mean that x has been rounded or approximated correctly to the nearest hundredth; that is,

$$x \approx 13.76 \quad \text{means} \quad 13.755 \leq x \leq 13.765$$

We say that x has four **significant digits** (1, 3, 7, and 6). If we round to $x \approx 14$, we say that x has two significant digits (1 and 4).

But how many significant digits are there in $y = 542,000,000$? Is this number correct only to the nearest million? We can use scientific notation to indicate the accuracy of a number by agreeing to include all signficant digits (and only significant digits) in the b part of $b(10^n)$. Thus, if we write $y = 5.42(10^8)$, we mean that the 2 is the last significant digit; that is, y is correct to the nearest million. But if we write $y = 5.42000(10^8)$ we mean that the third 0 is the last significant digit; that is, y is correct to the nearest thousand.

When doing arithmetic operations on several numbers, we cannot expect the answer to be more accurate than the numbers used in the calculation. Rules for rounding the answer are given here.

> ROUNDING TO SIGNIFICANT DIGITS
>
> 1. *Addition and subtraction.* Convert the numbers to decimal form and round the answer to the number of accurate decimal positions in the least accurate of the numbers used in the computation.
> 2. *Multiplication and division.* Round the answer to the least number of significant digits in the numbers used in the computation.

In item 1 of the preceding box we use *accurate decimal positions* while in item 2 we use significant digits. The difference is illustrated by writing

$$514.63 = 5.1463 \times 10^2$$

The number 514.63 is *accurate to two decimal places;* the number 5.1463×10^2 has five significant digits.

EXAMPLE 4

Consider the numbers

$$x \approx 2.78(10^{-2}) \quad \text{and} \quad y \approx 3.16455(10^3)$$

Find each value.

a. $x + y$ **b.** $y - x$ **c.** $x \cdot y$ **d.** $\dfrac{y}{x}$

Express each answer with the appropriate number of significant digits.

SOLUTION

a. In decimal form, we have

$$x \approx 0.0278 \quad \text{and} \quad y \approx 3164.55$$

Then $x + y \approx 3164.5778 \approx 3164.58.$
The answer must be rounded to 3164.58 since y is accurate only to the nearest hundredth.

b. $y - x \approx 3164.5222 \approx 3164.52$

Again, the answer must be rounded off to the nearest hundredth.

c. $x \cdot y \approx 2.78(10^{-2}) \cdot (3.16455)(10^3)$
   $= (2.78)(3.16455)(10^1)$
   $= 87.97449$
   $= 8.797449(10^1)$

But the answer must be expressed with only three significant digits since x was given with only three significant digits. Thus, we write

$$x \cdot y \approx 8.80(10^1) = 88.0$$

Writing 88.0 instead of 88 indicates three significant digits and also accuracy to one decimal place.

d. $\dfrac{y}{x} \approx \dfrac{3.16455(10^3)}{2.78(10^{-2})} = \dfrac{3.16455}{2.78}(10^5) \approx 1.138(10^5)$

Again this must be rounded:

$$\dfrac{y}{x} \approx 1.14(10^5)$$

Section 1.3 Exercises

Write the following using exponential notation.

1. $a \cdot a \cdot a \cdot a \cdot a$
2. $b \cdot b \cdot b$
3. $(x - y)(x - y)(x - y)(x - y)$
4. $(r + s + t)(r + s + t)(r + s + t)$
5. $\dfrac{(u + v)(u + v)(u + v)}{(w + z)(w + z)}$
6. $\dfrac{(m - n)(m - n)}{(a + b)(a + b)(a + b)(a + b)}$

Simplify the following, eliminating parentheses and negative exponents, if any.

7. $3^2 \cdot 3^3$
8. $3^{-2}(-3)^3$
9. $2^3 \cdot 2^4$
10. $3^2 \cdot 2^4$
11. $3^{-2}2^{-4}$
12. $\dfrac{5^5}{5^4}$
13. $\dfrac{5^5}{(-5)^{-4}}$
14. $\dfrac{(-3)^4}{(-3)^5}$
15. $\dfrac{(-2)^4}{(-2)^3}$
16. $\dfrac{-3^4}{(-3)^5}$
17. $\dfrac{-2^4}{(-2)^3}$
18. $(3^{-2})^3$
19. $(2^3)^{-2}$
20. $(-3)^3(-3)^2$
21. $(-2)^2(-2)^3$
22. $(-3^3)(-3^2)$
23. $(-2^2)(-2^3)$
24. $(-3^2)^3$
25. $(-2^3)^2$
26. $(-3^3)^2$
27. $(-2^2)^3$
28. $[(-3)^2]^3$
29. $[(-2)^3]^2$
30. $[(-3)^3]^2$
31. $[(-2)^2]^3$
32. $-[(3^2)^3]$
33. $y^5 y^{12}$
34. $t^{-7} \cdot t^5$
35. $x^{13} \cdot x^{17}$
36. $u^{-13} u^{-4}$
37. $\dfrac{a^{10}}{a^5}$
38. $\dfrac{b^{12}}{b^3}$
39. $\dfrac{xy^3}{x^2 y}$
40. $\dfrac{u^2 v}{uv^2}$
41. $\dfrac{(x^{-2}y)}{xy^{-2}}$
42. $(3x^2 y)(2y^3)$
43. $(2xy^2 z)(6x^3 yz^2)$
44. $(-2ax^{-2}b)(4a^{-1}b^{-2}x^{-2})$
45. $(7a^3 b)(-2a^{-7}b)$
46. $(t^2)^3$
47. $(x^6)^{-2}$
48. $(kL^2 m^3)^4$
49. $(xy^2 z)^3 (x^2 yz^2)^2$
50. $(x^2 y^3 z^5)^4 (z^3 x^2 y)^3$
51. $27x^{-2}(u^{-73} v^{2900} w^{163})^0$
52. $(2x^{-2}y)^{-3}$
53. $(a^{-3}b^{-2})^{-3}$
54. $(xy^2 z)^{-3}(x^{-2}yz^2)^{-2}$
55. $\dfrac{(x^2 y^{-4})^{-3}}{(x^{-3}y^{-2})^2}$
56. $[(ab^2 c)^2 (c^2 ba^3)^4]^3$
57. $[(x^{-2}y)^3(y^{-1}x^3)]^{-4}$
58. $\dfrac{(14s^{-2}t^{-3})}{(7s^{-4}t^{-4})}$
59. $\dfrac{(xy^2 z)^{-3}}{(x^{-2}yz^2)^{-2}}$
60. $\dfrac{(q^2 r^3)^4}{(qr^5)^6}$
61. $\left(\dfrac{2s^2 t^3}{4st^2}\right)^3$
62. $\left(\dfrac{3xy^2}{2z}\right)^3 \cdot \left(\dfrac{4x^2 z}{y^4}\right)^2$
63. $\dfrac{(r^2 s^{-2})^{-3}}{(r^{-3}s)^2}$
64. $[(2x^{-2}y)^{-3}(xy^2 z^{-3})^{-4}]^2$
65. $[(ab^{-2}c^{-1})^{-2}(c^{-2}ba^3)^4]^{-3}$
66. $\left[\dfrac{(3^{-2}rs^{-4})^2}{(2^2 r^{-2}s^2)^3}\right]^{-2}$
67. $\left[\dfrac{(x^2 y^{-4}z^3)^{-2}}{(x^{-5}y^3 z^2)^{-1}}\right]^{-2}$
68. $3x^{-1} - 2y^{-2}$
69. $x^{-2} + 2y^{-1}$

70. $\dfrac{2}{t^{-1}} + \dfrac{3t^{-1}}{t^{-2}}$

71. $\dfrac{2x^{-1}}{x^{-2}} + \dfrac{4x^{-3}}{x^{-5}}$

72. $xy^2(x^{-3} + y^{-2})$

73. $(ab)^2(2a^{-1} + b^{-1})$

Verify the following relationships.

74. $2^2 + 3^2 \neq 5^2$

75. $3^3 - 2^3 \neq (3-2)^3$

76. $2^2 - 3^2 \neq (-1)^2$

77. $3^2 + 3^2 \neq 6^2$

78. $2^2 + 2^3 \neq 2^5$

79. $(2+3)^{-2} \neq 2^{-2} + 3^{-2}$

80. $(x+y)^{-1} \neq x^{-1} + y^{-1}$

81. A googol is a 1 followed by 100 zeros, and a googolplex is a 1 followed by a googol of zeros. Express these numbers as powers of 10.

Express each of the following in decimal notation.

82. $2.634(10^{-7})$

83. $1.73215(10^5)$

84. $7.41532(10^8)$

85. $3.512(10^{-4})$

86. $5.5(10^{15})$, the surface of the earth in square feet.

87. $9.11(10^{-31})$, the mass of an electron in kilograms.

88. $1.67(10^{-27})$, the mass of a proton in kilograms.

89. $3(10^{10})$, the speed of light in centimeters per second.

Express each of the following in scientific notation.

90. 26

91. 2463

92. 230,000,000,000

93. 143.17

94. 0.023

95. 0.00132

96. 0.000341

97. 0.0000000000000712

Perform the following calculations by first converting the numbers to scientific notation. Express your answer in scientific notation and in terms of decimals.

98. $\dfrac{(2000)(0.00009)}{0.03}$

99. $\dfrac{(300,000)(0.0000006)}{(0.0000000015)(40,000)}$

100. $\dfrac{(250,000,000)(63,000)}{(0.0007)(0.09)(500)}$

101. $\dfrac{(0.000004)(600,000)}{(0.004)(0.003)}$

Calculator Exercises

Express each of the following in scientific notation.

102. The number of seconds in a life span of 70 years.

103. The surface area of the earth in acres and in hectares (1 acre = 43,560 square feet, 1 hectare = 10,000 square meters). See Exercise 86 and use 1 meter ≈ 39.37 inches.

Find the indicated value and express the answer with the appropriate number of significant digits.

104. $1.34(10^{-2}) + 1.82165(10^2)$

105. $3.148(10^{-3}) - 1.21876(10^{-2})$

106. $6.7328(10^1) - 4.9649(10^{-1})$

107. $2.793805(10^2) + 1.826745(10^3)$

108. $[3.841(10^5)][1.4756(10^7)]$

109. $[2.1775(10^4)][4.742(10^{-6})]$

110. $[8.7747(10^{-8})][7.699(10^7)]$

111. $\dfrac{[1.618(10^{-3})][2.68(10^4)]}{[6.5768(10^6)]}$

112. $\dfrac{[4.7001(10^4)][2.209(10^{-5})]}{[4.88(10^{-2})]}$

113. $\dfrac{[2.382(10^5)]}{[7.1784(10^{-3})][5.67258(10^{-2})]}$

114. Scientists tell us that light rays travel at the rate of 186,000 miles per second. Celestial distances are so great that astronomers use a light-year as the unit of measurement; 1 light-year is the distance traveled by a ray of light in 1 year.

 a. Use scientific notation to express the number of miles in 1 light-year.

 b. A furlong is $\tfrac{1}{8}$ mile; a fortnight is 14 days. Express the speed of light in furlongs per fortnight.

115. If the sun suddenly burned out, we would not know it for 8 minutes. How far is the sun from us? (See Exercise 114.)

116. Our solar system is located in a galaxy known as the Milky Way. The diameter of the Milky Way is 100,000 light-years. Its thickness is 10,000 light-years. Express these dimensions as miles in scientific notation.

117. It is possible to view certain events today that happened a very long time ago. For example, astronomers view events in the Andromeda Galaxy as they happened 750,000 years ago. How far away is the Andromeda Galaxy in miles? Write your answer in scientific notation.

Section 1.4
Radicals

Biophysicists tell us that the walking speed of an animal is related to the square root of the length of its legs. (With tongue in cheek, one might be tempted to say that the walking speed is related to the number of feet per leg.) With legs L feet long, an animal might be capable of walking $7\sqrt{L}/3$ miles per hour. According to this, a man with 36-inch legs should be capable of walking $7\sqrt{3}/3 \approx 4$ miles per hour.

Square roots (and other roots) occur in other areas of science and engineering as well as in biophysics. In this section, we shall consider sums, differences, products, and quotients of expressions involving roots.

An **nth root** of x is a number that when multiplied by itself n times yields x. (If $n = 2$ or 3, we say "square root" or "cube root," respectively.)

Every positive real number has two real square roots. For instance, the square roots of 9 are ± 3 since $3^2 = 9$ and $(-3)^2 = 9$. On the other hand, negative numbers have no real square roots since $a^2 \geq 0$ for every real number a.

The same comments hold regarding nth roots whenever n is even. For instance, 16 has two real fourth roots: 2 and -2. But -8 has no real square roots or fourth roots, since there is no real number x for which $x^2 = -8$ or $x^4 = -8$. -8 does have precisely one real cube root: -2, since $(-2)^3 = -8$. In fact, every real number has precisely one real nth root for each odd integer $n > 0$.

The symbol $\sqrt[n]{x}$ is used to denote the nth root of x (the positive root if there are two roots). Since there are two real nth roots of x when n is even and $x \geq 0$, the positive root, denoted by $\sqrt[n]{x}$, is called the **principal nth root.** The two real nth roots are then given by $\pm\sqrt[n]{x}$. In particular, $\sqrt[2]{x}$ is used to denote the positive (principal) square root of a nonnegative number x. (We generally write $\sqrt{\ }$ instead of $\sqrt[2]{\ }$, but the index n must be used for other roots such as the cube root of x, denoted by $\sqrt[3]{x}$.)

The principal square root of 9 is $\sqrt{9} = 3$. The two square roots of 9 are denoted $\sqrt{9} = 3$ and $-\sqrt{9} = -3$; $\sqrt{9} = 3$ just as if there were a plus sign in front of the radical.

In the notation $\sqrt[n]{x}$, the symbol $\sqrt{\ }$ is called a **radical,** n is called the **index** of the radical, and x is called the **radicand.**

DEFINITION

$$\sqrt[n]{x} = r \quad \text{means} \quad x = r^n$$

1. If n is even and
   a. $x \geq 0$, $\sqrt[n]{x}$ denotes the nonnegative real nth root of x;
   b. $x < 0$, $\sqrt[n]{x}$ is undefined in the set of real numbers.
2. If n is odd, $\sqrt[n]{x}$ denotes the unique real nth root of x.

EXAMPLE 1

Evaluate

a. $\sqrt{36}$ b. $\sqrt[3]{-27}$ c. $\sqrt{-25}$

SOLUTION

a. $\sqrt{36} = 6$ since $6^2 = 36$
b. $\sqrt[3]{-27} = -3$ since $(-3)^3 = -27$

c. $\sqrt{-25}$ **is not defined in the real number system** since there is no $x \in R$ for which $x^2 = -25$. ∎

RULES FOR RADICALS

RULE	EXAMPLE
1. $\sqrt[n]{x^n} = \begin{cases} \|x\| & \text{if } n \text{ is even} \\ x & \text{if } n \text{ is odd} \end{cases}$	1. $\begin{cases} \sqrt[4]{(-2)^4} = \sqrt[4]{16} = 2 = \|-2\| \\ \sqrt[3]{(-2)^3} = \sqrt[3]{-8} = -2 \end{cases}$
2. $\begin{cases} \textbf{a.}\ \sqrt[n]{a} \cdot \sqrt[n]{b} = \sqrt[n]{ab} \\ \textbf{b.}\ \dfrac{\sqrt[n]{a}}{\sqrt[n]{b}} = \sqrt[n]{\dfrac{a}{b}},\quad b \neq 0 \end{cases}$	2. $\begin{cases} \textbf{a.}\ \sqrt{4}\sqrt{9} = 2 \cdot 3 = 6 = \sqrt{36} = \sqrt{4 \cdot 9} \\ \textbf{b.}\ \dfrac{\sqrt{100}}{\sqrt{4}} = \dfrac{10}{2} = 5 = \sqrt{25} = \sqrt{\dfrac{100}{4}} \end{cases}$
3. $\sqrt[k]{\sqrt[n]{x}} = \sqrt[kn]{x}$	3. $\sqrt[3]{\sqrt[2]{64}} = \sqrt[3]{8} = 2 = \sqrt[6]{64}$
4. Only identical radicals can be collected.	4. $3\sqrt{2} + 5\sqrt{2} = 8\sqrt{2}$ but $\sqrt{2} + \sqrt{3} \neq \sqrt{5}$

These rules are valid whenever the roots in question are defined. Thus, for example, in Rule 2a we do *not* try to equate $\sqrt{(-3)(-12)} = \sqrt{36} = 6$ and $\sqrt{-3}\sqrt{-12}$ since $\sqrt{-3}$ and $\sqrt{-12}$ are not defined as real numbers.

We should emphasize Rule 1 when n is even, since the radical denotes the *positive* root when n is even.

$$\boxed{\sqrt{x^2} = |x|}$$

In particular

$$\sqrt{(-3)^2} = \sqrt{9} = 3 = |-3|$$

To establish Rule 2a, we write

$(\sqrt[n]{a}\sqrt[n]{b})^n = (\sqrt[n]{a})^n(\sqrt[n]{b})^n$ (Rule 3a, Section 1.3)

$\quad = ab$ (Definition of $\sqrt[n]{\ }$)

Thus $\sqrt[n]{a}\sqrt[n]{b} = \sqrt[n]{ab}$, since when $\sqrt[n]{a}\sqrt[n]{b}$ is raised to the nth power, the result is ab. Rule 2b can be established in a similar way.

To establish Rule 3, we write

$(\sqrt[k]{\sqrt[n]{x}})^{kn} = [(\sqrt[k]{\sqrt[n]{x}})^k]^n$ (Rule 2, Section 1.3)

$\quad = [\sqrt[n]{x}]^n$ (Definition of $\sqrt[k]{\ }$)

$\quad = x$ (Definition of $\sqrt[n]{\ }$)

Thus $\sqrt[k]{\sqrt[n]{x}} = \sqrt[kn]{x}$, since when $\sqrt[k]{\sqrt[n]{x}}$ is raised to the knth power, the result is x.

Section 1.4 Radicals 33

> **DEFINITION**
> A radical is said to be **simplified** if the following hold.
> 1. The radicand has no exponent $\geq$ the index of the radical.
> 2. There are no fractions under the radical sign.
> 3. There are no radicals in the denominator.

EXAMPLE 2 Simplify the following radicals.

a. $\sqrt{3}\sqrt{12}$ b. $\sqrt[3]{\dfrac{27}{8}}$ c. $\sqrt[3]{32}$

d. $\sqrt[6]{8}$ e. $3\sqrt{5} + 2\sqrt{5}$ f. $2\sqrt{12} - 3\sqrt{48}$

SOLUTION

a. $\sqrt{3}\sqrt{12} = \sqrt{3 \cdot 12}$ (Rule 2a)
$= \sqrt{36}$
$= 6$

b. $\sqrt[3]{\dfrac{27}{8}} = \dfrac{\sqrt[3]{27}}{\sqrt[3]{8}}$ (Rule 2b)
$= \dfrac{3}{2}$

c. $\sqrt[3]{32} = \sqrt[3]{8 \cdot 4}$
$= \sqrt[3]{8} \cdot \sqrt[3]{4}$ (Rule 2a)
$= 2\sqrt[3]{4}$

d. $\sqrt[6]{8} = \sqrt[2]{\sqrt[3]{8}}$ (Rule 3)
$= \sqrt{2}$

e. $3\sqrt{5} + 2\sqrt{5} = (3 + 2)\sqrt{5}$ (Distributive law)
$= 5\sqrt{5}$

f. $2\sqrt{12} - 3\sqrt{48} = 2\sqrt{4}\sqrt{3} - 3\sqrt{16}\sqrt{3}$ (Rule 2a)
$= 2 \cdot 2\sqrt{3} - 3 \cdot 4\sqrt{3}$
$= 4\sqrt{3} - 12\sqrt{3}$
$= (4 - 12)\sqrt{3}$ (Distributive law)
$= -8\sqrt{3}$ ∎

EXAMPLE 3 Simplify the following radicals.

a. $\sqrt{64x^2y^4z^5}$ b. $\sqrt[3]{-27x^3y^6z^4}$ c. $\sqrt[3]{8}\sqrt[4]{y^{12}}\sqrt{x^3}$

SOLUTION

a. $\sqrt{64x^2y^4z^5} = \sqrt{64}\sqrt{x^2}\sqrt{y^4}\sqrt{z^4}\sqrt{z}$ (Rule 2a)
$= 8|x|y^2z^2\sqrt{z}$ (Rule 1)

Note here that $\sqrt{x^2} = |x|$ but $\sqrt{y^4} = |y^2| = y^2$ and $\sqrt{z^4} = |z^2| = z^2$, since $y^2 \geq 0$ and $z^2 \geq 0$.

b. $\sqrt[3]{-27x^3y^6z^4} = \sqrt[3]{-27}\sqrt[3]{x^3}\sqrt[3]{y^6}\sqrt[3]{z^3}\sqrt[3]{z}$ (Rule 2a)
$= -3x\sqrt[3]{(y^2)^3}z\sqrt[3]{z}$ (Rules 1 and 2, Section 1.3)
$= -3xy^2z\sqrt[3]{z}$ (Rule 1)

c. $\sqrt[3]{8\sqrt[4]{y^{12}}\sqrt{x^3}} = \sqrt[3]{8}\sqrt[3]{\sqrt[4]{y^{12}}}\sqrt[3]{\sqrt{x^3}}$ (Rule 2a)
$= 2\sqrt[12]{y^{12}}\sqrt[6]{x^3}$ (Rule 3)
$= 2|y|\sqrt[2]{\sqrt[3]{x^3}}$ (Rules 1 and 3)
$= 2|y|\sqrt{x}$ (Definition of $\sqrt[3]{\ }$)

■

EXAMPLE 4

Simplify the following expressions.

a. $3\sqrt{2} + 4\sqrt{8} - \sqrt{32}$

b. $\dfrac{\sqrt[4]{32x^3} - \sqrt[4]{8x^5}}{\sqrt[4]{2x}}$

c. $(4\sqrt{5} - \sqrt{6})(2\sqrt{5} + 3\sqrt{6})$

SOLUTION

a. We cannot combine $\sqrt{2}$, $\sqrt{8}$, and $\sqrt{32}$ directly. (See Rule 4.) However, $\sqrt{8}$ and $\sqrt{32}$ can be expressed in terms of $\sqrt{2}$, as follows.

$3\sqrt{2} + 4\sqrt{8} - \sqrt{32} = 3\sqrt{2} + 4\sqrt{4}\sqrt{2} - \sqrt{16}\sqrt{2}$ (Rule 2a)
$= 3\sqrt{2} + 4 \cdot 2\sqrt{2} - 4\sqrt{2}$
$= (3 + 8 - 4)\sqrt{2}$ (Distributive law)
$= 7\sqrt{2}$

b. $\dfrac{\sqrt[4]{32x^3} - \sqrt[4]{8x^5}}{\sqrt[4]{2x}} = \dfrac{\sqrt[4]{32x^3}}{\sqrt[4]{2x}} - \dfrac{\sqrt[4]{8x^5}}{\sqrt[4]{2x}} = \sqrt[4]{\dfrac{32x^3}{2x}} - \sqrt[4]{\dfrac{8x^5}{2x}}$ (Rule 2b)

$= \sqrt[4]{16x^2} - \sqrt[4]{4x^4} = \sqrt[4]{16}\sqrt[4]{x^2} - \sqrt[4]{4}\sqrt[4]{x^4}$ (Rule 2a)
$= 2\sqrt[4]{\sqrt{x^2}} - \sqrt{\sqrt{4}}|x|$ (Rules 3 and 1)
$= 2\sqrt{|x|} - \sqrt{2}|x|$ (Rule 1)
$= 2\sqrt{x} - \sqrt{2x}$

Here the absolute value symbols can be removed from x since the statement of the problem presumes that $x \geq 0$; $\sqrt[4]{2x}$ is undefined for $x < 0$.

c. $(4\sqrt{5} - \sqrt{6})(2\sqrt{5} + 3\sqrt{6})$
$= 4\sqrt{5}(2\sqrt{5} + 3\sqrt{6}) - \sqrt{6}(2\sqrt{5} + 3\sqrt{6})$ (Distributive law)
$= 4 \cdot 2(\sqrt{5})^2 + 4 \cdot 3\sqrt{5}\sqrt{6} - 2\sqrt{6}\sqrt{5} - 3(\sqrt{6})^2$ (Distributive law)
$= 8 \cdot 5 + 12\sqrt{30} - 2\sqrt{30} - 3 \cdot 6$ (Definition of $\sqrt{\ }$ and Rule 2a)
$= 40 + (12 - 2)\sqrt{30} - 18$ (Distributive law)
$= 22 + 10\sqrt{30}$

When multiplying lengthy expressions involving radicals, it is helpful to keep track of the work by computing the product in the following format:

$$
\begin{array}{l}
4\sqrt{5} - \sqrt{6} \\
2\sqrt{5} + 3\sqrt{6} \\
\hline
8 \cdot 5 - 2\sqrt{5}\sqrt{6} \quad\longrightarrow\; \boxed{= 2\sqrt{5}(4\sqrt{5} - \sqrt{6})} \\
\quad + 12\sqrt{5}\sqrt{6} - 3 \cdot 6 \;\longrightarrow\; \boxed{= 3\sqrt{6}(4\sqrt{5} - \sqrt{6})} \\
\hline
40 + 10\sqrt{30} - 18 \\
= 22 + 10\sqrt{30}
\end{array}
$$

Rationalizing Denominators

When an expression involves radicals, it is sometimes preferable to rewrite the expression in a form with no radicals in the denominator. Frequently, multiplying by an appropriate factor will remove such radicals. Both numerator and denominator must be multiplied by the same factor to avoid changing the value of the given expression. Removing the radicals in a denominator by this process is known as **rationalizing** the denominator.

EXAMPLE 5 Simplify the following expressions and rationalize the denominator in each case.

a. $\dfrac{\sqrt{5}}{\sqrt{2}}$ **b.** $\dfrac{\sqrt[3]{7}}{\sqrt[3]{2}}$ **c.** $\dfrac{\sqrt{5} + \sqrt{2}}{\sqrt{10}}$

SOLUTION

a. Multiply numerator and denominator by $\sqrt{2}$ to obtain

$$\frac{\sqrt{5}}{\sqrt{2}} = \frac{\sqrt{5}\sqrt{2}}{\sqrt{2}\sqrt{2}} = \frac{\sqrt{10}}{2} \qquad \text{(Rule 2a)}$$

b. To eliminate $\sqrt[3]{2}$ in the denominator, we observe that $(\sqrt[3]{2})^3 = 2$. Thus we multiply numerator and denominator by $\sqrt[3]{2}\sqrt[3]{2} = \sqrt[3]{2 \cdot 2} = \sqrt[3]{4}$. The expression then simplifies as follows:

$$\frac{\sqrt[3]{7}}{\sqrt[3]{2}} = \frac{\sqrt[3]{7} \cdot \sqrt[3]{4}}{\sqrt[3]{2} \cdot \sqrt[3]{4}}$$

$$= \frac{\sqrt[3]{7 \cdot 4}}{\sqrt[3]{2 \cdot 4}} \qquad \text{(Rule 2a)}$$

$$= \frac{\sqrt[3]{28}}{\sqrt[3]{8}}$$

$$= \frac{\sqrt[3]{28}}{2}$$

c. $\dfrac{\sqrt{5} + \sqrt{2}}{\sqrt{10}} = \dfrac{\sqrt{5} + \sqrt{2}}{\sqrt{10}} \cdot \dfrac{\sqrt{10}}{\sqrt{10}}$

$$= \frac{\sqrt{50} + \sqrt{20}}{10} \qquad \text{(Distributive law and Rule 2a)}$$

$$= \frac{\sqrt{25}\sqrt{2} + \sqrt{4}\sqrt{5}}{10} \qquad \text{(Rule 2a)}$$

$$= \frac{5\sqrt{2} + 2\sqrt{5}}{10}$$

> **CALCULATOR COMMENTS**
>
> The $\sqrt{}$ key is used to find square roots on a calculator. On some calculators the $\sqrt{}$ key is pressed before the number; on others it is pressed after entering the number. To find $\sqrt{16.3}$, for example, we enter either
>
> $\boxed{\sqrt{}}\;\boxed{1}\;\boxed{6}\;\boxed{.}\;\boxed{3}\;\boxed{=}$
>
> or
>
> $\boxed{1}\;\boxed{6}\;\boxed{.}\;\boxed{3}\;\boxed{\sqrt{}}\;\boxed{=}$
>
> (depending on the calculator), and the display reads
>
> $\boxed{4.037325848}$
>
> If the number 16.3 represents a measurement correct to three significant digits, then the square root must also be rounded to three significant digits. We write
>
> $\sqrt{16.3} \approx 4.04$
>
> If several numbers are involved in the calculation, the result must be rounded to the number of significant digits in the number that has the fewest significant digits.
>
> Whenever the answer to a problem turns out to be $\sqrt{2}$ or $\sqrt{3}$, this represents the *exact* answer. The calculator display for
>
> $\sqrt{2} \approx \boxed{1.414213562}$
>
> is only an *approximation* to $\sqrt{2}$ within the limits of the calculator display.

Section 1.4 Exercises

Evaluate and/or simplify the following expressions.

1. $\sqrt{49}$
2. $\sqrt{625}$
3. $\sqrt{72}$
4. $\sqrt{675}$
5. $\sqrt[3]{125}$
6. $\sqrt[6]{64}$
7. $\sqrt[5]{-32}$
8. $\sqrt[4]{625}$
9. $\sqrt{8} \cdot \sqrt{2}$
10. $\dfrac{\sqrt{8}}{\sqrt{2}}$
11. $\dfrac{\sqrt{2}}{\sqrt[3]{8}}$
12. $\dfrac{\sqrt{125}}{\sqrt{5}}$
13. $\sqrt{3} \cdot \sqrt{27}$
14. $\sqrt[3]{27} \cdot \sqrt{9}$
15. $\dfrac{\sqrt[3]{81}}{\sqrt[3]{3}}$
16. $\dfrac{\sqrt[4]{16}}{\sqrt{4}}$
17. $\sqrt[4]{32}\sqrt[4]{8}$
18. $\sqrt[3]{9}\sqrt[3]{3}$
19. $\sqrt{\sqrt[4]{625}}$
20. $\sqrt{\sqrt[3]{729}}$
21. $\sqrt[3]{32}\sqrt{4}$
22. $\sqrt[4]{2}\sqrt{64}$
23. $\sqrt[4]{27}\sqrt{9}$
24. $\sqrt[3]{25}\sqrt{25}$
25. $\sqrt{9+16}$
26. $\sqrt{400+225}$
27. $\sqrt{9}+\sqrt{16}$
28. $\sqrt{400}+\sqrt{225}$
29. $\sqrt{169-25}$
30. $\sqrt{25-9}$
31. $\sqrt{169}-\sqrt{25}$
32. $\sqrt{25}-\sqrt{9}$
33. $\sqrt[3]{\dfrac{(270{,}000{,}000)(0.000008)}{1.25}}$
34. $\sqrt{\dfrac{(16{,}000{,}000)(0.000025)}{0.0004}}$
35. $\sqrt{\dfrac{(450{,}000)(150{,}000)}{300}}$
36. $\sqrt[3]{\dfrac{(9000)(3{,}000{,}000)}{8000}}$
37. $\sqrt[4]{81r^8 s^4}$
38. $\sqrt{16a^2 b^4}$
39. $\sqrt[3]{125 x^6 y^{12}}$
40. $\sqrt[4]{16 a^8 b^{12}}$
41. $\sqrt[3]{16 a^4 b}\,\sqrt[3]{4 a^2 b^5}$
42. $\sqrt{8 x^3 y}\,\sqrt{2xy^5}$

43. $\sqrt{3u^2v^3}\sqrt{27u^4v^5}$
44. $\sqrt[3]{16w^4z^2}\sqrt[3]{4w^2z^7}$
45. $\dfrac{\sqrt{3x^3y^3z}}{\sqrt{27xy^5z^3}}$
46. $\dfrac{\sqrt{8a^{10}b^2}}{\sqrt{2a^4b^6}}$
47. $\dfrac{\sqrt[4]{256u^{10}v^{20}}}{\sqrt{4u^3v^2}}$
48. $\dfrac{\sqrt[3]{27r^6s^{12}}}{\sqrt{16r^4s^8}}$
49. $\sqrt[3]{x^6\sqrt{27}}$
50. $\sqrt[4]{y^4\sqrt{256}}$
51. $\sqrt{125\sqrt{25x^4y^6}}$
52. $\sqrt{256\sqrt[3]{x^2}}$
53. $2\sqrt{7}+\sqrt{63}$
54. $3\sqrt{7}-\sqrt{28}$
55. $2\sqrt{27}-4\sqrt{3}+\sqrt{75}$
56. $\sqrt{45}+3\sqrt{5}-2\sqrt{20}$
57. $\sqrt{99}+\sqrt{44}-\sqrt{176}$
58. $2\sqrt{80}-3\sqrt{45}+5\sqrt{20}$
59. $5\sqrt{27}+\sqrt{75}-\sqrt{147}$
60. $3\sqrt{50}+\sqrt{98}-2\sqrt{18}$
61. $\sqrt[3]{24}-\sqrt[3]{3}$
62. $\sqrt[3]{54}-2\sqrt[3]{16}$
63. $2\sqrt[3]{81}-\sqrt[3]{-24}+4\sqrt[3]{3}$
64. $\sqrt[3]{135}+\sqrt[3]{-40}+4\sqrt[3]{5}$
65. $5\sqrt[3]{48}-4\sqrt[3]{6}+\sqrt[3]{162}$
66. $\sqrt[3]{-576}-\sqrt[3]{243}-\sqrt[3]{72}$
67. $3\sqrt{8}-\sqrt[3]{81}-\sqrt{128}+\sqrt[3]{375}$
68. $\sqrt[4]{576}-\sqrt{150}+\sqrt[6]{216}$
69. $(2\sqrt{5}+3\sqrt{3})(2\sqrt{5}-3\sqrt{3})$
70. $(8\sqrt{6}+\sqrt{15})(\sqrt{6}-2\sqrt{15})$
71. $(\sqrt{5}+\sqrt{20})^2$
72. $(\sqrt{2}+\sqrt{3}+\sqrt{5})(\sqrt{2}-\sqrt{3}-\sqrt{5})$
73. $(3\sqrt{6}-\sqrt{2}+\sqrt{3})(2\sqrt{6}+\sqrt{2}+\sqrt{3})$
74. $(\sqrt[3]{9}+\sqrt[3]{4})(\sqrt[3]{3}+\sqrt[3]{2})$
75. $(\sqrt[3]{9}+\sqrt[3]{6}+\sqrt[3]{4})(\sqrt[3]{3}-\sqrt[3]{2})$
76. $(\sqrt[3]{25}-\sqrt[3]{10}+\sqrt[3]{4})(\sqrt[3]{5}+\sqrt[3]{2})$
77. $\dfrac{\sqrt{18}+\sqrt{32}}{\sqrt{2}}$
78. $\dfrac{4\sqrt{18}-2\sqrt{12}}{\sqrt{3}}$
79. $\dfrac{\sqrt{2\sqrt{2}y^3}+\sqrt{8y^5}}{\sqrt[4]{2y^2}}$
80. $\dfrac{\sqrt{\sqrt{3}u^5}-\sqrt{3u^7}}{\sqrt[4]{3u^8}}$

Verify the following statements.

81. $\sqrt[3]{1+8}\neq\sqrt[3]{1}+\sqrt[3]{8}$
82. $\sqrt{(1+3)}\neq\sqrt{1}+\sqrt{3}$
83. $\sqrt{2^2+2^2}\neq 2+2$
84. $\sqrt{9-4}\neq\sqrt{9}-\sqrt{4}$

Rationalize the denominators in each of the following.

85. $\dfrac{\sqrt{4}}{\sqrt{3}}$
86. $\dfrac{\sqrt{5}}{\sqrt{2}}$
87. $\dfrac{\sqrt{2}}{\sqrt{3}}$
88. $\dfrac{\sqrt{4}}{\sqrt{6}}$
89. $\dfrac{\sqrt[3]{4}}{\sqrt[3]{3}}$
90. $\dfrac{\sqrt[3]{5}}{\sqrt[3]{4}}$
91. $\dfrac{\sqrt[4]{3}}{\sqrt[4]{2}}$
92. $\dfrac{\sqrt[4]{4}}{\sqrt[4]{3}}$
93. $\dfrac{\sqrt{3}-\sqrt{2}}{\sqrt{6}}$
94. $\dfrac{\sqrt{3}+\sqrt{5}}{\sqrt{15}}$
95. $\dfrac{\sqrt{7}+\sqrt{3}}{\sqrt{21}}$
96. $\dfrac{\sqrt{6}-\sqrt{2}}{\sqrt{12}}$
97. $\dfrac{\sqrt{5}-\sqrt{3}}{\sqrt{15}}$
98. $\dfrac{\sqrt{6}+\sqrt{3}}{\sqrt{18}}$

Calculator Exercises

Use a calculator to find the following values. Round the answer to the appropriate number of significant digits. Note that $\sqrt[4]{x}=\sqrt{\sqrt{x}}$.

99. $\sqrt{4.816(10^3)}$
100. $\sqrt{2.319(10^4)}$
101. $\sqrt{5.3812(10^5)}$
102. $\sqrt{2.89(10^6)}$
103. $\sqrt[4]{1.982(10^4)}$
104. $\sqrt[4]{3.931(10^3)}$
105. $\sqrt{1.54(10^2)\cdot[2.387(10^{-3})]}$
106. $\sqrt{3.4762(10^3)\cdot[1.208(10^{-2})]}$

107. If an animal with legs L feet long can walk $7\sqrt{L/3}$ miles per hour and Elmo the giraffe has legs 84 inches long, how fast can Elmo walk?

108. The number of strides per minute taken by a running animal with legs L feet long is approximately $99/\sqrt{L}$.
   a. Approximately how many strides per minute are taken by Elmo the giraffe (Exercise 107) when it runs?
   b. How many are taken by Wiener the dog if its legs are 4 inches long?

38 Chapter 1 Basic Algebra

109. The manufacturer of Green-Grass Pellets was recently found guilty of false advertising and required by the Federal Trade Commission to spend $20 million on commercial messages disclaiming earlier advertising. The manufacturer's advertising agency claims that the number of products sold is determined by

$$\frac{\sqrt{x} + \sqrt{y} + \sqrt{z}}{\sqrt{u} + \sqrt{v}}$$

where x = amount of money spent on television commercials, y = amount of money spent on radio commercials, z = amount of money spent on advertising in print, u = amount of money spent by the government contradicting the manufacturer's claims, and v = amount of money the manufacturer spends on disclaimer commercials. Determine the value of this expression if x = $50 million, y = $32 million, z = $18 million, u = $20 million, and v = $20 million.

110. The velocity of a wave on a banjo string is given by $\sqrt{T/\rho}$ meters per second, where T is the tension (in newtons, N) and ρ (the Greek letter rho) is the mass per unit length (in kilograms per meter, kg/m). Find the velocity if T = 48 N and ρ = 0.012 kg/m.

Section 1.5

Rational Exponents

In this section, we extend the definition of x^n to include rational (or fractional) exponents as well as integer exponents. That is, we wish to define $x^{m/n}$ for rational numbers m/n. We wish to preserve the rules for exponents (Section 1.3) so that they will apply to rational exponents also.

To preserve the power-of-a-power rule (Rule 2, Section 1.3), we must require that

$$(x^{1/n})^n = x^{n/n} = x^1 = x$$

That is, $x^{1/n}$ must be a number that when multiplied by itself n times yields x. Thus $x^{1/n}$ is an nth root of x. We define $x^{1/n}$ as the principal nth root of x whenever this nth root exists.

DEFINITION

$$x^{1/n} = \sqrt[n]{x} \quad \text{when } \sqrt[n]{x} \text{ exists}$$

EXAMPLE 1 Evaluate each of the following.

a. $16^{1/2}$ **b.** $(-8)^{1/3}$ **c.** $(-36)^{1/2}$

SOLUTION

a. $16^{1/2} = \sqrt{16} = 4$
b. $(-8)^{1/3} = \sqrt[3]{-8} = -2$
c. $(-36)^{1/2} = \sqrt{-36}$ is not defined in the real number system. ∎

We now define $x^{m/n}$ whenever $\sqrt[n]{x} = x^{1/n}$ exists.

DEFINITION

$$x^{m/n} = (x^{1/n})^m = (x^m)^{1/n}$$
$$= (\sqrt[n]{x})^m = \sqrt[n]{x^m} \quad \text{when } \sqrt[n]{x} \text{ exists}$$

Section 1.5 Rational Exponents 39

EXAMPLE 2   Evaluate each of the following.

a. $(-8)^{2/3}$ **b.** $4^{3/2}$ **c.** $[(-4)^2]^{1/4}$
d. $[(-4)^{1/4}]^2$ **e.** $(-4)^{2/4}$

SOLUTION   **a.** We evaluate $(-8)^{2/3}$ using both forms of the definition.

$$(-8)^{2/3} = [(-8)^{1/3}]^2 = (\sqrt[3]{-8})^2 = (-2)^2 = 4$$
$$(-8)^{2/3} = [(-8)^2]^{1/3} = (64)^{1/3} = \sqrt[3]{64} = 4$$

b. We also evaluate $4^{3/2}$ using both forms of the definition.

$$4^{3/2} = (\sqrt{4})^3 = 2^3 = 8$$
$$4^{3/2} = (4^3)^{1/2} = (64)^{1/2} = \sqrt{64} = 8$$

c. $[(-4)^2]^{1/4} = 16^{1/4} = \sqrt[4]{16} = 2$
d. $[(-4)^{1/4}]^2 = (\sqrt[4]{-4})^2$ is not defined in the real number system.
e. $(-4)^{2/4} = (\sqrt[4]{-4})^2$ is not defined since $\sqrt[4]{-4}$ does not exist in the real number system. ∎

CAUTION

Be careful when simplifying rational exponents.
    For instance

$$(-1)^{2/6} \neq (-1)^{1/3}$$

This unhappy situation occurs because

$$(-1)^{2/6} = (\sqrt[6]{-1})^2$$

does not exist, since $\sqrt[6]{-1}$ is not a real number; on the other hand,

$$(-1)^{1/3} = \sqrt[3]{-1} = -1$$

EXAMPLE 3   Rewrite each of the following in terms of radicals and/or integer exponents.

a. $(x^2 - y^2)^{1/2}$ **b.** $(27x^4)^{1/3}$

SOLUTION   **a.** $(x^2 - y^2)^{1/2} = \sqrt{x^2 - y^2}$

Note that this does *not* reduce to $x - y$; substitute $x = 3$, $y = 2$ to obtain

$$\sqrt{x^2 - y^2} = \sqrt{9 - 4} = \sqrt{5}$$

but

$$x - y = 3 - 2 = 1.$$

b. $(27x^4)^{1/3} = \sqrt[3]{27x^4}$
$$= \sqrt[3]{27}\sqrt[3]{x^3}\sqrt[3]{x}$$
$$= 3x\sqrt[3]{x}$$ ∎

40 Chapter 1 Basic Algebra

EXAMPLE 4 Rewrite each of the following in terms of rational exponents.

  a. $\sqrt[3]{x^{10}}$ **b.** $\sqrt[5]{(a+b)^2}$

SOLUTION

  a. $\sqrt[3]{x^{10}} = (x^{10})^{1/3} = x^{10/3}$

  b. $\sqrt[5]{(a+b)^2} = [(a+b)^2]^{1/5} = (a+b)^{2/5}$

  Note that this does *not* simplify to $a^{2/5} + b^{2/5}$; substitute $a = b = 1$ to obtain

  $$(a+b)^{2/5} = (2)^{2/5} = \sqrt[5]{2^2} = \sqrt[5]{4}$$

  but

  $$a^{2/5} + b^{2/5} = 1 + 1 = 2$$ ∎

As before, we define $x^{-m/n} = 1/x^{m/n}$. With these definitions it can be shown that the previously derived properties of exponents hold for rational exponents as well. The basic properties of exponents that include all the others are listed here for reference.

RULES FOR RATIONAL EXPONENTS

RULE	EXAMPLE
1. $x^a x^b = x^{a+b}$	1. $2^2 2^3 = 2^5$, since $4 \cdot 8 = 32$
2. $(x^a)^b = x^{ab}$	2. $(2^3)^2 = 2^6$, since $8^2 = 64 = 2^6$
3. $(xy)^a = x^a y^a$	3. $(2 \cdot 3)^2 = 2^2 3^2$, since $36 = 4 \cdot 9$
4. $x^0 = 1, \quad (x \neq 0)$	4. $10^0 = 1$
5. $x^{-a} = \dfrac{1}{x^a}, \quad (x \neq 0)$	5. $2^{-2} = \dfrac{1}{2^2}$

EXAMPLE 5 Simplify the following.

  a. $\left(-\dfrac{1}{8}\right)^{-2/3}$ **b.** $(-0.0000128)^{1/7}$

SOLUTION

  a. $\left(-\dfrac{1}{8}\right)^{-2/3} = \left(\sqrt[3]{-\dfrac{1}{8}}\right)^{-2}$ (Definition)

  $= \left(-\dfrac{1}{2}\right)^{-2}$

  $= \dfrac{1}{\left(-\dfrac{1}{2}\right)^2}$ (Rule 5)

  $= \dfrac{1}{\dfrac{1}{4}} = 1 \cdot \dfrac{4}{1} = 4$

b. $(-0.0000128)^{1/7} = [(-128) \cdot (10^{-7})]^{1/7}$
$= (-128)^{1/7}(10^{-1})$ (Rule 3)
$= -2 \cdot 10^{-1}$
$= -0.2$

EXAMPLE 6

Simplify the following expressions.

a. $(25a^{-4}b^6)^{-3/2}$

b. $\left(\dfrac{36x^{-2}y^{-8}}{9x^{-6}y^4}\right)^{1/2} \cdot \left(\dfrac{2xy^{-1}}{x^2y}\right)^{-1}$

SOLUTION

a. $(25a^{-4}b^6)^{-3/2} = (25)^{-3/2}(a^{-4})^{-3/2}(b^6)^{-3/2}$ (Rule 3)
$= (\sqrt{25})^{-3}(\sqrt{a^{-4}})^{-3}(\sqrt{b^6})^{-3}$ (Definition)
$= 5^{-3}(a^{-2})^{-3}|b^3|^{-3}$
$= \dfrac{a^6}{125|b|^9}$ (Rules 2 and 5)

b. $\left(\dfrac{36x^{-2}y^{-8}}{9x^{-6}y^4}\right)^{1/2} \cdot \left(\dfrac{2xy^{-1}}{x^2y}\right)^{-1} = \left(\dfrac{4x^{-2+6}}{y^{4+8}}\right)^{1/2} \cdot \left(\dfrac{2x}{x^2y^2}\right)^{-1}$ (Rule 5)

$= \sqrt{\dfrac{4x^4}{y^{12}}} \cdot \dfrac{x^2y^2}{2x}$ (Definition and Rule 5)

$= \dfrac{2x^2}{y^6} \cdot \dfrac{x^2y^2}{2x}$

$= \dfrac{x^3}{y^4}$

EXAMPLE 7

Population growth is sometimes said to be exponential. For instance, a certain bacteria colony may be observed to double its population every hour. If P denotes the initial population, then after 1 hour the population is $P \cdot 2$, and after 2 hours the population is $(P \cdot 2) \cdot 2 = P \cdot 2^2$. After n hours, the population is $P \cdot 2^n$. If 10,000 bacteria are present at noon, how many are present at 3:30 P.M.?

SOLUTION

The population n hours after noon is $10{,}000 \cdot 2^n$. Assuming the population grows steadily rather than by sudden large increases each hour, we may also assume that the population $3\frac{1}{2}$ hours after noon (at 3:30) is

$10{,}000(2^{3\frac{1}{2}}) = 10{,}000(2^3)(2^{1/2})$ (Rule 1)
$= 10{,}000 \cdot 8 \cdot \sqrt{2}$ (Definition)
$= 80{,}000\sqrt{2}$
$\approx 113{,}137$

> **CALCULATOR COMMENTS**
>
> Many hand-held calculators have a "power" key (a $\boxed{y^x}$ or $\boxed{x^y}$ key), which can be used to compute powers like $2^{3\frac{1}{2}}$. These calculators generally work as follows. To compute $2^{3\frac{1}{2}}$, for example,
>
> 1. Press $\boxed{2}$.
> 2. Press the power key $\boxed{y^x}$.
> 3. Press $\boxed{3}\,\boxed{.}\,\boxed{5}$ (for $3.5 = 3\frac{1}{2}$).
> 4. Finally, press $\boxed{=}$.
>
> This gives the answer $\boxed{11.314}$.
>
> When computing rational powers of numbers, the rules for rounding to significant digits are the same as for multiplication: the number of significant digits in the answer is determined by the least number of significant digits in the numbers used in the computation.

Section 1.5 Exercises

Rewrite each of the following in terms of radicals and/or integer exponents.

1. $x^{3/5}$
2. $y^{2/3}$
3. $(a^2 - b^2)^{1/4}$
4. $(u^2 + v^2)^{1/7}$
5. $(8ab)^{2/3}$
6. $(16rs)^{3/4}$
7. $(32u)^{3/5}$
8. $(64v)^{5/6}$

Rewrite each of the following in terms of rational exponents.

9. $\sqrt{x^2 + y^2}$
10. $\sqrt[3]{x^3 - y^3}$
11. $\sqrt[3]{(r^3 - s^3)^4}$
12. $\sqrt[4]{(a^4 + b^4)^3}$
13. $\sqrt[5]{u^7}$
14. $\sqrt[4]{r^9}$
15. $\sqrt[4]{v^{12}}$
16. $\sqrt[4]{a^8}$

Evaluate and/or simplify each of the following.

17. $25^{1/2}$
18. $36^{1/2}$
19. $16^{1/4}$
20. $81^{1/4}$
21. $(-243)^{1/5}$
22. $(-125)^{1/3}$
23. $4^{3/2}$
24. $9^{5/2}$
25. $(-8)^{5/3}$
26. $(-32)^{3/5}$
27. $(-27)^{-1/3}$
28. $(64)^{-5/6}$
29. $(27)^{-2/3}$
30. $(16)^{3/4}$
31. $(-27)^{2/3}$
32. $(-32)^{-6/5}$
33. $\left(\dfrac{1}{27}\right)^{-2/3}$
34. $\left(-\dfrac{1}{32}\right)^{-2/5}$
35. $\left(\dfrac{1}{81}\right)^{-3/4}$
36. $\left(\dfrac{1}{81}\right)^{1/4}$
37. $\left[\dfrac{1600(0.0025)}{900}\right]^{-3/2}$
38. $\left[\dfrac{(90{,}000)(3600)}{(0.0004)}\right]^{5/2}$
39. $(-2x^{3/2}y^{2/5})(4x^{3/2}y^{8/5})$
40. $(3r^{-3/2}s^{1/4})(r^{-1/2}s^{1/4})$
41. $(u^{1/3}v^{2/3}w^{5/3})^6$
42. $(x^{-1/3}y^{2/3}z^{-5/3})^6$
43. $(-x^{1/2}y^{-2/3})(3x^{-3/2}y^{2/3})$
44. $(x^{1/8}y^{1/4}z^2)^2$
45. $(p^{-1/8}q^{1/4}r^{-2})^{-2}$
46. $(s^{2/3}t^{-5/12}u^{-1/6})^{-3}$
47. $(a^{27}b^9z^3)^{2/3}$
48. $(a^8b^{-4}c^2)^{-3/2}$
49. $(x^{5/3}y^{-5/12}z^{5/6})^{3/5}$
50. $(x^{2/3}y^{-5/12}z^{-1/6})^{-3}$
51. $(x^{1/4}y^{3/8}z^{3/16})^{16}$
52. $\dfrac{(a^{-2/3}b^{1/3})^3}{(a^{1/4}b^{-1/8})^4}$

53. $\dfrac{(u^3v^4)^{-3}}{(v^4w^{-2}u^2)^{-2}}$

54. $\left(\dfrac{25a^{-4}b^{-6}c^2}{16x^{-2}y^6z^{-2}}\right)^{-3/2} \cdot \left(\dfrac{2a^2b^{-1}c^{-3}}{x^2y^{-2}z^3}\right)^{-1}$

Calculator Exercises

Round off the answers in each of the following to the appropriate number of significant digits.

55. $[6.79(10^3)]^{2/3}$ 56. $[1.46(10^2)]^{3/5}$

57. $[2.008(10^{-2})]^{5/4}$ 58. $[2.804(10^{-3})]^{7/3}$

59. $[1.44(10^5)\, 4.3282(10^{-2})]^{2/3}$

60. $[1.873(10^4)\, 2.0251(10^3)]^{3/7}$

61. $\left[\dfrac{4.1006(10^{-2})\, 1.681(10^3)}{2.82758(10^2)}\right]^{12/5}$

62. $\left[\dfrac{2.82(10^4)\, 1.1081(10^3)}{4.438164(10^{-2})}\right]^{7/2}$

63. The doubling time of a certain bacteria culture is 2 hours. If the culture starts with 5000 bacteria, how many will there be $5\frac{1}{2}$ hours later?

64. If the doubling time of a bacteria culture is 3 hours and there are 1000 present now, how many were there 1 hour ago?

65. a. If the gross national product was $2.5 trillion in 1980 and triples every 15 years, what would you expect it to be in the year 2000?
    b. What would it have been in 1970?

66. If the U.S. population doubles every 60 years and the 1980 population was estimated at 225 million people, what will the population be in the year 2000?

67. An insect doubles its weight every 48 hours and now weighs 4 milligrams.
    a. How much will it weigh in 12 hours?
    b. How much will it weigh in 3 days?

68. The half-life of radium is approximately 1600 years; that is, the quantity of radium in an ore sample decomposes to half its original amount in 1600 years. Two thousand units of radium are now present in a certain sample.
    a. How many units will be present 800 years from now?
    b. How many units will be present 400 years from now?
    c. How many units were present 800 years ago?
    d. How many units were present 400 years ago?

69. The temperature in a blast furnace is 2000°C. Cooling chambers are designed to permit molten metal to lose one quarter of its temperature (in °C) in 20 minutes. What will the temperature of the metal be after 45 minutes in the cooling chamber?

70. The rate at which crude oil flows from a tank depends on the weight or amount of oil remaining in the tank. If $\frac{1}{5}$ of the oil flows out every hour and 50,000 gallons remain in the tank after 3 hours, how much will be in the tank after 10 hours?

Section 1.6

Polynomials

Letters or symbols that stand for numbers are called **variables.** For example, in the expression $2x + 3$, x is a variable. We can substitute many numbers for x. If $x = 2$, the expression has a **value** of $2 \cdot 2 + 3 = 7$; if $x = -1$, its value is $2(-1) + 3 = 1$.

If we combine variables and numbers algebraically (by adding, subtracting, multiplying, dividing, raising to a power, or extracting roots) the result is called an **algebraic expression.** To distinguish the numbers from the variables, the numbers in an algebraic expression are called **constants.**

An algebraic expression formed without additions or subtractions is called a **term.** Thus

$$13x^2 \quad \text{and} \quad \dfrac{36\sqrt{x^2y^3}\,z}{\sqrt[3]{w}}$$

are terms; 13 and 36 are constants. Adding and/or subtracting terms yields another algebraic expression.

Terms which differ only by a constant multiple are called **like terms**. Thus

$$2x\sqrt{y} \quad \text{and} \quad 15x\sqrt{y}$$

are like terms but $2x\sqrt{y}$ and $2x^2y^2$ are not.

In this section, we work with algebraic expressions of the form

$$a_n x^n + a_{n-1} x^{n-1} + \cdots + a_1 x + a_0$$

where $a_n, \ldots, a_0$ are fixed constants, x is a variable, and n is a nonnegative integer. These expressions are called **polynomials** in one variable. The greatest exponent is called the **degree of the polynomial**. The constants $a_n, \ldots, a_0$ are called the **coefficients** of $x^n, \ldots, x^0$ respectively. The coefficient of the term of highest degree is called the **leading coefficient** of the polynomial. For example,

$x^3 - 3x^2 + 1$ is a polynomial of degree 3.   The coefficient of x^2 is -3.

$4x^2 + 6x - 2$ is a polynomial of degree 2.   The coefficient of x is 6.

$5x^{100} + 2x^{17}$ is a polynomial of degree 100.   The coefficient of x^{100} is 5; 5 is the leading coefficient.

A number by itself is called a **constant polynomial**. A constant polynomial $A \neq 0$ is said to have degree 0 since $A = Ax^0$. For example, the polynomial 7 is the same as $7x^0$. The polynomial identically 0 is called the **zero polynomial** and has no degree.

Expressions like

$$6\sqrt{x} + \frac{5}{x} + 2$$

and

$$7x^{-2} - 4x^2 + 15$$

are not polynomials, since $\sqrt{x} = x^{1/2}$, $5/x = 5x^{-1}$, and $6x^{-2}$ involve exponents which are not nonnegative integers.

A polynomial with one term is called a **monomial**. A polynomial with two terms is called a **binomial**. One with three terms is called a **trinomial**. Thus,

$6x^2$ is a monomial.

$6x^2 + 2$ is a binomial.

$6x^2 + 3x + 2$ is a trinomial.

Expressions of the form

$$a_{n,k} x^n y^k + \cdots + a_{m,l} x^m y^l$$

are called **polynomials in two variables**. Polynomials in three or more variables are defined similarly. The sum of the exponents of the variables in any term of a polynomial is the degree of that term; the highest degree of any of its terms is the degree of the polynomial. For instance,

$$3x^2 y + 2xy + 4x + 2y - 10$$

has degree 3, the degree of the term $3x^2 y$. The polynomial

$$29x^{100} y^{10} - 46 y^{60} x^3 + 20$$

has degree 110, the degree of the term $29x^{100} y^{10}$. The polynomial

$$6x^2 y z^2 + 6xy - 3z$$

has degree 5, the degree of the term $6x^2 y z^2$.

Addition and Subtraction of Polynomials

Since the letters in a polynomial represent numbers, we shall manipulate them in precisely the same way that we manipulate the real numbers. For example, the **distributive law** is used in the simple addition or subtraction of like terms:

$$3x + 2x = (3 + 2)x = 5x$$

To add or subtract expressions involving unlike terms, we also use the **associative and commutative laws** since these permit the rearrangement of terms in an expansion.

EXAMPLE 1

Find the sum and difference of the given polynomials.

$$2x^3 + 3x^2 + 4x - 1 \quad \text{and} \quad 6x^3 - x + 5$$

SOLUTION

Use the commutative and associative laws to collect like terms in the sum, as follows.

$$(2x^3 + 3x^2 + 4x - 1) + (6x^3 - x + 5)$$
$$= (2x^3 + 6x^3) + 3x^2 + (4x - x) - 1 + 5$$
$$= 8x^3 + 3x^2 + 3x + 4$$

Next, we subtract the second expression from the first.

$$(2x^3 + 3x^2 + 4x - 1) - (6x^3 - x + 5)$$
$$= (2x^3 + 3x^2 + 4x - 1) + [(-1)(6x^3 - x + 5)] \quad \text{(Property of subtraction)}$$
$$= (2x^3 + 3x^2 + 4x - 1) + (-6x^3 + x - 5) \quad \text{(Distributive law)}$$
$$= (2x^3 - 6x^3) + 3x^2 + (4x + x) - 1 - 5 \quad \text{(Collecting like terms)}$$
$$= -4x^3 + 3x^2 + 5x - 6$$

We see in the preceding example that polynomials are added or subtracted by merely adding or subtracting like terms. This shortcut is incorporated in the method of displaying the work in the familiar manner that is used for addition and subtraction of numbers. This so-called displayed form is illustrated in Example 2.

EXAMPLE 2

Rework Example 1 by writing one polynomial below the other in column formation.

SOLUTION

The addition is displayed as follows.

$$\begin{array}{r} 2x^3 + 3x^2 + 4x - 1 \\ +6x^3 \qquad\quad - x + 5 \\ \hline 8x^3 + 3x^2 + 3x + 4 \end{array}$$

To subtract we must remember that $-(a + b) = -a - b$. The subtraction is then displayed in two steps.

$$\begin{array}{r} 2x^3 + 3x^2 + 4x - 1 \\ -(6x^3 \qquad\quad - x + 5) \\ \hline \end{array} \quad = \quad \begin{array}{r} 2x^3 + 3x^2 + 4x - 1 \\ -6x^3 \qquad\quad + x - 5 \\ \hline -4x^3 + 3x^2 + 5x - 6 \end{array}$$

46 Chapter 1 Basic Algebra

When using the displayed form of subtraction, we generally do the transition mentally and immediately write the second step. ∎

Multiplication of Polynomials

Multiplication of polynomials depends on the distributive law and the laws of exponents. A simple addition problem remains after applying these laws. The *displayed form of multiplication* (illustrated in Example 3) is an excellent computational tool.

EXAMPLE 3 Find the product of $(3x^2 + x - 1)$ and $(2x^2 - 2x + 3)$.

SOLUTION We can use the distributive law to write

$$(3x^2 + x - 1)(2x^2 - 2x + 3)$$
$$= (3x^2 + x - 1)(2x^2) + (3x^2 + x - 1)(-2x) + (3x^2 + x - 1)(3)$$
$$= (6x^4 + 2x^3 - 2x^2) + (-6x^3 - 2x^2 + 2x) + (9x^2 + 3x - 3) \quad \text{(Distributive law)}$$
$$= 6x^4 - 4x^3 + 5x^2 + 5x - 3 \quad \text{(Collect like terms)}$$

We could also display the work as follows:

$$
\begin{array}{r}
3x^2 + x - 1 \\
2x^2 - 2x + 3 \\
\hline
6x^4 + 2x^3 - 2x^2 \longrightarrow = (3x^2 + x - 1)(2x^2) \\
- 6x^3 - 2x^2 + 2x \longrightarrow = (3x^2 + x - 1)(-2x) \\
+ 9x^2 + 3x - 3 \longrightarrow = (3x^2 + x - 1)(3) \\
\hline
6x^4 - 4x^3 + 5x^2 + 5x - 3
\end{array}
$$

∎

The following is a list of product expansions occurring so often in mathematics that each should be memorized. Each can be verified by simply carrying out the indicated multiplication. Their verifications are left as exercises.

COMMON EXPANSIONS

FORMULA	EXAMPLE
1. $(a + b)^2 = a^2 + 2ab + b^2$	1. $(x + 1)^2 = x^2 + 2x + 1$
2. $(a - b)^2 = a^2 - 2ab + b^2$	2. $(x - 1)^2 = x^2 - 2x + 1$
3. $(a + b)(a - b) = a^2 - b^2$	3. $(x + 1)(x - 1) = x^2 - 1$
4. $(a + b)^3 = a^3 + 3a^2b + 3ab^2 + b^3$	4. $(x + 1)^3 = x^3 + 3x^2 + 3x + 1$
5. $(a - b)^3 = a^3 - 3a^2b + 3ab^2 - b^3$	5. $(x - 1)^3 = x^3 - 3x^2 + 3x - 1$

Formulas 1 and 2 are sometimes combined into one as

$$(a \pm b)^2 = a^2 \pm 2ab + b^2$$

Similarly, formulas 4 and 5 are sometimes written together as

$$(a \pm b)^3 = a^3 \pm 3a^2b + 3ab^2 \pm b^3.$$

In any such double formula, the upper signs stay together and the lower signs stay together.

EXAMPLE 4 Find the following products by using formulas 1 through 5 in the list of common expansions.

a. $(2x^2 + 3y)^2$
b. $(3r - 2s)(3r + 2s)$
c. $(2u^2 - 3v)^3$

SOLUTION

a. We use formula 1 with $a = 2x^2$ and $b = 3y$.
$$(2x^2 + 3y)^2 = (2x^2)^2 + 2(2x^2)(3y) + (3y)^2$$
$$= 4x^4 + 12x^2y + 9y^2$$

b. This product fits formula 3 with $a = 3r$ and $b = 2s$.
$$(3r - 2s)(3r + 2s) = (3r)^2 - (2s)^2$$
$$= 9r^2 - 4s^2$$

c. We recognize this as the cube of a difference given by formula 5 with $a = 2u^2$ and $b = 3v$.
$$(2u^2 - 3v)^3 = (2u^2)^3 - 3(2u^2)^2(3v) + 3(2u^2)(3v)^2 - (3v)^3$$
$$= 8u^6 - 36u^4v + 54u^2v^2 - 27v^3$$

FOIL Expansion

If we expand the product of $(ax + b)$ and $(cx + d)$ by using the distributive law, we obtain
$$(ax + b)(cx + d) = acx^2 + adx + bcx + bd$$

Rather than memorizing this formula, you should memorize the process for obtaining such a product. To help with this, we name the terms as follows.

Last terms (L)   Inside terms (I)
$(ax + b)(cx + d)$ $(ax + b)(cx + d)$
First terms (F)   Outside terms (O)

Then
$$(ax + b)(cx + d) = acx^2 + adx + bcx + bd$$
$$ \text{F} \quad\;\; \text{O} \quad\;\; \text{I} \quad\;\; \text{L}$$

We call this the **FOIL method** of finding the product.

EXAMPLE 5 Use the FOIL method to find the product $(5w + 6)(3w - 4)$.

SOLUTION
$$(5w + 6)(3w - 4) = 5w \cdot 3w + 5w(-4) + 6 \cdot (3w) + 6(-4)$$
$$ \text{F} \quad\quad\;\; \text{O} \quad\quad\;\; \text{I} \quad\quad\; \text{L}$$
$$= 15w^2 - 20w + 18w - 24$$
$$= 15w^2 - 2w - 24$$

Rationalizing Denominators

The formula $(u + v)(u - v) = u^2 - v^2$ is useful in rationalizing denominators of the form $(\sqrt{a} + \sqrt{b})$ or $(\sqrt{a} - \sqrt{b})$; these latter two expressions are said to be **conjugates** of one another. To rationalize a denominator of this type, multiply by its conjugate. The denominator then becomes $a - b$ since

$$(\sqrt{a} + \sqrt{b})(\sqrt{a} - \sqrt{b}) = (\sqrt{a})^2 - (\sqrt{b})^2 = a - b$$

EXAMPLE 6

Rationalize the denominator in the expression $\dfrac{5}{\sqrt{2} + \sqrt{3}}$.

SOLUTION

$$\frac{5}{\sqrt{2} + \sqrt{3}} = \frac{5}{\sqrt{2} + \sqrt{3}} \cdot \frac{\sqrt{2} - \sqrt{3}}{\sqrt{2} - \sqrt{3}}$$
$$= \frac{5\sqrt{2} - 5\sqrt{3}}{2 - 3} = \frac{5\sqrt{2} - 5\sqrt{3}}{-1}$$
$$= 5\sqrt{3} - 5\sqrt{2}$$

Section 1.6 Exercises

Perform the indicated operations, collecting any like terms.

1. $(2x^2 + 3x - 2) + (4x^2 + 3x - 6)$
2. $(10y^5 + 3y^3 - 2y + 4)$
   $- (5y^4 - 3y^3 + 2y^2 - 4y - 2)$
3. $(6a^3 - 2a^2 - a + 4) + (3a^4 - 2a^3 - a^2 + a + 5)$
4. $(3u^4 - 10u^3 + 2u + 1) + (2u^3 - 3u^2 + 4)$
5. $(14t^7 - 20t^6 + 4t^2 - 2t + 8)$
   $- (17t^6 - 12t^5 + 5t^3 - 2t - 8)$
6. $(6b^4 - 4b^3 + 3b - 4) - (5b^4 - 4b^3 + 2b^2 - 3)$
7. $(3v^5 - 2v^4 + 3v^2 - 2v + 1)$
   $- (3v^4 - 2v^3 + 3v^2 - 4v + 6)$
8. $(2z^6 + 5z^4 + 3z^3 - 2z^2 + 5)$
   $+ (4z^6 - 2z^5 + z^4 - 2z^3 + z^2 - z - 5)$
9. $(3x^2 + 2x - 1) - (4x^2 - 3x + 4) + (2x^2 + x + 5)$
10. $(2u^2 - 5u + 6) + (3u^2 - 9u - 5) - (4u^2 + 3u + 8)$
11. $(5v^3 - 6v^2 + 3v - 4) + (5v^2 - 2v + 4)$
    $- (v^3 - v^2 + v + 3)$
12. $(4w^3 + 3w^2 + 5) - (2w^3 + 3w + 4)$
    $+ (w^2 + 7w - 8)$
13. $(18y^5 + 3y^4 - y^3 + 4) - (12y^5 + 6y^4 + 2y^2 + y)$
    $- (4y^4 + 3y^3 + 5y^2 - 6y + 4)$
14. $(8z^7 + 5z^6 - 8z^4 + 3z^3 + 2z - 4)$
    $- (2z^6 - 3z^5 + 2z^3 + z^2 + 5)$
    $- (3z^7 + 2z^5 + 7z^4 - z^2 + 5z + 6)$
15. $(x + y)(3y - 2x)$
16. $(u - 3v)(4u + v)$
17. $(2r + 5s)(6r + 3s)$
18. $(4w - 2z)(3w - 5z)$
19. $(z^2 + z - 1)(z + 3)$
20. $(v - 2)(v^2 + v + 1)$
21. $(r^2 - r + 1)(r^2 + 2r + 5)$
22. $(a^2 + 2a + 1)(a^2 - 3a - 2)$
23. $(2s^3 + 3s - 2)(s^2 + 4)$
24. $(t^3 + t^2 - t + 1)(t^2 - 1)$
25. $(t - 2)(t - 1)(t + 3)$
26. $(q + 3)(q - 2)(q - 4)$
27. $(p^2 - p - 1)(p + 3)(p - 2)$
28. $(3u^2 + 2u - 3)(2u - 1)(u + 2)$

29. Verify the formulas in the list of common expansions.

Use the common expansions or FOIL to find the following.

30. $(x + 4)^2$
31. $(4a - 2b)^2$
32. $(3u - 4v)^2$
33. $(2r + s)^2$
34. $(2a + 3b)^2$
35. $(3x - 2y)^2$
36. $(x - 2y)(x + 2y)$
37. $(2u - 3v)(2u + 3v)$
38. $(4t + 5u)(4t - 5u)$
39. $(6r + 5s)(6r - 5s)$
40. $(3j - 7m)(3j + 7m)$
41. $(9n - 3k)(9n + 3k)$
42. $(r - 3)^3$
43. $(5 + t)^3$
44. $(3x + 5y)^3$
45. $(2p - 3q)^3$
46. $(3a - 2b)^3$
47. $(4z + 5y)^3$
48. $(a - 3)(2a + 1)$
49. $(4y + 2)(3y - 1)$
50. $(7x - 2)(2x + 3)$
51. $(3y - 4)(2y + 3)$
52. $(5u + 1)(u - 2)$
53. $(2v + 7)(3v + 5)$
54. $(3r - 4)(2r + 1)$
55. $(5s + 2)(3s - 1)$
56. $(r - s + t)(r - s - t)$
57. $(x - y + z)(x + y - z)$
58. $(u + 2)(u - 2)(u^2 + 4)$
59. $(v - 3)(v + 3)(v^2 + 9)$

The formula $(a + b)(a - b) = (a^2 - b^2)$ can be used to perform seemingly complicated multiplications very quickly. For instance, $51 \cdot 49 = (50 + 1)(50 - 1) = 50^2 - 1^2 = 2500 - 1 = 2499$. Use this technique to evaluate the following.

60. $17 \cdot 13$
61. $75 \cdot 65$
62. $23 \cdot 37$
63. $18 \cdot 12$

64. A sporting goods manufacturer makes downhill skis, cross country skis, and water skis. Each pair of downhill skis sells for $120 and costs $78 to manufacture and distribute. Cross country skis sell for $75 and cost $43; water skis cost $55 and are sold for $95. If x downhill skis, y cross country skis, and z water skis are sold, algebraically express the following:
  a. the total sales receipts,
  b. the total expenses, and
  c. the total net profit.

Simplify each of the following and rationalize all denominators in your answers.

65. $\dfrac{\sqrt{2} - \sqrt{3}}{\sqrt{2} + \sqrt{3}}$
66. $\dfrac{2\sqrt{3} - \sqrt{8}}{\sqrt{3} - \sqrt{2}}$
67. $\dfrac{5\sqrt{2} - 2\sqrt{5}}{\sqrt{5} + 2\sqrt{2}}$
68. $\dfrac{2\sqrt{6} + \sqrt{8}}{\sqrt{2} - 2\sqrt{3}}$
69. $\dfrac{3}{\sqrt{x} - 3}$
70. $\dfrac{4}{\sqrt{x} + 2}$
71. $\dfrac{3}{\sqrt{x - 3}}$
72. $\dfrac{4}{\sqrt{x + 2}}$
73. $\dfrac{\sqrt{u}}{\sqrt{u} + 2}$
74. $\dfrac{\sqrt{r}}{\sqrt{r} - 3}$
75. $\dfrac{\sqrt{a} - \sqrt{b}}{\sqrt{a} + \sqrt{b}}$
76. $\dfrac{\sqrt{a} + \sqrt{b}}{\sqrt{a} - \sqrt{b}}$
77. $\dfrac{\sqrt{r + 1} + \sqrt{r - 1}}{\sqrt{r + 1} - \sqrt{r - 1}}$
78. $\dfrac{y^2(x - \sqrt{x^2 - y^2})}{x + \sqrt{x^2 - y^2}}$

Section 1.7

Factoring

Since $6 = 2 \cdot 3$, we say that 2 and 3 are factors of 6. In the same way, if a given polynomial is written as a product of other polynomials, then we call these other polynomials **factors** of the given polynomial and say that the given polynomial has been **factored** into the product of these. **Factoring** is the process of finding the factors of a given polynomial. Thus, for example, $(x + 1)$ and $(x - 1)$ are factors of $x^2 - 1$, since $x^2 - 1 = (x + 1)(x - 1)$.

Common Factors

Each of the common expansions of Section 1.6 can be used in reverse to yield a factoring formula. Before using one of these formulas, however, we should look for common factors. These are factors which appear in, or are common to, each of the terms in the expression. This common factor then becomes a factor of the polynomial by the distributive law.

50 Chapter 1 Basic Algebra

EXAMPLE 1   Express each of the following in terms of a common factor.

    a. $16x^2 + 8xy + 24x$
    b. $6ab^3 + 3a^2b^4 - 9a^2b^2$

SOLUTION   **a.** Note that $8x$ is a factor of each term in this expression. We use the distributive law in reverse to factor out $8x$.

$$16x^2 + 8xy + 24x = 8x \cdot 2x + 8x \cdot y + 8x \cdot 3$$
$$= 8x(2x + y + 3)$$

    b. We identify a common factor, $3ab^2$. Then we write

$$6ab^3 + 3a^2b^4 - 9a^2b^2 = 3ab^2 \cdot 2b + 3ab^2 \cdot ab^2 - 3ab^2 \cdot 3a$$
$$= 3ab^2(2b + ab^2 - 3a),$$

again by the distributive law. ∎

Difference of Squares

Formula 3 of the common expansions can be written in reverse as

$$\boxed{a^2 - b^2 = (a + b)(a - b)} \qquad (1)$$

EXAMPLE 2   Factor the following, using formula 1.

    a. $25 - u^2$
    b. $9x^2 - 25y^2$
    c. $81r^4 - 16s^4$
    d. $36k^2 - (m + 2n)^2$

SOLUTION   **a.** By formula 1, we have

$$25 - u^2 = 5^2 - u^2$$
$$= (5 + u)(5 - u)$$

    b. $9x^2 - 25y^2 = (3x)^2 - (5y)^2$
$$= (3x + 5y)(3x - 5y)$$

by formula 1 with $a = 3x$ and $b = 5y$.

    c. $81r^4 - 16s^4 = (9r^2 + 4s^2)(9r^2 - 4s^2)$ (Formula 1)

Now $9r^2 - 4s^2$ is also a difference of squares. Thus we write this as

$$81r^4 - 16s^4 = (9r^2 + 4s^2)(3r + 2s)(3r - 2s) \qquad \text{(Formula 1)}$$

    d. $36k^2 - (m + 2n)^2 = [6k + (m + 2n)][6k - (m + 2n)]$ (Formula 1)
$$= [6k + m + 2n][6k - m - 2n] \qquad ∎$$

In this section, we are interested only in factors that have integer coefficients. Thus we are not interested in factorizations like $x^2 - 3 = (x + \sqrt{3})(x - \sqrt{3})$.

Binomial Perfect Squares

Formulas 1 and 2 in the common expansions can be written in reverse as follows.

$$a^2 + 2ab + b^2 = (a + b)^2 \qquad (2)$$
$$a^2 - 2ab + b^2 = (a - b)^2 \qquad (3)$$

EXAMPLE 3 Use formulas 2 and 3 to factor the following.

a. $u^2 + 6u + 9$
b. $25x^2 - 20xy + 4y^2$

SOLUTION

a. $u^2 + 6u + 9 = u^2 + 2 \cdot u \cdot 3 + 3^2$
$= (u + 3)^2$ (Formula 2)

b. $25x^2 - 20xy + 4y^2 = (5x)^2 - 2(5x)(2y) + (2y)^2$
$= (5x - 2y)^2$

by Formula 3 with $a = 5x$ and $b = 2y$. ∎

FOIL Factoring

Not all polynomials of the form $Ax^2 + Bx + C$ are perfect squares that lend themselves to factoring by formulas 2 and 3. We can sometimes factor such expressions by interpreting the polynomial as a FOIL product

$$Ax^2 + Bx + C = (ax + b)(cx + d)$$

and working out the coefficients a, b, c, and d step-by-step. The procedure is illustrated in the following example. To begin the factoring process, we note that ac is the coefficient of x^2 and bd is the constant term.

EXAMPLE 4 Factor each of the following.

a. $u^2 - 5u + 4$
b. $z^2 + 4z - 5$
c. $2t^2 + 7t + 6$
d. $10x^2 - 13x - 3$

SOLUTION

a. Since the coefficient of u^2 is 1, we use the FOIL formula with $a = c = 1$. Then

$$(u^2 - 5u + 4) = (u + b)(u + d)$$

with $bd = 4$ and $(b + d) = -5$. Thus, b and d have the same sign since $bd > 0$; in fact, both must be negative since the coefficient of the middle term is negative. The factorizations of 4 are $1 \cdot 4$ and $2 \cdot 2$. Since $1 + 4 = 5$ and $2 + 2 \neq 5$, the only possibility is

$$(u^2 - 5u + 4) = (u - 4)(u - 1)$$

b. Again, write

$$(z^2 + 4z - 5) = (z + b)(z + d)$$

with $bd = -5$ and $(b + d) = 4$. This time b and d must have opposite signs since $bd < 0$; the positive term must be larger since $b + d > 0$. Since the only factorization of 5 is $1 \cdot 5$, we have

$$(z^2 + 4z - 5) = (z + 5)(z - 1)$$

c. We use the FOIL formula and try to write

$$(2t^2 + 7t + 6) = (2t + b)(t + d)$$

with $bd = 6$. Thus, b and d have the same sign since $bd > 0$. In fact, both must be positive since the coefficient of the middle term ($= 7$) is positive. The factorizations of 6 are $1 \cdot 6$ and $2 \cdot 3$. Trial and error leads to only one successful combination:

$$(2t^2 + 7t + 6) = (2t + 3)(t + 2)$$

d. Using the FOIL formula we factor $10x^2 - 13x - 3$ as follows:

$$(10x^2 - 13x - 3) = \begin{cases} (10x + b)(x + d) \\ \quad \text{or} \\ (2x + b)(5x + d) \end{cases}$$

with $bd = -3$. Thus, b and d have opposite signs, and numerical values of 1 and 3 suggest themselves. After trying the various possibilities, we find that

$$(10x^2 - 13x - 3) = (2x - 3)(5x + 1)$$

is the only possible factorization. ∎

Factoring by Grouping

When more than three terms appear in an expression to be factored, it sometimes helps to partition the expression into two or more smaller groups. This technique is known as **factoring by grouping.**

EXAMPLE 5 Factor the following expressions.

a. $6x^2 + 3xy + 8x + 4y$
b. $x^2 - y^2 + 2x^2y + 2xy^2$
c. $3x - 2y + xy - 6$
d. $x^2 + 2x + 1 - a^2$

SOLUTION

a. Group the first two and the last two terms and then factor as follows:

$$6x^2 + 3xy + 8x + 4y = (6x^2 + 3xy) + (8x + 4y)$$
$$= 3x(2x + y) + 4(2x + y)$$
$$= (3x + 4)(2x + y)$$

b. $x^2 - y^2 + 2x^2y + 2xy^2 = (x^2 - y^2) + (2x^2y + 2xy^2)$
$$= (x - y)(x + y) + 2xy(x + y) \qquad \text{(Formula 1)}$$
$$= (x - y + 2xy)(x + y)$$

c. After rearranging terms, we can again group and factor as follows.

$$3x - 2y + xy - 6 = 3x + xy - 2y - 6$$
$$= x(3 + y) - 2(y + 3)$$
$$= (x - 2)(y + 3)$$

d. $x^2 + 2x + 1 - a^2 = (x^2 + 2x + 1) - a^2$
$ = (x+1)^2 - a^2 \hfill \text{(Formula 2)}$
$ = (x + 1 + a)(x + 1 - a) \hfill \text{(Formula 1)}$ ∎

Sum and Difference of Cubes

The following formulas can be verified by multiplying the factors. The proofs are left as exercises.

$$a^3 + b^3 = (a + b)(a^2 - ab + b^2) \quad (4)$$
$$a^3 - b^3 = (a - b)(a^2 + ab + b^2) \quad (5)$$

EXAMPLE 6

Use formulas 4 and 5 to factor the following.

a. $w^3 - 27$
b. $27r^6 + 8s^3$

SOLUTION

a. $w^3 - 27 = w^3 - 3^3$
$ = (w - 3)(w^2 + 3w + 9) \hfill \text{(Formula 5)}$

b. Use formula 4 with $a = 3r^2$ and $b = 2s$:

$27r^6 + 8s^3 = (3r^2)^3 + (2s)^3 = (3r^2 + 2s)[(3r^2)^2 - (3r^2)(2s) + (2s)^2]$
$ = (3r^2 + 2s)(9r^4 - 6r^2 s + 4s^2)$ ∎

Factoring by Completing the Square

Certain expressions resemble the "perfect square" forms $a^2 \pm 2ab + b^2$ but they don't match it exactly. In this case we adjust the incomplete term to fit the form and then balance the expression by adding or subtracting a term so that the result is equal to the original expression. This process is called **completing the square**. *Sometimes* the result can then be factored, as illustrated in Example 5.

EXAMPLE 7

Factor the following expressions by first completing the square.

a. $x^4 - 13x^2 + 4$
b. $9x^4 + 15x^2 y^2 + 16y^4$

SOLUTION

a. This expression resembles the form $a^2 - 2ab + b^2$ with $a = x^2$ and $b = 2$ except that the middle term should be $2ab = 2 \cdot x^2 \cdot 2 = 4x^2$. Thus we rewrite the expression as

$x^4 - 13x^2 + 4 = (x^4 - 4x^2 + 4) - 9x^2$
$ = (x^2 - 2)^2 - (3x)^2 \hfill \text{(Formula 3)}$
$ = [(x^2 - 2) + 3x][(x^2 - 2) - 3x] \hfill \text{(Formula 1)}$
$ = [x^2 + 3x - 2][x^2 - 3x - 2]$

b. The expression $9x^4 + 15x^2 y^2 + 16y^4$ resembles the form $a^2 + 2ab + b^2$ except that $15x^2 y^2 \neq 2ab$. To match the form, the middle term should be

$2(3x^2)(4y^2) = 24x^2y^2$. Thus, we write

$$9x^4 + 15x^2y^2 + 16y^4 = (9x^4 + 24x^2y^2 + 16y^4) - 9x^2y^2$$
$$= (3x^2 + 4y^2)^2 - 9x^2y^2 \quad \text{(Formula 2)}$$
$$= (3x^2 + 4y^2 + 3xy)(3x^2 + 4y^2 - 3xy) \quad \text{(Formula 1)}$$

The factoring formulas are summarized here for reference.

FACTORING FORMULAS

FORMULA	EXAMPLE
1. $a^2 - b^2 = (a + b)(a - b)$	1. $x^2 - 1 = (x + 1)(x - 1)$
2. $a^2 + 2ab + b^2 = (a + b)^2$	2. $x^2 + 2x + 1 = (x + 1)^2$
3. $a^2 - 2ab + b^2 = (a - b)^2$	3. $x^2 - 2x + 1 = (x - 1)^2$
4. $a^3 + b^3 = (a + b)(a^2 - ab + b^2)$	4. $x^3 + 1 = (x + 1)(x^2 - x + 1)$
5. $a^3 - b^3 = (a - b)(a^2 + ab + b^2)$	5. $x^3 - 1 = (x - 1)(x^2 + x + 1)$

Section 1.7 Exercises

Factor the following as far as possible.

1. $6z + 3$
2. $8a + 4$
3. $25 - 50x$
4. $ax + b^2x^2$
5. $6y^2 + 3y - 9ay$
6. $8u^2 - 4u + 12bu$
7. $4r^3s^2 - 2rs^8 + 10r^4s^4 - 12r^2s^2$
8. $9x^2y^4 + 6x^2y^2 - 12x^4y^3 + 15xy^3$
9. $2u(r^2 + s) - v(r^2 + s)$
10. $3x(3m - 2x) + 2y(3m - 2x)$
11. $r^2 - 25$
12. $4u^2 - 25v^2$
13. $36 - s^2$
14. $9x^2 - 4y^2$
15. $36x^2 - 49y^2$
16. $64m^2 - 81n^2$
17. $256 - w^4$
18. $r^4 - 16s^4$
19. $u^8 - v^8$
20. $p^{16} - q^{16}$
21. $4u^2 - 12uv + 9v^2$
22. $36w^2 - 12wz + z^2$
23. $25r^2 + 20rs + 4s^2$
24. $m^2 + 6mk + 9k^2$
25. $9a^2 + 24ab + 16b^2$
26. $16r^2 + 24rs + 9s^2$
27. $4r^2 - 28rs + 49s^2$
28. $25p^2 - 20pq + 4q^2$
29. $25u^2 + 60uv + 36v^2$
30. $j^2 + 14jn + 49n^2$
31. $y^2 - 9y + 20$
32. $24b^2 - 6b - 18$
33. $t^2 - 14t + 24$
34. $6x^2 - xy - 2y^2$
35. $p^2 - p - 20$
36. $u^2 - 2uv - 8v^2$
37. $q^2 - 18qr + 72r^2$
38. $12u^2 - uv - v^2$
39. $4a^2 - 16a + 15$
40. $15m^2 + 14mn - 8n^2$
41. $w^3 - 27$
42. $8x^3 - 125y^3$
43. $8 - u^3$
44. $1000r^3 + 27s^3$
45. $s^3 + 8$
46. $64 + v^3$
47. $8t^3 + 27u^3$
48. $27a^3 - 64b^3$
49. $x^4 - 5x^2 + 4$
50. $w^4 + 8w^2 - 9$
51. $a^4 - 2a^2 + 1$
52. $r^4 - 7r^2 - 18$
53. $1 - 32x^4 + 256x^8$
54. $m^4 - 3m^2n^2 - 4n^4$
55. $p^6 - 64q^6$
56. $(u + v)^3 - 8$
57. $x^6 - 7x^3 - 8$
58. $81 + 24r^3 - r^6$

Factor the following by grouping.

59. $2ar + 3br + 4as + 6bs$
60. $12ab - 8av + 3ub - 2uv$
61. $x^2y^2 - 4y^2 + x^2 - 4$
62. $2s^2 + 6 - r^2s^2 - 3r^2$
63. $x^5 - x^3 + x^2 - 1$
64. $x^3 - y^3 + x^2 - y^2$
65. $x^2y^2 + 6xy^2 + 5y^2 - 9x^2 - 54x - 45$
66. $9r^2u^2 - r^2v^2 - 9s^2u^2 + s^2v^2$
67. $r^2 - 4r + 4 - s^2$
68. $t^2 - u^2 + 6u - 9$

Factor by completing the square.

69. $r^4 + 7r^2 + 16$
70. $k^4 + 4$
71. $u^8 + 4$
72. $t^4 - 3t^2 + 1$
73. $u^4 - 6u^2 + 1$
74. $r^4 - 3r^2s^2 + s^4$
75. $z^4 + 2z^2a^2 + 9a^4$
76. $p^4 + 3p^2q^2 + 4q^2$
77. $x^4 + x^2y^2 + y^4$
78. $256 - 17t^2 + t^4$

Use formulas 4 and 5 in the common expansions (Section 1.6) to factor the following.

79. $8a^3 + 36a^2b + 54ab^2 + 27b^3$
80. $8r^3 + 12r^2s + 6rs^2 + s^3$
81. $27 - 54x + 36x^2 - 8x^3$
82. $64 - 144s + 108s^2 - 27s^3$

83. Prove formulas 4 and 5 for the sum and difference of cubes by multiplying the factors.

84. The Easy-Chair Company finds that a webbing machine generates a monthly profit of $60 + 37t - 2t^2$ thousand dollars t months after its purchase. How long does the machine remain profitable? (*Hint:* Factor the profit expression and then determine which value of t yields a profit of 0.)

85. The altitude of a scientific weather rocket t seconds after it is launched is given by the polynomial $-16^2 + 56t + 120$ feet.

   a. How high is the rocket 3 seconds after firing?
   b. Factor the expression to determine when the rocket hits the ground (altitude = 0).

Section 1.8

Rational Expressions

In this section we work with **quotients of polynomials.** Just as a quotient of two integers is called a rational number, a quotient of two polynomials is called a **rational expression.** Examples of rational expressions include:

$$\frac{3x^2 + 2x - 4}{x^2}$$

$$\frac{2x^3 - 3x + 10}{3x^2 + 2}$$

$$\frac{2xy - 3}{4x^2 + 3xy^3 - 4y^2}$$

Whenever we write a rational expression, we must restrict the variables to only those values for which the denominator is not 0.

In Section 1.1 we listed the rules for algebraic operations on fractions. Since rational expressions represent numbers, the same rules must apply. Thus, in order to combine rational expressions, we use the familiar laws governing fractions.

Rule 1 for fractions (Section 1.1) says that

$$\boxed{\frac{k \cdot m}{k \cdot n} = \frac{m}{n}, \quad k \neq 0, n \neq 0} \tag{1}$$

This rule is used to **simplify** rational expressions by **canceling common factors** in the numerator and denominator. Since the numerator and denominator may be polynomials, we must be careful to cancel only complete factors and not individual terms. Of course, the respective polynomials must be factored before any cancellation can occur.

EXAMPLE 1

Simplify the following by canceling common factors.

a. $\dfrac{x^2 - x - 6}{x^2 - 9}$

b. $\dfrac{1 - x^2}{x^2 + x - 2}$

SOLUTION

a. Factor the numerator and the denominator to obtain

$$\dfrac{x^2 - x - 6}{x^2 - 9} = \dfrac{(x - 3)(x + 2)}{(x + 3)(x - 3)}$$

$$= \dfrac{x + 2}{x + 3} \qquad \text{(Rule 1)}$$

Since the original fraction is undefined when $x = \pm 3$, x is restricted to values other than ± 3. Thus $(x - 3) \neq 0$ and (1) does indeed apply.

b. $\dfrac{1 - x^2}{x^2 + x - 2} = \dfrac{(1 - x)(1 + x)}{(x + 2)(x - 1)}$

$$= \dfrac{-(x - 1)(1 + x)}{(x + 2)(x - 1)}$$

$$= -\dfrac{x + 1}{x + 2}$$

Note that $(1 - x)$ had to be converted to $-(x - 1)$, since only common factors can be canceled. ■

CAUTION

We emphasize again that only factors can be canceled. Do not cancel terms which are not factors.

a. $\dfrac{\overset{1}{x + 2}}{\underset{2}{4}} \neq \dfrac{x + 1}{2}$

b. $\dfrac{x^2 - x - 6}{x^2 + 3} \neq \dfrac{-x - 6}{3}$

Section 1.8 Rational Expressions

Rules 5 and 6 of Section 1.1 govern multiplication and division of rational expressions; they are presented here as a reminder.

$$\frac{a}{b} \cdot \frac{c}{d} = \frac{a \cdot c}{b \cdot d} \qquad (2)$$

$$\frac{a}{b} \div \frac{c}{d} = \frac{a}{b} \cdot \frac{d}{c} \qquad (3)$$

EXAMPLE 2 Perform the indicated operations and simplify the results.

a. $\dfrac{x^2 - x - 2}{x + 3} \cdot \dfrac{x^2 + 6x + 9}{x^2 - 4}$

b. $\dfrac{x^2 - 4}{x^2 + 5x + 6} \div \dfrac{x - 2}{x^2 + 7x + 12}$

SOLUTION

a. We simply multiply the numerators and multiply the denominators as with any fraction. The result is then simplified by canceling like terms in numerator and denominator as follows.

$$\frac{x^2 - x - 2}{x + 3} \cdot \frac{x^2 + 6x + 9}{x^2 - 4} = \frac{(x^2 - x - 2)(x^2 + 6x + 9)}{(x + 3)(x^2 - 4)}$$

$$= \frac{(x - 2)(x + 1)(x + 3)(x + 3)}{(x + 3)(x + 2)(x - 2)}$$

$$= \frac{(x + 1)(x + 3)}{(x + 2)}$$

$$= \frac{x^2 + 4x + 3}{x + 2}$$

b. As with ordinary division of fractions, we invert the divisor and multiply and then proceed as in (a). Note the factorization of polynomials in the numerator and denominator; this factorization enables us to see the common factors that can be canceled.

$$\frac{x^2 - 4}{x^2 + 5x + 6} \div \frac{x - 2}{x^2 + 7x + 12} = \frac{x^2 - 4}{x^2 + 5x + 6} \cdot \frac{x^2 + 7x + 12}{x - 2}$$

$$= \frac{(x + 2)(x - 2)(x + 3)(x + 4)}{(x + 2)(x + 3)(x - 2)}$$

$$= x + 4 \qquad \blacksquare$$

To add or subtract rational expressions, we find the **least common denominator (LCD)** of the expressions. As with numbers, the LCD is the **least common multiple (LCM)** of the denominators in question. To find the LCM of two or more polynomials, we first factor the polynomials. In the LCM we include each factor that appears in any of the given polynomials; for each factor we use the largest exponent that appears on that factor. For example, the LCM of

$$(x + 1)^2(x - 1) \quad \text{and} \quad (x + 1)(x - 1)^3(x + 2)$$

is

$$(x + 1)^2(x - 1)^3(x + 2)$$

Once the LCD is identified, each rational expression can be written in terms of the LCD by multiplying its numerator and denominator by the appropriate factor(s). Then we need only add or subtract the numerators as required. The result is used as the numerator over the LCD as denominator. For instance

$$\frac{x}{(x + 1)(x - 1)} + \frac{2}{(x + 1)(x - 1)} = \frac{x + 2}{(x + 1)(x - 1)}$$

and *not* simply $x + 2$! Of course, the final result should then be simplified by canceling common factors.

EXAMPLE 3

Perform the indicated operations and simplify the results.

a. $\dfrac{2x - 8}{x^2 - 1} + \dfrac{3}{x - 1} + \dfrac{5x^2}{x + 1}$

b. $\dfrac{2x - 1}{(x^2 - x - 6)} - \dfrac{2}{3 - x} - \dfrac{1}{x + 2}$

SOLUTION

a. Since $x^2 - 1 = (x - 1)(x + 1)$, the least common denominator is clearly $(x - 1)(x + 1)$. Hence, we write each fraction in terms of the LCD, add the resulting numerators, and simplify as follows.

$$\frac{2x - 8}{x^2 - 1} + \frac{3}{x - 1} + \frac{5x^2}{x + 1}$$

$$= \frac{2x - 8}{(x - 1)(x + 1)} + \frac{3(x + 1)}{(x - 1)(x + 1)} + \frac{5x^2(x - 1)}{(x + 1)(x - 1)}$$

$$= \frac{(2x - 8) + (3x + 3) + (5x^3 - 5x^2)}{(x + 1)(x - 1)}$$

$$= \frac{5x^3 - 5x^2 + 5x - 5}{(x + 1)(x - 1)} = \frac{5x^2(x - 1) + 5(x - 1)}{(x + 1)(x - 1)}$$

$$= \frac{(5x^2 + 5)(x - 1)}{(x + 1)(x - 1)}$$

$$= \frac{5(x^2 + 1)}{x + 1}$$

Note that we cannot cancel the 1s in $\dfrac{5(x^2 + 1)}{x + 1}$ to get $\dfrac{5x^2}{x}$ since the 1s in these places are not *factors* but just *terms within a factor*.

b. Writing $x^2 - x - 6 = (x - 3)(x + 2)$ and $(3 - x) = -(x - 3)$, we find the LCD is $(x - 3)(x + 2)$. We then proceed as follows.

Section 1.8 Rational Expressions

$$\frac{2x-1}{x^2-x-6} - \frac{2}{3-x} - \frac{1}{x+2}$$

$$= \frac{2x-1}{(x-3)(x+2)} - \frac{-2(x+2)}{(x-3)(x+2)} - \frac{1 \cdot (x-3)}{(x+2)(x-3)}$$

$$= \frac{2x-1+2(x+2)-(x-3)}{(x-3)(x+2)}$$

$$= \frac{2x-1+2x+4-x+3}{(x-3)(x+2)}$$

$$= \frac{3x+6}{(x-3)(x+2)}$$

$$= \frac{3\cancel{(x+2)}}{(x-3)\cancel{(x+2)}}$$

$$= \frac{3}{x-3} \quad \blacksquare$$

We summarize our discussion of rational expressions at this point.

COMBINING RATIONAL EXPRESSIONS

1. To multiply, multiply numerators and multiply denominators separately.
2. To divide, invert the divisor and multiply.
3. To add or subtract
   a. First find the least common denominator (LCD).
   b. Write each rational expression in terms of this LCD.
   c. Then, add or subtract numerators; keep the denominator.
4. Finally, simplify the results by canceling like factors in the numerator and denominator.

Certain relationships in science and medicine involve rational expressions, the numerators and denominators of which are also rational expressions. Just as complex fractions reduce to a simple fraction, these complex rational expressions will reduce to a simple rational expression.

EXAMPLE 4

Simplify the following complex rational expression.

$$\frac{x + \dfrac{1}{x} + 2}{x - \dfrac{1}{x}}$$

SOLUTION

Two solutions are shown to illustrate the various techniques that may be used. In the first solution the terms in the numerator are combined into a single rational expression. After doing the same in the denominator, we carry out the indicated division by inverting and multiplying:

$$\frac{x + \frac{1}{x} + 2}{x - \frac{1}{x}} = \frac{\left(\frac{x^2 + 1 + 2x}{x}\right)}{\left(\frac{x^2 - 1}{x}\right)}$$

It is helpful to use parentheses to keep the numerator and denominator separate since there are three fraction bars in the expression at this stage.

Next we invert the expression in the denominator and multiply.

$$= \frac{x^2 + 1 + 2x}{x} \cdot \frac{x}{x^2 - 1} \qquad \text{(Cancel common factors)}$$

$$= \frac{(x + 1)^2}{(x + 1)(x - 1)} \qquad \text{(Factor and cancel)}$$

$$= \frac{x + 1}{x - 1}$$

For an alternate solution, we find the LCD of the numerator and the denominator; the LCD is x in this expression. Then we multiply numerator and denominator by this LCD.

$$\frac{x + \frac{1}{x} + 2}{x - \frac{1}{x}} = \frac{x\left(x + \frac{1}{x} + 2\right)}{x\left(x - \frac{1}{x}\right)}$$

$$= \frac{x^2 + 1 + 2x}{x^2 - 1}$$

$$= \frac{(x + 1)^2}{(x + 1)(x - 1)} \qquad \text{(Factor and cancel)}$$

$$= \frac{x + 1}{x - 1} \qquad \blacksquare$$

A variation on the above example occurs in expressions with negative exponents.

EXAMPLE 5

Simplify the following expression and write the result using only positive exponents.

$$\frac{2u^{-1} + v^{-1}}{(uv)^{-1}}$$

SOLUTION

Use the definition of negative exponents to rewrite the expression as

$$\frac{2u^{-1} + v^{-1}}{(uv)^{-1}} = \frac{\frac{2}{u} + \frac{1}{v}}{\frac{1}{uv}}$$

Now multiply numerator and denominator by their LCD, uv, as in the alternate solution to Example 4.

$$= \frac{uv\left(\frac{2}{u} + \frac{1}{v}\right)}{uv\left(\frac{1}{uv}\right)}$$

$$= \frac{2v + u}{1} \qquad \text{(Distributive law)}$$

$$= 2v + u \qquad \blacksquare$$

EXAMPLE 6

The circulatory system in the human body is a maze of blood vessels connected in various ways. The blood vessels illustrated in Figure 1.14 are said to be connected in parallel. Each vessel offers a certain resistance to the flow of blood. Diseases like arteriosclerosis increase this resistance and either reduce the flow of blood, increase the blood pressure, or both. If the resistance of v_1 is x and the resistance of v_2 is y, the net resistance of the parallel network of Figure 1.14 is $\dfrac{1}{\frac{1}{x} + \frac{1}{y}}$ (x and y must be measured in the same units).

a. Simplify the expression for the resistance of a parallel network.

b. To see how pressure is reduced in parallel networks, let the resistance in v_1 be 10 and in v_2 be 5. Compute the resistance of the parallel network.

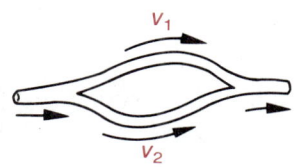

Figure 1.14

SOLUTION

a. $\dfrac{1}{\frac{1}{x} + \frac{1}{y}} = \dfrac{1}{\frac{y+x}{xy}} = \dfrac{xy}{x+y}$

b. Letting $x = 10$, $y = 5$, we have

$$\frac{1}{\frac{1}{x} + \frac{1}{y}} = \frac{xy}{x+y} = \frac{10 \cdot 5}{10 + 5} = \frac{50}{15} = 3\frac{1}{3}$$

The resulting resistance of the parallel network is less than that in either of the two separate branches. ∎

Polynomial Long Division

We have seen that some rational expressions can be simplified rather easily by factoring and applying the rules for manipulating fractions and the laws of exponents. However, other expressions are not so readily simplified because their factors are not readily apparent. For instance, it is not immediately obvious that

$$\frac{x^5 + 6x^4 - x^3 - 14x^2 + 18x + 8}{x^3 - 3x + 4}$$

can be simplified. For this kind of problem, the familiar procedure for long division of numbers can be adapted and used to advantage. In this particular problem we have

$$x^3 - 3x + 4 \overline{\smash{\big)}\, x^5 + 6x^4 - x^3 - 14x^2 + 18x + 8} \quad \begin{array}{c} x^2 + 6x + 2 \end{array}$$

```
                    x² + 6x  + 2
x³ − 3x + 4 ) x⁵ + 6x⁴ −  x³ − 14x² + 18x + 8
            −(x⁵       − 3x³ +  4x²)
              ─────────────────────────
                  6x⁴ + 2x³ − 18x²
               −(6x⁴       − 18x² + 24x)
                 ─────────────────────────
                        2x³        − 6x + 8
                      −(2x³        − 6x + 8)
                       ─────────────────────
                                            0
```

Thus,

$$\frac{x^5 + 6x^4 - x^3 - 14x^2 + 18x + 8}{x^3 - 3x + 4} = x^2 + 6x + 2$$

The long division can be checked by verifying that

$$(x^3 - 3x + 4)(x^2 + 6x + 2) = x^5 + 6x^4 - x^3 - 14x^2 + 18x + 8$$

LONG DIVISION OF POLYNOMIALS

1. Always write the polynomials with terms in order of *decreasing powers* of one of the variables. Then at each stage, only the first terms of the expressions need be examined.
2. Leave room for any missing powers of this variable in the dividend since they may appear in the intermediate stages of the long division.
3. Divide the leading term of the dividend by the leading term of the divisor and use the result for the next term in the quotient.
4. Multiply this result by the divisor and subtract the product from the dividend.
5. Repeat Steps 3 and 4 until the exponent of the variable (identified in Step 1) in the leading term of the dividend is less than the corresponding exponent in the leading term of the divisor.
6. Just as in long division of numbers, a remainder sometimes occurs when the process is terminated in Step 5.

EXAMPLE 7 Use the long-division algorithm on the indicated quotient.

$$\frac{6x^3 - x^4 - 14x - 8}{3x + x^2 - 4}$$

SOLUTION After rearranging the terms in decreasing powers of x and leaving space for an x^2 term, we can carry out the long division.

$$x^2 + 3x - 4 \overline{\smash{\big)}\begin{array}{r}-x^2 + 9x - 31 \\ -x^4 + 6x^3 + 0x^2 - 14x - 8\end{array}}$$

$$\begin{array}{r}
-(-x^4 - 3x^3 + 4x^2) \\ \hline
9x^3 - 4x^2 - 14x \\
-(9x^3 + 27x^2 - 36x) \\ \hline
-31x^2 + 22x - 8 \\
-(-31x^2 - 93x + 124) \\ \hline
115x - 132
\end{array}$$

Note the remainder of $115x - 132$. As with long division of numbers, we can write

$$\frac{-x^4 + 6x^3 - 14x - 8}{x^2 + 3x - 4} = -x^2 + 9x - 31 + \frac{115x - 132}{x^2 + 3x - 4}$$

Combining the terms on the right side of the equality according to the laws of fractions will verify the relationship. ■

Section 1.8 Exercises

Simplify the given expressions without resorting to long division.

1. $\dfrac{2x^3 - 6x^2 + 2x}{2x}$

2. $\dfrac{3y^4 + 9y^3 - 12y^2}{3y^2}$

3. $\dfrac{s^2 - 10s + 25}{s - 5}$

4. $\dfrac{t^2 + 18t + 81}{t + 9}$

5. $\dfrac{a^3 + 27}{a + 3}$

6. $\dfrac{b^3 - 64}{b - 4}$

7. $\dfrac{u^2 + 5u - 14}{u - 2}$

8. $\dfrac{2v^2 + 13v - 7}{2v - 1}$

9. $\dfrac{z^2 + 2z - 3}{z + 3}$

10. $\dfrac{6r^2 - 5r - 6}{2r - 3}$

In the following exercises, perform the indicated operations and simplify the results.

11. $\dfrac{m^2 + 2m}{m^2 - 3m} \cdot \dfrac{m^2 - 9}{m^2 - 4}$

12. $\dfrac{k^3 - 4k^2}{k + 5} \cdot \dfrac{k^3 - 25k}{k^2 - 16}$

13. $\dfrac{u^2 + 4u + 4}{4 - u^2} \cdot \dfrac{u - 2}{u^2 + 2u}$

14. $\dfrac{9 - t^2}{t - 3} \cdot \dfrac{t^2 - 3t}{t^2 - 6t + 9}$

15. $\dfrac{v^2 - 25}{v^2 - 16} \cdot \dfrac{v^2 + v - 12}{v^2 + 2v - 15}$

16. $\dfrac{s^2 - 2s - 3}{s^2 - 8s + 12} \cdot \dfrac{s^2 - 36}{s^2 - 9}$

17. $\dfrac{a^3 - b^3}{ab} \cdot \dfrac{a^2b^3}{(a^2 + ab + b^2)}$

18. $\dfrac{r^3 - 8}{r^3 + 8} \cdot \dfrac{(r - 1)(r^2 - 2r + 4)}{(r + 1)(r^2 + 2r + 4)}$

19. $\dfrac{x^2 - x - 2}{x + 2} \div \dfrac{x + 1}{x^2 - 4}$

20. $\dfrac{y^2 - 36}{y + 3} \div \dfrac{y + 6}{y^2 + 2y - 3}$

21. $\dfrac{u^2 + 5u + 6}{u^2 - 9} \div \dfrac{2 + u}{3 - u}$

22. $\dfrac{1 - 4w^2}{2w^2 - 3w + 1} \div \dfrac{4w^2 + 8w + 3}{2w^2 - 5w + 3}$

23. $\dfrac{x^2 - 25}{x^2 - 9} \div \dfrac{x^2 - x - 20}{x^2 + 7x + 12}$

24. $\dfrac{t^2 - 7t + 10}{t^2 - 6t - 16} \div \dfrac{t^2 - 11t + 18}{t^2 - 9t + 8}$

25. $\dfrac{x^3}{x^2 - xy + y^2} \cdot \dfrac{x}{x^2 - y^2} \div \dfrac{x^4}{x^3 + y^3}$

26. $\dfrac{18b^2 - 57b - 10}{18b^2 + 33b + 5} \div \dfrac{12b^2 - 47b + 45}{12b^2 + 5b - 72} \cdot \dfrac{6b^2 - 11b - 35}{6b^2 - 5b - 56}$

27. $\dfrac{a^2 - 2a}{2a + 4} \div \dfrac{a^2 - 4}{a^2 + 4a + 4} \div \dfrac{a^2 - 4a + 4}{2a^2 + 4a}$

28. $\left(\dfrac{u^2 - u - 2}{u + 1}\right) \div \left(\dfrac{u^2 - u - 20}{u^2 + 2u - 8} \div \dfrac{u^2 - 25}{u^2 + 5u}\right)$

29. $\dfrac{1}{t + 1} + \dfrac{1}{t}$

30. $\dfrac{1}{u + 1} - \dfrac{3}{u}$

31. $\dfrac{2}{w - 1} - \dfrac{1}{w}$

32. $\dfrac{2}{z} - \dfrac{1}{2z}$

33. $\dfrac{r}{2 - r} - \dfrac{r}{r - 2}$

34. $\dfrac{4s}{2s + 5} - \dfrac{6s}{3s - 1}$

35. $\dfrac{1}{x - 3} - \dfrac{x + 2}{x^2 - 3x}$

36. $\dfrac{3}{r + 3} + \dfrac{5}{r^2 - 9}$

37. $\dfrac{q}{9 - q^2} + \dfrac{3}{q^2 - 9}$

38. $\dfrac{m}{m^2 - 4} + \dfrac{2}{4 - m^2}$

39. $\dfrac{2u + v}{2v - u} + \dfrac{u + 2v}{v - 2u}$

40. $\dfrac{1 - k}{k^2 - 3k + 2} + \dfrac{2}{k + 1}$

41. $\dfrac{p + 3}{p^2 - p - 2} - \dfrac{1}{p - 2}$

42. $\dfrac{1}{c^2 - 4c + 3} - \dfrac{1}{c^2 - c - 6}$

43. $\dfrac{2a}{a - 2} - \dfrac{a^2 - 5}{a^2 - 4a + 4} - 1$

44. $\dfrac{3b}{b - 3} - \dfrac{2}{b^2 - 6b + 9} - 2$

45. $\dfrac{12p - 6q}{3p^2 - 3q^2} + \dfrac{5}{p + q} - \dfrac{1}{p - q}$

46. $\dfrac{3s + 4}{s^2 - 16} - \dfrac{s + 2}{s^2 - 5s + 4} + \dfrac{1}{s + 4}$

47. $\dfrac{3w^3}{w^3 + 27} - \dfrac{w}{w + 3} + \dfrac{2w^2}{w^2 - 3w + 9}$

48. $\dfrac{2t + 1}{4t^2 - 4t + 1} + \dfrac{1 - 18t}{16t^4 - 8t^2 + 1} - \dfrac{3 + t}{4t^2 + 4t + 1}$

49. $(t + 3)\left(\dfrac{1}{t} - \dfrac{1}{3}\right)$

50. $(x - y)\left(\dfrac{1}{x} + \dfrac{1}{y}\right)$

51. $\dfrac{s - \dfrac{9}{s}}{s - 3}$

52. $\dfrac{\dfrac{4}{t} - t}{t + 2}$

53. $\left(1 - \dfrac{y^2}{4x^2}\right) \div \left(\dfrac{y}{2x} - 1\right)$

54. $\left(\dfrac{v^2}{9u^2} - 1\right) \div \left(\dfrac{v}{3u} + 1\right)$

55. $\dfrac{\dfrac{1}{x} + \dfrac{1}{y}}{\left(\dfrac{1}{xy}\right)}$

56. $\dfrac{\dfrac{1}{u} - \dfrac{1}{v}}{\left(\dfrac{1}{uv}\right)}$

57. $\dfrac{\left(\dfrac{1}{r} + \dfrac{1}{s}\right)}{r + s}$

58. $\dfrac{\left(\dfrac{1}{a} - \dfrac{1}{b}\right)\left(\dfrac{1}{a - b}\right)}{\left(\dfrac{1}{a} + \dfrac{1}{b}\right)\left(\dfrac{1}{a + b}\right)}$

59. $\dfrac{3a}{3 - \dfrac{2}{a}} - \dfrac{9a^2 - \dfrac{2}{a} - 1}{9a + \dfrac{4}{a} - 12}$

60. $x - \dfrac{x}{x - \dfrac{1}{x}}$

61. $\dfrac{1 + \dfrac{1}{1 - \dfrac{1}{w}}}{1 - \dfrac{3}{1 - \dfrac{1}{w}}}$

62. $\dfrac{r + \dfrac{1}{1 + \dfrac{1}{1 + r}}}{r - \dfrac{1}{1 - \dfrac{5}{1 - r}}}$

63. $\dfrac{1 + m^{-1}}{1 - m^{-1}}$

64. $\dfrac{u + v^{-1}}{u - v^{-1}}$

65. $\dfrac{x^{-1} + y^{-2}}{x^{-2} + y^{-1}}$

66. $\dfrac{p^{-1} + q^{-1}}{(p + q)^{-1}}$

67. $\dfrac{rs^{-2} - sr^{-2}}{r^{-2} - s^{-2}}$

68. $\dfrac{xy^{-1} - yx^{-1}}{x^{-1} + y^{-1}}$

69. $\dfrac{1 - p^{-1}}{1 - p^{-2}}$

70. $\dfrac{1 + mk^{-1}}{1 - m^2k^{-2}}$

71. $\dfrac{(q + 1)^{-1} - q^{-1}}{(q + 1)^{-1}}$

72. $\dfrac{1 - (1 + z)^{-1}}{1 - (1 - z)^{-1}}$

Use long division on the following.

73. $\dfrac{12a^4 + 25a^2 + 42}{2a^2 + 3a + 4}$

74. $\dfrac{v^7 + 1}{v + 1}$

75. $\dfrac{2y^3 + 15y - 3y^2 - 1}{y^2 - 5}$

76. $\dfrac{7t^2 + t - t^3 - 3}{t + 3}$

77. $(x^4 + 2x^3 - 13x^2 - 14x + 24) \div (x + 2)$

78. $\dfrac{3s^3 - 2s + s^6 - 1}{s - 2}$

79. The Glossy Wax Company finds that its manufacturing costs for N quarts of wax are $\dfrac{(2N^2 + 7N + 6)}{(N + 2)}$ cents. The Environmental Protection Agency (EPA) and OSHA then require additional manufacturing safeguards and procedures that cost $\dfrac{(N + 2)}{3}$ cents for N quarts. Find an expression for the total cost of manufacturing N quarts and simplify this expression.

80. The yield of a certain plot of ground is generally $\dfrac{8}{(x + 4)}$ hundred bushels, where x is the number of rainless days in the growing season. An invasion of feasting beetles reduces the yield by $\dfrac{(2x + 3)}{[(x + 3)(x + 4)]}$ hundred bushels. What is the resulting yield?

81. The force exerted by two objects on one another is given by $\dfrac{6.67(10^{-11})m_1 m_2}{r^2}$, where m_1, m_2 are the respective masses of the objects in kilograms, and r is the distance between them in meters. Let three masses of 10 kilograms each be located as illustrated. Calculate the net force acting on each of the three masses.

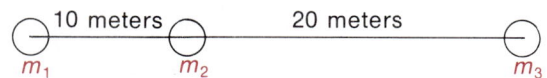

82. In the network of blood vessels shown in the figure, let x, y, and z denote the various resistances.

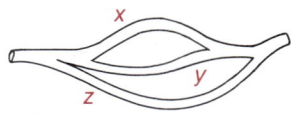

   a. Describe the net resistance as a rational expression. (See Example 6.)
   b. Find the net resistance if $x = 10$, $y = 10$, and $z = 20$.

83. In working with cameras and projectors, the distance of the image from the lens is given by the expression

$$\dfrac{1}{\dfrac{1}{x} - \dfrac{1}{y}}$$

where x is the focal length of the lens and y is the distance of the object or slide from the lens.

   a. Simplify this expression.
   b. Amy wishes to snap a picture of her current boyfriend standing 3 meters away. The lens on her camera has a focal length of 50 millimeters. How far must the lens be positioned from the film in order to focus the picture?
   c. Later Amy wishes to view the slide she took. She uses a projector with a fixed lens having a 10-centimeter focal length. The slides are positioned 10.2 centimeters behind the lens. Where should she place the screen?

Section 1.9

Complex Numbers

In the first section of this chapter we reviewed the operations on the real number system. At an early age, we became familiar with the "natural" or counting numbers and the fractions or rational numbers. The real numbers are more extensive than these, encompassing irrational numbers such as π as well. Every positive real number has two real square roots, but no negative real number has a real square root. This happens because

$$\sqrt{a} = b \quad \text{means} \quad a = b^2$$

and $b^2 \geq 0$ for every real number b, so a must be nonnegative (≥ 0) in order to have a real square root. For example

$$\sqrt{4} = 2 \quad \text{since} \quad 4 = 2^2 > 0$$

and

$$\sqrt{9} = 3 \quad \text{since} \quad 9 = 3^2 > 0$$

but
$$\sqrt{-8} \text{ is not a real number}$$
since $-8 < 0$.

We shall now extend the real numbers to complex numbers in order to generate square roots of negative numbers. Although the complex numbers may seem to be purely artificial mathematical constructions, they are indeed very useful in describing the behavior of certain electronic phenomena.

Our purpose here is to extend the real numbers to a larger number system in which square roots of negative numbers exist. If we agree to write

$$\sqrt{-a} = \sqrt{(-1) \cdot a} = \sqrt{-1}\sqrt{a}$$

for every $a > 0$, we see that by introducing a symbol i for $\sqrt{-1}$, we shall effectively introduce square roots for all negative numbers. Thus, for example,

$$\sqrt{-1} = i$$
$$\sqrt{-49} = \sqrt{-1}\sqrt{49} = 7i$$
$$\sqrt{-25} = \sqrt{-1}\sqrt{25} = 5i$$
$$\sqrt{-2} = \sqrt{-1}\sqrt{2} = \sqrt{2}i$$

DEFINITIONS
1. $i = \sqrt{-1}$
2. $i^2 = -1$
3. $\sqrt{-a} = \sqrt{a}\,i$ when $a \geq 0$.

CAUTION

The familiar rule $\sqrt{ab} = \sqrt{a}\sqrt{b}$ for positive real numbers does not necessarily hold for negative real numbers a and b. For instance

$$\sqrt{-4}\sqrt{-9} = 2i \cdot 3i = 6i^2 = -6$$

but

$$\sqrt{(-4)(-9)} = \sqrt{36} = 6$$

Historical Perspective

Just as the Pythagoreans had difficulty accepting numbers like $\sqrt{2}$, which were not rational, mathematicians of the seventeenth century (and many students today) found it difficult to accept numbers of the form $\sqrt{-4}$, which could not be real numbers. Descartes (1596–1650) called these numbers "imaginary," thereby giving them a permanent stigma. However, they are no more imaginary than any other mathematical abstraction.

Just as with positive real numbers, every negative real number $-x$ has two square roots given by $\pm\sqrt{-x}$ or $\pm i\sqrt{x}$. The notation $\sqrt{-x}$ denotes the **principal root** $i\sqrt{x}$.

In the real number system we must deal with numbers of the form $1 + \sqrt{2}$ and $7 - 3\sqrt{2}$; similarly, introduction of the number or symbol i means that we must consider numbers of the form $1 + i$ and $7 - 3i$.

DEFINITION

The system of **complex numbers** consists of numbers of the form

$$a + bi \qquad a, b \text{ real numbers}$$

The number i satisfies

$$i^2 = -1$$

The number a is called the **real part** and bi the **imaginary part** of the complex number $a + bi$. Addition and multiplication are defined according to the usual algebraic rules of combination, replacing i^2 with -1 whenever it appears:

$$(a + bi) + (c + di) = a + c + bi + di$$
$$= (a + c) + (b + d)i \qquad (1)$$
$$(a + bi)(c + di) = ac + bdi^2 + bci + adi$$
$$= (ac - bd) + (bc + ad)i \qquad (2)$$

It can be shown that the commutative, associative, and distributive laws also hold for complex numbers.

EXAMPLE 1

Simplify each of the following into the form $a + bi$.

a. $(5 + 3i) + (2 + 4i)$
b. $(5 + 3i) - (2 + 4i)$
c. $2(5 + 3i) - 3(2 - 4i)$
d. $(5 + 3i)(2 + 4i)$
e. $(2 + 4i)^2$

SOLUTION

a. $(5 + 3i) + (2 + 4i) = 7 + 7i$ (Formula 1)

b. $(5 + 3i) - (2 + 4i) = 5 + 3i - 2 - 4i$
$$= 3 - i \qquad \text{(Formula 1)}$$

c. $2(5 + 3i) - 3(2 - 4i) = 10 + 6i - 6 + 12i$
$$= 4 + 18i$$

d. $(5 + 3i)(2 + 4i) = (10 - 12) + (20 + 6)i$ (Formula 2)
$$= -2 + 26i$$

e. $(2 + 4i)(2 + 4i) = (4 - 16) + (8 + 8)i$ (Formula 2)
$$= -12 + 16i$$

We note that

$i^2 = -1$
$i^3 = i^2 i = (-1)i = -i$
$i^4 = i^2 i^2 = (-1)(-1) = 1$
$i^5 = i^4 i = 1 \cdot i = i$
$i^6 = i^4 i^2 = 1 \cdot i^2 = -1$

Further powers of i cycle through the values $-i$, 1, i, and -1 repeatedly. For example,

$i^{19} = i^{16} i^3 = (i^4)^4 i^2 i$
$= 1^4 \cdot (-1) \cdot i$
$= -i$

POWERS OF i

For each positive integer n,

$$i^n = \begin{cases} 1, & \text{if } n = 4k \\ -1, & \text{if } n = 4k + 2 \\ i, & \text{if } n = 4k + 1 \\ -i, & \text{if } n = 4k + 3 \end{cases}$$

where k is a nonnegative integer.

EXAMPLE 2 Simplify each of the following into the form $a + bi$.

a. i^{15}
b. i^{129}
c. $i^{27} - i^{36}$
d. $2i^5(4 + 3i) - 3i^3(2 - 4i)$

SOLUTION

a. $i^{15} = i^{12} i^3 = (i^4)^3 i^2 i = 1^3(-1)i = -i = 0 - i$

b. First divide 129 by 4: $129 = 4 \cdot 32 + 1$. Thus

$i^{129} = i^{128} i = (i^4)^{32} i = 1^{32} i = i = 0 + i$

c. $i^{27} - i^{36} = i^{24} i^3 - (i^4)^9$
$= (i^4)^6 i^3 - 1^9$
$= 1^6 i^2 i - 1$
$= (-1)i - 1$
$= -1 - i$

d. $2i^5(4 + 3i) - 3i^3(2 - 4i) = 2i^4 i(4 + 3i) - 3i^2 i(2 - 4i)$
$= 2i(4 + 3i) + 3i(2 - 4i)$
$= 8i + 6i^2 + 6i - 12i^2$
$= 8i - 6 + 6i + 12$
$= 6 + 14i$

Section 1.9 Complex Numbers   69

> **DEFINITION**
>
> Two complex numbers are **equal** if their real and imaginary parts are equal; that is,
>
> $$(a + bi) = (c + di) \quad \text{means} \quad a = c, b = d$$

Thus,

$$a + bi = 7 + 3i \quad \text{means} \quad a = 7, b = 3$$

Division

Division of complex numbers is accomplished in the same way that fractions with denominators of the form $a + b\sqrt{c}$ were simplified. Since the symbol $\sqrt{-1}$ now makes sense, we can simplify an expression of the form

$$\frac{2 + \sqrt{-1}}{3 + 4\sqrt{-1}}$$

by "rationalizing the denominator." The numerator and denominator must be multiplied by the **conjugate** of the denominator: $3 - 4\sqrt{-1}$.

$$\frac{2 + \sqrt{-1}}{3 + 4\sqrt{-1}} \cdot \frac{3 - 4\sqrt{-1}}{3 - 4\sqrt{-1}} = \frac{6 - 5\sqrt{-1} - 4(-1)}{9 - 16(-1)} = \frac{10 - 5\sqrt{-1}}{25}$$

$$= \frac{2}{5} - \frac{1}{5}i$$

> **DEFINITION**
>
> The **conjugate** of a complex number $z = a + bi$ is the complex number
>
> $$\bar{z} = a - bi$$

For example,

$$\overline{1 + 3i} = 1 - 3i$$
$$\overline{1 - 3i} = 1 - (-3)i = 1 + 3i$$

Note that the product of a complex number and its conjugate is always real and nonnegative.

$$(a + bi)\overline{(a + bi)} = (a + bi)(a - bi)$$
$$= a^2 - (bi)^2$$
$$= a^2 - b^2 i^2$$
$$= a^2 + b^2 \geq 0$$

Both a and b are real numbers.

70 Chapter 1 Basic Algebra

> **PRODUCT OF A COMPLEX NUMBER AND ITS CONJUGATE**
>
> For all complex numbers z, the product $z \cdot \bar{z}$ is a **nonnegative real number:**
>
> $$z\bar{z} \geq 0 \quad \text{for all complex numbers } z$$
>
> $$(a + bi)(a - bi) = a^2 + b^2 \geq 0 \tag{3}$$

It is this property that makes division by complex numbers possible.

> **DIVISION WITH COMPLEX NUMBERS**
>
> *To divide one complex number by another, write the indicated division as a "fraction" and multiply both numerator and denominator by the conjugate of the denominator.* The result is a new fraction with a positive real denominator. Write the result in the form $a + bi$ by dividing both parts of the numerator by the real denominator.

EXAMPLE 3 Simplify the following into the form $a + bi$.

a. $\dfrac{2}{3i}$ b. $\dfrac{-2}{3+i}$ c. $\dfrac{2+i}{3-2i}$ d. $\dfrac{2-i}{(1+i)(3+i)}$

SOLUTION

a. $\dfrac{2}{3i} = \dfrac{2}{3i} \cdot \dfrac{(-3i)}{(-3i)} = \dfrac{-6i}{9} = \dfrac{-2}{3}i$

b. $\dfrac{-2}{3+i} = \dfrac{-2}{3+i} \cdot \dfrac{3-i}{3-i}$

$= \dfrac{-6 + 2i}{3^2 + 1^2}$ (Formula 3)

$= \dfrac{-6 + 2i}{10}$

$= \dfrac{-3}{5} + \dfrac{1}{5}i$

c. $\dfrac{2+i}{3-2i} = \dfrac{2+i}{3-2i} \cdot \dfrac{3+2i}{3+2i}$

$= \dfrac{6 - 2 + 7i}{9 + 4}$ (Formulas 2 and 3)

$= \dfrac{4 + 7i}{13}$

$= \dfrac{4}{13} + \dfrac{7}{13}i$

d. $\dfrac{2-i}{(1+i)(3+i)} = \dfrac{2-i}{3-1+4i} = \dfrac{2-i}{2+4i}$ (Formula 2)

$= \dfrac{2-i}{2+4i} \cdot \dfrac{2-4i}{2-4i}$

$= \dfrac{4-4-10i}{4+16} = \dfrac{0-10i}{20}$ (Formulas 2 and 3)

$= 0 - \dfrac{1}{2}i$ ∎

EXAMPLE 4

Simplify the following into the form $a + bi$.

$$\dfrac{3}{2+i} - \dfrac{2}{1-i}$$

SOLUTION

To combine the fractions, we must have a common denominator.

$\dfrac{3}{2+i} - \dfrac{2}{1-i} = \dfrac{3}{2+i} \cdot \dfrac{1-i}{1-i} - \dfrac{2}{1-i} \cdot \dfrac{2+i}{2+i}$

$= \dfrac{3-3i}{2-i+1} - \dfrac{4+2i}{2-i+1}$

$= \dfrac{3-3i-4-2i}{3-i}$

$= \dfrac{-1-5i}{3-i}$

$= \dfrac{-1-5i}{3-i} \cdot \dfrac{3+i}{3+i}$

$= \dfrac{-3-i-15i+5}{3^2+1^2}$

$= \dfrac{2-16i}{10} = \dfrac{1}{5} - \dfrac{8}{5}i$ ∎

Section 1.9 Exercises

Simplify each of the following into a real or imaginary number.

1. $\sqrt{-9}$
2. $\sqrt{-16}$
3. $\sqrt{-36}$
4. $\sqrt{-64}$
5. $\sqrt{-121}$
6. $\sqrt{-81}$
7. $\sqrt{-5}\sqrt{-20}$
8. $\sqrt{-3}\sqrt{-12}$
9. $\sqrt{-2}\sqrt{-32}$
10. $\sqrt{-7}\sqrt{-28}$
11. $\dfrac{\sqrt{-12}}{\sqrt{-3}}$
12. $\dfrac{\sqrt{-20}}{\sqrt{-5}}$
13. $\dfrac{\sqrt{-28}}{\sqrt{-7}}$
14. $\dfrac{\sqrt{-32}}{\sqrt{-2}}$

Simplify each of the following into the form $a + bi$.

15. $(2+i) + (1-3i)$
16. $(5+12i) + (4+3i)$
17. $(3+i) + (2-5i)$
18. $(2+3i) + (4-5i)$
19. $(2+3i) - (-2-4i)$
20. $(1+3i) - (-2+5i)$
21. $(3-2i) - (5-4i)$
22. $2(2+5i) + 3i(3-2i)$

23. $i^2(10 + i) - 3(2 - 4i)$
24. $i(4 + 9i) - 2i^2(1 + 3i)$
25. $2i^3(3 + 2i) - 2i(3 + 2i)$
26. $(1 - 2i)^2 i^3$
27. $3i(1 + i) - 2i^3(1 - i)$
28. $(1 - i)(2 + i)$
29. $(1 - 4i)(3 - i)$
30. $(2 - i)(1 - 3i)$
31. $(3 - 4i)(3 + 4i)$
32. $(1 + 2i)(1 + i)i^2$
33. $(2 + i)^2$
34. $(1 - \sqrt{3}i)^2$
35. $(1 + 2i)^3$
36. $(1 + 2i)^2(1 - i)^3$
37. $(2 - i)^{30}(i^2 + 1)$
38. i^{39}
39. i^{632}
40. $i^{57} - i^{26}$
41. $i^{17} + i^{12} - i^3$
42. $3i^{23} - 2i^{42} + 3i^3$
43. $-2i^{63} - (2i)^3 + 2i^3$
44. $4i^{27} + 3i^7 - (5i)^2$

Find values of x and y which satisfy the following.

45. $x + 2i = 5 + yi$
46. $2x + 10i = 10 + 2yi$
47. $3x + 5iy = 15$
48. $2x + 5iy = 15i$
49. $-2x + 3yi = 8 - 21i$
50. $(x + 1) + 3(y - 1)i = 5 - 6i$

Find $\bar{z}$ for each of the following.

51. $2 - 5i$
52. $3 + 4i$
53. $1 + \sqrt{2}i$
54. $-3 - 2i$
55. $-5 - 3i$
56. $4 - 3i$
57. $-2 + 6i$
58. $-6 - 3i$
59. $\sqrt{-4} + \sqrt{-9}i$
60. $\sqrt{-25} - \sqrt{16}i^3$

Write each of the following in the form $a + bi$.

61. $\dfrac{-2}{1 + 3i}$
62. $\dfrac{3i}{1 + i}$
63. $\dfrac{-3i^3}{2 - i}$
64. $\dfrac{2 - i}{1 - 2i}$
65. $\dfrac{-2 + \sqrt{3}i}{2 - \sqrt{3}i}$
66. $\dfrac{3 - 4i}{1 + \sqrt{2}i}$
67. $\dfrac{2 - 3i}{3 + 2i}$
68. $\dfrac{3 + 2i}{2 - 3i}$
69. $\dfrac{1 + i}{1 - i}$
70. $\dfrac{4 - 3i}{3 + 4i}$
71. $\dfrac{5 + 3i}{2 - i}$
72. $\dfrac{1 + i}{(2 + i)^2}$
73. $\dfrac{1 + i}{(3 + i)(2 - i)}$
74. $\dfrac{2}{1 + i} - \dfrac{3}{i - 1}$
75. $\dfrac{1}{1 + i} - \dfrac{1}{1 - i}$
76. $\dfrac{2i}{4 + i} + \dfrac{3}{2 - i}$
77. $\dfrac{5}{i + 1} + \dfrac{7i}{2 + i}$
78. $\dfrac{2 + i}{2 - i} + \dfrac{1 + i}{1 - i}$
79. $\dfrac{1 - i}{1 + i} - \dfrac{1 + i}{1 - i}$
80. $\dfrac{3 + i}{2 - i} + \dfrac{2 + i}{1 - i}$
81. $\dfrac{-1 + 3i}{1 - 2i} + \dfrac{i - 1}{i + 1}$

Write $z = a + bi$, $w = c + di$ and show that the following formulas hold.

82. $\overline{z + w} = \bar{z} + \bar{w}$
83. $\overline{zw} = \bar{z}\bar{w}$
84. $\bar{z} = z$ if and only if z is a real number.
85. $\bar{\bar{z}} = z$
86. $\overline{\left(\dfrac{z}{w}\right)} = \dfrac{\bar{z}}{\bar{w}}$
87. $z + \bar{z}$ is always a real number.
88. $z + \bar{z} = 0$ if and only if the real part of z is 0.
89. $\overline{z - w} = \bar{z} - \bar{w}$
90. $\overline{az} = a\bar{z}$ for a a real number.

Section 1.10 Chapter Review

Terms and Concepts

Natural Number 1, 2, 3, ...
  1. Prime number 2, 3, 5, 7, 11, 13, ...
  2. Composite number $6 = 2 \cdot 3$
  3. Factor 1, 2, 3, 6 are factors of 6
  4. LCM LCM (4, 6) = 12
  5. GCD GCD (4, 6) = 2

Integer $\ldots, -2, -1, 0, 1, 2, \ldots$

Rational Number $\dfrac{m}{n}, n \neq 0$

Irrational Number $\sqrt{2}, \pi$

Real Number Includes all of the above

Complex Number $a + bi, i^2 = -1$

Complex Conjugate $\overline{a + bi} = a - bi$

Decimal Representation $0.67 = \dfrac{67}{100}$

Scientific Notation $12{,}340 = 1.234(10^4)$

Set $\{a, b, c\}$

Interval (a, b)

Polynomial $a_n x^n + a_{n-1} x^{n-1} + \cdots + a_1 x + a_0$

Rules and Formulas

Real Number Properties	See page 3.
Rules for Negative Numbers	See pages 4–5.
Rules for Fractions	See page 6.
Rules for Inequalities	See page 15.
Rules for Absolute Value	See page 18.
Rules for Exponents	See pages 23, 40.
Rules for Radicals	See page 32.

Set Operations

1. Intersection: $A \cap B$ $\{1, 2, 3\} \cap \{2, 3, 4\} = \{2, 3\}$
2. Union: $A \cup B$ $\{1, 2, 3\} \cup \{2, 3, 4\} = \{1, 2, 3, 4\}$
3. Difference: $A \setminus B$ $\{1, 2, 3\} \setminus \{2, 3, 4\} = \{1\}$

Hierarchy of Operations

1. Expressions in parentheses are to be evaluated before being combined with numbers outside the parentheses.

2. Exponentiations are done before other multiplications and divisions.

3. Multiplications and divisions are performed before additions and subtractions and in the order in which they appear from left to right.

4. Then additions and subtractions are done in the order in which they appear from left to right.

$(8 + 2) \div 2 + 6 \div 3 \cdot 2^2 + 9 \cdot (2 + 1)$
$= 10 \div 2 + 6 \div 3 \cdot 2^2 + 9 \cdot 3$

$= 10 \div 2 + 6 \div 3 \cdot 4 + 9 \cdot 3$

$= 5 + 2 \cdot 4 + 27 = 5 + 8 + 27$

$= 40$

Rounding to Significant Digits

1. Addition and subtraction.
Convert the numbers to decimal form and round the answer to the number of accurate decimal positions in the least accurate of the numbers in the computation.

$2.361 + 42.4 = 44.761 \approx 44.8$

2. Multiplication and division.
Round the answer to the least number of significant digits in the numbers used in the computation.

$(56.1)(382.4) = 5.61(10^1)[3.824(10^2)]$
$= (2.145264)10^4 \approx 2.15(10^4)$
$= 21{,}500$

Techniques

Factoring and Expanding

1. $a^2 - b^2 = (a + b)(a - b)$ $\qquad$ $x^2 - 9 = (x + 3)(x - 3)$
2. $a^2 \pm 2ab + b^2 = (a \pm b)^2$ $\qquad$ $x^2 - 6x + 9 = (x - 3)^2$
3. $a^3 \pm b^3 = (a \pm b)(a^2 \mp ab + b^2)$ $\qquad$ $x^3 + 1 = (x + 1)(x^2 - x + 1)$
4. $(a \pm b)^3 = a^3 \pm 3a^2b + 3ab^2 \pm b^3$ $\qquad$ $(x - 1)^3 = x^3 - 3x^2 + 3x - 1$
5. FOIL factoring $\qquad$ $2x^2 + x - 3 = (2x + 3)(x - 1)$
6. Grouping and taking out common factors

$$2ax + 3bx - 2ay - 3by = (2a + 3b)x - (2a + 3b)y$$
$$= (2a + 3b)(x - y)$$

7. Completing the square

$$x^4 + x^2 + 1 = (x^4 + 2x^2 + 1) - x^2$$
$$= (x^2 + 1)^2 - x^2$$
$$= [(x^2 + 1) + x][(x^2 + 1) - x]$$

Combining Rational Expressions

1. To multiply, multiply numerators and multiply denominators separately.

$$\frac{3}{x} \cdot \frac{2}{x + 1} = \frac{6}{x(x + 1)}$$

2. To divide, invert the divisor and multiply.

$$\frac{3}{x} \div \frac{6}{x} = \frac{3}{x} \cdot \frac{x}{6} = \frac{3x}{6x}$$

3. To add or subtract:
   a. First find the least common denominator (LCD).
   b. Write each rational expression in terms of this LCD.
   c. Then, add or subtract numerators.

$$\frac{3}{x} - \frac{2}{x + 1} = \frac{3(x + 1)}{x(x + 1)} - \frac{2x}{x(x + 1)} = \frac{x + 3}{x(x + 1)}$$

4. Simplify the results by canceling like factors in the numerator and denominator.

$$\frac{3x}{6x} = \frac{3x}{2 \cdot 3x} = \frac{1}{2}$$

Long Division of Polynomials

1. Always write the polynomial with terms in order of *decreasing powers* of one of the variables. Then at each stage, only the first terms of the expressions need be examined.

2. Leave room for any missing powers of x since they may appear in the intermediate stages of the long division.

3. Just as in long division of numbers, a remainder sometimes occurs. The remainder will be a polynomial whose degree is less than that of the divisor.

$$\begin{array}{r} x - 1 \\ x + 1 \overline{)\, x^2 + 1} \\ \underline{-(x^2 + x)} \\ -x \\ \underline{-(-x - 1)} \\ 2 = R \end{array}$$

Division by a Complex Number or by Radicals, Rationalizing

1. Complex number

$$\frac{1}{a + bi} = \frac{1}{a + bi} \cdot \frac{a - bi}{a - bi} = \frac{a - bi}{a^2 + b^2}$$

2. Radicals

$$\frac{1}{\sqrt{3} - \sqrt{2}} = \frac{1}{\sqrt{3} - \sqrt{2}} \cdot \frac{\sqrt{3} + \sqrt{2}}{\sqrt{3} + \sqrt{2}} = \frac{\sqrt{3} + \sqrt{2}}{3 - 2}$$
$$= \sqrt{3} + \sqrt{2}$$

Section 1.11 Supplementary Exercises

Factor the following into primes.

1. 76
2. 341
3. 5280

Find the LCM and GCF of the following sets of numbers.

4. 68, 76
5. 210, 1425
6. 18, 24, 40

Locate each of the following points on the number line.

7. 1.25
8. −3.63
9. 2.333…

State which properties of real numbers (commutative, associative, distributive, identity, or inverse) are used in the following equalities.

10. $(3 + 0) \cdot 7 = 7 \cdot 3$
11. $\dfrac{1}{5 + 3}(3 + 5) = 1$
12. $3 \cdot (-2 + 4) = (-2) \cdot 3 + 3 \cdot 4$
13. $6 \cdot (9 \cdot 1) = 9 \cdot 6$
14. $[3 \cdot (-6)] \cdot 7 = (-6)(3 \cdot 7)$

Reduce each of the following to lowest terms.

15. $\dfrac{21}{66}$
16. $\dfrac{96}{108}$
17. $\dfrac{-360}{144}$

Express each of the following as a single integer.

18. $2 + (-5)$
19. $(-3)(5)$
20. $8 - 2 \cdot 4 + (-10) - 7(6 - 8)$
21. $7 - 2[8 + 2 - (-9)] + 6$
22. $\left(\dfrac{1}{2}\right)[4 + 8(6 - 22)(-1)]$
23. $-2 + 3 \cdot 4 - 3 \div 6 - 8$
24. $-2 + 3 \cdot (4 - 3) \div (6 - 8)$
25. $(-2 + 3) \cdot 4 - (3 \div 6 - 8)$

Perform the indicated operations and express the answer in lowest terms.

26. $\dfrac{1}{4} + \dfrac{1}{3}$
27. $\dfrac{2}{3} - \dfrac{1}{5} + 3\dfrac{1}{2}$
28. $\dfrac{1}{2} + \dfrac{1}{3} - \dfrac{1}{4}$
29. $\dfrac{25}{28} \cdot \dfrac{14}{15} \cdot \dfrac{36}{10}$
30. $\dfrac{7}{9} \div \dfrac{-4}{15}$
31. $2\dfrac{1}{3} \div 4\dfrac{1}{2}$
32. $\left(\dfrac{1}{2} - \dfrac{2}{3}\right) \div \left(\dfrac{1}{3} - \dfrac{3}{4}\right)$
33. $\dfrac{7 + \dfrac{1}{6}}{\dfrac{2}{3} + \dfrac{5}{14}}$

Calculator Exercises

Be sure to enter data in such a way that the correct result appears.

34. a. $3 \div 13 \cdot 6$
    b. $3 \div (13 \cdot 6)$
35. a. $7 \div 11 \cdot 12$
    b. $7 \div (11 \cdot 12)$
36. a. $6 + 2 \div 5 - 1 \cdot 3 + 2$
    b. $(6 + 2) \div (5 - 1) \cdot (3 + 2)$
37. a. $6 \div 2 \cdot 5 + 1 \div 8 - 2$
    b. $6 \div (2 \cdot 5 + 1) \div (8 - 2)$
38. a. $12 \cdot 3 + 5 - 16 \div 8 + 2$
    b. $12 \cdot (3 + 5) - 16 \div (8 + 2)$

Let

$$A = \{5, 6, 7, 8, 9\}$$
$$B = \{2, 3, 4, 5, 6, 7\}$$
$$C = \{0, 2, 4, 6, 8, 10\}$$

Find the following sets.

39. $A \cup (B \cap C)$
40. $B \cap (A \cup C)$
41. $C \setminus (A \cap B)$
42. $(A \setminus B) \setminus C$
43. $A \setminus (B \setminus C)$

Round each of the numbers to the nearest thousandth.

44. $\dfrac{5}{7}$
45. $\dfrac{8}{9}$
46. $\dfrac{37}{13}$
47. $1.23434\ldots$
48. $2.506506\ldots$
49. $15.20652065\ldots$

Express an order relation ($<$, $=$, $>$) between the following numbers. A bar denotes a repeating block of digits.

50. $2, -4, -10, 5$
51. $\dfrac{2}{3}, \dfrac{3}{4}, \dfrac{5}{7}, \dfrac{7}{11}$
52. $0.\overline{204}, 0.2\overline{04}$

53. $13.24\overline{5}, 13.2\overline{45}, 13.\overline{245}$

54. $\dfrac{2}{3}, \dfrac{262}{393}$

55. $|-2|, -2, -|-3|, 4$

Use the inequality symbols to express the following.

56. x is greater than -5.

57. y is strictly between -2 and 4.

58. z is strictly between -3 and -5.

Sketch the following intervals on the real number line.

59. $(-3, 6]$ 60. $[2, 5]$ 61. $(1, \infty)$

Describe each of the following sets as an interval.

62. The set of all s such that $4 < s < 7$.

63. The set of all b such that $b \geq -3$.

64. The set of all u such that $|u| < 5$.

65. The set of all v such that $-4 \leq v < -3$.

66. The set of all w such that $7 \geq w \geq 2$.

67. $(-2, 3] \cap [2, 5)$

68. $[8, 12] \cup (2, 10)$ 69. $(-2, 3] \cup [2, 5)$

70. $(4, 7) \cap [6, 8]$ 71. $(4, 7) \cup [6, 8]$

72. $(-1, 5) \setminus [4, 6)$ 73. $(2, 8) \setminus (-5, 4]$

Evaluate the following expressions for the given values of a and b.

74. $|3x + 4y|, x = 2, y = -2$

75. $3|x| - 4|y|, x = -3, y = 2$

76. $|u - 3v| - |3u - v|, u = -4, v = 3$

77. $||r + 4s| - |2s - 3r||, r = -1, s = 2$

Write each of the following without using the absolute value symbols.

78. $|u - 3|, u < 3$ 79. $|2v + 6|, v < -3$

80. $|r^2 + 2|$ 81. $|r^2 - 4|, |r| \leq 2$

Find the distance between a and b as given.

82. $a = -3, b = -2$ 83. $a = 5, b = 10$

84. $a = 7, b = -6$ 85. $a = -3, b = 4$

Write each of the following using exponential notation.

86. $u \cdot u \cdot u \cdot u \cdot u \cdot u \cdot u$

87. $(r + t)(r + t)(r + t)$

88. $[(a - b)(a - b)(a - b)(a - b)] \div [(r + t)(r + t)(r + t)]$

Simplify the given expressions. Write the answers in terms of radicals or positive exponents if necessary. Rationalize any denominators in your answers.

89. $(2^3)^2$ 90. $\dfrac{(-3)^2}{(-3)^3}$

91. $\left(\dfrac{1}{16}\right)^{1/4}$ 92. $\left(\dfrac{1}{32}\right)^{-3/5}$

93. $\sqrt{16a^2}$ 94. $(2^{-2})^3$

95. $\dfrac{10^{20}}{10^{-4}}$ 96. $(81)^{3/4}$

97. $(-8)^{2/3}$ 98. $(25)^{-1/2}$

99. $\sqrt{144 + 25}$

100. $\sqrt[7]{0.0000128} \cdot \sqrt[5]{0.00243}$

101. $\dfrac{2}{\sqrt{7} - \sqrt{2}}$

102. $\dfrac{5}{\sqrt{5} - \sqrt{3}}$

103. $\dfrac{\sqrt{6} + \sqrt{3}}{\sqrt{6} - \sqrt{3}}$

104. $\sqrt[3]{54} + 3\sqrt[3]{24} + 2\sqrt[3]{250} - 5\sqrt[3]{-81}$

105. $(\sqrt[3]{9} - \sqrt[3]{6} + \sqrt[3]{4})(\sqrt[3]{2} + \sqrt[3]{3})$

106. $\sqrt[4]{64x^2y^6}$ 107. $(t^3)^2$

108. $\dfrac{x^{10}}{x^4}$ 109. $s^4 s^{-16}$

110. $(3x^{-2}y)(2y^{-3})$ 111. $(2x^2)^3$

112. $(ab^2c^3)^4$ 113. $[(x^2y)^3(yx^3)^2]^4$

114. $\dfrac{(x^2y^2)^3}{(x^3y)^2}$ 115. $\dfrac{x^{-10}}{x^{-3}}$

116. $\dfrac{xy^{-3}}{x^{-2}y}$

117. $(2x^{-1}y^2z^{-1})(6x^3y^{-1}z^2)$

118. $(ab^2c^{-3})^{-4}$ 119. $(uv^2w)^{-3}(u^{-2}vw^2)^{-2}$

120. $\left[\dfrac{(a^{-2}b^3)^{-4}}{(ab^{-5})^3}\right]^2$ 121. $\dfrac{(x-y)^{-2}}{x^{-2} - y^{-2}}$

122. $\sqrt[3]{125x^6y^{12}}$ 123. $\sqrt{8a^{10}b^2} \cdot \sqrt{2a^4b^6}$

124. $(a^{27}b^9z^3)^{2/3}$

125. $\dfrac{(a^{-3/2}y^{-1/4})^{-2}}{(y^{-2/3}z^{-5/6})^{-3}}$

126. $\left(\dfrac{25a^{-4}b^{-6}c^2}{16x^{-2}y^6z^{-2}}\right)^{-3/2} \cdot \left(\dfrac{2a^2b^{-1}c^{-3}}{x^2y^{-2}z^3}\right)^{-1}$

127. $(x^{3/4}y^{-3/8}z^{9/24})^{4/3}$

128. $\dfrac{x^2 - y^2}{\sqrt{x} + \sqrt{y}}$

Verify the following.

129. $2^3 + 4^3 \neq 6^3$

130. $1^{-3} + 2^{-3} \neq 3^{-3}$

131. $2^{-2} - 3^{-2} \neq (-1)^{-2}$

132. $(x + y)^{-2} \neq x^{-2} + y^{-2}$

Express each of the following in decimal notation.

133. $6.5(10^9)$

134. $4.2(10^{-6})$

135. $6.63(10^{-34})$

Express each of the following in scientific notation.

136. 16,700,000

137. 0.2372

138. 0.0000000019

139. 26.0014

Perform the following calculations by first converting the numbers to scientific notation. Express your answers in scientific notation and in terms of decimals.

140. $\dfrac{(0.000004)(1{,}200{,}000)}{0.0008}$

141. $\dfrac{(26{,}200{,}000)(0.00543)}{(3{,}570{,}000)}$

Find the indicated value and express the answer with the appropriate number of significant digits.

142. $2.513457(10^2) + 3.72023(10^{-1})$

143. $4.5731(10^{-3}) - 1.569(10^{-2})$

144. $[3.2178(10^{-2})][1.878(10^4)]$

145. $\dfrac{[3.531(10^3)][1.24649(10^{-5})]}{[1.5384(10^{-3})]}$

Perform the indicated operations, collecting like terms, if any.

146. $(r^2 + 3r - 2) + (3r^2 - 2r + 1)$

147. $(3r - 5)(2r + 7)$

148. $(2w + 3)(2w - 3)(4w^2 + 9)$

149. $\dfrac{3x^2}{x^3 - 8} - \dfrac{1}{x - 2} - \dfrac{3x}{x^2 + 2x + 4}$

150. $\dfrac{w^2 - 16}{w^2 - 9} \cdot \dfrac{w^2 + 5w + 6}{w^2 + 6w + 8}$

151. $\dfrac{y^2 + 14y - 15}{y^2 + 4y - 5} \div \dfrac{y^2 + 12y - 45}{y^2 + 6y - 27}$

152. $x - \dfrac{x^2}{x - \dfrac{x}{y - 2}}$

153. $\dfrac{u^2 - \dfrac{v^3}{u}}{\dfrac{v^2}{u} + v + u}$

154. $\dfrac{\dfrac{1}{u + v}}{\dfrac{1}{u^2} - \dfrac{1}{v^2}}$

155. $\dfrac{\dfrac{u}{v} - \dfrac{v}{u}}{u^2 - v^2}$

156. $\dfrac{1}{\dfrac{1}{x} + \dfrac{1}{y}}$

157. $\dfrac{x^{-1} - y^{-1}}{x^{-1} + y^{-1}}$

158. $\dfrac{xy^{-1} - yx^{-1}}{xy^{-1} + yx^{-1}}$

Factor the following as far as possible.

159. $16 + 12a$

160. $x^2 - 5x + 6$

161. $x^2 - x - 20$

162. $x^2 - 10x + 21$

163. $81 - a^4$

164. $25a^2 - 36b^2$

165. $x^8 - 256$

166. $64x^6 - y^6$

167. $4x^2 + 12xy + 9y^2$

168. $x^2 - 18xy + 81y^2$

169. $625 + 25z^2 + z^4$

170. $uv + 2v - 3u - 6$

171. $8 - 36y + 54y^2 - 27y^3$

172. $x^3 + 6x^2y + 12xy^2 + 8y^3$

173. $4x^2y^2 - 4x^2 - y^4 + y^2$

Use long division on the following.

174. $\dfrac{u^8 - 1}{u - 1}$

175. $\dfrac{4r^3 + 7r + 5}{1 + 2r}$

Simplify each of the following into the form a + bi.

176. $(3 + 2i) + (2 - 4i)$

177. $(1 + 2i) - (3 - i)$

178. $(4 + i)(2 - 4i)$

179. $3(2 - i) + 4i(2 + i)$

180. $(3 + i)^3$

181. i^{57}

182. $i^{54} + i^{27}$

183. $2i^{58} + 3i^{62} - 5i^2$

184. $\dfrac{3}{2+i}$

185. $\dfrac{5-2i}{1+3i}$

186. $\dfrac{-1-i}{-1+i}$

187. $\dfrac{1+i}{2+i} + \dfrac{1-i}{3-i}$

Find $\bar{z}$ for each of the following.

188. $7 - 3i$

189. $6 + 5i$

190. $\sqrt{-2} + \sqrt{-3}\,i$

191. $\sqrt{-16} + \sqrt{25}\,i^3$

192. The Soft-Eze Tissue Company distributes its paper to various retailers. Some vendors run specials of six rolls for $1.00. Others need more money to cover overhead expenses and prefer to sell four for $1.00. What is the maximum number of rolls the company should place in a package so that neither vendor need sell broken packages?

193. The highest U.S. city is Leadville, Colorado, with an elevation of 10,200 feet. The lowest U.S. town is Calipatria, California, which lies 183 feet below sea level. How much higher is Leadville than Calipatria?

194. An arena with an advertised seating capacity of 13,257 is filled except for three rows of 63 seats each. How many people are in attendance?

195. A dress pattern requires $2\tfrac{2}{3}$ yards of cotton and $1\tfrac{1}{8}$ yards of velvet. If cotton costs $5 per yard and velvet costs $12 per yard, what is the cost of the material needed to make the dress?

196. Suppose that a certain microorganism causes the human little finger to twitch. The number of people it has infected quintuples every week.
   a. If one person initially becomes infected, how many will have twitchy little fingers in 25 days?
   b. In 365 days?

197. Proxima Centauri is the star nearest to us other than our sun. When astronomers view Proxima Centauri by telescope, they see it as it was 4 years, 4 months ago. How far is Proxima Centauri from earth? (See Exercise 114, Section 1.3.)

198. a. If the half-life of radium is 1600 years and at at given time there are 100 units present, how many units are present 6400 years later?
   b. How many units were present 1600 years earlier?

199. If a ball is dropped from a height of 500 feet and returns to 80 percent of its original height after each bounce, how high does it bounce after hitting the pavement for the third time?

200. The "doubling time" of a certain bacteria culture is 2 hours. If the culture starts with 500 bacteria, how many will there be 6 hours later?

201. If the doubling time of a bacteria culture is known to be 3 hours and there are 1000 bacteria present now, how many were there 6 hours ago?

Equations and Algebraic Inequalities

2

A statement that two quantities or expressions are equal is called an **equation**. Examples of equations include

$$x + 5 = 0$$

and

$$t^2 + 3t + 7 = 10$$

Various numbers may be substituted for the letters or **variables** in an equation. The **solution** of the equation consists of those numbers that, when assigned to the variables, make the equation true. Such numbers are said to *satisfy* the equation. The numbers that satisfy an equation are also called **roots** of that equation. Since the values that satisfy an equation are not generally known beforehand, the variables are also called **unknowns**.

We can perform various algebraic manipulations on an equation until the solution is apparent. But to maintain the equality represented by the original equation, *whatever is done to one side of the equation must be done to the other.* This is analogous to a balanced seesaw: if the weight on one side is increased or decreased by a given amount, the other side must be changed by the same amount in order to maintain the balance. At the end of this chapter, we will apply the techniques and skills for solving equations to inequalities.

Certain equations, such as $3x - 2x = x$, are true for all values of x. Such equations are called **identities**. Others, such as $x + 5 = 0$, are true only for certain values of x. These are called **conditional equations**. Still other equations, such as $2x - 1 = 2x + 1$ are not true for any value of x; we call these **contradictions**. In this chapter, we work primarily with conditional equations in one variable.

If an equation can be written in the form polynomial = 0, then the degree of the polynomial is called the **degree of the equation**. Thus, $x + 5 = 0$ is an equation of degree 1; $t^2 + 3t + 7 = 10$ is an equation of degree 2. The numbers (not variables) in an equation are called **constants** or **coefficients**. For example, 3 is the coefficient of t in $t^2 + 3t + 7$.

The various techniques we will discuss for solving equations are all based on the following rule.

> If the same operations are performed on equal quantities, then the results must be equal.

In other words, whatever you do to one side of an equation, you must do to the other.

Section 2.1

First-degree Equations in One Variable

First-degree equations in one variable are also called **linear equations**. The procedure for solving linear equations is outlined in the following box.

SOLVING FIRST-DEGREE EQUATIONS IN ONE VARIABLE

1. Simplify the algebraic expressions.
2. Isolate the variable on one side of the equation by
   a. Adding or subtracting terms.
   b. Dividing by the coefficient of the variable.

EXAMPLE 1

Solve the following equation.
$$x - 3 = 5$$

SOLUTION

Here we add 3 to both sides of the equation:
$$x - 3 = 5$$
$$x - 3 + 3 = 5 + 3$$
$$x = 8$$

The number 8 is the only solution to this equation. ∎

In Example 1 there is only one number, 8, in the set of solutions. We say the **solution set** is $\{8\}$.

EXAMPLE 2

Solve the following equation.
$$3x = -6$$

SOLUTION

Here we divide both sides of the equation by 3:
$$3x = -6$$
$$\frac{3x}{3} = \frac{\overset{-2}{\cancel{-6}}}{\cancel{3}}$$
$$x = -2$$

The solution set is $\{-2\}$; -2 is the only solution. ∎

EXAMPLE 3

Solve the following equation.
$$5x = 2x + 12$$

Section 2.1 *First-degree Equations in One Variable* **81**

SOLUTION   In order to isolate x on one side of the equation, we subtract $2x$ from both sides:

$$5x = 2x + 12$$
$$5x - 2x = 2x + 12 - 2x$$
$$3x = 12$$
$$x = 4 \qquad \text{(Divide both sides by 3)}$$

The solution set is $\{4\}$; 4 is the only solution.

EXAMPLE 4   Solve the following equation.

$$7y + 6 = y + 6$$

SOLUTION
$$7y + 6 = y + 6$$
$$7y = y \qquad \text{(Subtract 6 from both sides)}$$
$$6y = 0 \qquad \text{(Subtract } y \text{ from both sides)}$$
$$y = 0 \qquad \text{(Divide both sides by 6)}$$

Thus, 0 is the only solution. The possibility of y being 0 prevented us from dividing both sides of the equation in step 2 by y. Had we attempted this division, we would have obtained the contradictory statement $7 = 1$.

There is only one number, 0, in the set of solutions; the solution set is $\{0\}$.

EXAMPLE 5   Solve the following equation.

$$3(x - 2) - 4(x - 3) = 8$$

SOLUTION   We first remove parentheses and collect like terms. Then we proceed as in Example 3 to isolate x on one side of the equation.

$$3(x - 2) - 4(x - 3) = 8$$
$$3x - 6 - 4x + 12 = 8 \qquad \text{(Remove parentheses)}$$
$$-x + 6 = 8 \qquad \text{(Collect terms)}$$
$$-x = 2 \qquad \text{(Subtract 6 from both sides)}$$
$$x = -2 \qquad \text{(Multiply both sides by } -1\text{)}$$

The only solution to the equation is -2. The solution set is $\{-2\}$.

EXAMPLE 6   Solve the following equation.

$$\frac{2x + 3}{5} - 10 = \frac{4 - 3x}{2}$$

SOLUTION   Our first task is to simplify the left side of this equation. We can do this and simultaneously eliminate the fractions if we multiply both sides by the LCD of the two fractions.

$$10 \cdot \left(\frac{2x+3}{5} - 10\right) = \left(\frac{4-3x}{2}\right) \cdot 10 \quad \text{(Multiply both sides by 10)}$$

$$\frac{\overset{2}{\cancel{10}} \cdot (2x+3)}{\cancel{5}} - 10 \cdot 10 = \frac{(4-3x)\overset{5}{\cancel{10}}}{\cancel{2}} \quad \text{(Cancellation and distributive laws)}$$

$$4x + 6 - 100 = 20 - 15x \quad \text{(Remove parentheses)}$$
$$4x - 94 = 20 - 15x \quad \text{(Collect like terms)}$$
$$19x - 94 = 20 \quad \text{(Add } 15x \text{ to both sides)}$$
$$19x = 114 \quad \text{(Add 94 to both sides)}$$
$$x = \frac{114}{19} \quad \text{(Divide both sides by 19)}$$
$$x = 6$$

The solution to the equation is 6; the solution set is $\{6\}$. ∎

Adding or subtracting the same term on both sides of an equation is sometimes called **transposition;** in effect, the given term is transposed to the other side of the equation, and its sign is changed. For instance, in Examples 4, 5, and 6 we could write

4. $7y + 6 = y + 6$
$7y = y + 6 - 6$
$7y = y$
$7y - y = 0$
$6y = 0$

5. $-x + 6 = 8$
$-x = 8 - 6$
$-x = 2$

6. $4x - 94 = 20 - 15x$
$19x - 94 = 20$
$19x = 114$

Equations Involving Fractions

Equations that have the same set of solutions are said to be **equivalent.** Thus

$$x - 3 = 4, \quad x = 7, \quad 2x = 14$$

are equivalent equations; $x - 3 = 4$ and $2x = 10$ are not equivalent.

The methods used in the preceding examples always yield equations equivalent to the original. These methods can be extended to solve more complex equations as well. We can certainly add or subtract the same term on both sides of the equation. We can likewise multiply both sides of an equation by the same nonzero value. However, if both sides of an equation are multiplied by 0, the result is

$$0 = 0$$

which is true regardless of the truth of the original equation for any given values of the variable. Thus we must guard against inadvertently multiplying both sides by 0; this can happen if we multiply by an expression involving the variable.

EXAMPLE 7

Solve the following equation.

$$\frac{2y-2}{y-2} - \frac{2}{y-2} = 5$$

SOLUTION

To eliminate the fractions, we multiply both sides of the equation by $(y-2)$.

$$(y-2)\left(\frac{2y-2}{y-2} - \frac{2}{y-2}\right) = 5(y-2)$$

$$\frac{(y-2)(2y-2)}{(y-2)} - \frac{2(y-2)}{(y-2)} = 5(y-2) \qquad \text{(Distributive law)}$$

$$2y - 2 - 2 = 5y - 10$$

$$2y - 5y = -10 + 4 \qquad \text{(Collect terms and transpose)}$$

$$-3y = -6$$

$$y = 2 \qquad \text{(Divide by } -3\text{)}$$

But the original equation is undefined for $y = 2$ since then the denominator, $(y-2)$, is 0. Thus 2 is unacceptable as a solution. We check by substituting $y = 2$ into the original equation:

$$\frac{2 \cdot 2 - 2}{(2-2)} - \frac{2}{2-2} \stackrel{?}{=} 5$$

$$\frac{2}{0} - \frac{2}{0} \stackrel{?}{=} 5$$

Since division by 0 is undefined, this equation is invalid.

We showed that 2 was the only possible solution and then we observed that 2 cannot be a solution. There are no solutions to this equation. The solution set is empty: $\emptyset$. ∎

Although it is always a good idea to check our solutions to any equation, the solutions *must* be checked when we multiply both sides of an equation by an expression involving the variable. Such multiplications might introduce new roots (which are not roots of the original equation). These new roots are called **extraneous roots**.

EXAMPLE 8

Solve the following equation.

$$\frac{x+9}{x+1} - \frac{x-1}{x-2} = \frac{4}{x+1}$$

SOLUTION

First we note that the LCD of these fractions is $(x+1)(x-2)$. To eliminate the fractions, we multiply both sides of the equation by this LCD:

$$(x+1)(x-2)\left(\frac{x+9}{x+1} - \frac{x-1}{x-2}\right) = \frac{4}{x+1}(x+1)(x-2)$$

$$\frac{(x+1)(x-2)(x+9)}{(x+1)} - \frac{(x+1)(x-2)(x-1)}{(x-2)} = \frac{4(x+1)(x-2)}{(x+1)}$$

$$\qquad\qquad\qquad\qquad\qquad\qquad\qquad\text{(Distributive law)}$$

$$(x^2 + 7x - 18) - (x^2 - 1) = 4x - 8$$

$$7x - 4x = -8 + 17 \qquad \text{(Collect terms and transpose)}$$

$$3x = 9$$

$$x = 3 \qquad \text{(Divide by 3)}$$

We note that the original equation is undefined only when $x = -1$ or $x = 2$. Thus 3 is a valid solution; we have not multiplied by 0. We check by substituting into the original equation:

$$\frac{3+9}{3+1} - \frac{3-1}{3-2} \stackrel{?}{=} \frac{4}{3+1}$$

$$\frac{12}{4} - \frac{2}{1} \stackrel{?}{=} \frac{4}{4}$$

$$3 - 2 \stackrel{?}{=} 1$$

$$1 \stackrel{\checkmark}{=} 1$$

Thus 3 satisfies the original equation. The solution set is {3}.

EXAMPLE 9

Solve the following equation.

$$(3x + 1)(2x - 3) = 6(x - 1)(x + 2)$$

SOLUTION

We first carry out the multiplication and then simplify as follows.

$$(3x + 1)(2x - 3) = 6(x - 1)(x + 2)$$
$$6x^2 - 7x - 3 = 6(x^2 + x - 2)$$
$$6x^2 - 7x - 3 = 6x^2 + 6x - 12$$
$$-7x - 6x = -12 + 3$$
$$-13x = -9$$
$$x = \frac{9}{13}$$

Section 2.1 Exercises

Solve the following equations.

1. $4x - 5 = 2x + 5$
2. $5 - 7z = 5z - 19$
3. $6 - 2t = 4t + 12$
4. $\frac{1}{5}u - 3 = 9 + \frac{1}{2}u$
5. $\frac{1}{3}v + 4 = 7 - \frac{1}{2}v$
6. $5y - 3y = 32 - 2y$
7. $6z + 4 = 10z - 20$
8. $2 + 7t + 3 = 9t + 6 + 4t$
9. $4u + 6u = 2u$
10. $3u + 21 = 0$
11. $4t - 3 + 2t = (8t + 5) - (2t + 8)$
12. $2(u - 1) + 3(2 - u) = (4 - u)$
13. $4x - 6 = 3x + 10$
14. $7(q - 2) = 6(q + 3)$
15. $4(x + 7) + 3(8 - 2x) = 0$
16. $3(3 - 2w) = 2(w - 5) + 3$
17. $4(p - 1) = 5 - 3(4p + 3)$
18. $(3s + 2) - (6s - 3) + 1 = 0$
19. $7(2 - y) - 26 = 2(8 - 7y)$
20. $4(3z - 2) + 1 = 6(1 + z) + 5$
21. $(4t + 3)3 - 4(1 - t) = 5$
22. $3(2x + 1) - 2(3x + 1) = 1$
23. $3(s + 3) + (s - 2) = 2(2s + 4) - 1$
24. $3(2x + 1) + 2(3x + 1) = 17$
25. $(2 - y) + 4(2y - 1) = 12$
26. $(r + 6) - (5 - 2r) + 2 = 0$
27. $4(2v + 3) - 2 = 5v + 4$
28. $\frac{3p + 4}{2} - 6 = \frac{1 - 3p}{5}$
29. $\frac{1 + u}{3} - \frac{u}{2} = \frac{u - 2}{7}$

30. $\dfrac{x+600}{3} - \dfrac{x}{2} = 100$

31. $\dfrac{2-w}{3} + \dfrac{w-2}{3} = 0$

32. $\dfrac{1+2x}{2} + \dfrac{2-x}{4} = \dfrac{3x+2}{6} + \dfrac{6x+8}{12}$

33. $\dfrac{2w-1}{3} + 2 = \dfrac{10+3w}{2}$

34. $\dfrac{q+1}{2} = \dfrac{2q-1}{5}$

35. $\dfrac{r+3}{4} = \dfrac{4-r}{3}$

36. $\dfrac{2s-5}{3} + \dfrac{s-4}{8} = \dfrac{-7}{12}$

37. $\dfrac{2-x}{7} = \dfrac{x}{2} - \dfrac{x+1}{3}$

38. $v - 2 = \dfrac{v+2}{6} - \dfrac{v-7}{3}$

39. $\dfrac{w+1}{6} + \dfrac{w-1}{4} + \dfrac{w-2}{3} + 3 = 0$

40. $\dfrac{3+2s}{3} + \dfrac{1-3s}{2} + 1 = 0$

41. $\dfrac{1}{2} - \dfrac{3}{x} = \dfrac{5}{x}$

42. $\dfrac{5}{2s+9} = \dfrac{1}{3-s}$

43. $\dfrac{3}{x-2} = \dfrac{5}{2x+7}$

44. $\dfrac{5}{4x+11} = \dfrac{3}{x+6}$

45. $\dfrac{3}{4s+2} = \dfrac{1}{2s+1}$

46. $\dfrac{z}{z-2} = \dfrac{z-3}{z+2}$

47. $\dfrac{x+2}{x-3} - \dfrac{8}{x-3} = \dfrac{2x}{x-3}$

48. $\dfrac{2q-2}{q-3} - 3 = \dfrac{4}{q-3}$

49. $\dfrac{3}{2t-1} - \dfrac{2}{6t-3} = \dfrac{1}{3}$

50. $\dfrac{u+4}{u+1} + \dfrac{u+1}{u+5} = 2$

51. $1 - \dfrac{3}{x+1} = \dfrac{3x}{x+1}$

52. $\dfrac{4}{y-12} = \dfrac{8}{2y-13} - \dfrac{3}{y-12}$

53. $w + \dfrac{3w+2}{w} = \dfrac{3w+2}{3}$

54. $2 - \dfrac{6}{3y+2} = \dfrac{9y}{3y+2}$

55. $\dfrac{6r-1}{12} + \dfrac{1}{r-1} = \dfrac{r-1}{2}$

56. $\dfrac{4v+5}{v+3} = 3 + \dfrac{v-1}{v}$

57. $2(y-3)(y-1) = (y-5)(2y-3)$

58. $(6x+5)(4x+7) = (3x+7)(8x+6)$

59. $7 + 6r(r-2) = (2r+1)(3r-5)$

60. $6t(t+8) + 96 = (3t+8)(2t+1)$

61. $(x-5)^2 = 4 + (x+5)^2$

62. $8(z-3)(z+1) = (2z+5)(4z+1)$

63. $z - 3i = 5i - z$

64. $z + 2i - 4 = 3iz - 2$

65. $z + 3i + 2 = 4iz - 6 + i$

66. $(z+2i)^2 = (z+3i)(z-3i)$

67. $(z-3i-2)(z+2i) = (z+i+1)(z-i)$

68. $(z+i+2)(z-i-2) = (z+3i)(z-2i)$

Section 2.2

Literal Equations and Word Problems

A **literal equation**, or **formula**, is an equation that relates two or more variables. Examples of literal equations include the formula for the circumference C and area A of a circle in terms of its radius r:

$$C = 2\pi r$$
$$A = \pi r^2$$

and the formula for the distance d traveled at a given rate r for a length of time t:

$$d = rt$$

Such formulas are not always used in their given forms. It is frequently useful to solve for a certain variable in terms of the others.

EXAMPLE 1

Express the radius r of a circle in terms of each of the following.

a. Its circumference C
b. Its area A

SOLUTION

a. Writing $C = 2\pi r$, we solve for r as follows:

$$2\pi r = C$$
$$r = \frac{C}{2\pi}$$

b. Writing $A = \pi r^2$, we solve for r as follows:

$$\pi r^2 = A$$
$$r^2 = \frac{A}{\pi}$$
$$r = \sqrt{\frac{A}{\pi}}$$

Here we take only the nonnegative square root, since a negative radius would be meaningless. ∎

EXAMPLE 2

If a, b, and c are related by the formula

$$\frac{a}{b+c} + \frac{b}{c} = 1$$

solve for each of the following.

a. a in terms of b and c
b. b in terms of a and c

SOLUTION

We begin by eliminating the fractions. To do so we multiply both sides of the equation by the LCD, $(b+c)c$:

$$(b+c)c\left(\frac{a}{b+c} + \frac{b}{c}\right) = 1 \cdot (b+c)c$$

$$\frac{\cancel{(b+c)} \cdot c \cdot a}{\cancel{(b+c)}} + \frac{(b+c) \cdot \cancel{c} \cdot b}{\cancel{c}} = (b+c)c$$

$$ca + b^2 + bc = bc + c^2 \qquad (1)$$

a. $ca = c^2 - b^2$

$$a = \frac{c^2 - b^2}{c}$$

b. $b^2 = c^2 - ca$

$$b = \pm\sqrt{c^2 - ca}$$

∎

EXAMPLE 3 The sales of thermal underwear in a region are related to the average winter temperature of the region: Low temperatures give high sales volume and vice versa. A simplified expression describing this behavior is $S = 150 - 2(t + 7)$, where S represents sales volume in hundred case lots, and t temperature in degrees Fahrenheit.

   a. At what temperature do sales drop to 0?
   b. If the manufacturing plant can produce only 200 lots annually, at what temperature is the capacity of the plant reached?
   c. Solve the given equation for t in terms of S.

SOLUTION

   a. The company sells no thermal underwear when
   $$150 - 2(t + 7) = 0$$
   Solving this equation for t, we obtain
   $$150 - 2t - 14 = 0$$
   $$136 - 2t = 0$$
   $$-2t = -136$$
   $$t = 68$$

   In regions with an average winter temperature of 68°F or above there would be no sales.

   b. On the other hand, the capacity of the plant is reached when 200 lots are sold annually:
   $$150 - 2(t + 7) = 200$$
   $$-2(t + 7) = 50$$
   $$t + 7 = -25$$
   $$t = -32$$

   In a region with an average winter temperature of $-32°F$, the plant would be operating at full capacity.

   c. Solving for t in terms of S, we have
   $$S = 150 - 2(t + 7)$$
   $$2(t + 7) = 150 - S$$
   $$t + 7 = \frac{150 - S}{2}$$
   $$= 75 - \frac{S}{2}$$
   $$t = 68 - \frac{S}{2}$$

Word Problems

Before a real-life problem can be solved mathematically, it must be translated into mathematical terms. Developing the ability to do this requires time and practice at increasing levels of difficulty. Solving word problems helps to develop this ability. You will probably find word problems challenging, mainly because you have not had much practice at solving them. As with most new skills, practice is the key to success. The steps outlined in the following box are essential for solving word problems.

> ### SOLVING WORD PROBLEMS
>
> 1. *Understand the problem.* Read the problem through quickly to pick up its general thrust. Then read it through again very carefully and completely, possibly several times.
> 2. *Make a sketch* if it will help you understand the problem.
> 3. *Identify the unknown quantities* that are to be determined.
> 4. *Name these quantities.* Use a variable such as x or y. Try to minimize the number of variables used. If you have drawn a sketch, label these quantities on the sketch.
> 5. *Express all known relationships mathematically.* Do so in terms of the variable names x, y, and so on.
> 6. *Reread the problem.* Check the relationships you have written down. Has all the information supplied in the problem been incorporated into these relationships?
> 7. *Solve the resulting equations.*
> 8. *Interpret your results* in terms of the original problem.
> 9. *Check your solutions* against the conditions of the problem and discard solutions which do not satisfy the problem as *originally* stated.

EXAMPLE 4

Monica has test scores of 72, 64, and 84 in her algebra class. What grade must she achieve on the final exam to obtain an 80 average if the final exam counts as two tests.

SOLUTION

Let x represent the desired final exam score. With the three tests already taken plus the final exam, which counts as two tests, Monica will have a total of $(72 + 64 + 84 + 2x)$ points on 5 tests. The average is to be 80:

$$\frac{72 + 64 + 84 + 2x}{5} = 80$$

$220 + 2x = 400$ (Collect terms and multiply both sides by 5)

$2x = 180$ (Subtract 220 from both sides)

$x = 90$ (Divide by 2)

Monica must get a 90 on the final exam in order to obtain an 80 average. ∎

EXAMPLE 5

Dean uses 160 feet of fence to enclose his rectangular garden plot. If the length is 20 feet more than the width, find the dimensions of the plot.

SOLUTION

We begin by letting x denote the width. Rather than introduce another variable for the length we note that the length is then $x + 20$. See Figure 2.1.

Figure 2.1

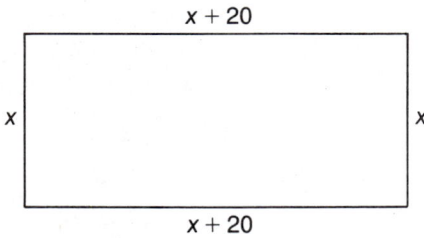

The perimeter is given by $2x + 2(x + 20)$. Thus we have the equation

$$2x + 2(x + 20) = 160$$
$$2x + 2x + 40 = 160 \qquad \text{(Distributive law)}$$
$$4x + 40 = 160 \qquad \text{(Collect terms)}$$
$$4x = 120 \qquad \text{(Subtract 40 from both sides)}$$
$$x = 30 \qquad \text{(Divide by 4)}$$
$$x + 20 = 50$$

The garden plot is 30 feet wide by 50 feet long. ∎

EXAMPLE 6

Ms. Pleiman sells an investment property and places the proceeds in certificates of deposit (CDs). The maximum current interest rate is paid by a 1-year certificate yielding 12 percent. She can lock in a 10 percent rate for $2\frac{1}{2}$ years. Thinking that interest rates may go down, she splits her money, depositing three times as much in the 12 percent account as in the 10 percent account. The yearly return on these deposits is $8280. How much did she realize from the sale of her property? How much did she invest in each account?

SOLUTION

Let

$$x = \text{amount (in dollars) invested at } 10\%$$
$$3x = \text{amount (in dollars) invested at } 12\%$$
$$4x = \text{amount (in dollars) realized from sale of the house}$$
$$(0.10)x = \text{yearly interest from the } 10\% \text{ investment}$$
$$(0.12)3x = \text{yearly interest from the } 12\% \text{ investment}$$

For each year's income, we have the following equation:

(Interest from 10% account) + (Interest from 12% account) = (Total interest)
$$(0.10)x \qquad + \qquad (0.12)3x \qquad = 8280.00$$

We then solve for the unknowns:

$$(0.10)x + (0.36)x = 8280.00$$
$$10x + 36x = 828{,}000 \qquad \text{(Multiply by 100)}$$
$$46x = 828{,}000$$
$$x = 18{,}000$$
$$3x = 54{,}000$$
$$4x = 72{,}000$$

Ms. Pleiman realized $72,000 from the sale of her property. She then deposited $18,000 at 10 percent interest and $54,000 at 12 percent interest. ∎

EXAMPLE 7

A chemist needs to mix 10 liters of a 20 percent acid solution with an 80 percent acid solution to obtain a mixture that will be a 30 percent acid solution. How many liters of the 80 percent acid solution must be used?

SOLUTION

Let

x = number of liters of 80 percent solution that must be added to 10 liters of 20 percent solution to yield a 30 percent solution

The total amount of acid in each container is shown in Figure 2.2.

Figure 2.2

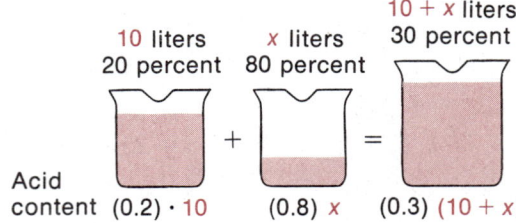

To find x, we must solve the equation

$$(0.2) \cdot 10 + (0.8)x = (0.3)(10 + x)$$
$$2 + 0.8x = 3 + 0.3x$$
$$0.5x = 1$$
$$5x = 10$$
$$x = 2$$

Thus, 2 liters of 80 percent solution must be added to 10 liters of 20 percent solution to yield a 30 percent solution. The formula for 30 percent solution is two parts 80 percent solution with ten parts 20 percent solution or "strong and weak acids are mixed in the ratio 1 to 5." ∎

EXAMPLE 8

The Vacation Realty Company plans to sell shares in a luxury resort condominium. Each share will entitle the shareholder to 1 week's use of the condominium each year. The company already has 50 prospective buyers. Knowing that not everyone will want to return to this particular spot every year, the company advises the prospective buyers to allow the project to be oversold. By allowing 25 additional shares to be sold, the original investors can save $1200 each. What is the total sale price of the condominium? What is the cost per share if 50 shares are sold? 75 shares?

SOLUTION

Let

P = total sales price (in dollars) of the condominium

$\dfrac{P}{50}$ = cost per share with 50 investors

$\dfrac{P}{75}$ = cost per share with 75 investors

With 75 investors, the cost per share is $1200 less than with 50 investors, which gives us the equation

(Cost with 50 investors) − $1200 = (Cost with 75 investors)

$$\frac{P}{50} - 1200 = \frac{P}{75}$$

This equation is solved for the P as follows.

$$150 \cdot \left(\frac{P}{50} - 1200\right) = \frac{P}{75} \cdot 150$$

$$3P - 180{,}000 = 2P$$

$$P = 180{,}000$$

$$\frac{P}{50} = \frac{180{,}000}{50} = 3600$$

$$\frac{P}{75} = \frac{180{,}000}{75} = 2400$$

The condominium costs $180,000; the costs per share are $3600 or $2400 depending on whether 50 or 75 shares are sold. ∎

EXAMPLE 9

A small plane can fly 585 nautical miles with the wind in the same time that it takes to fly 495 nautical miles against the wind. If the speed of the wind is 10 knots (nautical miles per hour), how fast does the plane fly when there is no wind?

SOLUTION

Let

x = speed of plane (no wind), in knots
$x + 10$ = speed of plane flying with the wind
$x - 10$ = speed of plane flying against the wind

We use the formula

distance = rate · time

or

$$d = rt$$

This formula can also be written as

$$r = \frac{d}{t} \quad \text{or} \quad t = \frac{d}{r}$$

It is convenient to display the given information in a table as follows.

	Distance	Rate	Travel time
With the wind	585	$x + 10$	$\dfrac{585}{x + 10}$
Against the wind	495	$x - 10$	$\dfrac{495}{x - 10}$

since $t = \dfrac{d}{r}$

Since the travel times are said to be the same, we set

$$\frac{585}{x+10} = \frac{495}{x-10}$$

$$(x+10)(x-10)\frac{585}{(x+10)} = \frac{495}{(x-10)}(x+10)(x-10)$$

$$585x - 5850 = 495x + 4950$$
$$(585 - 495)x = 5850 + 4950$$
$$90x = 10{,}800$$
$$x = 120$$

The speed of the plane is 120 knots.

EXAMPLE 10

A painter knows that her assistant can paint only half as much in a day as she can. She is planning to fire the assistant, but she must finish painting the houses in a certain real estate development by a given date or forfeit a bond. In the past 5 days they have painted three houses working together. How long would it take each woman working alone to paint a house?

SOLUTION

Let

t = number of days required for the painter to paint a house
$2t$ = number of days required for the assistant to paint a house
$\dfrac{1}{t}$ = number of houses painted in 1 day by the painter
$\dfrac{1}{2t}$ = number of houses painted in 1 day by the assistant
$\dfrac{1}{t} + \dfrac{1}{2t}$ = number of houses painted in 1 day by the two working together

Since they can paint three houses in 5 days working together, this yields the equation:

$$\left(\frac{1}{t} + \frac{1}{2t}\right)5 = 3$$

$$\frac{3}{2t} \cdot 5 = 3$$

$$\frac{5}{2t} = 1$$

$$5 = 2t$$

$$2\frac{1}{2} = t$$

The painter can paint a house in $2\frac{1}{2}$ days, whereas her assistant takes 5 days.

Section 2.2 Exercises

1. The formula for converting Celsius temperatures to Fahrenheit is $F = \frac{9}{5}C + 32$.
   a. Convert 77°F to Celsius.
   b. At what temperature do the Celsius and Fahrenheit temperature scales agree?
   c. Express the Celsius temperature C in terms of F.

2. The distance d traveled by a moving object equals its speed r times the length of time t it travels: $d = rt$.
   a. If Beth travels 2100 miles in 6 hours to visit her mother, how fast does the plane fly?
   b. Express r in terms of d and t.

3. The interest I earned on an investment equals the principal P invested multiplied by the interest rate r and the elapsed time t: $I = Prt$.
   a. If Mary earns $21 in 6 months at a rate of 6% per year, how much money has she invested?
   b. Express each of the variables P, r, t in terms of the others and I.

4. The total resistance in the parallel network of blood vessels pictured here is 5. If one of the branches has a resistance of 6, find the resistance of the other. (Hint: $\frac{1}{1/x + 1/6} = 5$; see Example 6, Section 1.8.)

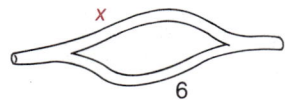

5. The formula $\frac{1}{x} + \frac{1}{y} = \frac{1}{r}$ arises in the study of resistance in a parallel network of blood vessels or electrical circuits. Solve this expression for y in terms of x and r.

6. The area A of a triangle with base length b and height h is given by $A = \frac{1}{2}bh$. Solve this equation for each of the following.
   a. b in terms of A and h
   b. h in terms of b and A

7. The surface area S and volume V of a sphere of radius r are given by $S = 4\pi r^2$ and $V = \frac{4}{3}\pi r^3$. Solve for r in terms of each of the following.
   a. S b. V

8. The volume V of a right circular cone of base radius r and altitude h is given by $V = \frac{1}{3}\pi r^2 h$. Solve this expression for each of the following.
   a. r b. h

9. The area A and perimeter P of a rectangle having length l and width w are given by $A = lw$ and $P = 2(l + w)$. Solve each of these expressions for each of the following.
   a. l b. w

10. Einstein's mass-energy equation is $E = mc^2$. Solve this expression for each of the following.
    a. m b. c

11. The taxes T on a given day's sales of x dollars at the XYZipper Outlet Store are given by
    $$T = \frac{x + 600}{3} - \frac{x}{2}$$
    a. If today's taxes are $100, what were today's sales?
    b. Express sales, x, in terms of taxes, T.

12. The equation for a certain parabola is $y = ax^2 + b$.
    a. Solve for x in terms of y, a, and b.
    b. Solve for a in terms of y, x, and b.

13. If $K = \frac{1}{2}mv^2$, solve for m in terms of K and v.

14. If $s = \frac{1}{2}gt^2$, solve for t in terms of s and g.

15. If $x^2 + y^2 = r^2$, solve for y in terms of x and r.

16. If $S = \frac{a}{1-r}$, solve for r in terms of S and a.

17. If $m = \frac{y-2}{x-1}$, solve for x in terms of m and y.

18. If $ax + by = c$, solve for x in terms of a, b, c, and y.

19. If $V = P + Prt$, solve for (a) P and (b) r in terms of the other variables.

20. If $A = 2\pi rh + 2\pi r^2$, solve for h in terms of the other variables.

21. If four test scores are 82, 78, 90, and 86, what grade must be achieved on the fifth test for an average of 85?

22. What grade must be achieved on the test in Exercise 21 if it counts as two test grades?

23. If 240 feet of fence are used to enclose a rectangular corral whose length is 30 feet less than twice its width, what are the dimensions of the corral.

24. Work Exercise 23 assuming 300 feet of fence are used and the length is 30 feet more than the width.

25. The manager of a bicycle shop knows she can sell 400 Reflecto handlebars if she charges $8 each. Suppose that for each 50¢ increase in price she loses 20 sales. At what selling price does she lose all sales?

26. If a television set sells for $450 and the markup is 30 percent of the selling price, how much did the dealer pay for the set?

27. If a television set sells for $468 and the markup is 30 percent of the dealer's cost, how much did he pay for the set?

28. On a recent visit to Las Vegas, Sandy doubled her money at the roulette wheel and then spent $300. She then tripled her money at the craps table and spent $540. Later she quadrupled her money at blackjack and spent $720. If she went home with $480 in her pocket, how much did she have when she arrived in Las Vegas?

29. If Todd spends one third of his money, loses two thirds of the remainder, and is left with $10, with how much did he start?

30. Jan has $6000 in a savings account. She withdraws $1000 to loan to a co-worker at twice the bank interest rate. Find the respective rates if the annual interest earned is $280.

31. Roger invested $6000 in a CD at 5 percent. He later found a better opportunity offering 8 percent. If he is now realizing 7 percent on his total investment, how much additional money did he invest at 8 percent?

32. Chuck has deposits of $12,000 and $9000 in separate CDs. The $9000 earns a $\frac{1}{2}$ percent higher interest rate but $120 per year less than the $12,000. Find the respective interest rates.

33. Mary recently moved to a colder climate and must strengthen the antifreeze in her car. Her radiator is now filled to capacity with 12 liters of a 40 percent solution. How much must be drained and replaced with pure antifreeze to obtain a 65 percent solution?

34. Thirty quarts of 20 percent medicinal saline solution is to be strengthened to 50 percent by replacing some of it with 80 percent solution. How much must be replaced?

35. The rangers at the Fish and Wildlife Service wish to spray the perimeter of Lake Mosquito in order to protect the campers and cabin dwellers from mosquitos. Last year they sprayed a solution consisting of 10 percent pesticide, which was ineffective. Because last year's price was so good they overstocked the 10 percent solution by 42 barrels. This year they have decided to use a 25 percent solution but wish to use their leftover 10 percent solution. A 60 percent solution is available. How many barrels of 60 percent solution must be added to their 42 barrels of 10 percent solution to get a 25 percent solution?

36. An oil mixture worth $1.17 a quart is to be made by adding oil worth 72¢ a quart to 100 quarts of oil priced at $1.35 a quart. How much must be added?

37. A group of 10 investors plans to construct an office building, but the cost is very high. If they can find 5 additional investors their individual investments will be reduced by $10,000 each. How much will the building cost assuming each must invest the same amount?

38. A major corporation plans to raise additional capital by issuing and selling stock. It can sell 3 million shares at a given price or 5 million shares at a price of $4 less per share. How much capital is it planning to raise?

39. Bill and Jane are moving to another city. Being a two-car family, they must each drive a car to their new home. Bill drives at 50 miles per hour but Jane drives 10 miles per hour faster. It takes Bill 1 hour longer than Jane to make the trip. Find their respective travel times and the distance between the cities.

40. A plane travels for 4 hours at a steady speed with a tail wind. On the return trip it must face the wind, thereby reducing its speed 100 miles per hour and increasing its time by 1 hour.
    a. Find the respective speeds for both the initial and return trips.
    b. What is the one-way distance involved?

41. Eric and Cindy are avid bicyclists living in cities 130 miles apart. Eric begins riding toward Cindy at noon traveling at 15 miles per hour and Cindy begins riding toward Eric at 2 P.M. traveling at 10 miles per hour.
    a. At what time will they meet?
    b. How far does each travel?

42. Renée starts on a hike at 3 P.M. walking at 4 miles per hour, expecting José to follow immediately. José delays $\frac{1}{2}$ hour but then jogs 6 miles per hour in order to catch up. How long does it take José to catch up with Renée?

43. Members of an outdoor club paddle a canoe in still water at 5 miles per hour. They traveled downstream one day for 24 miles and then camped overnight. Returning the next day, they noticed that after paddling the same amount of time, they still had 18 miles to go.
    a. What was the speed of the water in the stream? Assume the water speed was the same on both days.
    b. At this rate, how long will the return trip take them?

44. Tim can walk at 4 miles per hour or ride his bicycle at 12 miles per hour. It takes him 1 hour longer to walk to work than it does to ride.
    a. How far is his workplace from home?
    b. How long does it take him to walk to work and to ride to work?

45. If Mary can paint a room in 2 hours and Janice can do so in 3 hours, how long will it take them to paint a room working together? Assume that they do not get in each other's way or otherwise distract one another. Thus, the answer is *not* 5 hours.

46. If John can load a truck in 10 hours and Bill can load it in 15 hours, how long will it take them to load the truck working together?

47. At the poultry plant, Gert can clean 500 turkeys in half a day (4 hours), whereas Gail can do only 500 turkeys in a full 8-hour day. Working together, how long would it take them to clean 1000 turkeys?

48. Cindy can repair a VCR in 4 hours, but with Pam's help, it only takes 3 hours. At this rate, how long would it take Pam working alone to fix the VCR?

49. A magician asks a spectator to do the following:
    i. Think of a number.
    ii. Double it.
    iii. Add 4.
    iv. Divide by 2.
    v. Subtract the original number.

    The magician then states that the result is 2. Explain. (*Hint:* Let x denote the original number and then list the result of each action as it is taken.)

50. To conceal the trick of Exercise 49 from the audience, the magician instructs the next subject to perform the same steps, except that some other number is used in part **iii**. She knows that the final result will be half the number added in part **iii**. Explain.

Some mentalists like to give the impression that they have memorized the calendar and are capable of performing feats of lightning addition. Typical stunts are described in the exercises that follow. Each of three participants from the audience is asked to enclose a three-by-three block of nine dates in a square, as illustrated here. Explain each of the stunts described in the following three exercises.

13	14	15
20	21	22
27	28	29

51. The first participant is requested to add the numbers in the four corners of the square and state the result. The mentalist then recites all nine dates enclosed.

52. The next participant is asked to add the numbers in any two opposite corners of the square and state the result. The mentalist can then list all entries in the three-by-three block without even knowing which two corners were chosen!

53. The third participant is asked to indicate the smallest number in his chosen block. The mentalist immediately gives the total of all nine numbers in the block.

The following are variations of ancient puzzles. Find the solution to each.

54. Three honest women are to share equally in a gift from their mother. Confident of their honesty, the mother stacks the money on a table for them to pick up at their convenience. They arrive at separate times. Each thinks she is the first to arrive and takes only one third of the money that is on the table when she arrives. After they have what they consider to be their respective shares, the mother finds $800 left on the table. What was the total of her original gift?

55. A hungry man meets two friends with three sandwiches and five sandwiches, respectively. After seeing his plight, they agree to share the food equally three ways. The hungry man then pays his friends a total of 80¢. The two friends divide the money according to the amount of food that each gave the hungry man. How much money did each friend receive?

56. A messenger is sent from the rear to the front of a column of soldiers 3 kilometers long.
    a. If the column is marching forward at 5 kilometers per hour and the messenger travels at 25 kilometers per hour, how long does it take the messenger to deliver the message and return?
    b. What would be the corresponding time for a messenger delivering a message from front to rear and returning?

57. On the gravestone of Diophantus is a riddle that purports to indicate his age. The message is that he spent his life as follows:
    i. One sixth as a child
    ii. One twelfth as a youth
    iii. One seventh as an unmarried adult.

    Diophantus had a son who lived to be half as old as he. The son was born 5 years after Diophantus was married and died 4 years before Diophantus did. How old was Diophantus?

58. In a given room half the people are Caucasian, one fourth are black, one seventh are Chicano, and three are Oriental. How many people are in the room?

59. On a statue of Pallas Athene in Greece is engraved a message claiming the following of all the gold used to make the statue:
    i. Half came from Kariseus.
    ii. One eighth came from Thespian.
    iii. One tenth came from Solon.
    iv. One twentieth came from Themison.
    v. Nine units came from Aristodikos.

    How many units of gold are in the statue?

The following are "puzzle" problems that some people find interesting.

60. Two numbers differ by 3. Three times the larger number is 7 less than four times the smaller number. Find the numbers.

61. Find two consecutive positive even integers, the squares of which differ by 28.

62. Find three consecutive integers such that the sum of the first two is 2 more than the last.

63. Find three consecutive odd integers such that the product of the first two is 20 less than the product of the last two.

64. If a square is enlarged to a rectangle by lengthening one pair of sides 2 centimeters and the other pair of sides 3 centimeters, the area is increased by 36 square centimeters. Find the dimensions of the original square.

65. The border around a square picture is 2 inches wide and has an area of 56 square inches. Find the dimensions of the picture.

Section 2.3

Quadratic Equations

A second-degree equation in one variable is called a **quadratic equation**. For example,

$$x^2 - x - 12 = 0$$
$$2t^2 + 3t - 2 = 7t - 3$$
$$u^2 - 5u + 4 = 2u^2 + 3u + 2$$

Section 2.3 Quadratic Equations

are all quadratic equations. Quadratic equations have the general form
$$ax^2 + bx + c = 0 \qquad a \neq 0$$
or can be changed to this form.

Solution by Factoring

The method of factoring relies on the following property of the real number system.

> **ZERO PRODUCT PRINCIPLE**
>
> The product of two real numbers is 0 if and only if at least one of the factors is 0:
> $$uv = 0 \qquad \text{if and only if} \qquad u = 0 \text{ or } v = 0$$

This property is used to solve quadratic equations by factoring as follows.

> **SOLUTION BY FACTORING**
>
> If the quadratic equation can be factored into the form
> $$(ax + b)(cx + d) = 0$$
> then x satisfies the equation if and only if at least one of these factors is 0:
> $$ax + b = 0 \qquad \text{or} \qquad cx + d = 0$$
> $$x = -\frac{b}{a} \qquad \text{or} \qquad x = -\frac{d}{c}$$

EXAMPLE 1 Solve the following quadratic equation.
$$x^2 - x - 12 = 0$$

SOLUTION Factor the quadratic expression to obtain
$$(x - 4)(x + 3) = 0$$
This product is 0 if and only if
$$x - 4 = 0 \qquad \text{or} \qquad x + 3 = 0$$
Hence, 4 and -3 are the solutions to the equation. The solution set is $\{-3, 4\}$.

EXAMPLE 2 Solve the following equation.
$$4u^2 - 12u = -9$$

SOLUTION Before attempting to factor, we must put the equation in the form
$$4u^2 - 12u + 9 = 0$$

Then we observe that this quadratic expression is a perfect square: $(a - b)^2 = a^2 - 2ab + b^2$ with $a = 2u$ and $b = 3$. The solution is obtained as follows:

$$4u^2 - 12u + 9 = 0$$
$$(2u - 3)^2 = 0$$
$$2u - 3 = 0$$
$$u = \frac{3}{2}$$

The solution set is $\left\{\frac{3}{2}\right\}$. ∎

In Example 2, there is only one solution since the original expression was a perfect square; its factors were identical. The root $u = \frac{3}{2}$ is called a **double root** since it arises twice: Both factors are $(2u - 3)$.

EXAMPLE 3

Solve the following equation.

$$x^2 = 25$$

SOLUTION

If $x^2 = 25$, we simply take the square root of both sides to obtain

$$x = \pm 5$$

The solution set is $\{-5, 5\}$. ∎

We can solve any equation of the form $x^2 = a$ in a similar way.

THE SQUARE ROOT PROPERTY

If $a \geq 0$ and

$$x^2 = a \quad \text{or} \quad x^2 = -a$$

then the solutions are

$$x = \pm\sqrt{a} \quad \text{or} \quad x = \pm i\sqrt{a}$$

respectively.

Completing the Square

The method of completing the square is based on an extension of the square root property. If an equation has the form

$$(x - E)^2 = D, \quad D > 0$$

we take the square roots of both sides to obtain

$$x - E = \pm\sqrt{D}$$
$$x = E \pm \sqrt{D}$$

To complete the square of an expression of the form

$$x^2 + Bx = C$$

we add $\left(\dfrac{B}{2}\right)^2$ to both sides to obtain

$$x^2 + Bx + \left(\dfrac{B}{2}\right)^2 = C + \left(\dfrac{B}{2}\right)^2$$

Then let $D = C + \left(\dfrac{B}{2}\right)^2$ and factor the left side:

$$\left(x + \dfrac{B}{2}\right)^2 = D \qquad \text{(Formula 2, Section 1.7)}$$

We now outline the method of solving quadratic equations by completing the square.

COMPLETING THE SQUARE

To solve a quadratic equation $ax^2 + bx + c = 0$ ($a \neq 0$) by completing the square:

1. Transpose the constant term to the right side of the = sign and divide through by a if necessary to put the equation in the form

$$x^2 + Bx = C$$

2. Add $\left(\dfrac{B}{2}\right)^2$ to both sides of the equation to obtain the form

$$\left(x + \dfrac{B}{2}\right)^2 = D$$

3. Take the square root of both sides.
4. Isolate x on one side of the = sign.

EXAMPLE 4   Solve the following quadratic equation by completing the square.

$$x^2 - 4x - 5 = 0$$

SOLUTION

$$\begin{aligned}
x^2 - 4x &= 5 \\
x^2 - 4x + 4 &= 5 + 4 \qquad &\text{(Complete the square)} \\
(x - 2)^2 &= 9 \\
x - 2 &= \pm 3 \qquad &\text{(Take the square root of both sides)} \\
x &= 2 \pm 3
\end{aligned}$$

The solutions are $x = 2 + 3 = 5$

and $x = 2 - 3 = -1$.

The solution set is $\{5, -1\}$.

Chapter 2 Equations and Algebraic Inequalities

EXAMPLE 5 Complete the square to solve the equation.

$$2x^2 + x - 3 = 0$$

SOLUTION

$$2x^2 + x = 3 \quad \text{(Put the constant term on the right side)}$$

$$x^2 + \frac{1}{2}x = \frac{3}{2} \quad \text{(Divide through by 2 to set leading coefficient = 1)}$$

$$x^2 + \frac{1}{2}x + \left(\frac{1}{4}\right)^2 = \frac{3}{2} + \frac{1}{16} \quad \text{(Complete the square)}$$

$$\left(x + \frac{1}{4}\right)^2 = \frac{25}{16} \quad \text{(Factor the left side and simplify the right side)}$$

$$x + \frac{1}{4} = \pm\frac{5}{4} \quad \text{(Take the square root)}$$

$$x = -\frac{1}{4} \pm \frac{5}{4} \quad \text{(Isolate } x\text{)}$$

$$= \frac{4}{4}, -\frac{6}{4}$$

$$x = 1, -\frac{3}{2}$$

The solution set is $\left\{1, -\frac{3}{2}\right\}$. ∎

The Quadratic Formula

The quadratic formula enables us to solve any quadratic equation in one variable. It can be derived by completing the square in a general quadratic equation. If there is an x such that

$$ax^2 + bx + c = 0$$

then

$$ax^2 + bx = -c \quad \text{(Subtract } c \text{ from both sides)}$$

$$x^2 + \frac{b}{a}x = -\frac{c}{a} \quad \text{(Divide through by } a\text{)}$$

$$x^2 + \frac{b}{a}x + \left(\frac{b}{2a}\right)^2 = -\frac{c}{a} + \frac{b^2}{4a^2} \quad \text{(Complete the square)}$$

$$\left(x + \frac{b}{2a}\right)^2 = \frac{b^2 - 4ac}{4a^2} \quad \text{(Factor the left side and simplify the right side)}$$

Take the square root of both sides:

$$x + \frac{b}{2a} = \pm\frac{\sqrt{b^2 - 4ac}}{2|a|} \quad \text{(Recall } \sqrt{a^2} = |a|\text{)}$$

Since we are interested in both roots (those corresponding to the positive and negative signs on the right side), we can replace $|a|$ with a.

$$x = -\frac{b}{2a} \pm \frac{\sqrt{b^2 - 4ac}}{2a}$$

Each of the steps in this derivation is reversible. Thus we obtain the **quadratic formula.**

> ### THE QUADRATIC FORMULA
> The roots of a quadratic equation with real coefficients
> $$ax^2 + bx + c = 0 \qquad a \neq 0$$
> are given by
> $$x = \frac{-b + \sqrt{b^2 - 4ac}}{2a} \quad \text{and} \quad x = \frac{-b - \sqrt{b^2 - 4ac}}{2a}$$

The number $b^2 - 4ac$ is called the **discriminant** of the quadratic equation. It indicates the number and form of the roots:

> ### THE DISCRIMINANT
> If $ax^2 + bx + c = 0$; $a \neq 0$; a, b, and c are real numbers and
> i. $b^2 - 4ac > 0$, there are two distinct real roots.
> ii. $b^2 - 4ac = 0$, there is precisely one real root, $-\dfrac{b}{2a}$, which is called a double root.
> iii. $b^2 - 4ac < 0$, there are no real roots.

EXAMPLE 6 Solve the following equation by using the quadratic formula.
$$2x^2 + 5x - 3 = 0$$

SOLUTION We use $a = 2$, $b = 5$, and $c = -3$ in the quadratic formula to find

$$x = \frac{-5 \pm \sqrt{5^2 - 4 \cdot 2 \cdot (-3)}}{2 \cdot 2}$$

$$= \frac{-5 \pm \sqrt{25 + 24}}{4}$$

$$= \frac{-5 \pm 7}{4}$$

$$= \frac{-5 + 7}{4} \quad \text{or} \quad \frac{-5 - 7}{4}$$

$$= \frac{1}{2}, -3$$

The solutions are $\dfrac{1}{2}$ and -3.

Checking, we have the following:

$x = \frac{1}{2}$: $2\left(\frac{1}{2}\right)^2 + 5\left(\frac{1}{2}\right) - 3 = 2\left(\frac{1}{4}\right) + \frac{5}{2} - 3$

$= \frac{1}{2} + \frac{5}{2} - 3$

$= \frac{6}{2} - 3 \stackrel{\checkmark}{=} 0$

$x = -3$: $2(-3)^2 + 5(-3) - 3 = 2 \cdot 9 - 15 - 3$

$= 18 - 15 - 3$

$\stackrel{\checkmark}{=} 0$

The solution set is indeed $\left\{\frac{1}{2}, -3\right\}$. ∎

EXAMPLE 7   Solve the following quadratic equation.

$x^2 - 2x + 2 = 0$

SOLUTION   By the quadratic formula

$x = \dfrac{-(-2) \pm \sqrt{(-2)^2 - 4 \cdot 1 \cdot 2}}{2 \cdot 1}$

$= \dfrac{2 \pm \sqrt{4 - 8}}{2}$

$= \dfrac{2 \pm \sqrt{-4}}{2}$

$= \dfrac{2 \pm 2i}{2}$

$= 1 \pm i$

The roots are $1 + i$ and $1 - i$. There are no real roots. The solution set is $\{1 + i, 1 - i\}$. ∎

The roots in Example 7 are complex conjugates of one another: $1 \pm i$. Whenever the discriminant is negative ($b^2 - 4ac < 0$), there are no real roots; but there are two non-real roots and they are complex conjugates of one another. These non-real roots are given by

$x_1 = \dfrac{-b}{2a} + i\dfrac{\sqrt{|b^2 - 4ac|}}{2a}$ and $x_2 = \dfrac{-b}{2a} - i\dfrac{\sqrt{|b^2 - 4ac|}}{2a}$

Note that a given quadratic equation cannot have both real and non-real roots.

If a, b, and c are rational numbers and $b^2 - 4ac \geq 0$ is a perfect square (for example, 36), then the real roots are rational numbers; if $b^2 - 4ac > 0$ is not a perfect square (for example, 35), then the real roots are irrational.

EXAMPLE 8   Without solving the equation, classify the roots of each of the following.

a. $x^2 - 6x + 9 = 0$

b. $3x^2 + 2x - 1 = 0$

c. $x^2 + x + 5 = 0$
d. $2x^2 + 3x - 1 = 0$

SOLUTION

To classify the roots, we need only evaluate the discriminant, $b^2 - 4ac$.

a. $b^2 - 4ac = (-6)^2 - 4 \cdot 1 \cdot 9 = 36 - 36 = 0$

Thus there is only one real root; it is a double root. We can also verify this by factoring:

$x^2 - 6x + 9 = (x - 3)^2 = 0$ has the double root $x = 3$.

b. $b^2 - 4ac = 2^2 - 4 \cdot 3 \cdot (-1) = 4 + 12 = 16 > 0$

There are two real roots and these roots are rational since $16 \,(= 4^2)$ is a perfect square.

c. $b^2 - 4ac = 1^2 - 4 \cdot 1 \cdot 5 = 1 - 20 = -19 < 0$

Thus there are two non-real (complex conjugate) roots; there are no real roots.

d. $b^2 - 4ac = 3^2 - 4 \cdot 2 \cdot (-1) = 9 + 8 = 17 > 0$

There are two real roots; these roots are irrational since 17 is not a perfect square. ∎

EXAMPLE 9

If $x^2 - y^2 = xz$, solve for x in terms of y and z.

SOLUTION

We rewrite the equation as

$$x^2 - xz - y^2 = 0$$

and use the quadratic formula:

$$x = \frac{-(-z) \pm \sqrt{(-z)^2 - 4 \cdot 1 \cdot (-y^2)}}{2 \cdot 1}$$

$$= \frac{z \pm \sqrt{z^2 + 4y^2}}{2}$$
∎

Section 2.3 Exercises

Solve the following equations by factoring.

1. $x^2 - 4 = 0$
2. $u^2 - 9 = 0$
3. $x^2 + 4 = 0$
4. $u^2 + 9 = 0$
5. $s^2 + 6 = 5s$
6. $4y^2 - 4y + 1 = 0$
7. $z^2 + 4z + 4 = 0$
8. $4x^2 = 15 + 4x$
9. $v^2 - v = 2$
10. $w^2 + w = 20$
11. $6x^2 + 5x = 6$
12. $4y^2 + 3y = 1$
13. $9z^2 + 12z + 4 = 0$
14. $3s^2 = 10 + s$
15. $2t^2 + 9t = 5$
16. $4u^2 + 11u = 3$
17. $25v^2 + 1 = 10v$
18. $3y^2 = 4y + 4$
19. $2u^2 = 15 + 7u$
20. $3z^2 + 5z + 2 = 0$

Solve the following by completing the square.

21. $y^2 + 4y - 21 = 0$
22. $x^2 + 2x - 3 = 0$
23. $p^2 + 18p - 88 = 0$
24. $t^2 - 8t + 13 = 0$
25. $w^2 + 4w + 2 = 0$
26. $3s^2 + 12s - 15 = 0$
27. $2q^2 + 4q - 1 = 0$
28. $2u^2 - 6u + 1 = 0$

29. $z^2 - z - 72 = 0$
30. $r^2 + 3r + 5 = 0$
31. $r^2 - 4r - 12 = 0$
32. $u^2 + 4u - 2 = 0$
33. $2x^2 - 4x - 3 = 0$
34. $5p^2 - 10p + 3 = 0$
35. $4y^2 - 4y + 1 = 0$
36. $x^2 + 3x - 4 = 0$
37. $4v^2 - 4v - 3 = 0$
38. $3s^2 - s - 10 = 0$

76. $3z^2 - 4z + 2 = 0$
77. $r^2 - 8r + 16 = 0$
78. $3s^2 - 9s + 4 = 0$
79. $4t^2 - 20t + 25 = 0$
80. $u^2 - 2u - 2 = 0$
81. $8v^2 - 14v - 15 = 0$
82. $x^2 - x + 1 = 0$
83. $2y^2 - 17y + 36 = 0$

Use the quadratic formula to solve the following equations.

39. $x^2 + 3x - 4 = 0$
40. $y^2 - 3y - 4 = 0$
41. $9u^2 - 3u = 2$
42. $2z^2 + 9z = 5$
43. $4u^2 + 9 = 12u$
44. $25v^2 + 10v + 1 = 0$
45. $3p^2 + 5p + 2 = 0$
46. $x^2 + 2 = 2x$
47. $y^2 + 1 = y$
48. $z^2 + 9 = 4z$
49. $r^2 + 34 = 6r$
50. $4w^2 + 11w = 3$
51. $3z^2 = 4z + 4$
52. $(x-1)(x-2) + x = 0$
53. $5w^2 = 3w + 14$
54. $s^2 = 2s - 5$
55. $t^2 = 2t - 4$
56. $u^2 = 2u - 3$
57. $v^2 + 2v + 2 = 0$
58. $2w^2 - 2w + 1 = 0$
59. $p^2 + 8p + 25 = 0$
60. $6v^2 + v = 2$
61. $r^2 - 6r + 16 = 0$
62. $4y^2 + 3y = 1$
63. $3s^2 = 14s - 8$
64. $q^2 = 2q - 10$
65. $2x^2 - 2\sqrt{2}x + 3 = 0$
66. $y^2 + y + 1 = 0$
67. $p^2 + 2p + 5 = 0$
68. $u^2 + 4u + 5 = 0$
69. $w^2 + 4w + 13 = 0$
70. $2z^2 + 4z + 3 = 0$
71. $q^2 = 2q - 2$
72. $v^2 + 2v + 6 = 0$
73. $t^2 + 6t + 10 = 0$

Solve the following quadratic equations with complex coefficients by using the quadratic formula.

84. $2z^2 + 2iz - 13 = 0$
85. $3z^2 - 5iz + 2 = 0$
86. $2iz^2 - 5z - 3i = 0$
87. $-2iz^2 + 2z + 13i = 0$
88. $7z^2 - 4iz + 3 = 0$
89. $z^2 - 3iz + 4 = 0$

In each of the following, solve for the indicated variable in terms of the others.

90. $A = 2\pi rh + 2\pi r^2$, for r
91. $s = \frac{1}{2}gt^2 + vt$, for t
92. $x^2 + xy + y^2 = 0$, for y
93. $r^2 = 2rs + 3s$, for r
94. $(p - q)^2 + q^2 = 2qp$, for p
95. $(u + v)^2 + 2uv = v^2$, for u

Calculator Exercises

Use the quadratic formula and a calculator to find the roots of the following equations. Round the answers to two decimal places.

96. $1.6x^2 + 4.2x - 9.3 = 0$
97. $20x^2 + 79x + 36 = 0$
98. $12.76x^2 - 70.68x + 86.81 = 0$
99. $1.4x^2 - 6.7x - 9.2 = 0$

Without solving the quadratic equation, classify the roots in each of the following.

74. $2x^2 - 5x - 9 = 0$
75. $8y^2 - 30y + 27 = 0$

Section 2.4

Equations Solvable with Quadratic Techniques

Certain other kinds of equations can also be solved by using the techniques for solving quadratic equations. For instance, in an equation of the form

$$ax^{2k} + bx^k + c = 0$$

substitution of $u = x^k$ results in a quadratic equation. Such a substitution is illustrated in Examples 1 through 4 with k an integer and k a rational number.

Section 2.4 Equations Solvable with Quadratic Techniques 105

EXAMPLE 1 Solve the following equation.
$$x^4 - 2x^2 - 3 = 0$$

SOLUTION This is a quadratic equation in x^2. Setting $u = x^2$, this equation becomes

$$u^2 - 2u - 3 = 0$$
$$(u + 1)(u - 3) = 0 \qquad \text{(Factor)}$$
$$u = -1, 3$$

But recall that $u = x^2$ and it is x we wish to solve for; thus

$$x^2 = 3 \qquad \text{or} \qquad x^2 = -1$$
$$x = \pm\sqrt{3} \qquad \text{or} \qquad x = \pm i$$

We can substitute these values into the original equation to verify that they are all indeed solutions.

$(x = \pm 3)$: $(\pm\sqrt{3})^4 - 2(\pm\sqrt{3})^2 - 3 = 9 - 2 \cdot 3 - 3 \stackrel{\checkmark}{=} 0$
$(x = \pm i)$: $(\pm i)^4 - 2(\pm i)^2 - 3 = 1 - 2(-1) - 3 \stackrel{\checkmark}{=} 0$

The solution set is $\{\sqrt{3}, -\sqrt{3}, i, -i\}$. ■

EXAMPLE 2 Solve the following equation.
$$x^{2/3} + x^{1/3} - 2 = 0$$

SOLUTION This time we let $u = x^{1/3}$ and put the equation in the following form.

$$u^2 + u - 2 = 0 \qquad (x^{2/3} = [x^{1/3}]^2 = u^2)$$
$$(u + 2)(u - 1) = 0$$
$$u = -2, 1$$

But $u = x^{1/3} = \sqrt[3]{x}$; thus

$$u = \sqrt[3]{x} = -2 \qquad \text{or} \qquad u = \sqrt[3]{x} = 1$$
$$(\sqrt[3]{x})^3 = (-2)^3 \qquad\qquad (\sqrt[3]{x})^3 = 1^3$$
$$x = -8 \qquad \text{or} \qquad x = 1$$

We can check that each of these values is indeed a solution of the original equation. The solution set is $\{-8, 1\}$. ■

EXAMPLE 3 Solve the following equation.
$$2y^4 - 8y^2 + 3 = 0$$

SOLUTION We let $u = y^2$ and rewrite the equation as

$$2u^2 - 8u + 3 = 0$$

This quadratic expression will not factor but the quadratic formula yields

$$u = \frac{-(-8) \pm \sqrt{(-8)^2 - 4 \cdot 2 \cdot 3}}{2 \cdot 2}$$

$$= \frac{8 \pm \sqrt{64 - 24}}{4}$$

$$= \frac{8 \pm \sqrt{40}}{4} = \frac{8 \pm 2\sqrt{10}}{4}$$

$$y^2 = u = \frac{4 \pm \sqrt{10}}{2}$$

$$y = \pm\sqrt{u} = \pm\sqrt{\frac{4 \pm \sqrt{10}}{2}}$$

Note that $4 > \sqrt{10}$ so that $4 - \sqrt{10} > 0$. Thus both $\dfrac{4 + \sqrt{10}}{z}$ and $\dfrac{4 - \sqrt{10}}{z}$ have two square roots in the real number system.

Checking, we obtain

$$2\left(\pm\sqrt{\frac{4 \pm \sqrt{10}}{2}}\right)^4 - 8\left(\pm\sqrt{\frac{4 \pm \sqrt{10}}{2}}\right)^2 + 3$$

$$= 2\left(\frac{4 \pm \sqrt{10}}{2}\right)^2 - 8\left(\frac{4 \pm \sqrt{10}}{2}\right) + 3$$

$$= 2\left(\frac{16 \pm 8\sqrt{10} + 10}{4}\right) - 4(4 \pm \sqrt{10}) + 3$$

$$= 13 \pm 4\sqrt{10} - 16 \mp 4\sqrt{10} + 3 \stackrel{\checkmark}{=} 0$$

The solution set is $\left\{\sqrt{\dfrac{4 + \sqrt{10}}{2}}, \sqrt{\dfrac{4 - \sqrt{10}}{2}}, -\sqrt{\dfrac{4 + \sqrt{10}}{2}}, -\sqrt{\dfrac{4 - \sqrt{10}}{2}}\right\}$. ∎

EXAMPLE 4

Solve the following equation.

$$x + \sqrt{x} - 6 = 0$$

SOLUTION

We set $u = \sqrt{x}$ and obtain

$$u^2 + u - 6 = 0 \qquad\qquad (x = \sqrt{x}^2 = u^2)$$
$$(u + 3)(u - 2) = 0$$
$$u = -3, 2$$

But the symbol $\sqrt{x}$ stands for the positive square root of x. Thus, $u = \sqrt{x} = -3$ is impossible. Hence, we discard -3 and set

$$u = \sqrt{x} = 2$$

or

$$x = 4$$

We check the original equation.

$$x = 4: \qquad 4 + \sqrt{4} - 6 = 4 + 2 - 6 \stackrel{\checkmark}{=} 0$$

The number 4 is the (only) solution. The solution set is $\{4\}$. ∎

Equations Involving Radicals

Another type of problem that can lead to a quadratic equation involves radicals. Squaring both sides of the equation one or more times will eliminate the radicals. We begin by offering an alternate solution to Example 4.

ALTERNATE SOLUTION

Isolate the radical on the left side of the equation and rewrite the equation in the form

$$\sqrt{x} = 6 - x$$

Then square both sides of the equation to obtain

$$x = (6 - x)^2$$
$$= 36 - 12x + x^2$$

Collect all terms on the same side of the equation:

$$x^2 - 13x + 36 = 0$$
$$(x - 9)(x - 4) = 0$$
$$x = 9, 4$$

Checking in the original equation ($x + \sqrt{x} - 6 = 0$), we see that 4 is a legitimate solution, but for $x = 9$, we have

$$9 + \sqrt{9} - 6 = 9 + 3 - 6 = 12 - 6 \neq 0$$

Thus 4 is the only solution. ∎

In the alternate solution to Example 4, we obtained *apparent* roots 9 and 4. *The apparent root 9 is a root of the squared equation but not of the original equation.* Such a root is said to be **extraneous** and must be rejected.

EXTRANEOUS ROOTS

Multiplying an equation by an expression containing the variable (as when the equation is squared) can introduce **extraneous roots**. Roots obtained after such operations *must* be checked by substituting into the *original* equation.

EXAMPLE 5

Solve the equation

$$1 + \sqrt{2x + 1} = x$$

SOLUTION

Isolate the radical on one side of the equation:

$$\sqrt{2x + 1} = x - 1$$
$$2x + 1 = (x - 1)^2 \quad \text{(Square both sides)}$$
$$2x + 1 = x^2 - 2x + 1$$
$$0 = x^2 - 4x \quad \text{(Subtract } 2x + 1 \text{ from both sides)}$$
$$0 = x(x - 4)$$
$$x = 0, 4$$

Checking, we see that 4 is indeed a solution; but for $x = 0$,

$$1 + \sqrt{2 \cdot 0 + 1} = 1 + 1 \neq 0$$

The solution set is {4}.

EQUATIONS INVOLVING RADICALS

To solve an equation involving radicals

1. **Isolate the (a) radical** on one side of the equation.
2. **Remove the radical** by raising both sides of the equation to the power which equals the index of the radical.
3. **Simplify the results.**
4. **Repeat steps 1 and 2** until no more radicals remain in the equation.
5. **Solve the resulting equation** (which contains no radicals) using whatever other techniques are appropriate.
6. **Check** your solutions.

EXAMPLE 6 Solve the following equation.

$$\sqrt{x + 1} + \sqrt{2x - 5} = 3$$

SOLUTION First, we move one of the radicals to the other side of the equation:

$$\sqrt{2x - 5} = 3 - \sqrt{x + 1}$$
$$2x - 5 = 9 - 6\sqrt{x + 1} + (x + 1) \quad \text{(Square both sides)}$$
$$6\sqrt{x + 1} = 15 - x \quad \text{(Collect terms and isolate the radical)}$$
$$36(x + 1) = 225 - 30x + x^2 \quad \text{(Square both sides)}$$
$$0 = x^2 - 66x + 189 \quad \text{(Collect terms)}$$
$$0 = (x - 3)(x - 63)$$
$$x = 3, 63$$

If x satisfies the original equation, x will also satisfy the equation that results from multiplying by any expression. Hence, in this example we can say only that *if x is a solution*, then $x = 3$ or $x = 63$; that is, 3 and 63 are the only *possible* solutions. Checking the original equation shows us that 3 is the only solution.

$x = 3$: $\quad \sqrt{3 + 1} + \sqrt{2 \cdot 3 - 5} = \sqrt{4} + \sqrt{1} = 2 + 1 \stackrel{\checkmark}{=} 3$

$x = 63$: $\quad \sqrt{63 + 1} + \sqrt{2 \cdot 63 - 5} = \sqrt{64} + \sqrt{121} = 8 + 11 \neq 3$

The solution set is {3}.

The introduction of extraneous roots is illustrated very vividly by modifying Example 6 slightly, as in Example 7.

EXAMPLE 7

Solve the following equation.

$$\sqrt{x+1} + \sqrt{2x-5} = -3$$

SOLUTION

There are *no solutions* because the sum of two positive numbers cannot be negative. However, if we proceed as above, we again generate the equation $x^2 - 66x + 189 = 0$. (Try it!) Thus we obtain the *two* extraneous solutions $x = 3$ and $x = 63$, both of which we would have to reject.

EXAMPLE 8

Solve the following equation.

$$\sqrt[3]{x^2 - 36} = -3$$

SOLUTION

We cube (or raise to the third power) both sides of the equation:

$$x^2 - 36 = (-3)^3 = -27$$
$$x^2 = 36 - 27 = 9$$
$$x = \pm 3$$

Each of these values checks as a solution; the solution set is $\{3, -3\}$.

Equations Involving Fractions

To solve an equation involving rational expressions (fractions):

1. Multiply through by the LCD of the fractions in the equation.
2. Simplify and solve the resulting equation.
3. Check the solutions in the original equation.

Whenever the LCD involves the variable we *must* check to see that the apparent solutions satisfy the original equation since extraneous roots may be introduced. This is because we are multiplying by an expression involving the variable.

EXAMPLE 9

Solve the following equation.

$$\frac{x-1}{x+2} + \frac{x+7}{x+4} = 0$$

SOLUTION

Multiplying through by the LCD which is $(x+2)(x+4)$, we obtain

$$\frac{(x-1)(x+2)(x+4)}{(x+2)} + \frac{(x+7)(x+2)(x+4)}{(x+4)} = 0$$

$$(x-1)(x+4) + (x+7)(x+2) = 0$$
$$(x^2 + 3x - 4) + (x^2 + 9x + 14) = 0$$
$$2x^2 + 12x + 10 = 0$$
$$x^2 + 6x + 5 = 0$$
$$(x+1)(x+5) = 0$$
$$x = -1, -5$$

110 Chapter 2 Equations and Algebraic Inequalities

Checking $x = -1$, we have

$$\frac{(-1)-1}{(-1)+2} + \frac{(-1)+7}{(-1)+4} = \frac{-2}{1} + \frac{6}{3} = -2 + 2 \stackrel{\checkmark}{=} 0$$

Similarly, $x = -5$ can be shown to satisfy the original equation. The solutions are -1 and -5. The solution set is $\{-1, -5\}$.

Since we multiplied both sides of the original equation by $(x+2)(x+4)$, we *must* check to see that our apparent solutions satisfy the original equation; we have multiplied by an expression involving the variable. For instance, had we found -2 as a possible solution, we would have had to reject it because checking the original problem would have resulted in division by zero. ∎

EXAMPLE 10 Solve the following equation.

$$\frac{6x+2}{x^2-9} + \frac{4}{x-3} = 4$$

SOLUTION Since $x^2 - 9 = (x+3)(x-3)$, the LCD is $x^2 - 9$. Multiplying through by this LCD, we obtain:

$$(x^2-9)\left(\frac{6x+2}{x^2-9} + \frac{4}{x-3}\right) = 4(x^2-9)$$

$$\frac{(x^2-9)(6x+2)}{(x^2-9)} + \frac{4(x+3)(x-3)}{(x-3)} = 4(x^2-9) \quad (x^2-9) = (x+3)(x-3)$$

$$(6x+2) + (4x+12) = 4x^2 - 36$$
$$0 = 4x^2 - 10x - 50 \quad \text{(Collect terms)}$$
$$= 2x^2 - 5x - 25 \quad \text{(Divide by 2)}$$
$$= (2x+5)(x-5) \quad \text{(Factor)}$$
$$x = -\frac{5}{2}, 5$$

Since neither of these values ($x = -\frac{5}{2}$ or $x = 5$) causes $x^2 - 9$ to be 0, each of them checks as a solution. This happens because we did not multiply through by 0 when we multiplied by the LCD, $x^2 - 9$. Unfortunately there is no such shortcut to checking when we square both sides of an equation. ∎

EXAMPLE 11 Solve the following equation.

$$\frac{3}{x-2} - \frac{3}{(x-1)(x-2)} = \frac{2}{x-1}$$

SOLUTION Multiply through by the LCD, $(x-1)(x-2)$.

$$\frac{3(x-1)(x-2)}{(x-2)} - \frac{3(x-1)(x-2)}{(x-1)(x-2)} = \frac{2(x-1)(x-2)}{(x-1)}$$

$$3x - 3 - 3 = 2x - 4$$
$$x = 2$$

Checking, we see that $x = 2$ causes division by 0. Thus there are *no solutions* since 2 was the only *possible* solution. The solution set is empty. ∎

Section 2.4 Exercises

Find all real solutions to the following equations.

1. $x^4 + 5 = 6x^2$
2. $u^4 + 4u^2 = 12$
3. $2w^4 + 15w^2 = 8$
4. $y^4 + y^2 = 2$
5. $z^4 + 7 = 8z^2$
6. $w^4 - 5w^2 + 4 = 0$
7. $r^6 + 5 = 6r^3$
8. $s^6 + 4s^3 = 12$
9. $t^6 + 6t^3 + 9 = 0$
10. $9p^6 - 12p^3 + 4 = 0$
11. $t^6 + 3t^3 = 10$
12. $z^6 + 2z^3 = 15$
13. $q^8 + 8 = 6q^4$
14. $x^8 + 10x^4 = 24$
15. $3v^8 - 5v^4 + 2 = 0$
16. $6y^8 = 5y^4 + 6$
17. $z^{10} = z^5 + 6$
18. $3s^{10} + 8s^5 = 3$
19. $y^{12} = y^6 + 6$
20. $4t^{16} + 4t^8 = 15$
21. $x^4 + 2x^2 = 6$
22. $u^4 + u^2 = 1$
23. $v^4 + 7 = 6v^2$
24. $2r^4 + 2r^2 = 3$
25. $5w^4 + 3w^2 - 1 = 0$
26. $2t^4 = 4t^2 + 5$
27. $x^{2/3} + 6 = 5x^{1/3}$
28. $p^{2/3} + 4p^{1/3} + 4 = 0$
29. $r^{2/5} = r^{1/5} + 2$
30. $4w^{2/5} + 1 = 4w^{1/5}$
31. $9q^{-2} + 4 = 12q^{-1}$
32. $x^{-2} - 3x^{-1} + 2 = 0$

$\left(\text{Hint: Let } u = q^{-1} = \dfrac{1}{q}.\right)$

33. $2y^{-2} + 2 = 5y^{-1}$
34. $3v^{-2} + 8v^{-1} - 3 = 0$
35. $\dfrac{4}{u^4} - \dfrac{17}{u^2} + 4 = 0$
36. $\dfrac{3}{t^2} - \dfrac{11}{t} - 20 = 0$
37. $\dfrac{9}{z^4} - \dfrac{10}{z^2} + 1 = 0$
38. $\dfrac{6}{r^2} - \dfrac{5}{r} - 6 = 0$
39. $2w - 5\sqrt{w} = 3$
40. $x - 5\sqrt{x} + 6 = 0$
41. $w - \sqrt{w} - 2 = 0$
42. $t + 3\sqrt{t} - 10 = 0$
43. $2p + 3\sqrt{p} - 2 = 0$
44. $3q - 7\sqrt{q} - 6 = 0$
45. $4r - 4\sqrt{r} + 1 = 0$
46. $v - \sqrt{v} - 6 = 0$
47. $\sqrt{r} + \sqrt[4]{r} - 20 = 0$
48. $\sqrt[3]{s} + \sqrt[6]{s} - 6 = 0$
49. $\sqrt{z - 4} = 4 - z$
50. $4 + \sqrt{x + 2} = x$
51. $\sqrt{r + 2} = r - 10$
52. $\sqrt{4 - s} = 2 - s$
53. $\sqrt{2u + 9} = u - 3$
54. $\sqrt{3w + 1} = w - 3$
55. $5 + \sqrt{y - 3} = y$
56. $\sqrt{3s - 2} = 2 + \sqrt{s}$
57. $\sqrt{3w + 4} + 6 - \sqrt{w} = 0$
58. $\sqrt{r + 20} = \sqrt{r + 4} + 2$
59. $2 + \sqrt{q - 4} = \sqrt{2q - 1}$
60. $\sqrt{3x - 8} - \sqrt{5x - 11} = -1$
61. $\sqrt{w + 3} - \sqrt{11w + 58} + 5 = 0$
62. $\sqrt{p + 6} + \sqrt{p + 3} = \sqrt{5 - 2p}$
63. $\sqrt{v + 1} + \sqrt{v - 2} = \sqrt{2v + 3}$
64. $\sqrt[3]{w^2 - 1} = 2$
65. $\sqrt[3]{r^2 + 2} = 3$
66. $\sqrt[4]{y^3 + 9} = 1$
67. $\sqrt[4]{p^3 - 11} = 2$
68. $\dfrac{w - 1}{w + 1} + \dfrac{w + 1}{w - 1} = 0$
69. $\dfrac{v + 2}{v - 1} = \dfrac{v - 3}{2v - 2}$
70. $\dfrac{s + 3}{s - 1} + \dfrac{s + 1}{s - 8} = 0$
71. $\dfrac{3}{2y + 7} - 2y - 5 = 0$
72. $5 - \dfrac{24}{x + 3} = \dfrac{1}{x - 2}$
73. $\dfrac{3}{2x + 1} - 3 = \dfrac{2x}{3x - 1}$
74. $\dfrac{4}{y + 2} + \dfrac{2y}{y - 1} - 5 = 0$
75. $\dfrac{2}{z - 3} + 1 = \dfrac{6}{z - 8}$
76. $\dfrac{1}{3p - 4} + 3p - 2 = 0$
77. $\dfrac{45q}{3q - 4} - 1 = \dfrac{40q}{2q - 3}$
78. $\dfrac{2u}{3u + 2} - 3 = \dfrac{5u}{1 - 2u}$
79. $\dfrac{2}{r^2 - 3} + 1 = \dfrac{6}{r^2 - 8}$
80. $\dfrac{v^2 + 2}{v^2 - 3} = \dfrac{3v^2 - 6}{2v^2 - 7}$
81. $\dfrac{4}{r^2 + 2} + \dfrac{2r^2}{r^2 - 1} = 5$
82. $\dfrac{3}{u + 1} = 2\left[1 - \dfrac{1}{(u + 1)^2}\right]$
83. $\dfrac{w^3 - 3}{w^3 + 2} + \dfrac{w^3 + 1}{w^3 + 6} + 4 = 0$
84. $\dfrac{3}{2 - 3z} + \dfrac{1}{z - 1} + \dfrac{-1}{3z - 2} = 0$

Calculator Exercises

Find the real solutions to the following exercises, rounded to two decimal places.

85. $20x^4 + 13x^2 - 12 = 0$
86. $7u^4 - 12u^2 - 5 = 0$
87. $3v^{2/3} - 7v^{1/3} + 3 = 0$
88. $5y^{2/3} + 3y^{1/3} - 4 = 0$
89. $\sqrt[3]{p^2 - 5} = 16$
90. $\sqrt[4]{w^3 + 200} = 3.5$

Solve for the specified variable in terms of the others.

91. $\sqrt{x} + \sqrt{y} = \sqrt{a}$, for y
92. $\sqrt[3]{a} + \sqrt[3]{b} = \sqrt[3]{c}$, for a
93. $x^{2/3} + y^{2/3} = z^{2/3}$, for x
94. $p^{2/5} - q^{2/5} = A$, for q
95. $\sqrt{x^2 + y^2 + 1} = x + y$, for y
96. $\sqrt{u^2 - v^2 + 3} = u - v$, for v

Section 2.5

Applications of Quadratic Equations

Applications of any area of mathematics typically arise as word problems. In this section, we solve word problems which lead to quadratic equations. We summarize here the instructions for solving word problems as presented in Section 2.2

> SOLVING WORD PROBLEMS
> 1. Understand the problem; read it several times.
> 2. Make a sketch if it will help.
> 3. Identify the unknown quantities.
> 4. Name these quantities.
> 5. Express all known relationships mathematically.
> 6. Reread the problem; check your equations.
> 7. Solve the resulting equations.
> 8. Interpret the results.
> 9. Check the solutions.

EXAMPLE 1

A rectangular pool is 15 feet longer than it is wide and has a total area of 1000 square feet. Find the width and length of the pool.

SOLUTION

Label the rectangle as in Figure 2.3.

Figure 2.3

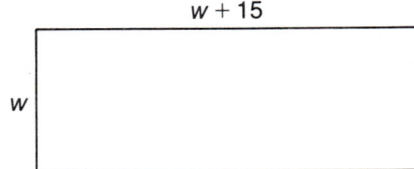

The area is given by

$$w(w + 15) = 1000$$
$$w^2 + 15w - 1000 = 0$$
$$(w + 40)(w - 25) = 0 \quad \text{(Factor)}$$
$$w = 25, -40$$
$$w + 15 = 40 \quad (= 25 + 15)$$

(We discard $w = -40$ since we must have $w > 0$.)

The dimensions of the pool are 25 feet by 40 feet; checking, we do have $25 \times 40 = 1000$.

EXAMPLE 2

A sidewalk is constructed diagonally across a rectangular park which is 700 yards longer than it is wide. If the sidewalk is 1300 yards long, what are the dimensions of the park?

SOLUTION

Label the dimensions (in hundreds of yards) as in Figure 2.4.

Figure 2.4

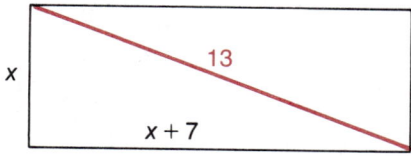

Then by the Pythagorean Theorem

$$x^2 + (x + 7)^2 = 13^2$$
$$x^2 + (x^2 + 14x + 49) = 169$$
$$2x^2 + 14x - 120 = 0 \qquad \text{(Collect terms)}$$
$$x^2 + 7x - 60 = 0 \qquad \text{(Divide through by 2)}$$
$$(x + 12)(x - 5) = 0 \qquad \text{(Factor)}$$
$$x = 5, -12$$
$$x + 7 = 12 \qquad (= 5 + 7)$$

(Again we discard $x = -12$ since we must have $x > 0$.) The dimensions of the park are 500 yards by 1200 yards.

EXAMPLE 3

The altitude h in feet of a projectile t seconds after firing is given by

$$h = -16t^2 + v_u t + h_0$$

where v_u is the *upward* component of its initial velocity in feet per second and h_0 is the altitude in feet from which it is fired. A rocket is launched from a hilltop 2400 feet above the desert with an initial upward velocity of 400 feet per second. When will it land on the desert? (See Figure 2.5.)

Figure 2.5

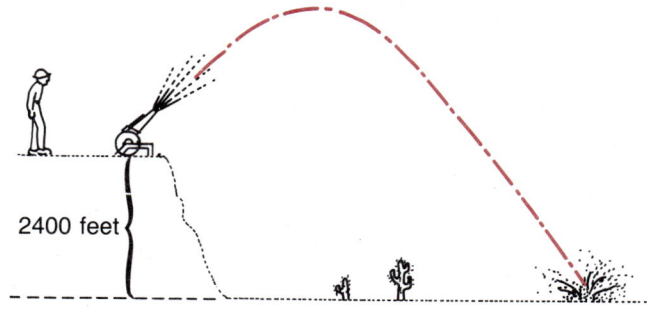

114 Chapter 2 Equations and Algebraic Inequalities

SOLUTION The altitude of the rocket t seconds after launching is

$$-16t^2 + 400t + 2400$$

When it lands, its altitude is zero. To determine the time when its altitude is zero, we must solve the quadratic equation

$$h = -16t^2 + 400t + 2400 = 0$$
$$-16(t^2 - 25t - 150) = 0 \quad \text{(Factor out } -16)$$
$$t^2 - 25t - 150 = 0 \quad \text{(Divide by } -16)$$
$$(t - 30)(t + 5) = 0 \quad \text{(Factor)}$$
$$t = 30, -5$$

The solution $t = -5$ represents the time 5 seconds *before* launching and is irrelevant. Thus, the rocket should land on the desert 30 seconds after it is fired. ∎

EXAMPLE 4

If the rocket in Example 3 is launched with an initial *downward* velocity of 80 feet per second, when does it land on the desert?

SOLUTION Since v_u in the formula given in Example 3 represents *upward* velocity, we must set $v_u = -80$ to represent the *downward* velocity. Then we solve as follows:

$$h = -16t^2 - 80t + 2400 = 0$$
$$-16(t^2 + 5t - 150) = 0$$
$$(t - 10)(t + 15) = 0$$
$$t = 10, -15$$

The rocket lands 10 seconds after firing (and *not* 15 seconds before firing). ∎

EXAMPLE 5

A rectangular piece of metal is to be used to make a box by cutting 4 inch squares from each of its corners and then folding up the sides. If the piece of metal is twice as long as it is wide and the volume of the box is 768 cubic inches, find the dimensions of the piece of metal.

SOLUTION We label the width and length of the metal as x and $2x$ respectively. See Figure 2.6.

Figure 2.6

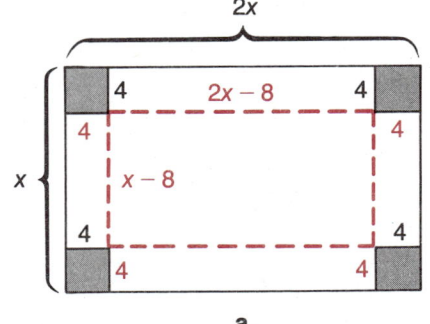
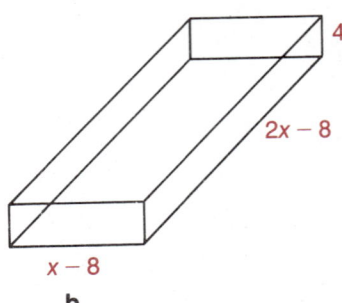

a.   b.

In Figure 2.6b, we see that the volume of the box is

$$V = 4(x - 8)(2x - 8) = \overset{192}{768}$$

$$(x - 8)2(x - 4) = \overset{96}{192}$$
$$x^2 - 12x + 32 = 96$$
$$x^2 - 12x - 64 = 0$$
$$(x - 16)(x + 4) = 0$$
$$x = 16, -4$$

We can discard $x = -4$ since we must have $x > 0$. The dimensions of the metal are 16 inches by 32 inches (x and $2x$).

EXAMPLE 6

Economists say that a given commodity is in equilibrium when supply equals demand. Suppose that when corn sells for $1.50 per bushel, there is a demand for 100,000 metric tons and a supply of only 50,000 metric tons. Also, suppose that each x¢ increase in the price per bushel stimulates an additional production of $5x^2$ metric tons and decreases the demand by $750x$ metric tons. What is the equilibrium price for corn?

SOLUTION

At a price of $(150 + x)$¢ per bushel,

Production $= 50,000 + 5x^2$
Demand $= 100,000 - 750x$

Equating these two for equilibrium:

$$50,000 + 5x^2 = 100,000 - 750x$$
$$5x^2 + 750x - 50,000 = 0$$
$$x^2 + 150x - 10,000 = 0$$
$$(x + 200)(x - 50) = 0$$
$$x = -200, 50$$

The value $x = -200$ represents a decrease of $2.00 in the price of corn to -50¢. Such a solution is meaningless. The value $x = 50$ makes sense. An equilibrium price for corn is $(150 + x)$¢ $= (150 + 50)$¢ $= \$2.00$ per bushel.

The supply and demand relationship is actually much more complicated than that given in Example 6. In this and other examples and exercises, simplified expressions are used so that the problem can be worked in this course.

EXAMPLE 7

The $60,000 cost of a charter flight must be shared equally by those who take the trip. A certain number of travelers have already expressed an interest. If 100 additional customers sign up, the cost to each of the original patrons will be reduced by $100. How many have already signed up?

116 Chapter 2 Equations and Algebraic Inequalities

SOLUTION

Let
$$x = \text{number of people already signed up}$$
$$\frac{60{,}000}{x} = \text{cost to each patron}$$
$$x + 100 = \text{total number of people if 100 more people sign up}$$
$$\frac{60{,}000}{x + 100} = \text{cost to each patron if 100 more people sign up}$$

Thus, we obtain the equation

(Original cost) − $100 = (Cost if 100 extra sign up)

$$\frac{60{,}000}{x} - 100 = \frac{60{,}000}{x + 100}$$

$$x(x + 100) \cdot \left[\frac{60{,}000}{x} - 100\right] = \left[\frac{60{,}000}{x + 100}\right] \cdot x(x + 100)$$

$$(x + 100)(60{,}000) - 100x(x + 100) = 60{,}000x$$

$$60{,}000x + 6{,}000{,}000 - 100x^2 - 10{,}000x = 60{,}000x$$

$$-100x^2 - 10{,}000x + 6{,}000{,}000 = 0$$

$$x^2 + 100x - 60{,}000 = 0$$

$$(x + 300)(x - 200) = 0$$

$$x = -300,\ 200$$

Thus, 200 people have already signed up. ∎

EXAMPLE 8

On an outing, Cindy's Explorer troop rowed upstream 60 miles and returned. The entire trip took 40 hours. If the stream is flowing at the rate of 2 miles per hour, how fast can Cindy's troop row in still water? How much more time was spent rowing upstream than downstream?

SOLUTION

Let
$$x = \text{rate in miles per hour the troop can row in still water}$$
$$x + 2 = \text{rate they travel downstream}$$
$$x - 2 = \text{rate they travel upstream}$$

Recall from Section 2.2 that

distance = rate · time

$$d = rt$$

	Distance	Rate	Time
Upstream	60	$x - 2$	$\dfrac{60}{x - 2}$
Downstream	60	$x + 2$	$\dfrac{60}{x + 2}$

$$t = \frac{d}{r}$$

We then obtain the equation:

(Time upstream) + (Time downstream) = (Total time)

$$\frac{60}{x-2} + \frac{60}{x+2} = 40$$

$$(x-2)(x+2)\left[\frac{60}{x-2} + \frac{60}{x+2}\right] = 40(x-2)(x+2)$$

$$60(x+2) + 60(x-2) = 40(x^2-4)$$

$$60x + 120 + 60x - 120 = 40x^2 - 160$$

$$-40x^2 + 120x + 160 = 0$$

$$x^2 - 3x - 4 = 0$$

$$(x-4)(x+1) = 0$$

$$x = 4, -1$$

The value $x = -1$ is meaningless in this problem since we must have $x \geq 0$. Thus, Cindy's troop can row 4 miles per hour in still water. They traveled 2 $(= x - 2)$ miles per hour upstream and 6 $(= x + 2)$ miles per hour downstream. The upstream trip took them 30 hours $[= 60/(x-2)]$ compared with 10 hours $[= 60/(x+2)]$ for the downstream trip. ∎

Section 2.5 Exercises

1. Find two consecutive positive even integers, the product of which is 168.

2. Find two consecutive positive even integers, the product of which is 528.

3. Find two consecutive positive integers the sum of whose squares is 113.

4. Find two consecutive positive integers, the sum of whose squares is 85.

5. Find the dimensions of a rectangular garden plot if it has an area of 600 square feet and can be surrounded with 100 feet of fence.

6. A fence 240 yards long is used to enclose a rectangular area of 3200 square yards. Find the dimensions of the rectangle.

7. Enclosing a certain rectangular kennel having an area of 1350 square meters requires 150 meters of fence. What are the dimensions of the kennel?

8. The perimeter of a rectangular region is 150 meters. Its area is 1250 square meters. Find the dimensions of the rectangle.

9. The diagonal distance across a rectangular playground is 29 feet. The playground is 1 foot longer than it is wide. Find the dimensions of the playground.

10. The hypotenuse of a right triangle is 17. The sum of its legs is 23. Find the length of each leg.

11. The length of a certain rectangle is 2 feet more than twice its width. Find the dimensions of the rectangle if its area is 60 square feet.

12. The length of a certain rectangle is 2 inches more than 3 times its width. If its area is 33 square inches, find the dimensions of the rectangle.

13. The Rescue Service wishes to send supplies by projectile to a mountaintop 3200 feet above their location. The projectile is fired with an initial upward velocity of 480 feet per second. When do the supplies land (from above) on the mountaintop? (See Example 3.)

14. A plane flying at an altitude of 1000 feet over the ocean ejects a canister downward with a velocity of 120 feet per second. How long does it take the canister to reach the ocean? (See Example 4.)

15. It is found that the concentration of a pollutant in a river decreases steadily as one moves downstream from the polluting source.

   a. If the concentration of the pollutant x kilometers from the source is $(10{,}000 - 3x - x^2/20)$ parts per million, how far must one be from the source to find water containing only 365 parts per million of the pollutant?

b. 9.8 parts per million?

16. Suppose that a lake covering x square acres can accommodate $(10{,}000x^2 + 45{,}000x)$ fish of various sizes and varieties. How many acres are then required to support one million fish?

17. For various diseases, the amount of serum injected in humans is computed in terms of body weight. If the proper dosage for a person weighing w kilograms is $(w^2 + 100w)$ milligrams, what size person requires an injection of 11.9 grams?

18. Lou's sporting goods store finds that sales receipts depend on the size of the showroom. Suppose that the store can sell $(100T^2 + 10T + 194)$ dollars worth of sporting goods per day with a showroom space of T thousand square feet. How large should the showroom be in order for $2000 worth to be sold each day?

19. If running at a speed of s miles per hour burns $(5s^2 + 20s + 25)$ calories per hour, how fast should one run in order to burn 505 calories in one hour?

20. Suppose that Flo's Fashions can sell $(596 - 4d - d^2)$ coats if the markup on each coat is $d.

    a. What should the markup be if Flo has 200 coats to sell?

    b. What if she has 375 coats to sell?

21. A white mat of uniform width is to be placed around a photograph which measures 5 inches by 7 inches. If the larger rectangular area is 63 square inches, find the width of the mat.

22. The length of a certain photograph exceeds its width by 3 inches. If a mat $1\frac{1}{2}$ inches wide is placed around the picture and the larger area is 154 square inches, find the dimensions of the picture.

23. A certain rectangular pool is twice as long as it is wide. A walkway 5 feet wide surrounds the pool. The total area (pool plus walkway) is 1000 square feet. Find the dimensions of the pool.

24. A rectangular garden plot 30 feet by 40 feet is surrounded by a walkway of uniform width. If the total area (garden plus walkway) is 1496 square feet, find the width of the walkway.

25. Mr. Rindler wishes to plant a border of flowers around the outside of his garden. He plans to use two thirds of his total available area for produce. How wide should his border be if his available space is a rectangle 50 feet by 60 feet?

26. The Great Hall in the Nature Museum consists of a 10-foot-wide walkway above an open display area. The open area (the smaller rectangle in the figure) is 50 percent longer than it is wide. What are the dimensions of the museum if 800 square yards of carpet are required to carpet the entire first floor together with the walkway above it?

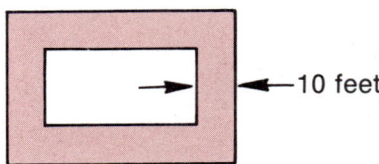

27. Dale is given instructions to construct an open metal box by cutting identical pieces from the four corners of a rectangular piece of metal. The box is to be 4 centimeters high and have a volume of 160 cubic centimeters. Its base is to be a rectangle 3 centimeters longer than it is wide. With what size piece of sheet metal should he begin?

28. Squares measuring 3 inches by 3 inches are cut from each corner of a rectangular piece of metal that is twice as long as it is wide. The sides are then folded up to make a box having a volume of 324 cubic inches. Find the dimensions of the metal.

29. According to Poiseuille's law, the rate at which fluid flows through an unobstructed pipe is given by $Q = kr^4$, where r is the radius of the pipe and k is a constant related to the viscosity of the fluid. A certain blood vessel has fatty deposits that reduce the flow to

$$Q = k(r^4 - ar^2) \text{ cubic centimeters per second}$$

where r is given in centimeters and $k = 120$, $a = \frac{1}{20}$ for blood. What radius must the blood vessel have in order to handle 6 cubic centimeters a second?

30. The cost of manufacturing a calculator increases directly with its complexity. On the other hand, more sophisticated units generate a larger market, thereby lowering the manufacturing cost per unit. Suppose that a calculator with n functions can profitably be sold at a price of $(2n - 6\sqrt{n})$ each.

How many functions can be expected in a calculator selling for $56?

31. Suppose that when fish oil extract sells for $50 per gallon, there is a demand for 50,000 gallons per week and a supply of 80,000 gallons per week. Also, suppose that if the price decreases x per gallon, the demand increases by $200x^2$ gallons while the supply decreases by $1000x$ gallons. What is the equilibrium price for fish oil extract?

32. The Zany auto parts store can sell 2000 "Pedal Power" units if the price is $8.00 per unit. Suppose that n price increases of 25¢ each decrease sales by $5n^2$ units. At what price do sales drop off to zero?

33. The manager of a toy store paid a total of $80 for some scooters. She was unable to sell three of them but made a profit of $6 on each of the others. Her gross receipts from those which she did sell also totaled $80. How many did she purchase?

34. Suppose that when the price of apples increases by $1 per bushel, the grower sells one less bushel than before for the same income of $30. What are the respective prices?

35. Find an integer that is 5 more than 24 times its reciprocal.

36. Find two positive integers such that one is 6 more than the other and their reciprocals sum to $\frac{1}{4}$.

37. Two blood vessels connected in parallel have a net resistance of 2. The resistance of one is 3 units more than the resistance of the other. What are the individual resistances of the blood vessels? (*Hint:* With resistances of r_1 and r_2, the net resistance is $\dfrac{1}{1/r_1 + 1/r_2}$. See Example 6, Section 1.8, and Exercise 4, Section 2.2.)

38. Suppose that an animal with legs L feet long walks at the rate of $7\sqrt{L}/3$ miles per hour and that Rapid Dancer's legs are exactly 3 feet longer than Rover's. If they are separated by 14 miles and walk toward each other, they will meet in 2 hours. How long are their legs?

39. Distance at sea is measured in nautical miles. Two ocean-going vessels leave port at noon traveling in perpendicular directions in the Atlantic Ocean. One travels 10 nautical miles per hour (knots) faster than the other. If the boats are 50 nautical miles apart at 1 P.M., find their respective rates of speed.

40. A passenger train heading north leaves the station at the same time as a freight train heading east. The passenger train travels 20 miles per hour more than twice as fast as the freight train. After 2 hours, they are 260 miles apart. How fast is each traveling?

41. Fred and his mother drive separate cars to a destination 150 miles away. Fred's shoes are heavier than his mother's so he drives 10 miles per hour faster than his mother and arrives $\frac{1}{2}$ hour sooner. How fast did each drive?

42. An outdoors enthusiast takes a boat 40 miles upstream and then back again in 15 hours. If the stream is flowing at 2 miles per hour, find the speed of the boat in still water. How long did each part of the trip take?

43. Making a round trip from Dayton to Chicago, a pilot faces a 30-mile-per-hour head wind one way and a 30-mile-per-hour tail wind on the return trip. The return trip takes 45 minutes less than the initial journey.
    a. If the distance between the two cities is 350 miles, what is the air-speed?
    b. How long did each part of the trip take?

44. Dan bid the labor for a remodeling job at $336. However, the job took 4 hours longer than he estimated. Consequently, he earned $2 an hour less than he had planned. What was the estimated time?

45. Ms. Ranly repaired a broken antique chair for $240. She earned 50¢ less per hour than she planned because the work took 2 hours longer than expected. In what amount of time had she planned to fix the chair?

46. A class decides to buy a calculator for $160. Had there been eight fewer members in the class, the cost per student would have risen by only $1. How many students are in the class?

47. Dan can build a garage in 14 days less time than Dale can. They complete a garage in 24 days when they work together. How long does it take each to build a garage working alone?

48. Todd can paint a barn in 6 days less than it takes Doug. Working together they paint the barn in 4

days. How long would it take each to paint the barn working alone?

49. A large pipe can fill a pool in 39 hours less than a small pipe can. If both pipes are running, it takes 40 hours to fill the pool. How long would it take each pipe alone to fill the pool?

50. According to the theory of relativity, the energy E, momentum p, and mass m of a particle are related to the speed of light c as follows:

$$E^2 = p^2c^2 + m^2c^4$$

Express c explicitly in terms of the other quantities.

Section 2.6

Proportionality; Variation

We noted earlier that the circumference and area of a circle of radius r are given as $C = 2\pi r$ and $A = \pi r^2$, respectively. We say that the circumference is directly proportional to r and the area is directly proportional to r^2. The concept of *proportionality* or *variation* is formalized as follows:

DEFINITIONS

If there is a constant $k \neq 0$ and an $r > 0$ such that one of the following relationships holds, then k is called a **constant of proportionality** and the terminology of proportion and variation is used as indicated.

1. $y = kx$ y **varies directly** as x, or y is **directly proportional** to x.
2. $y = \dfrac{k}{x}$ y **varies inversely** as x, or y is **inversely proportional** to x.
3. $y = kuv$ y **varies jointly** as u and v, or y is *directly proportional* to u and v.
4. $y = \dfrac{k}{uv}$ y is *inversely proportional* to u and v.

You should be able to expand on these notions of proportionality. For instance, each of the following can be written in the given form for some constant $k \neq 0$.

EXAMPLES	
y is directly proportional to x^4	$y = kx^4$
y is inversely proportional to x^2	$y = \dfrac{k}{x^2}$
y is directly proportional to u^2 and inversely proportional to v^3	$y = k\dfrac{u^2}{v^3}$
y is directly proportional to x^3 and w^2 and inversely proportional to $\sqrt[3]{u}$ and $v^{3/2}$	$y = k\dfrac{x^3 w^2}{\sqrt[3]{u}\, v^{3/2}}$

Section 2.6 *Proportionality; Variation*

In many attempts to develop a mathematical model for a physical phenomenon, the investigator has some idea regarding direct and inverse variation among variables. The task then is to determine the constant of proportionality k. This can be done by obtaining corresponding values of the variables and solving the proportionality equations for k. The procedure is illustrated in the following examples.

EXAMPLE 1

Let r vary directly as s^2 and assume $r = 50$ when $s = 5$.

a. Write r in terms of s.
b. Find r when $s = 12$.

SOLUTION

a. There is some constant $k \neq 0$ for which

$$r = ks^2$$

When $s = 5$, $r = 50$:

$$50 = k \cdot 5^2 = 25k$$
$$k = 2$$

Then $r = ks^2 = 2s^2$

b. When $s = 12$, we have

$$r = 2s^2 = 2 \cdot 12^2 = 2 \cdot 144$$
$$= 288$$

EXAMPLE 2

The volume of a sphere is directly proportional to the cube of its radius. If the volume is 36π when the radius is 3, write the volume in terms of the radius.

SOLUTION

There is some constant $k \neq 0$ such that

$$V = kr^3$$

When $r = 3$, $V = 36\pi$

$$36\pi = k \cdot 3^3 = 27k$$
$$k = \frac{36\pi}{27} = \frac{4\pi}{3}$$

Thus,

$$V = \frac{4\pi}{3} r^3$$

EXAMPLE 3

If y varies directly as x^2 and inversely as z and if $y = 50$ when $x = 5$ and $z = 1$, write y in terms of x and z.

SOLUTION

We have

$$y = k \frac{x^2}{z}$$

for some constant $k \neq 0$. The given conditions tell us that

$$50 = k \frac{5^2}{1}$$
$$= 25k$$

That is,
$$k = 2$$
so that
$$y = \frac{2x^2}{z}$$

The proportionality relationship alone sometimes gives information that can be used even if we do not know the constant of proportionality. This situation is illustrated in the following examples.

EXAMPLE 4 Suppose that the efficiency of the liver is inversely proportional to the square of the amount of alcohol in the bloodstream. What effect does doubling the alcohol content have on the liver's efficiency?

SOLUTION Letting E denote the efficiency of the liver and a denote the alcohol content, we have
$$E = \frac{k}{a^2}$$
for some constant k. Doubling a, the efficiency is reduced to
$$E^* = \frac{k}{(2a)^2} = \frac{k}{4a^2} = \frac{1}{4}\frac{k}{a^2}$$
$$= \frac{1}{4}E$$

Doubling the alcohol content reduces the efficiency of the liver to $\frac{1}{4}$ its original value.

EXAMPLE 5 Let y vary directly as u^2 and v and inversely as $\sqrt{w}$. If u is tripled, v is halved, and w is quadrupled, what happens to y?

SOLUTION We can write
$$y = \frac{ku^2v}{\sqrt{w}}$$

However, if u is tripled, v is halved, and w is quadrupled, the new value y^* is found as follows:
$$y^* = \frac{k(3u)^2\left(\frac{v}{2}\right)}{\sqrt{4w}} = \frac{k9u^2\left(\frac{v}{2}\right)}{2\sqrt{w}}$$
$$= \frac{k9u^2v}{4\sqrt{w}} = \frac{9}{4}\frac{ku^2v}{\sqrt{w}}$$
$$= \frac{9}{4}y$$

That is, y is multiplied by $\frac{9}{4}$ when the given changes are effected on u, v, and w.

Section 2.6 Exercises

1. Let y vary directly as x.
   a. If x is tripled, what happens to y?
   b. Write y in terms of x if $y = 2$ when $x = 10$.
   c. Find y when $x = 30$.

2. Let P vary inversely as q^2.
   a. If q is doubled, what happens to P?
   b. Write P in terms of q if $P = 3$ when $q = 3$.
   c. Find P when $q = 1$.

3. Let M vary directly as v^3.
   a. If v is halved, what happens to M?
   b. Write M in terms of v if $M = 500$ when $v = 5$.
   c. Find M when $v = 4$.

4. Let y vary directly as x and inversely as z^2 and w^3.
   a. If x, w, and z are all doubled, what happens to y?
   b. Write y in terms of x, w, and z if $y = 4$ when $x = 144$, $w = 2$, and $z = 3$.
   c. Under the conditions of (b), find y when $x = 162$, $z = 2$, and $w = 3$.

5. Let s be directly proportional to t^2 and inversely proportional to $\sqrt{u}$ and v^2.
   a. If t is doubled, u is multiplied by 9, and v is quadrupled, what happens to s?
   b. Write s in terms of t, u, and v if $s = 6$ when $t = 10$, $u = 4$, and $v = 5$.
   c. Under the conditions of (b), find s when $t = 2$, $u = 9$, and $v = 2$.

6. Let r vary directly as s^2 and inversely as h and l.
   a. If s, h, and l are all tripled, what happens to r?
   b. Write r in terms of s, h, and l if $r = 5$ when $s = 10$, $h = 2$, and $l = 10$.
   c. Under the conditions of (b), find r when $s = 20$, $h = 40$, and $l = 8$.

7. Let y be directly proportional to w and $\sqrt{x}$ and inversely proportional to z^3.
   a. If x is quadrupled, w is tripled, and z is halved, what happens to y?
   b. Write y in terms of w, x, and z if $y = 80$ when $x = 4$, $w = 128$, and $z = 4$.
   c. Under the conditions of (b), find y when $x = 25$, $w = 243$, and $z = 3$.

8. If the circumference of a circle is doubled, how is the area affected? (See Example 1 Section 2.2.)

9. If the perimeter of a square is tripled, how is the area affected? (See Exercise 9, Section 2.2.)

10. If the circumference of a great circle on a sphere is tripled, how is the surface area of the sphere affected? (See Exercise 7, Section 2.2.)

11. If the altitude of a triangle is doubled and its base is decreased to $\frac{1}{3}$ its original length, how is its area affected? (See Exercise 6, Section 2.2.)

12. At a fixed velocity the distance traveled by a moving object is directly proportional to the time of travel. Suppose the object travels 100 feet in 20 seconds.
    a. Write distance in terms of time.
    b. How far does the object travel in 5 seconds?

13. The distance required to stop a vehicle varies directly as the square of its speed. Suppose the stopping distance is 144 feet when the speed is 88 feet per second (60 miles per hour).
    a. Express stopping distance in terms of speed.
    b. Find the distance required to stop this vehicle if it is traveling 132 feet per second (90 miles per hour).

14. The time required for an automobile to travel a fixed distance is inversely proportional to its speed. Suppose $t = 3$ seconds when $r = 80$ feet per second.
    a. Express the time t in terms of the speed r.
    b. How long would it take to travel the distance if $r = 20$ feet per second?

15. The amount of fuel used by an automobile is directly proportional to the number of miles traveled. Suppose 5 gallons are required to travel 200 miles.
    a. Express the fuel requirement in terms of the distance traveled.
    b. How much fuel is required to travel 600 miles?

16. The demand for a certain item is inversely proportional to the square of the price. Suppose 500 units can be sold when the price is $20.
    a. Express the number that can be sold in terms of the price.
    b. How many units can be sold if the price is $10?

17. Sales tax is directly proportional to the cost of an item. Suppose the tax on an item costing $32 is $2.
    a. Express the tax in terms of the cost of an item.
    b. What is the sales tax on something that costs $240?
18. In physics, Hooke's law says that the force exerted by a spring is directly proportional to the amount it is stretched. Suppose a force of 100 pounds is required to stretch the spring 4 inches.
    a. Express the force in terms of the stretch.
    b. How many pounds of force are required to stretch the spring 25 inches?
19. The cost of manufacturing a certain item varies directly as the cube root of the number of units to be produced. Suppose it costs $9000 to produce 27,000 units.
    a. Express manufacturing cost in terms of the number to be produced.
    b. How much will it cost to produce 125,000 units?
20. The time it takes for a falling object to reach the ground is directly proportional to the square root of the altitude from which it falls. Suppose it takes 2 seconds for an object to fall 64 feet.
    a. Express time in terms of distance fallen.
    b. How long would it take the object to fall 400 feet?
21. The cost of having business cards printed varies directly as the square root of the number of cards ordered. Suppose it costs $40 to print 400 cards.
    a. Express the printing cost in terms of the number of cards ordered.
    b. What is the printing cost for an order of 1600 cards?
22. The intensity of illumination that an object receives from a light source varies inversely as the square of its distance from the source. If the distance from the source is tripled, how is the intensity of illumination affected?
23. The distance we can see to the horizon varies directly as the square root of our distance above the surface of the earth. Suppose a person on top of a building 300 feet high can see for 10 miles.
    a. Express the distance we can see to the horizon in terms of height above the earth's surface.
    b. How far can we see to the horizon from a plane flying at an altitude of 7500 feet?
24. The force of attraction between two objects varies directly as their masses and inversely as the square of the distance between them. If one of the masses is halved while the other is tripled and the distance between them shrinks to $\frac{1}{3}$ its original value, how is the force affected?
25. The interest I paid on a deposit is directly proportional to the principal P on deposit and the time t for which it earns interest. Suppose $I = \$200$ when $P = \$5000$ and $t = \frac{1}{2}$ year.
    a. Express I in terms of P and t.
    b. How much interest will be earned with a principal of $15,000 for 2 years under these conditions?
26. The kinetic energy of a moving object is directly proportional to its mass and the square of its velocity. If the mass is quadrupled, what adjustment in velocity must be made in order to maintain the same kinetic energy?
27. *The electrical current carried by a wire is directly proportional to the applied voltage and inversely proportional to the resistance of the wire.* If the voltage is doubled and the resistance is halved, how is the current flow affected?
28. *The resistance of a wire is directly proportional to its length and inversely proportional to the square of its cross-sectional radius.* If the diameter of the wire is tripled, how much longer can the length be without increasing the resistance?
29. a. *Using the information given in the preceding two problems*, indicate how the current flow, i, varies with respect to the length, l, of a wire, its radius, r, and the applied voltage, v.
    b. If 20 amps current must flow through 660 feet of wire and 220 volts are used if the wire has a radius $= \frac{1}{16}$ inch, how large a wire must be used if only 110 volts are available?
30. According to Charles' law in physics, the absolute temperature of an enclosed gas varies directly as the pressure and volume.
    a. How does the volume vary with respect to the temperature and pressure?
    b. How does the pressure vary with respect to the volume and temperature?

c. If the absolute temperature remains the same and the volume is halved, how is the pressure affected?

31. The load that can be safely carried by a steel beam is directly proportional to its width and the square of its depth and inversely proportional to its length.
   a. If the initial construction plans called for beams 20 feet long, 5 inches wide, and 12 inches high to support a load of 30 tons, express the safe load in terms of width, depth, and length (all in inches).
   b. If for some reason a supporting wall is eliminated so that the span must now be 80 feet rather than 20 feet, how much wider must the beams be in order to support the same weight, assuming the depth remains the same? How much deeper must the beams be in order to support the same weight, assuming the width remains the same?
   c. If steel is sold by weight and weight is proportional to width, depth, and length, is it cheaper for the contractor to buy wider or deeper beams in order to support the 30 tons over the longer span of 80 feet?

32. Certain science-fiction movies depict giant insects wreaking havoc on the human population. These movies are more fiction than science. A little mathematics should convince us that such giant insects would be so weak they wouldn't even be able to move their own limbs. Recall that the volume of a sphere of radius r is $4\pi r^3/3$ and the area of a circle of radius r is πr^2. It is reasonably accurate to say that the volume, and hence, the weight, of an animal or insect is directly proportional to the cube of its height or length; the cross-sectional area of its muscles varies directly as the square of its height or length. The weight the animal can lift is directly proportional to the cross-sectional area of its muscles. The strength of an animal is defined as

$$\text{Strength} = \frac{\text{Weight the animal can lift}}{\text{Weight of the animal}}$$

   a. Show that the strength s of an animal is inversely proportional to its height or length h: $s = k/h$ for some constant k; the constant k will depend on the type of animal.
   b. An average-sized ant 1 centimeter long can lift about three times its own weight; how strong would a human-sized ant 180 centimeters long be?
   c. An average-sized man 180 centimeters tall can lift about half his body weight; how strong would an ant-sized man 1 centimeter tall be?
   d. Whose body structure is inherently stronger, the human's or the ant's?

33. The rate at which an animal metabolizes energy is directly proportional to its surface area, which, in turn, varies directly as the square of its height or length. The oxygen for this metabolism is transported from the lungs to individual cells by the bloodstream. Thus, the rate of metabolism is directly proportional to the availability of oxygen, which, in turn, varies directly as the volume of blood pumped by each heartbeat and the rate at which the heart beats. The volume of the heart is directly proportional to the cube of the height or length of the animal.
   a. Show that the heart rate r of an animal is inversely proportional to its length h: $r = k/h$ for some constant k, depending on the type of animal.
   b. Compare the heart rates of each of the following.
      1. An 18-inch infant
      2. A 2-year-old who is 3 feet tall
      3. An adult 6 feet tall

Section 2.7

Algebraic Inequalities with One Variable

Since algebraic expressions represent numbers, the rules for order and absolute value discussed in Section 1.2 also hold for operations on such expressions.

Expressions of the form

$$2x^2 + 3x < 5$$
$$3y + 10 \leq 0$$
$$5z + 20z^3 > 16z^2$$

are called algebraic inequalities. The **solution** to an inequality consists of all values that, when assigned to the variable, make the inequality true. The following table reviews the rules for operations on inequalities.

RULES FOR INEQUALITIES

RULE	EXAMPLE
1. $a < b$ if and only if $b - a > 0$.	1. $2 < 5$ and $5 - 2 = 3 > 0$
2. If $a < b$ and $b < c$, then $a < c$.	2. $2 < 7$ and $7 < 9$, so $2 < 9$
3. If $a < b$, then $(a + c) < (b + c)$.	3. $2 < 7$ so $2 + 3 = 5 < 7 + 3 = 10$
4. If $a < b$, then $-a > -b$.	4. $2 < 5$ and $-2 > -5$
5. If $a < b$ and a. $x = 0$, then $ax = bx$. b. $x > 0$, then $ax < bx$. c. $x < 0$, then $ax > bx$.	5. $3 < 5$ and a. $3 \cdot 0 = 5 \cdot 0 = 0$ b. $3 \cdot 2 < 5 \cdot 2$, since $6 < 10$ c. $3(-2) > 5(-2)$, since $-6 > -10$
6. If $0 < a < b$, then $0 < \frac{1}{b} < \frac{1}{a}$.	6. Larger denominators make smaller fractions. $0 < 2 < 5$ and $0 < \frac{1}{5} < \frac{1}{2}$

In addition to these rules, we have the following properties of products, which are important in studying inequalities.

PRODUCT RULES

RULE	EXAMPLE
1. The product of two numbers is 0 if and only if at least one of the factors is 0.	1. If $(x - 2)(x - 3) = 0$, then $x = 2$ or $x = 3$
2. The product of two positive or two negative numbers is positive.	2. $2 \cdot 3 = 6$ $(-2)(-3) = 6$
3. The product of a negative and a positive number is negative.	3. $2(-3) = -6$

Linear Inequalities

In this section we restrict our attention to inequalities that can be solved by analyzing expressions of the form $ax + b$. These are called **linear inequalities**. As with first-degree equations in one variable, we work to isolate the variable on one side of the inequality.

Section 2.7 Algebraic Inequalities with One Variable **127**

> **SOLVING LINEAR INEQUALITIES**
>
> 1. Simplify the algebraic expressions.
> 2. Isolate the variable on one side of the inequality by
>   a. Adding or subtracting terms.
>   b. Multiplying or dividing by a non-zero constant. Note that multiplication or division by a negative number reverses the sense of the inequality.

EXAMPLE 1

Solve the inequality

$$10 - x > 2x - 2$$

and sketch the solution on the real number line.

SOLUTION

$10 - x > 2x - 2$
$10 - 3x > -2$ (Subtract $2x$ from both sides)
$-3x > -12$ (Subtract 10 from both sides)
$3x < 12$ (Reverse the inequality)
$x < 4$ (Divide both sides by 3)

Figure 2.7

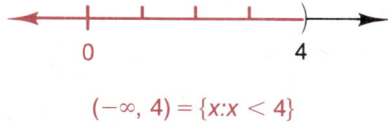

$(-\infty, 4) = \{x : x < 4\}$

The solution consists of the set of all $x < 4$ as shown in Figure 2.7. Recall from Section 1.2 that such a set is an interval, denoted by $(-\infty, 4)$. Also recall that the parenthesis next to the 4 indicates that 4 is not included in the set. Had 4 been included, we would have used a square bracket] and written $(-\infty, 4] = \{x : x \leq 4\}$. Here the solution set is the interval $(-\infty, 4) = \{x : x < 4\}$. See Figure 2.7. ∎

EXAMPLE 2

Solve the inequality

$$x - 16 \leq 3x - 4$$

and sketch the solution set on the real number line.

SOLUTION

$x - 16 \leq 3x - 4$
$-16 \leq 2x - 4$ (Subtract x from both sides)
$-12 \leq 2x$ (Add 4 to both sides)
$-6 \leq x$ (Divide by 2)

The solution set is the interval $\{x : x \geq -6\} = [-6, \infty)$. See Figure 2.8.

Figure 2.8

128 Chapter 2 *Equations and Algebraic Inequalities*

Multiple Inequalities

Problems often arise that require two or more inequalities to be satisfied simultaneously. Such multiple inequalities are treated as separate inequalities that are solved separately. Only values of the variable that satisfy each of the separate inequalities will satisfy the multiple inequality.

EXAMPLE 3

Solve the following double inequality.

$$2 - x < 3x + 2 \leq x + 4$$

SOLUTION

The notation indicates that two inequalities are to be satisfied simultaneously:

$$2 - x < 3x + 2 \quad \text{and} \quad 3x + 2 \leq x + 4$$

The first step is to solve these *separately* as follows.

The first inequality:

$$2 - x < 3x + 2$$
$-x < 3x$ (Subtract 2 from both sides)
$0 < 4x$ (Add x to both sides)
$0 < x$ (Divide by 4)

The solution set for the first inequality is the interval

$$S_1 = \{x : x > 0\} = (0, \infty)$$

The second inequality:

$$3x + 2 \leq x + 4$$
$3x \leq x + 2$ (Subtract 2 from both sides)
$2x \leq 2$ (Subtract x from both sides)
$x \leq 1$ (Divide by 2)

For the second inequality, the solution set is the interval

$$S_2 = \{x : x \leq 1\} = (-\infty, 1]$$

Since *both* inequalities must be satisfied simultaneously, the solution set S consists of those x's which are in both solution sets:

$$S = S_1 \cap S_2 = (0, \infty) \cap (-\infty, 1]$$
$$= (0, 1] = \{x : 0 < x \leq 1\}$$

See Figure 2.9.

Figure 2.9

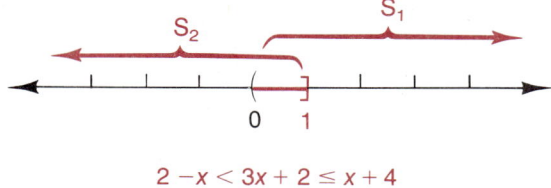

$2 - x < 3x + 2 \leq x + 4$

Or Inequalities

Some inequality conditions are stated in terms of two or more inequalities, only one of which need be satisfied. In this case the solution set consists of the *union* of the solution sets for the individual inequalities. The following example illustrates the situation.

EXAMPLE 4

Find the set of real numbers x which satisfies either

$$2x - 3 < 1 \quad \text{or} \quad 3x - 2 \geq 13$$

Section 2.7 Algebraic Inequalities with One Variable 129

SOLUTION First we solve the individual inequalities separately.

The first inequality:

$$2x - 3 < 1$$
$$2x < 4$$
$$x < 2$$

The solution set here is

$$S_1 = (-\infty, 2)$$

The second inequality:

$$3x - 2 \geq 13$$
$$3x \geq 15$$
$$x \geq 5$$

The solution set here is

$$S_2 = [5, \infty)$$

The problem requires only that at least one of the two inequalities is satisfied. Thus the solution set is

$$S = S_1 \cup S_2 = (-\infty, 2) \cup [5, \infty)$$

See Figure 2.10.

Figure 2.10

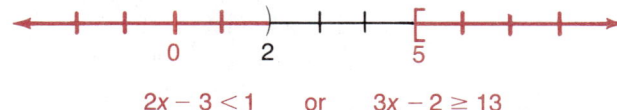

$2x - 3 < 1$ or $3x - 2 \geq 13$

AND VERSUS OR INEQUALITIES

For a problem involving two inequalities, solve each of the inequalities separately. Call the solution sets S_1, S_2 respectively.

1. If both the first *and* second inequalities must be satisfied, the solution set is $S_1 \cap S_2$.
2. If either the first inequality *or* the second inequality must be satisfied, the solution set is $S_1 \cup S_2$.

Higher-Order Inequalities

1. Write the inequality as an expression which is a product of linear factors (if possible) with 0 on the other side of the inequality symbol.
2. Find the points in which these factors become 0, and mark these on the real number line. The expression can change signs (+ or −) only at these points.
3. Choose a "test point" inside each interval determined by the points located in Step 2.
4. Determine the sign of the factors of the expression at each of these test points. An odd number of negative factors is required to make the product negative. Otherwise the product is positive (except at the 0-points).
5. All points in the interval containing a given test point either satisfy the inequality or not, depending on whether the test point does or does not.
6. The endpoints of these solution intervals are included for $\leq$ or $\geq$ inequalities; for $<$ or $>$ inequalities, they are not.

EXAMPLE 5

Solve the following inequality.

$$(2u - 6)(3u + 12) < 0$$

SOLUTION

This inequality is already expressed as a product of linear factors with 0 on the other side of the inequality symbol. These factors become 0 when

$$2u - 6 = 0 \quad \text{and when} \quad 3u + 12 = 0$$
$$2u = 6 \qquad\qquad\qquad 3u = -12$$
$$u = 3 \qquad\qquad\qquad u = -4$$

Figure 2.11

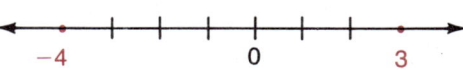

The expression can change signs only at these points. Now choose a test point inside each interval shown in Figure 2.11.

On the interval $(-\infty, -4)$, if we use -5 as a test point we have

$$2u - 6 = 2(-5) - 6 = -16 < 0$$
$$3u + 12 = 3(-5) - 6 = -3 < 0$$

Thus the sign of each factor is negative so that the product $(2u - 6)(3u + 12)$ is positive. We continue in this way with $(-4, 3)$ and $(3, \infty)$. It helps to collect this information into a table as shown here.

INTERVAL	TEST POINT	SIGN OF $(2u - 6)(3u + 12)$
$(-\infty, -4)$	-5	$(-) \cdot (-) = (+)$
$(-4, 3)$	0	$(-) \cdot (+) = (-)$
$(3, \infty)$	4	$(+) \cdot (+) = (+)$

Since we want $(2u - 6)(3u + 12) < 0$, we want its sign to be $(-)$.
The solution set is the interval $(-4, 3)$ as shown in Figure 2.12.

Figure 2.12

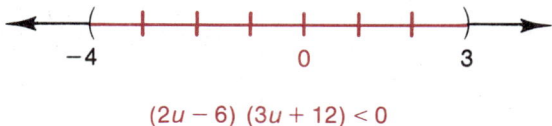

$(2u - 6)(3u + 12) < 0$ ∎

EXAMPLE 6

Solve the inequality

$$x^2 + x \geq 6$$

SOLUTION

First rewrite the inequality as

$$x^2 + x - 6 \geq 0$$
$$(x + 3)(x - 2) \geq 0$$

These factors become 0 when $x = -3$ and when $x = 2$ respectively.

Figure 2.13

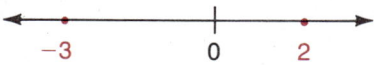

Choose a test point inside each interval shown in Figure 2.13 and proceed as follows.

INTERVAL	TEST POINT	SIGN OF $(x + 3)(x - 2)$
$(-\infty, -3)$	-4	$(-) \cdot (-) = (+)$
$(-3, 2)$	0	$(+) \cdot (-) = (-)$
$(2, \infty)$	3	$(+) \cdot (+) = (+)$

Since we want $x^2 + x - 6 \geq 0$, we want its sign to be $(+)$. Thus the solution set is

$$S = (-\infty, -3] \cup [2, \infty)$$

The endpoints are included here since this is a $\geq$ inequality. See Figure 2.14.

Figure 2.14

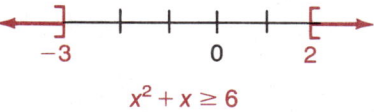

$x^2 + x \geq 6$

EXAMPLE 7 Solve the following inequality.

$$(x^2 - 7x + 12)(2x + 10) \leq 0$$

SOLUTION To solve this inequality, we factor the quadratic expression to obtain

$$(x - 4)(x - 3)(2x + 10) \leq 0$$

We note that the factors become 0 at $x = 4, 3, -5$, respectively. The expression can change sign only at these points. Choose a test point inside each interval determined by these points (see Figure 2.15).

Figure 2.15

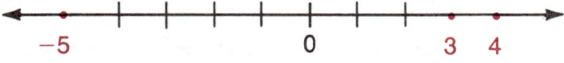

INTERVAL	TEST POINT	SIGN OF $(x - 4)(x - 3)(2x + 10)$
$(-\infty, -5)$	-6	$(-) \cdot (-) \cdot (-) = (-)$
$(-5, 3)$	0	$(-) \cdot (-) \cdot (+) = (+)$
$(3, 4)$	3.5	$(-) \cdot (+) \cdot (+) = (-)$
$(4, \infty)$	5	$(+) \cdot (+) \cdot (+) = (+)$

Since we want $(x - 4)(x - 3)(2x + 10) \leq 0$, we want its sign to be $(-)$.

132 Chapter 2 Equations and Algebraic Inequalities

The endpoints are included in the solution intervals here because of the ≤ inequality: the factors *are* permitted to be 0.

The solution set is $(-\infty, -5] \cup [3, 4]$ as sketched in Figure 2.16.

Figure 2.16

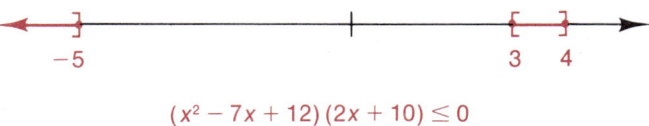

$(x^2 - 7x + 12)(2x + 10) \leq 0$ ∎

Fractional Inequalities

Although "cross-multiplying" is sometimes a quick procedure for solving fractional equations, this procedure should be avoided with inequalities, for multiplication by an expression involving the variable may reverse the inequality (if the multiplier is negative). Rather than consider additional cases (positive versus negative multipliers), it is best to proceed as follows.

> **INEQUALITIES INVOLVING FRACTIONS**
>
> 1. **Bring all terms to one side**, with 0 on the other side of the inequality symbol.
> 2. **Combine the fractions.**
> 3. **Simplify and factor.**
> 4. **Find the points in which these factors become 0** and mark these on the real number line. The expression can change sign only at these points.
> 5. **Choose a test point inside each interval determined by these points** and proceed as with higher-order inequalities. Be careful to exclude points from your solution if they make the denominator 0.
> 6. As with products, **an odd number of negative factors makes the quotient negative.** Otherwise, the quotient is positive (except at the 0-points).

EXAMPLE 8

Solve the inequality

$$\frac{x+2}{x-3} \geq 0$$

SOLUTION

First we find the zeros of the factors: $x + 2 = 0$ when $x = -2$, and $x - 3 = 0$ when $x = 3$. See Figure 2.17.

Figure 2.17

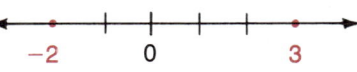

Section 2.7 Algebraic Inequalities with One Variable **133**

Then we proceed as follows.

INTERVAL	TEST POINT	SIGN OF $\dfrac{x+2}{x-3}$
$(-\infty, -2)$	-3	$\dfrac{(-)}{(-)} = (+)$
$(-2, 3)$	0	$\dfrac{(+)}{(-)} = (-)$
$(3, \infty)$	4	$\dfrac{(+)}{(+)} = (+)$

For the quotient to be positive, the factors must have the same sign. Thus the solution set S includes the intervals $(-\infty, -2]$ and $(3, \infty)$:

$$S = (-\infty, -2] \cup (3, \infty)$$

-2 is included in the solution set since the numerator may be 0 (a $\geq$ inequality), but 3 is not included in the solution set since division by 0 is not permitted. ∎

EXAMPLE 9 Solve the following inequality.

$$\frac{(x+1)^2(2x-6)}{x^2+1} < 0$$

SOLUTION Here we note that $x^2 + 1 > 0$ for all real numbers x. On the other hand, $2x - 6 = 0$ when $x = 3$, and $x + 1 = 0$ when $x = -1$.

Figure 2.18

INTERVAL	TEST POINT	SIGN OF $\dfrac{(x+1)^2(2x-6)}{x^2+1}$
$(-\infty, -1)$	-2	$\dfrac{(-)^2 \cdot (-)}{(+)} = (-)$
$(-1, 3)$	0	$\dfrac{(+)^2 \cdot (-)}{(+)} = (-)$
$(3, \infty)$	4	$\dfrac{(+)^2 \cdot (+)}{(+)} = (+)$

Note in the above that we must consider $(x + 1)^2$ and not just $(x + 1)$ in our sign analysis. The solution set is

$$S = (-\infty, -1) \cup (-1, 3)$$

as shown in Figure 2.19. We do not include -1 in the solution set since this is a $<$ inequality. ∎

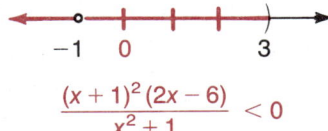

Figure 2.19

EXAMPLE 10 Solve the following inequality.

$$\frac{2}{t+3} \leq \frac{1}{t-4}$$

SOLUTION

$$\frac{2}{t+3} - \frac{1}{t-4} \leq 0 \qquad \text{(Bring all terms to one side)}$$

$$\frac{2(t-4) - 1(t+3)}{(t+3)(t-4)} \leq 0 \qquad \text{(Combine the fractions using the LCD, } (t+3)(t-4))$$

$$\frac{2t - 8 - t - 3}{(t+3)(t-4)} \leq 0$$

$$\frac{t - 11}{(t+3)(t-4)} \leq 0 \qquad \text{(Simplify)}$$

These factors become 0 when $t = 11, -3$, or 4 respectively.

Figure 2.20

INTERVAL	TEST POINT	SIGN OF $\dfrac{(t-11)}{(t+3)(t-4)}$
$(-\infty, -3)$	-4	$\dfrac{(-)}{(-)\cdot(-)} = (-)$
$(-3, 4)$	0	$\dfrac{(-)}{(+)\cdot(-)} = (+)$
$(4, 11)$	5	$\dfrac{(-)}{(+)\cdot(+)} = (-)$
$(11, \infty)$	12	$\dfrac{(+)}{(+)\cdot(+)} = (+)$

The solution set is

$$S = (-\infty, -3) \cup (4, 11]$$

The solution set includes 11 (a $\leq$ inequality) but not -3 or 4 since these would cause division by 0 (which is undefined). See Figure 2.21.

Figure 2.21

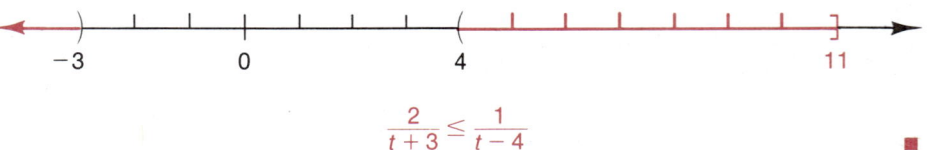

Section 2.7 Algebraic Inequalities with One Variable

CAUTION

Blindly cross-multiplying in Example 10 would have given

$$2(t-4) \le (t+3)$$
$$2t - 8 \le t + 3$$
$$t \le 11$$

But we saw that not every $t \le 11$ satisfies the inequality. Since multiplication by a negative number reverses the sense of an inequality, we *must consider the signs* of $(t+3)$ and $(t-4)$ when cross-multiplying. If we were careful in this respect, we could then arrive at the same solution as before.

EXAMPLE 11

Solve the following inequality.

$$\frac{3}{x} < \frac{6}{x+2} < \frac{2}{x-2}$$

SOLUTION

We must solve the two inequalities

$$\frac{3}{x} < \frac{6}{x+2} \quad \text{and} \quad \frac{6}{x+2} < \frac{2}{x-2}$$

The first inequality:

$$\frac{3}{x} - \frac{6}{x+2} < 0$$

$$\frac{3(x+2) - 6x}{x(x+2)} < 0$$

$$\frac{6 - 3x}{x(x+2)} < 0$$

These factors are 0 when $x = 2, 0, -2$, respectively. See Figure 2.22.

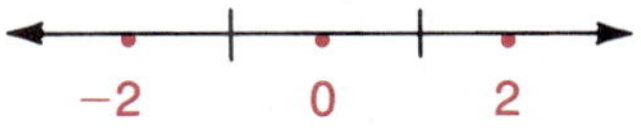

Figure 2.22

The second inequality:

$$\frac{6}{x+2} - \frac{2}{x-2} < 0$$

$$\frac{6(x-2) - 2(x+2)}{(x+2)(x-2)} < 0$$

$$\frac{4x - 16}{(x+2)(x-2)} < 0$$

These factors are 0 when $x = 4, -2, 2$, respectively. See Figure 2.23.

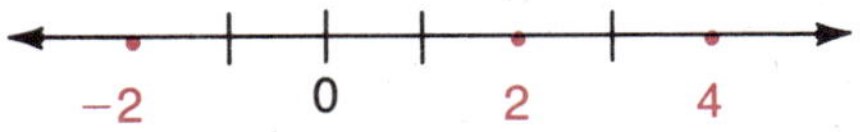

Figure 2.23

INTERVAL	TEST POINT	SIGN OF $\frac{6-3x}{x(x+2)}$
$(-\infty, -2)$	-3	$\frac{+}{(-) \cdot (-)} = (+)$
$(-2, 0)$	-1	$\frac{(+)}{(-) \cdot (+)} = (-)$
$(0, 2)$	1	$\frac{(+)}{(+) \cdot (+)} = (+)$
$(2, \infty)$	3	$\frac{(-)}{(+) \cdot (+)} = (-)$

INTERVAL	TEST POINT	SIGN OF $\frac{4x-16}{(x-2)(x+2)}$
$(-\infty, -2)$	(-3)	$\frac{(-)}{(-) \cdot (-)} = (-)$
$(-2, 2)$	0	$\frac{(-)}{(-) \cdot (+)} = (+)$
$(2, 4)$	3	$\frac{(-)}{(+) \cdot (+)} = (-)$
$(4, \infty)$	5	$\frac{(+)}{(+) \cdot (+)} = (+)$

The solution set here is

$$S_1 = (-2, 0) \cup (2, \infty)$$

See Figure 2.24.

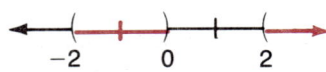

Figure 2.24

The solution set here is

$$S_2 = (-\infty, -2) \cup (2, 4)$$

See Figure 2.25.

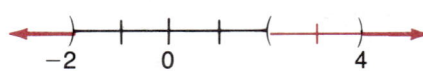

Figure 2.25

Since both inequalities must be satisfied the solution set is

$$S = S_1 \cap S_2 = (2, 4)$$

See Figure 2. 26.

Figure 2.26

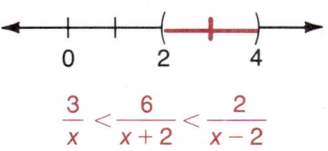

$$\frac{3}{x} < \frac{6}{x+2} < \frac{2}{x-2}$$

EXAMPLE 12 The manager of the Campus Bookstore can sell 4000 copies of "Passing Tests Made Easy" if he charges $8 per copy. For each 50¢ increase in price, he loses 100 sales. If each book costs the store $5, express the profits in terms of the selling price x. At what selling prices does the store make a profit?

SOLUTION   Let

$$x = \text{sales price in dollars}$$
$$x - 8 = \text{increase in price over } \$8$$
$$2(x - 8) = \text{number of 50¢ increases over } \$8$$
$$2(x - 8)100 = \text{number of sales lost}$$
$$4000 - 2(x - 8)100 = \text{resulting number of sales}$$
$$x - 5 = \text{profit per book}$$
$$[4000 - 2(x - 8)100](x - 5) = \text{total profit}$$

The bookstore makes a profit when

$$[4000 - 2(x - 8)100](x - 5) > 0$$

Since no profit is made if the book sells for no more than the store's purchase price, the manager will clearly set $x > 5$. Thus, he makes a profit when

$$4000 - 2(x - 8)100 > 0$$

since $(x - 5) > 0$ when $x > 5$. Solving this inequality, we have

$$4000 - 2(x - 8)100 > 0$$
$$4000 > 200(x - 8)$$
$$20 > x - 8$$
$$28 > x$$

The store makes a profit if the book sells for more than $5 and less than $28. ∎

Section 2.7 Exercises

Solve the following inequalities. Write the solutions in interval notation.

1. $2y - 16 > 0$
2. $4z + 20 \leq 0$
3. $3x - 2 < 10$
4. $3r + 15 \leq 0$
5. $3s - 9 > 0$
6. $3t - 2 > \dfrac{(10 - 2t)}{8}$
7. $u + 1 < 7u - 2$
8. $4w - 8 \geq w + 1$
9. $2p - 3 \geq 6p + 5$
10. $3v + 4 < 5v + 8$
11. $3 < 9 - 3z \leq 6$
12. $2 \leq \dfrac{4 - 2r}{3} \leq 8$
13. $-2 < 2y + 6 < 4$
14. $2 \leq 3v - 1 < 8$
15. $-11 \leq 1 - 2q \leq -5$
16. $8 \leq 5s - 7 \leq 25$
17. $h + 5 > 3 - 2h \geq h + 3$
18. $t - 2 \leq 3t - 1 \leq 2 - 3t$
19. $4u - 8 < u + 1 < 6u + 2$
20. $2w - 4 \geq 2 - 4w > w - 13$
21. $3x + 1 < 2$ or $2x - 1 \geq 5$
22. $1 - 4x \geq 9$ or $3x - 4 \geq 5$
23. $x + 3 \geq 2x - 1$ or $x + 5 \leq 2x - 4$
24. $5x + 3 \leq 2x - 6$ or $3x + 2 \leq 5x - 2$
25. $7 - 2x \leq x + 5$ or $2 - x \geq x + 1$
26. $5 - 3x \leq 2x + 10$ or $2 - 3x \leq 5 - 6x$
27. $(x - 3)(x + 2) < 0$
28. $(z + 5)(z - 1) > 0$
29. $(2s + 1)(3s - 6) \geq 0$
30. $(2v + 4)(3v - 1)(v - 7) < 0$
31. $(6 - w)(4 + 2w)(w - 5) > 0$
32. $(2x + 8)(3x - 12)(4 - x) \leq 0$
33. $(u - 3)(2 - u)(14u + 28) \geq 0$
34. $x^2 - x + 2 \leq 4$
35. $2y^2 + 3y - 2 \geq 0$
36. $8r^2 + 5r + 4 < 2r^2 - 5r - 8$
37. $(6t^2 + 6t - 36)(3t - 6) > 0$
38. $(s^2 + s - 2)(s + 3) \leq 0$
39. $(x^2 + 1)(x^2 - 1) \leq 0$
40. $(r^2 + 4)(r^2 - 3r - 10) \geq 0$
41. $(v - 4)^2(v^2 - 8v + 12) < 0$
42. $(y^2 - 2y + 1) > 0$
43. $(1 - w)^3(w^2 - w - 6) > 0$
44. $u^3 - 5u^2 + 6u < 0$
45. $(v^2 - v - 6)(-4v^2 + 8v - 3) < 0$
46. $\dfrac{1}{w - 2} \leq 0$
47. $\dfrac{2}{x + 4} \geq 0$
48. $\dfrac{-3}{y - 5} \leq 0$
49. $\dfrac{-4}{5 - r} \geq 0$
50. $\dfrac{2x - 6}{x + 4} \geq 0$
51. $\dfrac{x - 5}{8 - 4x} \leq 0$
52. $\dfrac{2x + 8}{3x - 9} < 0$
53. $\dfrac{5x + 15}{4x - 20} > 0$
54. $\dfrac{4}{r^2 + 2r - 15} \geq 0$
55. $\dfrac{2}{v^2 - v - 2} \leq 0$
56. $\dfrac{2x + 8}{x^2 - 9} \leq 0$
57. $\dfrac{w - 1}{w^2 - w - 2} \geq 0$
58. $\dfrac{(x - 2)(x + 3)}{(x + 1)(x - 4)} > 0$
59. $\dfrac{(y + 1)(y - 3)}{(y - 2)(y + 5)} < 0$
60. $\dfrac{(z - 3)(z - 1)}{(z + 2)(z + 4)} \leq 0$
61. $\dfrac{(w - 4)(w + 2)}{(w + 3)(w - 2)} \geq 0$
62. $\dfrac{2y + 8}{(y^2 - 4)(2y^2 + 5y - 3)} > 0$
63. $\dfrac{r^2 - 5r + 6}{(r + 4)(r^2 + 3r - 4)} \leq 0$
64. $\dfrac{(x - 2)^2(x + 2)}{(x - 3)} < 0$
65. $\dfrac{(x - 1)(2x + 4)^2}{(x + 1)} > 0$
66. $\dfrac{(3x + 6)(2x - 8)}{(x + 3)^2} > 0$
67. $\dfrac{(2x + 12)(x - 5)}{(3x - 12)^2} < 0$

138 Chapter 2 Equations and Algebraic Inequalities

68. $\dfrac{4}{y-2} \geq 2$

69. $\dfrac{9}{x-3} \leq 3$

70. $\dfrac{3}{z+1} < -1$

71. $\dfrac{6}{r+2} > 2$

72. $\dfrac{2x+3}{x} \leq 2$

73. $\dfrac{4x-1}{x} \geq 5$

74. $\dfrac{s-1}{s+2} < 2$

75. $\dfrac{2t+6}{6-2t} \geq 2$

76. $\dfrac{8}{x^2-12} \leq 2$

77. $\dfrac{18}{y^2-2} \geq 3$

78. $\dfrac{3}{r-1} < \dfrac{6}{r+1}$

79. $\dfrac{2}{6+s} \geq \dfrac{3}{s-3}$

80. $1 \leq \dfrac{2}{y-4} < 3$

81. $0 < \dfrac{1}{r-5} \leq \dfrac{3}{r+4}$

82. $-1 < \dfrac{1}{2z+1} \leq 2$

83. $\dfrac{2}{3s+6} < \dfrac{1}{s-7} \leq 0$

84. $\dfrac{1}{(t-2)(t-3)} \leq \dfrac{4}{t+2} \leq 2$

85. $\dfrac{1}{(u-1)(u-2)} \leq \dfrac{2}{(u+1)(u-2)} \leq \dfrac{2}{(u+3)(u-3)}$

86. Mary's Florist has fixed costs of $900 each month. For each $100 in retail sales, she has additional costs of $70. How much business must she have each month to make a profit?

87. A heart patient enters the hospital with a resting heart rate of 100 beats per minute. Her physician wants to decrease this rate medically to between 60 and 80 beats per minute. If 2 milligrams of medication per hour reduces the rate by 5 beats per minute, what should the prescription be?

88. If your grades on the first four tests in this course are 75, 95, 70, and 90, what grade must you receive on the fifth test in order to increase your average to 85 or better?

89. What would be the answer to Exercise 88 if the instructor is willing to drop your lowest grade?

90. A rectangle with one side of length 5 centimeters must have an area greater than 25 square centimeters and less than 75 square centimeters. What lengths are permitted for its other dimension?

91. A triangle with base of length 5 inches must have an area greater than 25 square inches and less than 75 square inches. What altitudes are permitted?

Section 2.8

Equations and Inequalities Involving Absolute Values

The absolute value of a real number represents its distance from 0 on the real number line. See Section 1.2. For example $|7| = 7$ and $|-7| = 7$. Thus for any *positive* real number a, there are *two* real numbers for which $|x| = a$; these are $x = a$ and $x = -a$.

EXAMPLE 1

Solve the following equations.

a. $|x| = 5$
b. $|x-2| = 3$
c. $|2x-3| = 5$

SOLUTION

a. If $|x| = 5$, then x can have only two values as a real number: $x = 5$ or $x = -5$. We write this as

$$x = \pm 5$$

The solution set is $\{5, -5\}$.

b. Here we treat $(x - 2)$ as a single real number. Then if $|x - 2| = 3$, this number must be ± 3 as in part (a). Thus if

$$|x - 2| = 3$$

then

$$x - 2 = \pm 3$$

That is,

$$x - 2 = 3 \quad \text{or} \quad x - 2 = -3$$

Finally,

$$\begin{aligned} x &= 2 + 3 &\text{or}&& x &= 2 - 3 \\ &= 5 &&& &= -1 \end{aligned}$$

We check $|5 - 2| = |3| = 3$ and $|-1 - 2| = |-3| = 3$. The solution set is $\{-1, 5\}$.

c. If

$$|2x - 3| = 5$$

then

$$2x - 3 = \pm 5$$
$$2x = 3 \pm 5$$

That is,

$$\begin{aligned} 2x &= 3 + 5 = 8 &\text{or}&& 2x &= 3 - 5 = -2 \\ x &= 4 &&& x &= -1 \end{aligned}$$

We check $|2 \cdot 4 - 3| = |8 - 3| = |5| = 5$ and $|2(-1) - 3| = |-2 - 3| = |-5| = 5$. The solution set is $\{4, -1\}$. ∎

EXAMPLE 2

Express each of the following in terms of absolute values.

a. The distance from v to 9 is 3.
b. The distance between y and 6 is less than 2.
c. u and -3 are at least 4 units apart.

SOLUTION

Since the distance between two numbers x and y on the real number line is $|x - y|$, we have the following.

a. $|v - 9| = 3$
b. $|y - 6| < 2$
c. $|u - (-3)| \geq 4$
  $|u + 3| \geq 4$ ∎

EXAMPLE 3

Solve the following absolute value equation.

$$|x - 2| = |6 - 2x|$$

SOLUTION   Since the two numbers $(x - 2)$ and $(6 - 2x)$ have the same absolute value, they are either equal or of opposite sign.

$$x - 2 = \pm(6 - 2x)$$

$$x - 2 = 6 - 2x \quad \text{or} \quad x - 2 = -(6 - 2x)$$
$$3x = 8 \qquad\qquad\qquad\quad = 2x - 6$$
$$x = \frac{8}{3} \qquad\qquad\qquad\quad -x = -4$$
$$\qquad\qquad\qquad\qquad\qquad x = 4$$

The equation is satisfied by $\frac{8}{3}$ or 4. The solution set is $\left\{\frac{8}{3}, 4\right\}$.   ■

EXAMPLE 4   Solve the following inequalities.

a. $|x| < 3$

b. $|x| > 3$

SOLUTION   **a.** Since $|x|$ represents the distance of x from 0 on the real number line, the solution to this inequality consists of all real numbers whose distance from 0 is less than 3: $|x - 0| = |x| < 3$. The two points whose distance from 0 equals 3 are 3 and -3. All points which are closer to the origin then satisfy $|x| < 3$. In Figure 2.27 we see that these are the points in the interval $(-3, 3)$.

Figure 2.27

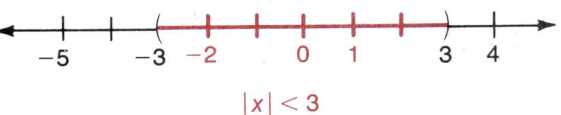

$|x| < 3$

To verify the above argument, we can check arbitrarily chosen points inside and outside this interval. For example $|-2| = 2 < 3$ and $|1| = 1 < 3$, but $|4| = 4 \not< 3$ and $|-5| = 5 \not< 3$. The solution set S is the interval $S = (-3, 3)$.

b. The numbers x that satisfy $|x| > 3$ are those whose distance from 0 is more than 3. These are illustrated in Figure 2.28.

Figure 2.28

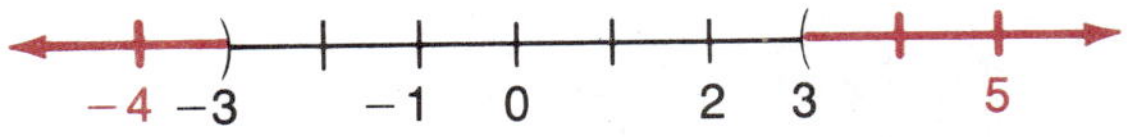

Checking various points, we have $|-4| = 4 > 3$ and $|5| = 5 > 3$, but $|-1| = 1 \not> 3$ and $|2| = 2 \not> 3$. The solution set S consists of the two intervals $(-\infty, -3)$ or $(3, \infty)$:

$$S = (-\infty, -3) \cup (3, \infty)$$ ■

In Example 4a we see that

$$|x| < 3 \quad \text{if and only if} \quad -3 < x < 3$$

(that is, $x > -3$ *and* $x < 3$), whereas in 4b,

$$|x| > 3 \quad \text{if and only if} \quad x < -3 \quad \text{or} \quad x > 3$$

For any *positive* real number a, we obtain similar statements:

$$|x| < a \quad \text{if and only if} \quad -a < x < a$$

Section 2.8 Equations and Inequalities Involving Absolute Values

(that is, $x > -a$ and $x < a$)

$$|x| > a \quad \text{if and only if} \quad x < -a \quad \text{or} \quad x > a$$

These properties are summarized in the following table.

PROPERTIES OF ORDER AND ABSOLUTE VALUE

EXPRESSION	SOLUTION	ILLUSTRATION
$|u| < a$	$-a < u < a$	$\xleftarrow{\quad\;(\;\;\;\;|\;\;\;\;)\;\;\;}\xrightarrow{}$ $-a$ $\;0\;$ a
$|u| = a$	$u = \pm a$	$\xleftarrow{\quad\;\bullet\;\;\;\;|\;\;\;\;\bullet\;\;\;}\xrightarrow{}$ $-a$ $\;0\;$ a
$|u| > a$	$u < -a \quad \text{or} \quad u > a$	$\xleftarrow{\quad\;)\;\;\;\;|\;\;\;\;(\;\;\;}\xrightarrow{}$ $-a$ $\;0\;$ a

EXAMPLE 5

Solve the following inequality.

$$|x - 3| < 5$$

SOLUTION

To solve the absolute value inequality $|x - 3| < 5$, we treat $|x - 3|$ as a single value and use these properties as follows.

$$|x - 3| < 5 \quad \text{means} \quad -5 < x - 3 < 5$$
$$-2 < \quad x \quad < 8 \quad \text{(Add 3 to all parts of the previous inequality)}$$

The solution set is the interval $(-2, 8)$. See Figure 2.29.

Figure 2.29

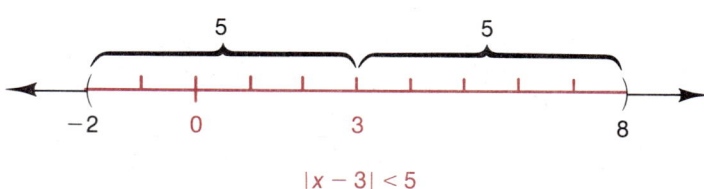

$|x - 3| < 5$

Note in Figure 2.29 that the solution does indeed consist of the numbers whose distance from 3 is less than 5. ∎

EXAMPLE 6

Solve the following inequality.

$$|2x + 1| < 3$$

SOLUTION

We know that $|2x + 1| < 3$ if and only if

$$-3 < 2x + 1 < 3$$
$$-4 < \quad 2x \quad < 2 \quad \text{(Subtract 1 from both sides)}$$
$$-2 < \quad x \quad < 1 \quad \text{(Divide by 2)}$$

$|2x + 1| < 3$

Figure 2.30

The solution set is the interval $(-2, 1)$. It is sketched in Figure 2.30. ∎

EXAMPLE 7

Solve the following inequality.

$$|1 - 2t| \geq 3$$

SOLUTION

Treat $(1 - 2t)$ as a single number.

$|1 - 2t| \geq 3$ means

$1 - 2t \leq -3$	or	$1 - 2t \geq 3$	
$-2t \leq -4$		$-2t \geq 2$	(Subtract 1 from both sides)
$2t \geq 4$		$2t \leq -2$	(Reverse the inequality)
$t \geq 2$	or	$t \leq -1$	(Divide by 2)

The solution set S consists of those numbers that are either ≥ 2 or ≤ -1:

$$S = (-\infty, -1] \cup [2, \infty)$$

The solution is sketched in Figure 2.31.

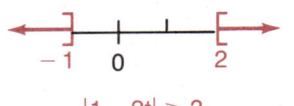

$|1 - 2t| \geq 3$

Figure 2.31

EXAMPLE 8

Solve the following inequalities.

a. $(|x| + 2)(x - 3) < 0$

b. $2 < |x| < 7$

SOLUTION

a. Note that $(|x| + 2) > 0$ for all real numbers x. Thus the indicated product is negative precisely when

$$x - 3 < 0$$
$$x < 3$$

The solution set is the interval

$$S = (-\infty, 3)$$

b. This double inequality represents the two inequalities

$$2 < |x| \quad \text{and} \quad |x| < 7$$

Thus, x must satisfy

$$[x < -2 \quad \text{or} \quad x > 2] \quad \text{and} \quad [-7 < x < 7]$$

Hence the solution set is

$$S = [(-\infty, -2) \cup (2, \infty)] \cap (-7, 7)$$
$$= (-7, -2) \cup (2, 7)$$

(Note that *or* corresponds to union and *and* corresponds to intersection of solution sets.) See Figure 2.32.

Figure 2.32

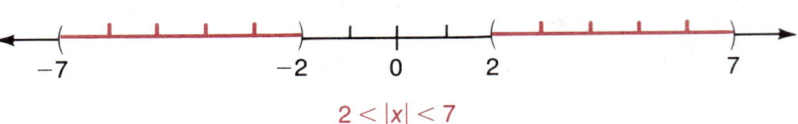

$2 < |x| < 7$

Section 2.8 *Equations and Inequalities Involving Absolute Values* **143**

EXAMPLE 9 Rancher MacDonald must build a corral for her cattle. She plans to use a rectangular canyon 100 feet wide. If she needs an area of approximately 40,000 square feet but will tolerate an error of 1000 square feet, how accurately must her overseer measure the length of the corral before building the fences?

SOLUTION Let x denote the length of the corral. Then $100x$ is its area. The area is to be within 1000 square feet of 40,000:

$$|100x - 40{,}000| < 1000$$
$$-1000 < 100x - 40{,}000 < 1000$$
$$39{,}000 < 100x < 41{,}000$$
$$390 < x < 410$$

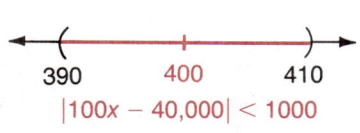

$|100x - 40{,}000| < 1000$

Figure 2.33

The length must be between 390 and 410 feet to satisfy Ms. MacDonald's requirements. Figure 2.33 shows that this is equivalent to $|x - 400| < 10$. The overseer should measure the length of the corral as 400 feet; she must be accurate to within 10 feet. ■

Section 2.8 Exercises

Write each of the following as an equation or inequality involving absolute values.

1. The distance between x and 4 is no more than 7.
2. y is less than two units away from -4.
3. r is exactly 2 units away from s.
4. u and t are 5 units apart.
5. w is at least 7 units away from -2.
6. The distance between z and h exceeds 6.

Solve the following equations and inequalities. Write the inequality solutions in interval notation.

7. $|x + 5| = 7$
8. $|y - 3| = 2$
9. $|2z + 4| = 10$
10. $|3r - 7| = 2$
11. $|2s - 10| = -5$
12. $|6t - 15| = 9$
13. $|a^2 - 2| = 2$
14. $|b^2 - 8| = 8$
15. $|c - 4| = c - 4$
16. $|3u - 9| = 3u - 9$
17. $|t - 3| = 3 - t$
18. $|4 - p| = p - 4$
19. $|x - 3| = |x - 7|$
20. $|y + 1| = |y - 9|$
21. $|2q - 3| = |8 - 3q|$
22. $|1 - 3r| = |2r - 2|$
23. $|y^2 + 1| = |3y^2 - 9|$
24. $|x^2 - 1| = |2x^2 - 7|$
25. $|2s - 1| < 7$
26. $|3v + 1| \leq 8$
27. $|11 - 3x| \geq 13$
28. $|2w - 5| > 11$
29. $|3y - 12| \geq 12$
30. $|2z + 5| < 3$
31. $|2r - 3| < -2$
32. $|t - 7| \geq 3$
33. $|2u - 5| > 5$
34. $\left|\dfrac{x - 2}{3}\right| < 4$
35. $\left|\dfrac{3q - 1}{2}\right| < 5$
36. $\left|\dfrac{2 - 3p}{4}\right| \geq 3$
37. $\left|\dfrac{3 - x}{4}\right| \geq 2$
38. $|2s + 3| < 9$
39. $|z + 5| < 2$
40. $|2s - 10| \geq 4$
41. $|2v - 3| \leq 3$
42. $|2t + 5| > 1$
43. $|1 - 3t| < 4$
44. $|4w - 7| \leq 5$
45. $|x^2 - 4|(x + 2) > 0$
46. $(x + 1)|x + 20| < 0$
47. $1 < |x| < 3$
48. $2 < |x| \leq 4$
49. $2 \leq |x - 1| < 5$
50. $1 < |x - 2| < 3$
51. $|x^2 - 4x + 4| \leq 1$
52. $|x^2 - 6x + 7| \leq 2$
53. $|w^2 - 2w - 4| < -1$
54. $-|x^2 + 6x - 7| > 2$

55. In order for Rare Oak acorns to germinate, the temperature must be within 5° of 27°C.

   a. Write an absolute value inequality that must be satisfied by the temperature T in order for germination to take place.

   b. Express the solution to this inequality as an interval.

56. Certain sensitive instruments must be shipped in a special cylinder. The circumference of the cylinder must be approximately 16π centimeters. An error of no more than $\pi/5$ centimeters from this figure will be tolerated. How accurately must the radius of the cylinder be determined? (*Hint:* The circumference of a circle of radius r is $2\pi r$ where π is a constant with value approximately 3.14159.)

57. The Gas-N-Power Utility Company estimates the monthly cost of heating or cooling a medium-sized home to be $3.75 times the difference between 60°F and the average temperature during the month. Ms. Walsh can afford at most $150 per month for heating and cooling.

   a. Write an absolute value inequality that must be satisfied by the temperature T if she is to be able to pay her utility bills.

   b. In what temperature range can she meet her obligation to the utility company? Express the solution set as an interval.

Section 2.9 Chapter Review

Terms and Concepts

Variable (or unknown) Represents a number (whose value is not specified).

Constant A fixed or specified number.

Equations The indicated equality of two quantities or expressions.
  1. Identities True for all values of the unknown.
  2. Conditional equations True for only certain values of the unknown.

Solution (or root) Values of the variable that make an equation or inequality true.

Proportionality
  1. y is directly proportional to x. $y = kx$
  2. y is inversely proportional to x. $y = k/x$
  3. y is jointly proportional to u and v. $y = kuv$

Rules and Formulas

Equations

$uv = 0$ if and only if $u = 0$ or $v = 0$.

The product of two positive or two negative numbers is positive.

The product of a positive and a negative number is negative.

Operations performed on one side of an equation must also be done on the other side.

If $(x - 2)(x - 3) = 0$, then $x = 2$ or $x = 3$.

$2 \cdot 3 = 6$ and $(-2)(-3) = 6$

$2 \cdot (-3) = -6$

Solution Techniques

First-Degree Equations in One Variable

  1. Simplify the algebraic expressions.
  2. Isolate the variable on one side of the equation by adding or subtracting terms and multiplying or dividing by a constant.

$3x + 4 - x - 10 = 0$
$2x - 6 = 0$
$2x = 6$
$x = 3$

Quadratic Equations

1. Factoring

$$x^2 - 4x + 3 = 0$$
$$(x - 3)(x - 1) = 0$$
$$x = 1, 3$$

2. Completing the square

$$x^2 - 2x - 1 = 0$$
$$x^2 - 2x + 1 = 2$$
$$(x - 1)^2 = 2$$
$$x - 1 = \pm\sqrt{2}$$
$$x = 1 \pm \sqrt{2}$$

3. Quadratic formula

If $ax^2 + bx + c = 0$, $(a \neq 0)$ then
$$x = \frac{-b \pm \sqrt{b^2 - 4ac}}{2a}$$

$$x^2 - 2x - 1 = 0$$
$$x = \frac{-(-2) \pm \sqrt{(-2)^2 - 4 \cdot 1 \cdot (-1)}}{2 \cdot 1}$$
$$= \frac{2 \pm 2\sqrt{2}}{2} = 1 \pm \sqrt{2}$$

Word Problems

1. Understand the problem; read it several times.
2. Make a sketch.
3. Identify the unknown quantities.
4. Name these quantities.
5. Express all known relationships mathematically.
6. Reread the problem to check your information.
7. Solve the resulting equations.
8. Interpret your results.
9. Check your solutions.

Equations Involving Fractions

1. Multiply through by the LCD.
2. Solve the resulting equation.

$$\frac{3}{x + 1} = \frac{2}{x}$$
$$3x = 2(x + 1)$$
$$3x = 2x + 2$$
$$x = 2$$

3. Check your answers in the original equation; you may have introduced extraneous roots.

$$\frac{3}{2 + 1} \stackrel{\checkmark}{=} \frac{2}{2}$$

Linear Inequalities

1. Simplify the algebraic expressions.
2. Isolate the variable on one side of the inequality. Note that the multiplication or division by a negative number reverses the inequality.

$$2x + 10 < 5x - 2$$
$$2x < 5x - 12$$
$$-3x < -12$$
$$x > 4$$

Higher-Order Inequalities

1. Write the inequality as an expression which is a product of linear factors (if possible) with 0 on the other side of the inequality symbol.

$$(2u - 6)(3u + 12) < 0$$

2. Find the points in which these factors become 0 and mark these on the real number line.

$$u = 3, -4$$

146 Chapter 2 *Equations and Algebraic Inequalities*

3. Determine the sign (+ or −) of this expression at a test point inside each interval determined by the points located in step 2.

4. If a test point satisfies the inequality, then so do all points in the open interval containing the test point; otherwise no point in the interval satisfies the inequality.

5. Include endpoints for $\leq$ or $\geq$ inequalities only.

Sign of $(2u - 6)(3u + 12)$

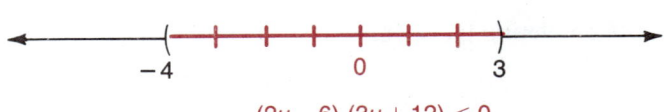

$(-)\cdot(-) = (+) \qquad (-)\cdot(+) = (-) \qquad (+)\cdot(+) = (+)$

$(2u - 6)(3u + 12) < 0$

Inequalities Involving Fractions

1. Bring all terms to one side.

2. Combine the fractions.

3. Simplify and factor.

4. Find the points in which these factors become 0 and mark them on the real number line.

5. Use test points as with higher order inequalities.

$$\frac{3}{x+1} < \frac{2}{x}$$

$$\frac{3}{x+1} - \frac{2}{x} < 0$$

$$\frac{3x - 2(x+1)}{x(x+1)} < 0$$

$$\frac{x - 2}{x(x+1)} < 0$$

Sign of $\dfrac{x-2}{x(x+1)}$

$\dfrac{(-)}{(-)\cdot(-)} = (-) \qquad \dfrac{(-)}{(-)\cdot(+)} = (+) \qquad \dfrac{(-)}{(+)\cdot(+)} = (-) \qquad \dfrac{(+)}{(+)\cdot(+)} = (+)$

$$\frac{3}{x+1} < \frac{2}{x}$$

Equations and Inequalities Involving Absolute Value

1. $|u| = a$ means $u = \pm a$

2. $|u| < a$ means $-a < u < a$

3. $|u| > a$ means $u > a$ or $u < -a$

$|x - 2| = 3$
$x - 2 = \pm 3$
$x = 2 \pm 3$
$x = 5, -1$

$|2x + 1| < 3$
$-3 < 2x + 1 < 3$
$-4 < 2x < 2$
$-2 < x < 1$

$|x - 1| > 2$
$x - 1 < -2 \quad$ or $\quad x - 1 > 2$
$x < -1 \quad$ or $\quad x > 3$

Section 2.10 Supplementary Exercises

Solve the following equations.

1. $3y - 10 = 6y + 20$
2. $3t - 5 = 3 - 5t$
3. $4x - 6 = 3x + 10$
4. $7(q - 2) = 6(q + 3)$
5. $7(w - 2) - 3(5 - w) = 11$
6. $\dfrac{r}{6} - 5 = \dfrac{r}{2} - 7$
7. $\dfrac{v}{3} - 2 = 3 - \dfrac{v}{2}$
8. $\dfrac{3p + 4}{2} - 6 = \dfrac{1 - 3p}{5}$
9. $\dfrac{1 + u}{3} - \dfrac{u}{2} = \dfrac{u - 2}{7}$

10. $\dfrac{z-3}{2} + \dfrac{z+3}{4} = \dfrac{8-z}{3} + 2$

11. $(x+1) + 3yi = 5 - 6i$

12. $2x + 3i = 8 + yi$

13. $x^2 - 6x + 9 = 0$

14. $y^2 - 10y + 21 = 0$

15. $3s^2 + 19s + 6 = 0$

16. $3r^2 - 8r - 3 = 0$

17. $x^2 + 2x + 2 = 0$

18. $t^2 - 5t + 6 = 0$

19. $r^2 - 4r - 12 = 0$

20. $2x^2 - 5x - 3 = 0$

21. $8 - 14t - 15t^2 = 0$

22. $2u^2 - 7u - 4 = 0$

23. $3t^2 - 10t + 3 = 0$

24. $4y^2 + 3y - 1 = 0$

25. $2x^2 + 2x + 5 = 0$

26. $w^2 + 2w + 10 = 0$

27. $2y^2 - 6y + 9 = 0$

28. $z^2 + 6z + 25 = 0$

29. $r^2 - 4r + 5 = 0$

30. $t^2 - 6t + 10 = 0$

31. $s^2 + 6s + 13 = 0$

32. $x^2 + 6x + 34 = 0$

33. $u^2 + 8u + 25 = 0$

34. $v^2 - 4v + 13 = 0$

35. $u^4 + 4u^2 - 12 = 0$

36. $s^6 + 10s^3 - 24 = 0$

37. $s^8 - 6s^4 + 8 = 0$

38. $v^{10} + 7v^5 + 12 = 0$

39. $4u^{12} - 3u^6 - 1 = 0$

40. $2x^{-2} + 4 = 7x^{-1}$

41. $6u^{-2} = 5u^{-1} - 1$

42. $p^{2/3} + 2p^{1/3} = 3$

43. $r^{2/3} - r^{1/3} = 2$

44. $q - \sqrt{q} - 2 = 0$

45. $\sqrt{z} - 3\sqrt[4]{z} + 2 = 0$

46. $3v + 11\sqrt{v} = 4$

47. $\sqrt{2t} - 5\sqrt{t} + \sqrt{18} = 0$

48. $\sqrt{27u} + 5\sqrt{u} = \sqrt{48}$

49. $\sqrt{2x-1} + \sqrt{x-1} = 5$

50. $\sqrt{2x-1} = 2x - 1$

51. $\sqrt{2-4u} + 3 = \sqrt{8u+5}$

52. $\sqrt{5y+6} - \sqrt{y+3} = 3$

53. $\sqrt{r-6} + \sqrt{2r+2} = \sqrt{3r+4}$

54. $\sqrt[3]{u^2 - 5} = 3$

55. $\sqrt[4]{t^2 + 12} = 2$

56. $\dfrac{x-1}{x+1} + \dfrac{5-2x}{x-1} = 0$

57. $\dfrac{6}{u+1} - \dfrac{3}{u+2} = 1$

58. $\dfrac{3}{t-2} - \dfrac{5}{2t-4} = \dfrac{2}{t+1}$

59. $\dfrac{3 - 2q^2}{q^2 + q - 2} + \dfrac{2q}{q+2} = \dfrac{5}{q-1}$

60. $\dfrac{w}{w^2 + w - 2} + \dfrac{2}{1 - w^2} = \dfrac{-2}{w^2 + 3w + 2}$

61. $(z + i + 1)^2 = (z+3)^2$

62. $(z - i + 3)(z + 2i) = (z - 3i)(z + i + 2)$

63. $z^2 - 2iz - 5 = 0$

64. $-iz^2 + 4z + 2i = 0$

65. $3iz^2 + 5z + 2i = 0$

66. $|z + 4| = 2$

67. $|u - 5| = -6$

68. $|x - 2| = |x + 4|$

69. $|y^2 - 4| = |y + 2|$

Solve the following inequalities. Write the solutions in interval notation.

70. $3r + 15 \le 0$

71. $7x + 10 \ge 3x - 2$

72. $5v - 2 < 3v - 8$

73. $(z+5)(z-1) > 0$

74. $(r-1)(r+4) \le 0$

75. $-2x^2 + 4x + 7 > 2x - 5$

76. $(t+1)(t-2)(2t-8) < 0$

77. $(u^2 - 1)(u^2 + u - 6) \ge 0$

78. $\dfrac{2}{v^2 - v - 2} < 0$

79. $\dfrac{x^2 - 2x - 8}{x+3} \ge 0$

80. $\dfrac{y^2 - 2y - 3}{y^2 + y - 2} < 0$

81. $-2 \le \dfrac{10 + 5s}{-5} \le 2$

82. $2v - 4 > 2 - 4v \ge v - 8$

83. $0 < \dfrac{1}{r-3} \le \dfrac{3}{r-5}$

84. $|5q + 4| \ge 6$

85. $|p + 4| \ge 2$

86. $|2s + 3| < 9$

87. $|3t + 1| < 4$

88. $|r + 3| < 4$

89. $\left|\dfrac{3-x}{4}\right| \ge 2$

90. $1 < |u| < 5$

91. $2 < |w| \le 4$

92. $3 \le |v - 1| < 6$

93. $1 < |z + 2| < 4$

94. If $A = \dfrac{x}{x+h}$, solve for h in terms of x and A.

95. In Exercise 94, solve for x in terms of h and A.

96. If $r^2 s = \dfrac{t+u}{v^2}$, solve for u in terms of the other variables.

97. In Exercise 96, solve for v in terms of the other variables.

98. The area of a square varies directly as the square of the lengths of its sides; its perimeter varies directly as the lengths of its sides.
    a. If the length of a side is tripled, what happens to the perimeter?
    b. If the length of a side is tripled, what happens to the area?

99. Let r vary directly as p^3 and s and inversely as the square root of q.
    a. If p is doubled, s is halved, and q is quadrupled, what happens to r?
    b. Write r in terms of p, q, and s if $r = 270$ when $p = 3$, $q = 4$, and $s = 2$.
    c. Under the conditions of (b), find r when $p = 1$, $q = 9$, and $s = 12$.

100. Suppose that the walking speed of an animal is directly proportional to the square root of its leg length.
    a. If a fully grown person with 30-inch legs walks at a rate of 4 miles per hour, how fast will a 2-year-old with 15-inch legs walk?
    b. How fast will a horse walk if its legs are 42 inches long?
    c. How fast will a giraffe walk if its legs are 5 feet long?

101. The cost of operating an automobile varies directly as the square root of its age and weight. If it costs 20¢ per mile to operate a 2-year-old 2000-pound automobile, how much will it cost to operate a 4-year-old 4000-pound car?

102. Because of fixed operating costs, the profit earned by an electronics supply firm varies directly as the square of the number of units of Super Stereos it sells and inversely as its utility costs. If it traditionally sells 50 percent more units in the winter when its utility costs are double those of the summer, compare its summer and winter profits.

103. To service a backwoods cabin, a utility company must run an electric cable through the forest to a pole along the highway. The direct distance from the cabin to the pole is
    i. 200 meters greater than the perpendicular distance from the cabin to the highway, and
    ii. 100 meters greater than the distance from that point on the highway to the pole.
    a. How far is the cabin from the highway?
    b. How much wire is needed if the cable is to be stretched directly from the cabin to this nearest pole?

104. Ten thousand dollars are invested in two CDs paying 6 percent and 8 percent interest, respectively. If the annual return on these investments is $660, find the amount invested at each rate. [*Hint:* Let the respective amounts be x and $(10,000 - x)$.]

105. Five small businesses plan to share in the purchase of a small computer. If they encourage three additional companies to join them, their individual investments will be decreased by $1200 each. How much does the computer cost?

106. Mary bought an order of roses for $1000. She can resell the roses at a profit of $25 per carton. But on delivery, four cartons were inadvertently left standing in the sun until they were ruined. The rest of the order was then sold for a net profit of $200. How many cartons of roses did she order?

107. Members of Friendship Fraternity plan to build a meeting house for $30,000. If they can find five more members, the cost per member will be reduced $500. How many members do they now have?

108. With a 15-mile-per-hour head wind, it takes a plane 1 hour, 40 minutes longer to travel 300 miles than with no wind. What is its airspeed?

109. A boat travels 15 miles upstream and then back again in 2 hours.
    a. If the stream is flowing at 4 miles per hour, find the speed of the boat in still water.
    b. How long did each part of the trip take?

110. Two kinds of industrial cleanser worth $1.20 a gallon and $1.80 a gallon, respectively, are mixed. If the mixture is worth $1.40 a gallon, what is the formula for the mixture? (*Hint:* Determine the relative proportions in a 120-gallon mixture.)

111. A farmhand is paid $5000 and a new automobile for labor in a given year. Had he worked only 7 months he would have received only $1000 in addition to his car. How much is the automobile worth?

112. To encourage quality work, the Plastics Molding Company pays Elizabeth 8¢ for each good case

she molds but fines her 5¢ for each defective one. After producing 2600 cases her net earnings were zero. How many bad cases did she mold?

113. A master plumber charges $18 an hour for his labor and $12 an hour for his apprentice's. On a given job the apprentice showed up 2 hours late for work one day. The total labor charges for this job were $276. How many hours did each spend on this job?

114. Art patched and painted Bill's hot rod for $360. Finishing the job in 12 hours less than estimated, he increased his wage by $1 per hour. What was the estimated time for this job?

115. Mr. Koesters has a field with dimensions 3 furlongs by 4 furlongs. After beginning to sow wild oats around the outside of the field, he decides to put corn in half the field. If he continues to sow oats around the outside until half the field is in oats, what are the dimensions of the inner plot of corn?

116. If a square is transformed into a rectangle by decreasing one pair of sides 1 inch and increasing the other pair of sides 2 inches, the area remains the same. Find the dimension of the original square.

117. A magician asks you to enclose a two-by-two block on a calendar and state the sum of the four entries. She then states the four entries. Explain. (*Hint:* See the note preceding Exercise 51, Section 2.2.)

118. Find two consecutive odd integers, the product of which is 22 more than the square of the first.

119. Find a positive integer that is 14 more than 51 times its reciprocal.

120. Find an integer that is 1 less than 56 times its reciprocal.

Graphs of Equations

3

A **graph** is a pictorial representation of a relationship between two or more quantities. Graphs are very efficient devices for communicating information regarding many aspects of daily living. For example, physicians use electrocardiograms to monitor heart function; stockbrokers plot stock prices to predict future trends; and economists plot new-car sales as an indicator of economic activity (Figure 3.1).

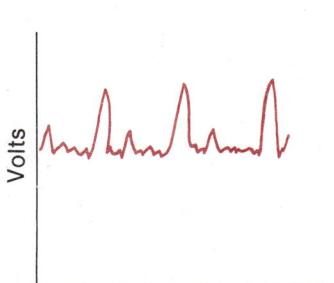
a. Electrocardiogram

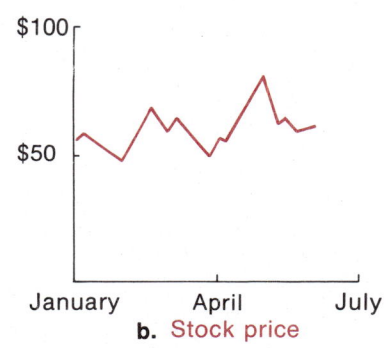
b. Stock price

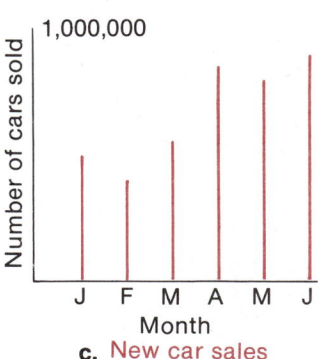
c. New car sales

Figure 3.1

Equations may also be used to describe everyday relationships. For instance, if you travel at a rate of 50 miles per hour, then the distance d traveled in t hours is given by $d = 50t$.

In this chapter we shall be concerned with the graphs of relationships that are expressed as an equation or inequality.

Section 3.1

Cartesian Coordinates; Distance and Circles

The graph of a relationship between two quantities P and Q generally includes a horizontal line and a vertical line. The horizontal line represents one of the quantities (say P) and the vertical line the other (say Q). Various values of P and Q are labeled on the horizontal and vertical lines, respectively. Then corresponding values are plotted on the graph. Thus in Figure 3.1b, for example, the horizontal line represents time (in months) and the vertical line represents price (in dollars).

To graph mathematical relationships, we place two perpendicular lines (to be called coordinate lines or **coordinate axes**) on a plane and use them in the same way. Unless otherwise specified, one of these lines is to be horizontal and the other vertical. The horizontal line is commonly called the **x-axis** and the vertical line is called the **y-axis;** however, other letters or symbols that are more appropriate to a given problem may also be used. The point O in which these two lines meet is

152 Chapter 3 Graphs of Equations

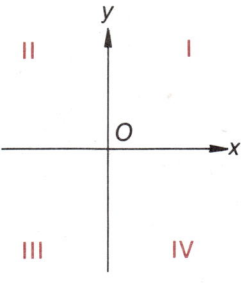

Figure 3.2

called the **origin.** The coordinate axes divide the plane into four sections, called **quadrants,** which are labeled I, II, III, and IV, as illustrated in Figure 3.2.

Each of the coordinate axes is considered a copy of the real number line (Section 1.1). It is customary to assign the positive direction to the right on the x-axis and upward on the y-axis. From each point P in the plane, we drop a perpendicular to each of the coordinate axes. The numbers corresponding to the points at which these perpendiculars meet the respective axes are called, respectively, the **x-coordinate** of P and the **y-coordinate** of P. If a is the x-coordinate of P and b is the y-coordinate of P, we say that P has coordinates (a, b). The context should prevent confusion of the coordinates (a, b) of a point P and the open interval (a, b).

The x-coordinate of a point indicates its distance to the right or left of the y-axis; the y-coordinate gives its distance above or below the x-axis. The coordinates of a number of points are illustrated in Figure 3.3. Note in particular that

Figure 3.3

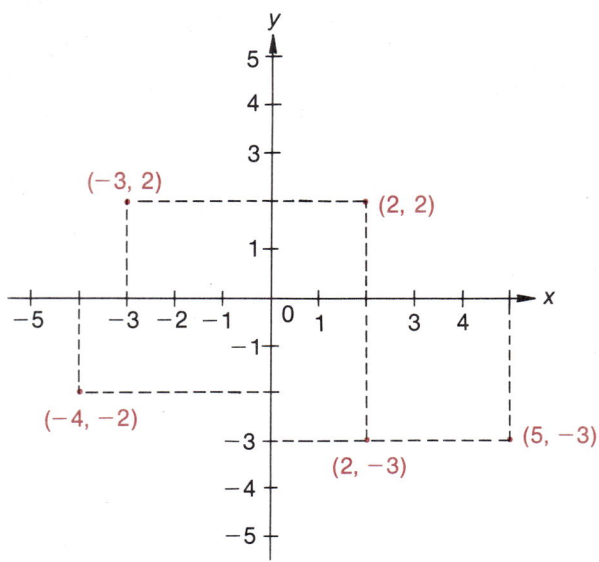

Historical Perspective

RENÉ DESCARTES
(1596–1650)

This system of plotting points is known as the **Cartesian coordinate system** in honor of René Descartes (1596–1650), the French mathematician and philosopher who is credited with its discovery and development. Descartes is said to have been inspired by watching a fly crawling near a corner on a ceiling. A plane on which is imposed a Cartesian coordinate system is called a **Cartesian plane** or **real plane.** Descartes introduced and developed the technique of studying geometry by analyzing coordinates of points and equations of figures. This branch of mathematics is now called **analytic geometry.**

$(2, -3) \neq (-3, 2)$ since $(2, -3)$ is in Quadrant IV, whereas $(-3, 2)$ is in Quadrant II. (See Figure 3.3.) Because of the importance of the order in which a and b are written, (a, b) is called an **ordered pair.**

We shall see how coordinates are used in three geometric problems: finding the midpoint of a line segment, computing distance, and describing a circle. If P has coordinates (a, b), we shall find it convenient to write $P = (a, b)$. However, if we are working with more than one coordinate system, we must then indicate in which system the coordinates (a, b) apply.

The Midpoint Formula

To obtain the coordinates of the midpoint $M = (x, y)$ of the line segment joining $P = (x_1, y_1)$ and $Q = (x_2, y_2)$, we observe that the x-coordinate of M appears to be halfway between x_1 and x_2 (see Figure 3.4):

$$x = x_1 + \frac{1}{2}(x_2 - x_1)$$
$$= \frac{x_1 + x_2}{2}$$

Similarly,

$$y = \frac{y_1 + y_2}{2}$$

Figure 3.4

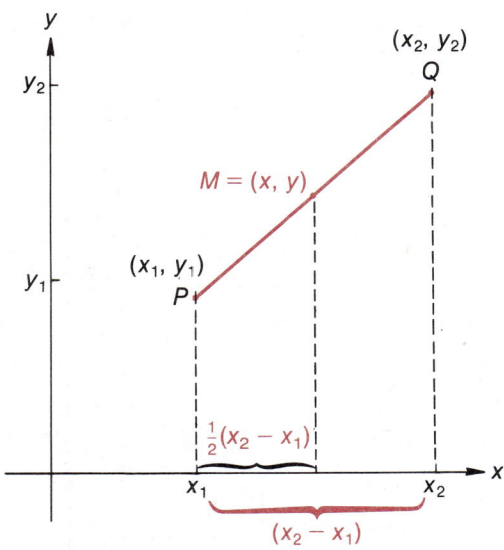

MIDPOINT FORMULA

The **midpoint** of the line segment joining $P = (x_1, y_1)$ and $Q = (x_2, y_2)$ has coordinates

$$\left(\frac{x_1 + x_2}{2}, \frac{y_1 + y_2}{2} \right)$$

EXAMPLE 1 Find the midpoint of the segment joining the points $(-2, 3)$ and $(6, -5)$.

SOLUTION The midpoint formula gives the coordinates of the midpoint as

$$\left(\frac{-2+6}{2}, \frac{3+(-5)}{2}\right) = \left(\frac{4}{2}, \frac{-2}{2}\right) = (2, -1)$$

The points $(-2, 3)$, $(6, -5)$, and $(2, -1)$ are shown in Figure 3.5.

Figure 3.5

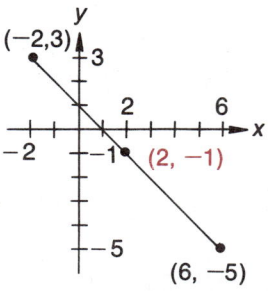

The Distance Formula

The distance between two points in a Cartesian plane is determined by their coordinates. If $P = (x_1, y_1)$ and $Q = (x_2, y_2)$ are two points in the plane, let $R = (x_2, y_1)$ as in Figure 3.6. The points P, Q, R then form a triangle with a right angle at R. Recall from Section 1.2 that the distance between two points x_1 and x_2 on the real number line is $|x_1 - x_2|$. If we let $d(A, B)$ denote the distance between two points A and B, the Pythagorean theorem tells us that

$$[d(P, Q)]^2 = [d(P, R)]^2 + [d(R, Q)]^2$$

Figure 3.6

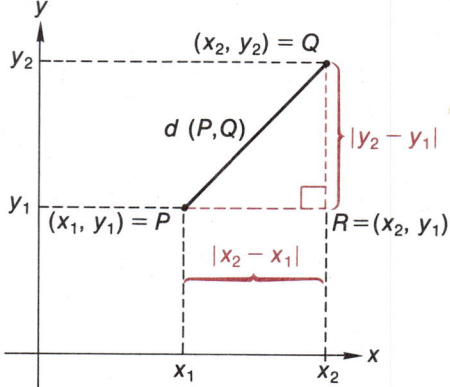

But $d(P, R) = |x_2 - x_1|$ and $d(R, Q) = |y_2 - y_1|$, as illustrated in Figure 3.6. Thus,

$$[d(P, Q)]^2 = |x_2 - x_1|^2 + |y_2 - y_1|^2$$
$$= (x_2 - x_1)^2 + (y_2 - y_1)^2 \qquad (a^2 = |a|^2)$$

Taking the square root of both sides of this equation, we obtain the *distance formula*.

DISTANCE FORMULA

The **distance** between two points $P = (x_1, y_1)$ and $Q = (x_2, y_2)$ in the real plane is given by

$$d(P, Q) = \sqrt{(x_2 - x_1)^2 + (y_2 - y_1)^2}$$

EXAMPLE 2 Find the distance between $(-2, 3)$ and $(4, -1)$.

SOLUTION
$$\begin{aligned} d[(-2, 3), (4, -1)] &= \sqrt{[4 - (-2)]^2 + [(-1) - 3]^2} \\ &= \sqrt{6^2 + (-4)^2} \\ &= \sqrt{52} \\ &= 2\sqrt{13} \end{aligned}$$

Circles

DEFINITION

A **circle** consists of those points whose distance from a given point A is some positive constant r. A is called the **center** and r the **radius** of the circle.

If $A = (a, b)$ is the center of a circle of radius r, then $P = (x, y)$ is on this circle if and only if $d(A, P) = r$ (see Figure 3.7):

$$d(A, P) = \sqrt{(x - a)^2 + (y - b)^2} = r \qquad (r > 0)$$

For $r > 0$, this latter equation is equivalent to

$$(x - a)^2 + (y - b)^2 = r^2$$

STANDARD EQUATION OF A CIRCLE

The circle of radius r centered at (a, b) has equation

$$(x - a)^2 + (y - b)^2 = r^2$$

In particular,

$$x^2 + y^2 = r^2$$

is the circle of radius r centered at the origin.

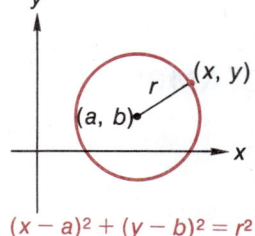

$(x - a)^2 + (y - b)^2 = r^2$

Figure 3.7

EXAMPLE 3 Write the equation of the circle of radius 5 centered at $(4, 3)$.

SOLUTION The given circle is just the locus (or set) of points lying 5 units away from $(4, 3)$. According to the preceding formula, the point (x, y) is on this circle if and only if

$$(x - 4)^2 + (y - 3)^2 = 25$$
$$x^2 - 8x + 16 + y^2 - 6y + 9 = 25$$
$$x^2 - 8x + y^2 - 6y = 0$$

EXAMPLE 4   Write the equation of the circle centered at $(-1, 2)$ that passes through $(-3, 4)$. Sketch the circle.

SOLUTION   If the circle passes through $(-3, 4)$, we can find the radius of the circle as the distance from the center to this point. Thus,

$$\begin{aligned} r &= d[(-1, 2), (-3, 4)] \\ &= \sqrt{[-3 - (-1)]^2 + (4 - 2)^2} \\ &= \sqrt{(-3 + 1)^2 + 2^2} \\ &= \sqrt{(-2)^2 + 2^2} \\ &= \sqrt{8} \end{aligned}$$

The equation of the circle is then

$$[x - (-1)]^2 + (y - 2)^2 = (\sqrt{8})^2$$
$$(x + 1)^2 + (y - 2)^2 = 8$$

The sketch is given in Figure 3.8.

Figure 3.8

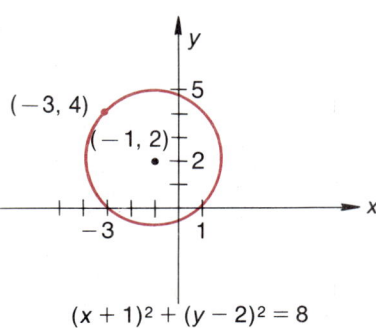

$(x + 1)^2 + (y - 2)^2 = 8$

EXAMPLE 5   Describe and sketch the set of points (x, y) satisfying

$$x^2 - 4x + y^2 + 6y - 12 = 0$$

SOLUTION   Rewriting the preceding equation as

$$(x^2 - 4x \quad) + (y^2 + 6y \quad) = 12$$

we complete the squares by adding 4 and 9 to both sides of the equation.

$$(x^2 - 4x + 4) + (y^2 + 6y + 9) = 12 + 4 + 9$$
$$(x - 2)^2 + (y + 3)^2 = 25$$
$$= 5^2$$

Thus, the given equation represents the circle of radius 5 centered at $(2, -3)$. Plotting the points located $r \,(= 5)$ units above, below, to the left, and to the right of the center $(2, -3)$ helps to sketch the circle. These points are $(2, 2)$, $(2, -8)$, $(-3, -3)$, and $(7, -3)$. See Figure 3.9.

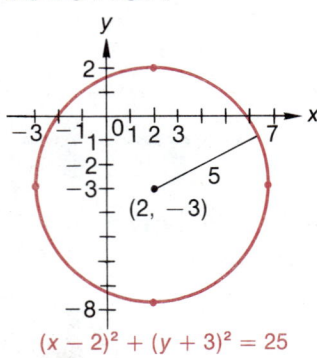

$(x - 2)^2 + (y + 3)^2 = 25$

Figure 3.9

Section 3.1 Cartesian Coordinates; Distance and Circles 157

EXAMPLE 6 Describe and sketch the set of points (x, y) satisfying each equation.

a. $x^2 - 4x + y^2 + 6y + 20 = 0$
b. $x^2 - 4x + y^2 + 6y + 13 = 0$

SOLUTION Each of these is like Example 5 except that the constant term -12 has been changed to 20 and 13, respectively. Thus we can complete the square in each and obtain the following.

a. $(x - 2)^2 + (y + 3)^2 = -20 + 4 + 9$
$= -7$

No point (x, y) can satisfy this equation, since $(x - 2)^2 \geq 0$ and $(y + 3)^2 \geq 0$, and so their sum cannot be less than 0.

b. $(x - 2)^2 + (y + 3)^2 = -13 + 4 + 9$
$= 0$

If either $(x - 2)^2 > 0$ or $(y + 3)^2 > 0$, their sum is >0. Thus we must have

$(x - 2)^2 = 0$ and $(y + 3)^2 = 0$
$x = 2$ $y = -3$.

Only the single point $(2, -3)$ satisfies the equation. ∎

By completing the square as in Examples 5 and 6, it can be shown that any equation of the form

$$\boxed{x^2 + ax + y^2 + by + c = 0}$$

represents either a circle or a single point or is not satisfied by any (x, y). This equation is called the **general equation** of a circle.

EXAMPLE 7 Describe the set of points satisfying the equation
$$3x^2 - 12x + 3y^2 + 18y - 36 = 0.$$

SOLUTION If we divide both sides of the equation by 3, we obtain
$$x^2 - 4x + y^2 + 6y - 12 = 0.$$

This is the same equation that was given in Example 5. It represents the circle of radius 5 centered at $(2, -3)$. See Figure 3.9. ∎

As in Example 7, any equation of the form
$$Bx^2 + Cx + By^2 + Dy + E = 0, \quad (B \neq 0)$$

can be converted into the general form of the equation for a circle by dividing

Section 3.1 Exercises

1. Find the coordinates of each of the points illustrated in the figure. Each grid mark represents one unit.

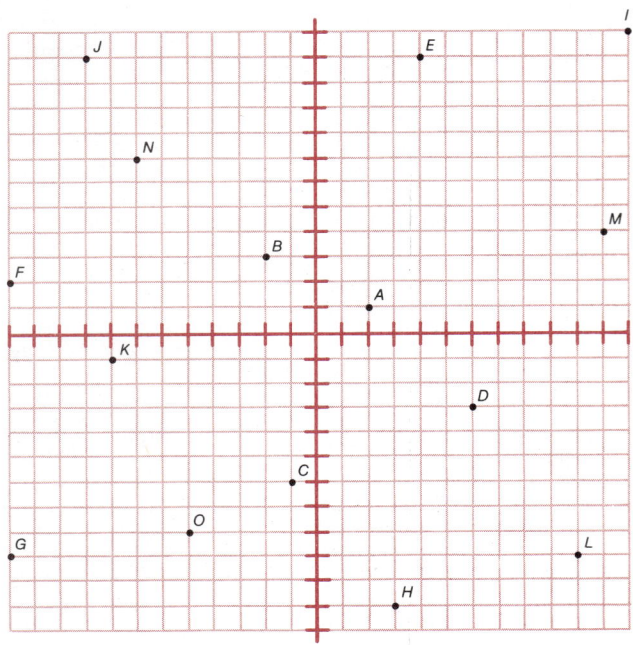

Plot the following points in the real plane and indicate the quadrant in which each point lies.

2. (1, 2)
3. (2, −1)
4. (−2, −1)
5. (3, −5)
6. (0, −1)
7. (5, 4)
8. (−1, 0)
9. (0, 1)
10. (10, 2)
11. (−3, 2)
12. (−2, 1)
13. (3, 0)
14. (−1, −3)
15. (−4, −2)
16. (−2, 0)

17. Road maps are commonly marked off in sections by a grid. We look for our destination in the index and read, for example, St. Henry, M11. St. Henry can then be found in block M11, as illustrated in the figure. Compile an index for the villages listed on the map in the figure.

Find the midpoints of the segments joining the following pairs of points.

18. (−3, 1), (1, 5)
19. (5, 2), (−1, −2)
20. (−3, 0), (−7, 0)
21. (1, 0), (0, 2)
22. (−4, 2), (2, −4)
23. (0, −2), (0, 6)
24. (3, 2), (1, 10)

Calculator Exercises

Find the distance between the following pairs of points,
  a. exactly, in terms of radicals if necessary.
  b. rounded to two decimal places.

25. (3, 10), (2, 1)
26. (−3, 1), (2, 4)
27. (−2, 5), (−3, −6)
28. (−2, 0), (5, 0)
29. (−2, −1), (−1, −2)
30. (4, 3), (6, −4)

31. $(0, -2), (0, -5)$
32. $(67, 32), (47, 42)$
33. $(-10, 5), (-10, -3)$
34. $(-10, 5), (6, 5)$
35. $(5, -2), (2, -5)$
36. $(-3, 0), (0, -4)$

A partial street naming scheme is to fix a common reference point (such as the intersection of Main St. and Central Ave.) and count "streets" east of this point and "avenues" north of this point. Then, for example, Fifth Street would run north and south and be five blocks east of this reference point.

37. With this street-labeling scheme, find the distance "as the crow flies" between the corner of 59th St. and 7th Ave. and the corner of 64th St. and 19th Ave., if each block is $\frac{1}{2}$ mile long and the streets form a rectangular grid.

38. A developer must construct a street running diagonally from the corner of 20th Ave. and 15th St. to the corner of 4th Ave. and 27th St. If the numbered streets and avenues form a rectangular grid and each block is $\frac{1}{4}$ mile long, how long is the new street?

39. Show that the points $(-2, 4), (0, -1)$, and $(3, 6)$ are vertices of an isoceles triangle. (*Hint:* Show that two sides have the same length.)

40. Show that $(-2, 2), (2, 0), (1, 3)$, and $(-1, -1)$ are vertices of a square.

41. Show that $(-2, 1), (4, 5), (2, 2)$ and $(0, 4)$ are vertices of a parallelogram. (*Hint:* Show that opposite sides have the same length.)

42. Show that the points $(2, 7), (-2, -1), (6, 5)$ are vertices of a right triangle. (*Hint:* Show that $a^2 + b^2 = c^2$.)

43. By finding the midpoints, show that the line segment from $(1, 3)$ to $(-1, -1)$ bisects the segment connecting $(2, 0)$ and $(-2, 2)$.

44. Locate a parallelogram on a convenient coordinate system and show that its diagonals bisect one another.

45. Use a convenient coordinate system to show that the midpoint of the hypotenuse of a right triangle is equidistant from its three vertices.

46. Locate a rectangle on a convenient coordinate system and show that its diagonals have the same length.

47. Write an equation (in x and y) that must be satisfied for $P = (x, y)$ to be on the perpendicular bisector of the segment joining $R = (1, 2)$ and $Q = (3, 1)$. (*Hint:* $d(P, R) = d(P, Q)$.)

48. Find all points with coordinates of the form $(x, -x)$ whose distance from $(2, -1)$ is 2.

49. Find all points with coordinates of the form $(2y, y)$ whose distance from $(-3, 2)$ is $\sqrt{10}$.

Sketch the circle represented by the given equation.

50. $x^2 + y^2 = 9$
51. $x^2 + y^2 = 36$
52. $(x - 3)^2 + (y - 1)^2 = 16$
53. $(x - 4)^2 + (y + 1)^2 = 100$
54. $(x - 2)^2 + (y + 2)^2 = 4$
55. $(x - 3)^2 + (y + 1)^2 = 81$
56. $(x - 3)^2 + (y - 2)^2 = 9$
57. $(x + 2)^2 + (y + 4)^2 = 4$

Write the general equation of the indicated circle.

58. Center $(-2, 5)$, radius 5
59. Center $(0, 4)$, radius 7
60. Center $(3, 2)$, radius 8
61. Center $(-1, -1)$, radius 1
62. Center $(3, 0)$, radius 6
63. Center $(4, 2)$ radius 3
64. Center $(2, -1)$, radius 4
65. Center $(-2, 3)$, radius 2
66. Center $(-1, -1)$, passing through $(0, -3)$
67. Center $(2, -4)$, passing through $(11, -16)$
68. Center $(-1, -3)$, passing through $(2, -5)$
69. Center $(-2, 6)$, passing through $(2, 9)$
70. Center $(2, 4)$, passing through $(5, 8)$
71. Center $(1, 5)$, passing through $(6, 17)$
72. Center $(3, -2)$, passing through $(4, -4)$
73. $(-2, 0)$ and $(-2, 6)$ are on opposite ends of a diameter.
74. $(1, 2)$ and $(5, 2)$ are at opposite ends of a diameter.

160 Chapter 3 Graphs of Equations

75. $(-17, -6)$ and $(31, 8)$ are at opposite ends of a diameter.
76. $(1, -2)$ and $(7, 6)$ are at opposite ends of a diameter.
77. Center $(1, 2)$, tangent to (that is, just touches) the y-axis.
78. Center $(-2, 4)$, tangent to the x-axis.

Describe the set of points $P = (x, y)$ *satisfying each of the following equations.*

79. $x^2 - 6x + y^2 + 2y - 15 = 0$
80. $x^2 - 10x + y^2 - 6y + 30 = 0$
81. $2x^2 + 8x + 2y^2 + 16y - 58 = 0$
82. $2x^2 + 4x + 2y^2 + 8y + 8 = 0$
83. $x^2 - 10x + y^2 + 6y + 30 = 0$
84. $x^2 - 4x + y^2 - 8y + 29 = 0$
85. $x^2 - 10x + y^2 + 10y + 25 = 0$
86. $3x^2 + 6x + 3y^2 - 12y + 6 = 0$
87. $3x^2 - 18x + 3y^2 + 6y + 30 = 0$
88. $x^2 - 12x + y^2 - 14y - 15 = 0$
89. $x^2 + 2x + y^2 - 4y - 4 = 0$
90. $x^2 - 4x + y^2 - 8y - 29 = 0$
91. $4x^2 - 16x + 4y^2 + 24y - 12 = 0$
92. $5x^2 - 30x + 105 = 40y - 5y^2$

Section 3.2

Graphs

In a graph indicating the relationship of energy demand to temperature (Figure 3.10), we see that daily energy use is 1.5 million kilowatt-hours when the temperature is 70°F. Five million kilowatt-hours are needed when the temperature is 100°F. The graph shows that energy use changes in relation to the temperature. The position of the curve above the x-axis at $x = T$ indicates the daily energy demanded when the temperature is T°F.

Figure 3.10

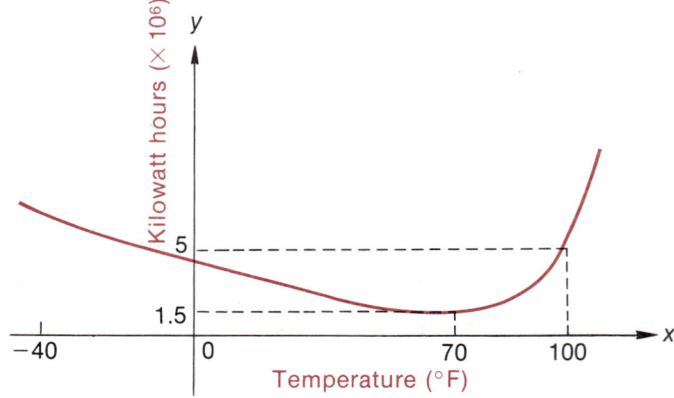

> ### THE GRAPH OF A RELATIONSHIP
>
> The **graph** of any relationship between x and y consists of those points (x, y) that satisfy the given relationship. The points in which the graph crosses the x-axis are called **x-intercepts**; **y-intercepts** are points in which the graph crosses the y-axis.

Note that the x-intercepts correspond to $y = 0$ (since $y = 0$ on the x-axis) and the y-intercepts correspond to $x = 0$ (since $x = 0$ on the y-axis).

Section 3.2 Graphs **161**

> PLOTTING POINTS
>
> An elementary approach to graphing a relationship is to
>
> 1. List corresponding x and y values in a table.
> 2. Plot these points on a coordinate system.
> 3. "Connect the dots" when enough points have been plotted to indicate a trend.
>
> Convenient points to plot are those corresponding to $x = 0$ and $y = 0$. These are the y- and x-intercepts of the graph, respectively.

EXAMPLE 1

Sketch the graph of the following relationship. Find the x- and y-intercepts.

$$2x - y + 1 = 0$$

SOLUTION

The graph consists of those points (x, y) that satisfy the equation $2x - y + 1 = 0$. Observe that y can be written explicitly in terms of x:

$$y = 2x + 1$$

In the table accompanying Figure 3.11, several values for x are listed with corresponding y values. For example,

if $x = 0,$ then $y = 2 \cdot 0 + 1 = 1$

and

if $x = 1,$ then $y = 2 \cdot 1 + 1 = 3$

The graph can then be sketched by plotting these points and connecting them as in Figure 3.11.

Figure 3.11

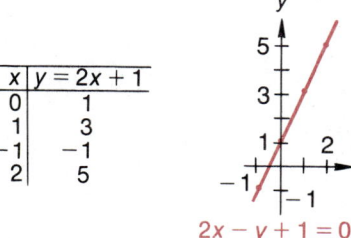

We have already found the y-intercept: $y = 1$ when $x = 0$. To find the x-intercept, we set $y = 0$:

$$2x + 1 = 0$$

$$x = -\frac{1}{2}$$

The graph crosses the x-axis at $x = -\frac{1}{2}$. ∎

We can indeed connect the dots to form a straight line in Example 1. However, we should be careful in this regard because a graph may contain bends and

162 Chapter 3 Graphs of Equations

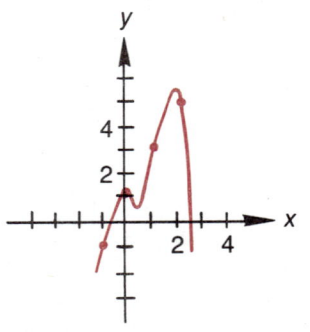

Figure 3.12

kinks that we may have missed. For example, we could also connect the dots corresponding to the values listed in the table accompanying Figure 3.11 as in Figure 3.12. Later we shall develop analytical methods to assist us in connecting the dots so as to ascertain the general shape of the graph. In the next section, we shall show that the graph in Example 1 is indeed a straight line and that, in fact,

---

THE GRAPH OF $ax + by + c = 0$

The graph of any equation of the form

$$ax + by + c = 0 \qquad a \text{ and } b \text{ not both } 0$$

is a straight line.

Consequently, the graph of any such equation can be sketched by plotting just two points and drawing the line that is determined by these two points.

EXAMPLE 2   Sketch the graph of each relationship.

a. $x = 0$ **b.** $y = 3$

SOLUTION

a. No restriction is placed on the y-coordinate. Any point of the form $(0, y)$ is on the graph; the graph is just the y-axis, as shown in Figure 3.13a.

b. The graph consists of all points of the form $(x, 3)$. This constitutes the line 3 units above and parallel to the x-axis, as shown in Figure 3.13b.

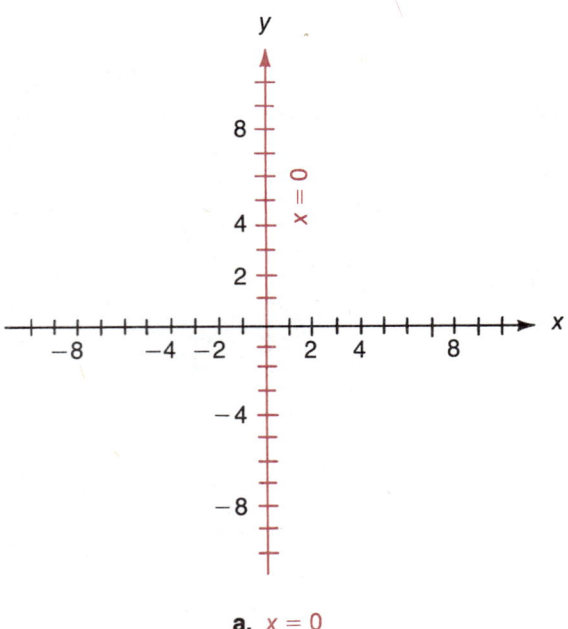
a. $x = 0$

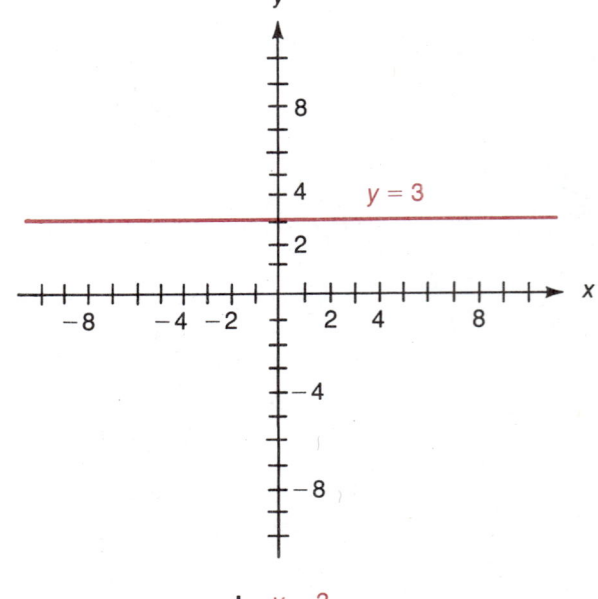
b. $y = 3$

Figure 3.13

Graphing Inequalities

To graph a relationship that is expressed as an inequality, first convert the inequality to an equation and sketch the graph of the equation. This graph will divide the plane into several regions. A point can be tested in each region to determine which regions consist of points satisfying the inequality.

Section 3.2 Graphs 163

EXAMPLE 3 Sketch the graph of the relationship

$$2x - y + 1 > 0$$

SOLUTION This relationship can be rewritten as $2x + 1 > y$ or

$$y < 2x + 1$$

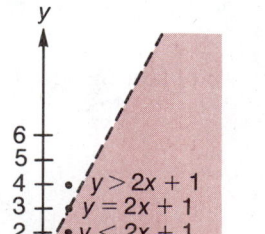

$y < 2x + 1$

Figure 3.14

In Example 1, the y-altitude was graphed as $y = 2x + 1$. Here, the y-altitude must be less than $2x + 1$; that is, this graph consists of all points below the graph of Example 1, as shown in Figure 3.14. The line is broken to indicate that it is *not* part of the solution set. For instance, considering $x = 1$, we have

$$3 = 2 \cdot 1 + 1 \quad \text{so} \quad (1, 3) \text{ is not on the graph}$$

Similarly,

$$4 > 2 \cdot 1 + 1 \quad \text{so} \quad (1, 4) \text{ is not on the graph}$$

but

$$2 < 2 \cdot 1 + 1 \quad \text{so} \quad (1, 2) \text{ is on the graph}$$

EXAMPLE 4 Sketch the graph of each of the following.

a. $|x| = 1$ **b.** $|x| > 1$

SOLUTION
a. Recall that $|x| = 1$ when $x = 1$ or $x = -1$. Since there is no restriction placed on y, the equations $x = 1$ and $x = -1$ represent the vertical lines shown in Figure 3.15a.

b. The lines in Figure 3.15a divide the x-y plane into three regions. We test a point in each of these regions to determine which regions consist of points satisfying $|x| > 1$. For instance,

$$(-2, 3): \quad |-2| = 2 > 1$$
$$(0, 2): \quad |0| = 0 \not> 1$$
$$(3, -1): \quad |3| = 3 > 1.$$

The graph appears as the shaded region in Figure 3.15b.

Figure 3.15

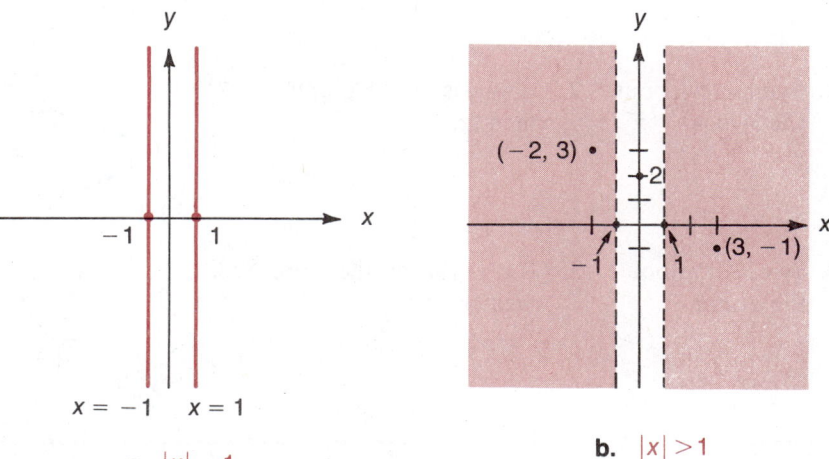

a. $|x| = 1$

b. $|x| > 1$

EXAMPLE 5 Sketch the graph of each relationship.

a. $y = x^2$ **b.** $x = y^2$

SOLUTION

a. Several points are found by listing in a table those y values corresponding to certain x values. We then plot these points and connect the dots as in Figure 3.16a.

b. If we simply interchange the roles of x and y in part (a), we obtain the graph of $x = y^2$, as shown in Figure 3.16b.

Figure 3.16

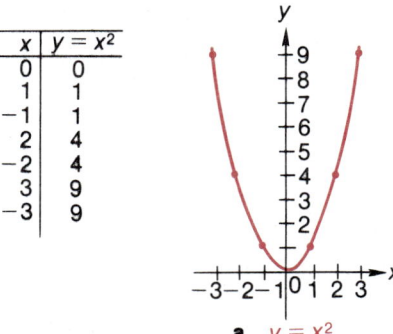

a. $y = x^2$

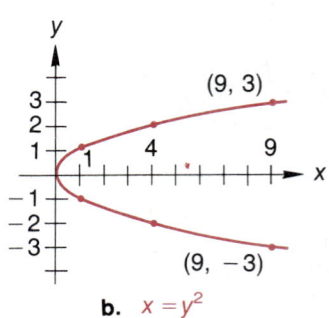

b. $x = y^2$

Graphs such as those illustrated in Figure 3.16 are called **parabolas**. The point $(0, 0)$, at which the parabola "turns around," is called its **vertex**.

The left and right halves of the graph in Figure 3.16a are mirror images of one another when reflected through the y-axis. Such a graph is said to be **symmetric** about the y-axis. Similarly, the graph in Figure 3.16b is said to be symmetric about the x-axis.

SYMMETRY

GEOMETRIC DESCRIPTION	ALGEBRAIC DESCRIPTION	ILLUSTRATION	TEST
1. Symmetry about the y-axis	1. $(-x, y)$ is on the graph when (x, y) is.	1.	1. Substitution of $-x$ for x leads to an equivalent equation.
2. Symmetry about the x-axis	2. $(x, -y)$ is on the graph when (x, y) is.	2.	2. Substitution of $-y$ for y leads to an equivalent equation.
3. Symmetry about the origin	3. $(-x, -y)$ is on the graph when (x, y) is.	3.	3. Substitution of $-x$ for x and $-y$ for y leads to an equivalent equation.

EXAMPLE 6

Sketch the graph of $x = y^3$.

SOLUTION

Substitution of $-x$ for x and $-y$ for y leads to

$$-x = (-y)^3 = -y^3$$
$$x = y^3 \qquad \text{(Multiply by } -1\text{)}$$

which is the original equation. Thus $(-x, -y)$ is on the graph whenever (x, y) is. The graph is symmetric about the origin. *Thus, we need plot only points corresponding to* $x \geq 0$ *and then extend the graph using this symmetry*. The graph and a small table of values appear in Figure 3.17. A calculator can be used to approximate the coordinates of additional points if desired.

Figure 3.17
$x = y^3$

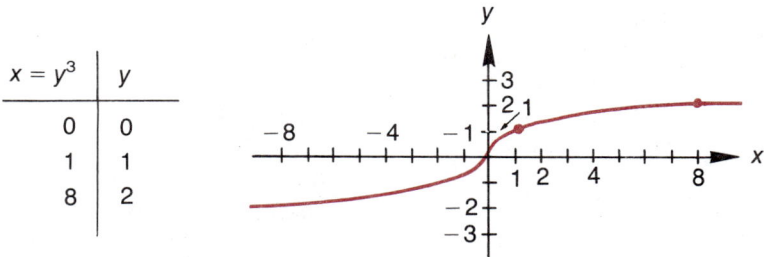

$x = y^3$	y
0	0
1	1
8	2

Note that $(-1, -1)$ and $(-8, -2)$ are on the graph; these are the points corresponding to $(1, 1)$ and $(8, 2)$, respectively, by the symmetry about the origin. ∎

Information that might otherwise be difficult to find can sometimes be obtained very easily from a graph. For example, the demand for and supply of a given product fluctuate depending on the price the product commands. The economy is in equilibrium when supply and demand are equal. Graphing the supply and demand curves helps to find this equilibrium point as illustrated in the following example.

EXAMPLE 7

A commercial bank finds that when its prime lending rate is 10% it has $10 million to lend and loan applications totaling $15 million. An increase of x percent in the prime rate increases the money supply by $x^2/5$ million dollars and decreases the loan demand to $15 \Big/ \left(1 + \dfrac{x}{10}\right)$ million dollars. Find the equilibrium interest rate at which supply and demand are equal.

SOLUTION

The money supply and loan demand at an interest rate of $(10 + x)$ percent are given, in millions of dollars, by

$$S(x) = 10 + \frac{x^2}{5}$$

and

$$D(x) = \frac{15}{1 + \dfrac{x}{10}}$$

166 Chapter 3 Graphs of Equations

respectively. To solve for x in the equation $S(x) = D(x)$ would be difficult. Thus, we graph the two curves separately; tables of values are included with Figure 3.18.

Figure 3.18

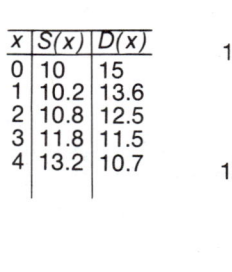
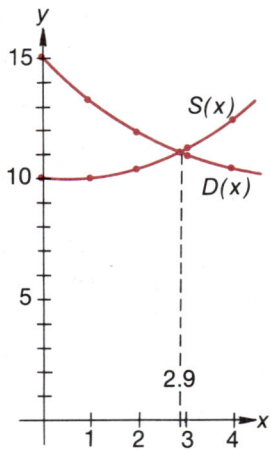

Note that different scales are used on the horizontal and vertical axes in Figure 3.18 in order to enlarge the graph. This enables us to estimate more accurately the equilibrium point x for which $S(x) = D(x)$. In this case, x is approximately 2.9. The bank should set its prime rate at $(10 + 2.9)\% = 12.9\%$. ■

Section 3.2 Exercises

Sketch the graphs of the following relationships. Indicate symmetry about the x-axis, the y-axis, or the origin.

1. $3x + y = 6$
2. $2x - 4y = 8$
3. $3x + y \leq 6$
4. $2x - 4y > 8$
5. $x = 2$
6. $y = 0$
7. $x \geq 2$
8. $y \leq 0$
9. $y = -1$
10. $x = -2$
11. $y < -1$
12. $x > -2$
13. $|y| = 4$
14. $|x| = 5$
15. $|y| \leq 4$
16. $|x| \geq 2$
17. $-1 < x < 2$, $y = 3$
18. $2 \leq y \leq 4$, $x = 1$
19. $0 \leq x \leq 1$, $0 \leq y \leq 2$
20. $1 < x < 3$, $2 \leq y \leq 4$
21. $1 \leq |x| \leq 3$, $1 < y < 2$
22. $1 \leq x \leq 3$, $2 < |y| < 4$
23. $y = x$
24. $x = 2y$
25. $y < x$
26. $x \leq 2y$
27. $y = |x|$
28. $x^2 = y^2$
29. $y \geq |x|$
30. $x^2 < y^2$
31. $y^2 = 4x^2$
32. $y - 4x^2 = 0$
33. $y^2 > 4x^2$
34. $y > 4x^2$
35. $x = y^4$
36. $x = 2y^2$
37. $x \geq y^4$
38. $x \geq 2y^2$
39. $y = -x^2$
40. $y = -2x^2$
41. $y \geq -x^2$
42. $y \geq -2x^2$
43. $y = \sqrt{x}$
44. $x = \sqrt{y}$
45. $y < \sqrt{x}$
46. $x > \sqrt{y}$
47. $x = \sqrt{1 - y}$
48. $y = \sqrt{1 + x}$
49. $x \geq \sqrt{1 - y}$
50. $y \leq \sqrt{1 + x}$
51. $y^2 = |x|$
52. $|x| = 2|y|$
53. $y^2 < |x|$
54. $|x| < 2|y|$
55. $y = |x^3|$
56. $y = x^4$

57. $y \geq |x|^3$
58. $y \leq x^4$
59. $|y| < |x|^3$
60. $|y| \leq x^4$

To sketch the graphs of the following relationships, first replace the inequality symbol with equality and sketch the resulting circle. Then use the distance formula to determine where the points satisfying the inequality are to be plotted.

61. $x^2 + y^2 < 9$
62. $x^2 + y^2 \geq 36$
63. $x^2 + 6x + y^2 - 4y + 9 < 0$
64. $x^2 + 6x \geq 4y - y^2 + 3$
65. $x^2 + y^2 + 30 \leq 10x + 6y$
66. $x^2 + 4x + y^2 > 29 - 8y$
67. $x^2 + y^2 < 12x + 14y + 15$

Calculator Exercises

Use a calculator to approximate the coordinates of a number of points on the graph of the given equation. Then sketch the graph.

68. $x = y^{3/4}$
69. $y = x^{2/5}$
70. $y = \sqrt[5]{(x+1)^2}$
71. $x = \sqrt[3]{(y-1)^2}$
72. $x = \dfrac{y^2}{1 - y^2}$
73. $y = \dfrac{x}{x+1}$
74. $y = \dfrac{x^4 + 2}{x^3}$
75. $x = \dfrac{y^3 - 1}{y^2}$

76. When wheat sells for $3.50 per bushel, there is a demand for 150 million metric tons and a supply of only 50 million metric tons. Increasing the price per bushel by x increases the supply to $\dfrac{(1 + 2x)}{(1 + x)} \times 10^8$ metric tons and *decreases the demand by* $x^2 \times 10^8$ metric tons. Sketch the supply and demand curves and use these curves to find an equilibrium price for wheat.

77. When eggs sell for 75¢ per dozen, there is a demand for 50,000 cases per week and a supply of 80,000 cases per week. If the price decreases x¢ per dozen, the demand increases by $1000x^2$ cases whereas the supply decreases by $5000x$ cases. Sketch the supply and demand curves to find an equilibrium price for eggs.

78. Suppose that a lake covering $10x$ acres can support enough fish to regenerate x^2 million fish each year. However, it also attracts fishermen in sufficient numbers to catch $5x$ million fish each year. Sketch the "regeneration" and "catching" curves in order to determine when the fish population remains steady.

79. Since 1959, the number of components in an integrated circuit (IC) has doubled each year. As a result, a higher percentage of the production must be scrapped as defective. When the increase in scrapping costs exceeds the savings achieved by further miniaturization, it is no longer profitable to develop more complex instruments.

    Suppose that for a certain calculator costing $100 to produce in 1980, the scrapping costs averaged $5 per calculator. If its production costs decrease by 25% each year thereafter, but the scrapping costs increase by 25% each year, use graphs to determine when the *increase* in scrapping costs will exceed the additional savings in production costs.

80. A forest fire starts 50 miles west and 40 miles north of Butler and fire advances at a rate of 2 miles per day in every direction. The rangers direct the local volunteers to dig a fire trench that is to stop the fire after 5 days. Sketch a map indicating the fire trench and the region to be sacrificed to the forest fire. How close does the fire come to Butler?

81. A prisoner escapes from the Placerville jail, but the only means of leaving the area is by foot or horseback. If the prisoner can average at most 10 miles per hour, where should the authorities place roadblocks that are operative 5 hours after the escape?

82. Chicago is located at a latitude of 41°52′28″ North and a longitude of 87°38′22″ West. Dayton's coordinates are 39°45′32″ North and 84°11′43″ West. Here, 1° represents about 69 miles, 60′ = 1°, and 60″ = 1′. The radar at the Chicago airport has a range of 200 miles, whereas Dayton's has a range of only 150 miles.

    a. Find the distance between Chicago and Dayton.
    b. Sketch the area covered by both Dayton's and Chicago's radar units.

168 Chapter 3 Graphs of Equations

83. Sketch the graph consisting of all points equidistant from (2, 4) and (4, 2). Use this distance relationship to find the equation represented by this graph.

84. Sketch the graph consisting of all points whose distance from the x-axis is twice the distance from the y-axis. Use this distance relationship to find the equation represented by this graph.

Section 3.3

Lines and Linear Equations

When a carpenter describes a roof as having a "5–12 pitch," he means that the roof rises 5 inches for each horizontal advance of 12 inches. He then knows that the roof will rise 5 feet for each horizontal advance of 12 feet; it will rise 10 feet above a horizontal advance of 24 feet, and so on.

In the same way, the vertical **rise** corresponding to a given horizontal advance, called the **run**, indicates the steepness of any nonvertical line. Corresponding angles of the three triangles in Figure 3.19 are equal as indicated. Such triangles are said to be **similar**. A theorem in geometry says that corresponding sides of similar triangles are proportional. Thus, in Figure 3.19,

$$\frac{y_2 - y_1}{x_2 - x_1} = \frac{y_2' - y_1'}{x_2' - x_1'} = \frac{y_2'' - y_1''}{x_2'' - x_1''}$$

That is, for a given line, the ratio rise/run is the same regardless of the location at which this computation is made.

Figure 3.19
RISE/RUN IS CONSTANT ON A GIVEN LINE

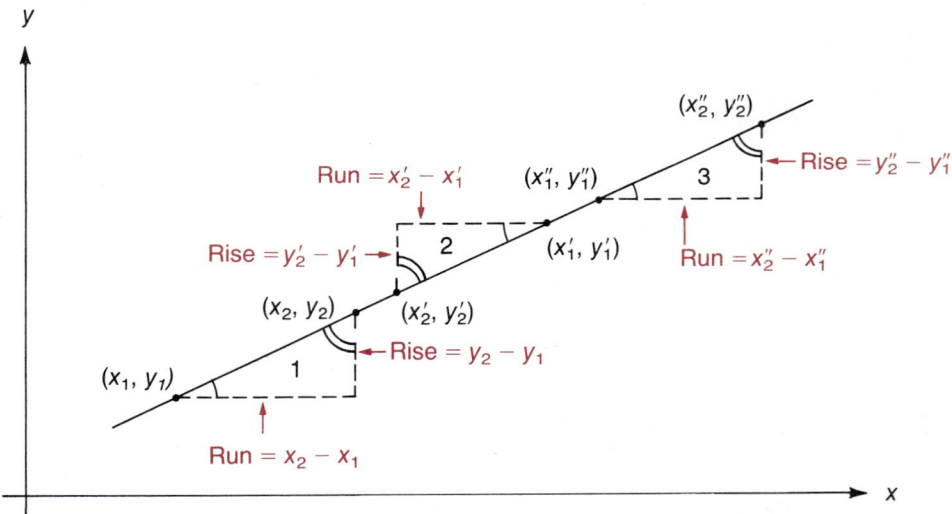

Slope

DEFINITION

The **slope** m of a nonvertical line l is given by

$$m = \frac{\text{rise}}{\text{run}} = \frac{y_2 - y_1}{x_2 - x_1} = \frac{y_1 - y_2}{x_1 - x_2}$$

for any two distinct points (x_1, y_1) and (x_2, y_2) on the line.

In triangle 2 of Figure 3.19, we see that the rise and/or run may be negative. If the line falls as we move to the right, a negative rise corresponds to a positive run and the ratio rise/run = slope is negative.

The slope of a line is a measure of its inclination or steepness. Lines having small positive or negative slope will rise or fall slowly as we move to the right, whereas lines having a larger slope will be very steep. Figure 3.20 shows lines of different slopes.

Figure 3.20
SLOPE IS A MEASURE OF "STEEPNESS"

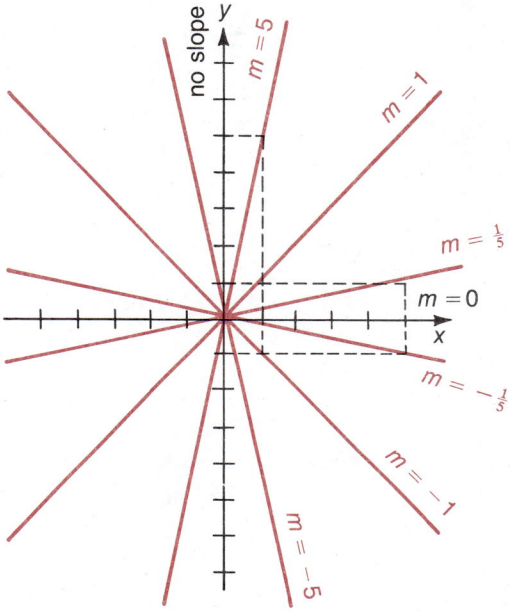

A line having 0 slope is horizontal; it does not rise or fall as we move to the right. Such lines can be described by equations of the form $y =$ constant as in Figure 3.13b of Section 3.2.

Note that slope is not defined for vertical lines. Vertical lines are parallel to the x-axis and thus are represented by equations of the form $x =$ constant. Were we to attempt to compute the slope from the given formula, we would be attempting to divide by 0. For instance, $(3, 2)$ and $(3, 6)$ are on the vertical line $x = 3$. The slope formula yields

$$m = \frac{y_2 - y_1}{x_2 - x_1} = \frac{6 - 2}{3 - 3} = \frac{4}{0}$$

an undefined quantity. Thus vertical lines do not have a slope. This property of *no slope* for vertical lines is in contrast to horizontal lines, which do have a slope, $m = 0$. For example, $(4, 5)$ and $(2, 5)$ are on the horizontal line $y = 5$. The slope formula for these points yields

$$m = \frac{5 - 5}{4 - 2} = \frac{0}{2} = 0.$$

170 Chapter 3 Graphs of Equations

VERTICAL AND HORIZONTAL LINES

1. A horizontal line with b as its y-intercept has equation $y = b$; it has slope $m = 0$.
2. A vertical line with a as its x-intercept has equation $x = a$; it has no slope.

See Figure 3.21.

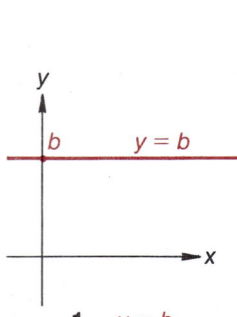
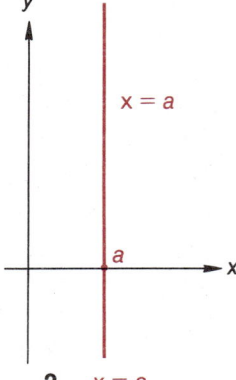

1. $y = b$
2. $x = a$

Figure 3.21

EXAMPLE 1

Sketch the line passing through $(-1, 4)$ and having slope $m = -2$.

SOLUTION

Since the slope $m = -2$ is negative, the vertical rise is negative when the horizontal advance is positive; that is, the line falls rather than rises as we move to the right. An advance of one unit to the right causes a vertical drop of two units. Thus, we sketch the line as in Figure 3.22.

Figure 3.22
THE LINE THROUGH $(-1, 4)$
WITH SLOPE -2

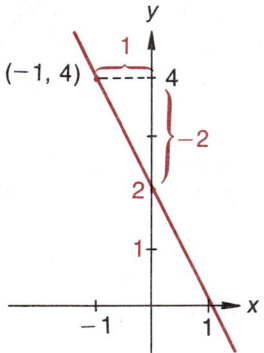

EXAMPLE 2

Find the slope of the line passing through each of the following pairs of points. Sketch the lines.

a. $(2, 3), (4, 5)$
b. $(-1, -2), (-1, 3)$
c. $(-1, 4), (3, 2)$
d. $(-4, -1), (6, -1)$

SOLUTION In each of these we let (x_1, y_1) and (x_2, y_2) be the first and second points listed and use the slope formula

$$m = \frac{y_2 - y_1}{x_2 - x_1}$$

a. $m = \dfrac{5-3}{4-2} = \dfrac{2}{2} = 1$

b. $m = \dfrac{3-(-2)}{(-1)-(-1)} = \dfrac{5}{0}$ is undefined; the line is vertical and has no slope.

c. $m = \dfrac{2-4}{3-(-1)} = \dfrac{-2}{4} = -\dfrac{1}{2}$

d. $m = \dfrac{-1-(-1)}{6-(-4)} = \dfrac{0}{10} = 0$; the line is horizontal.

The lines are sketched in Figure 3.23.

Figure 3.23

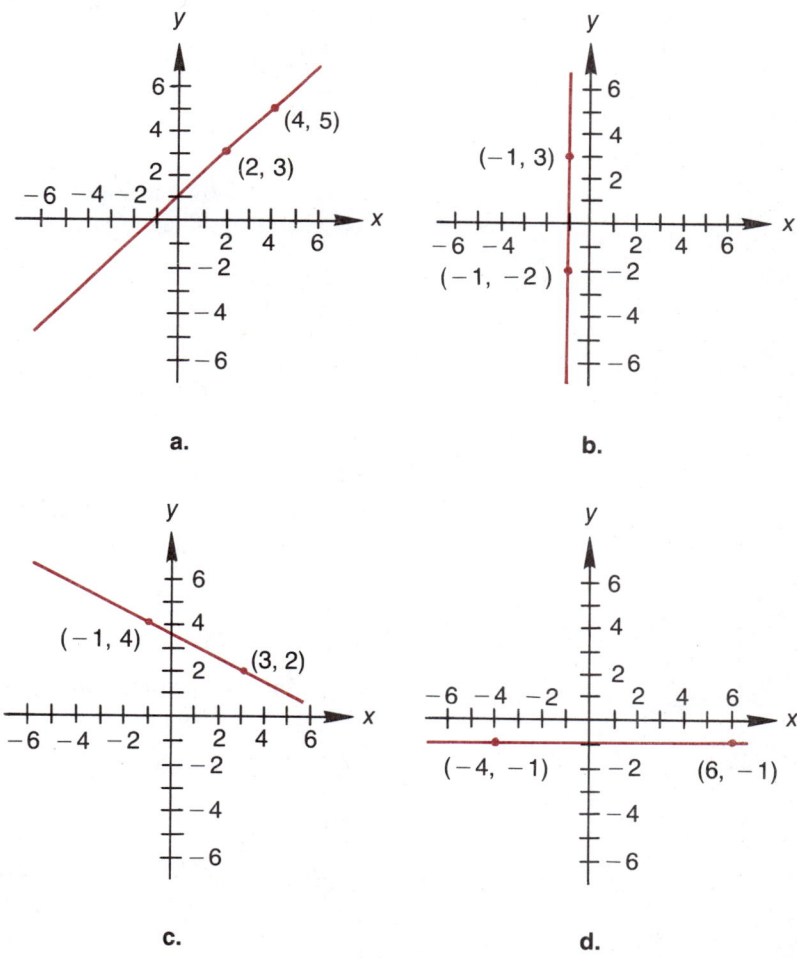

Vertical lines have equations of the form $x = $ constant. There are several forms for the equation of a nonvertical line. Each of these forms involves the slope of the line. In addition to the slope, we must also know at least one point on the line. If two points on the line are known, the slope can be computed from these points.

The Point-Slope Formula

Let (x_0, y_0), (x_1, y_1), and (x_2, y_2) be points on a nonvertical line as in Figure 3.24. The slope of the line, defined as rise/run, is given by

$$m = \frac{y_2 - y_1}{x_2 - x_1}$$

Figure 3.24

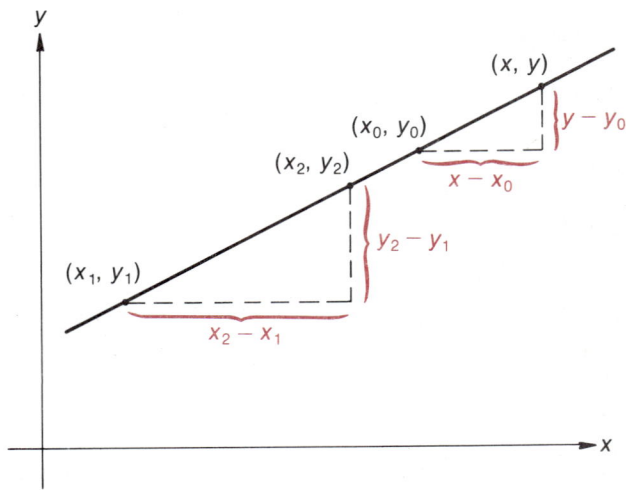

On the other hand, as shown in Figure 3.19, the slope must be the same regardless of where it is calculated. Thus,

$$m = \frac{y - y_0}{x - x_0}$$

for every point $(x, y) \neq (x_0, y_0)$ on this line. Multiplying this last equation by $(x - x_0)$, we see that *every* point on the line—(x_0, y_0) included—satisfies the equation

$$(y - y_0) = m(x - x_0)$$

Any point not on this line will lie at a different inclination to (x_0, y_0) and hence will not satisfy this equation.

POINT-SLOPE FORMULA

The slope of a nonvertical line may be computed as

$$m = \frac{y_2 - y_1}{x_2 - x_1}$$

for any two distinct points (x_1, y_1), (x_2, y_2) on the line.

The equation of the line passing through (x_0, y_0) with slope m is

$$(y - y_0) = m(x - x_0)$$

EXAMPLE 3

Sketch the line passing through $(2, -3)$ and $(-1, 6)$ and write its equation. Find the points at which it intersects the x- and y-axes, respectively.

SOLUTION

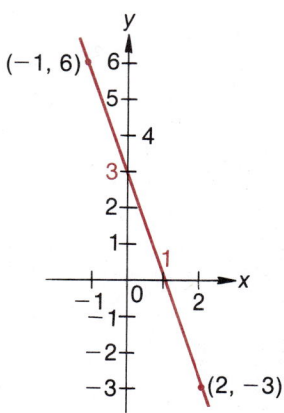

Figure 3.25

The line can be sketched by simply plotting the two given points as in Figure 3.25. Using the two given points, the slope of the line is found to be

$$m = \frac{-3-6}{2-(-1)} = \frac{-9}{3} = -3$$

Then, using the point-slope formula with the point $(2, -3)$ and $m = -3$, the equation of the line is

$$y - (-3) = -3(x - 2)$$
$$y + 3 = -3x + 6$$
$$y = -3x + 3$$

This line crosses the x-axis when $y = 0$:

$$0 = -3x + 3$$
$$x = 1$$

The x-intercept occurs at $x = 1$. The line crosses the y-axis when $x = 0$:

$$y = -3 \cdot 0 + 3$$
$$y = 3$$

The y-intercept occurs at $y = 3$. ∎

The Slope-Intercept Formula

If $y = b$ is the y-intercept of a given line, then the point $(0, b)$ is on the line (Figure 3.26).

Figure 3.26

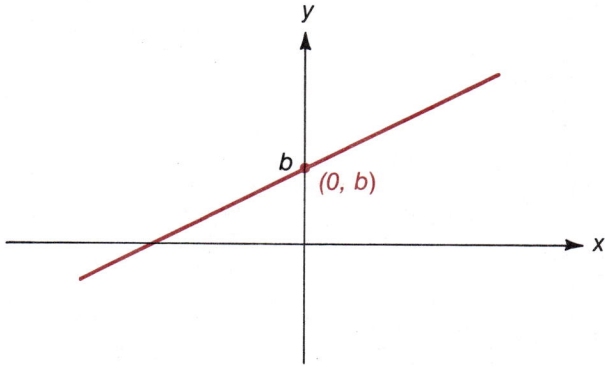

The point-slope formula then gives

$$(y - b) = m(x - 0) = mx$$
$$y = mx + b$$

Here m represents the slope of the line and b is its y-intercept. (Observe also that $y = b$ when we substitute $x = 0$ into this equation.)

SLOPE-INTERCEPT FORMULA

The equation of a line intersecting the y-axis at $y = b$ with slope m is

$$y = mx + b$$

---

174 Chapter 3 *Graphs of Equations*

In particular, when $b = 0$, the lines $y = mx$ pass through the origin. The lines from Figure 3.20 have the equations shown in Figure 3.27.

Figure 3.27

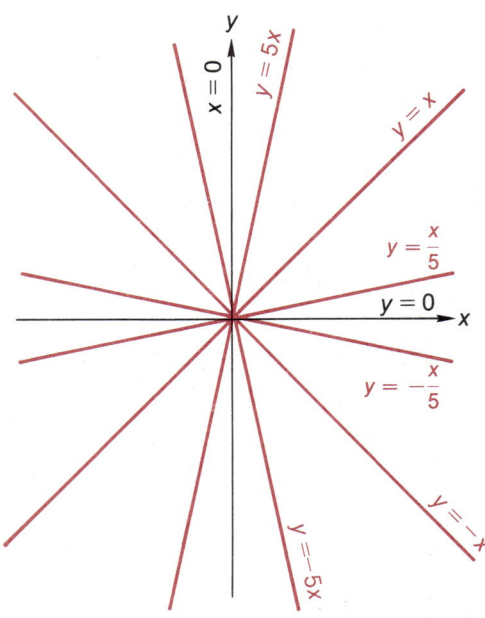

EXAMPLE 4

Sketch the line with slope $m = 3$ and y-intercept $b = -6$. Write its equation in slope-intercept form. Find its x-intercept.

SOLUTION

The slope-intercept form of the equation for this line is

$$y = 3x - 6$$

Its x-intercept occurs when $y = 0$:

$$0 = 3x - 6$$
$$x = 2$$

Consequently, the points $(0, -6)$ and $(2, 0)$ are on the line. The line can be sketched by plotting these two points and drawing a line through them. See Figure 3.28.

Since slope is defined as rise/run and $m = 3$, this means that rise = 3 when run = 1; that is, a vertical rise of 3 units corresponds to a horizontal run of 1 unit. Thus, we could also have moved 1 unit to the right and 3 units up from the point $(0, -6)$ to plot the point $(1, -3)$; these two points also determine the line.

For still another approach to graphing the line, observe that the y-coordinates on the line

$$y = 3x - 6$$

are exactly 6 units below the y-coordinates on the graph of

$$y = 3x$$

Thus we could sketch the line $y = 3x$ and then lower this graph 6 units. See Figure 3.28.

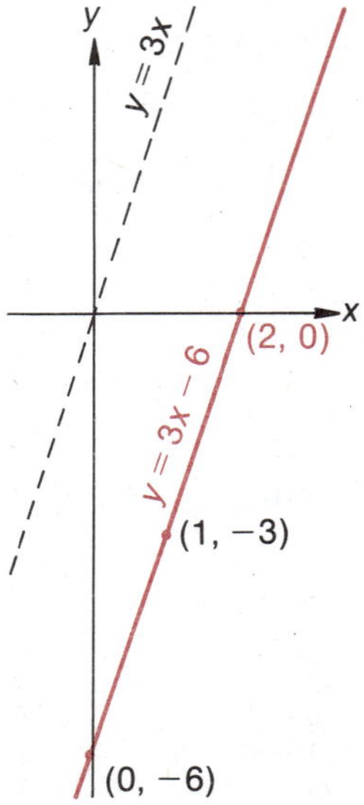

Figure 3.28

The General Equation of a Line

We have seen that every line has an equation as indicated here:

$y = mx + b$ if the line is not vertical

$x = a$ if the line is vertical

Each of these equations can be rewritten in the general form

$Ax + By + C = 0$ A, B not both 0

Conversely, every equation of this form represents a line. For such an equation can be rewritten in the form

$$y = -\frac{A}{B}x - \frac{C}{B} \quad \text{if } B \neq 0$$

or

$$x = -\frac{C}{A} \quad \text{if } B = 0$$

GENERAL EQUATION OF A LINE

The general form of the equation for a line is

$$Ax + By + C = 0 \quad A, B \text{ not both } 0$$

Such an equation is called a **linear equation**; if neither A nor B is zero in a linear equation, the equation is said to describe a **linear relationship** between x and y.

EXAMPLE 5

Sketch the line given by the following equation; find its slope and intercepts.

$$6x - 2y + 8 = 0$$

SOLUTION

Since the y-coefficient is not 0, we rewrite the equation as

$$2y = 6x + 8$$
$$y = 3x + 4$$

The line has slope $m = 3$ and y-intercept $b = 4$. Its x-intercept occurs when $y = 0$:

$$0 = 3x + 4$$
$$x = -\frac{4}{3}$$

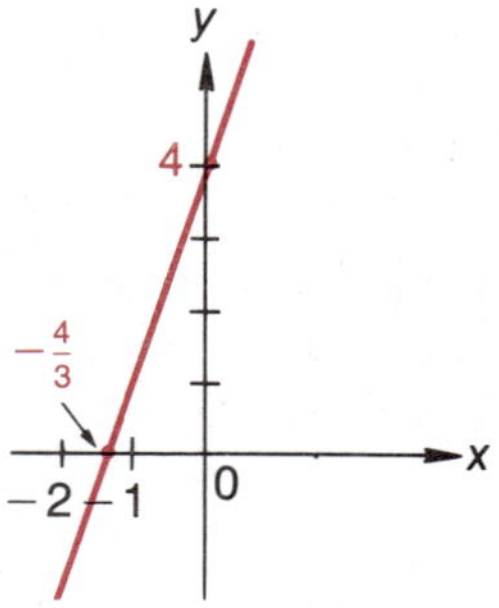

Figure 3.29 shows the line.

Figure 3.29

EXAMPLE 6

On the Fahrenheit temperature scale, water freezes at 32°F and boils at 212°F at sea level. On the Celsius scale, the freezing and boiling points are 0°C and 100°C, respectively. Assuming a linear relationship between the two scales, express Fahrenheit temperature in terms of Celsius temperature and vice versa. When do the two temperature scales have the same numerical value?

SOLUTION   If we assume a linear relationship, we have

$$F = aC + b$$

where F and C represent Fahrenheit and Celsius temperatures, respectively, and a and b are constants to be determined. At the freezing point $F = 32$ and $C = 0$.

$$32 = a \cdot 0 + b$$

Thus, $b = 32$ and $F = aC + 32$. At the boiling point $F = 212$ and $C = 100$:

$$212 = a \cdot 100 + 32$$
$$180 = a \cdot 100$$
$$a = \frac{180}{100} = \frac{9}{5}$$

Thus,

$$F = \frac{9}{5}C + 32 \tag{1}$$

Writing Celsius temperatures in terms of Fahrenheit, we have

$$\frac{9}{5}C = F - 32$$

$$C = \frac{5}{9}(F - 32)$$

The two scales have the same value when $F = C$. Substitute $F = C$ into (1):

$$C = \frac{9}{5}C + 32$$

$$-\frac{4}{5}C = 32$$

$$C = -\frac{5}{4} \cdot 32$$

$$= -40°$$

Thus, $-40°F = -40°C$. Figure 3.30 shows the graph. ■

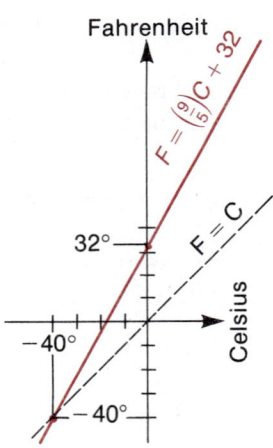

Figure 3.30

Parallel and Perpendicular Lines

Note that the line $y = mx + b$ can be obtained by simply raising or lowering the line $y = mx$. Several lines $y = mx + b$ are sketched in Figure 3.31 for a given m and various values of b. In particular, we note that nonvertical lines are parallel if and only if they have the same slope.

Figure 3.31

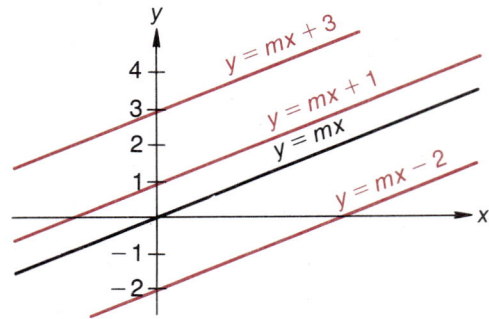

It is also possible to state a condition that determines whether two lines are perpendicular (see Exercise 120).

PARALLEL OR PERPENDICULAR LINES

Let l_1 and l_2 be two nonvertical lines having equations

$$y = m_1 x + b_1$$
$$y = m_2 x + b_2$$

respectively.

a. The lines are **parallel** if and only if

$$m_1 = m_2.$$

b. The lines are **perpendicular** if and only if

$$m_1 m_2 = -1$$

$\left(\text{equivalently, } m_2 = \dfrac{-1}{m_1}\right)$ or one line is horizontal and the other is vertical.

EXAMPLE 7

Write the equations of the lines through (1, 2) that are

a. Parallel
b. Perpendicular

to the line $3x + 9y - 18 = 0$.

SOLUTION

Solving for y in this equation, we obtain

$$9y = -3x + 18$$
$$y = -\frac{1}{3}x + 2$$

The slope of this line is $m_1 = -\dfrac{1}{3}$.

a. The slope of any line parallel to this line is also $m_1 = -\dfrac{1}{3}$; thus, the equation of the line through (1, 2) parallel to this line is

$$(y - 2) = -\frac{1}{3}(x - 1) \qquad \text{(Use the point-slope formula)}$$
$$3(y - 2) = -(x - 1)$$
$$3y - 6 = -x + 1$$
$$x + 3y - 7 = 0$$

b. The slope of any line perpendicular to this line has slope $m_2 = -1/m_1 = 3$. Hence, the perpendicular line has equation

$$(y - 2) = 3(x - 1)$$
$$y = 2 + 3x - 3$$
$$= 3x - 1$$
$$3x - y - 1 = 0 \qquad \blacksquare$$

The following table summarizes the forms for linear equations and shows their respective graphs.

LINES AND THEIR EQUATIONS

LINES	FORMULA	SKETCH
General form	$Ax + By + C = 0$, A, B not both 0	
Slope-intercept form	$y = mx + b$	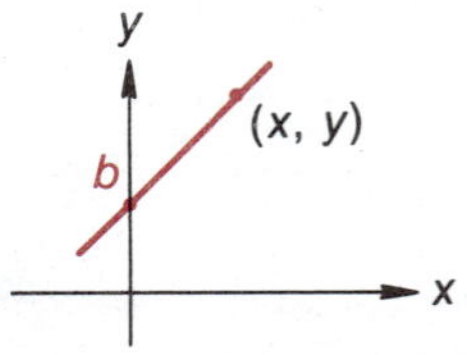
Point-slope form	$(y - y_1) = m(x - x_1)$	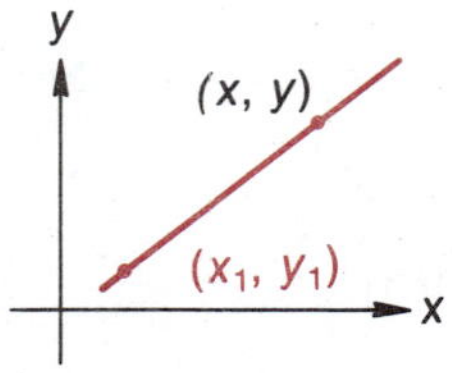
Parallel lines	$\left.\begin{array}{l} y = m_1 x + b_1 \\ y = m_2 x + b_2 \end{array}\right\} m_1 = m_2$	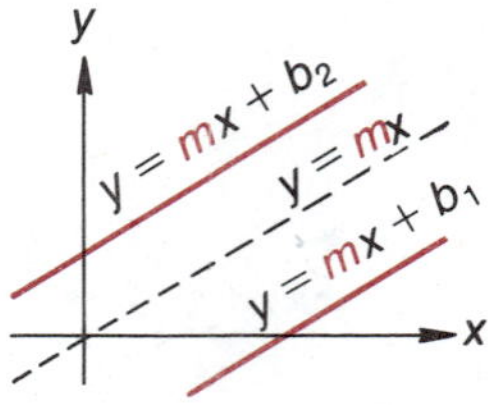
Perpendicular lines	$\left.\begin{array}{l} y = m_1 x + b_1 \\ y = m_2 x + b_2 \end{array}\right\} m_1 m_2 = -1$ or $\left.\begin{array}{l} x = a \\ y = b \end{array}\right\}$	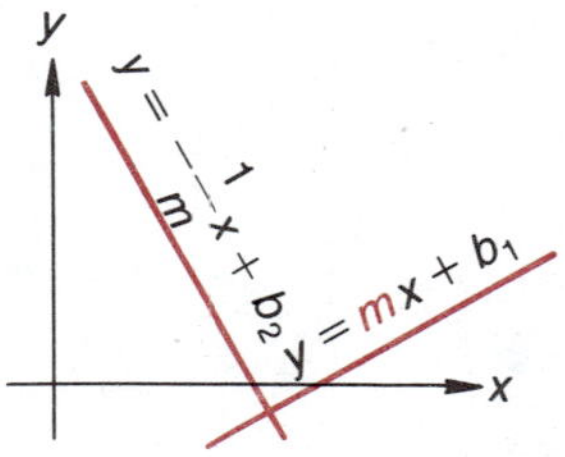

Section 3.3 Exercises

Sketch the line that passes through the given point and has the indicated slope.

1. $(1, 2)$, $m = 2$
2. $(2, 4)$, $m = 3$
3. $(-1, 3)$, $m = -3$
4. $(-3, 2)$, $m = -2$
5. $(-2, -1)$, $m = \frac{3}{2}$
6. $(-1, -2)$, $m = \frac{1}{2}$
7. $(2, -3)$, $m = -\frac{1}{2}$
8. $(3, -2)$, $m = -\frac{3}{2}$

For the line passing through the two points given in each exercise:
  a. *Find the slope.*
  b. *Write the equation in general form.*
  c. *Find the x- and y-intercepts.*

9. $(-4, 1), (-2, 3)$
10. $(1, 5), (4, 2)$
11. $(-3, 6), (-1, -2)$
12. $(2, -3), (2, 4)$
13. $(-1, 2), (-1, -4)$
14. $(1, 2), (3, 6)$
15. $(4, -2), (-1, 8)$
16. $(3, 4), (8, 4)$
17. $(-6, 1), (4, 1)$
18. $(-6, -4), (-3, -2)$
19. $(8, 2), (0, -2)$
20. $(5, -2), (-2, -2)$
21. $(3, 1), (3, -5)$
22. $(0, 9), (-5, 4)$
23. $(3, 7), (1, -7)$
24. $(2, -8), (3, 0)$
25. $(-4, -1), (6, -1)$
26. $(-2, -2), (-2, 7)$

Write the general equation of the line that passes through the given point with the indicated slope m. Find the x- and y-intercepts.

27. $(3, 5)$, $m = \frac{1}{3}$
28. $(-2, 8)$, $m = -10$
29. $(5, -3)$, $m = -1$
30. $(-1, 3)$, $m = -2$
31. $(-3, -1)$, $m = 0$
32. $(1, -2)$, $m = 0$
33. $(-3, -1)$, no slope
34. $(1, -2)$, no slope
35. $(4, -3)$, $m = -2$
36. $(3, 2)$, $m = -\frac{1}{3}$
37. $(-2, -5)$, $m = -\frac{1}{2}$
38. $(10, -5)$, $m = \frac{1}{5}$

Find the equation of the line with slope m and y-intercept b. Write the equation in general form.

39. $m = 3$, $b = 6$
40. $m = -1$, $b = 8$
41. $m = 2$, $b = -12$
42. $m = 6$, $b = 2$
43. $m = -10$, $b = -4$
44. $m = \frac{4}{3}$, $b = \frac{2}{3}$
45. $m = -5$, $b = 4$
46. $m = 5$, $b = -20$

Graph the following lines. Find their slopes and intercepts.

47. $x - 2y - 3 = 0$
48. $x + y = 2$
49. $3x - 2y + 12 = 0$
50. $-6x + 4y - 1 = 0$
51. $2x + y - 6 = 0$
52. $3x = 2$
53. $-4x - 8y + 16 = 0$
54. $2x - 3y + 6 = 0$
55. $16y = 32$
56. $4x + 7y + 28 = 0$
57. $3x - 6y + 6 = 0$
58. $10x = 100$

Determine whether each of the following pairs of lines are parallel, perpendicular, or neither.

59. $7x - 3y + 10 = 0$
    $6x + 14y + 5 = 0$
60. $2x - 3y + 4 = 0$
    $-16x + 24y - 2 = 0$
61. $3x - y - 1 = 0$
    $x - 3y - 1 = 0$
62. $x = 10$
    $y = -10$
63. $x - y = 6$
    $y - x = 3$
64. $x = 16$
    $x = -3$
65. $-2x - 2y + 3 = 0$
    $5x - 5y - 3 = 0$
66. $-10x - 5y + 3 = 0$
    $4x + 2y - 10 = 0$
67. $x - 9y + 6 = 0$
    $9x - y - 6 = 0$
68. $x - 9y - 6 = 0$
    $9x + y + 6 = 0$

Determine whether the given points are **collinear** *(lie on the same line).*

69. $(2, 0), (-6, -6), (6, 3)$
70. $(0, -2), (3, 7), (-2, -8)$
71. $(1, 1), (3, 7), (-2, -6)$
72. $(1, -1), (-2, 5), (5, -9)$
73. $(3, 8), (7, 10), (-9, 2)$
74. $(2, -3), (-2, 5), (5, -5)$

Write the equation of the line through the given point that is parallel to the given line and also of the line that is perpendicular to the given line.

75. $(-1, 3)$, $x = 5$
76. $(2, 0)$, $4x - 2y - 6 = 0$

77. $(-1, -2)$, $x + 3y - 2 = 0$

78. $(4, -1)$, $2x - 3y + 4 = 0$

79. $(0, -3)$, $2x + y - 5 = 0$

80. $(2, -11)$, $4x + 7y + 28 = 0$

81. $(6, 2)$, $y = 7$

82. $(100, -1)$, $3x - 9y + 6 = 0$

83. $(-1, 30)$, $x + y = 2$

84. $(1, 2)$, $x - y = 6$

85–102. Write the equation of the perpendicular bisector of the segment joining each pair of points given in Exercises 9–26, respectively.

103. Write the equation of the line tangent to the circle $x^2 + y^2 = 25$ at the point $(-4, 3)$. (A line is tangent to a circle if it is perpendicular to a line through the center of the circle at the point at which this line meets the circle.)

104. Write the equation of the line tangent to the circle $x^2 + y^2 = 169$ at the point $(5, -12)$. (See Exercise 103.)

105. Show that $x/a + y/b = 1$, $a \neq 0$, $b \neq 0$, is the equation of the line intersecting the x-axis at a and the y-axis at b. This equation is known as the **two-intercept form** of the equation of a line. (*Hint:* Rewrite the equation in general form to observe that it represents a line. Then find its x- and y-intercepts.)

Use Exercise 105 to write the general equation of the line with indicated x- and y-intercepts a and b, respectively. Find the slope of each line.

106. $a = 1$, $b = 2$
107. $a = 3$, $b = -5$
108. $a = 2$, $b = -4$
109. $a = -3$, $b = -2$
110. $a = 0$, $b = 2$
111. $a = 2$, $b = 1$

112. The Ahlers' house needs a new roof. The cost will depend on the pitch of the roof; a steep roof requires more equipment to keep the workers from falling off. If their house has dimensions as shown in the figure, what should they tell the contractor when he requests the pitch of their roof?

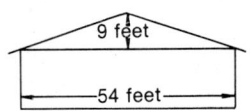

113. For tax purposes, certain property such as real estate, machinery, and equipment can be depreciated in several ways. One such method of depreciation is known as the straight-line method. In this method, the value of the property—for tax purposes—is reduced or depreciated by the same amount each year until the property has been depreciated to $0. If a $400,000 investment property is depreciated over 20 years by using the straight-line method, plot the remaining value of the property during each of these 20 years.

114. An automobile salesperson receives a monthly salary of $400 plus $100 for each car sold. Write an equation and sketch a graph that indicates her monthly income in terms of the number of cars she sells.

115. A car-rental agency charges $50 per day for a compact car and an additional 40¢ per mile driven. Write an equation and sketch a graph that indicates the total costs for a one-day rental in terms of the miles driven.

116. The RJ Calculator Company developed a new calculator at a cost of $50,000. In mass production it will cost $10 to produce each calculator. Environmental regulations then require an additional investment of $20,000 to bring the production line up to federal standards.
    a. Graph the cost of producing x calculators with and without the additional $20,000 investment.
    b. With the added investment, how many more units must the company sell before it begins to realize a profit if each unit sells for $15?

117. Each night that a downtown department store is open, the owners must pay $10,000 in additional salary and utility expenses. The markup on their merchandise equals the price they paid.
    a. Write an equation and sketch a graph that gives profit in terms of total sales for the night.
    b. How much business must they have each night to meet expenses?
    c. If the owners do not feel they can justify remaining open for less than $5000 profit per night, how much business must they have in order to justify the nighttime hours?

118. If (x_1, y_1), (x_2, y_2) are any two points on a nonvertical line, establish the **two-point formula** for the equation of the line:

$$\frac{y - y_1}{x - x_1} = \frac{y_2 - y_1}{x_2 - x_1}$$

119. If A and B are both 0 in the linear equation $Ax + By + C = 0$, show that the equation cannot represent a line. (*Hint:* Show that the set of points satisfying the equation is either empty or the entire plane depending on the value of C.)

120. The formula $m_1 m_2 = -1$ for perpendicular lines is established in this exercise. Let l_1 and l_2 be nonparallel lines neither of which is horizontal or vertical. These lines have nonzero slopes m_1 and m_2, respectively, and intersect at some point $P_0 = (x_0, y_0)$.

   a. Show that the line $x = x_0 + 1$ meets l_1 and l_2 at points

   $$P_1 = (x_1, y_1) = (x_0 + 1, y_0 + m_1)$$
   $$P_2 = (x_2, y_2) = (x_0 + 1, y_0 + m_2)$$

   respectively, as illustrated in the figure.

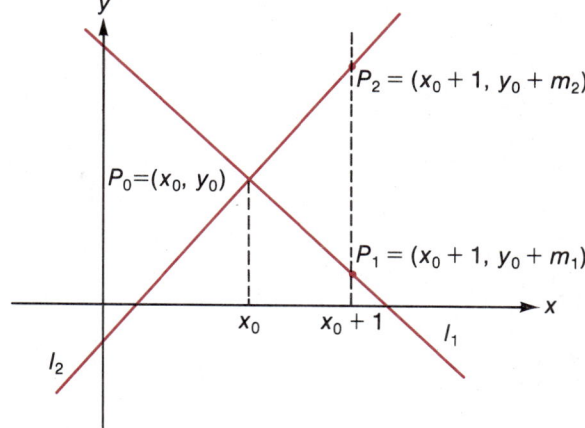

   b. Observe that the following statements are equivalent.
   1. l_1 is perpendicular to l_2
   2. $\Delta P_0 P_1 P_2$ is a right triangle with a right angle at P_0
   3. $[d(P_1, P_2)]^2 = [d(P_1, P_0)]^2 + [d(P_0, P_2)]^2$ (the Pythagorean formula).

   c. Expand statement (3) in terms of the coordinates of P_1, P_2, and P_0 and show that it is equivalent to $m_1 m_2 = -1$.

Section 3.4

Parabolas and Quadratic Equations

The U-shaped curve called a parabola was obtained in Section 3.2 as the graph of the relation $y = x^2$ (Example 5(a)). This parabola opens upward and is symmetric about the y-axis. Similarly, the graph of $x = y^2$ is a parabola that opens to the right rather than upward. It is symmetric about the x-axis. These parabolas are sketched in Figure 3.32.

Figure 3.32

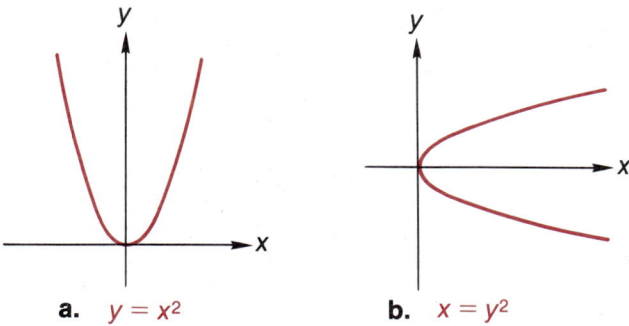

a. $y = x^2$ b. $x = y^2$

Parabolas can be used to describe the behavior of objects moving under the influence of gravity; a projectile such as a bullet or flare fired from a gun traces out a parabolic trajectory. Parabolas are also used in many light-focusing devices such as automobile headlights, beacons, and telescopes, as explained further on page 204.

By completing the square, any equation of the form

$$y = ax^2 + bx + c, \quad a \neq 0$$

or

$$x = Ay^2 + By + C, \quad A \neq 0$$

can be converted into the form

$$y - y_0 = a(x - x_0)^2 \quad \text{or} \quad x - x_0 = A(y - y_0)^2$$

respectively. If we let $x' = x - x_0$ and $y' = y - y_0$, these equations become, respectively,

$$y' = a(x')^2 \quad \text{or} \quad x' = A(y')^2$$

As in Figure 3.32, these represent parabolas in an x'-y' coordinate system. See Figure 3.33.

Figure 3.33

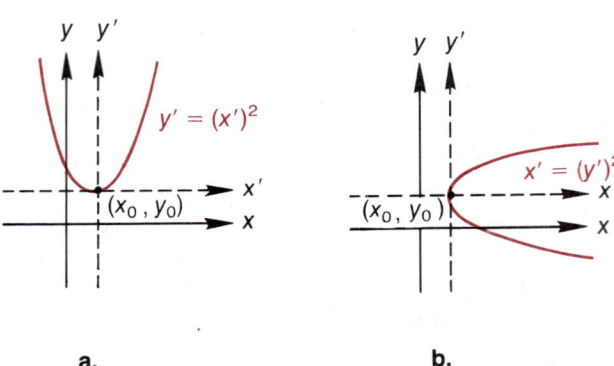

a.

b.

The parabolas will be symmetric about the y'-axis ($x' = 0$ or $x = x_0$) or x'-axis ($y' = 0$ or $y = y_0$), respectively. This **line of symmetry** is called the **axis** of the parabola.

For $a > 0$,

$$y - y_0 = a(x - x_0)^2 \geq 0$$
$$y \geq y_0 \qquad\qquad (y = y_0 \text{ when } x = x_0)$$

Thus the parabola "turns around" at the point (x_0, y_0); this point is called the **vertex** of the parabola.

For $A > 0$,

$$x - x_0 = A(y - y_0)^2 \geq 0$$
$$x \geq x_0 \qquad\qquad (x = x_0 \text{ when } y = y_0)$$

The vertex is again (x_0, y_0).

The cases $a < 0$ and $A < 0$ are similar, but then $y \leq y_0$ and $x \leq x_0$, respectively. Thus, (x_0, y_0) is the vertex in each case.

Section 3.4 Parabolas and Quadratic Equations   183

GRAPHING QUADRATIC EQUATIONS

The graph of a quadratic equation
$$y = ax^2 + bx + c, \quad a \neq 0 \quad \text{or} \quad x = Ay^2 + By + C, \quad A \neq 0$$
is a parabola. To sketch this graph:

1. Complete the square and write the equation in the alternate form
$$y - y_0 = a(x - x_0)^2 \quad \text{or} \quad x - x_0 = A(y - y_0)^2$$
2. The vertex of the parabola is at (x_0, y_0).
3. The axis or line of symmetry is
$$x = x_0 \quad \text{or} \quad y = y_0$$
4. The parabola opens

   upward if $a > 0$ to the right if $A > 0$

   downward if $a < 0$ to the left if $A < 0$

5. Plot additional points (a few, not many) to determine the shape of the parabola.
   a. Large (positive or negative) values of a or A yield a narrow parabola.
   b. Small values of a or A yield a wide parabola.
6. Determine the y-intercept(s) by setting $x = 0$ and the x-intercept(s) by setting $y = 0$.

See Figure 3.34.

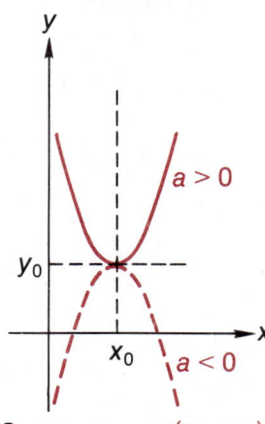
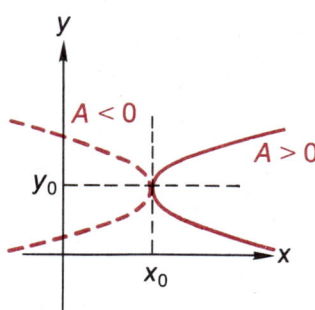

a. $y - y_0 = a(x - x_0)^2$ b. $x - x_0 = A(y - y_0)^2$

Figure 3.34

EXAMPLE 1

Sketch the graph of the equation
$$y = 2x^2 - 4x + 1$$
Find the vertex, the line of symmetry, and the x- and y-intercepts of the graph.

SOLUTION

We can write
$$y = 2(x^2 - 2x \quad) + 1$$

and then complete the square to obtain
$$y = 2(x^2 - 2x + 1) + 1 - 2 \cdot 1$$
$$y = 2(x - 1)^2 - 1$$
$$y - (-1) = 2(x - 1)^2$$

This matches the form $y - y_0 = a(x - x_0)^2$ with $a = 2$, $x_0 = 1$, $y_0 = -1$. Thus the graph is a parabola that

    i. Opens upward $(a = 2 > 0)$
    ii. Has vertex $(1, -1)$ $[=(x_0, y_0)]$
    iii. Has as its axis the line $x = 1$ $(x_0 = 1)$

To determine the shape of the parabola, we plot several points and require symmetry about the line $x = 1$. For $x = 0$, we have
$$y = 2 \cdot 0^2 - 4 \cdot 0 + 1 = 1$$

(Thus, by symmetry, $y = 1$ also when $x = 2$.) For $x = 3$, we have
$$y = 2 \cdot 3^2 - 4 \cdot 3 + 1 = 7$$

(By symmetry, $y = 7$ also when $x = -1$.)

The y-intercept occurs when $x = 0$; this was already found to be $y = 1$. The x-intercepts occur when $y = 0$:

$$2x^2 - 4x + 1 = 0$$

$$x = \frac{-(-4) \pm \sqrt{16 - 4 \cdot 2 \cdot 1}}{2 \cdot 2} \qquad \text{(Quadratic formula)}$$

$$= \frac{4 \pm \sqrt{8}}{4} = \frac{4 \pm 2\sqrt{2}}{4}$$

$$= 1 \pm \frac{\sqrt{2}}{2}$$

The graph is sketched in Figure 3.35.

Figure 3.35

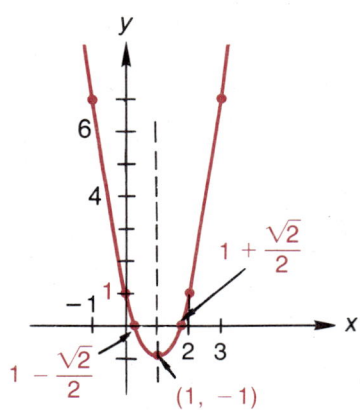

$$y = 2x^2 - 4x + 1 = 2(x - 1)^2 - 1$$

EXAMPLE 2   Graph the parabola represented by the following equation. Find its vertex, line of symmetry, and x- and y-intercepts.
$$x = -2y^2 - 8y - 9$$

SOLUTION

$$x = -2(y^2 + 4y \quad) - 9$$
$$= -2(y^2 + 4y + 4) - 9 + 8$$
$$= -2(y + 2)^2 - 1$$
$$x - (-1) = -2(y - [-2])^2$$

This matches the form $x - x_0 = A(y - y_0)^2$ with $A = -2$, $x_0 = -1$, $y_0 = -2$. The graph is a parabola that

    i. Opens to the left $(A = -2 < 0)$
    ii. Has vertex $(-1, -2)$ $[=(x_0, y_0)]$
    iii. Has as its axis the line $y = -2$ $(y_0 = -2)$

To determine the shape of the parabola, plot several additional points and require symmetry about the line $y = -2$.

    If $y = 0$,   then $x = -2 \cdot 0^2 - 8 \cdot 0 - 9 \quad = -9$

(By symmetry, $x = -9$ also when $y = -4$.)

    If $y = -1$,   then $x = -2(-1)^2 - 8(-1) - 9 = -3$

(By symmetry, $x = -3$ also when $y = -3$.) The graph is sketched in Figure 3.36.

Figure 3.36

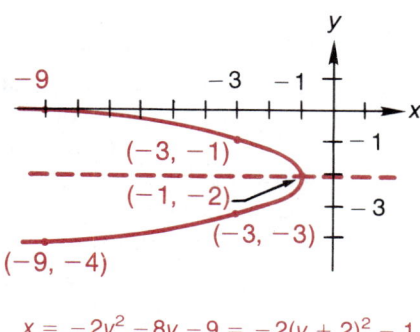

$$x = -2y^2 - 8y - 9 = -2(y + 2)^2 - 1$$

Since the parabola opens to the left and the vertex is at $(-1, -2)$, there can be no y-intercept. The x-intercept occurs when $y = 0$: $x = -9$. ∎

If we set $x = 0$ and use the quadratic formula to solve for y in Example 2, we find

$$-2y^2 - 8y - 9 = 0$$
$$y = \frac{-(-8) \pm \sqrt{(-8)^2 - 4(-2)(-9)}}{2(-2)}$$
$$= \frac{8 \pm \sqrt{64 - 72}}{-4}$$
$$= \frac{8 \pm \sqrt{-8}}{-4}$$
$$= \frac{8 \pm 2\sqrt{2}i}{-4}$$

This indicates that $x = 0$ for no real y values. There are no y-intercepts, confirming what we observed in Figure 3.36.

186   Chapter 3   Graphs of Equations

EXAMPLE 3   If the altitude of a toy rocket t seconds after launching is given by $h = 320t - 16t^2$ feet, find its maximum altitude. When does the rocket attain this maximum altitude? When does it hit the ground?

SOLUTION   If we label the horizontal and vertical axes as t and h, respectively, then the graph of the equation $h = 320t - 16t^2$ is a parabola opening downward ($a = -16 < 0$). We can answer the first two questions by finding the vertex of the parabola. Since

$$h = -16t^2 + 320t$$
$$= -16(t^2 - 20t\quad)$$
$$= -16(t^2 - 20t + 100) + 1600$$
$$= -16(t - 10)^2 + 1600$$
$$h - 1600 = -16(t - 10)^2$$

we see that the vertex of the parabola is the point (10, 1600). The maximum altitude of 1600 feet is attained 10 seconds after launching, as shown in Figure 3.37. Since $h = 0$ when $t = 0$ and the parabola is symmetric about the line $t = 10$, we see that the rocket hits the ground 20 seconds after launching. ∎

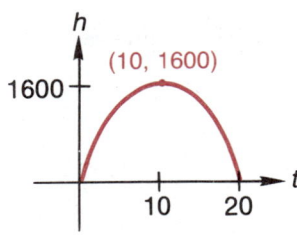

Figure 3.37

EXAMPLE 4   The Solar Oil Company has 1.6 million gallons of gasoline available for a holiday weekend. The usual price is $1.30 per gallon. For each 10,000 gallons that is held off the market, the selling price of the remainder can be increased by 1¢ per gallon. How much gasoline should the company withhold from the market in order to maximize its revenues?

SOLUTION   Let

$$x = \text{number of 10,000 gallon units withheld}$$
$$1{,}600{,}000 - 10{,}000x = \text{number of gallons sold}$$
$$130 + x = \text{cost per gallon in cents}$$

The revenue received is then

$$y = (1{,}600{,}000 - 10{,}000x)(130 + x)$$
$$= 10{,}000(160 - x)(130 + x)$$
$$= 10{,}000(20{,}800 + 30x - x^2)$$
$$= -10{,}000[(x^2 - 30x\quad) - 20{,}800]$$
$$= -10{,}000[(x^2 - 30x + 225) - 20{,}800 - 225]$$
$$= -10{,}000[(x - 15)^2 - 21{,}025]$$
$$= -10{,}000(x - 15)^2 + 2.1025[10^8]$$
$$y - 2.1025[10^8] = -10{,}000(x - 15)^2$$

The graph of this equation is a parabola opening downward ($a = -10{,}000 < 0$) with vertex at $(15, 2.1025(10^8))$, as shown in Figure 3.38. The maximum attainable revenue is $2,102,500. It is obtained by holding back 15 units of 10,000 gallons or 150,000 gallons. The remainder is then sold at $1.45 per gallon. ∎

Figure 3.38

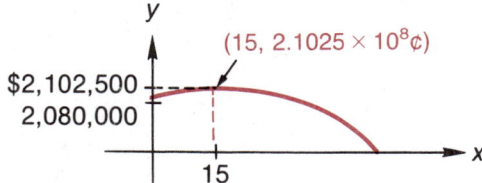

Quadratic Inequalities

Parabolas can also be used to solve quadratic inequalities in one variable. In Section 2.7, solutions to certain quadratic inequalities were found by factoring the quadratic expression and then analyzing the signs of its factors in various regions. The technique introduced here involves only finding the roots of the quadratic equation and looking at the sign of the x^2 coefficient.

SOLVING QUADRATIC INEQUALITIES BY GRAPHING

To solve a quadratic inequality

$$ax^2 + bx + c < 0, \quad a \neq 0$$

make a rough sketch of the parabola

$$y = ax^2 + bx + c$$

1. Find the roots of the quadratic equation

$$ax^2 + bx + c = 0$$

 The parabola crosses the x-axis at these points.
2. Determine from the sign of a whether the parabola opens upward or downward.
3. Roughly sketch the parabola; there is no need even to find the vertex.
4. Determine from this graph the region or regions in which the inequality is satisfied

This technique applies to $\leq, >, \geq$ inequalities also.

EXAMPLE 5 Solve the inequality $50x \leq x^2 + 600$.

SOLUTION First we rewrite the inequality as follows:

$$50x \leq x^2 + 600$$
$$-x^2 + 50x - 600 \leq 0$$

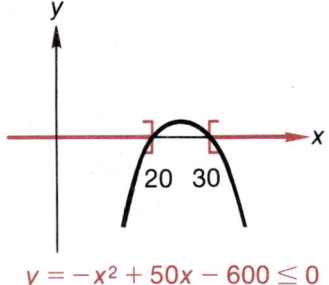

$y = -x^2 + 50x - 600 \leq 0$

Figure 3.39

The quadratic formula gives the roots of this quadratic as

$$x = \frac{-50 \pm \sqrt{2500 - 2400}}{-2}$$

$$= \frac{-50 \pm 10}{-2}$$

$$= 20, 30$$

Furthermore, the graph of $y = -x^2 + 50x - 600$ opens downward. Without even finding the vertex of this parabola, we note that $y \leq 0$ when $x \leq 20$ and when $x \geq 30$ (Figure 3.39). The solution set is

$$(-\infty, 20] \cup [30, \infty). \qquad \blacksquare$$

Section 3.4 Exercises

Graph each of the following parabolas. In each, find the vertex, the line of symmetry, and the intercepts.

1. $y = 2x^2$
2. $x = 3y^2$
3. $x = -3y^2$
4. $y = -2x^2$
5. $y = x^2 + 2$
6. $x = y^2 + 3$
7. $x = y^2 - 2$
8. $y = x^2 - 3$
9. $y = 1 - x^2$
10. $y = 2x^2 + 1$
11. $x = -3y^2 + 9$
12. $x = 1 - 2y^2$
13. $y = (x - 3)^2$
14. $y = (x + 2)^2$
15. $x = (y + 1)^2$
16. $x = (y - 4)^2$
17. $y = 2(x - 1)^2 - 1$
18. $y = -3(x + 1)^2 + 2$
19. $x = -3(y + 2)^2 + 4$
20. $x = 2(y - 2)^2 - 3$
21. $x = y^2 + 8y$
22. $x = -y^2 + 10y$
23. $y = x^2 + 6x$
24. $y = 12x - x^2$
25. $y = -x^2 + 6x - 7$
26. $x = y^2 - 4y + 2$
27. $x = 3y^2 + 6y + 4$
28. $y = -2x^2 + 4x - 6$
29. $y = 2x^2 + 12x + 16$
30. $x = y^2 - 4y + 4$
31. $x = -2y^2 + 4y - 3$
32. $y = x^2 + 6x + 5$
33. $x = -y^2 + 6y - 6$
34. $y = -2x^2 + 4x - 3$
35. $y = -2x^2 + 12x - 12$
36. $y = x^2 - 2x - 1$
37. $x = y^2 + 2y - 5$
38. $x = -2y^2 + 8y - 4$
39. $y = 3x^2 + 6x - 2$
40. $x = 3y^2 + 12y + 4$
41. $x = -y^2 - 2y + 3$
42. $y = -5x^2 + 10x + 4$
43. $y = 3x^2 - 6x + 2$
44. $x = -4y^2 + 4y + 2$

In the following problems, let h be the altitude in feet of a projectile t seconds after it is fired. Determine

a. *Its maximum altitude*
b. *The time at which it attains its maximum altitude*
c. *The time at which it hits the ground*

45. $h = 640t - 16t^2$
46. $h = 480t - 16t^2 + 2800$
47. $h = 640t - 16t^2 + 3600$
48. $h = 160t - 16t^2 + 1200$
49. $h = 320t - 16t^2 + 2000$

Solve the following inequalities. Write the solution in interval notation.

50. $-3x^2 - 8x + 3 \geq 0$
51. $2x^2 + x - 1 \leq 0$
52. $x^2 - 2x + 1 > 0$
53. $4x^2 - 12x + 5 < 0$
54. $x^2 + 5 < 2x$
55. $6x^2 + 4 \leq 14x$
56. $5x^2 > 12 + 4x$
57. $4x^2 - 4x + 1 \geq 0$
58. $2x^2 \geq 2x - 1$
59. $x^2 < 3x + 4$
60. $2x \leq x^2 - 2$
61. $3x^2 + 5x - 8 \leq 2x^2 + 6x + 4$
62. $2x^2 + 8x - 8 \leq x^2 + 2x - 1$

In the following, let $u = x^k$ and sketch a parabola in u and v in order to solve the inequalities.

63. $x^6 + 10x^3 - 24 \leq 0$
64. $x^4 - 5x^2 + 4 \geq 0$
65. $x - \sqrt{x} - 6 > 0$
66. $x - 5\sqrt{x} + 4 < 0$

67. The operator of a motorized bicycle business has determined that he can sell 40 bicycles per week when the price is $500 each. For each $50 increase in price, he sells two fewer units per week. What price will maximize weekly receipts?

68. The businessman in Exercise 67 sells his deluxe model for $800. At this price he can sell 20 units per week. Each $50 decrease in the sales price nets him two additional sales per week. What price will maximize weekly receipts?

69. The Red-White-Blue Paint Store generally sells 6000 gallons of paint each week at $15 per gallon. For each $1 decrease in the price per gallon, the store will sell an additional 1000 gallons each week. What sale price should the store advertise in order to maximize its income?

70. The Clean-Air Manufacturing Company can produce 4000 pollution-control units per week to sell for $100 each. For each additional 100 units produced each week, the sales costs are reduced by $1 per unit. At what production level is the total sales cost for the week maximized?

71. A sheet of metal 16 inches wide is to be used to make a trough by folding up the front and back sides by the same amount.
    a. What amount should be folded up in order to maximize the cross-sectional area of the trough?
    b. What is the maximum cross-sectional area?
    (*Hint:* Let x inches be folded up at the front and back. Then the cross-sectional area is $x(16 - 2x)$. Draw a picture.)

72. A rectangular corral is to be made by using 100 yards of fence to span three sides of the corral; the fourth side is a river. What dimensions of the corral will maximize the area under these conditions?

73. Answer Exercise 72 if the fence must span all four sides of the corral.

74. The Wonderful Gadget Company expects each salesperson to sell 100 gadgets each month. The seller is paid $10 for each unit sold up to 100 units. For each additional 10 units in a given month, a bonus of $2 per unit is payable on all units sold that month. The units are sold for $50 each. By selling how many gadgets does a given salesperson maximize the income he or she produces for the company in a given month?

75. Suppose that the cost of producing x widgets per day is $\$(2x^2 + 2400)$. These x widgets can be sold for $\$140x$. How many widgets can be produced at a profit?

76. As an incentive to its sales personnel, the Universal Motors Company determines the commission rate by the number of cars sold in a given month. If a certain salesperson sells x cars in a given month, her commission is $\$(80 + x)$ per car. She also earns a base salary of $225 per month. How many cars must she sell in order to earn at least $1650 this month?

Section 3.5

Central Ellipses and Hyperbolas (Optional)

(This section may be postponed with no loss of continuity. No later topics in this text depend on this material.)

An ellipse is a certain type of oval shaped curve. Hyperbolas have two distinct branches and are more difficult to describe geometrically. Orbiting bodies, including stars, planets, and artificial satellites, travel in elliptical paths. Television adventure stories have popularized the concept of locating an object by "triangulation." The hyperbola is the basis for the method of triangulation.

An ellipse or a hyperbola is defined as a **locus** (or set) of points satisfying a given condition; that is, it consists of precisely those points that satisfy the stated condition.

Ellipses

DEFINITION

An **ellipse** is the locus of points in a plane, the sum of whose distances from two fixed points remains constant. Each of the two fixed points is called a **focus** of the ellipse.

190   Chapter 3   Graphs of Equations

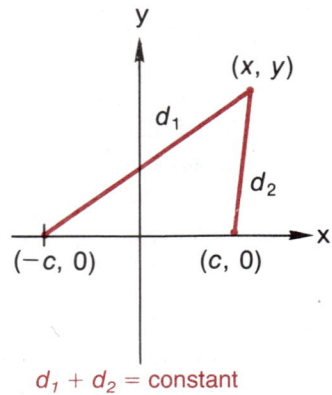

$d_1 + d_2$ = constant

Figure 3.40

To obtain the equation of an ellipse, we set up a coordinate system with the x-axis passing through the two foci. Place the y-axis midway between the foci. These foci then have coordinates $(\pm c, 0)$ for some $c > 0$ as indicated in Figure 3.40. Let (x, y) be an arbitrary point on the ellipse and label d_1, d_2 as in the figure. The condition stated then requires that $d_1 + d_2$ remain constant. It will be convenient to label this constant $2a$.

Using the distance formula to find d_1 and d_2, we find that the defining condition becomes

$$d_1 + d_2 = 2a$$
$$\sqrt{[x-(-c)]^2 + (y-0)^2} + \sqrt{(x-c)^2 + (y-0)^2} = 2a$$
$$\sqrt{(x+c)^2 + y^2} = 2a - \sqrt{(x-c)^2 + y^2}$$

Square both sides to obtain

$$x^2 + 2xc + c^2 + y^2 = 4a^2 - 4a\sqrt{(x-c)^2 + y^2} + x^2 - 2xc + c^2 + y^2$$
$$4a\sqrt{(x-c)^2 + y^2} = 4a^2 - 4xc$$
$$a\sqrt{(x-c)^2 + y^2} = a^2 - xc$$

Square both sides again:

$$a^2(x^2 - 2xc + c^2 + y^2) = a^4 - 2a^2xc + x^2c^2$$
$$a^2x^2 - x^2c^2 + a^2y^2 = a^4 - a^2c^2$$
$$x^2(a^2 - c^2) + a^2y^2 = a^2(a^2 - c^2)$$

Since the sum of the lengths of two sides of the triangle exceeds the length of the third side, we see in Figure 3.40 that $2a = d_1 + d_2 > 2c$ or $a > c$. Thus, we can let $b^2 = a^2 - c^2 > 0$. The preceding equation then becomes

$$x^2b^2 + a^2y^2 = a^2b^2$$

Dividing this equation by a^2b^2, we obtain the **standard equation of an ellipse.**

$$\boxed{\dfrac{x^2}{a^2} + \dfrac{y^2}{b^2} = 1}$$

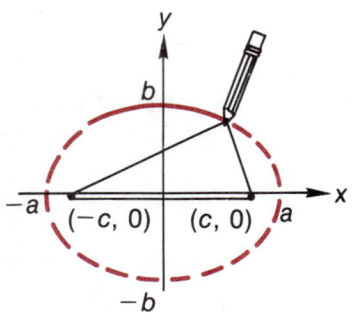

Figure 3.41
SKETCHING AN ELLIPSE

A convenient method of sketching an ellipse is to use a pencil, two tacks, and a piece of string. Place the two tacks at the focal points and knot the string, making sure that it is $2a = d_1 + d_2$ units longer than the distance between the tacks. Let the string guide the pencil and sketch the ellipse as indicated in Figure 3.41.

Alternately letting $y = 0$ and $x = 0$ in the equation for the ellipse, we see that it intersects the x- and y-axes at $(\pm a, 0)$ and $(0, \pm b)$, respectively. These points are called the **vertices** of the ellipse. The vertices and foci are related by the equation

$$a^2 - c^2 = b^2$$

or

$$a^2 = b^2 + c^2$$

This relationship is illustrated in Figure 3.42 along with the graph of the ellipse. The graph of any equation of the form

$$\dfrac{x^2}{a^2} + \dfrac{y^2}{b^2} = 1$$

Section 3.5 Central Ellipses and Hyperbolas

Figure 3.42
$\dfrac{x^2}{a^2} + \dfrac{y^2}{b^2} = 1, a > b$

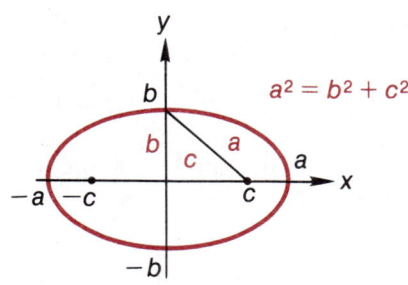

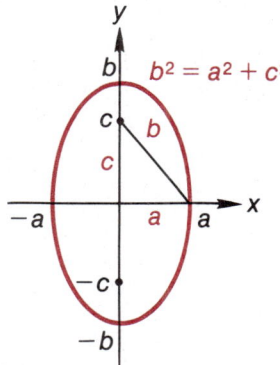

Figure 3.43
$\dfrac{x^2}{a^2} + \dfrac{y^2}{b^2} = 1, b > a$

is symmetric about the x-axis, the y-axis, and the origin (see Section 3.2). Thus, an ellipse has two axes of symmetry. The intersections of these axes of symmetry with the interior of the ellipse are called the **major** and **minor axes,** respectively, depending on which is larger. In Figure 3.42 with $a > b$, the major axis has length $2a$, whereas the length of the minor axis is $2b$. Half the major and minor axes are called the **semimajor** and **semiminor axes,** respectively. Of course, if $a = b$, the ellipse degenerates into a circle.

If $b > a$, the ellipse is elongated vertically. This means that the foci are on the y-axis at points $(0, \pm c)$. The relationship between vertices and foci illustrated in Figure 3.42 can be generally stated as *"use the semimajor axis as the hypotenuse of a right triangle in order to find the foci."* This property is illustrated in Figure 3.43 for an ellipse with $b > a$.

The point in which its major and minor axes intersect is called the **center** of the ellipse. In this section we shall treat only the **central ellipses:** ellipses centered at the origin with the x- and y-axes as the axes of symmetry.

CENTRAL ELLIPSES

Equation: $\dfrac{x^2}{a^2} + \dfrac{y^2}{b^2} = 1$

Vertices: $(\pm a, 0)$ and $(0, \pm b)$.

Elongation:
1. Horizontal if $a > b$
2. Vertical if $b > a$
3. Degenerates into a circle if $a = b$

Foci: $(\pm c, 0)$ or $(0, \pm c)$ as indicated in Figure 3.44.

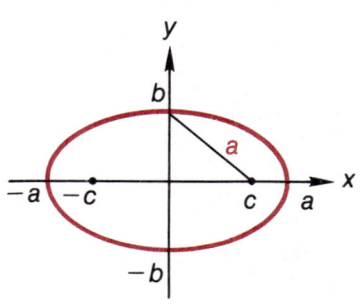 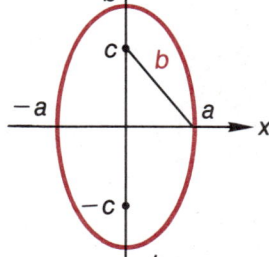

   a. $a^2 = b^2 + c^2, a > b$ **b.** $b^2 = c^2 + a^2, b > a$

Figure 3.44
$\dfrac{x^2}{a^2} + \dfrac{y^2}{b^2} = 1$

EXAMPLE 1

Sketch the ellipse given by the following equation, and label its vertices and foci.

$$4x^2 + 25y^2 = 100$$

SOLUTION

Dividing by 100, we obtain the standard equation of an ellipse:

$$\frac{x^2}{25} + \frac{y^2}{4} = 1$$

Its vertices are $(\pm 5, 0)$ and $(0, \pm 2)$. Use the semimajor axis $a = 5$ as the hypotenuse of a right triangle and the semiminor axis $b = 2$ as one of its legs. The foci are then found by using the Pythagorean formula on this triangle:

$$b^2 + c^2 = a^2$$
$$4 + c^2 = 25$$
$$c = \pm\sqrt{21}$$

The foci are located at $(\pm\sqrt{21}, 0)$ (see Figure 3.45).

Figure 3.45
$4x^2 + 25y^2 = 100$

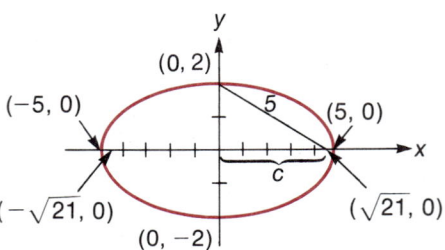

EXAMPLE 2

Find the equation and sketch the graph of the central ellipse passing through $(2, 0)$ and $(-1, 3\sqrt{3})$. Label its vertices and foci.

SOLUTION

The vertices of a central ellipse fall on the x- and y-axes. Thus, $(2, 0)$ is a vertex; that is, $a = 2$. The equation then has the form

$$\frac{x^2}{2^2} + \frac{y^2}{b^2} = 1$$

Since $(-1, 3\sqrt{3})$ is on the ellipse, it must be the case that

$$\frac{(-1)^2}{4} + \frac{(3\sqrt{3})^2}{b^2} = 1$$

Solving for b^2, we obtain $b^2 = 36$. Thus, the equation of the ellipse is

$$\frac{x^2}{4} + \frac{y^2}{36} = 1$$

Its vertices are $(\pm 2, 0)$ and $(0, \pm 6)$. Using 6 as the hypotenuse of a right triangle having one leg of length 2 as illustrated in Figure 3.46, we find the foci at $(0, \pm 4\sqrt{2})$:

$$2^2 + c^2 = 6^2$$
$$4 + c^2 = 36$$
$$c = \pm\sqrt{32} = \pm\sqrt{16 \cdot 2} = \pm 4\sqrt{2}$$

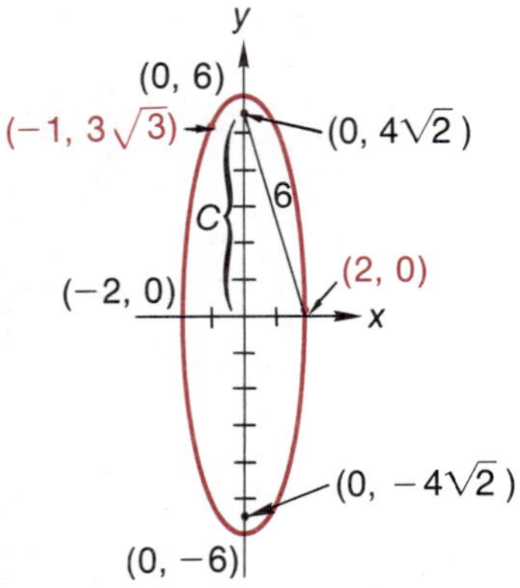

Figure 3.46
$\dfrac{x^2}{4} + \dfrac{y^2}{36} = 1$

Section 3.5 Central Ellipses and Hyperbolas

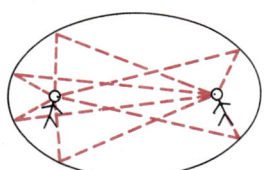

Figure 3.47
FOCAL PROPERTIES OF AN ELLIPSE

Ellipses have the property that a ray of light emanating from one focus is reflected to the other focus. This is the principle behind the so-called whispering galleries. One of these whispering galleries is the Rotunda of the Capitol Building in Washington, D.C. Two spots are marked on the floor. Two persons, one standing on each of these two spots, can whisper and hear each other quite plainly. The two spots are foci for the elliptical ceiling (see Figure 3.47).

Another use of the focal properties of an ellipse is found in the design of dental and surgical lights. The reflecting surface is elliptical, with the light source at one of its foci. The light then becomes concentrated at the other focus of the ellipse. By adjusting the position of the light so that the point of concentration is at the opening of the patient's mouth, several things happen:

1. The patient is not bothered by glare from the reflector.
2. A small area inside the patient's mouth is illuminated very brightly.
3. The dentist's hands do not interfere with the light.

Hyperbolas

The locus definition of a hyperbola is very similar to that of an ellipse except that the difference rather than the sum of distances is used.

> **DEFINITION**
>
> A **hyperbola** is the locus of points in a plane for which the difference of the distances to two fixed points remains constant. Each of the fixed points is called a **focus** of the hyperbola.

Historical Perspective

JOHANNES KEPLER
(1571–1630)

Tycho Brahe, who was court astronomer to Kaiser Rudolph II, collected and recorded very precise observations of astronomical phenomena for over 20 years. When Brahe died in 1601, his assistant, Johannes Kepler (1571–1630), inherited not only Brahe's position but also his massive collection of data. Poring over this data, Kepler was able to show that the orbit of a planet is not a circle (as was thought to be the case at that time) but rather an ellipse with the sun at one focus. Kepler actually formulated three "Laws of Planetary Motion" based on empirical observations. It was Issac Newton's attempts to prove these laws that eventually led to the development of calculus. An application of Kepler's laws closer to home tells us that the orbits of the moon and of satellites launched from earth are ellipses with the center of the earth at one focus (see Figure 3.48). However, these are not perfect ellipses because of perturbations caused by other bodies in the solar system.

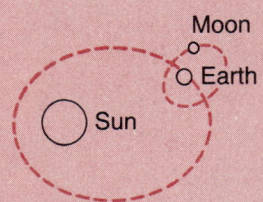

Figure 3.48
ELLIPTICAL ORBITS IN SPACE

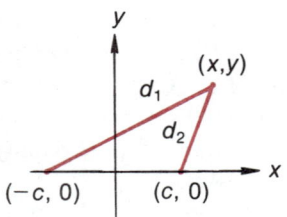

Figure 3.49
$|d_1 - d_2|$ = constant

In this context, the "difference" means "larger minus smaller." As with the ellipse, it is convenient to set up a coordinate system with one of the axes passing through the two foci. For the sake of the derivation, let the x-axis pass through the foci and place the y-axis midway between the foci. The foci then have coordinates $(\pm c, 0)$ for some $c > 0$. Let (x, y) be an arbitrary point on the hyperbola and label d_1, d_2 as in Figure 3.49.

The defining condition then requires that $|d_1 - d_2|$ remain constant. As before, label this constant as $2a$. Then

$$|d_1 - d_2| = 2a$$
$$d_1 - d_2 = \pm 2a$$
$$d_1 = d_2 \pm 2a$$

Using the distance formula to find d_1 and d_2, we can rewrite this equation as

$$\sqrt{[x-(-c)]^2 + (y-0)^2} = \sqrt{(x-c)^2 + (y-0)^2} \pm 2a \qquad (1)$$

As with the ellipse, repeated squaring operations reduce this equation to

$$x^2(a^2 - c^2) + a^2 y^2 = a^2(a^2 - c^2) \qquad (2)$$

Since the shortest distance between two points is a straight line, we see in Figure 3.49 that

$$d_2 + 2c > d_1$$
$$2c > d_1 - d_2 = 2a$$

Thus, $c > a$, and we can set $b^2 = c^2 - a^2 > 0$. Equation (2) then becomes

$$-b^2 x^2 + a^2 y^2 = -a^2 b^2, \qquad a^2 > 0, b^2 > 0$$

Dividing by $-a^2 b^2$, we obtain the **standard equation of a hyperbola with horizontal axis:**

$$\boxed{\dfrac{x^2}{a^2} - \dfrac{y^2}{b^2} = 1} \qquad (3)$$

Again, note the symmetry with respect to the x-axis, the y-axis, and the origin (see Section 3.2). Thus, in graphing this equation, we shall at first concern ourselves only with the first quadrant; that is, consider only $x \geq 0$ and $y \geq 0$. Note that there can be no y-intercept; setting $x = 0$ yields the equation $-y^2/b^2 = 1$, which has no real solutions. Setting $y = 0$, we obtain the x-intercepts $x = \pm a$. Equation (3) can be rewritten as

$$\dfrac{x^2}{a^2} = 1 + \dfrac{y^2}{b^2} \qquad (4)$$

In equation (4) we observe that

$$\dfrac{x^2}{a^2} \geq 1$$

or

$$x^2 \geq a^2$$

Hence, $x \geq a$ in the first quadrant. We could also solve equation (3) for y in terms of x to obtain

$$\frac{y^2}{b^2} = \frac{x^2}{a^2} - 1 = \frac{x^2 - a^2}{a^2}$$

$$y^2 = \frac{b^2}{a^2}(x^2 - a^2) = \frac{b^2}{a^2}x^2\left(1 - \frac{a^2}{x^2}\right)$$

Hence, in the first quadrant

$$y = \frac{b}{a}x\sqrt{1 - \frac{a^2}{x^2}} < \frac{b}{a}x \tag{5}$$

Now as x gets very large,

$$1 - \frac{a^2}{x^2} \approx 1$$

Equation (5) then indicates that for large x,

$$y \approx \frac{b}{a}x \quad \text{and} \quad y < \frac{b}{a}x$$

Thus, y approaches $(b/a)x$ from below as x gets large. Plotting additional points if necessary, we see finally that the graph of equation (3) in the first quadrant is as given in Figure 3.50.

Using the symmetry previously mentioned, we can then complete the graph as illustrated in Figure 3.51. It is called a **hyperbola**. For a hyperbola, the coordinates $(\pm c, 0)$ of the focus are related to a and b by

$$c^2 - a^2 = b^2$$

or

$$c^2 = a^2 + b^2$$

This relationship is also illustrated in Figure 3.51.

The lines $y = \pm(b/a)x$ are called **asymptotes** of the hyperbola. The points $(\pm a, 0)$ are its **vertices**. The point halfway between the vertices is called its **center**. As with the ellipses, we shall be concerned only with **central hyperbolas**: those with the origin as center and with the x- and y-axes as axes of symmetry.

Placing the foci on the y-axis at $(0, \pm c)$ simply interchanges the roles of x and y in the preceding derivation. Thus, the **standard equation of a hyperbola with vertical axis** is

$$\boxed{\frac{y^2}{b^2} - \frac{x^2}{a^2} = 1} \tag{6}$$

The focus is related to a and b in the same way as before:

$$c^2 = a^2 + b^2$$

Expressing y in terms of x in equation (6), we have

$$\frac{y^2}{b^2} = 1 + \frac{x^2}{a^2} \geq 1 \tag{7}$$

Thus,

$$y \geq b$$

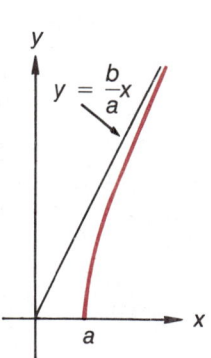

Figure 3.50

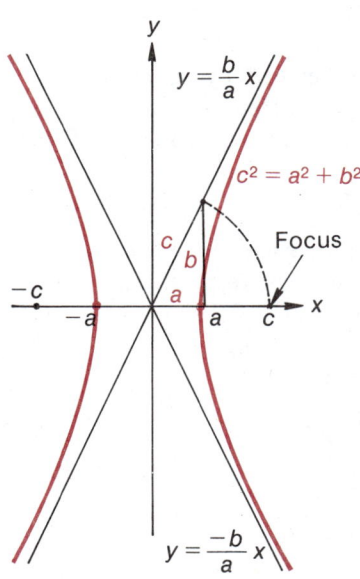

Figure 3.51
$\frac{x^2}{a^2} - \frac{y^2}{b^2} = 1$

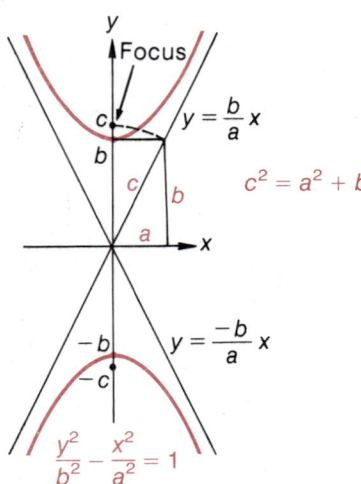

Figure 3.52

in the first quadrant; there is no x-intercept. Continuing from equation (7), we see that

$$\frac{y^2}{b^2} = \frac{a^2 + x^2}{a^2} = \frac{x^2}{a^2}\left(1 + \frac{a^2}{x^2}\right)$$

Thus,

$$y = \frac{bx}{a}\sqrt{1 + \frac{a^2}{x^2}} > \frac{bx}{a}$$

in the first quadrant. For large x,

$$y \approx \frac{bx}{a} \quad \text{and} \quad y > \frac{bx}{a}$$

That is, y approaches bx/a from above as x gets large. The graph of equation (6) is seen to be another hyperbola with the same asymptotes as equation (3). The vertices are now $(0, \pm b)$ and the hyperbola opens vertically. It is sketched and its foci are indicated in Figure 3.52. The hyperbolas given by equations (3) and (6) are said to be **conjugate** to one another.

CENTRAL HYPERBOLAS

The equations

$$\frac{x^2}{a^2} - \frac{y^2}{b^2} = 1 \qquad \frac{y^2}{b^2} - \frac{x^2}{a^2} = 1$$

represent conjugate hyperbolas. Each can be sketched by constructing a rectangle centered at the origin with horizontal sides of length $2a$ and vertical sides of length $2b$. The distance of the foci from the origin equals half the length of the diagonal of this rectangle:

$$c^2 = a^2 + b^2$$

The diagonals of the rectangle are asymptotes for each of these conjugate hyperbolas. The vertices occur at the intersection of the respective axes with the rectangle constructed as indicated in Figure 3.53.

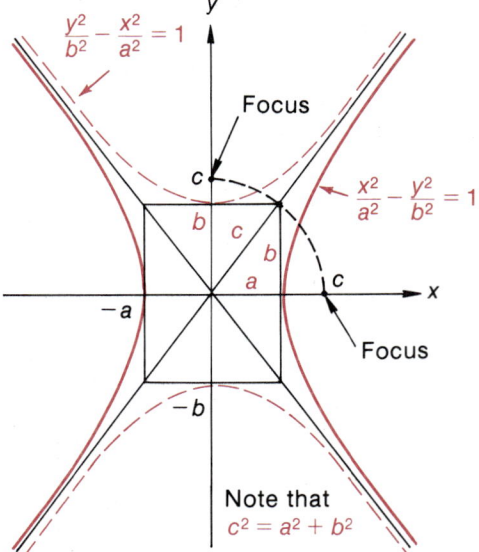

Note that $c^2 = a^2 + b^2$

Figure 3.53

EXAMPLE 3 Sketch the conjugate hyperbolas. Label the vertices and foci.

a. $\dfrac{x^2}{4} - \dfrac{y^2}{9} = 1$

b. $\dfrac{y^2}{9} - \dfrac{x^2}{4} = 1$

SOLUTION

In each of these hyperbolas, $a = 2$ and $b = 3$. Thus, the asymptotes are $y = \pm\tfrac{3}{2}x$. The vertices of the first hyperbola are $(\pm 2, 0)$; the vertices of the second are $(0, \pm 3)$. We sketch the rectangle centered at the origin and passing through these vertices. The diagonals of this rectangle are the asymptotes of both hyperbolas. It is then a simple matter to sketch the hyperbolas. The foci can be determined from the relationship $c^2 = 2^2 + 3^2 = 13$ (see Figure 3.54).

Figure 3.54

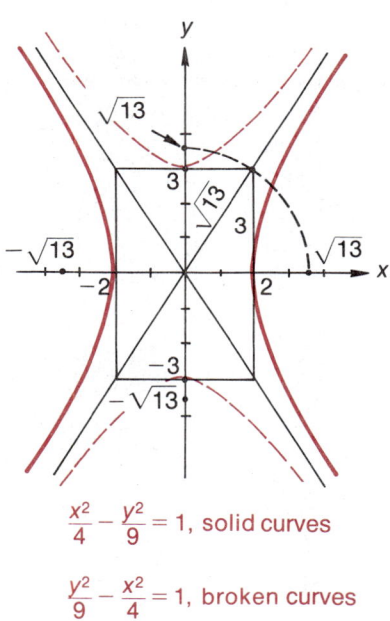

$\dfrac{x^2}{4} - \dfrac{y^2}{9} = 1$, solid curves

$\dfrac{y^2}{9} - \dfrac{x^2}{4} = 1$, broken curves

EXAMPLE 4 Find the equation and sketch the graph of the central hyperbola with vertices at $(\pm 5, 0)$ and the line $y = 2x$ as an asymptote.

SOLUTION

Note that $a = 5$ since $(\pm 5, 0)$ are vertices. The slopes of the asymptote lines are $\pm(b/a) = \pm b/5$. Since $y = 2x$ is an asymptote, we then have $b/5 = 2$ or $b = 10$. Since the vertices are on the x-axis, the equation of the hyperbola must then be

$$\dfrac{x^2}{25} - \dfrac{y^2}{100} = 1$$

It is graphed in Figure 3.55.

Figure 3.55

$\dfrac{x^2}{25} - \dfrac{y^2}{100} = 1$

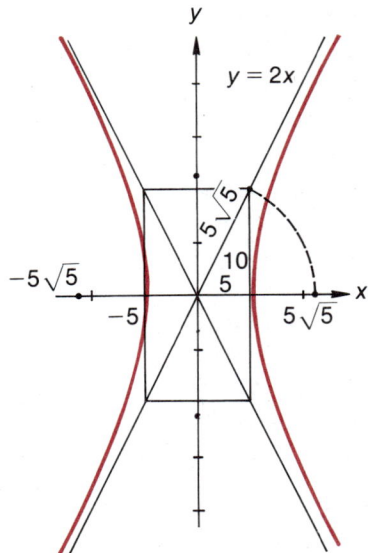

EXAMPLE 5

Find the equation and sketch the graph of the central hyperbola passing through $(1, 0)$ and $(2, 2\sqrt{3})$.

SOLUTION

Since this is a central hyperbola, $(1, 0)$ must be a vertex. Thus, the hyperbola opens horizontally and has the form

$$\frac{x^2}{a^2} - \frac{y^2}{b^2} = 1$$

Since $(1, 0)$ is a vertex, $a = 1$; the equation is

$$\frac{x^2}{1} - \frac{y^2}{b^2} = 1$$

or

$$x^2 - \frac{y^2}{b^2} = 1$$

Since $(2, 2\sqrt{3})$ is on the graph, we have

$$2^2 - \frac{(2\sqrt{3})^2}{b^2} = 1$$

$$4 - \frac{12}{b^2} = 1$$

From this we find

$$b^2 = 4$$

The equation of the hyperbola is

$$x^2 - \frac{y^2}{4} = 1$$

See Figure 3.56.

Figure 3.56
$x^2 - \dfrac{y^2}{4} = 1$

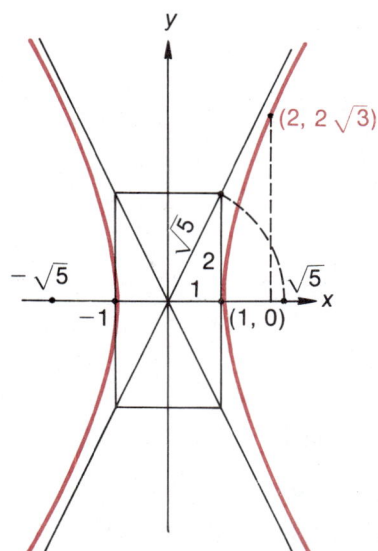

Section 3.5 Exercises

Sketch the following ellipses; indicate vertices and foci.

1. $\dfrac{x^2}{9} + \dfrac{y^2}{16} = 1$
2. $\dfrac{x^2}{16} + \dfrac{y^2}{9} = 1$
3. $\dfrac{x^2}{25} + \dfrac{y^2}{9} = 1$
4. $\dfrac{x^2}{25} + \dfrac{y^2}{16} = 1$
5. $x^2 + 9y^2 = 9$
6. $9x^2 + y^2 = 9$
7. $4x^2 + 9y^2 = 36$
8. $9x^2 + 4y^2 = 36$
9. $25x^2 + 36y^2 = 900$
10. $25x^2 + 9y^2 = 225$
11. $16x^2 + 4y^2 = 64$
12. $36x^2 + 81y^2 = 2916$
13. $169x^2 + 25y^2 = 4225$
14. $576x^2 + 625y^2 = 360{,}000$

Find the equation of the central ellipse satisfying the following conditions.

15. Vertices at (0, 2) and (−3, 0)
16. Vertical major axis of length 26, horizontal minor axis of length 24
17. Horizontal major axis of length 10, minor axis of length 6
18. Focus at (3, 0) and horizontal axis of length 10
19. Focus at (−12, 0), vertical axis of length 10
20. Focus at $(0, -\sqrt{5})$ and vertex at (0, 3)
21. Focus at (7, 0) and vertex at (0, −24)
22. Focus at (2, 0) and minor axis of length 4
23. Focus at (0, −12) and major axis of length 26
24. Vertex at (4, 0), passing through $\left(2, \dfrac{-3\sqrt{3}}{2}\right)$
25. Vertex at (0, 8) passing through $(-\sqrt{3}, 4)$
26. Passing through (0, 1) and $\left(1, \dfrac{\sqrt{2}}{2}\right)$
27. Passing through (2, 0) and $\left(1, \dfrac{\sqrt{15}}{2}\right)$

Sketch the graph of each of the given equations. In each case, write the equation of the conjugate hyperbola and also sketch its graph. Indicate vertices, foci, and asymptotes.

28. $\dfrac{x^2}{16} - \dfrac{y^2}{9} = 1$
29. $\dfrac{y^2}{16} - \dfrac{x^2}{9} = 1$
30. $\dfrac{x^2}{25} - \dfrac{y^2}{144} = 1$
31. $x^2 - y^2 = 9$
32. $9x^2 - y^2 = 9$
33. $x^2 - 4y^2 = 16$
34. $25y^2 - 4x^2 = 100$
35. $144y^2 - 25x^2 = 3600$
36. $16y^2 - 25x^2 = 400$
37. $y^2 - 4x^2 = 12$
38. $4x^2 - 9y^2 = 36$
39. $9x^2 - 4y^2 = 36$
40. $\dfrac{y^2}{49} - \dfrac{x^2}{576} = 1$
41. $64x^2 - 25y^2 = 1600$

Find the equation of the central hyperbola satisfying the following conditions.

42. Vertex $(0, 2)$, asymptote $y = 2x$

43. Vertex $(4, 0)$, asymptote $y = -\dfrac{3x}{4}$

44. Vertex $(16, 0)$, asymptote $y = \dfrac{5x}{4}$

45. Vertex $(0, -1)$, asymptote $y = -3x$

46. Vertex $(0, 2)$, passing through $(1, -3)$

47. Vertex $(3, 0)$, passing through $(-5, 8)$

48. Vertex $(-5, 0)$, passing through $(13, 36)$

49. Vertex $(0, -4)$, passing through $(1, -5)$

50. Vertex $(0, -3)$, focus $(0, 5)$

51. Vertex $(2, 0)$, focus $(-3, 0)$

52. Vertex $(5, 0)$, focus $(13, 0)$

53. Focus $(3, 0)$, asymptote $y = 2x$

54. Focus $(0, 2)$, asymptote $y = x$

55. Focus $(0, -2\sqrt{5})$, asymptote $x = 2y$

56. Sketch and write the equation of the locus of points the sum of whose distances from $(-2, 3)$ and $(4, 3)$ equals 10.

57. Sketch and write the equation of the locus of points the sum of whose distances from $(-1, 2)$ and $(1, 3)$ is 5.

58. Sketch and write the equation of the locus of points the difference of whose distances from $(1, 2)$ and $(1, 8)$ is 3.

59. Sketch and write the equation of the locus of points the difference of whose distances from $(3, -2)$ and $(-5, 4)$ is 8.

60. A race track is in the form of an ellipse with major and minor axes of length 500 and 140 meters, respectively. Posts are driven into the ground at the foci of this ellipse. A rope looped through Rapid Dancer's bridle is tied and draped over the posts as in Figure 3.41. How long must the rope be if Rapid Dancer is to be free to run around this track?

61. Bob's backyard is a rectangle 30 meters by 50 meters. He wants to let his dog exercise in the yard, but the neighbors get upset if the dog ventures onto their property. His solution is to drive two stakes into the ground and fasten a chain to both of these stakes after passing it through the dog's collar. Determine where the stakes should be located and how much chain is needed if the dog is to have a maximum exercise area without being able to encroach on neighboring property.

62. Two ranger stations 30 miles apart pick up an SOS broadcast from a hiker who has lost his way. It is determined that Ranger Alice receives the signal $\dfrac{1}{10,000}$ of a second later than Ranger Bob.

    a. Sketch a graph indicating the possible locations of the SOS broadcaster. (Radio waves travel at a speed of 186,000 miles per second.)

    b. Indicate how the location of the SOS broadcaster can be pinpointed if a third ranger also picks up the broadcast. (This is an example of "triangulation.")

Section 3.6

The Conic Sections and Translation (Optional)

(This section may be postponed with no loss of continuity. No later topics in this text depend on this material.)

You may be surprised to learn that some of the most important curves used by scientists and engineers to describe various natural phenomena can be obtained simply by slicing an ordinary object with which we are all familiar: the (ice cream) cone. These curves, called **conic sections,** are obtained by intersecting a plane and a double cone.

Only five general curves are generated in this way: lines, circles, ellipses, parabolas, and hyperbolas. Figure 3.57 illustrates how these conic sections are generated.

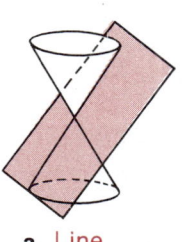 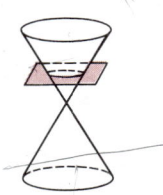 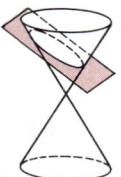 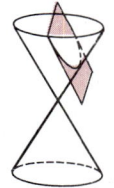 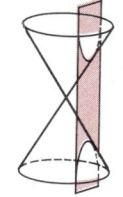 

a. Line   b. Circle   c. Ellipse   d. Parabola   e. Hyperbola   f. Intersecting lines (sometimes called a degenerate hyperbola)

Figure 3.57
THE CONIC SECTIONS

Locus

Each of the conic sections can be defined as a locus of points satisfying a given condition.

Circles

DEFINITION

A **circle** is the locus of points in a plane whose distance from a given point A is some positive constant r. A is called the **center** and r the **radius** of the circle.

The distance formula is used to write the equation of the circle of radius r centered at (x_0, y_0) as

$$(x - x_0)^2 + (y - y_0)^2 = r^2$$

Since circles were characterized in this way and studied extensively in Section 3.1, we shall not have any more to say about them in this section.

Lines

Since every point on the perpendicular bisector of a line segment is equidistant from the ends of that segment (see Figure 3.58), a line can be characterized as follows.

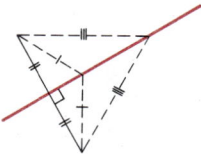

DEFINITION

A **line** is the locus of points in a plane equidistant from two given points.

Figure 3.58

EXAMPLE 1

Write the equation of the line representing the locus of points equidistant from $(1, 2)$ and $(3, 8)$.

SOLUTION

The line described must be the perpendicular bisector of the segment joining $(1, 2)$ and $(3, 8)$. Using the midpoint formula if necessary, we find the midpoint of this segment to be $(2, 5)$. The slope of the line joining the two points is

$$m_1 = \frac{8 - 2}{3 - 1} = \frac{6}{2} = 3$$

The slope of the perpendicular bisector is then

$$m_2 = -\frac{1}{m_1} = -\frac{1}{3}$$

Thus the equation of the line in question is

$$(y - 5) = -\frac{1}{3}(x - 2)$$ (Point-slope formula, Section 3.3)

This equation can be simplified into

$$3y - 15 = -x + 2$$
$$x + 3y = 17$$

See Figure 3.59.

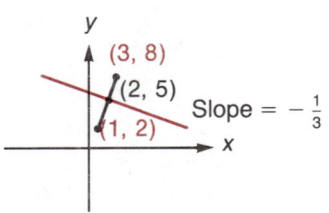

Figure 3.59

Parabolas

DEFINITION

A **parabola** is the locus of points in a plane equidistant from a given line and a given point; the line is called the **directrix,** and the fixed point is called the **focus** of the parabola.

To see why such a locus describes a parabola, set up a coordinate system as follows. Let the y-axis be obtained by dropping the perpendicular from the focus to the directrix. Then place the x-axis midway between the focus and directrix. In this coordinate system, the focus will have coordinates $(0, d)$, and the directrix is the line $y = -d$ for some number d. If (x, y) is on the described locus, and d_1, d_2 are labeled as in Figure 3.60, then

$$d_1 = d_2$$
$$\sqrt{(x - 0)^2 + (y - d)^2} = y + d$$

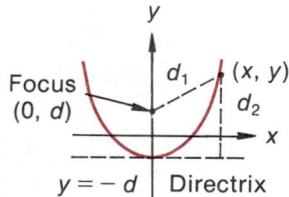

Figure 3.60

Squaring both sides yields

$$x^2 + y^2 - 2dy + d^2 = y^2 + 2dy + d^2$$
$$x^2 = 4dy$$
$$y = \frac{1}{4d}x^2$$

This last equation is the equation of a vertical parabola with its vertex at the origin. See Section 3.4. Similarly, the equation

$$x = \frac{1}{4D}y^2$$

represents a horizontal parabola with vertex at the origin, focus at $(D, 0)$, and directrix at $x = -D$.

The numbers d or D, respectively, can be used to determine the width of a parabola at its focus. Consider the equation

$$y = \frac{1}{4d}x^2$$

Letting $x = 2d,$ we see that the corresponding y value is

$$y = \frac{(2d)^2}{4d} = d$$

This relationship and its application to sketching the graph of a parabola is illustrated in Figure 3.61.

Section 3.6 The Conic Sections and Translation 203

PARABOLAS

The equation

$$y = \frac{1}{4d}x^2 \quad \text{or} \quad x = \frac{1}{4D}y^2$$

represents a parabola with vertex at the origin.

1. The parabola is

   vertical or horizontal

2. The focus is at

   $(0, d)$ or $(D, 0)$

3. The directrix is the line

   $y = -d$ or $x = -D$

4. The length of the

   horizontal or vertical

   chord through the focus is

   $4|d|$ or $4|D|$
   (half-length $= 2|d|$) (half-length $= 2|D|$)

5. The parabola opens

   upward if $d > 0$ to the right if $D > 0$
   downward if $d < 0$ to the left if $D < 0$

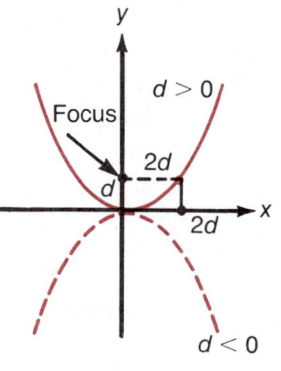
a. $y = \left(\frac{1}{4d}\right)x^2$

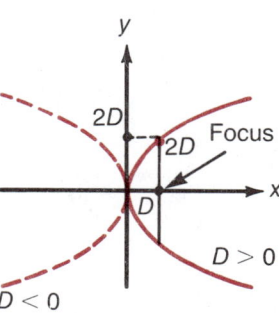
b. $x = \left(\frac{1}{4D}\right)y^2$

Figure 3.61

EXAMPLE 2 Sketch the graph of $y^2 = -12x$ and find its focus and directrix.

SOLUTION Writing the equation in the form $x = \frac{1}{4D}y^2$, we have

$$x = \frac{1}{-12}y^2 = \frac{1}{4(-3)}y^2$$

204 Chapter 3 *Graphs of Equations*

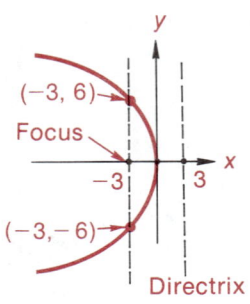

Figure 3.62
$y^2 = -12x$ or $x = -\frac{1}{12}y^2$

Thus $D = -3$. The graph is a horizontal parabola with vertex at the origin, focus at $(-3, 0)$, directrix at $x = -(-3) = 3$, and opening to the left. The vertical chord through the focus has length 12. (See Figure 3.62). ∎

The focus of a parabola is very important from a practical viewpoint. The line joining the focus and vertex of a parabola is called its **axis** (or line of symmetry; see Section 3.4). A light source situated at the focus of a parabola will have its rays reflected parallel to the principal axis of the parabola. This property is used in the design of automobile headlights and beacon lights. The property is used in reverse in designing parabolic-reflector telescopes; all light entering the parabolic reflector is reflected *into* the focus (see Figure 3.63).

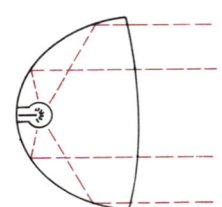 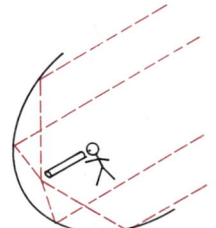

Figure 3.63

a. Automobile headlight b. Parabolic-reflector telescope

Ellipses and Hyperbolas

These were already characterized in terms of a locus in Section 3.5.

Translation of Axes

Until now we have been concerned only with the central conics. To discuss conics that are not centered at the origin, we must investigate the changes that are effected on an equation when the origin is moved. In this section, we shall only **translate** the origin—that is, move it to a new location in such a way that the axes are not stretched or rotated.

Consider now a plane on which are imposed two coordinate systems parallel to one another and with the same units as illustrated in Figure 3.64. Any point in this plane can then be given coordinates in each of the two coordinate systems. To relate the two coordinate systems, let the origin of the X'-Y' system have coordinates (h, k) in the X-Y system.

Let the coordinates of some point P be (x', y') and (x, y) in the X'-Y' and X-Y coordinate systems, respectively. We note in Figure 3.64 that

$$x = x' + h$$
$$y = y' + k$$

Solving for x', y' in terms of x and y, we obtain

$$x' = x - h$$
$$y' = y - k$$

Figure 3.64
PARALLEL COORDINATE SYSTEMS

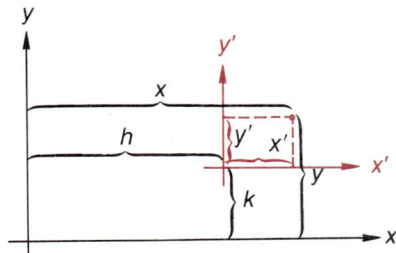

TRANSLATION OF AXES

If we know how to write the equation of a given figure when it is centered at the origin, then the following steps indicate how to write the equation of such a figure when it is centered at (h, k).

1. Introduce a new x'-y' coordinate system with origin at (h, k).
2. Write the equation of the curve in this new coordinate system.
3. Substitute

$$x' = x - h$$
$$y' = y - k$$

into the equation to obtain the equation in the original x-y coordinate system.

We have already seen these relationships in our study of lines, circles, and parabolas. See Figure 3.65. For instance, we should recognize the point-slope formula in Figure 3.65a.

Figure 3.65

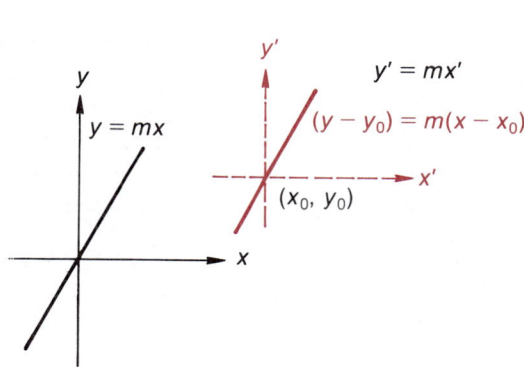

a. A line of slope m through (x_0, y_0)

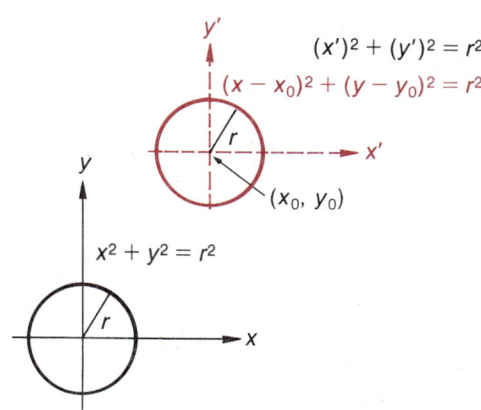

b. A circle of radius r centered at (x_0, y_0)

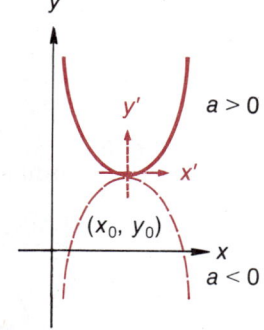

$y' = a(x')^2$
$y - y_0 = a(x - x_0)^2$

c. Vertical parabolas

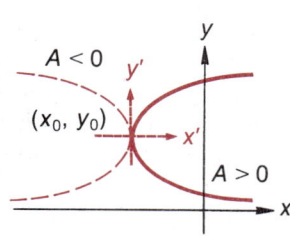

$x' = A(y')^2$
$x - x_0 = A(y - y_0)^2$

d. Horizontal parabolas

EXAMPLE 3

Sketch the following parabola; indicate its vertex, focus, directrix, and principal axis.

$$4y = x^2 - 6x + 17$$

SOLUTION

First complete the square in order to identify h and k:

$$4y = (x^2 - 6x \quad) + 17$$
$$= (x^2 - 6x + 9) + 17 - 9$$
$$= (x - 3)^2 + 8$$
$$4y - 8 = (x - 3)^2$$
$$4(y - 2) = (x - 3)^2$$

Thus

$$y - 2 = \frac{1}{4}(x - 3)^2$$

is the equation of the parabola. It opens upward with vertex at (3, 2). In the coordinate system centered at (3, 2), this equation becomes

$$y' = \frac{1}{4}(x')^2$$

If the focus is located d units above the vertex, its equation is

$$y' = \frac{1}{4d}(x')^2$$

Thus $d = 1$; that is, the focus is 1 unit above the vertex, at (3, 3). The directrix is then 1 unit below the vertex, at $y = 1$. The line $x = 3$ is the principal axis. The horizontal chord through the focus has length 4 (see Figure 3.66).

Figure 3.66

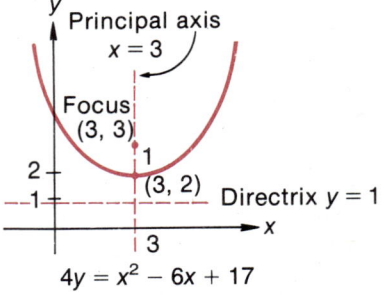

$$4y = x^2 - 6x + 17$$

Ellipses

If an ellipse is centered at (h, k) with semimajor and semiminor axes a and b, it is a simple matter to write its equation in a coordinate system centered at (h, k). The equation is

$$\frac{(x')^2}{a^2} + \frac{(y')^2}{b^2} = 1$$

Substituting the translation conditions into this equation yields

$$\frac{(x - h)^2}{a^2} + \frac{(y - k)^2}{b^2} = 1$$

as the equation of an ellipse centered at (h, k) with semimajor and semiminor axes a and b (see Figure 3.67).

Figure 3.67
$\frac{(x-h)^2}{a^2} + \frac{(y-k)^2}{b^2} = 1$

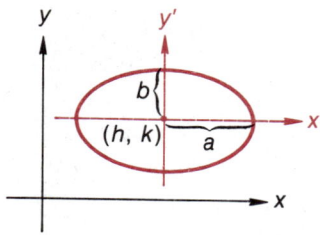

EXAMPLE 4

Sketch the ellipse with upper vertex at $(5, 9)$ and rightmost vertex at $(8, 7)$, assuming that its axes of symmetry are parallel to the x- and y-axes, respectively. Find its equation and foci.

SOLUTION

The vertices lie on the axes of symmetry. Accordingly, plot the given vertices and draw a vertical line through the upper vertex and a horizontal line through the rightmost vertex. The center of the ellipse is at the intersection of these axes of symmetry, at the point $(5, 7)$, as shown in Figure 3.68. The semimajor and semiminor axes are the distances from the center of the ellipse to the right (or left) vertex and to the top (or bottom) vertex; these are 3 and 2 respectively as shown in Figure 3.68. From this information we can write the equation of the ellipse as

$$\frac{(x-5)^2}{9} + \frac{(y-7)^2}{4} = 1$$

For the foci we use the equation $a^2 = b^2 + c^2$, since $a = 3 > b = 2$ (see Section 3.5). Thus,

$$3^2 = 2^2 + c^2$$
$$c = \sqrt{5}$$

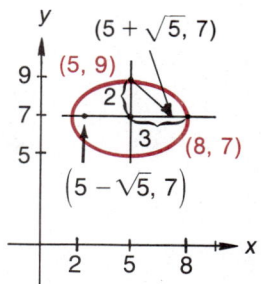

Figure 3.68

Thus, the foci are located at $(5 \pm \sqrt{5}, 7)$ (see Figure 3.68).   ∎

Hyperbolas

For a hyperbola centered at (h, k), we introduce a new coordinate system with origin at (h, k). In this new coordinate system, the equation of the hyperbola is

1. $\dfrac{(x')^2}{a^2} - \dfrac{(y')^2}{b^2} = 1$ if the hyperbola opens horizontally

or

2. $\dfrac{(y')^2}{b^2} - \dfrac{(x')^2}{a^2} = 1$ if the hyperbola opens vertically

In terms of the original variables, these equations become

1. $\dfrac{(x-h)^2}{a^2} - \dfrac{(y-k)^2}{b^2} = 1$

or

2. $\dfrac{(y-k)^2}{b^2} - \dfrac{(x-h)^2}{a^2} = 1$

Their graphs are shown in Figure 3.69.

Figure 3.69

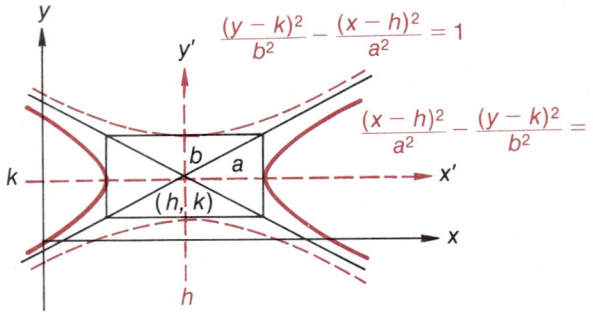

EXAMPLE 5

Sketch the graph of the following equation. Find its foci.

$$4x^2 - 9y^2 + 8x + 54y = 113$$

SOLUTION

Manipulate the equation into a more recognizable form by completing the squares.

$(4x^2 + 8x \quad) - (9y^2 - 54y \quad) = 113$ (Group common variables)

$4(x^2 + 2x \quad) - 9(y^2 - 6y \quad) = 113$ (Factor out coefficients of x^2 and y^2)

$4(x^2 + 2x + 1) - 9(y^2 - 6y + 9) = 113 + 4 - 81$ (Complete the squares)

$4(x + 1)^2 - 9(y - 3)^2 = 36$

$\dfrac{(x + 1)^2}{9} - \dfrac{(y - 3)^2}{4} = 1$ (Divide by 36)

The final form represents a hyperbola opening horizontally and centered at $(-1, 3)$ with $a = 3, b = 2$. It is sketched in Figure 3.70. To find its foci, we write $c^2 = a^2 + b^2$ (see Section 3.5).

$c^2 = 3^2 + 2^2 = 13$

$c = \sqrt{13}$

Thus, the foci are located at $(-1 \pm \sqrt{13}, 3)$.

Figure 3.70
$4x^2 - 9y^2 + 8x + 54y = 113$

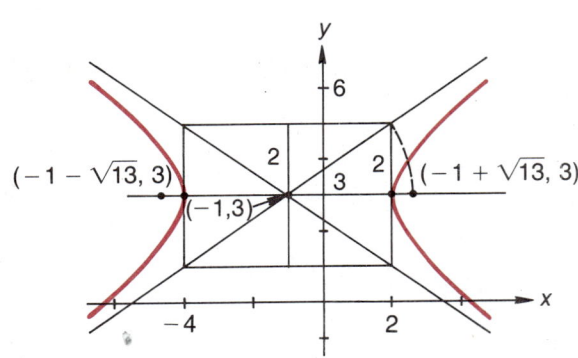

Section 3.6 Exercises

Write the equation for the locus of points equidistant from each of the following pairs of points.

1. (2, 0) and (0, 4)
2. (2, 3) and (4, 5)
3. (3, −1) and (6, 1)
4. (4, −2) and (−2, 6)
5. (−1, 1) and (−3, 3)
6. (−6, −1) and (0, −1)
7. (3, −3) and (3, 1)
8. (−3, 5) and (−4, −2)

Write the equation for the locus of points equidistant from the given point and the given line.

9. (1, 0) and $x = -1$
10. (−3, 0) and $x = 3$
11. (0, −5) and $y = 5$
12. (0, 2) and $y = -2$
13. (−2, 0) and $x = 2$
14. (4, 0) and $x = -4$
15. (0, 7) and $y = -7$
16. (0, −2) and $y = 2$
17. (5, 0) and $x = -5$
18. (−8, 0) and $x = 8$
19. (0, −6) and $y = 6$
20. (0, 10) and $y = -10$
21. (−2, −1) and $x = 6$
22. (0, 2) and $y = 8$
23. (2, 4) and $y = 6$
24. (−3, 2) and $y = 0$
25. (5, −7) and $x = 9$
26. (3, 2) and $x = -1$

Sketch the graph of each of the following parabolas. Label the vertex, focus, and directrix and indicate the length of the horizontal or vertical chord passing through the focus.

27. $y = \dfrac{x^2}{8}$
28. $x^2 = 12y$
29. $x^2 = -20y$
30. $x^2 = -4y$
31. $4x = y^2$
32. $y^2 = 16x$
33. $y^2 = -8x$
34. $y^2 = -20x$
35. $x^2 = -16y$
36. $x^2 = 24y$
37. $y^2 = 32x$
38. $y^2 = -28x$

Write the equations of the following parabolas.

39. Focus at (−1, 2), vertex at (1, 2)
40. Focus at (3, 1), directrix $y = 3$
41. Directrix $y = 3$, vertex (−2, 0)
42. Vertex (1, 2), focus at (1, 4)
43. Directrix $x = 0$, focus at (2, −1)
44. Vertex (2, 3), directrix $x = -2$

Write the equation of the ellipse satisfying the given conditions.

45. Lower vertex at (−2, −3), left vertex at (−7, 1)
46. Upper vertex at (6, 2), right vertex at (12, −1)
47. Left vertex at (1, 1), upper vertex at (3, 6)
48. Right vertex at (5, −2), lower vertex at (3, −7)
49. Upper vertex at (2, 4), right vertex at (4, 3)
50. Lower vertex at (5, −1), left vertex at (3, 5)

Write the equation of the hyperbola satisfying the given conditions. Do the same for its conjugate.

	OPENS	CENTERED AT	a	b
51.	Vertically	(−1, −4)	1	5
52.	Horizontally	(6, −9)	4	2
53.	Vertically	(8, 1)	10	6
54.	Horizontally	(2, −7)	4	6
55.	Horizontally	(−10, 8)	9	10
56.	Vertically	(9, −3)	7	5

Sketch the graphs of the given equations. Label the vertices and foci. If a parabola, indicate the directrix; if a hyperbola, indicate the asymptotes.

57. $2x^2 - 4x - y + 1 = 0$
58. $25x^2 + 9y^2 - 100x + 54y = 44$
59. $x^2 + 16y^2 + 4x - 32y + 16 = 0$
60. $9y^2 - 16x^2 + 64x + 54y + 161 = 0$
61. $2y^2 - 16y + x + 4 = 0$
62. $4y^2 - 9x^2 - 36x + 24y = 36$
63. $x^2 - 2x + y + 2 = 0$
64. $16y^2 + 4x^2 + 48y - 28x = 171$
65. $y^2 - x^2 + 2y - 4x - 7 = 0$
66. $x - 4y + 2y^2 + 2 = 0$
67. $4x^2 - 25y^2 + 16x + 150y = 109$
68. $16x^2 - 9y^2 + 64x - 18y - 89 = 0$

69. $4x^2 + 9y^2 + 16x - 36y + 16 = 0$

70. $5x^2 - 20x + y - 1 = 0$

71. $9x^2 + 4y^2 - 18x + 16y = 11$

72. Write the equation of the locus of points equidistant from the line $y = -3$ and the point $(2, -1)$. Sketch the graph.

73. Write the equation of the locus of points the sum of whose distances from $(2, 1)$ and $(8, 1)$ is 10. Sketch the graph.

74. Write the equation of the locus of points the difference of whose distances from $(2, -6)$ and $(2, 4)$ is 6. Sketch the graph.

Section 3.7 Chapter Review

Terms and Concepts

Cartesian Coordinates Enable us to plot points in an x-y plane.

Graph A pictorial representation of a relationship between two or more quantities. The graph of an equation or relationship between two variables x and y consists of those points (x, y) that satisfy the relationship.

Intercepts Points at which a graph meets the x- and y-axes

Circle The locus of points whose distance from a given point A is some positive constant r. A is called the **center** and r the **radius** of the circle.

Line The locus of points equidistant from two given points.

Parabola The locus of points equidistant from a given line and a given point; the line is called the **directrix**, and the fixed point is called the **focus** of the parabola.

Ellipse The locus of points the sum of whose distances from two fixed points remains constant. Each of the two fixed points is called a **focus** of the ellipse.

Hyperbola The locus of points for which the difference of the distances to two fixed points remains constant. Each of the fixed points is called a **focus** of the hyperbola.

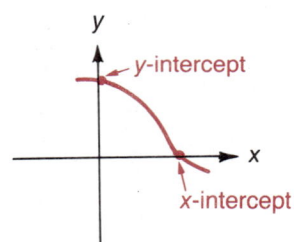

Symmetry

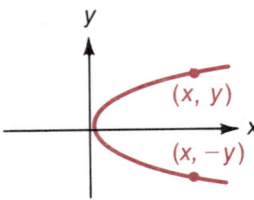

Symmetry about x-axis

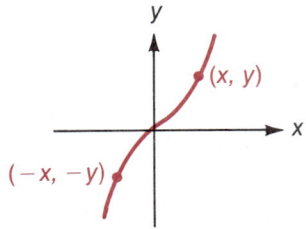

Symmetry about origin

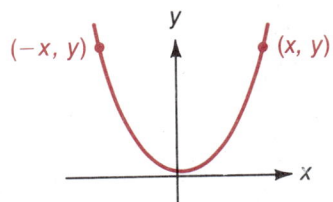

Symmetry about y-axis

Rules and Formulas

Midpoint Formula $M = \left(\dfrac{x_1 + x_2}{2}, \dfrac{y_1 + y_2}{2}\right)$ Midpoint of the line segment joining $P = (x_1, y_1)$ and $Q = (x_2, y_2)$

Distance Formula $d(P, Q) = \sqrt{(x_2 - x_1)^2 + (y_2 - y_1)^2}$ Distance between $P = (x_1, y_1)$ and $Q = (x_2, y_2)$

Circles

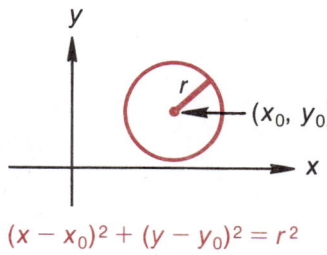

Center at (x_0, y_0), radius $= r$.

$$(x - x_0)^2 + (y - y_0)^2 = r^2$$

Lines

1. Slope $m = \dfrac{y_2 - y_1}{x_2 - x_1}$ $(x_1, y_1), (x_2, y_2)$ on a nonvertical line
   a. Horizontal lines — have slope $m = 0$
   b. Vertical lines — do not have a slope
2. Point-slope formula $(y - y_0) = m(x - x_0)$ slope $= m$, (x_0, y_0) on the line
3. Slope-intercept formula $y = mx + b$ slope $= m$, y-intercept $= b$
4. Vertical line formula $x = a$ $a = $ x-intercept
5. General formula $Ax + By + C = 0$ A, B not both 0
6. Parallel lines $m_1 = m_2$ slopes m_1, m_2
7. Perpendicular lines $m_1 m_2 = -1$ slopes m_1, m_2

Parabolas

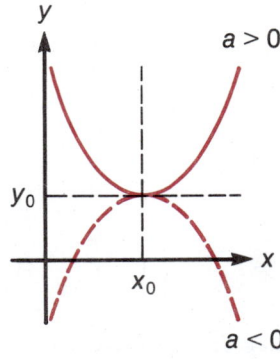

$y = ax^2 + bx + c$
$y - y_0 = a(x - x_0)^2$

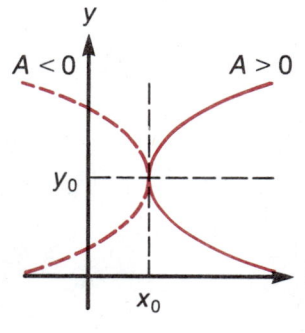

$x = Ay^2 + By + C$
$x - x_0 = A(y - y_0)^2$

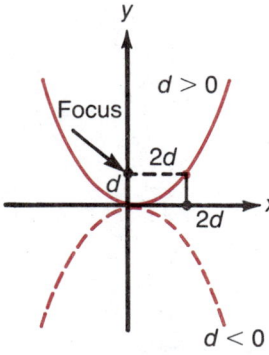

$y = \left(\dfrac{1}{4d}\right)x^2$

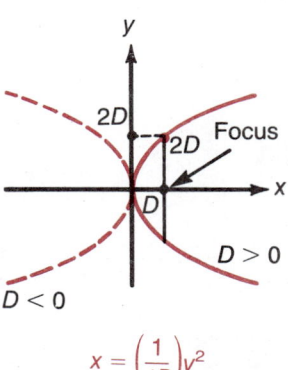

$x = \left(\dfrac{1}{4D}\right)y^2$

Ellipses

$$\dfrac{x^2}{a^2} + \dfrac{y^2}{b^2} = 1$$

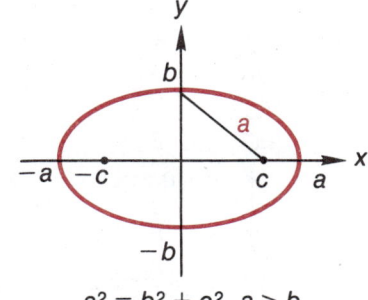

$a^2 = b^2 + c^2$, $a > b$

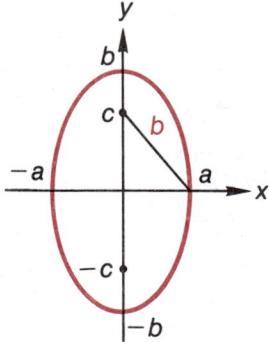

$b^2 = c^2 + a^2$, $b > a$

Hyperbolas

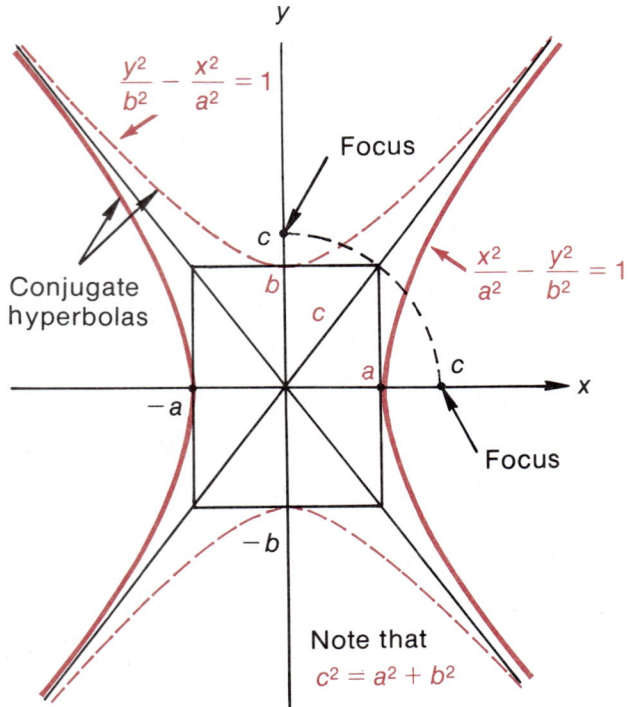

Techniques

Testing for Symmetry

1. Symmetry about the y-axis — Substitution of $-x$ for x leads to an equivalent equation.

2. Symmetry about the x-axis — Substitution of $-y$ for y leads to an equivalent equation.

3. Symmetry about the origin — Substitution of $-x$ for x and $-y$ for y leads to an equivalent equation.

Solving Quadratic Inequalities by Graphing

1. Find the roots of the equation $ax^2 + bx + c = 0$.
2. The corresponding parabola $y = ax^2 + bx + c$ crosses the x-axis at these points.
3. Determine from this parabola the region or regions in which the inequality is satisfied.

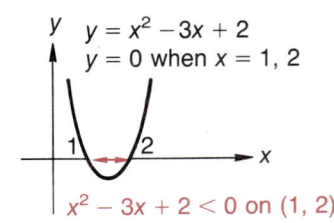

$y = x^2 - 3x + 2$
$y = 0$ when $x = 1, 2$

$x^2 - 3x + 2 < 0$ on $(1, 2)$

Graphing $Ax^2 + By^2 + Cx + Dy + E = 0$

1. Complete the squares in x and y.
2. Substitute $x' = x - h$, $y' = y - k$ and rewrite the equation in terms of x' and y'.
3. Sketch an x'-y' coordinate system centered at (h, k) on the x-y coordinate system.
4. Graph the equation in the x'-y' coordinate system.

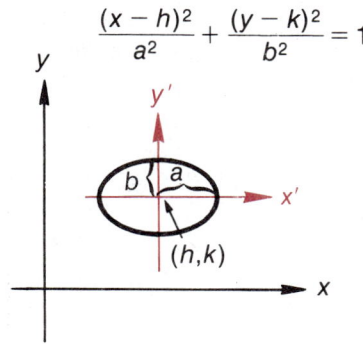

Section 3.8 Supplementary Exercises

For each of the following pairs of points.
  a. Plot them.
  b. Find the distance between them.
  c. Find the midpoint of the segment joining them.
  d. Write the equation of the line passing through them.
  e. Write the equation of the perpendicular bisector of the segment joining them.

1. $(-3, 0), (-7, 0)$
2. $(0, 2), (0, 5)$
3. $(6, 5), (4, 2)$
4. $(5, 2), (-1, -2)$
5. $(1, 0), (0, 2)$

Write the equation of the indicated circle.

6. Center $(1, 3)$, radius 2
7. Center $(-3, 4)$, radius 5
8. Center $(-4, -1)$, radius 3
9. Center $(4, 1)$, passing through $(6, 2)$
10. Center $(-10, 7)$, passing through $(-7, 10)$
11. Center $(-1, -2)$, passing through $(-3, -6)$
12. $(8, 26)$ and $(-6, -22)$ are at opposite ends of a diameter

In each of the following, determine whether the given points fall on a line.

13. $(-7, -6), (18, 4), (3, -2)$
14. $(-2, 6), (9, -2), (3, 2)$
15. $(-8, 3), (2, -1), (12, -5)$

In each of the following, find an equation of the line satisfying the stated conditions; m denotes slope, b = y-intercept, and a = x-intercept.

16. $m = 2, b = 6$
17. $m = -6, b = 3$
18. $m = -4, b = -2$
19. $m = 0$, passing through $(4, 2)$
20. $m = -\frac{1}{4}$, passing through $(-4, 3)$
21. $m = -\frac{1}{2}$, passing through $(-4, -5)$
22. $a = 6, b = 10$
23. $a = -3, b = 4$
24. $a = 0, b = 5$
25. Tangent to the circle $x^2 + y^2 = 25$ at the point $(3, 4)$
26. Tangent to the circle $x^2 + y^2 = 169$ at the point $(-12, 5)$

Determine whether the following pairs of lines are parallel, perpendicular, or neither.

27. $3x + 2y - 7 = 0$
    $9x + 6y + 3 = 0$
28. $3x - y + 2 = 0$
    $x + 3y - 2 = 0$
29. $11x - 3y + 5 = 0$
    $-22x + 6y - 3 = 0$
30. $7x - 3y - 4 = 0$
    $3x - 7y + 4 = 0$
31. $-4x + 2y - 6 = 0$
    $-3x - 6y = 0$

Write the equation of the line through the given point that is parallel to the given line and also of the line that is perpendicular to the given line.

32. $(2, 3)$, $2x + y - 5 = 0$
33. $(-1, 5)$, $y = -2$
34. $(2, 1)$, $3x - 9y - 6 = 0$
35. $(0, 7)$, $x = 3$
36. $(1, -2)$, $x - 2y - 3 = 0$

Graph the following lines. Find their slopes and intercepts.

37. $x - 5y - 10 = 0$
38. $2x + y + 8 = 0$
39. $4x = 6$
40. $10y = 0$
41. $3x - y = 9$
42. $-5x + 10y - 10 = 0$

Graph each of the following parabolas. In each, find the vertex, the line of symmetry, and the intercepts.

43. $y = -2x^2 - 1$
44. $x = 2y^2 - 4y$
45. $y = -2x^2 - 2x - 1$
46. $x = 2y^2 - 12y + 8$
47. $y = -2x^2 + 4x + 10$
48. $y = -x^2 + 2x - 2$
49. $x = y^2 - 2y + 6$

Write the equation of the indicated parabola.

50. Vertex at $(3, 4)$, directrix $y = 2$
51. Vertex at $(-2, -1)$, focus at $(2, -1)$
52. Focus at $(1, 3)$, directrix $y = 5$

Solve the following inequalities. Write the solution in interval notation.

53. $x^2 - x - 6 \leq 0$
54. $2x^2 + 4 < 6x$
55. $2x^2 - 6x - 10 > -x^2 + x + 10$
56. $4x^2 - 8x - 20 \geq 2x^2 - 10x + 4$
57. $x^4 - x^2 - 6 < 0$
58. $4x - 3\sqrt{x} - 1 \geq 0$

Sketch the graphs of the following relations.

59. $x^2 + y^2 = 4$
60. $x^2 + y^2 \leq 4$
61. $x = |y|$
62. $|x| = y^2$
63. $x < |y|$
64. $|x| < y^2$
65. $y^2 = 2x^2$
66. $y > 2x^2$
67. $x^2 - 4x + y^2 - 6y = 12$
68. $x^2 + 6x + y^2 + 10y + 30 = 0$
69. $x^2 + y^2 < 4x + 8y + 29$
70. $y^2 + 6y \geq 4x - x^2 + 12$
71. $x^2 - 10x + y^2 + 6y + 18 > 0$
72. $x^2 + 10y < 10x - y^2 - 25$

Write the equation of the locus of points equidistant from the given data.

73. Points $(-1, 3)$ and $(3, 3)$
74. Points $(2, 6)$ and $(6, 2)$
75. Points $(3, 4)$ and $(7, 2)$
76. Points $(-4, -6)$ and $(2, 4)$
77. Point $(2, -2)$ and the line $y = 4$
78. Point $(5, 2)$ and the line $x = 1$

Sketch the graphs of the following equations. Indicate vertices and foci. For hyperbolas, indicate the asymptotes, write the equation of the conjugate hyperbola, and sketch its graph also. If a parabola, also indicate the directrix.

79. $\dfrac{x^2}{36} + \dfrac{y^2}{49} = 1$
80. $x^2 + 4y^2 = 4$
81. $25x^2 + 16y^2 = 400$
82. $\dfrac{x^2}{169} + \dfrac{y^2}{144} = 1$
83. $49x^2 + 625y^2 = 30{,}625$
84. $\dfrac{x^2}{4} - \dfrac{y^2}{25} = 1$
85. $x^2 - 9y^2 = 9$
86. $\dfrac{y^2}{576} - \dfrac{x^2}{49} = 1$
87. $36y^2 - 9x^2 = 324$
88. $4y^2 - 3x^2 = 24$
89. $16x^2 - 25y^2 = 400$
90. $8x + 4y - y^2 = 12$
91. $9x^2 + 4y^2 + 36x - 24y + 36 = 0$
92. $4x^2 - 9y^2 + 24x + 36y = 144$
93. $4x^2 + 9y^2 + 16x - 18y = 11$
94. $x^2 + 6x = 4y - 1$
95. $y^2 - 4x^2 - 2y + 8x = 7$
96. $9x^2 - 4y^2 - 54x - 16y + 61 = 0$
97. $25x^2 + 9y^2 - 150x - 90y + 225 = 0$
98. $x^2 - 2x + y = 3$

Find the equation of the central ellipse satisfying the following conditions.

99. Focus at $(0, 3)$ and horizontal axis of length 8
100. Focus at $(0, 4)$ and minor axis of length 6
101. Vertex at $(-2, 0)$ and focus at $(\sqrt{3}, 0)$
102. Passing through $(0, 5)$ and $(\frac{12}{5}, 4)$
103. Vertex at $(5, 0)$ and passing through $(-4, \frac{39}{5})$

Find the equation of the central hyperbola satisfying the following conditions.

104. Vertex at $(0, 4)$, asymptote $y = 2x$
105. Vertex at $(-5, 0)$, asymptote $y = -\dfrac{4x}{5}$
106. Vertex at $(-2, 0)$, asymptote $x = -2y$
107. Vertex at $(0, 2)$, passing through $(2, 3)$
108. Vertex at $(0, 3)$, passing through $(-4, -6)$
109. Vertex at $(4, 0)$, passing through $(-8, 10)$
110. Vertex at $(2, 0)$, focus at $(-4, 0)$
111. Focus at $(0, \sqrt{10})$, asymptote $x = 3y$

Write the equation of the ellipse satisfying the given conditions.

112. Lower vertex at $(7, -7)$, left vertex at $(4, -2)$
113. Upper vertex at $(-3, 6)$, left vertex at $(-8, 2)$
114. Upper vertex at $(3, 3)$, right vertex at $(8, 0)$
115. Lower vertex at $(-1, -5)$, right vertex at $(4, 8)$

Write the equation of the hyperbola satisfying the given condition. Do the same for the conjugate hyperbola.

	OPENS	CENTERED AT	a	b
116.	Horizontally	$(-7, 2)$	8	6
117.	Vertically	$(3, 5)$	3	9
118.	Horizontally	$(-2, -1)$	2	3
119.	Vertically	$(4, -6)$	7	3

In the following problems, let h be altitude in feet of a projectile t seconds after it is fired. Determine
    a. *its maximum altitude,*
    b. *the time at which it attains its maximum altitude, and*
    c. *the time at which it hits the ground.*

120. $h = 480t - 16t^2$
121. $h = 32t - 16t^2 + 128$
122. $h = 800t - 16t^2 + 4400$

Functions and Their Graphs

4

In this chapter, we consider those relationships in which only *one* value of a second quantity corresponds to any given value of a first quantity. The second quantity is then said to be a **function** of the first. (A more formal definition of function is given in Section 4.1.) For example, if your car can travel 50 miles on a gallon of gas, you should be able to travel 100 miles on 2 gallons of gas, 250 miles on 5 gallons, and so on. The distance your automobile can travel without a fuel stop is uniquely *determined by*, or *is a function of*, the number of gallons its gas tank holds.

On the other hand, the relationship "y is a square root of x" does not describe a function; each positive real number has *two* square roots, not just one. If two quantities are related in some way that may or may not describe a function, we say that there is a **relation** between the two quantities.

In this chapter, we will also develop additional graphing techniques, the central idea being to modify and move simpler graphs in order to obtain the graphs of more complicated relations.

Section 4.1

Functions

Mail-order sales companies generally impose a shipping charge in addition to the cost of the item purchased. For instance, the cost of shipping an item from Chicago to Phoenix might be $1.10 for the first 4 pounds and 11¢ for each additional pound. The shipping charge on a 3-pound item is then $1.10, and the charge on a 6-pound item is $1.32. The shipping cost is expressed in terms of the weight of the item.

If we travel at a rate of 50 miles per hour, the relationship $D = 50t$ expresses distance in terms of travel time. The distance traveled depends on the amount of time spent traveling.

In each of these examples, a quantity—cost or distance—is *uniquely* determined by another quantity—the weight of an item or the travel time, respectively. These examples illustrate functional relationships or **functions.** Informally, a function is a rule that enables us to compute exactly one "new number" from each "given number." However, the concept of function is not limited to numbers. For instance, each house in a city has a unique street address; each person has a unique name at any given time (legally at least). Some teachers assign students certain seats for class. Each of these also represents a functional relationship: "street address" is a function of "house," "name" is a function of "person," and "seat location" is a function of "student in the class."

218 Chapter 4 Functions and Their Graphs

> **DEFINITION**
>
> 1. A **function** f from a set X to a set Y is a *correspondence*, or *rule*, that assigns to each element $x \in X$ a unique element $y \in Y$. We say *y is a function of* x and write $y = f(x)$.
> 2. The set of x values for which the function is defined is called the **domain** of the function f. The **range** of f is the set of all y values that can be obtained by considering all x values in the domain.
> 3. If y is a function of x, then y is called a **dependent variable** and x is called an **independent variable.**

You may find it helpful to visualize a function as a machine in which:

1. A given object is fed into the machine.
2. The machine performs some operations on the object.
3. The unique result then comes out of the machine.

See Figure 4.1.

Figure 4.1

The symbol $f(x)$ is read "f of x" or "function of x." If an equation or formula is used to express the relationship between x and y, then we find the value y of a function for some particular value of x by substituting that value of x into the functional equation.

EXAMPLE 1 Let $y = f(x) = x^2 + 2x - 3$. Find the values of this function for the given values of x.

a. $x = -1$
b. $x = 20$

SOLUTION We substitute -1 and 20, respectively, for x into the formula given for $f(x)$:

$$f(x) = x^2 + 2x - 3$$

a. $f(-1) = (-1)^2 + 2(-1) - 3 = -4$
b. $f(20) = (20)^2 + 2(20) - 3 = 437$

EXAMPLE 2 Consider an automobile getting 50 miles per gallon of gasoline and let

x = number of gallons of gasoline available
y = number of miles that can be traveled

a. Express y as a function of x.
b. Find the domain and range of this function.

SOLUTION **a.** The relationship between x and y is expressed by the equation

$$y = f(x) = 50x$$

b. The domain of this function is the set of all $x \geq 0$; we do not consider a negative amount of gasoline. The domain is the interval $[0, \infty)$. The range consists of all values of $y = 50x$ corresponding to $x \geq 0$. If

$$0 \leq x < \infty$$

then

$$0 \leq 50x < \infty$$

Thus the range consists of all $y \geq 0$. The range is also the interval $[0, \infty)$. ∎

Functions are frequently described by giving only the functional equation, such as $f(x) = \sqrt{4 - x^2}$, without indicating its domain or range. The domain is then implicitly taken to be the set of values of x for which the function or expression involving x is defined.

> **DETERMINING DOMAIN AND RANGE**
>
> If the domain of a function $y = f(x)$ is not specified, it is understood that the domain consists of all real numbers x for which the expression or equation yields a real number y; the range is then determined from these x values.

EXAMPLE 3 Find the implied domain and corresponding range for the following functions.

a. $f(x) = 3x - 2$
b. $f(x) = 10$

SOLUTION **a.** For each real number x, $f(x) = 3x - 2$ is a real number. Thus the domain of f is the entire set of real numbers $R = (-\infty, \infty)$. As x runs over the interval $(-\infty, \infty)$, so do the values of $f(x) = 3x - 2$. Thus its range is also $(-\infty, \infty)$.

b. Again the domain is $R = (-\infty, \infty)$. However, this time the range consists of just one value: 10. The range is $\{10\}$. ∎

EXAMPLE 4 Find the implied domain and corresponding range for the function defined by $y = f(x) = \sqrt{4 - x^2}$ if x and y are to be real numbers.

SOLUTION   In order for the square root to be a real number, we must have

$$0 \leq 4 - x^2$$
$$x^2 \leq 4 \quad \text{(Add } x^2 \text{ to both sides)}$$
$$|x| \leq 2 \quad \text{(Take the principal square root of both sides)}$$
$$-2 \leq x \leq 2$$

The domain of this function is the interval $[-2, 2]$. The corresponding functional values range between 0 (when $x = \pm 2$) and 2 (when $x = 0$). The range for this function consists of those y values satisfying $0 \leq y \leq 2$. The range is the interval $[0, 2]$. ■

In actual practice, various letters and symbols such as F, G, H, f, g, h, and θ are also used to denote functions. Letters other than x and y are used to denote independent and dependent variables, too. The letters used are not important. What matters is the functional relationship. Thus,

$$s = f(t) = t^2 - 1$$
$$y = g(x) = x^2 - 1$$
$$c = H(\$) = \$^2 - 1$$
$$\Delta = G(\square) = \square^2 - 1$$

all describe the same function; even though different symbols are used, exactly the same relationship or computational process is described. Each value of the independent variable is squared and then 1 is subtracted from the result. For this reason, the symbols used to describe the variables are sometimes called **dummy variables.** The *relationship* described by these symbols is what matters.

Most of the functions and relations we use will be described by an equation or formula. Sometimes more than one formula is needed to describe a function; different formulas may be necessary for various elements of the domain.

EXAMPLE 5   Consider the shipping cost example in the first paragraph of this section. Let

   $c = $ shipping cost, in dollars
   $w = $ weight of the item, in pounds

Express c as a function of w.

SOLUTION   The cost of shipping an item weighing 4 pounds or less is $1.10. Thus,

$$c = \$1.10 \quad \text{when} \quad 0 < w \leq 4$$

To ship more than 4 pounds, the cost is $1.10 plus 11¢ for each extra pound:

$$c = \$1.10 + 0.11(w - 4) \quad \text{when} \quad w > 4 \quad (w \text{ is a natural number})$$

We state these together as

$$c = f(w) = \begin{cases} \$1.10 & 0 < w \leq 4 \\ \$1.10 + 0.11(w - 4) & w > 4 \end{cases}, w \in N \quad ■$$

In Example 5, different formulas are used to define $f(w)$ on different intervals. Such a function is said to be defined **piecewise.**

Graphs and Functions

We have already studied two important classes of functions in detail:

Linear functions $\quad y = ax + b$

and

Quadratic functions $\quad y = ax^2 + bx + c$

When we purchase several identical items at a specified price per item, the total cost is a linear function of the number of items purchased. So also, the distance traveled at a constant speed is a linear function of the time of travel. Quadratic functions help to describe phenomena such as the effect of gravity on moving objects like baseballs.

Studying the graphs of various functions increases our understanding of the functions themselves. We have seen that the graph of a linear function is a straight line (Section 3.3); the graph of a quadratic function is a parabola (Section 3.4).

In Section 3.2, we sketched the graph of a relation by plotting those points that satisfied the relation. A function is a special kind of relation—one in which each value of a given variable determines precisely one value of another variable.

If y is a function of x, then each x value determines one and only one y value. This means that each vertical line intersects the graph in at most one point.

VERTICAL LINE TEST

A graph represents a function $y = f(x)$ if and only if each vertical line intersects the graph in at most one point. In this case, the graph consists of those points (x, y) whose altitude above or below the x-axis at each x is given by $y = f(x)$.

Similarly, a given y value determines precisely one x value when no two points on the graph lie on the same horizontal line.

HORIZONTAL LINE TEST

A graph determines x as a function of y if and only if each horizontal line intersects the graph in at most one point.

Figure 4.2 shows the graphs of some functions and relations.

Figure 4.2

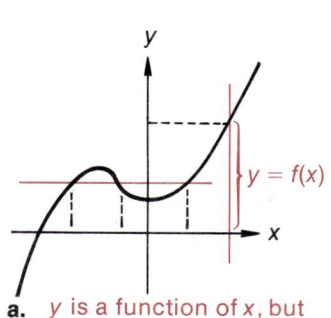
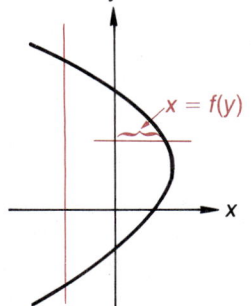
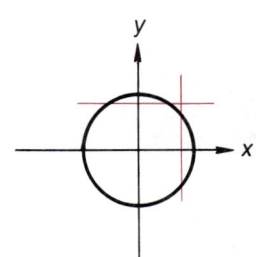

a. y is a function of x, but x is not a function of y.

b. y is a function of x, and x is a function of y.

c. x is a function of y, but y is not a function of x.

d. Neither x nor y is a function of the other.

EXAMPLE 6

Determine whether the following equation describes y as a function of x. Is x a function of y? If either is a function, sketch its graph and indicate its domain and range.

$$y - x^2 = 4$$

SOLUTION

Since $y = x^2 + 4$, we see that for each x, one and only one value of y can be computed. Thus, *y is indeed a function of x.* However,

$$x^2 = y - 4$$

so that

$$x = \pm\sqrt{y - 4}$$

For $y = 8$, note that $x = \pm 2$; thus, *x is not a function of y.* From Section 3.4, we know that the graph of $y = f(x) = x^2 + 4$ is a parabola opening upward with vertex at (0, 4) (see Figure 4.3). Using the horizontal line test on the graph also indicates that x is not a function of y. Note that $f(x) = x^2 + 4$ is defined for all x. Its domain is the entire real number line. On the other hand, $y = f(x) = x^2 + 4 \geq 4$; its range is the interval $4 \leq y < \infty$, or $[4, \infty)$.

Figure 4.3

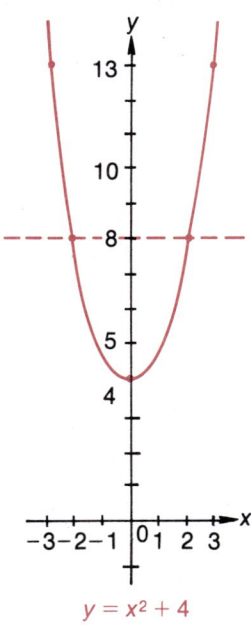

$y = x^2 + 4$

EXAMPLE 7

The Environmental Equipment Company has developed a new generator. The prototype model has cost the company $10 million in research, development, and manufacturing expenses. It will cost another $40 million to modify the manufacturing facility to make these generators. Then labor and materials for manufacturing and installing one of these generators will cost the company another $2 million. If each generator can be sold for $3 million, how many must the company sell before it begins to realize a profit?

SOLUTION

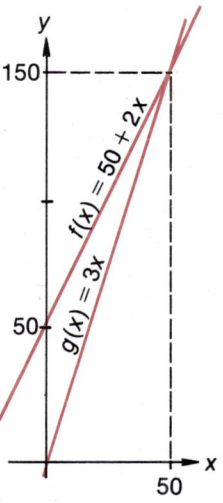

Figure 4.4

The total amount (in millions of dollars) that must be spent to produce x generators is

$$f(x) = 10 + 40 + 2x$$
$$= 50 + 2x$$

From the sale of these generators, the company will receive

$$g(x) = 3x$$

Each of these functions is graphed as a straight line in Figure 4.4. Income exceeds costs when the graph of g rises above the graph of f. This happens when $g(x) \geq f(x)$:

$$3x \geq 50 + 2x$$
$$x \geq 50 \quad \text{(Subtract } 2x \text{ from both sides)}$$

Thus, the company must sell at least 50 generators before it begins to make a profit. ∎

We noted earlier that not all functions are described by a single equation. In fact, a function is defined by *any* rule associating a value $f(x)$ to each x in the domain of the function. Of particular importance in calculus and more advanced courses are functions defined by several formulas, each applying in different intervals as illustrated in the next example.

EXAMPLE 8

Sketch the graph of the function defined piecewise by

$$f(x) = \begin{cases} x^2, & x < 0 \\ 1, & x = 0 \\ 2 - x, & x > 0 \end{cases}$$

SOLUTION

The graph of $y = x^2$ was sketched in Example 5a, Section 3.2. We shall use the portion of this graph corresponding to $x < 0$. It is easy to plot the value $y = 1$ when $x = 0$. The graph of $y = 2 - x$ is a line of slope -1 with $y = 0$ when $x = 2$. Thus, we sketch the portion of this line corresponding to $x > 0$. The complete graph appears in Figure 4.5. An open circle, ∘, is used to indicate that a given endpoint is *not* on the graph.

Figure 4.5

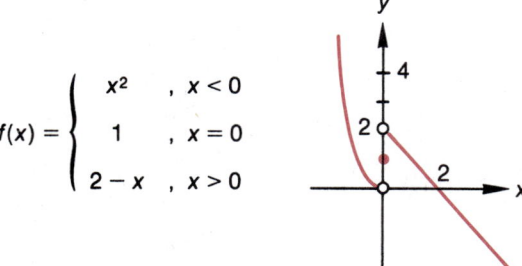

∎

Section 4.1 Exercises

Evaluate the following functions at the indicated points.

1. $g(y) = 5y - 6$; find $g(2)$, $g(3)$, $g(-3)$
2. $H(m) = 8 - 3m$; find $H(0)$, $H(4)$, $H\left(\frac{1}{3}\right)$
3. $f(t) = t^2 - 5t + 1$; find $f(2)$, $f(3)$, $f(-3)$
4. $N(z) = 5 + 3z - z^2$; find $N(3)$, $N(4)$, $N(-4)$
5. $\theta(b) = (b^2 + 2b - 3)$; find $\theta(1)$, $\theta(2^2)$, $[\theta(2)]^2$
6. $h(n) = n^3 - 2n^2 + 5n + 7$; find $h(0)$, $h(1)$, $h(-1)$
7. $f(x) = \dfrac{1+x}{1-x}$; find $f(0)$, $f(-1)$, $f(2)$
8. $T(z) = \dfrac{2z^2 - z - 1}{z + 2}$; find $T(1)$, $T(-3)$, $T\left(\dfrac{1}{2}\right)$
9. $r(u) = \sqrt{u^2 - 2u + 1}$; find $r(4)$, $r(-3)$, $r(2)$
10. $U(t) = \sqrt{t + 2} + \sqrt{t - 3}$; find $U(7)$, $U(3)$, $U(5)$

In each of the following, find the understood domain of the given function; if indicated, also find the range. Both domain and range are to be restricted to real numbers.

11. $f(x) = \sqrt{x - 1}$; range also
12. $H(s) = \sqrt{s^2 + 2s + 1} + 2$; range also
13. $F(v) = \dfrac{1}{v^2 - 4v + 4}$; range also
14. $p(t) = \dfrac{1}{t^2 - 4}$
15. $g(r) = \dfrac{\sqrt{25 - r^2}}{r + 2}$
16. $p(x) = \dfrac{\sqrt{16 - x^2}}{|x - 3|}$
17. $G(x) = \dfrac{\sqrt{x^2 - 4}}{\sqrt{9 - x^2}}$
18. $K(y) = \dfrac{2y - 3}{\sqrt{y + 1}}$
19. $R(x) = \dfrac{\sqrt{36 - x^2}}{x^2 - x - 2}$
20. $H(b) = \dfrac{3b - 2}{6b^2 + b - 1}$
21. $f(x) = \sqrt{x(x - 1)(x - 2)}$
22. $g(x) = \sqrt{(x + 3)(x + 1)(x - 1)}$
23. $h(x) = \sqrt{\dfrac{x - 1}{x - 2}}$
24. $F(x) = \sqrt{\dfrac{x - 3}{x(x + 1)}}$

Sketch the graphs of the following functions.

25. $f(x) = 2x + 1$
26. $g(x) = x^2 - 3$
27. $F(t) = \dfrac{1}{t}$
28. $H(v) = v^3$
29. $R(x) = |x^3|$
30. $T(x) = 1 - x^2$
31. $G(s) = s^2 + 1$
32. $H(u) = u^2 - 2u + 1$
33. $g(x) = x^2 + 2x + 2$
34. $f(x) = \sqrt{x}$
35. $F(x) = \sqrt{x - 1}$
36. $G(x) = \sqrt{x^2 + 1}$
37. $H(x) = \sqrt{x^2 - 1}$
38. $f(x) = \begin{cases} 1, & x \leq -1 \\ 2, & x > -1 \end{cases}$
39. $g(x) = \begin{cases} 0, & x \leq 0 \\ 1, & 0 < x < 1 \\ 2, & x \geq 1 \end{cases}$
40. $F(x) = \begin{cases} 2x + 1, & x \leq 1 \\ 2, & 1 < x \leq 2 \\ 1 - x, & x > 2 \end{cases}$
41. $G(t) = \begin{cases} t, & t < 0 \\ 1, & t = 0 \\ t, & 0 < t \leq 2 \\ 4 - t, & t > 2 \end{cases}$
42. $H(v) = \begin{cases} 1 - v, & v \leq 0 \\ v^2 + 1, & v > 0 \end{cases}$
43. $R(x) = \begin{cases} x^2, & x < 0 \\ -1, & x = 0 \\ \sqrt{x}, & x > 0 \end{cases}$
44. $Q(x) = \begin{cases} x, & x < 0 \\ 2, & x = 0 \\ \sqrt{x + 1}, & x > 0 \end{cases}$

In each of the following, determine whether y is a function of x and whether x is a function of y.

45. $x^2 = y - 1$
46. $2x - 3y = 4$
47. $x < 4y$
48. $y \leq 3x^2 + 2$
49. $x = y^2 + 4$
50. $y^2 = x^2 - 1$
51. $y = x^3$
52. $y^5 = x^2$
53. $y^2 + 2y - x = 4$
54. $x^2 + 6x + 2y = 5$

Which of the following graphs define y as a function of x? Which define x as a function of y?

55.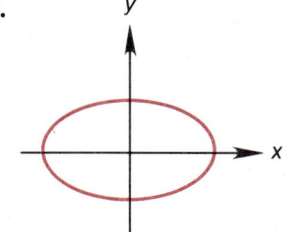

56.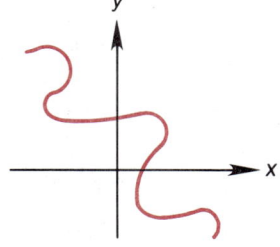

57.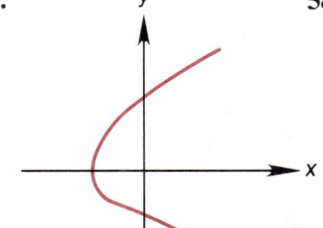

58.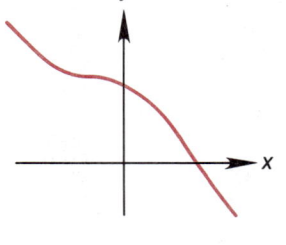

59.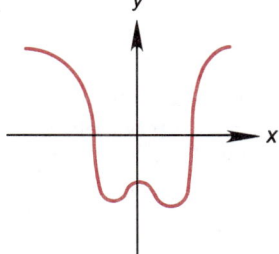

60.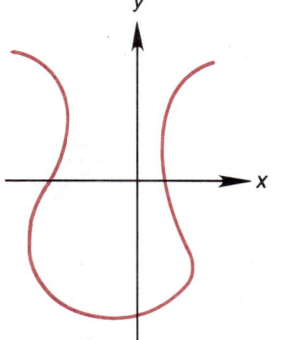

61.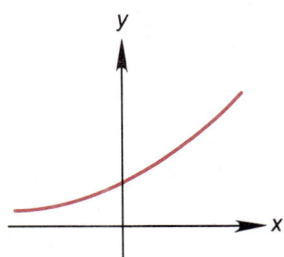

62.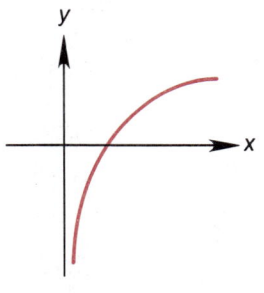

63.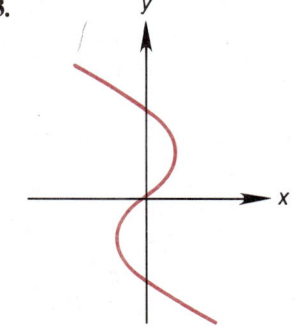

64.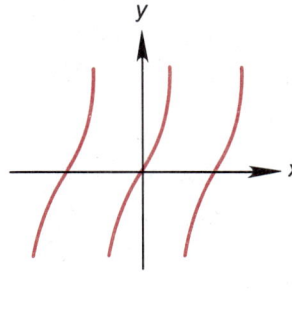

65.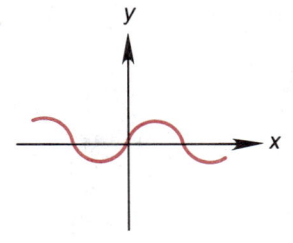

66.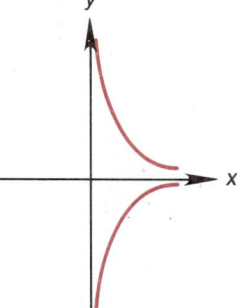

67.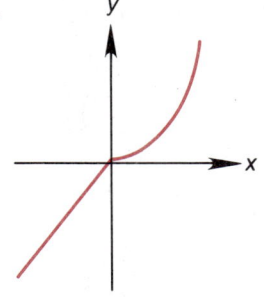

68.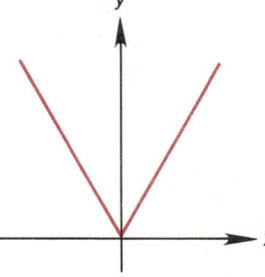

69.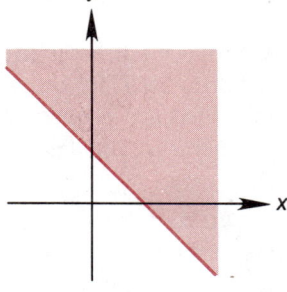

70. Ann drives her automobile at 55 miles per hour.
  a. Express the distance traveled as a function of time.
  b. How far does she travel in $2\frac{1}{2}$ hours?
  c. Describe the domain and range of this function.
  d. Sketch the graph of this function.

71. On a recent vacation trip, Bill traveled at a rate of 25 miles per hour for 3 hours while he enjoyed the beautiful sights. He then continued at a rate of 55 miles per hour.
  a. Express the distance he traveled as a function of time.
  b. Determine the time when he averages 45 miles per hour.
  c. How far did he travel in 8 hours?
  d. Describe the domain and range of this function.

72. A snowball of radius 10 inches is left in the sun. It melts at a rate that decreases its radius by 1 inch each hour.
    a. Express its radius, volume, and surface area as a function of time. (See Exercise 7, Section 2.2.)
    b. Find the domain and range of each of these functions.

73. A grocer must place an order for canned goods. The price is $2 per case for 9 or fewer cases, $1.75 per case for 10 to 99 cases, and $1.50 per case for 100 or more cases.
    a. Express the cost as a function of the number of cases ordered.
    b. Which costs more, 9 cases or 10?
    c. For what number of cases is it just as cheap to buy 100 cases even though that many are not needed.
    d. Describe the domain and range of this function.

74. The Gas and Electric Company estimates Ms. Walsh's monthly utility bill to be $3.75 times the absolute value of the difference between 60°F and the average temperature during the month.
    a. Express her estimated utility bill as a function of the average monthly temperature.
    b. What is her estimated bill when the average temperature is 80°F?
    c. What is her estimated bill when the average temperature is 20°F?
    d. Describe the domain and range of this function.
    e. Sketch the graph of this function.

75. The U-No-Pedal Moped Sales Company must pay $200 for each moped plus $200 per month to the manufacturer for advertising expenses. It sells each moped for $400, but this is offset by $4000 per month of overhead.
    a. Express expenses, income, and profit as functions of the number of mopeds sold in a given month.
    b. Determine the level of sales activity that generates a profit for the dealer.
    c. Describe the domain and range of the profit function.

76. A department store charges interest on credit card purchases as follows. The finance charge on a balance of $600 or less is $1\frac{1}{2}\%$ per month; but there is a minimum service charge of 75¢ on balances less than $50. Of course, no charge is levied when the balance is $0. On the balance that exceeds $600, the finance charge is 1% per month.
    a. Express the finance charge as a function of the outstanding balance.
    b. What is the finance charge on a balance of $750?
    c. What is the finance charge on a balance of 93¢?
    d. Describe the domain and range of this function.

77. For stability in a certain wildlife population, there must be at least 25 rabbits for each fox.
    a. Express this statement as a relation between the number of rabbits and the number of foxes.
    b. Does this relationship express either the number of rabbits or the number of foxes as a function of the other?

78. An advertising agency finds that each minute of commercial television time for advertising beer generates $2.6 million in sales for its client.
    a. If each commercial minute costs $100,000, express the net revenue from beer sales generated by television commercials as a function of the number of commercial minutes purchased.
    b. How much revenue is generated by a commerical blitz of 20 minutes in a given week?
    c. Describe the domain and range of this function.

79. A heart patient enters the hospital with a resting heart rate of 120 beats per minute. Administration of 2 milligrams of medication per hour reduces the rate by 5 beats per minute.
    a. Express the heart rate of this patient as a function of the amount of medication prescribed.
    b. What should the prescription be in order to reduce the rate to 70 beats per minute?

c. Assuming that a patient dies when the pulse rate reaches 10 beats per minute or 250 beats per minute, what is the domain of this function if the physician wants to keep the patient alive?

d. What is the range?

80. The altitude of a falling object t seconds after it is dropped from an altitude of h feet is given by $f(t) = h - 16t^2$.

a. Two girls decide to measure the height of a tall building by dropping a pebble from its top. It takes the pebble 10 seconds to reach the ground. How high is the skyscraper?

b. Find the domain and range of this function.

Section 4.2

Graphing Techniques

Often the graph of a seemingly complicated relation can be found by *scaling* (*stretching* or *compressing*), *reflecting*, and/or *translating* (*shifting*) a well-known graph. These techniques and others are developed and illustrated in this section.

Consider, for example, the function whose graph is sketched in Figure 4.6. Various modifications of this function are discussed in the following examples.

Figure 4.6

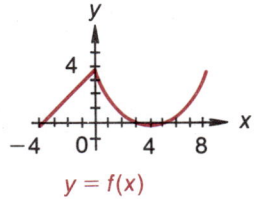

$y = f(x)$

Translation

EXAMPLE 1

With $y = f(x)$ as graphed in Figure 4.6, sketch the graph of each of the following.

a. $y = f(x) + 2$ and $y = f(x) - 2$

b. $y = f(x - 2)$ and $y = f(x + 2)$

SOLUTION

a. For $y = f(x) + 2$, the y-coordinate of each point in Figure 4.6 is increased two units; that is, the graph of $y = f(x)$ is moved two units **upward**. Similarly, the graph of $y = f(x) - 2$ would be obtained by moving the graph of $y = f(x)$ two units **downward** (see Figure 4.7).

Figure 4.7

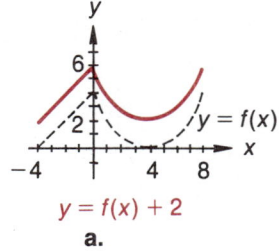

$y = f(x) + 2$

a.

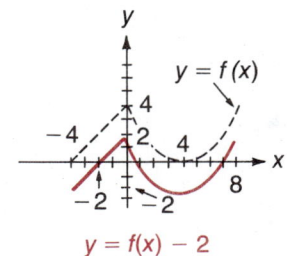

$y = f(x) - 2$

b.

b. For $y = f(x - 2)$, we plot

$y = f(-4)$ when $x = -2$
$y = f(0)$ when $x = 2$
$y = f(4)$ when $x = 6$
$y = f(8)$ when $x = 10$

Thus the graph of $y = f(x)$ is shifted two units to the right. In the same way we see that the graph of $y = f(x + 2)$ would be obtained by shifting the original graph two units to the left (see Figure 4.8).

Figure 4.8

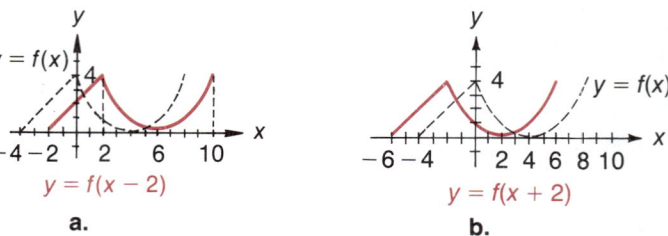

Modifications of the type discussed in Example 1 are called **translations**. These ideas are summarized in the table.

TRANSLATION

ALGEBRAIC CHANGE	CHANGE IN GRAPH
1. $y = f(x) + c$	1. The graph of $y = f(x)$ is moved vertically by c units. The graph moves up if $c > 0$ and down if $c < 0$ (see Figure 4.7).
2. $y = f(x - a)$	2. The graph of $y = f(x)$ is moved horizontally. The graph moves right if $a > 0$ and left if $a < 0$. Begin by plotting $y = f(0)$ when $x = a$ (see Figure 4.8).

Reflection

EXAMPLE 2   With $y = f(x)$ as graphed in Figure 4.6, sketch the graph of each of the following.

a. $y = -f(x)$ **b.** $y = f(-x)$

SOLUTION

a. The y-coordinate of each point on the graph of $y = f(x)$ changes sign; that is, the graph of $y = f(x)$ is reflected through the x-axis (see Figure 4.9).

b. We still plot $y = f(0)$ when $x = 0$. But we plot

$y = f(-4)$ when $x = 4$
$y = f(4)$ when $x = -4$
$y = f(8)$ when $x = -8$

The graph of $y = f(x)$ is reflected through the y-axis (see Figure 4.10).

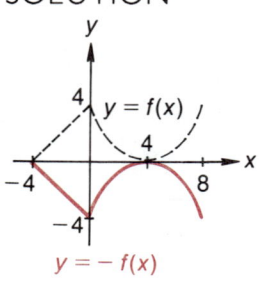

Figure 4.9

Figure 4.10

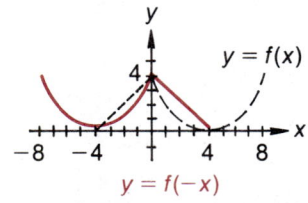

REFLECTION

ALGEBRAIC CHANGE	CHANGE IN GRAPH
1. $y = -f(x)$	1. The graph of $y = f(x)$ is reflected through the x-axis (see Figure 4.9).
2. $y = f(-x)$	2. The graph of $y = f(x)$ is reflected through the y-axis (see Figure 4.10).

Scaling

EXAMPLE 3 Again, with $y = f(x)$ as graphed in Figure 4.6, sketch the graph of each of the following.

a. $y = 2f(x)$ and $y = \dfrac{1}{2}f(x)$

b. $y = f(2x)$ and $y = f\left(\dfrac{1}{2}x\right)$

SOLUTION

a. For $y = 2f(x)$, the y-coordinate of each point on the graph of $y = f(x)$ is doubled. The graph in Figure 4.6 is stretched vertically. Similarly, the graph of $y = \frac{1}{2}f(x)$ would be obtained from the graph of $y = f(x)$ by vertical compression (see Figure 4.11).

Figure 4.11

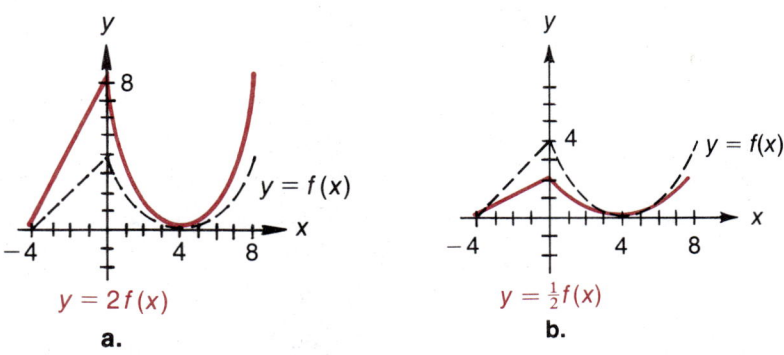

b. For $y = f(2x)$, observe that we must plot

$y = f(-4)$ when $x = -2$
$y = f(0)$ when $x = 0$
$y = f(4)$ when $x = 2$
$y = f(8)$ when $x = 4$

The graph of $y = f(x)$ is compressed in the horizontal direction.

Similarly, the graph of $y = f(\tfrac{1}{2}x)$ is obtained by stretching the graph of $y = f(x)$ in the horizontal direction. For $y = f(\tfrac{1}{2}x)$, we plot

$y = f(-4)$ when $x = -8$
$y = f(8)$ when $x = 16$

See Figure 4.12.

Figure 4.12

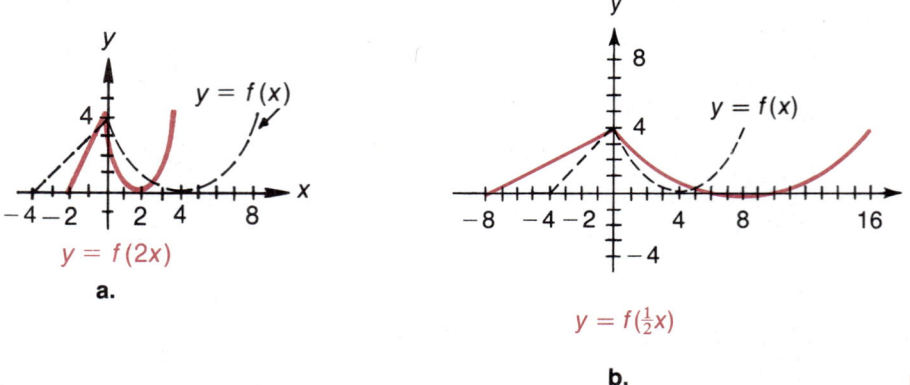

a.

$y = f(\tfrac{1}{2}x)$
b.

This example illustrated changes in the horizontal and vertical scales of a graph. The techniques are summarized in the table.

SCALING

ALGEBRAIC CHANGE	CHANGE IN GRAPH
1. $y = Af(x)$	1. A vertical stretch or compression is effected on the graph of $y = f(x)$. (See Figure 4.11.)
a. $\|A\| > 1$	a. Vertical stretch
b. $\|A\| < 1$	b. Vertical compression
2. $y = f(ax)$	2. A horizontal stretch or compression is effected on the graph of $y = f(x)$. (See Figure 4.12.)
a. $\|a\| > 1$	a. Horizontal compression
b. $\|a\| < 1$	b. Horizontal stretch

If $A < 0$ or $a < 0$, a reflection is also involved.

Note the differing stretch-compression behavior.

Stretch: $|A| > 1$ or $|a| < 1$
Compression: $|A| < 1$ or $|a| > 1$

Note that in all the changes discussed so far, changes in the independent variable cause horizontal modifications of the graph, whereas changes in the dependent variable cause vertical modifications.

The principles behind the techniques developed here can also be adapted to sketching graphs of relations giving x in terms of y. We shall do this in part (b) of the following example. In Example 4, we show essentially that the graph of any quadratic expression

$$y = ax^2 + bx + c, \quad a \neq 0$$

or

$$x = Ay^2 + By + C, \quad A \neq 0$$

is a parabola. We shall begin by completing the square as in Section 3.4, but now a basic graph will be modified and moved until we find the final graph.

EXAMPLE 4

Use the graphing techniques developed in this section to sketch the following graphs.

a. $y = 2x^2 - 4x + 3$
b. $x = y^2 + 2y - 1$

SOLUTION

a. Completing the square, we obtain

$$\begin{aligned} y &= 2x^2 - 4x + 3 \\ &= 2(x^2 - 2x \quad) + 3 \\ &= 2(x^2 - 2x + 1) + 3 - 2 \\ &= 2(x - 1)^2 + 1 \\ y - 1 &= 2(x - 1)^2 \end{aligned}$$

Then we sketch the following sequence of graphs (see Figure 4.13). Note that the vertex is (1, 1) in agreement with our observations in Section 3.4.

Figure 4.13

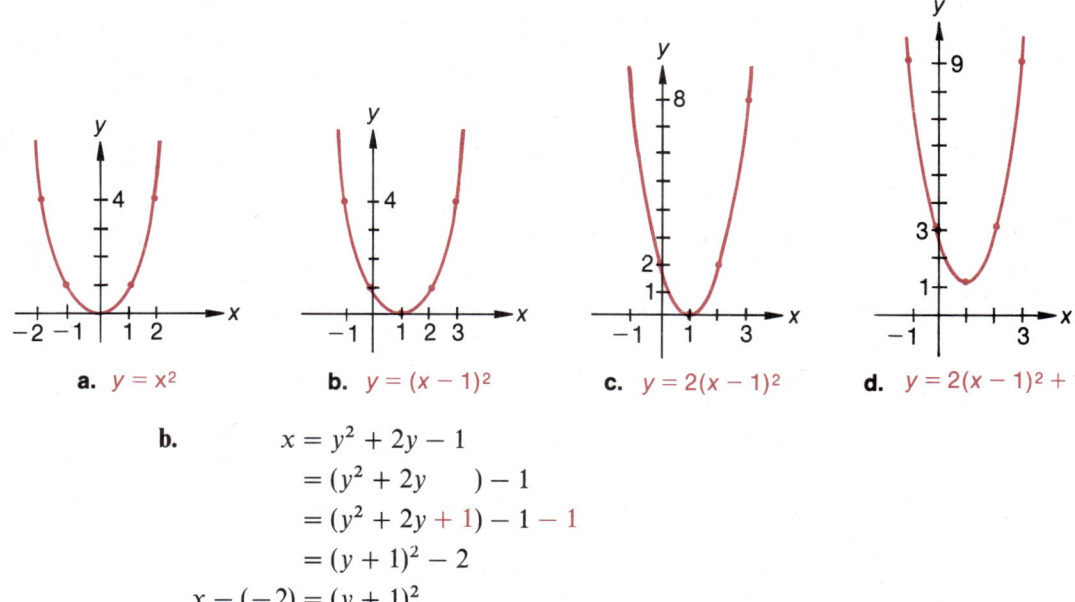

a. $y = x^2$
b. $y = (x - 1)^2$
c. $y = 2(x - 1)^2$
d. $y = 2(x - 1)^2 + 1$

b.
$$\begin{aligned} x &= y^2 + 2y - 1 \\ &= (y^2 + 2y \quad) - 1 \\ &= (y^2 + 2y + 1) - 1 - 1 \\ &= (y + 1)^2 - 2 \end{aligned}$$
$$x - (-2) = (y + 1)^2$$

In plotting the sequence of graphs that follow, we obtain Figure 4.14(b) from (a) by plotting $x = 0$ at $y = -1$ rather than at $y = 0$. Then (c) is obtained from (b) by subtracting 2 from each x-coordinate in (b); that is, the graph of (b) is moved two units to the left. Again, the vertex is $(-2, -1)$, in agreement with our observations in Section 3.4.

Figure 4.14

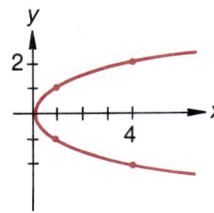
a. $x = y^2$

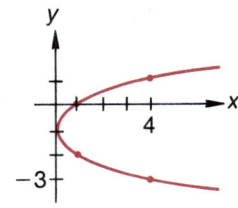
b. $x = (y + 1)^2$

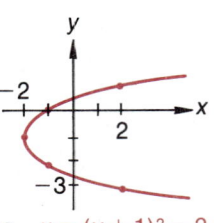
c. $x = (y + 1)^2 - 2$

EXAMPLE 5

With $y = f(x)$ as in Figure 4.6, sketch the graph of the following.
$$y = -2f(2x + 4) + 8$$

SOLUTION

In order to effect a horizontal translation, we must write $2x + 4$ as $2(x + 2)$. Thus we write
$$y = -2f[2(x + 2)] + 8$$

We then sketch the sequence of graphs shown in Figure 4.15. Each graph is obtained from its predecessor by using the techniques illustrated earlier.

Figure 4.15

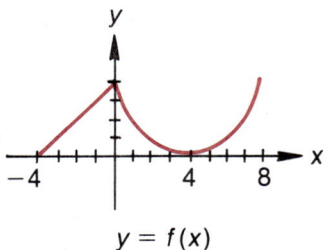
a. Original
$y = f(x)$

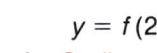
b. Scaling
$y = f(2x)$

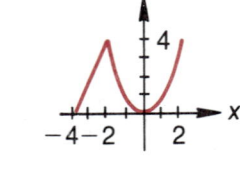
c. Translation
$y = f[2(x + 2)]$

Horizontal Modifications

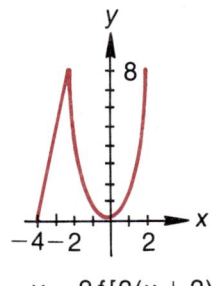
d. Scaling
$y = 2f[2(x + 2)]$

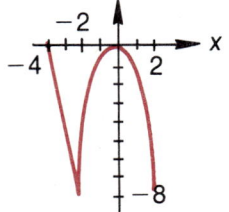
e. Reflection
$y = -2f[2(x + 2)]$

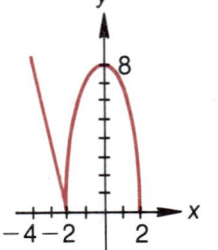
f. Translation
$y = -2f[2(x + 2)] + 8$

Vertical Modifications

It is important that the steps illustrated in parts (b) and (c) of Figure 4.15 not be reversed. Try it! That is, account for any horizontal stretch or compression before any horizontal shifting or translating.

> HORIZONTAL AND VERTICAL CHANGES
>
> Horizontal and vertical changes are independent of one another, but in either case, scaling and/or reflection must be taken care of before translation.

After finishing, it is a good idea to check the results at the endpoints. For instance, in Example 5, we have

$x = -4$: $\quad -2f(2x + 4) + 8 = -2f(-8 + 4) + 8$
$\qquad\qquad\qquad\qquad = -2f(-4) + 8 = -2 \cdot 0 + 8 = 8$

$x = 2$: $\quad -2f(2x + 4) + 8 = -2f(4 + 4) + 8 = -2f(8) - 8$
$\qquad\qquad\qquad\qquad = -2 \cdot 4 + 8 = 0$

These are in agreement with the values indicated on our graph (see Figure 4.15f).

Symmetry

The graphing techniques illustrated up to this point enable us to obtain new graphs from known graphs. Symmetry properties, on the other hand, enable us to obtain a complete graph from a portion of itself.

If the relationship between x and y is expressed as a function $y = f(x)$, then symmetry about the y-axis can be expressed as $f(-x) = f(x)$ and symmetry about the origin can be expressed as $f(-x) = -f(x)$.

DEFINITION

1. A function f for which

$$f(-x) = f(x)$$

is called an **even function**. Its graph is symmetric about the y-axis.

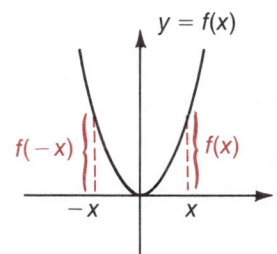

2. A function f for which

$$f(-x) = -f(x)$$

is called an **odd function**. Its graph is symmetric about the origin.

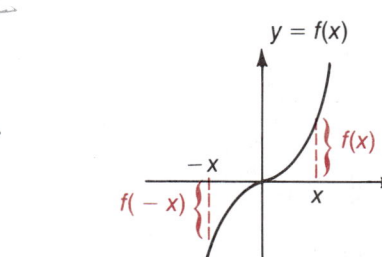

If f is an even function or an odd function, we need only sketch the branch of its graph corresponding to $x > 0$; this branch is then reflected through the y-axis or the origin, as the case may be. For example, if $y = f(x)$ is graphed as in

Figure 4.16a for $x > 0$, then it can be completed as in Figure 4.16b or c if it is even or odd, respectively. Of course a function could be neither even nor odd as in Figure 4.16d.

Figure 4.16

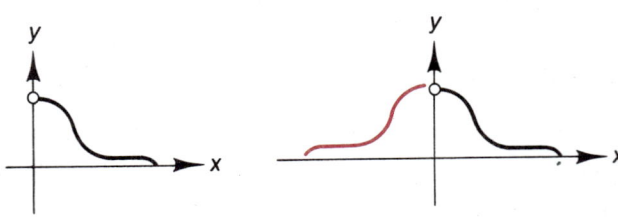

a. $y = f(x)$ b. $y = f(x)$, f is even: $f(-x) = f(x)$

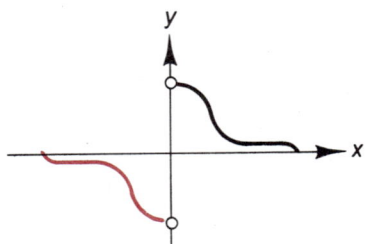 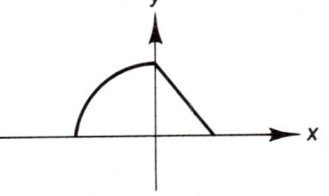

c. $y = f(x)$, f is odd: $f(-x) = -f(x)$ d. $y = f(x)$: f is neither even nor odd.

EVEN AND ODD FUNCTIONS

To determine whether a function defined by $y = f(x)$ is even or odd:

1. Replace x by $-x$.
2. The result will be equivalent to
   a. y (the original function) if the function is even, or
   b. replacing y by $-y$ if the function is odd.
3. Otherwise the function is neither even nor odd.

EXAMPLE 6

Determine whether the following functions are even, odd, or neither even nor odd.

a. $g(x) = x^3$
b. $h(x) = x + x^2$
c. $f(x) = x^2 - 4$

SOLUTION

a. Letting $y = g(x)$ and replacing x by $-x$, we have

$$g(-x) = (-x)^3 = -(x^3)$$
$$= -g(x) = -y$$

This is an odd function.

b. Let $y = h(x)$ and replace x by $-x$:

$$h(-x) = (-x) + (-x)^2$$
$$= -x + x^2 \neq \begin{cases} y = x + x^2 \\ -y = -x - x^2 \end{cases}$$

This function is neither even nor odd.

c. Letting $y = f(x)$ and replacing x by $-x$, we obtain
$$f(-x) = (-x)^2 - 4 = x^2 - 4$$
$$= f(x)$$
This is an even function. ∎

Graphing Absolute Values

The relation $f(x) = |x|$ defines an even function of x since $|-x| = |x|$. Its graph appears in Figure 4.17. To graph a relation involving absolute values, it helps to graph the part inside the absolute value symbols first. The absolute value operation then reflects any portion below the x-axis into the corresponding positive altitudes. This technique is illustrated in the next two examples.

Figure 4.17
$f(x) = |x|$

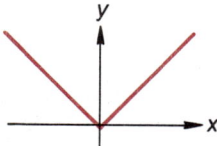

EXAMPLE 7

Sketch the graph of the following equation.
$$y = |12 - 3x| - 3$$

SOLUTION

The graph is sketched in the following stages.

a. First sketch the graph of that part of the function inside the absolute value symbol:
$$y = 12 - 3x$$
See Figure 4.18a.

b. Then reflect any negative y values into the corresponding positive values. This reflection of the negative y portion of the graph in Figure 4.18a through the x-axis yields the graph of
$$y = |12 - 3x|$$
See Figure 4.18b. Note that this graph is symmetric about the line $x = 4$.

c. Finally, a vertical translation gives us the graph of
$$y = |12 - 3x| - 3$$
See Figure 4.18c.

Figure 4.18

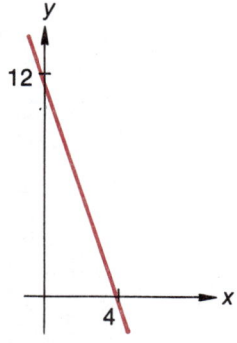
a. $y = 12 - 3x$

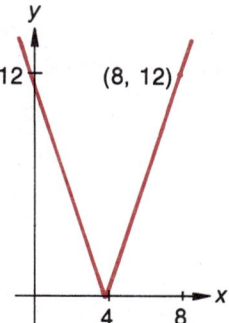
b. $y = |12 - 3x|$

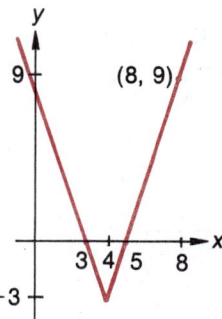
c. $y = |12 - 3x| - 3$ ∎

EXAMPLE 8

Mary is programming a computer to sketch a hungry fish chasing minnows. After much deliberation, she decides that the outline of the fish should satisfy the equation

$$|y^2 - 4y + 3| = x - 2$$

Sketch the fish.

SOLUTION

The equation can be rewritten as

$$x = |y^2 - 4y + 3| + 2$$

The part inside the absolute value can be written as

$$y^2 - 4y + 3 = (y^2 - 4y + 4) + 3 - 4$$
$$= (y - 2)^2 - 1$$

The graph of $x = y^2 - 4y + 3 = (y - 2)^2 - 1$ is a parabola opening to the right with vertex at $(-1, 2)$ and line of symmetry $y = 2$. See Figure 4.19a. Next, any negative x values are reflected into the corresponding positive x values to obtain the graph of

$$x = |y^2 - 4y + 3|$$

in Figure 4.19b. Finally, a horizontal translation yields the graph of

$$x = |y^2 - 4y + 3| + 2$$

in Figure 4.19c.

Figure 4.19

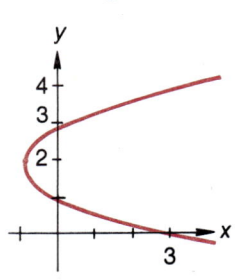

a. $x = y^2 - 4y + 3 = (y - 2)^2 - 1$

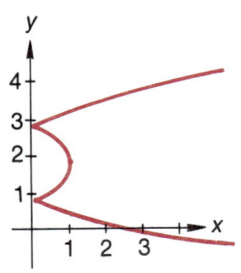

b. $x = |y^2 - 4y + 3|$

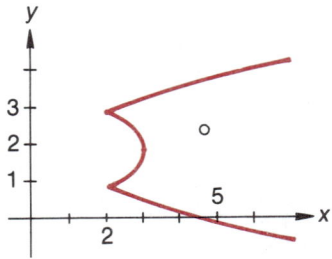

c. $x = |y^2 - 4y + 3| + 2$

Section 4.2 Exercises

1. Use the graph of $y = f(x)$ as given by the figure to sketch the graph of each of the following functions.

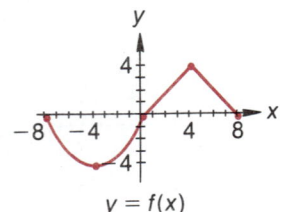

$y = f(x)$

a. $y = f(x - 4)$
b. $y = f(x + 4)$
c. $y = f(x) + 4$
d. $y = f(x) - 4$
e. $y = -3f(x)$
f. $y = \frac{1}{2}f(x)$
g. $y = f(2x)$
h. $y = f(-2x)$
i. $y = f(2x - 2)$
j. $y = f(-2x - 2)$
k. $y = f(-2x + 8)$
l. $y = f(\frac{1}{2}x - 4)$
m. $y = 3f(2x + 2)$
n. $y = -3f(-2x + 2) + 4$
o. $y = -\frac{1}{2}f(\frac{1}{2}x + 4) - 2$
p. $y = |f(x)|$
q. $y = f(|x|)$

2. Repeat Exercise 1 with the following figure.

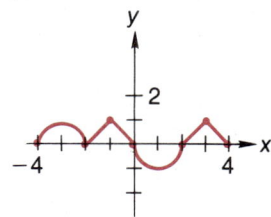

Use the techniques discussed in this section to sketch the graphs of the indicated functions.

3. **a.** $y = x$ **b.** $y = 2x$
   c. $y = -\frac{1}{2}x$ **d.** $y = x - 3$
   e. $y = 2x - 4$ **f.** $y = -3x + 1$
   g. $y = |x - 3|$ **h.** $y = |x + 2|$
   i. $y = |2x + 2|$ **j.** $y = |2x - 4|$
   k. $y = |2 - 4x|$ **l.** $y = |-3x + 1|$
   m. $y = |3x - 6|$

4. **a.** $y = x^2$ **b.** $y = 4x^2$
   c. $y = (x - 1)^2$ **d.** $y = (x + 2)^2$
   e. $y = -2(x + 1)^2$ **f.** $y = 3(x - 2)^2 - 3$
   g. $y = |3(x - 2)^2 - 3|$ **h.** $y = |-2(x + 1)^2 + 8|$
   i. $x = y^2$ **j.** $x = (y - 1)^2$
   k. $x = (y - 1)^2 - 1$ **l.** $x = |(y - 1)^2 - 1|$
   m. $x = 2|(y - 2)^2 - 4|$

5. **a.** $y = x^3$ **b.** $y = x^3/8$
   c. $y = (x + 1)^3$ **d.** $y = (x - 2)^3$
   e. $y = -\frac{1}{2}(x - 1)^3$ **f.** $y = \frac{1}{4}(x + 2)^3 + 2$
   g. $y = |x^3|$ **h.** $y = |(x + 1)^3|$
   i. $y = |x - 2|^3$ **j.** $y = |\frac{1}{4}(x + 2)^3 + 2|$
   k. $y = x^4$ **l.** $y = (x - 1)^4$
   m. $y = |(x - 1)^4 - 1|$

Sketch the graphs of the following relations.

6. $y = 3 - |1 - 2x|$
7. $y = |2x - 4| - 4$
8. $x = |y + 2| - 3$
9. $x = -|3y - 6| + 6$
10. $y = |x^2 - 2x|$
11. $y = 2 - |x^2 - 2x - 3|$
12. $x = -|y^2 - 4y + 1| + 2$
13. $x = |y^2 + 2y - 3| - 2$
14. $y = |-2x^2 + 8x - 4| - 4$
15. $x = |3y^2 + 6y - 2| - 2$

Extend each of the following graphs under the condition that it represents an **a.** even function; **b.** odd function.

16.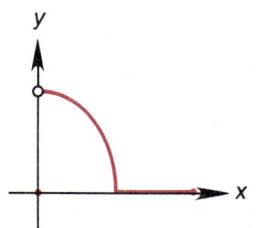

17.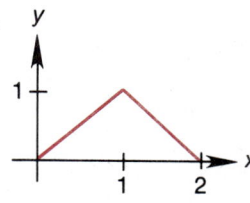

18.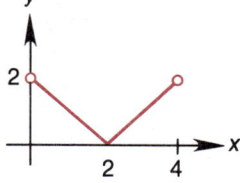

19.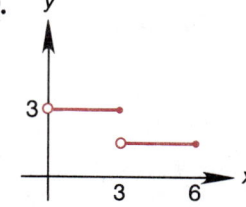

20.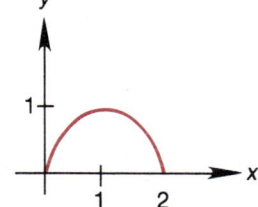

Indicate whether each of the following is even $[f(-x) = f(x)]$, odd $[f(-x) = -f(x)]$, or neither.

21. $y = 2x^2 + 3$
22. $y = |-2x^3|$
23. $y = \sqrt[3]{x}$
24. $y = \sqrt[5]{x^2}$
25. $y = \dfrac{1}{x + 1}, \quad x \neq -1$
26. $y = (x^4 + 1)\sqrt[3]{x}$
27. $y = |x^2 - 2x - 3| + 2$
28. $y = \left|2x^3 + 3x^5 - \dfrac{1}{x^7}\right|$
29. $y = 2x^3 + 3x - \dfrac{1}{x}$
30. $y = \dfrac{-2}{x^3}$
31. $y = \dfrac{x^2}{|x| + 1}$

32. Show that if f and g are even (or odd) functions, then the function h as defined is also an even (or odd) function
   a. $h(x) = f(x) + g(x)$

b. $h(x) = f(x) - g(x)$

c. $h(x) = cf(x)$ for some constant c.

33. Show that if f and g are both even functions or both odd functions, then the function h defined by $h(x) = f(x)g(x)$ is an even function.

34. Show that the product $h(x) = f(x)g(x)$ of an even function f and an odd function g is an odd function.

35. The French Party Supply Company's sales graph for 1985 is given in the figure. The company expects to double this sales volume by 1990. If the same monthly variations are experienced in 1990 as in 1985, sketch its projected sales graph for 1990.

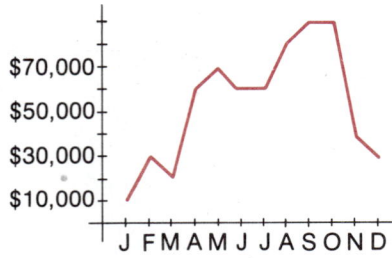

36. Suppose that the concentration of pollution in Indian Lake varies seasonally as sketched in the following graph. The Water Resources Division takes steps to combat these pollution levels and finds that its efforts reduce the pollution level to 50% of these levels.

Unforeseen reactions to their additives also delay the seasonal variations by 6 weeks. Sketch the resulting pollution-level curve.

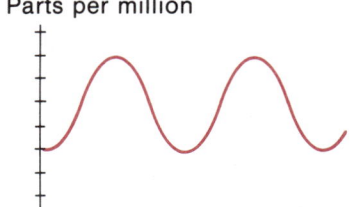

37. Injection of a certain drug into a laboratory animal cuts its heart rate in half as indicated in the graph. The antidote increases its heartbeat to its normal rate just as rapidly as the drug decreases it. Graph the heart rate of an animal from noon to 6 P.M. if the drug is administered at 1 P.M. and the antidote is administered at 4 P.M.

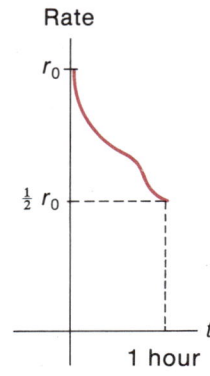

38. Can a function be *both* even and odd?

Section 4.3

Operations on Functions

We are often asked to combine several functions algebraically. For example, several different kinds of expenses contribute to the cost of operating an automobile. If the car consumes 11¢ worth of gas and oil for each mile driven, then the total fuel cost for driving x miles is

$$G(x) = 0.11x$$

Maintenance and tire costs might be estimated at 4¢ per mile. For x miles, these costs total

$$M(x) = 0.04x$$

Another major expense is insurance, which is a fixed fee, say $500 per year, regardless of the number of miles driven. Annual insurance expense is then given by the constant function

$$I(x) = 500$$

We shall simplify the problem by ignoring other expenses such as depreciation and interest; the total cost of driving the vehicle x miles per year is the sum of the preceding functions:

$$\begin{aligned} C(x) &= G(x) + M(x) + I(x) \\ &= 0.11x + 0.04x + 500 \\ &= 500 + 0.15x \end{aligned}$$

If f and g are both real-valued functions, we can clearly add, subtract, multiply, or divide the functional values. Thus, we make the following definitions.

DEFINITION

Let f and g both be real-valued functions with domains D_f and D_g respectively. Let $D = D_f \cap D_g$. Then the following functions are defined on D.

1. For a constant a, af is the function whose value at x is

   $$(af)(x) = a[f(x)]$$

   (This function is defined for all x in the domain of f.)

2. $f + g$ is the function whose value at x is

   $$(f + g)(x) = f(x) + g(x)$$

3. $f - g$ is the function whose value at x is

   $$(f - g)(x) = f(x) - g(x)$$

4. $f \cdot g$ is the function whose value at x is

   $$(f \cdot g)(x) = f(x) \cdot g(x)$$

5. f/g is the function whose value at x is

   $$\left(\frac{f}{g}\right)(x) = \frac{f(x)}{g(x)}$$

   $$D_{f/g} = D_f \cap D_g - \{x : g(x) = 0\}$$

   (The domain of $\dfrac{f}{g}$ must be restricted to preclude division by 0.)

EXAMPLE 1

Let

$$f(x) = 4x^3 - 2x$$
$$g(x) = 2x^2 - 1$$

Then,

$$(f + g)(x) = f(x) + g(x) = (4x^3 - 2x) + (2x^2 - 1)$$
$$= 4x^3 + 2x^2 - 2x - 1$$

$$(f - g)(x) = f(x) - g(x) = (4x^3 - 2x) - (2x^2 - 1)$$
$$= 4x^3 - 2x^2 - 2x + 1$$

$$(f \cdot g)(x) = f(x) \cdot g(x) = (4x^3 - 2x) \cdot (2x^2 - 1)$$
$$= 8x^5 - 8x^3 + 2x$$

$$\left(\frac{f}{g}\right)(x) = \frac{f(x)}{g(x)} = \frac{4x^3 - 2x}{2x^2 - 1} = \frac{2x(2x^2 - 1)}{(2x^2 - 1)}$$
$$= 2x \qquad \text{when } x \neq \frac{\pm 1}{\sqrt{2}}$$

$$(2f)(x) = 2f(x) = 2(4x^3 - 2x) = 8x^3 - 4x \qquad \blacksquare$$

Composition of Functions

As we indicated earlier, the value of a function $y = f(x)$ at any particular value of x, say, at $x = a$, is found by substituting a for x in the expression for $f(x)$. For the function

$$f(x) = 3x^2 - 2x + 1$$

we have

$$f(1) = 3 \cdot 1^2 - 2 \cdot 1 + 1 = 2$$

and

$$f(-\sqrt{2}) = 3(-\sqrt{2})^2 - 2(-\sqrt{2}) + 1 = 7 + 2\sqrt{2}$$

There is nothing to prevent us from substituting other unknown expressions for x also. Thus,

$$f(t) = 3t^2 - 2t + 1$$

and

$$f(t + 2) = 3(t + 2)^2 - 2(t + 2) + 1$$
$$= 3t^2 + 10t + 9$$

In fact, there is nothing to prevent us from substituting into the expression for $f(x)$ a number that is itself computed as a value for some other function. Thus, if $g(x) = 2x + 1$ and $f(x) = 3x^2 - 2x + 1$, we have

$$f[g(x)] = 3 \cdot [g(x)]^2 - 2[g(x)] + 1$$
$$= 3[2x + 1]^2 - 2[2x + 1] + 1$$
$$= 3[4x^2 + 4x + 1] - 4x - 2 + 1$$
$$= 12x^2 + 8x + 2$$

On the other hand,

$$g[f(x)] = 2[f(x)] + 1$$
$$= 2[3x^2 - 2x + 1] + 1$$
$$= 6x^2 - 4x + 3$$

Note that $f[g(x)] \neq g[f(x)]$. This procedure is important enough to warrant the following definition.

> **DEFINITION**
>
> For two functions f and g, the **composition** $g \circ f$ is defined as the function whose value at any x is given by
>
> $$(g \circ f)(x) = g[f(x)]$$
>
> In order for the composition $(g \circ f)(x)$ to be defined, x must lie in the domain for f; in turn, $f(x)$ must be permissible for substitution into the expression for g; that is, $f(x)$ must be in the domain of g.

The previous illustration showed that sometimes $(g \circ f)(x) \neq (f \circ g)(x)$; that is, $g \circ f \neq f \circ g$. *Composition of functions is not a commutative operation.* Considering function machines we should not even expect the composition of functions to be commutative, for if f paints an object red and g paints it grey, then

$(f \circ g)(x) = f[g(x)]$ yields red object

$(g \circ f)(x) = g[f(x)]$ yields grey object

The final color of the object depends only on the last paint job it received. The order in which the operations are performed can make a difference (Figure 4.20).

Figure 4.20
$f \circ g \neq g \circ f$

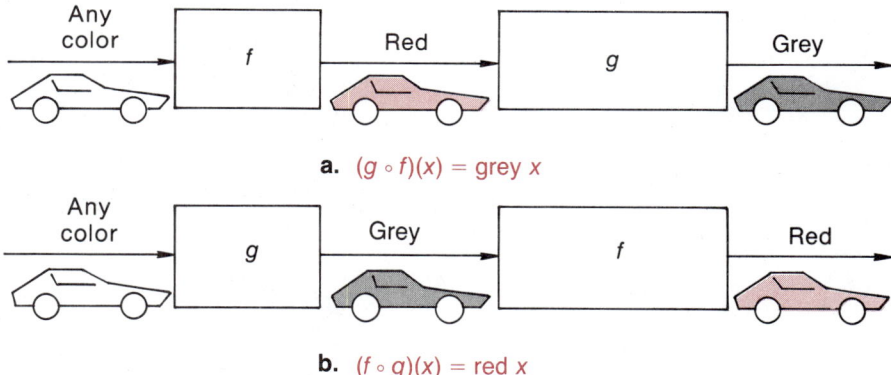

a. $(g \circ f)(x) = $ grey x

b. $(f \circ g)(x) = $ red x

EXAMPLE 2 Let $R(u) = u^2 + 1$ and $S(v) = v - 1$; find $(R \circ S)(t)$ and $(S \circ R)(t)$. When are they equal?

SOLUTION

$(R \circ S)(t) = R[S(t)] = R(t-1)$
$\qquad = (t-1)^2 + 1$ $\qquad$ (Substitute $u = t - 1$ into $R(u)$)
$\qquad = t^2 - 2t + 1 + 1$
$\qquad = t^2 - 2t + 2$

$(S \circ R)(t) = S[R(t)] = S(t^2 + 1)$
$\qquad = (t^2 + 1) - 1$ $\qquad$ (Substitute $v = t^2 + 1$ into $S(v)$)
$\qquad = t^2$

$(S \circ R)(t) = (R \circ S)(t)$ only when

$$t^2 = t^2 - 2t + 2$$
$$2t = 2$$
$$t = 1$$

$(R \circ S)(t) \neq (S \circ R)(t)$ except when $t = 1$.

EXAMPLE 3 Let $f(x) = \sqrt{x - 4}$ and $g(x) = x^2$.

a. Find an expression for $(f \circ g)(x)$ and find the domain for $f \circ g$.
b. Find an expression for $(g \circ f)(x)$ and find the domain for $g \circ f$.

SOLUTION

a. $(f \circ g)(x) = f[g(x)]$
$= \sqrt{g(x) - 4}$
$= \sqrt{x^2 - 4}$

Now $g(x)$ is defined for all x but $f[g(x)]$ is a real value only for

$$x^2 - 4 \geq 0$$
$$x^2 \geq 4$$
$$|x| \geq 2$$

Thus the domain for $f \circ g$ is

$$D_{f \circ g} = \{x : |x| \geq 2\} = (-\infty, -2] \cup [2, \infty)$$

b. $(g \circ f)(x) = g[f(x)]$
$= [f(x)]^2$
$= (\sqrt{x - 4})^2$
$= x - 4$

While it would appear from this result that $(g \circ f)(x) = x - 4$ is defined for all x, *this is not the case*. For $(g \circ f)(x) = g[f(x)]$ to be defined, x must first be in the domain of f. But $f(x) = \sqrt{x - 4}$ is a real number only for

$$x - 4 \geq 0$$
$$x \geq 4$$

Thus the domain of $g \circ f$ is

$$D_{g \circ f} = \{x : x \geq 4\} = [4, \infty)$$

EXAMPLE 4 Find functions f and g to describe the function h as $(g \circ f)(t)$.

$$h(t) = \sqrt[3]{(t + 1)^2}$$

SOLUTION Since $h(t) = \sqrt[3]{(t + 1)^2} = (t + 1)^{2/3}$, we could let $f(t) = (t + 1)$ and $g(u) = u^{2/3}$. Then,

$(g \circ f)(t) = g[f(t)] = g(t + 1)$
$= (t + 1)^{2/3}$

On the other hand, we could also let $f(t) = (t + 1)^2$ and $g(u) = \sqrt[3]{u}$. Then,

$(g \circ f)(t) = g[f(t)] = g[(t + 1)^2]$
$= \sqrt[3]{(t + 1)^2}$

EXAMPLE 5   Let $f(x - 3) = x^2 + 2x + 1$. Find each of the following.

a. $f(t)$ **b.** $f(x)$ **c.** $f(1)$ **d.** $f(x + 2)$

SOLUTION

a. To evaluate $f(t)$, we let $t = x - 3$ since f is described in the form $f(x - 3)$. Then $x = t + 3$ and substitution into the formula for $f(x - 3)$ yields

$$\underset{\downarrow}{x-3} \quad \underset{\downarrow}{x} \quad \underset{\downarrow}{x}$$
$$f(t) = (t + 3)^2 + 2(t + 3) + 1$$
$$= (t^2 + 6t + 9) + 2t + 6 + 1$$
$$= t^2 + 8t + 16$$

b. In part a, we found $f(t) = t^2 + 8t + 16$. But t is a dummy variable used to describe the function. Thus

$$f(x) = x^2 + 8x + 16$$

c. Substituting $x = 1$ into part b, we obtain

$$f(1) = 1^2 + 8 \cdot 1 + 16$$
$$= 25$$

d. Substituting $x + 2$ for either t or x in part b, we obtain

$$f(x + 2) = (x + 2)^2 + 8(x + 2) + 16$$
$$= (x^2 + 4x + 4) + 8x + 16 + 16$$
$$= x^2 + 12x + 36$$

EXAMPLE 6   The Environmental Protection Agency has determined that in a certain section of the country the average level of air pollution is $0.5\sqrt{P + 10{,}000}$ parts per million (ppm), where P is the population. The 1980 census predicts that the population t years after 1980 will be $7000 + 40t^2$.

a. Express the pollution level t years after 1980 as a composite function and reduce the composite function to a function of t.

b. What pollution level can be expected in 1990?

c. What pollution level can be expected in 2000?

SOLUTION

a. The pollution level is expressed as a function of population P as

$$f(P) = 0.5\sqrt{P + 10{,}000}$$

The population is expressed as a function of time t as

$$P(t) = 7000 + 40t^2$$

The pollution level t years after 1980 is given by the composite function

$$(f \circ P)(t) = f[P(t)]$$
$$= 0.5\sqrt{P(t) + 10{,}000}$$
$$= 0.5\sqrt{7000 + 40t^2 + 10{,}000}$$
$$= 0.5\sqrt{17{,}000 + 40t^2}$$

b. In 1990, $t = 10$ and the pollution level is estimated as

$$(f \circ P)(10) = 0.5\sqrt{17{,}000 + 40 \cdot 10^2}$$
$$= 0.5\sqrt{21{,}000}$$
$$\approx 72 \text{ ppm} \qquad \text{(Use a calculator)}$$

c. In the year 2000, the pollution level is predicted to be

$$(f \circ P)(20) = 0.5\sqrt{17{,}000 + 40 \cdot 20^2} \qquad (t = 20)$$
$$= 0.5\sqrt{33{,}000}$$
$$\approx 91 \text{ ppm}$$

■

Difference Quotient

In Figure 4.21, we see that the slope of the line joining the two points corresponding to x and $x + h$ on the graph of $y = f(x)$ is given by

$$m = \frac{y_2 - y_1}{x_2 - x_1} = \frac{f(x+h) - f(x)}{(x+h) - x} = \frac{f(x+h) - f(x)}{h}$$

This slope depends not only on x but also on h. The **difference quotient** for f is defined as

$$D(x, h) = \frac{f(x+h) - f(x)}{h}, \qquad \text{when } h \neq 0$$

It is a function of two variables: x and h.

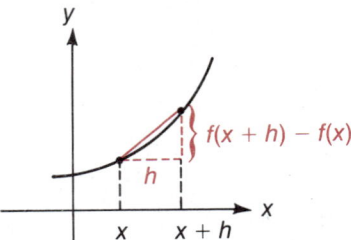

Figure 4.21

EXAMPLE 7 Find the difference quotient for $f(x) = x^2 + 2$.

SOLUTION
$$D(x, h) = \frac{f(x+h) - f(x)}{h} = \frac{[(x+h)^2 + 2] - [x^2 + 2]}{h}$$
$$= \frac{[(x^2 + 2xh + h^2) + 2] - [x^2 + 2]}{h}$$
$$= \frac{2xh + h^2}{h} = \frac{h(2x + h)}{h}$$
$$= 2x + h$$

■

Section 4.3 Exercises

In each of the following, find expressions for $f(x) + g(x)$, $f(x) - g(x)$, $f(x) \cdot g(x)$, $f(x)/g(x)$, and $af(x)$ for the indicated value of a.

1. $f(x) = x^2 + 5$, $g(x) = x^2 + x - 6$; $a = 2$
2. $f(s) = s^2 - 2s - 3$, $g(t) = 6 - t - t^2$, $a = -1$
3. $f(u) = u^3 - 3u$, $g(v) = 2v^2 + v - 6$; $a = 3$
4. $f(x) = x^3 + 3$, $g(y) = 2y^2 + y - 10$; $a = -2$
5. $f(a) = a^3 - a^2 - a + 1$, $g(r) = r^2 - 1$; $a = 4$

In each of the following, find the requested expressions.

6. Let $A(w) = 2w + 3$; find $A(x - 1)$, $A(2y - 1)$, $A(w - 2)$.

7. Let $f(x) = x + \dfrac{1}{x}$; find $f\left(\dfrac{1}{x}\right)$, $f(x^2)$, $f(2x)$.

8. Let $F(z) = z^2 + 6z + 20$; find $F(x)$, $F(2t)$, $F(z-1)$.

9. Let $g(t) = 3t^2 - t + 2$; find $g\left(\dfrac{1}{t}\right)$, $g(-x)$, $g(t+2)$.

10. Let $s(h) = 2h^2 + h - 1$; find $s(x-2)$, $s(2y-3)$, $s(h-1)$.

In each of the following
  a. Find expressions for $(f \circ g)(t)$ and $(g \circ f)(t)$;
  b. Find the domain for $f \circ g$ and $g \circ f$;
  c. Evaluate $(f \circ g)(a)$ and $(g \circ f)(a)$ for the given value of a.

11. $f(x) = 2x + 1$, $g(y) = 3 - y$; $a = -1$

12. $f(u) = u + 4$, $g(v) = 3v - 2$; $a = 2$

13. $f(x) = 2x$, $g(x) = 3x^2$; $a = -2$

14. $f(t) = 3t^2 + 2t + 1$, $g(u) = 2u$; $a = 0$

15. $f(s) = s^2 + s$, $g(s) = 3s - 2$; $a = 1$

16. $f(r) = r^2 - 4r + 1$, $g(r) = r + 2$; $a = -1$

17. $f(t) = t^3 + t + 1$, $g(r) = r - 2$; $a = 2$

18. $f(p) = p + 1$, $g(q) = q^3 - 1$; $a = -2$

19. $f(y) = 2y - y^2$, $g(x) = \dfrac{1}{x}$; $a = 4$

20. $f(v) = \dfrac{1}{v+1}$, $g(r) = \dfrac{1}{r}$; $a = 3$

21. $f(t) = t^2$, $g(u) = \sqrt{u}$; $a = -4$

22. $f(x) = \sqrt{x-1}$, $g(t) = t^2 + 1$; $a = -2$

23. $f(r) = \sqrt{r^2 - 1}$, $g(s) = \dfrac{1}{s}$; $a = 2$

24. $f(w) = \dfrac{1}{\sqrt{w-1}}$, $g(z) = z^2 + 1$; $a = 2$

25. $f(x) = -|x+1|$, $g(x) = x^2 - 1$; $a = -2$

26. $f(x) = x^2 + 2$, $g(x) = -|x-2|$; $a = -3$

Find functions f and g to describe the given function as g ∘ f.

27. $h(t) = (t-1)^2$

28. $F(x) = \sqrt{x+1}$

29. $u(n) = \dfrac{1}{(n+1)^2}$

30. $H(w) = (w^2 + 1)^{3/2}$

31. $G(u) = 3(u^3 + 2)^2 + 2(u^3 + 2) + 4$

32. $V(t) = 2(t^2 - 1)^3 + (t^2 - 1)^2 - (t^2 - 1) + 2$

Find the requested expressions.

33. Let $G(x - 2) = x^2 - 5x + 6$; find $G(t)$, $G(x)$, $G(0)$, $G(x+1)$.

34. Let $Y(t + 1) = t^2 + 5$; find $Y(x)$, $Y(t)$, $Y(2)$, $Y(t - 2)$.

35. Let $P(x + 1) = x^2 + 2x + 1$; find $P(t)$, $P(x)$, $P(2)$, $P(x+2)$.

36. Let $I(w + 1) = 3w^2 + 2w + 1$; find $I(u)$, $I(w)$, $I(2)$, $I(w-2)$.

37. Let $h(x + 2) = x^2 + 6x + 5$; find $h(t)$, $h(x)$, $h(4)$, $h(x-3)$.

The operations combining two different functions or dependent variables should not be confused with operations on independent variables. In general, there is no relationship between operations on independent variables and the corresponding operations on dependent variables. *Comparing the values requested in the following exercises should illustrate the situation sufficiently.*

38. Let $H(z) = z^2 - 10z + 11$; find and compare
  a. $H(1) + H(2)$ and $H(1 + 2)$
  b. $\dfrac{H(1)}{H(2)}$ and $H\left(\dfrac{1}{2}\right)$
  c. $H(2 - 1)$ and $H(2) - H(1)$

39. Let $L(s) = 3s - 4$; find and compare
  a. $L(2^2)$ and $[L(2)]^2$
  b. $L(\sqrt{4})$ and $\sqrt{L(4)}$
  c. $L(-1)$ and $-L(1)$

40. Let $G(x) = x^2 + 2x - 10$; find and compare
  a. $G(1 \cdot 2)$ and $G(1) \cdot G(2)$
  b. $G\left(\dfrac{1}{2}\right)$, $\dfrac{G(1)}{G(2)}$, and $\dfrac{1}{G(2)}$
  c. $G(2x)$ and $2G(x)$

41. Let $F(x) = x^2 + 1$; find and compare
  a. $F(1 + 1)$ and $F(1) + F(1)$
  b. $F(1 - 1)$ and $F(1) - F(1)$
  c. $F(1 \cdot 1)$ and $F(1) \cdot F(1)$

42. Let $P(t) = 8 + 2t$; find and compare
    a. $P(\sqrt{9})$ and $\sqrt{P(9)}$
    b. $P(3^2)$ and $[P(3)]^2$
    c. $P(3t)$ and $3P(t)$

Find the difference quotient for the following functions.

43. $f(x) = 3$
44. $g(x) = 4x - 1$
45. $h(x) = 2 - 3x$
46. $F(x) = 5x + 100$
47. $G(x) = 2x^2 - 1$
48. $H(x) = 3 - 4x^2$
49. $f(x) = x^2 + 2x - 5$
50. $g(x) = 3x^2 - 5x + 7$
51. $h(x) = 1 + 4x - 5x^2$
52. $F(x) = x^3$
53. $G(x) = x^4$
54. $H(x) = \dfrac{1}{x}$

55. An employee deposits a certain sum of money at 5% interest in the company credit union. In order to encourage thrift, the company announces a bonus equal to 7% of the value of each employee's savings account. Express both the interest and the bonus as functions of the value of the savings account. Express the net return as a combination of these two functions.

56. Viki paddles her canoe at a rate of 5 miles per hour. The stream is flowing at a rate of 2 miles per hour.
    a. Express the distance Viki rows *in the water* as a function of time.
    b. Express the distance the water flows as a function of time.
    c. If she paddles upstream, express her net distance traveled as a combination of these two functions.
    d. What if she rows downstream?

57. The efficiency of a jet engine is affected by the quality of fuel used; the plane can travel faster and longer on better grades of fuel. Suppose that with a tankful of fuel of quality q, the plane can travel $3q^2 + 2q$ miles per hour and can travel for $2q + 1$ hours. Express the distance the plane can travel on a tankful as a combination of these two functions of fuel quality.

58. The kinetic energy (K) of an object is given in terms of its mass m and velocity v as $K = mv^2/2$. Its momentum (M) is given by $M = mv$. If the kinetic energy at time t is $50t^2 + 50t - 300$ and the momentum at time t is $25t + 75$, express the velocity as a function of t by writing it as the quotient of two functions of t. (*Hint:* Show that $v = 2K/M$.)

59. A vacuum cleaner sales representative earns a commission consisting of 50% of the negotiated sales price. At the end of the year, the company gives to each representative a bonus of $1000 plus 10% of the commissions earned during the year.
    a. Express the representative's bonus as a function of the total amount of sales concluded by writing the bonus as a composition of two functions.
    b. Express the representative's total income for the year in terms of total sales.

60. Bertha deposits money in the bank at 6% interest but must then pay 20% of her interest income to the Internal Revenue Service.
    a. Express the amount of interest income she pays to the government in terms of her bank balance by writing it as a composition of two functions.
    b. What is her net gain from her savings account?

61. The rabbit population in the Midwest depends on the vegetation available as food. The available vegetation in turn depends on the amount of rainfall. The fox population depends on the number of rabbits. There are 20% as many foxes as rabbits and each rabbit requires 200 pounds of vegetation per year. Each inch of rainfall in Ogden County supplies 20 tons of excess vegetation, which is available to the rabbits. Express the number of foxes in Ogden County as a function of rainfall by writing it as a composition of three functions.

In the following, determine whether $f \circ g = g \circ f$.

62. Let f be the operation of combing your hair and g be the operation of washing your hair.

63. Let f be the operation of studying for an exam and g be the operation of taking the exam.

64. Let f be the operation of eating a steak and g the operation of grilling the steak.

65. Let f denote the operation of mowing the lawn and g the operation of sweeping the floor.

66. Let f denote buying a piece of property and g denote inspection of the property.

Section 4.4

Inverse Functions

When a dependent variable is given as a function of an independent variable, say, $s = f(t)$, it is sometimes necessary to determine which value or values of the independent variable t generate a certain value of the dependent variable s. For example, if the altitude in feet of a toy rocket t seconds after launching is given by

$$s = f(t) = 240t - 16t^2$$

we might want to know how long it takes the rocket to reach an altitude of 800 feet. Thus, for a known value of $s = 800$, we are to determine which values of t yield $s = f(t) = 800$. Solving for t, we obtain

$$800 = 240t - 16t^2$$
$$16t^2 - 240t + 800 = 0$$
$$t^2 - 15t + 50 = 0 \qquad \text{(Divide by 16)}$$
$$(t - 10)(t - 5) = 0$$
$$t = 5, 10$$

These answers tell us that 5 seconds after launching the rocket is 800 feet above the ground on its way up; 10 seconds after launching, it is again 800 feet above the ground, this time on its way down. In this example, t is not a function of s, since for some values of s there are two values of t related to s. In fact, for any nonnegative s value less than 900 there are exactly two values of t which yield that s value (see Figure 4.22).

In the preceding illustration, we saw that *even though s is a function of t, it does not necessarily follow that t is a function of s.* For t to be a function of s, each value of s must be related to only one t value; no two points of the graph can lie on the same horizontal line. Recall from Section 4.1 the horizontal line test.

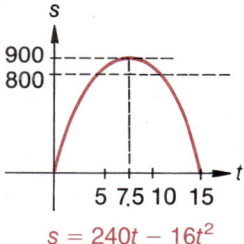

Figure 4.22

HORIZONTAL LINE TEST

A function $y = f(x)$ determines x as a function of y if and only if each horizontal line intersects its graph in at-most one point.

If each horizontal line intersects the graph of $y = f(x)$ in at most one point, then we cannot have $f(x_1) = f(x_2)$ for distinct values $x_1 \neq x_2$. Such functions are said to be *one-to-one*.

DEFINITION

A function f is said to be **one-to-one** if

$$f(x_1) \neq f(x_2) \qquad \text{whenever} \qquad x_1 \neq x_2$$

(for all x_1, x_2 in the domain of f).

248 Chapter 4 Functions and Their Graphs

EXAMPLE 1 Determine whether the given functions are one-to-one. Assume the domain of each is the set of real numbers.

a. $y = f(x) = 2x - 1$
b. $y = g(x) = x^2 + 2$

SOLUTION

a. We test to determine when we can have

$$f(x_1) = f(x_2)$$

If this happens, then

$$2x_1 - 1 = 2x_2 - 1$$
$$2x_1 = 2x_2$$
$$x_1 = x_2$$

We cannot have $f(x_1) = f(x_2)$ unless $x_1 = x_2$. Thus this function is one-to-one. In Figure 4.23a, we also see that no two distinct x values can yield the same $y = f(x)$ value, since each horizontal line crosses the graph in only one point.

b. Again we determine when we can have

$$g(x_1) = g(x_2)$$

This happens when

$$x_1^2 + 2 = x_2^2 + 2$$
$$x_1^2 = x_2^2$$
$$|x_1| = |x_2|$$

That is, if two numbers have the same absolute value, they yield the same $y = g(x)$ value. For instance

$$g(2) = 2^2 + 2 \quad = 4 + 2 = 6$$
$$g(-2) = (-2)^2 + 2 = 4 + 2 = 6$$
$$g(2) = g(-2)$$

This function is not one-to-one.

Figure 4.23

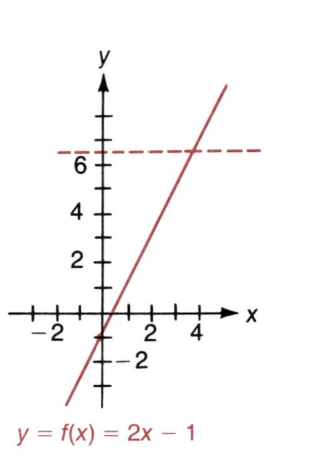

$y = f(x) = 2x - 1$

a.

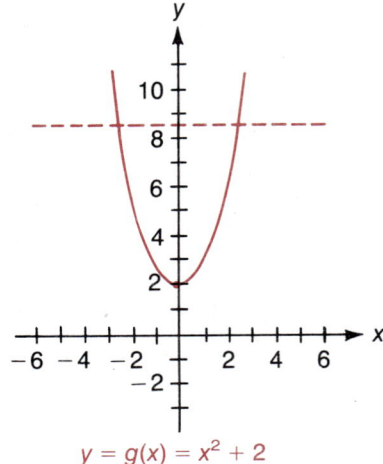

$y = g(x) = x^2 + 2$

b.

The graph in Figure 4.23b also shows that the function is not one-to-one since some horizontal lines intersect the graph in two points. ■

If a function f is one-to-one, then each $y = f(x)$ value is associated with exactly one x. Thus a given y in the range of f determines *exactly one x* in the domain of f; x is also a function of y.

DEFINITION

If v is a function of u given by $v = f(u)$ and this relationship also determines u as a function of v, the function that gives u in terms of v is called the **inverse** of the function f and is denoted by f^{-1}. We write $u = f^{-1}(v)$.

$$f^{-1}(v) \text{ is that } u \text{ such that } f(u) = v$$

The domain of f^{-1} is the range of f. The range of f^{-1} is the domain of f.

The -1 is used here solely to denote *inverse function*; f^{-1} *does not stand for the reciprocal,* $1/f$.

We can now state a more comprehensive version of the horizontal line test.

HORIZONTAL LINE TEST FOR f^{-1}

Let a function be defined by $y = f(x)$. Then the following statements are equivalent.

1. $y = f(x)$ also determines x as a function of y.
2. Each horizontal line intersects the graph in at most one point.
3. The function f is one-to-one.
4. The function f has an inverse, f^{-1}.

The function $s = f(t) = 240t - 16t^2$ that was graphed in Figure 4.22 does not have an inverse, t is not a function of s, the horizontal line at $s = 800$ intersects the graph in two points, and the function is not one-to-one, since $s(5) = s(10) = 800$.

As stated in the definition, the inverse function is obtained by expressing the independent variable in terms of the dependent variable. Thus if $v = f(u)$, we find $u = f^{-1}(v)$ by solving for u in terms of v.

FINDING f^{-1}

Let $v = f(u)$. Then find f^{-1} as follows.

1. Solve the equation $v = f(u)$ for u in terms of v.
2. a. If u is determined uniquely by v, set this expression equal to $f^{-1}(v)$: $u = f^{-1}(v)$.
   b. If u is not determined uniquely by v, then u is not a function of v and f^{-1} does not exist.
3. Since the variables used to define a function are dummy variables, any variable can be substituted for v in $f^{-1}(v)$. For instance, we can describe the inverse function as $f^{-1}(t)$ or $f^{-1}(x)$ if we wish.

EXAMPLE 2 In each of the following, determine whether the inverse function exists. If so, find $f^{-1}(x)$. Assume that the domain of each is the set of real numbers.

    a. $v = f(u) = 2u + 1$
    b. $w = f(z) = z^2 - 1$
    c. $y = f(x) = x^3 + 9$
    d. $s = f(r) = \dfrac{r+2}{r-2}, \quad r \neq 2$

SOLUTION   **a.** We solve for u in $v = f(u) = 2u + 1$.

$$2u + 1 = v$$
$$2u = v - 1$$
$$u = \frac{v-1}{2}$$

Thus, u is indeed a function of v:

$$u = f^{-1}(v) = \frac{v-1}{2}$$

Since v is a dummy variable, we can also write

$$f^{-1}(x) = \frac{x-1}{2}$$

b. Solve for z in $w = f(z) = z^2 - 1$:

$$z^2 - 1 = w \quad \text{(Note } w \geq -1\text{)}$$
$$z^2 = w + 1 \quad \text{(Note } w + 1 \geq 0\text{)}$$
$$z = \pm\sqrt{w+1}, \quad w \geq -1$$

Since each positive real number has two real square roots, z is not determined uniquely by w; z is not a function of w. The inverse function does not exist.

c. Solve for x in $y = f(x) = x^3 + 9$:

$$x^3 + 9 = y$$
$$x^3 = y - 9$$
$$x = \sqrt[3]{y-9}$$

Since each real number has a unique real cube root, x is a function of y:

$$x = f^{-1}(y) = \sqrt[3]{y-9}$$

Now we treat y as a dummy variable and write

$$f^{-1}(x) = \sqrt[3]{x-9}$$

d. Solve for r in terms of s:

$$s = \frac{r+2}{r-2}$$
$$(r-2)s = r + 2 \quad \text{(Multiply both sides by } (r-2)\text{)}$$
$$rs - 2s = r + 2$$

$$rs - r = 2s + 2$$
$$r(s - 1) = 2s + 2$$
$$r = \frac{2s + 2}{s - 1}, \quad s \neq 1$$

Then we write
$$r = f^{-1}(s) = \frac{2s + 2}{s - 1}, \quad s \neq 1$$

Using x as the variable we have
$$f^{-1}(x) = \frac{2x + 2}{x - 1}, \quad x \neq 1 \quad \blacksquare$$

The inverse of a function "undoes" the operation of the original function. From u we compute $v = f(u)$ and then $f^{-1}(v)$ takes us back to u. In the same way, f undoes the operation of f^{-1}: applying f to $u = f^{-1}(v)$ takes us back to v. This process is illustrated in Figure 4.24.

Figure 4.24

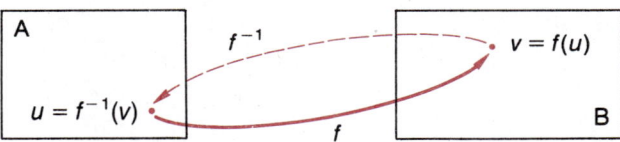

f^{-1}

Whenever f^{-1} exists,

1. $f^{-1}[f(t)] = t$ for all t in the domain of f
2. $f[f^{-1}(t)] = t$ for all t in the domain of f^{-1} ($=$ range of f)

EXAMPLE 3   For the functions given in Examples 2a, c, and d, verify that $f[f^{-1}(t)] = t$ and that $f^{-1}[f(t)] = t$ for all t in the domain of f^{-1} or f, respectively.

SOLUTION   In both 2a and 2c, the domain of f and of f^{-1} is the set of all real numbers.

a. In 2a,
$$f(u) = 2u + 1 \quad \text{and} \quad f^{-1}(v) = \frac{v - 1}{2}$$

Then
$$f[f^{-1}(t)] = 2f^{-1}(t) + 1$$
$$= 2\left(\frac{t - 1}{2}\right) + 1$$
$$= t$$

and
$$f^{-1}[f(t)] = \frac{f(t) - 1}{2}$$
$$= \frac{(2t + 1) - 1}{2}$$
$$= t$$

c. In 2c,
$$f(x) = x^3 + 9$$
and
$$f^{-1}(x) = \sqrt[3]{x - 9}$$
Then
$$f[f^{-1}(t)] = [f^{-1}(t)]^3 + 9$$
$$= [\sqrt[3]{t - 9}]^3 + 9$$
$$= [t - 9] + 9$$
$$= t$$
and
$$f^{-1}[f(t)] = \sqrt[3]{f(t) - 9}$$
$$= \sqrt[3]{(t^3 + 9) - 9}$$
$$= \sqrt[3]{t^3}$$
$$= t$$

d. In 2d,
$$f(r) = \frac{r + 2}{r - 2}$$
and
$$f^{-1}(s) = \frac{2s + 2}{s - 1}$$
Then
$$f[f^{-1}(t)] = \frac{f^{-1}(t) + 2}{f^{-1}(t) - 2}$$
$$= \frac{\frac{2t + 2}{t - 1} + 2}{\frac{2t + 2}{t - 1} - 2}$$
$$= \frac{\frac{2t + 2}{t - 1} + 2}{\frac{2t + 2}{t - 1} - 2} \cdot \frac{(t - 1)}{(t - 1)}$$
$$= \frac{2t + 2 + 2(t - 1)}{2t + 2 - 2(t - 1)}$$
$$= \frac{4t}{4}$$
$$= t$$

and

$$f^{-1}[f(t)] = \frac{2f(t) + 2}{f(t) - 1}$$

$$= \frac{2\frac{t+2}{t-2} + 2}{\frac{t+2}{t-2} - 1}$$

$$= \frac{2\frac{t+2}{t-2} + 2}{\frac{t+2}{t-2} - 1} \cdot \frac{(t-2)}{(t-2)}$$

$$= \frac{2(t+2) + 2(t-2)}{(t+2) - (t-2)}$$

$$= \frac{4t}{4}$$

$$= t$$

The property of *undoing* the effect of a given function characterizes the inverse of the function, as stated in the following box.

INVERSE FUNCTIONS

If f and g are functions with

$$\text{Domain } f = \text{range } g = A$$
$$\text{Domain } g = \text{range } f = B$$

and

$$g[f(x)] = x \quad \text{for every } x \text{ in } A$$
$$f[g(x)] = x \quad \text{for every } x \text{ in } B$$

(that is, each function undoes the other), then g and f are inverses of one another:

$$g = f^{-1} \quad \text{and} \quad f = g^{-1}$$

EXAMPLE 4

Show that the functions

$$f(x) = \frac{1}{x} - 1, \quad x \neq 0$$

and

$$g(x) = \frac{1}{x+1}, \quad x \neq -1$$

are inverses of one another by showing that

$$g[f(x)] = x \quad \text{for all } x \neq 0$$

and

$$f[g(x)] = x \quad \text{for all } x \neq -1$$

SOLUTION   Note that $f(x) \ne -1$, since $\dfrac{1}{x} \ne 0$ for $x \ne 0$. Thus $f(x)$ is in the domain of g and

$$g[f(x)] = \frac{1}{f(x) + 1} \quad \text{for all } x \ne 0$$

$$= \frac{1}{\left[\dfrac{1}{x} - 1\right] + 1}$$

$$= \frac{1}{\dfrac{1}{x}}$$

$$= x$$

Conversely, note that $g(x) \ne 0$ since $\dfrac{1}{x+1} \ne 0$ for $x \ne -1$. Thus $g(x)$ is in the domain of f and

$$f[g(x)] = \frac{1}{g(x)} - 1 \quad \text{for all } x \ne -1$$

$$= \frac{1}{\left[\dfrac{1}{x+1}\right]} - 1$$

$$= [x + 1] - 1$$

$$= x$$

Thus, g undoes the effect of f and f undoes the effect of g. The functions g and f are inverses of one another. ∎

If we know that the inverse function, f^{-1}, exists, we can use the property $f[f^{-1}(t)] = t$ to find f^{-1}. The procedure is illustrated in the following example.

EXAMPLE 5   Let $y = f(x) = 2x^2 + 1$ be defined only for $x \ge 0$.

   a. Show that the inverse function exists.

   b. Find $f^{-1}(t)$; $f^{-1}(x)$.

SOLUTION   **a.** Writing

$$y = 2x^2 + 1$$
$$= 2(x - 0)^2 + 1$$
$$y - 1 = 2(x - 0)^2$$

we see that its graph is a parabola with vertex at $(0, 1)$. (See Section 3.4.) But since the given function is defined only for $x \ge 0$, its graph is only the right-hand branch of this parabola, as shown in Figure 4.25.

Since each horizontal line meets the graph in at most one point, x can be computed in terms of y; f^{-1} exists.

b. The function is defined by

$$f(x) = 2x^2 + 1, \quad x \ge 0$$

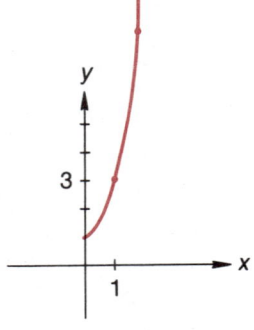

Figure 4.25
$y = 2x^2 + 1$, $x \ge 0$

Since $f[f^{-1}(t)] = t$, we write

$$f[f^{-1}(t)] = 2[f^{-1}(t)]^2 + 1 = t, \qquad f^{-1}(t) \geq 0$$
$$2[f^{-1}(t)]^2 = t - 1$$
$$[f^{-1}(t)]^2 = \frac{t-1}{2}$$
$$f^{-1}(t) = \sqrt{\frac{t-1}{2}}$$

We take only the positive square root here since we must have $f^{-1}(t) \geq 0$. The domain of f^{-1} is the range of f. Since $f(x) = 2x^2 + 1 \geq 1$,

$$\text{Domain}(f^{-1}) = \text{Range}(f) = [1, \infty)$$

Thus, f^{-1} is described by

$$f^{-1}(t) = \sqrt{\frac{t-1}{2}}, \qquad t \geq 1$$

Treating t as a dummy variable, we have

$$f^{-1}(x) = \sqrt{\frac{x-1}{2}}, \qquad x \geq 1 \qquad \blacksquare$$

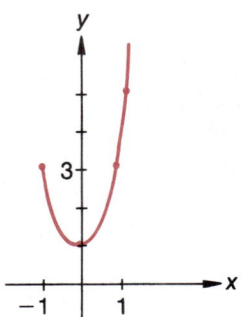

Figure 4.26
$y = 2x^2 + 1, \quad x \geq -1$

Had some $x < 0$ been allowed in the domain of the function in Example 5, we would not have obtained an inverse function. For instance, if the domain of f is expanded to $x \geq -1$, the graph becomes that of Figure 4.26. Since some horizontal line meets the graph in two points, x is then not a function of y.

In Figure 4.25 we see that $y = f(x)$ increases as x moves from left to right and thus f has an inverse.

DEFINITION

Let a function f be defined on an interval I and let x_1 and x_2 be in I.

1. f is **increasing** on I if $f(x_2) > f(x_1)$ when $x_2 > x_1$.
2. f is **decreasing** on I if $f(x_2) < f(x_1)$ when $x_2 > x_1$.
3. f is **constant** on I if $f(x_1) = f(x_2)$ for all x_1, x_2 in I.

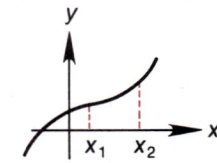 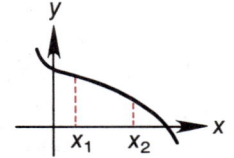 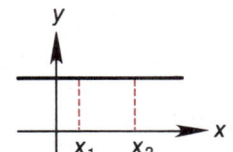

An increasing function   A decreasing function   A constant function

(f is said to be **nondecreasing** if $f(x_2) \geq f(x_1)$ when $x_2 > x_1$; **nonincreasing** is defined similarly.)

EXAMPLE 6

Determine the regions where the function whose graph is given in Figure 4.27 is

a. increasing
b. decreasing
c. constant
d. nondecreasing
e. nonincreasing

Figure 4.27

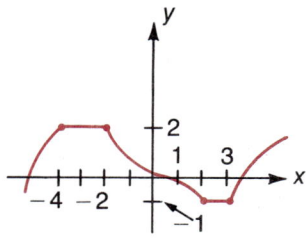

SOLUTION

The actual y values are not important; we must look for the trend in y values (increasing, decreasing, or constant) as x moves from left to right. The function is

a. increasing on $(-\infty, -4]$ and on $[3, \infty)$
b. decreasing on $[-2, 2]$
c. constant on $[-4, -2]$ and on $[2, 3]$
d. nondecreasing on $(-\infty, -2]$ and on $[2, \infty)$
e. nonincreasing on $[-4, 3]$

EXAMPLE 7

Determine where the function

$$f(t) = 2t + 1$$

is increasing and where it is decreasing.

SOLUTION

Consider $t_1 < t_2$ and determine the relationship between $f(t_1)$ and $f(t_2)$. Now if

$$t_1 < t_2$$

then

$$2t_1 < 2t_2 \qquad \text{(Rule 5b, Section 2.7)}$$

and

$$2t_1 + 1 < 2t_2 + 1 \qquad \text{(Rule 3, Section 2.7)}$$

That is,

$$f(t_1) < f(t_2)$$

This function is increasing on $(-\infty, \infty)$.

Section 4.4 Inverse Functions 257

EXAMPLE 8 Determine where the function
$$f(x) = x^2 - 2x + 4$$
is increasing and where it is decreasing.

SOLUTION Write
$$f(x) = (x^2 - 2x + 1) + 3$$
$$= (x - 1)^2 + 3$$

The graph of this function is a parabola opening upward with vertex at (1, 3). See Figure 4.28. Thus the function is decreasing on $(-\infty, 1]$ and increasing on $[1, \infty)$. ∎

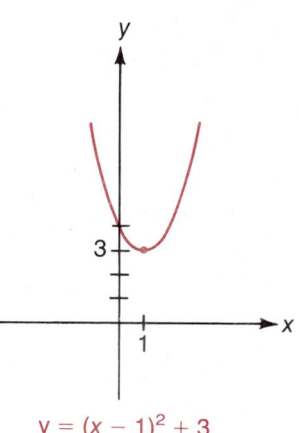

$y = (x - 1)^2 + 3$

Figure 4.28

The graph of an increasing (decreasing) function rises (respectively, falls) as x moves from left to right. The horizontal line test shows that an increasing or decreasing function has an inverse. The function graphed in Figure 4.26 has no inverse; it is neither an increasing nor a decreasing function, since its graph falls on the interval $[-1, 0]$ and rises on the interval $[0, \infty)$. However, if we restrict our attention to the interval $[0, \infty)$ on which f is increasing, the restricted function has an inverse; see Figure 4.25 again.

While increasing functions and decreasing functions do have inverses, a function need not be strictly increasing or decreasing in order to have an inverse. For example, the horizontal line test indicates that the function graphed in Figure 4.29 has an inverse.

Figure 4.29

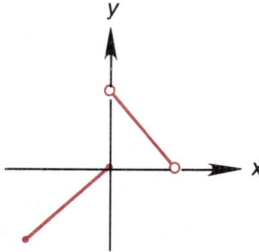

Graphing Inverse Functions

Let f be a function that has an inverse f^{-1}. If $v = f(u)$, then $u = f^{-1}(v)$. Specifically, if $f(2) = 4$, then $f^{-1}(4) = 2$. This means that if the point (2, 4) is on the graph of f, then (4, 2) is on the graph of f^{-1}. More generally, if (u, v) is on the graph of f, then (v, u) is on the graph of f^{-1}.

> **GRAPHING f^{-1}**
>
> The graph of f^{-1} can be obtained by interchanging the order of the coordinates of each point on the graph of f. That is, (b, a) is on the graph of f^{-1} if and only if (a, b) is on the graph of f.

Note that interchanging the order of the coordinates reflects the point through the line $y = x$; see Figure 4.30a. Figure 4.30b illustrates this procedure for sketching the graph of f^{-1} when the graph of f is given.

Figure 4.30

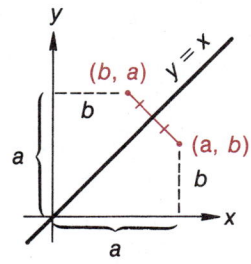
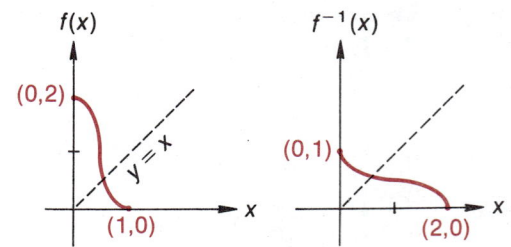

a. Reversing coordinates is equivalent to reflection through the line $y = x$.

b. Reflecting the graph of f through the line $y = x$ obtains the graph of f^{-1}.

EXAMPLE 9   If $y = f(x) = \sqrt{2x - 2}$ for $x \geq 1$, sketch the graphs of both the function and its inverse. Find the domain and range for both f and f^{-1}.

SOLUTION   Setting
$$y = \sqrt{2x - 2} \geq 0$$
we find
$$2x - 2 = y^2$$
Thus,
$$x = f^{-1}(y) = \frac{y^2 + 2}{2}, \qquad y \geq 0$$

To graph this functional relationship, we write
$$f^{-1}(x) = \frac{x^2 + 2}{2}, \qquad x \geq 0$$

(Remember, the variable used to describe the function is immaterial.)

In this case, it is more convenient to graph $y = f^{-1}(x)$ first and then "reflect through the diagonal" to obtain the graph of $y = f(x)$. Figure 4.31b shows the graph of $y = f^{-1}(x) = \frac{(x^2 + 2)}{2}$. Because of the restriction $x \geq 0$, we use only the right half of this graph. The graph of $y = f(x) = \sqrt{2x - 2}$ is then sketched in Figure 4.31a by reflecting b through the line $y = x$.

Figure 4.31

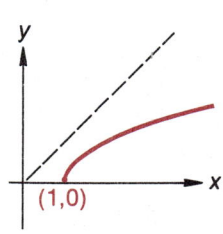
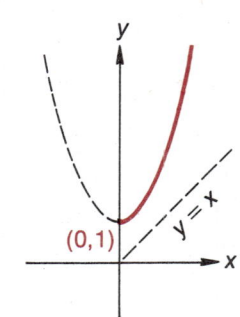

a. $y = f(x) = \sqrt{2x - 2}, \quad x \geq 1, y \geq 0$

b. $y = f^{-1}(x) = \frac{x^2 + 2}{2}, \quad x \geq 0, y \geq 1$

The domain of f consists of all $x \geq 1$; its range consists of all $y \geq 0$. The domain of f^{-1} consists of all $x \geq 0$; its range consists of all $y \geq 1$.

$$\text{Domain } f = [1, \infty), \qquad \text{Range } f = [0, \infty)$$
$$\text{Domain } f^{-1} = [0, \infty), \qquad \text{Range } f^{-1} = [1, \infty)$$

Section 4.4 Exercises

Which of the following functions are one-to-one? Let the domain be implied by the graph or formula.

1.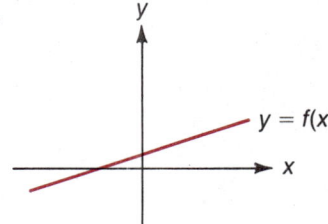

2.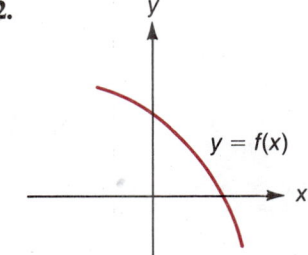

3.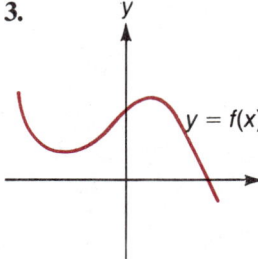

4.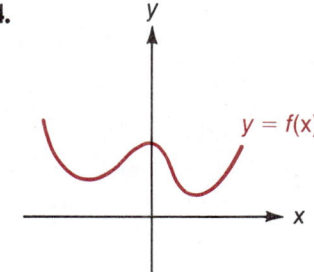

5. $f(x) = x + 3$

6. $g(x) = 2 - 5x$

7. $F(x) = \dfrac{1}{x}$

8. $G(x) = \dfrac{2}{x - 1}$

9. $h(x) = x^2 - 4$

10. $P(x) = 3 - 2x^2$

11. $f(x) = \sqrt{x - 1}$

12. $g(x) = \sqrt{x^2}$

13. $F(x) = |x + 2|$

14. $G(x) = \dfrac{1}{|x - 1|}$

15. $h(x) = 10$

16. $K(x) = 29$

17. $f(x) = \sqrt[3]{x^3 - 8}$

18. $g(x) = \sqrt[5]{x + 2}$

19. $F(x) = \dfrac{1}{x^2 - 1}$

20. $G(x) = \dfrac{3}{1 - 2x^2}$

In each of the following, determine whether an inverse function exists. If so, find it.

21. $f(x) = -4x + 1$

22. $H(u) = 2u - 8$

23. $g(v) = 32v^5$

24. $F(r) = \dfrac{1}{4r^4}, \quad r \neq 0$

25. $P(s) = 16s^4, \quad s \geq 0$

26. $F(x) = -\sqrt[3]{2x}$

27. $g(y) = \sqrt[6]{y^2}$

28. $Q(u) = \dfrac{2}{\sqrt[3]{3u}}, \quad u \neq 0$

29. $h(t) = 2t^2 - 1$

30. $s(t) = \sqrt{2t - 3}, \quad t \geq \dfrac{3}{2}$

31. $g(u) = \sqrt[3]{u^2 - 1}$

32. $G(x) = (x - 1)^2$

33. $s(x) = (x - 3)^2, \quad x \geq 3$

34. $s(t) = (t + 2)^2, \quad t \geq -2$

35. $F(x) = \dfrac{2x + 1}{x - 2}, \quad x \neq 2$

36. $G(s) = \dfrac{s + 1}{s - 1}, \quad s \neq 1$

37. $p(t) = \dfrac{3 - 2t}{2 - 3t}, \quad t \neq \dfrac{2}{3}$

38. $Q(r) = \dfrac{1 - r}{1 + r}, \quad r \neq -1$

Show that the following functions are inverses of one another by showing that $g[f(x)] = x$ and $f[g(x)] = x$ for all appropriate values of x.

39. $f(x) = -x; \quad g(x) = -x$

40. $f(x) = \dfrac{1}{x}, \quad x \neq 0; \quad g(x) = \dfrac{1}{x}, \quad x \neq 0$

41. $f(x) = x + 2; \quad g(x) = x - 2$

42. $f(x) = 2x + 2$; $g(x) = \dfrac{x-2}{2}$

43. $f(x) = \sqrt{x-1}$, $x \geq 1$; $g(x) = x^2 + 1$, $x \geq 0$

44. $f(x) = x^2 - 2x + 1$, $x \geq 1$; $g(x) = \sqrt{x} + 1$, $x \geq 0$

45. $f(x) = \dfrac{1}{x-1}$, $x \neq 1$; $g(x) = \dfrac{1+x}{x}$, $x \neq 0$

46. $f(x) = \dfrac{2}{x^2 - 1}$, $x > 1$; $g(x) = \sqrt{\dfrac{2+x}{x}}$, $x > 0$

47. $f(x) = \dfrac{x-2}{x+2}$, $x \neq -2$; $g(x) = \dfrac{2(x+1)}{1-x}$, $x \neq 1$

48. $f(x) = \dfrac{2x-1}{2x+1}$, $x \neq -\dfrac{1}{2}$; $g(x) = \dfrac{x+1}{2(1-x)}$, $x \neq 1$

Sketch the graph of f^{-1} for the given graph of f. Find the domain and range of both f and f^{-1}.

49.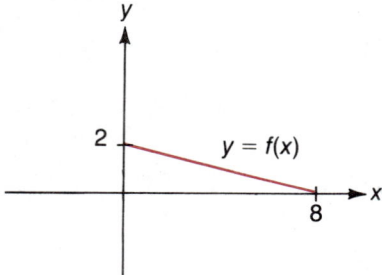

50.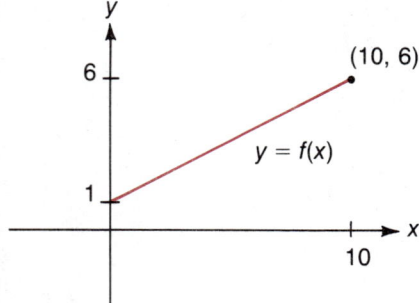

51.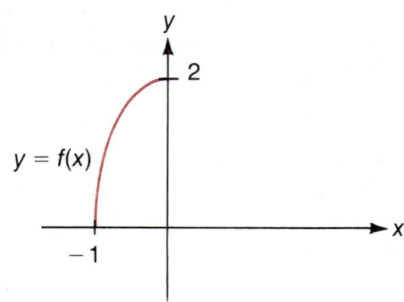

52.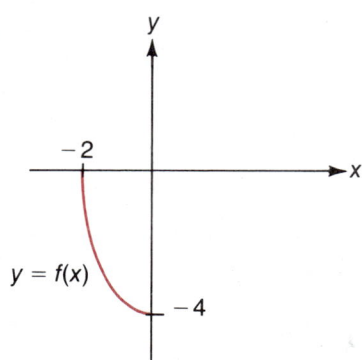

53.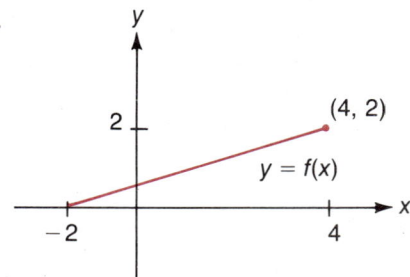

54.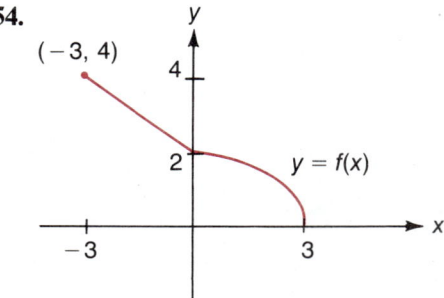

In each of the following
    a. Find the inverse function
    b. Show that $f[f^{-1}(t)] = t$
    c. Show that $f^{-1}[f(t)] = t$
    d. Sketch the graphs of both f and f^{-1}
    e. Find the domain and range of both f and f^{-1}

55. $f(x) = 3x - 2$

56. $f(x) = 6 - 2x$

57. $f(x) = -\sqrt{6 - 2x}$, $x \leq 3$

58. $f(x) = -\sqrt{x + 4}$, $x \geq -4$

59. $f(x) = \dfrac{1}{2x}$, $x \neq 0$

60. $f(x) = -\dfrac{1}{8x^3}$, $x \neq 0$

61. $f(x) = \sqrt[3]{8x}$

62. $f(x) = -\dfrac{2}{\sqrt[3]{8x}}$, $x \neq 0$

63. $f(x) = (x - 2)^2$, $x \geq 2$

64. $f(x) = -(x - 3)^2$, $x \geq 3$

The following are graphs of functions. Determine intervals where each is
  a. *increasing*
  b. *decreasing*
  c. *constant*
  d. *nondecreasing*
  e. *nonincreasing*

65.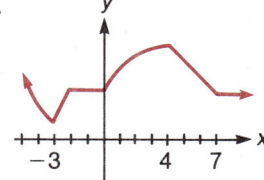

66.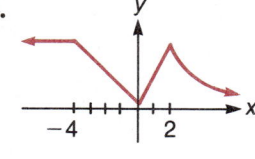

67.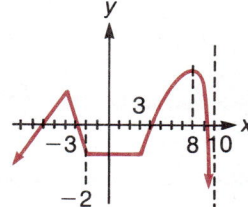

68.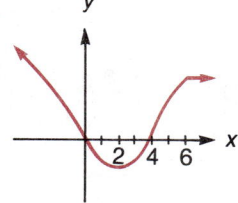

69.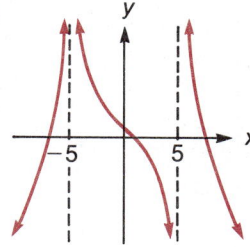

70.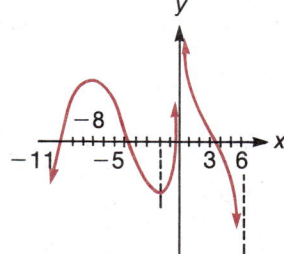

71.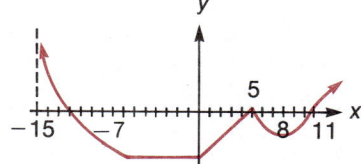

Determine the regions where the following functions are increasing and where they are decreasing.

72. $f(x) = 4 - 3x$

73. $g(u) = 2u + 6$

74. $f(t) = -\dfrac{2}{t}$

75. $h(v) = \dfrac{2}{(v^4 + v^2)}$

76. $F(w) = -w^2 - 2w + 5$

77. $G(t) = 3t^2 - 12t + 7$

78. $f(x) = \sqrt{x}$

79. $g(r) = |r| + r$

80. Show that the sum of two increasing (decreasing) functions is increasing (respectively, decreasing).

81. The function that expresses Fahrenheit temperature in terms of Celsius temperature is given by $F(C) = (\tfrac{9}{5})C + 32$. Find the inverse function expressing Celsius temperature in terms of Fahrenheit temperature.

82. The legal speed limit is now 88 kilometers per hour.
  a. Express the distance that can be traveled at this speed as a function of time.
  b. Find the inverse function expressing the time required for a given trip as a function of the distance involved.

83. The interest earned on an investment equals the principal multiplied by the interest rate and the elapsed time.
  a. At 5% interest, express the amount earned each year as a function of the amount invested.
  b. Find the inverse function expressing the investment required to earn a certain sum each year.

84. An object falling freely toward the earth will have fallen $16t^2$ feet t seconds after it is dropped.
  a. Express the distance fallen as a function of time.
  b. Find the inverse function expressing the time required to fall a certain distance.
  c. How long will it take to fall 40,000 feet?

85. The sales graph for the Energetic Battery Company is given.
  a. Can the Internal Revenue Service subpoena the company's records for the month in which it sold $300,000 worth of batteries? That is to say, is time a function of sales receipts?
  b. Are the sales receipts a function of time?

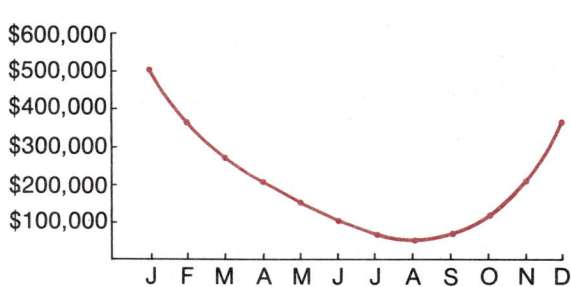

86. The average net yearly income per farm in Ohio is graphed here.
   a. During what periods did farm income increase?
   b. During what periods did farm income decrease?

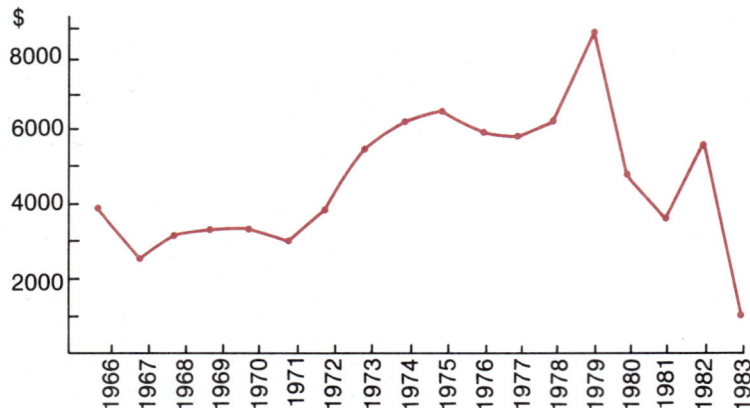

Section 4.5 Chapter Review

Terms and Concepts

Function $y = f(x)$ A correspondence or rule that assigns to each element x of some set X a unique element y of a set Y.

Domain The set X of x values for which a function is defined.

Range The set of all values $y = f(x)$ that can be obtained by considering all x values in the domain.

Independent Variable x
Dependent Variable y $\Big\}$ in $y = f(x)$

Linear Function $f(x) = ax + b$

Quadratic Function $f(x) = ax^2 + bx + c$

Even Function $f(-x) = f(x)$

Odd Function $f(-x) = -f(x)$

Operations on Functions

  Multiple of a Function $(af)(x) = a[f(x)]$

Sum of Functions $(f + g)(x) = f(x) + g(x)$

Difference of Functions $(f - g)(x) = f(x) - g(x)$

Product of Functions $(f \cdot g)(x) = f(x) \cdot g(x)$

Quotient of Functions $\left(\dfrac{f}{g}\right)(x) = \dfrac{f(x)}{g(x)}, \quad g(x) \neq 0$

Composition of Functions $(g \circ f)(x) = g[f(x)]$

Difference Quotient $\dfrac{f(x + h) - f(x)}{h}$

One-to-One Function $f(x_1) \neq f(x_2)$ when $x_1 \neq x_2$

Inverse of a Function $u = f^{-1}(v)$ means $v = f(u)$

Increasing Function $f(x_2) > f(x_1)$ when $x_2 > x_1$

Decreasing Function $f(x_2) < f(x_1)$ when $x_2 > x_1$

Constant Function $f(x_2) = f(x_1)$ for all x_1, x_2

Rules and Formulas

Vertical Line Test A graph represents y as a function of x if and only if each vertical line intersects the graph in at most one point.

Horizontal Line Test A graph determines x as a function of y if and only if each horizontal line intersects the graph in at most one point.

Section 4.5 Chapter Review

Horizontal Line Test for f^{-1}

If $y = f(x)$, the following are equivalent to each other.
1. $y = f(x)$ determines x as a function of y.
2. Each horizontal line intersects the graph in at most one point.
3. f is one-to-one.
4. f^{-1} exists.

Inverse Functions

$f^{-1}[f(t)] = t$ and $f[f^{-1}(t)] = t$

Finding f^{-1}

First Method

1. Solve $v = f(u)$ for u in terms of v. $\quad v = f(u) = 2u + 1$
2. a. If u is determined uniquely by v, then $u = f^{-1}(v)$. $\quad u = \dfrac{v-1}{2}$
   b. Otherwise f^{-1} does not exist.

   $f^{-1}(v) = \dfrac{v-1}{2}$

3. v is a dummy variable; change to $f^{-1}(t)$ or $f^{-1}(x)$ if desired. $\quad f^{-1}(x) = \dfrac{x-1}{2}$

Alternate Method

1. Verify that f^{-1} exists.
   a. Show that f is one-to-one, or
   b. Use the horizontal line test.
2. Solve the equation $f[f^{-1}(t)] = t$ for $f^{-1}(t)$.

$f(u) = 2u + 1$
$f[f^{-1}(t)] = 2[f^{-1}(t)] + 1 = t$
$2f^{-1}(t) = t - 1$
$f^{-1}(t) = \dfrac{t-1}{2}$

Graphing Techniques

Translation

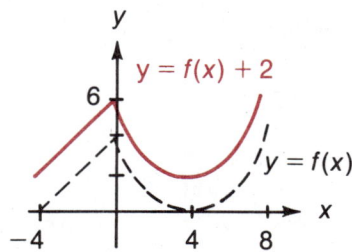

Vertical translation
$y = f(x) + c$

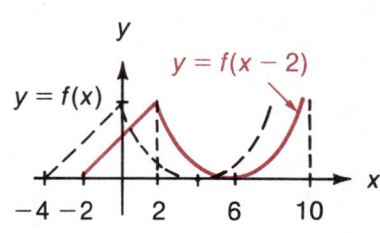

Horizontal translation
$y = f(x - a)$

Reflection

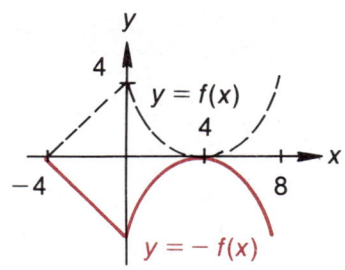

Vertical reflection
$y = -f(x)$

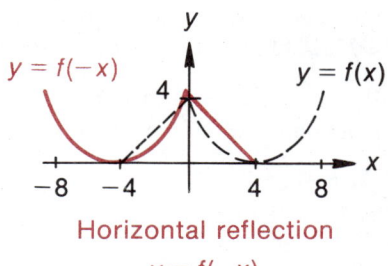

Horizontal reflection
$y = f(-x)$

264 Chapter 4 Functions and Their Graphs

Scaling

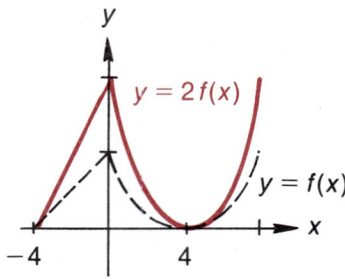
Vertical scaling
$y = Af(x)$

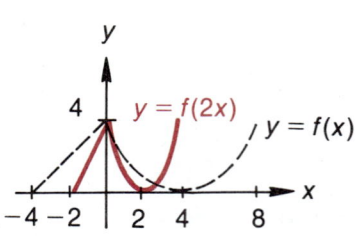
Horizontal scaling
$y = f(ax)$

Symmetry

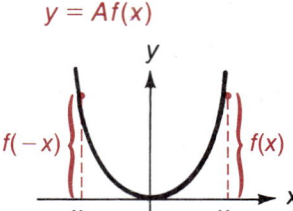
Symmetry about y-axis
(an even function)

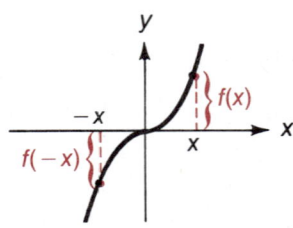
Symmetry about origin
(an odd function)

Absolute Values Any point with a negative y value is reflected through the x-axis to the point with the corresponding positive y value.

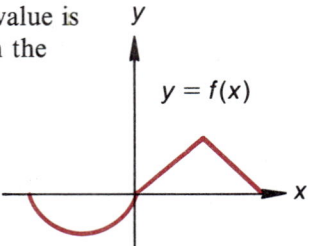
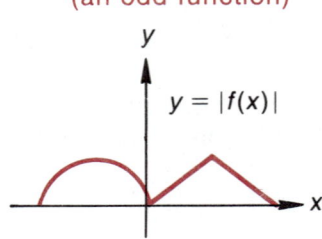

Inverse Functions The graph of f^{-1} can be obtained by interchanging the order of the coordinates of each point on the graph of f; that is, by reflecting the graph of f through the line $y = x$.

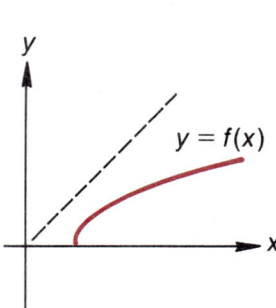
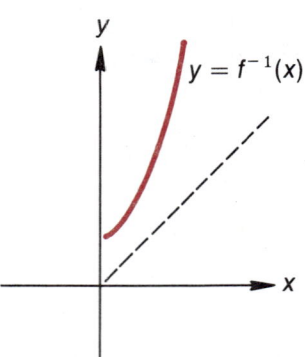

Section 4.6 Supplementary Exercises

Which of the following equations define y as a function of x? Which define x as a function of y?

1. $x - y^2 = 3x$
2. $2x^2 - 3y^2 = 4$
3. $x^3 = y^3$
4. $x^2 = y^2$

5. Which of the following graphs define y as a function of x? Which define x as a function of y?

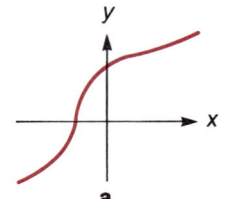
a.

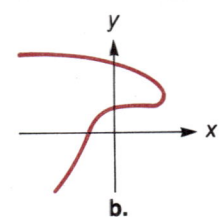
b.

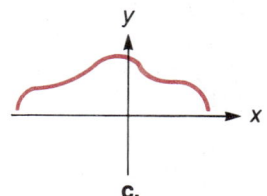
c.

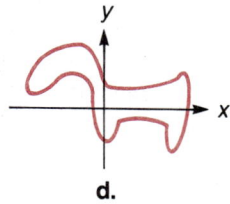
d.

In each of the following, find the requested expressions.

6. $y(x) = 3x - 1$; find $y(1)$, $y(4)$, $y(-4)$.
7. $g(y) = y^2 + 2y + 1$; find $g(5)$, $g(2)$, $g(-2)$.
8. $M(s) = 6s^3 + 3s - 5$; find $M(0)$, $M(1)$, $M(-1)$.
9. $f(v) = \dfrac{2v + 1}{4v - 1}$; find $f(0)$, $f(2)$, $f\left(\dfrac{1}{2}\right)$.
10. $J(c) = 3 + 2c$; find $J(a - 2)$, $J(2a + 1)$, $J(2c - 3)$.
11. $w(x) = x^2 + 2x + 3$; find $w(2x)$, $w(x^2)$, $w\left(\dfrac{1}{x}\right)$.
12. $p(r + 2) = r^2 - 1$; find $p(t)$, $p(r)$, $p(2)$, $p(x + 1)$.
13. $g(t + 5) = t^2 - 3t + 10$; find $g(x)$, $g(t)$, $g(1)$, $g(t - 1)$.
14. $H(s) = 2s - 5$; find and compare
    a. $H(1 + 1)$ and $H(1) + H(1)$
    b. $H(1 - 1)$ and $H(1) - H(1)$
    c. $H(1 \cdot 2)$ and $H(1) \cdot H(2)$
    d. $H\left(\dfrac{1}{2}\right)$, $\dfrac{H(1)}{H(2)}$, and $\dfrac{1}{H(2)}$
15. $w(k) = 3k + 4$; find and compare
    a. $w(2^2)$ and $[w(2)]^2$
    b. $w(\sqrt{4})$ and $\sqrt{w(4)}$
    c. $w(2x)$ and $2w(x)$
    d. $w(-2)$ and $-w(2)$

In the following exercises, find the understood domain of the given function. Both domain and range are to be restricted to real numbers.

16. $g(u) = \dfrac{u^2 - 2u + 1}{u + 2}$
17. $f(w) = \dfrac{\sqrt{w + 1}}{2w - 2}$
18. $f(x) = \dfrac{\sqrt{16 - x^2}}{\sqrt{x^2 - 1}}$
19. $G(v) = \dfrac{\sqrt{4 - v^2}}{\sqrt{v^2 - 16}}$

Sketch the graphs of the following functions.

20. $f(x) = \begin{cases} x, & x < 2 \\ 0, & x = 2 \\ 1, & x > 2 \end{cases}$

21. $g(x) = \begin{cases} x^2, & x < 1 \\ 0, & x = 1 \\ 2 - x^2, & x > 1 \end{cases}$

Use the graph of $y = f(x)$ as given after Exercise 26 to sketch the graph of each of the following.

22. $y = f(x - 3)$
23. $y = -2f(x)$
24. $y = f(x) - 2$
25. $y = f\left(\dfrac{x}{3}\right)$
26. $y = 2f(2x + 6) - 4$

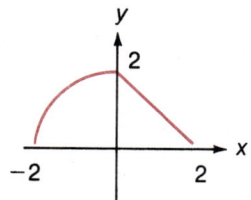

Use the techniques developed in this chapter to sketch the graphs of each of the following.

27. $y = 2x + 3$
28. $y = -(x - 2)^2$
29. $y = (2 - x)^3$
30. $y = 2(x + 1)^2 - 2$
31. $y = -\dfrac{1}{2}(x - 2)^3 + 3$
32. $y = |x + 2| - 2$
33. $y = |x + 1| - 1$
34. $y = 3 - |x - 1|$
35. $y = 2 - |2 - x|$
36. $x = |y^2 + 2y - 3|$
37. $y = -|x^2 - 4x + 1| + 3$
38. $y = |x^2 - 2x - 3| - 4$
39. Extend the given graph under the condition that it be
    a. Even   b. Odd

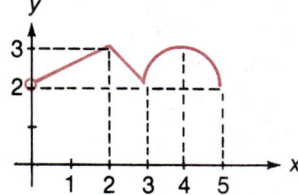

Indicate whether each of the following functions is even $[f(-x) = f(x)]$ or odd $[f(-x) = -f(x)]$ or neither.

40. $y = |x^2 + 2x + 1|$
41. $y = \dfrac{x^2 + 1}{\sqrt[3]{x}}$
42. $y = \sqrt[3]{x}\sqrt[5]{x}$

Find the difference quotient for the following functions.

43. $f(x) = -2$

44. $g(x) = 2x + 5$

45. $h(x) = 4x^2 - 2x + 1$

46. $F(x) = 2x^3 - x + 1$

47. $G(x) = \dfrac{1}{x^2}$

In each of the following, find expression for $(f + g)(x)$, $(f - g)(x)$, $(f \cdot g)(x)$, $(f/g)(x)$, $(f \circ g)(x)$, $(g \circ f)(x)$, and $(af)(x)$, for the indicated value of a. Also evaluate $(f \circ g)(a)$ and $(g \circ f)(a)$ for this a.

48. $f(x) = 2x - 3$, $g(t) = 5t - 3$, $a = -1$

49. $f(u) = 3u$, $g(v) = 3 - 2v$, $a = 2$

50. $f(r) = r^2 + r + 1$, $g(w) = 2 - 4w$, $a = 3$

51. $f(z) = z^2 - z - 1$, $g(a) = 2a + 1$, $a = -3$

52. $f(t) = 16t^2$, $g(s) = \dfrac{(s-1)}{(s+1)}$, $a = -2$

Find functions f and g to describe the given function as $g \circ f$.

53. $F(t) = \left(\dfrac{t+1}{t-1}\right)^2$

54. $G(x) = (x^2 - 1)^5 + 3(x^2 - 1)^3 + (x^2 - 1) + 2$

55. $V(u) = (u^2 + 2u + 5)^{2/3}$

Which of the following functions are one-to-one? Let the domain be implied by the graph or formula.

56.

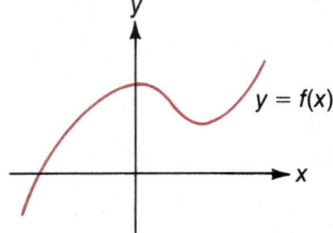

57.

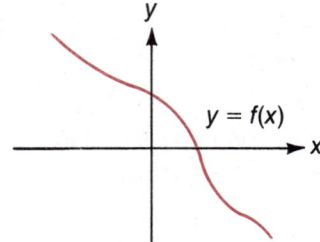

58. $f(x) = \sqrt{x - 2}$

59. $g(x) = \sqrt{x^2 - 4}$

60. $h(x) = x^4 + x^2 + 1$

61. $F(x) = x^3 + 8$

In each of the following, determine whether an inverse function exists. If so, find it.

62. $G(w) = \dfrac{3}{w^7}$, $w \neq 0$

63. $H(r) = r^2$

64. $F(v) = -3\sqrt[4]{v^2}$

65. $p(t) = 2\sqrt[7]{t}$

66. $f(x) = \dfrac{(2 - 3x)}{(x + 2)}$, $x \neq -2$

67. $g(u) = \dfrac{(1 + u)}{(1 - u)}$, $u \neq 1$

Show that f and g are inverses of one another by showing that $g[f(x)] = x$ and $f[g(x)] = x$ for all appropriate x.

68. $f(x) = x + 3$; $g(x) = x - 3$

69. $f(x) = \sqrt{x} - 1$, $x \geq 0$; $g(x) = x^2 + 2x + 1$, $x \geq -1$

70. $f(x) = \dfrac{2}{(x - 2)}$, $x \neq 2$; $g(x) = \dfrac{2(x+1)}{x}$, $x \neq 0$

71. $f(x) = \dfrac{(x+3)}{(x-1)}$, $x \neq 1$; $g(x) = \dfrac{(x+3)}{(x-1)}$, $x \neq 1$

Sketch the graph of f^{-1} for the given graph of f. Find the domain and range for both f and f^{-1}.

72.

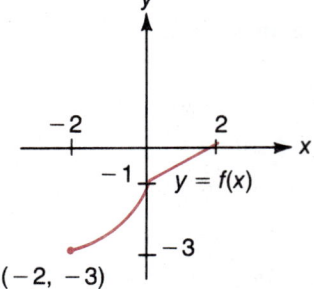

73.

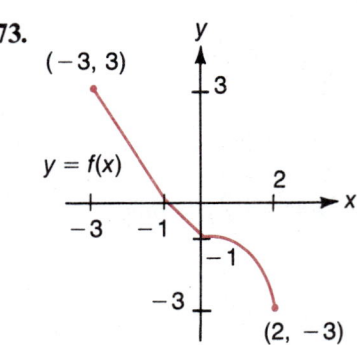

In each of the following
  a. *Find the inverse function*
  b. *Show that* $f[f^{-1}(t)] = t$
  c. *Show that* $f^{-1}[f(t)] = t$
  d. *Sketch the graphs of both f and* f^{-1}
  e. *Find the domain and range for both f and* f^{-1}

74. $f(x) = 1 - 4x$

75. $f(x) = (x - 1)^2, \quad x \leq 1$

76. $f(x) = 2\sqrt{x - 2}, \quad x \geq 2$

77. $f(x) = \dfrac{3}{x^5}, \quad x \neq 0$

78. $f(x) = -3\sqrt[4]{x}, \quad x \geq 0$

79. Determine where the function defined by the following graph is
  a. increasing
  b. decreasing
  c. constant
  d. nondecreasing
  e. nonincreasing

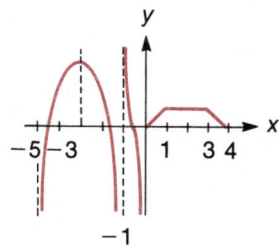

Determine the regions where the following functions are increasing and where they are decreasing.

80. $f(t) = 1 - 2t$

81. $g(x) = \dfrac{1}{x + 1}$

82. $h(r) = \dfrac{-2}{r^2 + |r| + 1}$

83. $f(x) = |x^2 - 2x - 3| + 2$

Polynomial and Rational Functions

5

A function that can be expressed as a polynomial is called a **polynomial function.** The general form of a polynomial function is $P(x) = a_n x^n + \cdots + a_1 x + a_0$ with $a_n \neq 0$ and n a nonnegative integer.

Examples of polynomial functions include the expressions for the volume and surface area of a sphere in terms of its radius x:

$$V(x) = \frac{4}{3}\pi x^3 \quad \text{and} \quad S(x) = 4\pi x^2$$

respectively.

A function that can be expressed as a quotient of two polynomials is called a **rational function:** $R(x) = P(x)/Q(x)$ where P and Q are polynomials. The domain of a rational function $P(x)/Q(x)$ must be restricted so that $Q(x) \neq 0$. The intensity of illumination of a lamp varies inversely as the square of the distance from the lamp: $I(x) = k/x^2$ and the current carried by an electrical wire is inversely proportional to the resistance of the wire: $C(x) = k/x$. Each of these relationships is an example of a rational function.

We have already studied the polynomial functions of degree ≤ 2 in detail, the linear functions $P(x) = ax + b$ and the quadratic functions $P(x) = ax^2 + bx + c$. We now examine polynomial functions of degree >2 and rational functions. In the final sections of this chapter, we shall graph these functions. Locating the x-intercepts will be a crucial step in our graphing efforts. Techniques for finding these intercepts will be described in the early sections of the chapter.

Section 5.1

Roots, Factors, and Synthetic Division

DEFINITION

If $P(r) = 0$, then r is called a **zero** or **root** of the polynomial $P(x)$.

The x-intercepts of a graph occur when $y = 0$; if $y = P(x) = 0$, then x is a root of $P(x)$. Thus the x-intercepts of the graph of a polynomial function $P(x)$ occur at its real roots. Most of the techniques for finding roots of polynomials will depend in some way on the division of one polynomial by another.

Remainder Theorem

If we use the long-division procedure (Section 1.8) to divide a polynomial $P(x)$ by $(x - a)$, where a is a given constant, we obtain a quotient $Q(x)$ (which is also a polynomial) and a remainder R, which is a constant; the remainder contains no x terms.

$$\frac{P(x)}{x-a} = Q(x) + \frac{R}{x-a}$$

$$P(x) = Q(x)(x-a) + R \qquad \text{(Multiply through by } x-a\text{)}$$

Evaluating the polynomial at $x = a$ yields

$$P(a) = Q(a)\underbrace{(a-a)}_{=0} + R$$

$$= R$$

($P(a)$, $Q(a)$ both exist since P and Q are polynomials.) The remainder is the value of the polynomial at $x = a$. This result is embodied in the next theorem.

REMAINDER THEOREM

If a polynomial $P(x)$ is divided by $(x - a)$, where a is a constant, the remainder R is the value of the polynomial at $x = a$: $R = P(a)$.

EXAMPLE 1 Check the validity of the remainder theorem by dividing $(x^3 + 2x^2 - 3x + 4)$ by $(x - 2)$. Then verify the formula $P(x) = Q(x)(x - a) + R$.

SOLUTION

$$\begin{array}{r}
x^2 + 4x + 5 \\
x - 2 \overline{\smash{)}\, x^3 + 2x^2 - 3x + 4} \\
\underline{-(x^3 - 2x^2)} \\
4x^2 - 3x \\
\underline{-(4x^2 - 8x)} \\
5x + 4 \\
\underline{-(5x - 10)} \\
14
\end{array}$$

Here we have

$$P(x) = x^3 + 2x^2 - 3x + 4$$
$$Q(x) = x^2 + 4x + 5$$
$$R = 14$$

The remainder is 14. Evaluating the polynomial at $x = 2$ also yields 14, as we see:

$$P(2) = 2^3 + 2 \cdot 2^2 - 3 \cdot 2 + 4 = 8 + 8 - 6 + 4$$
$$= 14 = R$$

To verify the formula $P(x) = Q(x)(x - a) + R$ (here $a = 2$), we write

$$\begin{aligned}
Q(x)(x-2) + R &= (x^2 + 4x + 5)(x-2) + 14 \\
&= x^3 + 4x^2 + 5x && [= (x^2 + 4x + 5)x] \\
& - 2x^2 - 8x - 10 && [= (x^2 + 4x + 5)(-2)] \\
& + 14 \\
&= x^3 + 2x^2 - 3x + 4 && \text{(Collect like terms)} \\
&= P(x)
\end{aligned}$$ ∎

Factor Theorem

Note that the remainder in a long-division problem $\frac{P(x)}{D(x)}$ is 0 if and only if the denominator $D(x)$ is a factor of the numerator $P(x)$. For if $\frac{P(x)}{D(x)} = Q(x)$, where $Q(x)$ is a polynomial, then $P(x) = D(x)Q(x)$. Combining this observation with the remainder theorem, we obtain the **factor theorem.**

> **FACTOR THEOREM**
>
> $(x - a)$ is a factor of a polynomial $P(x)$ if and only if $P(a) = 0$. That is, $(x - a)$ is a factor of $P(x)$ if and only if a is a root of $P(x)$.

EXAMPLE 2

Without carrying out the long division, show that $(x + 2)$ is a factor of

$$P(x) = (x^4 + 2x^3 - 13x^2 - 14x + 24)$$

SOLUTION

Since $(x + 2) = [x - (-2)]$, it follows that $(x + 2)$ is a factor of $P(x)$ if and only if the value of the polynomial at $x = -2$ is 0. Evaluating at $x = -2$ we obtain

$$P(-2) = (-2)^4 + 2(-2)^3 - 13(-2)^2 - 14(-2) + 24$$
$$= 16 - 16 - 52 + 28 + 24$$
$$= 0$$

$(x + 2)$ is indeed a factor of the given polynomial. This fact can also be verified by the long-division procedure if you wish to check it. ∎

EXAMPLE 3

Find a polynomial $P(x)$ of degree 3 having roots $x = 3$, $x = -2$, and $x = 4$ and satisfying $P(1) = 36$.

SOLUTION

By the factor theorem, $P(x)$ has factors $(x - 3)$, $[x - (-2)]$, and $(x - 4)$; that is, $P(x)$ has the form

$$P(x) = (x - 3)(x + 2)(x - 4)Q(x)$$

for some polynomial $Q(x)$. Since $P(x)$ has degree 3, $Q(x)$ must have degree 0, so $Q(x)$ is some constant A.

$$P(x) = A(x - 3)(x + 2)(x - 4)$$

Since $P(1) = 36$, we have

$$P(1) = A(1 - 3)(1 + 2)(1 - 4) = 36$$
$$A(-2)(3)(-3) = 36$$
$$18A = 36$$
$$A = 2$$

Finally,
$$P(x) = 2(x-3)(x+2)(x-4)$$
$$= 2x^3 - 10x^2 - 4x + 48 \qquad \blacksquare$$

Synthetic Division

To use the remainder and factor theorems, we first "guess" at a root $x = a$ of a polynomial $P(x)$ and then check to see whether $P(a) = 0$. If so, $x = a$ is a root and $x - a$ is a factor of $P(x)$. We then find the corresponding factor by using the long-division process to divide $P(x)$ by $(x - a)$.

The division of a polynomial by an expression of the form $(x - a)$ can be accomplished without writing out all the details of the long-division process. We can find and record the coefficients by position only, by using a process called **synthetic division**.

Consider, for example, the long division of $2x^4 + 22x - 9x^3 + 20$ by $(x - 3)$. As usual, we arrange the terms in decreasing order and include any missing terms by using 0 as their coefficients. This division is displayed on the left in the following illustration. This division is then rewritten in the middle display but with the variable suppressed; the circled entries are also suppressed since each one is simply a repetition of the entry directly above it. Finally, this format is condensed into the form on the right.

$$
\begin{array}{r}
2x^3 - 3x^2 - 9x - 5 \\
x - 3 \overline{\smash{\big)}\, 2x^4 - 9x^3 + 0x^2 + 22x + 20} \\
\underline{-(2x^4 - 6x^3)} \\
-3x^3 + 0x^2 \\
\underline{-(-3x^3 + 9x^2)} \\
-9x^2 + 22x \\
\underline{-(-9x^2 + 27x)} \\
-5x + 20 \\
\underline{-(-5x + 15)} \\
5 = R
\end{array}
$$

In the final display on the right, we can eliminate the "answer" row provided we include its leading 2 in the bottom row. The bottom row then includes both the quotient and the remainder. This is illustrated on the left in the following display.

Next we change the subtraction to addition by changing the signs of each entry in the second row. But the second row was found by computing $(-3)2 = -6$, $(-3)(-3) = 9, (-3)(-9) = 27$, and $(-3)(-5) = 15$. To maintain this method of obtaining the second row, we change the sign of the -3 as illustrated above on the right.

SYNTHETIC DIVISION

To divide a polynomial $P(x)$ by $(x - a)$:

1. Write $P(x)$ in order of decreasing powers of x, including 0s for any missing powers of x.
2. Place a outside the division symbol and the coefficients of $P(x)$ inside.
3. Fill in the second and third rows as indicated.

$$\frac{P(x)}{(x-a)} = Q(x) + \frac{R}{(x-a)}$$

4. Read the quotient and remainder from the last row, as indicated. If $P(x)$ has degree n, $Q(x)$ has degree $n - 1$.

Divide $P(x) = 2x^4 + 22x - 9x^3 + 20$ by $x - 3$.

1. $P(x) = 2x^4 - 9x^3 + 0x^2 + 22x + 20$

2. $3 \overline{)\, 2 \quad -9 \quad 0 \quad 22 \quad 20\,}$

3.
```
3 ) 2   -9    0    22    20
        6    -9   -27   -15
    ─────────────────────────
    2   -3   -9    -5     5 = R
```
Coefficients of $Q(x)$

4. $Q(x) = 2x^3 - 3x^2 - 9x - 5$
   $R = 5$

That is,

$$\frac{2x^4 + 22x - 9x^3 + 20}{x - 3} = 2x^3 - 3x^2 - 9x - 5 + \frac{5}{x - 3}$$

EXAMPLE 4 Use synthetic division to show that $x = 2$ is a root of $P(x) = 3x^5 - 8x^3 - 6x^2 + x - 10$. Factor $P(x)$ into $(x - 2)Q(x)$.

SOLUTION Using synthetic division to divide $P(x)$ by $(x - 2)$, we have the following display:

```
2 ) 3    0   -8   -6    1   -10
         6   12    8    4    10
    ───────────────────────────
    3    6    4    2    5     0 = R
```

Since the remainder is 0, $(x - 2)$ is indeed a factor of $P(x)$ and $P(2) = 0$ (by the factor theorem) as was to be shown. Furthermore, the last row represents the quotient:

$$\frac{3x^5 - 8x^3 - 6x^2 + x - 10}{x - 2} = 3x^4 + 6x^3 + 4x^2 + 2x + 5$$

Thus, we have the factorization

$$3x^5 - 8x^3 - 6x^2 + x - 10 = (x - 2)(3x^4 + 6x^3 + 4x^2 + 2x + 5)$$

By using synthetic division in Example 4, we were able to factor $P(x)$ almost as fast as we could have evaluated $P(2)$ to determine *if* $(x - 2)$ were a factor. Thus, *synthetic division is a shortcut to finding roots and simultaneously factoring out linear factors (x − a).*

Section 5.1 Exercises

Find a polynomial having degree n and satisfying the given conditions.

1. $n = 3$; roots 2, 1, −5; $P(0) = 20$
2. $n = 3$; roots −1, −2, −4; $P(0) = 8$
3. $n = 4$; roots 3, 5, −3, −5; $P(0) = -450$
4. $n = 4$; roots 2, 4, −2, −4; $P(0) = 128$
5. $n = 3$; roots −1, 2, 3; $P(1) = 16$
6. $n = 3$; roots 4, −2, 5; $P(1) = 144$
7. The graph of $P(x) = x^4 + ax^3 + bx^2 + cx + d$ intersects the x-axis at $x = 1, -2, 3, -4$. Find $P(0)$.
8. The graph of $P(x) = x^4 + ax^3 + bx^2 + cx + d$ intersects the x-axis at $x = -2, 2, 1, 3$. Find $P(4)$.
9. The graph of $P(x) = x^4 + ax^3 + bx^2 + cx + d$ intersects the x-axis at $x = -1, 2, -3$. If $P(4) = 70$, find the y-intercept.
10. The graph of $P(x) = x^4 + ax^3 + bx^2 + cx + d$ intersects the x-axis at $x = 1, -3, 4$. If $P(2) = -60$, find the y-intercept.

Use synthetic division and verify the remainder theorem in the following. Write the polynomial in the form Q(x)(x − a) + R.

11. $(4x^2 + 10x - 8) \div (x + 3)$
12. $(x^2 + 2x + 5) \div (x - 2)$
13. $(u^2 + 5u + 7) \div (u + 2)$
14. $(3x^3 + 2x^2 - x) \div (x - 2)$
15. $(t^3 - 2t^2 + 3t - 4) \div (t - 1)$
16. $(3u^3 + 2u^2 + u - 1) \div (u - 1)$
17. $(3y^3 - 4y^2 + 2y - 10) \div (y - 3)$
18. $(5x^4 + 3x^2 - 2) \div (x + 1)$
19. $(2t^4 - t^3 - 3t^2 - t - 10) \div (t + 2)$
20. $(4x^4 + 3x^3 + 2x^2 + x - 2) \div (x + 1)$
21. $(3w^4 + 6w^3 + 2w^2 - w - 3) \div (w + 2)$
22. $(x^4 - x^3 - x^2 + x) \div (x - 3)$
23. $(v^4 + 6v^3 + v^2 - 8) \div (v - 1)$
24. $(t^5 - 10t^3 + 7t + 6) \div (t - 3)$
25. $(x^6 + x^5 + x^2 + x + 1) \div (x + 1)$

Use the factor theorem to determine whether the second expression is a factor of the first; if so, use synthetic division to factor out the second expression.

26. $(r^3 + r^2 + r + 6)$, $(r + 2)$
27. $(3s^3 + 2s^2 - s - 4)$, $(s + 1)$
28. $(3t^3 + 2t^2 + t + 2)$, $(t + 1)$
29. $(2a^3 - a^2 - 6a - 3)$, $(a - 4)$
30. $(r^4 - 10r^2 + 10r - 1)$, $(r - 1)$
31. $(b^4 - 6b^3 + 6b^2 + 2b + 4)$, $(b - 2)$
32. $(2v^4 - v + 6)$, $(v - 5)$
33. $(x^5 - 32)$, $(x - 2)$
34. $(u^5 - 17u + 2)$, $(u - 2)$
35. $(w^5 - 10w^3 + 6w - 36)$, $(w + 2)$
36. $(z^5 + 3z^4 + 2z^2 - 4)$, $(z + 1)$
37. $(y^6 - 2y^4 + 10y - 12)$, $(y + 2)$
38. $(5u^6 - 7u^3 - 6u + 8)$, $(u - 1)$
39. $(p^6 + 2p^4 + 2p^3 + 30p - 20)$, $(p + 2)$
40. $(s^{100} + s^{67} + s^{13} + s^7 + 1)$, $(s + 1)$

Section 5.2

Real Roots of Polynomials

We found synthetic division, in conjunction with the factor and remainder theorems, to be useful in testing for roots of polynomials. In this section, we address the problem of determining possible roots to be tested.

Rational Roots

For a polynomial with integer coefficients, consideration of its first and last coefficients will limit appreciably the *rational* numbers, which need be considered as potential roots. For if $x = p/q$ is a rational root *in lowest terms* (that is, p and q have no common integer factors other than ± 1) of $P(x) = a_n x^n + a_{n-1} x^{n-1} + \cdots + a_1 x + a_0$, then

$$a_n \left(\frac{p}{q}\right)^n + a_{n-1} \left(\frac{p}{q}\right)^{n-1} + \cdots + a_1 \left(\frac{p}{q}\right) + a_0 = 0$$

Transposing the a_0 term and multiplying by q^n yields

$$a_n p^n + a_{n-1} p^{n-1} q + \cdots + a_1 p q^{n-1} = -a_0 q^n$$
$$p(a_n p^{n-1} + a_{n-1} p^{n-2} q + \cdots + a_1 q^{n-1}) = -a_0 q^n$$

This says that p is a factor of $-a_0 q^n$. But p/q was written in lowest terms. Thus, p cannot have any factors in common with q (or even q^n). This means that p must be a factor of a_0. In a similar fashion, it can be shown that q is a factor of a_n (see Exercise 34).

RATIONAL ROOT THEOREM

If a rational number p/q *in lowest terms* is a root of a polynomial $a_n x^n + \cdots + a_1 x + a_0$ with integer coefficients, then p is an integer factor of a_0 and q is an integer factor of a_n.

The rational root theorem does not indicate that any particular rational number is actually a root. It merely restricts the list of candidates from which we need search for rational roots.

EXAMPLE 1

Find all rational roots of

$$P(x) = 3x^2 - 2x - 1$$

SOLUTION

For $\dfrac{p}{q}$ to be a rational root in lowest terms,

p must be an integer factor of -1: $\quad p = \pm 1$
q must be an integer factor of 3: $\quad q = \pm 1, \pm 3$

Possible rational roots are

$$\frac{p}{q} = \pm 1, \pm \frac{1}{3}$$

Each of the possible roots a must be tested by one of the following.

  i. direct substitution of $x = a$ into $P(x)$ and evaluation of $P(a)$
  ii. polynomial long division by $(x - a)$
  iii. synthetic division by $(x - a)$

(since a is a root if and only if $(x - a)$ is a factor)

We never use polynomial long division for this purpose since synthetic division is more efficient. However, for $a = \pm 1$, substitution into $P(x)$ is also very efficient.

$a = 1$: $P(1) = 3 \cdot 1^2 - 2 \cdot 1 - 1 = 0$ 1 is a root

$a = -1$: $P(-1) = 3 \cdot (-1)^2 - 2 \cdot (-1) - 1 = 4 \neq 0$ -1 is *not* a root

For $a = \pm\frac{1}{3}$, we use synthetic division:

$a = \frac{1}{3}$:
$$\frac{1}{3} \overline{)\, 3 \quad -2 \quad -1\phantom{-\tfrac{4}{3}}}$$
$$\phantom{\frac{1}{3})\, 3 \quad} 1 \quad -\tfrac{1}{3}$$
$$\overline{\phantom{\frac{1}{3})\,} 3 \quad -1 \quad -\tfrac{4}{3}} = R = P(\tfrac{1}{3}) \qquad \text{(By the remainder theorem)}$$

Thus, $\frac{1}{3}$ is not a root since $P\left(\frac{1}{3}\right) \neq 0$

$a = -\frac{1}{3}$:
$$-\tfrac{1}{3} \overline{)\, 3 \quad -2 \quad -1}$$
$$\phantom{-\tfrac{1}{3})\,\, 3\,\,} -1 \quad 1$$
$$\overline{\phantom{-\tfrac{1}{3})\,} 3 \quad -3 \quad 0} = R = P(-\tfrac{1}{3}) \qquad \text{(By the remainder theorem)}$$

Thus $-\frac{1}{3}$ is also a root since $P\left(-\frac{1}{3}\right) = 0$. The rational roots are 1 and $-\frac{1}{3}$. ∎

We could have found the roots in Example 1 more easily by factoring:

$$P(x) = 3x^2 - 2x - 1 = 0$$
$$(3x + 1)(x - 1) = 0$$
$$3\left(x + \frac{1}{3}\right)(x - 1) = 0$$
$$x = -\frac{1}{3}, 1$$

In fact, for linear and quadratic polynomials, the techniques of this chapter are not normally used. The power and usefulness of the rational root theorem become apparent only when it is applied to higher-degree polynomials, as in Example 2.

The rational root theorem will identify ± 1 as possible roots for every polynomial

$$P(x) = a_n x^n + a_{n-1} x^{n-1} + \cdots + a_1 x + a_0$$

Testing $x = 1$, we obtain

$$P(1) = a_n + a_{n-1} + \cdots + a_1 + a_0$$

We thus obtain the following.

TESTING 1 AS A ROOT OF P(x)

If
$$P(x) = a_n x^n + a_{n-1} x^{n-1} + \cdots + a_1 x + a_0$$
then
$$P(1) = 0 \quad \text{if and only if} \quad a_n + a_{n-1} + \cdots + a_1 + a_0 = 0$$

That is, 1 is a root of a given polynomial if and only if the sum of its coefficients is 0.

EXAMPLE 2 Find all the rational roots of the following expression and factor out the corresponding linear factors.

$$P(x) = 3x^4 - 7x^3 - 3x^2 - 7x - 6$$

SOLUTION According to the rational root theorem, for p/q to be a rational root in lowest terms,

p must be an integer factor of -6: $p = \pm 1, \pm 2, \pm 3, \pm 6$
q must be an integer factor of 3: $q = \pm 1, \pm 3$

Thus, we obtain potential roots

$$\frac{p}{q} = \pm 1, \pm 2, \pm 3, \pm 6, \pm \frac{1}{3}, \pm \frac{2}{3}$$

To test 1, we check the sum of the coefficients:

$$P(1) = 3 - 7 - 3 - 7 - 6 = -20 \neq 0$$

Thus 1 is *not* a root. Substituting $x = -1$, we have

$$P(-1) = 3 + 7 - 3 + 7 - 6 = 8$$

Thus -1 is also *not* a root. To test the other potential roots, we use synthetic division.

$$x = 2: \quad \begin{array}{r|rrrrr} 2 & 3 & -7 & -3 & -7 & -6 \\ & & 6 & -2 & -10 & -34 \\ \hline & 3 & -1 & -5 & -17 & -40 \end{array} = R = P(2)$$

Thus 2 is not a root since $P(2) \neq 0$; similarly, it can be shown that -2 is not a root.

On the other hand, synthetic division by $(x - 3)$ yields

$$\begin{array}{r|rrrrr} 3 & 3 & -7 & -3 & -7 & -6 \\ & & 9 & 6 & 9 & 6 \\ \hline & 3 & 2 & 3 & 2 & 0 \end{array} = R = P(3) \quad \text{(By the remainder theorem)}$$

Thus, 3 is a root since $P(3) = 0$ and, by the synthetic division

$$P(x) = (x - 3)(3x^3 + 2x^2 + 3x + 2)$$

Any further roots of $P(x)$ must now *also* be roots of

$$Q(x) = 3x^3 + 2x^2 + 3x + 2$$

Applying the rational root theorem to $Q(x)$ still leaves $p/q = \pm \frac{1}{3}, \pm \frac{2}{3}$ as potential roots. We have now eliminated -3 and ± 6 from consideration in addition to the previously eliminated ± 1 and ± 2. Testing $x = -\frac{2}{3}$ as a root for $Q(x)$, we have

$$\begin{array}{r|rrrr} -\frac{2}{3} & 3 & 2 & 3 & 2 \\ & & -2 & 0 & -2 \\ \hline & 3 & 0 & 3 & 0 \end{array} = R = P(-\tfrac{2}{3}) \quad \text{(By the remainder theorem)}$$

Thus $-\frac{2}{3}$ is a root and by the synthetic division, we have

$$Q(x) = \left(x + \frac{2}{3}\right)(3x^2 + 3)$$

$$= \left(x + \frac{2}{3}\right)3(x^2 + 1)$$

Now $x^2 + 1$ has no real roots. Thus, 3 and $-\frac{2}{3}$ are the only rational (even real) roots of $P(x)$ and

$$P(x) = (x - 3)\underbrace{\left(x + \frac{2}{3}\right)3}_{\text{Multiply}}(x^2 + 1)$$

$$= (x - 3)(3x + 2)(x^2 + 1) \qquad \blacksquare$$

Descartes' Rule of Signs

René Descartes, who developed the Cartesian coordinate system, has also given us a rule to help in locating real roots of polynomials. We shall state his rule here without proof.

> **DESCARTES' RULE OF SIGNS**
>
> 1. The number of positive real roots of a polynomial $P(x)$ is either
>   a. Equal to the number of sign changes in $P(x)$
>   b. Less than this number of sign changes by a positive integer multiple of 2
> 2. The number of negative real roots is limited in the same way by the number of sign changes in $P(-x)$.
>
> In each of these computations, *the terms of the polynomial must be arranged in descending order* and *missing terms are not inserted*.

If $P(x) = (x - a)^k Q(x)$ and a is not a root of $Q(x)$, then a is said to be a **multiple root of order k** of the polynomial $P(x)$. Multiple roots of order k are counted as a root k times in Descartes' rule of signs.

EXAMPLE 3

Use Descartes' rule of signs to discuss the number of positive and negative roots of

$$P(x) = 4x^5 - 3x^3 - 2x^2 + x + 1$$

SOLUTION

There are two changes of signs in $P(x)$ as indicated here:

$$P(x) = 4x^5 - 3x^3 - 2x^2 + x + 1$$
$$\quad\quad\quad\;\; \underbrace{}_{+ \text{ to } -} \;\; \underbrace{}_{- \text{ to } +}$$

Thus $P(x)$ has two or zero positive roots. On the other hand,

$$P(-x) = 4(-x)^5 - 3(-x)^3 - 2(-x)^2 + (-x) + 1$$
$$= -4x^5 + 3x^3 - 2x^2 - x + 1$$
$$\underbrace{}_{- \text{ to } +} \underbrace{}_{+ \text{ to } -} \quad \underbrace{}_{- \text{ to } +}$$

has three sign changes. Thus, $P(x)$ has three or one negative roots. In particular, this polynomial has at least one real root. $\blacksquare$

Bounds on Real Roots

The synthetic division process can also be used to find ranges in which all real roots of a polynomial must lie.

RANGE LIMITATION TEST

Let $P(x)$ be a polynomial with real numbers as coefficients and positive leading coefficient. Use synthetic division to divide $P(x)$ by $(x - a)$.

1. If $a > 0$ and the third row of this synthetic division contains no negative entry, then all real roots of $P(x)$ are less than or equal to a.
2. If $a < 0$ and the signs alternate in the third row of this synthetic division, then all real roots of $P(x)$ are greater than or equal to a. (Here 0 may be considered to be either positive or negative—but not both in the same location—in order to make the signs alternate.)

EXAMPLE 4 Establish a range that includes all real roots of
$$P(x) = 3x^4 - 4x^3 - x^2 + 4x - 1$$

SOLUTION To make the range as small as possible, we start by trying $a = 1, 2,$ and so on. Synthetic division by $(x - 1)$ yields negative entries in the last row. On the other hand, for $(x - 2)$ we have

$$\begin{array}{r|rrrrr} 2 & 3 & -4 & -1 & 4 & -1 \\ & & 6 & 4 & 6 & 20 \\ \hline & 3 & 2 & 3 & 10 & 19 = R = P(2) \end{array}$$

There are no negative entries in the last row. Thus, all real roots of $P(x)$ are less than 2. (2 is not a root since $P(2) = 19 \neq 0$.)

To limit roots on the negative side, we begin with $a = -1$:

$$\begin{array}{r|rrrrr} -1 & 3 & -4 & -1 & 4 & -1 \\ & & -3 & 7 & -6 & 2 \\ \hline & 3 & -7 & 6 & -2 & 1 = R = P(-1) \end{array}$$

Since the signs alternate, all real roots are greater than -1. (-1 is not a root, since $P(-1) = 1 \neq 0$.)

We have established the range $-1 < r < 2$ for the real roots r of $P(x)$. The real roots of $P(x)$ all lie in the interval $(-1, 2)$. ∎

Irrational Roots

The range limitation test and Descartes' rule of signs apply to *all* real roots, rational and irrational. To locate irrational roots, we resort to the method of successive approximations. It is based on the simple property that *if $P(a)$ and $P(b)$ differ in sign, then $P(x) = 0$ for some x between a and b* as illustrated in Figure 5.1.

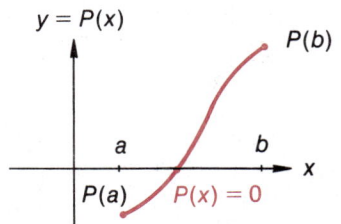

Figure 5.1

SUCCESSIVE APPROXIMATION OF ROOTS

1. Find two values $a, b,$ for which $P(a)$ and $P(b)$ differ in sign; there must be a root between a and b.
2. Break down the interval already obtained to isolate a root in a narrower interval.
3. Continue the process of refinement until a sufficient degree of accuracy is achieved.

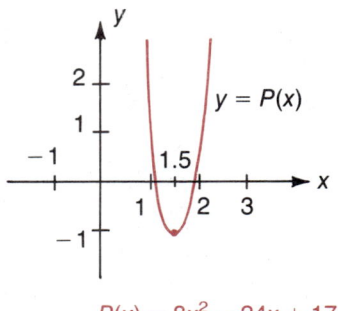

$P(x) = 8x^2 - 24x + 17$

Figure 5.2

Intervals containing roots will sometimes be missed if only integer values of x are tested. For instance, if $P(x) = 8x^2 - 24x + 17$, we have $P(1) = 1 > 0$ and $P(2) = 1 > 0$; yet there are two roots of $P(x)$ between 1 and 2. See Figure 5.2.

If we test $x = 1.5 = \frac{3}{2}$, we find $P(\frac{3}{2}) = -1 < 0$. Thus, there is at least one root in the interval (1, 1.5) and at least one root in the interval (1.5, 2). If not all roots are found to lie in certain intervals by testing integer endpoints, smaller intervals should be tested.

Needless to say, a calculator is essential for the method of successive approximations. In fact, a small computer would be even better as the number of computations required can be burdensome. In calculus, faster methods of locating roots are developed.

CALCULATOR COMMENTS

To evaluate a polynomial $P(x)$ at $x = a$, it is convenient to express the polynomial in telescoping form. The telescoping form of $P(x) = 5x^3 + 3x^2 + 6x + 1$ is given here.

$$5x^3 + 3x^2 + 6x + 1 = x[5x^2 + 3x + 6] + 1$$
$$= x[x(5x + 3) + 6] + 1$$

Then to evaluate $P(-2)$, we store $x = -2$ in memory:

$\boxed{-}\boxed{2}\boxed{\text{Sto}}$

($\boxed{\text{Sto}}$ denotes a memory or storage button.) Now we evaluate from the inside to the outside of the telescoping form. ($\boxed{\text{RCL}}$ denotes a memory recall key.) Press

$\boxed{5}\boxed{\times}\boxed{\text{RCL}}\boxed{+}\boxed{3}\boxed{=}$

$\boxed{\times}\boxed{\text{RCL}}\boxed{+}\boxed{6}\boxed{=}$

$\boxed{\times}\boxed{\text{RCL}}\boxed{+}\boxed{1}\boxed{=}$

and read the answer, $\boxed{-39} = P(-2)$.

EXAMPLE 5

Find all real roots of the following correct to the nearest tenth.

$$P(x) = 4x^5 - 2x^4 + 10x^3 - 5x^2 - 6x + 3$$

SOLUTION

In this case

$$P(-x) = -4x^5 - 2x^4 - 10x^3 - 5x^2 + 6x + 3$$

There are four sign changes in $P(x)$ and only one sign change in $P(-x)$. Thus Descartes' rule of signs indicates four, two, or zero positive (real) roots and exactly one negative (real) root.

Next, we limit the range in which the roots must lie. Synthetic division of $P(x)$ by $(x - 1)$ yields

$$\begin{array}{r|rrrrrr} 1) & 4 & -2 & 10 & -5 & -6 & 3 \\ & & 4 & 2 & 12 & 7 & 1 \\ \hline & 4 & 2 & 12 & 7 & 1 & 4 = R = P(1) \end{array}$$

There are no negative entries in the last row. Thus, all real roots are less than 1, ($P(1) \neq 0$).

Dividing $P(x)$ by $[x - (-1)]$ yields

$$
\begin{array}{r|rrrrrr}
-1 & 4 & -2 & 10 & -5 & -6 & 3 \\
   &   & -4 & 6 & -16 & 21 & -15 \\
\hline
   & 4 & -6 & 16 & -21 & 15 & -12 = R = P(-1)
\end{array}
$$

Since the signs alternate in the last row, all roots of $P(x)$ must be greater than $-1 (P(-1) \neq 0)$. All roots of $P(x)$ lie in the interval $(-1, 1)$. Testing the integers in this range by direct substitution or synthetic division, we have

$$P(-1) = -12$$
$$P(0) = 3$$
$$P(1) = 4$$

Since $P(-1) < 0$ and $P(1) > 0$, a root is indicated between -1 and 0, but we get no further information regarding the existence of roots between 0 and 1.

Possible rational roots are $p/q = \pm\frac{3}{4}, \pm\frac{1}{2}, \pm\frac{1}{4}$. The rational root theorem also gives $\pm 1, \pm 3$, and $\pm\frac{3}{2}$, but these lie outside the range $(-1, 1)$. Synthetic division can be used to eliminate all possibilities except $x = \frac{1}{2}$. For $x = \frac{1}{2}$ we have

$$
\begin{array}{r|rrrrrr}
\frac{1}{2} & 4 & -2 & 10 & -5 & -6 & -3 \\
 &   & 2 & 0 & 5 & 0 & -3 \\
\hline
 & 4 & 0 & 10 & 0 & -6 & 0 = R = P(\tfrac{1}{2})
\end{array}
$$

Thus, $\frac{1}{2}$ is a root, and by the synthetic division

$$P(x) = \left(x - \frac{1}{2}\right)(4x^4 + 10x^2 - 6)$$
$$= \left(x - \frac{1}{2}\right)2(2x^4 + 5x^2 - 3)$$
$$= (2x - 1)(2x^4 + 5x^2 - 3)$$

There must be at least one more positive root (the number of positive roots is zero, two, or four). Any further roots of $P(x)$ must be roots of

$$Q(x) = 2x^4 + 5x^2 - 3$$

But if r is a root of $Q(x)$, then $-r$ is also, since only even powers of x appear in $Q(x)$. Thus, there can be only one more positive root since there is only one negative root.

Using the method of successive approximations to find this positive root, we shall essentially halve the interval each time to minimize the computations necessary to close in on the root.

$$Q(0) = -3$$
$$Q(1) = 4 \qquad \text{(a root in } (0, 1))$$
$$Q(0.5) = -1.625 \qquad \text{(a root in } (0.5, 1))$$
$$Q(0.8) \approx 1.02 \qquad \text{(a root in } (0.5, 0.8))$$
$$Q(0.7) = -0.07 \qquad \text{(a root in } (0.7, 0.8))$$
$$Q(0.75) \approx 0.45$$

Thus, there is a root between 0.7 and 0.75. To the nearest tenth, the roots of $P(x)$ are $x = 0.5, 0.7$, and -0.7. There are no other real roots. Were we to use the techniques of Section 2.4, we would find the roots of $Q(x)$ to be $x = \pm\sqrt{2}/2 \approx \pm 0.707$.

Section 5.2 Exercises

Find all rational roots of the following and factor out the corresponding linear factor.

1. $v^3 - 3v^2 - 6v + 8$
2. $2w^3 + 3w^2 - 32w + 15$
3. $12s^3 + 16s^2 - 5s - 3$
4. $2t^3 + 5t^2 - 8t - 6$
5. $3u^3 + 7u^2 + 5u + 1$
6. $12y^4 - 4y^3 - 3y^2 + y$
7. $t^4 + 2t^3 - 2t^2 - 8t - 8$
8. $18z^3 + 21z^2 - 10z - 8$
9. Show that $\sqrt{2}$ is irrational by finding the rational roots of $P(x) = x^2 - 2$.
10. Show that $\sqrt{3}$ is irrational by finding the rational roots of $P(x) = x^2 - 3$.
11. Show that $\sqrt{5}$ is irrational by finding the rational roots of $P(x) = x^2 - 5$.
12. Show that $\sqrt{7}$ is irrational by finding the rational roots of $P(x) = x^2 - 7$.

Use Descartes' rule of signs to indicate the possible numbers of positive and negative roots of the following.

13. $x^3 + 3x^2 - 2x - 1$
14. $4s^4 - 2s^3 + s - 2$
15. $2u^5 + u^4 - u^3 - u^2 + 2u + 5$
16. $3y^6 - 10y^4 + 2y^2 - 3$
17. $x^5 + x^3 - x + 1$
18. $2t^4 + 3t^3 - 2t^2 + t - 1$
19. $v^7 - 10v^4 + 4v^3 - v + 3$
20. $-w^5 + w^4 - 4w^3 - 2w^2 + w + 1$

Use the range limitation test to find the smallest interval with integer endpoints containing all real roots of the following.

21. $u^4 + 4u^3 - 4u - 4$
22. $4s^4 + 2s^3 - s - 2$
23. $3x^6 + 10x^4 - 2x^2 - 3$
24. $t^4 - 4t^3 - 2t^2 + 12t - 3$
25. $2x^4 - 3x^3 + 2x^2 + x - 1$
26. $v^5 - v^4 - 4v^3 + 2v^2 - v + 1$
27. $2w^5 + 5w^4 - 2w - 5$
28. $-12y^4 + 4y^3 - 3y^2 + 2y - 5$

Show that the given polynomial has a root in the given interval.

29. $x^3 - x + 1$; $(-2, -1)$
30. $x^3 + x^2 + x - 1$; $(0, 1)$
31. $x^4 - x^2 + 4x - 2$; $(0, 1)$
32. $2x^4 + 3x^3 - x - 5$; $(1, 2)$
33. $x^5 + 2x^3 + x^2 + 1$; $(-1, 0)$
34. Finish the proof of the rational root theorem: show that q is a factor of a_n.

Calculator Exercises

Write each of the polynomials in telescoping form and evaluate at $x = a$.

35. $3x^3 + 8x^2 - 24x + 7$; $a = 2$
36. $4x^3 + 2x^2 - 3x + 4$; $a = -1$
37. $2x^4 + x^3 - 3x^2 - x + 4$; $a = 5$
38. $5x^4 - 4x^3 + x^2 - 5x - 5$; $a = 3$
39. $x^6 - x^5 + x^4 - x^3 + x^2 - x - 2$; $a = 3$
40. $x^7 - 2x^6 + x^5 - 3x^4 + x^3 - x^2 + 2x + 2$; $a = -2$

Estimate a real root of the following to within the nearest tenth and within the indicated interval.

41. $u^3 - u^2 - 6u + 2$; $(-3, -2)$
42. $v^3 + 5v^2 - 3$; $(-5, -4)$
43. $y^3 - 5y - 3$; $(-1, 0)$
44. $x^3 + 2x + 7$; $(-2, -1)$
45. $2x^4 - x^3 + 4x^2 + 6x - 4$; $(-2, -1)$

46. $t^3 + t + 1$; $(-1, 0)$

47. $s^3 + 5s^2 - 3$; $(-1, 0)$

48. $y^3 + 5y^2 + 4y + 5$; $(-5, -2)$

Find all real roots of the following. Estimate irrational roots to the nearest tenth.

49. $x^3 - 5x^2 + 2x + 12$
50. $s^3 - s^2 - 5s + 2$
51. $8x^3 + 12x^2 - 66x - 35$
52. $2x^4 - 3x^3 + 6x^2 + x - 15$
53. $t^3 - 2t + 7$
54. $u^3 + 3u^2 + 4u + 5$
55. $2y^5 + 5y^4 - 7y^3 - 2y^2 + 20y - 9$
56. $x^4 - 4x^3 + 3x^2 + 4x - 4$
57. $x^5 - x^4 - 5x^3 + 5x^2 + 6x - 6$
58. $8x^3 - 8x + 1$
59. $s^3 - 3s^2 - 4s + 13$

In the following, express your answers correct to 1 decimal place. Use the method of successive approximations.

60. A propane storage tank is to be constructed in the shape of a cylinder with hemispherical ends. If the tank is to be 20 feet long and have a volume of 1500 cubic feet, what should its radius be?

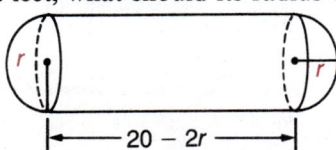

Hint: Let r denote the radius. The total volume in the two hemispheres is $\frac{4}{3}\pi r^3$; the volume of the cylindrical part is $\pi r^2(20 - 2r)$.

61. A spherical fuel tank having radius 20 feet is buried. A stick is to be calibrated to measure the amount of fuel in the tank. If the fuel in the tank is h feet deep, the volume of fuel is

$$\pi\left(20h^2 - \frac{h^3}{3}\right)$$

(when $h \geq 20$). How far from the bottom end of the stick should the mark be placed to indicate a volume of 25,000 cubic feet? (You may assume $h \geq 20$.)

62. Square corners are to be cut from a rectangular piece of metal 30 inches by 50 inches. Then the sides are to be folded up to make a box. What size squares should be cut from the corners to yield a box having a volume of 3000 cubic inches?

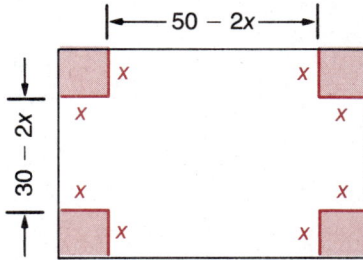

(*Hint:* Let x = side of square cut out. Then volume = $x(30 - 2x)(50 - 2x)$.)

63. A box is to have a square base and an altitude 2 inches longer than a side of the base. Find the dimensions of the box if it is to have a volume of 500 cubic inches. (Volume = length · width · height.)

Section 5.3

Complex Roots of Polynomials

The roots of a quadratic polynomial $P(x) = ax^2 + bx + c$ are given by the quadratic formula (Section 2.3) as

$$x = \frac{-b \pm \sqrt{b^2 - 4ac}}{2a}$$

The number $b^2 - 4ac$ is called the *discriminant* of $P(x) = ax^2 + bx + c$. (See Section 2.3.) It indicates the number and the nature of the roots of $P(x)$. There are

   i. Two real roots if $b^2 - 4ac > 0$.
  ii. Two nonreal roots if $b^2 - 4ac < 0$.
 iii. One (real) root $x_0 = -\dfrac{b}{2a}$ if $b^2 - 4ac = 0$.

284 Chapter 5 Polynomial and Rational Functions

In the latter case, the factor theorem tells us that

$$P(x) = (x - x_0)Q(x)$$

with $Q(x)$ of degree 1; that is, $Q(x) = ax - D = a(x - d)$ for some number d. But then $(x - d)$ is a factor, and d is a root of $P(x)$. As there is only one root x_0, this means that $d = x_0$ and

$$P(x) = ax^2 + bx + c = a(x - x_0)^2$$

Since the factor $(x - x_0)$ appears twice, x_0 is a root of multiplicity two; we say x_0 is a **double** root. Similarly, roots of multiplicity three are called **triple** roots, roots of multiplicity four, **quadruple** roots, and so on.

Fundamental Theorem of Algebra

Every second-degree polynomial having real coefficients has at least one complex root (real or nonreal); in fact, *it will have exactly two if we count double roots twice*. A similar statement holds for every polynomial having real coefficients. Although we are primarily interested in polynomials having real coefficients, the basic results of this section also apply to polynomials having complex coefficients and are stated in that form. The following theorem is stated without proof.

> **FUNDAMENTAL THEOREM OF ALGEBRA**
>
> Every nonconstant polynomial with complex coefficients has at least one complex root.

The remainder and factor theorems also hold for polynomials with nonreal roots as well as for those having real roots. Thus, if r_1 is a root of $P(x)$, then $P(r_1) = 0$, and by the factor theorem we can factor

$$P(x) = (x - r_1)Q(x)$$

Historical Perspective

CARL FRIEDRICH GAUSS
(1777–1855)

The first acceptable proof of the fundamental theorem of algebra was given by Carl Friedrich Gauss (1777–1855) in his doctoral dissertation, which he wrote at the age of 20. Gauss is an important figure in astronomy and electricity as well as mathematics. Legend has it that when Gauss was 3 years old, he discovered a bookkeeping error made by his father. Gauss is recognized as one of the three greatest mathematicians of all time, the other two being Archimedes and Isaac Newton.

If $P(x)$ has degree n, then $Q(x)$ has degree $n - 1$. The fundamental theorem of algebra can then be applied to $Q(x)$ to slice off another factor $(x - r_2)$. Continuing in this way we eventually write

$$P(x) = (x - r_1)(x - r_2) \cdots (x - r_n)A$$

for precisely n factors. We stop only when $Q(x)$ has degree 0; that is, $Q(x)$ is a constant. We thus make two observations.

ROOTS AND FACTORING OF POLYNOMIALS

Let $P(x)$ be a polynomial with complex coefficients.

1. $P(x)$ can be factored completely into linear factors:

$$P(x) = (x - r_1)(x - r_2) \cdots (x - r_n)A$$

where A is a constant (it is the leading coefficient of $P(x)$) and $r_1, \ldots, r_n$ are complex (real or nonreal) roots of $P(x)$.

2. $P(x)$ has precisely n roots; *some of these may be repeated roots.*

Conjugate Roots

If $z = a + bi$ is a root of $P(x) = a_n x^n + \cdots + a_1 x + a_0$ with real coefficients, then $\bar{z} = \overline{a + bi} = a - bi$ is also a root. For if

$$a_n z^n + \cdots + a_1 z + a_0 = 0$$

then

$$\overline{a_n z^n + \cdots + a_1 z + a_0} = \bar{0} = 0$$
$$\bar{a}_n \bar{z}^n + \cdots + \bar{a}_1 \bar{z} + \bar{a}_0 = 0 \quad \text{(The conjugate of a sum or product is the sum or product of the conjugates; see Exercises 82 and 83 in Section 1.9)}$$
$$a_n \bar{z}^n + \cdots + a_1 \bar{z} + a_0 = 0 \quad (a_k = \bar{a}_k \text{ since } a_k \text{ is real})$$

Hence, $\bar{z}$ is indeed a root. In this case, writing $z = a + bi$ and $\bar{z} = a - bi$, the factor theorem gives us

$$P(x) = (x - z)(x - \bar{z})Q(x)$$
$$= (x - [a + bi])(x - [a - bi])Q(x)$$
$$= (x^2 - [a + bi]x - [a - bi]x + [a + bi][a - bi])Q(x)$$
$$= (x^2 - 2ax + [a^2 + b^2])Q(x)$$

But $2a$ and $[a^2 + b^2]$ are real numbers. Thus, $P(x)$ has a real quadratic factor corresponding to the complex conjugate roots. We have obtained the following.

CONJUGATE ROOT THEOREM

The nonreal roots of a polynomial with real coefficients occur in conjugate pairs; if z is a root, then $\bar{z}$ is also a root. Every polynomial with real coefficients can be factored into real linear and quadratic factors.

The computations above have also indicated that

$$[x - (a + bi)][x - (a - bi)] = x^2 - 2ax + (a^2 + b^2)$$

EXAMPLE 1 Find a polynomial of lowest possible degree with real coefficients (leading coefficient of 1) and roots $2, -1, 1 + i, 2 - i$.

SOLUTION From the preceding observations, we know that there are at least two additional roots: $1 - i$ and $2 + i$. The factor theorem says that r is a root of $P(x)$ if and only if $(x - r)$ is a factor of $P(x)$. Thus, $(x - 2)$, $[x - (-1)]$, $[x - (1 + i)]$, $[x - (1 - i)]$, $[x - (2 - i)]$, $[x - (2 + i)]$ are all factors of $P(x)$. Writing $P(x)$ as a product of only these necessary factors, we obtain

$$P(x) = (x - 2)(x + 1)[x - (1 + i)][x - (1 - i)][x - (2 - i)][x - (2 + i)]$$
$$= (x - 2)(x + 1)(x^2 - 2x + 2)(x^2 - 4x + 5)$$
$$= x^6 - 7x^5 + 19x^4 - 21x^3 - 2x^2 + 26x - 20$$

This is the polynomial of lowest degree (with leading coefficient of 1) having the required roots, since any polynomial having these given roots must have the indicated factors. ∎

EXAMPLE 2 Find all roots of

$$P(x) = x^4 - 6x^3 + 10x^2 + 2x - 15$$

if it is known that $2 - i$ is a root. Factor $P(x)$ into real linear and quadratic factors.

SOLUTION Since $z = 2 - i$ is a root, then $\bar{z} = 2 + i$ is also a root and $[x - (2 - i)]$, $[x - (2 + i)]$ are factors of $P(x)$. Thus

$$P(x) = [x - (2 - i)][x - (2 + i)]Q(x)$$
$$= (x^2 - 4x + 5)Q(x)$$

To find $Q(x)$, we carry out the long division:

$$\begin{array}{r}
x^2 - 2x - 3 \\
x^2 - 4x + 5 \overline{\smash{)}\, x^4 - 6x^3 + 10x^2 + 2x - 15} \\
\underline{-(x^4 - 4x^3 + 5x^2)} \\
-2x^3 + 5x^2 \\
\underline{-(-2x^3 + 8x^2 - 10x)} \\
-3x^2 + 12x \\
\underline{-(-3x^2 + 12x - 15)} \\
0
\end{array}$$

Thus

$$P(x) = (x^2 - 4x + 5)(x^2 - 2x - 3)$$
$$= (x^2 - 4x + 5)(x - 3)(x + 1)$$

From this we find the additional roots of $P(x)$: $x = 3, -1$. The complete set of roots is $\{2 - i, 2 + i, 3, -1\}$. ∎

The task of finding real roots of polynomials was addressed in the preceding section. In Section 2.3, we saw that the quadratic formula can be used to find the real or complex roots of quadratic polynomials. Since any polynomial with real coefficients can be factored into real linear and quadratic factors, we can theoretically find all roots of any real polynomial. However, it sometimes requires a certain amount of ingenuity to find the quadratic factors. These correspond to the conjugate pairs of nonreal roots of the polynomial. Quadratic polynomials which have no real roots are said to be **irreducible.**

EXAMPLE 3

Find all the roots of the following polynomial and factor $P(x)$ into real linear and quadratic factors.
$$P(x) = 3x^6 + 5x^5 + x^4 + 5x^3 + x^2 + 5x - 2$$

SOLUTION

The only possible rational roots are ± 1, ± 2, $\pm \frac{1}{3}$, and $\pm \frac{2}{3}$. We find $P(1) = 18 \neq 0$ and $P(-1) = -12 \neq 0$. Testing $x = -2$ by synthetic division, we find

$$\begin{array}{r|rrrrrrr} -2) & 3 & 5 & 1 & 5 & 1 & 5 & -2 \\ & & -6 & 2 & -6 & 2 & -6 & 2 \\ \hline & 3 & -1 & 3 & -1 & 3 & -1 & 0 = R \end{array}$$

Thus, -2 is a root and
$$P(x) = (x + 2)Q(x)$$

where $Q(x) = 3x^5 - x^4 + 3x^3 - x^2 + 3x - 1$. Any remaining roots of $P(x)$ must be roots of $Q(x)$. The only possible rational roots of $Q(x)$ are $\pm \frac{1}{3}$ since ± 1 were previously eliminated. Testing $x = \frac{1}{3}$ we find

$$\begin{array}{r|rrrrrr} \frac{1}{3}) & 3 & -1 & 3 & -1 & 3 & -1 \\ & & 1 & 0 & 1 & 0 & 1 \\ \hline & 3 & 0 & 3 & 0 & 3 & 0 = R \end{array}$$

Hence, $x = \frac{1}{3}$ is a root and

$$P(x) = (x + 2)\left(x - \frac{1}{3}\right)(3x^4 + 3x^2 + 3)$$
$$= (x + 2)\underbrace{\left(x - \frac{1}{3}\right)3}_{\text{Multiply}}(x^4 + x^2 + 1)$$
$$= (x + 2)(3x - 1)(x^4 + x^2 + 1)$$

Now, $x^4 + x^2 + 1$ has no real roots since it is a sum of even powers of x. To determine its quadratic factors, we complete the square and write
$$x^4 + x^2 + 1 = (x^4 + 2x^2 + 1) - x^2$$
$$= (x^2 + 1)^2 - x^2$$
$$= (x^2 + 1 + x)(x^2 + 1 - x)$$

Thus,
$$P(x) = (x + 2)(3x - 1)(x^2 + x + 1)(x^2 - x + 1)$$

The nonreal roots can now be obtained from the quadratic formula by setting $x^2 + x + 1 = 0$ and $x^2 - x + 1 = 0$. They are

$$x = \frac{-1 \pm \sqrt{1-4}}{2} = \frac{-1 \pm \sqrt{3}i}{2}$$

and

$$x = \frac{1 \pm \sqrt{1-4}}{2} = \frac{1 \pm \sqrt{3}i}{2}$$

The complete list of six roots is

$$-2 \qquad \frac{1+\sqrt{3}i}{2} \qquad \frac{-1+\sqrt{3}i}{2}$$

$$\frac{1}{3} \qquad \frac{1-\sqrt{3}i}{2} \qquad \frac{-1-\sqrt{3}i}{2}$$

■

Section 5.3 Exercises

Find the polynomial of lowest degree with real coefficients, leading coefficient of 1, and the indicated roots.

1. $1, 2i$
2. $-1, -i$
3. $i - 1, i + 1$
4. $i, i - 2$
5. $1, 2, i$
6. $-2, 3, 1 + i$
7. $1, 2 - i, i$
8. $-1, i - 1, -i$
9. $2 + i, 1 + i, i$
10. $1 - i, 2, 2 + i$

Find all the roots of the following polynomials and factor them into real linear and quadratic factors.

11. $t^3 - t^2 - 4t - 6$, 3 is a root
12. $x^3 - 2x + 4$, -2 is a root
13. $w^4 - 3w^3 + 6w^2 - 12w + 8$, $-2i$ is a root
14. $y^4 - y^3 + 3y^2 - 9y - 54$, $3i$ is a root
15. $3u^4 + 16u^3 + 24u^2 - 44u - 39$, $2i - 3$ is a root
16. $y^5 - 2y^4 + 6y^3 + 24y^2 + 5y + 26$, $(2 - 3i)$ is a root
17. $z^4 - 10z^3 + 35z^2 - 50z + 34$, $(4 + i)$ is a root
18. $s^4 - 2s^3 + 4s^2 + 4s - 12$, $(1 + \sqrt{5}i)$ is a root
19. $x^4 - 16$
20. $u^4 - 81$
21. $y^4 - 9$
22. $z^4 - 4$
23. $x^3 - 1$
24. $v^3 - 8$
25. $r^3 + 8$
26. $s^3 + 1$
27. $x^4 + 4$ (*Hint:* Complete the square.)
28. $w^4 + 9$
29. $z^4 + 81$
30. $w^4 + 16$
31. $t^3 + 3t^2 + 25t + 75$
32. $u^3 - u^2 + 2$
33. $2w^3 - w^2 + 2w - 1$
34. $v^3 + 5v^2 + 4v - 10$
35. $4x^3 - 2x + 2$
36. $x^3 - x^2 - 4x - 6$

Section 5.4

Graphing Polynomial Functions

A polynomial function with real coefficients

$$y = P(x) = a_n x^n + a_{n-1} x^{n-1} + \cdots + a_1 x + a_0$$

can be graphed by making a table of values and plotting lots of points. But it soon becomes evident that a small computer or at least a programmable calculator would be an immense help for generating the values in the table. Even without resorting to such laborious methods, we can sometimes ascertain the *qualitative* nature of the function and its graph. That is to say, by plotting only a few selected points on the graph, we can sometimes sketch a rough graph that will indicate the behavior of the function.

We have already seen the graphs of the simple linear and quadratic polynomials $y = x$ and $y = x^2$ as illustrated in Figures 5.3a and 5.3b. We can graph $y = x^3$ and $y = x^4$ as in Figures 5.3c and 5.3d. If fact, the graph of $y = x^n$ resembles that of Figure 5.3c for n odd and 5.3d for n even.

Figure 5.3

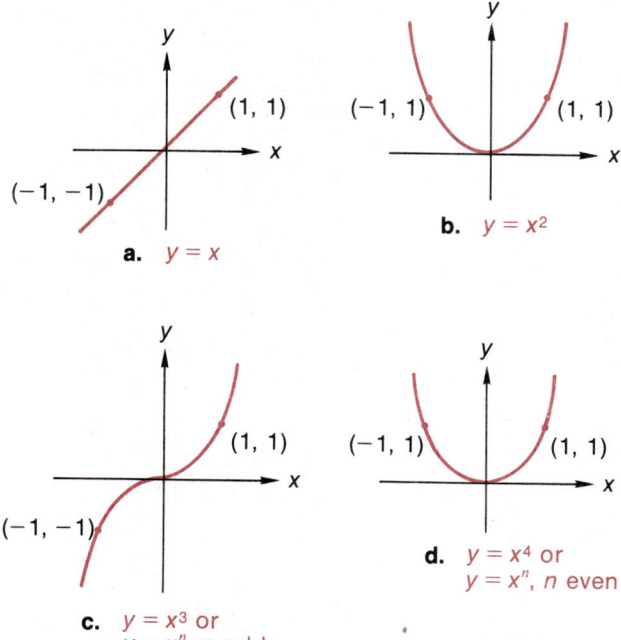

a. $y = x$

b. $y = x^2$

c. $y = x^3$ or $y = x^n$, n odd

d. $y = x^4$ or $y = x^n$, n even

THE BASIC POLYNOMIALS

The graph of $y = P(x) = x^n$, $n \geq 1$

1. Passes through $(0, 0)$ and $(1, 1)$
2. Is symmetric about
   a. The y-axis if n is even
   b. The origin if n is odd
3. If $n_1 < n_2$, then
   a. $x^{n_1} > x^{n_2}$ for $0 < x < 1$
   b. $x^{n_1} < x^{n_2}$ for $x > 1$

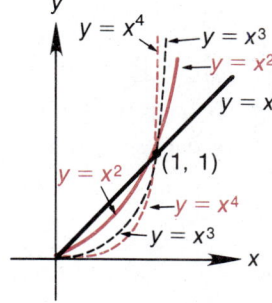

Figure 5.4

The graph of the polynomial

$$P(x) = x^2 - 2x + 2$$
$$= (x - 1)^2 + 1$$

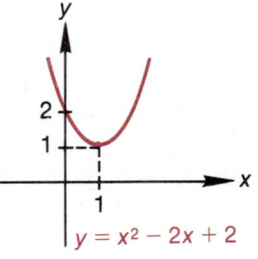

Figure 5.5

is given in Figure 5.5. Note that it does not meet the x-axis; $P(x) > 0$ for every real x. Finding the roots by the quadratic formula, we have

$$x = \frac{-(-2) \pm \sqrt{(-2)^2 - 4 \cdot 2}}{2}$$

$$= \frac{2 \pm \sqrt{-4}}{2} = \frac{2 \pm 2i}{2}$$

$$= 1 \pm i$$

The roots are nonreal. There could be no real roots since the graph does not cross the x-axis: $P(x) \neq 0$ for all real x. Any quadratic polynomial $P(x) = ax^2 + bx + c$ with real coefficients whose graph does not meet the x-axis will have only nonreal roots (see Figure 5.6).

Figure 5.6

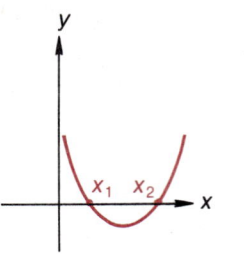
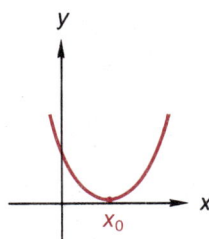
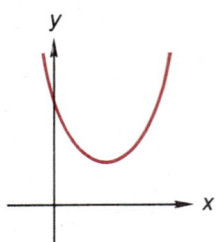

a. Two real roots: x_1 and x_2

b. One real root: x_0 (a double root), no nonreal roots

c. Two nonreal roots: no real roots

$y = ax^2 + bx + c,\ a \neq 0$

An effective tool in graphing a polynomial is the determination or estimation of its roots since the real roots are the x-intercepts of its graph. The fundamental theorem of algebra tells us that a polynomial of degree n has exactly n roots, some of which may be nonreal roots and/or repeated roots. Thus, the graph of a polynomial of degree n can have at most n x-intercepts; these intercepts correspond to its distinct real roots. Plotting these intercepts and a few points between and beyond the intercepts can already give us enough information to sketch a rough graph of the polynomial.

Behavior Between and Beyond Roots

We note in Figure 5.7 that if a polynomial $P(x)$ changes sign from $x = a$ to $x = b$, then there is a root or intercept between a and b. Thus, if all real roots are located, we know that $P(x)$ cannot change sign between two successive roots; for to do so would imply the existence of still another root. Hence, between successive real roots and beyond the largest and smallest real roots, the graph of the polynomial lies entirely above or below the x-axis. Evaluating the polynomial at a test point in each region indicates whether the graph in that region lies above or below the x-axis. Synthetic division and the remainder theorem can be used to facilitate these evaluations.

Figure 5.7

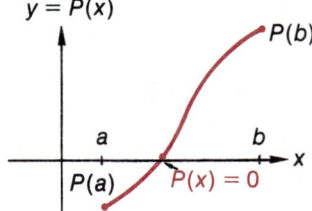

EXAMPLE 1

Sketch the following graphs.

a. $y = P(x) = x^3 - 2x^2 - x + 2$
b. $y = -P(x)$
c. $y = P(x) - 2$

SOLUTION

a. The possible rational roots of $P(x)$ are $x = \pm 1$ and $x = \pm 2$. Since the sum of the coefficients is $0 (1 - 2 - 1 + 2 = 0)$, 1 is a root. By substitution into $P(x)$, we can show that $P(-1) = 0$; -1 is a root. Synthetic division by $(x - 2)$ yields

$$\begin{array}{r|rrrr} 2) & 1 & -2 & -1 & 2 \\ & & 2 & 0 & -2 \\ \hline & 1 & 0 & -1 & 0 = R = P(2) \end{array}$$

Hence 2 is also a root.

There can be only three roots for this cubic polynomial; thus -1, 1, and 2 represent the complete set of roots. Plot these roots on the x-axis (Figure 5.8) and evaluate $P(x)$ at a test point inside each interval determined by these roots. (See Section 2.7.)

Figure 5.8

In the interval $(-\infty, -1)$ we use -2 as a test point and evaluate $P(-2)$ using synthetic division and the remainder theorem.

$$\begin{array}{r|rrrr} -2) & 1 & -2 & -1 & 2 \\ & & -2 & 8 & -14 \\ \hline & 1 & -4 & 7 & -12 = R = P(-2) \end{array}$$

Thus $P(x) < 0$ for all $x \in (-\infty, -1)$. The other values and signs in the following table are generated in a similar way.

INTERVAL	TEST POINT a	$P(a)$	SIGN OF $P(x)$ ON THE INTERVAL
$(-\infty, -1)$	-2	-12	$-$
$(-1, 1)$	0	2	$+$
$(1, 2)$	$\frac{3}{2}$	$-\frac{5}{8}$	$-$
$(2, \infty)$	3	8	$+$

The sign diagram in Figure 5.9 summarizes our discussion to this point.

Figure 5.9

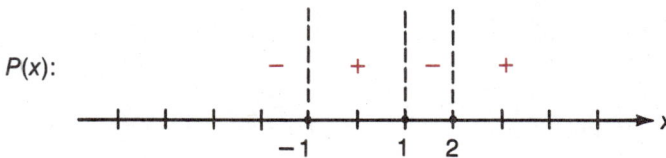

For relatively large (in magnitude) values of x, we write

$$P(x) = x^3 - 2x^2 - x + 2$$
$$= x^3\left(1 - \frac{2}{x} - \frac{1}{x^2} + \frac{2}{x^3}\right)$$

Note that all terms in the parentheses except the first "1" become smaller and smaller as x gets larger in magnitude. Thus,

$P(x) \approx x^3$ for x having large magnitude

In particular, $P(x)$ gets large (positively or negatively) as x does (positively or negatively, respectively). We write

$P(x) \to \infty$ as $x \to \infty$

and

$P(x) \to -\infty$ as $x \to -\infty$

We can now sketch a rough graph of $y = P(x)$.

1. Plot the roots of $P(x)$ on the x-axis:

    -1, 1, and 2

2. Plot the y-intercept:

    $P(0) = 2$

3. Plot the points used in the sign analysis:

    $(-2, -12)$, $(0, 2)$, $\left(\frac{3}{2}, \frac{5}{8}\right)$, and $(3, 8)$.

4. Join these points from left to right with a smooth curve. Beyond the roots, make sure

    $y \to \infty$ as $x \to \infty$

and

$y \to -\infty$ as $x \to -\infty$

See Figure 5.10a.

b. The graph of $y = -P(x)$ is just the reflection through the x-axis of the graph of $y = P(x)$ (see Figure 5.10b).

c. The graph of $y = P(x) - 2$ is obtained by lowering the first graph two units; it is sketched in Figure 5.10c. The x-intercepts can be found by observing that $P(x) - 2 = 0$ when

$$x^3 - 2x^2 - x = 0$$
$$x(x^2 - 2x - 1) = 0$$
$$x = 0, 1 \pm \sqrt{2} \qquad \text{(By the quadratic formula)}$$

Figure 5.10

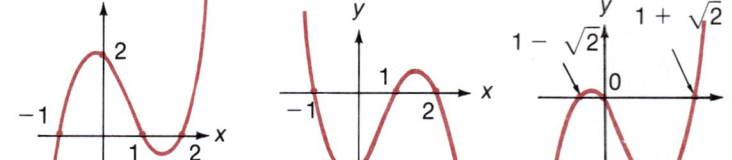

a. $y = P(x) = x^3 - 2x^2 - x + 2$ b. $y = -P(x)$ c. $y = P(x) - 2$

The leading term $a_n x^n$ is important in graphing polynomials. As we saw in Example 1a, the term $a_n x^n$ dominates the polynomial for large positive or negative values of x; the values of $a_n x^n$ tend to override any contribution by smaller powers of x. Thus, as $|x|$ gets large, $P(x)$ gets large and the sign of a_n determines whether these values are positive or negative. Specifically, for large positive x's, $P(x) > 0$ if $a_n > 0$ and $P(x) < 0$ if $a_n < 0$. For large negative x values, we note that for even n the extreme branches of the curve point in the same direction (up or down) whereas for odd n they point in opposite directions. For instance, if $a_n > 0$ and n is odd, then $a_n x^n < 0$ (and hence $P(x) < 0$) for large negative x.

Figure 5.10 illustrates the general shape of the graph of a third degree or "cubic" polynomial. Note that it has two **turnaround points,** at which the polynomial changes from increasing to decreasing, or vice versa. Precise location of such turnaround points requires methods of calculus and will not be addressed here. We will also not be concerned with how far a graph rises above or falls below the x-axis between roots of $P(x)$, except that we will plot the points used in the sign analysis as in Example 1a. Nor will we try to find additional wiggles in the graph between and beyond roots. These questions are all handled better using calculus.

A useful theorem concerning turnaround points is stated here without proof. It can be proved quite easily by using methods of calculus.

THEOREM   A polynomial of degree n can have at most $n - 1$ turnaround points.

This theorem gives us an insight into graphs of more general polynomials. For instance, typical fourth- and fifth-degree polynomials (quartics and quintics) are sketched in Figures 5.11 and 5.12, respectively.

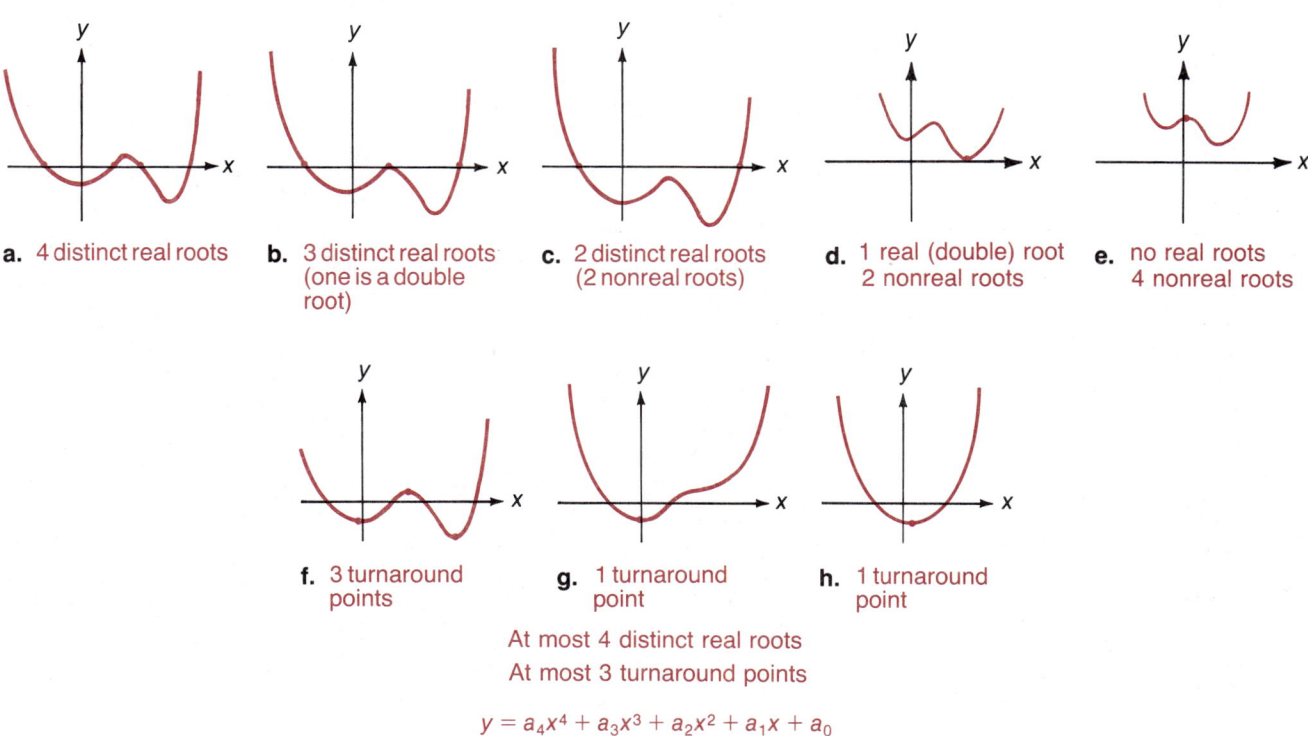

a. 4 distinct real roots

b. 3 distinct real roots (one is a double root)

c. 2 distinct real roots (2 nonreal roots)

d. 1 real (double) root 2 nonreal roots

e. no real roots 4 nonreal roots

f. 3 turnaround points

g. 1 turnaround point

h. 1 turnaround point

At most 4 distinct real roots
At most 3 turnaround points

$y = a_4 x^4 + a_3 x^3 + a_2 x^2 + a_1 x + a_0$

Figure 5.11

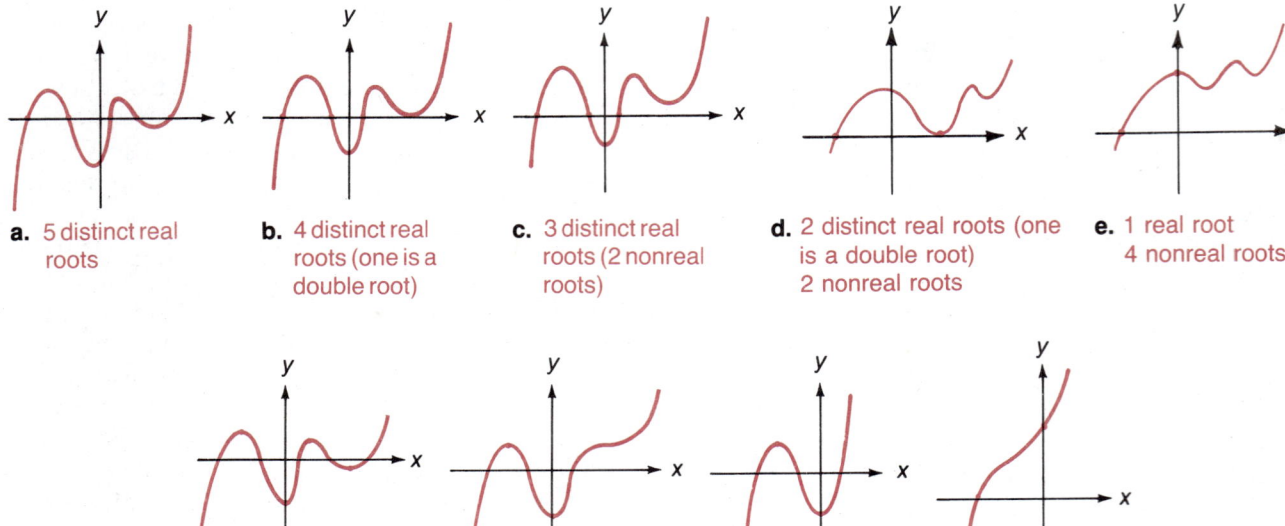

Figure 5.12

GRAPHING POLYNOMIALS

To sketch a rough graph of a polynomial

$$y = P(x) = a_n x^n + \cdots + a_1 x + a_0$$

1. Find the real roots of $P(x)$.
   a. There are at most n of these.
   b. These are the x-intercepts of the graph.
2. Evaluate $P(x)$ at a test point in each of the intervals determined by these roots.
3. Plot the corresponding points on the graph.
4. Within each interval determined by the roots of $P(x)$, the graph stays on the same side of the x-axis (above or below) as determined by the test point plotted there.
5. There are at most $n - 1$ turnaround points.
6. a. If $a_n > 0$, then $P(x) \to \infty$ as $x \to \infty$.
   b. If $a_n < 0$, then $P(x) \to -\infty$ as $x \to \infty$.
7. The extreme branches of the curve point
   a. In the same direction (up or down) if n is even.
   b. In the opposite direction if n is odd.
8. Draw a rough sketch satisfying the above conditions.

EXAMPLE 2

Sketch the graph of

$$y = P(x) = 2x^4 + 3x^3 - 12x^2 - 7x + 6$$

SOLUTION

The possible rational roots of $P(x)$ are $\pm 1, \pm 2, \pm 3, \pm 6, \pm \frac{1}{2}, \pm \frac{3}{2}$. The sum of the coefficients is not 0 ($2 + 3 - 12 - 7 + 6 = -8 \neq 0$); thus 1 is *not* a root of $P(x)$. To test $r = -1$, we use synthetic division to divide by $[x - (-1)]$:

$$\begin{array}{r|rrrrr} -1 & 2 & 3 & -12 & -7 & 6 \\ & & -2 & -1 & 13 & -6 \\ \hline & 2 & 1 & -13 & 6 & 0 = R = P(-1) \end{array}$$

Thus, $r = -1$ is a root and

$$P(x) = (x + 1)Q(x)$$

with

$$Q(x) = 2x^3 + x^2 - 13x + 6$$

The possible rational roots of $Q(x)$ are the same as for $P(x)$. We can verify by substitution into $Q(x)$ that -1 is not a root of $Q(x)$. Then synthetic division will show that 2 is a root of $Q(x)$ and that

$$Q(x) = (x - 2)(2x^2 + 5x - 3)$$

Finally,

$$\begin{aligned} P(x) &= (x + 1)Q(x) \\ &= (x + 1)(x - 2)(2x^2 + 5x - 3) \\ &= (x + 1)(x - 2)(2x - 1)(x + 3) \end{aligned}$$

The roots of $P(x)$ are $-1, 2, \frac{1}{2}, -3$. We then evaluate $P(x)$ at a test point inside each interval determined by these roots.

INTERVAL	TEST POINT a	$P(a)$	SIGN OF $P(x)$ ON THE INTERVAL
$(-\infty, -3)$	-4	162	$+$
$(-3, -1)$	-2	-20	$-$
$(-1, \frac{1}{2})$	0	6	$+$
$(\frac{1}{2}, 2)$	1	-8	$-$
$(2, \infty)$	3	120	$+$

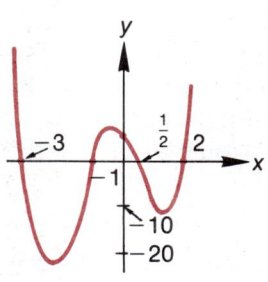

Figure 5.13
$y = P(x) = 2x^4 + 3x^3 - 12x^2 - 7x + 6$

Since the leading term is $2x^4$, we note that $P(x) \to \infty$ as $x \to \pm\infty$. Finally, we sketch a rough graph as in Figure 5.13. ∎

Section 5.4 Exercises

Use Figure 5.3c and d and the graphing techniques developed in Section 4.2 to graph the following polynomials. (Hint: See Examples 1b and c.)

1. $P(x) = x^4 - 1$
2. $P(x) = x^4 + 1$
3. $P(x) = x^4 + 2$
4. $P(x) = x^4 - 2$
5. $P(x) = 4 - x^4$
6. $P(x) = 1 - x^4$
7. $P(x) = x^3 + 8$
8. $P(x) = x^3 - 8$
9. $P(x) = x^3 - 1$
10. $P(x) = x^3 + 1$
11. $P(x) = 8 - x^3$
12. $P(x) = 1 - x^3$
13. $P(x) = (x - 1)^4$
14. $P(x) = (x + 1)^4$
15. $P(x) = (x + 2)^4$
16. $P(x) = (x - 2)^4$
17. $P(x) = (x - 2)^3$
18. $P(x) = (x - 1)^3$
19. $P(x) = (x + 1)^3$
20. $P(x) = (x + 2)^3$
21. $P(x) = -(x + 1)^3$
22. $P(x) = -(x + 2)^3$
23. $P(x) = (x - 1)^4 - 2$
24. $P(x) = (x + 1)^4 - 3$
25. $P(x) = 3 - (x + 2)^4$
26. $P(x) = 4 - (x - 2)^4$
27. $P(x) = (x - 2)^3 - 1$
28. $P(x) = (x - 1)^3 - 2$
29. $P(x) = 3 - (x + 1)^3$
30. $P(x) = 1 - (x + 2)^3$

Use the techniques developed in this section to sketch rough graphs of the following polynomials.

31. $y = (x - 1)(x - 2)(x - 3)$
32. $y = (x + 1)(x - 1)(2x - 1)$
33. $y = (3x - 2)(1 - 2x)(2x + 4)$
34. $y = (4x + 2)(2 - 4x)(x - 2)$
35. $y = (x + 1)(x - 2)(x + 3)(x - 4)$
36. $y = (x - 1)(x + 2)(x - 3)(x + 1)^2$
37. $y = (x^2 - 1)(x^2 - 4)$
38. $y = (x^2 - 9)(x^2 - 16)$
39. $y = (x^2 - 1)(x^2 - 4) - 4$
40. $y = (x^2 - 9)(x^2 - 16) - 2$
41. $y = 4 - (x^2 - 1)(x^2 - 4)$
42. $y = 2 - (x^2 - 9)(x^2 - 16)$
43. $y = (x - 1)^2(x - 2)^2$
44. $y = (x + 1)^2(x - 3)^2$
45. $y = (x - 1)^2(x - 2)^2 + 2$
46. $y = (x + 1)^2(x - 3)^2 + 2$
47. $y = 4 - (x - 1)^2(x - 2)^2$
48. $y = 1 - (x + 1)^2(x - 3)^2$
49. $y = x^3 - 4x$
50. $y = 4x^2 - x^3$
51. $y = x^3 - x^5$
52. $y = x^4 - 5x^2 + 4$
53. $y = x^3 - x^2 - 9x + 9$
54. $y = x^3 + 2x^2 - x - 2$
55. $y = 2x^4 + 4x^3 - 2x^2 - 4x$
56. $y = x^5 - x^4 - x^3 + x^2$
57. $y = x^4 + x^3 - 6x^2 - 4x + 8$
58. $y = x^4 - 5x^3 + 5x^2 + 5x - 6$
59. $y = x^4 + 7x^3 + 13x^2 - 3x - 18$
60. $y = x^5 + x^4 - 9x^3 - x^2 + 20x - 12$

61. Square corners are to be cut from a rectangular piece of metal 30 inches by 50 inches. Then the sides are to be folded up to make a box. If x = side of each square cut out, the volume of the resulting box is $V = P(x) = x(30 - 2x)(50 - 2x)$. See Exercise 62, Section 5.2.
    a. Sketch the graph of this $P(x)$.
    b. What interval of x values constitutes the domain of this volume function?
    c. Use a calculator to evaluate enough $P(x)$ values to estimate the x value that maximizes the volume. Estimate this x to the nearest tenth.

62. A propane storage tank is to be constructed in the shape of a cylinder with hemispherical ends. If the tank is to be 20 feet long and its radius is r, its volume is given by $V = P(r) = \frac{4}{3}\pi r^3 + \pi r^2(20 - 2r)$. See Exercise 60, Section 5.2.
    a. Sketch the graph of this $P(r)$.
    b. What interval of r values constitutes the domain of this volume function?
    c. What radius maximizes the volume?

Section 5.5

Rational Functions

A function such as

$$R(x) = \frac{2x - 1}{x^2 + 3} \quad \text{or} \quad S(x) = \frac{x^{10} + 10x - 3}{2x + 1}$$

which can be expressed as a quotient of two polynomials is called a **rational function**. The roots or zeros of its numerator and denominator are useful in graphing a rational function.

ROOTS OF $P(x)$ AND $Q(x)$

Let $R(x) = \dfrac{P(x)}{Q(x)}$ denote a rational function.

1. The roots of $Q(x)$ must be excluded from the domain of $R(x)$.
2. The roots of $R(x)$ are those roots of $P(x)$ that are not also roots of $Q(x)$; these are the x-intercepts of its graph.
3. The y-intercept of its graph is found by setting $x = 0$. However, if $Q(0) = 0$, there is no y-intercept because then 0 is not in the domain of $R(x)$.

Roots of the denominator that are not also roots of the numerator are called **poles** of the rational function. The graph cannot cross a vertical line at a pole since the pole is not in the domain of the function. We shall see that the graph does get arbitrarily close to such vertical lines, however, and that $y \to \pm\infty$ as this approach takes place.

ASYMPTOTES

1. If $f(x) \to \infty$ or $f(x) \to -\infty$ as $x \to a$, the line $x = a$ is called a **vertical asymptote** of the graph. Vertical asymptotes occur at the poles of a rational function.
2. If the points (x, y) on the graph get arbitrarily close to some curve or line as $x \to \infty$ or $x \to -\infty$ the curve or line is called an **asymptote** of the graph.
    a. If the asymptote is a horizontal line, it is called a **horizontal asymptote**.
    b. If the asymptote is a line with slope $m \neq 0$, it is called an **oblique asymptote**.
    c. If the asymptote is not a line, it is called a **curved asymptote**.

See Figure 5.14 for examples of asymptotes.

Figure 5.14

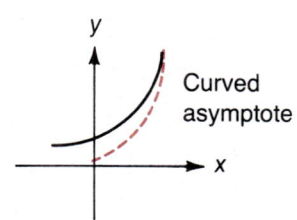
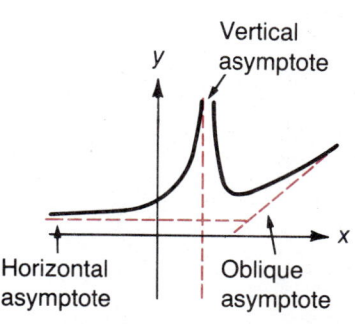

298 Chapter 5 *Polynomial and Rational Functions*

Except in the exercises whose numbers are in color, we will work only with straight line asymptotes (vertical, horizontal, or oblique) in this section.

EXAMPLE 1

Sketch the graph of the following rational functions.

a. $y = f(x) = \dfrac{1}{x}$ **b.** $y = g(x) = \dfrac{1}{x^2}$

c. $y = r(x) = \dfrac{1}{(x-2)}$ **d.** $y = s(x) = \dfrac{1}{(x+1)^2}$

SOLUTION

a. and b. We first sketch these two graphs for $x > 0$. The point $(1, 1)$ is on both graphs and for each

$$y \to 0 \quad \text{as} \quad x \to \infty$$
$$y \to \infty \quad \text{as} \quad x \to 0, x > 0$$

Thus the positive x and y axes are asymptotes for both these graphs. Plotting additional points corresponding to $x = \frac{1}{2}$ and $x = 2$ further determines the shape of these graphs as shown in the first quadrants of Figure 5.15a and b. These graphs are then completed, using the fact that a represents an odd function and b an even function. That is, a must be symmetric about the origin and b must be symmetric about the y-axis. See Figure 5.15a and b.

c. and d. The sketches in Figure 5.15c and d are obtained by shifting the first two graphs horizontally. (See Section 4.2.)

Figure 5.15

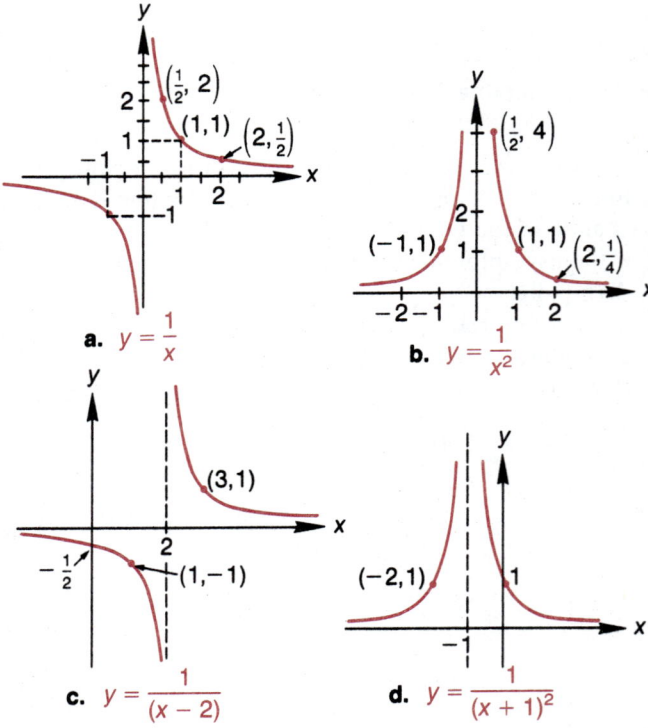

In describing asymptotes such as those that appear in Figure 5.15c, we write, for example, $y \to \infty$ as $x \to 2^+$; that is, y gets arbitrarily large as x approaches 2 from the right. Similarly, $y \to -\infty$ as $x \to 2^-$.

EXAMPLE 2

Sketch the graphs of the following rational functions.

a. $y = \dfrac{12}{(x^2 + 4)}$ **b.** $y = \dfrac{12}{(x^2 - 4)}$ **c.** $y = \dfrac{x}{(x^2 + 4)}$

SOLUTION

a. $y = \dfrac{12}{x^2 + 4}$ defines an even function; the graph will be symmetric about the y-axis. Note that $y = 3$ at $x = 0$; the y-intercept is $y = 3$. Also,

$$0 < y = \dfrac{12}{x^2 + 4} \leq 3 \quad \text{since} \quad x^2 + 4 \geq 4$$

There is no x-intercept, but note that $y \to 0$ as $x \to \pm\infty$. The graph is sketched in Figure 5.16.

Figure 5.16
$y = \dfrac{12}{x^2 + 4}$

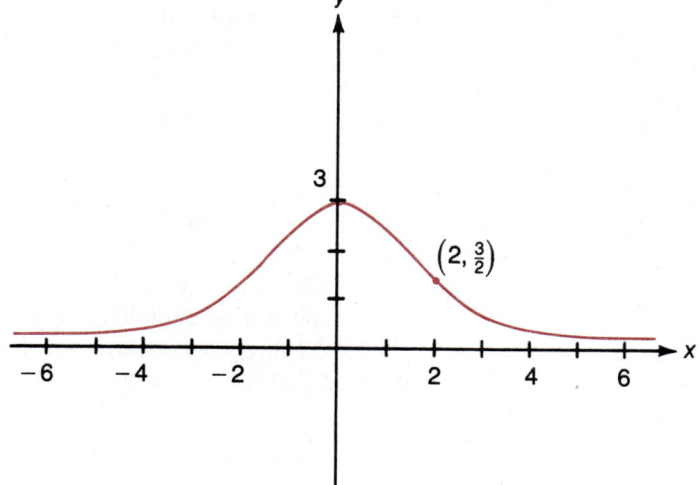

b. The graph of $y = \dfrac{12}{x^2 - 4}$ will also be symmetric about the y-axis (another even function). The y-intercept occurs when $x = 0$: $y = -3$. Again y is never 0 (no x-intercept) and $y \to 0$ as $x \to \pm\infty$. Writing

$$y = \dfrac{12}{(x^2 - 4)} = \dfrac{12}{(x - 2)(x + 2)}$$

we see that poles occur at $x = \pm 2$. The behavior of y near these poles is analyzed in the following chart. For instance, if x is slightly larger than 2, then $x - 2$ is a small positive number and $\dfrac{1}{(x - 2)}$ is a large positive number.

Then $y = \dfrac{1}{x-2} \cdot \dfrac{12}{x+2} \approx \dfrac{1}{x-2} \cdot \dfrac{12}{4}$; that is, y is large and positive. We say that $y \to +\infty$ as $x \to 2^+$. The other information in the chart is developed in a similar fashion. The graph is sketched in Figure 5.17.

Figure 5.17

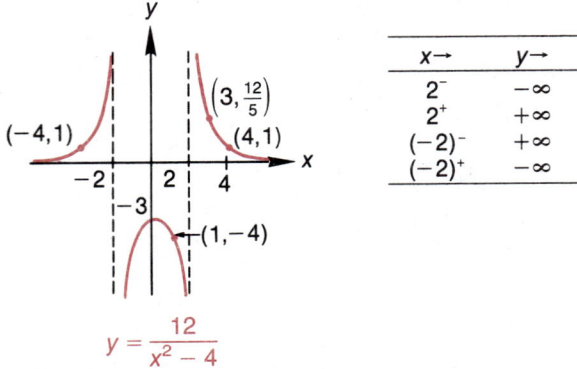

c. $y = \dfrac{x}{x^2 + 4}$ defines an odd function. The graph will be symmetric about the origin. This time $y = 0$ when $x = 0$. There are no other x-intercepts. $x^2 + 4 > 0$ for all real x, there are no poles and hence no vertical asymptotes. To analyze the behavior of this graph as x gets large, we write

$$y = \dfrac{x}{x^2 + 4} = \dfrac{1}{x + \dfrac{4}{x}} \qquad \text{(Divide numerator and denominator by } x\text{)}$$

For x large, $4/x$ becomes negligible in comparison to x, so $y \approx 1/x$; that is, $y \to 0$ as $x \to \pm\infty$.

This enables us to sketch a partial graph as indicated in Figure 5.18a. The graph can be completed by plotting several intermediate points (see Figure 5.18b). Determination of the exact location of the turnaround points (Section 5.4) requires methods of calculus.

Figure 5.18

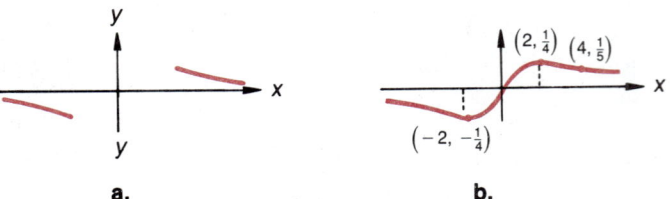

a. b.

As in Example 2c, we can show that for any *proper fraction* $y = \dfrac{P(x)}{Q(x)}$ (one in which degree $P(x) <$ degree $Q(x)$), $y \to 0$ as $x \to \pm\infty$. For instance, if

$$y = \dfrac{x^3 - 2x^2 + 3x + 4}{x^4 + 5x^3 + x - 2} = \dfrac{1 - \dfrac{2}{x} + \dfrac{3}{x^2} + \dfrac{4}{x^3}}{x + 5 + \dfrac{1}{x^2} - \dfrac{2}{x^3}} \qquad \text{(Divide numerator and denominator by } x^3\text{)}$$

then

$$y \approx \frac{1}{x+5} \quad \text{for } |x| \text{ large}$$

so that $y \to 0$ as $|x| \to \infty$.

If $y = R(x) = \dfrac{P(x)}{Q(x)}$ with degree $P(x) <$ degree $Q(x)$, then $y \to 0$ as $|x| \to \infty$.

It is not always the case that a root of the denominator indicates a vertical asymptote, as we will see in the following example. In this case the root of the denominator is not a pole.

EXAMPLE 3

Sketch the graph of the rational function

$$r(x) = \frac{x^2 - 5x + 6}{x - 2}$$

SOLUTION

Write

$$r(x) = \frac{x^2 - 5x + 6}{x - 2}$$
$$= \frac{(x - 2)(x - 3)}{(x - 2)}$$
$$= x - 3, \quad x \neq 2$$

Its graph is the straight line $y = x - 3$ with one point deleted since $x = 2$ is not in the domain of the function (see Figure 5.19). ∎

Figure 5.19
$y = \dfrac{x^2 - 5x + 6}{x - 2} = x - 3, \quad x \neq 2$

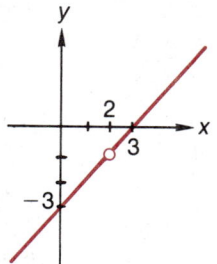

The situation illustrated in Example 3 occurs whenever numerator and denominator have a common root and hence a common factor, and this common factor cancels out of the denominator.

COMMON ROOTS IN $P(x)$ AND $Q(x)$

If the numerator and denominator of a rational function $R(x) = P(x)/Q(x)$ share a common root, r, and the multiplicity of r as a root of $Q(x)$ does not exceed the multiplicity of r as a root of $P(x)$, then this common root represents a missing point on the graph rather than an asymptote.

The following steps should be helpful in graphing rational functions.

> **GRAPHING RATIONAL FUNCTIONS** $R(x) = \dfrac{P(x)}{Q(x)}$
>
> 1. Check for symmetry (even or odd function).
> 2. Factor $P(x)$ and $Q(x)$ and find their roots.
> 3. Roots of $Q(x)$ are not in the domain of $R(x)$. Vertical asymptotes occur only at these roots.
> 4. The x-intercepts occur at the roots of $P(x)$ which are not also roots of $Q(x)$.
> 5. The y-intercept is $R(0)$; but if $Q(0) = 0$, there is no y-intercept.
> 6. Determine the behavior of $R(x)$ near the vertical asymptotes.
> 7. Find nonvertical asymptotes.
>   a. Use long division or synthetic division if necessary to write $R(x)$ in terms of a proper fraction:
> $$R(x) = A(x) + \frac{P_1(x)}{Q_1(x)}$$
>   with degree $P_1(x) <$ degree $Q_1(x)$.
>   b. Then $R(x) \to A(x)$ as $x \to \pm\infty$.
>   c. Determine whether these asymptotes are approached from above or below.
> 8. Determine if (and approximately where) the graph crosses the nonvertical asymptotes. The techniques for finding roots of polynomials may be useful here.
> 9. Use a calculator to plot additional points if necessary. Then, connect the dots and pieces to sketch a rough graph. Precisely locating turnaround points and other important points requires calculus and need not be done at this stage.

EXAMPLE 4

Sketch the graph of the following rational function.

$$y = \frac{x^2 - 5}{x^2 - 2x - 3}$$

SOLUTION

1. No symmetry is immediately obvious this time.
2. Now factor the denominator and write
$$\frac{x^2 - 5}{x^2 - 2x - 3} = \frac{x^2 - 5}{(x-3)(x+1)}$$
3. Vertical asymptotes are located at $x = 3$ and $x = -1$ since these are not roots of the numerator.
4. The x-intercepts are at $x = \pm\sqrt{5}$.
5. The y-intercept is $R(0) = \dfrac{-5}{(-3) \cdot 1} = \dfrac{5}{3}$.
6. Develop the following chart for behavior near the vertical asymptotes. For instance, if x is near 3, then $x^2 - 5 \approx 4$ and $x + 1 \approx 4$, but $x - 3$

is a small positive or negative number depending on whether $x > 3$ or $x < 3$, respectively. Thus the fraction is a large positive number when x is slightly larger than 3 and a large negative number when x is slightly less than 3.

$x \to$	$y \to$
3^-	$-\infty$
3^+	$+\infty$
$(-1)^-$	$-\infty$
$(-1)^+$	$+\infty$

7. Now write

$$\frac{x^2 - 5}{x^2 - 2x - 3} = \frac{(x^2 - 2x - 3) + 2x - 2}{x^2 - 2x - 3}$$

$$= 1 + \frac{2x - 2}{x^2 - 2x - 3}$$

$$= 1 + \frac{2(x - 1)}{(x - 3)(x + 1)}$$

(This could also have been obtained by long division)

Thus $y = 1$ is a horizontal asymptote. The proper fraction $\frac{2(x-1)}{(x-3)(x+1)}$ is greater than 0 for large positive x and less than 0 for large negative x. Thus

$$y \to 1^+ \quad \text{as} \quad x \to \infty$$
$$y \to 1^- \quad \text{as} \quad x \to -\infty$$

8. To find where the graph crosses the horizontal asymptote, we solve

$$y = \frac{x^2 - 5}{x^2 - 2x - 3} = 1$$

$$x^2 - 5 = x^2 - 2x - 3$$

$$2x = 2$$

$$x = 1$$

We see that $y = 1$ when $x = 1$; the graph crosses the horizontal asymptote at only one point, $x = 1$.

9. A calculator can be used to plot several additional points if desired. We thus obtain the graph sketched in Figure 5.20.

Figure 5.20

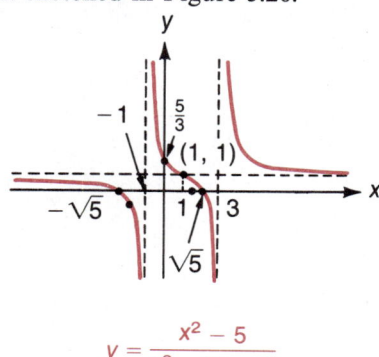

$$y = \frac{x^2 - 5}{x^2 - 2x - 3}$$

■

EXAMPLE 5

Sketch the graph of the following rational function.

$$y = \frac{x^2 - x - 1}{x - 2}$$

SOLUTION

1. Again, there is no obvious symmetry.
2. and 4. The quadratic formula can be used to find the zeros of the numerator of the rational function; these are the *x*-intercepts:

$$x = \frac{1 \pm \sqrt{1 + 4}}{2} = \frac{1 \pm \sqrt{5}}{2}$$

3. The line $x = 2$ is a vertical asymptote.
5. To find the *y*-intercept, we set $x = 0$; then $y = \frac{1}{2}$.
6. For x near 2, $x^2 - x - 1 \approx 1$, and $x - 2$ is a small positive or negative number depending on whether $x > 2$ or $x < 2$. Thus $y \to \infty$ as $x \to 2^+$ and $y \to -\infty$ as $x \to 2^-$.
7. We can use long division or synthetic division to write

$$y = \frac{x^2 - x - 1}{x - 2} = (x + 1) + \frac{1}{(x - 2)}$$

As $x \to \pm\infty$ the term $\frac{1}{(x-2)}$ becomes negligible so $y \approx (x + 1)$ for large $|x|$; the line $y = x + 1$ is an asymptote. If $x > 2$, then $y > x + 1$, and if $x < 2$, then $y < x + 1$.

8. It was shown in Step 7 that $y \neq x + 1$ for all $x \neq 2$ since $y = x + 1 + \frac{1}{x - 2}$.
9. The graph is sketched in Figure 5.21. It never crosses the asymptotes. ∎

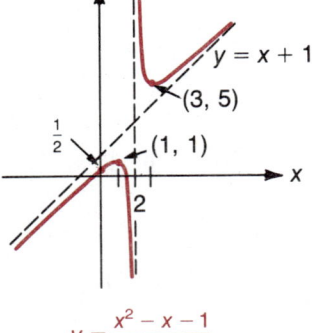

$$y = \frac{x^2 - x - 1}{x - 2}$$

Figure 5.21

In Example 5 with the degree of the numerator exceeding the degree of the denominator by 1, y is asymptotic to a linear function; $y \approx x + 1$ as $x \to \pm\infty$. Similarly, if the degree of the numerator exceeds the degree of the denominator by 2, y will be asymptotic to a quadratic function: $y \approx ax^2 + bx + c$ as $x \to \pm\infty$, and so on. See Exercises 39–42.

Section 5.5 Exercises

Sketch the graph of each of the following rational functions.

1. $y = \dfrac{1}{x - 3}$
2. $y = \dfrac{1}{(x - 3)^2}$
3. $y = \dfrac{-1}{x + 4}$
4. $y = \dfrac{2}{1 - x}$
5. $y = \dfrac{2}{1 - x}$
6. $y = \dfrac{2}{(1 - x)^4}$
7. $y = \dfrac{2}{(x - 1)^3}$
8. $y = \dfrac{4}{2x - 1}$
9. $y = \dfrac{3}{x^2 + 1}$
10. $y = \dfrac{5}{x^2 - 4x - 5}$
11. $y = \dfrac{3}{x^2 - 1}$
12. $y = \dfrac{2}{x^4 + 1}$
13. $y = \dfrac{2}{x^3 + 1}$
14. $y = \dfrac{x}{x^2 + 3}$
15. $y = \dfrac{2x}{4 - x^2}$
16. $y = \dfrac{2x + 4}{x^2 + 1}$
17. $y = \dfrac{x - 1}{2x^2 + x + 1}$
18. $y = \dfrac{3x + 6}{x^2 + 3x + 2}$

19. $y = \dfrac{x^2 - x - 2}{x + 1}$
20. $y = \dfrac{x^2 + 2x - 3}{x - 1}$
21. $y = \dfrac{x^2 - 3x + 2}{x - 2}$
22. $y = \dfrac{x - 4}{2x - 2}$
23. $y = \dfrac{4x + 2}{x}$
24. $y = \dfrac{2x}{x + 1}$
25. $y = \dfrac{x^2 - 1}{x^2 + 1}$
26. $y = \dfrac{x^2 - 3x + 2}{x^2 - 4}$
27. $y = \dfrac{x^2}{x^2 + x - 6}$
28. $y = \dfrac{x^3}{x^2 - 1}$
29. $y = \dfrac{x^2 + 2}{x - 1}$
30. $y = \dfrac{x^2 - x - 6}{x}$
31. $y = \dfrac{x^2 - 4x + 4}{x - 1}$
32. $y = \dfrac{x^3 - 4x}{x + 2}$
33. $y = \dfrac{x^3 - 2x + 1}{x - 1}$
34. $y = \dfrac{x^2 - 3x + 2}{x^2 - x}$
35. $y = \dfrac{x^2 - x - 2}{9x - x^3}$
36. $y = \dfrac{x^2 - x}{x^3 + x^2 - 2x}$
37. $y = \dfrac{x^2}{x^2 - 7x + 12}$
38. $y = \dfrac{x^2 - 1}{x^2 - 4}$
39. $y = \dfrac{x^3 - 2}{x + 1}$
40. $y = \dfrac{x^3 - 4x}{x - 1}$
41. $y = \dfrac{x^3}{x + 1}$
42. $y = \dfrac{x^3}{x - 2}$

43. The time T required to travel 1000 miles is inversely proportional to the speed:
$$T(x) = \dfrac{1000}{x}$$
  a. What is the domain of this function?
  b. Sketch the graph of this function.

44. The intensity I of illumination varies inversely as the square of the distance from the light source: $I(x) = 100/x^2$.
  a. What is the domain of this function?
  b. Sketch the graph of $I(x)$.

45. To operate a certain manufacturing facility, there are fixed costs of $10,000 and production costs of $20 per item.
  a. If x items are produced, what is the *total* manufacturing cost?
  b. What is the *average* manufacturing cost per item if 1000 units are produced?
  c. Express as a function $R(x)$ the *average* manufacturing cost per item if x units are produced.
  d. Graph the function $R(x)$ found in part c.

46. There is always a fixed demand for 5000 units of a certain medical product by those who require it for maintenance of life. The excess demand above this level is inversely proportional to the square of the price: $20,000/x^2$, where x is price per unit (in dollars).
  a. Express the total demand for this product as a function $R(x)$ of the price x.
  b. Graph the function $R(x)$ found in part a.

Section 5.6 Chapter Review

Terms and Concepts

Polynomial Function $P(x) = a_n x^n + \cdots + a_1 x + a_0$

Rational Function A quotient of polynomials:
$$R(x) = \dfrac{P(x)}{Q(x)}$$

Root r is a root of $P(x)$ if $P(r) = 0$

Asymptotes
  1. Vertical
     $f(x) \to \pm\infty$ as $x \to a$
  2. Horizontal
     $f(x) \to A$ as $x \to \pm\infty$
  3. Oblique
     Points on the graph approach a line as $x \to \pm\infty$

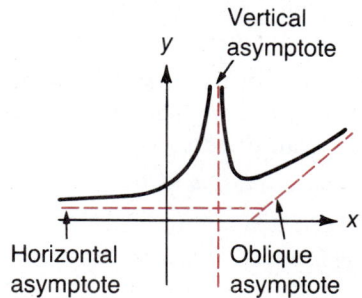

4. Curved
Points on the graph approach a curve (not a line) as $x \to \pm\infty$

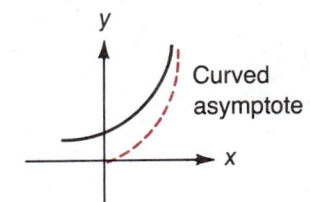

Rules and Formulas

Remainder Theorem If a polynomial is divided by $(x - a)$ where a is a constant, the remainder is the value of the polynomial at $x = a$.

Factor Theorem $(x - a)$ is a factor of a polynomial $P(x)$ if and only if $P(a) = 0$.

Rational Root Theorem If a rational number p/q in lowest terms is a root of a polynomial $a_n x^n + \cdots + a_1 x + a_0$ with integer coefficients, then p is an integer factor of a_0 and q is an integer factor of a_n.

Descartes' Rule of Signs

1. The number of positive real roots of a polynomial $P(x)$ is equal to the number of sign changes in $P(x)$ or less than the number of sign changes by a multiple of 2.

2. The number of negative real roots is limited in the same way by the number of sign changes in $P(-x)$.

Fundamental Theorem of Algebra Let $P(x)$ be a polynomial of degree n with real coefficients.

1. $P(x)$ can be factored completely into linear factors:
$$P(x) = (x - r_1)(x - r_2) \cdots (x - r_n) A$$
where A is real and $r_1, \ldots, r_n$ are complex (real or nonreal).

2. $P(x)$ has precisely n roots; some of these may be repeated roots.

3. The nonreal roots occur in conjugate pairs.

4. $P(x)$ can be factored into real linear and quadratic factors.

Techniques

Synthetic Division A short way of writing the steps in the long division of a polynomial by $x - a$.

```
2 ) 1  -6   15  -10
         2   -8  -14
    1  -4    7    4 = R
```

$$\frac{x^3 - 6x^2 + 15x - 10}{x - 2} = x^2 - 4x + 7 + \frac{4}{x - 2}$$

Range Limitation Test Let $P(x)$ be a polynomial with real coefficients and positive leading coefficient. Use synthetic division to divide $P(x)$ by $(x - a)$.

1. If $a > 0$ and the third row of this synthetic division contains no negative entry, then all real roots of $P(x)$ are less than or equal to a.

2. If $a < 0$ and the signs alternate in the third row of this synthetic division, then all real roots of $P(x)$ are greater than or equal to a.

Successive Approximation of Roots

1. Find two values a, b for which $P(a)$ and $P(b)$ differ in sign; there must be a root between a and b.

2. Break down the interval already obtained to isolate a root in a narrower interval.

3. Continue the process of refinement until a sufficient degree of accuracy is achieved.

Graphing Polynomial Functions

1. The basic polynomials $y = P(x) = x^n$, $n \geq 1$
   a. Pass through $(0, 0)$ and $(1, 1)$
   b. Are symmetric about
      i. The y-axis if n is even
      ii. The origin if n is odd

2. To graph $y = P(x) = a_n x^n + \cdots + a_1 x + a_0$,
   a. The real roots of $P(x)$ are the x-intercepts of the graph; there are at most n of these.
   b. Within each interval determined by these roots, the graph stays on the same side of the x-axis. Plot at least one test point inside each interval to determine whether $P(x) > 0$ or $P(x) < 0$ there.
   c. There are at most $n - 1$ turnaround points.
   d. If $a_n > 0$, then $P(x) \to \infty$ as $x \to \infty$; if $a_n < 0$, then $P(x) \to -\infty$ as $x \to \infty$.

e. The extreme branches of the graph point in the same direction if n is even but in the opposite direction if n is odd.

Graphing Rational Functions
1. Check for symmetry.
2. Find x- and y-intercepts.
3. Factor the denominator to find possible vertical asymptotes and analyze the behavior of the graph on either side of the vertical asymptotes.
4. Write $R(x) = A(x) + P_1(x)/Q_1(x)$ with degree $P_1(x) <$ degree $Q_1(x)$. Then $A(x)$ represents a nonvertical asymptote (horizontal, oblique, or curved). Determine whether this asymptote is approached from above or below as $x \to \infty$ and as $x \to -\infty$.
5. Determine if and where the graph crosses the nonvertical asymptotes.
6. Connect the pieces and dots to sketch a rough graph.

Section 5.7 Supplementary Exercises

Use synthetic division to verify the validity of the remainder theorem in each of the following. Write the polynomial in the form $Q(x)(x - a) + R$.

1. $(v^4 - 10v^2 + 25v - 2) \div (v + 4)$
2. $(5w^6 - 7w^3 - 6w + 8) \div (w - 1)$
3. $(2s^5 - s^3 + 3s - 2) \div (s + 1)$
4. $(3u^4 + 10u^3 - 4u + 5) \div (u + 4)$

Use the factor theorem to determine whether the second expression is a factor of the first; if so, use synthetic division to factor out the second expression.

5. $r^4 + 2r^2 - 3r + 5;\ (r - 3)$
6. $2t^5 - 5t^2 - t + 6;\ (t + 1)$
7. $s^5 + 4s^4 + s^3 + s - 1;\ (s + 1)$
8. $w^3 - 3w^2 + w + 1;\ (w - 2)$

Find all rational roots of the following and factor out the corresponding linear factors.

9. $3x^4 - 5x^3 + 4x^2 - 10x - 4$
10. $2y^4 + y^3 - 3y^2 + 2y - 2$
11. $u^4 - u^3 - u^2 + 3u - 6$
12. $12v^3 + 20v^2 + v - 3$

Use Descartes' rule of signs to indicate the possible numbers of positive and negative roots of the following.

13. $r^3 - 6r^2 + 11r - 6$
14. $3s^3 + 2s^2 - 7s + 6$
15. $t^5 - 3t^4 + 4t^3 - 4t^2 + 3t - 1$
16. $u^5 + u^4 - u^3 - u^2 + u + 5$

Use the range limitation test to find the smallest interval with integer endpoints containing all real roots of the following.

17. $2u^4 + 3u^3 + 2u^2 - u + 1$
18. $w^4 - 4w^3 + 4w + 4$
19. $2x^5 - 5x^4 - 2x + 5$
20. $t^7 + 10t^4 + 4t^3 - t - 3$

Show that the polynomial has a root in the given interval.

21. $2x^3 - 15x^2 + 27x - 10;\ (0, 1)$
22. $8x^3 - 6x^2 - 5x + 3;\ (-1, 0)$
23. $x^4 - 7x^3 + 10x^2 + 14x - 24;\ (-2, -1)$
24. $2x^4 - x^3 - 28x^2 + 30x - 8;\ (3, 4)$

Estimate a real root of the following to within the nearest tenth and within the indicated interval.

25. $r^3 - 2r^2 - r + 1;\ (-1, 0)$
26. $u^3 - 5u^2 + 4u - 5;\ (4, 5)$
27. $2s^4 + s^3 + 4s^2 - 6s - 4;\ (1, 2)$
28. $t^5 - t^4 + t^3 + 2t^2 - 2t + 2;\ (-2, -1)$

Find all real roots of the following; estimate all irrational roots to the nearest tenth.

29. $x^3 + 2x^2 - x - 1$
30. $t^4 - t^3 - 5t^2 + 3t + 6$
31. $3u^3 + u^2 - 8u + 2$
32. $v^4 + 5v^3 + 1$

Find the polynomial of lowest degree with real coefficients, leading coefficient of 1, and the indicated roots.

33. $-2, 3-i$

34. $(1+i), (2-i)$

35. $1, -1, 1-i$

36. $i, 1-i, 0, 1$

Find all the roots of the following and factor into real linear and quadratic factors.

37. $z^3 - 2z^2 + 4z - 8$

38. $2x^4 - 7x^3 + 11x^2 - 8x + 2$

39. $t^4 - t^3 + t^2 - t$

40. $2s^3 + s^2 + 2s + 1$; i is a root

41. $w^4 + 3w^3 + 3w^2 - 2$; $(-1-i)$ is a root

Factor the following by finding the roots and using the factor theorem.

42. $x^3 - 7x + 6$

43. $2w^3 - 12w^2 + 13w + 12$

44. $u^4 - 2u^3 + u - 2$

45. $v^4 + 2v^3 + v^2 - 2v - 2$

Sketch a rough graph of the following polynomials.

46. $y = x^6 - 1$

47. $y = 16 - (x-1)^4$

48. $y = x^3 + 3x^2 - x - 3$

49. $y = (x+1)(x-2)(x+4)(x-5)$

50. $y = x^4 + x^3 - 3x^2 - x + 2$

51. $y = 64 - x^6$

52. $y = 4x^2 - x^4$

53. $y = (x-2)(x+5)(x-3)$

54. $y = 16 - (x^2 - 4)(x-4)$

55. $y = x^4 - 13x^2 + 36$

Sketch the graph of each of the following functions.

56. $y = \dfrac{-2}{x+3}$

57. $y = \dfrac{2x}{x^2 + 2x + 1}$

58. $y = \dfrac{2x^2}{x^2 + 2x + 1}$

59. $y = \dfrac{x^2 + 5x + 4}{x+2}$

60. $y = \dfrac{4}{x^2 - 5x + 6}$

61. $y = \dfrac{x^2 + x - 6}{x+3}$

62. $y = \dfrac{x^2}{x+1}$

Exponential and Logarithmic Functions

6

In many natural phenomena, the rate at which something grows or decays depends on the amount of it that is present. The secretion of medications from the bloodstream by the kidneys is an example of this behavior. A typical situation is for half the medication to be removed every 4 hours (approximately). Thus if an initial dose of 400 milligrams (mg) of the medication is given at noon, then there are

$$400 \cdot \left(\frac{1}{2}\right) = 200 \text{ mg remaining at 4 P.M.}$$

$$400 \cdot \left(\frac{1}{2}\right) \cdot \left(\frac{1}{2}\right) = 100 \text{ mg remaining at 8 P.M.}$$

$$400 \cdot \left(\frac{1}{2}\right) \cdot \left(\frac{1}{2}\right) \cdot \left(\frac{1}{2}\right) = 50 \text{ mg remaining at midnight}$$

and so on. After n of these 4-hour periods, the amount of medication remaining in the bloodstream is given by $A(n) = 400 \cdot (\frac{1}{2})^n$. But this secretion does not occur suddenly every 4 hours on the hour. Rather, it is a *continuous* process. Thus, after $3\frac{1}{2}$ of these 4-hour periods (2 A.M.), the amount of medication remaining should be

$$A\left(3\frac{1}{2}\right) = 400 \cdot \left(\frac{1}{2}\right)^{3\frac{1}{2}} \approx 35.4 \text{ mg}$$

In fact, the amount of medication remaining in the bloodstream after t 4-hour periods have elapsed is expressed as

$$A(t) = 400 \cdot \left(\frac{1}{2}\right)^t$$

for *any* nonnegative value of t, not just for integer or rational values. Note that the variable t appears as an *exponent* in this expression.

This kind of function is called an **exponential function** and has many practical applications. Exponential functions are used in biology (population growth), physics (radioactive decay), and in modern banking for computations regarding continuously compounded interest.

To determine how long it takes the kidneys to eliminate 90% of the medication in the situation previously described, we must solve

$$400 \cdot \left(\frac{1}{2}\right)^t = 400 \cdot (0.10) = 40$$

or

$$\left(\frac{1}{2}\right)^t = \frac{40}{400} = \frac{1}{10}$$

That is, we must find t so that $(\frac{1}{2})^t$ yields a given value. The corresponding t value is called a **logarithm.** Thus a logarithm is really an exponent.

Logarithms were developed as a computational tool. They can be used to transform a multiplication or division problem into a simpler exercise in addition or subtraction. Although these computations are now performed by inexpensive, sophisticated hand calculators, logarithmic functions are still important and useful by virtue of their relationship to the exponential functions.

Section 6.1

Exponential Functions

In Chapter 1 we introduced integer exponents by defining $b^n = b \cdot b \cdots b$ (n times), $b^{-n} = 1/b^n$, and $b^0 = 1$. Rational exponents were introduced by defining $b^{1/n} = \sqrt[n]{b}$ and $b^{m/n} = (\sqrt[n]{b})^m$.

In this section we introduce numbers of the form b^x, *where* b *is a positive real number* and x *is an arbitrary real number.* Only positive b's are considered since, for example, $b^{1/2} = \sqrt{b}$ is not a real number when $b < 0$; for example, $\sqrt{-4} = 2i$. We shall now give meaning to numbers of the form $2^{\sqrt{3}}$, 3^π, $2^{-\sqrt{3}}$, and so on *but shall not consider numbers of the form* $(-2)^{\sqrt{3}}$.

If we want to get a smooth graph when we define b^x, then b^r must be close to b^x when r is a rational number close to x. This means that if we plot the values b^r for rational r, then b^x is "where it should be" on the graph.

To estimate $10^{\sqrt{2}}$, for example, we can write $\sqrt{2}$ in terms of its decimal expansion to as many places as desired: $\sqrt{2} = 1.414213562\ldots$. Whatever $10^{\sqrt{2}}$ is, it should be between $10^{1.4}$ and $10^{1.5}$, between $10^{1.41}$ and $10^{1.42}$, between $10^{1.414}$ and $10^{1.415}, \ldots,$ between $10^{1.414213}$ and $10^{1.414214}, \ldots$. A calculator yields the following estimate.

$$25.95452\ldots = 10^{1.414213} < 10^{\sqrt{2}} < 10^{1.414214} = 25.95458\ldots$$

The number $10^{\sqrt{2}}$ can be obtained to within any desired degree of accuracy by using sufficiently accurate rational approximations of $\sqrt{2}$. We can write $10^{\sqrt{2}} \approx 25.955$ and be assured of three-place accuracy.

With this interpretation of b^x, the earlier rules for exponents can be extended to include arbitrary real exponents.

RULES FOR EXPONENTS

For $b > 0$, $a > 0$, x and y real numbers

1. $b^x b^y = b^{x+y}$

2. $\dfrac{b^x}{b^y} = b^{x-y}$

3. $(b^x)^y = b^{xy}$ (Power of a power rule)

4. $b^0 = 1$

5. $b^1 = b$

6. $b^{-x} = \dfrac{1}{b^x}$

7. $(ab)^x = a^x b^x$ (Power of a product rule)

Section 6.1 Exponential Functions 311

For $b > 0$ and r an integer or rational number, $b^r > 0$. Thus $b^x > 0$ for all x. If $b > 1$, then $b, b^2, b^3, \ldots$ are increasing in magnitude; in fact, $b^x \to \infty$ as $x \to \infty$. But then $b^{-u} = 1/b^u \to 0$ as $u \to \infty$; that is, $b^x \to 0$ as $x \to -\infty$.

If $0 < b < 1$ we can write $a = 1/b > 1$. Then

$$b^{-t} = \frac{1}{b^t} = \left(\frac{1}{b}\right)^t = a^t \to \begin{cases} \infty & \text{as } t \to \infty \\ 0 & \text{as } t \to -\infty \end{cases}$$

Thus $b^x \to \infty$ as $x \to -\infty$ and $b^x \to 0$ as $x \to \infty$ when $0 < b < 1$.

VALUES OF b^x

For $b > 0$

1. $b^x > 0$ for all real x

2. If $b > 1$, then $b^x \to \begin{cases} \infty & \text{as } x \to \infty \\ 0 & \text{as } x \to -\infty \end{cases}$

3. If $0 < b < 1$, then $b^x \to \begin{cases} 0 & \text{as } x \to \infty \\ \infty & \text{as } x \to -\infty \end{cases}$

DEFINITION

If $b > 0$, $b \neq 1$, the function f defined by

$$f(x) = b^x \qquad x \text{ a real number}$$

is called an **exponential function** with **base** b.

The base $b = 1$ was excluded from the definition of an exponential function because it would simply yield the constant function $1^x \equiv 1$. (This is read "$1^x = 1$ for all x" or "1^x is **identically equal** to 1.")

EXAMPLE 1

Graph the following exponential functions

  a. $f(x) = 2^x$ **b.** $g(x) = \left(\frac{1}{2}\right)^x = 2^{-x}$ **c.** $h(x) = -2^x$

SOLUTION

Make a table of values, plot several points, and then "connect the dots" to sketch the graphs as illustrated in Figure 6.1. Note that the graph in part a "flattens" as x moves to the left and rises ever more rapidly as x moves to the right. The reverse holds true for part b.

x	2^x	$\left(\dfrac{1}{2}\right)^x$	-2^x
0	1	1	-1
1	2	$\frac{1}{2}$	-2
2	4	$\frac{1}{4}$	-4
3	8	$\frac{1}{8}$	-8
4	16	$\frac{1}{16}$	-16
-1	$\frac{1}{2}$	2	$-\frac{1}{2}$
-2	$\frac{1}{4}$	4	$-\frac{1}{4}$
-3	$\frac{1}{8}$	8	$-\frac{1}{8}$
-4	$\frac{1}{16}$	16	$-\frac{1}{16}$

Figure 6.1

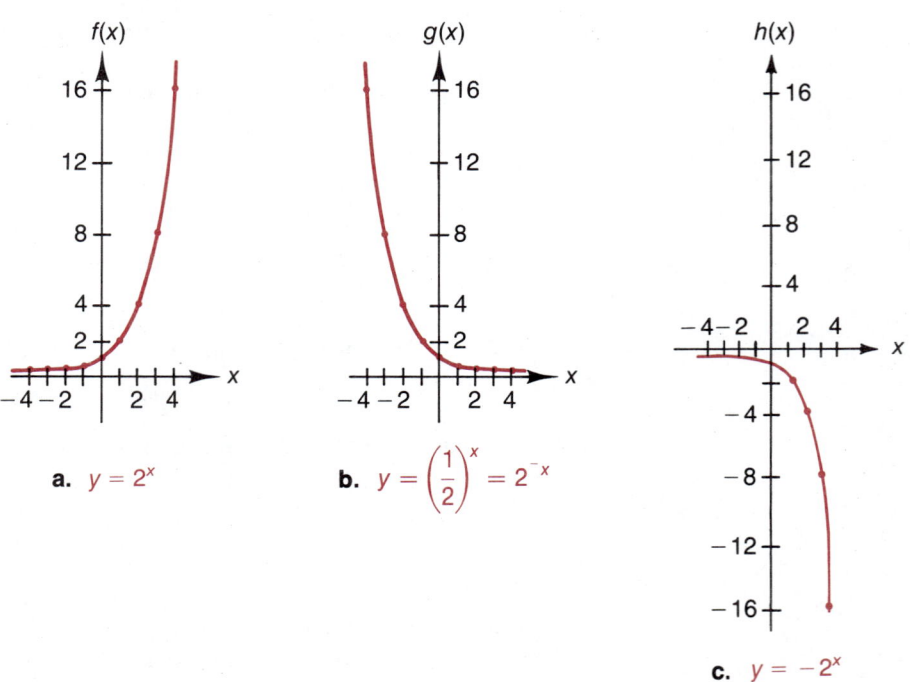

a. $y = 2^x$ b. $y = \left(\dfrac{1}{2}\right)^x = 2^{-x}$ c. $y = -2^x$

In part b of this example,

$$g(x) = \left(\dfrac{1}{2}\right)^x = \dfrac{1}{2^x} = 2^{-x} = f(-x)$$

Thus the graph of $y = (\frac{1}{2})^x$ could be obtained by reflecting the graph of $y = 2^x$ through the y-axis.

Similarly,

$$h(x) = -2^x = -f(x)$$

[Note that $-2^x = -(2^x) \neq (-2)^x$.] The graph in part c could be obtained by reflecting the graph of part a through the x-axis. ∎

Section 6.1 *Exponential Functions* **313**

GRAPHING $y = b^x$

The graphs of the exponential functions $y = b^x$ all pass through $(0, 1)$ and have the general shape that is indicated in Figure 6.2.

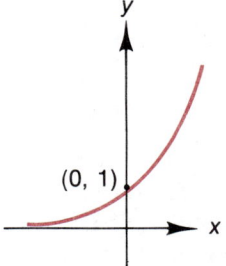
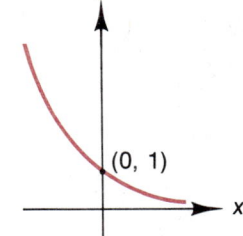

a. $y = b^x$, $b > 1$ **b.** $y = b^x$, $0 < b < 1$

Figure 6.2

EXAMPLE 2   Sketch the following graphs.

a. $y = 2^{|x|}$ **b.** $y = 2^{1-x}$

SOLUTION   **a.** Note that

$$y = 2^{|x|} = \begin{cases} 2^x, & x \geq 0 \\ 2^{-x} = \left(\dfrac{1}{2}\right)^x, & x \leq 0 \end{cases}$$

Thus this graph is obtained by using the right half ($x \geq 0$) of Figure 6.1a and the left half ($x \leq 0$) of Figure 6.1b. Or we could observe that this is an even function so that the graph must be symmetric about the y-axis. In any case, it is sketched as Figure 6.3a.

b. Write

$$y = 2^{1-x} = 2^{-(x-1)}$$
$$= \frac{1}{2^{x-1}}$$
$$= \left(\frac{1}{2}\right)^{x-1}$$
$$= g(x-1)$$

where g is the function of Example 1b. We obtain this graph by shifting the graph of Example 1b one unit to the right (see Figure 6.3b).

Another approach to this example is to write

$$y = 2^{1-x} = 2^1 2^{-x}$$
$$= 2 \cdot 2^{-x}$$
$$= 2g(x)$$

where g is again the function of Example 1b. Thus we could obtain this graph by vertically scaling (stretching) the graph of Example 1b. In either case we obtain Figure 6.3b. ∎

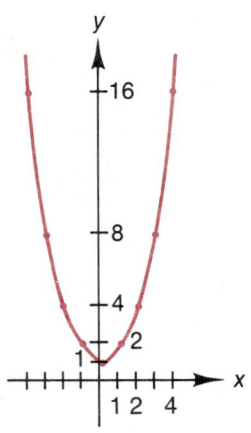

a. $y = 2^{|x|}$

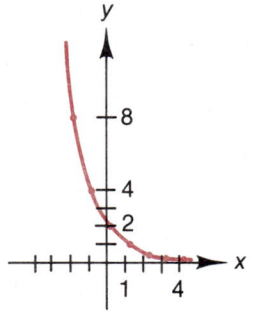

b. $y = 2^{1-x}$

Figure 6.3

Exponential Equations

The graphs of the exponential functions as illustrated in Figure 6.2 indicate that an exponential function with base b is increasing if $b > 1$ and decreasing if $b < 1$. (Recall a function is *increasing* if $f(x_2) > f(x_1)$ when $x_2 > x_1$; *decreasing* is defined in a similar fashion.) In particular, this means that if $0 < b \neq 1$ and $r \neq s$, then $b^r \neq b^s$. Alternatively, for $0 < b \neq 1$,

$$\text{if } b^r = b^s, \quad \text{then} \quad r = s$$

This property enables us to solve for x in equations that include x in the exponent. To solve such an equation, however, requires *expressing both sides of the equation in terms of the same base.* Later in the chapter we shall learn additional techniques to be used when no such common base is apparent.

EXAMPLE 3

Solve each of the following.

a. $3^x = 81$
b. $2^{3(1+x)} = 64$
c. $4^{2-v} = \dfrac{1}{8^{4/3}}$

SOLUTION

a. Since $81 = 3^4$, we write

$$3^x = 3^4$$

Thus

$$x = 4$$

b. Since $64 = 2^6$, we have

$$2^{3(1+x)} = 2^6$$

Thus

$$3(1 + x) = 6$$
$$1 + x = 2$$
$$x = 1$$

c. Both 4 and 8 are powers of 2; thus we rewrite the equation as follows:

$$4^{2-v} = \dfrac{1}{8^{4/3}}$$

$$(2^2)^{(2-v)} = \dfrac{1}{(2^3)^{4/3}}$$

$$2^{4-2v} = \dfrac{1}{2^4} = 2^{-4} \qquad \text{(Rule 3)}$$

$$4 - 2v = -4$$
$$-2v = -8$$
$$v = 4$$

Checking, we have

$$4^{(2-4)} = 4^{-2} = \dfrac{1}{4^2} = \dfrac{1}{16}$$

and
$$\frac{1}{8^{4/3}} = \frac{1}{(\sqrt[3]{8})^4} = \frac{1}{2^4} \stackrel{\checkmark}{=} \frac{1}{16}$$ ■

A variation on problems of this type occurs when the variable appears in the base rather than in the exponent:
$$u^r = A$$
Then
$$u = u^1 = (u^r)^{1/r} = A^{1/r}.$$

SOLVING EXPONENTIAL EQUATIONS

1. If the variable appears only in the exponent, express both sides in terms of the same base. Then the exponents must be equal.

    $2^x = 16 = 2^4$
    $x = 4$

2. If the variable appears only in the base, extract the base by raising both sides to the power which is the reciprocal of the exponent.

    $x^3 = 125$
    $(x^3)^{1/3} = 125^{1/3}$
    $x^1 = \sqrt[3]{125}$
    $x = 5$

EXAMPLE 4

Solve the following equations

a. $x^{5/3} = 32$
b. $(x - 2)^5 = 243$
c. $(x^2 + 5x)^{3/2} = 216$

SOLUTION

a. $(x^{5/3})^{3/5} = 32^{3/5}$
$\quad\quad x^1 = (\sqrt[5]{32})^3$
$\quad\quad x = 2^3 = 8$

Checking, we have
$8^{5/3} = (\sqrt[3]{8})^5 = 2^5 \stackrel{\checkmark}{=} 32$

b. $[(x - 2)^5]^{1/5} = 243^{1/5}$
$\quad\quad (x - 2)^1 = \sqrt[5]{243}$
$\quad\quad x - 2 = 3$
$\quad\quad\quad\quad x = 2 + 3 = 5$

Checking, we have
$(5 - 2)^5 = 3^5 \stackrel{\checkmark}{=} 243$

c. $[(x^2 + 5x)^{3/2}]^{2/3} = 216^{2/3}$
$\quad\quad (x^2 + 5x)^1 = (\sqrt[3]{216})^2$
$\quad\quad x^2 + 5x = 6^2 = 36$
$x^2 + 5x - 36 = 0$
$(x + 9)(x - 4) = 0$
$\quad\quad\quad\quad x = -9, 4$

316 Chapter 6 Exponential and Logarithmic Functions

Checking, we have

$$\begin{aligned} x=-9: & \quad [(-9)^2 + 5(-9)]^{3/2} = (81-45)^{3/2} \\ x=4: & \quad (4^2 + 5\cdot 4)^{3/2} = (16+20)^{3/2} \end{aligned} \Bigg\} = 36^{3/2} = (\sqrt{36})^3 = 6^3 \stackrel{\checkmark}{=} 216$$

The solution set is $\{-9, 4\}$. ∎

Functions of the form $g(x) = b^{-x}$ can be used to describe situations in which there is a rapid initial decrease that tends to level off. This situation is illustrated in the following example.

EXAMPLE 5   When an object is immersed in a temperature-controlled environment, it tends to take on the temperature of that environment. Specifically, Newton's law of cooling states that the temperature $T(t)$ of the object at time t after immersion is given by

$$T(t) = T_0 + (T_1 - T_0)b^{-at}$$

where T_0 is the temperature of the surrounding medium and T_1 is the initial temperature of the object; b and a are determined by properties of the object and its environment. Graph the temperature as a function of time for a piece of 340°F steel that is placed in 40°F water; the constants are $b = 2$, $a = 3$. Use the graph to estimate the time required to reduce the temperature of the steel to 150°F.

SOLUTION   The temperature function is given by

$$T(t) = 40° + (340° - 40°) \cdot 2^{-3t}$$
$$= 40 + 300 \cdot (2^{-3t})$$

The graph is sketched only for $t \geq 0$, since the object is presumed to be immersed at $t = 0$. The graph is best obtained in stages.

1. First, we graph $y = 2^{-3t}$. This graph is similar to Figure 6.1b except that the horizontal dimension is "compressed": At $t = 1$ we plot $2^{-3} = g(3)$, at $t = 2$ we plot $2^{-6} = g(6)$, and so on as in Figure 6.4a.
2. Next, we plot $y = 300 \cdot (2^{-3t})$ by multiplying each y-coordinate of Figure 6.4a by 300; see Figure 6.4b.
3. Finally, we raise the graph 40 units to obtain $y = 40 + 300 \cdot (2^{-3t})$, as in Figure 6.4c.

The graph should indicate that just before $t = \frac{1}{2}$, the temperature is reduced to 150°F. In fact, for $t = \frac{1}{2}$,

$$T\left(\frac{1}{2}\right) = 40 + 300(2^{-3/2})$$
$$= 40 + \frac{300}{2\sqrt{2}}$$
$$= 40 + \frac{300 \cdot \sqrt{2}}{2\sqrt{2} \cdot \sqrt{2}}$$
$$= 40 + \frac{300\sqrt{2}}{4}$$
$$= 40 + 75\sqrt{2}$$
$$\approx 146°F \qquad \text{(Use a calculator)}$$

We might estimate that $T(t) \approx 150°F$ when $t \approx 0.48$ or so. We will be able to find the exact solution later (Example 2, Section 6.3).

Figure 6.4

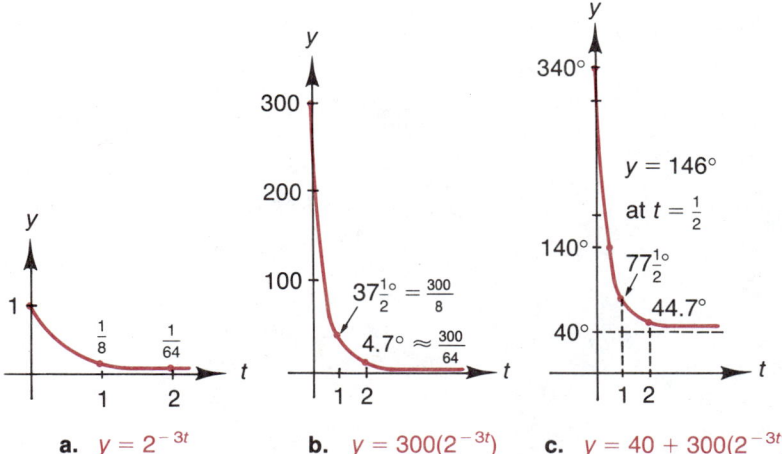

a. $y = 2^{-3t}$ b. $y = 300(2^{-3t})$ c. $y = 40 + 300(2^{-3t})$

The Number e

In calculus and in many of the applications of mathematics, it is convenient to use one particular irrational number as a base. This number is

$$e = 2.71828\ldots$$

The number e occurs naturally in the study of continuous growth processes such as bacteria growth (or growth of other populations), radioactive decay, and even continuously compounded interest. A discussion of compound interest and its relationship to the number e will be given in Section 6.5. The function $y = f(x) = e^x$ is called the **natural exponential function** and e is called the **natural exponential base**. The natural exponential function $f(x) = e^x$ is important in mathematical theory because of the nice properties it has in relation to calculus. The graph of $y = e^x$ has the same shape as the graph in Figure 6.2a since $e > 1$. The details are left to the exercises.

CALCULATOR COMMENTS

Most calculators with a "power" key also have an $\boxed{e^x}$ key or an $\boxed{\exp}$ key. To find e^3 on such a calculator

1. Press $\boxed{3}$.
2. Press $\boxed{e^x}$ or $\boxed{\exp}$.
3. Read the result $e^3 \approx \boxed{20.08553692}$ ....

Some calculators also have a $\boxed{10^x}$ key, which is used in the same way for base 10. For all other bases, the $\boxed{y^x}$ key must be used; note that it operates differently. You must

1. Enter the base y.
2. Press $\boxed{y^x}$.
3. Enter the exponent x.
4. Press $\boxed{=}$ and read the result.

Historical Perspective

LEONHARD EULER (1707–1783)

The symbol *e* was introduced by Leonhard Euler (1707–1783). Euler was one of the greatest mathematicians of his time and the most prolific mathematician of *all time*. His name occurs in every branch of mathematics. His life's work in mathematics comprises over 100 large volumes. In addition to the symbol *e*, he introduced the functional notation $f(x)$ that we use today. At an early age, he lost the sight of one eye, and 17 years before his death he became totally blind. Even so, he remained lively and cheerful and devoted to his family of 13 children. His powers of concentration were so great that even after going blind he produced volumes of important mathematics and could still perform mental calculations accurately to 15 figures.

Section 6.1 Exercises

Calculator Exercises (1–5)

Use a calculator with a power key to find an approximation, correct to the third decimal place, of each of the following.

1. $2^{\sqrt{3}}$, use $\sqrt{3} = 1.732050808\ldots$
2. $3^{\sqrt[3]{4}}$, use $\sqrt[3]{4} = 1.587401052\ldots$
3. $14^{\sqrt{5}}$, use $\sqrt{5} = 2.236067977\ldots$
4. $5^{\sqrt[3]{5}}$, use $\sqrt[3]{5} = 1.709975947\ldots$
5. $7^{\sqrt[3]{3}}$, use $\sqrt[3]{3} = 1.44224957\ldots$

Sketch the graph of the exponential functions f(x) =

6. 7^x
7. 3^x
8. $\left(\dfrac{1}{7}\right)^x$
9. $\left(\dfrac{1}{3}\right)^x$
10. 7^{-x}
11. 3^{-x}
12. $3 \cdot 7^{-x}$
13. $2 \cdot 3^{-x}$
14. $7^{|x|}$
15. $3^{|x|}$
16. $7^x - 1$
17. $3^x - 2$
18. 7^{x-1}
19. 3^{x-2}
20. $7^{|x-1|}$
21. $1 - 3^x$
22. $2 - 7^x$
23. 3^{1-x}
24. 7^{2-x}
25. $3^{|1-x|}$
26. 7^{x+1}
27. 3^{x+1}
28. $7^x + 1$
29. $3^x + 1$
30. $7 - 7^{x+1}$
31. $3 - 3^{x+1}$
32. $-3 \cdot 2^x$
33. 4^x
34. $\left(\dfrac{1}{4}\right)^x$
35. 4^{-x}
36. 2^{-2x}
37. $-4 \cdot 2^{-2x}$
38. 2^{2x}
39. 2^{2x-1}
40. 2^{x^2}
41. 2^{-x^2}
42. $2^{x/2}$
43. $2^{-|x|}$
44. $\dfrac{2}{(\sqrt{2})^x}$
45. $2 - 2^{-|x|}$
46. $3 - 2^{-x}$

Calculator Exercises (47–58)

Use a calculator to compile a table of values for e^x *and* e^{x^2} *(or you may use Table 3 at the end of this book). Then graph the following functions* f(x) =

47. e^x
48. e^{-x}
49. e^{x^2}
50. e^{-x^2}
51. e^{2x}
52. e^{x-2}
53. e^{2-x}
54. $e^{|x-2|}$
55. $e^{|x|}$

56. $e^{|x-1|}$ 57. $e^{|x|} - 1$ 58. $|e^x - 1|$

Solve the following equations.

59. $9^x = 3$ 60. $64^x = 4$ 61. $\left(\dfrac{1}{2}\right)^y = 8$

62. $\left(\dfrac{1}{3}\right)^v = 27$ 63. $27^r = 9$ 64. $125^s = 25$

65. $125^t = \dfrac{1}{25}$ 66. $27^u = \dfrac{1}{9}$ 67. $2^{2x} = 256$

68. $5^{8x} = \dfrac{1}{625}$ 69. $5^{9v} = 125^6$ 70. $4^{-3x} = 32$

71. $3^{-2w} = 81$ 72. $7^{10} = 49^y$ 73. $7^{4x} = 49^x$

74. $5^{y+1} = 125$ 75. $7^{3z} = 49^3$

76. $3^{2x+1} = 243$ 77. $25^{x+1} = 5^{3x}$

78. $10^{z-2} = 100^{2z-1}$ 79. $2^{x-3} = \dfrac{1}{16}$

80. $3^{u^2+1} = 9^{2.5}$ 81. $5^{2-3r} = \dfrac{1}{25^r}$

82. $3^{w^2+w} = (81)9^{w/2}$ 83. $7^{t^2-1} = 343$

84. $2^{3u+2} = 4^{u+2}$ 85. $2^{1-v^2} = \dfrac{2}{16^4}$

86. $27^{+3y} = 8^y \cdot 512$ 87. $81^{5-s} = 9^{s/2}$

88. $16^{2t-1} = 4^{2(t+1)}$ 89. $64^{2x/3} = \dfrac{1}{32}$

90. $81^{3x/5} = \dfrac{1}{27}$ 91. $u^{3/4} = 27$

92. $x^{5/6} = \dfrac{1}{32}$ 93. $(v+3)^{3/2} = 216$

94. $(w-20)^{4/3} = 625$ 95. $(x^2 - 6x)^{2/3} = 9$

96. $(r^2 - 18r)^{3/4} = 27$ 97. $(s^2 + 3s + 6)^{5/2} = 32$

98. $(t^2 + 5t + 20)^{3/4} = 8$

99. Bodies in cemeteries are usually buried 6 feet deep because at that depth the temperature remains fairly stable year round at 55°F. From the surface down to this depth, the temperature varies exponentially. This can be expressed by writing the temperature x feet below the surface as

$$T(x) = 55 + (T_s - 55)2^{-x}$$

where T_s denotes the average daily high temperature.

a. What is the temperature 6 feet down when the temperature generally reaches 105°F? $-50°F$?

b. Answer a for a depth of 1 foot and for a depth of 3 feet.

c. Graph temperature as a function of depth for $T_s = 105°F$ and for $T_s = -50°F$.

100. An injection of 30 milligrams of anesthetic per kilogram of body weight anesthetizes a patient for surgery. The body metabolizes this anesthetic so that it decreases exponentially after the injection. The amount remaining t minutes after the injection is given by

$$A(t) = A_0 \cdot 2^{-t/180}$$

where A_0 is the amount of the injection.

a. How much anesthetic must be injected to prepare an 80-kilogram patient for surgery?

b. After 1 hour, the effects of anesthesia have noticeably diminished. What should the amount of a second injection be to anesthetize the patient at the original level?

c. If the surgery lasts 5 hours and hourly injections are given as needed, sketch a curve indicating the amount of anesthetic in the patient at any given time t.

101. After an individual quits smoking, nicotine is gradually eliminated from the body. The nicotine level in the blood stream of a person who smokes n packs of cigarettes per day for t years is

$$N(t) = 500(1 - 2^{-nt})$$

and the nicotine level in the bloodstream q years after quitting is

$$N^*(q) = N_0 2^{-q}$$

where N_0 is the nicotine level at the time of quitting.

a. Sketch the graph of the nicotine level in the bloodstream of a person who smokes 2 packs per day for 10 years and then quits.

b. From the graph, estimate how long it takes to get down to the nicotine level achieved after 5 years of smoking.

c. Repeat b for the nicotine level after 1 year of smoking.

Section 6.2

Logarithmic Functions and their Graphs

In Example 5 of Section 6.1 we sketched a graph in order to estimate the time needed to cool an object to 150°F if its temperature at time t is given by

$$T(t) = 40 + 300(2^{-3t})$$

An exact determination of this time would require that we solve the following equation for t:

$$40 + 300(2^{-3t}) = 150$$
$$300(2^{-3t}) = 110$$
$$2^{-3t} = \frac{110}{300} = \frac{11}{30}$$
$$\left(\frac{1}{8}\right)^t = \frac{11}{30} \qquad \left[2^{-3t} = (2^{-3})^t = \left(\frac{1}{8}\right)^t\right]$$

We must find an exponent t such that b^t takes on a given value. This is precisely the task to be addressed in this section. Find an exponent u that will yield a given value for b^u.

Logarithms

Considering the graphs of the exponential functions $v = b^u$, we see that *for any $v > 0$ there is precisely one u for which $v = b^u$; this number u is called the **logarithm** of v to the **base** b and is denoted by $\log_b v$.* (See Figure 6.5 for the case $b > 1$.) Note that $\log_b v$ is defined only for $v > 0$ and $0 < b \neq 1$.

Figure 6.5

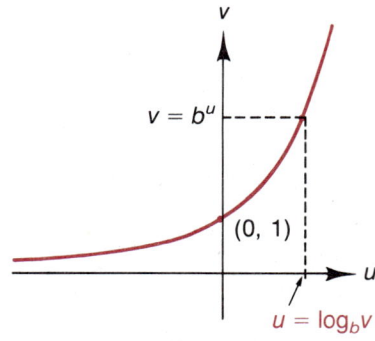

DEFINITION

For $v > 0$ and $0 < b \neq 1$,

$$u = \log_b v \quad \text{if and only if} \quad v = b^u$$

The function

$$f(x) = \log_b x, \quad x > 0$$

is called the **logarithmic function with base b.**

Section 6.2 *Logarithmic Functions and Their Graphs*

In words, **log$_b$ v** is the **exponent** *that must be used with base* b *to obtain the number* v. The following box shows some exponent-logarithm relationships.

---

1. $16 = 4^2$ $\log_4 16 = 2$
2. $16 = 2^4$ $\log_2 16 = 4$
3. $\dfrac{1}{25} = 5^{-2}$ $\log_5 \left(\dfrac{1}{25}\right) = -2$
4. $1 = 29^0$ $\log_{29} 1 = 0$
5. $\sqrt[4]{3} = 3^{1/4}$ $\log_3 \sqrt[4]{3} = \dfrac{1}{4}$

EXAMPLE 1

Find the following logarithms.

a. $\log_3 \sqrt{243}$ **b.** $\log_{25} 125$

SOLUTION

a. To find $\log_3 \sqrt{243}$, we must write $\sqrt{243}$ as a power of 3; if $\sqrt{243} = 3^u$, then $\log_3 \sqrt{243} = u$. Now

$$243 = 3^5$$
$$\sqrt{243} = \sqrt{3^5} = 3^{5/2}$$

Thus,

$$\log_3 \sqrt{243} = \frac{5}{2}$$

b. Here we must write 125 as a power of 25:

$$125 = 25 \cdot 5$$
$$= 25\sqrt{25} = 25^{3/2}$$

Thus,

$$\log_{25} 125 = \frac{3}{2}$$

■

Usually the logarithmic base is not specifically indicated when $b = 10$ or $b = e$. For base 10 we simply write log x and for the natural base e we write ln x:

$$\mathbf{log}\ x = \log_{10} x$$
$$\mathbf{ln}\ x = \log_e x$$

Log x is called the **common logarithm** of x and ln x is called the **natural logarithm** of x. Logarithm tables are available or calculators can be used to find common logarithms and natural logarithms. See Tables 1 and 2 at the end of this book.

From the definition of logarithmic functions, we find that

$$u = \log_b v \quad \text{and} \quad v = b^u$$

are equivalent statements. Substituting u and v from each of these into the other yields

$$u = \log_b b^u \quad \text{and} \quad v = b^{\log_b v}$$

respectively. If we let $f(x) = \log_b x$ and $g(x) = b^x$, we can rewrite this box as

$$f[g(u)] = u \quad \text{and} \quad g[f(v)] = v$$

That is, f and g are inverse functions of one another (see Section 4.4). Each "undoes" the effect of the other and returns the original argument.

INVERSE FUNCTIONS

The logarithmic and exponential functions with the same positive base

$$f(x) = \log_b x \quad \text{and} \quad g(x) = b^x$$

are inverses of one another.

In particular, the functions $f(x) = e^x$ and $g(x) = \ln x$ are inverses of one another as are $f(x) = 10^x$ and $g(x) = \log x$.

EXAMPLE 2

Use the inverse properties to evaluate each of the following.

a. $25^{\log_5 17}$ **b.** $\log_{16} 4^3$

SOLUTION

To use the inverse properties, each of these must be rewritten so that the exponential and logarithmic bases agree.

a. $25^{\log_5 17} = (5^2)^{\log_5 17}$

$\phantom{25^{\log_5 17}} = 5^{(\log_5 17) \cdot 2}$ (Rule 3, Section 6.1)

$\phantom{25^{\log_5 17}} = [5^{\log_5 17}]^2$ (Rule 3, Section 6.1)

$\phantom{25^{\log_5 17}} = 17^2$ (Inverse property)

$\phantom{25^{\log_5 17}} = 289$

b. $\log_{16} 4^3 = \log_{16}(\sqrt{16})^3$

$\phantom{\log_{16} 4^3} = \log_{16}(16^{1/2})^3$

$\phantom{\log_{16} 4^3} = \log_{16} 16^{3/2}$ (Rule 3, Section 6.1)

$\phantom{\log_{16} 4^3} = \dfrac{3}{2}$ (Inverse property) ■

Graphing Logarithmic Functions

Since the logarithmic and exponential functions are inverses of one another, the graph of a logarithmic function is the reflection of the corresponding exponential graph through the line $y = x$.

EXAMPLE 3 Sketch the graph of $f(x) = \log_2 x$.

SOLUTION The graph of the corresponding exponential function $y = 2^x$ is sketched as the broken curve in Figure 6.6. The graph of its inverse function, $f(x) = \log_2 x$, can be obtained by reflecting this graph through the line $y = x$ (or equivalently, interchanging the order of the coordinates on the broken line graph). See Section 4.4.

For verification, we list a table of certain values of $\log_2 x$.

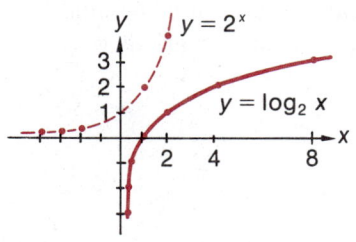

Figure 6.6
$y = \log_2 x$

x	1	2	$\frac{1}{2}$	4	$\frac{1}{4}$	8	$\frac{1}{8}$
$x = 2^?$	2^0	2^1	2^{-1}	2^2	2^{-2}	2^3	2^{-3}
$\log_2 x$	0	1	-1	2	-2	3	-3

EXAMPLE 4 Sketch the graph of $f(x) = \log_2(-x)$.

SOLUTION This graph can be obtained from the graph in Example 3 by reflecting it through the y-axis. For example, $f(-2) = \log_2 2$, $f(-4) = \log_2 4$, and so on. For $x \geq 0$, $f(x)$ does not exist. The graph appears in Figure 6.7.

Figure 6.7
$y = \log_2(-x)$

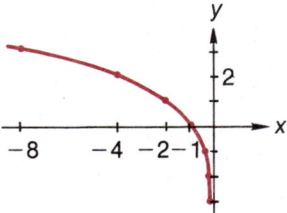

EXAMPLE 5 Sketch the graph of $g(x) = \log_{1/2} x$.

SOLUTION We could make a table of values and plot points, but observe that if

$$v = g(x) = \log_{1/2} x$$

then
$$x = \left(\frac{1}{2}\right)^v = \frac{1}{2^v}$$
$$= 2^{-v}$$
$$-v = \log_2 x$$
$$v = -\log_2 x$$

Thus
$$g(x) = -f(x)$$

where f is the function of Example 3. Consequently, this graph is obtained by reflecting the graph of Example 3 through the x-axis (see Figures 6.6 and 6.8). Positive y-coordinates in Figure 6.6 become negative for the graph of g and negative altitudes in Figure 6.6 become positive for Figure 6.8.

Figure 6.8

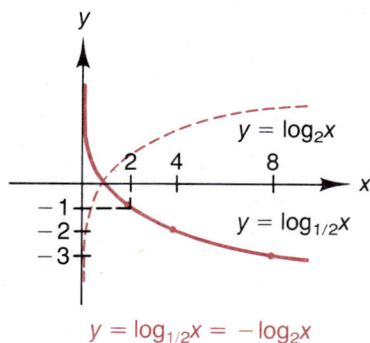

$y = \log_{1/2} x = -\log_2 x$

Figures 6.6 and 6.8 illustrate the typical shapes taken by graphs of logarithmic functions.

GRAPHING $y = \log_b x$

The graphs of the logarithmic functions $y = \log_b x$ all pass through $(1, 0)$ and have the general shape indicated in Figure 6.9.

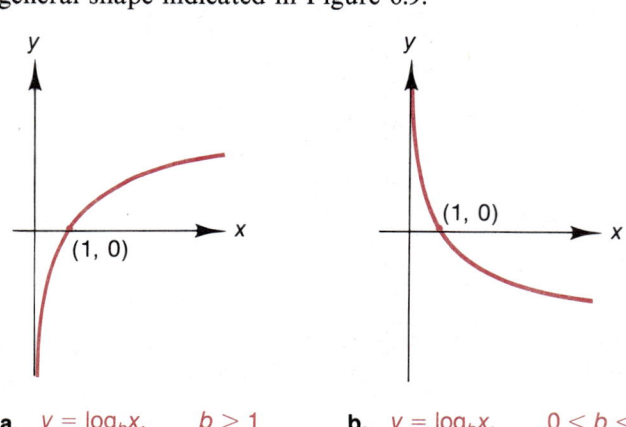

a. $y = \log_b x$, $b > 1$ b. $y = \log_b x$, $0 < b < 1$

Figure 6.9

We make the following observations from Figure 6.9.

> **VALUES OF $\log_b x$**
>
> 1. For $b > 1$, $\log_b x$ is an increasing function and
> $$\log_b x \to \begin{cases} \infty \text{ as } x \to \infty \\ -\infty \text{ as } x \to 0^+ \end{cases}$$
> 2. For $0 < b < 1$, $\log_b x$ is a decreasing function and
> $$\log_b x \to \begin{cases} -\infty \text{ as } x \to \infty \\ \infty \text{ as } x \to 0^+ \end{cases}$$

Compare Figures 6.2a and 6.9a. Note that for base $b > 1$, exponential functions increase much more rapidly than logarithmic functions do as $x \to \infty$.

EXAMPLE 6

Sketch the following graphs.

a. $y = \log_2|x|$ **b.** $y = \log_2|x - 1|$

SOLUTION

a. This function is defined for all $x \neq 0$ since $|x| > 0$ for all x. Note that here we have an even function so that the graph must be symmetric about the y-axis. Now $y = \log_2|x| = \log_2 x$ for $x > 0$, and this graph was sketched in Figure 6.6. Reflecting this portion of the graph through the y-axis, we obtain the complete graph as in Figure 6.10a.

b. The graph of $y = \log_2|x - 1|$ will have the same shape as the graph of $y = \log_2|x|$ with "1" playing the role of "0"; that is, we translate the graph of $y = \log_2|x|$ one unit to the right to obtain the graph of $y = \log_2|x - 1|$ as in Figure 6.10b.

Figure 6.10

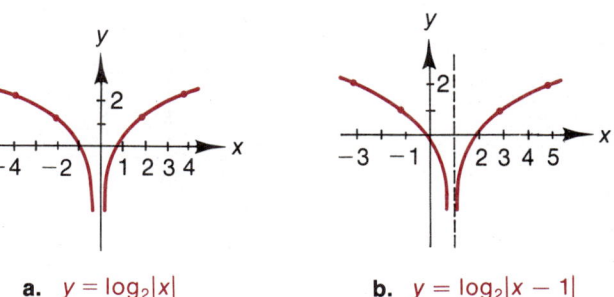

a. $y = \log_2|x|$ **b.** $y = \log_2|x - 1|$

EXAMPLE 7

As each day passes, students will remember less and less of today's lecture. With no further reinforcement, the percentage P of students who could recall the important features of today's lecture x days later decreases logarithmically. P might be given by

$$P(x) = 95 - 30 \log_2 x$$

a. After how many days do fewer than half the students recall the important features of the lecture?

b. After how many days will everyone have forgotten?

c. Graph this function.

SOLUTION

a. If only 50% of the students recall, then

$$50 = 95 - 30 \log_2 x$$
$$30 \log_2 x = 45$$
$$\log_2 x = \frac{45}{30} = \frac{3}{2}$$
$$x = 2^{3/2} = 2\sqrt{2} \approx 2(1.414)$$
$$\approx 2.828$$

After 3 days, fewer than half the students recall the important points of the lecture.

b. Everyone will have forgotten when

$$0 = 95 - 30 \log_2 x$$
$$30 \log_2 x = 95$$
$$\log_2 x = \frac{95}{30} = \frac{19}{6}$$
$$x = 2^{19/6}$$
$$\approx 8.98 \qquad \text{(Use a calculator)}$$

After 9 days, everyone has forgotten.

c. The graph is obtained from the graph of $\log_2 x$ (Figure 6.6) in three steps as indicated in Figure 6.11. Note that the graph does not extend beyond $x = 8.98$ since we cannot have $P(x) < 0$. ∎

Figure 6.11

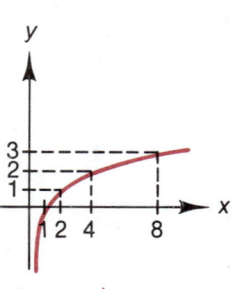
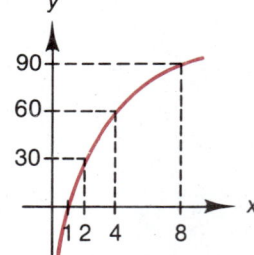

a. $y = \log_2 x$ b. $y = 30 \log_2 x$

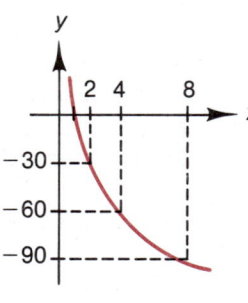

c. $y = -30 \log_2 x$ d. $y = -30 \log_2 x + 95$

Section 6.2 Exercises

Write the equivalent logarithmic form of each of the following.

1. $10^5 = 100{,}000$
2. $3^4 = 81$
3. $\dfrac{1}{125} = 5^{-3}$
4. $\dfrac{1}{64} = 2^{-6}$
5. $(0.001)^0 = 1$
6. $(15{,}631)^0 = 1$
7. $\sqrt[5]{32} = 2$
8. $\sqrt[3]{125} = 5$

Write the equivalent exponential form of each of the following.

9. $\log_3 9 = 2$
10. $\log_7 49 = 2$
11. $\log_4 8 = \dfrac{3}{2}$
12. $\log_{27} 81 = \dfrac{4}{3}$
13. $\log_{1/16} \dfrac{1}{4} = \dfrac{1}{2}$
14. $\log_{1/2} \dfrac{1}{8} = 3$
15. $\log_{1/6} 36 = -2$
16. $\log_{1/5} 125 = -3$
17. $\log_{16} \dfrac{1}{2} = -\dfrac{1}{4}$
18. $\log_{81} \dfrac{1}{9} = -\dfrac{1}{2}$

Find the indicated logarithms.

19. $\log_{10} 100{,}000{,}000$
20. $\log_2 256$
21. $\log_3 81$
22. $\log_{10}(0.000001)$
23. $\log_7 49$
24. $\log_{49} 7$
25. $\log_{25} 125$
26. $\log_{27} 243$
27. $\log_2 \dfrac{1}{4}$
28. $\log_{10} \sqrt{10{,}000}$
29. $\log_2 \sqrt{8}$
30. $\log_{10} \sqrt[5]{0.001}$
31. $\log_2 \sqrt[3]{256}$
32. $\log_2 (\sqrt[3]{2})^{1/5}$
33. $\log_3 \left[\left(\dfrac{1}{3}\right) \cdot (\sqrt[3]{3})\right]$
34. $\ln(\sqrt{e^3})$

Use the inverse properties to evaluate the following.

35. $5^{\log_5 16}$
36. $2^{\log_2 129}$
37. $\log_4 4^7$
38. $\log_3 (3^{163})$
39. $1000^{\log_{10} 3}$
40. $3^{\log_9 5}$
41. $\log_3(81^{-2})$
42. $\log_5 125^4$
43. $\log_{27} 3^{10}$
44. $\log_4(2^8)$
45. $\dfrac{\ln e^6}{3}$
46. $4e^{\ln \sqrt{e}}$

Sketch the graphs of the following functions f(x) =

47. $2 \log_2 x$
48. $\log_3 x$
49. $\log_2(1 - x)$
50. $\log_3 3x$
51. $|\log_2 x|$
52. $\log_3 \dfrac{x}{3}$
53. $|\log_2 |x||$
54. $\log_3(x - 1)$
55. $\log_2(x - 2)$
56. $\log_3(1 - x)$
57. $\log_2(2 - x)$
58. $\log_3 |x - 1|$
59. $\log_2 |x - 2|$
60. $|\log_3(x - 1)|$
61. $|\log_2(x - 2)|$
62. $|\log_3 |x - 1||$
63. $|\log_2 |x - 2||$
64. $\log_3(x - 2)$
65. $\log_2(x + 2)$
66. $|\log_3(x - 2)|$
67. $\log_2 |x + 2|$
68. $\log_3 |x - 2|$
69. $|\log_2(x + 2)|$
70. $|\log_3 |x - 2||$
71. $|\log_2 |x + 2||$

Calculator Exercises

Use a calculator or Tables 1 and 2 at the end of this book to compile a table of values for log x and ln x and then sketch the graphs of

72. $f(x) = \log x$
73. $f(x) = \ln x$
74. $g(x) = \log(-x)$
75. $h(x) = \ln|x|$
76. $F(x) = |\log x|$
77. $G(x) = \ln 2x$
78. $H(x) = \log(x - 1)$

79. Certain tasks involve the consecutive performance of several different steps. The time needed to learn these steps is a logarithmic function of the number of steps required. In a certain situation, this might be given as

$$T(t) = 302 - 50 \log_2(65 - t)$$

where $T(t)$ is the number of hours required to learn t steps.

   a. How many hours does it take to learn 64 steps? 1 step?
   b. Sketch the graph of this function.
   c. Express the number of steps that can be learned as a function of the time available.

80. Our earth depends on a layer of ozone in the atmosphere for protection from radiation. The strength of this ozone layer depends on many factors, one of which is the amount of forestland.

For the sake of argument, let

$$S(x) = \log_{10}(x + 1)^2$$

be the strength of this layer, where x represents the amount of forestland in billions of acres.

a. Sketch the graph of this function.

b. How does the strength of the ozone layer change if forest acreage is halved from 4 billion acres to 2 billion acres? (Use a calculator or Table 2 at the back of this book.)

c. If a 10% decrease in the strength of the ozone layer can cause serious problems, how much reduction in forest acreage from 4 billion acres can be permitted?

Section 6.3

Properties of Logarithms; Logarithmic Equations

The properties of exponents and the relationship between exponents and logarithms can sometimes be used to obtain information regarding logarithms. This occurs when the base b and the number x in $\log_b x$ are powers of the same number.

In Example 1 of Section 6.2, we found (a) $\log_3 \sqrt{243}$ and (b) $\log_{25} 125$. This was possible because (a) 243 is a power of the logarithmic base 3 and (b) 125 and 25 are both powers of the same number, 5. In the same way, we should be able to find $\log_{27} 9$, since 9 and 27 are both powers of 3: $9 = 3^2 = (\sqrt[3]{27})^2 = 27^{2/3}$. Thus $\log_{27} 9 = \tfrac{2}{3}$.

But in solving for t in $(\tfrac{1}{8})^t = \tfrac{11}{30}$ (see the introductory paragraph of Section 6.2), we are as yet unable to progress beyond $t = \log_{1/8}(\tfrac{11}{30})$, since no common base for $\tfrac{1}{8}$ and $\tfrac{11}{30}$ is readily apparent. Although logarithm tables or calculators can be used to find logarithms to the base e and to the base 10, the base $b = \tfrac{1}{8}$ is another matter entirely. To find a numerical value or estimate for t in this case requires other properties of logarithms.

Since logarithms are defined in terms of exponents, the properties of exponents translate into analogous properties of logarithms as indicated in the box.

PROPERTIES FOR EXPONENTS AND LOGARITHMS

$$0 < a \neq 1, \quad 0 < b \neq 1, \quad u > 0, \quad v > 0$$

Exponents		Logarithms	
Example	*Rule*	*Rule*	*Example*
$2^2 2^3 = 2^5$	E1. $b^x b^y = b^{x+y}$	L1. $\log_b(uv) = \log_b u + \log_b v$	$\log_2(4 \cdot 8) = \log_2 4 + \log_2 8 = 2 + 3 = 5$
$\dfrac{2^5}{2^2} = 2^3$	E2. $\dfrac{b^x}{b^y} = b^{x-y}$	L2. $\log_b\left(\dfrac{u}{v}\right) = \log_b u - \log_b v$	$\log_2\left(\dfrac{32}{4}\right) = \log_2 32 - \log_2 4 = 5 - 2 = 3$
$(2^2)^3 = 2^6$	E3. $(b^x)^r = b^{xr}$	L3. $\log_b(u^r) = r \log_b u$	$\log_2(4^3) = 3 \log_2 4 = 3 \cdot 2 = 6$
$2^0 = 1$	E4. $b^0 = 1$	L4. $\log_b 1 = 0$	$\log_2 1 = 0$
$2^1 = 2$	E5. $b^1 = b$	L5. $\log_b b = 1$	$\log_2 2 = 1$
$\dfrac{1}{2^2} = 2^{-2}$	E6. $\dfrac{1}{b^x} = b^{-x}$	L6. $\log_b\left(\dfrac{1}{u}\right) = -\log_b u$	$\log_2\left(\dfrac{1}{4}\right) = -\log_2 4 = -2$
$(2 \cdot 3)^2 = 2^2 \cdot 3^2$	E7. $(ab)^x = a^x b^x$	L7. $\log_b u = \dfrac{\log_a u}{\log_a b}$	$\log_4 16 = \dfrac{\log_2 16}{\log_2 4} = \dfrac{4}{2} = 2$
$3 = \log_2 2^3$	E8. $x = \log_b(b^x)$	L8. $u = b^{\log_b u}$	$3 = 2^{\log_2 3}$

To establish these properties, we let
$$x = \log_b u, \quad y = \log_b v$$
and convert these to exponential form:
$$u = b^x, \quad v = b^y$$
Then
$$uv = b^x y^y = b^{x+y} \tag{1}$$
$$\frac{u}{v} = \frac{b^x}{b^y} = b^{x-y} \tag{2}$$
$$u^r = (b^x)^r = b^{xr} \tag{3}$$
$$\frac{1}{u} = \frac{1}{b^x} = b^{-x} \tag{4}$$

Converting these statements from exponential to logarithmic form, we have

L1: $\log_b uv = x + y = \log_b u + \log_b v$ (From 1)

L2: $\log_b \dfrac{u}{v} = x - y = \log_b u - \log_b v$ (From 2)

L3: $\log_b u^r = rx = r \log_b u$ (From 3)

L6: $\log_b \dfrac{1}{u} = -x = -\log_b u$ (From 4)

Properties L4 and L5 are obtained by converting E4 and E5, respectively, to logarithmic form.

Properties E7 and L7 relate exponential and logarithmic functions having different bases a and b; L7 is called the *change-of-base formula* for logarithms. To establish L7, we again let
$$x = \log_b u$$
Then
$$u = b^x$$
and
$$\log_a u = x \log_a b \quad \text{(By L3)}$$
Thus we obtain L7:
$$\log_b u = x = \frac{\log_a u}{\log_a b}$$

Properties E8 and L8 were established in Section 6.2. For common and natural logarithms, Properties E8 and L8 take the form

$$\boxed{\begin{array}{ll} x = \log 10^x & x = \ln e^x \\ u = 10^{\log u} & u = e^{\ln u} \end{array}} \quad \text{and}$$

The preceding properties are used in the following examples. Note that the table in Example 1 gives logarithms only for prime numbers. The logarithm of any positive integer can be found in terms of the logarithms of its prime factors.

Chapter 6 Exponential and Logarithmic Functions

EXAMPLE 1 Consider the following table of logarithms to some base b.

n	$\log_b n$
2	0.3010
3	0.4771
5	0.6990
7	0.8451
11	1.0414

Use this table to estimate each of the following.

a. $\log_b 6$
b. $\log_b 7\sqrt{5}$
c. $\log_b \left(\dfrac{350}{\sqrt[10]{63}} \right)$

SOLUTION

a. $\log_b 6 = \log_b (2 \cdot 3)$
$ = \log_b 2 + \log_b 3$ (By Property L1)
$ \approx 0.3010 + 0.4771$
$ = 0.7781$

b. $\log_b 7\sqrt{5} = \log_b(7 \cdot 5^{1/2})$
$\phantom{\log_b 7\sqrt{5}} = \log_b 7 + \log_b 5^{1/2}$ (By Property L1)
$\phantom{\log_b 7\sqrt{5}} = \log_b 7 + \dfrac{1}{2}\log_b 5$ (By Property L3)
$\phantom{\log_b 7\sqrt{5}} \approx 0.8451 + \dfrac{1}{2}(0.6990)$
$\phantom{\log_b 7\sqrt{5}} = 0.8451 + 0.3495$
$\phantom{\log_b 7\sqrt{5}} = 1.1946$

c. $\log_b \left(\dfrac{350}{\sqrt[10]{63}} \right) = \log_b \dfrac{5 \cdot 7 \cdot 5 \cdot 2}{(7 \cdot 3^2)^{1/10}}$
$\phantom{\log_b \left(\dfrac{350}{\sqrt[10]{63}} \right)} = \log_b(5^2 \cdot 7 \cdot 2) - \log_b(7^{1/10} 3^{2/10})$ (By Property L2)
$\phantom{\log_b \left(\dfrac{350}{\sqrt[10]{63}} \right)} = 2\log_b 5 + \log_b 7 + \log_b 2 - \dfrac{1}{10}\log_b 7 - \dfrac{2}{10}\log_b 3$
     (By L1 and L3)
$\phantom{\log_b \left(\dfrac{350}{\sqrt[10]{63}} \right)} = 2\log_b 5 + \dfrac{9}{10}\log_b 7 + \log_b 2 - \dfrac{1}{5}\log_b 3$
$\phantom{\log_b \left(\dfrac{350}{\sqrt[10]{63}} \right)} \approx 2(0.6990) + 0.9(0.8451) + 0.3010 - 0.2(0.4771)$
$\phantom{\log_b \left(\dfrac{350}{\sqrt[10]{63}} \right)} = 2.36417$
$\phantom{\log_b \left(\dfrac{350}{\sqrt[10]{63}} \right)} \approx 2.3642$

Note that the result should be rounded off to four significant decimal places. (See Sections 1.1 and 1.3.)

CALCULATOR COMMENTS

Most scientific calculators are programmed to provide common and natural logarithms. These operations are generally provided through a $\boxed{\log}$ key and an $\boxed{\ln}$ key. On some calculators, these are one and the same key; if so, one of the operations must then be preceded by punching a $\boxed{\text{2nd}}$ key. See your calculator manual. To obtain logarithms on such a calculator:

1. Enter the number x, for example, 29.063, into the calculator:
   $\boxed{29.063}$

2. Press the $\boxed{\log}$ or $\boxed{\ln}$ key and read $\log x$ or $\ln x$ on the display:
   $$\log 29.063 \approx \boxed{1.463340442} \ldots$$
   or
   $$\ln 29.063 \approx \boxed{3.369465887} \ldots$$

EXAMPLE 2

Use the natural logarithm to find $\log_{1/8}\left(\frac{11}{30}\right)$.

SOLUTION

$$\log_{1/8}\left(\frac{11}{30}\right) = \frac{\log_e\left(\frac{11}{30}\right)}{\log_e\left(\frac{1}{8}\right)} \qquad \text{(By Property L7)}$$

$$= \frac{\ln 11 - \ln 30}{\ln 1 - \ln 8} \qquad \text{(By Property L2)}$$

$$\approx \frac{2.3979 - 3.4012}{0 - 2.0794} \qquad \text{(Use a calculator or Table 1)}$$

$$\approx 0.4825$$

Thus, in Example 5 of Section 6.1, the steel cools to 150°F when $t \approx 0.4825$. ∎

Logarithmic Equations

Science and engineering often require solving equations with an unknown quantity that appears in either an exponent or a logarithm. Several techniques can be used here. The simplest is to convert a logarithmic equation to an exponential equation.

EXAMPLE 3

If $\log_{10} 4x = 3$, find x.

SOLUTION

$$\log_{10} 4x = 3$$
$$4x = 10^3 = 1000$$
$$x = 250$$

Checking, we have

$$\log_{10} 4 \cdot 250 = \log_{10} 1000 = \log_{10} 10^3 \stackrel{\checkmark}{=} 3$$

EXAMPLE 4 Solve for v if
$$\log_6 4 + 2 \log_6 v = 2$$

SOLUTION
$$\log_6 4 + 2 \log_6 v = \log_6 4 + \log_6 v^2 \quad \text{(By L3)}$$
$$= \log_6 4v^2 \quad \text{(By L1)}$$

Thus
$$\log_6 4v^2 = 2$$
$$4v^2 = 6^2$$
$$= 36$$
$$v^2 = 9$$
$$v = \pm 3$$

But logarithms of negative numbers do not exist; hence the only solution is

$$v = 3$$

Checking, we have
$$\log_6 4 + 2 \log_6 3 = \log_6 4 + \log_6 3^2$$
$$= \log_6(4 \cdot 3^2) = \log_6 36$$
$$= \log_6 6^2 \stackrel{\checkmark}{=} 2$$

The extraneous root $v = -3$ arose when we converted $2 \log_6 v$ to $\log_6 v^2$. Since $v^2 \geq 0$, the form $\log_6 v^2$ no longer requires $v > 0$ but only $v \neq 0$. ∎

We observed earlier that the logarithmic and exponential functions are inverses of one another. A function that has an inverse must be one-to-one. (See Section 4.4.) Thus whenever an inverse function f^{-1} exists, $f(x_1) \neq f(x_2)$ for $x_1 \neq x_2$. Consequently, we have the following.

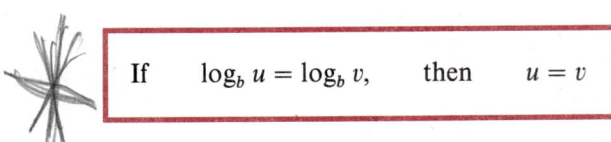

$$\text{If} \quad \log_b u = \log_b v, \quad \text{then} \quad u = v$$

EXAMPLE 5 Find w if
$$\log_b(3w + 2) + \log_b 4 = \log_b 64 + \log_b 2 - \log_b(3w - 2)$$

SOLUTION We can rewrite the equation as
$$\log_b(3w + 2) + \log_b(3w - 2) = \log_b 64 + \log_b 2 - \log_b 4$$
$$\log_b[(3w + 2)(3w - 2)] = \log_b \frac{64 \cdot 2}{4} \quad \text{(By L1 and L2)}$$
$$\log_b(9w^2 - 4) = \log_b 32$$

Then by the preceding property,

$$9w^2 - 4 = 32$$
$$9w^2 = 36$$
$$w^2 = 4$$
$$w = \pm 2$$

But if $w = -2$, then $3w + 2 = -4 < 0$, and there is no $\log_b(-4)$. Thus, the only solution is

$$w = 2$$

Checking, we have

$$\log_b(3 \cdot 2 + 2) + \log_b 4 = \log_b 8 + \log_b 4$$
$$= \log_b 8 \cdot 4 = \log_b 32$$

and

$$\log_b 64 + \log_b 2 - \log_b(3 \cdot 2 - 2) = \log_b(64 \cdot 2) - \log_b 4$$
$$= \log_b \frac{64 \cdot 2}{4}$$
$$\stackrel{\checkmark}{=} \log_b 32$$

Again, the extraneous root $w = -2$ arose when we combined logarithms to obtain an expression involving w^2 rather than w. ∎

EXAMPLE 6   Find t if $3^{t-1} = 2^{2t+4}$.

SOLUTION   Here we have an exponential equation involving different bases. We solve such equations by taking logarithms of both sides. Since common logarithms ($\log_{10}$) and natural logarithms (ln or $\log_e$) are readily available on calculators or from Tables 1 and 2 at the back of this book, we shall take $\log_{10}$ of both sides and then solve for t as indicated.

$$3^{t-1} = 2^{2t+4}$$
$$\log 3^{t-1} = \log 2^{2t+4}$$
$$(t-1)\log 3 = (2t+4)\log 2 \qquad \text{(By Property L3)}$$
$$t \log 3 - \log 3 = 2t \log 2 + 4 \log 2$$
$$t \log 3 - 2t \log 2 = 4 \log 2 + \log 3$$
$$t(\log 3 - 2 \log 2) = \log 2^4 + \log 3 \qquad \text{(By Property L3)}$$
$$t(\log 3 - \log 2^2) = \log 16 + \log 3 \qquad \text{(By Property L3)}$$
$$t \log\left(\frac{3}{2^2}\right) = \log 16 \cdot 3 \qquad \text{(By L1 and L2)}$$
$$t = \frac{\log 48}{\log\left(\frac{3}{4}\right)}$$
$$\approx \frac{1.6812}{-0.1249} \qquad \text{(Use a calculator or Table 2)}$$
$$\approx -13.46$$

Had we begun this problem by taking the natural logarithm (ln) of both sides, we would have obtained

$$t = \frac{\ln 48}{\ln\left(\frac{3}{4}\right)}$$

$$\approx \frac{3.8712}{-0.2877} \qquad \text{(Use a calculator or Table 1)}$$

$$\approx -13.46$$

The properties of logarithms can also be used as an aid in graphing logarithmic functions, as illustrated here.

EXAMPLE 7 Sketch the graph of $f(x) = \log_2(x - 1)^2$.

SOLUTION We note that $(x - 1)^2 \geq 0$ for all x; hence $\log_2(x - 1)^2$ exists for all $x \neq 1$. Now

$$\log_2(x - 1)^2 = \log_2|x - 1|^2$$
$$= 2 \log_2|x - 1|$$

We begin by graphing

$$y = \log_2|x| \qquad \text{(Figure 6.10a and Figure 6.12a)}$$

and then translate this graph one unit to the right to obtain the graph of

$$y = \log_2|x - 1| \qquad \text{(Figure 6.10b and Figure 6.12b)}$$

Finally, we double the y-coordinates of this second graph to obtain the graph of

$$y = 2 \log_2|x - 1| \qquad \text{(Figure 6.12c)}$$

Figure 6.12

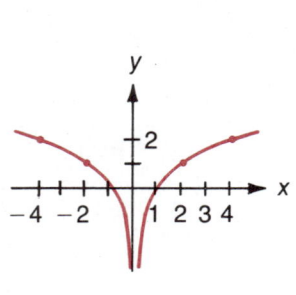

a. $y = \log_2|x|$

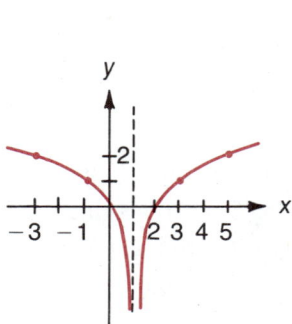

b. $y = \log_2|x - 1|$

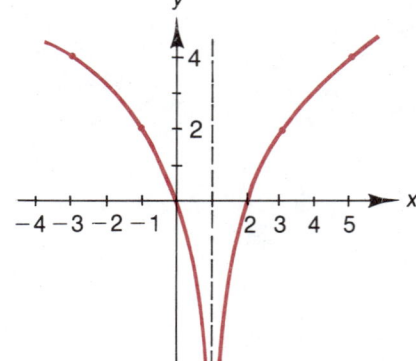

c. $y = \log_2(x - 1)^2 = 2 \log_2|x - 1|$

EXAMPLE 8 The basic unit of sound measurement is called a bel (in honor of Alexander Graham Bell). A decibel (one tenth of a bel) is the smallest *increase* detectable by the human ear. The ear does not respond directly to the intensity of the sound but rather to the logarithm of the intensity. The number of decibels difference in level between sound I_1 and I_2 is given by

$$D = 10 \log_{10} \frac{I_1}{I_2}$$

where I denotes the intensity or power of the sound. We say I_1 is twice as loud as I_2 if $D = 2$; if $D = 3$, I_1 is three times as loud as I_2, and so on. When we say that a given *sound is rated at a certain number of decibels,* it is being compared to the threshold of human hearing $I_0 = 10^{-16}$. Various decibel ratings are given in the table.

SOUND	INTENSITY I	NUMBER OF DECIBELS $\left(10 \log_{10} \dfrac{I}{I_0}\right)$
Threshold of hearing	I_0	0
Normal conversation	I_c	50
Street noise	I_s	70
Air hammer	I_h	90
Rock concert	I_r	110

a. How many times as loud as normal conversation is an air hammer? That is, find

$$D = 10 \log_{10} \frac{I_h}{I_c}$$

b. How much more intense or powerful is an air hammer than normal conversation? That is, find I_h/I_c.

SOLUTION

a. From the information given in the table, we have

$$10 \log_{10} \frac{I_h}{I_0} = 90 \quad \text{or} \quad \log_{10} \frac{I_h}{I_0} = 9$$

and

$$10 \log_{10} \frac{I_c}{I_0} = 50 \quad \text{or} \quad \log_{10} \frac{I_c}{I_0} = 5$$

From these values we obtain

$$10 \log_{10} \frac{I_h}{I_c} = 10 \log_{10} \left(\frac{I_h/I_0}{I_c/I_0}\right)$$

$$= 10 \left(\log_{10} \frac{I_h}{I_0} - \log_{10} \frac{I_c}{I_0}\right)$$

$$= 10(9 - 5) = 10 \cdot 4$$

$$= 40$$

The air hammer is 40 times as loud as normal conversation.

b. On the other hand,

$$\log_{10} \frac{I_h}{I_c} = 4$$

$$\frac{I_h}{I_c} = 10^4 = 10{,}000$$

The air hammer is 10,000 times as intense as normal conversation:

$$I_h = 10{,}000 I_c$$

Historical Perspective

JOHN NAPIER (1550–1617)

The incorporation of new technological developments into the daily operations of business, commerce, and science requires that computations be performed ever more rapidly and accurately. It is these demands that led to the development of the modern electronic computer and to the more recent development of sophisticated and inexpensive hand calculators and desk-top computers. Earlier demands led to the development of the slide rule. Even earlier, all calculations had to be done by hand. The development of our Hindu-Arabic system of numbers and decimal notation was a great improvement over hieroglyphics and Roman numerals; hand calculations were simplified immensely by the organization inherent in the decimal system.

In the early seventeenth century, John Napier developed a system of logarithms. The details of Napier's development need not concern us here, but his logarithm was

$$\log_{\text{nap}} y = 10^7 \log_{1/e}(y/10^7)$$

We do not use these logarithms today. Toward the end of his life, Napier collaborated with Henry Briggs. Together they decided that these logarithms were too cumbersome and developed the system of logarithms to the base 10.

To make these logarithms useful, extensive tables were needed. Developing these tables by hand was a very laborious task. Once the computations were completed and tabulated in a table of logarithms, the logarithms themselves became a very useful tool in simplifying calculations. In fact, they were so useful that they came to be known as **common logarithms.** Multiplying or dividing two numbers could then be accomplished by adding or subtracting their logarithms and then finding the number that had this value as its logarithm. Raising a number to a power or extracting a root was achieved by multiplying its logarithm by the power. By taking logarithms, we lower the level of difficulty of an operation by one notch: from multiplication to addition or from exponentiation to multiplication as indicated by Properties L1 and L3 and illustrated in Example 1.

In present times, however, even the simplest calculators can dispatch multiplication with ease. Calculators with a power key and/or root extraction capabilities are also very inexpensive. Thus tables of logarithms have almost become obsolete as a computational tool. The irony is that computing machines can easily generate logarithms but at the same time they render logarithms obsolete as a computational device.

Although we know logarithms as exponents, they were actually discovered before any use was made of exponential functions. Leonhard Euler (see page 318) first expressed logarithms in terms of exponents.

Section 6.3 Properties of Logarithms; Logarithmic Equations

> **CAUTION**
> Note that
> $$\log_b u + \log_b v \neq \log_b(u + v)$$
> For example,
> $$\log_2 4 + \log_2 8 \neq \log_2(4 + 8)$$
> since
> $$2 + 3 \neq 3.58\ldots = \log_2 12$$

Section 6.3 Exercises

Use the logarithm properties and the table given in Example 1 to find $\log_b$ of the following.

1. 40
2. $\dfrac{3}{25}$
3. $\dfrac{700}{49}$
4. $\dfrac{(30 \cdot 21)}{27}$
5. $\dfrac{(10^{-6} \cdot 4^2)}{7^3}$
6. $\sqrt{10}$
7. $\sqrt[3]{36}$
8. $\dfrac{1}{\sqrt[3]{25}}$
9. $\dfrac{\sqrt{6}}{14}$
10. $\sqrt[4]{15} \cdot \sqrt{12}$

Calculator Exercises (11–20)

Use common logarithms or natural logarithms and the change-of-base formula L7 to find the following. Use a calculator to do the computations and approximate the requested value.

11. $\log_3 7$
12. $\log_5 45$
13. $\log_3 72$
14. $\log_7 \dfrac{127 \cdot 243}{24}$
15. $\log_5 64\sqrt[3]{9}$
16. $\log_4 15$
17. $\log_{1/5} \dfrac{2}{3}$
18. $\log_{\sqrt{2}} 24$
19. $\log_{36} 100$
20. $\log_{3/2} \dfrac{1}{4}$

Use the properties of exponents and logarithms to express each of the following as a single logarithm.

21. $\log_b 100 - \log_b 25 + \log_b \dfrac{1}{4}$

22. $\dfrac{2}{3} \log_b 3 + \dfrac{1}{2} \log_b 6 - 2 \log_b 9$

23. $\log_b \dfrac{1}{4} - 3 \log_b 2 + \log_b \sqrt{16}$

24. $2 \log_b 3 + 3 \log_b 2 - \log_b \sqrt{9}$

25. $\log_2(2x + 1) + \log_2(2x - 1)$

26. $\log_3(y + 1) + \log_3(y - 1) - \log_3(y^2 - 1)$

27. $3 \log_3(3v - 9) - \log_3(v - 3)^3$

28. $4 \log_e \sqrt{u + 1} - \log_e(u + 1)$

29. $3 \log_b(u^2 - 1) - 3 \log_b(u - 1) - 5 \log_b(u + 1)$

30. $2 \log_b(v + 2) + 3 \log_b(v - 2) - 3 \log_b(v^2 - 4)$

Solve the following equations.

31. $\log_5 x = -3$
32. $\log_8 y = \dfrac{2}{3}$
33. $\log_{1/2}(-2v) = 4$
34. $\log_2(3r + 5) = 5$
35. $\log_3(u^2 + 2) = 3$
36. $\log_7 16\sqrt{t} = 0$
37. $\log_z 3 = \dfrac{1}{4}$
38. $\log_r 16 = 4$
39. $\log_t 1000 = -3$
40. $\log_t 9 = \dfrac{2}{3}$
41. $\log_k \dfrac{1}{32} = -\dfrac{5}{3}$
42. $\log_x 64 = \dfrac{3}{2}$
43. $\log_3(z + 1) = 2$
44. $\dfrac{1}{3} \log_3(2v + 1) = 1$
45. $2 \log_3 9w^2 = 8$
46. $\log_2(u - 1) + \log_2(u + 1) = 3$
47. $\log_{10}(2z - 4) - \log_{10}(z - 2) = 1$

48. $\log_2(2y-4) + \log_2(y-2) = 1$

49. $\log_2(-1-r) + \log_2(2-r) = 2$

50. $\log_3 2v + \log_3 v = \log_3 100 - \log_3 2$

51. $\log_2(2w+1) + \log_2(2w-1) = \log_2 11 + 2\log_2 3$

52. $\ln 6 + \ln 4 = 5\ln 2 + \ln 3 - \ln x$

53. $\log_3 2 - \log_3(-1-y) = \log_3(1-y)$

54. $4\log_8\sqrt{r+1} - \log_8(r+1) = \log_8 3 + \log_8 4$

55. $3\log_5 s^2 - 2\log_5 s^3 + \log_4(s+1) = \log_4 3$

56. $\log_b(1-z) - \log_b(-z-1) = \log_b 2$

57. $\log_b(x+2) + \log_b 3 = \log_b(2-x)$

58. $\log_b(4u+6) - \log_b 2 = \log_b(u+5)$

59. $\log_b(y+1) + \log_b(y-1) - 3\log_b 2 = 0$

60. $\log_b(v-3) + \log_b(v+3) - 4\log_b 2 = 0$

Calculator Exercises (61–70)

Solve the following equations explicitly and then use a calculator to approximate the solution.

61. $10^{3-y} = 3^{2-y}$

62. $2^{3-x} = 3^{2-x}$

63. $3^{2z-1} = 4^{z+5}$

64. $4^{2r+1} = 5^{r+2}$

65. $3^{2-t} = 2^{1+t}$

66. $5^{2s+1} = 4^{s+5}$

67. $7^{1-v} = 2^{1+v}$

68. $5^{u+4} = 3^{1-u}$

69. $11^{x-1} = 13^{x+2}$

70. $9^{2w} = 5^{4w}$

Sketch the following graphs: f(x) =

71. $\dfrac{1}{2}\log_2 4x$

72. $\log_2 x^2$

73. $\log_2(x+2)^2$

74. $\log_3(x-2)^2$

75. $\log_2\sqrt{x+2}$

76. $\log_3\sqrt{x-2}$

77. $\log_2 \dfrac{1}{(x+2)}$

78. $\log_3 \dfrac{1}{(x-2)}$

79. $\log_2 \dfrac{1}{|x+2|}$

80. $\log_3 \dfrac{1}{|x-2|}$

81. $\log_2 \dfrac{1}{x^2}$

82. $\log_{1/2} \dfrac{1}{x^2}$

83. $\log_{1/2} \dfrac{1}{x}$

84. $\log_2 \dfrac{1}{x}$

85. In a chain-letter scheme a letter contains five names. A new participant purchases a letter by mailing a $10 money order to the name at the top of the list, in the presence of the person selling the letter. The buyer is then instructed to remove the top name, place his name at the bottom, and make and sell five copies of the letter.

   a. What is the payoff if the chain is unbroken?

   b. How many steps can the chain take until the entire U.S. population (≈ 225 million) is involved?

   c. How many steps can the chain take until the population of the entire world (≈ 4 billion) is involved?

86. Betty wants to replace the antifreeze in her car. Her cooling system has a 16-liter capacity. When she drains the radiator, 4 liters remain in the engine block. How many times must she fill the radiator with fresh water, run the engine to mix the old coolant with the fresh water, and then redrain in order to have less than 5% polluted coolant in the final mixture?

87. How much louder is a rock concert than an air hammer? How much more intense? (See Example 8.)

88. The pH of a chemical substance is defined as

$$\text{pH} = \log_{10} \dfrac{1}{H}$$

where H is the concentration (moles/liter) of hydrogen ions in a water solution of the substance. Distilled water has a pH of 7. The substance is *acidic* if its pH is less than 7 and *basic* if its pH is greater than 7. Determine the pH of each of the substances listed below and state which are acidic.

SUBSTANCE	H
Beer	$3.16(10^{-5})$
Crackers	$3.16(10^{-8})$
Eggs	$1.58(10^{-8})$
Dill pickles	$3.98(10^{-4})$
Rhubarb	$7.08(10^{-4})$
Soft drinks	$1.12(10^{-3})$
Tuna	$9.55(10^{-7})$
Blood plasma	$3.98(10^{-8})$
Gastric contents	$9.77(10^{-3})$
Bile	$1.26(10^{-7})$

89. The exponential law of cooling can be used to determine the time of death for accident or murder victims. The temperature of a body t hours after death is

$$T(t) = T_0 + (T_1 - T_0)(0.97)^t$$

where T_0 is the air temperature and T_1 is the body temperature at the time of death. Assuming normal body temperature (98.6°F) and room temperature of 70°F, this formula becomes

$$T(t) = 70 + (98.6 - 70)(0.97)^t$$
$$= 70 + 28.6(0.97)^t$$

a. Express t as a logarithmic function of temperature.

b. At 10 P.M., John Doe was found murdered. His body temperature at that time was 82°F. When was he killed?

90. The Richter scale relates the magnitude of the shock waves of an earthquake to the energy released by the quake. If an earthquake releases an energy E (in watts per cubic centimeter), then the Richter scale measures a magnitude

$$M = \frac{\log_{10} E - 11.4}{1.5}$$

(The relationship is of the form $\log_{10} E = A + B \cdot M$, where A and B are still being refined. The values given here are those that were accepted in 1978.) One full unit difference on the Richter scale represents 10 times the amount of motion.

M	EFFECT
2	Can be felt
4.5	Slight damage
6	Moderate damage
8.5	Largest recorded

a. How many times as much energy is represented by 1 unit on the Richter scale? (Find the change in E when M is replaced by $M + 1$.)

b. How many times as much motion and how many times as much energy is represented by 2 units on the Richter scale?

c. Compare the 1964 Alaska earthquake ($M = 8.5$), the 1906 San Francisco quake ($M = 8.3$), and a 1975 Yellowstone Park quake ($M = 6.7$).

d. Express the energy released in terms of the magnitude M measured by the Richter scale.

Section 6.4

Using Tables (Optional)

The tables at the end of this book list values for e^x and for the natural and common logarithms. Since each of these tables is used in the same manner, we shall concentrate on only one (Table 2: Common Logarithms) in the following discussion.

These tables list values correct to four decimal places. The column headings are used for further refinement of the argument. For instance, in the portion of Table 2 that is reproduced in Figure 6.13 we see that $\log 2.37 \approx 0.3747$.

Figure 6.13

x	0	1	2	3	4	5	6	7	8	9
⋮	⋮	⋮	⋮	⋮	⋮	⋮	⋮	⋮	⋮	⋮
2.3	.3617	.3636	.3655	.3674	.3692	.3711	.3729	.3747	.3766	.3784
⋮	⋮	⋮	⋮	⋮	⋮	⋮	⋮	⋮	⋮	⋮

The process of **linear interpolation** is used to estimate values of $f(x)$ at an x-value between two of those listed in the table for f.

> **LINEAR INTERPOLATION**
>
> For x between x_1 and x_2, linear interpolation estimates $f(x)$ by adding to (or subtracting from) $f(x_1)$ that portion of the increase (or decrease) $[f(x_2) - f(x_1)]$ determined by the portion of the distance that x has traveled from x_1 to x_2.

EXAMPLE 1 Find log 2.346.

SOLUTION Use Table 2 to find log 2.34 and log 2.35. Then set up the following display.

$$
0.010 \left\{ 0.006 \left\{ \begin{array}{c|c} x & \log x \\ \hline 2.340 & 0.3692 \\ 2.346 & \\ 2.350 & 0.3711 \end{array} \right\} d \right\} \text{Increase} = 0.0019
$$

As x moves from 2.34 to 2.35, log x increases from 0.3692 to 0.3711; the increase of log x is 0.0019. Now 2.346 is

$$\frac{0.006}{0.010} = \frac{6}{10}$$

of the way from 2.34 to 2.35. Thus log 2.346 should be approximately $\frac{6}{10}$ of the way from log 2.34 to log 2.35; that is,

$$d \approx \frac{6}{10}(0.0019)$$
$$= (0.6)(0.0019) = 0.00114$$

Then

$$\log 2.346 = \log 2.34 + d$$
$$\approx 0.3692 + 0.00114$$
$$= 0.37034$$
$$\approx 0.3703 \qquad \blacksquare$$

In this example, we rounded 0.37034 to 0.3703 as the approximate value of log 2.346. The reason for this is that the values in the table are already rounded to four places and the linear interpolation procedure is also just an estimate of the intermediate value; hence we cannot expect to obtain logarithms correct to five places from a four-place table. For the same reasons, we interpolate to only one more decimal place in the argument; for example, we interpolate between 2.34 and 2.35 to 2.346 but not to 2.3463.

The relationship $d = \frac{6}{10}(0.0019)$ in Example 1 can be written in the form of a ratio or proportion:

$$d = \frac{6}{10}(0.0019)$$

$$\frac{d}{0.0019} = \frac{6}{10} = \frac{0.006}{0.010}$$

This ratio can also be obtained from the two sides of the display given at the beginning of the solution.

The tables can be used in reverse, too. Again, interpolation may be necessary.

EXAMPLE 2

Find x for which $\log x = 0.3741$.

SOLUTION

We bracket $\log x$ between two consecutive values in Table 2 as follows:

u	$\log u$
2.360	0.3729
x	0.3741
2.370	0.3747

$0.010 \begin{Bmatrix} d \begin{Bmatrix} 2.360 \\ x \end{Bmatrix} \end{Bmatrix}$ with differences 0.0012 and 0.0018.

From this display, we obtain the ratio

$$\frac{d}{0.010} \approx \frac{0.0012}{0.0018} = \frac{2}{3}$$

$$d \approx \frac{2}{3}(0.010)$$

That is, x is approximately $\frac{2}{3}$ of the way from 2.360 to 2.370:

$$x \approx 2.360 + d$$
$$\approx 2.360 + \frac{2}{3}(0.010)$$
$$= 2.360 + 0.00666\ldots$$
$$= 2.36666\ldots$$
$$\approx 2.367$$

Again, because of the rounding errors inherent in Table 2 and because of errors introduced in the interpolation process, we write $x \approx 2.367$ rather than $2.36666\ldots$. ∎

Any real number x is the $\log_b$ of precisely one real number u. The number u that has x as its $\log_b$ is called the **antilogarithm to the base b** of x; it is denoted antilog$_b$ x. In view of the equivalence of the statements

$$x = \log_b u \quad \text{and} \quad u = b^x$$

we see immediately that

$$\text{antilog}_b x = b^x$$

That is, the antilogarithm function is just the exponential function.

As with common logarithms, whenever the base $b = 10$, it will not be mentioned. Thus

$$\text{antilog } x = \text{antilog}_{10} x = 10^x$$

For instance, we see in Figure 6.13 (Table 2) that

$$10^{0.3747} = \text{antilog}(0.3747) \approx 2.37$$

In Section 1.3 we saw that any positive number x could be written in *scientific notation* as

$$x = b \cdot 10^n \qquad 1 \leq b < 10$$

Then we have

$$\log x = \log b + \log 10^n$$
$$= \log b + n$$

That is, the logarithm of any number can be expressed as an integer plus the log of some number between 1 and 10. For this reason, Table 2 lists values of $\log x$ only for x between 1 and 10.

The integer part of $\log x$ is called its **characteristic**; $\log b$ is called the **mantissa** of $\log x$. Note that the mantissa is a **positive decimal fraction**:

$$\log 1 \leq \log b < \log 10$$
$$0 \leq \log b < 1$$
$$\log b = 0.a_1 a_2 a_3 \ldots$$

The mantissa of log x depends only on the order of the digits in the decimal representation for x rather than on the location of the decimal point. Moving the decimal point only changes the characteristic of the logarithm. For example, if $\log 2.346 = 0.3703$, then

$$\log 2346 = \log 2.346(10^3)$$
$$= \log 2.346 + \log 10^3$$
$$\approx 0.3703 + 3$$
$$= 3.3703$$

and

$$\log 234600 = \log 2.346(10^5)$$
$$\approx 5.3703$$

Similarly

$$\log 0.002346 = \log 2.346(10^{-3})$$
$$= \log 2.346 + \log(10^{-3})$$
$$\approx 0.3703 - 3$$
$$= -2.6297$$

Here 0.3703 is the mantissa and -3 is the characteristic. The form -2.6297 does not display the characteristic and mantissa since $-2.6297 = -2 + (-0.6297)$; -0.6297 cannot be the mantissa since the mantissa must be greater than or equal to 0. For this reason, we usually leave this logarithm in the form

$$\log 0.002346 \approx 0.3703 - 3$$

when working with tables. However a calculator would give

$$\log 0.002346 \approx \boxed{-2.6297\ldots}$$

The characteristic is also useful in finding antilogarithms:

$$\text{antilog } 0.3703 \approx 2.346$$
$$\text{antilog } \mathbf{3}.3703 \approx 2346 \qquad (2.346) \qquad \text{3 places right}$$
$$\text{antilog } \mathbf{5}.3703 \approx 234600 \qquad (2.34600) \qquad \text{5 places right}$$
$$\text{antilog}(0.3703 - 3) \approx 0.002346 \qquad (002.346) \qquad \text{3 places left}$$

To convert a negative number like -2.6297 into a form that exhibits mantissa and characteristic, we *add and subtract a positive integer* as follows:

$$-2.6297 = 3 - 2.6297) - 3$$
$$= 0.3703 - 3$$

USING TABLE 2 TO FIND LOGS AND ANTILOGS

To find log x

1. Write x in scientific notation: $\qquad\qquad x = 2346$
   $$x = b \cdot 10^n \qquad\qquad\qquad\qquad = 2.346(10^3)$$

2. Find log b in Table 2; interpolate if necessary. $\qquad\qquad \log 2.346 \approx 0.3703$

3. $\log x = n + \log b$. $\qquad\qquad\qquad\qquad \log 2346 \approx 3.3703$

To find antilog y

1. Write y in a form that exhibits a positive decimal fraction (mantissa) and an integer (characteristic), which may be positive or negative.
   $$y = -2.6297$$
   $$= (3 - 2.6297) - 3$$
   $$= 0.3703 - 3$$

2. Find the antilog of the mantissa in Table 2. $\qquad$ antilog $0.3703 \approx 2.346$

3. Move the decimal point to the right (if characteristic > 0) or left (if characteristic < 0) the number of places indicated by the characteristic. $\qquad$ antilog $y \approx 0.002346$

EXAMPLE 3 Find $\log(0.00365)^4$.

SOLUTION
$$\log(0.00365)^4 = 4 \cdot \log(0.00365) \qquad\qquad \text{(By Property L3)}$$
$$= 4 \cdot \log[(3.65) \cdot 10^{-3}]$$
$$= 4(-3 + \log 3.65)$$
$$\approx 4(-3 + 0.5623) \qquad\qquad \text{(Table 2)}$$
$$= -12 + 2.2492$$
$$= -9.7508 \qquad (\text{or } 0.2492 - 10) \qquad\blacksquare$$

EXAMPLE 4 Find x if $\log x = -3.6259$

SOLUTION We write

$$\begin{aligned}\log x &= -3.6259 \\ &= (4 - 3.6259) - 4 \\ &= 0.3741 - 4\end{aligned}$$

Thus

$$x = \text{antilog}(0.3741 - 4)$$

Now

$$\text{antilog } 0.3741 \approx 2.367 \qquad \text{(Example 2)}$$

Then, moving the decimal point four places to the left, we have

$$x \approx 0.0002367 \qquad \blacksquare$$

Tables 1 and 3 (at the end of this book) need not be used in reverse, since each is, in effect, the reverse of the other.

As we mentioned earlier, logarithms are a useful computational tool in the absence of a sophisticated hand calculator. They can also be used to extend the capabilities of a simple four-function $(+, -, \times, \div)$ calculator to permit computation of roots and powers.

EXAMPLE 5 Use logarithms to calculate

$$\frac{(2.367)^{26} \cdot \sqrt[17]{0.0003471}}{\sqrt[5]{(6027)^{13}}}$$

SOLUTION Let x denote the value of the given expression. Then

$$\log x = 26 \cdot \log(2.367) + \frac{1}{17} \cdot \log(0.0003471) - \frac{13}{5} \cdot \log 6027$$

$$\approx 26(0.3742) + \frac{1}{17}(0.5404 - 4) - \frac{13}{5}(3.7801)$$

$$\approx 9.7292 + (-0.2035) - 9.8283$$

$$\approx -0.3026 = 0.6974 - 1$$

$$x \approx \text{antilog}(0.6974 - 1)$$

$$\text{antilog } 0.6974 \approx 4.982$$

$$x \approx 0.4982 \qquad \blacksquare$$

Section 6.4 Exercises

Use Table 2 (at the end of this book) to find the logarithms of the following numbers.

1. **a.** 123   **b.** 12,300   **c.** 0.00123

2. **a.** $(421)^4$ **b.** $(42100)^{-3}$ **c.** $(0.000421)^3$

3. **a.** $\sqrt[4]{0.0888}$ **b.** $(88.8)^3$ **c.** $(888,000)^{-6}$

4. **a.** $(105)^{63}$ **b.** $\sqrt{10,500}$ **c.** $(0.0105)^{-3}$

Section 6.5 Applications of Exponential and Logarithmic Functions   345

5. **a.** $\sqrt[3]{4720}$ **b.** $\dfrac{1}{\sqrt[3]{4720}}$ **c.** $\dfrac{1}{\sqrt[3]{47.2}}$

6. **a.** $(7270)^2$ **b.** $\dfrac{1}{\sqrt[3]{7270}}$ **c.** $(0.00727)^{1/3}$

7. **a.** $(1540)^{-2}$ **b.** $(0.00154)^2$ **c.** $\sqrt{154}$

8. **a.** $(0.0436)^{20}$ **b.** $(43{,}600)^{20}$ **c.** $\sqrt[3]{436}$

9. **a.** $(71{,}300)^3$ **b.** $(0.0713)^{1/3}$ **c.** $\dfrac{1}{\sqrt[3]{713}}$

10. **a.** $(306)^{18}$ **b.** $(3060)^{1.8}$ **c.** $(3.06)^{180}$

Use Table 2 to find x for log x as given in each of the following exercises.

11. **a.** 3.7168   **b.** 1.7168   **c.** 6.7168
12. **a.** 1.7642   **b.** 0.7642 − 5   **c.** 0.7642 − 3
13. **a.** −2.2197   **b.** −3.2197   **c.** −5.2197
14. **a.** 3.8681   **b.** 0.8681 − 3   **c.** −3.1319
15. **a.** 2.9058   **b.** 6.9058   **c.** 3.9058
16. **a.** −0.0640   **b.** −2.0640   **c.** −1.0640
17. **a.** −1.0405   **b.** −3.0405   **c.** 3.9595
18. **a.** 7.2148   **b.** 1.2148   **c.** 0.2148 − 3
19. **a.** 2.4281   **b.** 1.4281   **c.** 0.4281

20. **a.** 2.5403   **b.** 6.5403   **c.** −2.4597

Approximate the logarithms of the following numbers by linear interpolation from Table 2.

21. 2.631   22. 7.642
23. 676.7   24. 0.04586
25. 31,430   26. 642,700
27. 0.0006475   28. 0.000006175

Interpolate from Table 2 to find x if log x is given as follows.

29. 0.2230   30. 0.0921
31. −0.3430   32. 2.6704
33. −3.7568   34. 4.2516
35. −4.6824

Use logarithms from Table 2 to calculate the following expressions.

36. $(1237)(0.4945)(0.007654)$

37. $\dfrac{(839{,}200)(0.3044)}{0.06237}$

38. $\dfrac{\sqrt[3]{74.99}\,\sqrt{5042}}{\sqrt[3]{(6.718)^5}}$

39. $\dfrac{(3.01)^{15}(2.34)^{-5}}{(16.89)^6}$

40. $\dfrac{(4.579)^2\sqrt{82.71}}{\sqrt[3]{0.1224}}$

Section 6.5

Applications of Exponential and Logarithmic Functions

Exponential functions are useful in studying, among other things, population growth, compound interest, and radioactive decay. Because of their inverse relationship to exponential functions, logarithmic functions are often helpful in these situations as well. For example, determination of the time required to raise a certain sum of money in a compound interest situation requires logarithms.

The applications in this section illustrate bona fide uses of logarithms in the sense that one or more logarithms must be found in order to answer a certain question. Logarithms are not used here merely as a computational device.

Population Growth

EXAMPLE 1   A certain bacteria colony is known to double its population every hour; 10,000 bacteria are present at noon. Under these conditions, the number $f(x)$ of bacteria present x hours after noon is

$$f(x) = 10{,}000(2^x)$$

At what time will the population be 15,000?

SOLUTION   We must solve

$$10{,}000(2^x) = 15{,}000$$

$$2^x = \frac{15{,}000}{10{,}000} = \frac{3}{2}$$

$$x \log 2 = \log \frac{3}{2} \qquad \text{(L3, Section 6.3)}$$

$$x = \frac{\log \frac{3}{2}}{\log 2}$$

$$\approx \frac{0.1761}{0.3010}$$

$$\approx 0.5850$$

Now, 0.585 hours is approximately 35 minutes. Hence, the population is 15,000 at approximately 12:35 P.M. ∎

Compound Interest

Most banks now compound the interest on savings accounts. This means that the interest is paid into the account after a certain period of time; it then begins to earn interest, too.

i. Assume that a principal of P_0 dollars is invested at I percent interest per year and that the interest is to be compounded N times per year.

ii. Write $r = I/100$ (e.g., if $I = 5\%$, then $r = 0.05$).

iii. After one compounding period, or $1/N$th of a year, the interest payment is

$$P_0 \cdot r \cdot \left(\frac{1}{N}\right)$$

iv. The new principal is then

$$P = P_0 + P_0 \cdot r \cdot \left(\frac{1}{N}\right) = P_0\left(1 + \frac{r}{N}\right)$$

v. After another compounding period, this principal is multiplied by $[1 + (r/N)]$ again:

$$P = P_0\left(1 + \frac{r}{N}\right)\left(1 + \frac{r}{N}\right) = P_0\left(1 + \frac{r}{N}\right)^2$$

vi. After t years ($= Nt$ compounding periods), the principal increases to

$$P(t) = P_0\left(1 + \frac{r}{N}\right)^{Nt} = P_0\left[\left(1 + \frac{r}{N}\right)^N\right]^t$$

vii. **Continuous compounding** can be *approximated* by compounding over very small time intervals, such as every minute or every second. This yields a very large value of N. For instance, there are $N = 365 \cdot 24 \cdot 60 \cdot 60 = 31{,}536{,}000$ seconds in 1 year. As N gets larger and larger, the number $[1 + (r/N)]^N$ becomes a better and better estimate of e^r, where $e \approx 2.71828\ldots$ (see Exercise 40). Thus after t years of

Section 6.5 *Applications of Exponential and Logarithmic Functions* **347**

continuous compounding, we accumulate a principal of

$$P(t) \approx P_0 \left[\left(1 + \frac{r}{N}\right)^N \right]^t \quad \text{(For } N \text{ very large)}$$

$$\approx P_0 [e^r]^t = P_0 e^{rt}$$

COMPOUND INTEREST

Let an initial principal of P_0 be invested at I percent and let $r = I/100$. The principal resulting after compounding the interest for t years is given by

$$P(t) = \begin{cases} P_0 \left(1 + \dfrac{r}{N}\right)^{Nt} & \text{(compounding } N \text{ times per year)} \\ P_0 e^{rt} & \text{(compounding continuously)} \end{cases}$$

In the equation $P(t) = P_0 e^{rt}$, t may be any real number; it may even represent an irrational number of years. Most scientific calculators are programmed to compute the powers e^{rt} needed to make these calculations. These values can be estimated from Table 3 if you do not have such a calculator.

While continuous compounding may seem to be an artificial and complicated mathematical construction, it is really the "natural" form of compounding. For instance, the *growth in value* of a real estate investment depends on market prices and selling time. This growth in value may be considered as *interest* on the original investment. The rise in value is not recorded on an annual, monthly, or daily basis but is only realized at the time of sale. Yet this rise takes place steadily during the time of ownership; the interest, or growth in value, is compounded continuously. Whenever there is no outside "interest manager" to determine when interest should be paid, the interest compounds continuously.

EXAMPLE 2   If $500 is invested at 6 percent interest, how much capital is accumulated after 5 years in each case?

a. No compounding

b. Yearly compounding

c. Quarterly compounding

d. Daily compounding

e. Continuous compounding

Use the formulas given in the preceding box.

SOLUTION   Here we are given $P_0 = \$500$, $I = 6$, $r = 0.06$, and $t = 5$.

a. With no compounding, the 6% interest is paid for each of the 5 years. But this interest is never paid into the account to earn more interest. Thus the total principal after 5 years is

$$P(5) = \$500 + \$500 \cdot (0.06) \cdot 5 = \$650$$

b. Compounding yearly, we have $N = 1$ and

$$P(5) = \$500\left(1 + \frac{0.06}{1}\right)^{1 \cdot 5}$$
$$= \$500 \cdot (1.06)^5 \approx \$669.11$$

c. With quarterly compounding, we have $N = 4$ and

$$P(5) = \$500\left(1 + \frac{0.06}{4}\right)^{4 \cdot 5}$$
$$= \$500\left(1 + \frac{0.06}{4}\right)^{20} \approx \$673.43$$

d. Under daily compounding, $N = 365$ and

$$P(5) = \$500\left(1 + \frac{0.06}{365}\right)^{365 \cdot 5}$$
$$= \$500\left(1 + \frac{0.06}{365}\right)^{1825} \approx \$674.91$$

e. Under continuous compounding,

$$P(5) = \$500 e^{(0.06) \cdot 5} \approx \$500 \cdot (1.34986) = \$674.93$$

Lending institutions frequently advertise passbook savings accounts or certificates of deposit (CDs) paying a certain rate of interest, which is compounded regularly. Then they state an *effective* interest rate, which is slightly higher. This is called an APR for "annual percentage rate." This effective rate, or APR, is the interest rate that generates the same annual return with no compounding.

EXAMPLE 3

A passbook savings account pays 5% interest. Find the effective interest rates for quarterly and continuous compounding.

SOLUTION

With quarterly compounding ($N = 4$) for a full year ($t = 1$) at 5% ($r = 0.05$), an investment of P_0 dollars will grow to

$$P(1) = P_0\left(1 + \frac{0.05}{4}\right)^{4 \cdot 1} \approx 1.0509 P_0 = P_0 + (0.0509)P_0$$

The total increase in principal, or interest, earned during the year is about $(0.0509)P_0$. The effective interest *rate* is about 5.09%.

Compounding continuously for 1 year yields a principal of

$$P(1) = P_0(e^{0.05})^1 \approx 1.05127 P_0 = P_0 + (0.05127)P_0$$

The total interest earned during the year is about $(0.05127)P_0$; the effective interest rate is about 5.127%.

EXAMPLE 4

If we wish to double our investment in 5 years, what interest rate must we receive if the interest is compounded continuously?

SOLUTION

We use $t = 5$ in the formula for compound interest:

$$P(5) = P_0 e^{r \cdot 5}$$

Section 6.5 Applications of Exponential and Logarithmic Functions   349

To double our money in 5 years, we must have $P(5) = 2P_0$:

$$P_0 e^{5r} = 2P_0$$
$$e^{5r} = 2$$
$$5r = \ln 2$$
$$r = \frac{\ln 2}{5} \approx \frac{0.693}{2}$$
$$\approx 0.1386.$$

An interest rate of about 13.86% is required.

EXAMPLE 5 Find the lengths of time required to double an investment at 7% interest under the following compounding situations:

a. Quarterly
b. Daily
c. Continuous

SOLUTION We use $r = 0.07$. Then, given an initial principal P_0, we are to determine the time t that yields $P(t) = 2P_0$.

a. Compounding quarterly, we use $N = 4$ and solve

$$P(t) = P_0\left(1 + \frac{0.07}{4}\right)^{4t} = 2P_0$$
$$(1.0175)^{4t} = 2$$
$$4t \log(1.0175) = \log 2 \qquad \text{(L3, Section 6.3)}$$
$$t = \frac{\log 2}{4 \log(1.0175)}$$
$$\approx \frac{(0.30103)}{4(0.00753)}$$
$$\approx 9.99$$

Since the interest is paid only at the end of each quarter, it takes 10 years to double the investment at 7% compounded quarterly.

b. Compounding daily, we use $N = 365$ and solve

$$P(t) = P_0\left(1 + \frac{0.07}{365}\right)^{365 \cdot t} = 2P_0$$
$$365t \log\left(1 + \frac{0.07}{365}\right) = \log 2 \qquad \text{(L3, Section 6.3)}$$
$$t = \frac{\log 2}{365 \log\left(1 + \frac{0.07}{365}\right)}$$
$$\approx 9.903 \text{ years}$$

We can translate this into 9 years + 0.903 years, or 9 years and 330 days; this is the length of time needed to double an investment at 7% compounded daily.

c. Compounding continuously for t years should yield

$$P(t) = P_0(e^{0.07})^t = 2P_0$$
$$0.07t = \ln 2$$
$$t = \frac{\ln 2}{0.07}$$
$$\approx \frac{0.69315}{0.07}$$
$$\approx 9.902 \text{ years}$$

Continuous compounding doubles a 7% investment in 9.902 years, or 9 years, 329 days, and $5\frac{1}{2}$ hours.

Thus continuous compounding is almost indiscernible from daily compounding. Daily compounding, in turn, doubles a 7% investment only 35 days earlier than quarterly compounding. ∎

Radioactive Decay

Radioactive substances decay exponentially. If Q_0 represents an initial amount, the amount remaining at time t is given by the following law.

LAW OF EXPONENTIAL DECAY

$$Q(t) = Q_0 e^{-kt}, \quad \text{for some constant } k > 0$$

Half-life is an important physical concept relating to radioactivity. The **half-life** of a radioactive substance is the time in which a given amount of the substance will decay to half the original amount.

EXAMPLE 6

In the aftermath of the 1986 Chernobyl nuclear accident, neighboring residents were endangered by radioactive iodine (^{125}I), which has a half-life of 60 days.

a. Express the amount $I(t)$ of radioactive iodine present at any time t after an initial amount I_0 is present.

b. How long does it take for radioactive iodine to be reduced to 10% of its initial amount?

SOLUTION

a. Using the law of exponential decay on iodine, we have

$$I(t) = I_0 e^{-kt}$$

When $t = 60$ days, $I(t) = \frac{1}{2}I_0$:

$$I(60) = I_0 e^{-k \cdot 60} = \frac{1}{2}I_0$$
$$-60k = -\ln 2 \qquad \left(\ln \frac{1}{2} = -\ln 2\right)$$
$$k = \frac{\ln 2}{60}$$

Substituting this value of k into the initial equation, we obtain

$$I(t) = I_0 e^{-(\ln 2/60)t}$$
$$= I_0 (e^{\ln 2})^{-t/60}$$
$$= I_0 2^{-t/60}$$

where t is given in days.

b. To reduce $I(t)$ to 10% of its original amount I_0, we must have

$$I(t) = I_0 2^{-t/60} = (0.10) I_0$$

$$\frac{-t}{60} \ln 2 = \ln(0.10) \qquad \text{(L3, Section 6.3)}$$

$$t = \frac{-60 \ln(0.10)}{\ln 2}$$

$$\approx 199 \text{ days} \qquad \blacksquare$$

If a radioactive substance has a half-life H and an initial amount Q_0 is present, then as in Example 6a, the amount remaining at time t will be given by the following law.

HALF-LIFE DECAY LAW

$$Q(t) = Q_0 2^{-t/H}$$

Carbon Dating

EXAMPLE 7 When archaeologists discover a new site, they determine the approximate time during which its inhabitants lived by a process known as carbon-14 dating. Carbon-14 and carbon-12 exist in equilibrium in the atmosphere and in living organisms, the ratio of carbon-14 to carbon-12 being $(1.3)10^{-12}$. Carbon-14 decays with a half-life of 5730 years. As long as a being lives, it continues to replace its lost carbon-14 through interaction with the atmosphere by such activities as eating and breathing. Once it dies, however, its carbon-14 content continues to decay but can no longer be replaced. Its carbon-12 content remains the same.

Thus at the time of death the ratio of carbon-14 to carbon-12 begins to decrease with a half-life of 5730 years. Determine the age of a bone in which the ratio of carbon-14 to carbon-12 is $(3.25)10^{-13}$.

SOLUTION Let

$R(t)$ = ratio of carbon-14 to carbon-12 t years after death
$R_0 = (1.3)10^{-12}$ = ratio of carbon-14 to carbon-12 at time of death (and while living)

Then, by the half-life decay law, we have

$$R(t) = R_0 2^{-t/H} = (1.3)(10^{-12}) 2^{-t/5730}$$

To determine the age of a bone for which $R(t) = (3.25)10^{-13}$, we must solve

$$R(t) = (1.3)(10^{-12})2^{-t/5730} = (3.25)(10^{-13})$$

$$2^{-t/5730} = \frac{(3.25)(10^{-13})}{(1.3)(10^{-12})}$$

$$= \frac{3.25}{13}$$

$$\frac{-t}{5730} \ln 2 = \ln 3.25 - \ln 13 \qquad \text{(L3 and L2, Section 6.3)}$$

$$t = \frac{-5730(\ln 3.25 - \ln 13)}{\ln 2}$$

$$\approx 11{,}460$$

The bone is from an animal that has been dead for approximately 11,460 years. ∎

Section 6.5 Exercises
Calculator Exercises

1. A manufacturer of heavy duty trucks claims that if it produces T_0 trucks in a given year, then $T_0 \cdot 2^{(-0.012)t}$ of these trucks will be in service t years later.
   a. If 50,000 trucks were produced 10 years ago, how many should still be in service?
   b. After how many years are 90% of its trucks still serviceable?
   c. 75%?

2. The intensity of illumination (brightness) of a camping lantern is $I_0 \cdot 3^{(-0.05)t}$ lumens, where I_0 represents the initial brightness and t is the number of hours that fuel has been used from its tank.
   a. After 5 hours of operation, how will its brightness have been affected?
   b. After how many hours will the brightness have been halved?

3. A department store can make a yearly profit of $100,000 + (500,000)4^{(-0.2)n}$ on a popular gadget n years after the introduction of the gadget. How much profit does it make on these gadgets after they have been on the market 10 years?

4. If crops are not rotated and fertilized, the yield declines from year to year. If each succeeding year's crop will be only 80% that of the preceding year and Y_0 is the current yield, the yield t years from now is given by $Y(t) = Y_0(0.80)^t$. How long does it take for the yield to be halved?

5. A flu virus triples the number of its victims every 10 days. If it has infected 25,000 people now, the number of people infected t days from now is given by $I(t) = 25{,}000(3^{t/10})$
   a. How many will be infected 20 days from now?
   b. How many will be infected 25 days from now?

6. If the doubling time of a bacteria culture is known to be 3 hours, the population at time t is given by $P(t) = P_0 2^{t/3}$. There are 1000 bacteria present now ($t = 0$).
   a. When will the population reach 2500 members?
   b. When were only 100 members present?
   c. How many will there be 5 hours from now?
   d. How many were there 6 hours ago?

7. The doubling time of a bacteria culture is 2 hours. The population at time t is given by $P(t) = P_0 2^{t/2}$. The culture now ($t = 0$) consists of 100 bacteria.

a. How long will it take to reach 750 members?
b. When were there 10 bacteria present?
c. What is the population $5\frac{1}{2}$ hours from now?
d. How many were there 1 hour ago?

8. The population of the world has been doubling approximately every 35 years: $P(t) = P_0 2^{t/35}$. The population is now ($t = 0$) estimated to be about 4 billion.
   a. When will the population reach 5 billion?
   b. When will the population reach 6 billion?
   c. What will the population be 10 years from now?
   d. When was the population 3 billion?

9. The U.S. population doubles in about 60 years: $P(t) = P_0 2^{t/60}$. The current ($t = 0$) U.S. population is estimated to be 225 million. When will the U.S. population reach each value?
   a. 250 million
   b. 300 million
   c. 400 million
   d. What population can be projected for 25 years from now?

10. How quickly do prices double for each inflation rate?
    a. 3% compounded semiannually
    b. 5% compounded quarterly
    c. 6% compounded monthly
    d. 7% compounded continuously
    e. 9% compounded semiannually
    f. 12% compounded annually

11. In each of the following, find the effective annual interest rate, or APR.
    a. A passbook pays $5\frac{1}{2}$% compounded quarterly.
    b. A certificate of deposit (CD) pays 8% compounded semiannually.
    c. A CD pays 7% compounded daily.
    d. A CD pays 7% compounded continuously.
    e. A CD pays $7\frac{1}{2}$% compounded daily.

12. Five years ago Dan invested a certain sum of money at 7% interest compounded continuously. How much did he invest if his principal is now worth $30,000?

13. Paula invested a certain sum of money at 6% compounded continuously. Two years later, her investment was worth $10,000.
    a. How much is it worth 5 years after this?
    b. How much did she initially invest?

14. If Verna invests a given sum at 5% interest, how long does it take to triple her investment if the interest is compounded
    a. Not at all   b. Yearly   c. Daily
    d. Continuously

15. Dale invests $5000 and the interest is compounded quarterly. How long does it take to double his investment for each interest rate?
    a. 5%   b. 6%   c. 7%

16. If Bernice needs $7500 5 years from now and has $5000 to invest now, what interest rate compounded daily does she need?

17. If we need $10,000 10 years from now and have $5000 to invest now, what interest rate do we need if it is compounded continuously?

18. If Joan needs $5000 12 years from now and has $2000 to invest now, what interest rate compounded quarterly does she need?

(*The **present value** of a sum of money S to be received at a certain time in the future is the principal P, which if invested now would grow to S in that time.*)

19. a. What principal must be invested now at $6\frac{1}{2}$% interest compounded quarterly to achieve a value of $15,000 in 10 years?
    b. What sum is needed with continuous compounding?

20. a. What principal must be invested now at 7% interest compounded daily to achieve a value of $6000 in 5 years?
    b. What sum is needed with continuous compounding?

21. a. What principal must be invested now at 5% interest compounded semiannually to achieve a value of $2000 in 6 years?
    b. What sum is needed with continuous compounding?

22. a. What principal must be invested now at 6% interest compounded annually to achieve a value of $15,000 in 8 years?
    b. What sum is needed with continuous compounding?

23. In 1626 Peter Minuit purchased Manhattan Island from the Manhattan Indians for about $24.
    a. Had this money been invested at 6% interest compounded annually, what would the investment be worth in 1986?
    b. How about 4%?

24. The half-life of carbon-14 is approximately 5730 years. If 3 milligrams of carbon-14 are present in a fossil that is 3000 years old, how much carbon-14 was present in this fossil when it lived?

25. The half-life of radium is approximately 1600 years. Suppose 2000 units are present now.
    a. When were there 5000 units present?
    b. When will there be only 200 units present?
    c. How much will be present 50 years from now?

26. A given sample of radium consists of 100 units. (See Exercise 25.)
    a. When will it contain only 75 units?
    b. How much will there be 1000 years from now?
    c. When were 150 units present?
    d. When will one third of the original amount remain?

27. A tracer element (^{131}I) used in medical diagnostics has a half-life of 8 days. Let 400 units of this tracer be injected at time $t = 0$. How much remains after each time period?
    a. 1 day
    b. 4 days
    c. 12 days
    d. How long does it take to reduce the tracer to 10 units?

28. Another radioactive element released during the Chernobyl accident (see Example 6) was strontium-90 (^{90}Sr), which has a half-life of 29 years. How long does it take for ^{90}Sr to be reduced to each amount?
    a. 20% of its initial amount
    b. 5% of its initial amount

29. A mummy is found to have ratio $4.9(10^{-13})$ of carbon-14 to carbon-12. How old is the mummy? (See Example 7.)

30. Scientists can estimate the age of the earth by comparing the relative amounts of two kinds of uranium: ^{235}U and ^{238}U. ^{235}U has a half-life of $7.1(10^8)$ years, while the half-life of ^{238}U is $4.5(10^9)$ years. There is now 140 times as much ^{238}U in the earth as there is ^{235}U. If it is determined that ^{238}U was 3.2 times as plentiful as ^{235}U when the earth was formed, what is the age of the earth?

31. The charge on a certain capacitor has a half-life of 10 minutes. The charge now ($t = 0$) is 50 coulombs.
    a. When will the charge be 40 coulombs?
    b. When was the charge 60 coulombs?
    c. What will the charge be 2 minutes from now?
    d. What was the charge 5 minutes ago?

32. Water flows into a tank in such a way that its volume triples every hour: $W(t) = W_0 3^t$. The tank now ($t = 0$) contains 10 gallons.
    a. When will it contain 15 gallons?
    b. When did the tank contain only 5 gallons?
    c. How much will it contain 15 minutes from now?
    d. How long does it take for the volume to double?

33. The Goodyear blimp has a leak and 5% of its gas is escaping each minute: $V(t) = V_0(0.95)^t$. At present ($t = 0$), the blimp holds 500,000 cubic feet of gas.
    a. When will 250,000 cubic feet of gas remain in the blimp?
    b. When will 50,000 cubic feet of gas remain?
    c. When were 700,000 cubic feet present?
    d. How much was present $2\frac{1}{4}$ minutes ago?

34. The rate at which crude oil flows from a tank depends on the amount of fuel remaining in the tank; $\frac{1}{5}$ of the fuel flows out every 3 hours: $V(t) = V_0(0.80)^{t/3}$. Suppose 50,000 gallons of the oil are in the tank now ($t = 0$).
    a. How much will remain in the tank after 1 hour?
    b. After 5 hours?
    c. After 10 hours?
    d. How long will it take to reduce the oil to $\frac{1}{2}$ its original amount?

35. The temperature in a blast furnace is 2000°C. Cooling chambers are designed to permit molten metal to lose $\frac{1}{4}$ of its temperature (in °C) in 20 minutes: $T(t) = 2000(0.75)^{t/20}$.
   a. What will the temperature be after 45 minutes in the cooling chamber?
   b. After 2 hours?
   c. How long does it take to reduce the temperature to 30°C?

36. The value of a certain piece of machinery declines (depreciates) 20% each year; $V(t) = V_0(0.80)^t$. Suppose it is now ($t = 0$) worth $20,000.
   a. What will it be worth 5 years from now?
   b. 15 years from now?
   c. How long does it take to depreciate to a value of $1000?

37. The gross national product (GNP) of the United States has recently been growing at a rate so that it triples each 15 years. $P(t) = P_0 3^{t/15}$. The GNP was approximately $2.5 trillion in 1980.
   a. What is it expected to be in 1990?
   b. In 2000?
   c. What was the GNP in 1970?
   d. In 1960?
   e. How long does it take the GNP to double?

38. A municipal government finds that each time it raises taxes by 10%, 5% of its residents will move to the suburbs. If it raises taxes 10% each year for 10 years, how much will its population decrease in this time?

39. When a cold front moved in, Mr. Waters observed that the temperature (in °C) of the water in his pool decreased by 5% in 15 minutes. Assume that the temperature continues to drop at this rate and that the temperature is now 30°C.
   a. Write a formula for the temperature t minutes from now.
   b. When will the water freeze (0°C)?

40. To see that $(1 + r/N)^N \approx e^r$ for large N, write

$$\left(1 + \frac{r}{N}\right)^N = \left(1 + \frac{1}{N/r}\right)^N$$
$$= \left[\left(1 + \frac{1}{N/r}\right)^{N/r}\right]^r$$
$$= \left[\left(1 + \frac{1}{x}\right)^x\right]^r$$

for $x = N/r$. As N gets larger, x gets larger. Now use a calculator to evaluate $(1 + 1/x)^x$ for $x = 10^n$, $n = 1, 2, 3, 4, 5, 6$.

Section 6.6 Chapter Review

Terms and Concepts

Exponential Function $f(x) = b^x$ $(0 < b \ne 1)$

Natural Exponential Base $e = 2.71828 \cdots$

Logarithms $u = \log_b v$ if and only if $v = b^u$
    ($v > 0$ and $0 < b \ne 1$)

Logarithmic Function $f(x) = \log_b x$, $x > 0$ $(0 < b \ne 1)$

Rules and Formulas   ($0 < a \ne 1, 0 < b \ne 1, u > 0, v > 0$)

Exponents

1. $b^x b^y = b^{x+y}$ $2^2 2^3 = 2^5$

2. $\dfrac{b^x}{b^y} = b^{x-y}$ $\dfrac{2^5}{2^2} = 2^3$

3. $(b^x)^y = b^{xy}$ $(2^2)^3 = 2^6$

4. $b^0 = 1$ $2^0 = 1$

5. $b^1 = b$ $2^1 = 2$

6. $\dfrac{1}{b^x} = b^{-x}$ $\dfrac{1}{2^2} = 2^{-2}$

7. $(ab)^x = a^x b^x$ $(2 \cdot 3)^2 = 2^2 \cdot 3^2$

8. $x = \log_b(b^x)$ $3 = \log_2 2^3$

9. If $b^r = b^s$, then $r = s$.   If $2^x = 2^3$, then $x = 3$.

Logarithms

1. $\log_b(uv) = \log_b u + \log_b v$ $\log_2(2 \cdot 4) = \log_2 2 + \log_2 4 = 1 + 2 = 3$

2. $\log_b\left(\dfrac{u}{v}\right) = \log_b u - \log_b v$ $\log_2\left(\dfrac{4}{2}\right) = \log_2 4 - \log_2 2 = 2 - 1 = 1$

3. $\log_b(u^r) = r \log_b u$ $\log_2(4^2) = 2 \log_2 4 = 2 \cdot 2 = 4$

4. $\log_b 1 = 0$ $\log_2 1 = 0$

5. $\log_b b = 1$ $\log_2 2 = 1$

6. $\log_b\left(\dfrac{1}{u}\right) = -\log_b u$ $\log_2\left(\dfrac{1}{4}\right) = -\log_2 4 = -2$

7. $\log_b u = \dfrac{\log_a u}{\log_a b}$ $\log_4 16 = \dfrac{\log_2 16}{\log_2 4} = \dfrac{4}{2} = 2$

8. $u = b^{\log_b u}$ $3 = 2^{\log_2 3}$

9. If $\log_b u = \log_b v$, then $u = v$.   If $\log_2 x = \log_2 8$, then $x = 8$.

Exponential Growth $Q(t) = Q_0 e^{kt}, \quad k > 0$

Exponential Decay $Q(t) = Q_0 e^{-kt}, \quad k > 0$

Half-life Decay $Q(t) = Q_0 2^{-t/H}, \quad (H = \text{half-life})$

Compound Interest

1. Compounding N times per year $P(t) = P_0\left(1 + \dfrac{r}{N}\right)^{Nt}$

2. Continuous compounding $P(t) = P_0 e^{rt}$

Graphing Techniques

Exponential Functions

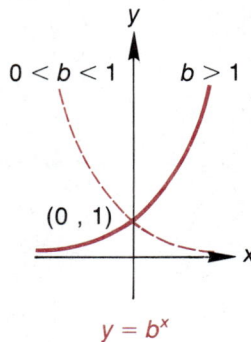

$y = b^x$

Logarithmic Functions

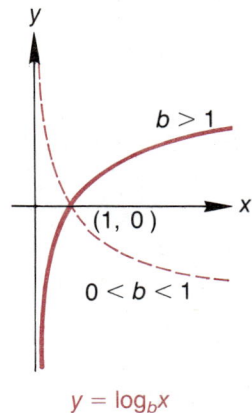

$y = \log_b x$

Section 6.7 Supplementary Exercises

Write the equivalent logarithmic form of each of the following.

1. $5^3 = 125$
2. $4^{-3} = \dfrac{1}{64}$
3. $16^0 = 1$
4. $3^4 = 81$

Write the equivalent exponential form of each of the following.

5. $\log_{10} 100 = 2$
6. $\log_{64} 4 = \dfrac{1}{3}$
7. $\log_{1/2} 8 = -3$
8. $\log_{1/3} 27 = -3$

Evaluate each of the following without using a calculator.

9. $7^{\log_7 132}$
10. $\ln(e^{\sqrt{4}})$
11. $49^{\log_7 5}$
12. $2^{\log_4 3}$
13. $\log_2 4^{-3}$
14. $\log_9 3^3$
15. $\log_9 81$
16. $\log_{16} 256$
17. $\log_{27} 3$
18. $\log_{16} \dfrac{1}{2}$
19. $\log_3 (\sqrt{3})^7$
20. $\log_7 \dfrac{1}{(\sqrt[4]{7})^5}$
21. $3e^{\ln 6}$
22. $5 \ln(e^{-3})$

Use the properties of exponents and logarithms to express each of the following as a single logarithm.

23. $2 \log_6 9 - 3 \log_6 \left(\dfrac{1}{3}\right) + 2 \log_6 \sqrt{12}$
24. $\log_2(2t - 4) - \log_2(t - 2)$
25. $\left(\log_b 27 + 3 \log_b \dfrac{1}{3}\right)(\log_{126} 4729 - \log_{372} 628)$
26. $(\log_b 16 - \log_b 2)\left(\log_b \dfrac{1}{4} + 2 \log_b 2\right)$

Use the logarithm properties and the table given in Example 1 of Section 6.3 to find each of the following.

27. $\log_b 10$
28. $\log_b \dfrac{49}{8}$
29. $\log_b (10^6)(14^5)$
30. $\log_b \sqrt[3]{4}$
31. $\log_b \sqrt{\dfrac{5}{48}}$
32. $\log_b \sqrt[5]{63}$

Use common logarithms or natural logarithms and the change-of-base formula L7 to find the following logarithms. You will also need a calculator or Tables 1 or 2.

33. $\log_5 \dfrac{2}{3}$
34. $\log_2 \dfrac{63}{\sqrt{250}}$
35. $\log_9 25$
36. $\log_{\sqrt{3}} 6$
37. $\log_6 21$
38. $\log_{21} 42$

Sketch the graphs of the following functions: $f(x) =$

39. 5^x
40. $\left(\dfrac{1}{5}\right)^x$
41. 5^{-x}
42. 5^{3+x}
43. 5^{2-x}
44. $1 - 5^x$
45. $5^{|x|}$
46. $5^{|x-2|}$
47. $1 - e^x$
48. $(-4) \cdot 5^{2-x}$

49. $\log_5 x$
50. $\log_{1/5} x$
51. $\log_2(x + 3)$
52. $\log_2|x + 3|$
53. $\log_2(x + 3)^2$
54. $\log_2 \sqrt{x + 3}$
55. $\log_2\left(\dfrac{1}{x + 3}\right)$
56. $|\log x|$
57. $|\log_2(-x)|$
58. $|\log_2|x||$

Solve the following equations.

59. $7^{s-1} = 343$
60. $3^{2s-4} = (81)^2$
61. $8^z = 16$
62. $\left(\dfrac{1}{4}\right)^y = 64$
63. $3^s = \dfrac{1}{243}$
64. $100^{2y+4} = 10^{3y}$
65. $9^{1-2w} = 27^{-2/3}$
66. $5^{4-r} = \dfrac{1}{25}$
67. $16^{5x/4} = 32$
68. $9^{7y/2} = \dfrac{1}{243}$
69. $(125)^{3u} = 5^{(3u)^2}$
70. $(x - 4)^{2/3} = 16$
71. $(r^2 - 14r)^{2/5} = 4$
72. $(u^2 + 6u + 24)^{3/4} = 8$
73. $\log_2 w = -4$
74. $\log_4 z = 3$
75. $\log_{27} r = \dfrac{4}{3}$
76. $\log_v 25 = \dfrac{1}{2}$
77. $\log_u\left(\dfrac{1}{64}\right) = -3$
78. $\log_y 343 = \dfrac{3}{2}$
79. $\log_2 4t^3 = 5$
80. $\log_3(2 - 5w) = 3$
81. $\log_6(2x + 6) = 2$
82. $3 \ln 2 - \ln 4 + 2 \ln x = 3 \ln 3 + \ln 2 - \ln 3$
83. $\log_2(2x - 4) + \log_2(x - 2) = 1$
84. $\log_{\sqrt{3}}(z + 1) - \log_{\sqrt{3}}(z - 1) = 2$
85. $\log_3(2z) - \log_3(z + 4) = \log_3(z - 4) - \log_3(z - 5)$
86. $\log_b(x + 2) + \log_b 3 = \log_b(x - 2)$
87. $\log_b(u - 4) + \log_b 2 = \log_b(u + 6)$
88. $\log_b(v + 1) + \log_b(v - 1) = \log_b 3$
89. $3^{3-t} = 2^{2-t}$
90. $5^{2x+1} = 4^{x+2}$
91. $2^{1-y} = 7^{1+y}$
92. $15^{u-1} = 12^{u+2}$

93. The half-life of the charge on a certain capacitor is 10 minutes. The charge is now 40 coulombs.

   a. What will the charge be 7 minutes from now?
   b. What was the charge 3 minutes ago?
   c. When will the charge be 10 coulombs?
   d. When was the charge 60 coulombs?

94. Water flows into a tank in such a way that its volume triples every hour: $W(t) = W_0 3^t$. The tank now contains 25 gallons.

   a. How much will it contain 4 hours, 15 minutes from now?
   b. How much did it contain 2 hours, 10 minutes ago?
   c. When will it contain 500 gallons?
   d. When did it contain 1 gallon?

95. Five years ago Mary invested some money at 7% interest compounded continuously.

   a. If her investment is now worth $32,000, how much did she invest initially?
   b. How much will her investment be worth 10 years from now?

In each of the following problems, let the principal P be invested at the annual interest rate I for n years. For each, find the resulting principal assuming

a. No compounding
b. Yearly compounding
c. Semiannual compounding
d. Quarterly compounding
e. Daily compounding
f. Continuous compounding

96. $P = \$100$, $I = 5\%$, $n = 100$ years
97. $P = \$1000$, $I = 6\%$, $n = 10$ years
98. $P = \$2000$, $I = 8\%$, $n = 5$ years
99. $P = \$4500$, $I = 6\tfrac{1}{4}\%$, $n = 5$ years
100. $P = \$5000$, $I = 7\tfrac{1}{2}\%$, $n = 5$ years

Optional Table Exercises.
Use Table 2 (at the end of this book) to find the logarithms of the following numbers

101. a. 26.4 b. 2640 c. 0.0264
102. a. $\sqrt{798}$ b. $\sqrt[3]{0.00798}$ c. $\sqrt[3]{0.0798}$
103. a. $(0.0631)^{-2}$ b. $(63.1)^{1/2}$ c. 6310^2

104. **a.** $(2060)^3$ **b.** $\dfrac{1}{(2060^{30})}$ **c.** $(2060)^{1/3}$

105. **a.** $(59600)^{10}$ **b.** $(0.00596)^{10}$ **c.** 596^{10}

Use Table 2 (at the end of this book) to find x for log x as given in each of the exercises that follow.

106. **a.** 7.6656 **b.** 3.6656 **c.** 2.6656
107. **a.** 6.3345 **b.** −3.6655 **c.** −2.6655
108. **a.** −1.1831 **b.** −0.1831 **c.** −3.1831
109. **a.** 6.9827 **b.** 3.9827 **c.** −2.0173
110. **a.** 2.9926 **b.** 0.9926 − 2 **c.** −2.0074

Approximate the logarithms of the following by linear interpolation in Table 2 (at the end of this book).

111. 5.064   112. 964.2   113. 0.01588
114. 1,001,000   115. 0.00003998

Interpolate in Table 2 to find x if log x is given here.

116. 0.9610   117. 7.9515   118. −3.1541
119. 3.7555   120. −2.5227

Use logs to calculate the following expressions.

121. $(0.0675)^{4.5}(423)^{3.7}$ 122. $\dfrac{(31.08)^2 \sqrt[3]{0.7926}}{(68.37)^3}$

The Trigonometric Functions

7

The word *trigonometry* means triangle (*trigon-*) measurement (*-metry*). Trigonometry deals with the measurement and relationships of the angles and sides of a triangle; it was developed to facilitate the computations required for surveying and navigation. Even today, surveying is essentially basic trigonometry.

The various ratios among the sides of a right triangle are given the names *sine, cosine, tangent, cotangent, secant,* and *cosecant*. These ratios are called the **trigonometric functions.** In addition to describing triangular relations, the trigonometric functions can also be used to describe certain phenomena (such as alternating electrical current and the oscillatory motion of a moving piston in an automobile engine) that are time-dependent and repetitive. Not only do sophisticated radar systems and automatic pilots use trigonometry to locate and direct a plane or ship, but the electronic systems within these units depend on properties of the trigonometric functions for their very development.

In many modern applications of the trigonometric functions, the connection between these functions and triangles is not evident. Since the trigonometric functions also have a close relationship to the circle and its properties, mathematicians today refer to the historical trigonometric functions as the **circular functions.** This terminology is used to emphasize the numerous and varied applications of these functions by downplaying their relation to triangles.

Section 7.1

Angles

Simply speaking, an **angle** consists of two **rays** or half-lines that are affixed to a common **endpoint.** Angles appear on everyday objects, such as on the face of a clock (the hands), at street corners and highway intersections, and on bridges and buildings (Figure 7.1).

Figure 7.1
ANGLES IN EVERYDAY OBJECTS

An angle is formed by rotating a ray or half-line $\overrightarrow{OA}$ about its endpoint O to some new position $\overrightarrow{OB}$. The resulting configuration is called $\angle AOB$ (read "angle AOB"); see Figure 7.2a. The point O is called the **vertex** of this angle; $\overrightarrow{OA}$ is called its **initial side;** and $\overrightarrow{OB}$ is called its **terminal side.** The same configuration is generated by rotating $\overrightarrow{OB}$ to $\overrightarrow{OA}$. This angle is illustrated in exactly the same way except that the rotation is in the opposite direction as shown in Figure 7.2b. In this case, $\overrightarrow{OB}$ is called the initial side and $\overrightarrow{OA}$ the terminal side of $\angle BOA$.

Figure 7.2

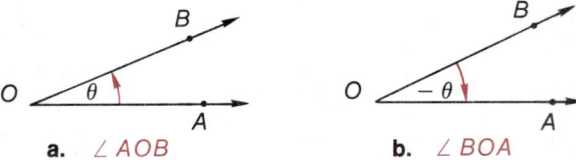

a. $\angle AOB$ b. $\angle BOA$

An angle formed by a counterclockwise rotation is called a **positive angle;** a **negative angle** is formed by a clockwise rotation. In Figure 7.2, $\angle AOB$ is a positive angle and $\angle BOA$ is a negative angle.

For simplicity, we frequently use Greek letters when referring to angles. For example, in Figure 7.2 we might refer to $\angle AOB$ simply as θ; $\angle BOA$, having the same size but opposite orientation, is then $-\theta$. In situations in which we are more concerned with the geometric configuration represented by an angle than with the direction of rotation used to generate the angle, we may refer to an angle by simply naming its vertex. For example, in a triangle, $\triangle ABC$, we may refer to the angles at vertices A, B, and C as $\angle A$, $\angle B$, and $\angle C$, respectively, as shown in Figure 7.3.

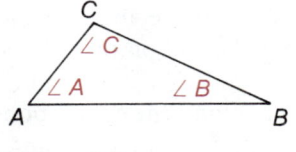

Figure 7.3

Angle Measurement

There are two important units used in measuring angles: **degrees** and **radians.** Each is based on preassigning a number to the angle formed by one complete counterclockwise revolution. Partial or multiple revolutions are then assigned a number in direct proportion to that part of a complete revolution. Although it is possible to assign any number to an angle formed by a complete revolution, in practice only two are used: 360 and 2π. The symbol π denotes the irrational number $\pi = 3.14159\ldots$; it is the ratio of the circumference of a circle to its diameter.

An angle generated by one complete counterclockwise revolution is said to measure 360 **degrees** (°); thus a half-revolution or **straight angle** measures 180° and a quarter revolution or **right angle** measures 90°. An angle of 1° represents $\frac{1}{360}$ of a complete revolution. Each degree is subdivided into 60 equal parts called **minutes** ('); each minute is, in turn, subdivided into 60 equal parts called **seconds** ("). Since there are 60 minutes in a degree, one-half a degree must equal 30 minutes; that is, $29\frac{1}{2}°$ equal $29°30'$. One-fourth of a degree is 15 minutes, two-thirds of a minute is 40 seconds.

On the other hand, the **radian measure** of one complete counterclockwise revolution is 2π; a straight angle or half a counterclockwise revolution then has radian measure π, and a right angle or positive quarter-revolution has radian measure $\pi/2$. The radian is not subdivided into named smaller units (see Figure 7.4).

Surveyors, navigators, and others who are primarily interested in triangle relationships use the degree measure of an angle. Scientists and engineers whose principal concern is with periodic or oscillatory phenomena generally use radian measure. Radian measure lends itself naturally to the study of time-dependent situations.

An angle having its vertex at the center of a circle is called a **central angle** of that circle. Figure 7.5 illustrates on a circle of radius 1 the degree and radian

Section 7.1 Angles   363

Figure 7.4
RADIAN AND DEGREE MEASURE

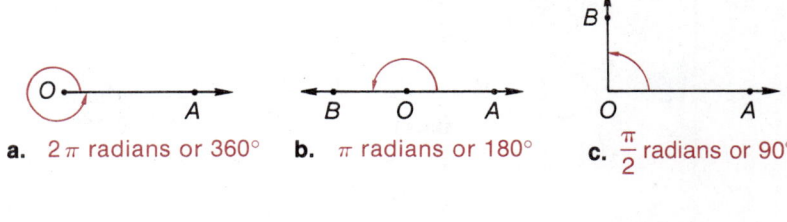

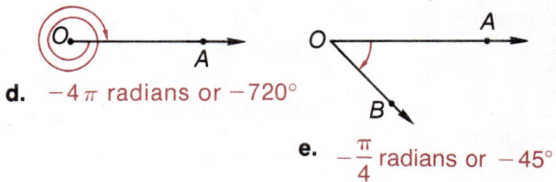

measures of a number of central angles generated by less than one counterclockwise revolution. The degree measure is indicated inside the circle; the radian measure is noted outside the circle.

Note that on a circle of radius 1, the radian measure of a central angle equals the length of its corresponding arc. The circumference of a circle of radius r is $C = 2\pi r$. Thus, for radius $r = 1$, 2π is the circumference of the circle and also the angular measure of a complete revolution; π is the radian measure of one-half a complete revolution and also one-half the total circumference; $2\pi/3$ represents one third a complete revolution and also one-third the total circumference; and so on.

Figure 7.5
DEGREES, RADIANS, AND ARC LENGTH ON A CIRCLE OF RADIUS 1

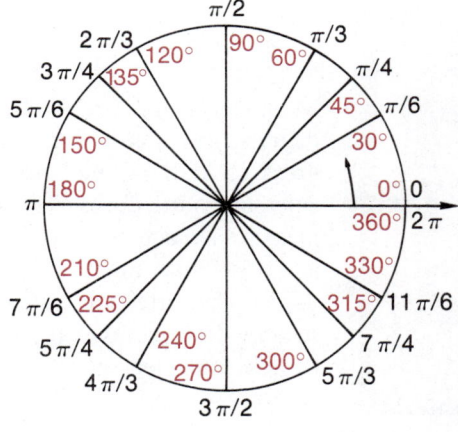

Degree-Radian Conversion

Since $1°$ represents $\frac{1}{360}$ of a complete revolution and a complete revolution is 2π radians, we have

$$1° = \frac{1}{360} \cdot 2\pi$$

$$= \frac{\pi}{180} \text{ radians}$$

Conversely, multiplying both sides of this equation by $180/\pi$, we obtain

$$\left(\frac{180}{\pi}\right)° = 1 \text{ radian}$$

DEGREE–RADIAN CONVERSION FORMULAS

$$1° = \frac{\pi}{180} \text{ radians} \qquad 1 \text{ radian} = \left(\frac{180}{\pi}\right)°$$

$$\approx 0.017 \text{ radian} \qquad \approx 57.3°$$

EXAMPLE 1

a. Convert 60° to radian measure.

b. Convert $\frac{\pi}{6}$ radians to degree measure.

SOLUTION

a. In Figure 7.5 we see that 60° = π/3 radians. This can be verified by the conversion formula since

$$60° = 60 \cdot (1°)$$
$$= 60 \cdot \left(\frac{\pi}{180} \text{ radians}\right) = \frac{\pi}{3} \text{ radians}$$

b. In a similar fashion, we see that

$$\frac{\pi}{6} \text{ radians} = \frac{\pi}{6} \cdot (1 \text{ radian}) = \frac{\pi}{6} \cdot \left(\frac{180}{\pi}\right)° = 30°$$

in agreement with the figure. You might also check these conversions on a hand calculator. ∎

CALCULATOR COMMENTS

Most calculators that include the trigonometric functions also have a built-in degree–radian conversion. Simply pushing one or more buttons converts degrees to radians and vice versa. One must be careful, however, for most calculators use decimal notation to represent an angle, whether that angle be measured in degrees or radians. Thus a calculator will ordinarily represent 7°45′ as 7.75°. Since most navigators and surveyors describe an angle in terms of degrees, minutes, and seconds, it is up to the user to convert minutes and seconds to decimal degrees and vice versa. On some calculators, this, too, can be accomplished simply by pushing a few keys.

EXAMPLE 2

Convert $\frac{5}{2}$ radians to degrees, minutes, and seconds.

SOLUTION

From the degree–radian conversion formulas, we have

$$\frac{5}{2} \text{ radians} = \frac{5}{2} \cdot (1 \text{ radian})$$
$$= \frac{5}{2} \cdot \left(\frac{180}{\pi}\right)°$$
$$= \frac{450°}{\pi}$$
$$\approx 143.24°$$

Now, each degree is 60 minutes. Thus

$$0.24° = (0.24) \cdot 1°$$
$$= (0.24) \cdot 60 \text{ minutes}$$
$$= 14.4 \text{ minutes}$$

Hence $\frac{5}{2}$ radians $\approx 143°14.4'$. But each minute equals 60 seconds. Hence

$$0.4' = (0.4) \cdot 1'$$
$$= (0.4) \cdot 60 \text{ seconds}$$
$$= 24 \text{ seconds}$$

Finally, $\frac{5}{2}$ radians $\approx 143°14'24''$.

EXAMPLE 3

Convert $67°32'16''$ to radians.

SOLUTION

Begin by converting $16''$ to minutes. One second is $\frac{1}{60}$ of a minute. Then

$$16'' = 16 \cdot (1 \text{ second})$$
$$= 16 \cdot \left(\frac{1}{60} \text{ minute}\right) = \frac{4}{15} \text{ minutes}$$
$$32'16'' = 32\frac{4}{15} \text{ minutes} = \frac{484}{15} \text{ minutes}$$

Now, each minute is $\frac{1}{60}$ of a degree and

$$32'16'' = \frac{484}{15} \text{ minutes} = \frac{484}{15} \cdot (1 \text{ minute})$$
$$= \frac{484}{15} \cdot \left(\frac{1}{60} \text{ degree}\right)$$
$$= \frac{121}{225} \text{ degrees}$$
$$\approx 0.54°$$

Finally,

$$67°32'16'' \approx 67.54° = (67.54) \cdot 1°$$
$$= (67.54) \cdot \frac{\pi}{180} \text{ radians}$$
$$\approx 1.18 \text{ radians}$$

EXAMPLE 4

How soon after 4 o'clock will the minute and hour hands coincide?

SOLUTION

Let θ denote the angle through which the minute hand travels until it meets the hour hand. Since the hour hand travels $\frac{1}{12}$ as fast, it moves through an angle of $\theta/12$. In Figure 7.6, we see that

$$\theta = 120 + \frac{\theta}{12} \text{ degrees}$$

Historical Perspective

In *The Exact Sciences of Antiquity*, Otto Nuegebauer suggests that the **degree** might be related to a large Babylonian unit of distance of about 7 miles. The time needed to travel this distance—about 2 hours—became for them a unit of time. One day corresponds to a complete revolution of the sky and takes 12 of these time periods. On the other hand, for smaller measurements, the Babylonians divided the large distance into 30 equal parts. Thus 1 day—or a complete revolution of the sky—consisted of $12 \cdot 30 = 360$ of the smaller units. This explanation relates the units of time, distance, and angle. As further evidence of this relationship between time and angles, we note that, just as with the hour, a degree is further subdivided into 60 minutes, and each minute is subdivided into 60 seconds.

Radian measure derives from the relationship between the circumference and radius of a circle. Apparently the ancients already knew or assumed that the ratio of circumference to diameter was the same for every circle. That ratio is given the name π: $C = \pi d = 2\pi r$. The radian measure of an angle is thus the length of the arc subtended by the angle on a circle of radius $r = 1$ centered at the vertex of the angle.

Any work done with the radian measure of an angle necessarily presupposes a good estimate of the value of π. Since π is an irrational number, we can at best approximate it in decimal notation. The Egyptians and Babylonians generally used $\pi \approx 3$; that is, they estimated the circumference of a circle as three times its diameter. There is evidence to suggest that they knew better, however. An old Babylonian tablet discovered in 1936 indicates that the Babylonians knew that $3\frac{1}{8} = 3.125$ was better than 3 as an estimate for π. The Rhind Papyrus indicates that the Egyptians sometimes used $4(\frac{8}{9})^2 = (\frac{4}{3})^4 \approx 3.1605$ as a better estimate for π.

Perhaps the ancients were compelled by their rulers to use $\pi \approx 3$ as the residents of Indiana were almost compelled to do in 1897. In that year, the Indiana House of Representatives, with the support of the State Superintendent of Instruction, passed a bill (H.B. No. 246) legislating $\pi = 3$. However, it generated so much publicity and ridicule in the newspapers that it never passed the Senate.

The Greeks had a very good estimate of π. By inscribing a regular polygon of 96 sides inside a circle and computing its circumference by geometric methods, Archimedes was able to show that

$$3\frac{10}{71} < \pi < 3\frac{1}{7}$$

The rational number $3\frac{1}{7} = \frac{22}{7}$ is the estimate of π most of us are familiar with from our elementary school days. Actually, Archimedes' technique can be used to find π to within any desired degree of accuracy. To nine decimal places, we have $\pi = 3.141592654\ldots$.

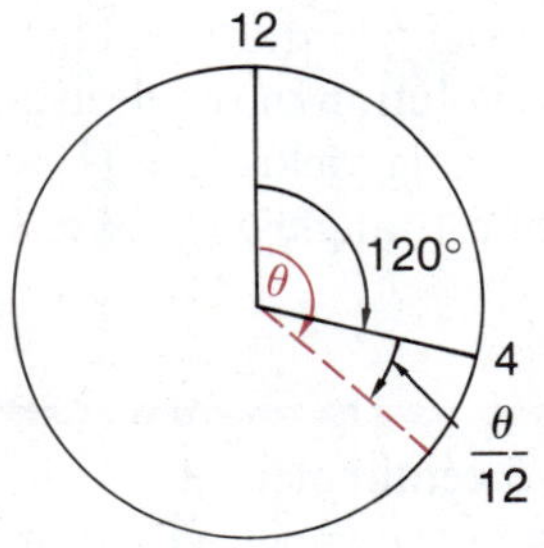

Solving for θ, we obtain

$$\frac{11}{12}\theta = 120$$

$$\theta = \frac{12 \cdot 120}{11} \text{ degrees}$$

Since 60 minutes of time represents a complete revolution (360°) of the minute hand, each minute of time represents 360°/60 = 6° in the angle θ. Letting *m* denote the time that has elapsed (in minutes), we have θ = 6*m* or

$$m = \frac{\theta}{6} = \frac{\overset{2}{\cancel{12}} \cdot 120}{\cancel{6} \cdot 11} = \frac{240}{11}$$

$$\approx 21.82 \text{ minutes}$$

$$\approx 21 \text{ minutes } 49 \text{ seconds}$$

The hands coincide at about 4:21:49. ∎

Figure 7.6

Sectors: Arc Length and Area

In the preceding example the size of the central angle θ determined the time that had elapsed. In a similar fashion the size of a central angle θ determines arc length and the area of a sector. In Figure 7.7, the ratio of the arc length *s* to the entire circumference $2\pi r$ should be the same as the ratio of the angle θ to the complete revolution 2π. Similarly, the ratio of the area *A* to the entire area πr^2 of the circle should also equal the ratio of the angle θ to the complete revolution 2π. That is,

$$\frac{s}{2\pi r} = \frac{\theta}{2\pi} \quad \text{and} \quad \frac{A}{\pi r^2} = \frac{\theta}{2\pi}$$

From these we obtain

$$s = 2\pi r \cdot \frac{\theta}{2\pi} = r\theta \qquad A = \pi r^2 \cdot \frac{\theta}{2\pi} = \frac{1}{2}r^2\theta$$

The first of these can be written as

$$\theta = \frac{s}{r}$$

That is, the radian measure of a central angle θ equals the length of the arc it subtends divided by the radius of the circle.

ARC LENGTH AND AREA OF A SECTOR

Let a central angle θ (*measured in radians*) subtend an arc of length *s* and an area *A* in a circle of radius *r*. Then

$$\theta = \frac{s}{r}$$

$$s = r\theta$$

$$A = \frac{1}{2}r^2\theta$$

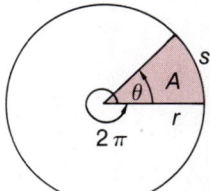

Figure 7.7
ARC LENGTH, AREA, AND RADIAN MEASURE

The angle θ must be measured in radians for these arc length and area computations since 2π was used as the measure of a complete revolution in the derivation of these formulas. Substituting $s = r\theta$ into the area formula yields $A = (\frac{1}{2})rs$; note the resemblance of this formula to that for the area of a triangle with base b and height h: $A = (\frac{1}{2})bh$.

EXAMPLE 5

a. For a central angle of $\pi/4$ radians in a circle of radius 5 centimeters, determine the length of the subtended arc and the area of the enclosed sector.

b. Repeat part a for a central angle of 30°.

SOLUTION

We use the formulas for arc length and area with $r = 5$ and $\theta = \pi/4$.

a. $s = r\theta$

$$= 5 \cdot \frac{\pi}{4}$$

$$\approx 3.93 \text{ centimeters}$$

and

$$A = \frac{1}{2}r^2\theta$$

$$= \frac{1}{2} \cdot 5^2 \cdot \frac{\pi}{4}$$

$$= \frac{25\pi}{8}$$

$$\approx 9.82 \text{ square centimeters}$$

b. An angle whose measure is given in degrees must be converted to radian measure before using the arc length and area formulas. Since $30° = \pi/6$ radians, we have

$s = r\theta$

$$= 5 \cdot \frac{\pi}{6}$$

$$\approx 2.62 \text{ centimeters}$$

and

$$A = \frac{1}{2}r^2\theta$$

$$= \frac{1}{2} \cdot 5^2 \cdot \frac{\pi}{6}$$

$$= \frac{25\pi}{12}$$

$$\approx 6.54 \text{ square centimeters}$$

EXAMPLE 6

As part of an advertising campaign, a painter is instructed to paint the exterior wall of a manufacturing plant. A circle is chosen to depict the company's niche in the marketplace. After appropriate consideration, the painter decides to use a circle having a radius of 6 meters. On this circle, the company's market share is represented by a sector having an arc length of 9 meters. To assist the painter in purchasing paint, find the area of this sector. Find the degree measure of its central angle.

SOLUTION   From the formula $\theta = s/r$ we obtain

$$\theta = \frac{9}{6} = \frac{3}{2}$$

as the *radian measure* of the central angle θ. The area of the enclosed sector is

$$A = \frac{1}{2}r^2\theta$$
$$= \frac{1}{2} \cdot 6^2 \cdot \frac{3}{2} \qquad (\theta = \frac{3}{2} \text{ radians})$$
$$= \frac{36 \cdot 3}{4} = 27 \text{ square meters}$$

Converting $\theta = \frac{3}{2}$ radians to degrees, we have

$$\theta = \frac{3}{2} \text{ (1 radian)}.$$
$$= \frac{3}{2} \cdot \left(\frac{180}{\pi}\right)^\circ = \left(\frac{270}{\pi}\right)^\circ \approx 85°56'37''$$

Section 7.1 Exercises

Sketch the angle with the given measure. If the measure is given in degrees, convert to radians and vice versa.

1. 75°
2. −15°
3. 105°
4. −120°
5. 240°
6. 285°
7. −255°
8. −487.5°
9. −540°
10. 585°
11. −780°
12. −735°
13. 810°
14. 1102.5°
15. 1350°
16. $-\frac{\pi}{6}$
17. 7π
18. $\frac{7\pi}{12}$
19. $-\frac{7\pi}{2}$
20. $-\frac{9\pi}{2}$
21. $\frac{8\pi}{3}$
22. $-\frac{15\pi}{4}$
23. $\frac{17\pi}{6}$
24. $-\frac{271\pi}{2}$
25. $-\frac{22\pi}{3}$
26. $\frac{29\pi}{6}$
27. $\frac{29\pi}{12}$
28. $-\frac{61\pi}{3}$
29. $-\frac{23\pi}{4}$
30. $\frac{67\pi}{4}$

Calculator Exercises (31–50)

Convert the given radian measure to degrees, minutes, and seconds.

31. 2
32. 5
33. 2.3
34. 6.2
35. 5.3
36. −4.7
37. −3.28
38. 1.54
39. 1.98
40. −5.42

Convert the given degree measure to radians correct to three decimal places.

41. 23°14'5''
42. 137°57'54''
43. −210°45'20''
44. 106°15'50''
45. 300°3'30''
46. −183°32'15''

47. $-78°27'45''$
48. $350°28'40''$
49. $250°42'5''$
50. $-155°15'10''$

For the central angle θ in a circle of radius r, determine the length of the subtended arc and the area of the enclosed sector in each of the following.

51. $r = 2$ centimeters, $\theta = \dfrac{5\pi}{12}$ radians
52. $r = 3$ inches, $\theta = \dfrac{2\pi}{3}$ radians
53. $r = 5$ meters, $\theta = \dfrac{3\pi}{4}$ radians
54. $r = 5$ feet, $\theta = \dfrac{19\pi}{12}$ radians
55. $r = 4$ centimeters, $\theta = 60°$
56. $r = 6$ inches, $\theta = 225°$
57. $r = 9$ meters, $\theta = 105°$
58. $r = 10$ yards, $\theta = 210°$

Two of the five quantities—radius, arc length, area of sector, radian measure of θ, degree measure of θ—are given for a central angle θ in a circle of radius r. From these two quantities, determine the other three in each of the following.

59. $s = 10$ centimeters, $r = 5$ centimeters
60. $s = 3$ feet, $A = 12$ square feet
61. $s = 6$ inches, $\theta = 2$ radians
62. $s = 5$ meters, $\theta = 120°$
63. $r = 5$ kilometers, $A = 50$ square kilometers
64. $\theta = \dfrac{7\pi}{4}$ radians, $A = 14\pi$ square centimeters
65. $\theta = 210°$, $A = \dfrac{21\pi}{4}$ square meters

66. Captain Ahab sights a whale at an angle of 15° to the right of his direction of travel. He also sights another ship at an angle of 60° to the right. What is the angle he sights between the ship and the whale?

67. "High noon" is the time when the sun is directly overhead.
   a. If sundown occurs at 9 P.M., what time is it when the sun makes an angle of 45° with the vertical?
   b. What time is it when the sun is 30° above the horizon?

68. A railroad roundhouse is being built to service five tracks. If the tracks are to be evenly distributed around a circle, what is the angle between any two adjacent tracks? Sketch the layout.

69. How soon after 7 o'clock does the minute hand coincide with the hour hand?

70. How soon after 9:30 will the minute hand overtake the hour hand?

71. How soon after 8 o'clock will the minute and hour hands form angles of the same size with the vertical?

72. An electromagnet is to be made by wrapping copper wire around an iron core. If 200 turns around the core are required and the diameter of the core is 2 centimeters, how much wire is needed? *Disregard the thickness of the wire.*

73. Sixty yards of thread are wrapped around a spool having a diameter of 1 inch.
   a. How many times does the thread wrap around the spool?
   b. What are the relative positions of the two ends of the thread?

   Again, disregard the thickness of the thread.

74. An expedition discovers a very tall tree. From a certain vantage point, the line of sight to the top of the tree makes an angle of 53°20' with the horizontal. Express this angle in radians.

75. A tunnel runs through Mount Insurmountable. From a neighboring peak, the angle between the lines of sight to the ends of this tunnel is measured as 1.13 radians. Convert this angle to degrees, minutes, and seconds.

76. A racetrack has the shape of a rectangle with a semicircle at each end. These semicircles have a radius of 20 meters. At each end of the track, a sector of the semicircle is to be planted with marigolds. If the length of the fence (arc) bounding each sector from the pavement is 25 meters, what is the total area to be planted in marigolds?

77. A manufacturing plant uses a belt to drive some of its machinery. Find the length of a belt connecting four pulleys as shown in the figure.

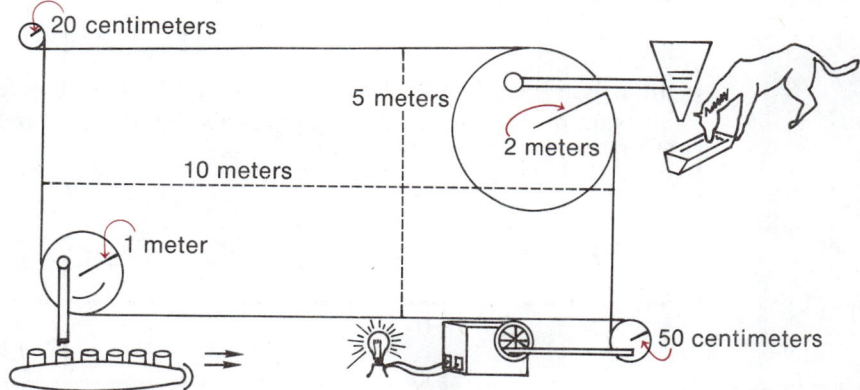

78. Eratosthenes (276–198 B.C.) was the first person to obtain a reasonably accurate estimate of the radius and circumference of the earth. On a given day at high noon, he observed that the sun cast no shadow into a well in the Egyptian city of Syene (modern-day Aswan). At the same time, the sun made an angle of 7.2° with the vertical in Alexandria, approximately 500 miles due north of Aswan. Use this information to estimate the radius and circumference of the earth. Assume the sun's rays are parallel.

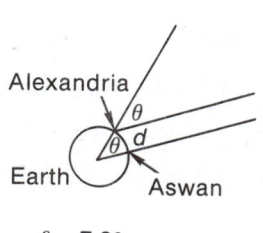

$\theta = 7.2°$
$d = 500$ miles

Section 7.2

Right Triangle Trigonometry

The sum of the angles of any triangle is 180°; each angle of a triangle must be less than 180°. The angles up to and including 180° are classified according to the following scheme.

DEFINITION	θ IN DEGREES	θ IN RADIANS	ILLUSTRATION
Acute angle	$0° < \theta < 90°$	$0 < \theta < \dfrac{\pi}{2}$	
Right angle	$90°$	$\dfrac{\pi}{2}$	
Obtuse angle	$90° < \theta < 180°$	$\dfrac{\pi}{2} < \theta < \pi$	
Straight angle	$180°$	π	

A triangle is called a **right triangle** if one of its angles is a right angle. Of course, the remaining angles must then be acute. The side opposite the right angle is called the **hypotenuse** of the right triangle; the other two sides are called its **legs**. The Pythagorean theorem relates the sides of a right triangle to one another.

> **PYTHAGOREAN THEOREM**
>
> In a right triangle with legs of length a and b and hypotenuse of length c, the sum of the squares of the legs equals the square of the hypotenuse:
>
> $$c^2 = a^2 + b^2$$

EXAMPLE 1 In a right triangle with hypotenuse of length $c = 13$ and one leg of length $a = 5$, find the length of the other leg.

SOLUTION By the Pythagorean theorem, the length b of the other leg satisfies the equation

$$a^2 + b^2 = c^2$$
$$5^2 + b^2 = 13^2$$
$$b^2 = 13^2 - 5^2 = 169 - 25$$
$$= 144$$
$$b = 12$$

We do not use $b = -12$ since we do not allow the hypotenuse to be negative. ∎

The trigonometric functions express relationships among the sides and angles of right triangles. These relationships are the essence of surveying and navigation. If θ is one of the acute angles in a right triangle as in Figure 7.8, then the trigonometric functions of θ are defined in terms of the lengths of the sides adjacent to or opposite the angle and the length of the hypotenuse as follows.

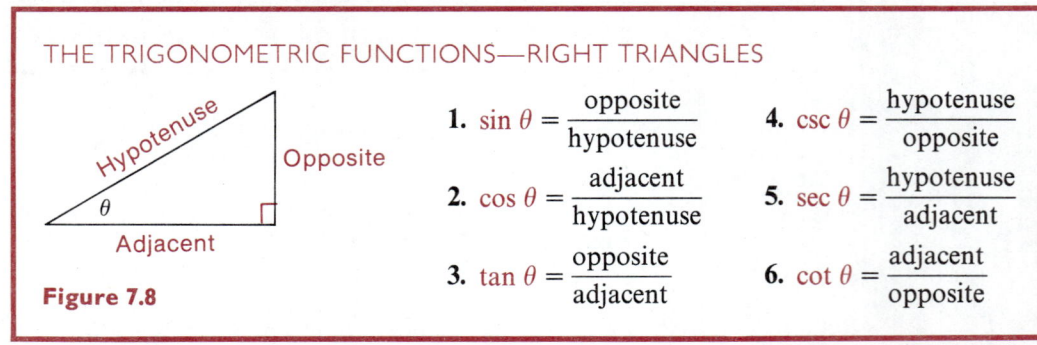

THE TRIGONOMETRIC FUNCTIONS—RIGHT TRIANGLES

1. $\sin \theta = \dfrac{\text{opposite}}{\text{hypotenuse}}$
2. $\cos \theta = \dfrac{\text{adjacent}}{\text{hypotenuse}}$
3. $\tan \theta = \dfrac{\text{opposite}}{\text{adjacent}}$
4. $\csc \theta = \dfrac{\text{hypotenuse}}{\text{opposite}}$
5. $\sec \theta = \dfrac{\text{hypotenuse}}{\text{adjacent}}$
6. $\cot \theta = \dfrac{\text{adjacent}}{\text{opposite}}$

Figure 7.8

The notations used for these functions are abbreviations for **sine** (sin), **cosine** (cos), **tangent** (tan), **cosecant** (csc), **secant** (sec), and **cotangent** (cot).

Two triangles in which corresponding angles have the same measure are said to be **similar**. An important property of similar triangles is that corresponding sides are proportional. In Figure 7.9, the measure of the remaining angle in each triangle is $180° - (90° + \theta)$; thus these triangles are similar. Consequently, $a/c = a'/c'$ so that $\sin \theta = a/c = a'/c'$ is the same using either triangle. The other trigonometric functions, being ratios, are also independent of the size of the triangle and depend only on the size of the angle.

Figure 7.9

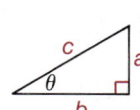 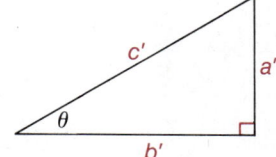

EXAMPLE 2 Evaluate the six trigonometric functions of the angle θ in the given triangle.

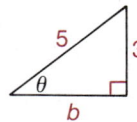

SOLUTION We can use the Pythagorean theorem to find b:

$$b^2 + 3^2 = 5^2$$
$$b = \sqrt{25 - 9} = 4$$

Thus,

$$\sin \theta = \frac{3}{5} \qquad \csc \theta = \frac{5}{3}$$
$$\cos \theta = \frac{4}{5} \qquad \sec \theta = \frac{5}{4}$$
$$\tan \theta = \frac{3}{4} \qquad \cot \theta = \frac{4}{3}$$

EXAMPLE 3 If θ is an acute angle for which $\cos \theta = \frac{2}{3}$, find the values of the other five trigonometric functions of θ.

SOLUTION If θ is an acute angle *of a triangle* and

$$\cos \theta = \frac{2}{3} = \frac{\text{adjacent}}{\text{hypotenuse}}$$

then, since the size of the triangle does not affect the value of $\cos \theta$, we can consider the triangle in Figure 7.10.

We can use the Pythagorean theorem to obtain

$$a = \sqrt{3^2 - 2^2} = \sqrt{5}$$

Then

$$\sin \theta = \frac{\sqrt{5}}{3} \qquad \csc \theta = \frac{3}{\sqrt{5}} = \frac{3\sqrt{5}}{5}$$
$$\cos \theta = \frac{2}{3} \qquad \sec \theta = \frac{3}{2}$$
$$\tan \theta = \frac{\sqrt{5}}{2} \qquad \cot \theta = \frac{2}{\sqrt{5}} = \frac{2\sqrt{5}}{5}$$

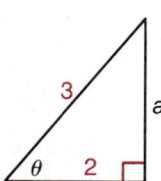

Figure 7.10

Reciprocals

In a right triangle with acute angle θ,

$$\sin \theta = \frac{\text{opposite}}{\text{hypotenuse}} \quad \text{and} \quad \csc \theta = \frac{\text{hypotenuse}}{\text{opposite}}$$

Thus $\sin \theta$ and $\csc \theta$ are reciprocals of one another:

$$\sin \theta = \frac{1}{\csc \theta} \quad \text{and} \quad \csc \theta = \frac{1}{\sin \theta}$$

Each of the following reciprocal relationships can be verified in a similar manner.

RECIPROCAL RELATIONSHIPS

$$\csc \theta = \frac{1}{\sin \theta} \qquad \sin \theta = \frac{1}{\csc \theta}$$

$$\sec \theta = \frac{1}{\cos \theta} \qquad \cos \theta = \frac{1}{\sec \theta}$$

$$\cot \theta = \frac{1}{\tan \theta} \qquad \tan \theta = \frac{1}{\cot \theta}$$

The two acute angles in a right triangle are complementary. More generally, any two angles α and β are

Complementary if $\alpha + \beta = 90° = \dfrac{\pi}{2}$ radians

Supplementary if $\alpha + \beta = 180° = \pi$ radians

We say the functions

sine	and	cosine
tangent	and	cotangent
secant	and	cosecant

are **cofunctions** of one another. A trigonometric function of an acute angle equals its cofunction of the complementary angle. We can easily verify this from the triangle relationships shown in Figure 7.11.

$$\sin \theta = \frac{a}{h} = \cos(90° - \theta) = \cos\left(\frac{\pi}{2} - \theta\right)$$

$$\cos \theta = \frac{b}{h} = \sin(90° - \theta) = \sin\left(\frac{\pi}{2} - \theta\right)$$

$$\tan \theta = \frac{a}{b} = \cot(90° - \theta) = \cot\left(\frac{\pi}{2} - \theta\right)$$

$$\cot \theta = \frac{b}{a} = \tan(90° - \theta) = \tan\left(\frac{\pi}{2} - \theta\right)$$

$$\sec \theta = \frac{h}{b} = \csc(90° - \theta) = \csc\left(\frac{\pi}{2} - \theta\right)$$

$$\csc \theta = \frac{h}{a} = \sec(90° - \theta) = \sec\left(\frac{\pi}{2} - \theta\right)$$

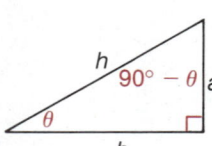

Figure 7.11
$T(\theta) = T^*(90° - \theta)$ FOR COFUNCTIONS T, T^* AND ACUTE ANGLE θ

Section 7.2 Right Triangle Trigonometry **375**

A triangle is **isosceles** if exactly two of its sides have the same length and **equilateral** if all three sides have the same length. The three angles of an equilateral triangle also have the same measure (which must be $\frac{180°}{3} = 60°$, or $\pi/3$ radians); the two angles opposite the equal sides of an isosceles triangle have the same measure. Thus in an isosceles *right* triangle, the two acute angles measure 45° $\frac{180° - 90°}{2}$, or $\pi/4$ radians.

EXAMPLE 4 Find $\sin\theta$, $\cos\theta$, and $\tan\theta$ for $\theta = 45°$, 30°, and 60°.

SOLUTION The triangle relationships allow us to calculate the requested values by geometric methods. We must construct a right triangle with an angle $\theta = \pi/4 = 45°$. Since one of the acute angles of the triangle is 45°, the other must also be 45° ($= 180° - 90° - 45°$). Then the sides opposite these equal angles must have the same length. Since the values of the trigonometric functions depend only on the angle and not on the size of the triangle, we can choose any convenient length for these sides. If we let the sides have length 1, then the hypotenuse has length $\sqrt{2}$ by the Pythagorean theorem. Figure 7.12 shows the triangle. From this figure, we see immediately that

$$\sin 45° = \frac{1}{\sqrt{2}} = \frac{\sqrt{2}}{2}$$

$$\cos 45° = \frac{1}{\sqrt{2}} = \frac{\sqrt{2}}{2}$$

$$\tan 45° = \frac{1}{1} = 1$$

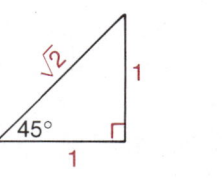

Figure 7.12

For 30° and 60°, consider an equilateral triangle having sides of length 2. Each angle of this triangle is 60° and the perpendicular from one vertex to the opposite side bisects that side as well as the vertex angle (see Figure 7.13). The Pythagorean formula can then be used to compute the altitude of this perpendicular as $\sqrt{3}$. Finally, the triangle relationships indicate that

$$\sin 60° = \frac{\sqrt{3}}{2} \qquad\qquad \sin 30° = \frac{1}{2}$$

$$\cos 60° = \frac{1}{2} \quad\text{and}\quad \cos 30° = \frac{\sqrt{3}}{2}$$

$$\tan 60° = \frac{\sqrt{3}}{1} = \sqrt{3} \qquad \tan 30° = \frac{1}{\sqrt{3}} = \frac{\sqrt{3}}{3}$$

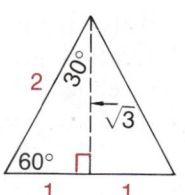

Figure 7.13

The values of the six trigonometric functions for the 30°, 45°, and 60° angles are compiled in Table 7.1. Rather than memorizing these values, it would be better to *compute* these values mentally when they are needed. Visualizing the triangles in Figures 7.12 and 7.13 should aid such mental computations.

Table 7.1
IMPORTANT ANGLES

DEGREES	RADIANS	$\sin \theta$	$\cos \theta$	$\tan \theta$	$\cot \theta$	$\sec \theta$	$\csc \theta$
30°	$\dfrac{\pi}{6}$	$\dfrac{1}{2}$	$\dfrac{\sqrt{3}}{2}$	$\dfrac{\sqrt{3}}{3}$	$\sqrt{3}$	$\dfrac{2\sqrt{3}}{3}$	2
45°	$\dfrac{\pi}{4}$	$\dfrac{\sqrt{2}}{2}$	$\dfrac{\sqrt{2}}{2}$	1	1	$\sqrt{2}$	$\sqrt{2}$
60°	$\dfrac{\pi}{3}$	$\dfrac{\sqrt{3}}{2}$	$\dfrac{1}{2}$	$\sqrt{3}$	$\dfrac{\sqrt{3}}{3}$	2	$\dfrac{2\sqrt{3}}{3}$

EXAMPLE 5   Find the exact value of

$$\frac{\tan \dfrac{\pi}{3} - \sin \dfrac{\pi}{4}}{\cos \dfrac{\pi}{4}}$$

SOLUTION   We can use the values in Table 7.1 to write

$$\frac{\tan \dfrac{\pi}{3} - \sin \dfrac{\pi}{4}}{\cos \dfrac{\pi}{4}} = \frac{\sqrt{3} - \dfrac{\sqrt{2}}{2}}{\dfrac{\sqrt{2}}{2}}$$

$$= \left(\sqrt{3} - \frac{\sqrt{2}}{2}\right) \cdot \frac{2}{\sqrt{2}}$$

$$= \left(\sqrt{3} - \frac{\sqrt{2}}{2}\right) \sqrt{2}$$

$$= \sqrt{6} - 1$$

Since the *exact* value was requested, we do not give a decimal approximation for $\sqrt{6}$. ∎

Except for a few special angles such as those just illustrated, evaluating the trigonometric functions can be difficult. Many calculators are programmed to provide approximate values of the sine, cosine, and tangent functions. Since the other trigonometric functions are reciprocals of these, they can also be evaluated on calculators of this type.

Using Tables (Optional)

Before the advent of calculators, tables provided approximate values for the trigonometric functions. This optional material on the use of tables is included for readers without access to a calculator. The values provided in the trigonometric tables at the back of this book are correct to four digits. (Tables referred to in this section are the trigonometric tables at the back of the book.)

Table 4 (at the back of the book) lists the values of the trigonometric functions for θ expressed in radians. The values are given for $0 \leq \theta \leq \pi/2$ radians in θ increments of 0.01 radians. Table 5 lists the values of the trigonometric functions for angles expressed in degrees and minutes. These values are given for $0° \leq \theta \leq 90°$ in steps of 10 minutes.

Tables 4 and 5 are set up somewhat differently from Tables 1, 2, and 3. Since there are six trigonometric functions to be evaluated, the columns are used to indicate the values of the six functions. For example, if we look for $\theta = 0.29$ radians in Table 4, we find

θ (radians)	$\sin \theta$	$\cos \theta$	$\tan \theta$	$\cot \theta$	$\sec \theta$	$\csc \theta$
.29	.2860	.9582	.2984	3.351	1.044	3.497

In particular, we note that $\sec(0.29 \text{ radians}) \approx 1.044$.

Note also that the left-hand column of Table 5 lists only those angles from $0°$ to $45°$, but the right-hand column lists angles from $90°$ down to $45°$. Because trigonometric tables are so extensive, Table 5 (at the back of the book) is condensed to provide all the necessary information in half the space. This can be done since $T(\theta) = T^*(90° - \theta)$ where T and T^* are cofunctions and θ is an acute angle.

To avoid computing $T(\theta) = T^*(90° - \theta)$ by hand, column labels are placed at the bottom of each column to relate to the angles listed in the right-hand column as illustrated in the following partial table. From this table we read

$$\sin 47°10' \approx 0.7333$$
$$\cos 77°30' \approx 0.2164$$

θ (DEGREES)	$\sin \theta$	$\cos \theta$	$\tan \theta$	$\cot \theta$	$\sec \theta$	$\csc \theta$	
0°00′	.0000	1.0000	.0000	—	1.000	—	90°00′
⋮	⋮	⋮	⋮	⋮	⋮	⋮	⋮
12°30′	.2164	.9763	.2217	4.511	1.024	4.620	77°30′
⋮	⋮	⋮	⋮	⋮	⋮	⋮	⋮
42°50′	.6799	.7333	.9271	1.079	1.364	1.471	47°10′
⋮	⋮	⋮	⋮	⋮	⋮	⋮	⋮
45°00′	.7071	.7071	1.0000	1.000	1.414	1.414	45°00′
	$\cos \theta$	$\sin \theta$	$\cot \theta$	$\tan \theta$	$\csc \theta$	$\sec \theta$	θ (DEGREES)

Values of the trigonometric functions for acute angles strictly between two listed angles are estimated by the process of **linear interpolation.** We use a straight line to approximate the graph between the closest two angle values.

> **LINEAR INTERPOLATION**
>
> For θ between θ_1 and θ_2, linear interpolation estimates $T(\theta)$ by adding to (or subtracting from) $T(\theta_1)$ that portion of the increase (or decrease) $[T(\theta_2) - T(\theta_1)]$ determined by the portion of the distance θ has traveled from θ_1 to θ_2.

EXAMPLE 6 Interpolate in Table 5 to find sin 23°12′.

SOLUTION Use Table 5 to find sin 23°10′ and sin 23°20′. Then set up the following display.

$$10'\left\{2'\left\{\begin{matrix}\theta & | & \sin\theta \\ 23°10' & | & 0.3934 \\ 23°12' & | & \\ 23°20' & | & 0.3961\end{matrix}\right\}d\right\}\text{Increase} = 0.0027$$

As in Section 6.4, we obtain the ratio

$$\frac{d}{0.0027} = \frac{2}{10} = 0.2$$

$$d = (0.2)(0.0027)$$
$$= 0.00054$$

Now, d represents the *increase* in the value of $\sin\theta$ as θ moves from 23°10′ to 23°12′. Thus

$$\sin 23°12' \approx \sin 23°10' + d$$
$$\approx 0.3934 + 0.00054$$
$$= 0.39394$$
$$\approx 0.3939 \qquad \blacksquare$$

In the preceding example the number 0.39394 is rounded to 0.3939 because the values in the table are already rounded to four places; the linear interpolation procedure is also just an estimate of the intermediate values. Thus we cannot expect to obtain five-place accuracy. For the same reason we interpolate only to the nearest minute; for example, we interpolate to 23°12′ but not to 23°12′40″. In Table 4 we might interpolate between 0.67 radians and 0.68 radians to 0.673 radians but not to 0.6728 radians.

Using a Calculator

Many calculators are programmed to evaluate at least the basic trigonometric functions: sine, cosine, and tangent. Pushing the reciprocal key then evaluates the other three trigonometric functions: cosecant, secant, and cotangent, respectively.

Most scientific calculators can work with both degrees and radians. Many even have built-in programs to convert between radians, decimal degrees, and the measure of an angle in degrees, minutes, and seconds. However, when evaluating the trigonometric functions for some θ, the angle must be expressed in either radians or decimal degrees. If 12.30 is entered as the input for evaluating the sine function at 12°30′, the calculator will evaluate $\sin 12.30° = \sin 12°18' \approx 0.2130$ rather than $\sin 12.5° = \sin 12°30' \approx 0.2164$. (Try it on yours.)

CALCULATOR COMMENTS

1. *When using a calculator to evaluate the trigonometric functions, make sure it is in the correct mode: degrees or radians; sin 5° is very much different from sin (5 radians).*

2. *Also be certain that your input is in a form the calculator understands. This generally necessitates that minutes and seconds be converted to decimal degrees; for example, 12°30′ must be entered as 12.5°.*

Section 7.2 Right Triangle Trigonometry

Interpolation is necessary when using the tables because there is a practical limit to the size of any table. However, calculators do *not* have tables of values stored internally. They are programmed to compute the required values from scratch each time. With a calculator, we need not be concerned with interpolation. The angle can be "fine tuned" to as many *decimal* places as the calculator permits.

EXAMPLE 7 Use a calculator to evaluate each of the trigonometric functions at $\theta = 1.063$ radians.

SOLUTION To save time, we enter 1.063 into memory. This is done by pushing the indicated keys in order.

$\boxed{1}\ \boxed{.}\ \boxed{0}\ \boxed{6}\ \boxed{3}\ \ \boxed{\text{STO}}$

The $\boxed{\text{STO}}$ key is the memory key; it may have other forms such as $\boxed{\text{MEM}}$ or $\boxed{\text{M}}$ on some calculators. *Make sure your calculator is in radian mode.* Recall the argument (1.063) from memory and evaluate the sine and cosecant as follows:

PRESS	READ ON DISPLAY
$\boxed{\text{RCL}}$	1.063
$\boxed{\text{SIN}}$	0.8738
$\boxed{1/x}$	1.144

The $\boxed{\text{RCL}}$ key is the recall key; $\boxed{1/x}$ is the reciprocal key. These, too, may have other forms on some calculators. From the displayed readings, we conclude that

$\sin(1.063 \text{ radians}) \approx 0.8738$

and

$\csc(1.063 \text{ radians}) \approx 1.144$

since $\csc \theta = 1/\sin \theta$. Use the calculator to evaluate the remaining functions as follows.

PRESS	READ ON DISPLAY	CONCLUDE THAT
$\boxed{\text{RCL}}$	1.063	
$\boxed{\text{COS}}$	0.4863	$\cos(1.063 \text{ radians}) \approx 0.4863$
$\boxed{1/x}$	2.0565	$\sec(1.063 \text{ radians}) \approx 2.0565$
$\boxed{\text{RCL}}$	1.063	
$\boxed{\text{TAN}}$	1.7970	$\tan(1.063 \text{ radians}) \approx 1.7970$
$\boxed{1/x}$	0.5565	$\cot(1.063 \text{ radians}) \approx 0.5565$

EXAMPLE 8
Evaluate the six trigonometric functions at $\theta = 62°32'16''$.

SOLUTION

Be careful! The input for the calculator must be in decimal degrees or radians. Converting θ to decimal degrees, we find

$$62°32'16'' = 62° + \left(32\frac{16}{60}\right)'$$
$$= \left[62 + \frac{32\frac{16}{60}}{60}\right]°$$
$$\approx 62.5378°$$

Enter 62.5378 into memory. *Make sure your calculator is in **degree** mode* and use it to evaluate the functions as follows.

PRESS	READ ON DISPLAY	CONCLUDE THAT
RCL	62.5378	
SIN	0.8873	$\sin \theta \approx 0.8873$
1/x	1.127	$\csc \theta \approx 1.127$
RCL	62.5378	
COS	0.4612	$\cos \theta \approx 0.4612$
1/x	2.168	$\sec \theta \approx 2.168$
RCL	62.5378	
TAN	1.924	$\tan \theta \approx 1.924$
1/x	0.5197	$\cot \theta \approx 0.5197$

Section 7.2 Exercises

Use the Pythagorean theorem to find the third side of a right triangle with the two given sides. Legs will be denoted by a *and* b; c *will denote the hypotenuse.*

1. $a = 7, b = 24$
2. $a = 9, b = 40$
3. $a = 3, c = 7$
4. $a = 5, c = 9$
5. $b = 4, c = 7$
6. $b = 7, c = 9$
7. $a = 15, b = 8$
8. $a = 20, b = 21$
9. $a = 2, c = 5$
10. $b = 3, c = 8$

11–20. Evaluate the six trigonometric functions for the acute angle adjacent to the smaller leg of the triangles given in Exercises 1–10.

21–30. Repeat Exercises 11–20 for the acute angle adjacent to the longer leg.

Let θ be an acute angle. Use the given value of the trigonometric function and the Pythagorean theorem to find the values of the other five trigonometric functions of θ.

31. $\sin \theta = \dfrac{1}{2}$
32. $\cos \theta = \dfrac{1}{3}$
33. $\tan \theta = \dfrac{3}{2}$
34. $\cot \theta = \dfrac{4}{3}$
35. $\sec \theta = \dfrac{5}{2}$
36. $\csc \theta = \dfrac{6}{5}$

37. $\cos\theta = \dfrac{3}{4}$

38. $\sin\theta = \dfrac{4}{7}$

39. $\cot\theta = \dfrac{\sqrt{2}}{3}$

40. $\tan\theta = \dfrac{\sqrt{3}}{5}$

41. $\csc\theta = \dfrac{7}{3}$

42. $\sec\theta = \dfrac{9}{2}$

Find the exact value for each of the following.

43. $\dfrac{\tan 30° - \sin 45°}{\cos 60°}$

44. $\dfrac{\cos 45° + \tan 60°}{\sin 30°}$

45. $\dfrac{2\sin\left(\dfrac{\pi}{4}\text{ rad}\right) - \cos\left(\dfrac{\pi}{6}\text{ rad}\right)}{\tan\left(\dfrac{\pi}{3}\text{ rad}\right)}$

46. $\dfrac{3\cos\left(\dfrac{\pi}{3}\text{ rad}\right) + 2\csc\left(\dfrac{\pi}{6}\text{ rad}\right)}{\sec\left(\dfrac{\pi}{4}\text{ rad}\right)}$

47. $\dfrac{\cot\left(\dfrac{\pi}{3}\text{ rad}\right) + \sin\left(\dfrac{\pi}{4}\text{ rad}\right)}{\tan\left(\dfrac{\pi}{6}\text{ rad}\right) + \cos\left(\dfrac{\pi}{4}\text{ rad}\right)}$

48. $\dfrac{\sec\left(\dfrac{\pi}{3}\text{ rad}\right) - \csc\left(\dfrac{\pi}{6}\text{ rad}\right)}{\tan\left(\dfrac{\pi}{6}\text{ rad}\right) + \cot\left(\dfrac{\pi}{4}\text{ rad}\right)}$

49. $\dfrac{\tan 30° + \sec 45°}{\cos 60° + \sin 45°}$

50. $\dfrac{\cos 45° - \csc 45°}{\tan 30° - \cot 45°}$

Table Exercises (Optional)

Use Table 4 or 5 (at the end of the book) to approximate the requested values.

51. cos(0.83 rad)
52. sin(1.06 rad)
53. tan(1.31 rad)
54. sec(0.37 rad)
55. sec 22°10′
56. csc 80°40′
57. cot 37°20′
58. cos 53°50′
59. sin(1.412 rad)
60. cos(0.165 rad)
61. tan(0.239 rad)
62. sec(0.934 rad)
63. csc(0.247 rad)
64. cot(0.985 rad)
65. sin 69°27′
66. tan 78°56′
67. cos 41°43′
68. sec 23°36′

Calculator Exercises

Evaluate all six trigonometric functions for each of the following angles.

69. 0.896 radians
70. 1.387 radians
71. 0.587 radians
72. 1.244 radians
73. 1.247 radians
74. 0.365 radians
75. 39°12′50″
76. 5°15′20″
77. 17°22′18″
78. 62°6′42″
79. 83°39′16″
80. 47°39′36″

Section 7.3

Solving Right Triangles

There are six parts to any triangle: three angles and three sides. If we know certain parts of a triangle, we can frequently determine the others on the basis of this limited information. For instance, if we know one acute angle α of a right triangle we can find the other angle $\beta = (180° - 90° - \alpha)$, or $\beta = (\pi - \pi/2 - \alpha)$. If we know two sides of a right triangle, we can find the third side by using the Pythagorean theorem. From these lengths we can find $\sin\theta$ for one of the acute angles. In this section we shall see how to use tables or a calculator "backward" to find θ from a known value for $\sin\theta$. In fact, if we know two parts of a right triangle in addition to its right angle, and at least one of these parts is a length, we will be able to find the remaining parts. When we have determined all parts of a triangle we say that we have *solved* the triangle.

EXAMPLE 1

Solve the right triangle satisfying the data given in Figure 7.14.

SOLUTION

We know that $\sin 30° = \frac{1}{2}$. (See Figure 7.13 or Table 7.1.) If we let h denote the length of the hypotenuse, we then have

$$\sin 30° = \frac{1}{2} = \frac{3}{h}$$

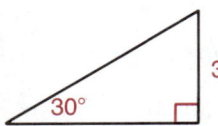

Figure 7.14

which we solve for h to obtain

$$h = 6$$

If b denotes the length of the other side of the triangle, the Pythagorean theorem yields

$$3^2 + b^2 = 6^2 = 36$$
$$b^2 = 36 - 3^2 = 27$$
$$b = 3\sqrt{3}$$

Finally, the other acute angle β must be

$$\beta = 180° - 90° - 30°$$
$$= 60°$$

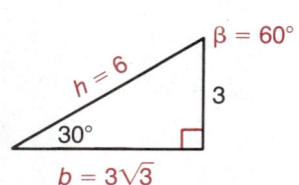

Figure 7.15

Figure 7.15 displays the solution. ∎

EXAMPLE 2

Solve the right triangle whose hypotenuse has length 10 if $\cos \alpha = \sqrt{2}/2$ for one of its acute angles α.

SOLUTION

We should recognize immediately that $\alpha = \pi/4$ radians. (See Figure 7.12 or Table 7.1.) But then the other acute angle $\beta = \pi - \pi/2 - \pi/4 = \pi/4$ radians. We have an isosceles right triangle, since the two acute angles have the same measure. If a denotes the length of each of the legs of this triangle, the Pythagorean theorem yields

$$a^2 + a^2 = 10^2$$
$$2a^2 = 100$$
$$a^2 = 50$$
$$a = 5\sqrt{2}$$

Figure 7.16

Figure 7.16 displays the solution. ∎

Using Tables In Reverse (Optional)

If the value of one of the trigonometric functions is known for some acute angle θ, Table 4 or 5 (at the end of the book) can be used to approximate θ. (You should not need to use Table 4 or 5 for any of the angles listed in Table 7.1, however!) Again, interpolation may be necessary. The following example illustrates the technique.

EXAMPLE 3

Find an acute angle θ for which $\cos \theta = 0.4377$.

SOLUTION

First we bracket the given value of $\cos \theta$ between two values that appear in Table 5 and display this information as follows.

Section 7.3 Solving Right Triangles

$$10'\left\{d\left\{\begin{array}{c}64°0'\\ \theta\\ 64°10'\end{array}\right.\right.\quad\begin{array}{c}\text{angle }\alpha\\ \hline 0.4384\\ 0.4377\\ 0.4358\end{array}\left.\begin{array}{c}\cos\alpha\\ \\ 0.0007\end{array}\right\}0.0026$$

As with earlier interpolation problems, we obtain

$$\frac{d}{10'} = \frac{0.0007}{0.0026} = \frac{7}{26}$$

$$d = \frac{7}{26}(10') \approx 2.69'$$

Thus

$$\theta \approx 64°0' + 2.69' = 64°2.69'$$

$$\approx 64°3'$$

Again, 2.69′ is rounded to 3′ because of the inaccuracies inherent in the table values that have already been rounded. Thus $\cos 64°3' \approx 0.4377$. ∎

Table 4 can be used in a similar fashion to determine θ from a known value of one of the trigonometric functions of θ.

Using Calculators in Reverse

A calculator can also be used to find θ from a known value for $\sin\theta$, $\cos\theta$, or $\tan\theta$. But since calculators do not include the secant, cosecant, or cotangent functions, the reciprocal relations must be used to convert values for these functions into values for the sine, cosine, or tangent functions before the angle θ can be determined. This process is illustrated in Example 4b.

Finding a θ that yields a given $T(\theta)$ is the *inverse* of computing with the function T. Thus the key for this operation is generally labeled INV, although some calculators use a 2ⁿᵈ F key. The INV or 2ⁿᵈ F key is punched before the respective function key SIN, COS, or TAN. With most calculators, the argument must be entered *before* punching these function keys.

EXAMPLE 4

Find an acute angle θ for which

a. $\cos\theta = \frac{1}{2}$

b. $\csc\theta = 3.124$

SOLUTION

a.

PRESS	READ ON DISPLAY	CONCLUDE THAT
2	2	
1/x	0.5	$\frac{1}{2} = 0.5$
INV or 2ⁿᵈ F		
COS	60 (if the calculator is in degree mode)	$\cos 60° = 0.5 = \frac{1}{2}$

This means that 60° is the only angle θ between 0° and 90° for which $\cos\theta = \frac{1}{2}$. If the calculator were in radian mode, it would have displayed 1.047197551 at the last stage of the computation. This is the decimal representation of $\pi/3$ within the limits of the calculator; however, it is not immediately recognizable as such. Unfortunately, a calculator cannot display the irrational number $\pi/3$ exactly.

b.

PRESS	READ ON DISPLAY	CONCLUDE THAT
3 . 1 2 4	3.124	
1/x	0.3201	$\sin\theta \approx 0.3201$
INV SIN or 2nd F SIN	18.66912	$\theta \approx 18.66912°$

We convert 18.66912° to degrees, minutes, and seconds as $\theta \approx 18°40'9''$. In radian mode we would have obtained $\theta \approx 0.3258$ radians. ∎

EXAMPLE 5 Solve the right triangle illustrated in Figure 7.17.

SOLUTION With the triangle labeled as in the figure, we have

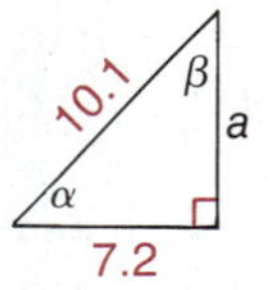

$$a = \sqrt{(10.1)^2 - (7.2)^2} \quad \text{(Pythagorean theorem)}$$
$$= \sqrt{50.17}$$
$$\approx 7.1$$

and

$$\cos\alpha = \frac{7.2}{10.1} \approx 0.7129$$

$$\alpha \approx 44.53° \quad \text{(Use a calculator or use Table 5 in reverse and interpolate.)}$$

$$\approx 44°32'$$

$$\beta = 90° - \alpha$$
$$\approx 90° - 44°32'$$
$$= 45°28'$$

Figure 7.17

The triangle is solved. ∎

EXAMPLE 6 Solve the right triangle, one of whose angles is 23° and for which the side opposite this acute angle has length 6.2.

SOLUTION Draw and label the triangle as in Figure 7.18. We have

$$\frac{b}{6.2} = \cot 23°$$
$$\approx 2.3559$$
$$b \approx (6.2)(2.3559)$$
$$\approx 14.6$$

$$\sin 23° = \frac{6.2}{h}$$

$$h = \frac{6.2}{\sin 23°}$$

$$\approx \frac{6.2}{0.3907}$$

$$\approx 15.9$$

$$\beta = 90° - 23°$$

$$= 67°$$

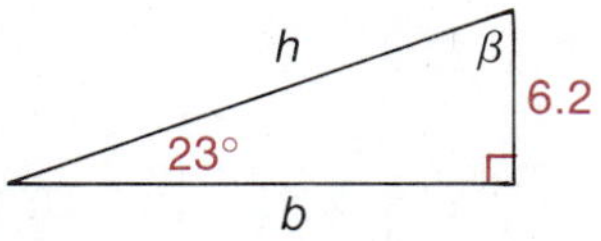

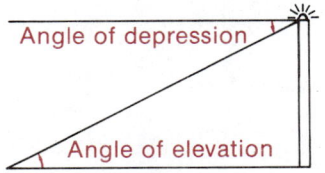

Figure 7.18

and the triangle is solved.

These triangular relationships are quite useful in surveying and in navigation. To determine the altitude of a cliff or a tree, we need only measure the angle to the top and the distance from the base; the tangent function can then be used to find the altitude. For problems of this type, angle of elevation and angle of depression are useful concepts. Sighting downward, the angle between the horizontal and the line of sight is called the **angle of depression;** sighting upward, the angle between the horizontal and the line of sight is called the **angle of elevation** (see Figure 7.19).

Figure 7.19

EXAMPLE 7

An expedition discovers a very tall tree and the members want to determine its height. A sextant and trigonometric tables are available. Standing 150 feet from the base of the tree, they measure the angle of elevation to the treetop as 53°20′. Find the height of the tree.

SOLUTION

Let a = height of the tree; then (see Figure 7.20)

$$\frac{a}{150} = \tan 53°20′$$

$$\approx 1.3432$$

$$a \approx (150)(1.3432)$$

$$\approx 201.5$$

The tree is approximately 201 feet 6 inches tall. In their journal, they would probably round this to 200 feet.

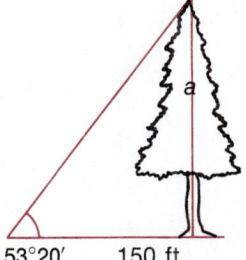

Figure 7.20

EXAMPLE 8

The next day the same expedition discovers a waterfall. The members would like to record in their journal the depth of the falls. Unfortunately, they are not standing level with the base of the falls. They measure the angle of elevation to the top of the falls as 72°10′ and the angle of elevation to the bottom of the falls as 32°40′. They have also determined that they are 300 feet from the area of the falls, as indicated in Figure 7.21. Find the depth of the falls.

SOLUTION

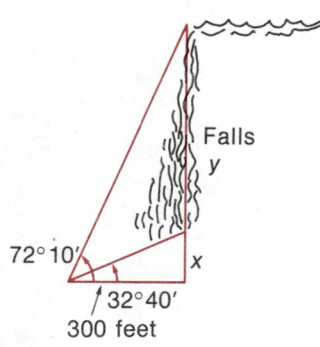

Figure 7.21

Let

x = altitude of the base of the falls
y = depth of the falls

Then

$$\frac{x+y}{300} = \tan 72°10' \approx 3.1084$$

$$x + y \approx 300(3.1084)$$
$$\approx 932.5$$

Similarly,

$$\frac{x}{300} = \tan 32°40' \approx 0.6412$$

$$x \approx 300(0.6412)$$
$$\approx 192.4$$

Finally,

$$y = (x + y) - x$$
$$\approx 932.5 - 192.4$$
$$\approx 740.1$$

The depth of the falls is approximately 740.1 feet. This would probably be rounded off to 740 feet for recording in their journal. ∎

EXAMPLE 9

The angle of elevation from John's house to the top of Shadow Mountain is 28°12'; from Jane's house, it is 47°38'. If Jane lives 6212 feet closer to the mountain than John, how tall is the mountain?

SOLUTION

Let h denote the height of the mountain and x the distance from Jane's house to a point directly below the peak as illustrated in Figure 7.22.

Figure 7.22

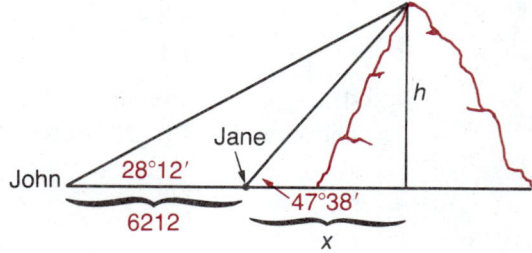

Then

$$\frac{h}{x} = \tan 47°38'$$

$$\frac{h}{x + 6212} = \tan 28°12'$$

From these we obtain

$$h = x \tan 47°38'$$
$$= (x + 6212) \tan 28°12'$$
$$= x \tan 28°12' + 6212 \tan 28°12'$$
$$x(\tan 47°38' - \tan 28°12') = 6212 \tan 28°12'$$
$$x = \frac{6212 \tan 28°12'}{(\tan 47°38' - \tan 28°12')}$$
$$\approx 5945.55 \qquad \text{(Use a calculator)}$$
$$h = x \tan 47°38'$$
$$\approx 6519 \text{ feet} \qquad \text{(Use a calculator)} \blacksquare$$

Section 7.3 Exercises

Solve the following triangles without using Tables or a calculator. Parts of the triangles are labeled as in the given figure.

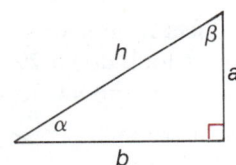

1. $\alpha = \frac{\pi}{4}, b = 5$
2. $\beta = \frac{\pi}{6}, h = 8$
3. $\beta = \frac{\pi}{3}, h = 12$
4. $\alpha = \frac{\pi}{4}, a = 3\sqrt{2}$
5. $\sin \beta = \frac{1}{2}, a = 3$
6. $\sin \alpha = \frac{\sqrt{2}}{2}, h = 6$
7. $\tan \beta = 1, h = 4\sqrt{2}$
8. $\tan \alpha = \sqrt{3}, h = 4$
9. $\sec \beta = 2, a = 3$
10. $\csc \beta = \sqrt{2}, h = 7$

Table Exercises (Optional)

Use Table 4 (at the end of the book) to find the acute angle θ that satisfies the given condition.

11. $\sin \theta = 0.0350$
12. $\cos \theta = 0.9260$
13. $\tan \theta = 0.1145$
14. $\sec \theta = 2.322$
15. $\cot \theta = 1.682$
16. $\csc \theta = 1.620$

Use Table 5 to find the acute angle θ that satisfies the given condition.

17. $\sin \theta = 0.0780$
18. $\tan \theta = 53$
19. $\csc \theta = 1.200$
20. $\cot \theta = 2.06$
21. $\cos \theta = 0.0911$
22. $\sec \theta = 1.300$

Calculator Exercises (23–42)

Find the acute angle θ in radians that satisfies the following.

23. $\sin \theta = 0.6723$
24. $\cos \theta = 0.1552$
25. $\sec \theta = 4.713$
26. $\tan \theta = 0.7348$
27. $\cos \theta = 0.7292$
28. $\sin \theta = 0.6204$
29. $\cot \theta = 72.37$
30. $\csc \theta = 30.00$
31. $\tan \theta = 4.650$
32. $\cot \theta = 2.707$

Find the acute angle θ in degrees that satisfies the following.

33. $\tan \theta = 4.832$
34. $\cos \theta = 0.6872$
35. $\cot \theta = 31.270$
36. $\sin \theta = 0.0341$
37. $\csc \theta = 1.313$
38. $\tan \theta = 0.0911$
39. $\sin \theta = 0.4295$
40. $\sec \theta = 23.74$
41. $\cos \theta = 0.0731$
42. $\cot \theta = 1.3502$

Use tables (at the back of the book) or a calculator in the following exercises.

In the following problems, parts of a right triangle are labeled as in the following illustration. Solve the triangles on the basis of the given information.

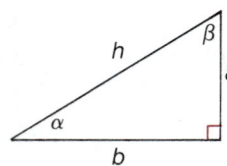

43. $a = 5.1$, $h = 6.3$
44. $b = 6.0$, $\alpha = 30°0'$
45. $h = 12.0$, $\alpha = 15°0'$
46. $b = 3.6$, $h = 4.2$
47. $a = 3.1$, $\beta = 37°36'$
48. $h = 22.0$, $\beta = 18°48'$
49. $a = 2.0$, $b = 9.0$
50. $b = 7.1$, $\beta = 26°0'$
51. $a = 7.0$, $\alpha = 12°6'$
52. $a = 14.0$, $\beta = 48°12'$
53. $b = 5.0$, $h = 14.6$
54. $h = 10.0$, $\alpha = 10°0'$
55. $a = 6.3$, $\alpha = 67°30'$
56. $a = 6.0$, $b = 7.0$

57. The pitch of a roof is defined as the vertical rise corresponding to a given horizontal run. Thus a 5-12 pitch indicates a vertical rise of 5 feet for each horizontal advance of 12 feet. What angle does the roofline make with the horizontal in this case?

58. What angle does a roof with a 4-12 pitch make with the horizontal? (See Exercise 57.)

59. The "grade" of a road is the vertical rise divided by the corresponding *road length*. If a highway has a 7% grade, what angle does it make with the horizontal?

60. A 1% grade is generally considered to be steep for railroads. What angle with the horizontal does this represent? (See Exercise 59.)

61. If a 30-foot ladder just reaches a roof 20 feet high, what angles does the ladder make with the ground?

62. If 100 meters of string are played out on a kite and the string makes an angle of 47° with the ground, how high is the kite?

63. A surveyor stands on a riverbank and sights to a point on the other side, which is opposite a point on side of the river 100 feet from the surveyor. The angle between the two lines of sight is 73°15'. How wide is the river?

64. A television antenna tower casts a shadow 10 meters long. From the tip of the shadow, the angle of inclination to the top of the tower is 52°.
   a. How long should the guy wires be if they will anchor the top of the tower to points 10 meters from the base of the tower?
   b. How tall is the tower?

65. Cindy is hiking on a cliff and sees a sign below that says "175 meters to base of cliff." She determines that her line of sight to the sign makes an angle of 58° with the horizontal. How high is the cliff?

66. If a man 6 feet tall casts a shadow 8 feet long when he stands 15 feet from a lamppost, how tall is the lamppost?

67. From a tourist viewpoint, Margaret Ann sees a river below. Her line of sight to the opposite side of the river makes an angle of 53°20' with the horizontal; the angle is 58°40' for this side of the river. If her view point is 3122 feet above the level of the river, how wide is the river?

68. Laara stands 10 feet from a wall in a public building. She must look down 29° to see the bottom of the wall and up 44° to see the top.
   a. How high is the room?
   b. How tall is Laara?

69. If a ship travels straight east for $2\frac{1}{2}$ hours at 100 nautical miles per hour (knots) and then veers 30° north of east for $1\frac{1}{2}$ hours at the same rate of speed, how far is the ship from its starting point?

70. A tree is spotted on top of a cliff. The angle of elevation to the bottom of the tree is 43°18'; the angle of elevation to the top of the tree is 54°6'. If the cliff is 320 feet high, how tall is the tree?

71. An exploring party wishes to determine the height of a mountain. Since the mountain is not a cliff, the party members cannot measure their distance from a point directly below the peak. Instead they measure the angle of inclination to the peak as 23°10'; then they move $1\frac{1}{2}$ miles (7920 feet) closer to the mountain and measure the angle of inclination as 34°50'. How high is the mountain?

72. The angle of inclination to the top of a water tower is 25°45'; 75 feet closer, the angle of elevation is 51°20'. How high is the tower?

73. Sighting directly along a highway in the valley, the angle of depression from a cliff to a certain mile marker is 32°17′; the angle of depression to next closer mile marker (assume exactly 5280 feet closer) is 46°52′. How high is the cliff?

74. The angle of depression from an air traffic control tower to the airport entrance gate is 6°22′. The angle to a point $\frac{1}{4}$ mile (1320 feet) closer is 52°56′. How high is the tower?

75. The *parsec* (parallax-second) is a large unit of distance used by astronomers. It is defined as that distance from which a parallax of 1 second would be observed between the earth and the sun (see the following figure).

    a. Taking the distance between the earth and the sun as 93,000,000 miles, express a parsec in

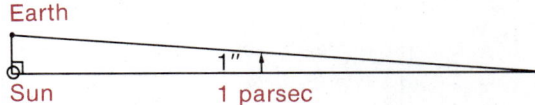

miles and in light-years; 1 light-year is the distance traveled by light in 1 year at 186,000 miles per second.

b. The distance from our solar system to the center of our Milky Way galaxy is on the order of 10,000 parsecs. What distance in miles does this represent?

c. What parallax is observed between the sun and the earth from the center of the galaxy?

Section 7.4

Trigonometric Functions of Arbitrary Angles and Real Numbers

In this section we extend the domain of the trigonometric functions to include all angles rather than just acute angles. We will need the following concepts in this section.

> **DEFINITION**
>
> 1. An angle $\angle AOB$ is in **standard position** on a coordinate system if its vertex O is at the origin and its initial ray $\overrightarrow{OA}$ coincides with the positive x-axis. See Figure 7.23a.
> 2. An angle is **in a specified quadrant** provided its terminal ray falls in that quadrant when the angle is placed in standard position. See Figure 7.23b.
> 3. Two angles are **coterminal** if their terminal rays coincide when they are both placed in standard position. See Figure 7.23c.

Figure 7.23

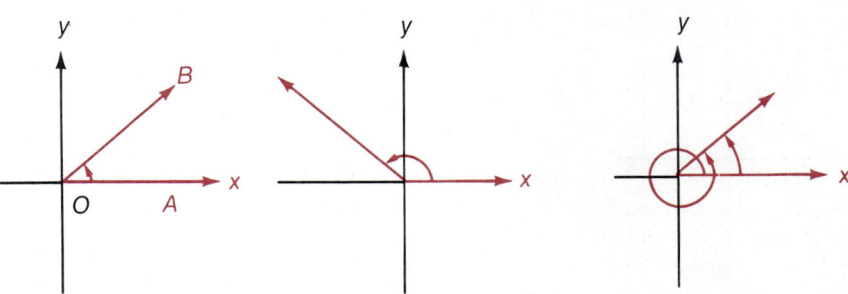

a. $\angle AOB$ in standard position   b. An angle in Quadrant II   c. Coterminal angles

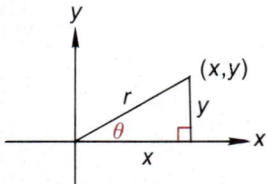

Figure 7.24

Placing an acute angle θ of a right triangle in standard position, we obtain the configuration illustrated in Figure 7.24. Letting (x, y) be the coordinates of the vertex indicated and r the distance of this vertex from the origin, we can write

$$\cos \theta = \frac{x}{r}, \quad \sin \theta = \frac{y}{r}, \quad \text{and so on.}$$

TRIGONOMETRIC FUNCTIONS—ARBITRARY ANGLES

To define the trigonometric functions for an arbitrary angle θ, we place it in standard position, choose any point $(x, y) \neq (0, 0)$ on its terminal ray, and find the distance r of this point from the origin (see Figure 7.25). The trigonometric functions of θ are then defined as follows.

1. $\sin \theta = \dfrac{y}{r}$ 4. $\csc \theta = \dfrac{r}{y}, \ y \neq 0$

2. $\cos \theta = \dfrac{x}{r}$ 5. $\sec \theta = \dfrac{r}{x}, \ x \neq 0$

3. $\tan \theta = \dfrac{y}{x}, \ x \neq 0$ 6. $\cot \theta = \dfrac{x}{y}, \ y \neq 0$

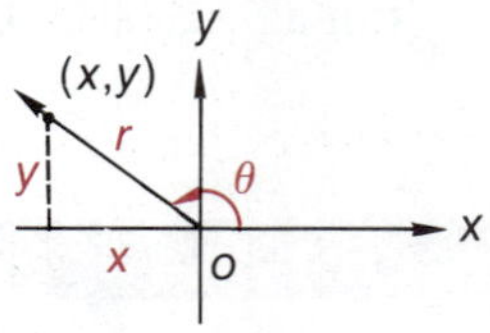

Figure 7.25

Again, consideration of similar triangles should convince us that the values of the trigonometric functions, being ratios, do not depend on r.

EXAMPLE 1

Evaluate the six trigonometric functions of the angle θ if the terminal side of θ passes through the point $(-3, 4)$ when θ is placed in standard position.

SOLUTION

Since $(-3, 4)$ is on the terminal side of the angle, we find its distance from the origin to be

$$r = \sqrt{(-3)^2 + 4^2} = 5$$

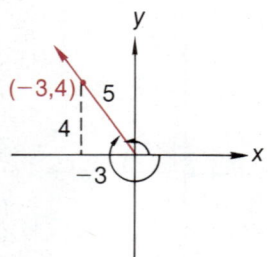

Figure 7.26

as shown in Figure 7.26. We see immediately that

$$\sin \theta = \frac{4}{5} \qquad \csc \theta = \frac{5}{4}$$

$$\cos \theta = \frac{-3}{5} = -\frac{3}{5} \qquad \sec \theta = \frac{5}{-3} = -\frac{5}{3}$$

$$\tan \theta = \frac{4}{-3} = -\frac{4}{3} \qquad \cot \theta = \frac{-3}{4} = -\frac{3}{4}$$

EXAMPLE 2

If θ is in standard position and its terminal side coincides with the line $2x + 3y = 0$ in the fourth quadrant, evaluate the six trigonometric functions at θ.

SOLUTION

First, we sketch the line $2x + 3y = 0$, as shown in Figure 7.27. The point $(3, -2)$ is on this line and in the fourth quadrant; hence $(3, -2)$ is on the terminal side of the angle. The distance of this point from the origin is

$$r = \sqrt{(3)^2 + (-2)^2}$$
$$= \sqrt{13}$$

Thus

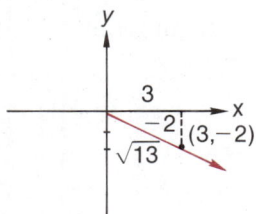

Figure 7.27

$$\sin \theta = \frac{-2}{\sqrt{13}} = -\frac{2}{\sqrt{13}} \qquad \csc \theta = \frac{\sqrt{13}}{-2} = -\frac{\sqrt{13}}{2}$$

$$\cos \theta = \frac{3}{\sqrt{13}} \qquad \sec \theta = \frac{\sqrt{13}}{3}$$

$$\tan \theta = \frac{-2}{3} = -\frac{2}{3} \qquad \cot \theta = \frac{3}{-2} = -\frac{3}{2}$$

Circular Functions: The Unit Circle

Since it does not matter which point we choose on the terminal ray of θ when we evaluate the trigonometric functions $T(\theta)$, we can choose the point which is at distance $r = 1$ from the origin. If $r = 1$, the values of $\cos \theta$ and $\sin \theta$ are the x- and y-coordinates, respectively, of the point in which the terminal ray of θ meets the circle of radius 1 centered at the origin. (See Figure 7.28.) This circle is called the **unit circle**.

Figure 7.28
$\sin \theta$ AND $\cos \theta$ ON THE UNIT CIRCLE

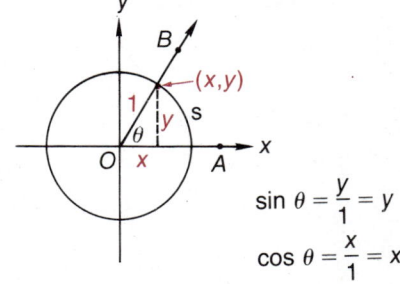

This relationship of the trigonometric functions to the unit circle is important for their applications to science and engineering. Mathematicians currently call these functions **circular functions** to indicate that many of their applications have little if any connection with triangles and to emphasize their relationship to the unit circle.

It is frequently useful to interpret the circular functions as being defined for all *real numbers* rather than just for *angles*. This interpretation is described as follows.

392 Chapter 7 The Trigonometric Functions

CIRCULAR FUNCTIONS OF REAL NUMBERS

1. Let s be any real number.
2. On the unit circle, start at the point $(1, 0)$ and mark off an arc length s in the counterclockwise direction (see Figure 7.29).
3. This generates a central angle θ which, by the arc length formula (Section 7.1), satisfies

$$s = r\theta = \theta \qquad \text{(Since } r = 1\text{)}$$

when θ is measured in radians; i.e., *a central angle of s radians is generated by the arc length s.* (Radian measure is in this sense a *natural* angle measure.)

4. Then $T(s)$ is defined as $T(s \text{ radians})$ for each real number s and each of the trigonometric or circular functions T.
5. In particular, note that *if $P(s) = (x, y)$ is the point on the unit circle corresponding to the arc length s, then $x = \cos s$ and $y = \sin s$* (see Figure 7.29).

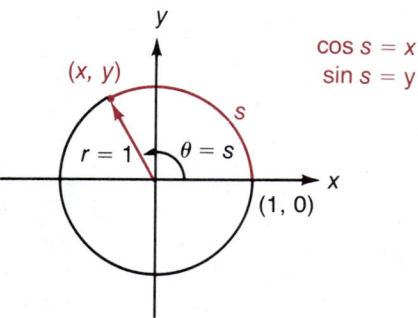

Figure 7.29
CIRCULAR FUNCTIONS ON THE UNIT CIRCLE

Figure 7.30

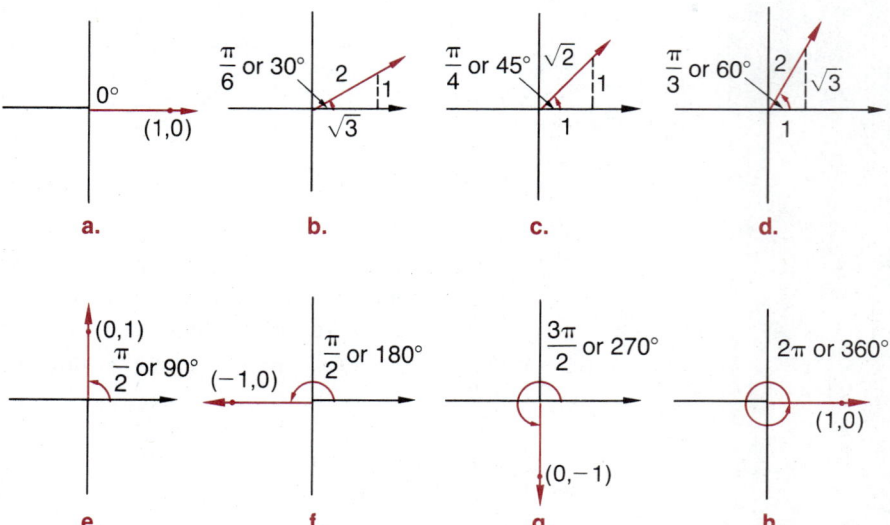

Table 7.2 is an extension of our earlier table of values of the trigonometric functions for certain angles (Table 7.1, Section 7.2). The values appearing in Table 7.2 are obtained from Figure 7.30 and can be used to evaluate the trigonometric functions of angles whose terminal sides are symmetric to the terminal sides of the given angles.

Table 7.2
IMPORTANT ANGLES

DEGREES	RADIANS	$\sin\theta$	$\cos\theta$	$\tan\theta$	$\cot\theta$	$\sec\theta$	$\csc\theta$
0°	0	0	1	0	—	1	—
30°	$\frac{\pi}{6}$	$\frac{1}{2}$	$\frac{\sqrt{3}}{2}$	$\frac{\sqrt{3}}{3}$	$\sqrt{3}$	$\frac{2\sqrt{3}}{3}$	2
45°	$\frac{\pi}{4}$	$\frac{\sqrt{2}}{2}$	$\frac{\sqrt{2}}{2}$	1	1	$\sqrt{2}$	$\sqrt{2}$
60°	$\frac{\pi}{3}$	$\frac{\sqrt{3}}{2}$	$\frac{1}{2}$	$\sqrt{3}$	$\frac{\sqrt{3}}{3}$	2	$\frac{2\sqrt{3}}{3}$
90°	$\frac{\pi}{2}$	1	0	—	0	—	1
180°	π	0	−1	0	—	−1	—
270°	$\frac{3\pi}{2}$	−1	0	—	0	—	−1
360°	2π	0	1	0	—	1	—

EXAMPLE 3

Evaluate each of the following.

a. $\sin\dfrac{\pi}{4}$

b. $\tan\dfrac{\pi}{3}$

SOLUTION

We use the definition of the circular functions of real numbers.

a. $\sin\dfrac{\pi}{4} = \sin\left(\dfrac{\pi}{4}\text{ radians}\right)$

$= \dfrac{\sqrt{2}}{2}$ (Table 7.2)

b. $\tan\dfrac{\pi}{3} = \tan\left(\dfrac{\pi}{3}\text{ radians}\right)$

$= \sqrt{3}$ (Table 7.2) ∎

EXAMPLE 4

Evaluate $\sin\theta$, $\cos\theta$, and $\tan\theta$ for $\theta = 5\pi/4$.

SOLUTION

Since $T(5\pi/4) = T(5\pi/4 \text{ radians})$ for each of the circular functions T, we work with the angle $5\pi/4$ radians. Note that $5\pi/4 = \pi + \pi/4$. Sketching $5\pi/4$ radians in standard position as in Figure 7.31, we observe its symmetry with the angle $\pi/4$ radians.

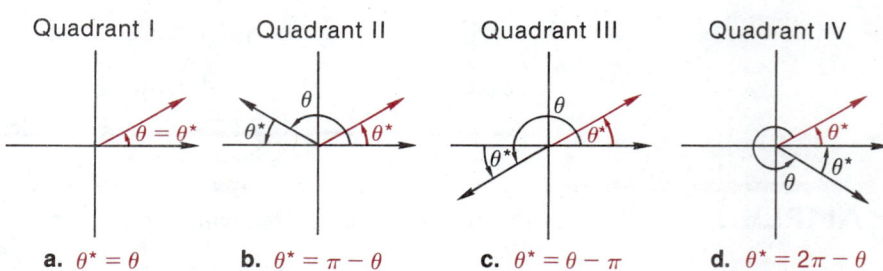

Thus with $x = -1$, $y = -1$, and $r = \sqrt{2}$, we obtain

$$\cos \frac{5\pi}{4} = \frac{x}{r} = \frac{-1}{\sqrt{2}} = -\frac{\sqrt{2}}{2}$$

$$\sin \frac{5\pi}{4} = \frac{y}{r} = \frac{-1}{\sqrt{2}} = -\frac{\sqrt{2}}{2}$$

$$\tan \frac{5\pi}{4} = \frac{y}{x} = \frac{-1}{-1} = 1$$

Figure 7.31

Reference Angles

In Example 4, $\pi/4$ radians is called a **reference angle** for $5\pi/4$ radians. The reference angle is an acute angle that can be used to find the values of the circular functions at the given angle. It is the *acute* angle that the terminal side of θ makes with the *x-axis* when θ is placed in standard position. Figure 7.32 illustrates reference angles θ^* for angles $\theta \in [0, 2\pi]$ whose terminal rays fall in any of the four quadrants.

Figure 7.32
REFERENCE ANGLES

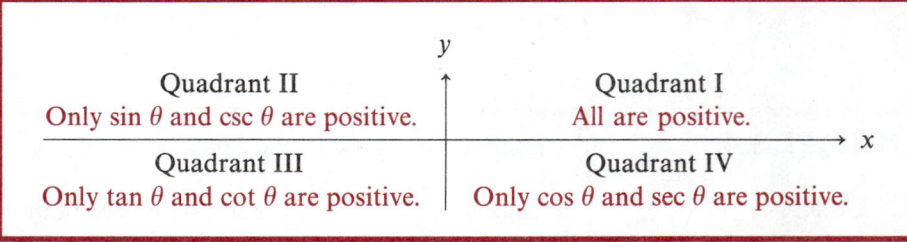

a. $\theta^* = \theta$ b. $\theta^* = \pi - \theta$ c. $\theta^* = \theta - \pi$ d. $\theta^* = 2\pi - \theta$

If θ is not in the range $0 \le \theta \le 2\pi$, then θ should first be replaced with a coterminal angle which is in this range. Then the reference angle may be found as in Figure 7.32.

It follows from the triangle relationships that $|T(\theta)| = T(\theta^*)$ for each of the trigonometric functions T when θ^* is the reference angle for θ. However, the appropriate $\pm$ sign for $T(\theta)$ must be determined in some other way, generally by considering the quadrant in which the terminal ray of θ falls when θ is placed in standard position. You should verify the validity of the following sign diagram for an angle θ in standard position whose terminal ray falls in the given quadrant.

Quadrant II Only $\sin \theta$ and $\csc \theta$ are positive.	Quadrant I All are positive.
Quadrant III Only $\tan \theta$ and $\cot \theta$ are positive.	Quadrant IV Only $\cos \theta$ and $\sec \theta$ are positive.

EXAMPLE 5   Find the values of sin θ, cos θ, and tan θ for θ = −22π/3.

SOLUTION   Again, $T(-22\pi/3) = T(-22\pi/3 \text{ radians})$ for each of the circular functions T. Note that

$$\frac{-22\pi}{3} = -7\frac{1}{3}\pi = -6\pi - \frac{4\pi}{3} = 3 \cdot (-2\pi) - \frac{4\pi}{3}$$

Since -2π represents a complete revolution, the terminal side of $-22\pi/3$ is the same as the terminal side of $-4\pi/3$. Since $-4\pi/3 = -\pi - \pi/3$, we find the reference angle for $-22\pi/3$ to be $\pi/3$, as illustrated in Figure 7.33. We then construct the right triangle with angle $\pi/3$ in standard position and having hypotenuse of length 2.

Figure 7.33

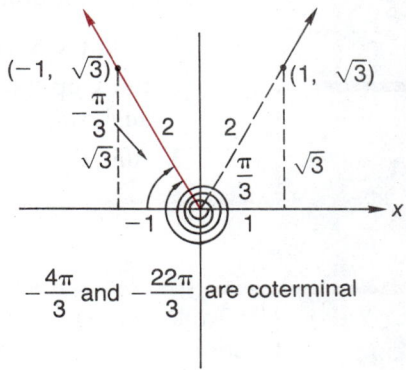

$-\frac{4\pi}{3}$ and $-\frac{22\pi}{3}$ are coterminal

Finally, symmetry is used to find the requested values as indicated in Figure 7.33. Note that cos θ and tan θ are negative, but sin θ is positive, since the terminal ray of θ falls in Quadrant II.

$$\cos \frac{-22\pi}{3} = -\frac{1}{2}$$

$$\sin \frac{-22\pi}{3} = \frac{\sqrt{3}}{2}$$

$$\tan \frac{-22\pi}{3} = \frac{\sqrt{3}}{-1} = -\sqrt{3}$$

EXAMPLE 6   If $\cos \theta = -\frac{5}{13}$ and $\sin \theta < 0$, in what quadrant does the terminal side of θ lie if θ is in standard position? Sketch the terminal side and evaluate tan θ.

SOLUTION   Place the angle in standard position. Let (x, y) be the coordinates of the point 13 units from the origin on the terminal side of the angle. Then $x = -5$ since

$$\cos \theta = \frac{-5}{13} = \frac{x}{r} = \frac{x}{13}$$

396 Chapter 7 The Trigonometric Functions

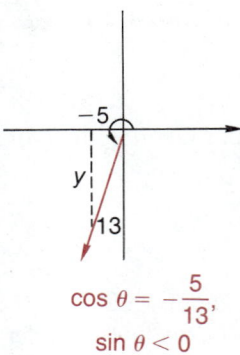

Figure 7.34

Since $x < 0$, the terminal side must lie in the second or third quadrant. But $\sin \theta = y/r < 0$ only when $y < 0$; thus the terminal side must lie in the third quadrant. To find y, we sketch the angle in standard position as in Figure 7.34 and use the Pythagorean theorem. Then

$$(-5)^2 + y^2 = 13^2$$
$$y^2 = 169 - 25$$
$$= 144$$
$$y = \pm 12$$

Since $y < 0$, we have $y = -12$. Thus $\tan \theta = \frac{-12}{-5} = \frac{12}{5}$. ∎

Using Tables for Arbitrary Angles (Optional)

To use tables for evaluating the circular functions at any angle θ:

1. Find the reference angle for θ.
2. Look up the values of the circular functions at the reference angle, interpolating if necessary.
3. Affix the appropriate $\pm$ signs.

EXAMPLE 7 Use Table 4 (at end of the book) to approximate the values of each of the circular functions at $\theta = 23$ radians. Do not interpolate. Use $\pi \approx 3.14$.

SOLUTION The angle 23 radians describes several counterclockwise revolutions. Each such revolution consists of $2\pi \approx 6.28$ radians. We can write

$$23 \approx 3 \cdot (6.28) + 3.14 + 1.02$$
$$\approx 3 \cdot (2\pi) + \pi + 1.02$$

The angle of 23 radians can now be sketched in standard position as in Figure 7.35a. We note in Figure 7.35b that the reference angle for 23 radians is 1.02 radians.

Figure 7.35

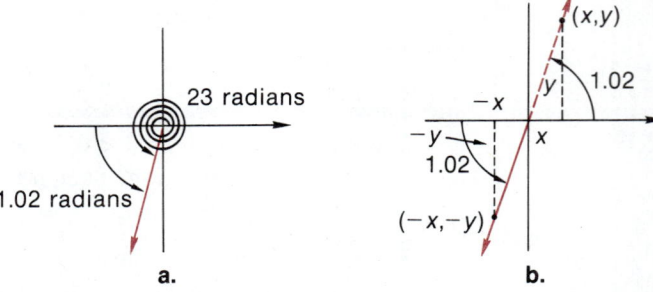

a.   b.

Only $\tan 23$ and $\cot 23$ will be positive since the terminal ray of 23 radians falls in the third quadrant. Using Table 4 for 1.02 radians and affixing the appropriate ± signs, we obtain

$$\sin 23 \approx -\sin 1.02 \approx -0.8521$$
$$\cos 23 \approx -\cos 1.02 \approx -0.5234$$
$$\tan 23 \approx \tan 1.02 \approx 1.628$$
$$\cot 23 \approx \cot 1.02 \approx 0.6142$$
$$\sec 23 \approx -\sec 1.02 \approx -1.911$$
$$\csc 23 \approx -\csc 1.02 \approx -1.174$$

EXAMPLE 8 Use Table 5 (at the end of the book) to find $\cos(-377°12')$.

SOLUTION To find the reference angle θ^* for $\theta = -377°12'$, we write

$$-377°12' = -360° - 17°12'$$
$$\theta^* = 17°12'$$

as indicated in Figure 7.36. We find the following values in Table 5 and display the information as shown.

	θ	$\cos \theta$	
10' { 2' {	17°10'	0.9555	} d } 0.0009
	17°12'		
	17°20'	0.9546	

Then

$$\frac{d}{0.0009} = \frac{2}{10} = 0.2$$

$$d = (0.2)(0.0009)$$
$$= 0.00018$$

Note that $\cos \theta$ decreases as θ moves from $17°10'$ to $17°20'$. Thus

$$\cos 17°12' \approx 0.9555 - d$$
$$\approx 0.9555 - 0.00018$$
$$= 0.95532$$
$$\approx 0.9553$$

(We cannot achieve five-place accuracy from a four-place table. One of the very-high-quality calculators gives $\cos 17°12' \approx 0.95527836$.) Finally, we note that $\cos \theta$ is positive when the terminal ray falls in the fourth quadrant. Then

$$\cos(-377°12') = \cos 17°12'$$
$$\approx 0.9553$$

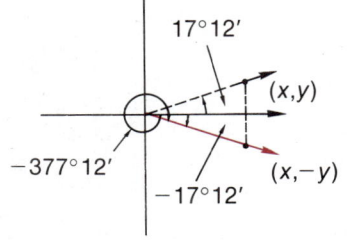

Figure 7.36

Using Calculators with Arbitrary Angles

To evaluate the trigonometric functions with a calculator, we need not be concerned with reference angles or interpolation. The angle entered into the computation does not have to lie in the range $0° \leq \theta \leq 90°$, or $0 \leq \theta \leq \pi/2$ radians. Just proceed in the same manner as with acute angles; the calculator is not concerned about the size or the sign of the angle. But as before, put the calculator in the proper mode (degrees or radians) and convert minutes and seconds to decimal degrees.

Section 7.4 Exercises

Evaluate each of the six trigonometric functions of θ if the indicated point is on the terminal side of θ when θ is placed in standard position.

1. $(-4, 0)$
2. $(0, -5)$
3. $(2, 1)$
4. $(-3, 2)$
5. $(-4, -5)$
6. $(2, -3)$

Evaluate each of the six circular functions of θ if the terminal side of θ coincides with the given line in the indicated quadrant when θ is placed in standard position.

7. $2x - y = 0$, Quadrant I
8. $5x - 12y = 0$, Quadrant III
9. $4x + 5y = 0$, Quadrant II
10. $4x + 3y = 0$, Quadrant IV
11. $x - 3y = 0$, Quadrant III
12. $2x + y = 0$, Quadrant II

13. Extend Table 7.2 of this section to include $\theta = 2\pi/3, 3\pi/4, 5\pi/6, 7\pi/6, 5\pi/4, 4\pi/3, 5\pi/3, 7\pi/4,$ and $11\pi/6$. (*Hint:* Place the angles in standard position and observe the symmetry of their terminal rays with those sketched in Figure 7.30.)

Evaluate the six circular functions at

14. $390°$
15. $-150°$
16. $-315°$
17. $420°$
18. $-570°$
19. $-405°$
20. $510°$
21. $600°$
22. $-750°$
23. $585°$

Evaluate the six circular functions at

24. 27π
25. $\dfrac{19\pi}{2}$
26. $\dfrac{155\pi}{4}$
27. -24π
28. $\dfrac{-91\pi}{3}$
29. $\dfrac{543\pi}{2}$
30. $\dfrac{59\pi}{6}$
31. $\dfrac{-53\pi}{4}$
32. $\dfrac{224\pi}{3}$
33. $\dfrac{-161\pi}{6}$

Find the exact value of each of the following.

34. $\dfrac{\sin 180° + \cos 180°}{\tan 120°}$

35. $\dfrac{\cos 270° - \sin 150°}{\csc(-120°)}$

36. $\dfrac{\sec \dfrac{5\pi}{3} + \cos \dfrac{7\pi}{3}}{\csc \dfrac{5\pi}{4}}$

37. $\dfrac{\cot \dfrac{7\pi}{4} - \sin\left(-\dfrac{5\pi}{3}\right)}{\cos \dfrac{13\pi}{3}}$

38. $\dfrac{\tan 150° - \cos 270°}{\sin(-120°) - \cot 135°}$

39. $\dfrac{\sec 120° + \cos(-60°)}{\csc 240° - \sin(-30°)}$

40. $\dfrac{\sin \dfrac{3\pi}{2} + \cos \pi}{\tan \dfrac{3\pi}{4} + \cot\left(-\dfrac{5\pi}{4}\right)}$

41. $\dfrac{\cos\left(-\dfrac{\pi}{4}\right) + \csc \dfrac{5\pi}{6}}{\sin \dfrac{3\pi}{4} - \sec\left(-\dfrac{\pi}{4}\right)}$

42. $\dfrac{\cot \dfrac{11\pi}{6} + \sin \dfrac{3\pi}{2}}{\tan\left(-\dfrac{3\pi}{4}\right) + \sec \dfrac{2\pi}{3}}$

43. $\dfrac{\tan\left(-\dfrac{7\pi}{4}\right) + \cos \dfrac{11\pi}{6}}{\sin\left(-\dfrac{3\pi}{2}\right) + \cos 4\pi}$

Using the given information, find the values of all six trigonometric functions of θ.

44. $\sin \theta = \dfrac{\sqrt{2}}{2}$, $\tan \theta < 0$

45. $\tan \theta = \dfrac{-4}{3}$, $\csc \theta < 0$

46. $\cos \theta = \dfrac{-1}{2}$, $\csc \theta > 0$

47. $\cot \theta = -3$, $\sin \theta < 0$

48. $\csc \theta = -2$, $\cos \theta < 0$

49. $\sec \theta = 3$, $\cot \theta > 0$

50. $\cot \theta = \sqrt{3}$, $\sin \theta > 0$

51. $\csc \theta = -3$, $\cos \theta < 0$

52. $\cos \theta = \dfrac{1}{5}$, $\sin \theta < 0$

53. $\sin \theta = \dfrac{-1}{3}$, $\tan \theta > 0$

Find ____ if θ is an ____ angle and ____.

54. $\sin \theta$ acute $\tan \theta = 6$

55. $\sec \theta$ obtuse $\csc \theta = \dfrac{3}{2}$

56. $\tan \theta$ obtuse $\sec \theta = \dfrac{-8}{3}$

57. $\cos \theta$ acute $\tan \theta = \dfrac{1}{3}$

58. $\csc \theta$ obtuse $\cos \theta = \dfrac{-2}{5}$

59. $\cot \theta$ acute $\sin \theta = \dfrac{2}{7}$

Table Exercises (Optional)

Use reference angles, Table 4 or 5 (at end of book), and interpolation if necessary to approximate the requested values. You may use $\pi \approx 3.142$.

60. $\sin(-18.412)$ 61. $\cos 16.5$

62. $\tan(-23.966)$ 63. $\sec(-22°10')$

64. $\csc(-100°40')$ 65. $\cot(637°20')$

66. $\sec 9.342$ 67. $\csc(-0.247)$

68. $\cot 19.846$ 69. $\sin 269°27'$

70. $\cos 141°43'$ 71. $\tan(-708°56')$

Calculator Exercises

Evaluate all six trigonometric functions for each of the following angles. Determine the quadrant in which the terminal ray falls when the angle is placed in standard position.

72. 2.036 radians   73. -4.899 radians

74. 5.876 radians   75. 17.438 radians

76. -0.247 radians   77. $125°10'14''$

78. $-139°12'50''$ 79. $632°6'42''$

80. $507°22'18''$ 81. $-147°39'36''$

Section 7.5

Elementary Properties of the Trigonometric Functions

The scientific applications of the trigonometric functions generally depend more on their analytic aspects than on the triangle relationships. In many cases, the graph of some modified trigonometric function conveys important information. The properties to be developed in this section will improve our understanding of the circular functions and later will help in graphing. Some of these properties can also be important to navigators and surveyors.

Referring to Figure 7.25 on page 390 we see that $\sin \theta = 0$ when the terminal ray falls on the x-axis ($y = 0$) and that $\cos \theta = 0$ when the terminal ray falls on the y-axis ($x = 0$). Thus

$$\sin \theta = 0 \quad \text{if and only if} \quad \theta = k\pi,$$
$$\cos \theta = 0 \quad \text{if and only if} \quad \theta = \frac{\pi}{2} + k\pi, \quad k \text{ an integer}$$

Cosine and sine are the two basic circular functions; the other four can be expressed in terms of these. For example, using the notation of Figure 7.25 we get

$$\tan \theta = \frac{y}{x} = \frac{y/r}{x/r} = \frac{\sin \theta}{\cos \theta}$$

when $\cos \theta \neq 0$. The other relationships in the following box can be established in a similar fashion. For each of the excluded angles, the given function is undefined.

RECIPROCAL RELATIONSHIPS

$$\csc \theta = \frac{1}{\sin \theta}, \quad \theta \neq k\pi \qquad \sin \theta = \frac{1}{\csc \theta}$$

$$\sec \theta = \frac{1}{\cos \theta}, \quad \theta \neq \frac{\pi}{2} + k\pi \qquad \cos \theta = \frac{1}{\sec \theta}$$

$$\cot \theta = \frac{\cos \theta}{\sin \theta} = \frac{1}{\tan \theta}, \quad \theta \neq k\pi \qquad \tan \theta = \frac{\sin \theta}{\cos \theta} = \frac{1}{\cot \theta}, \quad \theta \neq \frac{\pi}{2} + k\pi$$

k an integer

When an angle θ is placed in standard position, the terminal side of $\theta \pm 2\pi$ ($= \theta \pm 360°$) coincides with the terminal side of θ. Consequently,

$$T(\theta \pm 2\pi) = T(\theta \pm 360°) = T(\theta)$$

for each of the six circular functions T and every angle θ.

DEFINITION

Whenever any function f has the property that

$$f(t + c) = f(t)$$

for all t and some constant c, we say that f is **periodic**. The smallest positive constant c for which this property is true is called the **period** of f.

Of course, if f has period c, then

$$f(t + nc) = f(t)$$

for every integer n. For instance,

$$f(t + 2c) = f[(t + c) + c] = f(t + c) = f(t)$$

for all t.

We have just seen that each of the trigonometric functions is periodic. It can be shown that 2π or $360°$ is indeed the period of the sine and cosine functions and their reciprocals: cosecant and secant; see Exercise 89 at the end of this section. On the other hand, the tangent and cotangent functions have an even smaller

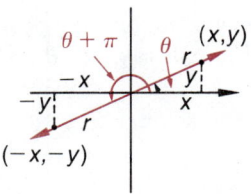

Figure 7.37

period. For if θ is any angle in standard position, the terminal side of $\theta + \pi$ is found by reflecting the terminal side of θ through the origin. Labeling as in Figure 7.37, we have

$$\tan \theta = \frac{y}{x} = \frac{-y}{-x} = \tan(\theta + \pi)$$

Since $\cot \theta = 1/\tan \theta$, we see that

$$\cot(\theta + \pi) = \frac{1}{\tan(\theta + \pi)} = \frac{1}{\tan \theta} = \cot \theta$$

It can be shown (see Exercise 90) that π is indeed the smallest c for which $\tan(\theta + c) = \tan \theta$ for all θ. Thus π or $180°$ is the period of these two functions.

PERIODICITY

sine, cosine, secant, cosecant are periodic with period 2π or $360°$ [e.g., $\sin(\theta + 2\pi) = \sin \theta$]

tangent, cotangent are periodic with period π or $180°$ [e.g., $\tan(\theta + \pi) = \tan \theta$]

EXAMPLE 1

A certain function whose graph is given in Figure 7.38 has period $= 3$. Extend the graph to the interval $[-4, 5]$.

Figure 7.38

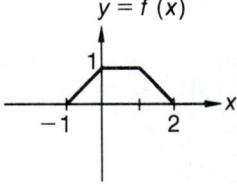

SOLUTION

Since the function is periodic, its graph simply repeats itself to the left and right as indicated in Figure 7.39.

Figure 7.39

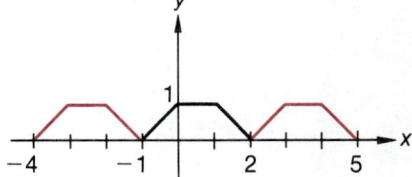

Symmetry Properties

In addition to their periodicity, other properties of the circular functions relating θ, $(\theta \pm \pi)$, and $(\theta \pm \pi/2)$ are also interesting. These properties all follow from the symmetry of terminal rays and are left to the exercises. Similar symmetry properties were used to develop the concept of "reference angle" discussed in Section 7.4.

Also dependent on the symmetry of terminal rays are properties relating positive and negative angles. Observe in Figure 7.40 that the terminal rays for θ and $-\theta$ are symmetric about the x-axis.

Labeling as in Figure 7.40, we make the following observations.

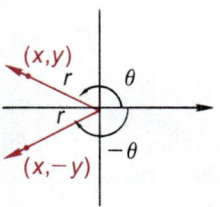

Figure 7.40

$$\cos(-\theta) = \frac{x}{r} = \cos \theta$$

$$\sin(-\theta) = \frac{-y}{r} = -\sin \theta$$

$$\tan(-\theta) = \frac{\sin(-\theta)}{\cos(-\theta)} = \frac{-\sin \theta}{\cos \theta} = -\tan \theta$$

From these relationships and their reciprocals, we see that each of the circular functions is either even $[f(-t) = f(t)]$ or odd $[f(-t) = -f(t)]$ (see Section 4.2).

SYMMETRY

$\left.\begin{array}{l}\text{cosine}\\\text{secant}\end{array}\right\}$ are even functions [e.g., $\cos(-\theta) = \cos \theta$]

$\left.\begin{array}{l}\text{sine}\\\text{cosecant}\\\text{tangent}\\\text{cotangent}\end{array}\right\}$ are odd functions [e.g., $\sin(-\theta) = -\sin \theta$]

EXAMPLE 2

If f is an odd function of period 4 and $f(-7) = 2$, find each value.

a. $f(1)$
b. $f(3)$
c. $f(4)$
d. $f(-t-8)$

SOLUTION

a. $f(1) = f(1 - 2 \cdot 4)$ (Period = 4)
   $= f(-7)$
   $= 2$

b. $f(3) = f(3 + 4)$ (Period = 4)
   $= f(7)$
   $= -f(-7)$ (f is odd)
   $= -2$

c. $f(4) = f(4 - 4)$ (Period = 4)
   $= f(0)$

But

$$f(0) = f(-0) = -f(0) \qquad (f \text{ is odd})$$

Thus

$$2f(0) = 0$$
$$f(0) = 0$$

Finally,

$$f(4) = f(0) = 0$$

d. $f(-t-8) = f(-t-8+2\cdot 4)$ (Period = 4)
$ = f(-t-8+8)$
$ = f(-t)$
$ = -f(t) \qquad (f \text{ is odd})$ ∎

Cofunction Properties

The trigonometric properties relating complementary acute angles also apply to arbitrary complementary angles. To show this, we sketch both θ and $\pi/2 - \theta$ in standard position on the same coordinate system. To sketch the angle $\pi/2 - \theta$ in standard position, we begin with the angle $\pi/2$ in standard position and then rotate the terminal ray θ radians in the *clockwise* direction from $\pi/2$ to $\pi/2 - \theta$ (see Figure 7.41).

Figure 7.41

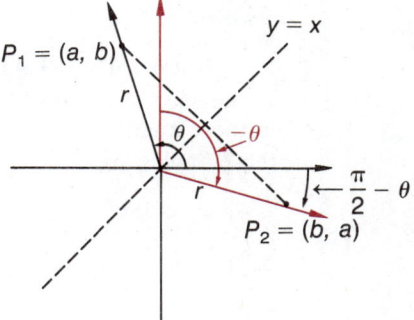

The terminal rays of θ and $\pi/2 - \theta$ are symmetric about the line $y = x$. In Figure 7.41, the coordinates of P_2 are obtained by reversing the order of the coordinates of P_1. Labeling $P_1 = (a, b)$ and $P_2 = (b, a)$, we have

$$\cos\left(\frac{\pi}{2} - \theta\right) = \frac{b}{r} = \sin\theta$$

$$\sin\left(\frac{\pi}{2} - \theta\right) = \frac{a}{r} = \cos\theta$$

Using these relationships, we can show (see Exercise 88) that for any two cofunctions T and T^* and any angle θ

$$T^*\left(\frac{\pi}{2} - \theta\right) = T(\theta)$$

Specifically,

$$\sin\left(\frac{\pi}{2} - \theta\right) = \cos\theta \qquad \csc\left(\frac{\pi}{2} - \theta\right) = \sec\theta$$

$$\cos\left(\frac{\pi}{2} - \theta\right) = \sin\theta \qquad \sec\left(\frac{\pi}{2} - \theta\right) = \csc\theta$$

$$\tan\left(\frac{\pi}{2} - \theta\right) = \cot\theta \qquad \cot\left(\frac{\pi}{2} - \theta\right) = \tan\theta$$

Symmetry properties are also useful when using tables or calculators in reverse to find θ from a known value of $T(\theta)$.

Using Tables in Reverse

To use Table 4 or 5 in reverse to find θ from a known value of $T(\theta)$, we must also use reference angles and work with the absolute value $|T(\theta)|$ since the tables are set up this way.

EXAMPLE 3 Find all angles θ between $0°$ and $360°$ for which $\cos\theta = -0.4377$.

SOLUTION First, we find an angle θ^* between $0°$ and $90°$ for which $\cos\theta^* = 0.4377$. This angle was found in Example 3 of Section 7.3; it is

$$\theta^* \approx 64°3'$$

But $\cos\theta = -0.4377$. Now $\cos\theta$ is negative when the terminal ray for θ falls in the second or third quadrant. We must find angles θ_1, θ_2 with terminal rays in these quadrants and having $64°3'$ as reference angle. In Figure 7.42 we see that these angles are

$$\theta_1 = 180° - 64°3' = 115°57'$$
$$\theta_2 = 180° + 64°3' = 244°3'$$

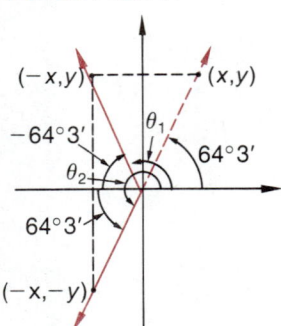

Figure 7.42

Using Calculators in Reverse

To use the calculator in reverse to find θ from a known value for $T(\theta)$ requires some care. For a given circular function T, many different θ's yield the same $T(\theta)$. But for a known value $T(\theta)$, the calculator will select just one θ that yields the given value.

FOR A KNOWN VALUE OF	THE CALCULATOR WILL SELECT θ SATISFYING	
$\sin\theta$	$-90° \leq \theta \leq 90°$	$-\dfrac{\pi}{2} \leq \theta \leq \dfrac{\pi}{2}$
$\cos\theta$	$0° \leq \theta \leq 180°$	$0 \leq \theta \leq \pi$
$\tan\theta$	$-90° < \theta < 90°$	$-\dfrac{\pi}{2} < \theta < \dfrac{\pi}{2}$

EXAMPLE 4

Find all angles θ between $0°$ and $360°$ for which $\sin\theta = 0.1357$.

SOLUTION

Enter 0.1357 into the display on the calculator. *Make sure the calculator is in degree mode.* Then press $\boxed{\text{INV}}$ $\boxed{\text{SIN}}$ (or $\boxed{\text{2nd F}}$ $\boxed{\text{SIN}}$). The resulting display will be approximately $\boxed{7.799099434}$. Since 0.1357 was presumably rounded to begin with, we round the answer to

$$\theta \approx 7.8° = 7°48'$$

But remember that the calculator always selects an acute angle when $\sin\theta > 0$. We must find any other requested θ's. Since $\sin\theta > 0$, the terminal ray of θ must fall in the first or second quadrant when θ is placed in standard position. In Figure 7.43 we can see that $\sin 172°12' = \sin 7°48'$. The requested angles are

$$\theta \approx 7°48' \quad \text{and} \quad \theta \approx 172°12' \quad \blacksquare$$

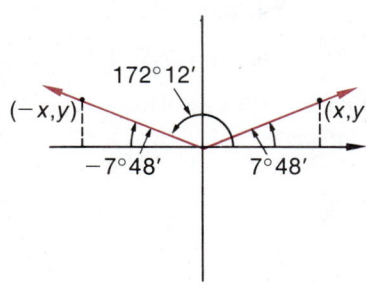

Figure 7.43

Section 7.5 Exercises

Extend to the indicated region the given portion of the graph of a function with the specified period.

1.

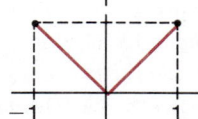

period = 2
a. extend to $[-3, 5]$
b. extend to $[-4, 6]$

2.

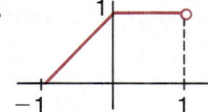

period = 2
a. extend to $[-3, 5]$
b. extend to $[-4, 8]$

3.

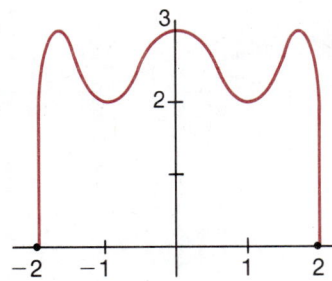

period = 4
a. extend to $[-6, 6]$
b. extend to $[-8, 9]$

4.

period = 4
a. extend to $[-6, 6]$
b. extend to $[-7, 9]$

5.

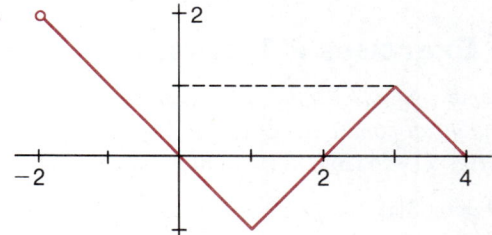

period = 6
a. extend to $[-8, 10]$
b. extend to $[-10, 15]$

If f is an odd function of period 3 and $f(1) = 10$, $f(\tfrac{1}{2}) = 4$, *find*

6. $f(0)$ **7.** $f(2)$

8. $f\left(\dfrac{5}{2}\right)$ **9.** $f(-2)$

10. $f(7)$ **11.** $f(9 - x)$

12. $f(-x - 15)$

If f is an even function of period 5 and $f(3) = 2$, $f(-4) = 1$, *find*

13. $f(-1)$ **14.** $f(2)$

15. $f(-6)$ **16.** $f(11)$

17. $f(-8)$ **18.** $f(15 - x)$

19. $f(-x + 25)$

If f is an even function of period 5 and f(2) = 2, f($\frac{1}{2}$) = 1, find

20. $f(-3)$ **21.** $f(3)$

22. $f\left(\frac{-9}{2}\right)$ **23.** $f\left(\frac{-11}{2}\right)$

24. $f(12)$ **25.** $f(20 - x)$

26. $f(-x + 65)$

If f is an odd function of period 4 and f(3) = 1, f($\frac{1}{3}$) = 2, find

27. $f(4)$ **28.** $f(1)$

29. $f(5)$ **30.** $f\left(\frac{11}{3}\right)$

31. $f\left(\frac{-13}{3}\right)$ **32.** $f(32 - x)$

33. $f(-x - 64)$ **34.** $f(3 - 4n)$, n an integer

Table Exercises (Optional)

Use Table 4 to find all angles between 0 and 2π that satisfy the given condition. Use $\pi \approx 3.142$. The results of Exercises 11–16, Section 7.3, may be helpful.

35. $\sin \theta = 0.0350$ **36.** $\cos \theta = -0.9260$

37. $\tan \theta = -0.1145$ **38.** $\sec \theta = 2.322$

39. $\cot \theta = -1.682$ **40.** $\csc \theta = -1.620$

Use Table 5 to find all angles between 0° and 360° that satisfy the given condition. The results of Exercises 17–22, Section 7.3, may be helpful.

41. $\sin \theta = 0.0780$ **42.** $\tan \theta = 53$

43. $\csc \theta = 1.200$ **44.** $\cot \theta = -2.06$

45. $\cos \theta = -0.0911$ **46.** $\sec \theta = -1.300$

Calculator Exercises (47–66)

Find all θ between 0 and 2π radians that satisfy the following. See Exercises 23–32, Section 7.3.

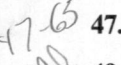

47. $\sin \theta = 0.6723$ **48.** $\cos \theta = 0.1552$

49. $\sec \theta = -4.713$ **50.** $\tan \theta = -0.7348$

51. $\cos \theta = 0.7292$ **52.** $\sin \theta = -0.6204$

53. $\cot \theta = -72.37$ **54.** $\csc \theta = 30.00$

55. $\tan \theta = 4.650$ **56.** $\cot \theta = -2.707$

Find all θ between 0° and 360° that satisfy the following. See Exercises 33–42, Section 7.3.

57. $\tan \theta = 4.832$ **58.** $\cos \theta = 0.6872$

59. $\cot \theta = 31.270$ **60.** $\sin \theta = 0.0341$

61. $\csc \theta = -1.313$ **62.** $\tan \theta = -0.0911$

63. $\sin \theta = -0.4295$ **64.** $\sec \theta = 23.74$

65. $\cos \theta = -0.0731$ **66.** $\cot \theta = -1.3502$

Refer to Figure 7.25 on page 390 to establish the following; k denotes an integer.

67. $\csc \theta = \dfrac{1}{\sin \theta}$, $\theta \neq k\pi$

68. $\sec \theta = \dfrac{1}{\cos \theta}$, $\theta \neq \dfrac{\pi}{2} + k\pi$

69. $\cot \theta = \dfrac{\cos \theta}{\sin \theta}$, $\theta \neq k\pi$

70. $\cot \theta = \dfrac{1}{\tan \theta}$, $\theta \neq k\pi$

Let θ be any angle in standard position. The terminal ray of $\theta \pm \pi$ is obtained by rotating the terminal ray of θ one half-revolution in either the clockwise or counterclockwise direction. The terminal rays for $\theta + \pi/2$ and $\theta - \pi/2$ lie halfway between the terminal rays for θ and $\theta \pm \pi$. Also note that the terminal ray for $-\theta$ is obtained by reflecting the terminal ray for θ through the x-axis. These properties are illustrated in the following figures.

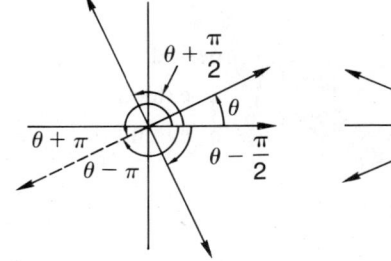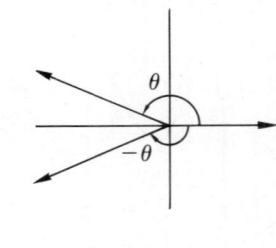

71. Exploit the symmetry properties evident in these relationships to evaluate $\sin(\theta \pm \pi)$ and $\cos(\theta \pm \pi)$ when θ is taken to be an angle in standard position having the given point on its terminal ray.

a. $(-4, 0)$ **b.** $(0, -5)$

c. $(2, 1)$ **d.** $(-3, 2)$

e. $(-4, -5)$ **f.** $(2, -3)$

72. Repeat Exercise 71 for $\theta + \dfrac{\pi}{2}$ in place of $\theta \pm \pi$.

73. Repeat Exercise 71 for $\theta - 90°$ in place of $\theta \pm \pi$.

74. Repeat Exercise 71 for $-\theta$ in place of $\theta \pm \pi$.

Evaluate all six trigonometric functions of the angle $(\pi - \theta)$ if θ is placed in standard position and

75. $(-2, 5)$ is on the terminal ray of θ.

76. $(-3, -4)$ is on the terminal ray of θ.

77. The terminal ray of θ coincides with the line $x - 4y = 0$ in the first quadrant.

78. The terminal ray of θ coincides with the line $3x + y = 0$ in the fourth quadrant.

79–82. Repeat Exercises 75–78 for $\pi/2 - \theta$ in place of $\pi - \theta$.

83. Use the symmetry of their terminal rays when the angles are placed in standard position to show that

$$\cos(\theta \pm \pi) = \cos(\theta \pm 180°)$$
$$= -\cos\theta$$
$$\sin(\theta \pm \pi) = \sin(\theta \pm 180°)$$
$$= -\sin\theta$$

and then show that

$$\sec(\theta \pm \pi) = \sec(\theta \pm 180°) = -\sec\theta$$
$$\csc(\theta \pm \pi) = \csc(\theta \pm 180°) = -\csc\theta$$

84. Show that

$$\cos(\pi - \theta) = \cos(180° - \theta)$$
$$= -\cos\theta$$
$$\sin(\pi - \theta) = \sin(180° - \theta)$$
$$= \sin\theta$$

and then show that

$$\sec(\pi - \theta) = \sec(180° - \theta) = -\sec\theta$$
$$\csc(\pi - \theta) = \csc(180° - \theta) = \csc\theta$$

(*Hint:* Use the properties established in Exercise 83 and the even-odd properties.)

85. Show that

$$\tan(180° - \theta) = \tan(\pi - \theta)$$
$$= -\tan\theta$$

and then show that

$$\cot(180° - \theta) = \cot(\pi - \theta) = -\cot\theta$$

86. Relate each of the trigonometric functions at $\theta + \pi/2$ to the same or another trigonometric function at θ.

87. Do the same for $\theta - \pi/2$.

88. Establish the cofunction properties:

$$\tan(90° - \theta) = \tan\left(\dfrac{\pi}{2} - \theta\right) = \cot\theta$$
$$\cot(90° - \theta) = \cot\left(\dfrac{\pi}{2} - \theta\right) = \tan\theta$$
$$\sec(90° - \theta) = \sec\left(\dfrac{\pi}{2} - \theta\right) = \csc\theta$$
$$\csc(90° - \theta) = \csc\left(\dfrac{\pi}{2} - \theta\right) = \sec\theta$$

89. Show that 2π or $360°$ is indeed the period of the sine, cosine, secant, and cosecant functions. [*Hint:* For the sine function you must find the smallest positive c for which $\sin(\theta + c) = \sin\theta$ for all θ. Consider $\theta = 0$ to show that $c = k\pi$ for k an integer. Then consider $\theta = \pi/2$ to show that $c = 2\pi$.]

90. Show that π or $180°$ is indeed the period of the tangent and cotangent functions.

Each of the following functions is periodic; find its period.

91. $k(v) = \sin 2v$

92. $g(u) = \cos 4u$

93. $F(y) = \cos 4y + \sin 2y$

94. $f(t) = \sin^2 t$

95. If f is an odd function of period 2, show that $f(n) = 0$ for every integer n.

Section 7.6

Graphing the Sine and Cosine Functions

In order to graph the circular functions, we plot their values at the "interesting" angles listed in Table 7.2 (Section 7.4): $\theta = 0$, $\pi/6$, $\pi/4$, $\pi/3$, $\pi/2$, π, $3\pi/2$, and 2π. Then we can use the properties of the circular functions to fill in the gaps and to extend the graph to include larger values of θ. We sketch these angles in standard position in the unit circle and label the corresponding points on the circle, as in Figure 7.44. Since $r = 1$,

$$x = \cos \theta$$
$$y = \sin \theta$$

Figure 7.44

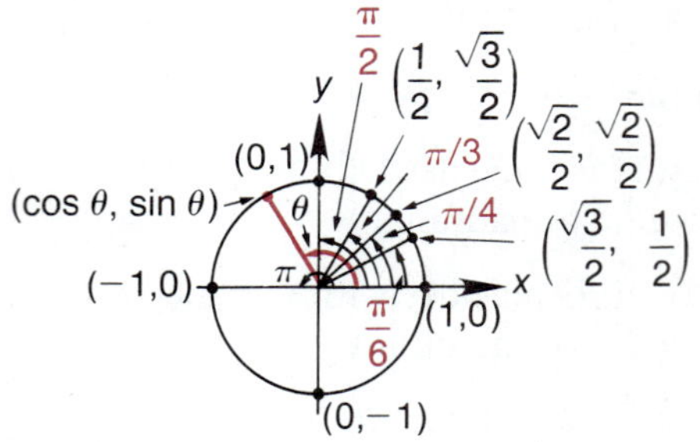

Graph of Sin θ

We note in Figure 7.44 that

1. $\sin 0 = 0$
2. $\sin \pi/2 = 1$
3. $\sin \theta$ increases from 0 to 1 as θ increases from 0 to $\pi/2$

This information enables us to sketch the graph of $\sin \theta$ for $0 \leq \theta \leq \pi/2$ (see Figure 7.45).

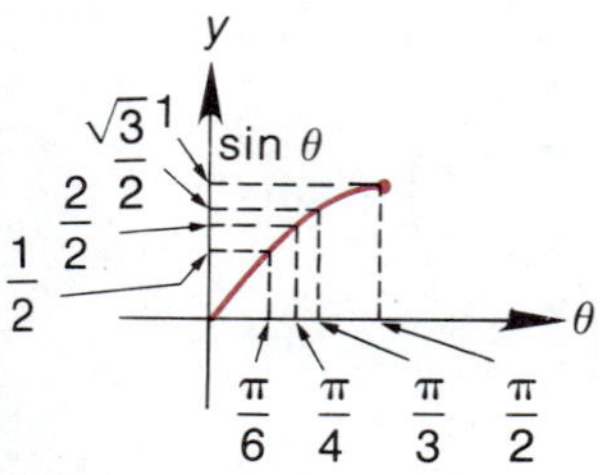

Figure 7.45
$y = \sin \theta, 0 \leq \theta \leq \dfrac{\pi}{2}$

Next, we observe in Figure 7.46a that

$$\sin\left(\frac{\pi}{2} + \alpha\right) = \sin\left(\frac{\pi}{2} - \alpha\right)$$

and extend this graph to $\pi/2 \leq \theta \leq \pi$ as in Figure 7.46b.

Figure 7.46

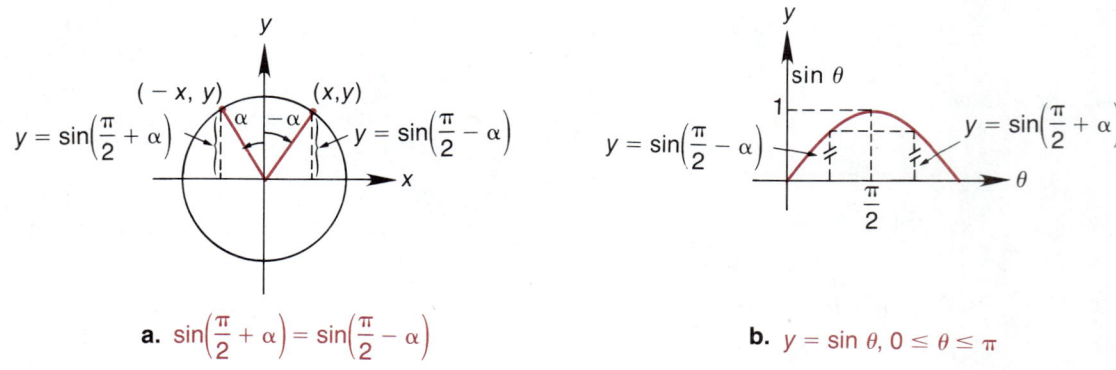

a. $\sin\left(\dfrac{\pi}{2} + \alpha\right) = \sin\left(\dfrac{\pi}{2} - \alpha\right)$

b. $y = \sin \theta, 0 \leq \theta \leq \pi$

Recall that $\sin(-\theta) = -\sin \theta$ in order to extend this graph to $-\pi \leq \theta \leq 0$ as in Figure 7.47.

We have now sketched the graph of $\sin \theta$ on an interval of length 2π: from $-\pi$ to π. Since $\sin \theta$ has period 2π, its graph now repeats itself over and over again to the left and to the right: $\sin(\theta + 2\pi) = \sin \theta$ for all θ (see Figure 7.48).

Figure 7.47

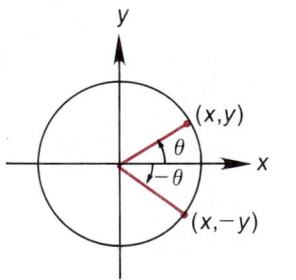

a. $\sin(-\theta) = -\sin\theta$

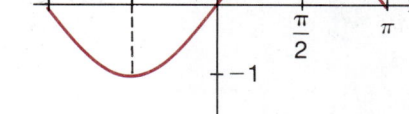

b. $y = \sin\theta, -\pi \leq \theta \leq \pi$

Figure 7.48
$y = \sin\theta$

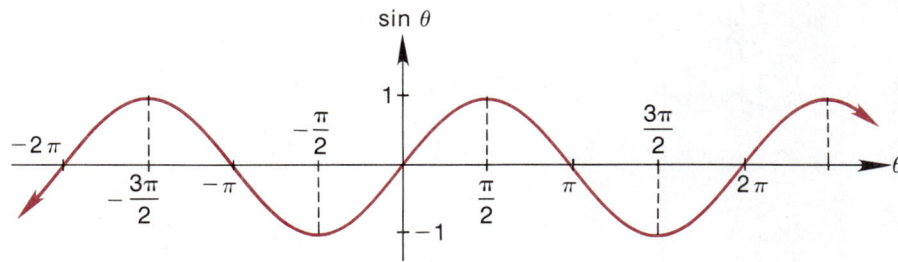

Graph of Cos θ

The graph of $y = \cos\theta$ can be obtained in a similar fashion. Observe in Figure 7.44 that

1. $\cos 0 = 1$
2. $\cos \pi/2 = 0$
3. $\cos\theta$ decreases from 1 to 0 as θ increases from 0 to $\pi/2$

The graph of $\cos\theta$ for $0 \leq \theta \leq \pi/2$ can be sketched as in Figure 7.49a. Next, observe in Figure 7.46a that $\cos(\pi/2 + \alpha) = -\cos(\pi/2 - \alpha)$. Then the graph of $\cos\theta$

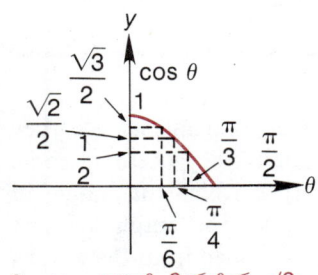

a. $y = \cos\theta, 0 \leq \theta \leq \pi/2$

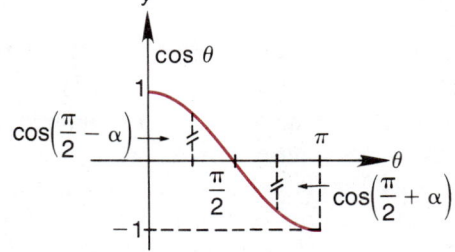

b. $y = \cos\theta, 0 \leq \theta \leq \pi$

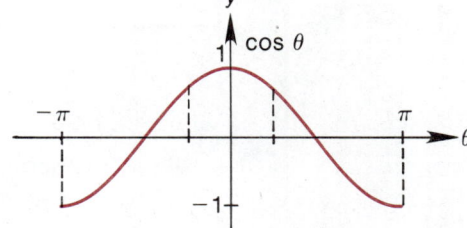

c. $y = \cos\theta, -\pi \leq \theta \leq \pi$

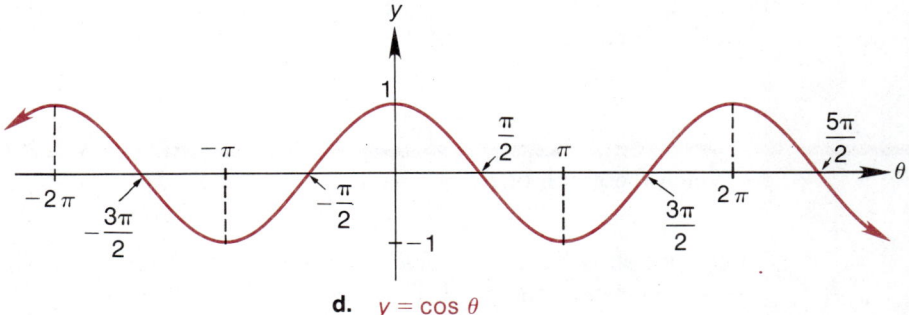

d. $y = \cos\theta$

Figure 7.49

can be extended to $\pi/2 \leq \theta \leq \pi$ as in Figure 7.49b. Finally, observe in Figure 7.47a that $\cos(-\theta) = \cos \theta$ and extend the graph of $\cos \theta$ to $-\pi \leq \theta \leq 0$ as in Figure 7.49c. The graph of $\cos \theta$ has now been sketched on an interval of length 2π. Since $\cos(\theta + 2\pi) = \cos \theta$, this graph now repeats itself to the left and to the right (see Figure 7.49d).

GRAPHS OF Sin θ AND Cos θ

1. Both $\sin \theta$ and $\cos \theta$ are periodic with period 2π.
2. Each of these functions has range $[-1, 1]$.
3. The basic period ($0 \leq \theta \leq 2\pi$) of each graph is included here. One complete period is called a **cycle**.

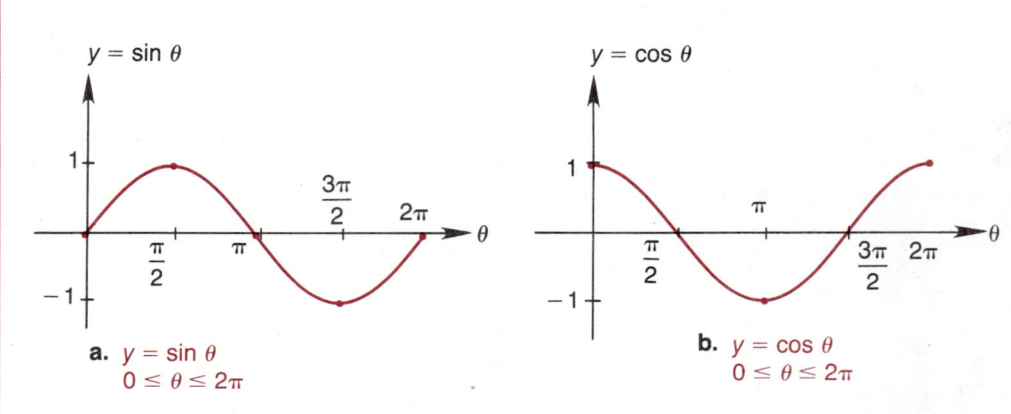

a. $y = \sin \theta$
$0 \leq \theta \leq 2\pi$

b. $y = \cos \theta$
$0 \leq \theta \leq 2\pi$

Figure 7.50

When the circular functions are encountered in engineering and science, they seldom appear in their pure form as $\sin \theta$ or $\cos \theta$. Usually they occur in some variation such as $2 \sin \theta$, $\cos 3t$, or even $7 \tan(2u + \pi/3)$ in which the "angle" θ is expressed as a function of time t or some other variable. The graphs of such functions are very helpful in solving problems. They are obtained from the graphs of the pure circular functions in a systematic manner. We rely heavily on the graphing techniques developed in Section 4.2: translation, reflection, and scaling (stretching or compressing). Unless otherwise indicated, the angle θ will be assumed to be expressed in radians.

EXAMPLE 1 Sketch the graph of $f(t) = 3 \sin t$.

SOLUTION This graph is basically a sine curve except that each y-coordinate on the curve $y = \sin t$ is multiplied by 3. The y-coordinates range between 3 and -3. This graph can be obtained by vertically scaling (see Section 4.2) the graph of $y = \sin t$, as shown in Figure 7.51.

Figure 7.51
$f(t) = 3 \sin t$

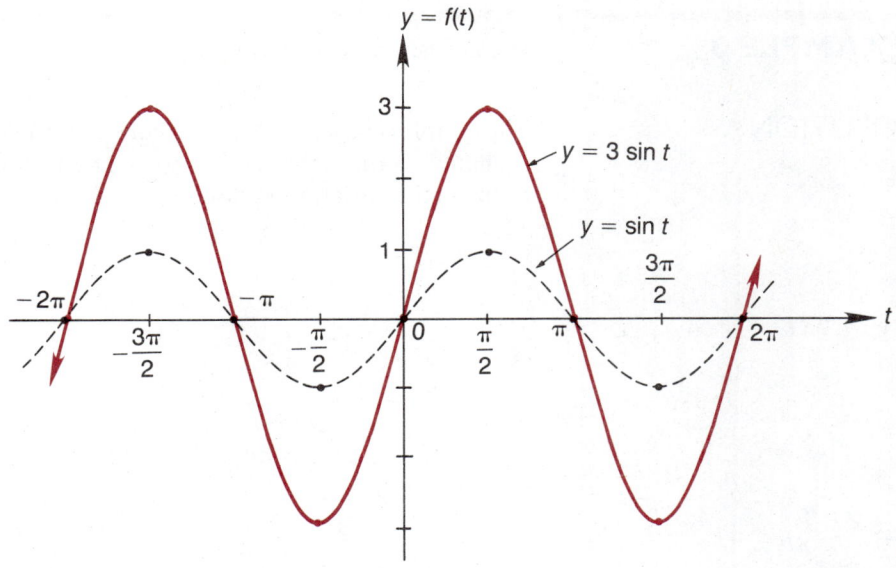

AMPLITUDE

The graphs of
$$y = A \sin \theta$$
and
$$y = A \cos \theta$$

have the same general shape as the basic sine and cosine curves, respectively, except that the range becomes $[-|A|, |A|]$. The number $|A|$ is called the **amplitude** of these functions and their graphs. If $A < 0$, the basic graph is also reflected through the x-axis.

EXAMPLE 2 Sketch the graph of $g(t) = |3 \sin t|$.

SOLUTION This graph is obtained by reflecting above the x-axis those portions of the graph in Figure 7.51 that lie below the x-axis. See Section 4.2. The graph is given in Figure 7.52.

Figure 7.52

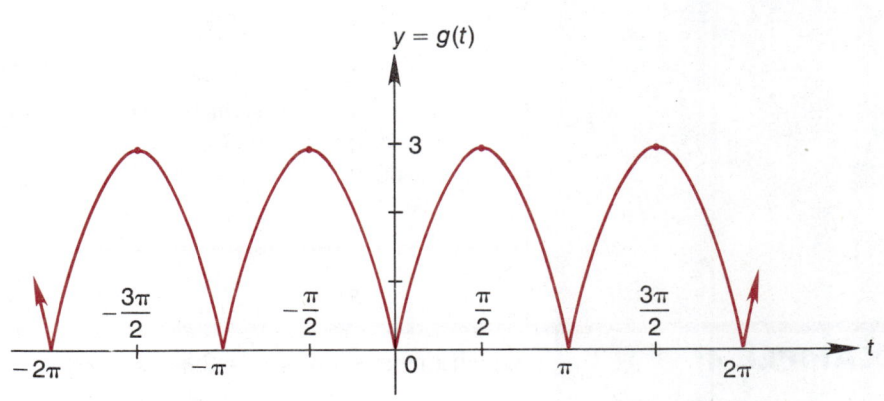

$g(t) = |3 \sin t|$

EXAMPLE 3

Sketch the graph of $f(x) = \cos(x - \pi)$.

SOLUTION

This graph is basically a cosine curve. However, the basic cosine curve $y = \cos x$ is shifted to the right π units; we plot cos 0 when $x = \pi$. (This is an example of horizontal translation as discussed in Section 4.2.) The graph is sketched in Figure 7.53.

Figure 7.53

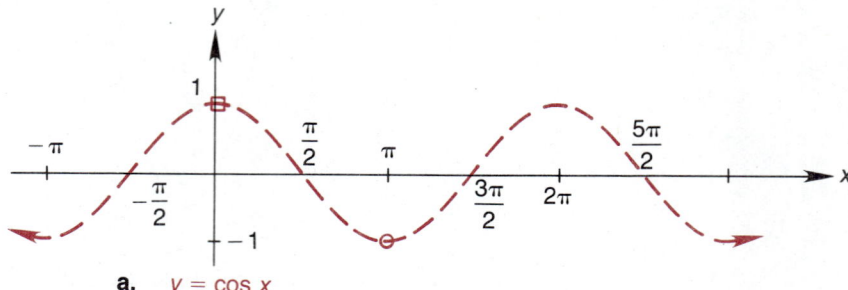

a. $y = \cos x$

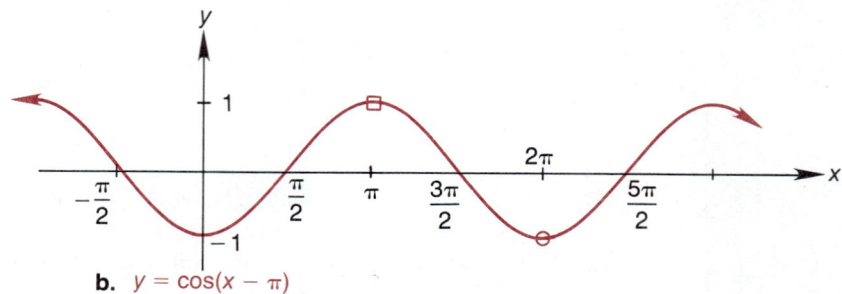

b. $y = \cos(x - \pi)$

PHASE SHIFT

The graphs of

$$y = \sin(x - c)$$

and

$$y = \cos(x - c)$$

have the same shape as the basic sine and cosine curves respectively. However, the basic graph is shifted horizontally. This horizontal displacement is called a **phase shift**. The basic graph is shifted to the right if $c > 0$ and to the left if $c < 0$.

EXAMPLE 4

Sketch the graph of $g(x) = 2 + \cos(x - \pi)$.

SOLUTION

This graph is obtained by translating the graph in Figure 7.53 two units upward. (See vertical translation in Section 4.2.) The graph is given in Figure 7.54.

Figure 7.54

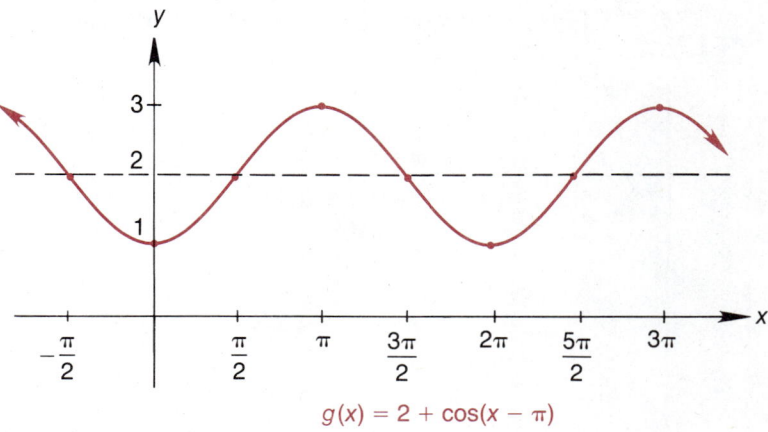

$g(x) = 2 + \cos(x - \pi)$

EXAMPLE 5 Sketch the graph of $g(t) = \cos 2t$.

SOLUTION Let $\theta = 2t$. We know that the cosine function is periodic with period 2π. This means that $\cos \theta$ runs through a complete period when $\theta = 2t$ runs from 0 to 2π. But if

$$0 \leq 2t \leq 2\pi$$

then

$$0 \leq t \leq \pi$$

That is, $\cos 2t$ runs through a complete period when t runs from 0 to π; $\cos 2t$ has period π. The graph has the same shape as the basic cosine curve but the graph of a complete period is "compressed" from $[0, 2\pi]$ to $[0, \pi]$, as illustrated in Figure 7.55.

Figure 7.55

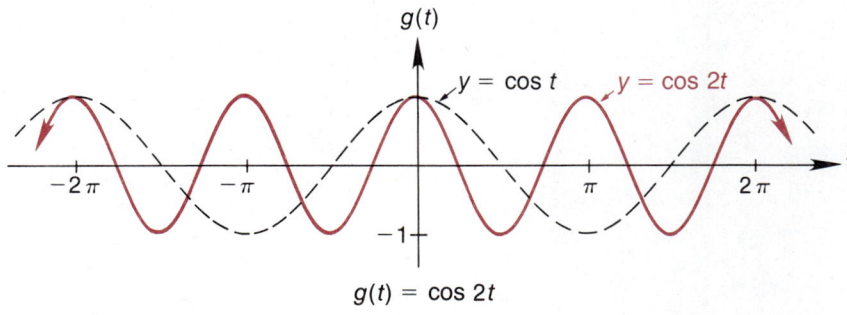

$g(t) = \cos 2t$

EXAMPLE 6 Sketch the graph of $g(t) = 2 \sin(-t/2)$.

SOLUTION Since sine is an odd function, we have

$$\begin{aligned} g(t) &= 2 \sin(-t/2) \\ &= -2 \sin(t/2) \end{aligned}$$

First we sketch the graph of $y = \sin t$ (Figure 7.56a). Then observe as in Example 5 that the graph of $y = \sin(t/2)$ is obtained by horizontally stretching the first graph. That is sin(*t*/2) runs through a complete period when

$$0 \le \frac{t}{2} \le 2\pi$$

or

$$0 \le t \le 4\pi$$

Figure 7.56

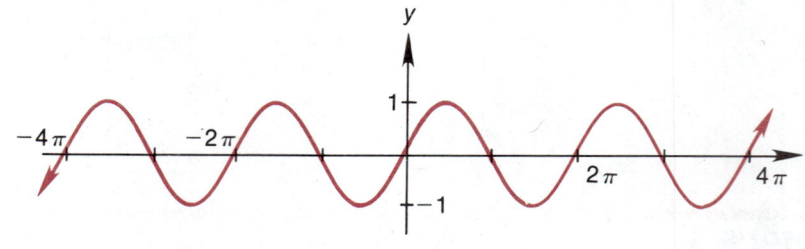

a. $y = \sin t$

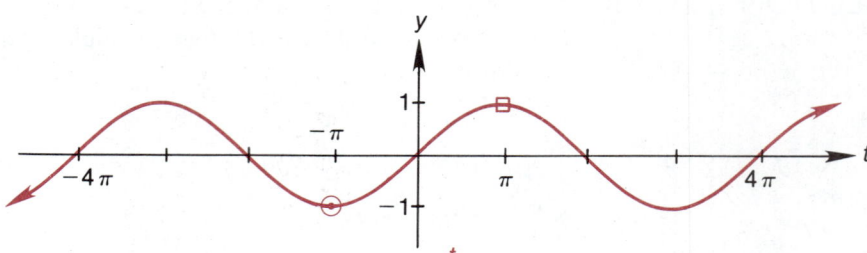

b. $y = \sin \dfrac{t}{2}$

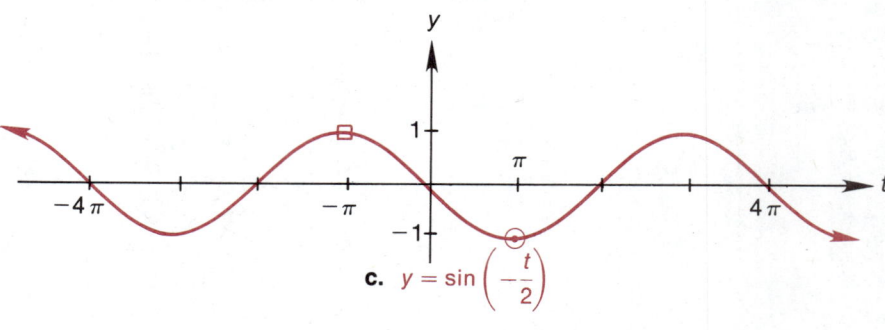

c. $y = \sin\left(-\dfrac{t}{2}\right)$

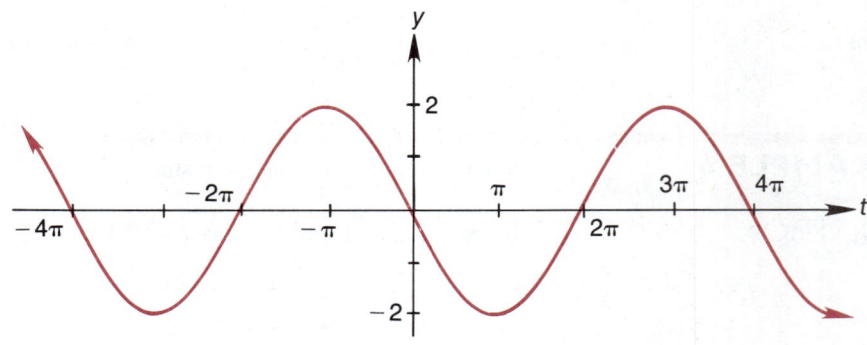

d. $y = 2\sin\left(-\dfrac{t}{2}\right) = -2\sin\left(\dfrac{t}{2}\right)$

Thus sin(*t*/2) has period 4π. See Figure 7.56b. Next we obtain the graph of $y = -\sin(t/2)$ by reflecting the graph of $y = \sin(t/2)$ through the *x*-axis, (as prescribed by the graphing techniques in Section 4.2). See Figure 7.56c. Finally, we double the *y*-coordinates of each point on this graph to obtain the graph of $y = 2\sin(-t/2) = -2\sin(t/2)$ as shown in Figure 7.5d. ∎

> ### PERIOD
> The graphs of
> $$y = \sin bt$$
> $$y = \cos bt \qquad (b > 0)$$
> have the same general shape as the basic sine and cosine curves respectively, except that the period is $2\pi/b$. If $b < 0$, use properties of odd or even functions respectively to express the function with a positive coefficient for *t*.

EXAMPLE 7

Sketch the graph of
$$f(t) = -\cos(2t - 3\pi)$$

SOLUTION

The graph of $y = \cos(2t - 3\pi)$ is a cosine curve that runs through its basic period when
$$0 \leq 2t - 3\pi \leq 2\pi$$
$$3\pi \leq 2t \leq 5\pi$$
$$\frac{3\pi}{2} \leq t \leq \frac{5\pi}{2}$$

The portion of the graph corresponding to this basic period is sketched in Figure 7.57a. The phase shift is $3\pi/2$; that is, we plot
$$\cos 0 = \cos(2t - 3\pi) \qquad \text{when} \qquad t = \frac{3\pi}{2}$$

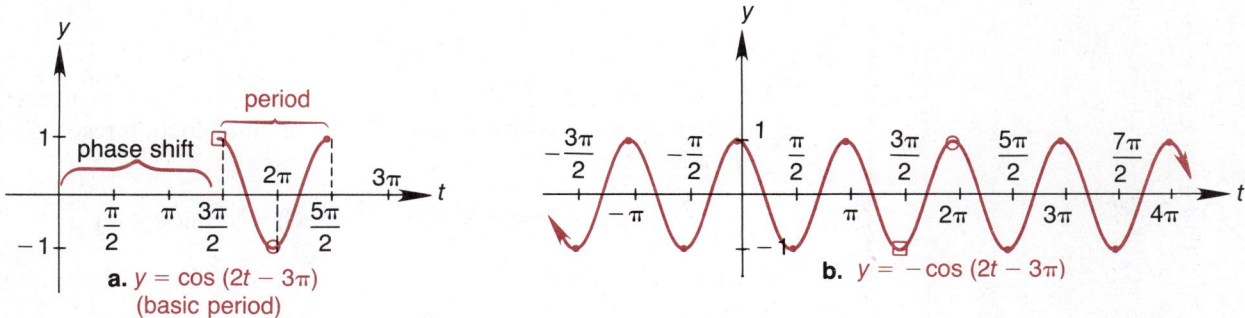

a. $y = \cos(2t - 3\pi)$ (basic period)

b. $y = -\cos(2t - 3\pi)$

Figure 7.57

The period is

$$\frac{5\pi}{2} - \frac{3\pi}{2} = \pi$$

The amplitude is $|A| = |-1| = 1$.

Finally, we reflect this portion of the graph of $y = \cos(2t - 3\pi)$ through the x-axis and then extend periodically to obtain the graph of $y = -\cos(2t - 3\pi)$. See Figure 7.57b. ∎

GRAPHING $y = A\cos(bt - c) + D$ **OR** $y = A\sin(bt - c) + D$

I. If $b > 0$:

1. Determine the period and phase shift of the graph by solving for t:

$$0 \le bt - c \le 2\pi$$

$$\frac{c}{b} \le t \le \frac{c}{b} + \frac{2\pi}{b}$$

  a. The period is

$$\frac{2\pi}{b}$$

  b. The phase shift is c/b. The graph shifts to the right if $c > 0$ and to the left if $c < 0$.

2. Horizontally scale the basic period ($0 \le \theta \le 2\pi$) of the cosine curve or sine curve respectively (see Figure 7.50) to fit the interval

$$\left[\frac{c}{b}, \frac{c}{b} + \frac{2\pi}{b} \right]$$

3. Vertically scale this curve to obtain amplitude $|A|$. If $A < 0$, the graph will also be reflected through the x-axis.
4. Extend this basic portion of the graph periodically.
5. Vertically translate the graph D units. The graph is moved upwards if $D > 0$ and downward if $D < 0$.

II. If $b < 0$:

1. Rewrite the expression with a positive coefficient for t by using the evenness or oddness of the cosine or sine function respectively. For instance,

$$\sin(-3t - \pi) = \sin(-[3t + \pi]) = -\sin(3t + \pi)$$

since $\sin(-\alpha) = -\sin \alpha$

2. Then follow the steps outlined in I.

Section 7.6 Graphing the Sine and Cosine Functions   417

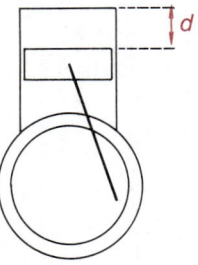

Figure 7.58

Many physical phenomena that vary periodically in time can be expressed in terms of sines and cosines. In mechanics and engineering, functions of the form $f(t) = A \sin(bt + c)$ or $g(t) = A \cos(bt + c)$ are used to describe oscillations and vibrations. A particle on a vibrating guitar string exhibits an important type of oscillatory phenomenon called **simple harmonic motion.** Seasonal changes can also be described by using such modified circular functions. Another example is given by a moving piston in an automobile engine. For instance, if a shaft is fastened near the circumference of a rotating wheel and to some other device, such as a piston in a cylinder as shown in Figure 7.58, the distance d is expressible in the form $A \sin(bt + c)$.

EXAMPLE 8

One periodic phenomenon that is familiar to everyone is the length of days. December 21 and June 21 are the shortest and longest days of the year, respectively. March 21 and September 21 are the equinoxes, in which there are 12 hours of daylight. The daylight time in between varies approximately like a sine function. If the shortest day has 9 hours of daylight and the longest day has 15 hours of daylight, we can sketch the graph indicating the number of hours of daylight as in Figure 7.59. Develop a formula for the function $D(t)$ that indicates the number of hours of daylight on day t.

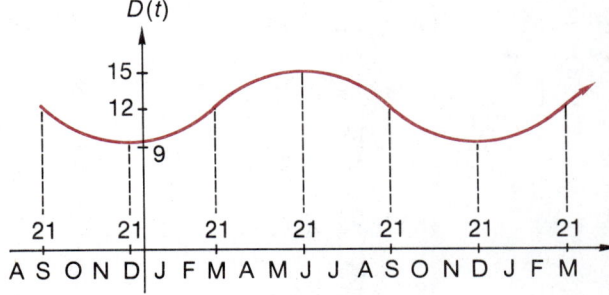

Figure 7.59

SOLUTION

The daylight function $D(t)$ is evidently periodic with a period of 365 days. To modify the basic sine curve to obtain a period of 365, recall that $\sin bt$ has period $2\pi/b$. Setting

$$365 = \frac{2\pi}{b}$$

we obtain

$$b = \frac{2\pi}{365}$$

If t is measured in days, $\sin(2\pi/365)t$ has period 365. In Figure 7.59 the point that corresponds to $\theta = 0$ on the basic sine curve occurs on March 21, the 80th day of the year. This indicates a phase shift of 80. Thus

$$\sin\left[\frac{2\pi}{365}(t - 80)\right]$$

includes the proper phase shift and period. The curve in Figure 7.59 oscillates between $y = 9$ and $y = 15$; it oscillates three units above and below $y = 12$. That is, its amplitude is $A = 3$. The daylight function $D(t)$ sketched in Figure 7.59 is expressed as

$$D(t) = 12 + 3 \sin\left[\frac{2\pi}{365}(t - 80)\right]$$

$$= 12 + 3 \sin\left[\frac{2\pi t}{365} - \frac{32\pi}{73}\right]$$

Section 7.6 Exercises

Sketch the graphs of the following functions.

1. $f(t) = -2 \sin t$
2. $f(t) = 5 \sin t$
3. $f(t) = 4 \cos t$
4. $f(t) = -3 \cos t$
5. $f(t) = -\frac{1}{2} \cos t$
6. $f(t) = \frac{1}{2} \sin t$
7. $f(t) = \cos \pi t$
8. $f(t) = \sin \pi t$
9. $f(t) = \sin 2t$
10. $f(t) = \cos 3t$
11. $f(t) = \sin\left(-\frac{\pi t}{2}\right)$
12. $f(t) = \cos\left(-\frac{\pi t}{4}\right)$
13. $f(t) = -2 \sin 3t$
14. $f(t) = -2 \cos(-3t)$
15. $f(t) = 3 \cos(-\pi t)$
16. $f(t) = 2 \sin\left(\frac{\pi t}{2}\right)$
17. $f(t) = 3 \sin(-\pi t)$
18. $f(t) = 2 \sin(-2t)$
19. $f(t) = \sin(t + \pi)$
20. $f(t) = \cos\left(t + \frac{\pi}{2}\right)$
21. $f(t) = \cos\left(t - \frac{3\pi}{2}\right)$
22. $f(t) = \sin\left(t - \frac{\pi}{2}\right)$
23. $f(t) = \cos(t + \pi)$
24. $f(t) = \sin(t - \pi)$
25. $f(t) = \sin\left(\frac{\pi}{2} - 3t\right)$
26. $f(t) = \cos\left(\frac{\pi}{3} - 2t\right)$
27. $f(t) = 2 \cos\left(t + \frac{\pi}{6}\right)$
28. $f(t) = -\cos\left(t + \frac{\pi}{4}\right)$
29. $f(t) = -\sin\left(t - \frac{\pi}{6}\right)$
30. $f(t) = 2 \sin\left(t - \frac{\pi}{2}\right)$
31. $f(t) = -2 \sin(t - \pi)$
32. $f(t) = 3 \cos(t + \pi)$
33. $f(t) = 2 \cos(3t - \pi)$
34. $f(t) = 3 \sin(2t + \pi)$
35. $f(t) = -2 \sin\left(\frac{t}{2} + \pi\right)$
36. $f(t) = 2 \cos\left(\frac{t}{3} - \frac{\pi}{3}\right)$
37. $f(t) = -3 \cos\left(\frac{t}{3} - \frac{\pi}{2}\right)$
38. $f(t) = -2 \sin\left(\frac{t}{2} - \frac{\pi}{4}\right)$
39. $f(t) = 2 \sin(\pi - 2t)$
40. $f(t) = -\cos\left(\frac{\pi}{2} - 2t\right)$
41. $f(t) = 2 - \sin 4t$
42. $f(t) = 1 + \cos 2t$
43. $f(t) = 1 + \cos\left(t + \frac{\pi}{2}\right)$
44. $f(t) = \sin\left(\pi - \frac{t}{2}\right) - 1$
45. $f(t) = 3 - 2 \cos(2t - \pi)$
46. $f(t) = 4 + 3 \sin\left(\frac{t}{2} - \frac{\pi}{4}\right)$
47. $f(t) = -2 + 2 \sin\left(\pi - \frac{t}{2}\right)$
48. $f(t) = -1 - 2 \cos(\pi - 2t)$

Review the techniques for graphing with absolute values (Section 4.2) and then sketch the graphs of the following functions.

49. $f(t) = |\sin t|$
50. $f(t) = |\cos t|$

51. $f(t) = \cos|t|$
52. $f(t) = \sin|t|$
53. $f(t) = -2|\cos(-3t)|$
54. $f(t) = 2|\sin 2t|$
55. $f(t) = 2 - |\sin 4t|$
56. $f(t) = |\cos 2t| - 1$
57. $f(t) = 2 - \sin|4t|$
58. $f(t) = (\cos|2t|) - 1$
59. $f(t) = 3\left|\sin\left|t - \dfrac{\pi}{4}\right|\right|$
60. $f(t) = -2\left|\cos\left|t + \dfrac{\pi}{2}\right|\right|$

61. "Biorhythm" experts tell us that each human being has three cycles: physical, emotional, and intellectual. The respective lengths of these cycles are 23, 28, and 33 days. Each cycle is represented by a sine curve and begins at birth. Times in which all three curves are above the horizontal axis are "good"; if all three curves are below this axis it is a "bad" time.
    a. Sketch portions of these three curves.
    b. For how many days after birth does a newborn have good days?
    c. How soon after birth do bad days begin?
    d. How long does the first bad period last?
    e. When does the next good period begin?

62. The moon has a period of 28 days. If a full moon occurs on January 15, sketch a curve indicating the portion of the moon that is lighted on a given night and express this in terms of the sine function.

63. The mean (or average) daily temperature is another periodic phenomenon. The lowest mean temperature in a particular town is 12°F and occurs on January 20. The highest mean temperature is 92°F and occurs on July 20. Assuming that the mean daily temperature varies like a cosine function,
    a. Sketch the graph of the average daily temperature.
    b. Express the mean daily temperature in terms of a cosine function.

64. Let the lowest mean temperature in a given city be 55°F occurring on January 10 and the highest mean temperature be 85°F occurring on July 10. Also assume that mean temperature varies like a sine function.
    a. Sketch the graph of the daily mean temperature.
    b. Express the daily mean temperature in terms of a sine function.

65. In a certain county, the average number of deaths per month is found to vary like a sine function between 60 (in August) and 100 (in February).
    a. Sketch a rough graph indicating monthly deaths.
    b. Express the average monthly death rate in terms of a sine function.

66. The flywheel illustrated in the figure has a diameter of 1 meter. It revolves clockwise, making three revolutions per minute. Express the distance d as a function of t if the given configuration is for $t = 0$. Sketch the graph of $d(t)$.

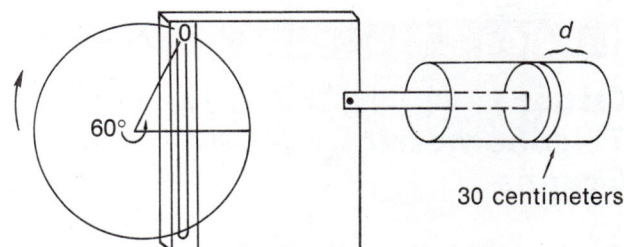

67. The electricity we use in our homes is called alternating current (ac). It gets its name because the voltage alternates between ± 120 volts. The effect is alternately to draw electrons from and return them to the earth, which is at "ground" or 0 voltage. The fluctuations are periodic and hence the voltage can be described in terms of a sine function. In the United States the current alternates at 60 cycles per second. *Sketch a graph that describes the voltage in this situation; you may use any phase shift you like since the shift will depend on when the generators are activated.* (When "phasing in" more generators to handle peak demands, power companies must "tune" them to be in phase with those already in operation.)

68. Suppose that the pollution level in a certain lake varies with the seasons according to the formula
$$P = 50 - 30\cos\left(\dfrac{2\pi t}{365}\right)$$

where t is given in days measured from the beginning of the year. A treatment program initiated by the Department of Wildlife reduces the pollution level by 50%. However, due to unforeseen reactions, the cycle is also delayed 30 days.

a. Write the formula for the pollution level after the treatment.

b. Sketch the graphs of the pollution level both before and after the treatment.

69. Both rabbits and foxes inhabit a certain wildlife area. As the rabbit population increases, the foxes have a plentiful food supply and increase their population. But when the fox population gets too large, too many rabbits are consumed. This results in a declining rabbit population followed by a declining fox population. When the fox population has decreased to a certain level, the rabbits can again increase and multiply. It takes some time before the foxes begin to increase their population, since their breeding stock is too low. Then the cycle repeats itself. Write equations and sketch graphs indicating the rabbit and fox populations under the following conditions.

   i. Both populations fluctuate on an 8-year cycle.
   ii. The fluctuations are approximately sinusoidal.
   iii. The rabbit population oscillates between 30,000 and 50,000 individuals.
   iv. The fox population oscillates between 2000 and 8000 individuals.
   v. The foxes reach their maximum population 1 year after the rabbits do.
   vi. There were 50,000 rabbits in 1980.

Section 7.7

Other Trigonometric Graphs

In this section, we discuss the graphs of the remaining four circular functions: tangent, cotangent, secant, and cosecant. Then we present the graphing techniques relating to damping and addition of ordinates.

Graph of $\tan \theta$

To graph the tangent function, we begin again by plotting the interesting values from Table 7.2.

θ	0	$\dfrac{\pi}{6}$	$\dfrac{\pi}{4}$	$\dfrac{\pi}{3}$	$\dfrac{\pi}{2}$
$\tan \theta$	0	$\dfrac{1}{\sqrt{3}} = \dfrac{\sqrt{3}}{3}$	1	$\sqrt{3}$	Undefined

Then observe that as θ increases from 0 and approaches $\pi/2$, $\sin \theta$ gets close to 1 and that $\cos \theta$ is positive and gets close to 0; hence the quotient $\tan \theta = \sin \theta / \cos \theta$ gets larger and larger. Using the notation of Section 5.5, we have

$$\tan \theta \to \infty \quad \text{as} \quad \theta \to \left(\dfrac{\pi}{2}\right)^-$$

"$\theta \to (\pi/2)^-$" means "θ approaches $\pi/2$ but $\theta < \pi/2$"; "$\tan \theta \to \infty$" means "$\tan \theta$ gets arbitrarily large."

This observation enables us to sketch the graph of $\tan\theta$ for $0 \leq \theta < \pi/2$ as in Figure 7.60a. Note that the graph approaches but never quite touches the vertical line at $\theta = \pi/2$. Such a line is an **asymptote** of the graph. Since $\tan(-\theta) = -\tan\theta$, this graph can be extended to $-\pi/2 < \theta \leq 0$ as in Figure 7.60b. Finally, since the tangent function has period π, we repeat this sketch over and over again as in Figure 7.60c. The graph has vertical asymptotes at $\pi/2 + k\pi$, k any integer.

Figure 7.60

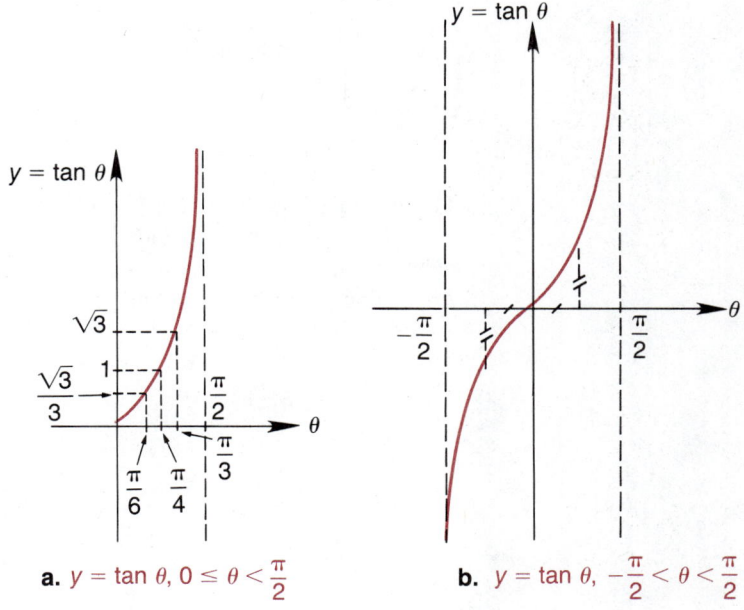

a. $y = \tan\theta$, $0 \leq \theta < \dfrac{\pi}{2}$

b. $y = \tan\theta$, $-\dfrac{\pi}{2} < \theta < \dfrac{\pi}{2}$

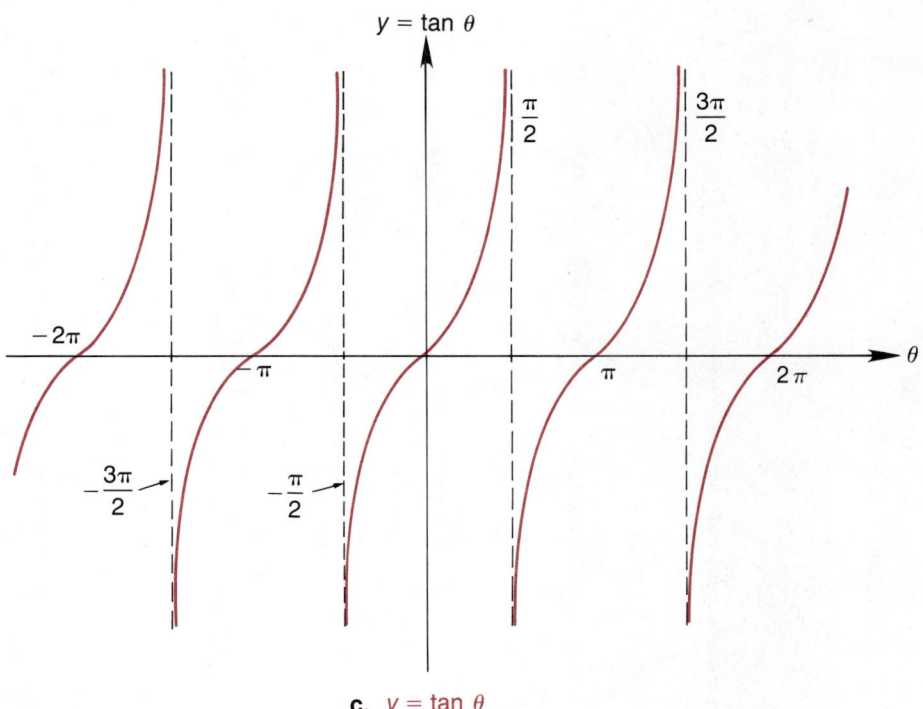

c. $y = \tan\theta$

Graph of cot θ

The interesting values from Table 7.2 are listed here.

θ	0	$\frac{\pi}{6}$	$\frac{\pi}{4}$	$\frac{\pi}{3}$	$\frac{\pi}{2}$
cot θ	Undefined	$\sqrt{3}$	1	$\frac{\sqrt{3}}{3}$	0

The graph of cot θ for $\theta \in (0, \pi/2]$ is sketched in Figure 7.61a. Since $\cot(-\theta) = -\cot \theta$, this graph is extended to $[-\pi/2, 0) \cup (0, \pi/2]$ in Figure 7.61b. The complete graph can now be obtained by observing that cot θ has period π. See Figure 7.61c. The graph of cot θ has vertical asymptotes at $\theta = k\pi$ for each integer k.

Figure 7.61

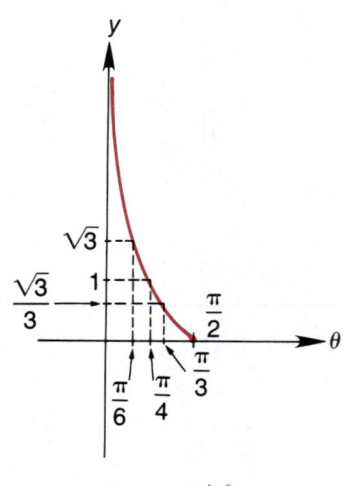

a. $y = \cot \theta$
$0 < \theta \leq \frac{\pi}{2}$

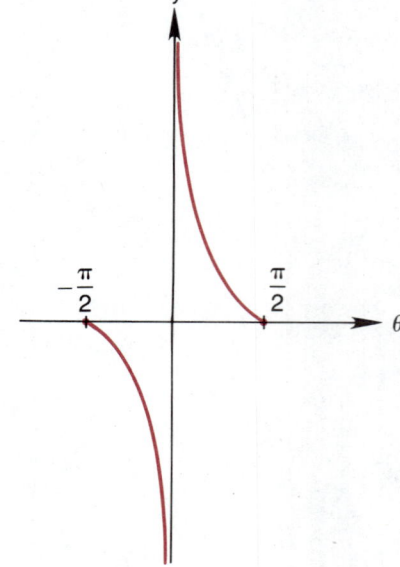

b. $y = \cot \theta$
$-\frac{\pi}{2} \leq \theta \leq \frac{\pi}{2}, \theta \neq 0$

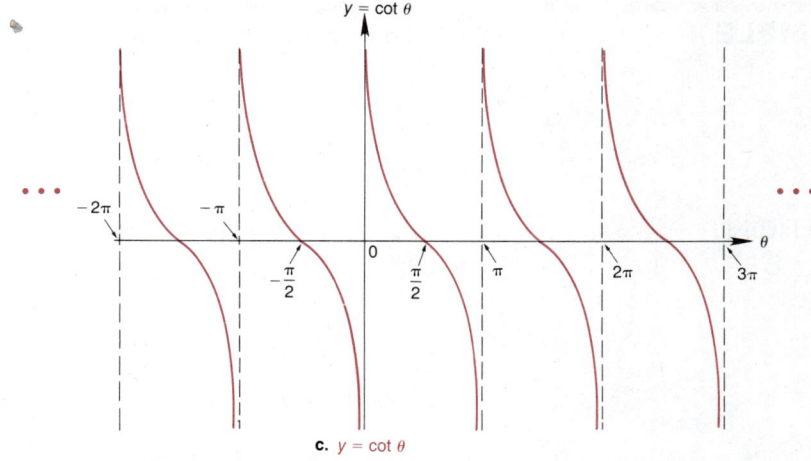

c. $y = \cot \theta$

GRAPHING tan θ AND cot θ

1. Both $\tan \theta$ and $\cot \theta$ are periodic with period π.
2. Each of these functions has range $(-\infty, \infty)$.
3. The graph corresponding to the basic period of each function is included here. See Figure 7.62. These are called the **basic branches** of the two graphs.

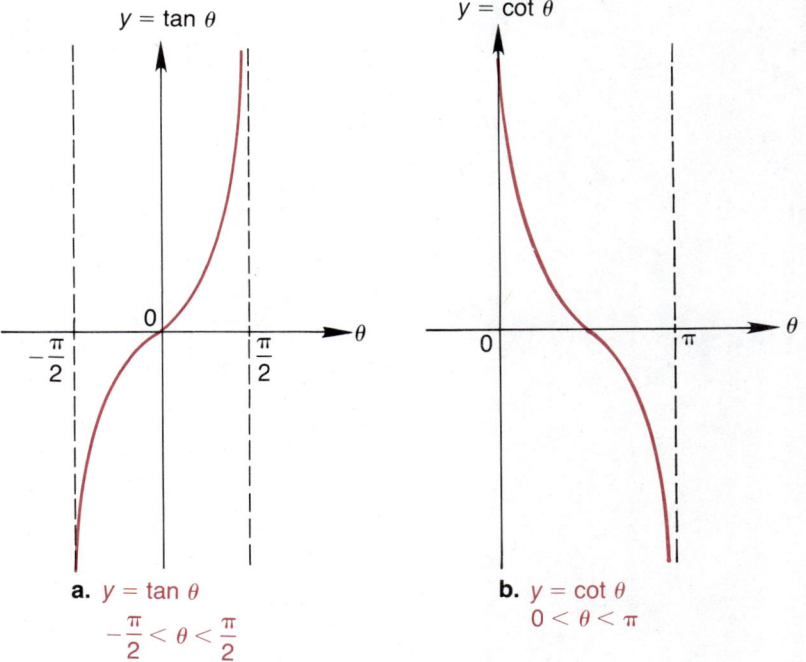

a. $y = \tan \theta$
$-\dfrac{\pi}{2} < \theta < \dfrac{\pi}{2}$

b. $y = \cot \theta$
$0 < \theta < \pi$

Figure 7.62

4. In each of these graphs, the branches are separated by vertical asymptotes at the values for which the function is undefined.
   a. $\tan \theta$ is undefined for $\theta = \pi/2 + k\pi$, k any integer.
   b. $\cot \theta$ is undefined for $\theta = k\pi$, k any integer.

EXAMPLE 1

Sketch the graphs of the following.

a. $y = \tan \dfrac{t}{2}$

b. $y = \cot 2t$

SOLUTION

a. Let $\theta = t/2$; $\tan \theta$ runs through a complete period when θ runs from $-\pi/2$ to $\pi/2$:

$$-\frac{\pi}{2} < \frac{t}{2} < \frac{\pi}{2}$$
$$-\pi < t < \pi$$

The basic branch of the tangent curve must be scaled horizontally to fit the interval $(-\pi, \pi)$. See Figure 7.63a. Then this basic part of the graph is extended periodically as in Figure 7.63b.

Figure 7.63

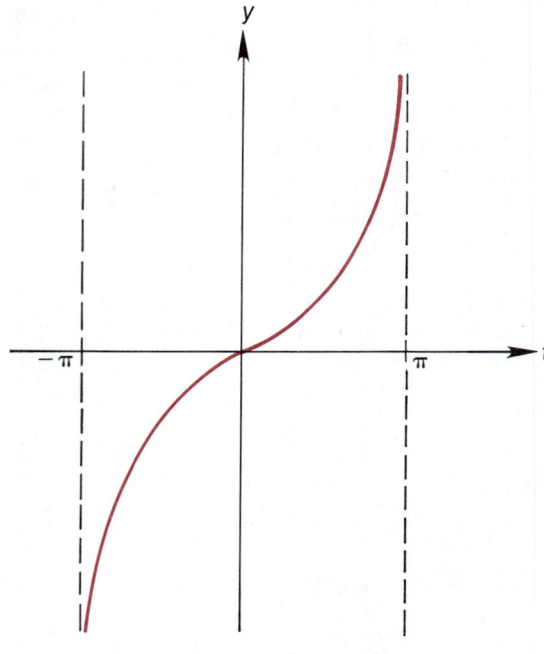

a. $y = \tan \dfrac{t}{2}$
$-\pi < t < \pi$

Figure 7.63 (Continued)

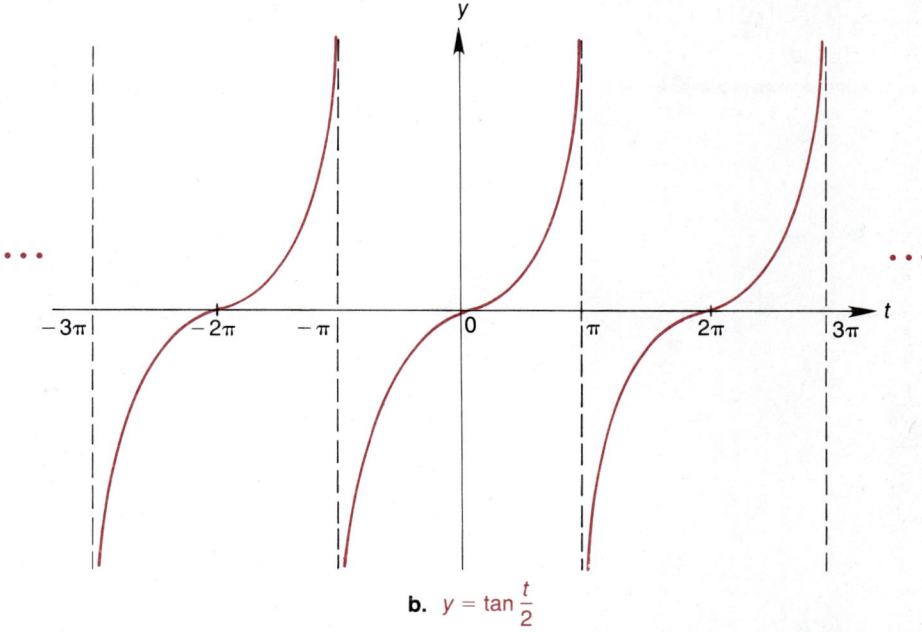

b. $y = \tan \dfrac{t}{2}$

b. The graph of cot $2t$ runs through a complete period when

$$0 < 2t < \pi$$
$$0 < t < \dfrac{\pi}{2}$$

This function has period $\pi/2$. The basic branch of the cotangent curve must be horizontally scaled to fit the interval $(0, \pi/2)$. See Figure 7.64a. Then this basic part of the graph is extended periodically as in Figure 7.64b.

Figure 7.64

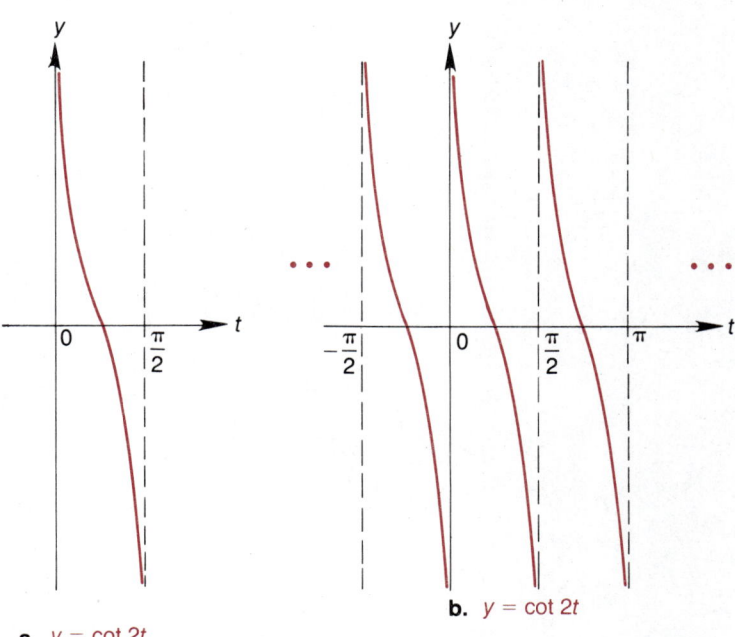

a. $y = \cot 2t$
$0 < t < \dfrac{\pi}{2}$

b. $y = \cot 2t$

Graphs of csc θ and sec θ

The graphs of csc θ and sec θ can be obtained by considering the graphs of sin θ and cos θ. Since

$$\csc \theta = \frac{1}{\sin \theta} \quad \text{and} \quad |\sin \theta| \leq 1$$

we have

$$|\csc \theta| \geq 1$$

As $\theta \to k\pi$, k an integer, $\sin \theta \to 0$, and then $\frac{1}{\sin \theta}$ gets larger and larger (positively or negatively). In fact

$$\csc \theta = \frac{1}{\sin \theta} \to \pm \infty \quad \text{as} \quad x \to k\pi, k \text{ an integer}$$

The graph of csc θ sketched in Figure 7.65a approaches but never contacts the

Figure 7.65

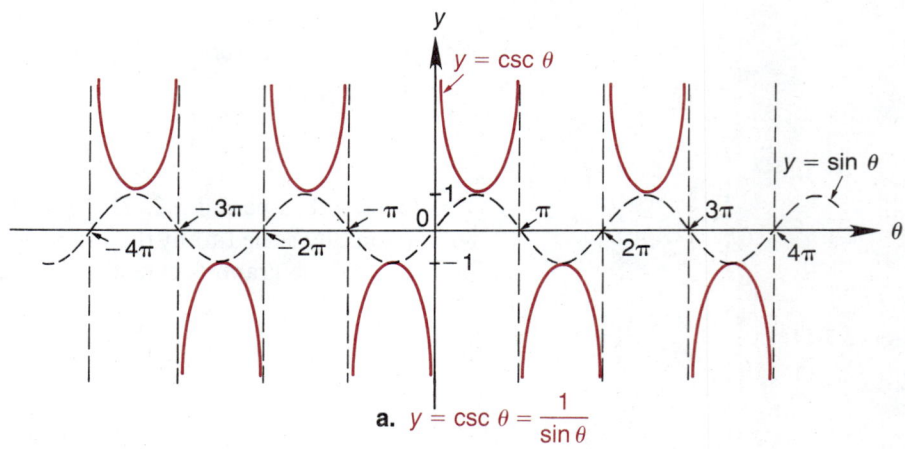

a. $y = \csc \theta = \frac{1}{\sin \theta}$

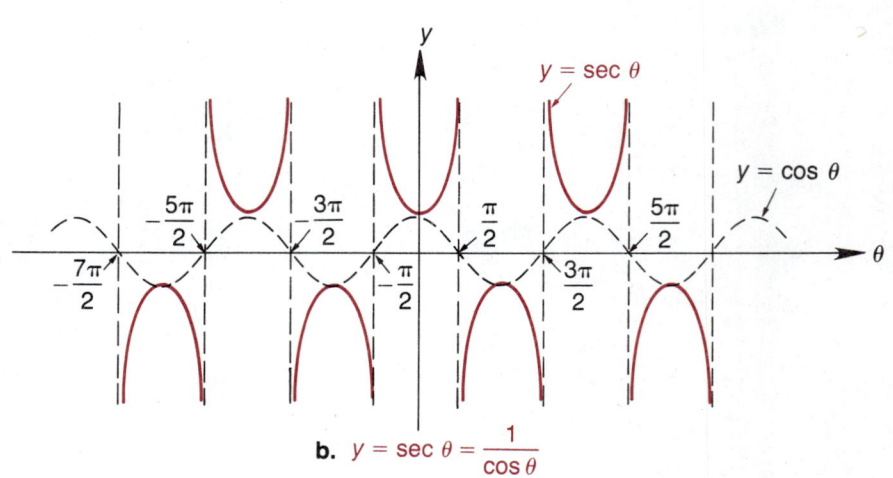

b. $y = \sec \theta = \frac{1}{\cos \theta}$

vertical lines at $\theta = 0, \pm \pi, \pm 2\pi, \ldots$. These are its asymptotes. The graph of $y = \sec \theta = \dfrac{1}{\cos \theta}$ is obtained in a similar fashion (Figure 7.65b).

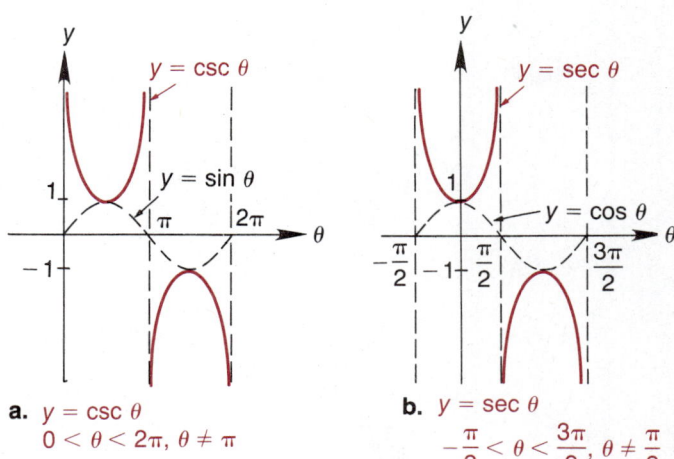

GRAPHING csc θ AND sec θ

1. Both $\csc \theta$ and $\sec \theta$ are periodic with period 2π.
2. Each of these functions has range $(-\infty, -1] \cup [1, \infty)$.
3. The graph corresponding to a basic period of each function is included here.

a. $y = \csc \theta$
$0 < \theta < 2\pi, \theta \neq \pi$

b. $y = \sec \theta$
$-\dfrac{\pi}{2} < \theta < \dfrac{3\pi}{2}, \theta \neq \dfrac{\pi}{2}$

Figure 7.66

EXAMPLE 2 Sketch the graph of $y = -3 \csc \dfrac{t}{2}$.

SOLUTION First we sketch the graph of $y = \csc \dfrac{t}{2}$. This graph runs through a complete period when $\dfrac{t}{2}$ runs from 0 to 2π:

$$0 < \dfrac{t}{2} < 2\pi, \quad \dfrac{t}{2} \neq \pi$$
$$0 < t < 4\pi, \quad t \neq 2\pi$$

Its period is 4π. The basic graph in Figure 7.66a must be scaled horizontally to fit the interval $(0, 4\pi)$ as shown in Figure 7.67a. Then the y-coordinates are multiplied by 3, as in Figure 7.67b. Finally, this graph is reflected through the x-axis and extended periodically to obtain the final graph $\left(y = -3 \csc \dfrac{t}{2}\right)$, as shown in Figure 7.67c.

Figure 7.67

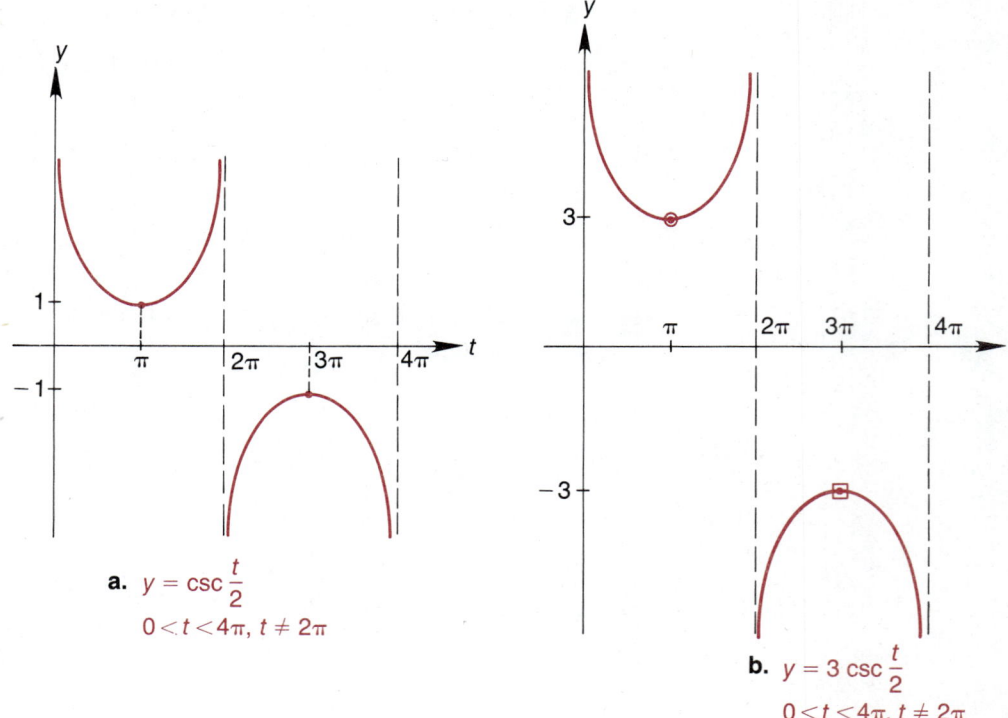

a. $y = \csc \dfrac{t}{2}$
$0 < t < 4\pi, t \neq 2\pi$

b. $y = 3 \csc \dfrac{t}{2}$
$0 < t < 4\pi, t \neq 2\pi$

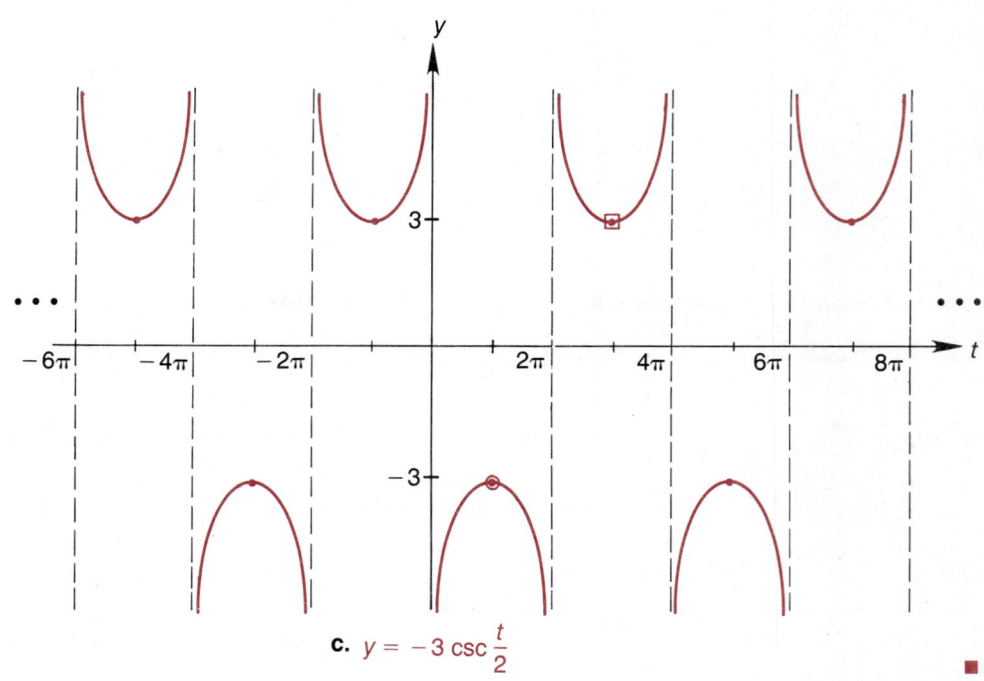

c. $y = -3 \csc \dfrac{t}{2}$

Graphing $A \cdot T(bx + c)$

We graph the general form

$$A \cdot T(bx + c)$$

of the other trigonometric functions (tan, cot, sec, and csc) in much the same way as we did for the sine and cosine functions. We begin with a basic period and then extend the graph.

EXAMPLE 3

Sketch the graph of

$$y = \sec(\pi - 2t)$$

SOLUTION

First obtain a positive coefficient for t as follows.

$$y = \sec(\pi - 2t) = \sec[-(2t - \pi)] = \sec(2t - \pi) \qquad [\sec(-\alpha) = \sec \alpha]$$

This graph runs through a complete period when $\theta = 2t - \pi$ runs from $-\pi/2$ to $3\pi/2$ (see Figure 7.66b):

$$-\frac{\pi}{2} < 2t - \pi < \frac{3\pi}{2}, \qquad 2t - \pi \neq \frac{\pi}{2}$$

$$\frac{\pi}{2} < 2t < \frac{5\pi}{2}, \qquad 2t \neq \frac{3\pi}{2}$$

$$\frac{\pi}{4} < t < \frac{5\pi}{4}, \qquad t \neq \frac{3\pi}{4}$$

The period is

$$\frac{5\pi}{4} - \frac{\pi}{4} = \pi$$

The basic graph in Figure 7.66b must be horizontally scaled to fit the interval $(\pi/4, 5\pi/4)$. See Figure 7.68a. The complete graph is obtained by extending this graph periodically; see Figure 7.68b.

Figure 7.68

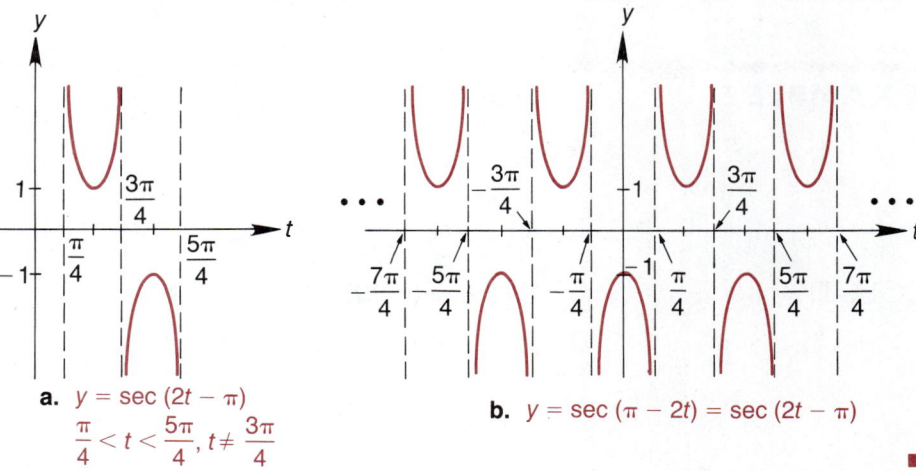

a. $y = \sec(2t - \pi)$
$\frac{\pi}{4} < t < \frac{5\pi}{4}, t \neq \frac{3\pi}{4}$

b. $y = \sec(\pi - 2t) = \sec(2t - \pi)$

Addition of Ordinates

When a function is expressed as a sum of two other functions, we sometimes graph the two parts on the same coordinate system and then add their ordinates (*y*-coordinates) to obtain the *y*-coordinate for the sum of the functions. This procedure is illustrated in the next example.

EXAMPLE 4

Sketch the graph of

$$f(x) = \sin x + \cos x$$

SOLUTION

First sketch the graphs of $y = \sin x$ and $y = \cos x$ on the same coordinate system. Then add their y-coordinates to obtain the y-coordinate at x for $f(x)$. Do this for enough points to detect the shape of the graph. Several such values are indicated on the graph, which is shown in Figure 7.69. The function f is periodic since

$$f(x + 2\pi) = \sin(x + 2\pi) + \cos(x + 2\pi)$$
$$= \sin x + \cos x$$
$$= f(x)$$

Figure 7.69

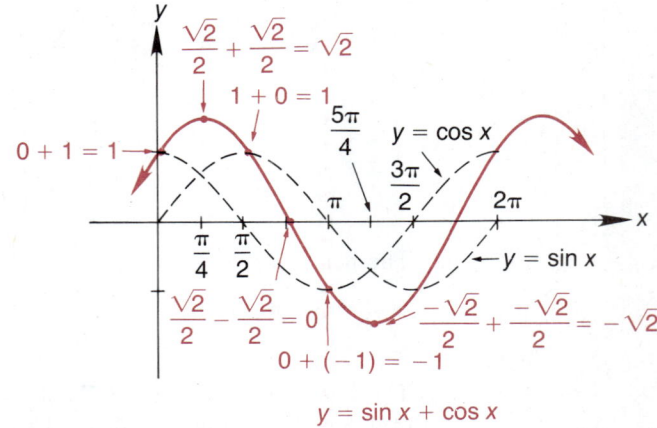

Damping

If the amplitude of an oscillatory phenomenon dies down, we say that the motion is **damped**. This situation is illustrated in the following example.

EXAMPLE 5

Sketch the graph of

$$y = 2^{-t} \cos t = \frac{\cos t}{2^t}$$

for $t \geq 0$.

SOLUTION

First observe that

$$|2^{-t} \cos t| = |2^{-t}||\cos t|$$
$$= 2^{-t}|\cos t|$$
$$\leq 2^{-t} \qquad \text{(Since } |\cos t| \leq 1\text{)}$$

Figure 7.70

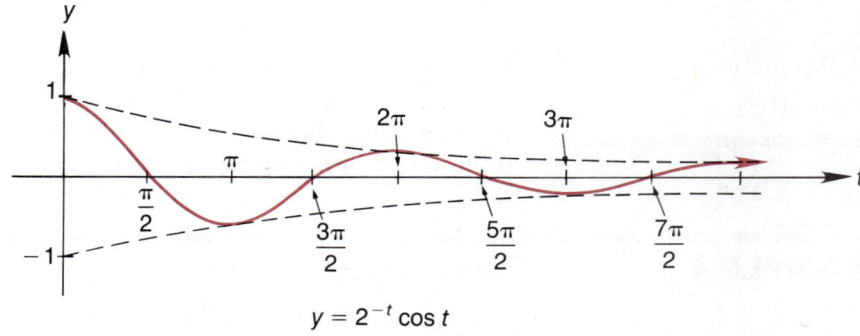

$y = 2^{-t} \cos t$

Section 7.7 Exercises

Sketch the graphs of the following functions.

1. $f(t) = -\tan t$
2. $f(t) = \frac{1}{2} \cot t$
3. $f(t) = 2 \sec t$
4. $f(t) = -\csc t$
5. $f(t) = \cot(-3t)$
6. $f(t) = \tan 2t$
7. $f(t) = \csc(-t)$
8. $f(t) = \sec(-t)$
9. $f(t) = \frac{1}{2} \tan 2t$
10. $f(t) = -\cot\left(\frac{t}{2}\right)$
11. $f(t) = -2 \sec 2t$
12. $f(t) = 2 \csc\left(-\frac{t}{2}\right)$
13. $f(t) = 1 - \cot(2t)$
14. $f(t) = \tan\left(\frac{t}{2}\right) - 1$
15. $f(t) = 2 \csc t - 2$
16. $f(t) = 2 - 3 \sec t$
17. $f(t) = \tan\left(t + \frac{\pi}{2}\right)$
18. $f(t) = \cot\left(\frac{\pi}{2} - t\right)$
19. $f(t) = \sec(\pi - t)$
20. $f(t) = \csc\left(t - \frac{\pi}{2}\right)$
21. $f(t) = 2 \tan\left(\frac{t}{2} + \frac{\pi}{3}\right)$
22. $f(t) = -\tan\left(\frac{\pi}{3} - 2t\right)$
23. $f(t) = -5 \cot\left(\frac{\pi}{3} - \frac{t}{2}\right)$
24. $f(t) = 2 \cot(2t - \pi)$
25. $f(t) = -\sec\left(2t + \frac{\pi}{2}\right)$
26. $f(t) = 3 \sec\left(\frac{\pi}{2} - 2t\right)$
27. $f(t) = 2 \csc(\pi - 2t)$
28. $f(t) = -\csc\left(\frac{t}{2} - \pi\right)$
29. $f(t) = \tan|t|$
30. $f(t) = \cot|t|$
31. $f(t) = |\tan t|$
32. $f(t) = |\cot t|$
33. $f(t) = \sec|t|$
34. $f(t) = \csc|t|$
35. $f(t) = |\sec t|$
36. $f(t) = |\csc t|$

Use the method of addition of ordinates to graph the following functions.

37. $f(t) = \sin t - \cos t$
38. $f(t) = 2 \cos t - \sin t$
39. $f(t) = \sin t + \sin 2t$
40. $f(t) = \cos 2t - \cos t$
41. $f(t) = 2 \sin t + \cos t$
42. $f(t) = \cos 2t + \cos t$

Use the technique of Example 5 to sketch the graph of the following functions.

43. $f(t) = t \sin t$
44. $f(t) = t \cos t$
45. $f(t) = e^{-t} \sin 2t$
46. $f(t) = e^{-t} \cos \frac{t}{2}$
47. $f(t) = \frac{t}{4} \sin(-2t)$
48. $f(t) = \frac{t}{2} \cos\left(\frac{t}{2}\right)$
49. $f(t) = t^2 \sin 4t$
50. $f(t) = t^2 \cos 2t$

51. Electric ranges and dryers require 240 volts of electricity. This is obtained by using two "hot" lines rather than a hot line and a ground. These two hot lines are out of phase in the sense that when one is at $+120$ volts, the other is at -120 volts. At these times, the difference in voltage is ± 240.

   a. Sketch a curve describing 240-volt current by analyzing two such 120-volt curves that are out of phase in this way.

   b. Express this voltage in terms of a sine function.

52. Two generators whose output voltages are expressed as sine waves have the same period and amplitudes of 120 volts and 60 volts, respectively. The generators are out of phase in the sense that the 60 volt generator hits its peak voltage when the 120-volt generator is at 0 voltage, after just having produced its peak voltage. Sketch the graph of the voltage produced by these two generators working together.

432 Chapter 7 The Trigonometric Functions

Section 7.8 Chapter Review

Terms and Concepts

Degree and Radian Measure of an Angle

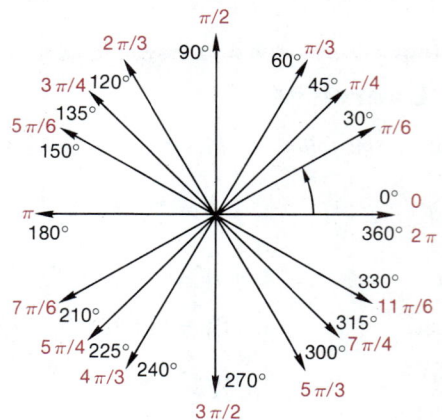

Acute Angle $\begin{cases} 0° < \theta < 90° \\ 0 < \theta < \dfrac{\pi}{2} \end{cases}$

Right Angle $\begin{cases} 90° \\ \dfrac{\pi}{2} \end{cases}$

Obtuse Angle $\begin{cases} 90° < \theta < 180° \\ \dfrac{\pi}{2} < \theta < \pi \end{cases}$

Straight Angle $\begin{cases} 180° \\ \pi \end{cases}$

Central Angle

Reference Angles

Quadrant I	Quadrant II	Quadrant III	Quadrant IV
a. $\theta^* = \theta$	b. $\theta^* = \pi - \theta$	c. $\theta^* = \theta - \pi$	d. $\theta^* = 2\pi - \theta$

The Trigonometric Functions

1. $\sin \theta = \dfrac{y}{r}$
2. $\cos \theta = \dfrac{x}{r}$
3. $\tan \theta = \dfrac{y}{x}, x \neq 0$
4. $\csc \theta = \dfrac{r}{y}, y \neq 0$
5. $\sec \theta = \dfrac{r}{x}, x \neq 0$
6. $\cot \theta = \dfrac{x}{y}, y \neq 0$

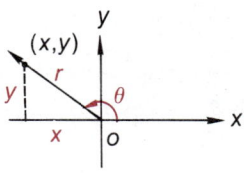

Circular Functions $T(s) = T(s \text{ radians})$ for each of the trigonometric functions T.

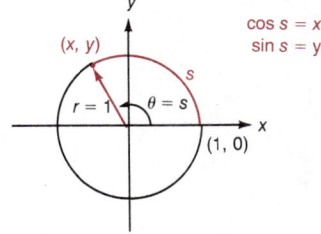

Right Triangle Trigonometry

1. $\sin \theta = \dfrac{\text{opposite}}{\text{hypotenuse}}$
2. $\cos \theta = \dfrac{\text{adjacent}}{\text{hypotenuse}}$
3. $\tan \theta = \dfrac{\text{opposite}}{\text{adjacent}}$
4. $\csc \theta = \dfrac{\text{hypotenuse}}{\text{opposite}}$
5. $\sec \theta = \dfrac{\text{hypotenuse}}{\text{adjacent}}$
6. $\cot \theta = \dfrac{\text{adjacent}}{\text{opposite}}$

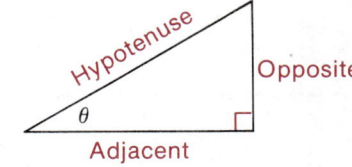

Rules and Formulas

Degree–Radian Conversion Formulas

1. $1° = \dfrac{\pi}{180}$ radians ≈ 0.017 radian
2. 1 radian $= \left(\dfrac{180}{\pi}\right)° \approx 57.3°$

Arc Length, Area, and Radian Measure

1. $\theta = \dfrac{s}{r}$
2. $s = r\theta$
3. $A = \left(\dfrac{1}{2}\right)r^2\theta$

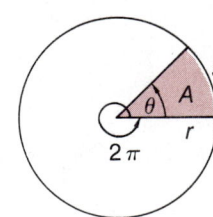

Pythagorean Theorem $c^2 = a^2 + b^2$

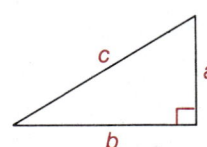

Chapter 7 The Trigonometric Functions

Important Values

DEGREES	RADIANS	$\sin\theta$	$\cos\theta$	$\tan\theta$	$\cot\theta$	$\sec\theta$	$\csc\theta$
0°	0	0	1	0	—	1	—
30°	$\dfrac{\pi}{6}$	$\dfrac{1}{2}$	$\dfrac{\sqrt{3}}{2}$	$\dfrac{\sqrt{3}}{3}$	$\sqrt{3}$	$\dfrac{2\sqrt{3}}{3}$	2
45°	$\dfrac{\pi}{4}$	$\dfrac{\sqrt{2}}{2}$	$\dfrac{\sqrt{2}}{2}$	1	1	$\sqrt{2}$	$\sqrt{2}$
60°	$\dfrac{\pi}{3}$	$\dfrac{\sqrt{3}}{2}$	$\dfrac{1}{2}$	$\sqrt{3}$	$\dfrac{\sqrt{3}}{3}$	2	$\dfrac{2\sqrt{3}}{3}$
90°	$\dfrac{\pi}{2}$	1	0	—	0	—	1
180°	π	0	−1	0	—	−1	—
270°	$\dfrac{3\pi}{2}$	−1	0	—	0	—	−1
360°	2π	0	1	0	—	1	—

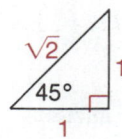
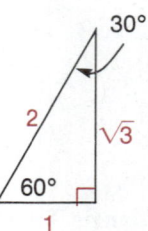

Reciprocal Properties (*k* denotes an integer)

1. $\csc\theta = \dfrac{1}{\sin\theta}, \quad \theta \neq k\pi$
2. $\sec\theta = \dfrac{1}{\cos\theta}, \quad \theta \neq \dfrac{\pi}{2} + k\pi$
3. $\cot\theta = \dfrac{\cos\theta}{\sin\theta} = \dfrac{1}{\tan\theta}, \quad \theta \neq k\pi$
4. $\sin\theta = \dfrac{1}{\csc\theta}$
5. $\cos\theta = \dfrac{1}{\sec\theta}$
6. $\tan\theta = \dfrac{\sin\theta}{\cos\theta} = \dfrac{1}{\cot\theta}, \quad \theta \neq \dfrac{\pi}{2} + k\pi$

Cofunction Properties

1. $\sin\left(\dfrac{\pi}{2} - \theta\right) = \cos\theta$
2. $\cos\left(\dfrac{\pi}{2} - \theta\right) = \sin\theta$
3. $\tan\left(\dfrac{\pi}{2} - \theta\right) = \cot\theta$
4. $\csc\left(\dfrac{\pi}{2} - \theta\right) = \sec\theta$
5. $\sec\left(\dfrac{\pi}{2} - \theta\right) = \csc\theta$
6. $\cot\left(\dfrac{\pi}{2} - \theta\right) = \tan\theta$

Periodicity

sine, cosine, secant, cosecant are periodic with period 2π or 360°
[e.g., $\sin(\theta + 2\pi) = \sin\theta$]

tangent, cotangent are periodic with period π or 180°
[e.g., $\tan(\theta + 180°) = \tan\theta$]

Symmetry

cosine, secant are even functions
[e.g., $\cos(-\theta) = \cos\theta$]

$\left.\begin{array}{l}\text{sine}\\ \text{cosecant}\\ \text{tangent}\\ \text{cotangent}\end{array}\right\}$ are odd functions [e.g., $\sin(-\theta) = -\sin\theta$]

Graphing Techniques

$y = \sin\theta$ (solid line)
$y = \csc\theta$ (broken line)

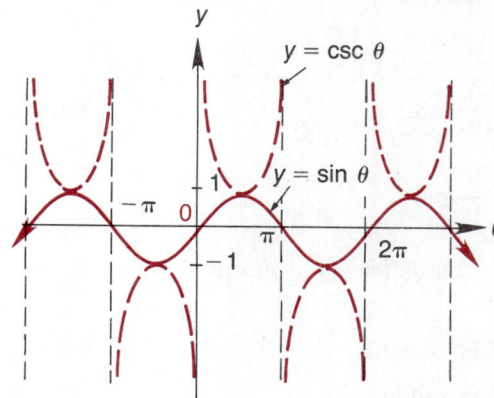

$y = \cos\theta$ (solid line)
$y = \sec\theta$ (broken line)

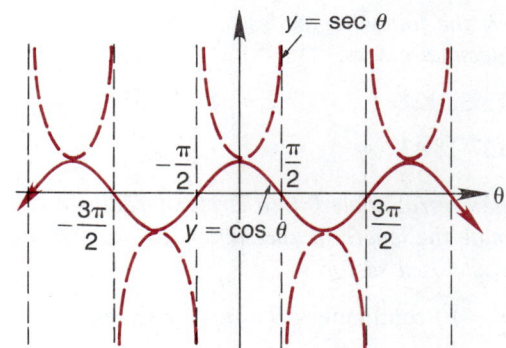

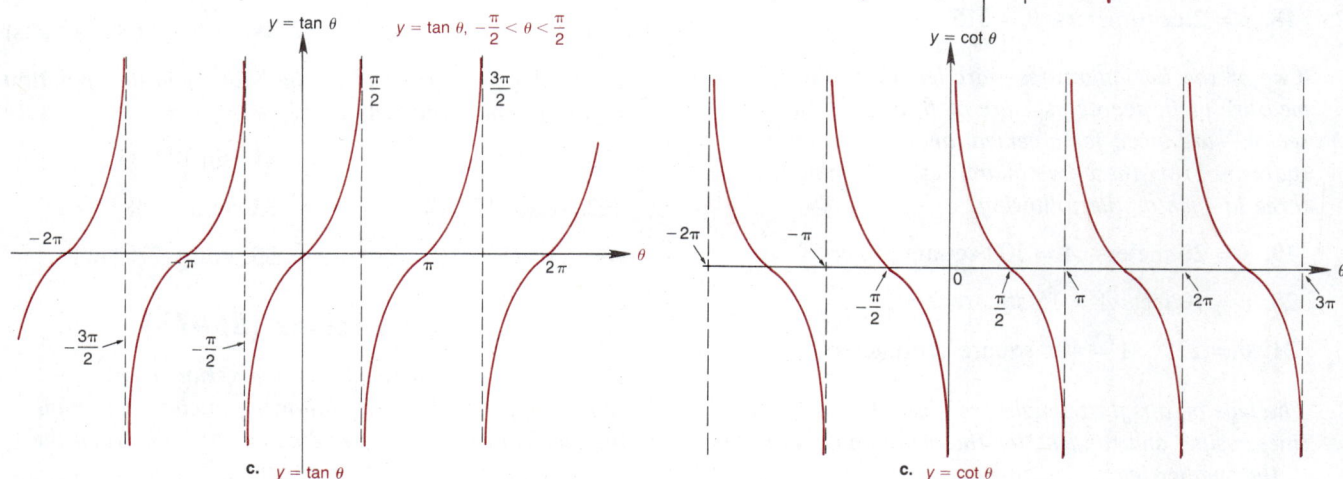

c. $y = \tan\theta$ c. $y = \cot\theta$

Period and Phase Shifts

For any of the trigonometric functions T, the graph of $T(bt - c)$ with $b > 0$ has the same basic shape as the graph of $T(\theta)$ but with a phase shift of c/b [plot $T(0)$ at $t = c/b$] and a period of $2\pi/b$ (π/b for tan and cot). If $b < 0$, use properties of even or odd functions to rewrite the expression with $b > 0$.

Section 7.9 Supplementary Exercises

Sketch in standard position the angle with the given measure. If the measure is given in degrees, convert to radians and vice versa.

1. 210°
2. −165°
3. −660°
4. 645°
5. 915°
6. $\pi/8$ rad
7. $\dfrac{23\pi}{12}$ rad
8. $\dfrac{-17\pi}{3}$ rad
9. $\dfrac{29\pi}{2}$ rad
10. $\dfrac{32\pi}{3}$ rad

Convert the following angles to degrees, minutes, and seconds.

11. 2.37 radians
12. −6.42 radians
13. 3.18 radians

Convert the following angles to radians correct to three decimal places.

14. 17°32′12″
15. −35°54′27″
16. 127°18′43″

For the central angle θ in a circle of radius r, determine the length of the subtended arc and the area of the enclosed sector.

17. $r = 10$ centimeters, $\theta = \pi/10$ radians
18. $r = 2$ centimeters, $\theta = 315°$

Two of the five quantities—arc length, radius, radian measure of θ, degree measure of θ, area of the sector—are given for a central angle θ in a circle of radius r. From these two quantities, determine the other three in each of the following.

19. $s = 20$ meters, $A = 100$ square meters
20. $r = 10$ feet, $A = 15$ square feet
21. $\theta = 225°$, $A = 40\pi$ square centimeters

The legs of a right triangle are given here, evaluate sine, cosine, and tangent for the acute angle adjacent to the smaller leg.

22. 2, 5
23. 6, 2
24. 4, 5
25. 2, 3
26. 1, 9

Solve each of the following right triangles on the basis of the given information: a *denotes the side opposite the acute angle α,* b *is the side opposite the acute angle β, and* h *is the hypotenuse.*

27. $a = 15.0$, $h = 36.0$
28. $h = 11.0$, $\alpha = 72°0′$
29. $a = 4.2$, $\beta = 53°18′$
30. $b = 4.7$, $\alpha = 60°0′$
31. $a = 5.0$, $b = 10.0$
32. $b = 2.3$, $\beta = 72°24′$
33. $h = 5.0$, $\beta = 83°54′$

Evaluate the six circular functions at the given angles. Do not use tables or a calculator.

34. −510°
35. −660°
36. 675°
37. 840°
38. 780°
39. 127π
40. $\dfrac{-639\pi}{2}$
41. $\dfrac{-195\pi}{4}$
42. $\dfrac{37\pi}{3}$
43. $\dfrac{-47\pi}{6}$

Table Exercises

Use Table 4 (at the end of the book) and interpolation to approximate the following. You may use $\pi \approx 3.142$.

44. csc(4.013 radians)
45. cos(6.582 radians)
46. sin(−0.732 radians)
47. tan(−1.632 radians)
48. sec(2.372 radians)
49. cot(−3.688 radians)

Use Table 5 (at the end of the book) and interpolation to approximate the following.

50. cos 322°48′
51. sin 147°39′
52. sec(−47°52′)
53. tan(−1683°38′)
54. csc 2127°44′
55. cot(−779°32′)

Calculator Exercises (56–73)

Use a calculator to evaluate all six trigonometric functions for each of the following angles. Determine the quadrant in which the terminal ray falls when the angle is placed in standard position.

56. 5.275 radians
57. 3.986 radians
58. −12.464 radians
59. −351°10′12″
60. 152°50′15″
61. −532°20′25″

Find all angles between 0 and 2π radians that satisfy the following.

62. $\sin \theta = 0.3342$
63. $\cos \theta = -0.7272$
64. $\cot \theta = -0.0314$
65. $\sec \theta = 3.102$
66. $\tan \theta = 2.147$
67. $\csc \theta = -10.37$

Find all angles θ between $0°$ and $360°$ that satisfy the following.

68. $\cos \theta = 0.2347$
69. $\tan \theta = -0.4182$
70. $\sec \theta = -7.243$
71. $\cot \theta = 0.8912$
72. $\sin \theta = -0.1043$
73. $\csc \theta = 6.348$

Evaluate each of the six trigonometric functions of θ in standard position if the indicated point is on the terminal side of θ.

74. $(0, 2)$
75. $(-1, 6)$
76. $(6, -3)$

Evaluate each of the six circular functions of θ in standard position if the terminal side of θ coincides with the given line in the indicated quadrant.

77. $4x - 3y = 0$, Quadrant I
78. $x - y = 0$, Quadrant III
79. $x + y = 0$, Quadrant II
80. $2x + y = 0$, Quadrant IV

81. Evaluate $\sin(\theta \pm \pi)$ and $\cos(\theta \pm \pi)$ if the given point is on the terminal ray of θ when θ is placed in standard position.
    a. $(0, 2)$
    b. $(-1, 6)$
    c. $(6, -3)$

82. Repeat Exercise 81 with $(\theta + 90°)$ in place of $(\theta \pm \pi)$.
83. Repeat Exercise 81 with $(\theta - \pi/2)$ in place of $(\theta \pm \pi)$.
84. Repeat Exercise 81 with $(-\theta)$ in place of $(\theta \pm \pi)$.
85. Relate each of the circular functions at $\theta + 3\pi/2$ to the same or another circular function at θ.
86. Do the same for $\theta - 3\pi/2$.
87. Evaluate all six circular functions of $\pi - \theta$ if θ is placed in standard position and
    a. $(2, 3)$ is on the terminal ray of θ
    b. $(4, -2)$ is on the terminal ray of θ
    c. The terminal ray of θ coincides with the line $2x + y = 0$ in the second quadrant
    d. The terminal ray of θ coincides with the line $2x - 3y = 0$ in the third quadrant.
88. Do the same for $\pi/2 - \theta$.

Using the given information, find the values of all six trigonometric functions of θ.

89. $\sin \theta = \frac{2}{3}$, $\tan \theta > 0$
90. $\cos \theta = -\frac{3}{5}$, $\csc \theta < 0$
91. $\csc \theta = -10$, $\sec \theta > 0$
92. $\sec \theta = -\frac{5}{3}$, $\tan \theta < 0$
93. $\tan \theta = 2$, $\cos \theta < 0$
94. $\cot \theta = \frac{3}{4}$, $\sec \theta > 0$

Find _____ if θ is an _____ angle and _____.

95. $\sin \theta$ obtuse $\cos \theta = -\frac{1}{5}$
96. $\sin \theta$ obtuse $\tan \theta = -2$
97. $\cot \theta$ acute $\sin \theta = \frac{3}{4}$
98. $\cos \theta$ acute $\csc \theta = \frac{11}{2}$

Each of the following functions is periodic; find its period.

99. $g(v) = 2 \sin 3v$
100. $H(u) = \cos \frac{u}{2}$
101. $F(w) = \cos 2w$
102. $f(t) = \tan \frac{t}{3}$

Extend the following graphs to the interval $[-4, 5]$ assuming each is periodic with period 3.

103.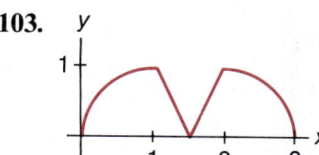

104.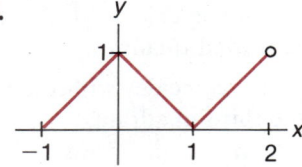

If f is an even function of period 3 and f(1) = 5, f(½) = −2, find

105. $f(2)$ **106.** $f(5/2)$

107–108. Rework the preceding two exercises if f is an odd function.

Sketch the graphs of the following functions f(t) =

109. $\sin 2\pi t$ **110.** $-2 \sec 2t$

111. $2 + \tan 2t$ **112.** $|\sin(-3t)|$

113. $\sin |2t|$

114. $\cot\left(\dfrac{\pi}{2} - t\right)$

115. $\tan\left(t - \dfrac{\pi}{2}\right)$

116. $\cos\left(t - \dfrac{\pi}{4}\right)$

117. $2 \sin\left(\dfrac{t}{3} - \dfrac{\pi}{3}\right)$

118. $4 \cot\left(3t + \dfrac{\pi}{2}\right)$

119. $\tan(\pi - 2t)$

120. $\csc\left(t + \dfrac{\pi}{6}\right)$

121. $3 \sin\left(\dfrac{5\pi}{3} - 2t\right)$

122. $2 + \cot(-t)$

123. $-|\cot(-t)|$

124. $\sin 2t + 2 \cos t$

125. $\cos 2t - 2 \sin t$

126. $2^{-t} \cos 2t$

127. $2^{-t} \sin 3t$

Trigonometric Identities and Equations; Inverse Trigonometric Functions

8

In Chapter 7, we discussed the circular, or trigonometric, functions and developed many of their properties. The relationship of the trigonometric functions to the unit circle establishes their periodicity. In this chapter we examine this relationship further in order to establish other relationships among the various trigonometric functions and to derive formulas for combinations of angles: 2α, $\alpha/2$, $\alpha \pm \beta$.

Those equations or relationships that hold for *all* angles in the common domain of the given functions are called **identities.** The identities permit us to rewrite certain trigonometric expressions in several equivalent forms. Consideration of more than one such representation can give helpful insights into phenomena such as radio and TV broadcasting and "beats" that are sometimes heard when two different tones are played simultaneously.

We also consider equations that are satisfied by only certain values of θ. Such equations arise, for example, in the design of electronic instruments, which are activated at a certain threshold level, and in studying the relationships among several oscillatory phenomena—biorhythms, for instance. Solving such equations will involve finding θ for which $T(\theta)$ takes on a given value (where T represents one of the trigonometric functions); we again encounter a need for "inverses" of the trigonometric functions.

Section 8.1

Elementary Identities

A given periodic relationship can be expressed in many different forms, which students frequently encounter when checking answers to textbook problems. The equivalent forms or relationships given in the exercises and examples of this section and the next are in themselves of minor significance; however, the exercise of establishing them should generate the important ability to manipulate the circular functions and to recognize equivalent forms.

In Chapter 7 we observed that

$$\sin(\theta + 2\pi) = \sin \theta$$

$$\tan \theta = \frac{\sin \theta}{\cos \theta}$$

$$\cot(-\theta) = -\cot \theta$$

$$\sin\left(\frac{\pi}{2} - \theta\right) = \cos \theta$$

$$\cos(\pi + \theta) = -\cos \theta$$

Each of these properties is an **identity**; *each holds for all values of θ for which the functions are defined.* Identities differ from **conditional equations,** which hold for certain values of θ but not for all values of θ.

EXAMPLE 1   Show that $\sin \theta + \cos \theta = 1$ is a conditional equation by exhibiting a value of θ that satisfies the equation and another value of θ that does not satisfy the equation.

SOLUTION $\theta = 0$ satisfies the equation:

$$\sin 0 + \cos 0 = 0 + 1 = 1$$

$\theta = \dfrac{\pi}{4}$ does not satisfy the equation:

$$\sin \frac{\pi}{4} + \cos \frac{\pi}{4} = \frac{\sqrt{2}}{2} + \frac{\sqrt{2}}{2} = \sqrt{2} \neq 1$$

Conditional equations are studied more extensively in Section 8.6. We are concerned primarily with identities in this and the next sections.

The basic trigonometric identities are those expressing the trigonometric functions in terms of sines and cosines. From these we obtain the reciprocal identities. These are listed here for reference.

FUNDAMENTAL IDENTITIES

1. $\csc \theta = \dfrac{1}{\sin \theta}$

2. $\sec \theta = \dfrac{1}{\cos \theta}$

3. $\tan \theta = \dfrac{\sin \theta}{\cos \theta}$

4. $\cot \theta = \dfrac{\cos \theta}{\sin \theta}$

5. $\sin \theta = \dfrac{1}{\csc \theta}$

6. $\cos \theta = \dfrac{1}{\sec \theta}$

7. $\cot \theta = \dfrac{1}{\tan \theta}$

8. $\tan \theta = \dfrac{1}{\cot \theta}$

EXAMPLE 2   Use the fundamental identities to verify that the following equation is an identity.

$$\cos \theta \csc \theta = \cot \theta$$

SOLUTION   We begin by converting the left side of the equation into an expression involving only sines and cosines. Then we identify the resulting expression with $\cot \theta$.

$$\cos \theta \csc \theta = \cos \theta \, \frac{1}{\sin \theta} \qquad \text{(By formula 1)}$$

$$= \frac{\cos \theta}{\sin \theta}$$

$$= \cot \theta \qquad \text{(By formula 4)}$$

Another important set of identities relates the arguments $-\theta$ and θ.

Section 8.1 Elementary Identities 441

> **EVEN–ODD IDENTITIES**
>
> 9. $\sin(-\theta) = -\sin\theta$ 12. $\csc(-\theta) = -\csc\theta$
> 10. $\cos(-\theta) = \cos\theta$ 13. $\sec(-\theta) = \sec\theta$
> 11. $\tan(-\theta) = -\tan\theta$ 14. $\cot(-\theta) = -\cot\theta$

EXAMPLE 3 Verify the identity

$$\sin(-\theta)\cos(-\theta)\tan\theta + (\sin\theta)^2 = 0$$

SOLUTION Again, convert to sines and cosines. Then use the even–odd identities.

$$\sin(-\theta)\cos(-\theta)\tan\theta + (\sin\theta)^2$$

$$= \sin(-\theta)\cos(-\theta)\frac{\sin\theta}{\cos\theta} + (\sin\theta)^2 \qquad \text{(By formula 3)}$$

$$= (-\sin\theta)(\cos\theta)\frac{\sin\theta}{\cos\theta} + (\sin\theta)^2 \qquad \text{(By formulas 9 and 10)}$$

$$= -(\sin\theta)^2 + (\sin\theta)^2$$

$$= 0 \qquad \blacksquare$$

To simplify writing $(\sin\theta)^2$, we write $\sin^2\theta$. In the same way, $\cos^2\theta = (\cos\theta)^2$, and so on.

Inscribing an angle θ in standard position in the unit circle, we see that the point in which the terminal ray meets the circle has coordinates $(\cos\theta, \sin\theta)$ as illustrated in Figure 8.1. Using the Pythagorean formula on the right triangle illustrated in the figure, we have $x^2 + y^2 = 1$ or

$$\cos^2\theta + \sin^2\theta = 1 \tag{15}$$

Figure 8.1

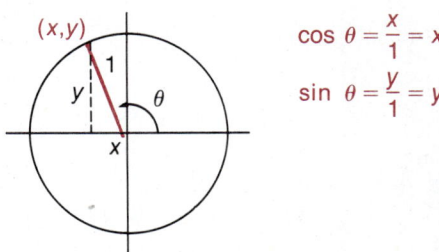

Dividing each term of equation (15) by $\cos^2\theta$ yields

$$\frac{\cos^2\theta}{\cos^2\theta} + \frac{\sin^2\theta}{\cos^2\theta} = \frac{1}{\cos^2\theta}$$

or

$$1 + \tan^2\theta = \sec^2\theta \tag{16}$$

Similarly, dividing each term of equation (15) by $\sin^2 \theta$, we obtain

$$\frac{\cos^2 \theta}{\sin^2 \theta} + \frac{\sin^2 \theta}{\sin^2 \theta} = \frac{1}{\sin^2 \theta}$$

or

$$\cot^2 \theta + 1 = \csc^2 \theta \qquad (17)$$

In equations (16) and (17), when either side of the equation is undefined, the other side is also undefined. Because of their dependence on the Pythagorean formula, these three identities are called the **Pythagorean identities;** they should be memorized.

PYTHAGOREAN IDENTITIES

$$\sin^2 \theta + \cos^2 \theta = 1 \qquad (15)$$
$$1 + \tan^2 \theta = \sec^2 \theta \qquad (16)$$
$$1 + \cot^2 \theta = \csc^2 \theta \qquad (17)$$

Alternate forms of (15) are

15a. $1 - \sin^2 \theta = \cos^2 \theta$
15b. $1 - \cos^2 \theta = \sin^2 \theta$

EXAMPLE 4 Verify the identity

$$(1 + \sin u)(1 - \sin u) = \frac{1}{\sec^2 u}$$

SOLUTION
$$\begin{aligned}(1 + \sin u)(1 - \sin u) &= 1 - \sin^2 u \\ &= \cos^2 u \qquad \text{(By formula 15a)} \\ &= \frac{1}{\sec^2 u} \qquad \text{(By formula 6)} \quad \blacksquare\end{aligned}$$

Trigonometric identities can be used to find values of the trigonometric functions when the value of some other function is known. This is illustrated in Example 5.

EXAMPLE 5 If θ is an acute angle and $\cos \theta = \frac{3}{7}$, find $\tan \theta$.

SOLUTION Formula 16 relates $\tan \theta$ and $\sec \theta \ (= 1/\cos \theta)$. First we write

$$\sec \theta = \frac{1}{\cos \theta} = \frac{7}{3}$$

Then

$$1 + \tan^2 \theta = \sec^2 \theta = \left(\frac{7}{3}\right)^2 \quad \text{(Formula 16)}$$

$$= \frac{49}{9}$$

$$\tan^2 \theta = \frac{49}{9} - 1 = \frac{40}{9}$$

$$\tan \theta = \pm\sqrt{\frac{40}{9}}$$

$$= \pm\frac{2\sqrt{10}}{3}$$

Since θ is an acute angle, $\tan \theta > 0$:

$$\tan \theta = \frac{2\sqrt{10}}{3}$$

Section 8.1 Exercises

Show that the following expressions are not identities, but exhibit at least one solution for each.

1. $\tan \theta + \cot \theta = 2$
2. $\sin \theta \cos \theta = \frac{1}{2}$
3. $\tan^2 \theta + \sin^2 \theta = 0$
4. $\sin^2 \theta - \cos^2 \theta = 0$
5. $\sin^2 \theta + \sec^2 \theta = 1$
6. $\sin x + \tan x = 0$
7. $\cos x - \sin x = 1$
8. $\cos^2 x + \tan^2 x = 1$
9. $\cot^2 x + \sin^2 x = 1$
10. $\sec^2 x + \tan^2 x = 3$

Convert the given expressions into sines and cosines and simplify.

11. $\sec t \cot t$
12. $\csc t \tan t$
13. $\sec^2 u \cos^2 u$
14. $\sec^2 u + \csc^2 u$
15. $\sec t + \tan t$
16. $\csc t + \cot t$
17. $\csc^2 v \tan^2 v$
18. $\sec t \csc t \tan t$
19. $\sec t \csc t \cot t$
20. $\csc^2 u - \sec^2 u + \tan^2 u$

Verify the following identities.

21. $\sec \theta = \tan \theta \csc \theta$
22. $\csc \alpha = \dfrac{\cot \alpha}{\cos \alpha}$
23. $\sin \beta \sec \beta = \tan \beta$
24. $\dfrac{\sin t}{\tan t} = \dfrac{1}{\sec t}$
25. $\cot u = \dfrac{\csc u}{\sec u}$
26. $\cos v \tan v = \sin v$
27. $\cos t + \tan t = \sin t(\cot t + \sec t)$
28. $\dfrac{\csc t}{\sec t} = \cot t$
29. $\cos(-x)\sec x = 1$
30. $\dfrac{\sec(-u)}{\csc(-u)} = -\tan u$
31. $\cos(-x)\tan(-x) + \sin x = 0$
32. $\tan(-\theta)\csc(-\theta) = \sec \theta$
33. $\sin(-\beta)\sec \beta = \tan(-\beta)$
34. $\sin(-t)\cos(-t)\cot t = -\cos^2 t$
35. $\dfrac{\csc(-\beta)}{\sec \beta} + \cot \beta = 0$
36. $\tan(-t)\csc(-t)\cos t = 1$
37. $\sin \alpha(\csc \alpha - \sin \alpha) = \cos^2 \alpha$
38. $2 - \sin^2 v = 1 + \cos^2 v$

39. $\sec^2 t - \tan^2 t = 1$

40. $\csc^2 u - \cot^2 u = 1$

41. $\dfrac{1 - \sin^2 v}{\sin^2 v} = \cot^2 v$

42. $\cos x(\sec x - \cos x) = \sin^2 x$

43. $\csc t(\csc t - \sin t) = \cot^2 t$

44. $\cos^2 u(\sec^2 u - 1) = \sin^2 u$

45. $1 + \tan^2(-\theta) = \sec^2 \theta$

46. $[1 - \cos^2(-\theta)]\csc^2 \theta = 1$

47. $\dfrac{\sin^2 t - 1}{\cos t} = -\cos t$

48. $\sin^2 t + \cos^2 t + \tan^2 t = \sec^2 t$

49. $(1 + \tan^2 t)\cos^2 t = 1$

50. $\cos \theta \cos(-\theta) - \sin(-\theta)\sin \theta = 1$

Use the identities in this section and the given value to find the requested value if the terminal ray for θ falls in the specified quadrant when θ is placed in standard position.

51. $\sin \theta = \dfrac{1}{4}$, quadrant II; find $\cot \theta$.

52. $\cos \theta = \dfrac{3}{4}$, quadrant IV; find $\tan \theta$.

53. $\tan \theta = 3$, quadrant III; find $\sec \theta$.

54. $\cot \theta = \dfrac{1}{2}$, quadrant I; find $\csc \theta$.

55. $\cot \theta = -2$, quadrant IV; find $\csc \theta$.

56. $\tan \theta = -5$, quadrant II; find $\sec \theta$.

57. $\cos \theta = -\dfrac{1}{3}$, quadrant II; find $\tan(-\theta)$.

58. $\sin \theta = -\dfrac{2}{3}$, quadrant III; find $\cot(-\theta)$.

59. $\tan \theta = 2$, quadrant I; find $\sec(-\theta)$.

60. $\cot \theta = -3$, quadrant II; find $\csc(-\theta)$.

Section 8.2

Verifying Identities

In this section we work with identities that require a little more algebraic manipulation than those of Section 8.1. The Pythagorean identities are very important and should be memorized.

All the trigonometric identities can be verified by appealing to the Pythagorean identities and/or the expressions of the other four circular functions in terms of sine and cosine. In verifying an identity, we can proceed in three ways.

METHODS FOR VERIFYING IDENTITIES

1. *Express each function in terms of sines and cosines and then show that the left and right sides of the equal sign become the same.* This method is certain to get results, although it can sometimes be tedious and time-consuming.

2. *Simplify both sides of the equal sign* (but not necessarily by changing everything into sines and cosines) *and show that the simplified expressions are equal.*

3. *Transform one side into the other side.* If one side is more complicated than the other, this is the side we work on. This is the fastest method once you have developed a certain skill in manipulating the circular functions.

Section 8.2 Verifying Identities

> **CAUTION**
>
> When verifying an identity, you must keep the expressions on the two sides of the equal sign separate. You should not multiply both sides of the equal sign by the same expression or add the same expression to both sides (even though these techniques are valid when solving equations). The reason for this is that you do not know beforehand that the two sides are really equal. You do not have an equation but rather a *claim* or *assertion* that the two expressions are equal. This claim must be verified.

EXAMPLE 1

Show that the following equation holds for all θ for which $\sec \theta$ is defined.

$$\cos \theta + \sin^2 \theta \sec \theta = \sec \theta$$

SOLUTION

Using method 1, we convert $\sec \theta$ to $\dfrac{1}{\cos \theta}$ and then show that the left and right sides of the equal sign become the same.

$$\cos \theta + \sin^2 \theta \sec \theta \stackrel{?}{=} \sec \theta$$

$$\cos \theta + \sin^2 \theta \sec \theta = \cos \theta + \sin^2 \theta \cdot \frac{1}{\cos \theta} \qquad\qquad \sec \theta$$

$$= \frac{\cos^2 \theta + \sin^2 \theta}{\cos \theta} \qquad\qquad\qquad\qquad \vdots$$

$$= \frac{1}{\cos \theta} \qquad\qquad\qquad\qquad = \frac{1}{\cos \theta} \qquad \text{(By formula 15, Section 8.1)}$$

The two sides of the equal sign reduce to the same expression. Since the steps on both sides are reversible, this verifies the identity. ∎

EXAMPLE 2

Show that the following is an identity.

$$(\tan \alpha)(1 + \cot^2 \alpha) = (\cot \alpha)(1 + \tan^2 \alpha)$$

SOLUTION

Using method 2, we simplify both sides of the equal sign separately. We use the Pythagorean identities $1 + \cot^2 \alpha = \csc^2 \alpha$ and $1 + \tan^2 \alpha = \sec^2 \alpha$.

$$(\tan \alpha)(1 + \cot^2 \alpha) \stackrel{?}{=} (\cot \alpha)(1 + \tan^2 \alpha)$$

$$(\tan \alpha)(1 + \cot^2 \alpha) = \tan \alpha \csc^2 \alpha \qquad\qquad (\cot \alpha)(1 + \tan^2 \alpha) = \cot \alpha \sec^2 \alpha$$

$$= \frac{\sin \alpha}{\cos \alpha} \cdot \frac{1}{\sin^2 \alpha} \qquad\qquad\qquad\qquad = \frac{\cos \alpha}{\sin \alpha} \cdot \frac{1}{\cos^2 \alpha}$$

$$= \frac{1}{\cos \alpha \sin \alpha} \qquad\qquad\qquad\qquad\qquad = \frac{1}{\sin \alpha \cos \alpha}$$

Hence $(\tan \alpha)(1 + \cot^2 \alpha) = \dfrac{1}{(\sin \alpha \cos \alpha)} = (\cot \alpha)(1 + \tan^2 \alpha)$. ∎

EXAMPLE 3

Verify the following identity.

$$(\sin t)(\cot t)(\sec t - 1) = 1 - \cos t$$

SOLUTION

Writing the left side in terms of sines and cosines we obtain the right side (a combination of methods 1 and 3).

$(\sin t)(\cot t)(\sec t - 1)$	$\overset{?}{=}$	$1 - \cos t$	
$(\sin t)(\cot t)(\sec t - 1) = \sin t \dfrac{\cos t}{\sin t}\left(\dfrac{1}{\cos t} - 1\right)$		$1 - \cos t$	(By formulas 4, 2, Section 8.1)
$= \cos t \left(\dfrac{1 - \cos t}{\cos t}\right)$		$\vdots$	
$= 1 - \cos t$		$= 1 - \cos t$	

EXAMPLE 4

Verify the following identity.

$$\frac{1 + \cos x}{\sin x} = \frac{\sin x}{1 - \cos x}$$

SOLUTION

We do *not* cross multiply here since we do not know that the two sides of the equal sign are really equal. Instead we must verify the equality by working with the two sides individually. We can convert the left side into the right side.

$\dfrac{1 + \cos x}{\sin x}$	$\overset{?}{=}$	$\dfrac{\sin x}{1 - \cos x}$	
$\dfrac{1 + \cos x}{\sin x} = \dfrac{1 + \cos x}{\sin x} \cdot \dfrac{1 - \cos x}{1 - \cos x}$		$\dfrac{\sin x}{1 - \cos x}$	
$= \dfrac{1 - \cos^2 x}{\sin x(1 - \cos x)}$			
$= \dfrac{\sin^2 x}{\sin x(1 - \cos x)}$		$\vdots$	(Formula 15b, Section 8.1)
$= \dfrac{\sin x}{1 - \cos x}$		$= \dfrac{\sin x}{1 - \cos x}$	

EXAMPLE 5

Verify the identity

$$\frac{\cos t + 1}{\tan^2 t} = \frac{\cos t}{\sec t - 1}$$

SOLUTION

Let's work on the right side this time. If we multiply by the conjugate of the denominator, we obtain $\sec^2 t - 1$, which is part of one of the Pythagorean identities.

$$\frac{\cos t + 1}{\tan^2 t} \stackrel{?}{=} \frac{\cos t}{\sec t - 1}$$

$$\frac{\cos t + 1}{\tan^2 t} \qquad \frac{\cos t}{\sec t - 1} = \frac{\cos t}{(\sec t - 1)} \frac{(\sec t + 1)}{(\sec t + 1)}$$

$$= \frac{\cos t(\sec t + 1)}{\sec^2 t - 1}$$

$$\vdots \qquad = \frac{\cos t(\sec t + 1)}{\tan^2 t} \qquad \text{(Formula 16, Section 8.1)}$$

$$= \frac{\cos t \left(\dfrac{1}{\cos t} + 1\right)}{\tan^2 t}$$

$$= \frac{1 + \cos t}{\tan^2 t} \qquad = \frac{1 + \cos t}{\tan^2 t}$$

∎

EXAMPLE 6

Find nonzero constants A and B for which the following expression becomes an identity.

$$A + B \cos^2 \theta = \sin^4 \theta - \cos^4 \theta$$

SOLUTION

We factor the right-hand side and then simplify the expression as follows.

$$A + B \cos^2 \theta = \sin^4 \theta - \cos^4 \theta$$
$$= (\sin^2 \theta - \cos^2 \theta)(\underbrace{\sin^2 \theta + \cos^2 \theta}_{=1}) \qquad \text{(Formula 15, Section 8.1)}$$
$$= \sin^2 \theta - \cos^2 \theta$$
$$= (1 - \cos^2 \theta) - \cos^2 \theta \qquad \text{(By formula 15b, Section 8.1)}$$
$$= 1 - 2 \cos^2 \theta$$

Clearly, $A = 1$, $B = -2$ are appropriate values; that is,

$$1 - 2 \cos^2 \theta = \sin^4 \theta - \cos^4 \theta$$

∎

Section 8.2 Exercises

Verify the following identities.

1. $(\sec^2 t - 1)\cos^2 t = \sin^2 t$
2. $\tan x(\cot x + \tan x) = \sec^2 x$
3. $(\cos x + \sin x)(\cot x + \tan x) = \csc x + \sec x$
4. $2 \csc t - \cot t \cos t = \csc t + \sin t$
5. $(1 + \csc u)(\sec u - \tan u) = \cot u$
6. $(1 - \sin v)(1 + \sin v)(1 + \tan^2 v) = 1$
7. $\tan \phi + \cot \phi = \csc \phi \sec \phi$
8. $\sec \alpha \cot \alpha - \csc \alpha \tan \alpha = \csc \alpha \sec \alpha(\cos \alpha - \sin \alpha)$
9. $\cos \beta \tan \beta - \sin^2 \beta = \sin \beta(1 - \sin \beta)$
10. $\sec^2 \gamma + \csc^2 \gamma = \sec^2 \gamma \csc^2 \gamma$
11. $(\sec \theta - 1)(\sin \theta + \cos \theta \sin \theta) = \tan \theta \sin^2 \theta$
12. $1 - 2 \cos^2 \phi = 2 \sin^2 \phi - 1$
13. $\sin^2 t \cot^2 t + \cos^2 t \tan^2 t = 1$
14. $\sec^2 u \csc^2 u = (\tan u + \cot u)^2$
15. $\cot v + \tan v = \sec^2 v \cot v$
16. $\tan^2 \theta - \sin^2 \theta = \tan^2 \theta \sin^2 \theta$
17. $\cot^2 \phi - \cos^2 \phi = \cot^2 \phi \cos^2 \phi$

18. $(1 + \sin^2 \alpha \sec^2 \alpha) = \sec^2 \alpha$

19. $(1 - \cos \beta)(1 + \sec \beta)(1 - \sin \beta)(1 + \csc \beta)$
    $= \sin \beta \cos \beta$

20. $\sec^2 s \cot^2 s - \csc^2 s \cos^2 s = 1$

21. $\sin^4 v + 2 \sin^2 v \cot^2 v - \cos^4 v = 1$

22. $(\sec u - \tan u)^2 + 2 \sin u \sec^2 u = 1 + 2 \tan^2 u$

23. $\csc^2 v = 2 \cos^2 v + \csc^2 v (\sin^4 v + \cos^4 v)$

24. $(1 + \csc \alpha)(\sec \alpha - \tan \alpha) = \cot \alpha$

25. $\sec^4 \beta (1 - \sin^4 \beta) = \sec^2 \beta + \tan^2 \beta$

26. $(\cos^4 \gamma - \sin^4 \gamma) = 1 - 2 \sin^2 \gamma$

27. $\csc^4 \theta - \cot^4 \theta = \csc^2 \theta + \cot^2 \theta$

28. $\dfrac{\sin \phi}{\sec \phi + 1} + \dfrac{\sin \phi}{\sec \phi - 1} = 2 \cot \phi$

29. $\dfrac{\sec t + \tan t}{\cos t - \tan t - \sec t} + \csc t = 0$

30. $\dfrac{1}{1 + \sin u} + \dfrac{1}{1 - \sin u} = 2 \sec^2 u$

31. $\dfrac{1 - \csc^2 v}{1 - \sec^2 v} = \cot^4 v$

32. $\dfrac{1 + \tan^2 \alpha}{\tan^2 \alpha} = \csc^2 \alpha$

33. $\dfrac{\tan \beta}{1 - \cot \beta} + \dfrac{\cot \beta}{1 - \tan \beta} = \tan \beta + \cot \beta + 1$

34. $(1 - \sin \gamma)(1 + \sin \gamma) = \dfrac{1}{1 + \tan^2 \gamma}$

35. $\dfrac{\cot \theta + \csc \theta}{\sin \theta + \tan \theta} = \cos \theta \csc^2 \theta$

36. $\dfrac{\sec \phi + \tan \phi}{\cos \phi + \cot \phi} = \sin \phi \sec^2 \phi$

37. $(\csc t - \cot t)^2 = \dfrac{1 - \cos t}{1 + \cos t}$

38. $\dfrac{1 + \sin u}{\cos u} + \dfrac{\cos u}{1 + \sin u} = 2 \sec u$

39. $\dfrac{\sin v + \cos v}{\cos^2 v} = \dfrac{1 - \tan^2 v}{\cos v - \sin v}$

40. $\dfrac{\tan \alpha + 1}{\tan \alpha - 1} = \dfrac{\sec \alpha + \csc \alpha}{\sec \alpha - \csc \alpha}$

41. $\dfrac{1 - \tan \beta}{1 + \tan \beta} = \dfrac{\cot \beta - 1}{\cot \beta + 1}$

42. $\dfrac{\cot t - 1}{\tan t - 1} = -\cot t$

In the following, find constants A, B, C that make the given expression an identity.

43. $(\cos^2 x - 1)(\tan^2 x + 1) = A + B \sec^2 x$

44. $\dfrac{\sin^2 \theta}{1 - \cos \theta} = A + B \cos \theta$

45. $\tan \phi \csc \phi \cos \phi + \cot \phi \sec \phi \sin \phi = A$

46. $\sec^4 \alpha - \tan^4 \alpha + \dfrac{2}{\sin^2 \alpha - 1} = A$

47. $\sin \beta \left(\dfrac{\cot \beta}{\sec \beta} + \csc \beta \right) = A + B \cos^2 \beta$

48. $\dfrac{\cos \gamma}{1 - \sin \gamma} - A \tan \gamma = \dfrac{1 - \sin \gamma}{\cos \gamma}$

49. $\dfrac{\tan v - \cot v}{\tan v + \cot v} = A + B \sin^2 v$

50. $(\tan \theta + \cot \theta)^2 \cdot \sin^2 \theta = A + B \tan^2 \theta$

51. $A \sin^4 u + B \sin^2 u \cos^2 u + C \cos^4 u = 1$

52. $\sec^4 t = A + B \tan^2 t + C \tan^4 t$

Section 8.3

Sum and Difference Formulas

There are certain relationships between the value of a circular function at a sum or difference of two independent variables [$(u + v)$ or $(u - v)$] and the values of circular functions at u and at v. We begin with the difference formula for the cosine; other sum and difference formulas can be derived from it.

Cosine Formulas

To obtain the formula for $\cos(u - v)$, where u and v are any two real numbers (or angles), we sketch arcs $\widehat{DA}$, $\widehat{DB}$, and $\widehat{DC}$ on the unit circle having lengths u, v, and $u - v$, respectively. These arcs, in turn, determine central angles in standard position having measure u, v, and $u - v$ radians, respectively. See Figure 8.2.

Figure 8.2

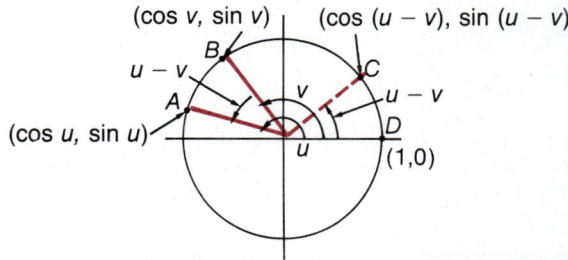

Since the circle has radius 1, the points A, B, C, and D have coordinates as indicated in Figure 8.2. (Also see Figure 8.1.) The arc length $\widehat{AB}$ is the same as the arc length $\widehat{CD}$. The lengths of the chords (line segments) $\overline{AB}$, $\overline{CD}$ are also equal; that is,

$$d(A, B) = d(C, D)$$

Substituting the coordinates of these points into the distance formula, we get

$$\sqrt{(\cos u - \cos v)^2 + (\sin u - \sin v)^2}$$
$$= \sqrt{[\cos(u - v) - 1]^2 - [\sin(u - v) - 0]^2}$$

Squaring both sides and expanding yields

$$\cos^2 u - 2(\cos u)(\cos v) + \cos^2 v + \sin^2 u - 2(\sin u)(\sin v) + \sin^2 v$$
$$= \cos^2(u - v) - 2\cos(u - v) + 1 + \sin^2(u - v)$$

Letting $\cos^2 \theta + \sin^2 \theta = 1$ whenever possible, this becomes

$$2 - 2\cos u \cos v - 2\sin u \sin v = 2 - 2\cos(u - v)$$

It follows that

$$\cos(u - v) = \cos u \cos v + \sin u \sin v \qquad (1a)$$

EXAMPLE I   Use formula 1a to find the exact value of $\cos \pi/12$.

SOLUTION   We can write $\pi/12 = \pi/3 - \pi/4$. Then

$$\cos \frac{\pi}{12} = \cos\left(\frac{\pi}{3} - \frac{\pi}{4}\right)$$

$$= \cos \frac{\pi}{3} \cos \frac{\pi}{4} + \sin \frac{\pi}{3} \sin \frac{\pi}{4} \qquad \text{(By formula 1a)}$$

$$= \frac{1}{2} \cdot \frac{\sqrt{2}}{2} + \frac{\sqrt{3}}{2} \cdot \frac{\sqrt{2}}{2}$$

$$= \frac{\sqrt{2} + \sqrt{6}}{4}$$

∎

Formula 1a leads in turn to a formula for $\cos(u + v)$ as follows.

$$\cos(u + v) = \cos[u - (-v)]$$
$$= \cos u \cos(-v) + \sin u \sin(-v)$$
$$= \cos u \cos v - \sin u \sin v$$

> **FORMULAS FOR** $\cos(u \mp v)$
>
> **1. a.** $\cos(u - v) = \cos u \cos v + \sin u \sin v$
> **b.** $\cos(u + v) = \cos u \cos v - \sin u \sin v$

EXAMPLE 2 Use formula 1b to find the exact value of $\cos 75°$.

SOLUTION
$$\cos 75° = \cos(45° + 30°)$$
$$= \cos 45° \cos 30° - \sin 45° \sin 30° \quad \text{(By formula 1b)}$$
$$= \frac{\sqrt{2}}{2}\frac{\sqrt{3}}{2} - \frac{\sqrt{2}}{2}\frac{1}{2}$$
$$= \frac{\sqrt{6} - \sqrt{2}}{4} \quad \blacksquare$$

Letting $u = \pi/2$ in formula 1a, we can verify again that
$$\cos\left(\frac{\pi}{2} - v\right) = \sin v$$

Then, letting $v = \pi/2 - t$ in this formula, it follows that
$$\sin\left(\frac{\pi}{2} - t\right) = \cos t$$

Sine Formulas

These two relationships can be used in conjunction with formula 1b to obtain the sum and difference formulas for the sine function as follows.

$$\sin(u - v) = \cos\left[\frac{\pi}{2} - (u - v)\right]$$
$$= \cos\left[\left(\frac{\pi}{2} - u\right) + v\right]$$
$$= \cos\left(\frac{\pi}{2} - u\right)\cos v - \sin\left(\frac{\pi}{2} - u\right)\sin v \quad \text{(By formula 1b)}$$
$$= \sin u \cos v - \cos u \sin v$$

We then find a formula for $\sin(u + v)$:
$$\sin(u + v) = \sin[u - (-v)]$$
$$= \sin u \cos(-v) - \cos u \sin(-v)$$
$$= \sin u \cos v + \cos u \sin v \quad \text{(By formulas 9, 10, Section 8.1)}$$

> **FORMULAS FOR** $\sin(u \mp v)$
>
> **2. a.** $\sin(u - v) = \sin u \cos v - \cos u \sin v$
> **b.** $\sin(u + v) = \sin u \cos v + \cos u \sin v$

Tangent Formulas

The sum and difference formulas for sine and cosine can be used to find sum and difference formulas for the tangent, as shown here.

$$\tan(u - v) = \frac{\sin(u - v)}{\cos(u - v)}$$

$$= \frac{\sin u \cos v - \cos u \sin v}{\cos u \cos v + \sin u \sin v}$$

To convert sines and cosines to tangents in this last expression, we divide both the numerator and the denominator by $\cos u \cos v$.

$$\tan(u - v) = \frac{\dfrac{\sin u \cos v}{\cos u \cos v} - \dfrac{\cos u \sin v}{\cos u \cos v}}{\dfrac{\cos u \cos v}{\cos u \cos v} + \dfrac{\sin u \sin v}{\cos u \cos v}}$$

$$= \frac{\tan u - \tan v}{1 + (\tan u)(\tan v)}$$

The formula for $\tan(u + v)$ is obtained in a similar fashion. (See Exercise 15.)

FORMULAS FOR $\tan(u \mp v)$

3. a. $\tan(u - v) = \dfrac{\tan u - \tan v}{1 + \tan u \tan v}$

b. $\tan(u + v) = \dfrac{\tan u + \tan v}{1 - (\tan u)(\tan v)}$

Although sum and difference formulas for cot, sec, and csc can be derived, the need for them seldom arises. Their verifications are left as exercises. (See Exercises 16–21.)

EXAMPLE 3 Evaluate $\sin t$ and $\tan t$ at $t = 17\pi/12$ by using the sum and difference formulas.

SOLUTION Write

$$\frac{17\pi}{12} = \frac{15\pi}{12} + \frac{2\pi}{12}$$

$$= \frac{5\pi}{4} + \frac{\pi}{6}$$

(See Figure 8.3.) Then

$$\sin \frac{17\pi}{12} = \sin\left(\frac{5\pi}{4} + \frac{\pi}{6}\right)$$

$$= \left(\sin \frac{5\pi}{4}\right)\left(\cos \frac{\pi}{6}\right) + \left(\cos \frac{5\pi}{4}\right)\left(\sin \frac{\pi}{6}\right) \qquad \text{(By formula 2b)}$$

$$= \frac{-\sqrt{2}}{2} \cdot \frac{\sqrt{3}}{2} + \frac{-\sqrt{2}}{2} \cdot \frac{1}{2}$$

$$= -\frac{\sqrt{2}}{4}(\sqrt{3} + 1)$$

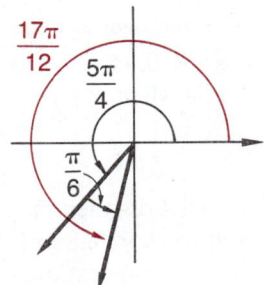

Figure 8.3

$$\tan\frac{17\pi}{12} = \tan\left(\frac{5\pi}{4} + \frac{\pi}{6}\right)$$

$$= \frac{\tan\frac{5\pi}{4} + \tan\frac{\pi}{6}}{1 - \left(\tan\frac{5\pi}{4}\right)\left(\tan\frac{5\pi}{4}\right)} \quad \text{(By formula 3b)}$$

$$= \frac{1 + \frac{1}{\sqrt{3}}}{1 - 1 \cdot \frac{1}{\sqrt{3}}}$$

$$= \frac{\sqrt{3} + 1}{\sqrt{3} - 1} \quad \text{(Multiply numerator and denominator by } \sqrt{3}\text{)}$$

$$= 2 + \sqrt{3} \quad \text{(Rationalize the denominator; see Section 1.6)} \quad \blacksquare$$

EXAMPLE 4 Find the exact value of

$$\frac{\tan\frac{7\pi}{12} - \tan\frac{5\pi}{12}}{1 + \tan\frac{7\pi}{12} \tan\frac{5\pi}{12}}$$

without using a calculator.

SOLUTION From formula 3b, we have

$$\frac{\tan\frac{7\pi}{12} - \tan\frac{5\pi}{12}}{1 + \tan\frac{7\pi}{12} \tan\frac{5\pi}{12}} = \tan\left(\frac{7\pi}{12} - \frac{5\pi}{12}\right)$$

$$= \tan\frac{\pi}{6}$$

$$= \frac{\sqrt{3}}{3} \quad \blacksquare$$

EXAMPLE 5 Evaluate $\sin(\alpha + \beta)$, $\cos(\alpha + \beta)$, and $\tan(\alpha + \beta)$ if $\sin\alpha = \frac{1}{3}$, $\cos\alpha < 0$, $\cot\beta = 3$, and $\sin\beta > 0$. In what quadrant does the terminal ray of $\alpha + \beta$ lie when $\alpha + \beta$ is placed in standard position?

SOLUTION If α and β are considered to be angles and are placed in standard position, then the terminal ray of α falls in the second quadrant since $\sin\alpha > 0$ and $\cos\alpha < 0$. Similarly, the terminal ray of β falls in the first quadrant. We obtain the configurations shown in Figure 8.4.

Figure 8.4

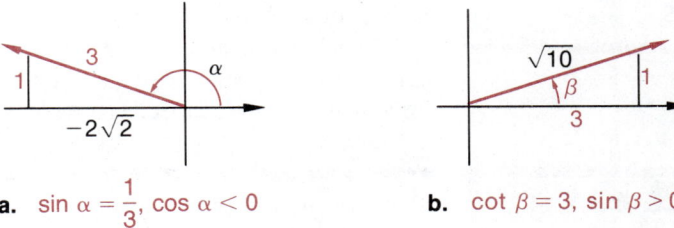

a. $\sin \alpha = \dfrac{1}{3}$, $\cos \alpha < 0$ b. $\cot \beta = 3$, $\sin \beta > 0$

The Pythagorean theorem can be used to find the horizontal leg of the triangle in Figure 8.4a and the hypotenuse in Figure 8.4b. These are $-2\sqrt{2}$ and $\sqrt{10}$, respectively. From these dimensions, we obtain

$$\sin \alpha = \frac{1}{3} \qquad \text{and} \qquad \sin \beta = \frac{1}{\sqrt{10}} = \frac{\sqrt{10}}{10}$$

$$\cos \alpha = \frac{-2\sqrt{2}}{3} \qquad\qquad\qquad \cos \beta = \frac{3}{\sqrt{10}} = \frac{3\sqrt{10}}{10}$$

$$\tan \alpha = \frac{1}{-2\sqrt{2}} = \frac{-\sqrt{2}}{4} \qquad\qquad \tan \beta = \frac{1}{3}$$

Finally,

$$\sin(\alpha + \beta) = \sin \alpha \cos \beta + \cos \alpha \sin \beta \qquad \text{(By formula 2b)}$$
$$= \frac{1}{3} \cdot \frac{3\sqrt{10}}{10} + \frac{-2\sqrt{2}}{3} \cdot \frac{\sqrt{10}}{10}$$
$$= \frac{\sqrt{10}}{10}\left(1 - \frac{2\sqrt{2}}{3}\right) > 0$$

$$\cos(\alpha + \beta) = \cos \alpha \cos \beta - \sin \alpha \sin \beta \qquad \text{(By formula 1b)}$$
$$= \frac{-2\sqrt{2}}{3} \cdot \frac{3\sqrt{10}}{10} - \frac{1}{3} \cdot \frac{\sqrt{10}}{10}$$
$$= \frac{-\sqrt{10}}{10}\left(2\sqrt{2} + \frac{1}{3}\right) < 0$$

$$\tan(\alpha + \beta) = \frac{\tan \alpha + \tan \beta}{1 - \tan \alpha \tan \beta} \qquad \text{(By formula 3b)}$$
$$= \frac{-\dfrac{\sqrt{2}}{4} + \dfrac{1}{3}}{1 - \left(-\dfrac{\sqrt{2}}{4} \cdot \dfrac{1}{3}\right)} = \frac{-3\sqrt{2} + 4}{12 + \sqrt{2}}$$
$$= \frac{(4 - 3\sqrt{2})(12 - \sqrt{2})}{144 - 2} = \frac{54 - 40\sqrt{2}}{142} \qquad \text{(Rationalize the denominator)}$$
$$= \frac{27 - 20\sqrt{2}}{71}$$

Since $\sin(\alpha + \beta) > 0$ and $\cos(\alpha + \beta) < 0$, the terminal ray of $\alpha + \beta$ will fall in the second quadrant when $\alpha + \beta$ is placed in standard position. ■

Identities

In Sections 8.1 and 8.2 we discussed identities involving the circular functions, all of which were evaluated simultaneously at the same u. The addition formulas derived in this section are also identities; they hold for all values of the variables for which they are defined. These identities, in turn, are the basis for other identities.

EXAMPLE 6 Show for all α, β that

$$\frac{\sin(\alpha + \beta)}{\cos \alpha \cos \beta} = \tan \alpha + \tan \beta$$

SOLUTION
$$\frac{\sin(\alpha + \beta)}{\cos \alpha \cos \beta} = \frac{\sin \alpha \cos \beta + \cos \alpha \sin \beta}{\cos \alpha \cos \beta} \quad \text{(By formula 2b)}$$

$$= \frac{\sin \alpha \cos \beta}{\cos \alpha \cos \beta} + \frac{\cos \alpha \sin \beta}{\cos \alpha \cos \beta}$$

$$= \tan \alpha + \tan \beta \qquad \blacksquare$$

Graphing $A \cos \alpha + B \sin \alpha$ (Optional)

Engineers frequently encounter expressions of the form

$$A \cos \alpha + B \sin \alpha$$

in which *both sine and cosine are evaluated at the same α*. In its given form, the expression is difficult to graph. However, such expressions can be rewritten in terms of a single sine or cosine function but with a phase shift on the angle. The procedure is illustrated in Example 7.

EXAMPLE 7
a. Express $\sqrt{3} \cos \alpha - \sin \alpha$ in terms of a single cosine function.
b. Sketch its graph.
c. Use the expression in part a to evaluate $\sqrt{3} \cos \dfrac{\pi}{12} - \sin \dfrac{\pi}{12}$.

SOLUTION

a. We sketch the right triangle having legs of length $\sqrt{3}$ and 1 (see Figure 8.5). Then $\theta = \pi/6$ or $30°$ and

$$\sqrt{3} \cos \alpha - \sin \alpha = 2\left(\frac{\sqrt{3}}{2} \cos \alpha - \frac{1}{2} \sin \alpha\right)$$

$$= 2(\cos \theta \cos \alpha - \sin \theta \sin \alpha)$$

$$= 2 \cos(\alpha + \theta) \quad \text{(By formula 1b)}$$

$$= 2 \cos\left(\alpha + \frac{\pi}{6}\right)$$

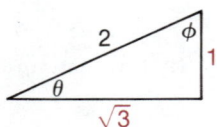

Figure 8.5

Figure 8.6
$y = \sqrt{3} \cos \alpha - \sin \alpha = 2 \cos\left(\alpha + \dfrac{\pi}{6}\right)$

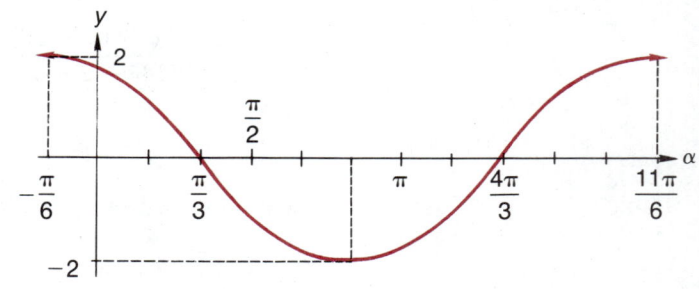

b. The graph of $y = 2\cos(\alpha + \pi/6)$ is found by moving the basic cosine curve to the left $\pi/6$ and doubling its amplitude (see Figure 8.6).

c.
$$\sqrt{3}\cos\frac{\pi}{12} - \sin\frac{\pi}{12} = 2\cos\left(\frac{\pi}{12} + \frac{\pi}{6}\right)$$
$$= 2\cos\frac{\pi}{4}$$
$$= 2\left(\frac{\sqrt{2}}{2}\right)$$
$$= \sqrt{2}$$

Had we used angle ϕ rather than θ in Figure 8.5 in the preceding example, we would have obtained $\phi = \pi/3$ and

$$\sqrt{3}\cos\alpha - \sin\alpha = 2(\sin\phi\cos\alpha - \cos\phi\sin\alpha)$$
$$= 2\sin(\phi - \alpha) \qquad \text{(Formula 2a)}$$
$$= 2\sin\left(\frac{\pi}{3} - \alpha\right)$$
$$= -2\sin\left(\alpha - \frac{\pi}{3}\right)$$

This graph would have involved a reflection as well as a phase shift from the basic sine curve.

CONVERTING $(A\cos\alpha + B\sin\alpha)$ to $C\cos(\alpha - \theta)$ or $C\sin(\alpha + \phi)$

$$\cos\theta = \frac{A}{\sqrt{A^2 + B^2}} = \sin\phi$$

$$\sin\theta = \frac{B}{\sqrt{A^2 + B^2}} = \cos\phi$$

$$A\cos\alpha + B\sin\alpha = \sqrt{A^2 + B^2}\left(\frac{A}{\sqrt{A^2 + B^2}}\cos\alpha + \frac{B}{\sqrt{A^2 + B^2}}\sin\alpha\right)$$

$$= \begin{cases} \sqrt{A^2 + B^2}(\cos\theta\cos\alpha + \sin\theta\sin\alpha) \\ \sqrt{A^2 + B^2}(\sin\phi\cos\alpha + \cos\phi\sin\alpha) \end{cases}$$

$$= \begin{cases} \sqrt{A^2 + B^2}\cos(\alpha - \theta) & \text{(By formula 1a)} \\ \sqrt{A^2 + B^2}\sin(\alpha + \phi) & \text{(By formula 2b)} \end{cases}$$

Section 8.3 Exercises

Use the sum or difference formulas to show the following.

1. $\cos\left(\dfrac{\pi}{2} - \theta\right) = \sin\theta$

2. $\sin\left(\dfrac{\pi}{2} - \theta\right) = \cos\theta$

3. $\tan\left(\dfrac{\pi}{2} - \theta\right) = \cot\theta$

4. $\cos(\theta + 2\pi) = \cos\theta$

5. $\sin(\theta + 2\pi) = \sin\theta$

6. $\tan(\theta + \pi) = \tan\theta$

7. $\tan(\theta - \pi) = \tan\theta$
8. $\cos(\theta - \pi) = -\cos\theta$
9. $\cos(\theta + \pi) = -\cos\theta$
10. $\sin(\theta + \pi) = -\sin\theta$
11. $\sin(\theta - \pi) = -\sin\theta$
12. $\cos(\pi - \theta) = -\cos\theta$
13. $\sin(\pi - \theta) = \sin\theta$
14. $\tan(\pi - \theta) = -\tan\theta$

Verify the following sum and difference formulas.

15. $\tan(u + v) = \dfrac{\tan u + \tan v}{1 - \tan u \tan v}$

16. $\cot(u - v) = \dfrac{\cot u \cot v + 1}{\cot v - \cot u}$

17. $\cot(u + v) = \dfrac{\cot u \cot v - 1}{\cot v + \cot u}$

18. $\sec(u + v) = \dfrac{\sec u \sec v}{1 - \tan u \tan v}$

19. $\sec(u - v) = \dfrac{\sec u \sec v}{1 + \tan u \tan v}$

20. $\csc(u - v) = \dfrac{\csc u \csc v}{\cot v - \cot u}$

21. $\csc(u + v) = \dfrac{\csc u \csc v}{\cot v + \cot u}$

Write $2\theta = (\theta + \theta)$ and use the sum formulas developed in Exercises 15, 17, 18, and 21 to develop formulas for the following.

22. $\sec 2\theta$
23. $\csc 2\theta$
24. $\cot 2\theta$
25. $\tan 2\theta$

Use the sum or difference formulas to evaluate the sine, cosine, and tangent of the following angles exactly, without using a calculator.

26. $\dfrac{\pi}{12}$ $\left(\text{use } \dfrac{\pi}{12} = \dfrac{\pi}{4} - \dfrac{\pi}{6}\right)$

27. $\dfrac{5\pi}{12}$ $\left(\text{use } \dfrac{5\pi}{12} = \dfrac{\pi}{4} + \dfrac{\pi}{6}\right)$

28. $\dfrac{7\pi}{12}$ $\left(\text{use } \dfrac{7\pi}{12} = \dfrac{5\pi}{6} - \dfrac{\pi}{4}\right)$

29. $\dfrac{13\pi}{12}$ $\left(\text{use } \dfrac{13\pi}{12} = \dfrac{5\pi}{6} + \dfrac{\pi}{4}\right)$

30. $\dfrac{19\pi}{12}$ $\left(\text{use } \dfrac{19\pi}{12} = \dfrac{4\pi}{3} + \dfrac{\pi}{4}\right)$

31. $\dfrac{23\pi}{12}$ $\left(\text{use } \dfrac{23\pi}{12} = \dfrac{13\pi}{6} - \dfrac{\pi}{4}\right)$

32. $165°$ (use $165° = 120° + 45°$)

33. $105°$ (use $105° = 135° - 30°$)

Use the sum or difference formulas to evaluate the following exactly, without using a calculator.

34. $\cos\dfrac{8\pi}{15}\cos\dfrac{2\pi}{15} - \sin\dfrac{8\pi}{15}\sin\dfrac{2\pi}{15}$

35. $\cos\dfrac{13\pi}{15}\cos\dfrac{\pi}{5} + \sin\dfrac{13\pi}{15}\sin\dfrac{\pi}{5}$

36. $\sin\dfrac{\pi}{8}\cos\dfrac{3\pi}{8} + \cos\dfrac{\pi}{8}\sin\dfrac{3\pi}{8}$

37. $\sin\dfrac{3\pi}{8}\cos\dfrac{7\pi}{8} - \cos\dfrac{3\pi}{8}\sin\dfrac{7\pi}{8}$

38. $\dfrac{\tan\dfrac{17\pi}{20} + \tan\dfrac{3\pi}{20}}{1 - \tan\dfrac{17\pi}{20}\tan\dfrac{3\pi}{20}}$

39. $\dfrac{\tan\dfrac{23\pi}{30} - \tan\dfrac{\pi}{10}}{1 + \tan\dfrac{23\pi}{30}\tan\dfrac{\pi}{10}}$

40. $\cos\dfrac{\pi}{15}\cos\dfrac{9\pi}{10} + \sin\dfrac{\pi}{15}\sin\dfrac{9\pi}{10}$

41. $\sin 25° \cos 145° - \cos 25° \sin 145°$

42. $\dfrac{\tan 105° - \tan 75°}{1 + \tan 105° \tan 75°}$

Use the given values of the indicated circular functions at α and β to evaluate sin, cos, and tan at $\alpha + \beta$ and $\alpha - \beta$. Use the signs of the sine and cosine functions to indicate the quadrants in which the terminal rays of $\alpha + \beta$ and $\alpha - \beta$ lie when they are placed in standard position.

43. $\sin\alpha = -\dfrac{3}{5}$, $\cos\alpha > 0$; $\cos\beta = \dfrac{1}{2}$, $\sin\beta > 0$

44. $\sec \alpha = \dfrac{5}{4}$, $\tan \alpha < 0$; $\sin \beta = \dfrac{\sqrt{3}}{2}$, $\cos \beta < 0$

45. $\tan \alpha = -\dfrac{\sqrt{5}}{2}$, $\sin \alpha > 0$; $\sec \beta = -3$, $\csc \beta > 0$

46. $\cot \alpha = \sqrt{3}$, $\sin \alpha < 0$; $\tan \beta = \dfrac{-3}{4}$, $\sec \beta > 0$

47. $\sin \alpha = \dfrac{1}{4}$, $\cos \alpha > 0$; $\csc \beta = 10$, $\tan \beta > 0$

48. $\sec \alpha = 2$, $\csc \alpha > 0$; $\cos \beta = -\dfrac{1}{2}$, $\csc \beta < 0$

49. $\cot \alpha = \dfrac{3}{4}$, $\cos \alpha < 0$; $\cos \beta = -\dfrac{\sqrt{3}}{3}$, $\cot \beta > 0$

50. $\cot \alpha = 2$, $\sin \alpha < 0$; $\csc \beta = -7$, $\cos \beta < 0$

Use the sum or difference formulas to verify the following identities.

51. $\tan(\theta + \phi)\tan(\theta - \phi) = \dfrac{\tan^2 \theta - \tan^2 \phi}{1 - \tan^2 \theta \tan^2 \phi}$

52. $\dfrac{\sin(\theta + \phi)}{\sin(\theta - \phi)} = \dfrac{\tan \theta + \tan \phi}{\tan \theta - \tan \phi}$

53. $\cos(u + v)\cos(u - v) = \cos^2 u - \sin^2 v$

54. $\sin u \sin(u + v) + \cos u \cos(u + v) = \cos v$

55. $\sin\left(t + \dfrac{\pi}{6}\right) - \sin\left(t - \dfrac{\pi}{6}\right) = \cos t$

56. $2\sin\left(u + \dfrac{\pi}{4}\right)\sin\left(u - \dfrac{\pi}{4}\right) = \sin^2 u - \cos^2 u$

57. $2\cos\left(v + \dfrac{\pi}{4}\right)\sin\left(v + \dfrac{\pi}{4}\right) = \cos^2 v - \sin^2 v$

Optional Exercises

Write each of the following expressions in terms of a single cosine function and sketch the graph of each.

58. $\cos \theta + \sin \theta$

59. $\sin \theta - \cos \theta$

60. $\cos \theta + \sqrt{3} \sin \theta$

61. $\cos \theta - \sqrt{3} \sin \theta$

62. $\sin \theta - \sqrt{3} \cos \theta$

63–67. Write each of the expressions in Exercises 58–62 in terms of a single sine function.

Use the expressions derived in Exercises 58–67 to find

68. $\cos \dfrac{\pi}{12} + \sin \dfrac{\pi}{12}$

69. $\sin \dfrac{7\pi}{12} - \cos \dfrac{7\pi}{12}$

70. $\cos \dfrac{13\pi}{12} + \sqrt{3} \sin \dfrac{13\pi}{12}$

71. $\cos 75° - \sqrt{3} \sin 75°$

72. $\sin 285° - \sqrt{3} \cos 285°$

Section 8.4

Multiple Angles; Half-angles

The sum formulas can be used to find formulas for multiple angles by writing $2\alpha = \alpha + \alpha$, $3\alpha = 2\alpha + \alpha$, and so on. As with the addition formulas, we discuss multiple angle formulas only for the sine, cosine, and tangent functions. Letting $\beta = \alpha$ in the formulas for $(\alpha + \beta)$, we obtain the following.

Double Angles

$$\sin(\alpha + \alpha) = \sin \alpha \cos \alpha + \cos \alpha \sin \alpha \quad \text{(By formula 2b, Section 8.3)}$$
$$= 2 \sin \alpha \cos \alpha$$

$$\cos(\alpha + \alpha) = \cos \alpha \cos \alpha - \sin \alpha \sin \alpha \quad \text{(By formula 1b, Section 8.3)}$$
$$= \cos^2 \alpha - \sin^2 \alpha$$

$$\tan(\alpha + \alpha) = \dfrac{\tan \alpha + \tan \alpha}{1 - \tan \alpha \tan \alpha} \quad \text{(By formula 3b, Section 8.3)}$$
$$= \dfrac{2 \tan \alpha}{1 - \tan^2 \alpha}$$

The identities $\cos^2 \alpha = 1 - \sin^2 \alpha$ and $\sin^2 \alpha = 1 - \cos^2 \alpha$ are used to obtain the alternate forms for $\cos 2\alpha$ in the following box.

DOUBLE-ANGLE FORMULAS

$$\sin 2\alpha = 2 \sin \alpha \cos \alpha \tag{1}$$

$$\cos 2\alpha = \cos^2 \alpha - \sin^2 \alpha \tag{2}$$

$$= 1 - 2 \sin^2 \alpha \tag{2a}$$

$$= 2 \cos^2 \alpha - 1 \tag{2b}$$

$$\tan 2\alpha = \frac{2 \tan \alpha}{1 - \tan^2 \alpha} \tag{3}$$

These double-angle formulas can be used in conjunction with the sum formulas to develop triple- and quadruple-angle formulas as follows.

EXAMPLE 1 Develop formulas for $\sin 3\alpha$ and $\sin 4\alpha$.

SOLUTION

$\sin 3\alpha = \sin(2\alpha + \alpha)$
$= (\sin 2\alpha)(\cos \alpha) + (\cos 2\alpha)(\sin \alpha)$ (By formula 2b, section 8.3)
$= (2 \sin \alpha \cos \alpha)(\cos \alpha) + (\cos^2 \alpha - \sin^2 \alpha)(\sin \alpha)$ (By formulas 1 and 2)
$= 2(\sin \alpha)(\cos^2 \alpha) + (\cos^2 \alpha)(\sin \alpha) - (\sin^3 \alpha)$
$= 3 \cos^2 \alpha \sin \alpha - \sin^3 \alpha$

$\sin 4\alpha = \sin(2 \cdot 2\alpha)$
$= 2 \sin 2\alpha \cos 2\alpha$ (By formula 1)
$= 2(2 \sin \alpha \cos \alpha)(\cos^2 \alpha - \sin^2 \alpha)$ (By formulas 1 and 2)
$= 4 \sin \alpha \cos \alpha (\cos^2 \alpha - \sin^2 \alpha)$
$= 4(\sin \alpha \cos^3 \alpha - \cos \alpha \sin^3 \alpha)$ ∎

EXAMPLE 2 If $\sin \alpha = \frac{1}{4}$ and the terminal ray lies in the second quadrant when α is placed in standard position, find $\sin 2\alpha$, $\cos 2\alpha$, and $\tan 2\alpha$ by using the double-angle formulas. In what quadrant is the terminal ray of 2α located?

SOLUTION Sketching α in standard position, we obtain the configuration shown in Figure 8.7. From the Pythagorean theorem, the horizontal leg of the triangle is found to be $-\sqrt{15}$. Thus

$$\cos \alpha = \frac{-\sqrt{15}}{4} \quad \text{and} \quad \tan \alpha = \frac{1}{-\sqrt{15}} = \frac{-\sqrt{15}}{15}$$

Finally,

$\sin 2\alpha = 2 \sin \alpha \cos \alpha$ (By formula 1)
$= 2 \cdot \frac{1}{4} \cdot \left(\frac{-\sqrt{15}}{4}\right)$
$= -\frac{\sqrt{15}}{8}$

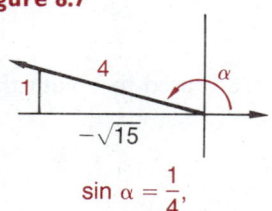

Figure 8.7
$\sin \alpha = \frac{1}{4}$, terminal ray in second quadrant

$$\cos 2\alpha = \cos^2 \alpha - \sin^2 \alpha \quad \text{(By formula 2)}$$
$$= \frac{15}{16} - \frac{1}{16} = \frac{7}{8}$$
$$\tan 2\alpha = \frac{2 \tan \alpha}{1 - \tan^2 \alpha} \quad \text{(By formula 3)}$$
$$= \frac{\frac{-2\sqrt{15}}{15}}{1 - \frac{1}{15}} = -\frac{\sqrt{15}}{7}$$

Since $\sin 2\alpha < 0$ and $\cos 2\alpha > 0$, the terminal ray of 2α lies in the fourth quadrant. ∎

Identities

EXAMPLE 3 Verify the identity
$$\frac{1 - \tan^2 t}{1 + \tan^2 t} = \cos 2t$$

SOLUTION In this example it is quickest to express the left side in terms of sines and cosines and manipulate it into the right side. Thus we have

$$\frac{1 - \tan^2 t}{1 + \tan^2 t} = \frac{1 - \frac{\sin^2 t}{\cos^2 t}}{1 + \frac{\sin^2 t}{\cos^2 t}}$$

$$= \frac{\cos^2 t - \sin^2 t}{\cos^2 t + \sin^2 t} \quad \text{(Multiply numerator and denominator by } \cos^2 t\text{)}$$

$$= \cos^2 t - \sin^2 t \quad (\cos^2 t + \sin^2 t = 1)$$

$$= \cos 2t \quad \text{(By formula 2)} \quad \blacksquare$$

EXAMPLE 4 Verify the identity
$$\frac{1 - \cos 2\theta}{\sin 2\theta} = \tan \theta$$

SOLUTION Use the double-angle formulas on the left side to convert it into the right side.

$$\frac{1 - \cos 2\theta}{\sin 2\theta} = \frac{1 - (\cos^2 \theta - \sin^2 \theta)}{2 \sin \theta \cos \theta} \quad \text{(By formulas 1 and 2)}$$

$$= \frac{1 - \cos^2 \theta + \sin^2 \theta}{2 \sin \theta \cos \theta}$$

$$= \frac{\sin^2 \theta + \sin^2 \theta}{2 \sin \theta \cos \theta} \quad (1 - \cos^2 \theta = \sin^2 \theta)$$

$$= \frac{2 \sin^2 \theta}{2 \sin \theta \cos \theta}$$

$$= \frac{\sin \theta}{\cos \theta}$$

$$= \tan \theta \quad \blacksquare$$

Half-angle Formulas

To obtain half-angle formulas, we rewrite double-angle formulas 2a and 2b in the form

$$\sin^2 \beta = \frac{1 - \cos 2\beta}{2} \quad \text{and} \quad \cos^2 \beta = \frac{1 + \cos 2\beta}{2}$$

respectively. For $\beta = \alpha/2$, these become

$$\sin^2\left(\frac{\alpha}{2}\right) = \frac{1 - \cos \alpha}{2} \quad \text{and} \quad \cos^2\left(\frac{\alpha}{2}\right) = \frac{1 + \cos \alpha}{2}$$

$$\sin \frac{\alpha}{2} = \pm\sqrt{\frac{1 - \cos \alpha}{2}} \qquad (4)$$

$$\cos \frac{\alpha}{2} = \pm\sqrt{\frac{1 + \cos \alpha}{2}} \qquad (5)$$

CAUTION

Note the uncertainty of the $\pm$ sign in $\sin \alpha/2$ and $\cos \alpha/2$. Unlike the sum, difference, and multiple-angle formulas, these half-angle formulas do not determine the sign at $\alpha/2$. It must be determined in some other way, for example, by locating the quadrant in which the terminal ray of $\alpha/2$ lies. Locating this quadrant can itself be an ambiguous task if we know only the location of the terminal ray of α without knowing α. For instance, if the terminal ray of α coincides with its initial ray, it could be that $\alpha = 0, \pm 2\pi, \pm 4\pi, \ldots$. If $\alpha = 2\pi$, then $\alpha/2 = \pi$ and its terminal ray lies on the negative x-axis; on the other hand, if $\alpha = 4\pi$, then $\alpha/2 = 2\pi$ and its terminal ray lies on the positive x-axis (see Figure 8.8).

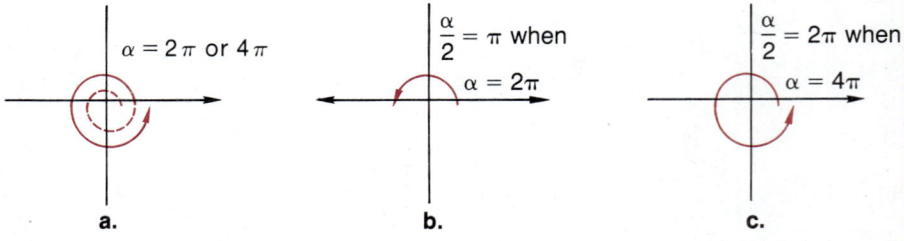

a. $\alpha = 2\pi$ or 4π

b. $\frac{\alpha}{2} = \pi$ when $\alpha = 2\pi$

c. $\frac{\alpha}{2} = 2\pi$ when $\alpha = 4\pi$

Figure 8.8 THE INHERENT AMBIGUITY OF THE $\pm$ SIGN IN THE FORMULAS FOR $\sin \frac{\alpha}{2}$ AND $\cos \frac{\alpha}{2}$

Surprisingly enough, this ambiguity of signs vanishes with $\tan \alpha/2$. To find $\tan \alpha/2$, we write

$$\tan\frac{\alpha}{2} = \frac{\sin\alpha/2}{\cos\alpha/2}$$

$$= \pm\frac{\sqrt{\frac{1-\cos\alpha}{2}}}{\sqrt{\frac{1+\cos\alpha}{2}}} \qquad \text{(By formulas 4 and 5)}$$

$$= \pm\sqrt{\frac{1-\cos\alpha}{1+\cos\alpha}}$$

$$= \pm\sqrt{\frac{(1-\cos\alpha)^2}{1-\cos^2\alpha}} \quad \text{or} \quad \pm\sqrt{\frac{1-\cos^2\alpha}{(1+\cos\alpha)^2}}$$

$$= \pm\sqrt{\frac{(1-\cos\alpha)^2}{\sin^2\alpha}} \quad \text{or} \quad \pm\sqrt{\frac{\sin^2\alpha}{(1+\cos\alpha)^2}}$$

$$= \pm\frac{1-\cos\alpha}{\sin\alpha} \quad \text{or} \quad \pm\frac{\sin\alpha}{1+\cos\alpha}$$

It can be verified that $\tan\alpha/2$ and $\sin\alpha$ have the same sign; (see Exercise 52). Since $|\cos\alpha| \leq 1$, $1 \pm \cos\alpha \geq 0$. Thus we can write

$$\boxed{\tan\frac{\alpha}{2} = \frac{1-\cos\alpha}{\sin\alpha} = \frac{\sin\alpha}{1+\cos\alpha}} \qquad (6)$$

EXAMPLE 5

Use the half-angle formulas to find $\sin\theta$, $\cos\theta$, and $\tan\theta$ for $\theta = \pi/12$.

SOLUTION

Each of the requested values will be positive since $\pi/12$ is an acute angle.

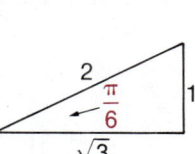

Figure 8.9

$$\sin\frac{\pi}{12} = \sin\frac{\pi/6}{2}$$
$$= \sqrt{\frac{1-\cos\pi/6}{2}} \qquad \text{(By formulas 4 and 5)}$$
$$= \sqrt{\frac{1-\sqrt{3}/2}{2}} \qquad \text{(See Figure 8.9)}$$
$$= \sqrt{\frac{2-\sqrt{3}}{4}}$$
$$= \frac{\sqrt{2-\sqrt{3}}}{2}$$

$$\cos\frac{\pi}{12} = \cos\frac{\pi/6}{2}$$
$$= \sqrt{\frac{1+\cos\pi/6}{2}}$$
$$= \sqrt{\frac{1+\sqrt{3}/2}{2}}$$
$$= \sqrt{\frac{2+\sqrt{3}}{4}}$$
$$= \frac{\sqrt{2+\sqrt{3}}}{2}$$

$$\tan\frac{\pi}{12} = \tan\frac{\pi/6}{2}$$
$$= \frac{1-\cos\pi/6}{\sin\pi/6} \qquad \text{(By formula 6)}$$
$$= \frac{1-\sqrt{3}/2}{1/2} \qquad \text{(See Figure 8.9)}$$
$$= 2 - \sqrt{3}$$

∎

EXAMPLE 6

If $\sin \alpha = \frac{2}{3}$ and $\pi/2 < \alpha < \pi$, find $\sin \alpha/2$, $\cos \alpha/2$, and $\tan \alpha/2$.

SOLUTION

If
$$\frac{\pi}{2} < \alpha < \pi$$
then
$$\frac{\pi}{4} < \frac{\alpha}{2} < \frac{\pi}{2}$$

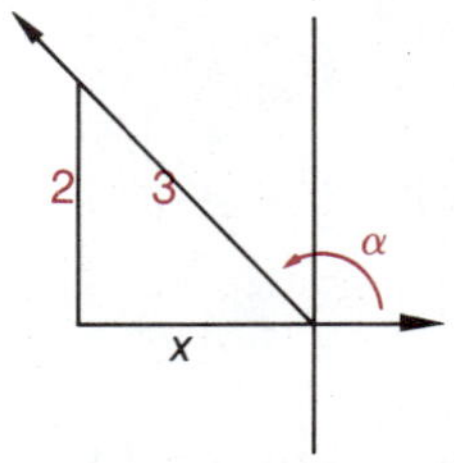

Hence the trigonometric functions of $\alpha/2$ are all positive. To use the half-angle formulas, we must find $\cos \alpha$. Sketch α as in Figure 8.10 and use the Pythagorean theorem to find $x = -\sqrt{5}$. Thus $\cos \alpha = -\sqrt{5}/3$; then

$$\sin \frac{\alpha}{2} = \sqrt{\frac{1 - \cos \alpha}{2}} \qquad \text{(By formulas 4 and 5)} \qquad \cos \frac{\alpha}{2} = \sqrt{\frac{1 + \cos \alpha}{2}}$$

$$= \sqrt{\frac{1 - (-\sqrt{5}/3)}{2}} \qquad \text{(See Figure 8.10)} \qquad = \sqrt{\frac{1 + (-\sqrt{5}/3)}{2}}$$

$$= \sqrt{\frac{3 + \sqrt{5}}{6}} \qquad\qquad\qquad\qquad\qquad\qquad\qquad = \sqrt{\frac{3 - \sqrt{5}}{6}}$$

$$\tan \frac{\alpha}{2} = \frac{1 - \cos \alpha}{\sin \alpha} \qquad \text{(By formula 6)}$$

$$= \frac{1 - (-\sqrt{5}/3)}{2/3} \qquad \text{(See Figure 8.10)}$$

$$= \frac{3 + \sqrt{5}}{2}$$

Figure 8.10

Identities

EXAMPLE 7

Verify the identity
$$\tan \frac{\theta}{2} = \csc \theta - \cot \theta$$

SOLUTION

$$\tan \frac{\theta}{2} = \frac{1 - \cos \theta}{\sin \theta} \qquad \text{(By formula 6)}$$

$$= \frac{1}{\sin \theta} - \frac{\cos \theta}{\sin \theta}$$

$$= \csc \theta - \cot \theta$$

EXAMPLE 8

Show that
$$\tan \frac{t}{2} + \cot \frac{t}{2} = 2 \csc t$$

SOLUTION

$$\tan\frac{t}{2} + \cot\frac{t}{2} = \tan\frac{t}{2} + \frac{1}{\tan\frac{t}{2}}$$

$$= \frac{\tan^2\frac{t}{2} + 1}{\tan\frac{t}{2}} = \frac{\sec^2\frac{t}{2}}{\tan\frac{t}{2}} \qquad (1 + \tan^2\theta = \sec^2\theta)$$

$$= \frac{\frac{1}{\cos^2\frac{t}{2}}}{\left(\sin\frac{t}{2}\right)/\left(\cos\frac{t}{2}\right)}$$

$$= \frac{1}{\left(\sin\frac{t}{2}\right)\left(\cos\frac{t}{2}\right)}$$

$$= \frac{2}{2\sin\frac{t}{2}\cos\frac{t}{2}}$$

$$= \frac{2}{\sin\left(2\cdot\frac{t}{2}\right)} = \frac{2}{\sin t} \qquad \text{(By formula 1)}$$

$$= 2\csc t \qquad\blacksquare$$

Historical Perspective

Trigonometry is as old as recorded history. Ancient documents as far back as the Rhind Papyrus and Babylonian tablets contain trigonometric computations. Hipparchus (~140 B.C.) is credited with developing a table of chords. This table is, for computational purposes, equivalent to a table of sines.

Later, Ptolemy (~150 A.D.) refined this table to develop a table of chord values equivalent to a sine table for angles between 0° and 90° in increments of 15′. In order to develop his table, Ptolemy used trigonometric formulas for chords which today are seen to be equivalent to the modern trigonometric formulas for $\sin(u + v)$, $\sin(u - v)$, and $\sin(\theta/2)$.

Section 8.4 Exercises

Use the sum formulas and the double-angle formulas to develop formulas for

1. $\cos 3\theta$
2. $\tan 3\theta$
3. $\cos 4\theta$
4. $\tan 4\theta$

Use the given information to find sine, cosine, and tangent of 2θ. Indicate the quadrant in which the terminal ray of 2θ lies when 2θ is placed in standard position.

5. $\sin\theta = -\frac{1}{3}, \quad \tan\theta < 0$

6. $\cos \theta = \dfrac{3}{5}$, $\tan \theta > 0$

7. $\sec \theta = -\dfrac{5}{4}$, $\sin \theta > 0$

8. $\tan \theta = -\dfrac{5}{12}$, $\cos \theta < 0$

9. $\cot \theta = \dfrac{3}{4}$, $\cos \theta < 0$

10. $\csc \theta = -10$, $\tan \theta > 0$

Use the half-angle formulas to find exact values for the sine, cosine, and tangent of the given angles.

11. $\dfrac{\pi}{8}$
12. $\dfrac{5\pi}{8}$
13. $\dfrac{9\pi}{8}$
14. $\dfrac{7\pi}{8}$
15. $\dfrac{5\pi}{12}$
16. $\dfrac{11\pi}{12}$
17. $\dfrac{19\pi}{12}$
18. $\dfrac{13\pi}{12}$
19. $105°$
20. $67.5°$

Use the given information to find sine, cosine, and tangent of $\alpha/2$.

21. $\tan \alpha = \dfrac{\sqrt{5}}{2}$, $0 < \alpha < \pi$

22. $\cot \alpha = \sqrt{3}$, $\pi < \alpha < 2\pi$

23. $\cos \alpha = \dfrac{1}{4}$, $2\pi < \alpha < 3\pi$

24. $\sec \alpha = 2$, $3\pi < \alpha < 4\pi$

25. $\sin \alpha = -\dfrac{1}{3}$, $\dfrac{5\pi}{2} < \alpha < \dfrac{7\pi}{2}$

26. $\csc \alpha = -7$, $\dfrac{3\pi}{2} < \alpha < \dfrac{5\pi}{2}$

27. $\cos \alpha = -\dfrac{1}{2}$, $3\pi < \alpha < 4\pi$

28. $\tan \alpha = -5$, $\dfrac{5\pi}{2} < \alpha < \dfrac{7\pi}{2}$

29. $\sec \alpha = -3$, $0 < \alpha < \pi$

30. $\cot \alpha = \dfrac{3}{4}$, $4\pi < \alpha < 5\pi$

Use the double-angle formulas to verify the following identities.

31. $\dfrac{\sin^2 2t}{(1 + \cos 2t)^2} = \sec^2 t - 1$

32. $\cos 4u = 1 + 8(\cos^4 u - \cos^2 u)$

33. $\cot v - \tan v = 2 \cot 2v$

34. $\dfrac{\sin 4x}{\sin 2x} = 2 \cos 2x$

35. $2 \csc 2\theta = \sec \theta \csc \theta$

36. $4 \cos^2 \alpha \sin^2 \alpha + \cos^2 2\alpha = 1$

37. $\cot \beta \sin 2\beta = 1 + \cos 2\beta$

38. $\tan 2w + \sec 2w = \dfrac{\cos w + \sin w}{\cos w - \sin w}$

39. $(1 + \tan^2 t) \sin 2t = 2 \tan t$

40. $\dfrac{\sin 2\theta}{\sin \theta} - \dfrac{\cos 2\theta}{\cos \theta} = \sec \theta$

41. $\dfrac{\cos 2u}{1 - \sin 2u} = \dfrac{1 + \tan u}{1 - \tan u}$

42. $\dfrac{1 + \cos 2\theta}{\sin 2\theta} = \cot \theta$

43. $\dfrac{2 \cot \alpha}{\csc^2 \alpha - 2} = \tan 2\alpha$

Use the half-angle formulas to verify the following.

44. $(1 + \cos t) \csc t = \cos \dfrac{t}{2}$

45. $\dfrac{\sin^2 \dfrac{u}{2} - \cos^2 \dfrac{u}{2}}{\sin u} + \cot u = 0$

46. $\dfrac{\tan \dfrac{v}{2} + 1}{\tan \dfrac{v}{2} - 1} = \dfrac{\cos v}{\sin v - 1}$

47. $(1 - \cos \alpha)^2 + \dfrac{(\sin 2\alpha)^2}{4(1 + \cos \alpha)^2} = \tan^2 \dfrac{\alpha}{2}$

48. $\dfrac{\sin^2 \beta}{2 \cos^2 \dfrac{\beta}{2}} = 1 - \cos \beta$

49. $\dfrac{2 \sin^2 \dfrac{\phi}{2} - 1}{\sin \phi} + \cot \phi = 0$

50. $\cot \dfrac{\theta}{2} = \csc \theta + \cot \theta$

51. $\dfrac{1 + \tan^2 \dfrac{u}{2}}{1 - \tan^2 \dfrac{u}{2}} = \sec u$

52. Show that $\tan \alpha/2$ and $\sin \alpha$ have the same sign for all α. (*Hint:* Consider successive intervals for α such as $0 \le \alpha \le \pi$, $\pi \le \alpha \le 2\pi$, $2\pi \le \alpha \le 3\pi$, and so on. For negative α, use the fact that both $\sin \theta$ and $\tan \theta$ are odd functions.)

Section 8.5

Product–Sum Formulas

Although the formulas derived in this section are not used as often as those in the preceding section, they are useful especially in electronic communications and in the study of resonance in vibrations. Understanding these relationships and their derivation can be very helpful.

The formulas relate the product of two sines, two cosines, or a sine and a cosine to a sum of sines and cosines. One set of formulas converts products into sums; the other converts sums to products. We first consider the product-to-sum formulas.

PRODUCT-TO-SUM FORMULAS

$$\sin \alpha \cos \beta = \dfrac{1}{2}[\sin(\alpha + \beta) + \sin(\alpha - \beta)] \qquad (1)$$

$$\cos \alpha \cos \beta = \dfrac{1}{2}[\cos(\alpha + \beta) + \cos(\alpha - \beta)] \qquad (2)$$

$$\sin \alpha \sin \beta = \dfrac{1}{2}[\cos(\alpha - \beta) - \cos(\alpha + \beta)] \qquad (3)$$

The first of these can be verified by considering the formulas for $\sin(\alpha - \beta)$ and $\sin(\alpha + \beta)$:

$\sin(\alpha - \beta) = \sin \alpha \cos \beta - \cos \alpha \sin \beta$
$\sin(\alpha + \beta) = \sin \alpha \cos \beta + \cos \alpha \sin \beta$ (Formulas 2a, 2b, Section 8.3)

Adding these yields

$\sin(\alpha + \beta) + \sin(\alpha - \beta) = 2 \sin \alpha \cos \beta$

Finally, division by 2 yields formula 1.
Similarly, we write

$\cos(\alpha - \beta) = \cos \alpha \cos \beta + \sin \alpha \sin \beta$
$\cos(\alpha + \beta) = \cos \alpha \cos \beta - \sin \alpha \sin \beta$ (Formulas 1a, 1b, Section 8.3)

Adding these formulas yields

$\cos(\alpha + \beta) + \cos(\alpha - \beta) = 2 \cos \alpha \cos \beta$

whereas subtraction of the second from the first yields

$\cos(\alpha - \beta) - \cos(\alpha + \beta) = 2 \sin \alpha \sin \beta$

Division by 2 then yields formulas 2 and 3.

EXAMPLE 1

Write the following as sums of sines or cosines.

a. $\sin 2\alpha \cos 3\alpha$
b. $\sin 2\alpha \sin 4\alpha$
c. $\cos^2 \alpha$

SOLUTION

Using the product-to-sum formulas, we have the following.

a. $\sin 2\alpha \cos 3\alpha = \dfrac{1}{2}[\sin(2\alpha + 3\alpha) + \sin(2\alpha - 3\alpha)]$ (By formula 1)

$= \dfrac{1}{2}[\sin 5\alpha + \sin(-\alpha)]$

$= \dfrac{1}{2}(\sin 5\alpha - \sin \alpha)$

b. $\sin 2\alpha \sin 4\alpha = \dfrac{1}{2}[\cos(2\alpha - 4\alpha) - \cos(2\alpha + 4\alpha)]$ (By formula 3)

$= \dfrac{1}{2}[\cos(-2\alpha) - \cos 6\alpha]$

$= \dfrac{1}{2}(\cos 2\alpha - \cos 6\alpha)$

c. $\cos^2 \alpha = \cos \alpha \cos \alpha$

$= \dfrac{1}{2}[\cos(\alpha + \alpha) + \cos(\alpha - \alpha)]$ (By formula 2)

$= \dfrac{1}{2}(\cos 2\alpha + \cos 0)$

$= \dfrac{1}{2}(\cos 2\alpha + 1)$ ■

EXAMPLE 2

Use the product-to-sum formulas to evaluate the following.

a. $\left(\sin \dfrac{\pi}{4}\right)\left(\cos \dfrac{\pi}{12}\right)$

b. $\left(\cos \dfrac{3\pi}{4}\right)\left(\cos \dfrac{5\pi}{12}\right)$

SOLUTION

a. $\left(\sin \dfrac{\pi}{4}\right)\left(\cos \dfrac{\pi}{12}\right) = \dfrac{1}{2}\left[\sin\left(\dfrac{\pi}{4} + \dfrac{\pi}{12}\right) + \sin\left(\dfrac{\pi}{4} - \dfrac{\pi}{12}\right)\right]$ (By formula 1)

$= \dfrac{1}{2}\left(\sin \dfrac{\pi}{3} + \sin \dfrac{\pi}{6}\right)$

$= \dfrac{1}{2}\left(\dfrac{\sqrt{3}}{2} + \dfrac{1}{2}\right)$

$= \dfrac{\sqrt{3} + 1}{4}$

b. $\left(\cos\dfrac{3\pi}{4}\right)\left(\cos\dfrac{5\pi}{12}\right) = \dfrac{1}{2}\left[\cos\left(\dfrac{3\pi}{4}+\dfrac{5\pi}{12}\right)+\cos\left(\dfrac{3\pi}{4}-\dfrac{5\pi}{12}\right)\right]$ (By formula 2)

$= \dfrac{1}{2}\left(\cos\dfrac{14\pi}{12}+\cos\dfrac{4\pi}{12}\right)$

$= \dfrac{1}{2}\left(\cos\dfrac{7\pi}{6}+\cos\dfrac{\pi}{3}\right)$

$= \dfrac{1}{2}\left(-\dfrac{\sqrt{3}}{2}+\dfrac{1}{2}\right)$

$= \dfrac{(1-\sqrt{3})}{4}$ ∎

The preceding product-to-sum formulas can be used to develop formulas to perform the reverse task: writing a sum of sines or cosines as a product of sines and/or cosines. For if we let $u = (\alpha + \beta)/2$, $v = (\alpha - \beta)/2$, then $u + v = \alpha$ and $u - v = \beta$. Using the product-to-sum formulas, we obtain the following:

$2\sin u \cos v = \sin(u+v) + \sin(u-v)$ (By formula 1)

or

$2\sin\dfrac{\alpha+\beta}{2}\cos\dfrac{\alpha-\beta}{2} = \sin\alpha + \sin\beta$

Each of the sum-to-product formulas listed here can be obtained similarly.

SUM-TO-PRODUCT FORMULAS

$$\sin\alpha + \sin\beta = 2\sin\left(\dfrac{\alpha+\beta}{2}\right)\cos\left(\dfrac{\alpha-\beta}{2}\right) \quad (4)$$

$$\sin\alpha - \sin\beta = 2\cos\left(\dfrac{\alpha+\beta}{2}\right)\sin\left(\dfrac{\alpha-\beta}{2}\right) \quad (5)$$

$$\cos\alpha + \cos\beta = 2\cos\left(\dfrac{\alpha+\beta}{2}\right)\cos\left(\dfrac{\alpha-\beta}{2}\right) \quad (6)$$

$$\cos\alpha - \cos\beta = -2\sin\left(\dfrac{\alpha+\beta}{2}\right)\sin\left(\dfrac{\alpha-\beta}{2}\right) \quad (7)$$

EXAMPLE 3   Write the following as a product of sines and/or cosines.

a. $\sin 2\alpha + \sin 3\alpha$ **b.** $\sin 3\alpha - \sin \alpha$
c. $\cos 4\alpha + \cos \alpha$ **d.** $\cos 3\alpha - \cos 2\alpha$

SOLUTION

a. $\sin 2\alpha + \sin 3\alpha = 2\sin\dfrac{2\alpha+3\alpha}{2}\cos\dfrac{2\alpha-3\alpha}{2}$ (By formula 4)

$= 2\sin\dfrac{5\alpha}{2}\cos\dfrac{-\alpha}{2}$

$= 2\sin\dfrac{5\alpha}{2}\cos\dfrac{\alpha}{2}$

b. $\sin 3\alpha - \sin \alpha = 2 \cos \dfrac{3\alpha + \alpha}{2} \sin \dfrac{3\alpha - \alpha}{2}$ (By formula 5)

$= 2 \cos \dfrac{4\alpha}{2} \sin \dfrac{2\alpha}{2}$

$= 2 \cos 2\alpha \sin \alpha$

c. $\cos 4\alpha + \cos \alpha = 2 \cos \dfrac{4\alpha + \alpha}{2} \cos \dfrac{4\alpha - \alpha}{2}$ (By formula 6)

$= 2 \cos \dfrac{5\alpha}{2} \cos \dfrac{3\alpha}{2}$

d. $\cos 3\alpha - \cos 2\alpha = -2 \sin \dfrac{3\alpha + 2\alpha}{2} \sin \dfrac{3\alpha - 2\alpha}{2}$ (By formula 7)

$= -2 \sin \dfrac{5\alpha}{2} \sin \dfrac{\alpha}{2}$ ∎

EXAMPLE 4 Use the sum-to-product formulas to evaluate the following.

a. $\sin \dfrac{\pi}{4} + \sin \dfrac{\pi}{12}$

b. $\cos 225° + \cos 105°$

SOLUTION

a. $\sin \dfrac{\pi}{4} + \sin \dfrac{\pi}{12} = 2 \sin \dfrac{\pi/4 + \pi/12}{2} \cos \dfrac{\pi/4 - \pi/12}{2}$ (By formula 4)

$= 2 \sin \dfrac{\pi}{6} \cos \dfrac{\pi}{12}$

$= 2 \cdot \dfrac{1}{2} \cdot \cos \dfrac{\pi}{12}$

$= \cos \dfrac{\pi}{12}$

$= \sqrt{\dfrac{1 + \cos \pi/6}{2}}$ (By formula 5, Section 8.4)

$= \sqrt{\dfrac{1 + \sqrt{3}/2}{2}}$

$= \dfrac{\sqrt{2 + \sqrt{3}}}{2}$

b. $\cos 225° + \cos 105° = 2 \cos \dfrac{225° + 105°}{2} \cos \dfrac{225° - 105°}{2}$ (By formula 6)

$= 2 \cos 165° \cos 60°$

$= \cos 165°$ ($\cos 60° = \tfrac{1}{2}$)

$= -\sqrt{\dfrac{1 + \cos 330°}{2}}$ (By formula 5, Section 8.4)

$= -\sqrt{\dfrac{1 + \sqrt{3}/2}{2}}$

$= -\dfrac{\sqrt{2 + \sqrt{3}}}{2}$ ∎

Identities

EXAMPLE 5 Verify the following identity.

$$\frac{\cos 2t - \cos 4t}{\sin 2t + \sin 4t} = \tan t$$

SOLUTION Using formula 7 on the numerator and formula 4 on the denominator, we obtain

$$\frac{\cos 2t - \cos 4t}{\sin 2t + \sin 4t} = \frac{-2 \sin\left(\frac{2t+4t}{2}\right) \sin\left(\frac{2t-4t}{2}\right)}{2 \sin\left(\frac{2t+4t}{2}\right) \cos\left(\frac{2t-4t}{2}\right)}$$

$$= \frac{-\sin(-t)}{\cos(-t)}$$

$$= \frac{\sin t}{\cos t} \qquad \text{(Formulas 9, 10, Section 8.1)}$$

$$= \tan t \qquad \blacksquare$$

Applications to Graphing

A certain type of oscillatory damping known as *modulation* is important in electronic communications theory. The "AM" in AM radio broadcasting stands for *amplitude modulation*. Amplitude modulation occurs when two sine curves or two cosine curves are summed. The sum-to-product formulas can be used to express these sums in a form that can be graphed using the damping technique of Section 7.7.

EXAMPLE 6 Sketch the graph of the following.

$$f(t) = \sin 2t + \sin 4t$$

SOLUTION We could sketch $\sin 2t$ and $\sin 4t$ and then add the y-coordinates of the points on these graphs (the addition-of-ordinates method of Section 7.7) but this task is time consuming and requires great precision. On the other hand, we can use the sum-to-product Formula 4 to write

$$f(t) = \sin 2t + \sin 4t = 2 \sin \frac{2t+4t}{2} \cos \frac{2t-4t}{2}$$

$$= 2 \sin 3t \cos t \qquad [\cos(-t) = \cos t]$$

In order to graph this product we proceed as with damped vibration (Section 7.7), except that now the damping curve is also periodic. We note that $\sin 3t$ has period $2\pi/3$, but $\cos t$ has period 2π. The $\sin 3t$ curve oscillates more rapidly than the $\cos t$ curve and the graph of f lies between the curves $y = 2 \cos t$ and $y = -2 \cos t$ since

$$|f(t)| = |2 \sin 3t \cos t|$$
$$\leq 2|\cos t|$$

Hence we sketch the curves $y = \pm 2 \cos t$ as the modulating or damping curves and modulate the amplitude of the sin $3t$ curve to fit this region. It helps to plot the critical values for which

$$\sin 3t = 0 \qquad \left(t = 0, \frac{\pi}{3}, \frac{2\pi}{3}, \pi, \frac{4\pi}{3}, \frac{5\pi}{3}, 2\pi\right)$$

$$\cos t = 0 \qquad \left(t = \frac{\pi}{2}, \frac{3\pi}{2}\right)$$

and

$$\sin 3t = \pm 1 \qquad \left(t = \frac{\pi}{6}, \frac{\pi}{2}, \frac{5\pi}{6}, \frac{7\pi}{6}, \frac{3\pi}{2}, \frac{11\pi}{6}\right)$$

Since sin $3t$ is multiplied by 2 cos t, we must be careful to reflect the sine curve through the t-axis when cos $t < 0$. We then obtain the graph as shown in Figure 8.11.

Figure 8.11
$f(t) = 2 \sin 3t \cos t = \sin 2t + \sin 4t$

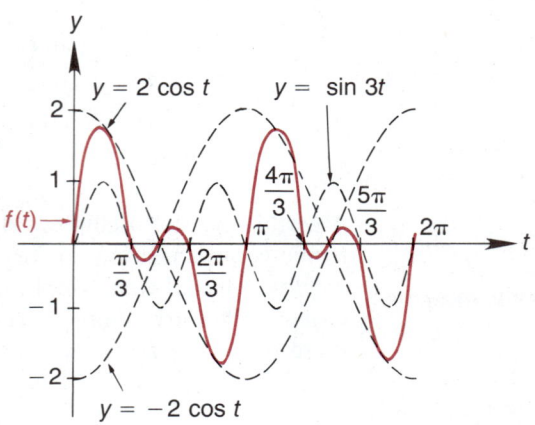

EXAMPLE 7   Sketch the graph of
$$f(t) = \cos 3t + \cos t$$

SOLUTION   We use the sum-to-product Formula 6 to write

$$f(t) = \cos 3t + \cos t$$
$$= 2 \cos\left(\frac{4t}{2}\right) \cos\left(\frac{2t}{2}\right)$$
$$= 2 \cos 2t \cos t$$

Again cos t has the longer period, so we sketch 2 cos t as the modulating curve and modulate cos $2t$ to lie between 2 cos $2t$ and $-2 \cos 2t$. To help in plotting the graph, we first plot the critical values where

$$\cos t = 0 \qquad \left(t = \frac{\pi}{2}, \frac{3\pi}{2}\right)$$

$$\cos 2t = 0 \qquad \left(t = \frac{\pi}{4}, \frac{3\pi}{4}, \frac{5\pi}{4}, \frac{7\pi}{4}\right)$$

and
$$\cos 2t = \pm 1 \quad \left(t = 0, \frac{\pi}{2}, \pi, \frac{3\pi}{2}, 2\pi\right)$$

The graph appears in Figure 8.12.

Figure 8.12
$f(t) = 2 \cos 2t \cos t = \cos 3t + \cos t$

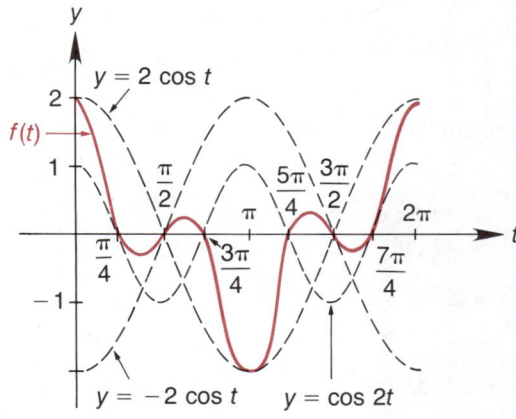

Section 8.5 Exercises

1. Verify sum-to-product formulas 5, 6, and 7.

Express each of the following as a sum or difference of sines or cosines.

2. $\sin \theta \sin 2\theta$
3. $\sin 4\alpha \cos 7\alpha$
4. $\sin 4\beta \cos 3\beta$
5. $\cos 10\gamma \cos 4\gamma$
6. $\cos 2x \cos x$
7. $\cos 10v \sin 4v$
8. $\sin 7w \cos 4w$
9. $\cos 3\phi \cos 4\phi$
10. $\sin 3t \cos 5t$
11. $\sin 4u \sin 7u$

Evaluate the following by using the product-to-sum formulas.

12. $\cos \dfrac{\pi}{4} \cos \dfrac{\pi}{12}$
13. $\sin \dfrac{3\pi}{4} \sin \dfrac{5\pi}{12}$
14. $\cos \dfrac{5\pi}{8} \sin \dfrac{\pi}{8}$
15. $\cos \dfrac{3\pi}{8} \sin \dfrac{\pi}{8}$
16. $\sin \dfrac{5\pi}{4} \cos \dfrac{7\pi}{12}$
17. $\sin \dfrac{3\pi}{4} \sin \dfrac{5\pi}{12}$
18. $\sin \dfrac{3\pi}{8} \cos \dfrac{7\pi}{8}$
19. $\cos^2 \dfrac{3\pi}{8}$
20. $\cos 75° \cos 105°$
21. $\sin 105° \cos 165°$

Express each of the following as a product of sines and/or cosines.

22. $\sin 2\alpha - \sin \alpha$
23. $\sin \beta + \sin 3\beta$
24. $\cos 4\gamma - \cos \gamma$
25. $\sin 4\theta - \sin \theta$
26. $\cos \phi - \cos 3\phi$
27. $\cos 2\eta + \cos 4\eta$
28. $\sin t + \sin 4t$
29. $\sin 4u - \sin 2u$
30. $\cos 2v - \cos 3v$
31. $\cos 3w + \cos w$
32. $\sin x - \sin 4x$
33. $\cos 3z - \cos z$

Evaluate the following by using the sum-to-product formulas.

34. $\cos \dfrac{5\pi}{12} + \cos \dfrac{\pi}{12}$
35. $\sin \dfrac{5\pi}{12} + \sin \dfrac{\pi}{12}$
36. $\cos \dfrac{7\pi}{12} - \cos \dfrac{\pi}{12}$
37. $\sin \dfrac{7\pi}{12} - \sin \dfrac{11\pi}{12}$
38. $\sin \dfrac{7\pi}{12} - \sin \dfrac{\pi}{12}$
39. $\cos \dfrac{\pi}{12} - \cos \dfrac{5\pi}{12}$
40. $\cos \dfrac{5\pi}{12} - \cos \dfrac{11\pi}{12}$
41. $\sin \dfrac{11\pi}{12} - \sin \dfrac{5\pi}{12}$
42. $\sin \dfrac{7\pi}{12} + \sin \dfrac{13\pi}{12}$
43. $\cos \dfrac{5\pi}{12} + \cos \dfrac{13\pi}{12}$
44. $\sin 75° - \sin 15°$
45. $\cos 15° - \cos 105°$

Verify the following identities.

46. $\dfrac{\sin 2x + \sin 6x}{\cos 2x + \cos 6x} = \tan 4x$

47. $\dfrac{\cos 7x + \cos 3x}{\sin 7x - \sin 3x} = \cot 2x$

48. $\dfrac{\cos 3x - \cos 5x}{\cos 3x + \cos 5x} = \tan x \tan 3x$

49. $\dfrac{\sin 5x - \sin x}{\cos 5x + \cos x} = \tan 2x$

50. $\dfrac{\cos 5x + \cos 7x}{\cos 5x - \cos 7x} = \cot 6x \cot x$

51. $\dfrac{\sin 3x + \sin 5x}{\cos 3x - \cos 5x} = \cot x$

52. $\dfrac{\sin 6x - \sin 4x}{\sin 6x + \sin 4x} = \cot 5x \tan x$

53. $\dfrac{\cos 2x - \cos 8x}{\sin 2x + \sin 8x} = \tan 3x$

54. $\dfrac{\sin 3u + \sin u}{\sin 3u - \sin u} = \tan 2u \cot u$

55. $\dfrac{\cos 4x + \cos 2x}{\sin 4x + \sin 2x} = \cot 3x$

Use the sum-to-product formulas to sketch the graphs of $f(t) =$

56. $\sin t + \sin 3t$
57. $\cos 4t - \cos 2t$
58. $\cos 5t + \cos 3t$
59. $\sin 10t - \sin 6t$
60. $\sin 9t - \sin 3t$
61. $\cos 2t + \cos 4t$
62. $\cos 2t - \cos 6t$
63. $\sin 10t + \sin 6t$
64. $\cos 3t + \cos t$
65. $\sin 5t - \sin 3t$

66. When two simultaneous vibrations (sine waves) have periods which are close in value but not equal, the phenomenon of *beats* can occur. Consider, for example, two pieces of metal on an automobile, which vibrate when the engine runs at a certain speed. If the two vibrations are described by

$$f(t) = \sin 50\pi t$$
$$g(t) = \sin 48\pi t$$

use the sum-to-product formulas to graph

$$f(t) + g(t)$$

Note how the amplitude of the vibration increases and decreases. With sound we would notice this phenomenon as alternating loudness and quiet.

Section 8.6

Trigonometric Equations

A **trigonometric equation** is an equation or formula relating trigonometric functions at one or more angles or arguments. An **identity** is an equation that holds for all values of the argument for which all the functions involved are defined. So far in this chapter we have been concerned with various trigonometric identities, but not all trigonometric equations are identities. In this section, we discuss solutions of equations which are satisfied by only certain values of the argument (independent variable); these are called conditional equations. These equations can sometimes be simplified by using certain trigonometric identities as well as the periodicity and reference angle relationships.

EXAMPLE I

Solve the equation

$$2 \tan x - 3 = \tan(-x)$$

SOLUTION

Since tangent is an odd function, we have $\tan(-x) = -\tan x$. Thus the equation becomes

$$2 \tan x - 3 = -\tan x$$
$$3 \tan x = 3$$

Then
$$\tan x = 1$$
$$x = \frac{\pi}{4}$$
is the only solution in the interval $[0, \pi]$. Since tangent has period π, other solutions occur when
$$x = \frac{\pi}{4} + k\pi, \quad k \text{ an integer}$$
The solution set is
$$S = \left\{\frac{\pi}{4} + k\pi : k \in \mathbf{Z}\right\}$$
See Figure 8.13.

Figure 8.13

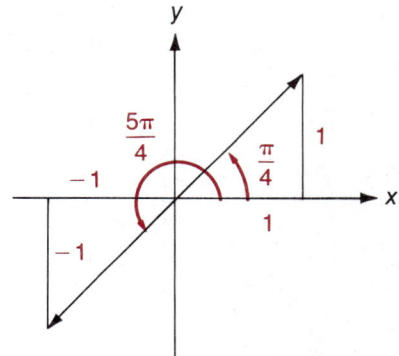

■

EXAMPLE 2

Find all θ for which the following is true.
$$2\cos^2 \theta - 11 \cos \theta - 6 = 0$$

SOLUTION

We recognize this as a quadratic equation in $\cos \theta$. Accordingly, let $w = \cos \theta$ and rewrite the equation as
$$2w^2 - 11w - 6 = 0$$
Factoring, we obtain
$$(2w + 1)(w - 6) = 0$$
$$w = -\frac{1}{2}, 6$$
But $w = \cos \theta \neq 6$ for any θ since $|\cos \theta| \leq 1$; thus $w = -\frac{1}{2}$ is the only possible solution. We must then find all θ for which
$$w = \cos \theta = -\frac{1}{2}$$
In Figure 8.14 we find solutions
$$\theta = \frac{2\pi}{3}, \frac{4\pi}{3} \quad \text{when } 0 \leq \theta \leq 2\pi$$

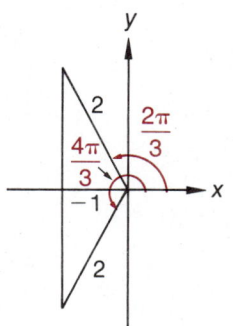

Figure 8.14

Since the period for cosine is 2π, the solutions are all those angles of the form

$$\theta = \frac{2\pi}{3} + 2k\pi \quad \text{or} \quad \theta = \frac{4\pi}{3} + 2k\pi$$

for any integer k. The solution set is

$$S = \left\{\frac{2\pi}{3} + 2k\pi : k \in \mathbb{Z}\right\} \cup \left\{\frac{4\pi}{3} + 2k\pi : k \in \mathbb{Z}\right\}$$

■

EXAMPLE 3   Find all θ in the interval $[0, 2\pi]$ for which the following is true.

$$\cos^2 \theta + 2 \sin \theta = 1$$

SOLUTION   This equation becomes a quadratic equation in $\sin \theta$ after substituting the identity

$$\cos^2 \theta = 1 - \sin^2 \theta$$

We can then simplify and factor as follows.

$$(1 - \sin^2 \theta) + 2 \sin \theta = 1$$
$$-\sin^2 \theta + 2 \sin \theta = 0$$
$$\sin \theta (-\sin \theta + 2) = 0$$

Since $\sin \theta \neq 2$, the equation is satisfied if and only if $\sin \theta = 0$; that is, the solutions in $[0, 2\pi]$ are

$$\theta = 0, \pi, 2\pi$$

The solution set is $S = \{0, \pi, 2\pi\}$.   ■

EXAMPLE 4   Two wheels are geared so that one rotates twice as fast as the other, both in the counterclockwise direction. Each wheel, in turn, is connected to a metal plate with a slit in it as shown in Figure 8.15a. When the holes are aligned, a light beam triggers a sensor which emits a beep. During a single revolution of the slower wheel, how many times is the beep heard? When does this happen?

Figure 8.15

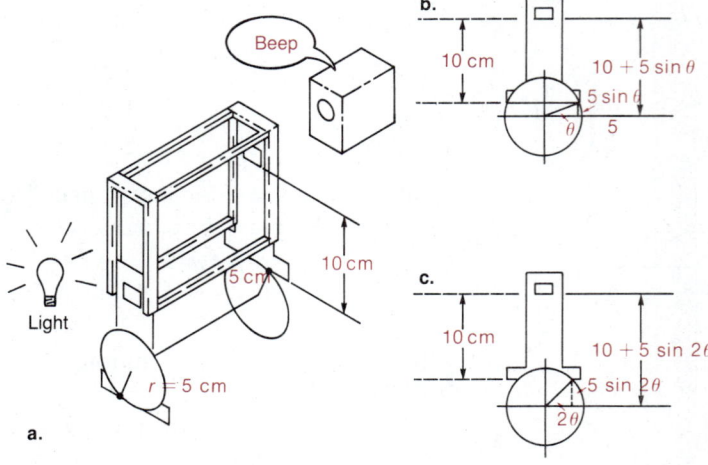

SOLUTION   Place the centers of the wheels at the origin, as in Figure 8.15b and c. The drive pins, which are near the rims of the wheels, should each be placed on the x-axis initially. The faster wheel rotates through 2θ radians when the slower one rotates through θ radians. Then the altitudes of the two slits are $10 + 5 \sin \theta$ and $10 + 5 \sin 2\theta$, respectively. (See Figure 8.15b and c). The slits coincide when

$$10 + 5 \sin 2\theta = 10 + 5 \sin \theta$$
$$\sin 2\theta = \sin \theta$$

The double-angle formula is useful here; when we write $\sin 2\theta = 2 \sin \theta \cos \theta$, the preceding equation becomes

$$2 \sin \theta \cos \theta = \sin \theta$$
$$2 \sin \theta \cos \theta - \sin \theta = 0$$
$$\sin \theta (2 \cos \theta - 1) = 0$$

The solutions to this equation occur when

$$\sin \theta = 0 \quad \text{or} \quad \cos \theta = \frac{1}{2}$$

The solutions in the interval $[0, 2\pi]$ are

$$\theta = 0, \pi, 2\pi \quad \text{or} \quad \theta = \frac{\pi}{3}, \frac{5\pi}{3}$$

The solution set is $\left\{ 0, \pi, 2\pi, \dfrac{\pi}{3}, \dfrac{5\pi}{3} \right\}.$

If the wheels are synchronized at $\theta = 0$, the slits are aligned three more times during one revolution of the slower wheel (when $\theta = \pi/3, \pi,$ and $5\pi/3$) and the wheels are again synchronized when the slower wheel completes the revolution ($\theta = 2\pi$). Thus the sensor beeps four times during every revolution of the slower wheel. The beeps occur in a triple burst and a single shot: ..., beep, beep, beep, , beep,   , beep, beep, beep,   , beep,   , beep, beep, beep. ... The triple burst occurs at $\theta = -\pi/3, 0, \pi/3,$ and the single shot occurs at $\theta = \pi$. ∎

EXAMPLE 5

Find those vaues of t in the interval $[0, 2\pi]$ that satisfy the following equation.

$$\cos 2t = \sin t$$

SOLUTION   Use the double-angle formula $\cos 2t = 1 - 2 \sin^2 t$ to rewrite the equation as

$$1 - 2 \sin^2 t = \sin t$$
$$-2 \sin^2 t - \sin t + 1 = 0$$
$$-(2 \sin t - 1)(\sin t + 1) = 0$$
$$\sin t = \frac{1}{2}, -1$$
$$t = \frac{\pi}{6}, \frac{5\pi}{6}, \frac{3\pi}{2}$$

The solution set is

$$S = \left\{\frac{\pi}{6}, \frac{5\pi}{6}, \frac{3\pi}{2}\right\}$$

See Figure 8.16.

Figure 8.16

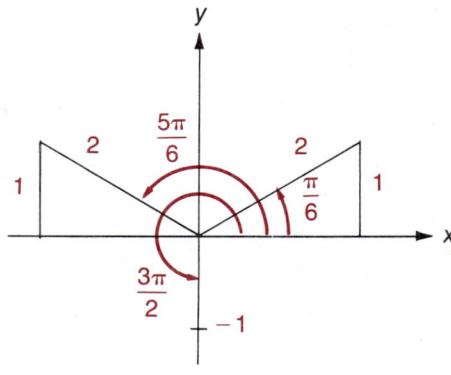

EXAMPLE 6

Solve the following trigonometric equation.

$$2 \sin 2x + \cos 2x = 2(\sin 2x)(\cos 2x) + 1$$

SOLUTION

We need not use the double-angle formulas here since each argument is written as $2x$. Bringing each term over to the left side, we rewrite the equation as

$$2 \sin 2x + \cos 2x - 2(\sin 2x)(\cos 2x) - 1 = 0$$

Regrouping and factoring yields

$$(2 \sin 2x - 2 \sin 2x \cos 2x) + (\cos 2x - 1) = 0$$
$$2 \sin 2x(1 - \cos 2x) - (1 - \cos 2x) = 0$$
$$(2 \sin 2x - 1)(1 - \cos 2x) = 0$$
$$\sin 2x = \frac{1}{2}, \quad \cos 2x = 1$$

Now $\sin 2x = \frac{1}{2}$ when

$$2x = \begin{cases} \dfrac{\pi}{6} + 2k\pi \\ \text{or} \\ \dfrac{5\pi}{6} + 2k\pi \end{cases} \quad k \text{ an integer}$$

that is, when

$$x = \begin{cases} \dfrac{\pi}{12} + k\pi \\ \text{or} \\ \dfrac{5\pi}{12} + k\pi \end{cases} \quad k \text{ an integer}$$

Also,
$$\cos 2x = 1 \quad \text{when} \quad 2x = 2k\pi, \quad k \text{ an integer}$$
that is, when
$$x = k\pi, \quad k \text{ an integer}$$
The solutions are all those numbers of the form
$$x = \frac{\pi}{12} + k\pi, \quad \frac{5\pi}{12} + k\pi, \quad k\pi$$
for any integer k. The solution set is
$$S = \left\{\frac{\pi}{12} + k\pi : k \in \mathbf{Z}\right\} \cup \left\{\frac{5\pi}{12} + k\pi : k \in \mathbf{Z}\right\} \cup \{k\pi : k \in \mathbf{Z}\}$$

EXAMPLE 7 Find all u in the interval $[0, 2\pi]$ for which $\tan^4 u = 9$.

SOLUTION Taking the square root of both sides, we obtain
$$\tan^2 u = 3 \qquad \text{(Since } a^2 \geq 0 \text{ for all real numbers } a\text{)}$$
Again, taking square roots, we have
$$\tan u = \pm\sqrt{3}$$
$$u = \frac{\pi}{3}, \frac{2\pi}{3}, \frac{4\pi}{3}, \frac{5\pi}{3}$$

The solution set is
$$S = \left\{\frac{\pi}{3}, \frac{2\pi}{3}, \frac{4\pi}{3}, \frac{5\pi}{3}\right\}$$

See Figure 8.17.

Figure 8.17

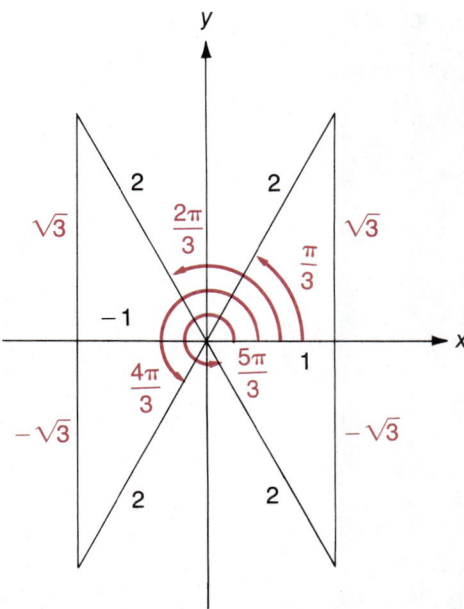

EXAMPLE 8

Find all $x \in [0, 2\pi]$ for which the following is true.

$$\sin^2 x + \sin x = 1$$

Approximate the answers to three decimal places.

SOLUTION

Letting $w = \sin x$ and subtracting 1 from both sides of the equation, we have

$$w^2 + w - 1 = 0$$

Since the quadratic expression does not factor, we use the quadratic formula to obtain

$$w = \frac{-1 \pm \sqrt{1+4}}{2}$$

$$= \frac{-1 \pm \sqrt{5}}{2}$$

Since $(-1 - \sqrt{5})/2 < -1$, the only possible value for $\sin x$ is

$$\sin x = w = \frac{-1 + \sqrt{5}}{2}$$

We use a calculator to find

$$x \approx 0.6662\ldots$$

as the only solution in the interval $[0, \pi/2]$. In $[0, 2\pi]$, the solutions are

$$0.6662\ldots \quad \text{and} \quad \pi - 0.6662\ldots \approx 2.475$$

The (approximate) solution set is

$$S = \{0.666, 2.475\}$$

Section 8.6 Exercises

Solve for x in each of the following equations.

1. $2 \sin x - 1 = 0$
2. $1 + \tan x = 2$
3. $1 - \cot x = 0$
4. $4 + 5 \tan x = 9$
5. $6 \sin \theta + 3 = 0$
6. $2 \sin x + \sqrt{3} = 0$
7. $3 + 2 \csc x = \csc x + 5$
8. $2 \cos x - \sqrt{3} = 0$
9. $\tan^2 x - 1 = 0$
10. $3 \tan^2 x = 1$
11. $\sec^2 x - 2 = 0$
12. $4 \cos^2 x - 3 = 0$
13. $\cos 2x = 0$
14. $\sin 2x = 1$
15. $\tan 3x = 1$
16. $\cos 2x = -\frac{\sqrt{3}}{2}$
17. $2 \cos 2x + \sqrt{3} = 0$
18. $2 \sin 2x + \sqrt{2} = 0$
19. $2 \sin \frac{x}{2} = -\sqrt{3}$
20. $2 \cos \frac{x}{2} = \sqrt{3}$
21. $\cos \frac{x}{2} = 1$
22. $\sin \frac{x}{2} = -1$
23. $(2 \cos x - 1)(2 \sin x + 4) = 0$
24. $(1 - \sin x)(2 - \cos x) = 0$
25. $(1 - \tan x)(1 - \sin x) = 0$
26. $\sin x(1 - 2 \cos x) = 0$
27. $(\sqrt{3} - \tan x)(\sqrt{3} - 2 \cos x) = 0$
28. $(2 \cos x - \sqrt{3})(\cos x - 1) = 0$
29. $\tan^2 x = \tan x$
30. $\cos^2 x - \cos x = 0$
31. $2 \sin x \cos x = \cos x$
32. $\cot x \cos x - \cot x = 0$

33. $\cos^2 x \sin x = \sin x$ 34. $\tan^3 x = \dfrac{1}{3} \tan x$

35. $\sin^3 x - \sin x = 0$ 36. $2 \cos^2 x - \cos x = 0$

Find all x in the interval $[0, 2\pi]$ that satisfy the given equations.

37. $\csc^2 x + \csc x = 2$ 38. $2 \sin^2 x - \sin x = 1$

39. $2 \cos^2 x + \cos x = 1$

40. $2 \cos^2 x - 7 \cos x + 3 = 0$

41. $\sin^2 x + \sin x = 2$

42. $2 \cos^2 x - 5 \cos x - 3 = 0$

43. $2 \sin^2 x - 3 \sin x + 1 = 0$

44. $\cot^2 x - 2 \cot x + 1 = 0$

45. $\cos^2 x - 2 \cos x + 1 = 0$

46. $2 \cos^2 x + 3 \cos x + 1 = 0$

47. $2 \cos^2 x - \sin x = 1$ 48. $\sin^2 x + 2 \cos x = 1$

49. $\sec^2 x = 2 \tan x$ 50. $2 \sin^2 x + \cos x = 2$

51. $4 \cos^2 x + 1 = 8 \sin x$ 52. $\csc^2 x + 2 \cot x = 0$

53. $\tan^2 x + 2 \sec x + 1 = 0$

54. $\cot^2 x + \csc x = 1$ 55. $2 \cos^2 2x = \cos 2x$

56. $2 \sin^2 2x + \sin 2x - 1 = 0$

57. $\tan^2 4x - 2 \tan 4x + 1 = 0$

58. $\csc^2 2x + \cot 2x = 1$

59. $2 \sin 2x + \tan 2x = 0$

60. $\cos x + \sin x \cot x = 3$

61. $3 \sin x + 2 \tan^2 x = 6 + \tan^2 x \sin x$

62. $2 \cos x \cot x + \cot x - 2 \cos x - 1 = 0$

63. $\cos^2 x + \cos x \tan x = \sin x$

64. $\sin 2x = 3 \sin x$ 65. $4 \sin 3x \cos 3x = 1$

66. $\cos 2x - 3 \sin x + 1 = 0$

67. $4 \sin^2 x + 4 \cos 2x = 1$

68. $\cos 2x + \sin^2 x = 1$

69. $\cos 2x + \sin^2 x = 0$ 70. $\sec^4 x = 4$

Square both sides of the following equations before attempting to solve them.

71. $\sin x + \cos x = 1$ 72. $\cos x - \sin x = 1$

73. $\sin 2x - \cos 2x = 1$ 74. $\sin \dfrac{x}{2} + \cos \dfrac{x}{2} = 1$

Calculator Exercises (75–84)

Approximate the solutions of the following equations to three decimal places. List only the solutions in the interval $[0, 2\pi]$.

75. $6 \sin^2 x + 13 \sin x + 6 = 0$

76. $\cos^2 x - \cos x = 20$

77. $\cos^2 t + 4 \cos t + 1 = 0$

78. $\sin^2 x - 2 \sin x - 1 = 0$

79. $\tan^2 x - 3 \tan x + 2 = 0$

80. $\tan^2 x + 8 \tan x + 13 = 0$

81. $3 \csc^2 x + 5 \csc x - 12 = 0$

82. $4 \sec^2 x + 4 \sec x - 3 = 0$

83. $3 \sec^2 x = 4 - 2 \tan x$

84. $\csc^2 \theta - 4 = 2 \cot \theta$

85. Rework Example 4 of this section under the assumption that the wheels are rotating in opposite directions.

86. Rework Example 4 of this section if the slower wheel has radius 10 centimeters, all other conditions remaining the same.

Section 8.7

The Inverse Circular Functions

As we have seen, problems do arise in which we must find θ when we know $T(\theta)$ for one of the trigonometric functions T. Since the circular functions are periodic, there are infinitely many θ's that yield a given value for any of these functions. For instance, $0 = \sin 0 = \sin \pi = \sin 2\pi$, and so on. Strictly speaking, the circular functions do not have inverses. However, if we restrict our attention to a particular region, we can frequently find exactly one θ in that region that yields the given value.

The Inverse Sine Function

Considering the graph of the sine function (Figure 8.18a), we see that all its values are attained exactly once in the interval $-\pi/2 \leq \theta \leq \pi/2$ and that the sine function is strictly increasing on this interval. For any value v between -1 and 1 inclusive, there is precisely one θ in the interval $-\pi/2 \leq \theta \leq \pi/2$ for which $v = \sin \theta$. This means that if we restrict our attention to this interval, the restricted function $v = \sin \theta$ has an inverse (see Figure 18b). This restricted function will be denoted by **Sin**, with a capital S. Thus

$$\text{Sin } \theta = \sin \theta \qquad -\frac{\pi}{2} \leq \theta \leq \frac{\pi}{2}$$

Figure 8.18

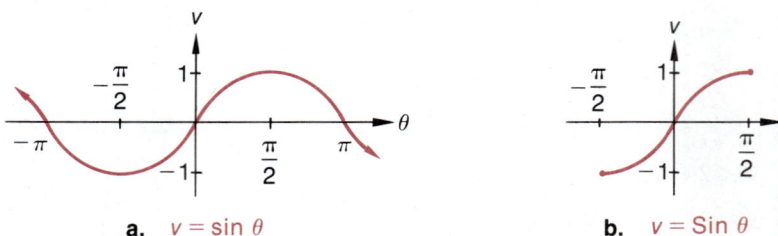

a. $v = \sin \theta$ \qquad b. $v = \text{Sin } \theta$

The inverse of the restricted function Sin is called the **arcsine function** and is denoted by Sin^{-1}. (Other texts may use the notation $\sin^{-1}$, arcsin, or Arcsin for the inverse sine function Sin^{-1}.) As with any inverse function, $\theta = \text{Sin}^{-1} x$ means $x = \text{Sin } \theta$. Thus

$$\boxed{\theta = \text{Sin}^{-1} x \quad \text{means} \quad \left(x = \sin \theta \quad \text{and} \quad -\frac{\pi}{2} \leq \theta \leq \frac{\pi}{2}\right)}$$

> **CAUTION**
>
> Do not confuse the notations $(\sin x)^{-1}$ and $\text{Sin}^{-1} x$. $\text{Sin}^{-1} x$ denotes the inverse sine function but
>
> $$(\sin x)^{-1} = \frac{1}{\sin x} \neq \text{Sin}^{-1} x$$

EXAMPLE 1

Find each value.

a. $\text{Sin}^{-1} \dfrac{1}{2}$ \qquad b. $\text{Sin}^{-1} \left(-\dfrac{\sqrt{2}}{2}\right)$ \qquad c. $\text{Sin}^{-1} \dfrac{\sqrt{3}}{2}$

d. $\text{Sin}^{-1}(-1)$ \qquad e. $\text{Sin}^{-1} 0$

SOLUTION

a. $\text{Sin}^{-1} \dfrac{1}{2} = \dfrac{\pi}{6}$ \qquad since \qquad $\sin \dfrac{\pi}{6} = \dfrac{1}{2}$

b. $\text{Sin}^{-1}\left(-\dfrac{\sqrt{2}}{2}\right) = -\dfrac{\pi}{4}$ \qquad since \qquad $\sin\left(-\dfrac{\pi}{4}\right) = -\dfrac{\sqrt{2}}{2}$

c. $\text{Sin}^{-1} \dfrac{\sqrt{3}}{2} = \dfrac{\pi}{3}$ since $\sin \dfrac{\pi}{3} = \dfrac{\sqrt{3}}{2}$

d. $\text{Sin}^{-1}(-1) = -\dfrac{\pi}{2}$ since $\sin -\dfrac{\pi}{2} = -1$

e. $\text{Sin}^{-1} 0 = 0$ since $\sin 0 = 0$ ∎

The graph of $y = \text{Sin}^{-1} x$ is given in Figure 8.19. Note that $-\pi/2 \leq \text{Sin}^{-1} x \leq \pi/2$.

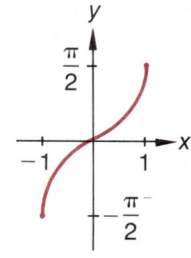

THE ARCSINE FUNCTION

Domain $= [-1, 1]$

Range $= \left[-\dfrac{\pi}{2}, \dfrac{\pi}{2}\right]$

$y = \text{Sin}^{-1} x = \arcsin x$

Figure 8.19

CALCULATOR COMMENTS

The inverse sine function, Sin^{-1}, is programmed into many calculators. There is no Sin^{-1} key, however. The inverse sine function is obtained by pressing the $\boxed{\text{INV}}$ $\boxed{\text{SIN}}$ keys in succession. (On some calculators, Sin^{-1} is obtained by pressing $\boxed{\text{2nd F}}$ $\boxed{\text{SIN}}$ in succession.) The input must be between -1 and 1; pressing the $\boxed{\text{INV}}$ $\boxed{\text{SIN}}$ (or $\boxed{\text{2nd F}}$ $\boxed{\text{SIN}}$) keys then yields an output

1. Between $-\pi/2$ and $\pi/2$ if the calculator is in radian mode.
2. Between $-90°$ and $90°$ if the calculator is in degree mode.

When reading the results, it is best to verify that the calculator is in the proper mode (degrees or radians).

EXAMPLE 2

Find $\text{Sin}^{-1}(-\tfrac{1}{4})$.

CALCULATOR SOLUTION

$\text{Sin}^{-1}(-\tfrac{1}{4})$ is the angle θ between $-\pi/2$ and $\pi/2$ for which $\sin \theta = -\tfrac{1}{4}$. On the calculator, this is found by entering -0.25 and pressing $\boxed{\text{INV}}$ $\boxed{\text{SIN}}$. The result is -0.253 radians, or $-14°29'$ ($= -14.48°$).

TABLE SOLUTION

Without a calculator, we would first find the reference angle θ^* for which $\sin \theta^* = \tfrac{1}{4}$ (since negative values do not appear in Table 4 or 5 at the end of the book).

482 Chapter 8 Trigonometric Identities and Equations; Inverse Trigonometric Functions

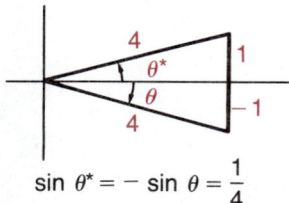

Figure 8.20

Using Table 4 or Table 5 in reverse, we find

$$\theta^* = 0.253, \quad \text{or} \quad 14°29'$$

(Interpolation is necessary.) But since $\sin \theta < 0$ and $-\pi/2 \leq \theta \leq \pi/2$, we then obtain θ from this reference angle by setting $\theta = -\theta^*$ (see Figure 8.20):

$$\theta = -0.253 \text{ radians}, \quad \text{or} \quad -14°29' \qquad \blacksquare$$

The Inverse Cosine Function

Considering the graph of $v = \cos \theta$, we see that only the positive values of $\cos \theta$ are attained in the interval $[-\pi/2, \pi/2]$; see Figure 8.21a. Furthermore, each of the values except $\cos 0 = 1$ is assumed twice on this interval. However, each of the values of the cosine function is attained exactly once in the interval $0 \leq \theta \leq \pi$. Hence we define

$$\text{Cos } \theta = \cos \theta, \quad 0 \leq \theta \leq \pi$$

See Figure 8.21b.

Figure 8.21

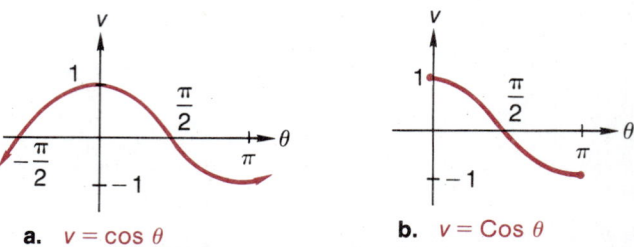

a. $v = \cos \theta$ **b.** $v = \text{Cos } \theta$

This gives a strictly decreasing function which then has an inverse, called the **arccosine function** and denoted Cos^{-1}. The graph of Cos^{-1} appears in Figure 8.22. Observe as with Sin^{-1} that

$$\theta = \text{Cos}^{-1} x \quad \text{means} \quad (x = \cos \theta \text{ and } 0 \leq \theta \leq \pi)$$

THE ARCCOSINE FUNCTION

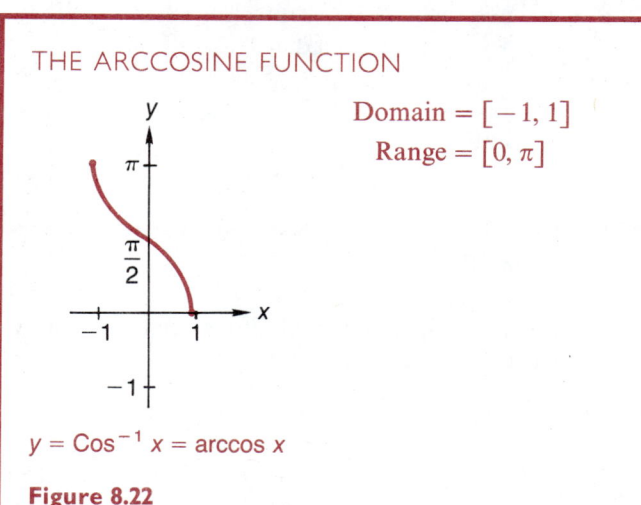

Domain = $[-1, 1]$
Range = $[0, \pi]$

$y = \text{Cos}^{-1} x = \arccos x$

Figure 8.22

EXAMPLE 3

Find the following.

a. $\operatorname{Cos}^{-1} \dfrac{\sqrt{2}}{2}$ b. $\operatorname{Cos}^{-1}\left(\dfrac{-\sqrt{3}}{2}\right)$ c. $\operatorname{Cos}^{-1}\left(\dfrac{-1}{2}\right)$

d. $\operatorname{Cos}^{-1}(-1)$ e. $\operatorname{Cos}^{-1} 0$

SOLUTION

a. $\operatorname{Cos}^{-1} \dfrac{\sqrt{2}}{2} = \dfrac{\pi}{4}$ since $\cos \dfrac{\pi}{4} = \dfrac{\sqrt{2}}{2}$

b. $\operatorname{Cos}^{-1}\left(\dfrac{-\sqrt{3}}{2}\right) = \dfrac{5\pi}{6}$ since $\cos \dfrac{5\pi}{6} = \dfrac{-\sqrt{3}}{2}$

c. $\operatorname{Cos}^{-1}\left(\dfrac{-1}{2}\right) = \dfrac{2\pi}{3}$ since $\cos \dfrac{2\pi}{3} = \dfrac{-1}{2}$

d. $\operatorname{Cos}^{-1}(-1) = \pi$ since $\cos \pi = -1$

e. $\operatorname{Cos}^{-1} 0 = \dfrac{\pi}{2}$ since $\cos \dfrac{\pi}{2} = 0$

CALCULATOR COMMENTS

The Cos^{-1} function is also programmed into many calculators. Again, the input must be between -1 and 1; pressing the [INV] [COS] (or [2nd F] [COS]) keys then yields an output

1. Between 0 and π if the calculator is in radian mode.
2. Between $0°$ and $180°$ if the calculator is in degree mode.

EXAMPLE 4

Find $\operatorname{Cos}^{-1}(-\tfrac{1}{3})$.

CALCULATOR SOLUTION

Entering $-1 \div 3$ and pressing [INV] [COS] (or [2nd F] [COS]) on the calculator yields 1.911 radians, or $109°28'$ ($=109.47°$).

TABLE SOLUTION

Let
$$\theta = \operatorname{Cos}^{-1}\left(-\dfrac{1}{3}\right)$$
Then
$$\cos \theta = -\dfrac{1}{3} \quad \text{and} \quad 0 \leq \theta \leq \pi$$

Since $\cos \theta < 0$, it must be the case that $\pi/2 \leq \theta \leq \pi$. We find the reference angle $\theta^* = \pi - \theta$, as illustrated in Figure 8.23. Then
$$\cos \theta^* = -\cos \theta = \dfrac{1}{3}$$

Interpolating in Tables 4 or 5 (at the back of the book), we find
$$\theta^* \approx 1.231 \quad \text{or} \quad \theta^* \approx 70°32'$$

Finally,
$$\theta = \pi - \theta^* \approx 1.911 \quad \text{or} \quad \theta = 180° - \theta^* \approx 109°28'$$

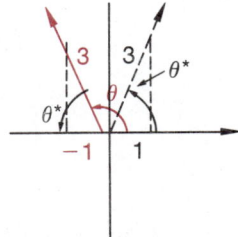

Figure 8.23
$\cos \theta^* = -\cos \theta = \dfrac{1}{3}$

EXAMPLE 5

Evaluate $\sin[\text{Cos}^{-1}(-\frac{1}{2})]$.

CALCULATOR SOLUTION

Treating this as a key-punching exercise on a calculator, we enter -0.5 and then press $\boxed{\text{INV}}$ (or $\boxed{\text{2nd F}}$) $\boxed{\text{COS}}$ $\boxed{\text{SIN}}$ to obtain 0.866 as the result rounded to three decimal places.

ALGEBRAIC SOLUTION

Without a calculator, we let $\theta = \text{Cos}^{-1}(-\frac{1}{2})$ and sketch θ in standard position, noting that $0 \leq \theta \leq \pi$; see Figure 8.24. Then use the Pythagorean theorem to find the altitude $a = \sqrt{3}$. Finally,

$$\sin\left[\text{Cos}^{-1}\left(-\frac{1}{2}\right)\right] = \sin \theta = \frac{\sqrt{3}}{2}$$

is the exact solution.

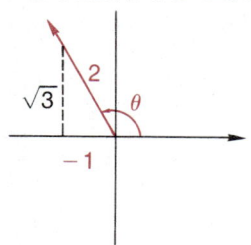

Figure 8.24
$\theta = \text{Cos}^{-1}\left(-\frac{1}{2}\right)$

CALCULATOR COMMENTS

Example 5 illustrates one of the disadvantages in using calculators: the output is always in decimal notation. The calculator displays $\sqrt{3}/2$ as ≈ 0.866. This difficulty is even more pronounced when using the calculator in radian mode to evaluate the inverse trigonometric functions. For example, the calculator displays $\text{Cos}^{-1}(-\frac{1}{2}) \approx 2.09$; one does not immediately recognize the "nice" angle $2\pi/3$ from this display. To recognize "nice" angles, it is better to put the calculator in degree mode when evaluating inverse circular functions. In degree mode, the calculator displays $\text{Cos}^{-1}(-\frac{1}{2}) = 120$.

EXAMPLE 6

Show for each $x \in [-1, 1]$ that $\sin[\text{Cos}^{-1} x] = \sqrt{1 - x^2}$.

SOLUTION

$\theta = \text{Cos}^{-1} x$ is between 0 and π. Sketching θ in standard position as in Figure 8.25, we see immediately that $\sin \theta = \sqrt{1 - x^2}/1 = \sqrt{1 - x^2}$.

Figure 8.25
$\theta = \text{Cos}^{-1} x$

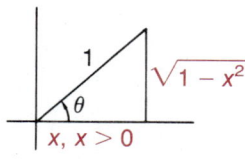
a.

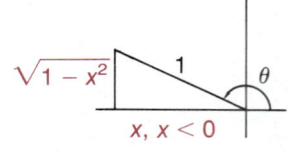
b.

Trigonometric Equations

EXAMPLE 7

Solve the following trigonometric equation.

$$\text{Cos}^{-1} \frac{3}{5} + \text{Sin}^{-1} \frac{5}{13} = \text{Sin}^{-1} x$$

Section 8.7 The Inverse Circular Functions 485

SOLUTION

Three angles are involved in this equation:

$$\alpha = \text{Cos}^{-1}\frac{3}{5} \qquad \beta = \text{Sin}^{-1}\frac{5}{13} \qquad \theta = \text{Sin}^{-1} x$$

Observe that $\alpha + \beta = \theta$ and

$$\cos \alpha = \frac{3}{5} \qquad \sin \beta = \frac{5}{13} \qquad x = \sin \theta$$

The angles α and β are sketched in standard position in Figure 8.26. Both have their terminal sides in Quadrant I since their arguments, $\frac{3}{5}$ and $\frac{5}{13}$, are positive. The Pythagorean theorem is used to find the third side of each triangle in Figure 8.26. We then use the addition formula and Figure 8.26 to obtain

$$\begin{aligned} x = \sin \theta &= \sin(\alpha + \beta) \\ &= \sin \alpha \cos \beta + \cos \alpha \sin \beta \quad \text{(By formula 2b, Section 8.3)} \\ &= \frac{4}{5} \cdot \frac{12}{13} + \frac{3}{5} \cdot \frac{5}{13} \\ &= \frac{63}{65} \end{aligned}$$

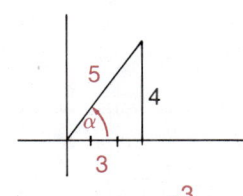
a. $\alpha = \text{Cos}^{-1}\frac{3}{5}$

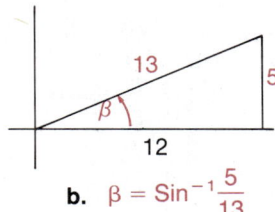
b. $\beta = \text{Sin}^{-1}\frac{5}{13}$

Figure 8.26

EXAMPLE 8

Solve the following trigonometric equation.

$$\text{Sin}^{-1} 2x + 2 \text{Cos}^{-1} x = \pi$$

SOLUTION

Let $\alpha = \text{Sin}^{-1} 2x$ and $\beta = \text{Cos}^{-1} x$; then $\alpha + 2\beta = \pi$ and

$$\sin \alpha = 2x \qquad -\frac{\pi}{2} \leq \alpha \leq \frac{\pi}{2}$$

$$\cos \beta = x \qquad 0 \leq \beta \leq \pi$$

Sketch α and β in standard position but keep in mind that x could be negative. Note that the third side of each triangle indicated in Figure 8.27 is given by the positive radical even if $x < 0$.

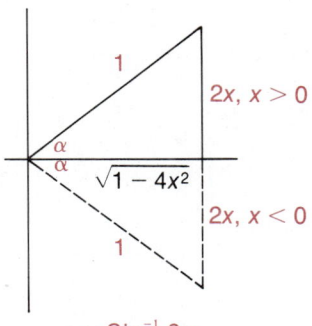
a. $\alpha = \text{Sin}^{-1} 2x$

Then observe that

$$\begin{aligned} \sin \pi &= \sin(\alpha + 2\beta) \\ 0 &= \sin \alpha \cos 2\beta + \cos \alpha \sin 2\beta \quad \text{(By formula 2b, Section 8.3)} \\ &= \sin \alpha (\cos^2 \beta - \sin^2 \beta) + \cos \alpha (2 \sin \beta \cos \beta) \\ &\quad \text{(By formulas 2 and 1, Section 8.4)} \\ &= 2x(x^2 - [1 - x^2]) + \sqrt{1 - 4x^2} \, (2 \cdot \sqrt{1 - x^2} \cdot x) \quad \text{(See Figure 8.27)} \\ &= 2x(2x^2 - 1) + \sqrt{1 - 4x^2} \, 2x\sqrt{1 - x^2} \\ &= 2x(2x^2 - 1 + \sqrt{1 - 4x^2}\sqrt{1 - x^2}) \end{aligned}$$

Hence

$$x = 0 \quad \text{or} \quad 2x^2 - 1 = -\sqrt{1 - 4x^2}\sqrt{1 - x^2}$$

Squaring both sides of this latter equation, we have

$$4x^4 - 4x^2 + 1 = 1 - 5x^2 + 4x^4$$

$$x^2 = 0$$

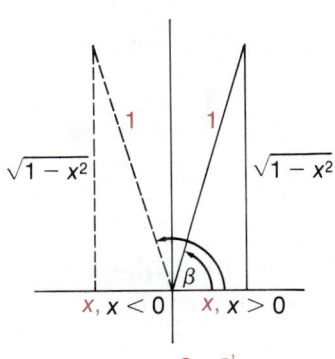
b. $\beta = \text{Cos}^{-1} x$

Figure 8.27

Thus $x = 0$ is the only possible solution.

Checking, we have

$$\text{Sin}^{-1} 2 \cdot 0 + 2 \text{Cos}^{-1} 0 = 0 + 2 \cdot \frac{\pi}{2} \stackrel{\checkmark}{=} \pi$$

The solution set is $\{0\}$. ∎

CAUTION

Even though

$$\sin(\text{Sin}^{-1} x) = x \quad \text{and} \quad \cos(\text{Cos}^{-1} x) = x$$

it is not necessarily the case that $\text{Sin}^{-1}(\sin x) = x$. For if $x = 2\pi$, then $\sin 2\pi = 0$ and $\text{Sin}^{-1}(\sin 2\pi) = \text{Sin}^{-1} 0 = 0$. However, $\text{Sin}^{-1}(\sin x) = x$ if x is already in the appropriate range, namely, x between $-\pi/2$ and $\pi/2$. Similarly, $\text{Cos}^{-1}(\cos x) = x$ only when x is between 0 and π.

$$\text{Sin}^{-1}(\sin x) = x \quad \text{only when} \quad x \in \left[-\frac{\pi}{2}, \frac{\pi}{2}\right]$$

$$\text{Cos}^{-1}(\cos x) = x \quad \text{only when} \quad x \in [0, \pi]$$

EXAMPLE 9 Find each value.

a. $\text{Sin}^{-1}\left(\sin \frac{5\pi}{4}\right)$ b. $\sin\left[\text{Sin}^{-1}\left(\frac{-\sqrt{2}}{2}\right)\right]$

c. $\text{Cos}^{-1}\left(\cos \frac{5\pi}{4}\right)$ d. $\cos\left[\text{Cos}^{-1}\left(\frac{-\sqrt{2}}{2}\right)\right]$

SOLUTION

a. $\text{Sin}^{-1}\left(\sin \frac{5\pi}{4}\right) = \text{Sin}^{-1}\left(\frac{-\sqrt{2}}{2}\right) = -\frac{\pi}{4} \neq \frac{5\pi}{4}$

b. $\sin\left[\text{Sin}^{-1}\left(\frac{-\sqrt{2}}{2}\right)\right] = \sin\left(-\frac{\pi}{4}\right) = \frac{-\sqrt{2}}{2}$

c. $\text{Cos}^{-1}\left(\cos \frac{5\pi}{4}\right) = \text{Cos}^{-1}\left(\frac{-\sqrt{2}}{2}\right) = \frac{3\pi}{4} \neq \frac{5\pi}{4}$

d. $\cos\left[\text{Cos}^{-1}\left(\frac{-\sqrt{2}}{2}\right)\right] = \cos \frac{3\pi}{4} = \frac{-\sqrt{2}}{2}$ ∎

Graphing

EXAMPLE 10 Sketch the graph of the function defined by $y = 3 \text{Sin}^{-1} \frac{x}{4}$.

SOLUTION Proceed as in Sections 7.6 and 7.7 with the graphs of modified trigonometric functions. First sketch $y = \text{Sin}^{-1} x$. Note that $\text{Sin}^{-1} x/4$ is defined for

$$-1 \leq \frac{x}{4} \leq 1 \quad \text{or} \quad -4 \leq x \leq 4$$

The graph of $\text{Sin}^{-1} x/4$ is obtained by "stretching" the graph of $\text{Sin}^{-1} x$ horizontally. Finally, the y-coordinate of each point on this graph is multiplied by 3 (see Figure 8.28).

Figure 8.28

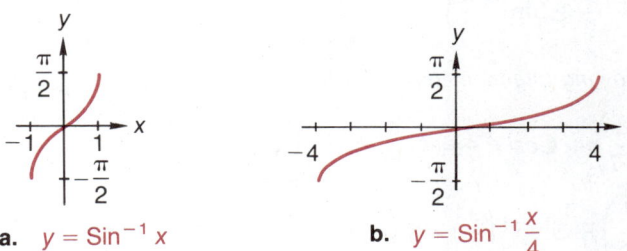

a. $y = \text{Sin}^{-1} x$ b. $y = \text{Sin}^{-1} \dfrac{x}{4}$

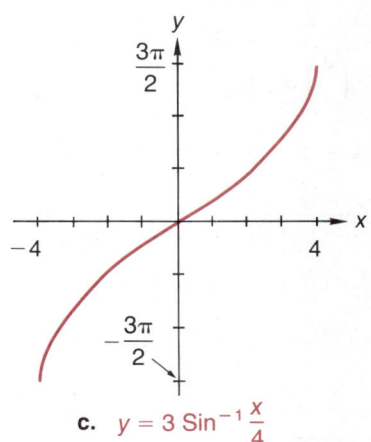

c. $y = 3 \text{Sin}^{-1} \dfrac{x}{4}$

Section 8.7 Exercises

Find the following without using tables or a calculator.

1. $\text{Cos}^{-1}\left(-\dfrac{\sqrt{2}}{2}\right)$
2. $\text{Sin}^{-1}\left(-\dfrac{1}{2}\right)$
3. $\text{Sin}^{-1}\left(\dfrac{\sqrt{2}}{2}\right)$
4. $\text{Cos}^{-1}\left(\dfrac{1}{2}\right)$
5. $\text{Cos}^{-1}\left(\dfrac{\sqrt{3}}{2}\right)$
6. $\text{Sin}^{-1}\left(-\dfrac{\sqrt{3}}{2}\right)$

Calculator or Table Exercises (7–16)

Use a calculator or Table 4 (at the end of the book) to express the following in radians correct to three decimal places.

7. $\text{Cos}^{-1}(0.9200)$
8. $\text{Sin}^{-1}(-0.1343)$
9. $\text{Sin}^{-1}(0.6830)$
10. $\text{Sin}^{-1}(0.4932)$
11. $\text{Cos}^{-1}(-0.4344)$
12. $\text{Sin}^{-1}(-0.8360)$
13. $\text{Sin}^{-1}(-0.2135)$
14. $\text{Cos}^{-1}(-0.3240)$
15. $\text{Cos}^{-1}(-0.9930)$
16. $\text{Cos}^{-1}(0.8200)$

Use the properties of right triangles to find exact values for each of the following. If you have access to a calculator, use it only to check your answers.

17. $\tan\left[\text{Sin}^{-1}\left(-\dfrac{3}{5}\right)\right]$
18. $\csc\left[\text{Cos}^{-1}\left(\dfrac{2}{3}\right)\right]$
19. $\sec\left[\text{Sin}^{-1}\left(\dfrac{2}{9}\right)\right]$
20. $\sin\left[\text{Cos}^{-1}\left(-\dfrac{4}{11}\right)\right]$
21. $\sin\left[2\,\text{Cos}^{-1}\left(\dfrac{1}{3}\right)\right]$
22. $\cot\left[\text{Sin}^{-1}\left(-\dfrac{1}{5}\right)\right]$
23. $\sin\left(\text{Cos}^{-1}\dfrac{-5}{13} - \text{Sin}^{-1}\dfrac{4}{5}\right)$
24. $\cos\left[\left(\dfrac{1}{2}\right)\text{Sin}^{-1}\left(\dfrac{3}{5}\right)\right]$

25. $\tan\left(2\,\text{Cos}^{-1}\dfrac{4}{5} + \text{Sin}^{-1}\dfrac{12}{13}\right)$

26. $\sin\left(\text{Cos}^{-1}\dfrac{2}{3} + \text{Sin}^{-1}\dfrac{2}{3}\right)$

Solve the following trigonometric equations.

27. $\text{Cos}^{-1}\left(\dfrac{8}{17}\right) + \text{Cos}^{-1}\dfrac{4}{5} = \text{Cos}^{-1} x$

28. $\text{Sin}^{-1}\left(\dfrac{7}{25}\right) + \text{Sin}^{-1}\left(\dfrac{3}{5}\right) = \text{Cos}^{-1} x$

29. $\text{Cos}^{-1}\left(\dfrac{4}{5}\right) - \text{Sin}^{-1}\left(-\dfrac{5}{13}\right) = \text{Sin}^{-1} x$

30. $\text{Sin}^{-1}\left(\dfrac{12}{13}\right) - \text{Cos}^{-1}\left(\dfrac{4}{5}\right) = \text{Sin}^{-1} x$

31. $2\,\text{Sin}^{-1} x - \text{Cos}^{-1} x = \pi$

32. $\text{Cos}^{-1} x - \text{Sin}^{-1} 2x = \dfrac{\pi}{2}$

Establish the following for all x in $[-1, 1]$.

33. $\sin(\text{Cos}^{-1} x) \geq 0$

34. $\cos(\text{Sin}^{-1} x) \geq 0$

35. $\text{Cot}[\text{Cos}^{-1} x] = \dfrac{x}{\sqrt{1 - x^2}}$

36. $\sec(\text{Cos}^{-1} x) = \dfrac{1}{x}$

37. $\csc(\text{Sin}^{-1} x) = \dfrac{1}{x}$

38. $\cos(\text{Sin}^{-1} x) = \sqrt{1 - x^2}$

39. $\tan(\text{Cos}^{-1} x) = \dfrac{\sqrt{1 - x^2}}{x}$

40. $\csc(\text{Cos}^{-1} x) = \dfrac{1}{\sqrt{1 - x^2}}$

41. $\cot(\text{Sin}^{-1} x) = \dfrac{\sqrt{1 - x^2}}{x}$

42. $\tan(\text{Sin}^{-1} x) = \dfrac{x}{\sqrt{1 - x^2}}$

Find the indicated values in each of the following. Do not use a calculator.

43. $\text{Cos}^{-1}\left[\cos\left(\dfrac{-45\pi}{4}\right)\right]$

44. $\text{Sin}^{-1}\left[\sin\left(\dfrac{37\pi}{3}\right)\right]$

45. $\text{Sin}^{-1}\left[\sin\left(\dfrac{-16\pi}{3}\right)\right]$

46. $\text{Cos}^{-1}\left[\cos\left(-\dfrac{\pi}{3}\right)\right]$

47. $\text{Sin}^{-1}\left[\sin\left(\dfrac{2\pi}{3}\right)\right]$

48. $\text{Cos}^{-1}\left[\cos\left(\dfrac{26\pi}{3}\right)\right]$

Sketch the graphs of the following functions.

49. $y = \text{Sin}^{-1} x + \dfrac{\pi}{2}$

50. $s = \dfrac{\pi}{2} - \text{Cos}^{-1} r$

51. $y = 3\,\text{Cos}^{-1}\left(\dfrac{x}{2}\right)$

52. $v = -2\,\text{Sin}^{-1} 3u$

53. $s = 4\,\text{Cos}^{-1} 2r$

54. $w = \left(-\dfrac{1}{2}\right)\text{Cos}^{-1}\left(\dfrac{y}{3}\right)$

55. $\theta = 3\,\text{Sin}^{-1} 2\phi$

56. $w = 2\,\text{Cos}^{-1}(x + 1)$

57. $y = -3\,\text{Sin}^{-1}(t - 2)$

58. $\phi = |\text{Sin}^{-1} v|$

59. $w = \text{Cos}^{-1} |x|$

60. $\beta = \text{Cos}^{-1}|r| - \dfrac{\pi}{2}$

61. $\alpha = \text{Cos}^{-1}(|s| - 1)$

62. $\theta = |\text{Sin}^{-1}(2u + 4)|$

63. $v = \left|\text{Cos}^{-1}(t + 1) - \dfrac{\pi}{2}\right|$

64. a. Sketch the graphs of $\text{Sin}^{-1} x$ and $(\text{Cos}^{-1} x - \pi/2)$ on the same coordinate system.

b. Show that
$$\text{Sin}^{-1} x + \text{Cos}^{-1} x = \dfrac{\pi}{2}$$

Section 8.8

More Inverse Trigonometric Functions

The remaining circular functions must also be restricted before an inverse can be found. Continuing in the same vein as with Sin^{-1} and Cos^{-1}, we define

$$\text{Tan}\,\theta = \tan\theta \qquad -\dfrac{\pi}{2} < \theta < \dfrac{\pi}{2}$$

$$\text{Cot}\,\theta = \cot\theta \qquad 0 < \theta < \pi$$

These restricted functions assume each value precisely once and hence have inverses. Their graphs and the graphs of their inverses are given in Figure 8.29.

Figure 8.29

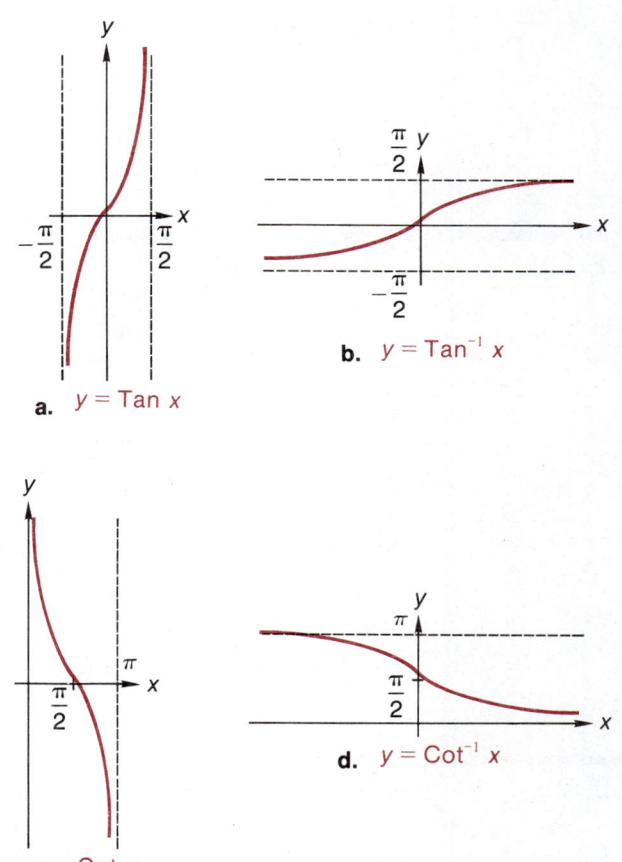

a. $y = \text{Tan } x$

b. $y = \text{Tan}^{-1} x$

c. $y = \text{Cot } x$

d. $y = \text{Cot}^{-1} x$

We have the usual properties of inverse functions:

1. $y = \text{Tan}^{-1} x$ means $\left(x = \tan y \quad \text{and} \quad -\dfrac{\pi}{2} < y < \dfrac{\pi}{2} \right)$

   Domain $= (-\infty, \infty)$

   Range $= \left(-\dfrac{\pi}{2}, \dfrac{\pi}{2} \right)$

2. $y = \text{Cot}^{-1} x$ means $(x = \cot y \quad \text{and} \quad 0 < y < \pi)$

   Domain $= (-\infty, \infty)$

   Range $= (0, \pi)$

EXAMPLE 1

Find each value.

a. $\text{Tan}^{-1}\sqrt{3}$

b. $\text{Cot}^{-1}(-1)$

SOLUTION

a. $\text{Tan}^{-1}\sqrt{3} = \dfrac{\pi}{3}$ since $\tan\dfrac{\pi}{3} = \sqrt{3}$

b. $\text{Cot}^{-1}(-1) = \dfrac{3\pi}{4}$ since $\cot\dfrac{3\pi}{4} = -1$

EXAMPLE 2

Evaluate $\sec[\text{Cot}^{-1}(\tfrac{3}{2})]$.

SOLUTION

Let $\theta = \text{Cot}^{-1}\tfrac{3}{2}$; then $\cot\theta = \tfrac{3}{2}$ and $0 < \theta < \pi$. The angle θ can be sketched in standard position as in Figure 8.30. Use the Pythagorean theorem to find the hypotenuse ($=\sqrt{13}$) of the triangle in Figure 8.30. From the figure, we observe that

$$\sec\left(\text{Cot}^{-1}\frac{3}{2}\right) = \sec\theta = \frac{\sqrt{13}}{3}$$

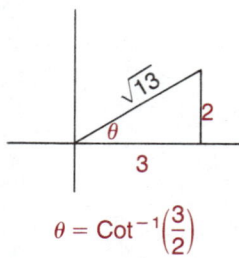

$\theta = \text{Cot}^{-1}\left(\dfrac{3}{2}\right)$

Figure 8.30

EXAMPLE 3

Evaluate $\cos[\text{Tan}^{-1}(\tfrac{1}{2}) + \text{Cot}^{-1}(-\tfrac{1}{2})]$.

SOLUTION

Let

$$\alpha = \text{Tan}^{-1}\left(\frac{1}{2}\right) \quad \text{and} \quad \beta = \text{Cot}^{-1}\left(-\frac{1}{2}\right)$$

Then

$$\tan\alpha = \frac{1}{2} \quad \text{and} \quad 0 < \alpha < \frac{\pi}{2}$$

$$\cot\beta = -\frac{1}{2} \quad \text{and} \quad \frac{\pi}{2} < \beta < \pi$$

These angles are sketched in Figure 8.31. Now

$$\cos\left[\text{Tan}^{-1}\left(\frac{1}{2}\right) + \text{Cot}^{-1}\left(-\frac{1}{2}\right)\right] = \cos(\alpha + \beta)$$

$$= \cos\alpha\cos\beta - \sin\alpha\sin\beta \quad \text{(By formula 1b, Section 8.3)}$$

$$= \frac{2}{\sqrt{5}} \cdot \frac{-1}{\sqrt{5}} - \frac{1}{\sqrt{5}} \cdot \frac{2}{\sqrt{5}} \quad \text{(See Figure 8.31)}$$

$$= \frac{-4}{5}$$

Figure 8.31

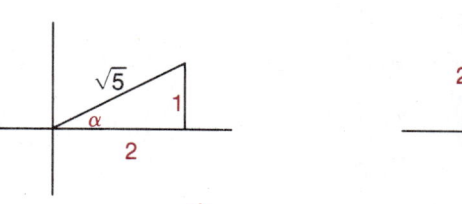
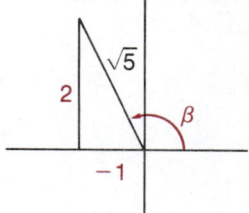

a. $\alpha = \text{Tan}^{-1}\left(\frac{1}{2}\right)$ b. $\beta = \text{Cot}^{-1}\left(-\frac{1}{2}\right)$ ∎

To complete our study of the inverse circular functions, we define

$$\text{Sec } \theta = \sec \theta \qquad 0 \le \theta \le \pi$$

$$\text{Csc } \theta = \csc \theta \qquad -\frac{\pi}{2} \le \theta \le \frac{\pi}{2}$$

These restricted functions also have inverses. The respective graphs appear in Figure 8.32. Note that both Sec^{-1} and Csc^{-1} are defined only for $|x| \ge 1$. A further investigation of Sec^{-1} and Csc^{-1} is left to the exercises.

Figure 8.32

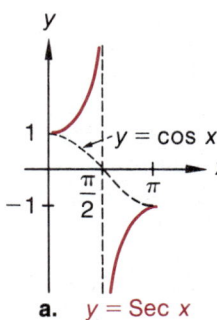
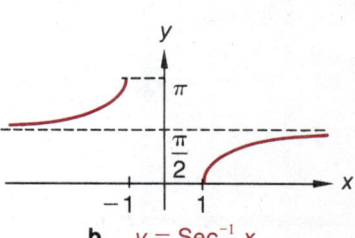

a. $y = \text{Sec } x$ b. $y = \text{Sec}^{-1} x$

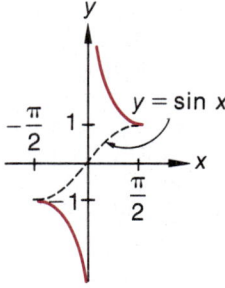
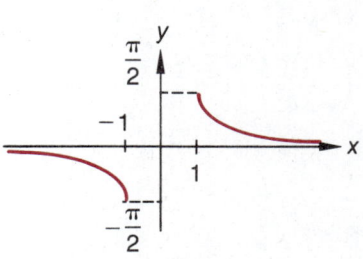

c. $y = \text{Csc } x$ d. $y = \text{Csc}^{-1} x$

CALCULATOR COMMENTS Tan^{-1} is also programmed into many calculators, but Cot^{-1}, Sec^{-1}, and Csc^{-1} are not programmed. Nevertheless, these three inverse functions can be found with a calculator by working with the inverses of their reciprocal functions, as illustrated in the next example.

EXAMPLE 4

Find $\text{Cot}^{-1}(-0.7536)$

SOLUTION

Let
$$\theta = \text{Cot}^{-1}(-0.7536)$$
Then
$$\cot \theta = -0.7536 \quad \text{and} \quad \frac{\pi}{2} < \theta < \pi$$
Thus
$$\tan \theta = \frac{1}{\cot \theta} = \frac{1}{-0.7536}$$
$$\approx -1.3270$$

Find the reference angle $\theta^* = \pi - \theta$ (see Figure 8.33). Then
$$\tan \theta^* = 1.3270$$
$$\theta^* = \text{Tan}^{-1}(1.3270)$$
$$\approx 0.925\ldots$$

Finally,
$$\theta = \pi - \theta^*$$
$$\approx 3.14159\ldots - 0.925\ldots$$
$$\approx 2.217$$

rounded to three decimal places.
Note in particular that $\theta \neq \text{Tan}^{-1}(-1.3270) \approx -0.925$.

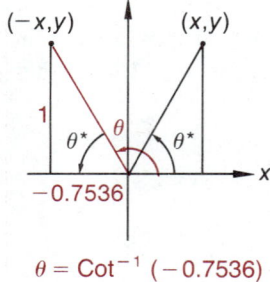

Figure 8.33

Section 8.8 Exercises

Find the following without using tables or a calculator.

1. $\text{Tan}^{-1}(-\sqrt{3})$
2. $\text{Cot}^{-1}(-\sqrt{3})$
3. $\text{Cot}^{-1}\left(\dfrac{1}{\sqrt{3}}\right)$
4. $\text{Tan}^{-1}\left(\dfrac{-1}{\sqrt{3}}\right)$
5. $\text{Sec}^{-1} 2$
6. $\text{Csc}^{-1}\left(\dfrac{-2\sqrt{3}}{3}\right)$
7. $\text{Csc}^{-1}(-2)$
8. $\text{Sec}^{-1}(-\sqrt{2})$

Use a calculator or Table 4 (at end of the book) to express the following in radians correct to three decimal places.

9. $\text{Tan}^{-1}(0.2500)$
10. $\text{Cot}^{-1}(0.8020)$
11. $\text{Sec}^{-1}(-16)$
12. $\text{Csc}^{-1}(2.5)$
13. $\text{Cot}^{-1}(-5)$
14. $\text{Tan}^{-1}(2)$
15. $\text{Csc}^{-1}(-3)$
16. $\text{Sec}^{-1}(-2.4)$

Use the properties of right triangles to evaluate the following. If you have access to a calculator, use it only to check your answers.

17. $\cos\left[\text{Tan}^{-1}\left(\dfrac{1}{3}\right)\right]$
18. $\sin[\text{Tan}^{-1}(6)]$
19. $\sin[\text{Sec}^{-1}(-4)]$
20. $\sec[\text{Cot}^{-1}(-2)]$
21. $\cos[\text{Csc}^{-1}(30)]$
22. $\tan\left[\text{Cot}^{-1}\left(-\dfrac{3}{8}\right)\right]$
23. $\tan[\text{Sec}^{-1}(-3)]$
24. $\cot[\text{Csc}^{-1}(-5)]$
25. $\sin[\text{Sec}^{-1}(2) - \text{Tan}^{-1}(2)]$
26. $\sin\left[\dfrac{1}{2}\text{Tan}^{-1}\left(\dfrac{24}{7}\right)\right]$
27. $\cos[2\,\text{Sec}^{-1}(3)]$
28. $\cos[\text{Cot}^{-1}(3) - \text{Tan}^{-1}(2)]$

29. $\tan\left[\operatorname{Csc}^{-1}\left(\frac{3}{2}\right) + \operatorname{Cot}^{-1}\left(\frac{2}{3}\right)\right]$

30. $\sin[\operatorname{Tan}^{-1}(2) - \operatorname{Sec}^{-1}(2)]$

Find the requested values in each of the following.

31. $\operatorname{Cot}^{-1}\left[\cot\left(-\frac{\pi}{4}\right)\right]$ 32. $\operatorname{Tan}^{-1}\left[\tan\left(\frac{16\pi}{3}\right)\right]$

Solve the following equations.

33. $2\operatorname{Tan}^{-1} x + \operatorname{Cos}^{-1} x = \frac{\pi}{2}$

34. $2\operatorname{Csc}^{-1} x - \operatorname{Sec}^{-1} x = \pi$

35. $\operatorname{Sec}^{-1}\left(\frac{5}{4}\right) - \operatorname{Csc}^{-1}\left(-\frac{13}{5}\right) = \operatorname{Sin}^{-1} x$

36. $\operatorname{Tan}^{-1}\left(\frac{4}{3}\right) - \operatorname{Cot}^{-1}(2) = \operatorname{Cos}^{-1} x$

Establish the following relationships for x in the appropriate range.

37. $\cot(\operatorname{Tan}^{-1} x) = \frac{1}{x}$ 38. $\tan(\operatorname{Cot}^{-1} x) = \frac{1}{x}$

39. $\sin(\operatorname{Csc}^{-1} x) = \frac{1}{x}$ 40. $\cos(\operatorname{Sec}^{-1} x) = \frac{1}{x}$

41. $\cos(\operatorname{Tan}^{-1} x) = \frac{1}{\sqrt{1 + x^2}}$

42. $\csc(\operatorname{Tan}^{-1} x) = \frac{\sqrt{1 + x^2}}{x}$

43. $\sec(\operatorname{Cot}^{-1} x) = \frac{\sqrt{1 + x^2}}{x}$

44. $\sin(\operatorname{Cot}^{-1} x) = \frac{1}{\sqrt{1 + x^2}}$

Sketch the graphs of the following functions.

45. $y = \operatorname{Tan}^{-1} 2x$ 46. $v = \operatorname{Cot}^{-1}\left(\frac{u}{3}\right)$

47. $s = \frac{\pi}{2} - \operatorname{Tan}^{-1} r$ 48. $\theta = 3\operatorname{Sec}^{-1}\left(\frac{u}{2}\right)$

49. $y = -2\operatorname{Tan}^{-1}(r - 1)$ 50. $w = |\operatorname{Csc}^{-1}(t + 1)|$

51. $u = 2\operatorname{Sec}^{-1}|v|$ 52. $y = \left|\operatorname{Sec}^{-1} x - \frac{\pi}{2}\right|$

Sketch two graphs on the same coordinate system and add their ordinates to verify the following.

53. $\operatorname{Tan}^{-1} x + \operatorname{Cot}^{-1} x = \frac{\pi}{2}$

54. $\operatorname{Sec}^{-1} x + \operatorname{Csc}^{-1} x = \frac{\pi}{2}$

Section 8.9 Chapter Review

Terms and Concepts

Identity An equation true for all values of the unknown(s)

$x^2 = x \cdot x$

Conditional Equation An equation true for only certain values of the unknown(s)

$2x = 6$

$y = \operatorname{Sin}^{-1} x$ means $\left(x = \sin y \text{ and } -\frac{\pi}{2} \leq y \leq \frac{\pi}{2}\right)$

$y = \operatorname{Csc}^{-1} x$ means $\left(x = \csc y \text{ and } -\frac{\pi}{2} \leq y \leq \frac{\pi}{2}\right)$

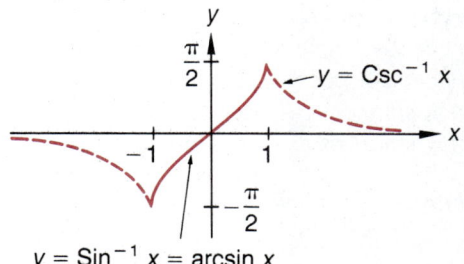

$y = \operatorname{Sin}^{-1} x = \arcsin x$

$y = \text{Cos}^{-1} x$ means $(x = \cos y \text{ and } 0 \leq y \leq \pi)$
$y = \text{Sec}^{-1} x$ means $(x = \sec y \text{ and } 0 \leq y \leq \pi)$

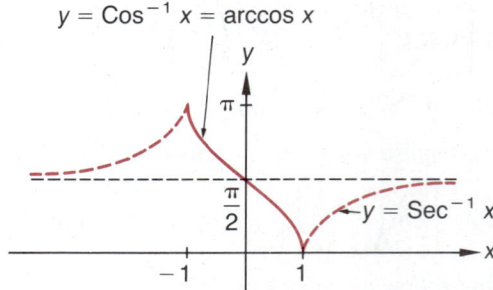

$y = \text{Tan}^{-1} x$ means $\left(x = \tan y \text{ and } -\dfrac{\pi}{2} < y < \dfrac{\pi}{2}\right)$
$y = \text{Cot}^{-1} x$ means $(x = \cot y \text{ and } 0 < y < \pi)$

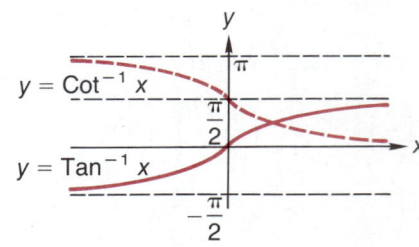

Rules and Formulas

Fundamental Identities

1. $\csc\theta = \dfrac{1}{\sin\theta}$
2. $\sec\theta = \dfrac{1}{\cos\theta}$
3. $\tan\theta = \dfrac{\sin\theta}{\cos\theta}$
4. $\cot\theta = \dfrac{\cos\theta}{\sin\theta}$
5. $\sin\theta = \dfrac{1}{\csc\theta}$
6. $\cos\theta = \dfrac{1}{\sec\theta}$
7. $\cot\theta = \dfrac{1}{\tan\theta}$
8. $\tan\theta = \dfrac{1}{\cot\theta}$

Even–Odd Identities

9. $\sin(-\theta) = -\sin\theta$
10. $\cos(-\theta) = \cos\theta$
11. $\tan(-\theta) = -\tan\theta$
12. $\csc(-\theta) = -\csc\theta$
13. $\sec(-\theta) = \sec\theta$
14. $\cot(-\theta) = -\cot\theta$

Pythagorean Identities

15. $\sin^2\theta + \cos^2\theta = 1$
16. $1 + \tan^2\theta = \sec^2\theta$
17. $1 + \cot^2\theta = \csc^2\theta$

Sum and Difference Formulas

1a. $\cos(u - v) = \cos u \cos v + \sin u \sin v$
 b. $\cos(u + v) = \cos u \cos v - \sin u \sin v$
2a. $\sin(u - v) = \sin u \cos v - \cos u \sin v$
 b. $\sin(u + v) = \sin u \cos v + \cos u \sin v$
3a. $\tan(u - v) = \dfrac{\tan u - \tan v}{1 + \tan u \tan v}$
 b. $\tan(u + v) = \dfrac{\tan u + \tan v}{1 - \tan u \tan v}$

Double-angle and Half-angle Formulas

1. $\sin 2\alpha = 2\sin\alpha\cos\alpha$
2. $\cos 2\alpha = \cos^2\alpha - \sin^2\alpha = 1 - 2\sin^2\alpha = 2\cos^2\alpha - 1$
3. $\tan 2\alpha = \dfrac{2\tan\alpha}{1 - \tan^2\alpha}$
4. $\sin\dfrac{\alpha}{2} = \pm\sqrt{\dfrac{1 - \cos\alpha}{2}}$
5. $\cos\dfrac{\alpha}{2} = \pm\sqrt{\dfrac{1 + \cos\alpha}{2}}$
6. $\tan\dfrac{\alpha}{2} = \dfrac{1 - \cos\alpha}{\sin\alpha} = \dfrac{\sin\alpha}{1 + \cos\alpha}$

Product-to-Sum Formulas

1. $\sin\alpha\cos\beta = \dfrac{1}{2}[\sin(\alpha + \beta) + \sin(\alpha - \beta)]$
2. $\cos\alpha\cos\beta = \dfrac{1}{2}[\cos(\alpha + \beta) + \cos(\alpha - \beta)]$
3. $\sin\alpha\sin\beta = \dfrac{1}{2}[\cos(\alpha - \beta) - \cos(\alpha + \beta)]$

Sum-to-Product Formulas

1. $\sin \alpha + \sin \beta = 2 \sin \dfrac{\alpha + \beta}{2} \cos \dfrac{\alpha - \beta}{2}$

2. $\sin \alpha - \sin \beta = 2 \cos \dfrac{\alpha + \beta}{2} \sin \dfrac{\alpha - \beta}{2}$

3. $\cos \alpha + \cos \beta = 2 \cos \dfrac{\alpha + \beta}{2} \cos \dfrac{\alpha - \beta}{2}$

4. $\cos \alpha - \cos \beta = -2 \sin \dfrac{\alpha + \beta}{2} \sin \dfrac{\alpha - \beta}{2}$

Techniques

Verifying Identities

1. Express each function in terms of sines and cosines and then show that the left and right sides of the equal sign become the same.

2. Simplify both sides of the equal sign and show that the simplified expressions are equal.

3. Transform one side into the other side.

Section 8.10 Supplementary Exercises

Show that the following expressions are not identities but exhibit at least one solution for each.

1. $\tan^2 \theta - \cot^2 \theta = 0$ 2. $\cos^2 \theta + \csc^2 \theta = 1$

Verify the following identities.

3. $\sin t \cot t = \cos t$

4. $\cos u \csc u = \cot u$

5. $\tan \alpha = \dfrac{\sec \alpha}{\csc \alpha}$

6. $\sec^2 \beta - \csc^2 \beta = \tan^2 \beta - \cot^2 \beta$

7. $\csc v - \sin v = \csc v \cos^2 v$

8. $\tan \theta \sin \theta + \cos \theta = \sec \theta$

9. $(\cot \gamma + 1)(\sin \gamma - \cos \gamma) = \sin \gamma - \cos \gamma \cot \gamma$

10. $\csc \phi - \cot \phi \cos \phi = \sin \phi$

11. $(1 - 2\cos^2 t)\sec t \csc t = \tan t - \cot t$

12. $\sin u \sec^2 u - \sin u = \cos u \tan^3 u$

13. $(1 + \sec v)(1 - \cos v) = \sec v \sin^2 v$

14. $(1 + \sec \alpha) = \csc \alpha (\sin \alpha + \tan \alpha)$

15. $(\sec \beta + \tan \beta)(1 - \sin \beta) = \cos \beta$

16. $\dfrac{\cos \gamma - \sin \gamma}{\cos \gamma + \sin \gamma} = \dfrac{1 - \tan \gamma}{1 + \tan \gamma}$

17. $\cos^2 \theta - \sin^2 \theta = \dfrac{1 - \tan^2 \theta}{1 + \tan^2 \theta}$

18. $\dfrac{1 - \sin^2 \phi}{\sin^2 \phi} = \cot^2 \phi$

19. $\cot t + \csc t = \dfrac{\sin t}{1 - \cos t}$

20. $\dfrac{1 - \sin u}{\cos u} = \dfrac{\cos u}{1 + \sin u}$

21. $\dfrac{\sin v}{\sec v} = \dfrac{1}{\tan v + \cot v}$

22. $(\tan \alpha - \sec \alpha)^2 = \dfrac{1 - \sin \alpha}{1 + \sin \alpha}$

23. $\dfrac{1 + \sin^2 \beta \sec^2 \beta}{1 + \cos^2 \beta \csc^2 \beta} = \tan^2 \beta$

24. $\sin(\alpha + \beta)\sin(\alpha - \beta) = \sin^2 \alpha - \sin^2 \beta$

25. $\cos \beta \sin(\alpha - \beta) + \sin \beta \cos(\alpha - \beta) = \sin \alpha$

26. $\cos\left(t - \dfrac{\pi}{6}\right) - \cos\left(t + \dfrac{\pi}{6}\right) = \sin t$

27. $2 \cos\left(t + \dfrac{\pi}{4}\right) \cos\left(t - \dfrac{\pi}{4}\right) = \cos^2 t - \sin^2 t$

28. $\csc 2t + \cot 2t = \cot t$

29. $\sin^2 2u - \cos^2 2u + \cos 4u = 0$

30. $(\sin v + \cos v)^2 = 1 + \sin 2v$

31. $\cos^4 \theta - \sin^4 \theta = \cos 2\theta$

32. $\tan \phi \sin 2\phi = 2 \sin^2 \phi$

33. $\dfrac{2 \tan(t/2)}{1 + \tan^2 (t/2)} = \sin t$

34. $\dfrac{2 \tan t}{\tan t + \sin t} = \sec^2\left(\dfrac{t}{2}\right)$

35. $\tan \dfrac{t}{2} = \csc t - \cot t$

36. $\tan\left(\dfrac{t}{2} + \dfrac{\pi}{4}\right) - \tan\left(\dfrac{t}{2} - \dfrac{\pi}{4}\right) = 2 \sec t$

37. $\tan\left(\dfrac{t}{2} + \dfrac{\pi}{4}\right) + \tan\left(\dfrac{t}{2} - \dfrac{\pi}{4}\right) = 2 \tan t$

38. $2 \cos\left(t + \dfrac{\pi}{4}\right) \sin\left(t - \dfrac{\pi}{4}\right) = \sin 2t - 1$

39. $2 \sin\left(t + \dfrac{\pi}{4}\right) \cos\left(t - \dfrac{\pi}{4}\right) = 1 + \sin 2t$

Find constants A and B that make the given expression an identity.

40. $(\sin \theta + \cos \theta)^2 = A + B \sin \theta \cos \theta$
41. $(\tan \phi + \cot \phi)^2 = A \sec^2 \phi + B \csc^2 \phi$
42. $\dfrac{1}{1 - \cos \alpha} + \dfrac{1}{1 + \cos \alpha} = A + B \cot^2 \alpha$
43. $\sin^2 \beta \cot^2 \beta + \cos^2 \beta \tan^2 \beta = A$

Use the sum or difference formulas to evaluate sine, cosine, and tangent of each of the following.

44. $\dfrac{\pi}{12}$, use $\dfrac{\pi}{12} = \dfrac{\pi}{3} - \dfrac{\pi}{4}$.

45. $\dfrac{5\pi}{12}$, use $\dfrac{5\pi}{12} = \dfrac{2\pi}{3} - \dfrac{\pi}{4}$.

46. $\dfrac{7\pi}{12}$, use $\dfrac{7\pi}{12} = \dfrac{\pi}{3} + \dfrac{\pi}{4}$.

47. $\dfrac{13\pi}{12}$, use $\dfrac{13\pi}{12} = \dfrac{4\pi}{3} - \dfrac{\pi}{4}$.

48. $\dfrac{19\pi}{12}$, use $\dfrac{19\pi}{12} = \dfrac{11\pi}{6} - \dfrac{\pi}{4}$.

49. $\dfrac{23\pi}{12}$, use $\dfrac{23\pi}{12} = \dfrac{5\pi}{3} + \dfrac{\pi}{4}$.

50. $\dfrac{11\pi}{12}$, use $\dfrac{11\pi}{12} = \dfrac{7\pi}{6} - \dfrac{\pi}{4}$.

Use the half-angle formulas to evaluate sine, cosine, and tangent of the given angles.

51. $\dfrac{7\pi}{8}$
52. $\dfrac{13\pi}{8}$
53. $\dfrac{11\pi}{8}$
54. $\dfrac{17\pi}{12}$
55. $\dfrac{13\pi}{12}$
56. $\dfrac{23\pi}{12}$

Use the given information to find sine, cosine, and tangent of $\alpha/2$.

57. $\cos \alpha = \dfrac{1}{5}$, $\dfrac{5\pi}{2} < \alpha < 4\pi$
58. $\sec \alpha = 10$, $2\pi < \alpha < 3\pi$
59. $\sin \alpha = -\dfrac{\sqrt{3}}{3}$, $4\pi < \alpha < \dfrac{11\pi}{2}$
60. $\cot \alpha = 2$, $\dfrac{3\pi}{2} < \alpha < \dfrac{5\pi}{2}$
61. $\tan \alpha = -\dfrac{3}{4}$, $\pi < \alpha < 2\pi$

Use the given information to find sine, cosine, and tangent of 2θ. Indicate the quadrant in which the terminal ray of 2θ lies when 2θ is placed in standard position.

62. $\sin \theta = -\dfrac{3}{5}$, $\cos \theta > 0$
63. $\tan \theta = -\dfrac{3}{4}$, $\sec \theta > 0$
64. $\cos \theta = \dfrac{1}{2}$, $\sin \theta > 0$

Use the given values of the indicated circular functions at α and β to evaluate sine, cosine, and tangent at $(\alpha + \beta)$ and $(\alpha - \beta)$. Determine the quadrant in which the terminal rays of $(\alpha + \beta)$ and $(\alpha - \beta)$ lie when they are placed in standard position.

65. $\sin \alpha = -\dfrac{1}{3}$, $\tan \alpha < 0$; $\tan \beta = -5$, $\sin \beta > 0$
66. $\cos \alpha = \dfrac{1}{5}$, $\sin \alpha > 0$; $\cos \beta = \dfrac{3}{5}$, $\tan \beta > 0$
67. $\tan \alpha = \dfrac{8}{15}$, $\sec \alpha > 0$; $\sin \beta = \dfrac{3}{5}$, $\sec \beta > 0$
68. $\sec \alpha = \dfrac{5}{4}$, $\sin \alpha > 0$; $\sec \beta = \dfrac{25}{24}$, $\cot \beta < 0$

Write each of the following in terms of a single cosine function and sketch its graph.

69. $\cos \theta - \sin \theta$
70. $\sqrt{3} \cos \theta + \sin \theta$

Write each of the following in terms of a single sine function and sketch its graph.

71. $\sqrt{3} \sin \theta + \cos \theta$
72. $\sin \theta + \cos \theta$

Use the expressions derived in Exercises 69–72 to find each of the following.

73. $\cos \dfrac{5\pi}{12} - \sin \dfrac{5\pi}{12}$

74. $\sqrt{3} \cos \dfrac{11\pi}{12} + \sin \dfrac{11\pi}{12}$

75. $\sqrt{3} \sin \dfrac{7\pi}{12} + \cos \dfrac{7\pi}{12}$

76. $\sin \dfrac{13\pi}{12} + \cos \dfrac{13\pi}{12}$

Express each of the following as the sum or difference of sines or cosines.

77. $\sin 3\alpha \sin 5\alpha$

78. $\sin \beta \cos 2\beta$

79. $\cos 3\gamma \cos 5\gamma$

80. $\sin 10\theta \cos 4\theta$

81. $\cos 3\phi \sin 5\phi$

Evaluate the following by using the product-to-sum formulas.

82. $\cos \dfrac{7\pi}{8} \sin \dfrac{13\pi}{8}$

83. $\sin \dfrac{5\pi}{12} \sin \dfrac{13\pi}{12}$

84. $\cos \dfrac{5\pi}{4} \sin \dfrac{7\pi}{12}$

85. $\cos \dfrac{3\pi}{4} \cos \dfrac{7\pi}{12}$

86. $\sin^2 \dfrac{\pi}{12}$

Express each of the following as the product of sines and/or cosines.

87. $\sin 2t + \sin 4t$

88. $\sin u - \sin 3u$

89. $\cos 4v - \cos 2v$

90. $\sin 3\alpha - \sin 2\alpha$

91. $\cos 2\gamma + \cos 3\gamma$

92. $\cos \beta - \cos 4\beta$

Evaluate the following by using the sum-to-product formulas.

93. $\sin \dfrac{\pi}{12} + \sin \dfrac{5\pi}{12}$

94. $\cos \dfrac{11\pi}{12} + \cos \dfrac{7\pi}{12}$

95. $\cos \dfrac{13\pi}{12} - \cos \dfrac{7\pi}{12}$

96. $\sin \dfrac{7\pi}{12} + \sin \dfrac{11\pi}{12}$

97. $\sin \dfrac{19\pi}{12} - \sin \dfrac{13\pi}{12}$

98. $\cos \dfrac{19\pi}{12} + \cos \dfrac{25\pi}{12}$

Use the sum-to-product formulas to verify the following identities.

99. $\dfrac{\cos 2x - \cos 4x}{\sin 4x - \sin 2x} = \tan 3x$

100. $\dfrac{\cos 2x - \cos 4x}{\sin 4x + \sin 2x} = \tan x$

101. $\dfrac{\sin 4x - \sin 10x}{\cos 10x - \cos 4x} = \cot 7x$

102. $\dfrac{\sin 8x + \sin 4x}{\cos 8x + \cos 4x} = \tan 6x$

Use the sum-to-product formulas to sketch the graphs of the following.

103. $f(t) = \sin 3t - \sin t$

104. $f(t) = \cos 2t + \cos 4t$

105. $f(t) = \cos 5t - \cos 3t$

106. $f(t) = \sin 5t + \sin 3t$

Find all $x \in [0, 2\pi]$ that satisfy the following equations.

107. $\sin x - \cos x = 0$

108. $\sec^2 x = 4$

109. $3 \tan 2x = \sqrt{3}$

110. $3 \cos^2 x = \sin^2 x$

111. $(\sqrt{3} \csc x - 2)(\csc x - 2) = 0$

112. $(\sqrt{2} \sin x - 1)(2 \cos x + 1) = 0$

113. $\csc x \sec x - 2 \sec x = 0$

114. $\cos 2x(\sec 2x + 2) = 0$

115. $4 \sin^2 x - 3 \sin x = 0$

116. $5 \cos^2 x - 2 \cos x = 0$

117. $2 \cos^2 x - \cos x = 1$

118. $4 \cos^2 x - 4 \cos x + 1 = 0$

119. $\csc^2 x - \csc x - 2 = 0$

120. $\tan^2 x = 5 \tan x - 4$

121. $\cot^2 x = 3 + 2 \cot x$

122. $2 \cos^2 x + 7 \sin x = 5$

123. $2 \sin^2 x - 3 \cos x = 3$

124. $\tan^2 x + \sec x = 1$

125. $\cot^2 x + 3 \csc x = 3$

126. $2 \sec^2 x - \tan x = 2$

127. $\csc^2 x = 4 \cot x - 2$

128. $2 \sin^2 2x = \sin 2x + 1$

129. $2 \cot^2 3x + 1 + \csc 3x = 0$

130. $2 \cos 2x + 3 \tan 2x = 0$

131. $\cot x = 4 \cos^2 x \cot x$

132. $\tan x \sec x - 4 = 2 \tan x - 2 \sec x$

133. $\sin 2x = \cos x$

134. $\cos 2x = \cos^2 x$

135. $\cos^2 x = \cos 2x + 1$

136. $4 \sin^2 x - 5 \cos x \tan x = 6$

137. $2 \sin 2x + 1 = 2 \cos x + 2 \sin x$

Find the following without using tables or a calculator.

138. $\text{Tan}^{-1}\left(\dfrac{1}{\sqrt{3}}\right)$

139. $\text{Cos}^{-1}\left(-\dfrac{\sqrt{2}}{2}\right)$

140. $\text{Sin}^{-1}\left(-\dfrac{\sqrt{3}}{2}\right)$

141. $\text{Cos}^{-1}\left(\dfrac{1}{2}\right)$

142. $\text{Tan}^{-1}(-\sqrt{3})$

143. $\text{Cot}^{-1}(\sqrt{3})$

144. $\text{Sec}^{-1}\left(\dfrac{2\sqrt{3}}{3}\right)$

145. $\text{Csc}^{-1}(\sqrt{2})$

Use a calculator or Table 4 (at the end of the book) to express the following in radians correct to three decimal places.

146. $\text{Tan}^{-1}(-0.3692)$

147. $\text{Cot}^{-1}(0.6666)$

148. $\text{Sin}^{-1}(-0.6666)$

149. $\text{Sin}^{-1}(-0.3802)$

150. $\text{Cos}^{-1}(0.8537)$

151. $\text{Sec}^{-1}(-2.1)$

152. $\text{Csc}^{-1}(1.5)$

153. $\text{Sin}^{-1}(0.9554)$

154. $\text{Cos}^{-1}(-0.4626)$

Use the properties of right triangles to evaluate the following. If you have access to a calculator, use it only to check your answers.

155. $\sin[\text{Cot}^{-1}(10)]$

156. $\csc\left[\text{Cos}^{-1}\left(-\dfrac{2}{5}\right)\right]$

157. $\tan\left[\text{Sin}^{-1}\left(\dfrac{2}{7}\right)\right]$

158. $\cot\left[\text{Sin}^{-1}\left(-\dfrac{3}{4}\right)\right]$

159. $\cos[\text{Tan}^{-1}(4)]$

160. $\cot[\text{Sec}^{-1}(-5)]$

161. $\sin[2 \text{ Cot}^{-1}(2)]$

162. $\cos\left[2 \text{ Cos}^{-1}\left(\dfrac{4}{5}\right)\right]$

163. $\cos\left[\text{Cos}^{-1}\left(\dfrac{1}{4}\right) + \text{Cos}^{-1}\left(\dfrac{3}{4}\right)\right]$

164. $\sin\left[\text{Cos}^{-1}\left(\dfrac{1}{2}\right) + \text{Sin}^{-1}\left(\dfrac{3}{5}\right)\right]$

165. $\cos\left[\text{Tan}^{-1}\left(\dfrac{1}{3}\right) + \text{Sec}^{-1}(3)\right]$

Evaluate the following.

166. $\text{Sin}^{-1}\left(\sin \dfrac{3\pi}{4}\right)$

167. $\text{Cos}^{-1}\left[\cos\left(-\dfrac{2\pi}{3}\right)\right]$

168. $\text{Tan}^{-1}\left[\tan\left(\dfrac{3\pi}{4}\right)\right]$

Solve the following equations.

169. $\text{Csc}^{-1}\left(\dfrac{25}{7}\right) - \text{Sec}^{-1}\left(\dfrac{5}{4}\right) = \text{Cos}^{-1} x$

170. $\text{Sin}^{-1}\left(\dfrac{5}{13}\right) - \text{Cos}^{-1}\left(\dfrac{3}{5}\right) = \text{Cos}^{-1} 2x$

171. $\text{Cos}^{-1} 2x - \text{Sin}^{-1} x = \dfrac{\pi}{2}$

172. $2 \text{ Cos}^{-1} x + \text{Sin}^{-1} x = \dfrac{\pi}{2}$

Sketch the graphs of the following equations.

173. $y = 3 \text{ Cos}^{-1}\left(\dfrac{x}{3}\right)$

174. $y = \left(\dfrac{1}{2}\right) \text{Sin}^{-1} 2x$

175. $y = \text{Tan}^{-1}\left(\dfrac{x}{2}\right)$

176. $y = \pi - 2 \text{ Sin}^{-1} x$

177. $y = \text{Sin}^{-1}|x|$

178. $y = \text{Sin}^{-1}(|x| - 1) + \dfrac{\pi}{2}$

Establish the following relationships for x in the appropriate range.

179. $\sec(\text{Sin}^{-1} x) = \dfrac{1}{\sqrt{1 - x^2}}$

180. $\sin(\text{Tan}^{-1} x) = \dfrac{x}{\sqrt{1 + x^2}}$

181. $\sec(\text{Tan}^{-1} x) = \sqrt{1 + x^2}$

182. $\cot(\text{Cos}^{-1} x) = \dfrac{x}{\sqrt{1 - x^2}}$

Further Applications of Trigonometry

Trigonometric functions can be used to "solve" a right triangle, provided enough information is given (Section 7.3). In the first two sections of this chapter we develop triangle relationships even further in order to solve **oblique triangles**—triangles that do not contain a right angle.

We then include a section on **vectors,** which have direction as well as magnitude. Since directions—and hence angles—are involved, trigonometry will be useful in these discussions. Vectors are useful in many scientific and engineering situations. For instance, the navigator of a plane or ship must account for the movement of the air or water when setting a direction; vectors facilitate these computations. Vectors also help to determine the strength needed in building braces and in certain machinery parts.

Complex numbers, having two distinct components (a real part and an imaginary part), are plotted on a coordinate plane. By characterizing a complex number in terms of its direction and distance from the origin, we are able to formulate a technique for finding the nth roots of a real or complex number.

Finally, the technique of expressing a complex number in terms of its direction and distance from the origin is adapted to represent ordered pairs (a, b) of real numbers in the Cartesian coordinate plane.

Section 9.1

The Law of Sines

In Chapter 7 we discussed the solution of right triangles. These techniques will now be extended to those triangles that do not contain a right angle, the so-called **oblique triangles.** Of course, any oblique triangle can be analyzed by considering the two right triangles formed by dropping a perpendicular from one vertex to the other side or its extension as illustrated in Figure 9.1. This technique was used in certain examples and exercises in Chapter 7. In this and the next sections, however, we develop triangle-solving methods that do not require using these associated right triangles. The computations involve only known parts of the given triangle. Of course, the new procedures will be developed by appealing to these associated right triangles. (A calculator is recommended for calculations in this and the next sections).

Figure 9.1

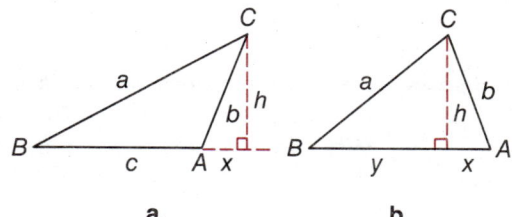

a.   b.

Consider any oblique triangle with parts labeled as in Figure 9.1. Since at most one angle of a triangle can be obtuse, there are essentially only two cases. These can be illustrated as

a. $A > 90°$

b. All angles are acute.

In each case, a perpendicular is dropped to side BA or its extension; the various lengths are then labeled as in the figure. In (a), we must use the fact that $\sin A = \sin(180° - A)$. Then we observe in both (a) and (b) that

$$\sin A = \frac{h}{b} \quad \text{and} \quad \sin B = \frac{h}{a}$$

Thus,

$$b \sin A = h = a \sin B$$

Finally,

$$\frac{\sin A}{a} = \frac{\sin B}{b}$$

This relationship holds even if A is a right angle. For in this case $\sin A = \sin 90° = 1$ and $\sin B = b/a$ (see Figure 9.2). Thus $(\sin B)/b = 1/a = (\sin A)/a$ as before.

Had we initially chosen to work with C instead of B, we would similarly have obtained

$$\frac{\sin A}{a} = \frac{\sin C}{c}$$

regardless of whether A is an acute, right, or obtuse angle.

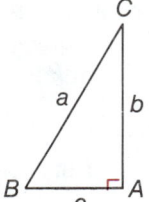

Figure 9.2

LAW OF SINES

In *any* triangle with sides a, b, c opposite angles A, B, C, respectively

$$\frac{\sin A}{a} = \frac{\sin B}{b} = \frac{\sin C}{c}$$

In order to solve a triangle using the law of sines, we must know one part in each of the fractions appearing in the law of sines and both numerator and denominator in at least one. Sometimes, however, even with less information we can determine enough additional information to continue the solution of the triangle. For instance, if we know two angles, we can find the third angle since the sum of the three angles must be 180°. Then, specifying the length of one side completely determines the triangle. This situation is illustrated in the following example.

EXAMPLE 1

Two observers at locations 6.3 kilometers apart wish to pinpoint the location of a ship in the fog. The observers A and B and the ship C form a triangle. The angle at A is 14°10′; the angle at B is 67°40′. How far is the ship from each of the observers?

SOLUTION   Make a rough sketch of the situation, as in Figure 9.3, labeling a, b, c as the sides opposite angles A, B, C, respectively. Then

$$c = 6.3 \text{ kilometers}$$

and

$$C = 180° - (14°10') - (67°40') = 98°10' \qquad (A + B + C = 180°)$$

Figure 9.3

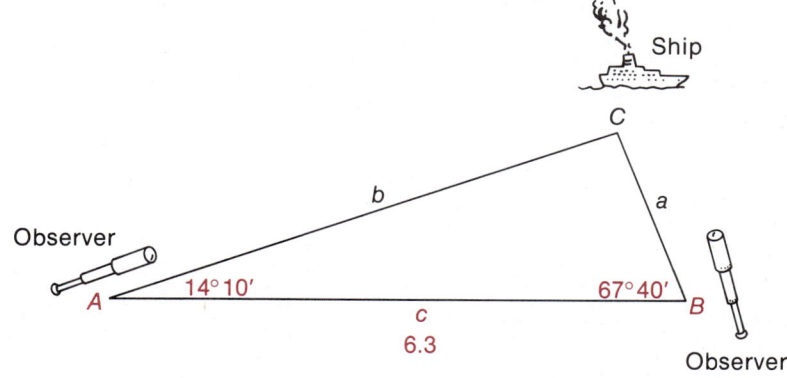

By the law of sines

$$\frac{\sin A}{a} = \frac{\sin C}{c}$$

$$a = \frac{c \sin A}{\sin C}$$

$$\approx \frac{(6.3)(0.2447)}{0.9899}$$

$$\approx 1.6$$

Similarly,

$$b = \frac{c \sin B}{\sin C} \qquad \text{(Law of sines)}$$

$$\approx \frac{(6.3)(0.9250)}{0.9899}$$

$$\approx 5.9$$

The triangle is solved. The ship is approximately 5.9 kilometers from A and 1.6 kilometers from B as illustrated in Figure 9.3. ∎

Congruent Triangles

Two triangles are said to be **congruent** provided their six corresponding parts (three sides and three angles), respectively, have the same measure. With **similar** triangles, only the three corresponding angles must have the same measure. Similar triangles need not be congruent; doubling all three sides of a triangle changes none of its angles.

If the corresponding parts indicated in any of the following situations have the same measure, then the triangles are congruent.

1. Two angles and the included side (ASA).
2. Two angles and a nonincluded side (AAS).
3. Three sides (SSS).
4. Two sides and the included angle (SAS).

Knowledge of its three angles is never sufficient to solve a triangle since similar triangles need not be congruent. However, if one side is also given, then precisely one of these similar triangles is singled out. The law of sines can be used to solve any triangle in which two (hence all three) angles and any side are given. (See 1 and 2 in the preceding box.)

The Ambiguous Case

If only one angle of a triangle is known, the law of sines can sometimes be used if enough sides are also known. Ambiguities might arise, however, from the fact that an angle θ must then be determined from $\sin \theta$. Since $\sin \theta = \sin(180° - \theta)$, there may be two possibilities as illustrated in the next example.

EXAMPLE 2 Solve the triangle with $C = 53°20'$, $a = 11.2$, $c = 9.3$.

SOLUTION

$$\sin A = \frac{a \sin C}{c} \qquad \text{(Law of sines)}$$

$$\approx \frac{(11.2)(0.8021)}{9.3}$$

$$\approx 0.9660$$

Using a calculator or Table 5 (at the end of the book) in reverse, we find $\sin \theta \approx 0.9660$ when $\theta \approx 75°1'$; but $\sin(180° - \theta) = \sin \theta$. We have two possibilities:

$$A \approx 75°1' \quad \text{or} \quad A \approx 104°59'$$

a. If $A \approx 75°1'$, then

$$B \approx 180° - (75°1') - (53°20') \qquad (A + B + C = 180°)$$

$$\approx 51°39'$$

and

$$b = \frac{c \sin B}{\sin C} \qquad \text{(Law of sines)}$$

$$\approx \frac{(9.3)(0.7842)}{0.8021}$$

$$\approx 9.1$$

b. If $A \approx 104°59'$, then

$$B \approx 180° - 104°59' - 53°20' \qquad (A + B + C = 180°)$$

$$\approx 21°41'$$

and

$$b = \frac{c \sin B}{\sin C} \qquad \text{(Law of sines)}$$

$$\approx \frac{(9.3)(0.3695)}{0.8021}$$

$$\approx 4.3$$

We have two possible solutions as shown in Figure 9.4.

Figure 9.4

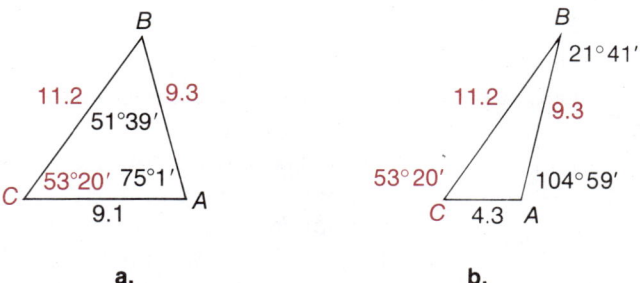

a.   b.

Example 2 illustrates the ambiguity that may arise in trying to solve a triangle when two of its sides and a nonincluded angle (SSA) are known. *This ambiguity is not a defect in the law of sines but rather is inherent in the concept of "triangle."* That is, given A and b, various situations could occur depending on the length a, as illustrated in Figures 9.5 and 9.6.

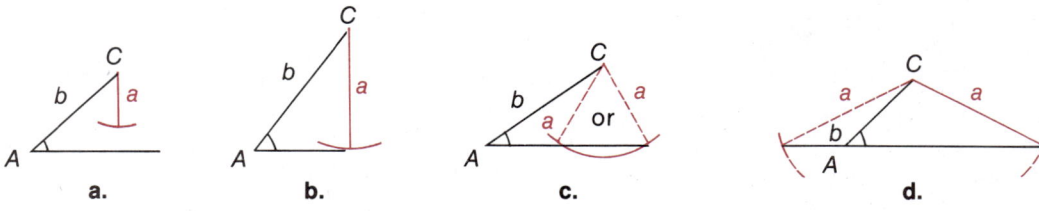

Figure 9.5
SIDE B AND ACUTE ANGLE A

Figure 9.6
SIDE B AND OBTUSE ANGLE A

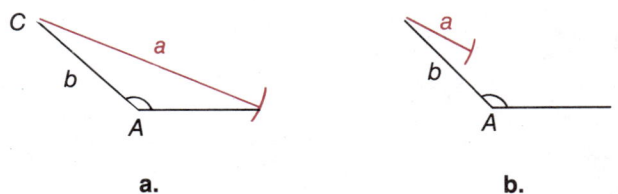

The ambiguous case occurs when A is acute and a is longer than the perpendicular dropped from C but shorter than b, as illustrated in Figure 9.5(c). Figures 9.5a and 9.6b illustrate the case when the data given are fictitious since the supposed side a is not long enough to close up the "triangle." If A is obtuse, there can be no triangle unless $a > b$ [Figure 9.6]. In any case, if $a \geq b$, there can be no ambiguity (Figures 9.5b, d, and 9.6a).

EXAMPLE 3 Solve the triangle with $A = 42°50'$, $b = 25$, $a = 10$.

SOLUTION

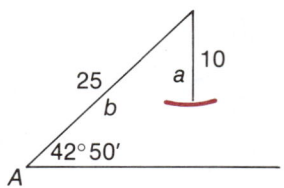

$$\sin B = \frac{b \sin A}{a} \qquad \text{(Law of sines)}$$

$$\approx \frac{25(0.6799)}{10}$$

$$\approx 1.70 > 1$$

This is an impossibility since $|\sin \theta| \leq 1$. There is no such triangle. See Figure 9.7. ∎

Figure 9.7

EXAMPLE 4 A small town is separated from the local power plant by moutainous terrain and several lakes. Until now, electrical power has been routed through a nearby city. The recent development of a stronger wire permits a direct line to be constructed. Sighting from the town, the angle between the city and the power plant is $77°10'$. The distance between the city and the town is 123 kilometers. The distance from the power plant to the city is 156 kilometers. What is the distance "as the crow flies" between the town and the power plant?

SOLUTION Let the triangle formed by the town, the city, and the power plant be labeled with vertices T, C, and P, respectively; see Figure 9.8. Let the opposite sides be labeled t, c, and p in accordance with the convention adopted earlier. Then $T = 77°10'$, $t = 156$ kilometers, and $p = 123$ are the data given.

Figure 9.8

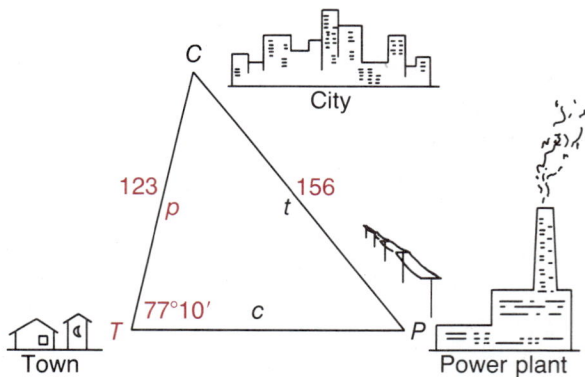

We are requested to find c, the distance between the town (T) and the power plant (P). Before we can do this, we must first find C; this can be done only after finding P by the law of sines.

$$\sin P = \frac{p \sin T}{t} \approx \frac{123(0.9750)}{156} \qquad \text{(Law of sines)}$$

$$\approx 0.7688$$

$$P \approx 50°15' \quad \text{or} \quad 129°45'$$

But if $P = 129°45'$, then $P + T > 180°$, an impossibility. That is, we must have

$P \approx 50°15'$; it follows that

$$C = 180° - T - P \qquad (T + P + C = 180°)$$
$$\approx 180° - (77°10') - (50°15')$$
$$= 52°35'$$

Finally,

$$c = \frac{t \sin C}{\sin T} \qquad \text{(Law of sines)}$$
$$\approx \frac{156(0.7942)}{0.9750}$$
$$\approx 127$$

The direct distance between the power plant and the town is approximately 127 kilometers. ∎

EXAMPLE 5

a. Solve the triangle with $A = 127°$, $b = 16$, $a = 10$.
b. Do the same with $a = 26$.

SOLUTION

a. Since angle A is obtuse, side a must be the longest side. But $a < b$; there is no triangle satisfying these conditions with $a = 10$ (see Figure 9.9). If we fail to realize this and attempt to use the law of sines, however, we have

$$\sin B = \frac{b \sin A}{a}$$
$$\approx \frac{16(0.7986)}{10}$$
$$\approx 1.28 > 1$$

also an impossibility!

Figure 9.9

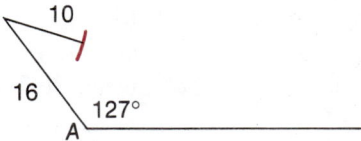

b. On the other hand, if $a = 26$, then

$$\sin B = \frac{b \sin A}{a} \qquad \text{(Law of sines)}$$
$$\approx \frac{16(0.7986)}{26}$$
$$\approx 0.4914$$
$$B \approx 29°26' \qquad (B \text{ is acute since } A \text{ is obtuse.})$$

Then we find

$$C \approx 180° - 127° - 29°26' \qquad (A + B + C = 180°)$$
$$= 23°34'$$

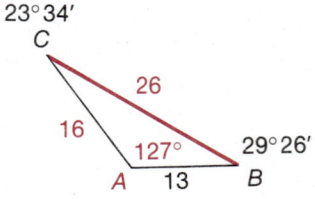

Figure 9.10

and

$$c = \frac{a \sin C}{\sin A} \qquad \text{(Law of sines)}$$

$$\approx \frac{26(0.3998)}{0.7986}$$

$$\approx 13$$

(See Figure 9.10.) ∎

> **SOLVING TRIANGLES–LAW OF SINES**
>
> For solving a triangle the law of sines is helpful in two situations:
>
> 1. *Two angles and any side are known* (AAS or ASA); these data always yield a unique triangle.
> 2. *Two sides and a nonincluded angle are known* (SSA); these data may yield two, one, or no triangles, however, depending on the length of the side opposite the angle. This is called the ambiguous case.
>
> No other situations lend themselves to solution by the law of sines.

As mentioned earlier, to solve a triangle we must know at least one side, since similar triangles need not be congruent. One side together with two angles then completely determine the triangle.

> **CAUTION**
>
> 1. If only one angle is known and its opposite side is not known, the law of sines will not yield a solution of the triangle. (Try it!)
> 2. Likewise, knowledge of all three sides of a triangle does not contain enough information to yield a solution of a triangle by using the law of sines.

The two situations described in the Caution box are somewhat disconcerting in view of the side-angle-side (SAS) and side-side-side (SSS) congruence properties of triangles. Even though the triangles may be completely determined by these data, the law of sines cannot solve such triangles. The law of cosines is developed in the next section to cover these latter two situations.

Section 9.1 Exercises

Solve the triangles with the following data.

1. $A = 36°40'$, $B = 67°20'$, $c = 15.0$
2. $a = 7.0$, $b = 15.0$, $A = 72°0'$
3. $A = 57°20'$, $B = 52°10'$, $a = 13.1$
4. $b = 9.0$, $a = 7.0$, $B = 29°0'$
5. $c = 15.0$, $b = 22.0$, $C = 33°0'$

6. $a = 12.0, b = 36.0, A = 122°0'$
7. $a = 42.0, b = 21.0, A = 112°0'$
8. $B = 126°10', C = 17°12', a = 5.3$
9. $a = 6.1, c = 11.3, C = 42°10'$
10. $c = 14.6, a = 3.1, A = 12°10'$
11. $c = 32.0, b = 27.0, B = 154°0'$
12. $B = 113°0', C = 26°10', c = 27.2$
13. $A = 43°0', B = 86°0', c = 22.0$
14. $b = 18.1, c = 22.1, B = 39°20'$
15. $B = 12°0', C = 15°0', a = 5.0$
16. $a = 72.3, c = 43.1, C = 131°0'$
17. $b = 32.0, c = 46.0, B = 15°0'$
18. $a = 54.0, c = 11.6, C = 12°20'$
19. $b = 32.1, c = 46.2, C = 143°0'$
20. $A = 43°20', C = 62°5', a = 14.6$
21. $A = 111°0', C = 40°0', c = 7.0$
22. $a = 47.2, b = 47.5, A = 83°30'$
23. $A = 67°10', B = 100°0', c = 12.0$
24. $A = 68°10', B = 42°50', a = 31.2$
25. $b = 19.7, c = 28.1, B = 57°40'$

26. A surveyor wishes to determine the height of a mountain. At a given position he measures the angle of elevation to the top as 43°40'. He then moves 1000.0 feet farther from the mountain and measures the angle of elevation to the top as 38°12'. How high is the mountain?

27. The Goodyear blimp hovers over a city. Observers 1 mile (5280 feet) apart at opposite ends of the city with the blimp between them measure the respective angles of elevation to the blimp as 23°0' and 52°0'. How high in feet is the blimp?

28. A timber buyer wishes to determine the height of a tree that is tilted 10°0' from the vertical. When the sun is aligned so that the tree leans directly away from the sun, she measures the shadow as 185.0 feet long and the angle of elevation of the line from the tip of the shadow to the tree top as 23°0'. How high is the tree?

29. On a slope of 45°0', a tower makes a shadow 120.0 between the slope and the line of sight) is 67°0'. from the tip of the shadow (not the angle between the slope and the line of sight) is 67°0'. How high is the tower if the shadow falls
    a. directly uphill?
    b. directly downhill?

30. A tunnel is to be constructed through one of the peaks in the Rocky Mountains from Apex (A) to Bounty (B). An observation tower is located at point (C). Angle BAC measures 132°0' and ∠ACB measures 23°0'. If the distance from A to C is 2 miles (10,560 feet), how long will the tunnel be?

31. In order to measure the length of a ski slope, a surveyor measures the angle of elevation of the slope, which is 29°0'. At a distance of 450.0 feet from the foot of the slope, the angle of elevation to the top is 19°0'. How long is the slope?

32. In order to measure the distance between two points A and B on opposite sides of a river, a surveyor takes readings at A and at another point C on the same side of the river as A. He measures ∠BAC = 63°0', ∠ACB = 87°0', and the distance from A to C is 320.0 feet. How far apart are A and B?

33. When traveling toward a given city at 55 miles per hour, a clever student of trigonometry measures the angle of elevation to the tallest building in the city as 3°0'. One minute later, the angle of elevation is 5°0'.
    a. How far is the student from the city?
    b. How tall is the building?

34. A ship traveling at 150.0 miles per hour sights an island in a direction making an angle of 42°0' with its direction of travel. One-half hour later the angle is 78°0'. How close will the ship come to the island if it does not change its direction?

35. Cities A and B are separated by a lake. City C is off to the side in such a way that straight highways connect C with A and C with B. The distance from A to C is 100.0 miles and the distance from B to C is 75.0 miles, making the trip from A to B 175.0 miles. If ∠BAC = 33°0', what is the distance from A to B "as the crow flies"?

Section 9.2

The Law of Cosines

$\cos \alpha = x/r$;
$\sin \alpha = y/r$;

$x = r \cos \alpha$
$y = r \sin \alpha$

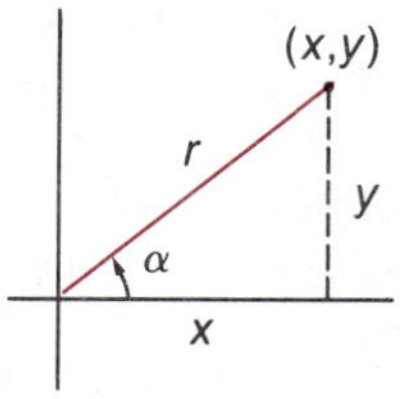

Figure 9.11

Figure 9.12

As we mentioned in the last section, the law of sines is not very useful for solving triangles when two sides and either the included angle or the remaining side are known—the so-called side-angle-side (SAS) or side-side-side (SSS) information. The law of cosines covers these two situations.

Specifically, the law of cosines relates an angle of any triangle to the three sides of that triangle. To avoid tying the results to a particular vertex of a triangle, we label one of its angles θ with adjoining sides u, v, and opposite side w. Although several derivations of this relationship are possible, the simplest stems from the distance formula.

Before we begin, recall that if the line joining a point (x, y) in the plane to the origin makes an angle α with the positive x-axis, then $x = r \cos \alpha$ and $y = r \sin \alpha$, where r is the distance of the point from the origin (see Figure 9.11). If the triangle is located with θ in standard position, its vertices can then be labeled as in Figure 9.12, which shows θ as (a) acute or (b) obtuse.

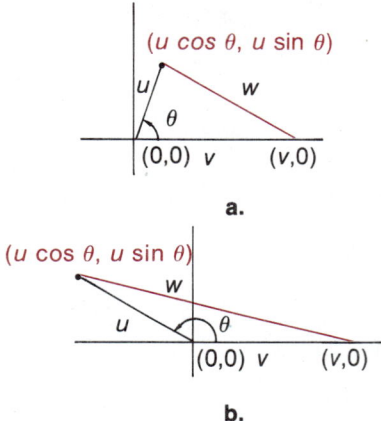

According to the distance formula,

$$w = d[(u \cos \theta, u \sin \theta), (v, 0)]$$
$$= \sqrt{(u \cos \theta - v)^2 + (u \sin \theta - 0)^2}$$
$$= \sqrt{u^2 \cos^2 \theta - 2uv \cos \theta + v^2 + u^2 \sin^2 \theta}$$
$$= \sqrt{u^2(\cos^2 \theta + \sin^2 \theta) + v^2 - 2uv \cos \theta}$$
$$= \sqrt{u^2 + v^2 - 2uv \cos \theta} \qquad (\cos^2 \theta + \sin^2 \theta = 1)$$
$$w^2 = u^2 + v^2 - 2uv \cos \theta \qquad \text{(Square both sides)}$$

This relationship is called the *law of cosines*.

LAW OF COSINES

In a given triangle, if sides u and v are adjacent to angle θ and w is opposite θ, then

$$w^2 = u^2 + v^2 - 2uv \cos \theta$$

The law of cosines lends itself to the solution of a triangle for which we have SSS or SAS data. The SAS data can be labeled u, θ, and v, enabling us to find the third side w. We then have SSS data, which can be used to find the other angles *with no indeterminancy* since the sought-after angle is acute or obtuse, depending on whether its cosine is positive or negative.

Incidentally, the Pythagorean theorem is a special case of the law of cosines. For if $\theta = 90°$, then $\cos \theta = 0$ and u, v are the legs of the right triangle with w as the hypotenuse; the law of cosines then reads $w^2 = u^2 + v^2$. We can also conclude that *a triangle is a right triangle if and only if the Pythagorean relation holds.* To see this, note that if it is not a right triangle, then $\cos \theta \neq 0$ and $w^2 = u^2 + v^2 - 2uv \cos \theta \neq u^2 + v^2$.

EXAMPLE 1

Solve the triangle with $a = 12.7$, $b = 22.1$, $C = 58°30'$.

SOLUTION

Draw a rough sketch of the triangle, as in Figure 9.13. By the law of cosines

$$c^2 = a^2 + b^2 - 2ab \cos C$$
$$\approx (12.7)^2 + (22.1)^2 - 2(12.7)(22.1)(0.5225)$$
$$\approx 161.29 + 488.41 - 293.30$$
$$\approx 356.40$$
$$c \approx 18.9$$

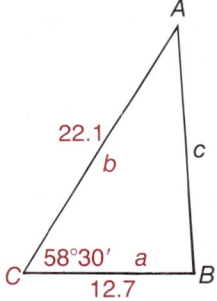

Figure 9.13

Now, knowing three sides and an angle, we could proceed with the law of cosines or the law of sines to find the other angles. Using the law of cosines to find A, we have

$$a^2 = b^2 + c^2 - 2bc \cos A$$
$$2bc \cos A = b^2 + c^2 - a^2$$
$$\cos A = \frac{b^2 + c^2 - a^2}{2bc}$$
$$\approx \frac{(22.1)^2 + (18.9)^2 - (12.7)^2}{2(22.1)(18.9)}$$
$$\approx 0.8192 \qquad \text{(Use a calculator)}$$
$$A \approx 35°0' \qquad \text{(Use a calculator or Table 5)}$$

Finally,

$$B = 180° - A - C$$
$$\approx 180° - 35° - 58°30' \qquad (A + B + C = 180°)$$
$$= 86°30' \qquad \blacksquare$$

EXAMPLE 2

Two roads lead out of a village in a remote mountain area. These roads go in the same direction but one climbs to a mine; the other descends to a city. The distance from the village to the city is 15.0 miles; the distance from the village to the mine is 3.0 miles. Sighting from the village, the angle between the city and the mine is $32°0'$. How far is it from the city to the mine? Sighting from the mine, what is the angle between the village and the city?

SOLUTION   The village, the city, and the mine form a triangle, the vertices of which we shall label as V, C, and M, respectively; see Figure 9.14. The given data then translate into $m = 15.0$, $c = 3.0$, and $V = 32°0'$.

Figure 9.14

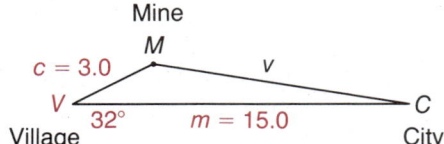

We are asked to find v and $\angle M$. By the law of cosines

$$v^2 = m^2 + c^2 - 2mc \cos V$$
$$\approx 225 + 9 - 2 \cdot 15 \cdot 3 \cdot (0.8480)$$
$$\approx 234 - 90 \cdot (0.8480)$$
$$\approx 157.68$$
$$v \approx 12.6$$

The distance between the city and the mine is approximately 12.6 miles. Since m is the longest side, M is the only possible obtuse angle. Accordingly, the acute angle C can be found by the law of sines:

$$\sin C = \frac{c \sin V}{v}$$
$$\approx \frac{3 \cdot (0.5299)}{12.6}$$
$$\approx 0.1262$$
$$C \approx 7°15'$$

Then

$$M \approx 180° - 32° - 7°15' \qquad (V + C + M = 180°)$$
$$\approx 140°45' \qquad\blacksquare$$

EXAMPLE 3   Solve the triangle with $a = 12.0$, $b = 20.0$, $c = 25.0$.

SOLUTION C is the only possible obtuse angle since c is the longest side. For this reason, we use the law of cosines to find C first.

$$c^2 = a^2 + b^2 - 2ab \cos C$$
$$\cos C = \frac{a^2 + b^2 - c^2}{2ab}$$
$$\approx \frac{144 + 400 - 625}{2 \cdot 12 \cdot 20}$$
$$\approx \frac{-81}{480}$$
$$\approx -0.1688$$
$$C \approx 99°43'$$

Then, by the law of sines

$$\sin B = \frac{b \sin C}{c} \approx \frac{20(0.9857)}{25}$$

$$\approx 7.885$$

$$B \approx 52°3' \qquad (B \text{ must be acute})$$

Finally,

$$A \approx (180° - 52°3' - 99°43') \qquad (A + B + C = 180°)$$
$$= 28°14'$$

Figure 9.15 shows the triangle.

Figure 9.15

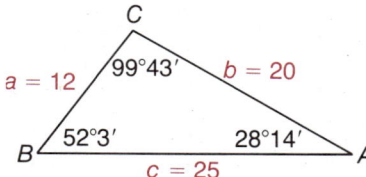

SOLVING TRIANGLES

1. If three sides are given (*SSS*), use the *law of cosines*.
2. If two sides and an angle are given, use the
   a. *Law of cosines* if the angle is included between the two sides (*SAS*).
   b. *Law of sines* if the angle lies opposite one of the sides (*SSA*); recall that this is the ambiguous case.
3. If one side and two angles are known (*ASA or AAS*), the *law of sines* yields a unique solution.

Together, the law of sines and law of cosines enable us to solve any solvable triangle since (a) two side lengths (SS), (b) two or three angles (AA or AAA), or (c and d) one side and one angle do not determine a unique triangle (see Figure 9.16).

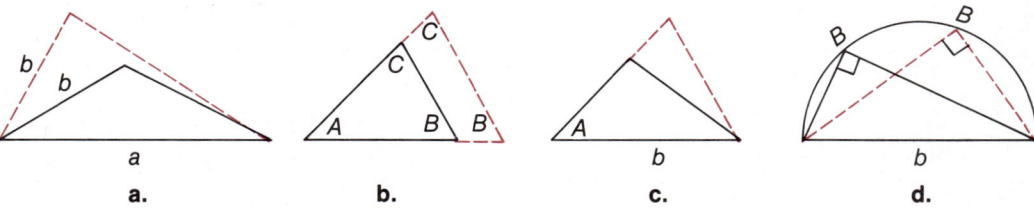

Figure 9.16

Area

The following formula for finding the area of any triangle can be established by using the law of cosines. The steps leading to a proof are outlined in Exercise 26 of this section.

> **HERON'S AREA FORMULA**
>
> Let a triangle have sides of length a, b, c and let $s = \frac{1}{2}(a + b + c)$; s denotes the semiperimeter. Then the area of the triangle is
>
> $$\text{Area} = \sqrt{s(s-a)(s-b)(s-c)}$$

EXAMPLE 4   Find the area of the triangle having sides of length 3, 4, 5.

SOLUTION   The semiperimeter is $s = \frac{1}{2}(3 + 4 + 5) = 6$. Using Heron's formula

$$\begin{aligned}\text{Area} &= \sqrt{6(6-3)(6-4)(6-5)} \\ &= \sqrt{6 \cdot 3 \cdot 2 \cdot 1} \\ &= \sqrt{36} \\ &= 6\end{aligned}$$

You may have recognized that the 3-4-5 triangle is a right triangle, since $3^2 + 4^2 = 5^2$ (see Example 2, Section 7.2). The sides of length 3 and 4 are perpendicular and we may consider one side to be the base and the other the height of the triangle. The area could then also be computed as

$$\begin{aligned}\text{Area} &= \frac{1}{2} \cdot \text{base} \cdot \text{height} \\ &= \frac{1}{2} \cdot 3 \cdot 4 \\ &= 6\end{aligned}$$

in agreement with Heron's formula. ∎

Section 9.2 Exercises

Solve the triangles satisfying the following conditions.

1. $b = 15.6$, $c = 10.5$, $A = 62°0'$
2. $a = 33.0$, $b = 20.0$, $c = 30.0$
3. $a = 55.2$, $c = 32.5$, $B = 100°20'$
4. $a = 17.7$, $b = 3.0$, $C = 25°10'$
5. $a = 12.2$, $b = 37.4$, $c = 42.1$
6. $a = 22.0$, $b = 55.0$, $c = 40.0$
7. $a = 53.7$, $c = 44.2$, $B = 58°40'$
8. $a = 28.0$, $c = 21.0$, $B = 21°10'$
9. $a = 24.0$, $b = 17.0$, $c = 20.0$
10. $a = 190.0$, $b = 95.0$, $C = 112°20'$
11. $a = 5.0$, $b = 12.0$, $c = 11.0$
12. $b = 20.0$, $c = 30.0$, $A = 104°30'$
13. $a = 206.0$, $b = 213.0$, $C = 53°40'$
14. $a = 100.0$, $c = 73.8$, $B = 27°53'$
15. $a = 7.0$, $b = 9.0$, $c = 13.0$

16. To measure the length of a proposed tunnel, a surveyor positions himself where he can see the suggested locations for the ends of the tunnel. The distance from his location to the tunnel ends is 13.0 and 17.0 kilometers, respectively. If the angle between the two lines of sight is 83°0', find the length of the proposed tunnel.

17. If two trucks leave a given city on highways making an angle of 132°0' with one another, traveling at 45.0 and 55.0 miles per hour,

respectively, what is the distance between the trucks 2 hours later?

18. Two cities A and B lie on opposite sides of a lake. A third city C is located at one end of the lake. If the distance from A to C is 10.0 miles and the distance from B to C is 15.0 miles and $\angle BCA = 12°0'$, what is the distance by boat from A to B?

19. Of three towns, A and B are on the same side of a river and C is on the opposite side. Towns B and C are connected by a bridge 5.0 miles long; the distance from A to B is 3.0 miles.
    a. If $\angle ABC = 110°0'$, how far is it from A to C?
    b. How wide is the river?

20. If the distance from A to B is 16.0 miles, from B to C is 23.0 miles, and from C to A is 30.0 miles, how close does one come to B when traveling from C to A?

21. A triangular plot of land has sides of length 700.0 feet, 800.0 feet, and 1000.0 feet. How many acres are contained in this plot? (One acre = 43,560 square feet.)

22. A mine shaft drops vertically downward 950.0 meters to a large cavern. A village lies 1625.0 meters downhill from the entrance to the shaft at an angle of depression of $57°0'$. A proposed tunnel will go directly from the village to the cavern.
    a. At what angle to the hillside must this tunnel be dug?
    b. How long will the tunnel be?

23. Presently the only means of communication between High Mesa and Red Gorge is by Pony Express through Moonlight City. It is 115.0 miles between High Mesa and Moonlight City and 212.0 miles from Moonlight City to Red Gorge. The angle between the two routes is $23°20'$. How much travel time could be saved if a direct route could be opened between High Mesa and Red Gorge assuming the horses average 28.0 miles per hour?

24. Two planes are observed at distances of 8.3 and 9.2 kilometers from the airport control tower. The angle between the lines of sight is $2°43'$. How close are the planes to one another?

25. A valley is nestled between three mountains. The valley is basically triangular with sides of length 27.0 kilometers, 13.0 kilometers, and 22.0 kilometers. Irrigation will increase the yield of the valley by 2.6 metric tons per hectare. How much will the yield of the valley be increased by irrigation? (One hectare = 10,000 square meters.)

26. Work through the following steps to verify Heron's area formula for a triangle with sides of lengths a, b, c.
    i. Show that the area can be written as $\frac{1}{2}bc \sin A$.
    ii. Show that $\sin A = \sqrt{(1 + \cos A)(1 - \cos A)}$.
    iii. Use the law of cosines to substitute for $\cos A$ in (ii).
    iv. Simplify the resulting expression for the area to obtain Heron's formula. (*Hint:* $(a + b - c)/2 = (a + b + c - 2c)/2 = (a + b + c)/2 - c = s - c$.)

Use Heron's formula to find the area of the following triangles.

27. $a = 8.0$ inches, $b = 3.0$ inches, $c = 7.0$ inches

28. $a = 4.5$ centimeters, $b = 5.2$ centimeters, $c = 6.3$ centimeters

29. $a = 8$ feet 6 inches, $b = 16$ feet 4 inches, $c = 15$ feet 9 inches

30. $a = 6.2$ miles, $b = 3.1$ miles, $c = 5.7$ miles

Section 9.3

Vectors

In our discussion of signed numbers and absolute value, we saw that in many cases the magnitude alone is not sufficient to describe a property; bank balances of $200 and −$200 are considerably different. With physical phenomena we must consider even more than just the positive and negative directions. For example, if a force is exerted on an object, we do not know which way the object will move until we know the direction in which the force is applied. To pilots, the direction of wind

514 Chapter 9 *Further Applications of Trigonometry*

velocity is just as important as its magnitude. Mathematicians have developed the concept of "vector" to describe such phenomena.

A **vector**—a quantity such as force, velocity, or acceleration—has both magnitude and direction. Vectors are distinguished from purely numerical quantities, which are called **scalars**—quantities such as temperature, volume, and area.

It is convenient to use directed line segments or arrows to represent vectors. The orientation of the arrow indicates the direction of the vector and the length of the arrow indicates its magnitude. If the arrow is drawn from A to B, the vector is denoted $\overrightarrow{AB}$; A is called the **initial point** or "tail," and B the **terminal point** or "tip" of the vector.

A vector has *only* magnitude and direction. It does not have a position, strictly speaking; that is, a vector is free to roam as long as its magnitude and direction are not changed. In Figure 9.17 $\overrightarrow{AB}$ and $\overrightarrow{CD}$ represent the same vector since they have the same magnitude and direction. In this case, we write $\overrightarrow{AB} = \overrightarrow{CD}$.

Figure 9.17

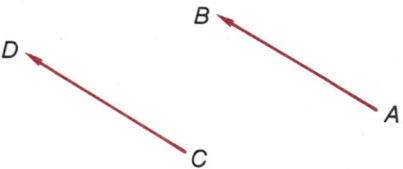

Frequently, it will be convenient to denote a vector by using only one letter since its position is irrelevant. In these cases, a bold-faced letter will be used. For instance, **F** is often used to denote a force vector. $|\mathbf{F}|$ or $|\overrightarrow{AB}|$ will denote the **magnitude** or *length* of the vector. Vectors of length 1 are called **unit vectors**. The magnitude of a velocity vector is called **speed.**

EXAMPLE 1

Draw an arrow representing the vector of magnitude 2 in the direction $\pi/3$ radians above the horizontal. Locate this vector with its initial point at the following coordinates and label the corresponding terminal points.

a. $(0, 0)$
b. $(1, 2)$
c. $(-1, 1)$

SOLUTION

The requested vectors are sketched in Figure 9.18 and labeled accordingly.

Figure 9.18

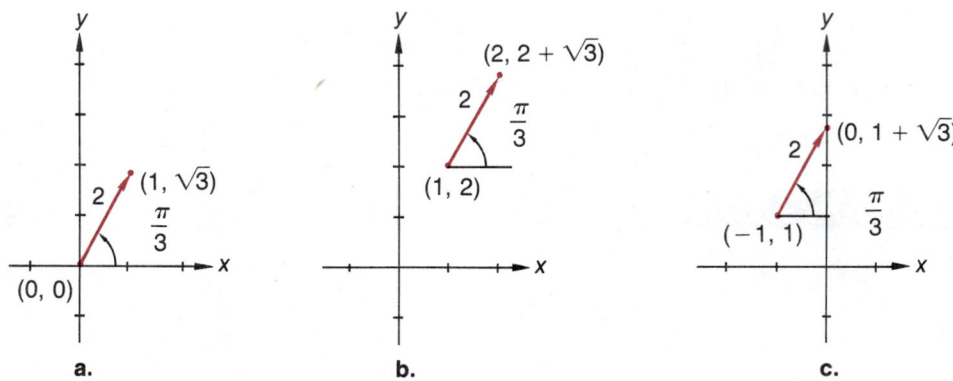

Section 9.3 Vectors 515

Vector Addition

If an object moves from *A* to *B* and then from *B* to *C*, the net result is that it has moved from *A* to *C*. We write $\overrightarrow{AB} + \overrightarrow{BC} = \overrightarrow{AC}$ as illustrated in Figure 9.19. This statement is the basis for the definition of vector addition.

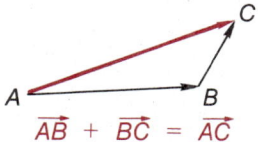

$\overrightarrow{AB} + \overrightarrow{BC} = \overrightarrow{AC}$

Figure 9.19

> **DEFINITION**
>
> To form the sum **U** + **V** of two vectors **U** and **V** in arrow form, join the tip of **U** to the tail of **V**; **U** + **V** is represented by the arrow from the tail of **U** to the tip of **V**.
>
>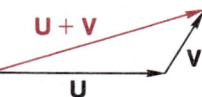

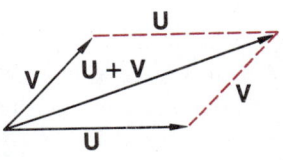

Figure 9.20

The preceding definition of addition is permitted since the vector **V** is free to be moved as long as its orientation and magnitude are not changed. This freedom of movement also yields the **parallelogram law of addition:** **U** + **V** is a diagonal of the parallelogram having sides **U** and **V** as illustrated in Figure 9.20.

Vector addition is useful in determining the net velocity or net force resulting from a combination of two or more velocities or forces. This net result is called the **resultant.**

EXAMPLE 2   A plane is oriented in a northeasterly direction and maintaining an air speed of 120 knots (nautical miles per hour) (1 nautical mile ≈ 1.15 land miles). The wind is blowing due east at 30 knots. Find the resultant velocity of the plane with respect to the ground.

SOLUTION   Velocity is a vector since direction as well as magnitude is important. In the absence of any air currents, the plane would travel 120 nautical miles northeast in 1 hour. The effect of the wind is to move the plane 30 nautical miles farther east. This can be illustrated in vector form as in Figure 9.21. The velocity vector for the plane in still air is denoted by **V**, **W** is the wind velocity, and **R** is the resultant velocity of the plane: **R** = **V** + **W**. To describe **R**, we must determine its magnitude and its

Figure 9.21

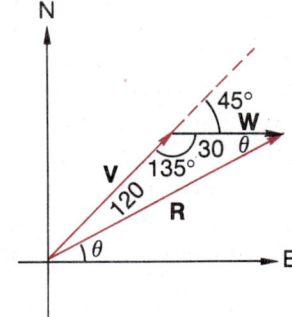

direction. The magnitude of **R** can be determined from the law of cosines:

$$|\mathbf{R}|^2 = 30^2 + 120^2 - 2 \cdot 30 \cdot 120 \cdot (\cos 135°)$$
$$= 900 + 14{,}400 - 7200 \cdot \left(-\frac{\sqrt{2}}{2}\right)$$
$$= 15{,}300 + 3600\sqrt{2}$$
$$\approx 20{,}391$$
$$|\mathbf{R}| \approx 143$$

The law of sines can then be used to determine θ:

$$\frac{\sin \theta}{120} = \frac{\sin 135°}{|\mathbf{R}|}$$
$$\sin \theta \approx 120 \cdot \frac{\sqrt{2}/2}{143}$$
$$\approx 0.5934$$
$$\theta \approx 36°24'$$

The resultant velocity of the plane is approximately 143 knots in a direction 36°24' north of due east (E36°24'N). Converting to land miles per hour the speed of the plane is approximately $(143) \cdot (1.15) \approx 164$ miles per hour. ∎

Forces are also combined by using vector addition. To visualize this, imagine forces acting independently for the same length of time. The first force generates a certain displacement, which is followed by a further displacement generated by the second force. The net displacement is the vector sum of the two displacements; the resultant force is the vector sum of the two individual forces.

Scalar Multiples

If **V** denotes a vector, then 2**V** denotes a vector having the same direction as **V** but twice as its magnitude; −**V** denotes a vector with direction opposite to that of **V** and having the same magnitude as **V**.

> **DEFINITION**
>
> For any real number a and any vector **V**, a**V** is defined as follows.
>
> 1. $|a\mathbf{V}| = |a|\,|\mathbf{V}|$; that is, the magnitude of **V** is multiplied by $|a|$.
> 2. The direction of a**V** is the same as the direction of **V** if $a > 0$ and opposite to that of **V** if $a < 0$.
>
> a**V** is called a **scalar multiple** of **V**.

According to the preceding definition, 0**V** has length 0; it is called the **zero vector** and denoted by **0**. Any direction can be associated with the zero vector; if we travel a 0 distance, it makes no difference in what direction we make this hypothetical journey. The zero vector is the only vector that does not have a unique direction.

The following are properties of addition and multiplication by a scalar.

ADDITION

PROPERTY	ILLUSTRATION
1. Commutativity: $U + V = V + U$	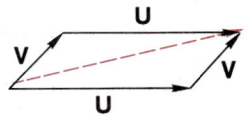
2. Associativity: $U + (V + W) = (U + V) + W$	
3. Zero vector: $U + 0 = U$	
4. Opposite vector: $U + (-U) = 0$	

MULTIPLICATION BY A SCALAR

PROPERTY	ILLUSTRATION
1. $a(U + V) = aU + aV$	 $2U + 2V = 2(U + V)$ (By similar triangles)
2. $aU + bU = (a + b)U$	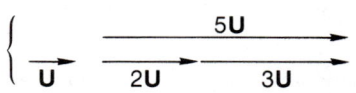
3. $a(bU) = (ab)U$	
4. $1 \cdot U = U$	No change in magnitude or direction is generated.
5. $0 \cdot U = 0 = a \cdot 0$	All have magnitude 0 and direction is then immaterial.
6. $-U = (-1)U$	Only the direction of U is changed (reversed) in each case.

Vector Subtraction

As in any algebraic system, subtraction is defined in terms of addition. Specifically, $U - V$ is that vector W that when added to V yields U:

$$U - V = W \quad \text{means} \quad U = V + W$$

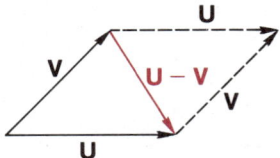

Figure 9.22

Subtraction can also be accomplished by using a "parallelogram law"; **U** + **V** and **U** − **V** are the two diagonals of the parallelogram having **U** and **V** as sides. Note in Figure 9.22 that for the diagonal labeled **U** − **V**, we do indeed have **V** + (**U** − **V**) = **U**.

We obtain the usual properties relating subtraction and negatives.

PROPERTY	ILLUSTRATION
1. **U** + (−**V**) = **U** − **V**	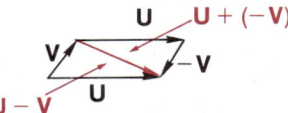
2. −(**U** + **V**) = −**U** − **V**	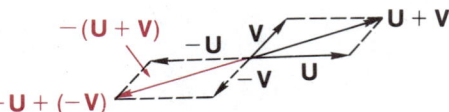
3. −(−**U**) = **U**	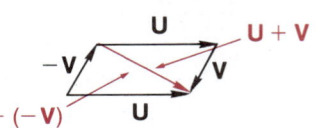
4. **U** − (−**V**) = **U** + **V**	

EXAMPLE 3

With vectors **A**, **B**, and **C** as shown in Figure 9.23 sketch each of the following.

a. **A** − (**B** + **C**)

b. (**A** − **B**) + **C**

c. $2\mathbf{A} - \frac{1}{2}\mathbf{B}$

Figure 9.23

SOLUTION Each of the sketches in Figure 9.24 should be self-explanatory.

Figure 9.24

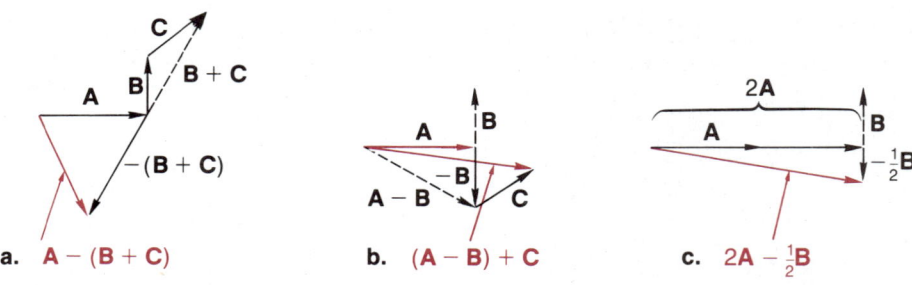

a. **A** − (**B** + **C**)   b. (**A** − **B**) + **C**   c. $2\mathbf{A} - \frac{1}{2}\mathbf{B}$

EXAMPLE 4

Use vectors to show that the diagonals of a parallelogram bisect each other.

SOLUTION

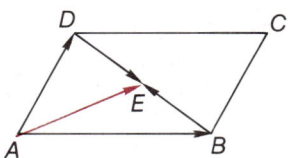

Figure 9.25

Let E be the midpoint of diagonal $\overline{DB}$ in a parallelogram having successive vertices $ABCD$ as sketched in Figure 9.25.
Then

$$\overrightarrow{AE} = \overrightarrow{AD} + \overrightarrow{DE} \qquad (1)$$
$$= \overrightarrow{AB} + \overrightarrow{BE} \qquad (2)$$

But

$$\overrightarrow{BE} = -\overrightarrow{DE}$$

Substituting this into 2 we have

$$\overrightarrow{AE} = \overrightarrow{AB} - \overrightarrow{DE} \qquad (3)$$

Adding corresponding sides of equations 1 and 3, we obtain

$$2\overrightarrow{AE} = \overrightarrow{AB} + \overrightarrow{AD}$$
$$= \overrightarrow{AB} + \overrightarrow{BC}$$
$$= \overrightarrow{AC}$$

$$\overrightarrow{AE} = \frac{1}{2}\overrightarrow{AC}$$

That is, the midpoint E of $\overline{DB}$ is also the midpoint of $\overline{AC}$. ∎

Coordinate Vectors

When working in a Cartesian coordinate system, we call the unit vectors in the positive x- and y-directions **i** and **j**, respectively, as illustrated in Figure 9.26(a). Any vector **V** in the Cartesian plane can be described in terms of these. Simply place the vector so that its initial point is at the origin. If its terminal point then falls on (x, y), it can be expressed as $\mathbf{V} = x\mathbf{i} + y\mathbf{j}$; see Figure 9.26b. It follows from the Pythagorean theorem than $|\mathbf{V}| = \sqrt{x^2 + y^2}$. When **V** is expressed as $x\mathbf{i} + y\mathbf{j}$, x and y are called the **i** and **j** **components** of **V**.

Figure 9.26

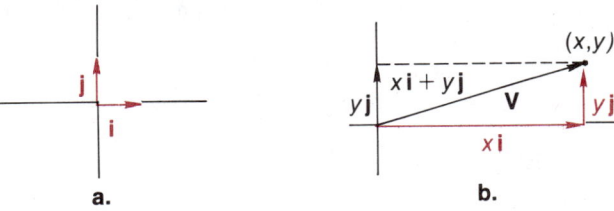

If two vectors are expressed in terms of components, the commutative and associative laws simplify addition, subtraction, and multiplication by a scalar. Specifically,

$$(x\mathbf{i} + y\mathbf{j}) + (w\mathbf{i} + z\mathbf{j}) = (x + w)\mathbf{i} + (y + z)\mathbf{j}$$
$$(x\mathbf{i} + y\mathbf{j}) - (w\mathbf{i} + z\mathbf{j}) = (x - w)\mathbf{i} + (y - z)\mathbf{j}$$
$$a(x\mathbf{i} + y\mathbf{j}) = (ax)\mathbf{i} + (ay)\mathbf{j}$$

EXAMPLE 5

Let $A = (2, -3)$ and $B = (5, 4)$.

a. Express $\overrightarrow{AB}$ in terms of **i** and **j**.

b. Find the magnitude of $\overrightarrow{AB}$.

c. Find a unit vector in the direction of $\overrightarrow{AB}$.

SOLUTION

a. In Figure 9-27, we see that we can travel from A to B by traveling 3 units in the **i** direction and then 7 units in the **j** direction:
$$\overrightarrow{AB} = 3\mathbf{i} + 7\mathbf{j}$$

b. $|\overrightarrow{AB}| = \sqrt{3^2 + 7^2} = \sqrt{9 + 49} = \sqrt{58}$

c. To obtain a unit vector in the direction of $\overrightarrow{AB}$, we must divide the length of AB by $\sqrt{58}$. The unit vector is
$$\mathbf{U} = \frac{1}{\sqrt{58}} \overrightarrow{AB} = \frac{3}{\sqrt{58}}\mathbf{i} + \frac{7}{\sqrt{58}}\mathbf{j}$$

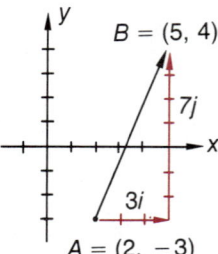

Figure 9.27

EXAMPLE 6

Use vectors to develop the formula for the midpoint of a line segment in the Cartesian plane.

SOLUTION

Figure 9.28

Consider $P = (x_1, y_1)$ and $Q = (x_2, y_2)$, as shown in Figure 9.28.

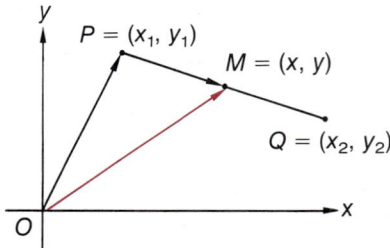

Let the midpoint $M = (x, y)$. Then as in Example 5, we can show that
$$\overrightarrow{PQ} = (x_2 - x_1)\mathbf{i} + (y_2 - y_1)\mathbf{j}$$
In terms of vectors, we now have
$$\overrightarrow{OM} = \overrightarrow{OP} + \frac{1}{2}\overrightarrow{PQ}$$

$$x\mathbf{i} + y\mathbf{j} = (x_1\mathbf{i} + y_1\mathbf{j}) + \frac{1}{2}[(x_2 - x_1)\mathbf{i} + (y_2 - y_1)\mathbf{j}]$$
$$= \left[x_1 + \frac{1}{2}(x_2 - x_1)\right]\mathbf{i} + \left[y_1 + \frac{1}{2}(y_2 - y_1)\right]\mathbf{j}$$
$$= \frac{1}{2}(x_1 + x_2)\mathbf{i} + \frac{1}{2}(y_1 + y_2)\mathbf{j}$$

The coordinates of the midpoint are

$$x = \frac{1}{2}(x_1 + x_2) \qquad y = \frac{1}{2}(y_1 + y_2)$$

The elementary vector operations can be accomplished by working with the **i** and **j** components of the vectors. For this reason, we introduce a shorthand notation that indicates only the components of the vector:

$$x\mathbf{i} + y\mathbf{j} = \langle x, y \rangle$$

As discusssed previously, we then have

1. $\langle x, y \rangle + \langle w, z \rangle = \langle x + w, y + z \rangle$
2. $\langle x, y \rangle - \langle w, z \rangle = \langle x - w, y - z \rangle$
3. $a\langle x, y \rangle = \langle ax, ay \rangle$

EXAMPLE 7

Let $\mathbf{A} = \langle 2, 2 \rangle$ and $\mathbf{B} = \langle 1, -3 \rangle$. Sketch each of the following and find the magnitude of each.

a. $\mathbf{A} + \mathbf{B}$
b. $3\mathbf{A} - 2\mathbf{B}$

SOLUTION

a. $\mathbf{A} + \mathbf{B} = \langle 2, 2 \rangle + \langle 1, -3 \rangle = \langle 2 + 1, 2 - 3 \rangle$
$= \langle 3, -1 \rangle$
$= 3\mathbf{i} - \mathbf{j}$
$|\mathbf{A} + \mathbf{B}| = \sqrt{3^2 + (-1)^2}$
$= \sqrt{10}$

b. $3\mathbf{A} - 2\mathbf{B} = 3\langle 2, 2 \rangle - 2\langle 1, -3 \rangle = \langle 3 \cdot 2, 3 \cdot 2 \rangle - \langle 2 \cdot 1, 2 \cdot (-3) \rangle$
$= \langle 6, 6 \rangle - \langle 2, -6 \rangle = \langle 6 - 2, 6 - (-6) \rangle$
$= \langle 4, 12 \rangle$
$= 4\mathbf{i} + 12\mathbf{j}$
$|3\mathbf{A} - 2\mathbf{B}| = \sqrt{4^2 + 12^2}$
$= \sqrt{160}$
$= 4\sqrt{10}$

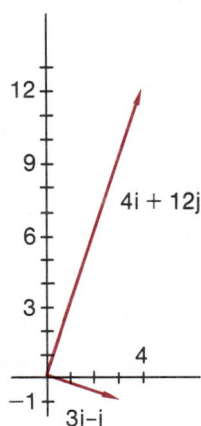

Figure 9.29

Each of these vectors is sketched in Figure 9.29.

Section 9.3 Exercises

Given the three vectors **U**, **V**, *and* **W** *shown in the figure, construct the requested vector.*

1. $\mathbf{U} - (\mathbf{V} + \mathbf{W})$

2. $2\mathbf{U} + \frac{1}{2}(\mathbf{V} - \mathbf{W})$

The following figure is to be used for Exercise 3–6.

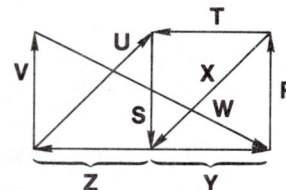

3. Write **U** in terms of **T**, **X**, and **Z**.

4. Write **Y** in terms of **R**, **S**, **V**, **W**, and **Z**.

5. Find $\mathbf{T} - \mathbf{X}$.

6. Find $\mathbf{V} + \mathbf{S} + \mathbf{U}$.

7. If $\overrightarrow{AD}$ is the median drawn from A to the midpoint of side $\overline{BC}$ in triangle ABC, show that $\overrightarrow{AD} = \frac{1}{2}(\overrightarrow{AB} + \overrightarrow{AC})$.

8. Suppose that the sides of a closed polygon represent vectors arranged tip to tail. What is the sum of these vectors?

9. Show that the line joining the midpoints of any two sides of a triangle is parallel to and half as long as the third side.

A regular hexagon is formed by six vectors of equal length oriented in a counter-clockwise fashion. If **A** *and* **B** *are two adjacent sides, find the following in terms of* **A** *and* **B**.

10. Each of the other sides.

11. The radius vectors from the center to each vertex.

12. All diagonal vectors connecting the vertices.

13. Let A, B, and C be the vertices of a triangle and let O be some other point. Let $\overrightarrow{AD}$ be the median drawn from A to the midpoint of side $\overline{BC}$.

    a. Find $\overrightarrow{AD}$ in terms of $\overrightarrow{OA}$, $\overrightarrow{OB}$, and $\overrightarrow{OC}$.

    b. Find $\overrightarrow{OA} + (2/3)\overrightarrow{AD}$ in terms of $\overrightarrow{OA}$, $\overrightarrow{OB}$, and $\overrightarrow{OC}$.

    c. What happens if the roles of A, B, and C are interchanged in the preceding?

    d. What geometric theorem has been established?

14. Let E, F, G, and H be the midpoints of the line segments $\overline{AB}$, $\overline{BC}$, $\overline{CD}$, and $\overline{DA}$, respectively. Show that

    a. $\overrightarrow{EF} = \frac{1}{2}\overrightarrow{AB} + \frac{1}{2}\overrightarrow{BC}$

    b. $\overrightarrow{HG} = \frac{1}{2}\overrightarrow{AD} + \frac{1}{2}\overrightarrow{DC}$

    c. $\overrightarrow{EF} = \overrightarrow{HG}$

    Express this as a geometric theorem.

15. Illustrate geometrically that $-(\mathbf{U} - \mathbf{V}) = -\mathbf{U} + \mathbf{V}$.

16. Use properties of a triangle to show that $|\mathbf{U} + \mathbf{V}| \leq |\mathbf{U}| + |\mathbf{V}|$.

In each of the following, express $\overrightarrow{AB}$ *in terms of* **i** *and* **j**.

17. $A = (1, 3)$, $B = (2, -1)$

18. $A = (2, 3)$, $B = (0, 0)$

19. $A = (4, 2)$, $B = (5, -1)$

20. $A = (3, -4)$, $B = (4, 1)$

In the following, find the vector $\overrightarrow{OC}$, *where* O *is the origin and* C *is the indicated point.*

21. C is the midpoint of the segment joining $(3, -4)$ and $(-1, 2)$

22. C is the midpoint of the segment joining $(1, 5)$ and $(-2, 1)$

23. C is the point one fourth of the way from $(3, 2)$ to $(4, -3)$

24. C is the point two fifths of the way from $(2, -1)$ to $(8, 6)$

Determine whether the following points are collinear by computing vectors from one to another.

25. $A = (2, 5)$, $B = (0, 1)$, and $C = (3, 7)$

26. $A = (1, 2)$, $B = (5, 4)$, and $C = (-2, -1)$

Let $\mathbf{A} = \langle 0, -1 \rangle$, $\mathbf{B} = \langle 3, -2 \rangle$, $\mathbf{C} = \langle 1, -2 \rangle$. *Find the given vector.*

27. $\mathbf{A} + \mathbf{B}$

28. $\mathbf{A} - \mathbf{B}$

29. $2\mathbf{A} + 3\mathbf{C}$ **30.** $3\mathbf{C} - \mathbf{A}$

31–34. Do the same for $\mathbf{A} = \langle 4, -2 \rangle$, $\mathbf{B} = \langle 3, 4 \rangle$, $\mathbf{C} = \langle -2, 3 \rangle$.

35–38. Do the same for $\mathbf{A} = 3\mathbf{j} - 2\mathbf{i}$, $\mathbf{B} = \mathbf{i} - 4\mathbf{j}$, $\mathbf{C} = 2\mathbf{i} + \mathbf{j}$.

Find the magnitude of each of the following.

39. $\mathbf{i} + \mathbf{j}$ **40.** $5\mathbf{i} - 12\mathbf{j}$

41. $\cos\theta\,\mathbf{i} + \sin\theta\,\mathbf{j}$ **42.** $\langle 5, 1 \rangle$

43. $\langle 8, -1 \rangle$ **44.** $\langle 3, -4 \rangle$

Find a unit vector in the direction of each of the given vectors.

45. $4\mathbf{i} + 3\mathbf{j}$ **46.** $7\mathbf{i} + 24\mathbf{j}$

47. $\mathbf{i} + \mathbf{j}$ **48.** $\langle 2, -2 \rangle$

49. $\langle 5, 12 \rangle$ **50.** $\langle \sqrt{3}, 1 \rangle$

Let $\mathbf{A} + \mathbf{B} = 4\mathbf{i} + 2\mathbf{j}$ and $\mathbf{A} = 3\mathbf{j} - 2\mathbf{i}$.

51. Find the magnitude of $\mathbf{B}$.

52. Find the magnitude of $\mathbf{A} - \mathbf{B}$.

If $|\mathbf{U}| = 2$, find the following.

53. $|3\mathbf{U}|$ **54.** $|-4\mathbf{U}|$

55. A plane is flying in a southwesterly direction with an airspeed of 250 knots. The wind is blowing E30°S (30° South of due East, see Example 2) at 20 knots. Describe the speed and direction of the flight with respect to the ground.

56. A pilot wishes to fly in a southwesterly direction. He will fly with an airspeed of 250 knots and the wind is blowing E30°S at 20 knots.
   a. In what direction should he orient his plane?
   b. What is his resulting speed with respect to the ground?

57. A rowboat is being rowed directly across stream at a rate of 15 feet per second. The stream is flowing at a rate of 5 feet per second. If the stream is 750 feet wide, how far downstream will the boat have drifted before reaching the opposite side?

58. A river is flowing at a rate of 5 feet per second. In what direction must Les row his boat if he wishes to travel directly across the river and he rows at a rate of 15 feet per second?

59. A 500-pound weight is supported as illustrated in the figure.

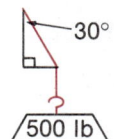

a. Find the tension in the rope.

b. What is the force exerted on the horizontal bracket?

(*Hint:* The vertical component of the force vector parallel to the rope must be 500 pounds.)

60. The weight of Exercise 59 is now supported as indicated in the figure.

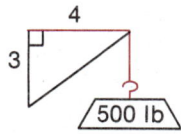

a. Find the tension in the rope.

b. Find the compression force in the angled bracket.

(*Hint:* Again, the vertical component of the force parallel to the bracket must be 500 pounds.)

61. A 3000-pound automobile is parked on a driveway inclined 7° above the horizontal.
   a. What force is needed to keep the automobile from rolling downhill?
   b. What is the force exerted perpendicular to the driveway?

Section 9.4

The Complex Plane and Polar Representation

Since two real numbers are needed to determine a complex number, one for the real part and one for the imaginary part, complex numbers can be plotted in the Cartesian plane. For this purpose we call the horizontal axis (or *x*-axis) the **real axis,** and the vertical axis (or *y*-axis) the **imaginary axis,** or **i axis.** The coordinate system consisting of the real and *i*-axes is called the **complex plane.** The complex number $x + iy$ is plotted as the point (x, y) in the complex plane. Various complex numbers are plotted in the complex plane in Figure 9.30.

Figure 9.30

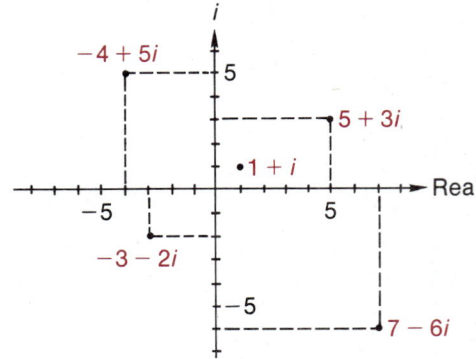

For any real number $x = x + 0i$, the absolute value of x is the distance of x from 0 on the real-axis. Similarly, for a complex number z, we define $|z|$ as the distance of z from the origin in the complex plane, as shown in Figure 9.31.

DEFINITION

For $z = x + iy$,

$$|z| = \sqrt{x^2 + y^2}$$

$|z|$ is called the **absolute value** or the **modulus** of z.

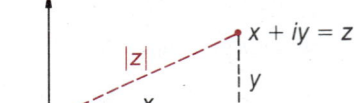

Figure 9.31

EXAMPLE 1   Find the modulus of the following complex numbers.

a. $z = 1 + i$
b. $z = -4 + 5i$
c. $z = -3 - 2i$

SOLUTION
a. $|1 + i| = \sqrt{1^2 + 1^2} = \sqrt{2}$
b. $|-4 + 5i| = \sqrt{(-4)^2 + 5^2} = \sqrt{16 + 25} = \sqrt{41}$
c. $|-3 - 2i| = \sqrt{(-3)^2 + (-2)^2} = \sqrt{9 + 4} = \sqrt{13}$ ∎

Before proceeding any further, we observe that the conjugate $\bar{z}$ of z is the reflection of z through the real axis, since

$$\overline{x + iy} = x - iy$$

Also note that $-z$ is the reflection of z through the origin. Then
$$|z| = |\bar{z}| = |-z| = |-\bar{z}| = |\overline{-z}|$$
(See Figure 9.32.)

Figure 9.32

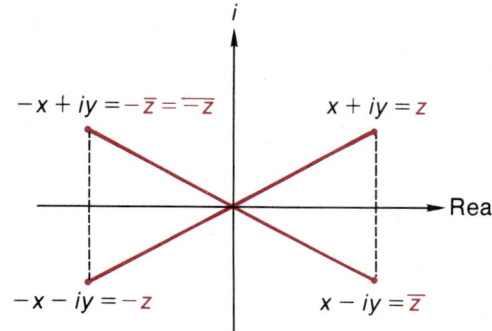

Note that if $z = x + iy$, then
$$|z| = \sqrt{x^2 + y^2}$$
$$= \sqrt{(x+iy)(x-iy)} = \sqrt{z\bar{z}}$$

$$\boxed{|z| = \sqrt{z\bar{z}}} \tag{1}$$

With this characterization of $|z|$, we can verify that two of the properties of the absolute value function (of real numbers) hold for the modulus of a complex number as well, namely,

$$\boxed{\begin{aligned} |z_1 z_2| &= |z_1||z_2| \\ \left|\frac{z_1}{z_2}\right| &= \frac{|z_1|}{|z_2|} \end{aligned}} \tag{2} \tag{3}$$

These properties are established as follows.

$$\begin{aligned}
|z_1 z_2| &= \sqrt{(z_1 z_2)(\overline{z_1 z_2})} & \text{(By Equation 1)} \\
&= \sqrt{z_1 \bar{z}_1 z_2 \bar{z}_2} & \text{(Exercise 83, Section 1.9)} \\
&= \sqrt{z_1 \bar{z}_1} \sqrt{z_2 \bar{z}_2} & \text{(Since } z\bar{z} \geq 0) \\
&= |z_1||z_2| & \text{(By Equation 1)} \\
\left|\frac{z_1}{z_2}\right| &= \sqrt{\left(\frac{z_1}{z_2}\right) \cdot \overline{\left(\frac{z_1}{z_2}\right)}} & \text{(By Equation 1)} \\
&= \sqrt{\frac{z_1}{z_2} \cdot \frac{\bar{z}_1}{\bar{z}_2}} & \text{(Exercise 86, Section 1.9)} \\
&= \frac{\sqrt{z_1 \bar{z}_1}}{\sqrt{z_2 \bar{z}_2}} \\
&= \frac{|z_1|}{|z_2|} & \text{(By Equation 1)}
\end{aligned}$$

A third property characteristic of the $|x|$ symbol for real numbers is the *triangle inequality*:

$$|x_1 + x_2| \le |x_1| + |x_2|$$

This property also holds for the complex modulus $|z|$. To see why this is so, consider the illustration in Figure 9.33 with

$$z_1 = a_1 + ib_1$$
$$z_2 = a_2 + ib_2$$
$$z_1 + z_2 = (a_1 + a_2) + i(b_1 + b_2)$$

and

$$|z_1| = \sqrt{a_1^2 + b_1^2}$$
$$|z_2| = \sqrt{a_2^2 + b_2^2}$$
$$|z_1 + z_2| = \sqrt{(a_1 + a_2)^2 + (b_1 + b_2)^2}$$

Figure 9.33
$|z_1 + z_2| \le |z_1| + |z_2|$

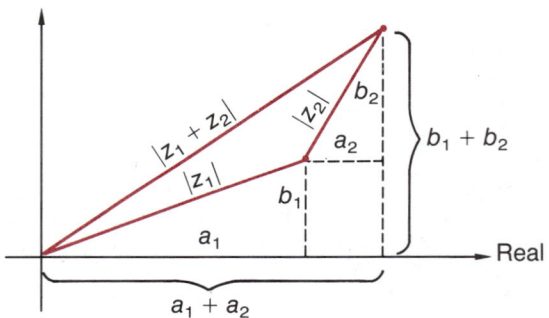

The length of any side of a triangle is less than or equal to the sum of the lengths of the other two sides. (The shortest distance between two points is a straight line.) Thus

$$|z_1 + z_2| \le |z_1| + |z_2| \tag{4}$$

for every two complex numbers z_1 and z_2.

Polar Representation of a Complex Number

Until now, we have located points in the plane by referring to their rectangular or Cartesian coordinates (x, y). On the other hand, it is sometimes advantageous to describe a complex number in terms of its distance from the origin and the angle that the ray from the origin through the point makes with the positive x-axis. This angle is called its **argument.** The argument of z is denoted **arg(z).** If $r = |z|$ and $\theta = \arg(z)$, the rectangular form $z = x + iy$ can be rewritten in terms of r and θ as illustrated in the following box.

POLAR REPRESENTATION OF A COMPLEX NUMBER

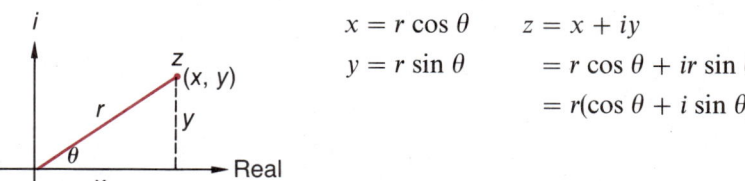

$$x = r \cos \theta \qquad z = x + iy$$
$$y = r \sin \theta \qquad\; = r \cos \theta + ir \sin \theta$$
$$\qquad\qquad\qquad = r(\cos \theta + i \sin \theta)$$

Figure 9.34

The expression $z = r(\cos \theta + i \sin \theta)$ is called the **polar form** or **polar representation** of the complex number z. As we observed, r must be the absolute value or modulus of z. The angle θ in the polar form of a complex number is not uniquely determined since $\theta \pm 2k\pi$ will also do (k any integer). The expression $(\cos \theta + i \sin \theta)$ is sometimes abbreviated as cis θ.

NOTATIONAL SHORTHAND

$$\cos \theta + i \sin \theta = \text{cis } \theta$$

EXAMPLE 2 Write each of the given complex numbers in the form $a + bi$ and then plot them in the complex plane.

a. $2 \text{ cis } \dfrac{\pi}{2}$

b. $4 \text{ cis}\left(\dfrac{-\pi}{4}\right)$

c. $2 \text{ cis}\left(\dfrac{2\pi}{3}\right)$

d. $4 \text{ cis}\left(\dfrac{-5\pi}{6}\right)$

SOLUTION

a. $2 \text{ cis } \dfrac{\pi}{2} = 2\left(\cos \dfrac{\pi}{2} + i \sin \dfrac{\pi}{2}\right)$
$= 2(0 + i)$
$= 0 + 2i$

b. $4 \text{ cis}\left(\dfrac{-\pi}{4}\right) = 4\left(\cos \dfrac{-\pi}{4} + i \sin \dfrac{-\pi}{4}\right)$
$= 4\left(\cos \dfrac{\pi}{4} - i \sin \dfrac{\pi}{4}\right)$
$= 4\left(\dfrac{\sqrt{2}}{2} - i\dfrac{\sqrt{2}}{2}\right)$
$= 2\sqrt{2} - 2\sqrt{2}i$

c. $2 \operatorname{cis} \dfrac{2\pi}{3} = 2\left(\cos \dfrac{2\pi}{3} + i \sin \dfrac{2\pi}{3}\right)$

$\qquad = 2\left(-\dfrac{1}{2} + i \dfrac{\sqrt{3}}{2}\right)$

$\qquad = -1 + \sqrt{3}\, i$

d. $4 \operatorname{cis}\left(-\dfrac{5\pi}{6}\right) = 4\left(\cos \dfrac{-5\pi}{6} + i \sin \dfrac{-5\pi}{6}\right)$

$\qquad = 4\left(\cos \dfrac{5\pi}{6} - i \sin \dfrac{5\pi}{6}\right)$

$\qquad = 4\left(\dfrac{-\sqrt{3}}{2} - i \dfrac{1}{2}\right)$

$\qquad = -2\sqrt{3} - 2i$

The points are plotted in Figure 9.35.

Figure 9.35

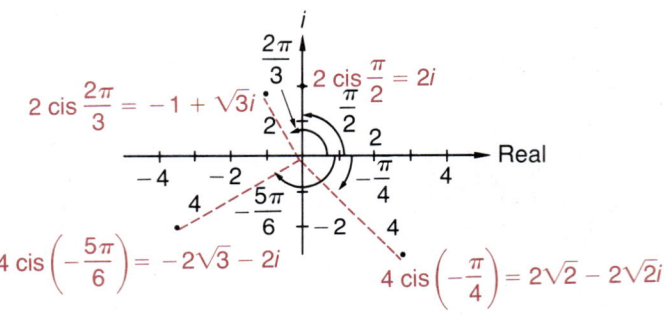

To determine the argument θ of a complex number $z = x + iy$, observe in Figure 9.34 and 9.36 that $\tan \theta = y/x$. But

$$\theta = \operatorname{Tan}^{-1}\left(\dfrac{y}{x}\right) \qquad \text{only when} \qquad x > 0$$

since $-\pi/2 < \operatorname{Tan}^{-1} t < \pi/2$ for all t. If $x < 0$, then $-z = -x - iy$ and the argument for $-z$ is $\operatorname{Tan}^{-1}(-y/-x) = \operatorname{Tan}^{-1}(y/x)$ since then $-x > 0$. The arguments for z and $-z$ differ by π since they are reflections of one another through the origin. That is,

$$\theta = \operatorname{Tan}^{-1}\left(\dfrac{y}{x}\right) + \pi \qquad \text{for } x < 0$$

(See Figure 9.36.)

$\tan \theta = \dfrac{y}{x}$

$\theta = \arg(z)$

$\quad = \pi + \arg(-z)$

$\quad = \pi + \operatorname{Tan}^{-1}\left(\dfrac{y}{x}\right) \qquad$ when $x < 0$

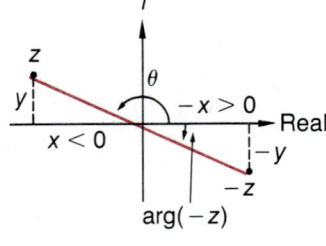

Figure 9.36

Section 9.4 The Complex Plane and Polar Representation 529

If $z = x + iy$, then
$$z = r(\cos\theta + i\sin\theta)$$
where
$$r = \sqrt{x^2 + y^2}$$
and
$$\theta = \begin{cases} \text{Tan}^{-1}\left(\dfrac{y}{x}\right) & \text{when } x > 0 \\ \pi + \text{Tan}^{-1}\left(\dfrac{y}{x}\right) & \text{when } x < 0 \\ \dfrac{\pi}{2} & \text{when } x = 0 \text{ and } y > 0 \\ \dfrac{-\pi}{2} & \text{when } x = 0 \text{ and } y < 0 \\ \text{any number} & \text{when } x = y = 0 \end{cases}$$

When $x = y = 0$, then
$$z = 0 = 0(\cos\theta + i\sin\theta)$$
for *any* θ as indicated in the preceding box.

EXAMPLE 3   Plot each of the given points in the complex plane and write each in polar form.
   a. $2 + 2i$
   b. $-3 + 3\sqrt{3}i$
   c. $-2\sqrt{3} - 2i$
   d. $2 - 2i$

SOLUTION   Each of the points is plotted in Figure 9.37. The respective arguments can be found as in the preceding discussion; these arguments are indicated in Figure 9.37 and are used to find the polar form of each number.

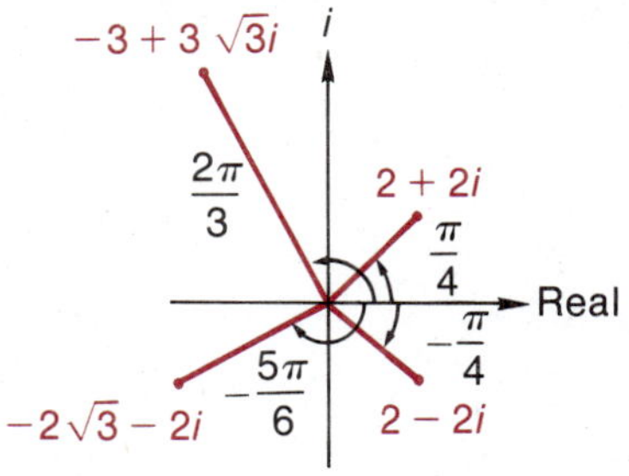

a. $2 + 2i = \sqrt{4 + 4}\left(\cos\dfrac{\pi}{4} + i\sin\dfrac{\pi}{4}\right)$

$\qquad\quad\; = 2\sqrt{2}\,\text{cis}\left(\dfrac{\pi}{4}\right)$

b. $-3 + 3\sqrt{3}i = \sqrt{9 + 27}\left(\cos\dfrac{2\pi}{3} + i\sin\dfrac{2\pi}{3}\right)$

$\qquad\qquad\quad\; = 6\,\text{cis}\left(\dfrac{2\pi}{3}\right)$

c. $-2\sqrt{3} - 2i = \sqrt{12 + 4}\left(\cos\dfrac{7\pi}{6} + i\sin\dfrac{7\pi}{6}\right)$ $\left(\text{We could also use } \dfrac{-5\pi}{6}\right)$

$\qquad\qquad\quad\; = 4\,\text{cis}\left(\dfrac{7\pi}{6}\right)$

Figure 9.37

d. $2 - 2i = \sqrt{4+4}\left(\cos\dfrac{7\pi}{4} + i\sin\dfrac{7\pi}{4}\right)$ $\left(\text{We could also use } \dfrac{-\pi}{4}\right)$

$\phantom{\textbf{d.} \ 2 - 2i \ } = 2\sqrt{2}\operatorname{cis}\left(\dfrac{7\pi}{4}\right)$ ∎

The following theorem makes "raising to a power" and "taking roots" easy tasks when a number is written in polar form.

THEOREM 9.1 If two complex numbers are written in polar form as

$$z_1 = r_1 \operatorname{cis}\theta_1$$
$$z_2 = r_2 \operatorname{cis}\theta_2$$

then

$$z_1 z_2 = r_1 r_2 \operatorname{cis}(\theta_1 + \theta_2)$$

and if $z_2 \neq 0$, then

$$\dfrac{z_1}{z_2} = \dfrac{r_1}{r_2}\operatorname{cis}(\theta_1 - \theta_2)$$

PROOF

$$\begin{aligned}z_1 z_2 &= r_1(\cos\theta_1 + i\sin\theta_1)r_2(\cos\theta_2 + i\sin\theta_2)\\ &= r_1 r_2[(\cos\theta_1\cos\theta_2 - \sin\theta_1\sin\theta_2) + i(\sin\theta_1\cos\theta_2 + \cos\theta_1\sin\theta_2)]\\ &= r_1 r_2[\cos(\theta_1 + \theta_2) + i\sin(\theta_1 + \theta_2)] \quad \text{(By formulas 1b, 2b Section 8.3)}\end{aligned}$$

To verify the quotient rule, observe that when $z_2 \neq 0$, z_1/z_2 is defined. Let

$$w = \dfrac{z_1}{z_2}$$

Then

$$|w| = \left|\dfrac{z_1}{z_2}\right| = \dfrac{|z_1|}{|z_2|} = \dfrac{r_1}{r_2}$$

It remains to find a polar angle α for w. Write w in polar form as

$$w = \dfrac{r_1}{r_2}\operatorname{cis}\alpha$$

It follows from the product rule applied to

$$z_1 = wz_2$$

that

$$r_1 \operatorname{cis}\theta_1 = \dfrac{r_1}{r_2} r_2 \operatorname{cis}(\alpha + \theta_2)$$

Now, $\alpha = \theta_1 - \theta_2$ satisfies this relationship. That is, $\theta_1 - \theta_2$ is *a* polar angle for w and

$$\dfrac{z_1}{z_2} = w = \dfrac{r_1}{r_2}\operatorname{cis}(\theta_1 - \theta_2)$$

Of course, other polar angles could also be used. ∎

Geometrically, Theorem 9.1 says that *two complex numbers can be multiplied by multiplying their moduli and adding their arguments; division can be accomplished by dividing their moduli and subtracting their arguments:*

POLAR FORM OF MULTIPLICATION AND DIVISION

$$(r_1 \text{ cis } \theta_1)(r_2 \text{ cis } \theta_2) = r_1 r_2 \text{ cis}(\theta_1 + \theta_2)$$

$$\frac{r_1 \text{ cis } \theta_1}{r_2 \text{ cis } \theta_2} = \frac{r_1}{r_2} \text{ cis}(\theta_1 - \theta_2), \qquad r_2 \neq 0$$

See Figure 9.38.

Figure 9.38

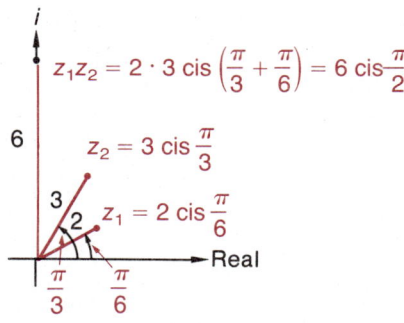

EXAMPLE 4

Use the polar form and Theorem 9.1 to find $z_1 z_2$ and z_1/z_2 for the following pairs of numbers.

a. $z_1 = -4 - 4\sqrt{3}i, \quad z_2 = 2\sqrt{3} + 2i$

b. $z_1 = 9\sqrt{3} + 9i, \quad z_2 = -3 + 3i$

SOLUTION

a. The points are plotted in Figure 9.39 in order to find their moduli and arguments. Thus we see that

$$z_1 = \sqrt{16 + 48}\left(\cos\frac{4\pi}{3} + i\sin\frac{4\pi}{3}\right)$$

$$= 8 \text{ cis } \frac{4\pi}{3}$$

$$z_2 = \sqrt{12 + 4}\left(\cos\frac{\pi}{6} + i\sin\frac{\pi}{6}\right)$$

$$= 4 \text{ cis } \frac{\pi}{6}$$

Figure 9.39

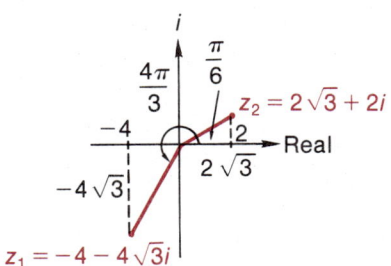

Then

$$z_1 z_2 = \left(8 \operatorname{cis} \frac{4\pi}{3}\right) \cdot \left(4 \operatorname{cis} \frac{\pi}{6}\right)$$

$$= 8 \cdot 4 \operatorname{cis}\left(\frac{4\pi}{3} + \frac{\pi}{6}\right) = 32 \operatorname{cis}\left(\frac{3\pi}{2}\right) \quad \text{(Theorem 9.1)}$$

$$= 32\left(\cos \frac{3\pi}{2} + i \sin \frac{3\pi}{2}\right) = 32(0 - i)$$

$$= -32i$$

and

$$\frac{z_1}{z_2} = \frac{8 \operatorname{cis} 4\pi/3}{4 \operatorname{cis} \pi/6}$$

$$= \frac{8}{4} \operatorname{cis}\left(\frac{4\pi}{3} - \frac{\pi}{6}\right) = 2 \operatorname{cis}\left(\frac{7\pi}{6}\right) \quad \text{(Theorem 9.1)}$$

$$= 2\left(\cos \frac{7\pi}{6} + i \sin \frac{7\pi}{6}\right)$$

$$= 2\left(\frac{-\sqrt{3}}{2} - i\frac{1}{2}\right)$$

$$= -\sqrt{3} - i$$

These results can be checked by using the earlier methods of computation (Section 1.9):

$$(-4 - 4\sqrt{3}i)(2\sqrt{3} + 2i) = (-8\sqrt{3} + 8\sqrt{3}) + (-24 - 8)i$$
$$= -32i$$

$$\frac{-4 - 4\sqrt{3}i}{2\sqrt{3} + 2i} = \frac{-4 - 4\sqrt{3}i}{2\sqrt{3} + 2i} \cdot \frac{2\sqrt{3} - 2i}{2\sqrt{3} - 2i}$$

$$= \frac{-8\sqrt{3} - 8\sqrt{3} + i(-24 + 8)}{12 + 4}$$

$$= \frac{-16\sqrt{3} - 16i}{16}$$

$$= -\sqrt{3} - i$$

b. Plotting the points as indicated in Figure 9.40, we see that

$$z_1 = \sqrt{9^2 \cdot 3 + 9^2}\left(\cos \frac{\pi}{6} + i \sin \frac{\pi}{6}\right)$$

$$= 18 \operatorname{cis}\left(\frac{\pi}{6}\right)$$

Figure 9.40

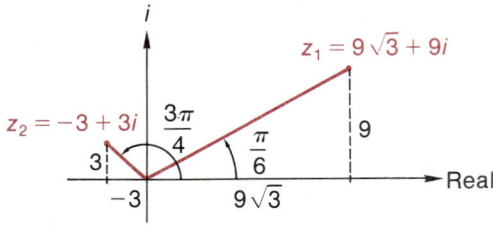

$$z_2 = \sqrt{9+9}\left(\cos\frac{3\pi}{4} + i\sin\frac{3\pi}{4}\right)$$

$$= 3\sqrt{2}\operatorname{cis}\left(\frac{3\pi}{4}\right)$$

Then

$$z_1 z_2 = 18 \cdot 3\sqrt{2}\operatorname{cis}\left(\frac{\pi}{6} + \frac{3\pi}{4}\right) \qquad \text{(Theorem 9.1)}$$

$$= 54\sqrt{2}\operatorname{cis}\left(\frac{11\pi}{12}\right)$$

$$= 54\sqrt{2}\cos\frac{11\pi}{12} + i\,54\sqrt{2}\sin\frac{11\pi}{12}$$

and

$$\frac{z_1}{z_2} = \frac{18}{3\sqrt{2}}\operatorname{cis}\left(\frac{\pi}{6} - \frac{3\pi}{4}\right) \qquad \text{(Theorem 9.1)}$$

$$= 3\sqrt{2}\operatorname{cis}\left(-\frac{7\pi}{12}\right)$$

$$= 3\sqrt{2}\cos\frac{7\pi}{12} - i\,3\sqrt{2}\sin\frac{7\pi}{12} \qquad \blacksquare$$

Section 9.4 Exercises

Plot the following points in the complex plane and write a polar form for each.

1. $6 + 6\sqrt{3}i$
2. $-3 - 3\sqrt{3}i$
3. $4 + 4i$
4. -4
5. $4\sqrt{3} + 4i$
6. $\sqrt{2} - \sqrt{2}i$
7. -3
8. $5i$
9. $-2 - 2i$
10. $-\sqrt{3} - i$
11. $-5\sqrt{3} + 5i$
12. 6
13. $2i$
14. $4\sqrt{3} + 4i$
15. 4
16. $-2\sqrt{2} + 2\sqrt{2}i$
17. $2 - 2\sqrt{3}i$
18. $-3i$
19. $-4i$
20. $-2 + 2\sqrt{3}i$

Plot the following points in the complex plane and write in the form a + bi.

21. $4\operatorname{cis}\dfrac{7\pi}{6}$
22. $4\operatorname{cis}\left(\dfrac{-\pi}{2}\right)$
23. $2\operatorname{cis}\left(\dfrac{3\pi}{2}\right)$
24. $2\operatorname{cis}\left(-\dfrac{7\pi}{6}\right)$
25. $2\operatorname{cis}\left(\dfrac{3\pi}{4}\right)$
26. $\sqrt{2}\operatorname{cis}\left(\dfrac{-5\pi}{4}\right)$
27. $10\operatorname{cis} 0$
28. $5\operatorname{cis}\pi$
29. $2\sqrt{2}\operatorname{cis}\left(\dfrac{-3\pi}{4}\right)$
30. $6\operatorname{cis}\left(\dfrac{-2\pi}{3}\right)$
31. $5\operatorname{cis}\dfrac{5\pi}{3}$
32. $2\operatorname{cis}\left(\dfrac{5\pi}{2}\right)$
33. $2\operatorname{cis}\left(\dfrac{-11\pi}{6}\right)$
34. $4\operatorname{cis}\dfrac{\pi}{6}$
35. $3\operatorname{cis}\dfrac{\pi}{2}$
36. $2\sqrt{2}\operatorname{cis}\dfrac{7\pi}{4}$
37. $2\operatorname{cis}\left(\dfrac{-\pi}{3}\right)$
38. $4\operatorname{cis}\dfrac{7\pi}{3}$
39. $16\operatorname{cis} 180°$
40. $3\operatorname{cis} 90°$

For each of the following pairs of complex numbers, find $|z_1|$, $|z_2|$, $|z_1 z_2|$, $|z_1/z_2|$, and $|z_1 + z_2|$.

41. $z_1 = 1 + i,\ z_2 = 1 - i$
42. $z_1 = 3 + i,\ z_2 = 1 - 3i$

43. $z_1 = -1 + i$, $z_2 = 2 + i$

44. $z_1 = 1 + 2i$, $z_2 = 3i$

45. $z_1 = 2 + i$, $z_2 = 3 + 4i$

46. $z_1 = -3 - 2i$, $z_2 = -8 - 4i$

47. $z_1 = 3 - 3i$, $z_2 = 2 - i$

48. $z_1 = 1 + i$, $z_2 = 2 - i$

49. $z_1 = 3 - 5i$, $z_2 = 2 - 2i$

50. $z_1 = 4 - 4i$, $z_2 = -3 - i$

Use the polar form to evaluate $z_1 z_2$, z_1/z_2 and z_2/z_1 for the following pairs of numbers.

51. $z_1 = 1 + i$, $z_2 = 1 - i$

52. $z_1 = \sqrt{3} + i$, $z_2 = 1 + \sqrt{3}i$

53. $z_1 = -1 + i$, $z_2 = \sqrt{3} + i$

54. $z_1 = \sqrt{2} + \sqrt{2}i$, $z_2 = 3i$

55. $z_1 = 2 + 2\sqrt{3}i$, $z_2 = 4\sqrt{3} + 4i$

56. $z_1 = -2\sqrt{3} - 2i$, $z_2 = -8\sqrt{3} - 8i$

57. $z_1 = 3 - 3i$, $z_2 = \sqrt{3} - i$

58. $z_1 = 1 + i$, $z_2 = \sqrt{3} - i$

59. $z_1 = 2\sqrt{3} - 2i$, $z_2 = 2 - 2i$

60. $z_1 = 4 - 4i$, $z_2 = -\sqrt{3} - i$

61. Show that $\dfrac{1}{z} = \dfrac{\bar{z}}{|z|^2}$.

62. Show that $|z_1 - z_2|$ is the distance between z_1 and z_2 in the complex plane.

Section 9.5

Powers and Roots of Complex Numbers

In this section we discuss the procedure for finding the roots of a complex number. Since an **nth root** of z is a complex number w for which $w^n = z$, this procedure depends on the method of "raising to a power." The polar form of a complex number is useful here. The fundamental result is known as DeMoivre's theorem, which is a consequence of Theorem 9.1.

DEMOIVRE'S THEOREM If a complex number is written in polar form as

$$z = r \text{ cis } \theta$$

then

$$z^n = r^n \text{ cis } n\theta$$

for every integer $n \geq 0$. If $z \neq 0$, the result also holds for negative integers n.

PROOF

As with real numbers, we define $z^0 = 1$ for every complex number z. The relationship holds for $n = 0$ since

$$r^0(\cos 0\theta + i \sin 0\theta) = 1(1 + 0i)$$
$$= 1$$
$$= z^0$$

For $n = 1$, $z^1 = r^1 \text{cis}(1 \cdot \theta) = r \text{ cis } \theta$ is just the polar form of z. For $n = 2$, Theorem 9.1 indicates that

$$z^2 = (r \text{ cis } \theta)(r \text{ cis } \theta)$$
$$= r^2 \text{ cis}(\theta + \theta)$$
$$= r^2 \text{ cis } 2\theta$$

For $n = 3$,

$$\begin{aligned}
z^3 &= z^2 \cdot z \\
&= (r^2 \text{ cis } 2\theta)(r \text{ cis } \theta) \\
&= r^3 \text{ cis}(2\theta + \theta) \quad &\text{(Theorem 9.1)} \\
&= r^3 \text{ cis } 3\theta
\end{aligned}$$

Continuing in this fashion, we can verify the relationship for every positive integer n.

When $z \neq 0$ and $n < 0$, we write $n = -k$ and

$$\begin{aligned}
z^n = z^{-k} &= \frac{1}{z^k} = \frac{1}{(r \text{ cis } \theta)^k} \\
&= \frac{1 \cdot \text{cis } 0}{r^k \text{ cis } k\theta} \quad &\text{(cis } 0 = 1) \\
&= \frac{1}{r^k} \text{cis}(0 - k\theta) \quad &\text{(Theorem 9.1)} \\
&= r^{-k} \text{cis}(-k\theta) \\
&= r^n \text{ cis } n\theta
\end{aligned}$$
∎

DeMoivre's theorem says that *to raise a complex number to an integral power, raise its modulus to that power and multiply its argument by that power:*

$$\boxed{(r \text{ cis } \theta)^n = r^n \text{ cis } n\theta}$$

(See Figure 9.41.)

Figure 9.41

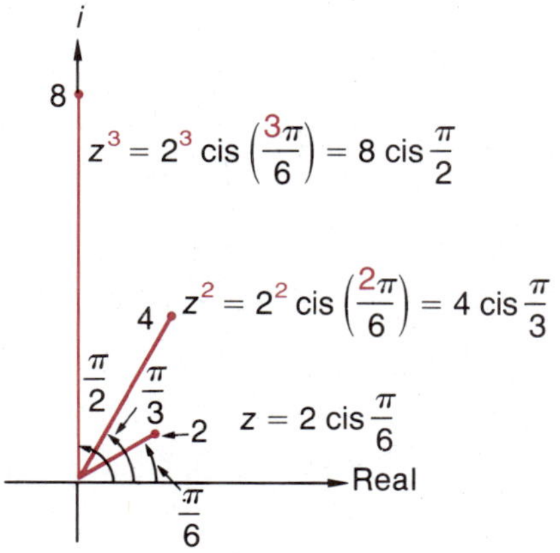

$z^3 = 2^3 \text{ cis}\left(\frac{3\pi}{6}\right) = 8 \text{ cis}\frac{\pi}{2}$

$z^2 = 2^2 \text{ cis}\left(\frac{2\pi}{6}\right) = 4 \text{ cis}\frac{\pi}{3}$

$z = 2 \text{ cis}\frac{\pi}{6}$

EXAMPLE 1

Use DeMoivre's theorem to find

a. $(1 + \sqrt{3}i)^6$
b. $1/(1 + \sqrt{3}i)^3$

SOLUTION

The polar form of $1 + \sqrt{3}i$ is exhibited in Figure 9.42. Thus, $1 + \sqrt{3}i = 2 \text{ cis}\frac{\pi}{3}$.

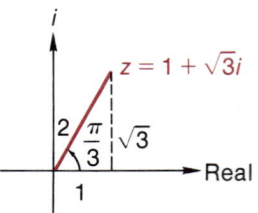

Figure 9.42

a. DeMoivre's theorem then indicates that

$$(1 + \sqrt{3}i)^6 = \left(2 \text{ cis } \frac{\pi}{3}\right)^6$$
$$= 2^6 \left[\text{cis } 6\left(\frac{\pi}{3}\right)\right]$$
$$= 2^6 \text{ cis } 2\pi$$
$$= 2^6(\cos 2\pi + i \sin 2\pi)$$
$$= 2^6(1 + 0i)$$
$$= 64$$

b. $\dfrac{1}{(1 + \sqrt{3}i)^3} = (1 + \sqrt{3}i)^{-3} = \left(2 \text{ cis } \dfrac{\pi}{3}\right)^{-3}$

$$= 2^{-3} \text{ cis}\left(-\frac{3\pi}{3}\right) = 2^{-3} \text{ cis}(-\pi) \quad \text{(DeMoivre's theorem)}$$
$$= 2^{-3}[\cos(-\pi) + i \sin(-\pi)]$$
$$= 2^{-3}(-1)$$
$$= -\frac{1}{8}$$

See Figure 9.43.

Figure 9.43

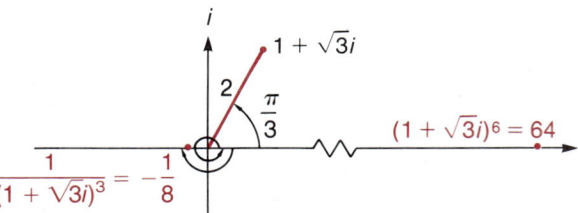

nth Roots

If n is a positive integer, the nth roots of a complex number z are those complex numbers w for which $w^n = z$. Writing z and w in polar form

$$z = r \text{ cis } \theta \qquad w = |w| \text{ cis } \phi$$

we use DeMoivre's theorem to write

$$w^n = z$$

as

$$|w|^n \text{ cis } n\phi = r \text{ cis } \theta$$

These two polar forms must then represent the same point in the complex plane; that is,

$$w^n = z \quad \text{if and only if} \quad \begin{cases} |w|^n = r \\ \text{and} \\ n\phi = \theta + 2k\pi \end{cases} \quad \text{for some integer } k$$

Equivalently, $w = |w| \text{ cis } \phi$ is an nth root of $z = r \text{ cis } \theta$ if and only if

$$\begin{cases} |w| = \sqrt[n]{r} \\ \text{and} \\ \phi = \dfrac{\theta}{n} + \dfrac{2k\pi}{n} \end{cases} \quad \text{for some integer } k$$

There are evidently many *n*th roots all located on the circle of radius $\sqrt[n]{r}$ centered at the origin. Labeling the roots as

$$w_k = \sqrt[n]{r}\, \text{cis}\left(\frac{\theta}{n} + \frac{2k\pi}{n}\right)$$

we see that the roots $w_0, w_1, w_2, \ldots, w_{n-1}$ are distinct but that all other w_k's coincide with one of these. For instance, when $k = n$

$$\text{cis}\left(\frac{\theta}{n} + \frac{2n\pi}{n}\right) = \text{cis}\left(\frac{\theta}{n} + 2\pi\right) = \text{cis}\left(\frac{\theta}{n}\right)$$

Then, $w_n = w_0$ since sine and cosine both have period 2π. Similarly $w_{n+1} = w_1$, $w_{-1} = w_{n-1}$, and so forth (see Figure 9.44).

Figure 9.44

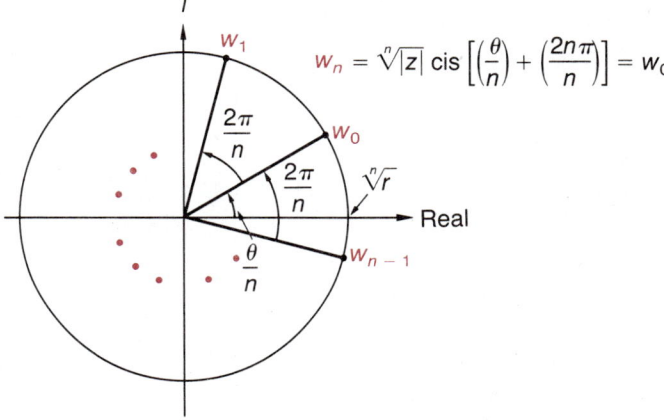

ROOTS OF A COMPLEX NUMBER

If *n* is a positive integer and *z* is a nonzero complex number with polar form $z = r\,\text{cis}\,\theta$, there are exactly *n* *n*th roots of *z*. These are given by

$$w_k = \sqrt[n]{r}\, \text{cis}\left(\frac{\theta}{n} + \frac{2k\pi}{n}\right) \qquad \text{for } k = 0, 1, \ldots, n-1$$

w_0 will be called the **principle *n*th root** of *z*.

FINDING THE *n*th ROOTS OF A COMPLEX NUMBER

To find the *n* *n*th roots of a complex number $z = r\,\text{cis}\,\theta$:

1. **Find the *n*th root of its modulus r.**
2. **Divide its argument θ by n.**
3. **The principal *n*th root is $\sqrt[n]{r}\,\text{cis}\left(\dfrac{\theta}{n}\right)$.**
4. **The others are distributed evenly on the circle of radius $\sqrt[n]{r}$ centered at the origin $\left(\text{at angular intervals of } \dfrac{2\pi}{n}\right)$.**

A circle centered at the origin in the complex plane will be called a **polar circle**.

EXAMPLE 2

Use the polar form to find the three cube roots of -8.

SOLUTION

Plotting -8 in the complex plane as in Figure 9.45, we see that its polar form is

$$-8 = 8 \text{ cis } \pi$$

Figure 9.45
THE CUBE ROOTS OF -8.

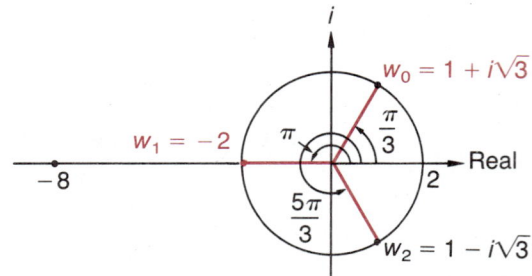

Its cube roots are $w_k = \sqrt[3]{8} \text{ cis}\left(\dfrac{\pi}{3} + \dfrac{2k\pi}{3}\right)$ for $k = 0, 1, 2$. These are

$$w_0 = 2 \text{ cis } \dfrac{\pi}{3}$$
$$= 2\left[\cos \dfrac{\pi}{3} + i \sin \dfrac{\pi}{3}\right]$$
$$= 2\left[\dfrac{1}{2} + i\dfrac{\sqrt{3}}{2}\right]$$
$$= 1 + i\sqrt{3}$$

$$w_1 = 2 \text{ cis}\left(\dfrac{\pi}{3} + \dfrac{2\pi}{3}\right)$$
$$= 2 \text{ cis } \pi$$
$$= 2[\cos \pi + i \sin \pi]$$
$$= 2(-1 + 0i)$$
$$= -2$$

$$w_2 = 2 \text{ cis}\left(\dfrac{\pi}{3} + \dfrac{4\pi}{3}\right)$$
$$= 2 \text{ cis } \dfrac{5\pi}{3}$$
$$= 2\left[\cos \dfrac{5\pi}{3} + i \sin \dfrac{5\pi}{3}\right]$$
$$= 2\left[\dfrac{1}{2} - i\dfrac{\sqrt{3}}{2}\right]$$
$$= 1 - i\sqrt{3}$$

See Figure 9.45.

EXAMPLE 3

Find the 6 sixth roots of 64.

SOLUTION

We know that $2^6 = 64$. Thus 2 is *one* sixth root of 64. The 6 distinct sixth roots are then equally spaced (separated by $\pi/3$ radians or 60°) on the polar circle of radius 2. They are illustrated and labeled in Figure 9.46.

Figure 9.46
THE SIXTH ROOTS OF 64.

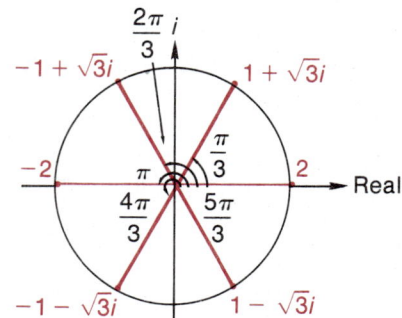

EXAMPLE 4

Find the cube roots of $1 + i$ in polar form.

SOLUTION

The polar form of $1 + i$ is $\sqrt{2}$ cis $\pi/4$. Its principal cube root is $\sqrt[6]{2}$ cis $\pi/12$ while the other two are equally spaced (or $2\pi/3$ radians apart) on the polar circle of radius $\sqrt[6]{2}$. The respective roots are

$$w_0 = \sqrt[6]{2}(\cos \pi/12 + i \sin \pi/12)$$

$$w_1 = \sqrt[6]{2}\left(\cos \frac{3\pi}{4} + i \sin \frac{3\pi}{4}\right) \qquad \left(\frac{\pi}{12} + \frac{2\pi}{3} = \frac{3\pi}{4}\right)$$

$$= \sqrt[6]{2}\left(-\frac{\sqrt{2}}{2} + i\frac{\sqrt{2}}{2}\right)$$

$$w_2 = \sqrt[6]{2}\left(\cos \frac{17\pi}{12} + i \sin \frac{17\pi}{12}\right) \qquad \left(\frac{\pi}{12} + \frac{4\pi}{3} = \frac{17\pi}{12}\right)$$

(See Figure 9.47). ∎

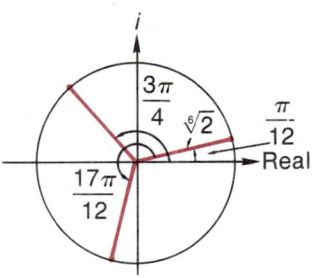

Figure 9.47
THE CUBE ROOTS OF $1 + i$.

For any negative real number $-r$, write $-r = r$ cis π. The square roots of $-r$ are then given by

$$\sqrt{r} \operatorname{cis}\left(\frac{\pi}{2} + \frac{2k\pi}{2}\right) \qquad k = 0, 1$$

They are

$$w_0 = \sqrt{r} \operatorname{cis} \frac{\pi}{2} \qquad \text{and} \qquad w_1 = \sqrt{r} \operatorname{cis} \frac{3\pi}{2}$$
$$= i\sqrt{r} \qquad\qquad\qquad\qquad = -i\sqrt{r}$$

Again, every negative real number $-r$ has exactly two square roots given by $i\sqrt{r}$ and $-i\sqrt{r}$ as discussed in Section 1.9. Most authors avoid the use of the symbol $\sqrt[n]{z}$ since there are n nth roots of a complex number z. When it is used, however, it is generally taken to mean the *principal nth root of z*. For example, $\sqrt{-1} = i$, $\sqrt{-4} = 2i$, and $\sqrt{-9} = 3i$, in agreement with the use of this notation ($\sqrt{}$) in Chapters 1 and 2. A problem remains, however, in using the symbol $\sqrt{z}$ since the rule $\sqrt{ab} = \sqrt{a}\sqrt{b}$ for positive real numbers does not necessarily hold for complex numbers a and b (see Section 1.9).

Historical Perspective

Abraham DeMoivre (1667–1754) is well known for his works in probability, statistics, and actuarial science as well as for his contributions regarding the relationship between trigonometry and complex numbers. While the formula

$$(\cos \theta + i \sin \theta)^n = \cos n\theta + i \sin n\theta$$

may or may not have originated with DeMoivre, he was certainly aware of it for n being a positive integer.

DeMoivre was a contemporary and a close friend of Issac Newton. In his later years, DeMoivre is said to have noticed that each day he required 15 minutes more sleep. When his sleep requirements reached 24 hours per day, he died.

Section 9.5 Exercises

Use DeMoivre's theorem to evaluate the following.

1. $(2 + 2i)^8$
2. $(\sqrt{3} + i)^7$
3. $(1 - i)^5$
4. $\dfrac{1}{(1 + i)^6}$
5. $(-1 - i)^5$
6. $\dfrac{1}{(\sqrt{3} + i)^3}$
7. $(-1 - i)^{10}$
8. $\left[\sqrt{2}\left(\cos\dfrac{2\pi}{5} - i\sin\dfrac{2\pi}{5}\right)\right]^5$
9. $\dfrac{1}{\left(\dfrac{1 + i\sqrt{3}}{2}\right)^{20}}$
10. $(\sqrt{3} - i)^3$
11. $\left[\sqrt{3}\left(\cos\dfrac{\pi}{8} + i\sin\dfrac{\pi}{8}\right)\right]^{10}$
12. $\dfrac{1}{(1 - i)^3}$
13. $\dfrac{1}{(2 + 2\sqrt{3}i)^3}$
14. $(1 - \sqrt{3}i)^{-4}$
15. $\dfrac{1}{(-i - 1)^5}$

Use DeMoivre's theorem and Theorem 9.1 (Section 9.4) to evaluate the following.

16. $\dfrac{(\sqrt{3} + i)^3}{(1 - i)^3}$
17. $(1 + i)^3(1 - i)^5$
18. $\dfrac{(1 - i)^6}{(1 + i)^4}$
19. $\left(\dfrac{1 + i\sqrt{3}}{2}\right)^5\left(\dfrac{1 + i}{\sqrt{2}}\right)^6$
20. $\left(\dfrac{1 + i}{\sqrt{2}}\right)^{50}\left(\dfrac{1 + i\sqrt{3}}{\sqrt{2}}\right)^{20}$
21. $(\sqrt{3} + i)^4(\sqrt{3} - i)^3$
22. $\dfrac{(1 + i)^3}{(1 - i)^5}$
23. $\dfrac{(\sqrt{3} + i)^3}{(1 - i\sqrt{3})^3}$
24. $\left(\dfrac{-1 - \sqrt{3}i}{2}\right)^7\left(\dfrac{-1 + \sqrt{3}i}{2}\right)^5$
25. $\dfrac{(1 + \sqrt{3}i)^4}{(\sqrt{3} - i)^3}$

Find the nth roots of the following.

26. $1, n = 6$
27. $8, n = 3$
28. $27, n = 3$
29. $16, n = 4$
30. $-16, n = 4$
31. $9, n = 2$
32. $81, n = 4$
33. $-1, n = 4$
34. $-64, n = 6$
35. $-27, n = 3$
36. $-1, n = 6$
37. $i, n = 2$
38. $i, n = 3$
39. $-i, n = 3$
40. $8i, n = 3$
41. $-4i, n = 2$
42. $\sqrt{2} + \sqrt{2}i, n = 2$
43. $2 + 2\sqrt{3}i, n = 2$
44. $1 - \sqrt{3}i, n = 2$
45. $-8 + 8\sqrt{3}i, n = 4$
46. $-2 + 2\sqrt{3}i, n = 2$
47. $-2 - 2\sqrt{3}i, n = 2$
48. $-8 - 8\sqrt{3}i, n = 4$
49. $-4\sqrt{2} - 4\sqrt{2}i, n = 3$
50. $-4\sqrt{2} + 4\sqrt{2}i, n = 3$

51. Use DeMoivre's theorem to obtain formulas for $\cos 3\theta$ and $\sin 3\theta$ in terms of $\sin \theta$ and $\cos \theta$. *Hint:* Expand $(\cos \theta + i \sin \theta)^3$.

52. Use DeMoivre's theorem to obtain formulas for $\cos 4\theta$ and $\sin 4\theta$ in terms of $\sin \theta$ and $\cos \theta$. *Hint:* Expand $(\cos \theta + i \sin \theta)^4$.

Solve the given equations and give the solution in the form $a + bi$ or r cis θ.

53. $z^4 - 16 = 0$
54. $z^4 + i = 0$
55. $z^3 + 1 = 0$
56. $z^3 - 64i = 0$
57. $z^5 - 243 = 0$
58. $z^5 + 32 = 0$
59. $z^4 - 8 + 8i\sqrt{3} = 0$
60. $z^4 + 32 - 32i\sqrt{3} = 0$

Section 9.6
Polar Coordinates

We have generally been locating points in the plane by referring to their rectangular or Cartesian coordinates (x, y). An exception to this practice was our use of a polar representation $r \operatorname{cis} \theta$ for complex numbers $z = x + iy$. Any point (x, y) can be located in a similar fashion by specifying its distance from the origin and the angle that the ray from the origin through the point makes with the positive x-axis.

In this section the origin is called the **pole** and the positive x-axis is called the **polar axis**. A ray originating at the origin O is called a **polar ray**. The angle that this ray makes with the positive x-axis is called a **polar angle** (see Figure 9.48a).

Figure 9.48
POLAR COORDINATES.

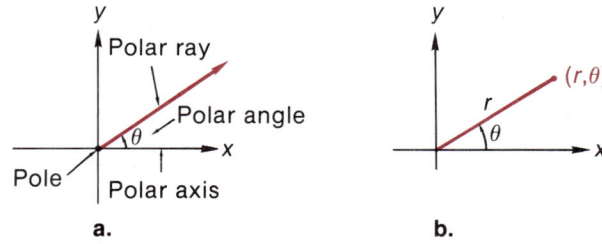

If a point P is located r units from the origin on a polar ray making an angle θ with the polar axis, then (r, θ) are called **polar coordinates** for P (see Figure 9.48b). The angle θ may be given in radians or degrees; the essential feature is that we be able to locate P with the given information r and θ. Of course, the polar coordinates of a point are not unique since (r, θ) and $(r, \theta \pm 2k\pi)$ specify the same point.

If a point has polar coordinates (r, θ) with $r < 0$, the point is located in the negative direction on the polar ray making an angle θ with the polar axis; that is, it is located on the ray opposite to the one specified by the angle θ. (*Note:* We did not allow $r < 0$ in the polar representation of a complex number.)

If there is any danger of confusion between polar and rectangular coordinates we use subscripts to distinguish between the two:

$(r, \theta)_p$ designates polar coordinates
$(x, y)_r$ designates rectangular coordinates

Polar coordinates of a number of points are indicated in Figure 9.49. Note that $(-1, \pi/4) = (1, 5\pi/4)$ and $(-2, 120°) = (2, -60°)$ in the sense that they represent the same points.

In general we have

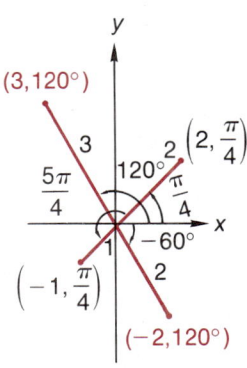

Figure 9.49

$$(-r, \theta) = \begin{cases} (r, \theta \pm \pi) & \theta \text{ in radians} \\ (r, \theta \pm 180°) & \theta \text{ in degrees} \end{cases}$$

The relationship between polar coordinates and rectangular coordinates is indicated in Figure 9.50. Note that we claim that $\theta = \operatorname{Tan}^{-1}(y/x)$ only when $x > 0$ (Figure 9.50a). This is because $\operatorname{Tan}^{-1}(t)$ is an angle between $-\pi/2$ and $\pi/2$. To describe points (x, y) with $x < 0$, we could either let $r = -\sqrt{x^2 + y^2}$ or $\theta = [\operatorname{Tan}^{-1}(y/x)] \pm \pi$ as indicated in Figure 9.50b. Points (x, y) with $x = 0$ have polar coordinates $(y, \pi/2)$ where y can be either positive or negative.

Figure 9.50

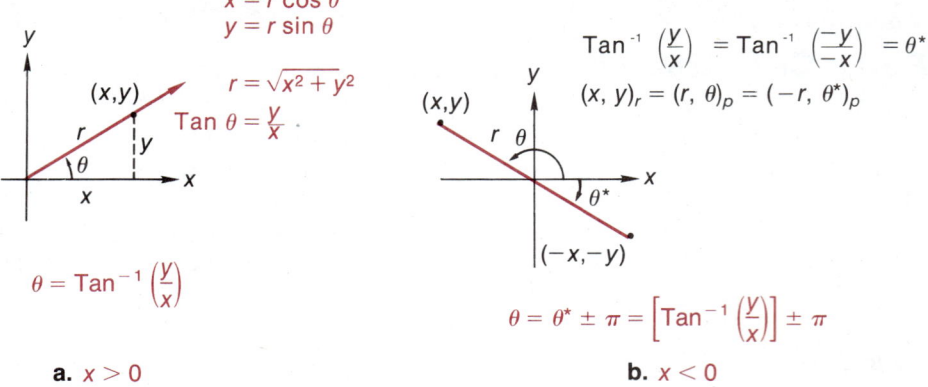

a. $x > 0$ **b.** $x < 0$

EXAMPLE 1

Locate the following polar points in the plane and find their rectangular coordinates.

a. $\left(2, \dfrac{-\pi}{3}\right)$ **b.** $\left(-2, \dfrac{\pi}{3}\right)$ **c.** $(3, 135°)$

SOLUTION

Each of the points is plotted and then the conversion formulas are used as indicated (see Figure 9.51).

Figure 9.51

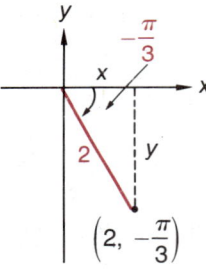

$x = 2 \cos\left(-\dfrac{\pi}{3}\right) = 2 \cdot \left(\dfrac{1}{2}\right) = 1$

$y = 2 \sin\left(-\dfrac{\pi}{3}\right)$

$\quad = 2 \cdot \left(-\dfrac{\sqrt{3}}{2}\right) = -\sqrt{3}$

a. $\left(2, -\dfrac{\pi}{3}\right)_p = (1, -\sqrt{3})_r$

$x = (-2) \cos\left(\dfrac{\pi}{3}\right)$

$\quad = (-2) \cdot \dfrac{1}{2} = -1$

$y = (-2) \sin\left(\dfrac{\pi}{3}\right)$

$\quad = (-2) \cdot \left(\dfrac{\sqrt{3}}{2}\right) = -\sqrt{3}$

b. $\left(-2, \dfrac{\pi}{3}\right)_p = (-1, -\sqrt{3})_r$

$x = 3 \cos 135°$

$\quad = 3\left(\dfrac{-\sqrt{2}}{2}\right) = \dfrac{-3\sqrt{2}}{2}$

$y = 3 \sin 135° = 3 \dfrac{\sqrt{2}}{2} = \dfrac{3\sqrt{2}}{2}$

c. $(3, 135°)_p = \left(\dfrac{-3\sqrt{2}}{2}, \dfrac{3\sqrt{2}}{2}\right)_r$

EXAMPLE 2

Locate the following points in the plane and find three sets of polar coordinates for each, one with $r < 0$.

a. $(3\sqrt{3}, 3)$ **b.** $(-6, 8)$

SOLUTION

Each of the points is plotted and the conversion formulas are used as indicated (see Figure 9.52).

Figure 9.52

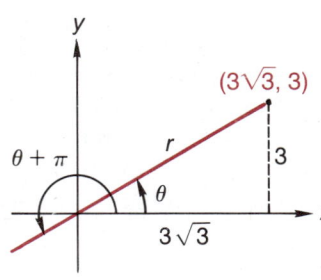

$$r = \sqrt{(3\sqrt{3})^2 + 3^2} = 6$$
$$\tan\theta = \frac{3}{(3\sqrt{3})} = \frac{1}{\sqrt{3}}$$
$$\theta = \text{Tan}^{-1}\left(\frac{1}{\sqrt{3}}\right) = \frac{\pi}{6} \text{ or } 30°$$

a. $(3\sqrt{3}, 3)_r = \left(6, \frac{\pi}{6}\right)_p$

$= \left(6, -\frac{11\pi}{6}\right)_p$

$= \left(-6, \frac{7\pi}{6}\right)_p$

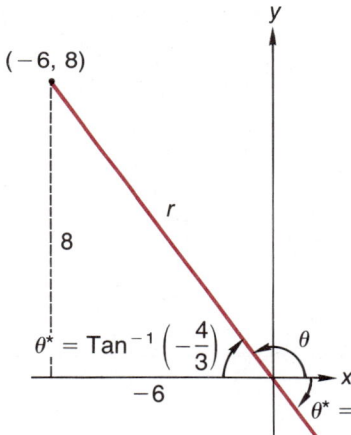

$$r = \sqrt{(-6)^2 + 8^2} = 10$$
$$\tan\theta = \frac{8}{-6} = -\frac{4}{3}$$

b. $(-6, 8)_r = \left(10, \text{Tan}^{-1}\left[-\frac{4}{3}\right] + \pi\right)_p$

$= \left(10, \text{Tan}^{-1}\left[-\frac{4}{3}\right] - \pi\right)_p$

$= \left(-10, \text{Tan}^{-1}\left[-\frac{4}{3}\right]\right)_p$

■

Polar Curve Sketching

Graphs of certain equations involving r and θ can be sketched in polar coordinates just as graphs of equations in x and y were sketched by using Cartesian coordinates. For instance, $\theta = \pi/6$ represents the line sketched in Figure 9.53(a); there is no restriction placed on r. Similarly, $r = 2$ represents the circle of radius 2 centered at the origin since there is no restriction placed on θ; see Figure 9.53(b).

Figure 9.53

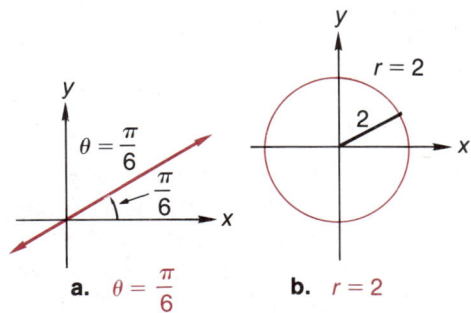

a. $\theta = \frac{\pi}{6}$ **b.** $r = 2$

EXAMPLE 3

Sketch the curve represented by $r = 1 - \sin\theta$.

SOLUTION

We begin with $\theta = 0$; then $r = 1$. As θ increases to $\pi/2$, $\sin\theta$ increases to 1 and r decreases to 0. Then as θ increases to π, $\sin\theta$ decreases to 0 and r increases to 1 again. As θ increases to $3\pi/2$, $\sin\theta$ decreases to -1 and r increases to 2. Finally, as θ increases to 2π, $\sin\theta$ increases to 0 and r decreases to 1. Then the curve repeats itself. This analysis is summarized in the following table. Several other specific points are plotted and then the curve is sketched as in Figure 9.54.

	θ	$\sin\theta$	$r = 1 - \sin\theta$
	0	0	1
①	↓	↓	↓
	$\pi/2$	1	0
②	↓	↓	↓
	π	0	1
③	↓	↓	↓
	$3\pi/2$	-1	2
④	↓	↓	↓
	2π	0	1

Figure 9.54
$r = 1 - \sin\theta$

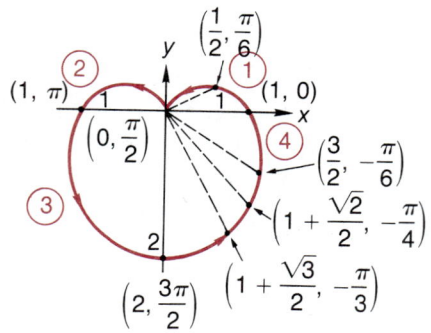

The heart-shaped curve in Figure 9.54 is called a **cardioid**. We note that it is symmetric about the y-axis. Had we noted this symmetry initially, we would only have had to sketch the curve for $-\pi/2 \leq \theta \leq \pi/2$ and then reflect that part of the curve through the y-axis. The following observations regarding symmetry may be helpful.

SYMMETRY TESTS

a.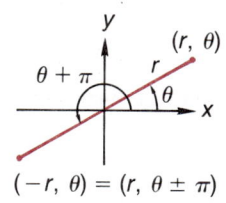

$(-r, \theta) = (r, \theta \pm \pi)$

Symmetry about the origin. r can be replaced by $-r$ and/or θ can be replaced by $\theta \pm \pi$ without changing the equation; that is, $(-r, \theta)$ and/or $(r, \theta \pm \pi)$ is on the curve whenever (r, θ) is.

b.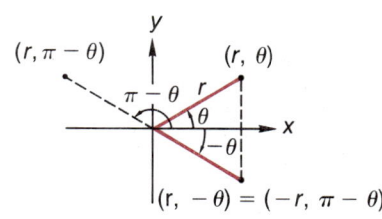

$(r, -\theta) = (-r, \pi - \theta)$

Symmetry about the x-axis. θ can be replaced by $-\theta$ and/or r, θ can be replaced by $-r, \pi - \theta$ without changing the equation; that is, $(r, -\theta)$ and/or $(-r, \pi - \theta)$ is on the curve whenever (r, θ) is.

c.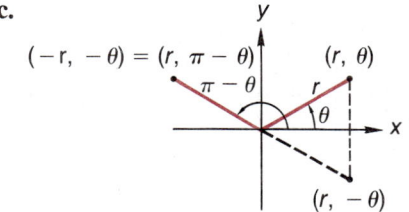

$(-r, -\theta) = (r, \pi - \theta)$

Symmetry about the y-axis. θ can be replaced by $\pi - \theta$ and/or r, θ can be replaced by $-r, -\theta$ without changing the equation; that is, $(r, \pi - \theta)$ and/or $(-r, -\theta)$ is on the curve whenever (r, θ) is.

In Example 3, we note that $1 - \sin(\pi - \theta) = 1 - \sin \theta$, indicating symmetry about the y-axis by symmetry test c.

EXAMPLE 4

Sketch the curve $r = 2 \cos 3\theta$.

SOLUTION

Since $\cos 3(-\theta) = \cos 3\theta$, this curve is symmetric about the x-axis. Initially, we need only sketch the curve for $0 \leq \theta \leq \pi$. Symmetry about the x-axis will then determine the curve for $-\pi \leq x \leq 0$. We generate the table as described in Example 3 and plot several intermediate points. The curve is sketched in Figure 9.55.

Figure 9.55
$r = 2 \cos 3\theta$

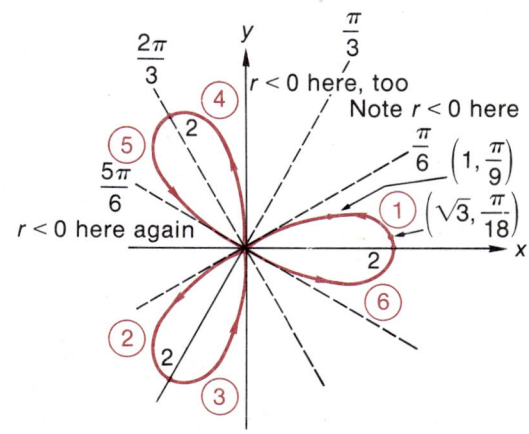

The curve must be symmetric about the x-axis. Reflecting the given curve through the x-axis simply duplicates the same curve. Even though we have only sketched the curve for $0 \leq \theta \leq \pi$, we have the complete graph. This curve is called a **three-leaved rose**.

	θ	3θ	$\cos 3\theta$	$r = 2\cos 3\theta$
	0	0	1	2
①	↓	↓	↓	↓
	$\pi/6$	$\pi/2$	0	0
②	↓	↓	↓	↓
	$\pi/3$	π	-1	-2
③	↓	↓	↓	↓
	$\pi/2$	$3\pi/2$	0	0
④	↓	↓	↓	↓
	$2\pi/3$	2π	1	2
⑤	↓	↓	↓	↓
	$5\pi/6$	$5\pi/2$	0	0
⑥	↓	↓	↓	↓
	π	3π	-1	-2

EXAMPLE 5 Sketch the curve represented by $r^2 = 9\sin 2\theta$.

SOLUTION This curve is symmetric about the origin since $(-r, \theta)$ is on the curve whenever (r, θ) is. We also observe that $\sin 2(\theta + \pi) = \sin(2\theta + 2\pi) = \sin 2\theta$. We need only plot the curve for $r > 0$ and $0 \leq \theta \leq \pi$ and then reflect this part of the curve through the origin. We generate the table and sketch the curve as before. Note that when $\sin 2\theta < 0$, there are no points on the curve since $r^2 \geq 0$.

	θ	2θ	$\sin 2\theta$	$|r| = 3\sqrt{\sin 2\theta}$
	0	0	0	0
①	↓	↓	↓	↓
	$\pi/4$	$\pi/2$	1	3
②	↓	↓	↓	↓
	$\pi/2$	π	0	0
③	↓	↓	↓	
	$3\pi/4$	$3\pi/2$	-1	There are no r values in this range since $\sin 2\theta < 0$.
④	↓	↓	↓	
	π	2π	0	0

The illustration in Figure 9.56a is only a partial graph. It can be completed by symmetrically extending it through the origin, as we observed above. The complete graph is given in Figure 9.56b. It is known as a **lemniscate**.

Figure 9.56

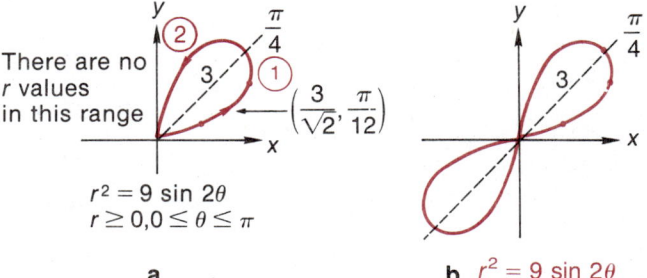

a.

b. $r^2 = 9 \sin 2\theta$

EXAMPLE 6

Sketch the curve $r = 2 \cos \theta$.

SOLUTION

Although we could proceed as in the preceding examples, another technique works well here. Multiplying both sides of this equation by r, we obtain

$$r^2 = 2r \cos \theta$$

This can be converted to a rectangular equation by setting $r^2 = x^2 + y^2$ and $x = r \cos \theta$:

$$x^2 + y^2 = 2x$$
$$x^2 - 2x + y^2 = 0$$
$$x^2 - 2x + 1 + y^2 = 0 + 1$$
$$(x - 1)^2 + y^2 = 1$$

This rectangular equation represents the circle of radius 1 centered at (1, 0) (see Figure 9.57).

Figure 9.57
$r = 2 \cos \theta$

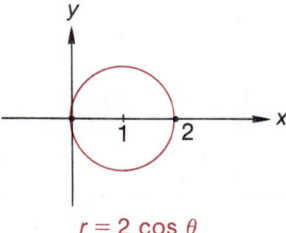

$r = 2 \cos \theta$

The technique used in Example 6 can be used to convert any polar equation of the form

$$r = A \cos \theta \quad \text{or} \quad r = B \sin \theta$$

into a rectangular equation which represents a circle passing through the origin. The center of the circle is on the x-axis (at $A/2$) for $r = A \cos \theta$ and on the y-axis (at $B/2$) for $r = B \sin \theta$.

Section 9.6 Exercises

Locate the following polar points in the plane and find their rectangular coordinates.

1. $\left(4, \dfrac{\pi}{3}\right)$

2. $\left(3, \dfrac{\pi}{2}\right)$

3. $\left(-2, \dfrac{\pi}{4}\right)$

4. $\left(6, \dfrac{7\pi}{4}\right)$

5. $(16, \pi)$

6. $\left(-4, \dfrac{11\pi}{6}\right)$

548 Chapter 9 Further Applications of Trigonometry

7. $\left(4, \dfrac{3\pi}{2}\right)$

8. $(10, 0)$

9. $\left(-6, \dfrac{2\pi}{3}\right)$

10. $\left(2, \dfrac{3\pi}{4}\right)$

11. $\left(-4, \dfrac{\pi}{2}\right)$

12. $(-1, \pi)$

13. $\left(-2, \dfrac{7\pi}{6}\right)$

14. $\left(2, -\dfrac{3\pi}{2}\right)$

15. $(4, 210°)$

16. $(-3, 270°)$

Locate the following rectangular points in the plane and give three sets of polar coordinates for each, one with r < 0.

17. $(2, 0)$

18. $(5, 5)$

19. $(-2, -2)$

20. $(-5\sqrt{3}, 5)$

21. $(-\sqrt{3}, -1)$

22. $(2, -2)$

23. $(-3, -3\sqrt{3})$

24. $(-3, 0)$

25. $(0, 3)$

26. $(-1, \sqrt{3})$

27. $(-4, 4)$

28. $(4\sqrt{3}, 4)$

29. $(0, -4)$

30. $(2, -2\sqrt{3})$

31. $(6, 6\sqrt{3})$

32. $(-4, 4\sqrt{3})$

Sketch the following curves in polar coordinates.

33. $r = 3 - 3\cos\theta$

34. $r = -2 - 2\cos\theta$

35. $r = 2 + 2\sin\theta$

36. $r = 1 + 2\sin\theta$

37. $r = 1 + 2\cos\theta$

38. $r = -2 + 4\cos\theta$

39. $r = 2 - 4\sin\theta$

40. $r = 5 - 4\sin\theta$

41. $r = 4 - \sin\theta$

42. $r = 4 - 3\cos\theta$

43. $r^2 = -25\sin 2\theta$

44. $r^2 = 4\cos 2\theta$

45. $r^2 = \cos 3\theta$

46. $r = 3\sin 2\theta$

47. $r = -2\cos 4\theta$

48. $r = -\sin 5\theta$

49. $r = 5\sin 3\theta$

50. $r = 2\sin^2 \dfrac{\theta}{2}$

51. $r = \sin \dfrac{\theta}{2}$

52. $r = 1 + \cos \dfrac{\theta}{2}$

Convert the following polar equations to rectangular form as in Example 6 and then sketch the graph.

53. $r = 4\sin\theta$

54. $r = -6\cos\theta$

55. $r = -8\sin\theta$

56. $r = 3\csc\theta$

57. $r = 4\sec\theta$

58. $r^2 \sin 2\theta = 2$ (*Hint:* $\sin 2\theta = 2\sin\theta\cos\theta$)

Section 9.7 Chapter Review

Terms and Concepts

Vector A quantity having both magnitude and direction.

Magnitude of a Vector For $\mathbf{v} = x\mathbf{i} + y\mathbf{j}$, $|\mathbf{v}| = \sqrt{x^2 + y^2}$

Modulus of a Complex Number For $z = x + iy$, $|z| = \sqrt{x^2 + y^2}$

Polar Representation of a Complex Number

$z = x + iy$
$ = r\cos\theta + ir\sin\theta$
$ = r(\cos\theta + i\sin\theta) = r\,\text{cis}\,\theta$

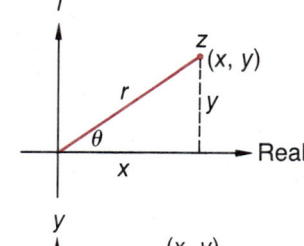

$x = r\cos\theta$
$y = r\sin\theta$

$r = \sqrt{x^2 + y^2} = |z|$
$\tan\theta = \dfrac{y}{x}$

Polar Coordinates

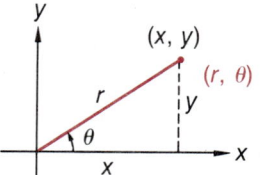

$x = r\cos\theta$
$y = r\sin\theta$

$r = \sqrt{x^2 + y^2}$
$\tan\theta = \dfrac{y}{x}$

Rules and Formulas

Law of Sines $\dfrac{\sin A}{a} = \dfrac{\sin B}{b} = \dfrac{\sin C}{c}$

Law of Cosines $a^2 = b^2 + c^2 - 2bc \cos A$

Heron's Area Formula Area $= \sqrt{s(s-a)(s-b)(s-c)}$
where $s = \tfrac{1}{2}(a+b+c)$

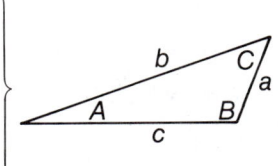

Vectors

Polar Multiplication and Division

$(r_1 \text{ cis } \theta_1)(r_2 \text{ cis } \theta_2) = r_1 r_2 \text{ cis}(\theta_1 + \theta_2)$

$\dfrac{r_1 \text{ cis } \theta_1}{r_2 \text{ cis } \theta_2} = \dfrac{r_1}{r_2} \text{ cis}(\theta_1 - \theta_2)$

De Moivre's Theorem $(r \text{ cis } \theta)^n = r^n \text{ cis } n\theta$

Roots of Complex Numbers

There are n nth roots of a complex number $z = x + iy$; these are given by

$w_k = \sqrt[n]{r} \text{ cis}\left(\dfrac{\theta}{n} + \dfrac{2k\pi}{n}\right)$ for $k = 0, 1, \ldots, n-1$

Techniques

Solving Triangles
1. Right triangles—must know two sides or one side and an acute angle.
2. Law of sines—must know two angles and any side (AAS or ASA) or two sides and the nonincluded angle (SSA); this latter case may not completely determine the triangle.
3. Law of cosines—must know two sides and the included angle (SAS) or three sides (SSS).

Graphing in Polar Coordinates

1. Symmetry about the origin. $(-r, \theta)$ and/or $(r, \theta \pm \pi)$ are on the curve whenever (r, θ) is.

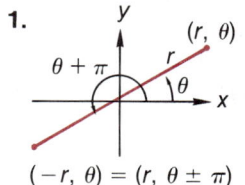

2. Symmetry about the x-axis. $(r, -\theta)$ and/or $(-r, \pi - \theta)$ are on the curve whenever (r, θ) is.

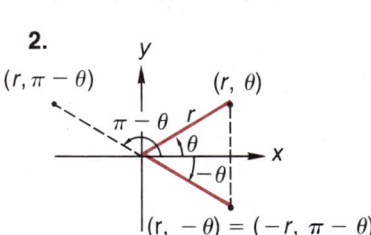

3. Symmetry about the y-axis. $(r, \pi - \theta)$ is on the curve and/or $(-r, -\theta)$ are on the curve whenever (r, θ) is.

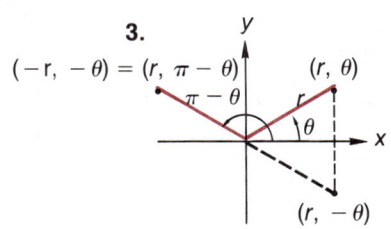

Section 9.8 Supplementary Exercises

Solve the triangle, given the following data.

1. $B = 42°0'$, $C = 63°0'$, $a = 19.0$
2. $a = 10.8$, $b = 22.1$, $A = 102°0'$
3. $A = 50°0'$, $B = 49°0'$, $a = 6.0$
4. $a = 13.0$, $b = 24.0$, $B = 137°0'$
5. $b = 36.0$, $c = 5.0$, $C = 16°0'$
6. $b = 42.3$, $c = 31.2$, $B = 74°30'$
7. $A = 46°0'$, $C = 93°0'$, $b = 6.7$
8. $a = 14.0$, $c = 10.0$, $C = 27°0'$
9. $a = 6.0$, $b = 2.5$, $C = 136°30'$
10. $a = 30.0$, $b = 50.0$, $c = 45.0$
11. $a = 5.0$, $b = 7.0$, $c = 10.0$
12. $b = 11.5$, $c = 14.0$, $A = 49°10'$
13. $a = 35.0$, $c = 42.1$, $A = 84°10'$
14. $a = 52.6$, $c = 22.0$, $A = 162°0'$
15. $B = 10°0'$, $C = 20°0'$, $b = 4.0$
16. $c = 12.7$, $a = 18.3$, $A = 61°20'$
17. $a = 42.4$, $b = 43.1$, $A = 75°0'$
18. $A = 26°20'$, $B = 83°10'$, $c = 10.0$
19. $B = 36°10'$, $A = 132°0'$, $b = 12.7$
20. $a = 7.0$, $b = 6.0$, $c = 5.0$
21. $c = 10.0$, $a = 80.0$, $B = 60°0'$
22. $b = 10.0$, $c = 7.0$, $A = 132°0'$

Use Heron's formula to find the area of the given triangles.

23. $a = 10.0$ centimeters, $b = 15.0$ centimeters, $c = 20.0$ centimeters
24. $a = 5.0$ feet, $b = 7.0$ feet, $c = 9.0$ feet
25. $a = 3.2$ miles, $b = 2.7$ miles, $c = 5.6$ miles

In each of the following, find
    a. $A + B$ b. $A - B$
    c. $A - 2B$ d. $3B - 2A$

26. $A = 2i - 3j$
    $B = i + 2j$
27. $A = 3i + 2j$
    $B = i - 2j$
28. $A = \langle 3, 4 \rangle$
    $B = \langle 2, 2 \rangle$

Find the magnitude of the given vector.

29. $-5i - 12j$ 30. $-2i + 3j$ 31. $\langle 9, -2 \rangle$

Find a unit vector in the same direction as the given vector.

32. $\langle 3, -6 \rangle$ 33. $\langle 3, -4 \rangle$ 34. $2i + 2\sqrt{3}j$

35. Find the vector from $(1, 2)$ to the midpoint of the segment joining $(-1, 0)$ and $(5, 6)$.

36. If A, B, and C are three points, express $\overrightarrow{BC}$ in terms of $\overrightarrow{AB}$ and $\overrightarrow{AC}$. What vector is $\overrightarrow{BA} - \overrightarrow{BC}$?

37. Show that the line segment joining the midpoints of the nonparallel sides of a trapezoid is parallel to the bases and half as long as the sum of the two base lengths.

38. Show that if a line divides two sides of a triangle proportionately, then it is parallel to the third side.

Plot the following points in the complex plane. Convert from rectangular to polar form or vice versa.

39. $3i$ 40. 2
41. $5 + 5i$ 42. $-1 + \sqrt{3}i$
43. $2\sqrt{3} - 2i$ 44. $3 \text{ cis } 60°$
45. $2 \text{ cis}\left(\dfrac{5\pi}{4}\right)$ 46. $\text{cis } \pi$
47. $4 \text{ cis}\left(\dfrac{\pi}{2}\right)$ 48. $\text{cis}\left(\dfrac{7\pi}{4}\right)$

In each of the following, find $|z_1|$, $|z_2|$, $|z_1 z_2|$, $|z_1/z_2|$, and $|z_1 + z_2|$.

49. $z_1 = 1$, $z_2 = 1 - i$
50. $z_1 = 6 + 3i$, $z_2 = 2 + 2i$
51. $z_1 = 3 + 9i$, $z_2 = -3 + 3i$
52. $z_1 = 1 - i$, $z_2 = 2 + i$
53. $z_1 = 2 + 2i$, $z_2 = 1 - i$

Use polar representations to evaluate $z_1 z_2$ and z_1/z_2 for the following pairs of complex numbers.

54. $z_1 = 1$, $z_2 = 1 - i$
55. $z_1 = 6 + 6\sqrt{3}i$, $z_2 = 2 + 2i$

56. $z_1 = 9\sqrt{3} + 9i$, $z_2 = -3\sqrt{3} + 3i$
57. $z_1 = 1 - i$, $z_2 = \sqrt{3} + i$
58. $z_1 = 2 + 2\sqrt{3}i$, $z_2 = 1 - i$

Use DeMoivre's theorem to evaluate the following.

59. $(-1 + i)^5$
60. $(1 - i)^{-4}$
61. $(-1 - \sqrt{3}i)^4$
62. $\dfrac{1}{(1 + i)^{10}}$
63. $\left(\dfrac{-1 + \sqrt{3}i}{2}\right)^{17}$

Use DeMoivre's theorem and Theorem 9.1 to evaluate the following.

64. $\left(\dfrac{\sqrt{3} - i}{1 + i}\right)^3$
65. $\left(\dfrac{1 - i\sqrt{3}}{2}\right)^2 \left(\dfrac{-1 + i}{2}\right)^3$
66. $\left(\dfrac{-1 + i}{\sqrt{2}}\right)^{10} \left(\dfrac{-1 + i\sqrt{3}}{2}\right)^8$
67. $\dfrac{(1 - i)^3}{(1 + i)^5}$
68. $\dfrac{(1 - \sqrt{3}i)^3}{(\sqrt{3} + i)^4}$

Find the n nth roots of the following.

69. -16, $n = 2$
70. -64, $n = 3$
71. i^3, $n = 2$
72. $4i$, $n = 2$
73. 256, $n = 8$
74. $-8i$, $n = 3$
75. $1 + \sqrt{3}i$, $n = 2$
76. $-1 + \sqrt{3}i$, $n = 2$
77. $-8 + 8\sqrt{3}i$, $n = 4$
78. $7 + 24i$, $n = 2$

Solve the given equation and give the solution in the form a + bi or r cis θ.

79. $z^3 - 27i = 0$
80. $z^5 - 32 = 0$

Locate the following polar points in the plane and find their rectangular coordinates.

81. $\left(-6, -\dfrac{3\pi}{4}\right)$
82. $\left(-6, \dfrac{4\pi}{3}\right)$
83. $\left(8, \dfrac{5\pi}{6}\right)$
84. $(5, 405°)$
85. $(4, -30°)$

Locate the following rectangular points in the plane and give three sets of polar coordinates for each, one with r < 0.

86. $(2, 2\sqrt{3})$
87. $(-4, -4)$
88. $(-3\sqrt{3}, -3)$
89. $(-2, 0)$
90. $(0, -2)$

Sketch the following curves in polar coordinates.

91. $r = 2 - 2 \sin \theta$
92. $r = 2 + 2 \cos \theta$
93. $r = 1 - 2 \sin \theta$
94. $r = 4 + 5 \cos \theta$
95. $r = 3 - 2 \cos \theta$
96. $r = 4 + 3 \sin \theta$
97. $r^2 = 4 \sin 2\theta$
98. $r^2 = -16 \cos 2\theta$
99. $r = 4 \sin 2\theta$
100. $r = \sin 3\theta$
101. $r = \cos \dfrac{\theta}{2}$
102. $r = 1 + \sin \dfrac{\theta}{2}$

Convert the following to rectangular form and then sketch the graph.

103. $r = -2 \sin \theta$
104. $r = 2 \sec \theta$
105. $r^2 \sin 2\theta = -4$

Systems of Equations and Inequalities

10

Scientists, engineers, economists, and politicans use equations to describe various phenomena. Separate equations must be used to describe several constraints, which may be in effect *simultaneously*.

For instance, in selecting a variety of foods to satisfy certain dietary requirements, a dietitian will find that from a given food group, each serving of a particular food contains certain amounts of vitamins A and B and that each serving of another food contains different amounts of these vitamins. The total number of units of vitamins A and B contained in a monthly diet depends on the number of times the various foods supplying these vitamins are served. For example, if carrots are served x times per month and each serving provides 8000 International Units (IU) of vitamin A, while spinach is served y times per month with each serving providing 4000 IU of vitamin A, the total number of units of vitamin A provided by these two foods each month is $8000x + 4000y$. To satisfy the monthly requirement for vitamin A (say 150,000 IU) with only these two foods requires

$$8000x + 4000y = 150{,}000$$

Of course, consideration of more foods introduces additional variables. Each dietary requirement results in a similar equation. To provide a balanced diet, all of these equations must then be satisfied simultaneously.

A collection of equations, all of which are to be satisfied simultaneously, is called a **system of equations**. A **system of inequalities** is a collection of inequalities, all of which must be satisfied simultaneously.

$$(1) \begin{cases} 2x + 3y - 4z = 0 \\ 2x^2 - 4y^2 + 2 = 0 \\ 9z - 3w + 4y = 0 \end{cases} \qquad (2) \begin{cases} x + 3y < 10 \\ 2xy + 4 \geq -2 \end{cases}$$

A system of equations   A system of inequalities

DEFINITION

The **solution** to a system of equations or inequalities consists of all values of the variables that satisfy each of the equations or inequalities in the system.

Section 10.1

Two Equations in Two Unknowns

In the first part of this section we discuss systems of linear equations in two variables: systems involving equations of the form

$$ax + by = c \qquad a, b, c \text{ constants}, \qquad a, b \text{ not both } 0$$

Toward the end of the section, we will study nonlinear systems.

554 Chapter 10 Systems of Equations and Inequalities

Graphing

The graph of a linear equation in two variables is a straight line (see Section 3.3). Thus the solution to a system of two linear equations in two variables consists of the points which the lines represented by these equations have in common. Three situations could occur.

1. If the lines intersect in precisely one point, there is precisely one solution: the system is said to be **consistent.**
2. If the lines are parallel, there are no solutions; the system is said to be **inconsistent.**
3. If the lines coincide, every point on this common line is a solution; the system is said to be **dependent.**

These situations are illustrated in Figure 10.1.

Figure 10.1

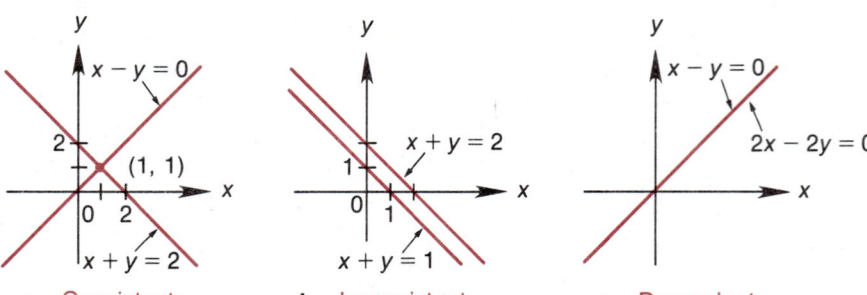

a. Consistent   b. Inconsistent   c. Dependent

Recall that two lines have the same slope if and only if they are parallel or coincide. A system of two linear equations in two unknowns is consistent if and only if the associated lines have different slopes.

EXAMPLE 1

Determine whether the following systems are consistent, inconsistent, or dependent. If the system is consistent, estimate the solution from the graphs.

a. $\begin{cases} 3x - 2y = 1 \\ -6x + 4y = -1 \end{cases}$ b. $\begin{cases} x - y = 4 \\ 2y - 2x = -8 \end{cases}$ c. $\begin{cases} x + 2y = 3 \\ x + 4y = 6 \end{cases}$

SOLUTION

a. The equations may be rewritten as

$$\begin{cases} 2y = 3x - 1 \\ 4y = 6x - 1 \end{cases} \quad \text{or} \quad \begin{cases} y = \dfrac{3}{2}x - \dfrac{1}{2} \\ y = \dfrac{3}{2}x - \dfrac{1}{4} \end{cases}$$

The latter version indicates parallel lines each having slope $m = \frac{3}{2}$; the system is inconsistent. The solution set is empty.

b. This system is equivalent to

$$\begin{cases} y = x - 4 \\ 2y = 2x - 8 \end{cases} \quad \text{or} \quad \begin{cases} y = x - 4 \\ y = x - 4 \end{cases}$$

The lines coincide; the system is dependent. The solution set is the entire line $\{(x, y): y = x - 4\}$.

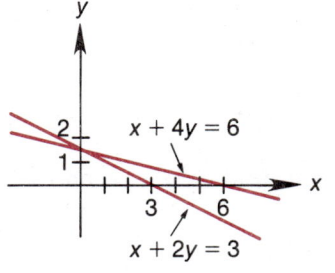

Figure 10.2

c. Again, rewrite the system in a form that indicates the slope of each line:

$$\begin{cases} 2y = -x + 3 \\ 4y = -x + 6 \end{cases} \quad \text{or} \quad \begin{cases} y = -\frac{1}{2}x + \frac{3}{2} \\ y = -\frac{1}{4}x + \frac{3}{2} \end{cases}$$

The slopes are $m_1 = -\frac{1}{2}$ and $m_2 = -\frac{1}{4}$, respectively. The lines are not parallel. The system is consistent; it has precisely one solution. The lines are sketched in Figure 10.2. The intersection of these lines gives us the solution

$$x = 0, \quad y = \frac{3}{2}$$

Substituting these values into the original equations verifies the solution:

$$0 + 2 \cdot \frac{3}{2} \stackrel{\checkmark}{=} 3$$

$$0 + 4 \cdot \frac{3}{2} \stackrel{\checkmark}{=} 6$$

The solution set is $\{(0, \frac{3}{2})\}$. ∎

The Substitution Method

It is sometimes difficult to estimate the solution to a system of equations with any great degree of precision by using a graph. On the other hand, one of the equations in a linear system can be used to express some variable in terms of the others. Substituting this expression into the other equation(s) simplifies the problem by reducing the number of variables. This technique is called the **substitution method of solution**. It is especially useful in solving systems of two equations in two unknowns.

EXAMPLE 2

Solve the following systems.

a. $\begin{cases} x - y = 1 \\ 2x + 3y = 7 \end{cases}$ b. $\begin{cases} x - y = 1 \\ 2x - 2y = 2 \end{cases}$ c. $\begin{cases} 2x + 2y = 4 \\ x + y = 1 \end{cases}$

SOLUTION

a. In order for the first equation to be satisfied, we must have

$$x = y + 1$$

Substituting $(y + 1)$ for x into the second equation gives

$$2(y + 1) + 3y = 7$$
$$5y + 2 = 7$$
$$5y = 5$$
$$y = 1$$

Hence $y = 1$ is the only possible y value that can satisfy this system of equations. But then

$$x = y + 1 = 1 + 1 = 2$$

The solution to the system is $x = 2$, $y = 1$, or

$$(x, y) = (2, 1)$$

We check the solution by substituting $x = 2$, $y = 1$ into the original equations:
$$\begin{cases} 2 - 1 \stackrel{\vee}{=} 1 \\ 2 \cdot 2 + 3 \cdot 1 \stackrel{\vee}{=} 7 \end{cases}$$

The solution set is $\{(2, 1)\}$.

b. Again, we have $x = y + 1$ from the first equation. Substituting, the second equation becomes
$$2(y + 1) - 2y = 2$$
$$2 = 2$$

No restrictions are placed on y. The only restriction on the variables is that $x = y + 1$. The system is dependent. The solution set is the entire line $\{(x, y) : x = y + 1\}$.

c. Use the second equation to write $y = 1 - x$ and substitute into the first equation:
$$2x + 2(1 - x) = 4$$
$$2 = 4$$

If these equations are both satisfied, then $2 = 4$. This cannot be; thus both equations cannot be satisfied simultaneously. There are no solutions. The solution set is empty.

The graphs are given in Figure 10.3.

Figure 10.3

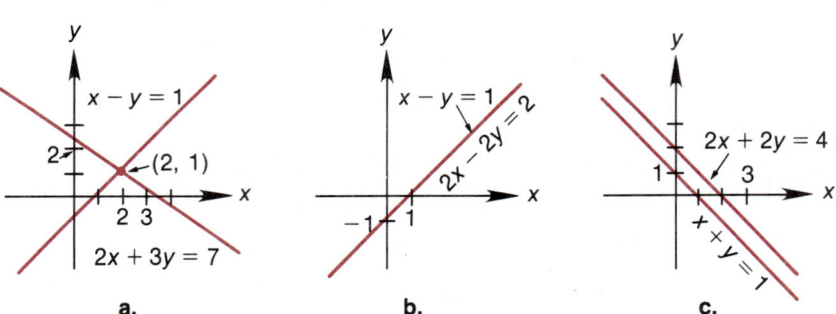

a.   b.   c.

EXAMPLE 3

Recall that for complex numbers, $a + bi = c + di$ if and only if $a = c$ and $b = d$. With this in mind, find real numbers x and y satisfying
$$x(1 - 2i) + y(1 + i) = 7 + i$$

SOLUTION

$$x(1 - 2i) + y(1 + i) = 7 + i$$
$$x - 2xi + y + yi = 7 + i$$
$$x + y + (-2x + y)i = 7 + i$$

This leads to the system
$$\begin{cases} x + y = 7 \\ -2x + y = 1 \end{cases}$$

which we can solve to obtain $x = 2$, $y = 5$. The solution set is $\{(2, 5)\}$.

EXAMPLE 4   Mixing a given quantity of 30% silver alloy with a quantity of 90% silver alloy yields a 54% silver alloy. However, the mixture would have been only 50% silver had 20 units less of the 90% alloy been used. How many units of each alloy were used?

SOLUTION   In solving any word problem, we begin by labeling the quantities we are expected to find. Thus we let

x = number of units of 30% alloy

y = number of units of 90% alloy

The situation is illustrated in Figure 10.4. The quantity of silver in each "container" is listed below each.

Figure 10.4

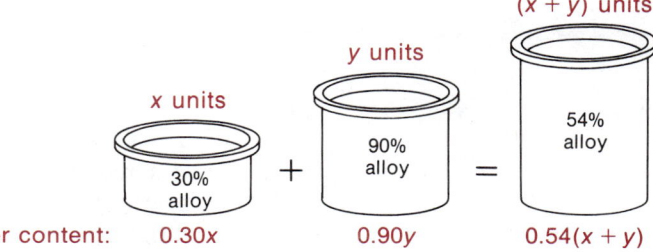

Thus we obtain the equation

$$0.30x + 0.90y = 0.54(x + y) \tag{1}$$

Had $(y - 20)$ units of the second alloy been used, computation of the corresponding amounts yields the equation

$$0.30x + 0.90(y - 20) = 0.50[x + (y - 20)] \tag{2}$$

We thus have a system of equations

$$\begin{cases} 0.30x + 0.90y = 0.54(x + y) & (1) \\ 0.30x + 0.90(y - 20) = 0.50(x + y - 20) & (2) \end{cases}$$

Multiplying each equation by 100, collecting terms, and simplifying, we can obtain the equivalent system

$$\begin{cases} -2x + 3y = 0 \\ -x + 2y = 40 \end{cases}$$

The second equation yields

$$x = 2y - 40$$

The first equation then becomes

$$-2(2y - 40) + 3y = 0$$
$$-y = -80$$
$$y = 80$$

Finally

$$x = 2y - 40$$
$$= 2 \cdot 80 - 40$$
$$= 120$$

Thus 120 units of 30% alloy and 80 units of 90% alloy were used.

Checking, we have

$$(0.30)120 + (0.90)80 \stackrel{?}{=} (0.54)(120 + 80)$$
$$36 + 72 \stackrel{?}{=} (0.54)200$$
$$108 \stackrel{\checkmark}{=} 108$$

and

$$(0.30)120 + (0.90)(80 - 20) \stackrel{?}{=} (0.50)(120 + 80 - 20)$$
$$36 + (0.90)(60) \stackrel{?}{=} (0.50)180$$
$$36 + 54 \stackrel{\checkmark}{=} 90$$

Nonlinear Systems

An engineer, scientist, or other user of mathematics will often encounter systems in which some or all the equations are nonlinear. An example of such a system might be

$$\begin{cases} x^2 + y^2 = 25 \\ 3x - y = 5 \end{cases}$$

For example, a frequent task is to find the points in which a certain line meets a given circle; this necessitates the solution of a system consisting of a linear equation and a second-degree equation, as given above.

Systems involving the logarithmic and exponential functions are also commonplace in the workaday world.

The substitution method can be used on nonlinear systems also. Note that with nonlinear systems it is possible to have a solution consisting of several points.

EXAMPLE 5   Determine where the line $x + y = 2$ meets the parabola $y = x^2$.

SOLUTION   We must solve the system

$$\begin{cases} x + y = 2 \\ y = x^2 \end{cases}$$

From the first equation, we find

$$y = 2 - x$$

and substitute this into the second equation

$$2 - x = x^2$$
$$0 = x^2 + x - 2$$
$$= (x + 2)(x - 1)$$
$$x = -2, 1$$

From $y = 2 - x$, we find

$y = 4$ when $x = -2$
$y = 1$ when $x = 1$

The points of intersection are $(-2, 4)$ and $(1, 1)$, as illustrated in Figure 10.5
The solution set for the system of equations is $\{(-2, 4), (1, 1)\}$.

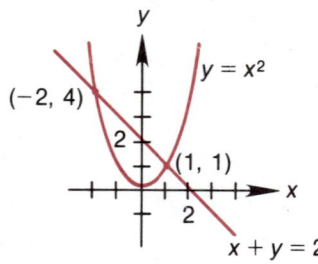

Figure 10.5

EXAMPLE 6

Find the points in which the line $3x - y = 5$ meets the circle $x^2 + y^2 = 25$.

SOLUTION

At the points (x, y) where the line meets the circle, each of the equations

$$\begin{cases} x^2 + y^2 = 25 \\ 3x - y = 5 \end{cases}$$

must be satisfied. We can use the second equation to find

$$y = 3x - 5$$

and substitute this into the first equation:

$$x^2 + (3x - 5)^2 = 25$$
$$x^2 + 9x^2 - 30x + 25 = 25$$
$$10x^2 - 30x = 0$$
$$10x(x - 3) = 0$$
$$x = 0, 3$$

Since $y = 3x - 5$, we find

$$y = -5 \quad \text{when} \quad x = 0$$
$$y = 4 \quad \text{when} \quad x = 3$$

The points of intersection are $(0, -5)$ and $(3, 4)$; see Figure 10.6.

Figure 10.6

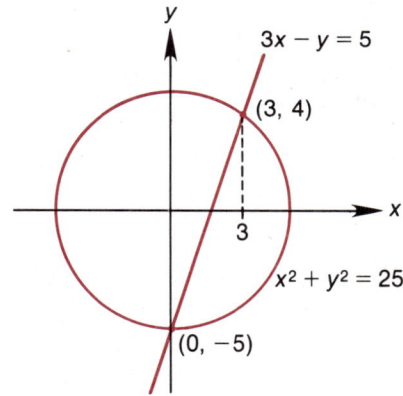

The solution set for the system of equations is $\{(0, -5), (3, 4)\}$. ∎

EXAMPLE 7

Determine where the parabola $y = x^2$ meets the circle $x^2 + y^2 = 2$.

SOLUTION

We must solve the system

$$\begin{cases} y = x^2 \\ x^2 + y^2 = 2 \end{cases}$$

Substituting $y = x^2$ into the second equation, we obtain

$$x^2 + (x^2)^2 = 2$$
$$(x^2)^2 + x^2 - 2 = 0$$

Use the quadratic formula to find

$$x^2 = \frac{-1 \pm \sqrt{1+8}}{2}$$
$$= \frac{-1 \pm 3}{2}$$

But $x^2 \geq 0$; hence

$$x^2 = \frac{-1+3}{2} = 1$$
$$x = \pm 1$$

Since $y = x^2$, the solutions are

(1, 1)   and   (−1, 1)

See Figure 10.7.

These are the points in which the parabola meets the circle. The solution set for the system of equations is $\{(1, 1), (-1, 1)\}$. ∎

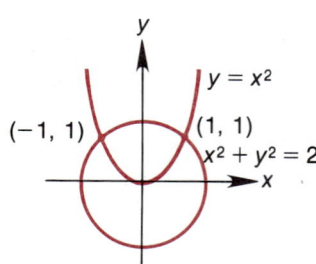

Figure 10.7

Section 10.1 Exercises

For each of the following systems
  a. *Sketch the graphs.*
  b. *Indicate whether the system is consistent, inconsistent, or dependent.*
  c. *If consistent, estimate the solution from the graphs.*
  d. *Use the substitution method to solve the system exactly.*

1. $\begin{cases} x - y = 2 \\ x + 2y = -1 \end{cases}$
2. $\begin{cases} 14r + 6s = 8 \\ 7r + 3s = 5 \end{cases}$
3. $\begin{cases} p - 5q = 7 \\ 3p + 4q = 2 \end{cases}$
4. $\begin{cases} u - 3v = 2 \\ 7u - 21v = 14 \end{cases}$
5. $\begin{cases} t - 4u = 24 \\ 2t + 3u = -7 \end{cases}$
6. $\begin{cases} 2x + y = 4 \\ x - 3y = 2 \end{cases}$
7. $\begin{cases} 2t - 6u = 8 \\ 3t - 9u = 9 \end{cases}$
8. $\begin{cases} 4p - 3q = 5 \\ 4p - 9q = -1 \end{cases}$
9. $\begin{cases} 2y + z = 1 \\ 4y + 2z = 2 \end{cases}$
10. $\begin{cases} 4v + 2w = 3 \\ 10v + 5w = 8 \end{cases}$
11. $\begin{cases} x + w = 2 \\ 3x + 2w = 1 \end{cases}$
12. $\begin{cases} 4t + 3x = 3 \\ 8t + 6x = 6 \end{cases}$
13. $\begin{cases} 2u + v = 6 \\ u + 2v = 0 \end{cases}$
14. $\begin{cases} 3y - 4u = 5 \\ y - 3u = -5 \end{cases}$
15. $\begin{cases} 5v + 7w = 3 \\ 10v + 14w = 6 \end{cases}$
16. $\begin{cases} 13t + 8y = 6 \\ 7t + 6y = 10 \end{cases}$
17. $\begin{cases} u + v = 0 \\ 2u - 3v = 5 \end{cases}$
18. $\begin{cases} 7x + 6y = 3 \\ 35x + 30y = 15 \end{cases}$
19. $\begin{cases} w - 2z = 5 \\ 2w - 3z = 8 \end{cases}$
20. $\begin{cases} v + 5w = -1 \\ v - 3w = 7 \end{cases}$
21. $\begin{cases} 2t - 3u = 15 \\ 6t - 9u = 5 \end{cases}$
22. $\begin{cases} 3y - 2z = 3 \\ 5y - z = -2 \end{cases}$
23. $\begin{cases} 3x - 2y = 4 \\ 4x - 6y = 2 \end{cases}$
24. $\begin{cases} 4r - s = 5 \\ 4r - 3s = -17 \end{cases}$
25. $\begin{cases} 3v - 2r = 5 \\ 4v - 5r = 2 \end{cases}$

Solve for x and y in the following where $x, y \in R$.

26. $(x - y) + (x + y)i = 1 + 7i$
27. $(x - yi) + (y + xi) = 3 - i$
28. $y + xi = 3i(y + 1) + x + 1$
29. $x(1 + 2i) + y(3i - 1) = 1 + 7i$
30. $x(1 + 2i) + y(3i - 4) = 20 + 7i$
31. $4x(1 + i) - y(1 + 3i) = 5 - 17i$
32. $(x + yi)(3i - 2) = 1 + 5i$

Solve the following nonlinear systems.

33. $\begin{cases} x - y = 3 \\ x - y^2 = 1 \end{cases}$
34. $\begin{cases} x^2 - y = 4 \\ x + y = 0 \end{cases}$

35. $\begin{cases} -x + y = 4 \\ x + y^2 = 2 \end{cases}$
36. $\begin{cases} x^2 + y = 4 \\ 2x + y = 1 \end{cases}$

37. $\begin{cases} x - y^2 = 1 \\ x - 3y = 11 \end{cases}$
38. $\begin{cases} x + 2y = 0 \\ x - y^2 = 1 \end{cases}$

39. $\begin{cases} x + 2y = 7 \\ x^2 + 4y^2 = 25 \end{cases}$
40. $\begin{cases} x + 3y = 2 \\ x^2 - 2y^2 = -1 \end{cases}$

41. $\begin{cases} x^2 + y^2 = 25 \\ x - 2y = 10 \end{cases}$
42. $\begin{cases} 2x^2 + 3y^2 = 5 \\ x + y = 2 \end{cases}$

43. $\begin{cases} x^2 + y^2 = 8 \\ x + y = 4 \end{cases}$
44. $\begin{cases} 2x + 3y = 0 \\ x^2 - y^2 = -4 \end{cases}$

45. $\begin{cases} x^2 + y^2 = 2 \\ x - 2y = -1 \end{cases}$
46. $\begin{cases} x^2 + y^2 = 5 \\ x + y = 1 \end{cases}$

47. $\begin{cases} x^2 + y^2 = 6 \\ x + y^2 = 0 \end{cases}$
48. $\begin{cases} x^2 + y = 0 \\ x^2 + y^2 = 2 \end{cases}$

49. $\begin{cases} xy = -1 \\ 4x + 2y = 7 \end{cases}$
50. $\begin{cases} xy = -6 \\ x + y = 5 \end{cases}$

51. $\begin{cases} xy = 8 \\ 2x + 3y = -16 \end{cases}$
52. $\begin{cases} xy = 4 \\ x - 2y = 7 \end{cases}$

53. $\begin{cases} x^2 - 2xy + y^2 = 1 \\ x + y = 1 \end{cases}$
54. $\begin{cases} 2x^2 - 2xy + y^2 = 8 \\ 2x + y = 4 \end{cases}$

55. $\begin{cases} x + y = 3 \\ x - 4y = y^2 + 7 \end{cases}$
56. $\begin{cases} x - y = 4 \\ x^2 + y^2 + 4y = 0 \end{cases}$

57. $\begin{cases} x^2 + y^2 + 4y = 0 \\ x + 2y = -2 \end{cases}$
58. $\begin{cases} 2x - y = 2 \\ x + 6y + y^2 = -9 \end{cases}$

59. $\begin{cases} x^2 + y^2 = 12 \\ x^2 + y^2 = -4 \end{cases}$
(Substitute for x^2 or y^2.)

60. $\begin{cases} x^2 + y^2 = 10 \\ x^2 - 2y^2 = -17 \end{cases}$
61. $\begin{cases} \log 2x + y = \log 2 \\ 2 \log x - y = 0 \end{cases}$

62. $\begin{cases} x + \log_2(y + 1) = 3 \\ x - \log_2(y + 1) = -1 \end{cases}$

63. $\begin{cases} e^{2x} + ye^x = 2 \\ 3e^x - y = 0 \end{cases}$
64. $\begin{cases} x - 2^y = 4 \\ x + 2^y = 8 \end{cases}$

65. The equation for converting Celsius temperatures to Fahrenheit temperatures is $F = (\frac{9}{5})C + 32$. One method for finding if and when the two temperature scales have the same numerical values is to solve the system of equations

$$\begin{cases} F = \frac{9}{5}C + 32 \\ F = C \end{cases}$$

a. Graph the two equations on the same coordinate system.
b. Solve the system of equations by the method of substitution.
c. Check your results on the graph.

66. When corn sells for $1.50 per bushel, there is a demand for 100,000 metric tons and a supply of only 50,000 metric tons. Each penny increase in the price per bushel stimulates an additional production of 8000 metric tons and decreases the demand by 2000 metric tons.

a. If equilibrium occurs when supply equals demand, write a system of equations whose solution indicates the equilibrium price for corn.
b. Sketch the lines represented by these equations.
c. Solve the system by the method of substitution.
d. Check your results on the graph.

67. A tortoise travels at a rate of 1 mile in 5 hours; the hare can travel 5 miles in 1 hour.

a. If the tortoise has an 8-hour head start, write a system of equations whose solution indicates when the hare passes the tortoise and how far each has traveled when this happens.
b. Solve the system of equations by the method of substitution.

68. How old are Karen and Bea if 4 years ago Karen was seven times as old as Bea and 4 years from now Karen will be only three times as old as Bea?

69. Four years ago Viki was 22 years younger than her mother. If Viki's mother is now 8 years more than twice as old as Viki, how old are Viki and her mother now?

70. Find a two-digit number such that the sum of the digits is 10 and the square of the tens digit is 2 more than the units digit.

71. The tens digit of a two-digit number exceeds the units digit by 3. If the digits are reversed, the result is 1 less than half the original. Find the number. *Hint:* If the tens and units digits are t and u respectively, the number is $10t + u$. For example $27 = 10 \cdot 2 + 7$.

72. Mr. Rammel invests a certain sum of money at 7% interest and another sum at 6% to achieve

an annual yield of $628. Were he to interchange his investments, his annual interest would decrease by $8. What sums did he invest at each rate of interest?

73. An 18% acid solution is mixed with a 27% acid solution to yield a 22% acid solution. Had 15 units more of the 18% solution been used, the resulting mixture would have been 21% acid. How many units of each solution were used?

74. Find the points in which the line $y = 2x$ meets the circle of radius $\sqrt{5}$ centered at $(1, 2)$.

75. Find the points in which the line passing through the origin with slope $m = 7$ meets the circle of radius 5 centered at $(5, 10)$.

76. Determine where the parabola $y = 2x^2$ meets the circle $x^2 + y^2 = 5$.

77. Wildlife experts are treating two problems simultaneously in a river. One chemical is supposed to purify the water; the other is supposed to promote growth of fish foods. Because of adverse reactions between the two chemicals, the relative amounts must be precisely controlled. The product of the amounts must be 168 and the sum of the squares of the amounts must be 340. What treatment levels are permissible?

78. The product of two nonnegative numbers is 65. The sum of their squares is 194. Find the numbers.

Section 10.2

The Elimination Method of Solution

The arithmetic involved in the substitution method frequently leads to messy fractions and becomes tedious. On the other hand, sketching the graphs represented by a system of equations and "guesstimating" the coordinates of their intersections can be very inaccurate. In this section, we introduce another method for solving systems of equations. Known as the **elimination method**, it yields precise results and the manipulations are sometimes easier than those of the substitution method. We shall first illustrate this technique on systems of two equations in two unknowns.

Systems with Two Variables

EXAMPLE 1 Solve the following system by using the two equations to eliminate one of the variables.

$$\begin{cases} x - y = 1 \\ 2x + 3y = 7 \end{cases}$$

SOLUTION Multiply the first equation by 3 and then add like terms of the corresponding members of the two equations to eliminate y:

$$\begin{cases} 3x - 3y = 3 \\ 2x + 3y = 7 \end{cases}$$
$$\overline{5x = 10}$$
$$x = 2$$

Substituting $x = 2$ into

$$x - y = 1$$

we obtain

$$2 - y = 1$$
$$-y = -1$$
$$y = 1$$

The solution set is {(2, 1)}. (This agrees with the solution we found by substitution in Example 2a of the preceding section.)

EXAMPLE 2

Solve the following system.

$$\begin{cases} x + 2y = 4 \\ 5x + 10y = 10 \end{cases}$$

SOLUTION

Multiply the first equation by 5 and then subtract corresponding members of the second equation from the first (to eliminate x):

$$\begin{array}{r} 5x + 10y = 20 \\ 5x + 10y = 10 \\ \hline 0 = 10 \end{array}$$

Since $0 \neq 10$, there can be no solutions. The solution set is empty. Of course, this is reasonable since the two equations represent distinct parallel lines.

Systems with Three Variables

Like two equations in two unknowns, larger systems may be

1. Consistent: have precisely one solution.
2. Inconsistent: have no solutions.
3. Dependent: have infinitely many solutions.

It can be shown that an equation of the form

$$ax + by + cz = d \qquad a, b, c \text{ not all } 0$$

represents a plane in three-dimensional space. Nevertheless, these are called linear equations. In general, *an equation involving* n *variables is said to be* **linear** *if it has the form*

$$ax_1 + bx_2 + \cdots + cx_n = d, \qquad a, b, \ldots, c \text{ not all } 0$$

The numbers $a, b \ldots, c$ are called **coefficients** of $x_1, x_2, \ldots, x_n$, respectively.

One linear equation in three unknowns represents a plane and hence has infinitely many solutions. A system of two such equations then represents the intersection of two planes. (See Figure 10.8.) This intersection could be (a) a line, (b) empty (if the planes are parallel), or (c) a plane (if the planes coincide).

Figure 10.8

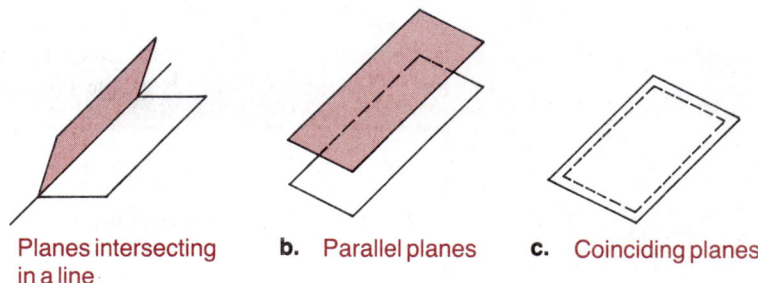

a. Planes intersecting in a line   b. Parallel planes   c. Coinciding planes

In any case, the solution to a system of *two* linear equations in *three* unknowns, being the intersection of two planes, cannot be a single point. In fact, *any linear system with fewer equations than unknowns never has precisely one solution; it is necessarily dependent or inconsistent.*

If we have three equations in three unknowns, we are essentially intersecting a third plane with one of the configurations in Figure 10.8. If two or more of these planes coincide, one of the configurations in Figure 10.8 results. But if no two planes coincide, the possibilities are illustrated in Figure 10.9.

Figure 10.9

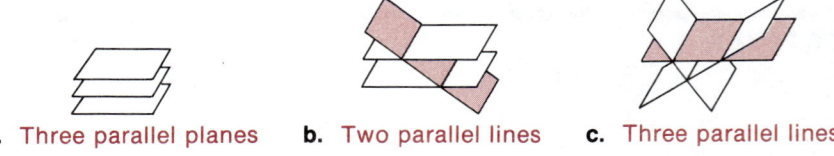

a. Three parallel planes   b. Two parallel lines   c. Three parallel lines

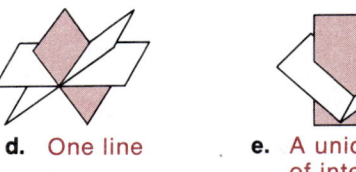

d. One line   e. A unique point of intersection

Figures 10.8(b) and 10.9(a), (b), and (c) represent inconsistent systems. Dependent systems are illustrated in Figure 10.8(a), (c) and 10.9(d). Only Figure 10.9(e) represents a consistent system.

A linear system with more equations than unknowns could also be consistent, dependent, or inconsistent. To visualize this, consider a system of four linear equations in three variables. Each equation represents a plane. If the first three equations have a unique solution, then the fourth equation may or may not represent a plane through this point; such a system may have exactly one or no solutions. For a dependent system, consider

$$\begin{cases} 2x + 4y + 4z = 14 \\ x + 2y + 2z = 7 \\ 2x - 2y - 5z = -1 \\ 4x - 4y - 10z = -2 \end{cases}$$

The first two equations represent a single plane (hence the same solutions), as do the last two. Figure 10.8(a) results.

The substitution method can be used on systems involving more than two unknowns or equations, but the computations get to be very messy for larger systems. The elimination method is advantageous when more than two equations are present. While we can proceed somewhat informally with two equations in two unknowns, we shall formalize the elimination procedure for larger systems by replacing a given system of equations with an **equivalent** system: a system that has the same solution set as the given system.

PRINCIPLES OF THE ELIMINATION METHOD

The following operations on a system of equations yield an equivalent system:

1. Interchanging equations.
2. Multiplying both sides (or members) of an equation by a nonzero constant.
3. Adding a nonzero multiple of one equation to another equation.

These operations are performed on a system of equations to eliminate one or more of the variables from one or several equations.

If two equations in a system are interchanged, the list of equations that must be satisfied is not changed; only the order of the listing has been changed. Regarding the second operation, it should be clear, for example, that

$$x + y = 2$$
$$2x + 2y = 4$$

represent the same line and hence provide the same solutions in a system. Concerning the third, consider

$$\begin{cases} x - y = 2 \\ 2x + 3y = -1 \end{cases} \quad \text{and} \quad (1)$$

$$\begin{cases} x - y = 2 \\ 3(x - y) + 2x + 3y = -1 + 3 \cdot 2 \end{cases} \quad (2)$$

System (2) is obtained by adding three times the first equation to the second equation of system (1). The two systems have the same solutions, since $3(x - y) = 3 \cdot 2$ in either case.

A compact notation, or "record-keeping system," for the steps used in the elimination method is introduced in the next example. In this notation, E_1 stands for equation 1, and so on.

EXAMPLE 3

Solve the following system of equations.

$$\begin{cases} 2x + y - z = 5 \\ x + y + z = 1 \\ -x + 2y + 2z = -4 \end{cases}$$

SOLUTION

1. Begin by obtaining a coefficient of 1 for x in the first equation. This is most easily accomplished by interchanging the first two equations. We shall keep track of our work by writing $E_1 \leftrightarrow E_2$ on the side. This yields the equivalent system:

$$\begin{cases} x + y + z = 1 \\ 2x + y - z = 5 \\ -x + 2y + 2z = -4 \end{cases} \quad (E_1 \leftrightarrow E_2)$$

2. Next, we use the first equation to eliminate x from the remaining equations. We do this by subtracting twice the first equation from the second and then adding the first equation to the third. We keep track of our work by writing $E_2 - 2E_1 \to E_2$ and $E_3 + E_1 \to E_3$ on the side. This results in the equivalent system:

$$\begin{cases} x + y + z = 1 \\ -y - 3z = 3 \quad (E_2 - 2E_1 \to E_2) \\ 3y + 3z = -3 \quad (E_3 + E_1 \to E_3) \end{cases}$$

3. Next, we obtain a coefficient of 1 for y in the second equation; we can do this by multiplying the second equation by -1. We keep track of our work by writing $(-1)E_2 \to E_2$ on the side. This results in the equivalent system:

$$\begin{cases} x + y + z = 1 \\ y + 3z = -3 \quad [(-1)E_2 \to E_2] \\ 3y + 3z = -3 \end{cases}$$

4. **Now use the second equation to eliminate y from the third equation** by subtracting 3 times the second equation from the third. We keep track of our work by writing $E_3 - 3E_2 \to E_3$ on the side. This yields the equivalent system:

$$\begin{cases} x + y + z = 1 \\ \phantom{x + {}} y + 3z = -3 \\ \phantom{x + y + {}} -6z = 6 \end{cases} \quad (E_3 - 3E_2 \to E_3)$$

5. **Next, we obtain a coefficient of 1 for z in the last equation** by multiplying this last equation by $-\frac{1}{6}$. We keep track of our work by writing $(-\frac{1}{6})E_3 \to E_3$ on the side. This yields the equivalent system:

$$\begin{cases} x + y + z = 1 \\ \phantom{x + {}} y + 3z = -3 \\ \phantom{x + y + {}} z = -1 \end{cases} \quad \left(-\frac{1}{6}E_3 \to E_3\right)$$

6. We can now read part of the solution:

$$z = -1$$

Substituting this value into the second equation yields:

$$y + 3(-1) = -3$$
$$y = 0$$

Substituting $z = -1$ and $y = 0$ into the first equation then yields:

$$x + 0 + (-1) = 1$$
$$x = 2$$

We obtain the solution $x = 2$, $y = 0$, $z = -1$ or

$$(x, y, z) = (2, 0, -1)$$

We check by substituting these values into each of the original equations:

$$\begin{cases} 2 \cdot 2 + \phantom{2 \cdot {}} 0 - \phantom{2 \cdot {}} (-1) \stackrel{\checkmark}{=} 5 \\ \phantom{2 \cdot {}} 2 + \phantom{2 \cdot {}} 0 + \phantom{2 \cdot {}} (-1) \stackrel{\checkmark}{=} 1 \\ -2 + 2 \cdot 0 + 2 \cdot (-1) \stackrel{\checkmark}{=} -4 \end{cases}$$

The solution set consists of the single point $\{(2, 0, -1)\}$. ∎

It is always a good idea to check your solutions to a system of equations. If the solution does not check out, the side notes can be used to retrace your steps in an attempt to find the error.

The procedure used in Step 6 of the last example is called **back substitution**. The value of z obtained from the last equation is substituted back into the second equation to obtain y. These values of y and z are then substituted back into the first equation to obtain x.

The first nonzero term of an equation is called its **leading term**: its coefficient is called the leading coefficient. The system of equations obtained in Step 5 of the previous example is said to be in *echelon form*.

Section 10.2 The Elimination Method of Solution 567

> **DEFINITION**
>
> A system of equations is said to be in **echelon form** if
>
> 1. The variables are written in the same order in each equation.
> 2. The leading coefficient of each nonzero equation is 1.
> 3. The leading term of each nonzero equation (other than the first) is to the right of the leading term of its preceding equation.
> 4. Equations of the form $0 = 0$ appear last.

Our approach to the elimination method can now be summarized as follows.

> **THE ELIMINATION METHOD OF SOLUTION**
>
> 1. Write the variables in the same order in each equation.
> 2. Interchange equations or multiply the first equation by a nonzero constant so that the leftmost leading term appears in the first equation and has coefficient 1 there.
> 3. Eliminate the corresponding variable from the following equations by adding or subtracting multiples of the first equation from these.
> 4. Repeat Steps 2 and 3 on the following equations.
> 5. Continue in this way until the system of equations is in echelon form.
> 6. Use back substitution to solve for the variables.

EXAMPLE 4 Solve the system
$$\begin{cases} 2x + 4y + 4z = 14 \\ 2x - 2y - 5z = -1 \\ x + 2y + 2z = 7 \end{cases}$$

SOLUTION The system is equivalent to
$$\begin{cases} x + 2y + 2z = 7 \\ 2x - 2y - 5z = -1 \quad (E_1 \leftrightarrow E_3) \\ 2x + 4y + 4z = 14 \end{cases}$$

Next we eliminate x from the second and third equations:
$$\begin{cases} x + 2y + 2z = 7 \\ -6y - 9z = -15 \quad (E_2 - 2E_1 \to E_2) \\ 0 = 0 \quad\quad\quad\quad (E_3 - 2E_1 \to E_3) \end{cases}$$

Dividing the new second equation by -6 yields the desired echelon form:

$$\begin{cases} x + 2y + 2z = 7 \\ \quad y + \dfrac{3}{2}z = \dfrac{5}{2} \quad \left(-\dfrac{1}{6}E_2 \to E_2\right) \\ \quad \quad 0 = 0 \end{cases}$$

The equation $0 = 0$ places no restriction on the variables; we essentially have a system of two equations in three unknowns. There cannot be a unique solution. Here we have infinitely many solutions. The last nonzero equation can be used to write y in terms of z:

$$y = \frac{5}{2} - \frac{3}{2}z$$

This y-value is then substituted into the first equation:

$$x + 2\left(\frac{5}{2} - \frac{3}{2}z\right) + 2z = 7$$
$$x + 5 - 3z + 2z = 7$$
$$x = z + 2$$

For *any* value of z, these corresponding x and y values yield a solution to the system. The solution consists of all points of the form

$$\left(z + 2, \frac{5 - 3z}{2}, z\right) \quad \text{for any real number } z$$

The solution set is $\left\{\left(z + 2, \dfrac{5 - 3z}{2}, z\right) : z \text{ any real number}\right\}$. ∎

EXAMPLE 5 Solve the system

$$\begin{cases} x - 2y - 4z = 7 \\ 2x - 3y + z = 5 \\ 3x - 4y + 6z = 2 \end{cases}$$

SOLUTION Using the elimination method, we have

$$\begin{cases} x - 2y - 4z = 7 \\ \quad y + 9z = -9 \quad (E_2 - 2E_1 \to E_2) \\ \quad 2y + 18z = -19 \quad (E_3 - 3E_1 \to E_3) \end{cases}$$

$$\begin{cases} x - 2y - 4z = 7 \\ \quad y + 9z = -9 \\ \quad \quad 0 = -1 \quad (E_3 - 2E_2 \to E_3) \end{cases}$$

Since the last equation can never be satisfied, there are no solutions to this system of equations; the solution set is empty. ∎

EXAMPLE 6 A lawn company has three kinds of fertilizer labeled as grades A, B, and C, respectively.

Grade A contains 40% plant food and 40% weed killer.
Grade B contains 60% plant food and 20% weed killer.
Grade C contains 50% plant food and 24% weed killer.

In mixing a ton (2000 pounds) of fertilizer, how many pounds of each must be used to achieve a mixture containing $52\frac{1}{2}\%$ plant food and 26% weed killer?

SOLUTION

Let

a = amount of A used in the mixture
b = amount of B used in the mixture
c = amount of C used in the mixture

Then

$a + b + c$ = total weight of the mixture
$(0.4)a + (0.6)b + (0.5)c$ = total amount of plant food in the mixture
$(0.4)a + (0.2)b + (0.24)c$ = total amount of weed killer in the mixture

In 2000 pounds of mixture containing $52\frac{1}{2}\%$ plant food and 26% weed killer, there are $(0.525)2000 = 1050$ pounds of plant food and $(0.26)2000 = 520$ pounds of weed killer. Thus we obtain the equations

$$\begin{cases} a + b + c = 2000 \\ (0.4)a + (0.6)b + (0.5)c = 1050 \\ (0.4)a + (0.2)b + (0.24)c = 520 \end{cases}$$

We solve this system as follows:

$$\begin{cases} a + b + c = 2000 \\ 4a + 6b + 5c = 10500 \quad (10E_2 \to E_2) \\ 40a + 20b + 24c = 52000 \quad (100E_3 \to E_3) \end{cases}$$

$$\begin{cases} a + b + c = 2000 \\ 2b + c = 2500 \quad (E_2 - 4E_1 \to E_2) \\ -20b - 16c = -28000 \quad (E_3 - 40E_1 \to E_3) \end{cases}$$

$$\begin{cases} a + b + c = 2000 \\ b + \frac{1}{2}c = 1250 \quad \left(\frac{1}{2}E_2 \to E_2\right) \\ 20b + 16c = 28000 \quad (-E_3 \to E_3) \end{cases}$$

$$\begin{cases} a + b + c = 2000 \\ b + \frac{1}{2}c = 1250 \\ 6c = 3000 \quad (E_3 - 20E_2 \to E_3) \end{cases}$$

Without even finalizing the echelon form, we obtain

$c = 500$

We then use back substitution to obtain

$b + \frac{1}{2} \cdot 500 = 1250$

$b = 1000$

and
$$a + 1000 + 500 = 2000$$
$$a = 500$$

Thus, 500 pounds of grade A, 1000 pounds of grade B, and 500 pounds of grade C are used to obtain the appropriate mixture in a ton of fertilizer.

It can be verified that these amounts do satisfy the conditions stated in the problem. ∎

Section 10.2 Exercises

Solve the following systems of equations by the elimination method.

1. $\begin{cases} t - u = -2 \\ t + 2u = 1 \end{cases}$

2. $\begin{cases} v + w = -1 \\ v - 2w = 8 \end{cases}$

3. $\begin{cases} 2w + z = 1 \\ 2w - z = 3 \end{cases}$

4. $\begin{cases} 8r + 2s = 3 \\ 32r + 8s = 12 \end{cases}$

5. $\begin{cases} 3x + 5y = 1 \\ x - 7y = 9 \end{cases}$

6. $\begin{cases} 6u - 5v = 10 \\ 9u - 4v = -13 \end{cases}$

7. $\begin{cases} 4x - z = 5 \\ 4x + 3z = 17 \end{cases}$

8. $\begin{cases} x - y = 1 \\ 6x + 5y = 28 \end{cases}$

9. $\begin{cases} 3y + 2z = 6 \\ 6y + 4z = 12 \end{cases}$

10. $\begin{cases} z - 2u = 10 \\ -5z + 10u = 3 \end{cases}$

11. $\begin{cases} 5t - 6v = 7 \\ 6t - 7v = 9 \end{cases}$

12. $\begin{cases} 9w + 4x = -1 \\ 3w + 2x = 1 \end{cases}$

13. $\begin{cases} 2r - 3s = 7 \\ 3r - 4s = 8 \end{cases}$

14. $\begin{cases} 7r - 6y = -10 \\ 3r + 2y = 14 \end{cases}$

15. $\begin{cases} 2v - 2r = 4 \\ 5v - 5r = 5 \end{cases}$

16. $\begin{cases} 3w - v = 7 \\ 2w + 4v = 0 \end{cases}$

17. $\begin{cases} 2x - y + z = -5 \\ x - 2y - 3z = 6 \\ x + y - 2z = 1 \end{cases}$

18. $\begin{cases} 2u - v + 2w = 5 \\ -u + v + 3w = 6 \\ 2u - 2v - w = 8 \end{cases}$

19. $\begin{cases} a - b + c = -1 \\ b + 2c = -6 \\ 3a + 2b + c = 3 \end{cases}$

20. $\begin{cases} p + 2q - 3r = 10 \\ p + q - r = 4 \\ 2p - q + 2r = 6 \end{cases}$

21. $\begin{cases} t - 2u - 3v = 4 \\ 2t - 3u + v = 7 \\ 2t - 4u - 6v = -1 \end{cases}$

22. $\begin{cases} p - 2q + r = 8 \\ 5p - q + 3r = 6 \\ 3p - 6q + 3r = 14 \end{cases}$

23. $\begin{cases} x - 3y + z = -9 \\ -2x + y + 3z = 8 \\ x - 5y + 3z = -13 \end{cases}$

24. $\begin{cases} t - v - w = 0 \\ 2t + v - 2w = 0 \\ t - 2v - w = 0 \end{cases}$

25. $\begin{cases} y - 2z + w = 5 \\ 3y + 2z - 2w = -13 \\ 2y - z + 3w = 11 \end{cases}$

26. $\begin{cases} 3t - 2z + p = 4 \\ 6t - 4z + 2p = 8 \\ 9t - 6z + 3p = 12 \end{cases}$

27. $\begin{cases} 3x + 4y + z = 6 \\ x + 3y - 2z = -7 \\ 2x + y + z = 1 \end{cases}$

28. $\begin{cases} 8a + 2b + 5c = 4 \\ 3a - b + 2c = 7 \\ a - b - 4c = -5 \end{cases}$

29. $\begin{cases} r - 3s + t = 3 \\ r + s + 3t = 3 \\ 2r + s + 2t = -1 \end{cases}$

30. $\begin{cases} 2x + 4y + 3z = 4 \\ x + 2y + z = 1 \\ 3x - y - z = 2 \end{cases}$

31. $\begin{cases} 2r + 3s - t = 5 \\ 2r - 6s - 10t = -4 \end{cases}$

32. $\begin{cases} 4v - 5w + 3x = 3 \\ 3v - 4w + 2x = 1 \end{cases}$

33. $\begin{cases} 3x - 4y = 5 \\ x - 3y = 10 \\ 2x + 5y = 4 \end{cases}$

34. $\begin{cases} r + s = 2 \\ 3r + 2s = 1 \\ 5r + 4s = 5 \end{cases}$

35. $\begin{cases} p - 2q + 3r = 0 \\ 2p + 3q - 2r = 1 \\ 2p + q - r = 2 \\ p + q - r = 1 \end{cases}$

36. $\begin{cases} u + v - w = 3 \\ 2u - v + w = 0 \\ 2u + v + 3w = 0 \\ u - 2v - 2w = 0 \end{cases}$

37. Decreasing the length of a rectangle by 3 inches and increasing its width by 2 inches decreases its area by 16 square inches. On the other hand, increasing its length by 4 inches and decreasing its width by 2 inches increases its area by 32 square inches. Find the dimensions of the rectangle.

38. Jean has $2.35 in dimes and quarters. The number of quarters is five less than twice the

number of dimes. How many dimes and how many quarters does she have?

39. If 5 pounds of fudge and 7 pounds of mints cost $5.62, whereas 8 pounds of fudge and 8 pounds of mints cost $8.48, find the price of each per pound.

40. Eileen flew 300 miles into the city in 90 minutes and back in 3 hours. What was her airspeed and what was the wind velocity?

41. Two years ago Laara's mother was eight times as old as Laara. Eight years from now her mother will be only three times as old as Laara. What are their ages now?

42. Mr. Frilling rents three farms paying $40, $30, and $25 per acre, respectively. The total rent for the $25/acre farm is one-half the total combined rent for the other two farms. If his total rental fee is $15,000 and the total acreage is 480, how much is his rental fee for each farm?

43. Clara has $20,000 invested, some at 4%, some at 5%, and some at 8%. She collects $1250 interest yearly. She has twice as much invested at 8% as at 5%. How much is invested at each rate of interest?

44. A fertilizer company has solutions containing 20%, 30%, and 70% nitrogen, respectively. Because of other characteristics of the liquids, the company wants half of any mixture to consist of the 30% solution. How many pounds of each solution should be used to make 1 ton of fertilizer containing 50% nitrogen?

45. The sum of the digits of a three-digit number is 12. Reversing the digits increases the number by 99. The tens digit is 2 more than the sum of the units and hundreds digits. Find the number. (*Hint:* A three-digit number like 684 is computed as $684 = 6 \cdot 100 + 8 \cdot 10 + 4$; in general, if h = hundreds digit, t = tens digit, u = units digit, then $100h + 10t + u$ = the number. Reversing the order of the digits yields the number $100u + 10t + h$.)

46. A three-digit number is 12 less than 25 times the sum of its digits. Reversing the digits increases the number by 99. The tens digit is 1 more than the sum of the hundreds and units digits. Find the number.

47. Doris spent $3.76 for 14 stamps in 16¢, 20¢, and 60¢ denominations. If she purchased twice as many 16¢ stamps as 60¢ stamps, find how many stamps of each denomination she purchased.

Section 10.3

Nonlinear Systems

There is no "cookbook" approach to the solution of nonlinear systems. In Section 10.1, the substitution method was used on nonlinear systems. The principles of the elimination method may also be used on a nonlinear system to obtain an equivalent system. However, these operations are generally not sufficient to yield a solution to a nonlinear system. Generally, one must also use special properties of the particular situation. About the closest thing to a set of rules for solving nonlinear systems is the following.

> **NONLINEAR SYSTEMS**
> 1. Look for an equation from which to make a substitution and decrease the size of the system by eliminating a variable.
> 2. Use the elimination method to whatever extent possible.
> 3. Call upon your knowledge and past experience to progress to a solution if possible.

With these admonitions and suggestions we shall work several examples.

EXAMPLE 1 Solve the system of equations
$$\begin{cases} x^2 - y^2 = 21 \\ x^2 + y^2 = 29 \end{cases}$$

SOLUTION We use the elimination method to solve for x^2 and y^2 first. Then we find x and y. We begin by adding corresponding members of the two equations to eliminate y^2.

$$\begin{cases} x^2 - y^2 = 21 \\ x^2 + y^2 = 29 \end{cases}$$
$$\begin{aligned} 2x^2 &= 50 \\ x^2 &= 25 \\ x &= \pm 5 \end{aligned}$$

Substituting $x^2 = 25$ into the second equation, we have
$$25 + y^2 = 29$$
$$y^2 = 4$$
$$y = \pm 2$$

We can check to see that all of these solutions satisfy each of the original equations. The solution set is
$$S = \{(5, 2), (5, -2), (-5, 2), (-5, -2)\}. \qquad \blacksquare$$

In Example 2, we rework Example 7 of Section 10.1 by using the principles of the elimination method.

EXAMPLE 2 Determine where the parabola $y = x^2$ meets the circle $x^2 + y^2 = 2$.

SOLUTION We must solve the system
$$\begin{cases} y = x^2 \\ x^2 + y^2 = 2 \end{cases}$$

To illustrate the use of the elimination method, we could rewrite the system as
$$\begin{cases} x^2 - y = 0 \\ x^2 + y^2 = 2 \end{cases}$$

This system is equivalent to
$$\begin{cases} x^2 - y = 0 \\ y^2 + y = 2 \end{cases} \qquad (E_2 - E_1 \to E_2)$$

Solving for y in the second equation:
$$y^2 + y - 2 = 0$$
$$(y + 2)(y - 1) = 0$$
$$y = -2, 1$$

Since $y = x^2 \geq 0$, the only possible y solution is

$$y = 1$$

Then from the first equation

$$x^2 = y = 1$$
$$x = \pm 1$$

Again, we find the solutions

$$(1, 1) \quad \text{and} \quad (-1, 1) \qquad \blacksquare$$

EXAMPLE 3 The Soviet KGB wishes to eliminate an illicit broadcaster. On the basis of the strength of the radio signals transmitted, they determine that it is located 2 kilometers from KGB station A and approximately $\sqrt{10}$ kilometers from station B. If A is 2 kilometers west and 1 kilometer north of the Kremlin and B is 1 kilometer east and 4 kilometers north of the Kremlin, where should they search for the illicit broadcaster?

SOLUTION Set up a north-south, east-west coordinate system with the Kremlin at the origin. Then A and B are situated as indicated in Figure 10.10, Let T, the illicit broadcaster, have coordinates (x, y) on this coordinate system. Points A and B have coordinates $(-2, 1)$ and $(1, 4)$, respectively. The distances are

$$d(A, T) = \sqrt{(x + 2)^2 + (y - 1)^2} = 2$$
$$d(B, T) = \sqrt{(x - 1)^2 + (y - 4)^2} = \sqrt{10}$$

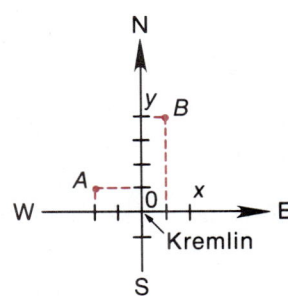

Figure 10.10

Squaring both sides of these equations we obtain

$$\begin{cases}(x + 2)^2 + (y - 1)^2 = 4 \\ (x - 1)^2 + (y - 4)^2 = 10\end{cases}$$

Expand the squared terms and rewrite the system as

$$\begin{cases}x^2 + 4x + 4 + y^2 - 2y + 1 = 4 \\ x^2 - 2x + 1 + y^2 - 8y + 16 = 10\end{cases}$$

Collect terms:

$$\begin{cases}x^2 + y^2 + 4x - 2y = -1 \\ x^2 + y^2 - 2x - 8y = -7\end{cases}$$

Subtract the first equation from the second:

$$\begin{cases}x^2 + y^2 + 4x - 2y = -1 \\ -6x - 6y = -6\end{cases} \quad (E_2 - E_1 \to E_2)$$

From the second equation, we find

$$y = 1 - x$$

which we substitute into the first:

$$x^2 + (1 - x)^2 + 4x - 2(1 - x) = -1$$

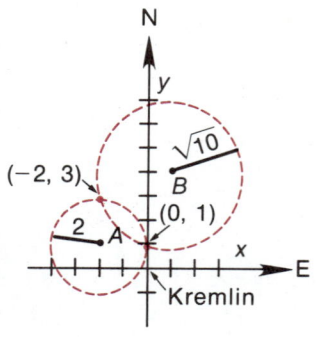

Figure 10.11

This is just a quadratic equation in x. It can be simplified as follows:

$$x^2 + (1 - 2x + x^2) + 4x - 2 + 2x = -1$$
$$2x^2 + 4x = 0$$
$$2x(x + 2) = 0$$

Its solutions are found to be $x = 0$ and $x = -2$. Substituting these values into the relationship $y = 1 - x$, we find the solutions to the system:

$$(0, 1) \quad \text{and} \quad (-2, 3)$$

There are only two possible locations for the illicit broadcaster. It is either directly 1 kilometer north of the Kremlin or 2 kilometers west and 3 kilometers north (see Figure 10.11).

Since T is on the circle of radius 2 centered at A and also on the circle of radius $\sqrt{10}$ centered at B, we have in actuality found the intersection of these two circles. ∎

EXAMPLE 4

Solve the system

$$\begin{cases} \log_3(x + 1) + y^2 = 5 \\ \log_3(x - 1) - y^2 = -4 \end{cases}$$

SOLUTION

Adding the two equations, we obtain the following

$$\log_3(x + 1) + \log_3(x - 1) = 5 - 4$$
$$\log_3(x + 1)(x - 1) = 1 \quad \text{(L1, Section 6.3)}$$
$$(x + 1)(x - 1) = 3^1$$
$$x^2 - 1 = 3$$
$$x^2 = 4$$
$$x = \pm 2$$

Since $\log_3(-2 - 1) = \log_3(-3)$ does not exist, $x = 2$ is the only x solution. Substituting this into either of the original equations, we obtain (from the second equation)

$$\log_3(2 - 1) - y^2 = -4$$
$$\log_3(1) - y^2 = -4$$
$$0 - y^2 = -4$$
$$y^2 = 4$$
$$y = \pm 2$$

The solution set S consists of two points $(x, y) = (2, 2)$ and $(2, -2)$.

$$S = \{(2, 2), (2, -2)\}$$
∎

EXAMPLE 5

Ron and Ken working together can paint a house in 45 hours. But after 30 hours Ron is stricken with "painter's elbow" and quits. Ken keeps working and finally finishes the job 40 hours later. How long would it have taken each of them to paint the house alone?

SOLUTION Let

x = number of hours needed for Ron to paint the house alone

y = number of hours needed for Ken to paint the house alone

Then Ron paints $1/x$ of the house in 1 hour, and Ken paints $1/y$ of the house in 1 hour. But working together they paint $\frac{1}{45}$ of the house in 1 hour. It is helpful to display this information as follows:

	Number of Hours Needed to Paint Entire House	Portion of House Painted in One Hour
Ron	x	$\dfrac{1}{x}$
Ken	y	$\dfrac{1}{y}$
Ron and Ken together	45	$\dfrac{1}{45}$

This translates into the equation.

$$\frac{1}{x} + \frac{1}{y} = \frac{1}{45} \qquad (1)$$

After 30 hours, Ron has painted $30 \cdot (1/x)$ of the house. Ken works a total of 70 hours, thereby painting $70 \cdot (1/y)$ of the house. After this time the entire house is painted. We thus write the equation

$$30 \cdot \frac{1}{x} + 70 \cdot \frac{1}{y} = 1 \qquad (2)$$

where the 1 indicates that the entire house is painted. These two equations give us the system

$$\begin{cases} \dfrac{1}{x} + \dfrac{1}{y} = \dfrac{1}{45} & (1) \\ 30 \cdot \dfrac{1}{x} + 70 \cdot \dfrac{1}{y} = 1 & (2) \end{cases}$$

which can be treated as a system in the variables $1/x$ and $1/y$. The elimination method yields

$$\begin{cases} \dfrac{1}{x} + \dfrac{1}{y} = \dfrac{1}{45} \\ 40 \cdot \dfrac{1}{y} = 1 - \dfrac{30}{45} = \dfrac{1}{3} \end{cases} \qquad (E_2 - 30E_1 \to E_2)$$

From the second equation, we find

$$\frac{1}{y} = \frac{1}{120}$$

Substitution of this value into the first equation then gives

$$\frac{1}{x} = \frac{1}{72}$$

Thus $x = 72$ and $y = 120$. Ron can paint the house in 72 hours and Ken can paint the house in 120 hours without any help.

Again, verify this result by checking the original statement. ∎

Section 10.3 Exercises

Solve the following systems of equations.

1. $\begin{cases} \dfrac{1}{x} - \dfrac{1}{y} = 2 \\ \dfrac{1}{x} + \dfrac{2}{y} = -1 \end{cases}$

2. $\begin{cases} \dfrac{1}{p} - \dfrac{5}{q} = 7 \\ \dfrac{3}{p} + \dfrac{4}{q} = 2 \end{cases}$

3. $\begin{cases} \dfrac{2}{t} + \dfrac{3}{u} = -7 \\ \dfrac{1}{t} - \dfrac{4}{u} = 24 \end{cases}$

4. $\begin{cases} \dfrac{2}{w} + \dfrac{3}{x} = 1 \\ \dfrac{1}{w} + \dfrac{1}{x} = 2 \end{cases}$

5. $\begin{cases} \dfrac{2}{u} + \dfrac{1}{v} = 6 \\ \dfrac{1}{u} + \dfrac{2}{v} = 0 \end{cases}$

6. $\begin{cases} \dfrac{2}{r} - \dfrac{3}{s} = 5 \\ \dfrac{1}{r} + \dfrac{1}{s} = 0 \end{cases}$

7. $\begin{cases} x^2 - y^2 = 5 \\ x^2 + 2y^2 = 17 \end{cases}$

8. $\begin{cases} x^2 - 3y^2 = 1 \\ 3x^2 + 2y^2 = 14 \end{cases}$

9. $\begin{cases} 3x^2 + 2y^2 = 62 \\ 4x^2 - y^2 = 68 \end{cases}$

10. $\begin{cases} 3x^2 - 2y^2 = 12 \\ x^2 + y^2 = 4 \end{cases}$

11. $\begin{cases} 9x^2 + 4y^2 = 72 \\ 25x^2 + 16y^2 = 244 \end{cases}$

12. $\begin{cases} 4x^2 - 3y^2 = 1 \\ 3x^2 + 2y^2 = 5 \end{cases}$

13. $\begin{cases} \log x - \log y = -2 \\ \log x + 2 \log y = 1 \end{cases}$

14. $\begin{cases} 4 \log x - \log y = 5 \\ 4 \log x + 3 \log y = 17 \end{cases}$

15. $\begin{cases} 3e^x - 2e^y = 4 \\ 4e^x - 6e^y = 2 \end{cases}$

16. $\begin{cases} e^x - e^y = 1 \\ 6e^x + 5e^y = 28 \end{cases}$

17. $\begin{cases} x^2 + y = -1 \\ x + y = -3 \end{cases}$

18. $\begin{cases} -2x + y = 3 \\ x^2 - y = 0 \end{cases}$

19. $\begin{cases} x - y^2 - 2y = 1 \\ 3x + y = 1 \end{cases}$

20. $\begin{cases} x^2 + 4x + y = -5 \\ -x + y = -1 \end{cases}$

21. $\begin{cases} x^2 + xy + y^2 = 2 \\ x^2 + y^2 = 4 \end{cases}$

22. $\begin{cases} 2x^2 + xy = 4 \\ 3x^2 - 5xy = 6 \end{cases}$

23. $\begin{cases} \log_2(x + 4) + y^3 = 11 \\ \log_2(x - 3) - y^3 = -8 \end{cases}$

24. $\begin{cases} r - \log_4(3s + 2) = -3 \\ 2r + \log_4(s - 1)^2 = -3 \end{cases}$

25. $\begin{cases} \log_2(7 - p) + q^2 = \log_2 5 \\ \log_2(2 + p) - q^2 = 2 \end{cases}$

26. $\begin{cases} \sqrt{u} + \log_2(3 - v) = 0 \\ \sqrt{u} - \log_2(2 + v) = -2 \end{cases}$

27. $\begin{cases} 3y - 2^{2x} = 2 \\ y - 2^x = 0 \end{cases}$

28. $\begin{cases} 3^{2v} - r^2 = 1 \\ 3^v + r^2 = 1 \end{cases}$

29. $\begin{cases} 2^{2x} + 3 \cdot 2^x = 28 \\ 3 \cdot 2^x + 2^{x+1} = 20 \end{cases}$

30. $\begin{cases} 3 \cdot 3^{2x} - 2 \cdot 3^x = 225 \\ 2 \cdot 3^{2x} - 3^x = 153 \end{cases}$

31. $\begin{cases} 4^x - (-2)^x = 72 \\ 2^{2x} + 2(-2)^x = 48 \end{cases}$

32. $\begin{cases} 2^{x+1} + 2^x = 6 \\ 2^{x+1} - 2^x = 2 \end{cases}$

33. $\begin{cases} 4 \cos x - y = 0 \\ y \cos x = 3 \end{cases}$

34. $\begin{cases} 2 \sin x \cos y = 1 \\ \sin^2 x + \cos^2 y = 1 \end{cases}$

35. $\begin{cases} 4 x \cos y = 3 \\ x - 3 \cos y = 0 \end{cases}$

36. $\begin{cases} 3 \tan x \sec x = 2 \\ \tan x + \sec x = \sqrt{3} \end{cases}$

Solve for real numbers x and y in the following.

37. $x^2 + 3 + (x + 1)i = y + 2x + yi$

38. $x(1 + i) - y(5 + i) = y^2 + 3(2 + i)$

39. $(2x + yi)(3x - yi) = 2xy + 9 + xi$

40. $(x - yi)(x + 2yi) = x^2 + y^2 + 2xi + 4$

41. Find two integers whose sum is 17 and for which the sum of their reciprocals is $\frac{17}{72}$.

42. If 3 is added to both numerator and denominator of a fraction, its value is increased by 25%. On the other hand, subtracting 1 from both numerator and denominator decreases its value by 25%. Find the fraction.

43. Two pipes with different capacities are used to fill a pool. If both are working, the pool can be filled in 20 hours. After 8 hours the first well runs dry; the second continues pumping for another 27 hours in order to fill the pool. How many hours would it take each of them individually to fill the pool?

44. Joseph and Adolph can empty a truck in 12 hours working together. But their schedules do not coincide. Joseph works for 4 hours alone, after which Adolph empties the truck in 18 hours. How many hours would it have taken each of them to empty the truck alone?

Section 10.4
Matrix Methods

The solution to a system of equations depends on the coefficients rather than on the specific names given to the variables. For instance, we would solve the two systems

$$\begin{cases} x + y = 2 \\ x - y = 0 \end{cases} \quad \begin{cases} u + v = 2 \\ u - v = 0 \end{cases}$$

in exactly the same way. The sequence of steps to be used in the elimination method is determined by considering the coefficients of the various variables. In fact, we can solve a system of equations by the elimination method without ever writing the variables, provided we keep track of the coefficients in the proper manner.

DEFINITION

An **m × n matrix** (where $m, n \in N$) is a rectangular array of real numbers:

$$M = \begin{bmatrix} a_{11} & a_{12} & \cdots & a_{1j} & \cdots & a_{1n} \\ a_{21} & a_{22} & \cdots & a_{2j} & \cdots & a_{2n} \\ \vdots & \vdots & & \vdots & & \vdots \\ a_{i1} & a_{i2} & \cdots & a_{ij} & \cdots & a_{in} \\ \vdots & \vdots & & \vdots & & \vdots \\ a_{m1} & a_{m2} & \cdots & a_{mj} & \cdots & a_{mn} \end{bmatrix}$$

$[a_{i1}, a_{i2}, \ldots, a_{ij}, \ldots, a_{in}]$ is the **ith row** of M, and

$$\begin{bmatrix} a_{1j} \\ a_{2j} \\ \vdots \\ a_{ij} \\ \vdots \\ a_{mj} \end{bmatrix}$$ is the **jth column** of M.

a_{ij} is the **element** in the ith row, jth column of M. If $m = n$, M is called a **square matrix**.

For any system of equations, such as Example 3 of Section 10.2, we consider its **coefficient matrix** and **augmented matrix,** as described here:

$$\begin{cases} 2x + y - z = 5 \\ x + y + z = 1 \\ -x + 2y + 2z = -4 \end{cases} \qquad \begin{bmatrix} 2 & 1 & -1 \\ 1 & 1 & 1 \\ -1 & 2 & 2 \end{bmatrix} \qquad \left[\begin{array}{ccc|c} 2 & 1 & -1 & 5 \\ 1 & 1 & 1 & 1 \\ -1 & 2 & 2 & -4 \end{array}\right]$$

System of equations   Coefficient matrix   Augmented matrix

We say that the coefficient matrix is **augmented** with the column

$$\begin{bmatrix} 5 \\ 1 \\ -4 \end{bmatrix}$$

The vertical broken line in the augmented matrix is optional. It merely indicates the location of the equal signs (=) in the corresponding equations.

The steps in our solution of this example are repeated here. Included alongside the equivalent systems are their respective augmented matrices. The recordkeeping notation of Section 10.2 is adapted by using R_i to denote the ith row. Then $(R_2 - 2R_1) \to R_2$ means 2 times the first row is subtracted from the second row; this becomes the new second row.

$$\begin{cases} 2x + y - z = 5 \\ x + y + z = 1 \\ -x + 2y + 2z = -4 \end{cases} \qquad \left[\begin{array}{ccc|c} 2 & 1 & -1 & 5 \\ 1 & 1 & 1 & 1 \\ -1 & 2 & 2 & -4 \end{array}\right]$$

$(E_1 \leftrightarrow E_2)$
$$\begin{cases} x + y + z = 1 \\ 2x + y - z = 5 \\ -x + 2y + 2z = -4 \end{cases} \qquad \left[\begin{array}{ccc|c} 1 & 1 & 1 & 1 \\ 2 & 1 & -1 & 5 \\ -1 & 2 & 2 & -4 \end{array}\right] \qquad (R_1 \leftrightarrow R_2)$$

$(E_2 - E_1 \to E_2)$
$(E_3 + E_1 \to E_3)$
$$\begin{cases} x + y + z = 1 \\ -y - 3z = 3 \\ 3y + 3z = -3 \end{cases} \qquad \left[\begin{array}{ccc|c} 1 & 1 & 1 & 1 \\ 0 & -1 & -3 & 3 \\ 0 & 3 & 3 & -3 \end{array}\right] \qquad \begin{array}{c} (R_2 - R_1 \to R_2) \\ (R_3 + R_1 \to R_3) \end{array}$$

$[(-1)E_2 \to E_2]$
$$\begin{cases} x + y + z = 1 \\ y + 3z = -3 \\ 3y + 3z = -3 \end{cases} \qquad \left[\begin{array}{ccc|c} 1 & 1 & 1 & 1 \\ 0 & 1 & 3 & -3 \\ 0 & 3 & 3 & -3 \end{array}\right] \qquad [(-1)R_2 \to R_2]$$

$(E_3 - 3E_2 \to E_3)$
$$\begin{cases} x + y + z = 1 \\ y + 3z = -3 \\ -6z = 6 \end{cases} \qquad \left[\begin{array}{ccc|c} 1 & 1 & 1 & 1 \\ 0 & 1 & 3 & -3 \\ 0 & 0 & -6 & 6 \end{array}\right] \qquad (R_3 - 3R_2 \to R_3)$$

$\left(-\dfrac{1}{6}E_3 \to E_3\right)$
$$\begin{cases} x + y + z = 1 \\ y + 3z = -3 \\ z = -1 \end{cases} \qquad \left[\begin{array}{ccc|c} 1 & 1 & 1 & 1 \\ 0 & 1 & 3 & -3 \\ 0 & 0 & 1 & -1 \end{array}\right] \qquad \left(-\dfrac{1}{6}R_3 \to R_3\right)$$

Each succeeding matrix is obtained from its predecessor in exactly the same way that succeeding equivalent systems are. That is to say, a linear system can be solved by manipulating its augmented matrix in the same way that we manipulate the system itself. This method of solution is called the **matrix form** of the elimination method.

> **PRINCIPLES OF THE MATRIX FORM OF THE ELIMINATION METHOD**
>
> *If* M *is the augmented matrix for a system of equations* S, *then, since a row of* M *corresponds to an equation in* S, *the following operations on* M *yield the matrix for a system of equations equivalent to the original system* S:
>
> 1. Interchanging two rows.
> 2. Multiplying any row by a nonzero constant.
> 3. Adding a nonzero multiple of one row to another.

EXAMPLE 1 Use the matrix form of the elimination method to solve the system of equations

$$\begin{cases} x + y = 2 \\ x - y = 0 \end{cases}$$

SOLUTION The augmented matrix for this system of equations is:

$$\begin{bmatrix} 1 & 1 & | & 2 \\ 1 & -1 & | & 0 \end{bmatrix}$$

Subtract the first row from the second:

$$\begin{bmatrix} 1 & 1 & | & 2 \\ 0 & -2 & | & -2 \end{bmatrix} \quad [(R_2 - R_1) \to R_2]$$

Multiply the last row by $-\frac{1}{2}$:

$$\begin{bmatrix} 1 & 1 & | & 2 \\ 0 & 1 & | & 1 \end{bmatrix} \quad \left(-\frac{1}{2}R_2 \to R_2\right)$$

This last matrix corresponds to the system

$$\begin{cases} x + y = 2 \\ \phantom{x + {}} y = 1 \end{cases}$$

Back substitution yields

$$x + 1 = 2$$
$$x = 1$$

The solution is $(x, y) = (1, 1)$. The solution set is $\{(1, 1)\}$. ∎

A matrix that represents a system of equations in echelon form is also said to have echelon form. Again, the first nonzero element of any row of a matrix is called the **leading element** of that row.

> **DEFINITION**
>
> A matrix is said to have **echelon form** if
>
> 1. The leading element in each nonzero row is 1.
> 2. The leading element of each nonzero row (other than the first) lies to the right of the leading element of its preceding row.
> 3. Rows consisting entirely of zeros appear last.

EXAMPLE 2

Use the matrix form of the elimination method to solve the following system.

$$\begin{cases} 3x - 4y - 2z = 3 \\ x + 2y + 2z = 0 \\ 2x + 2y + 5z = -2 \end{cases}$$

SOLUTION

We form the augmented matrix for the system and convert it to echelon form as follows.

$$\begin{bmatrix} 3 & -4 & -2 & | & 3 \\ 1 & 2 & 2 & | & 0 \\ 2 & 2 & 5 & | & -2 \end{bmatrix}$$

$$\begin{bmatrix} 1 & 2 & 2 & | & 0 \\ 3 & -4 & -2 & | & 3 \\ 2 & 2 & 5 & | & -2 \end{bmatrix} \quad (R_1 \leftrightarrow R_2)$$

$$\begin{bmatrix} 1 & 2 & 2 & | & 0 \\ 0 & -10 & -8 & | & 3 \\ 0 & -2 & 1 & | & -2 \end{bmatrix} \quad \begin{array}{l}(R_2 - 3R_1 \to R_2) \\ (R_3 - 2R_1 \to R_3)\end{array}$$

$$\begin{bmatrix} 1 & 2 & 2 & | & 0 \\ 0 & 1 & \frac{4}{5} & | & -\frac{3}{10} \\ 0 & -2 & 1 & | & -2 \end{bmatrix} \quad \left(-\frac{1}{10}R_2 \to R_2\right)$$

$$\begin{bmatrix} 1 & 2 & 2 & | & 0 \\ 0 & 1 & \frac{4}{5} & | & -\frac{3}{10} \\ 0 & 0 & \frac{13}{5} & | & -\frac{13}{5} \end{bmatrix} \quad (R_3 + 2R_2 \to R_3)$$

$$\begin{bmatrix} 1 & 2 & 2 & | & 0 \\ 0 & 1 & \frac{4}{5} & | & -\frac{3}{10} \\ 0 & 0 & 1 & | & -1 \end{bmatrix} \quad \left(\frac{5}{13}R_3 \to R_3\right)$$

This echelon form corresponds to the system of equations:

$$\begin{cases} x + 2y + 2z = 0 \\ y + \frac{4}{5}z = -\frac{3}{10} \\ z = -1 \end{cases}$$

Back substitution yields

$$y + \frac{4}{5}(-1) = -\frac{3}{10}$$

$$y = -\frac{3}{10} + \frac{4}{5} = \frac{5}{10}$$

$$y = \frac{1}{2}$$

and

$$x + 2 \cdot \frac{1}{2} + 2(-1) = 0$$

$$x = 1$$

The solution is $(x, y, z) = \left(1, \frac{1}{2}, -1\right)$. The solution set is $S = \left\{\left(1, \frac{1}{2}, -1\right)\right\}$. ∎

It is generally advantageous to continue generating zeros in the augmented matrix even after echelon form is reached. Done appropriately, the values of each of the variables can then be read from the matrix; back substitution becomes unnecessary.

DEFINITION

A matrix is said to be in **reduced row echelon form** if

1. It is in echelon form.
2. Each column which contains the leading element of a certain row contains only zeros in its other positions.

THE MATRIX METHOD OF SOLVING A SYSTEM OF EQUATIONS

1. Write the variables in the same order in each equation.
2. Extract the augmented matrix by suppressing the variables and replacing the equal signs with a broken vertical line. If desired, you may label the columns of the matrix corresponding to the respective variables.
3. Use the principles of the elimination method to convert the augmented matrix to one of these forms:
   a. Echelon form
   b. Reduced row echelon form
4. a. If the matrix is in echelon form, use back substitution to find the solution.
   b. If the matrix is in reduced row echelon form, read off the solution.

EXAMPLE 3 In the solution of Example 2, we found the echelon form of the augmented matrix of the system of equations. Convert this echelon form to reduced row echelon form and find the solution of the system from this form.

SOLUTION The echelon form for the system of Example 2 was

$$\begin{bmatrix} 1 & 2 & 2 & | & 0 \\ 0 & 1 & \dfrac{4}{5} & | & -\dfrac{3}{10} \\ 0 & 0 & 1 & | & -1 \end{bmatrix}$$

We convert this to reduced row echelon form as follows.

$$\begin{bmatrix} 1 & 2 & 0 & | & 2 \\ 0 & 1 & 0 & | & \dfrac{1}{2} \\ 0 & 0 & 1 & | & -1 \end{bmatrix} \quad \begin{array}{l}(R_1 - 2R_3 \to R_1) \\ \left(R_2 - \dfrac{4}{5}R_3 \to R_2\right)\end{array}$$

$$\begin{bmatrix} 1 & 0 & 0 & | & 1 \\ 0 & 1 & 0 & | & \dfrac{1}{2} \\ 0 & 0 & 1 & | & -1 \end{bmatrix} \quad (R_1 - 2R_2 \to R_1)$$

This matrix corresponds to the system

$$\begin{cases} x & = 1 \\ y & = \dfrac{1}{2} \\ z & = -1 \end{cases}$$

which gives us the solution $(x, y, z) = \left(1, \dfrac{1}{2}, -1\right)$ as in Example 2. No matter what solution technique is used, we should verify the solution by substitution into each of the original equations. That is left to the reader. ∎

EXAMPLE 4 Solve the following system by converting its augmented matrix to reduced row echelon form.

$$\begin{cases} 3w - 4x - 2y + z = 11 \\ 2w - 2x - 3y + 2z = 13 \\ 2w - 2x + 2y + z = 5 \\ 2w + 4x - y + 3z = 14 \end{cases}$$

SOLUTION The augmented matrix for the system is

$$\begin{array}{cccc} w & x & y & z \end{array}$$
$$\begin{bmatrix} 3 & -4 & -2 & 1 & | & 11 \\ 2 & -2 & -3 & 2 & | & 13 \\ 2 & -2 & 2 & 1 & | & 5 \\ 2 & 4 & -1 & 3 & | & 14 \end{bmatrix}$$

Rather than introduce fractions immediately, we find that we can obtain $a_{11} = 1$

by subtracting the second row from the first:

$$\begin{bmatrix} 1 & -2 & 1 & -1 & | & -2 \\ 2 & -2 & -3 & 2 & | & 13 \\ 2 & -2 & 2 & 1 & | & 5 \\ 2 & 4 & -1 & 3 & | & 14 \end{bmatrix} \quad (R_1 - R_2 \to R_1)$$

Then we generate zeros as the other elements in the first column:

$$\begin{bmatrix} 1 & -2 & 1 & -1 & | & -2 \\ 0 & 2 & -5 & 4 & | & 17 \\ 0 & 2 & 0 & 3 & | & 9 \\ 0 & 8 & -3 & 5 & | & 18 \end{bmatrix} \quad \begin{array}{l} (R_2 - 2R_1 \to R_2) \\ (R_3 - 2R_1 \to R_3) \\ (R_4 - 2R_1 \to R_4) \end{array}$$

We can also generate zeros in all but the a_{22} position of column 2 without resorting to fractions:

$$\begin{bmatrix} 1 & 0 & -4 & 3 & | & 15 \\ 0 & 2 & -5 & 4 & | & 17 \\ 0 & 0 & 5 & -1 & | & -8 \\ 0 & 0 & 17 & -11 & | & -50 \end{bmatrix} \quad \begin{array}{l} (R_1 + R_2 \to R_1) \\ \\ (R_3 - R_2 \to R_3) \\ (R_4 - 4R_2 \to R_4) \end{array}$$

Now we notice $a_{34} = -1$. If we interchange rows 3 and 4, we can obtain $a_{44} = 1$:

$$\begin{bmatrix} 1 & 0 & -4 & 3 & | & 15 \\ 0 & 2 & -5 & 4 & | & 17 \\ 0 & 0 & 17 & -11 & | & -50 \\ 0 & 0 & 5 & -1 & | & -8 \end{bmatrix} \quad (R_3 \leftrightarrow R_4)$$

$$\begin{bmatrix} 1 & 0 & -4 & 3 & | & 15 \\ 0 & 2 & -5 & 4 & | & 17 \\ 0 & 0 & 17 & -11 & | & -50 \\ 0 & 0 & -5 & 1 & | & 8 \end{bmatrix} \quad [(-1)R_4 \to R_4]$$

Next we use the $a_{44} = 1$ position to generate zeros in the rest of the fourth column:

$$\begin{bmatrix} 1 & 0 & 11 & 0 & | & -9 \\ 0 & 2 & 15 & 0 & | & -15 \\ 0 & 0 & -38 & 0 & | & 38 \\ 0 & 0 & -5 & 1 & | & 8 \end{bmatrix} \quad \begin{array}{l} (R_1 - 3R_4 \to R_1) \\ (R_2 - 4R_4 \to R_2) \\ (R_3 + 11R_4 \to R_3) \end{array}$$

Dividing row 3 by -38, we obtain $a_{33} = 1$. We then use the $a_{33} = 1$ position to generate zeros in the rest of the third column:

$$\begin{bmatrix} 1 & 0 & 11 & 0 & | & -9 \\ 0 & 2 & 15 & 0 & | & -15 \\ 0 & 0 & 1 & 0 & | & -1 \\ 0 & 0 & -5 & 1 & | & 8 \end{bmatrix} \quad \left(-\frac{1}{38}R_3 \to R_3\right)$$

$$\begin{bmatrix} 1 & 0 & 0 & 0 & | & 2 \\ 0 & 2 & 0 & 0 & | & 0 \\ 0 & 0 & 1 & 0 & | & -1 \\ 0 & 0 & 0 & 1 & | & 3 \end{bmatrix} \quad \begin{array}{l} (R_1 - 11R_3 \to R_1) \\ (R_2 - 15R_3 \to R_2) \\ \\ (R_4 + 5R_3 \to R_4) \end{array}$$

Finally, dividing the second row by 2 yields the reduced row echelon form:

$$\begin{array}{cccc} w & x & y & z \end{array}$$
$$\left[\begin{array}{cccc|c} 1 & 0 & 0 & 0 & 2 \\ 0 & 1 & 0 & 0 & 0 \\ 0 & 0 & 1 & 0 & -1 \\ 0 & 0 & 0 & 1 & 3 \end{array}\right] \quad \left(\frac{1}{2}R_2 \to R_2\right)$$

Converting back to the system of equations, we obtain the solution:

$$\begin{cases} w & = 2 \\ x & = 0 \\ y & = -1 \\ z = 3 \end{cases}$$

Checking the solution, we have

$$\begin{cases} 3 \cdot 2 - 4 \cdot 0 - 2(-1) + 3 \stackrel{\checkmark}{=} 11 \\ 2 \cdot 2 - 2 \cdot 0 - 3(-1) + 2 \cdot 3 \stackrel{\checkmark}{=} 13 \\ 2 \cdot 2 - 2 \cdot 0 + 2(-1) + 3 \stackrel{\checkmark}{=} 5 \\ 2 \cdot 2 + 4 \cdot 0 - (-1) + 3 \cdot 3 \stackrel{\checkmark}{=} 14 \end{cases}$$

The solution is $(w, x, y, z) = (2, 0, -1, 3)$. The solution set is $S = \{(2, 0, -1, 3)\}$. ∎

Checking the results is especially important when long, drawn-out computations are involved in finding the solution. Had an error been indicated when checking the solution in Example 4, the records that we kept along the way $(R_2 - 2R_1 \to R_2, R_1 + R_2 \to R_1$, and so on) would be invaluable in tracking down the error or errors.

There are generally many sequences of steps that will convert a matrix to reduced row echelon form. By being judicious in our selection of steps, we were able to avoid the introduction of fractions in Example 4. While this cannot be done in all cases, careful selection of steps can often reduce the number of fractions with which we must work to solve a system of equations.

EXAMPLE 5

Use the matrix method to solve the following system.

$$\begin{cases} x - 2y + z + w = 4 \\ 2x + y - 2z - w = 3 \\ -x - 8y + 7z + 5w = 13 \\ 3x + y - 2z + 2w = 6 \end{cases}$$

SOLUTION

Form the augmented matrix and generate zeros below the first element in column 1 as follows:

$$\left[\begin{array}{cccc|c} 1 & -2 & 1 & 1 & 4 \\ 2 & 1 & -2 & -1 & 3 \\ -1 & -8 & 7 & 5 & 13 \\ 3 & 1 & -2 & 2 & 6 \end{array}\right]$$

$$\begin{bmatrix} 1 & -2 & 1 & 1 & | & 4 \\ 0 & 5 & -4 & -3 & | & -5 \\ 0 & -10 & 8 & 6 & | & 17 \\ 0 & 7 & -5 & -1 & | & -6 \end{bmatrix} \quad \begin{array}{l} (R_2 - 2R_1 \to R_2) \\ (R_3 + R_1 \to R_3) \\ (R_4 - 3R_1 \to R_4) \end{array}$$

Again, trying to postpone the introduction of fractions as long as possible, we use the $a_{22} = 5$ position to eliminate $a_{32} = -10$:

$$\begin{bmatrix} 1 & -2 & 1 & 1 & | & 4 \\ 0 & 5 & -4 & -3 & | & -5 \\ 0 & 0 & 0 & 0 & | & 7 \\ 0 & 7 & -5 & -1 & | & -6 \end{bmatrix} \quad (R_3 + 2R_2 \to R_3)$$

We can stop here. The third row represents the equation

$$0 = 7$$

That is, *if* the original equations are satisfied, then $0 = 7$. This means there are no solutions. The solution set is empty. ∎

EXAMPLE 6

Use the matrix method to solve the following system.

$$\begin{cases} x + y + z + 3u + 2v = 4 \\ 4x + 3y - z + 8u + v = 2 \\ -2x - y + 2z - 3u + 2v = 3 \end{cases}$$

SOLUTION

Form the augmented matrix and convert it to reduced row echelon form as follows.

$$\begin{array}{ccccc} x & y & z & u & v \end{array}$$

$$\begin{bmatrix} 1 & 1 & 1 & 3 & 2 & | & 4 \\ 4 & 3 & -1 & 8 & 1 & | & 2 \\ -2 & -1 & 2 & -3 & 2 & | & 3 \end{bmatrix}$$

$$\begin{bmatrix} 1 & 1 & 1 & 3 & 2 & | & 4 \\ 0 & -1 & -5 & -4 & -7 & | & -14 \\ 0 & 1 & 4 & 3 & 6 & | & 11 \end{bmatrix} \quad \begin{array}{l} (R_2 - 4R_1 \to R_2) \\ (R_3 + 2R_1 \to R_3) \end{array}$$

$$\begin{bmatrix} 1 & 1 & 1 & 3 & 2 & | & 4 \\ 0 & 1 & 5 & 4 & 7 & | & 14 \\ 0 & 1 & 4 & 3 & 6 & | & 11 \end{bmatrix} \quad [(-1)R_2 \to R_2]$$

$$\begin{bmatrix} 1 & 0 & -4 & -1 & -5 & | & -10 \\ 0 & 1 & 5 & 4 & 7 & | & 14 \\ 0 & 0 & -1 & -1 & -1 & | & -3 \end{bmatrix} \quad \begin{array}{l} (R_1 - R_2 \to R_1) \\ (R_3 - R_2 \to R_3) \end{array}$$

$$\begin{bmatrix} 1 & 0 & -4 & -1 & -5 & | & -10 \\ 0 & 1 & 5 & 4 & 7 & | & 14 \\ 0 & 0 & 1 & 1 & 1 & | & 3 \end{bmatrix} \quad [(-1)R_3 \to R_3]$$

$$\begin{bmatrix} 1 & 0 & 0 & 3 & -1 & | & 2 \\ 0 & 1 & 0 & -1 & 2 & | & -1 \\ 0 & 0 & 1 & 1 & 1 & | & 3 \end{bmatrix} \quad \begin{array}{l} (R_1 + 4R_3 \to R_1) \\ (R_2 - 5R_3 \to R_2) \end{array}$$

The matrix is now in reduced row echelon form. It corresponds to the system of equations

$$\begin{cases} x + 3u - v = 2 \\ y - u + 2v = -1 \\ z + u + v = 3 \end{cases}$$

The best we can do is to express x, y, and z in terms of u and v:

$$x = 2 - 3u + v$$
$$y = -1 + u - 2v$$
$$z = 3 - u - v$$

The solution is $(x, y, z, u, v) = (2 - 3u + v, -1 + u - 2v, 3 - u - v, u, v)$ for arbitrary u and v. The solution set is

$$S = \{(2 - 3u + v, -1 + u - 2v, 3 - u - v, u, v) : u \in \mathbf{R}, v \in \mathbf{R}\}$$ ∎

EXAMPLE 7

Find the equation of the circle passing through the three points $(2, 2)$, $(-3, 7)$, $(1, 5)$; find its radius and center.

SOLUTION

Recall that the general equation of a circle (Section 3.1) is

$$x^2 + y^2 + ax + by + c = 0$$

If this circle is to pass through the given points, the following equations must be satisfied.

$(2, 2)$: $4 + 4 + 2a + 2b + c = 0$
$(-3, 7)$: $9 + 49 - 3a + 7b + c = 0$
$(1, 5)$: $1 + 25 + a + 5b + c = 0$

Collecting terms results in the following:

$$\begin{cases} 2a + 2b + c = -8 \\ -3a + 7b + c = -58 \\ a + 5b + c = -26 \end{cases}$$

Thus what originally seems to be a system of quadratic equations is in actuality a linear system, since we are seeking a, b, and c. This system can be solved in the usual way to find

$$a = 6 \qquad b = -4 \qquad c = -12$$

The equation of this circle is

$$x^2 + y^2 + 6x - 4y - 12 = 0$$

Completing the squares will recast this equation in a form that indicates the radius and center of the circle:

$$(x^2 + 6x) + (y^2 - 4y) = 12$$
$$(x^2 + 6x + 9) + (y^2 - 4y + 4) = 12 + 9 + 4$$
$$(x + 3)^2 + (y - 2)^2 = 25$$

The circle is centered at $(-3, 2)$ and has radius 5. ∎

The matrix method is especially useful for large systems because it cuts down on the writing needed to solve a problem. Another advantage is that the numbers are easier to see when the variables are suppressed. Since only the coefficients are used and the names of the variables are suppressed, this method lends itself to machine computation. It is very easy to program a computer to solve a system of equations by using the matrix method. Generally, the machine is instructed to put the matrix (or system of equations) in echelon form or reduced row echelon form.

Section 10.4 Exercises

Solve the following systems by the matrix method.

1. $\begin{cases} z - 4t = 7 \\ z + 2t = 1 \end{cases}$

2. $\begin{cases} 7r + 6t = 10 \\ 13r + 8t = 6 \end{cases}$

3. $\begin{cases} 6w - 2y = -5 \\ w + 7y = 1 \end{cases}$

4. $\begin{cases} 2s + 3t = 7 \\ s - 4t = -24 \end{cases}$

5. $\begin{cases} 6x - 4y = 7 \\ 5x - 3y = 6 \end{cases}$

6. $\begin{cases} 3u - 2v = 4 \\ 2u - v = 3 \end{cases}$

7. $\begin{cases} r - s = 2 \\ 2r - 3s = -1 \end{cases}$

8. $\begin{cases} a - 5b = 8 \\ 2a + 3b = -10 \end{cases}$

9. $\begin{cases} 3p + 2q = 8 \\ 2p + q = 5 \end{cases}$

10. $\begin{cases} 2s - 3t = 8 \\ 5s + 2t = 1 \end{cases}$

11. $\begin{cases} 2a + 5b = -2 \\ a - 3b = 10 \\ 3a + 4b = 4 \end{cases}$

12. $\begin{cases} 3x - y = -4 \\ 2x + 3y = 7 \\ -x + 2y = 7 \end{cases}$

13. $\begin{cases} 2x - 2y - z = -7 \\ -x + y + 3z = 6 \\ 2x - y + 2z = 5 \end{cases}$

14. $\begin{cases} 3a + 2b + c = 1 \\ a - b + c = 0 \\ b + 2c = -2 \end{cases}$

15. $\begin{cases} 2r + s + t = 0 \\ r - 3s - 2t = 0 \\ r - s - 2t = 0 \end{cases}$

16. $\begin{cases} 2u - v + w = -3 \\ u - 2v - 3w = 4 \\ u + v - 2w = 5 \end{cases}$

17. $\begin{cases} 2r - s - 3t = 0 \\ r - 5s - 6t = -9 \\ r - 2s - 3t = -3 \end{cases}$

18. $\begin{cases} x + y - z = 4 \\ 2x - y + 2z = 6 \\ x + 2y - 3z = 10 \end{cases}$

19. $\begin{cases} p - 2q + r = 8 \\ 2p - q + 3r = 11 \\ 3p + 2q - 2r = 3 \end{cases}$

20. $\begin{cases} 2u - v + w = 8 \\ -u + 2v + w = -1 \\ u + v + 2w = 7 \end{cases}$

21. $\begin{cases} 3x - y + 2z = 3 \\ 2x + 3y + z = -1 \end{cases}$

22. $\begin{cases} t - u + 2v = 7 \\ t - 2u + 3v = 11 \end{cases}$

23. $\begin{cases} 3r - 2s + 5t = 12 \\ 2r - 5s + 3t = 4 \\ r + 3s + 2t = 8 \end{cases}$

24. $\begin{cases} x - 2y - 3z = 3 \\ 2x + y - z = 1 \\ x + y + z = -4 \end{cases}$

25. $\begin{cases} u - 2v - w = 0 \\ 2u + v - w = 0 \\ 2u + v - 3w = 0 \end{cases}$

26. $\begin{cases} -x + 2y + z = 8 \\ 2x - y + z = -1 \\ x + y + 2z = 7 \end{cases}$

27. $\begin{cases} 3x + 4y + z = 0 \\ 5x + 3y + z = 1 \\ x - 4y - 3z = 5 \\ 2x - y + 4z = -5 \end{cases}$

28. $\begin{cases} x + y - z = 1 \\ 2x + 3y - 2z = 1 \\ x - 2y + 3z = 0 \\ 2x + y - z = 2 \end{cases}$

29. $\begin{cases} r - 2s - t - 3u = -9 \\ r + s - t = 0 \\ 3r + 4s + u = 6 \\ 2s - 2t + u = 3 \end{cases}$

30. $\begin{cases} -x + y + z - 2w = 1 \\ x - y + 2z - w = -1 \\ 3x + y + 2z - 2w = -2 \\ -2x - 2y + z + 2w = 3 \end{cases}$

31. $\begin{cases} 2r - s + t + u = 1 \\ r + 3s - 2t + 5u = 0 \\ 3r - 2s + 4t - u = 0 \\ -r + s - 3t + 6u = 1 \end{cases}$

32. $\begin{cases} 3u - 14v + 4w + 2x = -1 \\ -2u + 13v + 2w - 2x = 3 \\ u - 5v - 2w + 2x = 2 \\ u - 11v - 4w + 3x = -2 \end{cases}$

33. $\begin{cases} -a + 3b + 2c - d = -6 \\ 3a - 2b + c - 6d = 2 \\ -2a + b + 2c = -2 \\ a - 3b + c - 4d = 4 \end{cases}$

34. $\begin{cases} 2x + y - 2z = 1 \\ -x - 2y + z - 5t = -4 \\ -x - y - 2z + t = 2 \\ 3x + y + 3z - 10t = -1 \end{cases}$

35. $\begin{cases} 2r + s - t = 3 \\ r - 3s + 2t = -4 \\ 3r + s - 3t + u = 1 \\ r + 2s - 4t - u = -2 \end{cases}$

36. $\begin{cases} -2u + v - x = -6 \\ u + 2w - 3x = 4 \\ -2v + 3w - 6x = -2 \\ -u + 3v + w - 2x = 0 \end{cases}$

37. $\begin{cases} x + y - z + w = -2 \\ 2x - y + 2z - w = 7 \\ -x + 2y + z + 2w = -1 \end{cases}$

38. $\begin{cases} x - 3y + z - 6w = 6 \\ -2x + 3y + 8w = -5 \\ 5x - 3y - z - 12w = 6 \end{cases}$

39. $\begin{cases} 4u - 3v + w + 2x + 2y = -4 \\ u - 2v + 2w + 2x - 3y = -8 \\ -2u + v - w + x = 8 \\ -u - 2v + 2w + 3x - 7y = -8 \\ 2u - v - 2w - x - 5y = -4 \end{cases}$

40. $\begin{cases} r + s + t + u + 2v = 3 \\ 3r + s - t - v = 0 \\ 2r - 2s - t + 2u + v = 4 \\ 4r - s - 2t - u - 2v = 0 \\ 2r - 3s + t - u + 4v = 6 \end{cases}$

In each of the following, find the equation of the circle passing through the three points. Identify its center and radius.

41. (5, 5), (4, 6), (1, 7)

42. (−2, −16), (3, 9), (10, 2)

43. (0, 0), (48, 14), (17, −17)

44. (1, 7), (−2, 8), (−6, 6)

45. (−6, −4), (2, −16), (19, 1)

46. (4, −20), (22, 4), (21, −3)

47. Old MacDonald has three farms, which he rents to his neighbors. His holdings total 960 acres. The farms rent for $80, $60, and $50 per acre, respectively. His combined income from the higher-priced rentals exactly doubles his income from the $50-per-acre farm. If his total income is $60,000, how large is each farm?

48. Sam Showplace uses three brands of combination weed killer and plant food on his lawn. Super Feeder brand is 80% plant food, 10% weed killer, and 10% filler material. Weed-Go brand contains 30% plant food, 33% weed killer, and 37% filler. In bulk (no label), the mixture is 60% plant food, 12% weed killer, and 28% filler. How should Sam prepare his mixture in order to obtain 65% plant food, 16% weed killer, and 19% filler? (*Hint:* See Example 6, Section 10.2.)

49. Three pipes are used to fill a pool. Pipe 1 and pipe 2 both run for 2 hours; then these are turned off and the third is turned on for 1 hour and the pool is filled. Last week someone turned pipe 1 and pipe 2 off after 1 hour and then turned pipe 3 on. Pipe 3 then took 5 hours and 30 minutes to finish filling the pool. Yesterday all three were turned on simultaneously. After 30 minutes, pipe 3 stopped. After another 30 minutes, pipe 2 stopped. Two hours later the pool was filled. How long would it take each pipe alone to fill the pool? (*Hint:* See Example 5, Section 10.3.)

50. Dan, Paul, and Les can build a garage in 24 hours. However, they cannot always work together. If Les works alone for 54 hours, Dan and Paul can finish the job in 15 hours. If Dan works alone for 17 hours, Paul and Les can finish the job in 30 hours. How long would it take each one alone to build a garage? (*Hint:* See Example 5, Section 10.3.)

51. Henry lost $56 to the "one-armed-bandits" in Las Vegas. He lost three times as much on the half-dollar machines as on the quarter and nickel machines put together. The total number of coins lost was 224. How many nickels, quarters, and halves did he lose?

52. On her paper route, Cindy collected $5.20 in nickels, dimes, and quarters in a total of 33 coins. The number of quarters is one more than twice the number of nickels. How many of each type of coin does she have?

53. Kevin has 13 coins in nickels, dimes, and quarters totaling $1.45. If he has one more dime than quarters, find how many coins of each type he has.

54. Tara has $2.00 in nickels, dimes, and quarters. She has 14 coins in all, and there are twice as many dimes as nickels. How many of each type of coin does she have?

55. One number is 9 more than twice another number, whereas twice the other number is 5 more than three times the first. What are the two numbers?

56. A two-digit number is 2 more than six times the sum of the digits. If the digits are reversed the number is decreased by 18. What is the number? (*Hint:* See Exercise 71, Section 10.1.)

57. Of three numbers, the third is 1 less than the sum of the other two. The third is also 8 more than twice the second. The first is 1 less than three times the difference of the last two (third minus second). Find the numbers.

58. Find a three-digit number whose digits add up to 11 and whose tens digit is both twice the units digit and three times the hundreds digit.

59. Find a three-digit number whose units digit is 1 less than its tens digit; the sum of its digits is 10; if the digits are reversed, the new number is 297 more than the original. (*Hint:* See Exercise 45, Section 10.2.)

60. The units digit of a three-digit number is 1 more than the sum of the other two digits. If the digits are reversed, the number is increased by 792. If the hundreds and tens digits are interchanged, the original number is increased by 540. Find the number. (*Hint:* See Exercise 45, Section 10.2.)

Section 10.5

Determinants

In the preceding section, we indicated that the matrix of a system of equations can be used to solve the system; using the matrix eliminates the need to write down the variables repeatedly. There is another way in which the matrix can be used to find the solution. Central to this method of solution is the concept of **determinant** of a matrix. In this section, we shall be working solely with "square" matrices. An $n \times n$ matrix or determinant is said to have **order** n.

The determinant of a square matrix will be defined inductively: that is, a determinant of order $(n + 1)$ will be defined in terms of determinants of order n. The definitions may seem somewhat arbitrary at first, but we shall show that determinants arise naturally in the solution of linear systems.

> **DEFINITION**
>
> The **determinant of a** 1×1 **matrix** $A = [a_{11}]$ is *defined* as
>
> $$\det A = \det[a_{11}] = |A| = a_{11}$$

We generally use the vertical bars to denote the determinant of a matrix, but if there is the danger of confusion with the absolute value symbol, as there is with the 1×1 matrix, we use the alternative symbol det.

> **DEFINITION**
>
> The **determinant of a** 2×2 **matrix** $A = \begin{bmatrix} a_{11} & a_{12} \\ a_{21} & a_{22} \end{bmatrix}$ is *defined* as
>
> $$\det A = \det \begin{bmatrix} a_{11} & a_{12} \\ a_{21} & a_{22} \end{bmatrix} = \begin{vmatrix} a_{11} & a_{12} \\ a_{21} & a_{22} \end{vmatrix} = a_{11}a_{22} - a_{12}a_{21}$$

EXAMPLE 1

a. $\begin{vmatrix} 1 & 2 \\ 3 & 4 \end{vmatrix} = 1 \cdot 4 - 2 \cdot 3 = -2$

b. $\begin{vmatrix} -2 & -4 \\ 5 & 3 \end{vmatrix} = (-2) \cdot 3 - (-4) \cdot 5 = 14$

Before defining determinants of higher order, we introduce the concepts of *minor* and *cofactor*.

> **DEFINITION**
>
> For any square matrix $A = (a_{ij})$
>
> 1. The **minor** M_{ij} of the element a_{ij} is the determinant of the matrix that remains when the ith row and the jth column of the matrix are removed.
> 2. The **cofactor** of the element a_{ij} is
>
> $$A_{ij} = (-1)^{i+j} M_{ij}$$

EXAMPLE 2

Let $M = \begin{bmatrix} 1 & 2 & 0 \\ 3 & -1 & -2 \\ 4 & -3 & 5 \end{bmatrix}$. Find each of the following.

a. M_{11} and A_{11}
b. M_{13} and A_{13}
c. M_{32} and A_{32}

SOLUTION

a. $\begin{bmatrix} 1 & 2 & 0 \\ 3 & -1 & -2 \\ 4 & -3 & 5 \end{bmatrix}$ $M_{11} = \begin{vmatrix} -1 & -2 \\ -3 & 5 \end{vmatrix} = -5 - 6 = -11$

$A_{11} = (-1)^{1+1} M_{11} = (-1)^2(-11) = 1 \cdot (-11) = -11$

b. $\begin{bmatrix} 1 & 2 & 0 \\ 3 & -1 & -2 \\ 4 & -3 & 5 \end{bmatrix}$ $M_{13} = \begin{vmatrix} 3 & -1 \\ 4 & -3 \end{vmatrix} = -9 + 4 = -5$

$A_{13} = (-1)^{1+3} M_{13} = (-1)^4(-5) = 1 \cdot (-5) = -5$

c. $\begin{bmatrix} 1 & 2 & 0 \\ 3 & -1 & -2 \\ 4 & -3 & 5 \end{bmatrix}$ $M_{32} = \begin{vmatrix} 1 & 0 \\ 3 & -2 \end{vmatrix} = -2$

$A_{32} = (-1)^{3+2} M_{32} = (-1)^5(-2) = (-1)(-2) = 2$ ∎

An easy way to keep track of the appropriate sign $(-1)^{i+j}$ to use with the cofactor $A_{ij} = (-1)^{i+j} M_{ij}$ is to impose a checkerboard pattern of + and − signs on the matrix:

$$\begin{bmatrix} + & - & + & - & + \\ - & + & - & + & - \\ + & - & + & - & + \\ - & + & - & + & - \\ + & - & + & - & + \end{bmatrix}$$

> **DEFINITION**
>
> A **3 × 3 determinant** is defined in terms of 2 × 2 determinants:
> $$\begin{vmatrix} a_{11} & a_{12} & a_{13} \\ a_{21} & a_{22} & a_{23} \\ a_{31} & a_{32} & a_{33} \end{vmatrix} = a_{11}A_{11} + a_{12}A_{12} + a_{13}A_{13}$$
> $$= a_{11}M_{11} - a_{12}M_{12} + a_{13}M_{13}$$

Since other methods of evaluating determinants will be given later, this particular evaluation will be called *expansion by the first row*.

Note that for $A = \begin{bmatrix} a_{11} & a_{12} & a_{13} \\ a_{21} & a_{22} & a_{23} \\ a_{31} & a_{32} & a_{33} \end{bmatrix}$

$$\begin{aligned} |A| &= a_{11}A_{11} + a_{12}A_{12} + a_{13}A_{13} = a_{11}M_{11} - a_{12}M_{12} + a_{13}M_{13} \\ &= a_{11}(a_{22}a_{33} - a_{23}a_{32}) - a_{12}(a_{21}a_{33} - a_{23}a_{31}) + a_{13}(a_{21}a_{32} - a_{22}a_{31}) \\ &= a_{11}a_{22}a_{33} - a_{11}a_{23}a_{32} - a_{12}a_{21}a_{33} + a_{12}a_{23}a_{31} \\ &\quad + a_{13}a_{21}a_{32} - a_{13}a_{22}a_{31} \end{aligned} \quad (1)$$

We can show that
$$a_{21}A_{21} + a_{22}A_{22} + a_{23}A_{23} = |A|$$
and that
$$a_{13}A_{13} + a_{23}A_{23} + a_{33}A_{33} = |A|$$
by working the left side of each of these equations down to the value given in (1). That is, a 3 × 3 determinant can be evaluated by expanding by the second row or expanding by the third column. In fact, expanding by any row or column will yield the same value.

> The determinant of a 3 × 3 matrix A can be evaluated by either of the following methods.
>
> 1. Expanding by any row:
> $$|A| = a_{i1}A_{i1} + a_{i2}A_{i2} + a_{i3}A_{i3}$$
> 2. Expanding by any column:
> $$|A| = a_{1j}A_{1j} + a_{2j}A_{2j} + a_{3j}A_{3j}$$

EXAMPLE 3

Let $A = \begin{bmatrix} 1 & 2 & 0 \\ 3 & -1 & -2 \\ 4 & -3 & 5 \end{bmatrix}$. Evaluate $|A|$ by

a. Expanding by the first row.

b. Expanding by the second column.

SOLUTION

a. $\begin{vmatrix} 1 & 2 & 0 \\ 3 & -1 & -2 \\ 4 & -3 & 5 \end{vmatrix} = 1\begin{vmatrix} -1 & -2 \\ -3 & 5 \end{vmatrix} - 2\begin{vmatrix} 3 & -2 \\ 4 & 5 \end{vmatrix} + 0\begin{vmatrix} 3 & -1 \\ 4 & -3 \end{vmatrix}$

$= 1(-5 - 6) - 2(15 + 8) + 0$

$= -57$

b. In expanding by the second column, note that the alternating signs associated with the cofactors begin with a negative sign: $(-1)^{1+2} = (-1)^3 = -1$.

$\begin{vmatrix} 1 & 2 & 0 \\ 3 & -1 & -2 \\ 4 & -3 & 5 \end{vmatrix} = a_{12}A_{12} + a_{22}A_{22} + a_{32}A_{32}$

$= (-1)^{1+2}a_{12}M_{12} + (-1)^{2+2}a_{22}M_{22} + (-1)^{3+2}a_{32}M_{32}$

$= (-1) \cdot 2 \cdot \begin{vmatrix} 3 & -2 \\ 4 & 5 \end{vmatrix} + (+1) \cdot (-1) \cdot \begin{vmatrix} 1 & 0 \\ 4 & 5 \end{vmatrix}$

$+ (-1) \cdot (-3) \cdot \begin{vmatrix} 1 & 0 \\ 3 & -2 \end{vmatrix}$

$= -2(15 + 8) - (5 - 0) + 3(-2 - 0)$

$= -57$ ∎

The determinants of order 4 are defined in terms of determinants of order 3; then determinants of order 5 are defined in terms of determinants of order 4 and so on.

DEFINITION

$\begin{vmatrix} a_{11} & \cdots & a_{1n} \\ \vdots & & \vdots \\ a_{n1} & \cdots & a_{nn} \end{vmatrix} = a_{11}A_{11} + a_{12}A_{12} + \cdots + a_{1n}A_{1n}$

$= a_{11}M_{11} - a_{12}M_{12} + \cdots + (-1)^{1+n}a_{1n}M_{1n}$

The evaluation of fourth-order or larger determinants can be very laborious and time-consuming. In addition, the number of operations that are necessary for the evaluation is very conducive to arithmetical and/or sign errors. Therefore, we shall discuss several properties of determinants that simplify their evaluation. Formal proofs will not be given for determinants of arbitrary order; we shall only establish the validity of the properties for second- or third-order determinants and leave the verification for higher orders to a course in linear algebra. As with third-order determinants, we are not restricted to expansion by the first row.

PROPERTY I

A determinant of order n can be evaluated by either of the following methods.

1. Expanding by any row:

$$|A| = a_{i1}A_{i1} + a_{i2}A_{i2} + \cdots + a_{in}A_{in}$$

2. Expanding by any column:

$$|A| = a_{1j}A_{1j} + a_{2j}A_{2j} + \cdots + a_{nj}A_{nj}$$

For example, expanding by the fourth row of a fifth-order determinant, we have

$$\begin{vmatrix} a_{11} & \cdots & \cdots & \cdots & a_{15} \\ a_{21} & \cdots & \cdots & \cdots & \cdots \\ a_{31} & \cdots & \cdots & \cdots & \cdots \\ a_{41} & a_{42} & a_{43} & a_{44} & a_{45} \\ a_{51} & \cdots & \cdots & \cdots & a_{55} \end{vmatrix} = a_{41}A_{41} + a_{42}A_{42} + a_{43}A_{43} + a_{44}A_{44} + a_{45}A_{45}$$

$$= -a_{41}M_{41} + a_{42}M_{42} - a_{43}M_{43} + a_{44}M_{44} - a_{45}M_{45}$$

Note that the signs alternate when the cofactors are converted to minors; the sign in front of $a_{ij}M_{ij}$ is just $(-1)^{i+j}$.

EXAMPLE 4

Let $A = \begin{bmatrix} 1 & 2 & 3 & 4 \\ -1 & -3 & 0 & 2 \\ 0 & 1 & -4 & -1 \\ 2 & 0 & 1 & 5 \end{bmatrix}$. Evaluate $|A|$ by

a. Expanding across the second row.

b. Expanding down the third column.

SOLUTION

a. In expanding across the second row, note that the alternating signs associated with the cofactors begin with the negative sign: $(-1)^{2+1} = (-1)^3 = (-1)$.

$$\begin{vmatrix} 1 & 2 & 3 & 4 \\ -1 & -3 & 0 & 2 \\ 0 & 1 & -4 & -1 \\ 2 & 0 & 1 & 5 \end{vmatrix} = a_{21}A_{21} + a_{22}A_{22} + a_{23}A_{23} + a_{24}A_{24}$$

$$= -a_{21}M_{21} + a_{22}M_{22} - a_{23}M_{23} + a_{24}M_{24}$$

$$= -(-1)\begin{vmatrix} 2 & 3 & 4 \\ 1 & -4 & -1 \\ 0 & 1 & 5 \end{vmatrix} + (-3)\begin{vmatrix} 1 & 3 & 4 \\ 0 & -4 & -1 \\ 2 & 1 & 5 \end{vmatrix}$$

$$- 0\begin{vmatrix} 1 & 2 & 4 \\ 0 & 1 & -1 \\ 2 & 0 & 5 \end{vmatrix} + 2\begin{vmatrix} 1 & 2 & 3 \\ 0 & 1 & -4 \\ 2 & 0 & 1 \end{vmatrix}$$

(Note that the term $a_{23} = 0$ means that we need not evaluate the 3×3 minor M_{23}. With this observation, we shall evaluate the other 3×3 minors by expanding with rows or columns containing a zero as indicated.)

$$= \left[0\begin{vmatrix} 3 & 4 \\ -4 & -1 \end{vmatrix} - 1\begin{vmatrix} 2 & 4 \\ 1 & -1 \end{vmatrix} + 5\begin{vmatrix} 2 & 3 \\ 1 & -4 \end{vmatrix} \right]$$

$$- 3\left[-0\begin{vmatrix} 3 & 4 \\ 1 & 5 \end{vmatrix} + (-4)\begin{vmatrix} 1 & 4 \\ 2 & 5 \end{vmatrix} - (-1)\begin{vmatrix} 1 & 3 \\ 2 & 1 \end{vmatrix} \right]$$

$$+ 2\left[1\begin{vmatrix} 1 & -4 \\ 0 & 1 \end{vmatrix} - 0\begin{vmatrix} 2 & 3 \\ 0 & 1 \end{vmatrix} + 2\begin{vmatrix} 2 & 3 \\ 1 & -4 \end{vmatrix} \right]$$

$$= [-(-2-4) + 5(-8-3)] - 3[(-4)(5-8) + (1-6)]$$

$$+ 2[1(1-0) + 2(-8-3)]$$

$$= [6 - 55] - 3[12 - 5] + 2[1 - 22]$$

$$= -49 - 21 - 42 = -112$$

b. In expanding by the third column, note that the alternating signs associated with the cofactors begin with the $+$ sign: $(-1)^{1+3} = (-1)^4 = (+1)$.

$$\begin{vmatrix} 1 & 2 & 3 & 4 \\ -1 & -3 & 0 & 2 \\ 0 & 1 & -4 & -1 \\ 2 & 0 & 1 & 5 \end{vmatrix} = a_{13}A_{13} + a_{23}A_{23} + a_{33}A_{33} + a_{43}A_{43}$$

$$= a_{13}M_{13} - a_{23}M_{23} + a_{33}M_{33} - a_{43}M_{43}$$

$$= 3\begin{vmatrix} -1 & -3 & 2 \\ 0 & 1 & -1 \\ 2 & 0 & 5 \end{vmatrix} - 0\begin{vmatrix} 1 & 2 & 4 \\ 0 & 1 & -1 \\ 2 & 0 & 5 \end{vmatrix}$$

$$+ (-4)\begin{vmatrix} 1 & 2 & 4 \\ -1 & -3 & 2 \\ 2 & 0 & 5 \end{vmatrix} - 1\begin{vmatrix} 1 & 2 & 4 \\ -1 & -3 & 2 \\ 0 & 1 & -1 \end{vmatrix}$$

$$= 3\left[-0\begin{vmatrix} -3 & 2 \\ 0 & 5 \end{vmatrix} + 1\begin{vmatrix} -1 & 2 \\ 2 & 5 \end{vmatrix} - (-1)\begin{vmatrix} -1 & -3 \\ 2 & 0 \end{vmatrix} \right]$$

$$- 4\left[2\begin{vmatrix} 2 & 4 \\ -3 & 2 \end{vmatrix} - 0\begin{vmatrix} 1 & 4 \\ -1 & 2 \end{vmatrix} + 5\begin{vmatrix} 1 & 2 \\ -1 & -3 \end{vmatrix} \right]$$

$$- \left[1\begin{vmatrix} -3 & 2 \\ 1 & -1 \end{vmatrix} - (-1)\begin{vmatrix} 2 & 4 \\ 1 & -1 \end{vmatrix} + 0\begin{vmatrix} 2 & 4 \\ -3 & 2 \end{vmatrix} \right]$$

$$= 3[(-5 - 4) + (0 + 6)] - 4[2(4 + 12) + 5(-3 + 2)]$$
$$- [(3 - 2) + (-2 - 4)]$$
$$= 3[-9 + 6] - 4[32 - 5] - [1 - 6]$$
$$= -9 - 108 + 5 = -112 \qquad \blacksquare$$

In Example 4, we saw that zeros in a matrix facilitate the evaluation of its determinant. We saw in Section 10.4 that the principles of the elimination method can be used to generate zeros in a matrix. Thus we shall determine the effect of the principles of the elimination method on the value of a determinant (Properties 3, 4, and 5). Property 2 is given so that the principles of the elimination method can be applied to columns as well as to rows.

The **main diagonal** of a matrix runs from its upper left to lower right corners. Reflecting through this diagonal changes rows to columns and vice versa; this is called **transposing** the matrix. In Exercise 71 at the end of this section, you will be asked to show that transposing a third-order matrix does not affect the value of its determinant. For a second-order determinant, this is clear:

$$\begin{vmatrix} a & b \\ c & d \end{vmatrix} = ad - bc = \begin{vmatrix} a & c \\ b & d \end{vmatrix}$$

In fact, this property holds for determinants of any order.

PROPERTY 2

Transposing a matrix (reflecting through its main diagonal) does not change the value of its determinant.

EXAMPLE 5

$$\begin{vmatrix} 2 & -1 & 0 \\ 3 & 5 & 1 \\ 6 & -2 & 4 \end{vmatrix} = 2 \begin{vmatrix} 5 & 1 \\ -2 & 4 \end{vmatrix} - (-1) \begin{vmatrix} 3 & 1 \\ 6 & 4 \end{vmatrix} + 0 \begin{vmatrix} 3 & 5 \\ 6 & -2 \end{vmatrix}$$

(Expand by the first row)

$$= 2 \cdot (20 + 2) + 1 \cdot (12 - 6) + 0 = 50$$

$$\begin{vmatrix} 2 & 3 & 6 \\ -1 & 5 & -2 \\ 0 & 1 & 4 \end{vmatrix} = 2 \begin{vmatrix} 5 & -2 \\ 1 & 4 \end{vmatrix} - (-1) \begin{vmatrix} 3 & 6 \\ 1 & 4 \end{vmatrix} + 0 \begin{vmatrix} 3 & 6 \\ 5 & -2 \end{vmatrix}$$

(Expand by the first column)

$$= 2 \cdot (20 + 2) + 1 \cdot (12 - 6) + 0 = 50 \quad \blacksquare$$

In light of Property 2, *any* statement we make later regarding rows of a determinant must also hold for columns, because columns become rows when the matrix is transposed. Column operations on the original matrix become row operations on the transposed matrix.

PROPERTY 3

Interchanging two rows or two columns of a determinant changes its sign.

We shall confirm this property for third-order determinants. Interchange the first and second rows of

$$\begin{vmatrix} a_1 & b_1 & c_1 \\ a_2 & b_2 & c_2 \\ a_3 & b_3 & c_3 \end{vmatrix} \quad \text{to get} \quad \begin{vmatrix} a_2 & b_2 & c_2 \\ a_1 & b_1 & c_1 \\ a_3 & b_3 & c_3 \end{vmatrix}$$

Expand this latter determinant by its first column.

$$\begin{vmatrix} a_2 & b_2 & c_2 \\ a_1 & b_1 & c_1 \\ a_3 & b_3 & c_3 \end{vmatrix} = a_2 \begin{vmatrix} b_1 & c_1 \\ b_3 & c_3 \end{vmatrix} - a_1 \begin{vmatrix} b_2 & c_2 \\ b_3 & c_3 \end{vmatrix} + a_3 \begin{vmatrix} b_2 & c_2 \\ b_1 & c_1 \end{vmatrix}$$

$$= -\left[a_1 \begin{vmatrix} b_2 & c_2 \\ b_3 & c_3 \end{vmatrix} - a_2 \begin{vmatrix} b_1 & c_1 \\ b_3 & c_3 \end{vmatrix} + a_3(b_1 c_2 - b_2 c_1) \right]$$

$$= -\left[a_1 \begin{vmatrix} b_2 & c_2 \\ b_3 & c_3 \end{vmatrix} - a_2 \begin{vmatrix} b_1 & c_1 \\ b_3 & c_3 \end{vmatrix} + a_3 \begin{vmatrix} b_1 & c_1 \\ b_2 & c_2 \end{vmatrix} \right]$$

$$= - \begin{vmatrix} a_1 & b_1 & c_1 \\ a_2 & b_2 & c_2 \\ a_3 & b_3 & c_3 \end{vmatrix}$$

(From the expansion of this final determinant by its first column)

EXAMPLE 6

$$\begin{vmatrix} 2 & -1 \\ 3 & 5 \end{vmatrix} = 10 - (-3) = 13$$

$$\begin{vmatrix} 3 & 5 \\ 2 & -1 \end{vmatrix} = -3 - 10 = -13$$

∎

A consequence of Property 3 is that *if two rows or two columns of a matrix are identical, its determinant must be 0.* This is so because interchanging these two rows or two columns must change the sign of the determinant; yet the determinant must be unchanged because the matrix is exactly the same as it was before the interchange. Thus $|M| = -|M|$ for such a matrix. But the only number that equals its own negative is 0. Thus $|M| = 0$ when two columns or two rows of M are identical.

PROPERTY 4

If two rows or two columns of a matrix are identical, its determinant is 0.

EXAMPLE 7

$$\begin{vmatrix} 2 & 3 & 1 \\ 2 & 3 & 1 \\ 5 & 4 & -2 \end{vmatrix} = 5\begin{vmatrix} 3 & 1 \\ 3 & 1 \end{vmatrix} - 4\begin{vmatrix} 2 & 1 \\ 2 & 1 \end{vmatrix} + (-2)\begin{vmatrix} 2 & 3 \\ 2 & 3 \end{vmatrix}$$

$$= 5(3-3) - 4(2-2) - 2(6-6)$$

$$= 0$$

∎

PROPERTY 5

Multiplying each entry of any row or column of a matrix by the same number k multiplies its determinant by k; equivalently, a common factor of each element in a given row or column can be factored out of the determinant.

To verify this property for third-order determinants, note that

$$\begin{vmatrix} a_1 & kb_1 & c_1 \\ a_2 & kb_2 & c_2 \\ a_3 & kb_3 & c_3 \end{vmatrix} = kb_1 A_{12} + kb_2 A_{22} + kb_3 A_{32}$$

$$= k(b_1 A_{12} + b_2 A_{22} + b_3 A_{32})$$

$$= k\begin{vmatrix} a_1 & b_1 & c_1 \\ a_2 & b_2 & c_2 \\ a_3 & b_3 & c_3 \end{vmatrix}$$

EXAMPLE 8

$$\begin{vmatrix} 2 & 7(-1) \\ 3 & 7 \cdot 5 \end{vmatrix} = 2 \cdot 7 \cdot 5 - 7(-1) \cdot 3$$

$$= 7[2 \cdot 5 - (-1) \cdot 3]$$

$$= 7 \begin{vmatrix} 2 & -1 \\ 3 & 5 \end{vmatrix}$$

> **PROPERTY 6**
>
> Adding a multiple of any row (or column) to another row (column, respectively) does not change the value of the determinant.

This property is illustrated in third-order determinants by adding k times row 2 to row 3 and expanding by the new row 3 to obtain

$$\begin{vmatrix} a_1 & b_1 & c_1 \\ a_2 & b_2 & c_2 \\ a_3 + ka_2 & b_3 + kb_2 & c_3 + kc_2 \end{vmatrix}$$

$$= (a_3 + ka_2)M_{31} - (b_3 + kb_2)M_{32} + (c_3 + kc_2)M_{33}$$

$$= a_3 M_{31} - b_3 M_{32} + c_3 M_{33} + k(a_2 M_{31} - b_2 M_{32} + c_2 M_{33})$$

$$= \begin{vmatrix} a_1 & b_1 & c_1 \\ a_2 & b_2 & c_2 \\ a_3 & b_3 & c_3 \end{vmatrix} + k \begin{vmatrix} a_1 & b_1 & c_1 \\ a_2 & b_2 & c_2 \\ a_2 & b_2 & c_2 \end{vmatrix} \quad \text{(Note: Two rows are equal: this latter determinant is 0.)}$$

$$= \begin{vmatrix} a_1 & b_1 & c_1 \\ a_2 & b_2 & c_2 \\ a_3 & b_3 & c_3 \end{vmatrix}$$

EXAMPLE 9

$$\begin{vmatrix} 2 + 7(-1) & -1 \\ 3 + 7 \cdot 5 & 5 \end{vmatrix} = [2 + 7(-1)] \cdot 5 - [3 + 7 \cdot 5](-1)$$

$$= [2 \cdot 5 - 3(-1)] + 7[(-1) \cdot 5 - 5(-1)]$$

$$= \begin{vmatrix} 2 & -1 \\ 3 & 5 \end{vmatrix} + 7 \underbrace{\begin{vmatrix} -1 & -1 \\ 5 & 5 \end{vmatrix}}_{= 0}$$

$$= \begin{vmatrix} 2 & -1 \\ 3 & 5 \end{vmatrix}$$

We shall now illustrate the use of these properties in evaluating determinants. The procedure is to work on one row or column until all but one of its entries is 0; then expand by this row or column. We shall use the notation $R_1 - 3R_4 \to R_1$ from the elimination and matrix methods to indicate the operations we perform on the determinant; C_1 denotes column 1 and so on.

EXAMPLE 10

$$\begin{vmatrix} 4 & 1 & 3 & 1 \\ 3 & 4 & -2 & -2 \\ -2 & 3 & 1 & 5 \\ 2 & -2 & 1 & 3 \end{vmatrix}$$

↑ Use this entry to generate 0s in its column.

$$= \begin{vmatrix} -2 & 7 & 0 & -8 \\ 7 & 0 & 0 & 4 \\ -4 & 5 & 0 & 2 \\ 2 & -2 & 1 & 3 \end{vmatrix} \quad \begin{array}{l} (R_1 - 3R_4 \to R_1) \\ (R_2 + 2R_4 \to R_2) \\ (R_3 - R_4 \to R_3) \end{array}$$

↑ Expand by this column.

$$= -1 \begin{vmatrix} -2 & 7 & -8 \\ 7 & 0 & 4 \\ -4 & 5 & 2 \end{vmatrix}$$

$$= (-1) \begin{vmatrix} -2 & 7 & 2 \cdot (-4) \\ 7 & 0 & 2 \cdot 2 \\ -4 & 5 & 2 \cdot 1 \end{vmatrix}$$

↑ Factor 2 out of this column.

$$= (-1) \cdot 2 \begin{vmatrix} -2 & 7 & -4 \\ 7 & 0 & 2 \\ -4 & 5 & 1 \end{vmatrix} \leftarrow \text{Use this 1 to generate 0s in its row.}$$

$(C_1 + 4C_3 \to C_1)$
$(C_2 - 5C_3 \to C_2)$

$$= (-2) \begin{vmatrix} -18 & 27 & -4 \\ 15 & -10 & 2 \\ 0 & 0 & 1 \end{vmatrix}$$

← Expand by this row.

$$= (-2) \cdot 1 \cdot \begin{vmatrix} -18 & 27 \\ 15 & -10 \end{vmatrix} = -2 \begin{vmatrix} 9 \cdot (-2) & 9 \cdot 3 \\ 5 \cdot 3 & 5 \cdot (-2) \end{vmatrix}$$

Factor 9 and 5 out of rows 1 and 2, respectively.

$$= (-2) \cdot 9 \cdot 5 \begin{vmatrix} -2 & 3 \\ 3 & -2 \end{vmatrix}$$

$$= -90(4 - 9)$$

$$= 450$$

Section 10.5 Exercises

For the given matrix, find the requested minors and cofactors.

1. $\begin{vmatrix} -1 & 2 & -2 \\ 0 & 3 & 1 \\ 4 & -3 & 5 \end{vmatrix}$
   a. M_{11} and A_{11}
   b. M_{23} and A_{23}
   c. M_{31} and A_{31}

2. $\begin{vmatrix} 1 & 3 & -1 \\ 3 & 0 & -2 \\ 4 & -2 & 4 \end{vmatrix}$
   a. M_{21} and A_{21}
   b. M_{13} and A_{13}
   c. M_{32} and A_{32}

3. $\begin{vmatrix} 2 & 7 & -3 & 6 \\ 1 & 0 & 5 & 1 \\ 6 & 2 & -6 & 2 \\ 7 & 0 & 3 & 7 \end{vmatrix}$
   a. M_{12} and A_{12}
   b. M_{24} and A_{24}
   c. M_{43} and A_{43}

4. $\begin{vmatrix} 6 & 4 & 7 & 0 \\ 13 & 0 & 6 & 4 \\ -8 & -1 & -3 & 2 \\ 8 & 0 & 6 & 5 \end{vmatrix}$
   a. M_{41} and A_{41}
   b. M_{34} and A_{34}
   c. M_{12} and A_{12}

Evaluate the following determinants.

5. $\begin{vmatrix} 1 & 3 \\ -1 & 2 \end{vmatrix}$

6. $\begin{vmatrix} 1 & 2 \\ -1 & 1 \end{vmatrix}$

7. $\begin{vmatrix} 1 & 6 \\ 3 & -2 \end{vmatrix}$

8. $\begin{vmatrix} 3 & -5 \\ 8 & -2 \end{vmatrix}$

9. $\begin{vmatrix} 1 & 5 \\ 2 & -3 \end{vmatrix}$

10. $\begin{vmatrix} 3 & -5 \\ 8 & 1 \end{vmatrix}$

11. $\begin{vmatrix} 1 & -2 \\ -7 & 6 \end{vmatrix}$

12. $\begin{vmatrix} 2 & 4 \\ 7 & -9 \end{vmatrix}$

13. $\begin{vmatrix} -1 & 4 \\ 10 & -8 \end{vmatrix}$

14. $\begin{vmatrix} -6 & 3 \\ 5 & -8 \end{vmatrix}$

Without evaluating the determinants, explain why the following are true.

15. $\begin{vmatrix} 1 & 2 & 3 \\ 0 & 0 & 1 \\ 4 & 5 & 6 \end{vmatrix} = -\begin{vmatrix} 1 & 2 \\ 4 & 5 \end{vmatrix}$

16. $\begin{vmatrix} -1 & 4 & 0 \\ 5 & 3 & 0 \\ 2 & 1 & 4 \end{vmatrix} = 4\begin{vmatrix} -1 & 4 \\ 5 & 3 \end{vmatrix}$

17. $\begin{vmatrix} -1 & 2 & 4 \\ 3 & 1 & -2 \\ 5 & 0 & 7 \end{vmatrix} = \begin{vmatrix} -1 & 3 & 5 \\ 2 & 1 & 0 \\ 4 & -2 & 7 \end{vmatrix}$

18. $\begin{vmatrix} 5 & 3 & 7 \\ -2 & 6 & -4 \\ 2 & 1 & 0 \end{vmatrix} = \begin{vmatrix} 5 & -2 & 2 \\ 3 & 6 & 1 \\ 7 & -4 & 0 \end{vmatrix}$

19. $\begin{vmatrix} 1 & 2 & 3 \\ -1 & 0 & 4 \\ 5 & -2 & 1 \end{vmatrix} = -\begin{vmatrix} 2 & 1 & 3 \\ 0 & -1 & 4 \\ -2 & 5 & 1 \end{vmatrix}$

20. $\begin{vmatrix} 1 & 2 & 3 \\ -1 & 0 & 4 \\ 5 & -2 & 1 \end{vmatrix} = -\begin{vmatrix} 1 & 2 & 3 \\ 5 & -2 & 1 \\ -1 & 0 & 4 \end{vmatrix}$

21. $\begin{vmatrix} 1 & 3 & 5 \\ 4 & 0 & 6 \\ 1 & 3 & 5 \end{vmatrix} = 0$

22. $\begin{vmatrix} 1 & 0 & 1 \\ 3 & 2 & 3 \\ -1 & 4 & -1 \end{vmatrix} = 0$

23. $\begin{vmatrix} 1 & 3 & 2 \\ 2 & -1 & 5 \\ 4 & -2 & 2 \end{vmatrix} = 2\begin{vmatrix} 1 & 3 & 2 \\ 2 & -1 & 5 \\ 2 & -1 & 1 \end{vmatrix}$

24. $\begin{vmatrix} 1 & 3 & 3 \\ 2 & 6 & 4 \\ -1 & 9 & 7 \end{vmatrix} = 3\begin{vmatrix} 1 & 1 & 3 \\ 2 & 2 & 4 \\ -1 & 3 & 7 \end{vmatrix}$

25. $\begin{vmatrix} 1 & 2 & -1 \\ 3 & 1 & 4 \\ 2+2 & 3+4 & 5-2 \end{vmatrix} = \begin{vmatrix} 1 & 2 & -1 \\ 3 & 1 & 4 \\ 2 & 3 & 5 \end{vmatrix}$

26. $\begin{vmatrix} 3+3 & 2 & 1 \\ -1+3 & 4 & 1 \\ 2-6 & -3 & -2 \end{vmatrix} = \begin{vmatrix} 3 & 2 & 1 \\ -1 & 4 & 1 \\ 2 & -3 & -2 \end{vmatrix}$

27. $\begin{vmatrix} -1 & 2 & 4 \\ 3 & 1 & -2 \\ 5 & 0 & 7 \end{vmatrix} = \begin{vmatrix} 4 & -1 & 2 \\ -2 & 3 & 1 \\ 7 & 5 & 0 \end{vmatrix}$

28. $\begin{vmatrix} -1 & 2 & 4 \\ 3 & 1 & -2 \\ 5 & 0 & 7 \end{vmatrix} = \begin{vmatrix} 5 & 0 & 7 \\ -1 & 2 & 4 \\ 3 & 1 & -2 \end{vmatrix}$

29. $\begin{vmatrix} 1 & 3 & 3 \\ 2 & 6 & 4 \\ -1 & 9 & 7 \end{vmatrix} = 2 \cdot 3 \cdot \begin{vmatrix} 1 & 1 & 3 \\ 1 & 1 & 2 \\ -1 & 3 & 7 \end{vmatrix}$

30. $\begin{vmatrix} -12 & 6 & 9 \\ 8 & 1 & -2 \\ -12 & -1 & 4 \end{vmatrix} = 3 \cdot 4 \cdot \begin{vmatrix} -1 & 2 & 3 \\ 2 & 1 & -2 \\ -3 & -1 & 4 \end{vmatrix}$

31. $\begin{vmatrix} 1+6-4 & 2+8+2 & 1+2-6 \\ 3 & 4 & 1 \\ 2 & -1 & 3 \end{vmatrix} = \begin{vmatrix} 1 & 2 & 1 \\ 3 & 4 & 1 \\ 2 & -1 & 3 \end{vmatrix}$

32. $\begin{vmatrix} 1 & 3-2+2 & 2 \\ 2 & -1-4+5 & 5 \\ 1 & -2-2+3 & 3 \end{vmatrix} = \begin{vmatrix} 1 & 3 & 2 \\ 2 & -1 & 5 \\ 1 & -2 & 3 \end{vmatrix}$

Use the properties of determinants to assist in evaluating the following.

33. $\begin{vmatrix} 1 & 6 & 3 \\ 2 & -2 & 9 \\ 8 & 3 & 4 \end{vmatrix}$

34. $\begin{vmatrix} 1 & -2 & 5 \\ 2 & 3 & 1 \\ 4 & -1 & -6 \end{vmatrix}$

35. $\begin{vmatrix} 1 & -1 & -3 \\ 2 & -1 & 2 \\ -2 & 2 & 1 \end{vmatrix}$

36. $\begin{vmatrix} 6 & 2 & 7 \\ 11 & 4 & 9 \\ 2 & 6 & 4 \end{vmatrix}$

37. $\begin{vmatrix} 2 & 0 & 4 \\ 4 & 2 & 2 \\ 1 & 6 & -1 \end{vmatrix}$

38. $\begin{vmatrix} 4 & -1 & 3 \\ -8 & -3 & -1 \\ 6 & 4 & -1 \end{vmatrix}$

39. $\begin{vmatrix} 2 & 1 & -3 \\ -1 & 1 & -2 \\ 1 & -2 & 1 \end{vmatrix}$

40. $\begin{vmatrix} -2 & 2 & -3 \\ 2 & 1 & -1 \\ -1 & 3 & 2 \end{vmatrix}$

41. $\begin{vmatrix} 2 & 11 & 6 \\ 4 & 9 & 7 \\ 3 & 2 & 1 \end{vmatrix}$

42. $\begin{vmatrix} 4 & 2 & -5 \\ 6 & 3 & 0 \\ 7 & -1 & 1 \end{vmatrix}$

43. $\begin{vmatrix} 2 & 1 & 1 \\ 1 & 1 & -2 \\ -1 & 1 & -1 \end{vmatrix}$

44. $\begin{vmatrix} -2 & 1 & 2 \\ 1 & -3 & 2 \\ 2 & -1 & -1 \end{vmatrix}$

45. $\begin{vmatrix} 3 & 2 & 5 \\ 4 & 3 & -1 \\ -1 & -1 & -2 \end{vmatrix}$

46. $\begin{vmatrix} 2 & 1 & 3 \\ 1 & 5 & -2 \\ 4 & -6 & -1 \end{vmatrix}$

47. $\begin{vmatrix} 3 & -2 & 1 & 5 \\ 2 & 3 & 4 & -2 \\ 1 & -2 & 3 & 1 \\ -2 & 4 & 1 & 3 \end{vmatrix}$

48. $\begin{vmatrix} 3 & -1 & -1 & 2 \\ 3 & 1 & 2 & 1 \\ -1 & 3 & 2 & -1 \\ 1 & 2 & -1 & -1 \end{vmatrix}$

49. $\begin{vmatrix} -1 & -2 & 1 & 3 \\ 3 & 1 & -3 & -2 \\ -3 & -1 & 2 & 1 \\ 2 & 2 & 1 & 1 \end{vmatrix}$

50. $\begin{vmatrix} 2 & 1 & 2 & 1 \\ 1 & 2 & -1 & 0 \\ -1 & 1 & 2 & 2 \\ 2 & -1 & 1 & 1 \end{vmatrix}$

51. $\begin{vmatrix} 3 & 3 & 2 & -1 \\ 1 & 1 & 2 & -2 \\ -3 & -2 & 1 & 1 \\ -2 & -1 & 1 & 3 \end{vmatrix}$

52. $\begin{vmatrix} 4 & 3 & 2 & 1 \\ 2 & -1 & 0 & 3 \\ 1 & 2 & -2 & -1 \\ 2 & 1 & -1 & 2 \end{vmatrix}$

53. $\begin{vmatrix} 3 & 1 & -1 & -6 \\ 2 & 3 & -1 & 1 \\ 5 & -3 & 3 & 2 \\ -1 & -2 & -4 & 3 \end{vmatrix}$

54. $\begin{vmatrix} 2 & -1 & -2 & 0 \\ 5 & 1 & 4 & 3 \\ 4 & 2 & -2 & 1 \\ 0 & 3 & 3 & 1 \end{vmatrix}$

55. $\begin{vmatrix} 2 & 0 & 2 & 0 & 0 \\ 1 & -1 & 0 & 2 & -1 \\ 0 & -2 & 1 & 1 & 1 \\ 1 & 1 & 0 & -2 & -1 \\ 0 & 2 & 1 & -1 & 1 \end{vmatrix}$

56. $\begin{vmatrix} 0 & -1 & 0 & -1 & 0 \\ 1 & 0 & 0 & 1 & 0 \\ -1 & 0 & 1 & 0 & -1 \\ 0 & 1 & 0 & -1 & 0 \\ -1 & 0 & -1 & 0 & 1 \end{vmatrix}$

57. $\begin{vmatrix} 1 & 0 & 0 & 2 & 0 & 1 \\ 0 & 0 & -1 & 0 & 1 & 0 \\ -1 & 1 & 0 & 1 & 0 & 0 \\ 0 & 1 & 0 & 0 & 0 & 2 \\ 2 & 0 & 1 & 1 & 0 & 0 \\ 0 & 0 & 1 & 0 & 1 & 0 \end{vmatrix}$

58. $\begin{vmatrix} 1 & 0 & 2 & 0 & 3 & 0 \\ 0 & 1 & -1 & 0 & 0 & -1 \\ 2 & 3 & 1 & 0 & 4 & 1 \\ 1 & -1 & 2 & 2 & 3 & 1 \\ 0 & 0 & 1 & 1 & 0 & 1 \\ 0 & 3 & 1 & -1 & 0 & 2 \end{vmatrix}$

The equation of the line through two points (x_1, y_1) and (x_2, y_2) can be written as

$$\begin{vmatrix} x & y & 1 \\ x_1 & y_1 & 1 \\ x_2 & y_2 & 1 \end{vmatrix} = 0$$

Use this form to obtain the equation of the line through the following pairs of points.

59. (1, 3) and (2, 4)

60. (−5, 1) and (3, 7)

61. (2, 5) and (4, −3)

62. (−1, −3) and (−6, −4)

63. (3, −2) and (−5, 6)

64. (−2, −1) and (3, 8)

The area of a triangle with vertices at (a_1, b_1), (a_2, b_2), and (a_3, b_3) is the absolute value of

$$\begin{vmatrix} a_1 & b_1 & 1 \\ a_2 & b_2 & 1 \\ a_3 & b_3 & 1 \end{vmatrix}$$

Find the area of each triangle having the indicated vertices.

65. $(1, 2), (-1, 3), (8, 4)$

66. $(1, 5), (3, 7), (6, 0)$

67. $(3, 1), (1, -2), (2, 3)$

68. $(3, -8), (12, 6), (-9, 7)$

69. $(-1, 7), (3, -9), (8, -2)$

70. $(2, -7), (1, 8), (0, 3)$

71. Show that reflecting a third-order matrix through its main diagonal does not affect its determinant.

72. Show that a matrix that has a row or column consisting entirely of zeros must have determinant 0.

Section 10.6

Cramer's Rule

In a system of two linear equations in two unknowns

$$\begin{cases} a_1 x + b_1 y = c_1 \\ a_2 x + b_2 y = c_2 \end{cases}$$

we can eliminate y if we multiply the first equation by b_2 and the second by b_1 and subtract:

$$\begin{array}{r} b_2 a_1 x + b_2 b_1 y = b_2 c_1 \\ -(b_1 a_2 x + b_1 b_2 y = b_1 c_2) \\ \hline (b_2 a_1 - b_1 a_2) x \qquad\qquad = (b_2 c_1 - b_1 c_2) \end{array}$$

The quantities in parentheses are the values of 2×2 determinants. We can rewrite this last equation as

$$\begin{vmatrix} a_1 & b_1 \\ a_2 & b_2 \end{vmatrix} x = \begin{vmatrix} c_1 & b_1 \\ c_2 & b_2 \end{vmatrix}$$

If we eliminate x rather than y from the original system of equations, we obtain in the same way

$$\begin{vmatrix} a_1 & b_1 \\ a_2 & b_2 \end{vmatrix} y = \begin{vmatrix} a_1 & c_1 \\ a_2 & c_2 \end{vmatrix}$$

If $\begin{vmatrix} a_1 & b_1 \\ a_2 & b_2 \end{vmatrix} \neq 0$, the solution (x, y) is then given as

$$x = \frac{\begin{vmatrix} c_1 & b_1 \\ c_2 & b_2 \end{vmatrix}}{\begin{vmatrix} a_1 & b_1 \\ a_2 & b_2 \end{vmatrix}} \qquad y = \frac{\begin{vmatrix} a_1 & c_1 \\ a_2 & c_2 \end{vmatrix}}{\begin{vmatrix} a_1 & b_1 \\ a_2 & b_2 \end{vmatrix}}$$

Here the denominator

$$D = \begin{vmatrix} a_1 & b_1 \\ a_2 & b_2 \end{vmatrix}$$

is the determinant of the coefficient matrix for the original system of equations. The numerators used to find x and y are

$$D_x = \begin{vmatrix} c_1 & b_1 \\ c_2 & b_2 \end{vmatrix} \quad \text{and} \quad D_y = \begin{vmatrix} a_1 & c_1 \\ a_2 & c_2 \end{vmatrix}$$

respectively. D_x is obtained from D by replacing the coefficients a_1 and a_2 of x with the constants c_1 and c_2 from the right side of the equal signs in the original system; D_y is obtained from D by replacing the coefficients b_1, b_2 of y with c_1, c_2. The solution technique just described is called Cramer's rule.

> **CRAMER'S RULE**
>
> Let D, D_x, and D_y be as defined earlier for a system of two equations in two unknowns x and y. If $D \neq 0$, there is a unique solution to the system given by
>
> $$x = \frac{D_x}{D}, \quad y = \frac{D_y}{D}$$

EXAMPLE 1 Use Cramer's rule to solve the system

$$\begin{cases} 2x - 3y = 4 \\ 5x - y = -2 \end{cases}$$

SOLUTION The determinant of the coefficient matrix is

$$D = \begin{vmatrix} 2 & -3 \\ 5 & -1 \end{vmatrix} = -2 + 15 = 13$$

Replacing the first and second columns respectively with the column of constants $\begin{pmatrix} 4 \\ -2 \end{pmatrix}$, we obtain

$$D_x = \begin{vmatrix} 4 & -3 \\ -2 & -1 \end{vmatrix} = -4 - 6 = -10$$

and

$$D_y = \begin{vmatrix} 2 & 4 \\ 5 & -2 \end{vmatrix} = -4 - 20 = -24$$

Cramer's rule gives the solution

$$x = \frac{D_x}{D} = \frac{-10}{13} \quad y = \frac{D_y}{D} = \frac{-24}{13}$$

This can be checked by substitution into the original system of equations. The solution set is $S = \left\{ \left(-\frac{10}{13}, -\frac{24}{13} \right) \right\}$. ■

Cramer's rule can be generalized to systems of n equations in n unknowns. Put such a system in the form required for solution by the matrix method; that is, the variables should be aligned in columns on the left side of the equal signs and the constant terms should be on the right side of the equal signs.

> **CRAMER'S RULE (General Form)**
>
> In a system of n linear equations in n unknowns
>
> $$x_1, x_2, \ldots, x_n$$
>
> let $D =$ the determinant of the coefficient matrix
>
> $D_{x_i} =$ the determinant that results when the column of coefficients of x_i (in D) is replaced by the column of constants from the right side of the equal signs
>
> 1. If $D \neq 0$, there is a unique solution to the system given by
>
> $$x_1 = \frac{D_{x_1}}{D}, \quad x_2 = \frac{D_{x_2}}{D}, \ldots, \quad x_n = \frac{D_{x_n}}{D}$$
>
> 2. If $D = 0$, the system is dependent or inconsistent. Use the elimination or matrix methods for further clarification of the situation.

We shall not prove statement (2) of Cramer's rule. However, in light of (1) and (2), we can state the following.

> A system of n linear equations in n unknowns has a *unique solution* if and only if *the determinant of its coefficient matrix is not* 0.

EXAMPLE 2

Use Cramer's rule to solve the system

$$\begin{cases} 6x + y - z = 1 \\ -2x + 3y + 4z = 4 \\ 9x + 2y - 2z = -1 \end{cases}$$

SOLUTION

We will use the note-taking technique of the past several sections to keep track of operations on columns and rows when we simplify determinants for evaluation. The determinants in Cramer's rule are then evaluated as indicated.

$$x = \frac{D_x}{D} = \frac{\begin{vmatrix} 1 & 1 & -1 \\ 4 & 3 & 4 \\ -1 & 2 & -2 \end{vmatrix}}{\begin{vmatrix} 6 & 1 & -1 \\ -2 & 3 & 4 \\ 9 & 2 & -2 \end{vmatrix}} = \frac{\begin{vmatrix} 1 & 0 & 0 \\ 4 & -1 & 8 \\ -1 & 3 & -3 \end{vmatrix}}{\begin{vmatrix} 0 & 1 & 0 \\ -20 & 3 & 7 \\ -3 & 2 & 0 \end{vmatrix}}$$

$(C_2 - C_1 \to C_2)$
$(C_3 + C_1 \to C_3)$

(Expand each by the first row)

$(C_3 + C_2 \to C_3)$
$(C_1 - 6C_2 \to C_1)$

$$= \frac{1 \cdot \begin{vmatrix} -1 & 8 \\ 3 & -3 \end{vmatrix}}{(-1)\begin{vmatrix} -20 & 7 \\ -3 & 0 \end{vmatrix}} = \frac{3 - 24}{(-1)(21)} = \frac{-21}{-21} = 1$$

$$y = \frac{D_y}{D} = \frac{\begin{vmatrix} 6 & 1 & -1 \\ -2 & 4 & 4 \\ 9 & -1 & -2 \end{vmatrix}}{D}$$

Expand by the second column.

$$= \frac{\begin{vmatrix} 6 & 1 & -1 \\ -26 & 0 & 8 \\ 15 & 0 & -3 \end{vmatrix}}{-21} \quad \begin{array}{l} (R_2 - 4R_1 \to R_2) \\ (R_3 + R_1 \to R_3) \end{array}$$

Note that $D = -21$ was evaluated in the computation for x.

$$= \frac{(-1)\begin{vmatrix} -26 & 8 \\ 15 & -3 \end{vmatrix}}{-21} = \frac{-(78 - 120)}{-21}$$

$$= \frac{42}{-21} = -2$$

$$z = \frac{D_z}{D} = \frac{\begin{vmatrix} 6 & 1 & 1 \\ -2 & 3 & 4 \\ 9 & 2 & -1 \end{vmatrix}}{D}$$

Expand by the third column.

$$= \frac{\begin{vmatrix} 6 & 1 & 1 \\ -26 & -1 & 0 \\ 15 & 3 & 0 \end{vmatrix}}{-21} \quad \begin{array}{l} (R_2 - 4R_1 \to R_2) \\ (R_3 + R_1 \to R_3) \end{array}$$

$$= \frac{1 \cdot \begin{vmatrix} -26 & -1 \\ 15 & 3 \end{vmatrix}}{-21} = \frac{-78 + 15}{-21} = \frac{-63}{-21} = 3$$

The solution is $x = 1$, $y = -2$, $z = 3$. Verify it. ∎

Although Cramer's rule is quite useful in solving small systems, its use is impractical for large systems because of the large number of high-order determinants that must be evaluated. For large systems, the matrix method is probably the most efficient method of solution. However, for small systems, especially 2×2 systems, Cramer's rule is very fast.

As we mentioned earlier, it is easy to program a computer to solve a system of equations by using its matrix. These programs are very efficient and easy to write. In contrast, a computer can also be programmed to solve a system of equations by using Cramer's rule, but these programs are very inefficient and difficult to write.

Section 10.6 Exercises

Use Cramer's rule to solve the following systems of equations.

1. $\begin{cases} x - y = 3 \\ 3x - 2y = 1 \end{cases}$

2. $\begin{cases} 5x - 3y = 19 \\ 2x - 4y = 16 \end{cases}$

3. $\begin{cases} t + 2u = 0 \\ 3t + 2u = 5 \end{cases}$

4. $\begin{cases} 2x + y = 6 \\ x - y = 3 \end{cases}$

5. $\begin{cases} u - 2v = -3 \\ 2u - 3v = 1 \end{cases}$

6. $\begin{cases} 7x + 3y = 4 \\ 6x + 3y = 2 \end{cases}$

7. $\begin{cases} w + z = 0 \\ 2w - 3z = 5 \end{cases}$

8. $\begin{cases} 4x + 3y = -8 \\ 3x + 2y = -5 \end{cases}$

9. $\begin{cases} t - 2y = 5 \\ 2t - 3y = 8 \end{cases}$

10. $\begin{cases} 4x - 5y = -1 \\ 8x - 7y = -5 \end{cases}$

11. $\begin{cases} 2p - 5q = 2 \\ 4p + 3q = 1 \end{cases}$

12. $\begin{cases} 3x - 2y = 1 \\ 5x - 3y = -8 \end{cases}$

13. $\begin{cases} 3r - 5s = -3 \\ 2r + 4s = 2 \end{cases}$

14. $\begin{cases} 4x - 3y = 12 \\ 2x - y = 7 \end{cases}$

15. $\begin{cases} x - y = 2 \\ x + 2y = 1 \end{cases}$

16. $\begin{cases} 3x - 7y = 16 \\ 4x + 5y = 10 \end{cases}$

17. $\begin{cases} r - 4u = -1 \\ 3r + 2u = 2 \end{cases}$

18. $\begin{cases} x - 4z = 24 \\ 2x + 7z = 8 \end{cases}$

19. $\begin{cases} x + y = 0 \\ y + z = -1 \\ -x + 2z = 1 \end{cases}$

20. $\begin{cases} x + y - 3z = 1 \\ y + z = -2 \\ -2x + 3y = 4 \end{cases}$

21. $\begin{cases} 2x - y + z = -5 \\ x - 2y - 3z = -8 \\ x + y - 2z = 1 \end{cases}$

22. $\begin{cases} u - 2v + w = 8 \\ 5u - v + 3w = 6 \\ 3u - 6v + 3w = 14 \end{cases}$

23. $\begin{cases} 2r - s + 2t = 5 \\ -r + s + 3t = 6 \\ 2r - 2s - t = -7 \end{cases}$

24. $\begin{cases} 3p + 2q + 5r = 12 \\ 2p - 5q + 3r = 4 \\ p + 3q + 2r = 8 \end{cases}$

25. $\begin{cases} x - y + z = 0 \\ y + 2z = -2 \\ 3x + 2y + z = 1 \end{cases}$

26. $\begin{cases} w - 2x - 3y = 3 \\ 2w + x - y = 1 \\ w + x + y = -4 \end{cases}$

27. $\begin{cases} 3w + 4z + u = 0 \\ 5w + 3z + u = 1 \\ w - 4z - 3u = 5 \end{cases}$

28. $\begin{cases} 2x + 3y - 2z = 1 \\ 2x + y - z = 2 \\ x + y - z = 1 \end{cases}$

29. $\begin{cases} r - 2s + t = 2 \\ 3r + t = 0 \\ -s + u = 1 \\ r + s - t + u = 1 \end{cases}$

30. $\begin{cases} u + 2v = 0 \\ 2x + 3y + v = 2 \\ x + 2y - u = 0 \\ y + u = 0 \end{cases}$

31. $\begin{cases} -a + b + c - 2d = 1 \\ a - b + 2c - d = -1 \\ 3a + b + 2c - 2d = -2 \\ -2a - 2b + c + 2d = 3 \end{cases}$

32. $\begin{cases} r - 2s - t - 3u = -9 \\ r + s - t = 0 \\ 3r + 4s + u = 6 \\ 2s - 2t + u = 3 \end{cases}$

33. The larger of two numbers is 1 more than 3 times the smaller. The smaller plus twice the larger equals 30. Find the numbers.

34. A two-digit number is 1 less than 5 times the sum of its digits. If the digits are reversed the number is increased by 18. Find the number. (See Exercise 71, Section 10.1.)

35. Mr. Huber invests money in two institutions at 8% and 5% interest, respectively, and earns $1130 annually. He has $1000 more than twice the smaller investment invested at 8%. How much is deposited in each bank?

36. Cindy has $5.15 in nickels and dimes. The number of nickels is 8 more than 3 times the number of dimes. What is the number of each type of coin she has?

37. Two years from now Gideon's father will be 4 times as old as Gideon. Sixteen years from now the father will be twice as old as Gideon. How old are they now?

Section 10.7

Partial Fractions (Optional)

We obtain the sum of rational expressions

$$\frac{2}{x - 1} + \frac{1}{x + 1} - \frac{1}{(x + 1)^2}$$

by writing each fraction in terms of the LCD and adding numerators:

$$\frac{2(x+1)^2}{(x-1)(x+1)^2} + \frac{(x-1)(x+1)}{(x-1)(x+1)^2} - \frac{(x-1)}{(x-1)(x+1)^2}$$

$$= \frac{2(x^2+2x+1)+(x^2-1)-(x-1)}{(x-1)(x+1)^2}$$

$$= \frac{3x^2+3x+2}{(x-1)(x+1)^2}$$

It is sometimes useful (especially in calculus courses) to reverse this process: separate the combined form into the separate fractions with which we started. This process is called **partial fractions decomposition.**

PARTIAL FRACTIONS DECOMPOSITION

1. If the degree of the numerator is not less than that of the denominator, carry out the long division to obtain a proper fraction.
2. Factor the denominator as far as possible; linear and irreducible quadratic factors should be the result. (See Chapter 5.)
3. For each linear factor $(ax+b)$, consider a fraction of the form $\frac{A}{ax+b}$.
4. For each linear factor raised to a power $(ax+b)^k$, consider fractions of the form $\frac{A_1}{(ax+b)}, \frac{A_2}{(ax+b)^2}, \dots, \frac{A_k}{(ax+b)^k}$.
5. For each irreducible quadratic factor (ax^2+bx+c), consider a fraction of the form $\frac{Ax+B}{ax^2+bx+c}$.
6. For each irreducible quadratic factor raised to a power $(ax^2+bx+c)^k$, consider fractions of the form $\frac{A_1 x + B_1}{ax^2+bx+c}, \dots, \frac{A_k x + B_k}{(ax^2+bx+c)^k}$.

Linear Factors

EXAMPLE 1 Find the partial fractions decomposition of

$$\frac{7-x}{x^2+x-2}$$

SOLUTION We factor the denominator and write

$$\frac{7-x}{x^2+x-2} = \frac{7-x}{(x-1)(x+2)}$$

Then, applying Rule 3 for partial fractions decomposition, we try

$$\frac{7-x}{(x-1)(x+2)} = \frac{A}{x-1} + \frac{B}{x+2}$$

To solve for A and B, we begin by recombining the fractions:

$$= \frac{A(x+2)}{(x-1)(x+2)} + \frac{B(x-1)}{(x-1)(x+2)}$$

$$= \frac{A(x+2) + B(x-1)}{(x-1)(x+2)}$$

For these two fractions to be equal, their numerators must be equal, since their denominators are:

$$7 - x = A(x+2) + B(x-1)$$
$$= Ax + 2A + Bx - B$$
$$= (A+B)x + (2A-B)$$

This equality holds when the respective coefficients agree:

$$7 - x = (A+B)x + (2A-B)$$

x: $\quad A + B = -1$
constant: $\quad 2A - B = 7$

Adding these two equations, we obtain

$$3A = 6$$
$$A = 2$$

Substituting $A = 2$ into $A + B = -1$ yields

$$2 + B = -1$$
$$B = -3$$

The partial fractions decomposition is

$$\frac{7-x}{x^2+x-2} = \frac{2}{x-1} - \frac{3}{x+2}$$

The result can be checked by combining the rational expressions on the right. ∎

Repeated Linear Factors

EXAMPLE 2   Find the partial fractions decomposition of

$$\frac{x^3 - 3x^2 + 5x - 2}{x^3 - 2x^2 - 4x + 8}$$

SOLUTION   First, we carry out the long division to obtain a proper fraction:

$$\frac{x^3 - 3x^2 + 5x - 2}{x^3 - 2x^2 - 4x + 8} = 1 + \frac{-x^2 + 9x - 10}{x^3 - 2x^2 - 4x + 8}$$

Then we factor the denominator:

$$x^3 - 2x^2 - 4x + 8 = x^2(x-2) - 4(x-2)$$
$$= (x^2 - 4)(x - 2)$$
$$= (x+2)(x-2)^2$$

Thus we consider the following partial fractions decomposition of the proper fraction using Rule 4

$$\frac{-x^2 + 9x - 10}{(x + 2)(x - 2)^2} = \frac{A}{x + 2} + \frac{B}{x - 2} + \frac{C}{(x - 2)^2}$$

The procedure is to put these fractions back together and solve for A, B, and C.

$$\frac{-x^2 + 9x - 10}{(x + 2)(x - 2)^2} = \frac{A(x - 2)^2 + B(x + 2)(x - 2) + C(x + 2)}{(x + 2)(x - 2)^2}$$

Again, this equality holds when the numerators agree:

$$-x^2 + 9x - 10 = A(x - 2)^2 + B(x + 2)(x - 2) + C(x + 2)$$

An alternative to equating respective coefficients (as in Example 1) is to substitute various values of x into this equation. The values $x = 2$ and $x = -2$ will cause certain terms to be 0:

$x = 2$: $\quad A(2 - 2)^2 + B(2 + 2)(2 - 2) + C(2 + 2) \quad = -2^2 + 9 \cdot 2 - 10$

$$4C = 4$$
$$C = 1$$

$x = -2$: $\quad A(-2 - 2)^2 + B(-2 + 2)(-2 - 2) + C(-2 + 2) = -(-2)^2 + 9 \cdot (-2) - 10$

$$16A = -32$$
$$A = -2$$

Other convenient values of x to substitute into the equation are $x = 0, 1, -1$. We try $x = 0$.

$x = 0$: $\quad A(0 - 2)^2 + B(0 + 2)(0 - 2) + C(0 + 2) = -0 + 9 \cdot 0 - 10$

(We already know $A = -2$ and $C = 1$.)

$$(-2) \cdot 4 - 4B + 1 \cdot 2 = -10$$
$$-4B = -4$$
$$B = 1$$

Thus the partial fractions decomposition is

$$\frac{x^3 - 3x^2 + 5x - 2}{x^3 - 2x^2 - 4x + 8} = 1 + \frac{-x^2 + 9x - 10}{(x + 2)(x - 2)^2}$$

$$= 1 + \frac{-2}{x + 2} + \frac{1}{x - 2} + \frac{1}{(x - 2)^2} \quad \blacksquare$$

Both Linear and Quadratic Factors

EXAMPLE 3

Find the partial fractions decomposition of

$$\frac{4x^2 + 7x - 3}{2x^3 + x^2 - 6x - 3}$$

SOLUTION

First we factor the denominator:

$$2x^3 + x^2 - 6x - 3 = x^2(2x + 1) - 3(2x + 1)$$
$$= (x^2 - 3)(2x + 1)$$

Then for the partial fractions decomposition, we try

$$\frac{4x^2 + 7x - 3}{(x^2 - 3)(2x + 1)} = \frac{Ax + B}{x^2 - 3} + \frac{C}{2x + 1}$$

$$= \frac{(Ax + B)(2x + 1) + C(x^2 - 3)}{(x^2 - 3)(2x + 1)}$$

Equate numerators:

$$(Ax + B)(2x + 1) + C(x^2 - 3) = 4x^2 + 7x - 3$$

We use the method of substituting x values to find A, B, and C. Since one of the factors is $2x + 1$, we shall begin with $x = -\frac{1}{2}$.

$x = -\frac{1}{2}$: $\quad 0 + C\left[\left(-\frac{1}{2}\right)^2 - 3\right] = 4\left(-\frac{1}{2}\right)^2 + 7\left(-\frac{1}{2}\right) - 3$

$$-\frac{11}{4}C = -\frac{11}{2}$$

$$C = 2$$

$x = 0$: $\quad (A \cdot 0 + B)(2 \cdot 0 + 1) + C(0 - 3) = 4 \cdot 0^2 + 7 \cdot 0 - 3$

$$B + 2(-3) = -3 \qquad (C = 2)$$

$$B = 3$$

$x = 1$: $\quad (A \cdot 1 + B)(2 \cdot 1 + 1) + C(1^2 - 3) = 4 \cdot 1^2 + 7 \cdot 1 - 3$

$$(A + 3) \cdot 3 + 2(-2) = 4 + 7 - 3 \qquad (B = 3, C = 2)$$

$$3A + 9 - 4 = 8$$

$$3A = 3$$

$$A = 1$$

Thus the partial fractions decomposition is

$$\frac{4x^2 + 7x - 3}{2x^3 + x^2 - 6x - 3} = \frac{x + 3}{x^2 - 3} + \frac{2}{2x + 1} \qquad \blacksquare$$

Repeated Quadratic Factors

EXAMPLE 4 Find the partial fractions decomposition of

$$\frac{3x^4 - 2x^3 + 7x^2 - x + 1}{(x - 1)(x^2 + 1)^2}$$

SOLUTION We already have a proper fraction with denominator factored. Thus we proceed with the decomposition.

$$\frac{3x^4 - 2x^3 + 7x^2 - x + 1}{(x - 1)(x^2 + 1)^2}$$

$$= \frac{A}{x - 1} + \frac{Bx + C}{x^2 + 1} + \frac{Dx + E}{(x^2 + 1)^2}$$

$$= \frac{A(x^2 + 1)^2 + (Bx + C)(x - 1)(x^2 + 1) + (Dx + E)(x - 1)}{(x - 1)(x^2 + 1)^2}$$

Equating numerators, we obtain

$$3x^4 - 2x^3 + 7x^2 - x + 1$$
$$= A(x^4 + 2x^2 + 1) + B(x^4 - x^3 + x^2 - x) + C(x^3 - x^2 + x - 1)$$
$$+ D(x^2 - x) + E(x - 1)$$
$$= x^4(A + B) + x^3(-B + C) + x^2(2A + B - C + D)$$
$$+ x(-B + C - D + E) + (A - C - E)$$

In this example, we choose again to illustrate the method of equating corresponding coefficients:

$$\begin{aligned} x^4: & \quad A + B & = 3 \\ x^3: & \quad -B + C & = -2 \\ x^2: & \quad 2A + B - C + D & = 7 \\ x: & \quad -B + C - D + E & = -1 \\ \text{constant}: & \quad A \quad - C \quad - E & = 1 \end{aligned}$$

Solving the system, we obtain

$$A = 2, \quad B = 1, \quad C = -1, \quad D = 1, \quad E = 2$$
$$\frac{3x^4 - 2x^3 + 7x^2 - x + 1}{(x - 1)(x^2 + 1)^2} = \frac{2}{x - 1} + \frac{x - 1}{x^2 + 1} + \frac{x + 2}{(x^2 + 1)^2}$$

∎

Section 10.7 Exercises

Find the partial fractions decomposition of the following

1. $\dfrac{8x - 9}{(x - 3)(x + 2)}$

2. $\dfrac{3x + 4}{x(x + 2)}$

3. $\dfrac{x + 2}{(x + 1)(x - 2)}$

4. $\dfrac{4x + 6}{(x + 3)(2x + 1)}$

5. $\dfrac{2x^2 - 4x - 1}{x^2 - x - 6}$

6. $\dfrac{2x + 11}{x^2 + x - 6}$

7. $\dfrac{5x - 1}{x^2 - x - 2}$

8. $\dfrac{x}{x^2 - 5x + 6}$

9. $\dfrac{x^2 - 3x - 4}{x(x + 4)(x - 2)}$

10. $\dfrac{4x^2 - 7x - 12}{x(x + 1)(x - 2)}$

11. $\dfrac{4x^2 - 3x + 2}{x^2(x - 1)}$

12. $\dfrac{x^2 + x - 3}{x^2(x + 3)}$

13. $\dfrac{2}{x(x + 2)^2}$

14. $\dfrac{4x^2 - 10x + 3}{x(x - 1)^2}$

15. $\dfrac{3x^2 + 12x}{(x - 2)(x + 1)^2}$

16. $\dfrac{2x^2 - 11x + 11}{(x - 1)(x - 2)^2}$

17. $\dfrac{2x^4 - 5x^3 - 3x^2 + 13x - 12}{x^3 - 3x^2}$

18. $\dfrac{2x^2 + 7x + 7}{(x - 1)(x + 1)^2}$

19. $\dfrac{-2x^2 + 5x - 4}{(x - 4)(x - 2)^2}$

20. $\dfrac{3x^2 + 4x - 1}{(2x + 3)(x + 2)^2}$

21. $\dfrac{x - 2}{x(x^2 - 2)}$

22. $\dfrac{x^2 - 2x + 1}{x(x^2 + 1)}$

23. $\dfrac{3x^2 + 2x + 2}{(x + 1)(x^2 + 2)}$

24. $\dfrac{2x^2 + 4x - 2}{(x - 1)(x^2 + 1)}$

25. $\dfrac{4x^2 + 7x + 4}{(x + 2)(x^2 + 2)}$

26. $\dfrac{2x^4 + x^3 + 6x^2 + 3x + 5}{x^3 + 3x}$

27. $\dfrac{x + 11}{(x + 2)(x^2 + 2x + 2)}$

28. $\dfrac{x^3 - x^2 - 4x - 4}{(x - 1)(x^2 + x + 2)}$

29. $\dfrac{5x^2 + 6x + 3}{(x + 1)(x^2 + 2x + 2)}$

30. $\dfrac{5x^3 + 6x^2 + x + 2}{x(x + 1)(x^2 + 2)}$

31. $\dfrac{1}{x(x^2 + 1)^2}$

32. $\dfrac{x^3}{(x^2 + 2)^2}$

33. $\dfrac{3x^4 + 12x^2 - 3x + 9}{x(x^2 + 3)^2}$

34. $\dfrac{x^4 + 2x^2 - x - 1}{x(x^2 + 1)^2}$

35. $\dfrac{x^3 + 2x^2 + 4}{x(x^2 + 2)^2}$

36. $\dfrac{x^4 + 4x^2 - x}{(x - 1)(x^2 + 1)^2}$

37. $\dfrac{-2x^3 - 4x - 3}{(x-1)(x^2+2)^2}$

38. $\dfrac{x^3 + 2x - 1}{(x+1)(x^2+1)^2}$

39. $\dfrac{x^4 + 3x^3 - 3x^2 - 2x + 12}{(x-2)(x^2+2)^2}$

40. $\dfrac{2x^6 + 4x^5 + 9x^4 + 16x^3 + 12x^2 + 17x + 6}{(x+2)(x^2+2)^2}$

Section 10.8
The Algebra of Matrices

In Sections 10.4 and 10.5, we saw that matrices can be used to solve a system of equations. Matrices are also useful in other situations. Thus it is appropriate to develop some rules concerning the algebraic manipulation of matrices. Three algebraic operations are defined for matrices:

1. The sum $A + B$ and difference $A - B$ of two matrices A and B.
2. The product (scalar multiplication) kA of a real number k times a matrix A.
3. The product (matrix multiplication) AB of two matrices A and B. (Quotients are not defined, however.)

The notation a_{ij} denotes the element in the ith row, jth column of the matrix A. We shall use the shorthand notation $A = (a_{ij})$ for the matrix

$$A = \begin{bmatrix} a_{11} & \cdots & a_{1j} & \cdots & a_{1n} \\ \vdots & & \vdots & & \vdots \\ a_{i1} & \cdots & a_{ij} & \cdots & a_{in} \\ \vdots & & \vdots & & \vdots \\ a_{m1} & \cdots & a_{mj} & \cdots & a_{mn} \end{bmatrix}$$

Equality

DEFINITION

Two matrices $A = (a_{ij})$, $B = (b_{ij})$ are said to be **equal** ($A = B$) if

1. They are of the same shape (say both are $m \times n$).
2. Each pair of corresponding elements are equal:

$$a_{ij} = b_{ij} \quad \text{for all } i \text{ and } j \ (1 \le i \le m, 1 \le j \le n)$$

EXAMPLE 1

Find the values of each of the variables that will make the matrix equation true.

$$\begin{bmatrix} 1 & p & 3 \\ u & -1 & v \end{bmatrix} = \begin{bmatrix} x & 2 & z \\ 5 & r & 4 \end{bmatrix}$$

SOLUTION

Both matrices are 2×3 (2 rows, 3 columns). For corresponding elements to be equal, we have

$x = 1$
$p = 2$
$z = 3$
$u = 5$
$r = -1$
$v = 4$

Addition

DEFINITION

To add (or subtract) two matrices $A = (a_{ij})$, $B = (b_{ij})$,

1. They must have the same shape (have the same number of rows and the same number of columns).
2. Then add (or subtract) each pair of corresponding elements:
   a. $(a_{ij}) + (b_{ij}) = (a_{ij} + b_{ij})$
   b. $(a_{ij}) - (b_{ij}) = (a_{ij} - b_{ij})$

EXAMPLE 2 Let $A = \begin{bmatrix} 1 & -2 & -1 \\ 0 & 3 & 1 \end{bmatrix}$, $B = \begin{bmatrix} 3 & 1 & -4 \\ -3 & -2 & 1 \end{bmatrix}$. Find

a. $A + B$ b. $A - B$

SOLUTION

a. $\begin{bmatrix} 1 & -2 & -1 \\ 0 & 3 & 1 \end{bmatrix} + \begin{bmatrix} 3 & 1 & -4 \\ -3 & -2 & 1 \end{bmatrix} = \begin{bmatrix} 1+3 & -2+1 & -1-4 \\ 0-3 & 3-2 & 1+1 \end{bmatrix}$

$= \begin{bmatrix} 4 & -1 & -5 \\ -3 & 1 & 2 \end{bmatrix}$

b. $\begin{bmatrix} 1 & -2 & -1 \\ 0 & 3 & 1 \end{bmatrix} - \begin{bmatrix} 3 & 1 & -4 \\ -3 & -2 & 1 \end{bmatrix}$

$= \begin{bmatrix} 1-3 & -2-1 & -1-(-4) \\ 0-(-3) & 3-(-2) & 1-1 \end{bmatrix}$

$= \begin{bmatrix} -2 & -3 & 3 \\ 3 & 5 & 0 \end{bmatrix}$ ∎

Scalar Multiplication

To multiply a matrix by a real number (called **scalar multiplication**), multiply each of its elements by that number:

DEFINITION

$$k(a_{ij}) = (ka_{ij})$$

EXAMPLE 3 For $A = \begin{bmatrix} 1 & 2 & -1 \\ 0 & 3 & 2 \end{bmatrix}$, find $2A$.

SOLUTION

$2\begin{bmatrix} 1 & 2 & -1 \\ 0 & 3 & 2 \end{bmatrix} = \begin{bmatrix} 2 \cdot 1 & 2 \cdot 2 & 2 \cdot (-1) \\ 2 \cdot 0 & 2 \cdot 3 & 2 \cdot 2 \end{bmatrix} = \begin{bmatrix} 2 & 4 & -2 \\ 0 & 6 & 4 \end{bmatrix}$ ∎

A matrix, all of whose elements are 0, is called a **zero matrix**, denoted by **0**. For example,

$$\mathbf{0} = \begin{bmatrix} 0 & 0 & 0 \\ 0 & 0 & 0 \end{bmatrix}$$

is the 2×3 zero matrix.

Since addition of matrices and multiplication of a matrix by a real number are defined in terms of corresponding entries and these entries are real numbers, many of the properties of real numbers carry over to matrices. Specifically, if A, B, and C are matrices and k, l are real numbers, the following properties hold.

> **PROPERTIES OF MATRIX ADDITION AND SCALAR MULTIPLICATION FOR MATRICES A, B, AND C, AND $k, l \in R$**
>
> 1. $A + B = B + A$
> 2. $A + (B + C) = (A + B) + C$
> 3. $A + 0 = A$
> 4. $k(A + B) = kA + kB$
> 5. $(k + l)A = kA + lA$
> 6. $k(lA) = (kl)A$
> 7. $A + (-1)A = A - A = 0$
>   (where **0** is the zero matrix)

In light of Property 7, $(-1)A$ is called the additive inverse of A and is denoted by $-A$: $(-1)A = -A$. It then follows that

$$A + (-B) = A + (-1)B = A - B$$

EXAMPLE 4

Find $2A - 3B$ if

$$A = \begin{bmatrix} 1 & -1 \\ 3 & 1 \\ 2 & -4 \end{bmatrix} \quad \text{and} \quad B = \begin{bmatrix} -1 & 1 \\ 2 & 4 \\ 1 & -3 \end{bmatrix}$$

SOLUTION

$$2A - 3B = 2\begin{bmatrix} 1 & -1 \\ 3 & 1 \\ 2 & -4 \end{bmatrix} - 3\begin{bmatrix} -1 & 1 \\ 2 & 4 \\ 1 & -3 \end{bmatrix}$$

$$= \begin{bmatrix} 2 & -2 \\ 6 & 2 \\ 4 & -8 \end{bmatrix} - \begin{bmatrix} -3 & 3 \\ 6 & 12 \\ 3 & -9 \end{bmatrix} = \begin{bmatrix} 5 & -5 \\ 0 & -10 \\ 1 & 1 \end{bmatrix} \quad \blacksquare$$

Matrix Multiplication

Matrix multiplication is *not* defined in terms of corresponding entries. Rather, a different kind of product of matrices is found to be useful. Consider the following situation, for example.

Assume that the Apex Company produces compressors and fans on three machines: a press, a molder, and a welder. Each compressor requires 2 hours on the press, 1 hour on the molder, and $\frac{1}{2}$ hour on the welder. Each fan requires 1 hour on the press, $1\frac{1}{2}$ hours on the molder, and $\frac{3}{4}$ hour on the welder. This information can be stored in a matrix A as illustrated here.

$$A = \begin{bmatrix} 2 & 1 \\ 1 & \frac{3}{2} \\ \frac{1}{2} & \frac{3}{4} \end{bmatrix} \begin{matrix} \text{Press} \\ \text{Molder} \\ \text{Welder} \end{matrix}$$

with columns labeled Compressor and Fan.

If a customer orders 12 compressors and 8 fans, we store this information in a matrix B:

$$B = \begin{bmatrix} 12 \\ 8 \end{bmatrix} \begin{matrix} \text{Compressors} \\ \text{Fans} \end{matrix}$$

Placing these matrices side by side, we use these figures to find the total number of hours required on each machine to fill these orders as follows:

$$\text{Press} \begin{bmatrix} 2 & 1 \\ 1 & \frac{3}{2} \\ \frac{1}{2} & \frac{3}{4} \end{bmatrix} \begin{bmatrix} 12 \\ 8 \end{bmatrix} \qquad (\text{Press hours} = 2 \cdot 12 + 1 \cdot 8 = 32)$$

$$\text{Molder} \begin{bmatrix} 2 & 1 \\ 1 & \frac{3}{2} \\ \frac{1}{2} & \frac{3}{4} \end{bmatrix} \begin{bmatrix} 12 \\ 8 \end{bmatrix} \qquad \left(\text{Molder hours} = 1 \cdot 12 + \frac{3}{2} \cdot 8 = 24\right)$$

$$\text{Welder} \begin{bmatrix} 2 & 1 \\ 1 & \frac{3}{2} \\ \frac{1}{2} & \frac{3}{4} \end{bmatrix} \begin{bmatrix} 12 \\ 8 \end{bmatrix} \qquad \left(\text{Welder hours} = \frac{1}{2} \cdot 12 + \frac{3}{4} \cdot 8 = 12\right)$$

We store this information in a matrix as

$$C = \begin{bmatrix} 32 \\ 24 \\ 12 \end{bmatrix} \begin{matrix} \text{Press} \\ \text{Molder} \\ \text{Welder} \end{matrix}$$

If 1 hour of operation costs \$20 on the press, \$40 on the molder, and \$30 on the welder, we store this information as

$$D = \begin{bmatrix} 20 & 40 & 30 \end{bmatrix}$$

The total production cost to the company for producing these 12 compressors and 8 fans is given by

$$\begin{bmatrix} 20 & 40 & 30 \end{bmatrix} \begin{bmatrix} 32 \\ 24 \\ 12 \end{bmatrix} \qquad (\text{Cost} = 20 \cdot 32 + 40 \cdot 24 + 30 \cdot 12 = \$1960)$$

We now formalize these kinds of computations in our definition of matrix multiplication.

> **DEFINITION**
>
> If $A = (a_{ij})$ is an $m \times n$ matrix and $B = (b_{ij})$ is an $n \times k$ matrix (where $m, n, k \in N$), then
>
> $$AB = C = (c_{ij})$$
>
> is an $m \times k$ matrix whose element in row i, column j is given by
>
> $$c_{ij} = a_{i1}b_{1j} + a_{i2}b_{2j} + \cdots + a_{in}b_{nj}$$

Thus, to form the product $C = AB$ of two matrices, the number of columns of A must equal the number of rows of B. To find the entry in row i, column j of the product $C = AB$ of two matrices, we multiply corresponding entries in row i of A and column j of B. The sum of these values is the c_{ij} element of $C = AB$.

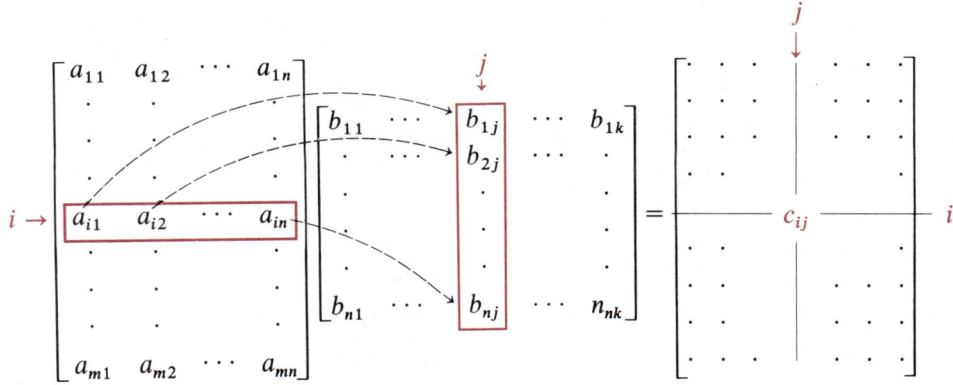

EXAMPLE 5

Find AB if

$$A = \begin{bmatrix} -1 & 1 \\ 2 & 4 \\ 3 & -1 \end{bmatrix} \quad \text{and} \quad B = \begin{bmatrix} 2 & 0 & 1 & 3 \\ -1 & 2 & -2 & 4 \end{bmatrix}$$

SOLUTION

Let

$$AB = \begin{bmatrix} -1 & 1 \\ 2 & 4 \\ 3 & -1 \end{bmatrix} \begin{bmatrix} 2 & 0 & 1 & 3 \\ -1 & 2 & -2 & 4 \end{bmatrix} = C$$

The product should have 3 rows and 4 columns. The c_{13} entry is found as illustrated.

$$\begin{bmatrix} \boxed{-1 \quad 1} \\ 2 \quad 4 \\ 3 \quad -1 \end{bmatrix} \begin{bmatrix} 2 & 0 & \boxed{1} & 3 \\ -1 & 2 & \boxed{-2} & 4 \end{bmatrix} = \begin{bmatrix} \cdot & \cdot & c_{13} & \cdot \\ \cdot & \cdot & \cdot & \cdot \\ \cdot & \cdot & \cdot & \cdot \end{bmatrix}$$

where $c_{13} = (-1) \cdot 1 + 1 \cdot (-2) = -3$. Similarly, the c_{22} computation is illustrated.

$$\begin{bmatrix} -1 & 1 \\ \boxed{2 \quad 4} \\ 3 & -1 \end{bmatrix} \begin{bmatrix} 2 & \boxed{0} & 1 & 3 \\ -1 & \boxed{2} & -2 & 4 \end{bmatrix} = \begin{bmatrix} \cdot & \cdot & -3 & \cdot \\ \cdot & c_{22} & \cdot & \cdot \\ \cdot & \cdot & \cdot & \cdot \end{bmatrix}$$

where $c_{22} = 2 \cdot 0 + 4 \cdot 2 = 8$. The other entries are computed in a similar fashion to yield

$$\begin{bmatrix} -1 & 1 \\ 2 & 4 \\ 3 & -1 \end{bmatrix} \begin{bmatrix} 2 & 0 & 1 & 3 \\ -1 & 2 & -2 & 4 \end{bmatrix} = \begin{bmatrix} -3 & 2 & -3 & 1 \\ 0 & 8 & -6 & 22 \\ 7 & -2 & 5 & 5 \end{bmatrix}$$ ∎

Note in Example 5 that we cannot even form a product BA since the shapes do not match up; B has 4 columns and A has only 3 rows. Even if the shapes do match up, it is not necessarily true that $AB = BA$. For instance, let

$$A = \begin{bmatrix} 1 & -1 \\ 0 & 1 \end{bmatrix} \quad \text{and} \quad B = \begin{bmatrix} 2 & 1 \\ 1 & 0 \end{bmatrix}$$

Then

$$AB = \begin{bmatrix} 1 & -1 \\ 0 & 1 \end{bmatrix} \begin{bmatrix} 2 & 1 \\ 1 & 0 \end{bmatrix} = \begin{bmatrix} 1 & 1 \\ 1 & 0 \end{bmatrix}$$

and

$$BA = \begin{bmatrix} 2 & 1 \\ 1 & 0 \end{bmatrix} \begin{bmatrix} 1 & -1 \\ 0 & 1 \end{bmatrix} = \begin{bmatrix} 2 & -1 \\ 1 & -1 \end{bmatrix} \neq AB$$

These are clearly not the same. Nevertheless, matrix multiplication is associative and distributive whenever the shapes match up so that the products can be formed. We mention these properties here without proof.

PROPERTIES OF MATRIX MULTIPLICATION

For matrices A, B, and C,

1. $A(BC) = (AB)C$
2. $A(B + C) = AB + AC$
3. $(A + B)C = AC + BC$
4. AB is not necessarily the same as BA; AB and BA may be equal in special cases.

Inverse of a Matrix

For any size n, the symbol I denotes the $n \times n$ matrix with $a_{ii} = 1$ and $a_{ij} = 0$ for $i \neq j$. For example,

$$I_2 = \begin{bmatrix} 1 & 0 \\ 0 & 1 \end{bmatrix}, \quad I_5 = \begin{bmatrix} 1 & 0 & 0 & 0 & 0 \\ 0 & 1 & 0 & 0 & 0 \\ 0 & 0 & 1 & 0 & 0 \\ 0 & 0 & 0 & 1 & 0 \\ 0 & 0 & 0 & 0 & 1 \end{bmatrix}$$

It can be verified that

$$AI_n = A \quad \text{and} \quad I_n B = B$$

for any matrices A and B for which these products are defined; that is, A has n columns and B has n rows.

I_n is called the $n \times n$ **identity matrix**. For some (but not all) $n \times n$ matrices A, there is a matrix B for which $AB = BA = I_n$. When such a matrix exists, it is called the **inverse** of A and is denoted by A^{-1}:

$$AA^{-1} = A^{-1}A = I_n$$

The principles of the elimination method can be used to determine whether A^{-1} exists and to find A^{-1}.

MATRIX INVERSE

To find the inverse of an $n \times n$ matrix A,

1. Augment A with the appropriate identity matrix I_n:

$$[A \mid I_n]$$

The result is an $n \times 2n$ matrix.

2. Use the principles of the elimination method on the rows of this augmented matrix to convert A to reduced row echelon form.

3. a. If the reduced row echelon form of A is the identity I_n, then the original I_n will have been converted to A^{-1}:

$$[I_n \mid A^{-1}]$$

b. If a row of zeros results in the reduced row echelon form of A, then A has no inverse.

EXAMPLE 6

Either find A^{-1} or show that A^{-1} does not exist for the following matrices.

a. $A = \begin{bmatrix} 2 & -1 \\ -5 & 3 \end{bmatrix}$ b. $A = \begin{bmatrix} 1 & -1 & 2 \\ -3 & 2 & 1 \\ 8 & -6 & 2 \end{bmatrix}$

SOLUTION

a. Augment A with I_2 and use the steps of the elimination method to convert A to reduced row echelon form as follows.

$$\begin{bmatrix} 2 & -1 & \vdots & 1 & 0 \\ -5 & 3 & \vdots & 0 & 1 \end{bmatrix} \quad [A \mid I_2]$$

Trying to avoid fractions as long as possible, we obtain a 1 in the a_{22} position by adding twice row 1 to row 2:

$$\begin{bmatrix} 2 & -1 & \vdots & 1 & 0 \\ -1 & 1 & \vdots & 2 & 1 \end{bmatrix} \quad (R_2 + 2R_1 \to R_2)$$

Then obtain 0 in the a_{12} position:

$$\begin{bmatrix} 1 & 0 & | & 3 & 1 \\ -1 & 1 & | & 2 & 1 \end{bmatrix} \quad (R_1 + R_2 \to R_1)$$

Finally, obtain 0 in the a_{21} position:

$$\begin{bmatrix} 1 & 0 & | & 3 & 1 \\ 0 & 1 & | & 5 & 2 \end{bmatrix} \quad (R_2 + R_1 \to R_2)$$

$\underbrace{}_{I_2} \quad \underbrace{}_{A^{-1}}$

Thus $A^{-1} = \begin{bmatrix} 3 & 1 \\ 5 & 2 \end{bmatrix}$. Checking, we find

$$AA^{-1} = \begin{bmatrix} 2 & -1 \\ -5 & 3 \end{bmatrix} \begin{bmatrix} 3 & 1 \\ 5 & 2 \end{bmatrix} = \begin{bmatrix} 1 & 0 \\ 0 & 1 \end{bmatrix} \stackrel{\checkmark}{=} I_2$$

Similarly, we can check that $A^{-1}A = I_2$.

b. $\begin{bmatrix} 1 & -1 & 2 & | & 1 & 0 & 0 \\ -3 & 2 & 1 & | & 0 & 1 & 0 \\ 8 & -6 & 2 & | & 0 & 0 & 1 \end{bmatrix}$ $[A \mid I_3]$

$\begin{bmatrix} 1 & -1 & 2 & | & 1 & 0 & 0 \\ 0 & -1 & 7 & | & 3 & 1 & 0 \\ 0 & 2 & -14 & | & -8 & 0 & 1 \end{bmatrix}$ $\begin{array}{l}(R_2 + 3R_1 \to R_2) \\ (R_3 - 8R_1 \to R_3)\end{array}$

$\begin{bmatrix} 1 & 0 & -5 & | & -2 & -1 & 0 \\ 0 & -1 & 7 & | & 3 & 1 & 0 \\ 0 & 0 & 0 & | & -2 & 2 & 1 \end{bmatrix}$ $\begin{array}{l}(R_1 - R_2 \to R_1) \\ \\ (R_3 + 2R_2 \to R_3)\end{array}$

The 0 row indicates that A^{-1} *does not exist*. We need not continue with the row reduction. ∎

Solving Systems of Equations Using A^{-1}

The inverse of a matrix can be used to solve a system of equations. To illustrate this, consider the following system:

$$\begin{cases} 2x - y = 1 \\ -5x + 3y = -2 \end{cases}$$

Rewrite this system in matrix form as

$$\begin{bmatrix} 2 & -1 \\ -5 & 3 \end{bmatrix} \begin{bmatrix} x \\ y \end{bmatrix} = \begin{bmatrix} 2x - y \\ -5x + 3y \end{bmatrix} = \begin{bmatrix} 1 \\ -2 \end{bmatrix}$$

Letting

$$A = \begin{bmatrix} 2 & -1 \\ -5 & 3 \end{bmatrix} \quad X = \begin{bmatrix} x \\ y \end{bmatrix} \quad B = \begin{bmatrix} 1 \\ -2 \end{bmatrix}$$

we have

$$AX = B$$

If A^{-1} exists, then

$$A^{-1}AX = A^{-1}B$$
$$IX = A^{-1}B$$
$$X = A^{-1}B$$

Thus if A^{-1} is known, the solution can be computed directly as $A^{-1}B$. In practice, though, finding A^{-1} requires as much work as solving the system by the matrix or elimination methods. However, if basically the same system must be solved repeatedly but with different B columns, this can be a relatively efficient technique.

EXAMPLE 7

Solve the system
$$2x - y = b_1$$
$$-5x + 3y = b_2$$

with

a. $\begin{bmatrix} b_1 \\ b_2 \end{bmatrix} = \begin{bmatrix} 1 \\ -2 \end{bmatrix}$ **b.** $\begin{bmatrix} b_1 \\ b_2 \end{bmatrix} = \begin{bmatrix} 2 \\ 3 \end{bmatrix}$ **c.** $\begin{bmatrix} b_1 \\ b_2 \end{bmatrix} = \begin{bmatrix} -1 \\ 1 \end{bmatrix}$

SOLUTION

Writing this system in the form $AX = B$, we note that A is the matrix of Example 6a. Thus $A^{-1} = \begin{bmatrix} 3 & 1 \\ 5 & 2 \end{bmatrix}$, $X = A^{-1}B$, and we proceed as follows:

a. $\begin{bmatrix} x \\ y \end{bmatrix} = A^{-1}B = \begin{bmatrix} 3 & 1 \\ 5 & 2 \end{bmatrix} \begin{bmatrix} 1 \\ -2 \end{bmatrix} = \begin{bmatrix} 1 \\ 1 \end{bmatrix}$ $x = 1,\quad y = 1$

b. $\begin{bmatrix} x \\ y \end{bmatrix} = A^{-1}B = \begin{bmatrix} 3 & 1 \\ 5 & 2 \end{bmatrix} \begin{bmatrix} 2 \\ 3 \end{bmatrix} = \begin{bmatrix} 9 \\ 16 \end{bmatrix}$ $x = 9,\quad y = 16$

c. $\begin{bmatrix} x \\ y \end{bmatrix} = A^{-1}B = \begin{bmatrix} 3 & 1 \\ 5 & 2 \end{bmatrix} \begin{bmatrix} -1 \\ 1 \end{bmatrix} = \begin{bmatrix} -2 \\ -3 \end{bmatrix}$ $x = -2,\quad y = -3$

Each of these solutions checks in the original system(s). Verify them! ∎

Historical Perspective

ARTHUR CAYLEY
(1821–1895)

Arthur Cayley (1821–1895) was a prominent English mathematician. He developed and studied matrices to assist him in his work on analytic geometry (such as elliptical orbits); he was one of the first mathematicians to study matrices. In 1858 he published his *Memoirs on the Theory of Matrices,* which presented his new and revolutionary ideas. His discovery of the laws of matrix manipulation, especially the noncommutativity of multiplication ($AB \neq BA$) opened up the broad new area of modern abstract algebra.

Section 10.8 Exercises

Find the values of the variables that satisfy the matrix equations.

1. $\begin{bmatrix} 1 & x \\ y & 2 \end{bmatrix} = \begin{bmatrix} u & 3 \\ 4 & v \end{bmatrix}$

2. $\begin{bmatrix} 1 & -1 & r \\ s & t & 3 \end{bmatrix} = \begin{bmatrix} p & q & -2 \\ 1 & 5 & n \end{bmatrix}$

3. $\begin{bmatrix} a & -1 \\ 2 & b \\ -3 & c \end{bmatrix} = \begin{bmatrix} 5 & d \\ e & -6 \\ f & 7 \end{bmatrix}$

4. $\begin{bmatrix} z & x \\ 0 & y \\ 1 & -1 \end{bmatrix} = \begin{bmatrix} -3 & 0 \\ u & 2 \\ v & w \end{bmatrix}$

Let $A = \begin{bmatrix} 1 & 2 \\ -1 & 3 \end{bmatrix}$, $B = \begin{bmatrix} 0 & 1 \\ 4 & -1 \end{bmatrix}$, $C = \begin{bmatrix} 2 & 5 \\ -1 & -3 \end{bmatrix}$. Find

5. $A + B$
6. $A - C$
7. $4A + 2B$
8. $5B - 3C$
9. AB
10. BA
11. $A(B - C)$
12. $AB + AC$
13. $(2A - B)C$
14. $AC - BC$
15. $(AB)C$
16. $A(BC)$

Let $A = \begin{bmatrix} 1 & 2 & -1 \\ 3 & 0 & 1 \end{bmatrix}$, $B = \begin{bmatrix} 2 & 1 \\ 1 & 3 \\ 4 & 0 \end{bmatrix}$, $C = \begin{bmatrix} -1 & 0 \\ -3 & 2 \\ 5 & -2 \end{bmatrix}$.

Find

17. $B - C$
18. $3C + 4B$
19. AB
20. AC
21. BA
22. CA
23. $A(B + C)$
24. $(B - C)A$
25. $2BA + 3CA$
26. $2AC - 4AB$

Let $A = \begin{bmatrix} 1 & 0 & -1 & 2 \\ 0 & 3 & 4 & 1 \\ 1 & -1 & 2 & -2 \\ -3 & -1 & 0 & 5 \end{bmatrix}$,

$B = \begin{bmatrix} 1 & -1 & 1 & 2 \\ 0 & 3 & 4 & 0 \end{bmatrix}$,

and $C = \begin{bmatrix} -2 & 0 & 1 & 4 \\ 3 & 1 & 2 & 5 \end{bmatrix}$. Find

27. BA
28. CA
29. $B - 2C$
30. $3BA - 2CA$
31. $B(2A) + C(3A)$
32. $(C + B)A$
33. AA
34. Find a matrix M so that $BA + M = CA$.

For the following matrices either find the inverse or show that no inverse exists.

35. $\begin{bmatrix} 2 & 5 \\ 1 & 3 \end{bmatrix}$
36. $\begin{bmatrix} 2 & 3 \\ -4 & -6 \end{bmatrix}$
37. $\begin{bmatrix} 1 & -1 \\ -6 & 7 \end{bmatrix}$
38. $\begin{bmatrix} 2 & -1 \\ 3 & -1 \end{bmatrix}$
39. $\begin{bmatrix} -2 & 8 \\ 3 & -12 \end{bmatrix}$
40. $\begin{bmatrix} 3 & 7 \\ 1 & 2 \end{bmatrix}$
41. $\begin{bmatrix} 10 & 4 & 1 \\ 5 & 1 & 1 \\ 7 & 2 & 1 \end{bmatrix}$
42. $\begin{bmatrix} -3 & -3 & 8 \\ 5 & -1 & -2 \\ 4 & 1 & -5 \end{bmatrix}$
43. $\begin{bmatrix} 2 & 1 & -1 \\ 1 & 1 & -1 \\ -1 & -2 & 3 \end{bmatrix}$
44. $\begin{bmatrix} 2 & 1 & -1 \\ 3 & 2 & -2 \\ 1 & 2 & -3 \end{bmatrix}$

Write the following systems in the form $AX = B$ and use A^{-1} to solve the system. A^{-1} will have been found in Exercises 35 to 44.

45. $\begin{cases} 2x + 5y = 2 \\ x + 3y = 3 \end{cases}$
(Hint: See Exercise 35.)

46. $\begin{cases} 2x - y = 3 \\ 3x - y = -2 \end{cases}$
(Hint: See Exercise 38.)

47. $\begin{cases} x - y = -1 \\ -6x + 7y = 2 \end{cases}$
(*Hint:* See Exercise 37.)

48. $\begin{cases} 3x + 7y = 4 \\ x + 2y = 1 \end{cases}$
(*Hint:* See Exercise 40.)

49. $\begin{cases} 10x + 4y + z = 1 \\ 5x + y + z = -1 \\ 7x + 2y + z = 1 \end{cases}$
(*Hint:* See Exercise 41.)

50. $\begin{cases} 2x + y - z = -2 \\ 3x + 2y - 2z = 1 \\ x + 2y - 3z = -1 \end{cases}$
(*Hint:* See Exercise 44.)

51. $\begin{cases} 2x + y - z = 0 \\ x + y - z = -2 \\ -x - 2y + 3z = 1 \end{cases}$
(*Hint:* See Exercise 43.)

The following information is used in the remaining exercises.

The Robinson Manufacturing Company produces bicycles, mopeds, and tricycles on four machines: a stamper, a press, a welder, and a painter. Each bicycle requires 2 hours on the stamper, 4 on the press, 1 on the welder, and 1 on the painter. Each moped requires 1, 2, 3, and 2 hours on these machines, respectively, and each tricycle requires 2, 1, 1, and 3 hours respectively. It costs $20 an hour to operate the stamper, $15 an hour for the press, $25 an hour for the welder, and $10 an hour for the painter.

52. **a.** Express the total production cost for each product as a product of two matrices.
    b. What are these total costs?

53. **a.** Express the total number of hours required on each machine to produce 10 bicycles, 5 mopeds, and 20 tricycles as a product of two matrices.
    b. How many hours are required on each machine?

54. **a.** Express the total cost of producing the items in Exercise 53 as a product of three matrices.
    b. What is this total cost?

55. The company constructs another factory for which the hourly costs are $10 for the stamper, $12 for the press, $20 for the welder, and $10 for the painter. Production times are not affected.
    a. If 40% of its products are manufactured in the new plant, express the average production cost for each item in matrix terms.
    b. What are these average costs?

56. **a.** If 10 bicycles, 20 mopeds, and 30 tricycles are manufactured at the original plant and 30 bicycles, 15 mopeds, and 20 tricycles are produced at the new plant, express the total production cost in matrix terms.
    b. What is the total production cost?

Section 10.9

Systems of Inequalities

Sometimes the constraints describing a situation result in a system of inequalities rather than in a system of equations. We have already discussed several procedures for solving a system of equations; among these were Cramer's rule and the elimination method. Unfortunately, there are no correspondingly slick techniques available for solving a system of inequalities.

In a system of inequalities, each inequality must be solved separately, generally by graphing its solution. The points that satisfy the system of inequalities are precisely those that lie on *each* of the individual graphs. The elimination and substitution methods are frequently useful in finding the points in which two boundary curves for the solutions meet.

Although this work can be extended to more than two variables, it can be very difficult to visualize the solutions with three variables and is impossible with four or more variables. For this reason, we shall work with only two variables.

> ### SYSTEMS OF INEQUALITIES IN TWO VARIABLES
>
> To solve a system of inequalities in two variables:
>
> 1. Replace the inequality symbols with = signs.
> 2. Sketch the resulting graphs. Use a solid line if the original inequality was $\leq$ or $\geq$; use a broken line if the original was $<$ or $>$.
> 3. The solution to a given inequality consists of one of the regions determined by the graph of the corresponding equality. A broken line would not be included in the solution set; a solid line would be included.
> 4. The solution set will consist of the region(s) common to all the regions determined in Step 3.
> 5. If possible, find the vertices (corner points) of the solution set.

EXAMPLE 1

Solve the following system of inequalities.

$$\begin{cases} 3x - y \leq 12 \\ 2x - 4y > -12 \end{cases}$$

SOLUTION

To illustrate the procedure just described, we shall sketch the graph of each inequality independently.

1. Rewrite

    $$3x - y \leq 12$$

    as

    $$y \geq 3x - 12$$

    and sketch the line $y = 3x - 12$. The solution to the first inequality consists of all points on or above this line as indicated in Figure 10.12(a). For instance, (5, 0) lies below the line and does not satisfy the inequality, (5, 5) satisfies the inequality and lies above the line, and (5, 3) is on the edge of the solution region.

2. Then write the second inequality $2x - 4y > -12$ as

    $$4y < 2x + 12$$

    $$y < \left(\frac{1}{2}\right)x + 3$$

    The solution to this inequality consists of all points below the line $y = (\frac{1}{2})x + 3$ as indicated in Figure 10.12b. For instance, (4, 4) lies below the line and satisfies the inequality, (4, 7) lies above the line and does not satisfy the inequality, and (4, 5) lies at the edge of the solution region but does not satisfy the inequality. The line in Figure 10.12(b) is broken to indicate that it is not part of the solution; the solid line is used in Figure 10.12(a) to indicate that it is part of the solution.

Figure 10.12

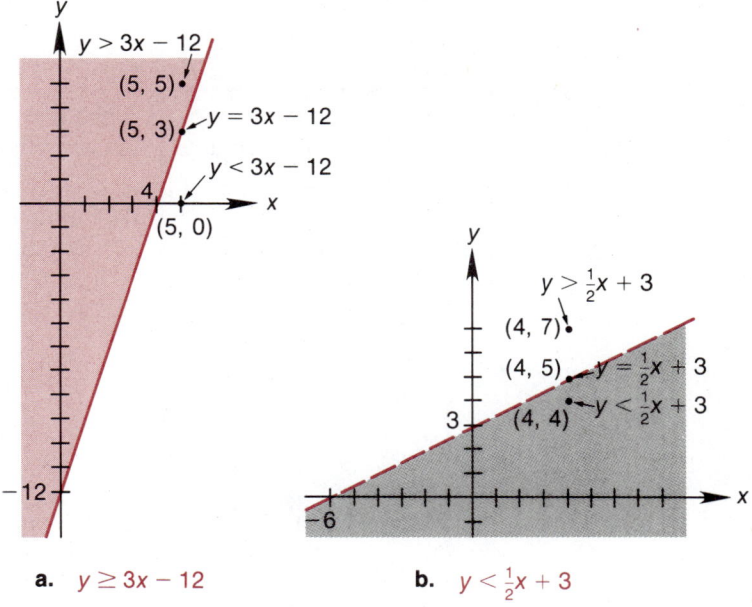

a. $y \geq 3x - 12$

b. $y < \frac{1}{2}x + 3$

Finally, both inequalities are satisfied in the darker shaded region that is sketched in Figure 10.13. The point at which the boundary lines meet can be found by the elimination or substitution methods or even by Cramer's rule.

Figure 10.13
$$\begin{cases} 3x - y \leq 12 \\ 2x - 4y > -12 \end{cases}$$

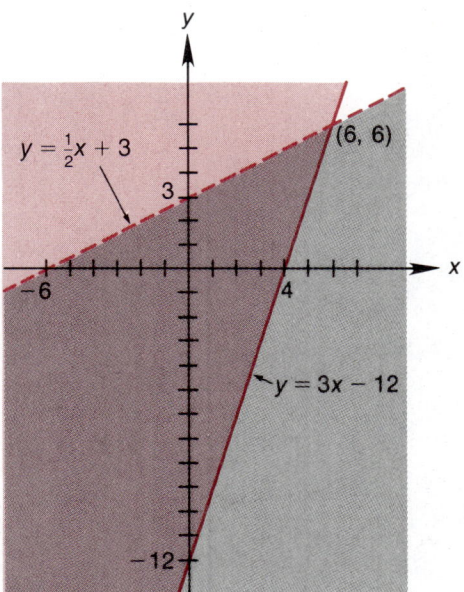

As in parts 1 and 2 of the solution to Example 1, we make the following observations regarding *individual* inequalities.

INEQUALITY SOLUTION SETS

If an inequality can be expressed as	Then the solution set consists of	
1. $y < f(x)$	1. Points *below* the graph of $y = f(x)$	
2. $y > f(x)$	2. Points *above* the graph of $y = f(x)$	
3. $x < g(y)$	3. Points to the left of the graph of $x = g(y)$	
4. $x > g(y)$	4. Points to the right of the graph of $x = g(y)$.	

Systems of inequalities do not always appear originally as "systems." For instance, in the following example, the inequality statement appears on a single line, yet is indeed a system of inequalities. In Example 3, a single $\geq$ is used; yet its solution also requires consideration of a system of inequalities. Any inequality involving an absolute value is also a system in disguise, as we shall see in Example 4.

EXAMPLE 2

Graph the following inequality.

$$3x + y < x - y \leq 2x + 3y$$

SOLUTION

Since both inequalities must be satisfied, we actually have a system of inequalities:

$$\begin{cases} 3x + y < x - y \\ x - y \leq 2x + 3y \end{cases}$$

Simplifying each of these expressions yields the equivalent system:

$$\begin{cases} y < -x \\ y \geq -\frac{1}{4}x \end{cases}$$

The solution to the first inequality consists of the region *below* the line $y = -x$. The solution to the second consists of the points *above or on* the line $y = -\frac{1}{4}x$. Each of these is then graphed and the solution is found as indicated by the darker shaded region in Figure 10.14c.

Figure 10.14

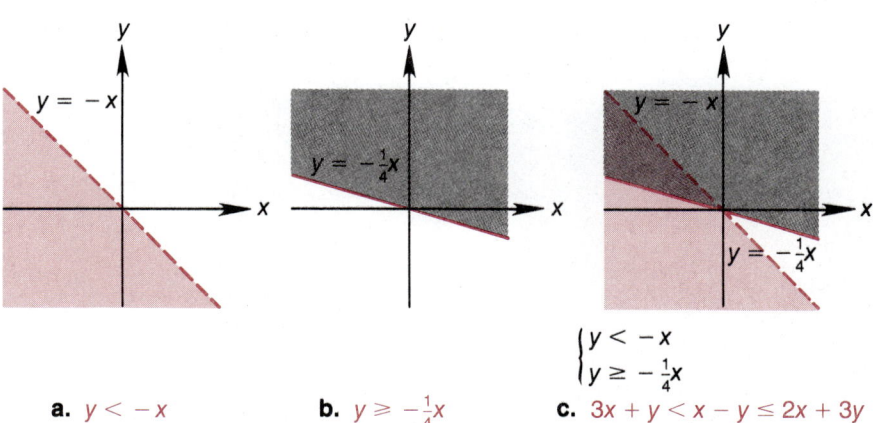

a. $y < -x$ **b.** $y \geq -\frac{1}{4}x$ **c.** $3x + y < x - y \leq 2x + 3y$

EXAMPLE 3

Graph the following inequality.

$$(x - 2y)(y - 2x) \geq 0$$

SOLUTION

In order for the product to be nonnegative, both factors must have the same sign.

a. If both factors are nonnegative, then

$$\begin{cases} x - 2y \geq 0 \\ y - 2x \geq 0 \end{cases} \quad \text{equivalently} \quad \begin{cases} y \leq \frac{1}{2}x \\ y \geq 2x \end{cases}$$

This solution is sketched as the darker shaded region in Figure 10.15.

Figure 10.15

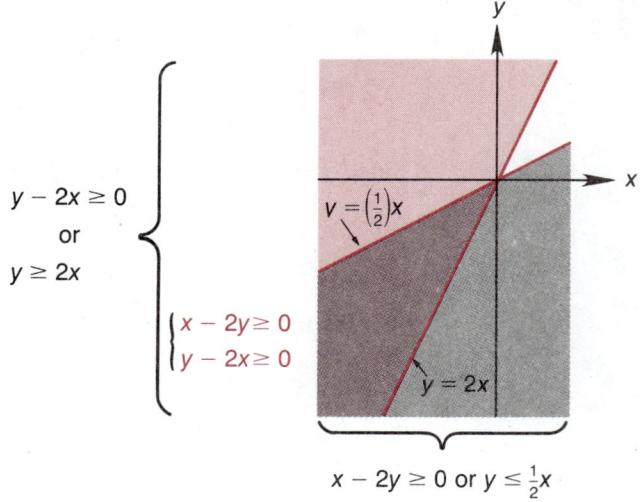

b. If both factors are nonpositive, then

$$\begin{cases} x - 2y \leq 0 \\ y - 2x \leq 0 \end{cases} \quad \text{equivalently} \quad \begin{cases} y \geq \frac{1}{2}x \\ y \leq 2x \end{cases}$$

This solution is sketched as the darker shaded region in Figure 10.16.

Figure 10.16

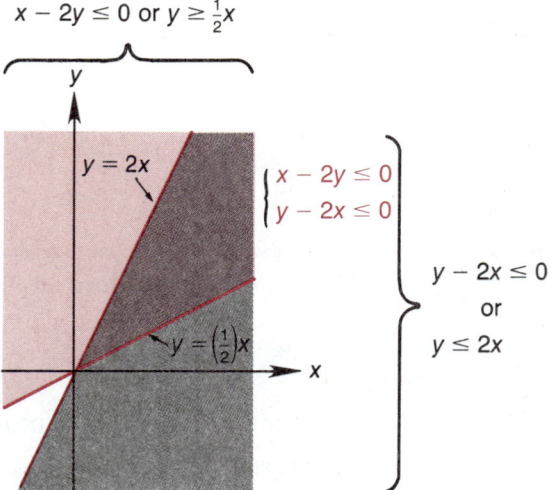

The original inequality is satisfied in both cases. Thus its solution is as illustrated in Figure 10.17.

Figure 10.17
$(x - 2y)(y - 2x) \geq 0$

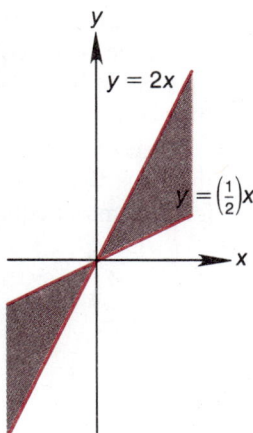

EXAMPLE 4

Graph the following inequality.

$$|y - x| \leq 2$$

SOLUTION

This inequality is equivalent to

$$-2 \leq y - x \leq 2 \qquad \text{(Section 2.8)}$$

Since two inequalities are expressed here, this is actually a system of inequalities:

$$\begin{cases} y - x \geq -2 \\ y - x \leq 2 \end{cases}$$

This is equivalent to the system

$$\begin{cases} y \geq x - 2 \\ y \leq x + 2 \end{cases}$$

Its solution is sketched as the darker shaded region in Figure 10.18.

Figure 10.18

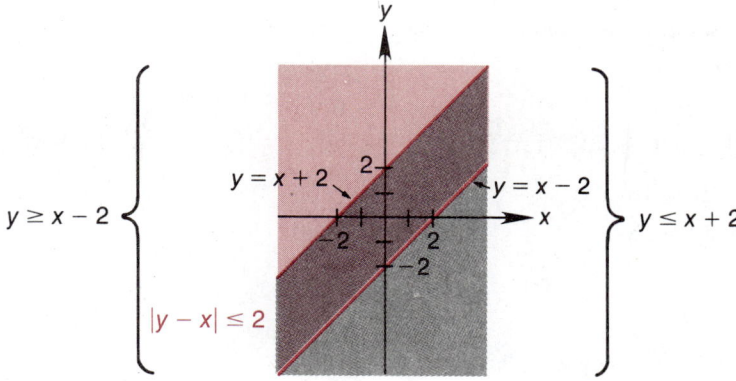

EXAMPLE 5

Solve the following system of inequalities.

$$\begin{cases} y > x^2 + 1 \\ y < x + 3 \end{cases}$$

SOLUTION

Even though one of these inequalities is quadratic, the technique is the same: graph the two inequalities and find the common solution. The solution to the first inequality consists of the region *above* the parabola $y = x^2 + 1$. The solution to the second inequality consists of the region *below* the line $y = x + 3$. The darker shaded region in Figure 10.19 represents the points satisfying *both* inequalities and hence is the solution set for the system of inequalities. The techniques developed for systems of equations are used to find the "corner points" of the region.

Figure 10.19
$\begin{cases} y > x^2 + 1 \\ y < x + 3 \end{cases}$

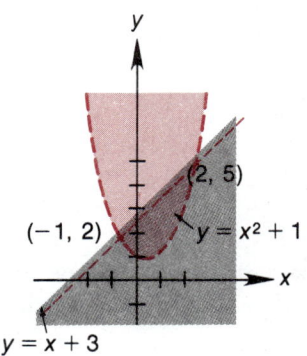

Section 10.9 Exercises

Graph the solution sets to the inequalities in Exercises 1–10. Note that these are not systems of inequalities.

1. $y > 1 - x$
2. $y < x + 2$
3. $x \leq y - 1$
4. $x \geq 2 - 2y$
5. $2x - y < 3$
6. $x + 3y \geq -3$
7. $y \geq x^2 - 1$
8. $y < 2 - x^2$
9. $x \leq 4 - y^2$
10. $x > y^2 - 9$

Graph the solution sets for the following systems of inequalities.

11. $\begin{cases} x - y > 0 \\ 3x - y \leq 0 \end{cases}$
12. $\begin{cases} x + y > 0 \\ x - 2y < 0 \end{cases}$
13. $\begin{cases} 3x - 2y \leq 16 \\ 2x + 3y > 18 \end{cases}$
14. $\begin{cases} 5x + 3y \leq 60 \\ 2x + 4y \geq 40 \end{cases}$
15. $\begin{cases} x + 2y < 2 \\ 2y - x \geq 4 \end{cases}$
16. $\begin{cases} y - x \geq 1 \\ y + x < 3 \end{cases}$
17. $2x - 3y < 3x + 4y \leq 3y - 2x$
18. $x - 2y < 3x - y \leq x + y$
19. $4x + y + 4 < 2x + 3y - 4 \leq x + 2y + 6$
20. $2x + y - 2 \leq x - y + 5 < 8 - 2x - y$
21. $(y - 3x)(2x - y) \geq 0$
22. $(x + y)(x - 2y) \leq 0$
23. $|2x + y| \leq 3$
24. $|x + 2y| < 4$
25. $|x - 2y - 6| < 4$
26. $|x + 6y - 8| \leq 5$
27. $|2x + y + 1| \leq 2$
28. $|3x - y + 2| < 3$
29. $\begin{cases} 2x^2 + y < 2 \\ x - y < 1 \end{cases}$
30. $\begin{cases} 4x^2 - 8x + 8 < y \\ y - x > 8 \end{cases}$
31. $\begin{cases} x^2 - y > 1 \\ y - x \leq 1 \end{cases}$
32. $\begin{cases} x^2 + y \geq 1 \\ x + y > 1 \end{cases}$
33. $\begin{cases} 2x^2 - y \leq 1 \\ 2x^2 + y \leq 1 \end{cases}$
34. $\begin{cases} 3x^2 - 4y \leq 0 \\ 4y - x^2 \leq 8 \end{cases}$
35. $\begin{cases} x^2 - y < 2 \\ x^2 \leq 2y \end{cases}$
36. $\begin{cases} x^2 + y > 1 \\ y \leq x^2 \end{cases}$
37. $\begin{cases} x^2 + y^2 < 25 \\ x + y \geq 1 \end{cases}$
38. $\begin{cases} x^2 + y^2 \geq 1 \\ x + y < 1 \end{cases}$
39. $\begin{cases} x^2 + y^2 < 9 \\ 2x^2 + 5y < 0 \end{cases}$
40. $\begin{cases} x^2 + y^2 \leq 16 \\ x^2 - y \geq 5 \end{cases}$
41. $\begin{cases} 2x + y \geq 2 \\ y \leq 2 \\ x \leq 1 \end{cases}$
42. $\begin{cases} 3x + 4y \leq 12 \\ x \geq 0 \\ y \geq 0 \end{cases}$
43. $\begin{cases} x - y \geq 0 \\ x + y \geq 2 \\ x \leq 4 \end{cases}$
44. $\begin{cases} x + y \geq 1 \\ x - y \leq 1 \\ y \leq 3 \end{cases}$

45. $\begin{cases} y > 2^x \\ y > 2^{-x} \end{cases}$
46. $\begin{cases} y > 2^x \\ y < 2^{-x} \end{cases}$

47. $\begin{cases} y \leq \log_2 x \\ x \geq 2 \\ y \geq 0 \end{cases}$
48. $\begin{cases} y \geq \log_2 x \\ y \leq x \\ y \geq 0 \end{cases}$

49. $|x - y||2x - y| \leq 0$

50. $|x + y - 3||2x - y + 4| \leq 0$

51. $|x + y - 1||x - y - 1| \leq 0$

52. The Universal Recycling Company manufactures a Tin-Can Crusher. The byproducts from manufacturing this item are, in turn, recycled and used to make Bottle Crushers. Limitations on the amount of material available require that the number of Bottle Crushers produced daily is no more than the square of the number of Tin-Can Crushers produced during that day. The Fizz Cola Company has succeeded in lobbying the state legislature to forbid the production of more than 90 crushers per day. What legal production schedules are available to this company?

53. Mr. Doerger is treating his meadow with two products. One is designed to decrease the growth of clover; the other is designed to increase the growth of alfalfa. Unfortunately, the products react with one another to kill wildlife in the area.
    Suppose that the adverse reactions are held to a minimum if the sum of the squares of the relative amounts per acre does not exceed 625 and the product of these amounts does not exceed 168. Sketch a graph to illustrate the relative treatment levels that are permissible.

54. Doris has two part-time jobs; one pays $6 per hour and the other pays $5 per hour. If she wants to earn at least $180 per week, sketch a graph indicating the ways in which she could divide her time between the two jobs.

55. Injecting a patient with x milliliters of a certain medication results in undesirable toxins in the amount of $x^2 + 3x$ micrograms above the patient's existing level of 2 micrograms. All this toxin must be destroyed by administering at least as much antitoxin as there is toxin in the patient. Sketch a graph indicating the permissible combinations of medication and antitoxin that can be given to this patient.

Section 10.10

Linear Programming

Linear programming is a branch of mathematics that has become increasingly popular in recent years. When one's choices are constrained, linear programming can be very useful in deciding which of various options to choose for maximum benefit. For example, a plant manager might use linear programming to schedule production to maximize profit. A dietitian could use it to purchase foods at the least cost that will satisfy certain dietary requirements. Although some very sophisticated techniques have been developed to solve linear programming problems, problems in two variables can be solved by graphing the solution to a system of linear inequalities.

This solution technique will be developed in the following example. Since the solution is rather lengthy, appropriate steps will be identified.

EXAMPLE 1 A calculator company manufactures two types of calculators: the basic and the scientific. Each basic requires 10 minutes in the electronics prefabrication area; each scientific requires 20 minutes of electronic prefabrication. For the final assembly and packaging, each basic requires 6 minutes; each scientific requires only 4 minutes since they are more completely assembled on the electronics assembly line. The company is able to produce only 500 plastic calculator bodies per day. There are only 150 work-hours available per day in the electronics prefabrication area and only 40 work-hours per day in the final assembly and packaging area. If each basic yields a profit of $4 and each scientific yields a profit of $5, how many of each type should the company produce daily in order to maximize its profits?

SOLUTION

Step 1. *Read the problem carefully.*

Step 2. As with any word problem, we must *label what we are to find.* Thus we let

x = number of basics produced per day

y = number of scientifics produced per day

Step 3. *Translate the given information into mathematical statements.* Since each basic requires 10 minutes and each scientific requires 20 minutes in the electronics prefabrication area, we shall use a daily total of $10x + 20y$ minutes for labor in the electronics prefabrication room. But there are only 150 hours = 9000 minutes available. This constraint translates into the inequality

$$10x + 20y \leq 9000 \tag{1}$$

In a similar fashion, the assembly time requirements and the limitations on available assembly time translate into the inequality

$$6x + 4y \leq 2400 \tag{2}$$

since 40 hours = 2400 minutes.

Finally, the plastic production capacity limits the daily production to at most 500 units:

$$x + y \leq 500 \tag{3}$$

The only other constraints are

$$x \geq 0 \tag{4}$$
$$y \geq 0 \tag{5}$$

since we cannot produce a negative number of units.

Our objective is to maximize the total profit by choosing to produce the "right" number of each type. Producing x basics and y scientifics at profits of $4 and $5, respectively, yields a total profit (in dollars) of

$$P = 4x + 5y \tag{6}$$

We now collect the preceding mathematical statements into the following statement of our problem.

MATHEMATICAL FORMULATION OF EXAMPLE I

Maximize $\quad P = 4x + 5y$

subject to $\quad \begin{cases} 10x + 20y \leq 9000 \\ 6x + 4y \leq 2400 \\ x + y \leq 500 \\ x \geq 0 \\ y \geq 0 \end{cases}$

Section 10.10 Linear Programming **631**

> ### LINEAR PROGRAMMING PROBLEMS
>
> The mathematical formulation of Example 1 is that of a typical linear programming problem. Its essential features are that
>
> 1. All expressions are linear: $ax + by$.
> 2. The constraints are inequalities: $\leq$, $\geq$.
> 3. The variables must be nonnegative.
> 4. Something is to be maximized or minimized.
>
> In a linear programming problem, that which is to be maximized or minimized is called the **objective function**. A point that optimizes (maximizes or minimizes as the case may be) the objective function is called an **optimal solution** to the problem. A point that satisfies the constraints of the problem but does not necessarily optimize the objective function is called a **feasible solution.**

We now continue with the solution of Example 1.

Step 4. *Find the feasible solutions.* The techniques of the preceding section can be used to sketch the set of feasible solutions to this problem as illustrated in Figure 10.20.

Figure 10.20

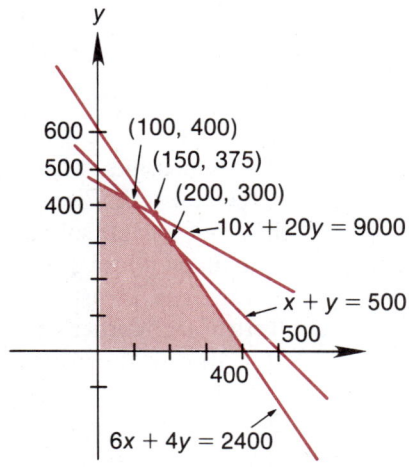

Step 5. *Analyze the values of the objective function* on the set of feasible solutions. In this problem the objective function is profit $P = 4x + 5y$. Lines representing various levels of profit as sketched in Figure 10.21. These lines are parallel since each has slope $m = -\frac{4}{5}$:

$$y = -\frac{4}{5}x + \frac{P}{5}$$

Figure 10.21

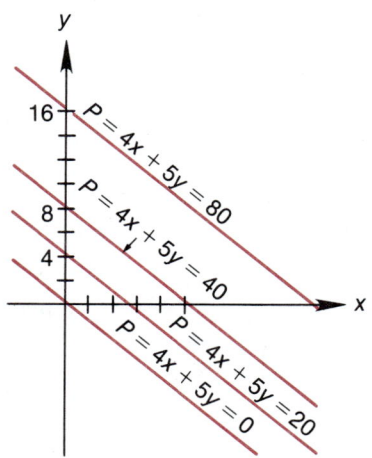

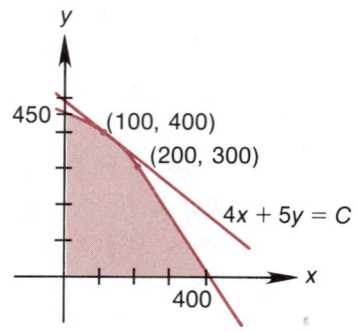

Figure 10.22

Note that each of the parallel lines $4x + 5y =$ constant represents a fixed level of profit. As such a "profit" line is moved upward, it represents a greater profit.

Step 6. *Find a point that maximizes the profit.* We can find the maximum profit level by moving the "profit" line as high as possible without leaving the set of feasible solutions. This is seen to be the line that passes through the point (100, 400), since any higher line would not contain any feasible solutions; any lower line would represent a lower profit (see Figure 10.22).

Thus the company should produce 100 basic and 400 scientific calculators daily in order to maximize its profit. Its maximum daily profit is then

$$P_{max} = 4 \cdot 100 + 5 \cdot 400$$
$$= \$2400$$

∎

Sometimes, when the slope of the profit line is very close to the slope of one of the constraint lines, it may be difficult to ascertain when the profit line leaves the set of feasible solutions on its way upward. However, we need not be especially meticulous about sketching the profit line because, as the profit line is moving upward and is just ready to leave the feasible set, it must be passing through a "vertex" or "corner" of the feasible set.

In case the profit coincides with an edge of the constraint set as it is ready to leave the constraint set, it will be passing through two vertices. In this case, there will be many optimal solutions; nevertheless, at least one of the optimal solutions occurs at a vertex.

> **LOCATION OF OPTIMAL SOLUTIONS**
>
> The optimal solution to a linear programming problem must occur at a vertex of the feasible region. If two vertices give this same optimal value, then all points on the segment joining these two points give the same optimal value.

Thus rather than sketching the profit lines, we need only check the profit at each vertex and choose that vertex that yields the maximum profit. For instance, in Example 1 the vertices of the feasible set are $v_1 = (0, 0)$, $v_2 = (0, 450)$,

$v_3 = (100, 400)$, $v_4 = (200, 300)$, and $v_5 = (400, 0)$. The corresponding profits are

$$P_1 = 4 \cdot 0 + 5 \cdot 0 = 0$$
$$P_2 = 4 \cdot 0 + 5 \cdot 450 = 2250$$
$$P_3 = 4 \cdot 100 + 5 \cdot 400 = 2400$$
$$P_4 = 4 \cdot 200 + 5 \cdot 300 = 2300$$
$$P_5 = 4 \cdot 400 + 5 \cdot 0 = 1600$$

The maximum profit of $2400 occurs at $v_3 = (100, 400)$.

Our approach to linear programming problems is summarized as follows.

SOLVING A LINEAR PROGRAMMING PROBLEM

1. *Read the problem* carefully from beginning to end.
2. *Label the quantities* that you are expected to find.
3. *Translate* the given information *into mathematical statements.* This should result in an objective function and a set of constraints described by a system of linear inequalities.
4. *Find the* set of *feasible solutions* by sketching the solution to the system of inequalities.
5. *Evaluate the objective function* at each vertex of the set of feasible solutions.
6. *Choose a (the) vertex* that optimizes the objective function.

The method we have described here generalizes to three variables but requires three-dimensional figures. Since we cannot visualize more than three dimensions, this graphical method breaks down when there are four or more unknowns. There is an abstract algebraic formulation of this procedure known as the **simplex method,** which can then be used. The simplex method is beyond the scope of this book, however.

EXAMPLE 2 In order to treat a certain deficiency, Dr. Quincy prescribes a monthly diet containing at least 60 units of carbohydrate, 40 units of fat, and 45 units of protein. The patient is being treated in Metropolitan Hospital, which is on an austerity budget. Metropolitan seeks to minimize the cost of the food. Providing a variety of foods can be very costly; thus, it purchases only two types of foods: a "Nutritious Mixture" and a "Tasty Feeder." Each unit of Mixture contains 2 units of carbohydrate, 1 unit of fat, 1 unit of protein, and costs $6. Each unit of Feeder contains 2 units of protein, 1 unit of fat, 1 unit of carbohydrate, and costs $9. What combination of Mixture and Feeder should the hospital use to compose a diet of minimum cost that satisfies the requirements of the prescription?

SOLUTION Let

x = number of units of Mixture
y = number of units of Feeder

to be used monthly in the patient's diet. This combination yields

$2x + y$ units of carbohydrate
$x + 2y$ units of protein
$x + y$ units of fat

at a cost of $C = 6x + 9y$ dollars. Hence we are to

> Minimize $\quad C = 6x + 9y$
>
> subject to $\quad \begin{cases} 2x + y \geq 60 \\ x + 2y \geq 45 \\ x + y \geq 40 \\ x \geq 0 \\ y \geq 0 \end{cases}$

The set of feasible solutions is shown in Figure 10.23.

Figure 10.23

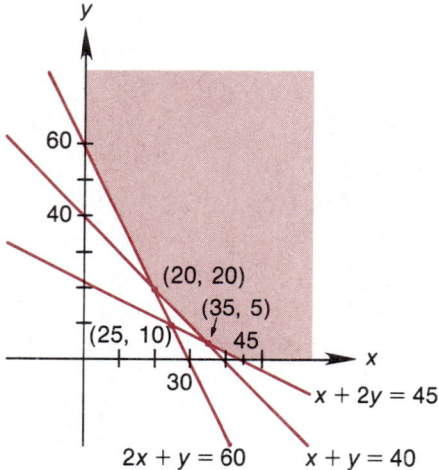

This time the lines of the form $6x + 9y =$ constant consist of (x, y)'s that yield the same *cost*. Since we wish to *minimize* cost, we move the cost line down as far as possible without leaving the feasible set. This line will pass through one of the vertices of the feasible set when it has reached this stage. Thus an optimal solution again occurs at one of the vertices. In the chart below we list the vertices and the corresponding costs.

VERTEX (x, y)	COST $(6x + 9y)$
(0, 60)	540
(20, 20)	300
(35, 5)	255
(45, 0)	270

Thus the minimum cost occurs at vertex (35, 5). The institution should purchase 35 units of Mixture and 5 units of Feeder in order to minimize its cost at $255. ∎

Section 10.10 Exercises

Exercises 1–20 are linear programming problems that have already been translated into mathematical statements. Solve them.

1. Maximize $\quad 5x + 3y$

   subject to $\quad \begin{cases} 2x + 5y \leq 300 \\ x + y \leq 90 \\ 0 \leq x \leq 70 \\ y \geq 0 \end{cases}$

2. Maximize $\quad 10x + 5y$

   subject to $\quad \begin{cases} 4x + 3y \leq 60 \\ x + 3y \leq 30 \\ x \geq 0 \\ y \geq 0 \end{cases}$

3. Minimize $\quad 5x + 4y$

   subject to $\quad \begin{cases} 5x + 6y \geq 90 \\ 5x + 3y \geq 60 \\ x \geq 0 \\ y \geq 0 \end{cases}$

4. Minimize $\quad x + y$

   subject to $\quad \begin{cases} 2x + y \geq 5 \\ x + 2y \geq 4 \\ x \geq 0 \\ y \geq 0 \end{cases}$

5. Maximize $\quad x + 2y$

   subject to $\quad \begin{cases} 3x + 4y \leq 200 \\ 2x + y \leq 100 \\ x \geq 0 \\ y \geq 0 \end{cases}$

6. Maximize $\quad x + 3y$

   subject to $\quad \begin{cases} 2x + y \leq 20 \\ x + y \leq 16 \\ x \geq 0 \\ y \geq 0 \end{cases}$

7. Minimize $\quad 2x + 3y$

   subject to $\quad \begin{cases} 5x + 2y \geq 20 \\ x + y \geq 7 \\ x \geq 0 \\ y \geq 0 \end{cases}$

8. Minimize $\quad 10x + 3y$

   subject to $\quad \begin{cases} x + y \geq 5 \\ 2x + y \geq 8 \\ x \geq 0 \\ y \geq 0 \end{cases}$

9. Maximize $\quad 4x + 5y$

   subject to $\quad \begin{cases} x + 3y \leq 9 \\ x + y \leq 5 \\ x \geq 0 \\ y \geq 0 \end{cases}$

10. Maximize $\quad 5x + 7y$

    subject to $\quad \begin{cases} 2x + 3y \leq 24 \\ x + y \leq 10 \\ x \geq 0 \\ y \geq 0 \end{cases}$

11. Minimize $\quad 7x + 4y$

    subject to $\quad \begin{cases} 3x + y \geq 6 \\ x + 5y \geq 8 \\ x + y \geq 4 \\ x \geq 0 \\ y \geq 0 \end{cases}$

12. Minimize $\quad 2x + 3y$

    subject to $\quad \begin{cases} x + 3y \geq 6 \\ 3x + y \geq 9 \\ 2x + y \geq 7 \\ x \geq 0 \\ y \geq 0 \end{cases}$

13. Maximize $\quad 5x + y$

    subject to $\quad \begin{cases} 10x + y \leq 175 \\ 5x + 2y \leq 125 \\ x + 2y \leq 105 \\ x \geq 0 \\ y \geq 0 \end{cases}$

14. Maximize $\quad 5x + 3y$

    subject to $\quad \begin{cases} 3x + 4y \leq 26 \\ 2x + y \leq 14 \\ x + 2y \leq 12 \\ x \geq 0 \\ y \geq 0 \end{cases}$

15. Minimize $3x + 10y$

subject to $\begin{cases} 5x + 2y \geq 24 \\ 2x + 3y \geq 24 \\ x + 3y \geq 18 \\ x \geq 0 \\ y \geq 0 \end{cases}$

16. Minimize $6x + 4y$

subject to $\begin{cases} 11x + 5y \geq 425 \\ x + y \geq 55 \\ 2x + 6y \geq 170 \\ x \geq 0 \\ y \geq 0 \end{cases}$

17. Maximize $2x + 9y$

subject to $\begin{cases} x + y \leq 48 \\ 5x + 2y \leq 180 \\ x + 2y \leq 80 \\ x \geq 0 \\ y \geq 0 \end{cases}$

18. Maximize $12x + 7y$

subject to $\begin{cases} x + 2y \leq 120 \\ x + y \leq 70 \\ 2x + y \leq 120 \\ x \geq 0 \\ y \geq 0 \end{cases}$

19. Maximize $4x + 5y$

subject to $\begin{cases} 2x + 3y \leq 180 \\ 3x + 4y \leq 250 \\ x + y \leq 75 \\ x \geq 0 \\ y \geq 0 \end{cases}$

20. Minimize $x + y$

subject to $\begin{cases} 2x + y \geq 70 \\ 2x + 3y \geq 120 \\ 3x + 2y \geq 130 \\ x \geq 0 \\ y \geq 0 \end{cases}$

Convert the following linear programming problems into mathematical form and then solve them.

21. The Swifty Bicycle Company makes two models, the Alpha and the Beta. Each Alpha requires 1 hour of manufacturing time and 2 hours of assembly time and yields a profit of $2. Each Beta requires 2 hours manufacturing time and 1 hour assembly time and yields a profit of $2.50.
    a. If 50 bikes can be manufactured simultaneously and both the manufacturing and the assembly areas operate only 12 hours per day, how many of each model should be produced daily so as to maximize profits?
    b. What is the maximum profit?

22. Repeat Exercise 21 with the following changes: Each Alpha requires 3 hours of manufacturing time, $\frac{1}{2}$ hour of assembly time, and yields $5 profit. Each Beta yields a $2 profit.

23. Change Exercise 22 to have each bicycle yielding the same profit of $3.

24. A farmer is planning to raise oats and corn. Each acre of oats yields a profit of $40; each acre of corn yields a profit of $60. To prepare the soil and plant the crops, she rents three machines, M_1, M_2, and M_3, for 70, 40, and 90 hours, respectively. Each acre of oats requires 2 hours of M_1's time and 1 hour each from M_2 and M_3; each acre of corn requires 3 hours of M_3's time and 1 hour each from M_1 and M_2.
    a. How many acres of each crop should she plant in order to maximize her profit?
    b. What is the maximum profit?

25. A truck manufacturer makes both light-duty pickups and heavy-duty semis. Each semi requires 5 worker-days' labor in the manufacturing area and 4 worker-days in the painting and finishing area and yields a profit of $300. Each pickup requires 3 worker-days in manufacturing and 2 days in finishing and yields a profit of $150. Enough people have been hired to provide 210 worker-days' labor in manufacturing and 160 worker-days' labor in the finishing area per week.
    a. How many of each type of truck should be produced in order to maximize the company's profits?
    b. What is the maximum profit each week?

26. Do Exercise 25 with the profit figures interchanged.

27. A dietitian must supply a diet including at least 36 units of protein, 24 units of carbohydrate, and 16 units of fat. These requirements can be met with a mixture of two foods, F_1 and F_2, costing $20 and $16 per unit, respectively. Each unit of F_1

supplies 9 units of protein, 3 units of carbohydrate, and 1 unit of fat. Each unit of F_2 supplies 2 units of protein, 2 of carbohydrate, and 2 of fat.

   a. What mixture of these two foods should be used in order to minimize the cost of supplying the given dietary requirements?

   b. What is this minimum cost?

28. Repeat Exercise 27 with F_1 costing $30 per unit and F_2 costing $20 per unit.

29. Repeat Exercise 27 with the cost of F_2 reduced to $12 per unit.

30. The Sunbright Mining Company owns two mines that produce silver and lead. The company has contracted to supply 3200 pounds of silver to the United Ring Company and 5000 pounds of lead to the Acme Plumbing Company. The first mine produces 40 pounds of silver and 100 pounds of lead per day. The second produces 80 pounds of silver and 50 pounds of lead per day.

   a. If it costs $150 per day to operate the first mine and $200 per day to operate the second, how many days should each mine be operated in order to fulfill the contract at least expense?

   b. What is the total cost of operating the mines?

31. Repeat Exercise 30 with the mining costs reversed.

32. A bank with $10 million in deposits must allocate its resources to loans and bond purchases. Its profit on loans is 6% of the amount loaned, whereas its net return on the bonds it purchases is only 4%. The Federal Reserve Board requires the bank to keep at least 20% of its deposits on hand as a cash reserve; this amount cannot be invested in either loans or bonds. The bank wants to invest at least 70% of its deposits with no more than 65% of its deposits on loan.

   a. What amount should it invest in bonds and in loans in order to maximize its return?

   b. What is its maximum return?

33. A contractor builds houses and apartments. He makes a profit of $6000 on each apartment unit and $8000 on each house he builds. He has enough financial backing to build 240 total residences. However, he has only 320 worker-months of labor available. Each home requires 2 worker-months and each apartment requires 1 worker-month of labor.

   a. How many homes and how many apartments should he build so as to maximize his profits?

   b. What is his maximum profit?

34. Repeat Exercise 33 with the profit figures reversed.

35. The Super Suds Soap Company must allocate part of its advertising budget to print and part to radio and television. It has $6 million to spend. Its market analysts have determined that each $1 spent on print advertising eventually returns $30; each $1 spent on radio and television commercials returns $40. The company president wants to spend no more than $4 million on printed advertising and no more than $5 million on radio and television commericals.

   a. How should the company distribute its advertising dollars in order to maximize its return?

   b. What is its maximum return?

36. Repeat Exercise 35 if the expected returns of $30 and $40 are reversed.

37. Laara likes to mix two kinds of breakfast cereal: Crunchy Bits and Fluffy Flakes. Each unit of Crunchy Bits supplies 12 units of vitamin A, 2 units of vitamin B, and 4 units of B_{12} and costs 40¢. Each unit of Fluffy Flakes supplies 4 units of A, 2 units of B, and 20 units of B_{12} and costs 20¢.

   a. To supply at least 48 units of A, 16 units of B, and 80 units of B_{12} at minimal cost, how many units of each cereal should Laara's parents buy?

   b. What is this minimal cost?

38. Repeat Exercise 37 with the following changes. Each unit of Crunchy Bits supplies 5, 3, and 5 units of A, B, and B_{12}, respectively, and each unit of Fluffy Flakes supplies 2, 2, and 1 units of A, B, and B_{12}, respectively. The pediatrician recommends a total of 60, 40, and 35 units of these vitamins in the same order.

39. Hilda's Ice Cream Company manufactures two kinds of ice cream: plain, which sells for $3/gallon, and extra creamy, which sells for $3.30/gallon. Each gallon of plain is made from 2 quarts of milk and 2 quarts of cream; each gallon of extra creamy is made from 4 quarts of cream. Currently on hand there are 1000 gallon containers, 1600 quarts of milk, and 3200 quarts of cream.

a. How many gallons of each type should be produced in order to collect the most money?

b. How much is then collected?

40. A toy manufacturer makes widgits and gadgets by using three machines: a mold, a lathe, and a sprayer. Each widgit requires 18 minutes mold time, 18 minutes lathe time, 6 minutes spraying time and yields $1. Each gadget requires 6 minutes mold time, 9 minutes lathe time, 12 minutes spraying time, and yields a profit of $1.50. The mold is available for 45 hours per week, the lathe for 48 hours, and the sprayer for 40.

a. What production should be scheduled each week in order to maximize revenues?

b. What is the maximum income?

Section 10.11 Chapter Review

Terms and Concepts

System of Equations

$$\begin{cases} 2x + y = 3 \\ x - y = 4 \end{cases}$$

1. Consistent — Exactly one solution.
2. Inconsistent — No solution.
3. Dependent — Infinitely many solutions.
4. Echelon Form

$$\begin{cases} x + 2y + 3z = 4 \\ y - 2z = 1 \\ z = 5 \end{cases}$$

System of Inequalities

$$\begin{cases} 2x + y \le 3 \\ x - y > 4 \end{cases}$$

Matrix

$$\begin{bmatrix} 1 & -1 & 2 \\ 2 & 3 & 1 \end{bmatrix}$$

1. Echelon Form

$$\begin{bmatrix} 1 & 2 & 3 & 4 \\ 0 & 1 & 5 & 6 \\ 0 & 0 & 1 & 7 \end{bmatrix}$$

2. Reduced Row Echelon Form

$$\begin{bmatrix} 1 & 0 & 0 & 8 \\ 0 & 1 & 0 & 9 \\ 0 & 0 & 1 & 5 \end{bmatrix}$$

Determinant

$$\begin{vmatrix} a & b \\ c & d \end{vmatrix} = ad - bc$$

1. Minor

$$\begin{bmatrix} 1 & 2 & 3 \\ 4 & 5 & 6 \\ 7 & 8 & 9 \end{bmatrix}, \quad M_{12} = \begin{vmatrix} 4 & 6 \\ 7 & 9 \end{vmatrix}$$

2. Cofactor

$$A_{ij} = (-1)^{i+j} M_{ij}$$

Identity Matrix I

$AI = IA = A$

$$I = \begin{bmatrix} 1 & 0 & \cdots & 0 \\ 0 & 1 & \cdots & 0 \\ \vdots & \vdots & & \vdots \\ 0 & \cdots & 0 & 1 \end{bmatrix}$$

Inverse of a Matrix

$AA^{-1} = A^{-1}A = 1$

$$\underbrace{\begin{bmatrix} 1 & 1 \\ 1 & 2 \end{bmatrix}}_{A} \underbrace{\begin{bmatrix} 2 & -1 \\ -1 & 1 \end{bmatrix}}_{A^{-1}} = \begin{bmatrix} 1 & 0 \\ 0 & 1 \end{bmatrix} = \begin{bmatrix} 2 & -1 \\ -1 & 1 \end{bmatrix} \begin{bmatrix} 1 & 1 \\ 1 & 2 \end{bmatrix}$$

Rules and Formulas

Matrix Algebra

1. To add or subtract matrices, add or subtract corresponding entries.

$$(a_{ij}) + (b_{ij}) = (a_{ij} + b_{ij})$$

2. To multiply a matrix by a number, multiply each entry of the matrix by that number.

$$k(a_{ij}) = (k \cdot a_{ij})$$

3. AB is not necessarily the same as BA; AB and BA may be equal in special cases.

$$\begin{bmatrix} 1 & 2 \\ -1 & 1 \end{bmatrix} \begin{bmatrix} 2 & 3 \\ 1 & 0 \end{bmatrix} = \begin{bmatrix} 4 & 3 \\ -1 & -3 \end{bmatrix}$$

$$\begin{bmatrix} 2 & 3 \\ 1 & 0 \end{bmatrix} \begin{bmatrix} 1 & 2 \\ -1 & 1 \end{bmatrix} = \begin{bmatrix} -1 & 7 \\ 1 & 2 \end{bmatrix}$$

Finding A^{-1}

1. Augment A with the identity.

$$[A \mid I]$$

2. Use the principles of the elimination method on the rows of this augmented matrix to convert A to reduced row echelon form.

3. **a.** If the reduced row echelon form of A is the identity I, then the original I will have been converted to A^{-1}.

$$[I \mid A^{-1}]$$

   b. If a row of zeros results in the reduced row echelon form of A, then A has no inverse.

Determinants

1. A determinant can be evaluated by expanding
   a. By any row
   b. By any column
2. Transposing a matrix (reflecting through its main diagonal) does not change the value of its determinant.
3. Interchanging two rows or two columns of a determinant changes its sign.
4. If two rows or two columns of a matrix are identical, its determinant must be 0.
5. Multiplying each element of any row or column of a matrix by the same number k multiplies its determinant by k; equivalently, a common factor of each element in a given row or column can be factored out of the determinant.
6. Adding a multiple of any row (or column) to another row (column, respectively) does not change the value of the determinant.

Partial Fractions $\quad \dfrac{1}{(x-1)^2(x^2+1)^2} = \dfrac{A}{x-1} + \dfrac{B}{(x-1)^2} + \dfrac{Cx+D}{x^2+1} + \dfrac{Ex+F}{(x^2+1)^2}$

Solution Techniques

Systems of Equations

1. Graphing — Find the point(s) common to the graphs of each equation in the system.

2. Substitution — Use one of the equations to express one of the variables in terms of the others and substitute this expression into the other equations.

3. Elimination

Perform the following operations to convert the system of equations to echelon form.
  a. Interchange equations.
  b. Multiply an equation by a nonzero constant.
  c. Add a nonzero multiple of one equation to another equation.

Then use back substitution to solve for the variables.

4. Matrix (a condensed form of elimination)

  a. Write the variables in the same order in each equation.
  b. Extract the augmented matrix by suppressing the variables and replacing the equal signs with a broken vertical line.
  c. Use the principles of the elimination method to convert the augmented matrix to either of the following.
      i. Echelon form.
      ii. Reduced row echelon form.
  d. i. If the matrix is in echelon form, use back substitution to find the solution.
     ii. If the matrix is in reduced row echelon form, read off the solution.

5. Matrix Inverse

If $AX = B$, then $X = A^{-1}B$.

6. Cramer's Rule

In a system of n linear equations in n unknowns

$$x_1, \ldots, x_n$$

let D = the determinant of the coefficient matrix and D_{x_i} = the determinant that results when the column of coefficients of x_i (in D) is replaced by the column of constants from the right side of the equal signs.

  a. If $D \neq 0$, there is a unique solution to the system given by

  $$x_i = \frac{D_{x_i}}{D}, \quad \text{for } i = 1, \ldots, n$$

  b. If $D = 0$, the system is dependent or inconsistent. Use the elimination or matrix method for further clarification of the situation.

Systems of Inequalities in Two Variables

1. Replace the inequality symbols with equal signs.
2. Sketch the resulting graphs.
3. The solution to a given inequality consists of one of the regions determined by the graph of the corresponding equality.

4. The solution set will consist of the region(s) common to all of the regions determined in step 3.

Linear Programming

1. *Read the problem* carefully from beginning to end.
2. *Label the quantities* that you are expected to find.
3. *Translate the given information into mathematical statements.* This should result in an objective function and a set of constraints described by a system of linear inequalities.
4. *Find the set of feasible solutions* by sketching the solution to the system of inequalities.
5. *Evaluate the objective function* at each vertex of the set of feasible solutions.
6. *Choose a (the) vertex* that optimizes the objective function.

Section 10.12 Supplementary Exercises

Solve the following systems of equations by the method of substitution.

1. $\begin{cases} y + 4z = 11 \\ 5y + 3z = 4 \end{cases}$
2. $\begin{cases} w - 3v = 8 \\ w + 2v = 3 \end{cases}$
3. $\begin{cases} 4x - 2y = 1 \\ 10x - 5y = 3 \end{cases}$
4. $\begin{cases} u + 2v = 3 \\ u - 2v = 15 \end{cases}$
5. $\begin{cases} 4s + v = 24 \\ 7s - 3v = 4 \end{cases}$
6. $\begin{cases} 2r + 3s = 1 \\ 8r + 12s = 4 \end{cases}$
7. $\begin{cases} 3p - 5q = 5 \\ 5p - 3q = 3 \end{cases}$
8. $\begin{cases} 5x + 3w = 19 \\ 3x + 4w = 7 \end{cases}$
9. $\begin{cases} 9x + 6y = 3 \\ 12x + 8y = 4 \end{cases}$
10. $\begin{cases} x - y^2 = 1 \\ x - 3y = 11 \end{cases}$
11. $\begin{cases} x + y^2 = 4 \\ x - 2y = 1 \end{cases}$
12. $\begin{cases} x^2 + y = 1 \\ 2x + y = 1 \end{cases}$
13. $\begin{cases} x - y = 2 \\ x^2 - y^2 = -4 \end{cases}$
14. $\begin{cases} x^2 + y^2 = 5 \\ 4x - 3y = 2 \end{cases}$

Solve the following systems by the elimination method.

15. $\begin{cases} 2x - 3y = 8 \\ x - 2y = 5 \end{cases}$
16. $\begin{cases} 2r - 3s = 5 \\ r + s = 5 \end{cases}$
17. $\begin{cases} 3t - 4u = -6 \\ 5t + 3u = 19 \end{cases}$
18. $\begin{cases} 12v + 11w = 8 \\ 36v + 33w = 20 \end{cases}$
19. $\begin{cases} z + 3y = 14 \\ 7z + 5y = 2 \end{cases}$
20. $\begin{cases} x - y + 2z = 5 \\ x - 2y + 3z = -1 \end{cases}$
21. $\begin{cases} a + 2b + c = 5 \\ 3a - 4b - 7c = 5 \\ 2a + b - 3c = 5 \end{cases}$
22. $\begin{cases} u - 2v + w = 1 \\ 2u - 4v + 3w = 4 \\ -2u + 8v - w = -1 \end{cases}$
23. $\begin{cases} 3r - s + t = 2 \\ 2r - 4s - t = 8 \\ r + s + t = -2 \end{cases}$
24. $\begin{cases} x + 4y = 11 \\ 5x + 3y = 4 \\ 17x + 34y = 85 \end{cases}$
25. $\begin{cases} 10u + 2v + 3w = -4 \\ -3u - 5v + 2w = 13 \\ 2u - 2v + 5w = 0 \end{cases}$
26. $\begin{cases} t - u + 5v = 2 \\ t + 5u - 7v = -10 \\ 3t - 2u + 3v = 4 \end{cases}$
27. $\begin{cases} 2x + 5z = 1 \\ y + 3z = 1 \\ x + y + 3z = 9 \end{cases}$
28. $\begin{cases} x^2 - 4y^2 = 4 \\ 9x^2 + 16y^2 = 140 \end{cases}$
29. $\begin{cases} 4x^2 + 3y^2 = 16 \\ 3x^2 - 2y^2 = -5 \end{cases}$
30. $\begin{cases} 5x^2 + 3y^2 = 17 \\ 3x^2 - 2y^2 = -5 \end{cases}$
31. $\begin{cases} \dfrac{1}{x} - \dfrac{2}{y} = 5 \\ \dfrac{2}{x} - \dfrac{3}{y} = 8 \end{cases}$
32. $\begin{cases} \dfrac{2}{x} + \dfrac{1}{y} = -3 \\ \dfrac{1}{x} - \dfrac{3}{y} = 2 \end{cases}$
33. $\begin{cases} \dfrac{3}{x} - \dfrac{2}{y} = 4 \\ \dfrac{2}{x} - \dfrac{3}{y} = 1 \end{cases}$
34. $\begin{cases} \dfrac{3}{x} - \dfrac{2}{y} = 5 \\ \dfrac{4}{x} - \dfrac{5}{y} = 2 \end{cases}$
35. $\begin{cases} \log|x - 1| + 2y = 1 \\ \log|x + 1| - 2y = -1 \end{cases}$

Solve the following systems by the matrix method.

36. $\begin{cases} 3r + s + 4t = -4 \\ 5r - s + 3t = 6 \\ r - 2s + t = 4 \end{cases}$

37. $\begin{cases} 2x - 4y - 6z = -1 \\ 2x - 3y + z = 7 \\ x - 2y - 3z = 4 \end{cases}$

38. $\begin{cases} 3u - 2v + 5w = 12 \\ 2u - 5v + 3w = 4 \\ u + 3v + 2w = 8 \end{cases}$

39. $\begin{cases} a - 2b - c = 0 \\ 2a + b - 2c = 0 \\ a - b - c = 0 \end{cases}$

40. $\begin{cases} 9u - 6v + 3w = 12 \\ 6u - 4v + 2w = 8 \\ 3u - 2v + w = 4 \end{cases}$

41. $\begin{cases} r - 2s - 3t = 3 \\ 2r + s - t = 1 \\ r + s + t = -4 \end{cases}$

42. $\begin{cases} x - 2y - 2z = 0 \\ 2x + y + 3z = 0 \\ 2x - y + z = 0 \\ x + y - z = 3 \end{cases}$

43. $\begin{cases} x + y + 4z = 0 \\ x + y + w = 5 \\ x + 2y - 2w = -6 \\ y + 2z + w = 3 \end{cases}$

44. $\begin{cases} t - 3v + w = -14 \\ t + u + v + w = 10 \\ t - 3u - v = -7 \\ 2u + v + w = 11 \end{cases}$

45. $\begin{cases} 2p - q - r + 2s = 8 \\ 4p + q + 5r + 2s = 6 \\ p + q + r + s = 1 \\ 3p + 2q - 2r - s = -3 \end{cases}$

Use Cramer's rule to solve the following systems.

46. $\begin{cases} 2x + y = 4 \\ -x + 3y = 2 \end{cases}$

47. $\begin{cases} u + 5v = 1 \\ u - 3v = 7 \end{cases}$

48. $\begin{cases} 2r + s = 6 \\ r + 2s = 0 \end{cases}$

49. $\begin{cases} t + 4u = 11 \\ 5t + 3u = 4 \end{cases}$

50. $\begin{cases} 5x + 3y = 17 \\ 3x + 4y = 6 \end{cases}$

51. $\begin{cases} x - 2y + z = 5 \\ 3x + 3y - 2z = -6 \\ 2x - y + 3z = 11 \end{cases}$

52. $\begin{cases} 3p - 4q + r = 3 \\ p + 3q + 2r = -4 \\ 2p - q - r = -1 \end{cases}$

53. $\begin{cases} 8t + 2u + 5v = 4 \\ 3t - u + 2v = 7 \\ t - u - 4v = -5 \end{cases}$

54. $\begin{cases} x - 3y + z = 3 \\ x + y + 3z = 3 \\ 2x + y + 2z = -1 \end{cases}$

55. $\begin{cases} 3u + 4v + 4w + 2x = 7 \\ -2u - 3v + 2w - 2x = -2 \\ 4u + 5v - 2w - 2x = 14 \\ u - v - 4w - 3x = 0 \end{cases}$

Evaluate the following determinants.

56. $\begin{vmatrix} 4 & -9 \\ -6 & 11 \end{vmatrix}$

57. $\begin{vmatrix} 3 & 2 & -4 \\ 1 & -3 & 1 \\ 4 & -1 & -3 \end{vmatrix}$

58. $\begin{vmatrix} 3 & 2 & -1 \\ 2 & -1 & 1 \\ 1 & 1 & -2 \end{vmatrix}$

59. $\begin{vmatrix} 2 & -2 & -1 \\ -1 & -3 & 2 \\ 2 & 4 & 6 \end{vmatrix}$

60. $\begin{vmatrix} -1 & 1 & -2 \\ -1 & -1 & -1 \\ 2 & 1 & -1 \end{vmatrix}$

61. $\begin{vmatrix} 3 & -2 & 2 \\ -1 & -4 & 5 \\ 4 & -2 & 3 \end{vmatrix}$

62. $\begin{vmatrix} 2 & 4 & 1 \\ 4 & 2 & -1 \\ 0 & -1 & -3 \end{vmatrix}$

63. $\begin{vmatrix} 1 & 1 & 0 & 2 \\ 2 & 2 & 1 & -1 \\ -1 & 1 & 2 & 1 \\ 1 & 2 & -1 & 2 \end{vmatrix}$

64. $\begin{vmatrix} 3 & -4 & 4 & -2 \\ 4 & 6 & 1 & 1 \\ -2 & 2 & 3 & 5 \\ 2 & 2 & -2 & 3 \end{vmatrix}$

65. $\begin{vmatrix} 4 & -4 & -1 & -1 \\ 2 & -2 & -1 & 2 \\ 2 & -6 & 2 & -1 \\ 6 & 2 & 1 & 3 \end{vmatrix}$

Write the equation of the circle passing through the given points. Find its center and radius.

66. (0, 0), (3, 1), and (7, −1)

67. (11, 2), (3, 14), and (−7, −10)

68. (−26, −10), (22, 4), and (5, 21)

Find the partial fractions decomposition of each of the following.

69. $\dfrac{x+3}{(x+2)(x+1)}$

70. $\dfrac{4x-2}{(x+1)(2x-3)}$

71. $\dfrac{2x-2}{x(x+1)^2}$

72. $\dfrac{-x^2+6x+6}{(x-2)(x^2+3)}$

73. $\dfrac{2x^4+4x^3+9x^2+7x+2}{(x+1)(x^2+x+1)^2}$

74. $\dfrac{2x^5+x^4-x^3-x^2+7x+2}{x^3-x}$

Let $A = \begin{bmatrix} 1 & 3 & 2 \\ 1 & 0 & 4 \end{bmatrix}$, $B = \begin{bmatrix} 1 & 2 \\ -1 & 3 \\ 4 & 0 \end{bmatrix}$, $C = \begin{bmatrix} 1 & 4 \\ 3 & 1 \\ 2 & 0 \end{bmatrix}$,

$D = \begin{bmatrix} -1 & 2 \\ 3 & 1 \end{bmatrix}$, and $E = \begin{bmatrix} 1 & 0 & 1 \\ 3 & 1 & -2 \\ 1 & 0 & 4 \end{bmatrix}$. *Find*

75. AB
76. $A(B+2C)$
77. ABD
78. BDA
79. AEC
80. AEB
81. EBA
82. $ECDA$
83. $BDAE$

Find the inverse for each of the following matrices.

84. $\begin{bmatrix} 5 & -2 \\ -3 & 1 \end{bmatrix}$

85. $\begin{bmatrix} 2 & -1 \\ 5 & -3 \end{bmatrix}$

86. $\begin{bmatrix} 8 & 3 & -1 \\ 3 & 1 & -1 \\ 6 & 2 & -1 \end{bmatrix}$

87. $\begin{bmatrix} 1 & -3 & 2 \\ -1 & 4 & -2 \\ 1 & -3 & 3 \end{bmatrix}$

Write the following systems in the form $AX = B$ *and use* A^{-1} *to solve the system.* A^{-1} *will have been found in the preceding exercises.*

88. $\begin{cases} 5x - 2y = -1 \\ -3x + y = 1 \end{cases}$

89. $\begin{cases} 2x - y = 1 \\ 5x - 3y = 2 \end{cases}$

90. $\begin{cases} 8x + 3y - z = 0 \\ 3x + y - z = 1 \\ 6x + 2y - z = 3 \end{cases}$

91. $\begin{cases} x - 3y + 2z = -1 \\ -x + 4y - 2z = 0 \\ x - 3y + 3z = -2 \end{cases}$

Graph the solutions for the following inequalities.

92. $\begin{cases} y - x > 1 \\ y + x < 1 \end{cases}$

93. $\begin{cases} y - 2x < 0 \\ y + x > 0 \end{cases}$

94. $|3x - 4y + 6| < 3$

95. $\begin{cases} x^2 + y^2 < 4 \\ x < y \end{cases}$

96. $\begin{cases} x^2 - 6y + y^2 < 0 \\ y - x \geq 3 \end{cases}$

97. $\begin{cases} x^2 + y^2 - 4y < 0 \\ x^2 + y^2 - 4x \geq 0 \end{cases}$

98. $\begin{cases} x^2 - 6x + y^2 + 5 < 0 \\ x^2 + y^2 - 6x - 4y + 12 < 0 \end{cases}$

99. $\begin{cases} x^2 - 4x + y^2 < 0 \\ y^2 < 3x - 6 \end{cases}$

100. $\begin{cases} y^2 - 5x - 50 \leq 0 \\ x^2 + 10x + y^2 < 0 \end{cases}$

101. Find two numbers whose sum is 90 and whose difference is 36.

102. A number is 2 less than half of another number. The second number is 7 less than three times the first. Find the numbers.

103. A 60-mile boat trip upriver takes 12 hours, and the return trip takes only 4 hours. What are the speeds of the boat and of the current?

104. The tens digit of a two-digit number is twice the units digit. If the digits are reversed and the result is doubled, the final result is 9 more than the original. Find the number. (*Hint:* See Exercise 71, Section 10.1.)

105. The sum of the digits of a three-digit number is 1 less than 3 times its hundreds digit. The number is 8 more than 111 times its hundreds digit. The tens digit is 3 less than the sum of the hundreds and units digits. Find the number. (*Hint:* See Exercise 45, Section 10.2.)

106. Cindy has 42¢ in pennies, nickels, and dimes. If the number of dimes is 1 less than the number of nickels, and she has a total of twelve coins, find the number of each type of coin she has.

107. Mrs. Schoen can buy 15 pounds of pork, 15 pounds of fish, and 20 pounds of steak for $102.50. She could also buy $7\frac{1}{2}$ pounds of pork, 30 pounds of fish, and 15 pounds of steak for $108.75. Or, she could buy $22\frac{1}{2}$ pounds of pork, 15 pounds of fish, and $7\frac{1}{2}$ pounds of steak for $82.50. What is the price per pound of each of these foods?

108. Ken, Les, and Roger can paint a garage in 4 hours working together. But Ken and Roger quit after 2 hours; Les continues to work and finishes the job in $4\frac{1}{2}$ more hours. If Roger had quit after 3 hours, Ken and Les could finish in $1\frac{1}{2}$ more hours. How long would it take each one alone to paint the garage? (*Hint:* See Example 5, Section 10.3)

Solve the following linear programming problems.

109. Maximize $\quad 10y + 15x$

subject to $\quad \begin{cases} y \leq 2x \\ x + y \leq 90 \\ y \geq 10 \end{cases}$

110. Maximize $\quad 4x + 9y$

subject to $\quad \begin{cases} x + 2y \leq 14 \\ 2x + 5y \leq 30 \\ x \geq 0 \\ y \geq 0 \end{cases}$

111. Minimize $\quad 5x + 10y$

subject to $\quad \begin{cases} x + 2y \geq 80 \\ x + y \geq 50 \\ 5x + 2y \geq 150 \\ x \geq 0 \\ y \geq 0 \end{cases}$

112. Minimize $\quad 150x + 200y$

subject to $\quad \begin{cases} 10x + y \geq 10 \\ x \leq y + 5 \\ x \geq 0 \\ y \geq 0 \end{cases}$

113. Minimize $\quad 6x + 3y$

subject to $\quad \begin{cases} x + 3y \geq 34 \\ x + y \geq 14 \\ x \geq 4 \\ y \geq 0 \end{cases}$

114. Maximize $\quad 20x + 32y$

subject to $\quad \begin{cases} x + y \leq 210 \\ x + 2y \leq 400 \\ 9x \leq 180 + 10y \\ x \geq 0 \\ y \geq 0 \end{cases}$

115. Maximize $\quad 10x + 20y$

subject to $\quad \begin{cases} 7x + 4y \leq 200 \\ 3x + y \leq 60 \\ x \geq 0 \\ y \geq 0 \end{cases}$

116. Maximize $\quad 20x + 14y$

subject to $\quad \begin{cases} 8x + 40 \geq 7y \\ 2x + y \geq 10 \\ 4x \leq 40 + y \\ x + y \leq 20 \\ y \geq 0 \end{cases}$

117. Minimize $\quad 4y + 2x$

subject to $\quad \begin{cases} 4y + 2 \leq x \\ x + 2y \geq 5 \\ y \geq 0 \end{cases}$

118. Minimize $\quad x - y$

subject to $\quad \begin{cases} 8x + 5y \geq 230 \\ 6x \leq 160 + 5y \\ 10y \leq x + 120 \\ x \geq 0 \\ y \geq 0 \end{cases}$

Mathematical Induction, Progressions, and Probability

11

Have you ever wondered how a banker computes the monthly payment required on a loan or how pension planners compute the premium needed to generate a given retirement income? Each of these computations is made by using formulas established by **mathematical induction.** (Actually, the banker probably looks up these values in a table that was compiled by using these formulas.)

Mathematical induction is generally applied to situations that progress in a step-by-step fashion. For instance, the amount still owed on a loan is reduced systematically after each payment.

Mathematical induction involves the discovery of a certain pattern in the first several stages of a continuing process. Then this pattern is established for the remaining stages by showing that the validity of the pattern or formula at a given stage forces its validity at the next stage.

We continue to deal with orderly patterns of computation by establishing various counting formulas regarding the number of combinations and rearrangements of a given collection. These formulas are in turn useful in determining the likelihood or **probability** that certain events will happen. Industrial concerns determine probabilities of failure for their products in order to develop proper warranty policies. Police work can sometimes rely on probability, for instance, in that it is very unlikely that two identical cars from different states have the same license number and are on the same street at the same time. At the gambling tables at Las Vegas, Reno, and Atlantic City, which attract millions of people each year, the odds in a betting situation are determined by the laws of probability.

Section 11.1

Mathematical Induction

We have already encountered the need to establish the truth of certain statements for every positive integer n. For instance, in the discussion of exponents we stated that $(xy)^n = x^n y^n$. In the section on compound interest (Section 6.5), we found a formula for the value of an investment after n compounded interest payments. And in Chapter 9 we established the formula $[r \text{ cis } \theta]^n = r^n \text{ cis } n\theta$. We were able to convince ourselves of the validity of each of these formulas without too much trouble. However, there are situations in which a formula is desired for every positive integer n but its form and its verification are in no way obvious.

The **principle of mathematical induction** is based on the axioms of the system of natural numbers. We shall only state and use this principle; its proof is not within the scope of this book. However, its statement is so simple and the principle itself seems so natural that we should have no qualms regarding its use.

> **PRINCIPLE OF MATHEMATICAL INDUCTION**
>
> Suppose that for each positive integer n, there is a corresponding statement S_n. If
>
> 1. S_1 is true, and
> 2. The truth of S_k implies the truth of S_{k+1} for every k
>
> then S_n is true for every positive integer n.

If conditions 1 and 2 are satisfied, then S_1 is true, and S_2 is true because S_1 is true. S_3 is true because S_2 is true, S_4 is true because S_3 is true, S_5 is true because S_4 is true, and so on.

Here we see the resemblance of mathematical induction to the domino principle. For if dominos are lined up so that when any one topples, it topples the next one, the entire line can be toppled simply by pushing the first domino. In the same way, if statements are lined up so that when any one is true the next one is true, the entire list can be verified simply by showing that the first statement is true (see Figure 11.1).

Figure 11.1

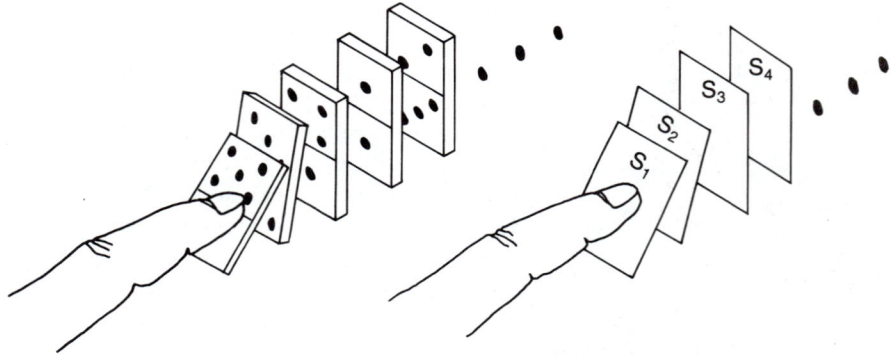

a. The domino principle   b. Mathematical induction

The analogy with the domino principle is carried further as follows.

DOMINO PRINCIPLE	MATHEMATICAL INDUCTION
1. If we do not push the first domino, the dominoes do not fall.	1. If S_1 is not true, then the induction fails.
2. If some domino does not reach the next one, then not all of the dominoes will fall, even though the first one is pushed.	2. If some S_j does not imply S_{j+1} (say S_{10} does not imply S_{11}), the induction fails.

The applications of the principle of mathematical induction are straightforward; we just verify conditions 1 and 2; that is,

1. Verify that S_1 is true.
2. Show that the truth of S_k implies the truth of S_{k+1}.

EXAMPLE 1

Use mathematical induction to prove that the sum of the first n natural numbers is $\frac{n(n+1)}{2}$; that is

$$1 + 2 + 3 + \cdots + n = \frac{n(n+1)}{2}$$

for $n = 1, 2, \ldots$. For instance, when $n = 4$, $1 + 2 + 3 + 4 = \frac{4(4+1)}{2} = 10$.

SOLUTION

Let S_n be the statement $1 + 2 + \cdots + n = \frac{n(n+1)}{2}$.

Step 1. Observe that S_1 is true since

$$1 = \frac{1(1+1)}{2}$$

Step 2. Next we assume the truth of S_k and try to show that then S_{k+1} must be true. Assume that

$$1 + \cdots + k = \frac{k(k+1)}{2}$$

Then try to show that

$$1 + \cdots + k + (k+1) = \frac{(k+1)[(k+1)+1]}{2}$$

Now

$$1 + \cdots + k + (k+1) = (1 + \cdots + k) + (k+1)$$
$$= \frac{k(k+1)}{2} + (k+1) \quad \text{(By the assumption on } S_k\text{)}$$
$$= \frac{k(k+1) + 2(k+1)}{2}$$
$$= \frac{(k+1)(k+2)}{2}$$
$$= \frac{(k+1)[(k+1)+1]}{2}$$

as was to be shown.

Thus S_n is true for every n. ∎

EXAMPLE 2

Use mathematical induction to prove that the sum of the first n perfect squares is $\frac{n(n+1)(2n+1)}{6}$; that is

$$1^2 + 2^2 + \cdots + n^2 = \frac{n(n+1)(2n+1)}{6}$$

for $n = 1, 2, \ldots$. For instance, when $n = 3$,

$$1^2 + 2^2 + 3^2 = \frac{3(3+1)(2 \cdot 3 + 1)}{6} = 14$$

SOLUTION

Let S_n be the statement in question.

Step 1. S_1 is true since

$$1^2 = \frac{1 \cdot (1+1)(2 \cdot 1 + 1)}{6}$$
$$= \frac{1 \cdot 2 \cdot 3}{6} = 1$$

Step 2. Next, we assume that S_k is true and try to prove S_{k+1}; that is, we assume

$$1^2 + 2^2 + \cdots + k^2 = \frac{k(k+1)(2k+1)}{6}$$

Then for S_{k+1},

$$1^2 + 2^2 + \cdots + k^2 + (k+1)^2 = (1^2 + \cdots + k^2) + (k+1)^2$$
$$= \frac{k(k+1)(2k+1)}{6} + (k+1)^2$$
$$= \frac{k(k+1)(2k+1) + 6(k+1)^2}{6}$$
$$= \frac{(k+1)[k(2k+1) + 6(k+1)]}{6}$$
$$= \frac{(k+1)(2k^2 + 7k + 6)}{6}$$
$$= \frac{(k+1)(k+2)(2k+3)}{6}$$
$$= \frac{(k+1)[(k+1)+1][2(k+1)+1]}{6}$$

Thus S_{k+1} is true provided S_k is true.

Hence S_n is true for every n. ∎

EXAMPLE 3

The tower of Hanoi, an ancient puzzle, consists of three pegs, or "towers," on one of which are stacked discs of decreasing size, as shown in Figure 11.2. The problem is to *move the discs one at a time from one peg to another, never placing a larger disc above a smaller one,* in such a way that the entire stack is finally located on another peg. Determine the minimum number of steps that are required to achieve this task if the tower consists of n discs.

Figure 11.2
THE TOWER OF HANOI

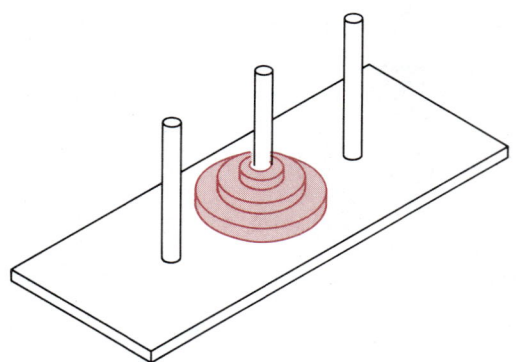

SOLUTION

This is a more difficult problem in that we must not only verify the answer by using mathematical induction, but we must also find the answer! In determining the correct answer, we shall use the inductive technique—we shall first study a tower with one disc, then a tower with two discs, and so on until we discern a clue. We shall let M_n be the solution for a tower of n discs.

For a tower with one disc, it takes only one move to complete the task. Thus $M_1 = 1$.

For a tower with two discs, we move the top disc to one vacant peg, the bottom disc to the other vacant peg; finally, place the small disc over the larger for a total of three moves as illustrated in Figure 11.3. Thus $M_2 = 3$.

Figure 11.3

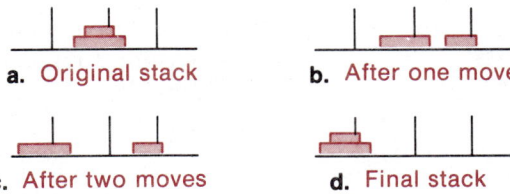

a. Original stack b. After one move

c. After two moves d. Final stack

For a tower with three discs, we observe that the bottom disc cannot be removed until the top two are stacked on another peg; that is, we must first work the Hanoi puzzle for $n = 2$. This takes $M_2 = 3$ moves. Then the bottom disc is moved to the vacant peg. Finally, the top two discs must be moved again to sit above the bottom disc; the Hanoi puzzle for $n = 2$ must be solved again.

This involves another M_2 moves. Hence

$$M_3 = M_2 + 1 + M_2 = 3 + 1 + 3 = 7$$

It appears that the discussion for $n = 3$ is perfectly general; that is, *to solve the puzzle with $k + 1$ discs* we must

1. Solve the puzzle with k discs.
2. Move the bottom peg.
3. Solve the puzzle with k discs again.

We need only reread the preceding discussion (for a tower of three discs) with k and $k + 1$ in place of 2 and 3, respectively. Hence

$$M_{k+1} = M_k + 1 + M_k$$
$$= 2M_k + 1$$

Since $M_1 = 1$, $M_2 = 3$, $M_3 = 7$, we conjecture $M_n = 2^n - 1$. Let S_n be the statement: $M_n = 2^n - 1$.

Step 1. Observe that S_1, S_2, and S_3 are true.

Step 2. Assume S_k is true:

$$M_k = 2^k - 1$$

and try to prove S_{k+1}. Now

$$M_{k+1} = 2M_k + 1$$
$$= 2(2^k - 1) + 1 \qquad \text{(By the assumption on } S_k\text{)}$$
$$= 2 \cdot 2^k - 2 + 1$$
$$= 2^{k+1} - 1$$

Thus *the least* number of moves required to solve a Hanoi puzzle with n discs is $2^n - 1$. With the currently marketed version having 8 discs, $2^8 - 1 = 255$ moves are required. ∎

If care is not taken in showing that the truth of S_k really does imply the truth of S_{k+1} for all k, some silly statements can seemingly be "proved" by using mathematical induction. For instance, Charles the Charlatan, who is color-blind, says that "all birds are the same color." In attempting to prove this statement, he reasons as follows:

1. In any set of birds consisting of only one bird, all birds (there is only one) are the same color.

2. Assume that in any set of birds consisting of k birds, all birds are the same color. Consider any set of $k + 1$ birds as shown in Figure 11.4. The first k birds are the same color; the last k birds are the same color. Thus all $k + 1$ birds are the same color.

Figure 11.4

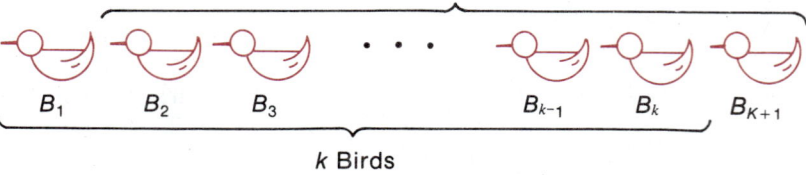

k Birds

3. Thus Charles concludes that all birds are the same color.

Evidently something is wrong with Charles' reasoning since the conclusion is obviously false. In Step 2 (Figure 11.4), we note that Charles is depending on the overlapping part of the two sets he considered, each of which contains k birds; $B_2, \ldots, B_k$ are in both these sets. Then B_1 and B_{k+1} are both the same color as these; hence they are all the same color. His argument breaks down when $k = 1$ since then the two sets do not overlap (Fig. 11.5). Thus S_1 does not imply S_2 although *if* S_2 were true, then S_3 would be true, and so on.

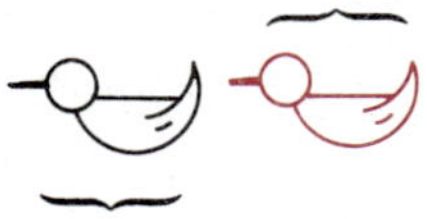

Figure 11.5

For Example 4, we use the following definition.

> **DEFINITION**
>
> If a and b are positive integers and there is a positive integer q such that $a = b \cdot q$, then we say that *a is divisible by b*.

EXAMPLE 4 Show that $6^n - 1$ is divisible by 5 for $n = 1, 2, 3, \ldots$.

SOLUTION **Step 1.** Observe that the statement is true for $n = 1$ since
$$6^1 - 1 = 5 = 5 \cdot 1$$

Step 2. If the statement is true for $n = k$, then there is a positive integer q such that
$$6^k - 1 = 5 \cdot q$$

Then for $n = k + 1$, we have

$$\begin{aligned}6^{k+1} - 1 &= (6^k \cdot 6) - 1 \\ &= 6^k(5 + 1) - 1 \\ &= 6^k \cdot 5 + (6^k - 1) \\ &= 6^k \cdot 5 + 5 \cdot q \quad \text{(By the induction assumption)} \\ &= 5(6^k + q) \\ &= 5 \cdot Q \quad \text{(For } Q = 6^k + q\text{)}\end{aligned}$$

Thus $6^{k+1} - 1$ is divisible by 5 provided $6^k - 1$ is divisible by 5. The truth of the statement for $n = k$ implies its truth for $n = k + 1$.

By mathematical induction, the statement is true for every $n = 1, 2, \ldots$. ∎

Section 11.1 Exercises

Use mathematical induction to establish the following formulas.

1. $1 + 8 + 16 + \cdots + 8(n-1) = (2n-1)^2$

2. $2^2 + 4^2 + 6^2 + \cdots + (2n)^2 = \dfrac{2n(n+1)(2n+1)}{3}$

3. $1^3 + 2^3 + 3^3 + \cdots + n^3 = \left[\dfrac{n(n+1)}{2}\right]^2$

4. $2^3 + 4^3 + \cdots + (2n)^3 = 2[n(n+1)]^2$

5. $2 + 4 + \cdots + 2^n = 2(2^n - 1)$

6. $4 + 16 + \cdots + 4^n = \dfrac{4(4^n - 1)}{3}$

7. $\dfrac{1}{2} + \dfrac{1}{4} + \cdots + \dfrac{1}{2^n} = 1 - \dfrac{1}{2^n}$

8. $1 + x + \cdots + x^n = \dfrac{1 - x^{n+1}}{1 - x}, x \neq 1$

9. $1 \cdot 2 + 2 \cdot 3 + \cdots + n(n+1) = \dfrac{n(n+1)(n+2)}{3}$

10. $1 \cdot 3 + 2 \cdot 5 + \cdots + n(2n+1) = \dfrac{n(n+1)(4n+5)}{6}$

11. $1 \cdot 3 + 3 \cdot 5 + \cdots + (2n-1)(2n+1)$
$= \dfrac{n(4n^2 + 6n - 1)}{3}$

12. $1 \cdot 2 + 2 \cdot 4 + \cdots + n \cdot 2^n = (n-1)2^{n+1} + 2$

13. $1 \cdot 2 \cdot 3 + 2 \cdot 3 \cdot 4 + \cdots + n(n+1)(n+2)$
$= \dfrac{n(n+1)(n+2)(n+3)}{4}$

14. $\dfrac{1}{1 \cdot 2} + \dfrac{1}{2 \cdot 3} + \cdots + \dfrac{1}{n(n+1)} = \dfrac{n}{n+1}$

15. $\dfrac{1}{1 \cdot 2 \cdot 3} + \dfrac{1}{2 \cdot 3 \cdot 4} + \cdots + \dfrac{1}{n(n+1)(n+2)}$
$= \dfrac{n(n+3)}{4(n+1)(n+2)}$

Establish the following statements using mathematical induction.

16. $4^n - 1$ is divisible by 3.

17. $5^n - 1$ is divisible by 4.

18. $9^n - 1$ is divisible by 4.

19. $2^{2n+1} + 1$ is divisible by 3.

20. $2^{[2n]} - 1$ is divisible by 3.

21. $n^3 + 6n^2 + 2n$ is divisible by 3.

22. $n(n^2 + 5)$ is divisible by 6.

23. $a^n - b^n$ is divisible by $a - b$, $a \neq b$.

24. $a^{2n} - b^{2n}$ is divisible by $a + b$, $a \neq -b$.

25. $a^{2n-1} + b^{2n-1}$ is divisible by $a + b$, $a \neq -b$.

26. For any two real or complex numbers a, b, show by mathematical induction that for every positive integer n
   a. $(ab)^n = a^n b^n$
   b. $\left(\dfrac{a}{b}\right)^n = \dfrac{a^n}{b^n}$, when $b \neq 0$

27. Assume that a principal of P_0 dollars grows to $P_1 = P_0(1 + r)$ after one compounding period.

Use mathematical induction to show that after n compounding periods, the principal P_0 grows to

$$P_n = P_0(1 + r)^n$$

28. **a.** What is the total number of gifts "my true love gave to me" during the 12 days of Christmas? (1 gift the first day, $1 + 2$ gifts the second day, $1 + 2 + 3$ gifts the third day, and so on.)

   b. How many gifts will "my true love give to me" during the n days of Christmas? (*Hint:* The formulas in Example 1 and Exercise 9 are useful.)

29. Advertising agencies frequently sponsor contests to promote certain products. One such contest requires the participants to count the number of ways in which the message "Our product is best" can be found in the pyramid shown in the figure. Use mathematical induction to find the correct solution for the given pyramid. (*Hint:*

Consider "messages" of increasing length

```
      S       E
   T, STS,   ESE
           ESTSE
```

as indicated.

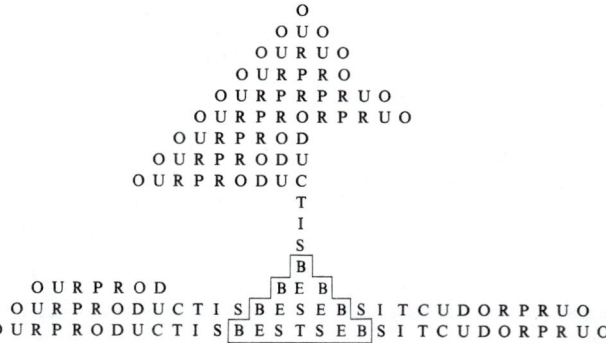

30. Use mathematical induction to establish DeMoivre's theorem for $n \geq 0$:

$$(r \text{ cis } \theta)^n = r^n \text{ cis } n\theta$$

Section 11.2

Sequences and Progressions

We have already considered the numbers $1, 2, 3, \ldots$, their reciprocals $1, \frac{1}{2}, \frac{1}{3}, \ldots$, and the inverse powers of two: $1, \frac{1}{2}, \frac{1}{4}, \frac{1}{8}, \ldots$. Each of these is an example of sequence. Intuitively, a **sequence** is an infinite list of objects or numbers in which there is a first object in the list, a second, and so on. Formally a sequence is a function whose domain is the set of natural numbers. For such a function a, we consider $a(1), a(2), a(3), \ldots$. In working with sequences, however, we usually dispense with functional notation and instead write $a(n) = a_n$; a_n is called the **nth term** of the sequence. We list the terms of the sequence as

$$a_1, a_2, a_3 \ldots$$

EXAMPLE 1

Write out the first five terms of the sequences whose nth term is

a. $a(n) = a_n = \dfrac{1}{n + 1}$ **b.** $b(n) = b_n = \dfrac{n^2}{2n + 1}$

SOLUTION

We are to write out the respective terms for $n = 1, \ldots, 5$. These are:

a. $a_1 = \dfrac{1}{1 + 1} = \dfrac{1}{2}$

$a_2 = \dfrac{1}{2 + 1} = \dfrac{1}{3}$

$a_3 = \dfrac{1}{3 + 1} = \dfrac{1}{4}$

$a_4 = \dfrac{1}{4 + 1} = \dfrac{1}{5}$

$a_5 = \dfrac{1}{5 + 1} = \dfrac{1}{6}$

b. $b_1 = \dfrac{1^2}{2 \cdot 1 + 1} = \dfrac{1}{3}$

$b_2 = \dfrac{2^2}{2 \cdot 2 + 1} = \dfrac{4}{5}$

$b_3 = \dfrac{3^2}{2 \cdot 3 + 1} = \dfrac{9}{7}$

$b_4 = \dfrac{4^2}{2 \cdot 4 + 1} = \dfrac{16}{9}$

$b_5 = \dfrac{5^2}{2 \cdot 5 + 1} = \dfrac{25}{11}$ ∎

Arithmetic Progressions

Of particular interest are the sequences that form arithmetic and geometric progressions.

DEFINITION

A sequence is an **arithmetic progression** if the difference between any two successive terms is constant; that is, there is some **common difference** d such that

$$a_{n+1} - a_n = d \quad \text{for } n = 1, 2, \ldots$$

In an arithmetic progression, $a_{n+1} = a_n + d$; an arithmetic progression is built up by adding the same d to any term to get the next term. Thus, if we know any term and the common difference, we can find any other term since

$$a_2 = a_1 + d$$
$$a_3 = a_2 + d = a_1 + 2d$$
$$a_4 = a_3 + d = a_1 + 3d$$
$$\vdots$$

*n*th TERM OF AN ARITHMETIC PROGRESSION

$$a_n = a_{n-1} + d = a_1 + (n-1)d, \quad \text{for } n = 1, 2, \ldots \quad (1)$$

This formula can be established by mathematical induction; see Exercise 63.

EXAMPLE 2 Write out the first four terms and the twelfth term of the arithmetic progression whose first term is 2 and whose common difference is 3.

SOLUTION The first four terms are

$$a_1 = 2$$
$$a_2 = a_1 + d = 2 + 3 = 5$$
$$a_3 = a_2 + d = 5 + 3 = 8$$
$$a_4 = a_3 + d = 8 + 3 = 11$$

The twelfth term is

$$a_{12} = a_1 + 11d$$
$$= 2 + 11 \cdot 3$$
$$= 35$$

EXAMPLE 3   If the fifteenth term of an arithmetic progression is 59 and its twenty-seventh term is 107, find its tenth term.

SOLUTION   The given information yields the following system of equations.

$$\begin{cases} a_{15} = a_1 + 14d = 59 \\ a_{27} = a_1 + 26d = 107 \end{cases}$$

Subtracting corresponding members of the first equation from the second, we obtain

$$12d = 48$$
$$d = 4$$

Then

$$a_1 + 14 \cdot 4 = 59$$
$$a_1 = 3$$

Finally,

$$a_{10} = a_1 + 9d$$
$$= 3 + 9 \cdot 4$$
$$a_{10} = 39$$

The terms of an arithmetic progression that appear between two given terms a and b are called **arithmetic means** of a and b. If there is only one term, it is the arithmetic mean; if there are two, they are called the two arithmetic means; n terms are called n arithmetic means.

EXAMPLE 4   Insert three arithmetic means between 10 and 22.

SOLUTION   We want to form an arithmetic progression of the form

$$\ldots 10, \underline{10 + d, 10 + 2d, 10 + 3d,} 22, \ldots$$

Thus

$$22 = 10 + 4d$$

or

$$d = 3$$

The arithmetic means requested are 13, 16, 19.

Geometric Progressions

DEFINITION

A **geometric progression** is a sequence in which each successive term is obtained by multiplying the given term by the same constant; that is, there is some number r called the **common ratio** such that

$$a_{n+1} = r \cdot a_n \quad \text{for } n = 1, 2, 3, \ldots$$

If the first term of a geometric sequence is a and the common ratio is r, the sequence has the form

$$a = ar^0, ar, ar^2, ar^3, \ldots$$

Thus

nth TERM OF A GEOMETRIC PROGRESSION

$$a_n = ar^{n-1}, \quad \text{for } n = 1, 2, \ldots \tag{2}$$

This formula too can be established by mathematical induction; see Exercise 64.

EXAMPLE 5

Find the first six terms and the twelfth term of the geometric progression whose first term is 2 and whose common ratio is $\frac{1}{2}$.

SOLUTION

Any given term is multiplied by $\frac{1}{2}$ to get the next term. Thus the first six terms are

$$2, 1, \frac{1}{2}, \frac{1}{4}, \frac{1}{8}, \frac{1}{16}$$

The twelfth term is

$$a_{12} = a_1 r^{11}$$
$$= 2 \cdot \left(\frac{1}{2}\right)^{11} = \frac{1}{2^{10}}$$
$$= \frac{1}{1024}$$

As with arithmetic progressions, if we know any two terms of a geometric progression we can find the others. The technique is illustrated in the following example.

EXAMPLE 6

If the fourth term of a geometric progression is 24 and the seventh term is 192, find the second term.

SOLUTION

Again, the given information yields a system of equations.

$$\begin{cases} a_4 = a_1 r^3 = 24 \\ a_7 = a_1 r^6 = 192 \end{cases}$$

Then, dividing corresponding members of these equations, we have

$$\frac{a_7}{a_4} = \frac{a_1 r^6}{a_1 r^3} = \frac{192}{24}$$

$$r^3 = 8$$

$$r = 2$$

Now,

$$24 = a_1 r^3 = a_1 \cdot 8$$

Hence

$$a_1 = 3$$

Finally,

$$a_2 = a_1 r = 3 \cdot 2$$

$$a_2 = 6$$

The terms of a geometric progression that appear between two given terms a and b are called **geometric means** of a and b. If there is only one term, it is called the geometric mean of a and b; if there are n terms, they are called the n geometric means.

EXAMPLE 7   Insert three geometric means between 6 and 486.

SOLUTION   We are seeking part of a geometric progression of the form

$$\ldots 6, 6r, 6r^2, 6r^3, 486, \ldots$$

Thus

$$6r^4 = 486$$

$$r^4 = 81$$

$$r = \pm 3$$

There are two sets of *three geometric means* between 6 and 486. They correspond to $r = 3$ and $r = -3$. The geometric means are

for $r = 3$:   18, 54, 162

for $r = -3$: $-18, 54, -162$

Applications

We have already seen examples of everyday applications of arithmetic and geometric progressions. In Chapter 6 we studied exponential growth and decay. If a given population P_0 doubles every hour, the hourly population figures form a geometric progression with first term P_0 and common ratio $r = 2$. For if the initial population is P_0, then 1 hour later the population is $P_1 = 2P_0$; still another hour later it is $P_2 = 4P_0$. After n hours, the population is $P_n = 2^n P_0$. Listing the population at each hour yields the geometric progression

$$P_0, 2P_0, 4P_0, 8P_0, \ldots$$

We also studied compound interest in Chapter 6. If 5% interest is compounded annually on an investment of P_0 dollars, the value of the investment after n years was seen to be $P_n = P_0 \cdot (1.05)^n$. Thus the yearly values of the investment form a geometric progression with first term P_0 and common ratio $r = 1.05$:

$$P_0, P_0(1.05), P_0(1.05)^2, \ldots, P_0(1.05)^n, \ldots$$

On the other hand, if the interest is not compounded, we receive interest each year of $(0.05)P_0$:

$$P_n = P_{n-1} + (0.05)P_0$$

This gives an arithmetic progression with $a_1 = P_0$ and $d = (0.05)P_0$:

$$P_0, \underbrace{P_0 + 0.05P_0}_{P_1}, \underbrace{P_0 + 2(0.05P_0)}_{P_2}, \ldots, \underbrace{P_0 + n(0.05)P_0}_{P_n}, \ldots$$

EXAMPLE 8

Mr. Jaworski is enclosing a stairway with poles as indicated in Figure 11.6. The length of the shortest pole is 30 inches; the longest is 8 feet. The span is 12 feet. If he places a pole every 6 inches, how long should each pole be?

Figure 11.6

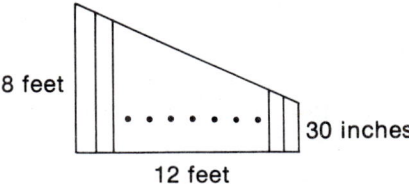

SOLUTION

The length of the poles increases steadily as we move from right to left. Thus the lengths form an arithmetic progression. In 12 feet there are twenty-four 6-inch steps to be taken. Thus there are twenty-three arithmetic means to be placed *between* 30 and 96 $(= 12 \cdot 8)$ inches. We form an arithmetic progression

$$30, 30 + d, 30 + 2d, \ldots, 30 + 23d, 30 + 24d = 96$$

From the last relation we obtain

$$30 + 24d = 96$$
$$24d = 66$$
$$d = \frac{66}{24} = \frac{11}{4}$$

$$d = 2\frac{3}{4} \text{ inches}$$

The lengths of the poles are, in inches,

$$30, 32\frac{3}{4}, 35\frac{1}{2}, 38\frac{1}{4}, 41, 43\frac{3}{4}, 46\frac{1}{2}, 49\frac{1}{4}, 52, 54\frac{3}{4}, 57\frac{1}{2}, 60\frac{1}{4}, 63,$$

$$65\frac{3}{4}, 68\frac{1}{2}, 71\frac{1}{4}, 74, 76\frac{3}{4}, 79\frac{1}{2}, 82\frac{1}{4}, 85, 87\frac{3}{4}, 90\frac{1}{2}, 93\frac{1}{4}, 96$$

Section 11.2 Exercises

Write out the first five terms and the tenth term of the sequences whose nth term is given.

1. $3n$
2. $2n - 1$
3. $2n + 3$
4. $5n + 3$
5. $2n^2$
6. $3n^2 + 2$
7. $1 + (-1)^n$
8. $1 + 2^n$
9. $(-2)^n$
10. $1 + \left(-\dfrac{1}{2}\right)^n$

Write out the first four terms and the eighth term of the sequences whose nth term is given.

11. $\dfrac{n-1}{n+1}$
12. $\dfrac{n}{2n-1}$
13. $\dfrac{n+1}{n+2}$
14. $\dfrac{n(n+1)}{2}$
15. $\dfrac{3}{n+5}$
16. $\dfrac{1}{n(n+1)}$
17. $\dfrac{2n+3}{2n-3}$
18. $\dfrac{1}{2n^2}$
19. $\dfrac{(-1)^n}{n^3}$
20. $\dfrac{n}{n^2+1}$

Write out the first five terms of the sequences defined as follows.

21. $a_1 = -2, a_n = -a_{n-1}$
22. $a_1 = 6, a_n = a_{n-1} + 1$
23. $a_1 = 1, a_n = \dfrac{na_{n-1}}{2}$
24. $a_1 = 2, a_n = \dfrac{1}{a_{n-1}}$
25. $a_1 = 2, a_n = (a_{n-1} - 1)^2$
26. $a_1 = 1, a_2 = 1, a_n = (a_{n-1} + a_{n-2})$

Write out the first five terms and the fifteenth term of the arithmetic progression with the following data.

27. $a_1 = 2, d = 3$
28. $a_1 = 3, d = -2$
29. $a_1 = -5, d = 2$
30. $a_4 = 15, d = 6$
31. $a_2 = 20, d = -2$
32. $a_{12} = -40, d = -3$

Find the requested term of an arithmetic progression given the following information.

33. $a_4 = -5, a_{23} = 52$, find a_{30}.
34. $a_{53} = 108, a_{86} = 174$, find a_{100}.
35. $a_{14} = 0, a_{10} = -12$, find a_{21}.
36. $a_{16} = 53, a_{23} = 74$, find a_6.
37. $a_6 = -9, a_{14} = -25$, find a_4.
38. $a_7 = 0, a_{10} = -6$, find a_3.

Insert the requested number of arithmetic means between the following numbers.

39. 7 means between 6 and 30
40. 3 means between -6 and 14
41. 6 means between 20 and 55
42. 4 means between -8 and 17
43. 7 means between 13 and 49
44. 5 means between 16 and 31

Write out the first four terms and the eighth term of the geometric progression with the following data.

45. $a_1 = 2, r = 3$
46. $a_1 = 48, r = \dfrac{1}{2}$
47. $a_3 = 48, r = 4$
48. $a_3 = \dfrac{1}{250}, r = \dfrac{1}{5}$
49. $a_7 = 18, r = \sqrt{3}$
50. $a_{12} = 512, r = 2$

Find the requested term of a geometric progression given the following information.

51. $a_3 = 12, a_6 = 96$, find a_{10}
52. $a_5 = 45, a_{13} = 3645$, find a_8
53. $a_4 = 81, a_8 = 9$, find a_2
54. $a_9 = 48, a_3 = 6$, find a_{11}
55. $a_4 = -12, a_7 = 96$, find a_2
56. $a_5 = 18, a_8 = -486$, find a_3

Insert the requested number of geometric means between the following numbers.

57. 5 terms between 21 and 567
58. 7 terms between 15 and 1215

59. 2 terms between 18 and -486

60. 3 terms between 648 and 8

61. 9 terms between 2048 and 2

62. 4 terms between -24 and 768

63. Use mathematical induction to establish Formula (1) for the nth term of an arithmetic progression.
$$a_n = a_1 + (n-1)d$$

64. Use mathematical induction to establish Formula (2) for the nth term of a geometric progression.
$$a_n = ar^{n-1}$$

65. Bob borrows $1000 interest-free from his father in order to buy a car. He agrees to repay his father $200 in 1 month and $50 per month thereafter.
  a. List the amount still owed to his father after each payment.
  b. Do these amounts form an arithmetic or geometric progression?

66. Sue has been jogging for 1 month and can now run 1 mile comfortably.
  a. She wishes to increase her distance by $\frac{1}{10}$ of a mile every other day. Write a sequence indicating the distance she runs every day for the next 2 weeks.
  b. She wishes to increase her distance by 10% every other day. Write a sequence indicating the distance she runs every day for the next 2 weeks.

67. The altitude of an object t seconds after it falls from a height of h_0 feet is $h_0 - 16t^2$ feet.
  a. Write an expression for the nth term of a sequence that indicates the altitude after each second.
  b. Write an expression for the nth term of a sequence that indicates the distance fallen *during each second*.
  c. Show that the sequence in part b forms an arithmetic progression.
  d. How far does an object fall during the tenth second? the twentieth second?

68. A certain ball bounces to three-quarters the height from which it falls. If it is initially dropped from a building 120 meters high, write the altitudes to which it rises on the first four bounces. Do these numbers form an arithmetic or geometric progression?

69. Suppose that an automobile loses 25% of its value each year.
  a. List the yearly values of a car that initially sells for $8000.
  b. What is its value after 10 years?
  c. Do these numbers form an arithmetic or geometric progression?

70. Betty wishes to flush her car radiator but cannot completely drain the engine block. The cooling system holds 16 liters, but she can drain only 12 liters from the radiator. She dilutes the old solution by successively filling the radiator with water, running the engine to mix the old solution with the water, and then draining the radiator again.
  a. How much of the old solution remains after each of the first five successive draining operations?
  b. Do these numbers form an arithmetic or geometric progression?

Section 11.3

The Summation Notation

Although it is a simple matter to write down short sums, very long sums can be cumbersome. Lengthy sums appear often enough that it is convenient to have a shorthand notation to describe them. The **summation symbol**, $\sum$, is used for this purpose.

> **DEFINITION**
> For $k, l \in N$ and $k < l$,
> $$\sum_{j=k}^{l} a_j = a_k + a_{k+1} + \cdots + a_l$$

In the preceding definition j is called the **index of summation.** It is a *dummy variable* in the sense that $\sum_{j=k}^{l} a_j = \sum_{n=k}^{l} a_n$ since both represent $a_k + \cdots + a_l$. Also, k and l are called the **lower** and **upper limits,** respectively, of the summation.

EXAMPLE 1

Evaluate $\sum_{j=2}^{5} (2j + 1)$.

SOLUTION

$$\sum_{j=2}^{5} (2j + 1) = (2 \cdot 2 + 1) + (2 \cdot 3 + 1) + (2 \cdot 4 + 1) + (2 \cdot 5 + 1)$$
$$= 5 + 7 + 9 + 11$$
$$= 32$$

EXAMPLE 2

Write $\sum_{j=10}^{27} 2^j$ in expanded form.

SOLUTION

$$\sum_{j=10}^{27} 2^j = 2^{10} + 2^{11} + \cdots + 2^{26} + 2^{27}$$

EXAMPLE 3

Write $(3 + 8 + 13 + 18 + 23)$ in shorthand form by using the summation symbol.

SOLUTION

There are many ways of writing a solution to this problem. Two of these are given here:

$$3 + 8 + 13 + 18 + 23 = \sum_{j=0}^{4} (3 + 5j)$$
$$= \sum_{j=1}^{5} (-2 + 5j)$$

We have already seen several situations in which the summation symbol could have been used to advantage. For instance, the formulas in Examples 1 and 2 of Section 11.1 are rewritten here.

$$1 + 2 + \cdots + n = \frac{n(n + 1)}{2}$$

$$\boxed{\sum_{j=1}^{n} j = \frac{n(n + 1)}{2}} \qquad (1)$$

$$1^2 + 2^2 + \cdots + n^2 = \frac{n(n + 1)(2n + 1)}{6}$$

$$\boxed{\sum_{j=1}^{n} j^2 = \frac{n(n + 1)(2n + 1)}{6}} \qquad (2)$$

The summation symbol is also useful in computing the sum of a finite number of successive terms in an arithmetic or geometric progression. Before discussing these applications, several elementary properties of the $\sum$ symbol should be established.

The symbol $\sum_{j=1}^{n} C$ denotes the addition of C to itself repeatedly from step 1 to step n. Thus

$$\sum_{j=1}^{n} C = \underset{j=1}{C} + \underset{j=2}{C} + \cdots + \underset{j=n}{C} = nC$$

$$\boxed{\sum_{j=1}^{n} C = nC} \qquad (3)$$

In general,

$$\boxed{\sum_{j=k}^{l} C = (l - k + 1)C} \qquad (3')$$

Because of the "rearrangement" properties of addition (the commutative and associative laws), we obtain

$$(a_k + b_k) + \cdots + (a_l + b_l) = (a_k + \cdots + a_l) + (b_k + \cdots + b_l)$$

That is,

$$\boxed{\sum_{j=k}^{l} (a_j + b_j) = \sum_{j=k}^{l} a_j + \sum_{j=k}^{l} b_j} \qquad (4)$$

Similarly, we could obtain

$$\boxed{\sum_{j=k}^{l} (a_j - b_j) = \sum_{j=k}^{l} a_j - \sum_{j=k}^{l} b_j} \qquad (4')$$

The distributive law becomes the *factorable constant* property. A constant multiplier can be factored through the $\sum$:

$$Ca_k + \cdots + Ca_l = C(a_k + \cdots + a_l)$$

That is,

$$\boxed{\sum_{j=k}^{l} (Ca_j) = C\left(\sum_{j=k}^{l} a_j\right)} \qquad (5)$$

Although "factorable constant" very aptly describes this property, its correct mathematical name is the property of **homogeneity.**

Finally, we observe that two $\sum$'s can be linked into one if their indices match up correctly. For instance,

$$\sum_{j=2}^{5} a_j + \sum_{6}^{10} a_j = (a_2 + \cdots + a_5) + (a_6 + \cdots + a_{10})$$

$$= \sum_{j=2}^{10} a_j$$

In general,

$$\sum_{j=k}^{l} a_j + \sum_{j=l+1}^{m} a_j = \sum_{j=k}^{m} a_j \qquad (6)$$

Arithmetic Progressions

Now to address the task of adding a finite number of successive terms of an arithmetic progression we consider

$$\ldots \underbrace{a_N, a_N + d, a_N + 2d, \ldots, a_N + (n-1)d}_{n \text{ terms}}, \ldots$$

The sum of n successive terms beginning at a_N is

$$a_N + (a_N + d) + \cdots + [a_N + (n-1)d]$$

$$= \sum_{j=0}^{n-1} (a_N + jd)$$

$$= \sum_{j=0}^{n-1} a_N + \sum_{j=0}^{n-1} jd \qquad \text{(Formula 4)}$$

$$= \sum_{j=0}^{n-1} a_N + d \sum_{j=0}^{n-1} j \qquad \text{(Formula 5)}$$

$$= \sum_{j=0}^{n-1} a_N + d \sum_{j=1}^{n-1} j \qquad (j=0 \text{ does not add to the sum})$$

$$= na_N + d\frac{(n-1)n}{2} \qquad \text{(Note that the last } j \text{ is } n-1; \text{ then use formulas 3' and 1)}$$

$$= \frac{n}{2}[2a_N + (n-1)d]$$

$$= \frac{n}{2}\{a_N + [a_N + (n-1)d]\}$$

This result can be stated as follows.

SUM OF AN ARITHMETIC PROGRESSION

The sum of n successive terms of an arithmetic progression equals $\dfrac{n}{2}$ times the sum of the first and last terms used:

$$a_1 + \cdots + a_n = \frac{n}{2}[a_1 + a_n] \qquad \text{for } n = 1, 2, \ldots \qquad (7)$$

Mathematical induction can be used to establish this formula; see Exercise 61.

EXAMPLE 4   Evaluate the sums given in

a. Example 1
b. Example 3

SOLUTION   a. Note that the terms of Example 1 form an arithmetic progression with $d = 2$. Thus applying formula 7,

$$\sum_{j=2}^{5} (2j + 1) = 5 + 7 + 9 + 11$$
$$= \frac{4}{2}(5 + 11) \qquad \text{(There are four terms)}$$
$$= 32$$

b. In Example 3, we also have an arithmetic progression, this time with $d = 5$.

$$\underbrace{3 + 8 + 13 + 18 + 23}_{\text{5 terms}} = \frac{5}{2}(3 + 23)$$
$$= 65$$

EXAMPLE 5   Find the sum of the odd integers between 50 and 490.

SOLUTION   We are to find

$$51 + 53 + \cdots + 487 + 489$$

These terms form an arithmetic progression with $d = 2$, 51 as the first term, and 489 as the last term:

$$489 = a_n = 51 + (n - 1)2 \qquad \text{(Formula 1, Section 8.2)}$$

Solving this for n, we obtain $n = 220$: $51 = a_1$ and $489 = a_{220}$. The sum of these 220 terms is then obtained by using Formula 7:

$$a_1 + \cdots + a_{220} = \frac{220}{2}(a_1 + a_{220})$$
$$= \frac{220}{2}(51 + 489) = 110 \cdot 540$$
$$= 59{,}400$$

Knowing the sum of an arithmetic progression can be useful in many everyday situations, as illustrated in the following example.

EXAMPLE 6   When booking a performance in an arena, the promoter wants to know the number of seats that are available in a given section. The section is wedge-shaped with 15 seats in the first row and each succeeding row having two additional seats. If there are 20 rows in the section, what is the total number of seats in the section?

SOLUTION There are 15 seats in the first row; each succeeding row has two additional seats. Thus the twentieth row has $(15 + 19 \cdot 2)$ seats. The total number of seats is

$$15 + 17 + 19 + \cdots + (15 + 19 \cdot 2) = \frac{20}{2}[15 + (15 + 19 \cdot 2)] \quad \text{(Formula 7)}$$
$$= 10(30 + 38)$$
$$= 680 \text{ seats} \quad \blacksquare$$

Geometric Progressions

In order to find the sum of the first n terms of a geometric progression with common ratio $r \neq 1$, we write

a. $S_n = a + ar + \cdots + ar^{n-1}$

Then

b. $rS_n = ar + \cdots + ar^n$

Subtracting corresponding members of both sides of equations a and b, we get

$$S_n(1 - r) = a - ar^n = a(1 - r^n)$$

and

$$S_n = \frac{a(1 - r^n)}{1 - r}$$

This observation is stated as follows:

SUM OF A GEOMETRIC PROGRESSION

$$a + ar + \cdots + ar^{n-1} = \sum_{j=0}^{n-1} ar^j = a\left(\frac{1 - r^n}{1 - r}\right), \quad r \neq 1 \quad (8)$$

This formula too can be established by mathematical induction; see Exercise 62.

EXAMPLE 7 Evaluate the sum in Example 2.

SOLUTION
$$\sum_{j=10}^{27} 2^j = 2^{10} + 2^{11} + \cdots + 2^{27}$$
$$= 2^{10}(1 + 2 + \cdots + 2^{17})$$
$$= 2^{10} \sum_{j=0}^{17} 2^j$$
$$= 2^{10}\left(\frac{1 - 2^{18}}{1 - 2}\right) \quad \text{(Formula 8)}$$
$$= 2^{10}(2^{18} - 1)$$
$$= 268{,}434{,}432 \quad \text{(From a calculator)} \quad \blacksquare$$

The formula for the sum of a geometric progression is useful in financial situations such as the study of investments, annuities, insurance, and mortgage payments. This formula is also useful in scientific or engineering situations as indicated in the following example.

EXAMPLE 8

If a ball falls from an altitude of 200 feet and has the property that it rebounds to three-quarters of its original altitude after each fall, how far has it traveled by the time it hits the ground for the tenth time?

SOLUTION

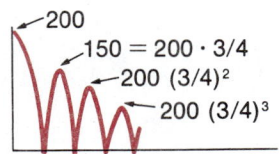

Figure 11.7

Figure 11.7 shows the situation. Since the ball initially falls 200 feet and then *rises and falls* nine times, the total distance traveled is

$$200 + 2 \cdot \left(200 \cdot \frac{3}{4}\right) + 2\left[200 \cdot \left(\frac{3}{4}\right)^2\right] + \cdots + 2\left[200 \cdot \left(\frac{3}{4}\right)^8\right] + 2\left[200\left(\frac{3}{4}\right)^9\right]$$

$$= 200 + \sum_{n=1}^{9} 2 \cdot 200 \cdot \left(\frac{3}{4}\right)^n$$

To use Formula 8, we need $\sum_{n=0}^{9}$ instead of $\sum_{n=1}^{9}$. This form is obtained by adding and subtracting the term corresponding to $\left(\frac{3}{4}\right)^0$. Then we continue as follows.

$$= 200 + \sum_{n=1}^{9} 2 \cdot 200 \left(\frac{3}{4}\right)^n$$

$$= 200 + \sum_{n=0}^{9} 2 \cdot 200 \left(\frac{3}{4}\right)^n - 2 \cdot 200 \qquad \left[\text{Since } \left(\frac{3}{4}\right)^0 = 1\right]$$

$$= -200 + 2 \cdot 200 \sum_{n=0}^{9} \left(\frac{3}{4}\right)^n \qquad \text{(Formula 5)}$$

$$= -200 + 400 \, \frac{1 - (3/4)^{10}}{1 - (3/4)} \qquad \text{(Formula 8)}$$

$$= -200 + 1600 \left[1 - \left(\frac{3}{4}\right)^{10}\right]$$

$$\approx -200 + 1510 \qquad \text{(From a calculator)}$$

$$= 1310 \text{ feet} \qquad \blacksquare$$

EXAMPLE 9

Evaluate $\sum_{n=1}^{10} (2n + 3n^2 + 2^n)$.

SOLUTION

$$\sum_{n=1}^{10} (2n + 3n^2 + 2^n)$$

$$= \sum_{n=1}^{10} 2n + \sum_{n=1}^{10} 3n^2 + \sum_{n=1}^{10} 2^n \qquad \text{(Formula 4)}$$

$$= 2\sum_{n=1}^{10} n + 3\sum_{n=1}^{10} n^2 + \sum_{n=0}^{10} 2^n - 2^0 \qquad \text{(Use Formula 5 twice; then add and subtract the term } 2^0\text{)}$$

$$= 2\,\frac{10 \cdot 11}{2} + 3\,\frac{10 \cdot 11 \cdot 21}{6} + \frac{1 - 2^{11}}{1 - 2} - 1 \qquad \text{(Use Formulas 1, 2, and 8)}$$

$$= 110 + 1155 + 2047 - 1$$

$$= 3311 \qquad \blacksquare$$

Geometric Series

If $-1 < r < 1$, we note that as n gets larger and larger, the values of $|r^n|$ get smaller and smaller. Using the notation introduced in Sections 5.4 and 5.5, we write

$$r^n \to 0 \qquad \text{as} \qquad n \to \infty \qquad \text{when } |r| < 1$$

If we now consider the sum

$$S_n = a + ar + \cdots + ar^{n-1} = \sum_{j=0}^{n-1} ar^j = a\frac{1-r^n}{1-r}$$

we see that

$$S_n \longrightarrow a\frac{1-0}{1-r} = \frac{a}{1-r} \quad \text{as} \quad n \to \infty$$

This means that if we add more and more terms to a geometric progression, the sum S_n gets closer and closer to $\frac{a}{1-r}$.

DEFINITION

A **geometric series** is the indicated nonterminating sum of the terms of a geometric progression:

$$\sum_{j=0}^{\infty} ar^j = a + ar + \cdots + ar^n + \cdots$$

If $|r| < 1$, the sum of the geometric series is

$$\sum_{j=0}^{\infty} ar^j = \frac{a}{1-r} \tag{9}$$

EXAMPLE 10

Find the sum of the following geometric series.

a. $3 + \frac{3}{2} + \frac{3}{4} + \frac{3}{8} + \cdots$ **b.** $\sum_{j=1}^{\infty} \left(\frac{1}{3}\right)^j$

SOLUTION

a. We can write this series in summation notation as $\sum_{j=0}^{\infty} 3\left(\frac{1}{2}\right)^j$. Then, using Formula 9, we have

$$\sum_{j=0}^{\infty} 3\left(\frac{1}{2}\right)^j = \frac{3}{1-\frac{1}{2}} = \frac{3}{\frac{1}{2}} = 6$$

b. To match the form of Formula 9, we must include the $j = 0$ term.

$$\sum_{j=1}^{\infty} \left(\frac{1}{3}\right)^j = \sum_{j=0}^{\infty} \left(\frac{1}{3}\right)^j - \left(\frac{1}{3}\right)^0$$

$$= \sum_{j=0}^{\infty} 1 \cdot \left(\frac{1}{3}\right)^j - 1$$

$$= \frac{1}{1-\frac{1}{3}} - 1 \quad \text{(Formula 9)}$$

$$= \frac{1}{\frac{2}{3}} - 1$$

$$= \frac{3}{2} - 1$$

∎

Section 11.3 Exercises

Expand and evaluate the following.

1. $\sum_{n=1}^{4} \frac{1}{n}$
2. $\sum_{k=2}^{3} k(-1)^k$
3. $\sum_{n=1}^{3} (2n-1)(2n)$
4. $\sum_{n=3}^{5} n^2(-1)^n$
5. $\sum_{n=2}^{6} (n^3 - n)$
6. $\sum_{j=2}^{7} j(j+2)$
7. $\sum_{k=1}^{3} \frac{1}{k(k+1)}$
8. $\sum_{n=3}^{5} \frac{2^n}{2n}$

Write the following in summation form.

9. $5 + 7 + 9 + 11 + 13$
10. $1 + 4 + 9 + 16 + \cdots + 81 + 100$
11. $\frac{1}{2} + \frac{1}{4} + \frac{1}{6} + \frac{1}{8} + \frac{1}{10}$
12. $3 + 9 + 27 + 81 + 243$
13. $20 + 30 + 40 + 50 + 60 + 70$
14. $2 \cdot 3 + 3 \cdot 4 + 4 \cdot 5 + 5 \cdot 6$
15. $2^2 + 3^3 + 4^4 + \cdots + 99^{99}$
16. $1 - 2 + 3 - 4 + 5 - 6 + 7$

Evaluate the following by using the results of this section.

17. $\sum_{k=1}^{10} (3k+1)$
18. $\sum_{i=0}^{6} (2+3i)$
19. $\sum_{k=2}^{80} (1+4k)$
20. $\sum_{j=1}^{15} (2j-1)$
21. $\sum_{n=15}^{25} (3n-1)$
22. $\sum_{n=1}^{15} (n^2 + 2n)$
23. $\sum_{n=1}^{10} (2n-1)(2n)$
24. $\sum_{n=1}^{10} (n-2)(n-3)$
25. $\sum_{n=1}^{14} (n^2 + 3n + 2)$
26. $\sum_{n=0}^{5} 5 \cdot \left(\frac{1}{2}\right)^n$
27. $\sum_{n=0}^{5} 2 \cdot 3^n$
28. $\sum_{n=1}^{7} 16 \cdot \left(\frac{1}{2}\right)^n$
29. $\sum_{k=1}^{4} 3 \cdot 4^k$
30. $\sum_{n=1}^{10} (n^2 - 4n + 3^n)$
31. $\sum_{k=1}^{5} (3k + 3^k \cdot 4 + 4k^2)$
32. $\sum_{l=1}^{20} [3 \cdot 2^l + l(l+3) + 4]$

Find the sum of the first n terms of the arithmetic progressions that satisfy the following.

33. $a_1 = 3, d = 4, n = 20$
34. $a_1 = -2, d = 3, n = 16$
35. $a_1 = 12, a_2 = 16, n = 8$
36. $a_1 = 5, a_2 = 7, n = 14$
37. $a_3 = 11, a_{15} = 35, n = 10$
38. $a_5 = 24, a_{20} = 69, n = 12$

Find the sum of the first n terms of the geometric progressions that satisfy the following.

39. $a = 2, r = 3, n = 5$
40. $a = 3, r = 2, n = 7$
41. $a = 5, r = -2, n = 7$
42. $a = 4, r = -3, n = 4$

Find the sum of the given geometric series.

43. $2 + \frac{2}{3} + \frac{2}{9} + \frac{2}{27} + \cdots$
44. $3 + \frac{3}{4} + \frac{3}{16} + \frac{3}{64} + \cdots$
45. $1 - \frac{1}{2} + \frac{1}{4} - \frac{1}{8} + \frac{1}{16} - \cdots$
46. $1 - \frac{1}{3} + \frac{1}{9} - \frac{1}{27} + \frac{1}{81} - \cdots$
47. $5 + \frac{5}{6} + \frac{5}{36} + \frac{5}{216} + \cdots$
48. $3 - \frac{3}{5} + \frac{3}{25} - \frac{3}{125} + \cdots$
49. $\sum_{j=0}^{\infty} 2\left(-\frac{1}{4}\right)^j$
50. $\sum_{j=0}^{\infty} 5\left(\frac{1}{7}\right)^j$
51. $\sum_{j=1}^{\infty} 3\left(\frac{2}{3}\right)^j$
52. $\sum_{j=1}^{\infty} 4\left(-\frac{2}{5}\right)^j$

53. A grocer stacks cans in a pyramid with 30 cans on the bottom row and 1 fewer can on each higher row until the peak has just 1 can. How many cans are in the pyramid?

54. Barrels are stacked in a brewery in a pyramid fashion with 50 barrels on the bottom row, 1 fewer barrel in each successive row, and a total height of 10 rows. How many barrels are there?

55. Karen deposits $1 on May 1, $2 on May 2, ... $31 on May 31. How many dollars does she deposit in May?

56. Rob deposits 1¢ on May 1, 2¢ on May 2, 4¢ on May 3, and so on, doubling the deposit each day. How much does he deposit in May?

57. The annual premium on a certain insurance policy is 2% of the value of an article. If a given article depreciates at the rate of 10% per year so that its value at the end of a given year is 90% of its value at the beginning of that year, what is the total of the insurance premiums for a period of 10 years?

58. Fred receives a chain letter including a list of five names. He is requested to send $1.00 to the name at the top of the list, remove this name, and add his name to the bottom of the list. He is then supposed to mail 10 copies of the new letter with the same instructions.
   a. How much money will he eventually receive if the chain is unbroken?
   b. How many steps can this chain take before it will have involved the entire U.S. population of about 250 million people?
   c. How many steps can this chain take before it will have involved the entire world population of about 4 billion people?

59. So-called pyramid selling schemes frequently become popular. There are many variations on these schemes but common to all of them is an emphasis on "building a sales organization" rather than actually selling products. One such plan requires an investment of $10,000. Then, for each person you convince to join your organization, you collect $10,000. You keep $500 of this and turn the rest over to your "sponsor." This sponsor then keeps $250 and passes the rest up to his or her sponsor, and so on, until the chain reaches the company president. The company expects each participant to enlist 5 people into the sales organization: Assume each participant enlists 5 people.
   a. How many echelons must be built below you before you recoup your initial investment?
   b. Through how many echelons can this pyramid proceed before including the entire available adult U.S. population of about 100 million people?

60. Number the squares on a checkboard from 1 to 64 and then place one dot on square 1, two dots on square two, four dots on square 3, and so on.
   a. How many dots are required on square 64?
   b. How many dots are required in total?
   c. Answer the preceding questions if three dots are placed on square 3, four dots on square 4, and so on.

61. Use mathematical induction to establish Formula 7 for the sum of an arithmetic progression:
$$a_1 + \cdots + a_n = \frac{n}{2}[a_1 + a_n]$$

62. Use mathematical induction to establish Formula 8 for the sum of a geometric progression:
$$a + ar + \cdots + ar^{n-1} = a\left(\frac{1-r^n}{1-r}\right), \quad r \neq 1$$

Section 11.4

The Binomial Theorem

Expanding a binomial such as $(a + b)^{10}$ by repeated multiplication can be a very laborious task and provides many opportunities for error. The binomial theorem provides a straightforward way of writing such expansions directly without repeated multiplications. For a clue regarding a pattern for repeated multiplication of a binomial, consider the following:

$$(a + b)^0 = 1$$
$$(a + b)^1 = a + b$$
$$(a + b)^2 = a^2 + 2ab + b^2$$
$$(a + b)^3 = a^3 + 3a^2b + 3ab^2 + b^3$$
$$(a + b)^4 = a^4 + 4a^3b + 6a^2b^2 + 4ab^3 + b^4$$
$$(a + b)^5 = a^5 + 5a^4b + 10a^3b^2 + 10a^2b^3 + 5ab^4 + b^5$$

We can make several observations about the expansion of $(a+b)^n$.

1. The first term is $a^n = a^n b^0$.
2. The last term is $b^n = a^0 b^n$.
3. All terms are of the form $a^k b^l$ with $k + l = n$.
4. In moving from left to right to successive terms, the power of a decreases by 1 and the power of b increases by 1.
5. There are $n + 1$ terms in all.

The coefficients of $a^k b^l$ in the expansion of $(a + b)^n$ are called **binomial coefficients.** If we extract the binomial coefficients from the preceding expansions we obtain **Pascal's triangle** as shown here.

PASCAL'S TRIANGLE

$$
\begin{array}{c}
1 \\
1 \quad 1 \\
1 \quad 2 \quad 1 \\
1 \quad 3 \quad 3 \quad 1 \\
1 \quad 4 \quad 6 \quad 4 \quad 1 \\
1 \quad 5 \quad 10 \quad 10 \quad 5 \quad 1 \\
1 \quad 6 \quad 15 \quad 20 \quad 15 \quad 6 \quad 1 \\
1 \quad 7 \quad 21 \quad 35 \quad 35 \quad 21 \quad 7 \quad 1
\end{array}
$$

In Pascal's triangle the leading entry and the final entry in each row is a 1. Inside each row, a given entry is the sum of the two entries above it. For instance, consider the rows corresponding to $(a + b)^4$ and $(a + b)^5$.

$$
\begin{array}{ccccccccc}
1 & & 4 & & 6 & & 4 & & 1 \\
1 & 5 & & 10 & & 10 & & 5 & & 1
\end{array}
$$

The binomial coefficients appearing in the expansion of $(a + b)^n$ are found in the corresponding row of Pascal's triangle; the second entry in that row will be n.

We can use Pascal's triangle to read off

$$(a + b)^7 = a^7 + 7a^6 b + 21a^5 b^2 + 35a^4 b^3 + 35a^3 b^4 + 21a^2 b^5 + 7ab^6 + b^7$$

EXAMPLE 1

Use Pascal's triangle to expand the following.

a. $(x + 2y)^4$ **b.** $(x - 2)^5$

SOLUTION

a. According to Pascal's triangle, the coefficients for $n = 4$ are 1, 4, 6, 4, 1 in that order. Thus

$$(x + 2y)^4 = 1 \cdot x^4 + 4x^3(2y) + 6x^2(2y)^2 + 4x(2y)^3 + 1 \cdot (2y)^4$$
$$= x^4 + 8x^3 y + 24x^2 y^2 + 32xy^3 + 16y^4$$

b. The coefficients for $n = 5$ are 1, 5, 10, 10, 5, 1. Thus
$$\begin{aligned}(x - 2)^5 &= [x + (-2)]^5 \\ &= 1 \cdot x^5 + 5x^4(-2) + 10x^3(-2)^2 + 10x^2(-2)^3 \\ &\quad + 5x(-2)^4 + 1 \cdot (-2)^5 \\ &= x^5 - 10x^4 + 40x^3 - 80x^2 + 80x - 32\end{aligned}$$
∎

Although Pascal's triangle is quite useful for small n, it could be quite a chore to generate 100 rows of the triangle in order to find some coefficient in $(a + b)^{100}$. A formula for the binomial coefficients is certainly desirable. It will be helpful to introduce the following notation:

DEFINITION

$$n! = 1 \cdot 2 \cdot \cdots \cdot n \quad \text{(for } n \in N\text{)}$$
$$0! = 1$$
$$1! = 1$$

The ! symbol is read "**factorial**." The expression **0!** is seldom used; its notation is introduced merely to simplify writing certain expressions later. Also note the following property of factorials.

$$\boxed{n!(n + 1) = (n + 1)!}$$

EXAMPLE 2   Evaluate the following factorial expressions.

a. $\dfrac{5!}{3!}$ **b.** $\dfrac{5!}{2!(5 - 2)!}$

SOLUTION

a. $\dfrac{5!}{3!} = \dfrac{1 \cdot 2 \cdot 3 \cdot 4 \cdot 5}{1 \cdot 2 \cdot 3} = 4 \cdot 5 = 20$

b. $\dfrac{5!}{2!(5 - 2)!} = \dfrac{5!}{2!3!} = \dfrac{1 \cdot 2 \cdot 3 \cdot 4 \cdot 5}{(1 \cdot 2)(1 \cdot 2 \cdot 3)} = \dfrac{20}{2} = 10$ ∎

The factorial notation helps in writing a formula for the binomial coefficients. These coefficients are embodied in the binomial theorem.

THE BINOMIAL THEOREM

$$(a + b)^n = \sum_{l=0}^{n} \left(\frac{n!}{l!(n - l)!} a^{n-l} b^l \right)$$

PROOF   It is a simple matter to verify that the binomial theorem holds for small values of n, say, for $n = 1, 2, 3$. We then assume the formula holds for $n = k$ and attempt to establish it for $n = k + 1$. (That is, we prove this formula by mathematical induction.) For $n = k$, we have

$$(a + b)^k = \sum_{l=0}^{k} \frac{k!}{l!(k-l)!} a^{k-l} b^l$$

Then

$$(a + b)^{k+1} = (a + b)(a + b)^k$$

$$= (a + b) \left[\sum_{l=0}^{k} \frac{k!}{l!(k-l)!} a^{k-l} b^l \right]$$

$$= a \sum_{l=0}^{k} \frac{k!}{l!(k-l)!} a^{k-l} b^l + b \sum_{l=0}^{k} \frac{k!}{l!(k-l)!} a^{k-l} b^l$$

$$= \sum_{l=0}^{k} \frac{k!}{l!(k-l)!} a^{k+1-l} b^l + \sum_{l=0}^{k} \frac{k!}{l!(k-l)!} a^{k-l} b^{l+1}$$

$$= a^{k+1} + \sum_{l=1}^{k} \frac{k!}{l!(k-l)!} a^{k+1-l} b^l + \sum_{l=0}^{k-1} \frac{k!}{l!(k-l)!} a^{k-l} b^{l+1} + b^{k+1}$$

Now, replace the (dummy) index of summation l with $l - 1$ in the second summation.

$$= a^{k+1} + \sum_{l=1}^{k} \frac{k!}{l!(k-l)!} a^{k+1-l} b^l + \sum_{l=1}^{k} \frac{k!}{(l-1)!(k-l+1)!} a^{k-l+1} b^l + b^{k+1}$$

$$= a^{k+1} + \sum_{l=1}^{k} \left[\frac{k!}{l!(k-l)!} + \frac{k!}{(l-1)!(k-l+1)!} \right] a^{k+1-l} b^l + b^{k+1}$$

$$= a^{k+1} + \sum_{l=1}^{k} \left[\frac{k!(k-l+1)}{l!(k-l)!(k-l+1)} + \frac{k! \, l}{(l-1)!(k-l+1)! \, l} \right] a^{k+1-l} b^l + b^{k+1}$$

$$= a^{k+1} + \sum_{l=1}^{k} \frac{k!(k-l+1+l)}{l!(k+1-l)!} a^{k+1-l} b^l + b^{k+1} \quad [(k-l)!(k-l+1) = (k-l+1)! \text{ and } (l-1)! \, l = l!]$$

$$= a^{k+1} + \sum_{l=1}^{k} \frac{(k+1)!}{l!(k+1-l)!} a^{k+1-l} b^l + b^{k+1} \quad [k!(k+1) = (k+1)!]$$

$$= \sum_{l=0}^{k+1} \frac{(k+1)!}{l!(k+1-l)!} a^{k+1-l} b^l$$

Thus the truth of the formula for $n = k$ implies its truth for $n = k + 1$. Since the formula holds for $n = 1, 2$, and 3, mathematical induction then tells us that it holds for all positive integer powers n. ∎

From the binomial theorem, we note the following.

The terms in the expansion of $(a + b)^n$ are of the form $c_{jk} a^j b^k$ with

1. $j + k = n$
2. $c_{jk} = \dfrac{n!}{j! k!}$

The kth term in this expansion is the term involving b^{k-1}.

Thus, in $(a + b)^2 = a^2 + 2ab + b^2$, a^2 is the first term; $2ab$ is the second term; and b^2 is the third term.

EXAMPLE 3   Use the binomial theorem to find the following.

a. The third term of $(x - 3)^{15}$
b. The fifth term of $(x + 2y)^{10}$

SOLUTION

a. The third term of $(x - 3)^{15} = [x + (-3)]^{15}$ is the one involving $x^{13}(-3)^2$:

$$\frac{15!}{2!(15-2)!} x^{13}(-3)^2 = \frac{15!}{2!13!} x^{13}(-3)^2$$

$$= \frac{(13!) \cdot 14 \cdot 15}{2 \cdot (13!)} x^{13} \cdot 9$$

$$= \frac{14 \cdot 15}{2} \cdot 9x^{13}$$

$$= 945x^{13}$$

b. The fifth term of $(x + 2y)^{10}$ is the one involving $x^6(2y)^4$:

$$\frac{10!}{4!6!} x^6 \cdot (2y)^4 = \frac{(6!) \cdot 7 \cdot 8 \cdot 9 \cdot 10}{(4!)6!} x^6(2y)^4$$

$$= \frac{7 \cdot 8 \cdot 9 \cdot 10}{2 \cdot 3 \cdot 4} \cdot 16x^6y^4$$

$$= 3360x^6y^4$$

EXAMPLE 4   Use the binomial theorem to evaluate $(99)^4$.

SOLUTION

$$(99)^4 = (100 - 1)^4 = [100 + (-1)]^4$$

$$= 100^4 + \frac{4!}{1!3!} 100^3(-1) + \frac{4!}{2!2!} 100^2(-1)^2$$

$$+ \frac{4!}{3!1!} 100(-1)^3 + \frac{4!}{4!0!} (-1)^4$$

$$= (100)^4 - 4(100)^3 + 6(100)^2 - 4(100) + 1$$

$$= 100,000,000 - 4,000,000 + 60,000 - 400 + 1$$

$$= 96,059,601$$

Section 11.4 Exercises

Evaluate the following factorial expressions.

1. $6!$
2. $9!$
3. $8!$
4. $10!$
5. $4!6!$
6. $3!7!$
7. $3!8!$
8. $4!4!$
9. $\dfrac{9!}{5!}$
10. $\dfrac{8!}{3!}$
11. $\dfrac{3!2!}{4}$
12. $\dfrac{5!6!}{3 \cdot 5}$
13. $\dfrac{6!}{4!3!}$
14. $\dfrac{8!}{5!4!}$

15. $\dfrac{10!}{7! \cdot 8 \cdot 9}$

16. $\dfrac{12!}{8! \cdot 9 \cdot 10}$

17. $\dfrac{100!\,3!}{99!\,4!}$

18. $\dfrac{150!\,74!}{75!\,149!}$

Use Pascal's triangle to expand the following.

19. $(x + 1)^4$
20. $(t - 3)^5$
21. $(r - s)^{10}$
22. $(u + v)^9$
23. $(4x - y)^3$
24. $(2x + y)^6$
25. $(v - w^2)^5$
26. $(3a - 5b)^4$
27. $(2x + 3y^2)^4$
28. $(u^2 - 3v)^4$
29. $(a^3 - 2b^2)^6$
30. $(x^2 - 3y^2)^5$
31. $\left(x + \dfrac{1}{x}\right)^6$
32. $\left(\dfrac{3}{u} - \dfrac{u}{3}\right)^5$
33. $(t - u^{-2})^8$
34. $(r^{-2} + r^2)^7$
35. $\left(2p - \dfrac{2}{q^2}\right)^5$
36. $\left(\dfrac{1}{a^2} + 3b\right)^6$
37. $\left(x - \dfrac{1}{\sqrt{x}}\right)^5$
38. $\left(\sqrt{u} - \dfrac{1}{u}\right)^6$

Use the binomial theorem to evaluate the following.

39. 11^5
40. 9^5
41. $(101)^4$
42. $(98)^4$
43. $(9.9)^3$
44. $(1.02)^5$
45. $(0.995)^3$
46. $(1.1)^6$

Use the binomial theorem to find the indicated term of the following.

47. $(2a - b)^7$, fifth term
48. $(x + y)^{17}$, fourth term
49. $(2v + w)^{20}$, eighteenth term
50. $(r + 2s)^{20}$, third term
51. $(x^2 + y)^{15}$, eleventh term
52. $(u^2 - 2v)^{10}$, fourth term
53. $(3a - 2b^2)^5$, fourth term
54. $\left(t + \dfrac{1}{t}\right)^{12}$, eighth term
55. $\left(x - \dfrac{3}{y}\right)^{12}$, seventh term
56. $\left(\dfrac{p^2}{q} - \dfrac{2}{p}\right)^{10}$, fourth term
57. Find the term involving x^6 in the expansion of $(x^2 - y^2)^7$.
58. Find the term involving v^8 in the expansion of $(u^3 + v^2)^8$.
59. Find the constant term in the expansion of $\left(r - \dfrac{1}{r}\right)^{12}$.
60. Find the constant term in the expansion of $\left(s^2 + \dfrac{1}{s}\right)^{15}$.
61. Find the term involving x^{12} in the expansion of $(x + \sqrt[3]{x})^{30}$.
62. Find the term involving u^7 in the expansion of $\left(u - \dfrac{2}{u^2}\right)^{40}$.
63. Find the first four terms in the expansion of $(2x + 3y)^{15}$.
64. Find the first four terms in the expansion of $(s - 3t)^{30}$.
65. Find the last four terms in the expansion of $(2x + 3y)^{15}$.
66. Find the last four terms in the expansion of $(s - 3t)^{30}$.

Section 11.5

Permutations and Combinations

Whenever a system can assume several different characteristics, in various combinations with one another, it can sometimes be a time-consuming process to enumerate all possible configurations for the system. Accordingly, some kind of systematic counting procedure should be developed to determine the total number of such configurations.

674 Chapter 11 Mathematical Induction, Progressions, and Probability

EXAMPLE 1 A certain racing bicycle can be purchased in various configurations:

    i. Wheels may be either metric size (these are commonly called "700 c" and measure about 70 centimeters in diameter) or English size (27 inches in diameter).

    ii. Brakes may be either center pull or side pull.

    iii. Tires may be bald, treaded, or ribbed.

How many configurations are possible for this bicycle?

SOLUTION We find it convenient to display the information in a **tree diagram.** Let the wheels be denoted by m or E for metric or English, respectively. Similarly, let the brakes be denoted by c or s and the tires, b, t, or r.

When we start to select the components of our bicycle, we have two choices for the wheels.

```
         m
Start<
         E
```

Once the wheels are chosen, we may choose either center or side pull brakes

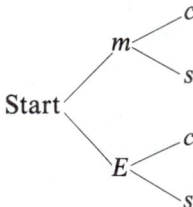

Now, for each of these selections, we can choose one of three kinds of tires.

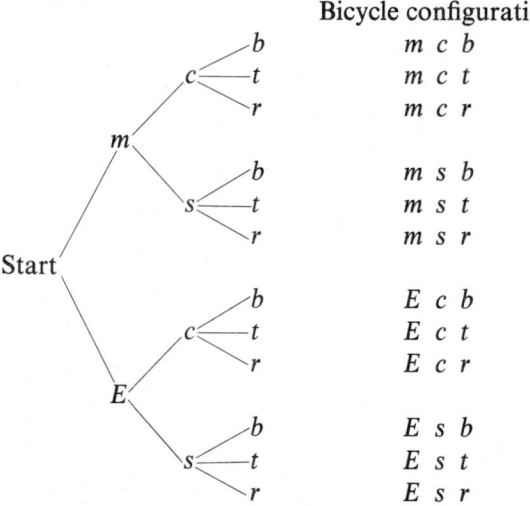

There are, in total, 12 different configurations denoted by $m\,c\,b$ (for metric wheels, center-pull brakes, bald tires), $m\,c\,t$, and so on. ■

Consider another simple—yet important—situation from everyday life: classification of human blood types.

EXAMPLE 2

Three substances, called antigens, affect the blood donor–recipient relationship. The medical profession calls these antigens A, B, and Rh. Any, all, or none of these may be present in the blood of a given individual. How many blood types can be classified by using these three antigens?

SOLUTION

The presence or absence of Rh is denoted by Rh+ or Rh−, respectively. Then, using a tree diagram, we classify the blood types as follows.

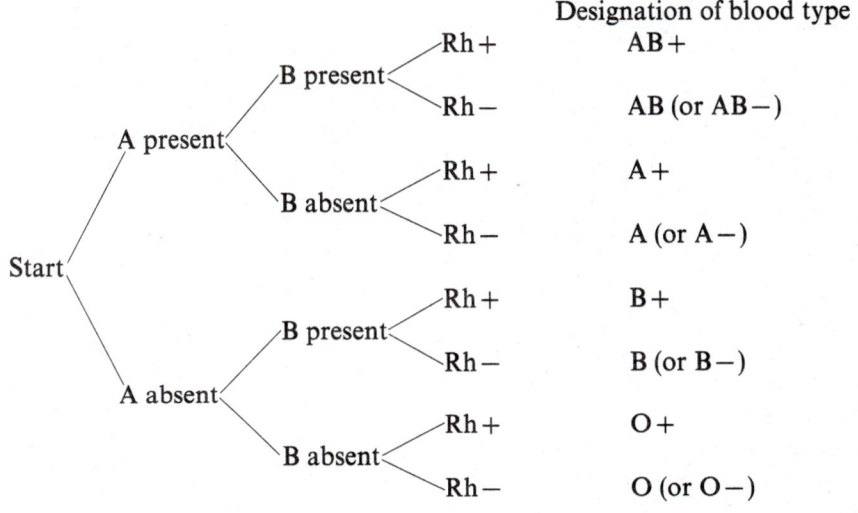

There are eight different blood types.

In the following example, we begin to develop a counting procedure that does not rely on tree diagrams.

EXAMPLE 3

A manufacturing concern developing a top-secret new product uses three of the letters A, B, C, D, E to label the versions it is currently testing.

a. If no letter may appear more than once on a given label, how many different labels are possible?
b. If letters may be repeated, how many different labels are possible?

SOLUTION

a. If only one letter were allowed, then just five labels would be possible. If two letters are to be used, then once the first letter is chosen, it can be followed by any one of the four remaining letters. Since there are five possible choices for the first letter and each of these can then be followed by four possible second letters, there are

$$5 \cdot 4 = 20$$

possible two-letter labels:

AB	BA	CA	DA	EA
AC	BC	CB	DB	EB
AD	BD	CD	DC	EC
AE	BE	CE	DE	ED

For a three-letter label, only three letters remain after the choice of the first two letters. Thus each of the preceding two-letter labels can be followed by any one of three remaining letters; for example, we have

ABC
ABD
ABE

Consequently, there are

$5 \cdot 4 \cdot 3 = 60$

possible three-letter labels if no letter can appear more than once on a given label.

b. If repetitions of letters are permitted, there will be five choices for the second letter. This yields

$5 \cdot 5 = 25$

two-letter labels. Then any one of five letters could be used in the third position, yielding

$5 \cdot 5 \cdot 5 = 125$

three-letter labels if repetitions are permitted. ∎

These observations are embodied in the following principle.

FUNDAMENTAL COUNTING PRINCIPLE

Suppose that n events are to take place and that

1. The first event can result in N_1 different outcomes.
2. After the first event, there remain N_2 possible outcomes for the second event.
⋮
n. After the preceding $(n - 1)$ events, there remain N_n possible outcomes for the nth event.

Then the total number of possible outcomes from the sequence of n events is the product

$$N_1 \cdot N_2 \cdots N_n$$

In Example 1, the events are the selection of wheels ($N_1 = 2$), brakes ($N_2 = 2$), and tires ($N_3 = 3$). We obtained

$N_1 \cdot N_2 \cdot N_3 = 2 \cdot 2 \cdot 3 = 12$

possible configurations.

In Example 2, the events were the determination of the presence or absence of the three antigens A, B, Rh ($N_1 = N_2 = N_3 = 2$). There were

$N_1 \cdot N_2 \cdot N_3 = 2 \cdot 2 \cdot 2 = 8$

possible blood types.

In Example 3, the events were the selection of the first, second, and third letters on the labels:

3a. $N_1 = 5, N_2 = 4, N_3 = 3$

$N_1 \cdot N_2 \cdot N_3 = 5 \cdot 4 \cdot 3 = 60$ possible labels

3b. $N_1 = N_2 = N_3 = 5$

$N_1 \cdot N_2 \cdot N_3 = 5 \cdot 5 \cdot 5 = 125$ possible labels

EXAMPLE 4 An automobile-license numbering system currently used by several states consists of 3 letters followed by 3 digits. Find the total number of distinct license numbers of this type if

a. Repetitions are permitted.
b. Repetitions are not permitted.
c. Letters may not be repeated, digits may be repeated, but 0 may not be the first digit.

SOLUTION

a. There are 26 choices for each of the three letters and 10 choices for each of the three digits. Thus the total number of distinct license numbers is

$26 \cdot 26 \cdot 26 \cdot 10 \cdot 10 \cdot 10 = 17{,}576{,}000$

b. If no repetitions are permitted, then there are

26 choices for the first letter
25 choices for the second letter
24 choices for the third letter
10 choices for the first digit
9 choices for the second digit
8 choices for the third digit

This gives a total of

$26 \cdot 25 \cdot 24 \cdot 10 \cdot 9 \cdot 8 = 11{,}232{,}000$

distinct license numbers.

c. The differences between this case and b are that only 9 choices are permitted for the first digit and 10 choices are available for each of the remaining digits. Thus the total is

$26 \cdot 25 \cdot 24 \cdot 9 \cdot 10 \cdot 10 = 14{,}040{,}000$ ■

Permutations

DEFINITION

Let S be a collection of n objects or symbols. An arrangement of r of these objects ($r \leq n$) in which

1. Order is important
2. Repetitions are not permitted

is called a **permutation** of S. The total number of permutations of n objects taken r at a time is denoted by $_nP_r$.

To find the total number of permutations of n objects taken r at a time, observe that there are

n ways to choose the first object
$n - 1$ ways to choose the second object
$n - 2$ ways to choose the third object
$\vdots$
$n - (r - 1) = n - r + 1$ ways to choose the rth object

Thus by the fundamental counting principle, there are

$$n \cdot (n - 1) \cdot (n - 2) \cdots (n - r + 1)$$

such permutations. Another way to write this is to observe that

$$n(n - 1)(n - 2) \cdots (n - r + 1) = \frac{n(n - 1) \cdots (n - r + 1)[(n - r) \cdots 2 \cdot 1]}{[(n - r) \cdots 2 \cdot 1]}$$

$$= \frac{n!}{(n - r)!}$$

PERMUTATION FORMULA

The total number of permutations of n objects taken r at a time is

$$_nP_r = n(n - 1) \cdots (n - r + 1)$$

$$= \frac{n!}{(n - r)!}$$

In particular, the number of permutations in which all n objects are used is

$$_nP_n = n(n - 1) \cdots 2 \cdot 1$$
$$= n!$$

(Recall from Section 8.4 that $0! = 1$ by definition. Thus $n! = n!/(n - n)!$ in agreement with the first formula.)

EXAMPLE 5 On Competition Day at Southdale Elementary School, one of the events is a softball game between the fifth- and sixth-grade classes. If 15 persons in the sixth grade are interested in playing softball, in how many ways can a 9-person team be formed?

SOLUTION Since position is important and no person can play two positions simultaneously, the number is $_nP_r$ with $n = 15$, $r = 9$:

$$_{15}P_9 = \frac{15!}{9!} = \frac{15 \cdot 14 \cdot \cdots \cdot 10 \cdot 9 \cdot \cdots \cdot 2 \cdot 1}{9 \cdot 8 \cdot \cdots \cdot 2 \cdot 1}$$

$$= 15 \cdot 14 \cdot 13 \cdot 12 \cdot 11 \cdot 10$$

$$= 3{,}603{,}600$$

Combinations

In many types of arrangements, the order is immaterial. For instance, in a bridge hand only the hand of 13 cards is important. The order in which these 13 cards are received does not matter.

> **DEFINITION**
>
> Let S be a collection of n objects or symbols. A collection of r of these objects in which
>
> 1. Order is irrelevant
> 2. Repetitions are not permitted
>
> is called a **combination** from S. The total number of combinations of n objects taken r at a time is denoted $_nC_r$.

The total number of permutations of n objects taken r at a time is $_nP_r$. If no repetitions are allowed but the order of the listing is immaterial, there are $_nC_r$ combinations of these n objects taken r at a time. But the r objects in a given combination can be arranged in $_rP_r = r!$ ways; that is, each combination represents $r!$ permutations. Hence

$$_nC_r \cdot r! = {_nP_r}$$

or

$$_nC_r = \frac{_nP_r}{r!} = \frac{n!}{(n-r)!r!}$$

> **COMBINATION FORMULA**
>
> The total number of combinations of n objects taken r at a time is
>
> $$_nC_r = \frac{n!}{(n-r)!r!}$$

Note that the numbers $_nC_r$ have appeared earlier as the binomial coefficients; see the binomial theorem, Section 11.4.

EXAMPLE 6 How many different bridge hands of 13 cards can be dealt from a 52-card deck?

SOLUTION The total number of hands of 13 cards that can be dealt *in a given order* from a 52-card deck is $_{52}P_{13}$; the total number of distinct hands, ignoring the order in which the individual cards are received, is $_{52}C_{13}$. Once the player picks up her hand, she can rearrange the cards any way she chooses. Thus the same hand could be received in $_{13}P_{13} = 13!$ different ways. Hence each hand represents 13! different permutations or ordered hands. Since there are $_{52}C_{13}$ hands and $_{52}P_{13}$ permutations, we have

$$_{52}C_{13} \cdot 13! = {_{52}P_{13}}$$

or

$$_{52}C_{13} = \frac{_{52}P_{13}}{13!}$$

$$= \frac{52!}{39!13!}$$

$$= \frac{52 \cdot 51 \cdots 40 \cdot [39 \cdots 2 \cdot 1]}{[39 \cdot 38 \cdots 2 \cdot 1][13 \cdot 12 \cdots 2 \cdot 1]}$$

$$= \frac{52 \cdot 51 \cdots 41 \cdot 40}{13 \cdot 12 \cdots 2 \cdot 1}$$

$$\approx (6.35)10^{11} \qquad \text{(Use a calculator)} \quad \blacksquare$$

EXAMPLE 7 In panel tests of consumer products, the technique of paired comparisons has seen a certain amount of success. In this technique each product is paired with each of the other products and a direct comparison of the two products is made. With a sample consisting of 6 different products, how many paired comparisons must be made?

SOLUTION We must find the total number of combinations of 6 objects taken 2 at a time (i.e., in pairs). This is just

$$_6C_2 = \frac{6!}{(6-2)!2!} = \frac{6!}{4!2!} = \frac{6 \cdot 5 \cdot 4 \cdot 3 \cdot 2 \cdot 1}{(4 \cdot 3 \cdot 2 \cdot 1)(2 \cdot 1)} = \frac{6 \cdot 5}{2} = 15 \quad \blacksquare$$

Distinguishable Permutations (Optional)

Still another counting problem arises if certain objects in an arrangement cannot be distinguished from others. This happens, for example, if the objects to be arranged are coins not all of different denominations (or dates or mints).

EXAMPLE 8 Find the total number of ways in which three Kennedy half-dollars and five Anthony dollars can be arranged in a line if coins of like denomination cannot be distinguished from one another.

SOLUTION Temporarily attach labels K_1, K_2, K_3 and A_1, A_2, A_3, A_4, A_5 to the halves and dollars, respectively. Since there are 8 objects, these can be arranged in $_8P_8 = 8!$ different ways. If we remove the labels from the halves, then for any given arrangement of the Anthony dollars, there are $_3P_3 = 3!$ ways of arranging the Kennedy half-dollars, and we will not be able to distinguish one such arrangement from another. Thus there are only $8!/3!$ distinct arrangements after the labels are removed from the half-dollars. If we now remove the labels from the dollars, then for any three positions occupied by the half-dollars, there will be $_5P_5 = 5!$ ways of placing the dollars in the remaining 5 positions. Thus there are only $\frac{8!/3!}{5!}$, or $\frac{8!}{3!5!}$, distinct arrangements of the coins when all labels are removed.

$$\frac{8!}{3!5!} = \frac{5! \cdot 6 \cdot 7 \cdot 8}{3!5!} = \frac{6 \cdot 7 \cdot 8}{2 \cdot 3} = 56$$

The same answer could have been found by viewing this problem as the assignment of 3 coins (Kennedy half-dollars) to 8 positions: $_8C_3 = \dfrac{8!}{3!5!}$. ∎

> **DEFINITION**
>
> Consider a collection of objects, some of which are identical. Two permutations that differ only in the arrangement of identical objects are said to be **indistinguishable** permutations. Otherwise the permutations are said to be **distinguishable.**

Reasoning as in the solution of Example 8, we can find the number of distinguishable permutations of a given collection, as stated next.

> **DISTINGUISHABLE PERMUTATION FORMULA**
>
> For a collection of objects having
>
> n_1 identical objects of a given type
> n_2 identical objects of another type
> $\vdots$
> n_k identical objects of still another type
>
> for a total of $n = n_1 + n_2 + \cdots + n_k$ objects, the number of distinguishable permutations of these objects is
>
> $$\frac{n!}{n_1! n_2! \cdots n_k!}$$

EXAMPLE 9 How many different "words," or letter patterns, can be obtained by rearranging the letters in *MENOMENEE*?

SOLUTION There are 2 *M*'s, 4 *E*'s, 2 *N*'s, and 1 O, for a total of 9 letters. The number of distinguishable permutations is

$$\frac{9!}{2!4!2!1!} = \frac{4! \cdot 5 \cdot 6 \cdot 7 \cdot 8 \cdot 9}{2 \cdot 4! \cdot 2} = 3780$$

∎

Counting problems appear in a wide variety of situations. These can range from those that are merely interesting, such as puzzles and games, to very important practical problems that occur in the design of computer systems and telephone-switching networks. A whole branch of mathematics called "combinatorics" is devoted essentially to the development of counting techniques for increasingly complex situations.

Circular Permutations (Optional)

In this section we have illustrated a variety of counting problems. The techniques discussed here are by no means exhaustive but they are basic in that many more sophisticated counting techniques rely on these principles of permutations and combinations. One topic we have not yet discussed is that of **circular permutations:** arranging objects in a circle rather than a line. An application of the principles presented earlier allows us to handle these also as illustrated in the following example.

EXAMPLE 10 In how many ways can nine colors be arranged on a color wheel?

SOLUTION If we mark a first position on the wheel, there are $_9P_9 = 9!$ such arrangements. But it really does not matter which position was marked as first; only the circular order matters. Since there are 9 positions which could have been marked first, there are only $9!/9 = 8! = 40{,}320$ arrangements.

On the other hand, if the color wheel is transparent so that the colors can be viewed from both sides, then we cannot distinguish one arrangement from that obtained by flipping the wheel over. In this case, there are $8!/2 = 20{,}160$ arrangements. ∎

CALCULATOR COMMENTS Some calculators have a factorial key (!), while others do not. Even with calculators that have a factorial key, it is advisable to cancel as much as possible with large factorials, since the capacity of the calculator or its display can easily be exceeded. For instance, a calculator cannot evaluate $\dfrac{100!}{98!}$ unless it is simplified as

$$\frac{100!}{98!} = \frac{98! \cdot 99 \cdot 100}{98!} = 99 \cdot 100 = 9900$$

Section 11.5 Exercises

Evaluate the following expressions.

1. $_3C_2$
2. $_5C_4$
3. $_5C_5$
4. $_8C_8$
5. $_{12}C_9$
6. $_{10}C_4$
7. $_{20}C_{16}$
8. $_{15}C_{12}$
9. $_7C_1$
10. $_9C_1$
11. $_{50}C_2$
12. $_{60}C_{56}$
13. $_{36}C_{35}$
14. $_{40}C_{39}$
15. $_4P_3$
16. $_6P_4$
17. $_{13}P_9$
18. $_{11}P_5$
19. $_{30}P_{28}$
20. $_{40}P_{36}$
21. $_{15}P_{12}$
22. $_{20}P_{16}$
23. $_{60}P_{56}$
24. $_{50}P_2$

25. **a.** How many 7-digit telephone numbers are possible if 0 is not permitted as a first digit?

b. How many 10-digit telephone numbers (area code and number) are possible if 0 is not permitted as the first digit of either the area code or the number?

26. Compare the possible numbers of 5-digit and nine-digit ZIP codes if 0 is permitted as the first digit.

27. In how many ways could you answer a true-false test consisting of 10 questions?

28. In how many ways could you answer a multiple choice (A, B, C, D, E) test having 10 questions?

29. How many different meals can be planned by using 3 meats, 4 vegetables, 2 breads, 4 fruits,

and 3 drinks if each meal is to contain exactly 1 item from each group?

30. An automobile can be purchased in 3 body types, 8 colors (1 or 2 of which can be applied to any given car), 3 engines, 2 transmissions, and 2 "rear ends." In how many different configurations can this automobile be purchased?

31. In selecting skiing equipment, Laara has narrowed her options down to 3 skis, 2 bindings, 2 poles, and 3 kinds of boots. How many options does she have?

32. a. Using at least 1 and up to 10 different ingredients, how many different combination pizzas can be made?
    b. What if 3 of these ingredients (crust, cheese, sauce) must be used?

33. A builder provides 5 basic house plans with choice of 3 carpets in each of 5 rooms, colonial or contemporary trim, 0 to 5 appliances (2 brands each), and 2 paint colors in each of 5 rooms. How many options does a buyer have?

34. Computers operate in the binary system—switches are either on or off. How many switches are necessary to distinguish
    a. Digits 0 through 9?
    b. Letters *a* through *z*?
    c. Letters and digits?
    d. Both upper- and lower-case letters?
    e. Both upper- and lower-case letters and digits?
    f. Both upper- and lower-case letters, digits and up to 20 punctuation and mathematics symbols?

35. In how many ways can 12 jurors be seated in the jury box?

36. In how many ways can a president, vice-president, secretary, and treasurer be elected
    a. From a membership of 15 people?
    b. From a membership of 7 males, 8 females if the president must be female and the secretary must be male?

37. In how many ways can 6 candidates be arranged on a ballot?

38. In how many ways can 6 candidates for county commission and 4 candidates for city council be arranged on a ballot if the candidates for a given office must be grouped together?

39. In how many ways can 4 flags be arranged on a pole if all flags are of a different color?

40. The presidents and sales managers of 4 companies are to make presentations. In how many ways can these presentations be arranged if
    a. All the presidents must speak before the sales managers?
    b. Each president is followed by his or her sales manager?

41. Jane has received invitations to interview 6 companies for employment. In how many ways can she order these interviews if
    a. No restrictions are placed on the order of the interviews?
    b. The 6 companies are located in 2 cities, 3 in each, and only 1 trip can be made to each city?
    c. The 6 companies are located in 3 cities, 2 in each, and only 1 trip can be made to each city?

42. How many distinct lines can be drawn through 10 points, no 3 of which fall on the same line?

43. How many distinct circles can be drawn through 10 points, no 3 of which lie on the same line? (*Hint:* Any 3 points which do not lie on the same line determine a unique circle.)

44. Semaphore or flag-signaling is based on holding a flag in each hand and then positioning each flag at 1 of 8 positions around a circle as indicated in the sketch. If identical flags are used and the flags must occupy different positions, how many different signals can be sent?

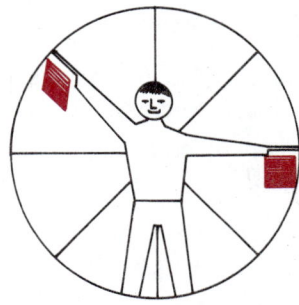

45. In how many ways can 12 players be split into 2 teams of 6?

46. How many different "euchre" hands of 5 cards can be dealt from a 24-card deck?

47. How many different (euchre) hands are there if it is recognized that the declaration of 1 of the 4 suits as "trump" affects the value of a hand and trump is always declared? (*Hint:* See Exercise 46.)

48–49. Answer Exercises 46–47 for the entire deal of 4 players.

50. In how many ways can a jury of 12 people be chosen from a pool of 20 people?

51. In how many ways can an extended jury of 13 people (12 jurors and 1 alternate) be chosen from a pool of 20 people? Assume that the order of selection of the 12 jurors is immaterial but that the 13th juror selected is designated as the alternate.

52. Rework Exercise 51 under the assumption that 13 jurors are chosen and then any 1 of these is later designated as the alternate.

53. On a test of 10 questions, how many ways are there to choose the questions to answer if
  a. Any 2 can be omitted?
  b. Any 2 of the last 5 can be omitted?
  c. Seven problems must be worked including 3 of the last 5?

54. a. How many different numbers can be formed from the digits in 10,632?
  b. How many of these are even numbers?
  c. How many are odd numbers?
  d. How many of the numbers are greater than 5000?
  e. How many are less than 3500?

55. If 10 points are placed on a circle and adjacent points are connected to form a polygon, how many distinct diagonals does this polygon have?

56. a. How many different dominoes ranging from double blank to double 6 are there?
  b. How many different sums can be formed on dominoes?

57. How many different combinations can be obtained from rolling
  a. Three dice (6 sides)?
  b. Three tetrahedra (4 sides, choose the number on the bottom)?
  c. Three dodecahedra (12 sides)?

58. In a family of 6 girls and 5 boys, how many different family picture arrangements are possible if the parents alone occupy the first row, the girls occupy the second row, and the boys occupy the last row?

59. In how many ways can 3 history, 6 math, 4 sociology, and 5 science books be arranged on a shelf if books on the same subject are to be shelved together?

60. Determine the number of ways in which 30 cars can be arranged in a funeral procession, assuming the first 3 cars are specified in a given order, the next group of 5 is identified but no order within this group is required, then 3 groups of 4 are specified where the groups are to follow in a given order but order within each group is arbitrary, and the remaining 10 cars are to follow in any order.

61. How many different 5-card poker hands can be dealt from a 52-card deck if the first 2 cards are dealt face down (order is immaterial) and the last 3 are dealt face up (order affects the betting)?

62. a. In how many ways can 12 members of a ski club ride in 3 distinct cars (4 persons each) if seating within each car is ignored?
  b. What if each owner rides in his or her own car?

63. In how many ways can 16 people be split into
  a. Four bridge "tables" of 4 persons each? Ignore the distinction between tables and the division of tables into teams.
  b. Eight bridge teams of 2 persons each?

64. In how many ways can union- and management-negotiating teams be seated on opposite sides of a table if each team consists of 5 members?

65. In how many ways can a group of 12 horses be subdivided into
  a. Four teams of 3?
  b. One team of 6 and two teams of 3?

66. Answer Exercise 65 if the teams are assigned to distinct jobs.

67–68. Answer Exercises 65–66 if 2 of the horses cannot be made to cooperate on the same team.

69. Three salespeople from the Safer-Alarm Company attend a convention. Twelve presentations are of interest to them but each can attend only 4. In how many ways can this be done assuming

  a. Three presentations are given simultaneously at 8, 10, 12, and 2 respectively?
  b. Each can attend any 4?

70. a. How many different 4-person committees can be formed from a group of 5 men and 3 women?
  b. Answer a if at least 1 committee member must be female.
  c. Answer a if the committee must have at least 1 male and 2 females.
  d. Answer a if the committee must be sexually balanced.
  e. Answer a if Lois must be a member of the committee.

The following exercises involve distinguishable permutations or circular permutations.

71. How many different numbers can be formed from the digits in 1961?

72. a. How many different numbers can be formed from the digits in 355,662?
  b. How many of these are even numbers?
  c. How many are odd numbers?
  d. How many of these numbers are greater than 650,000?
  e. How many are less than 243,450?

73. How many different "words," or letter patterns, can be formed from the letters in each word?
  a. COMMITTEE   b. SUCCESS
  c. BANANA   d. TALLAHASSEE
  e. SCISSORS   f. AUTOMOBILE

74. Rework Exercise 73 under the restriction that the second and fourth letters must be vowels.

75. How many different patterns can be formed by equally spacing the symbols $+, -, \cdot, \times, \div,$ and : around the edge of a disc that is
  a. Opaque?   b. Transparent?

76. In how many ways can 10 keys be arranged
  a. In a key case?
  b. On a circular ring?
  c. On a circular ring attached to a chain?

77. A triominoe is a triangle with a number ranging from zero to five at each of its vertices. How many different triominoes ranging from triple 0 to triple 5 are there?

78. In how many ways can 16 people be assigned seats at 4 tables (4 persons per table) if
  a. Table assignment is important but at each table only the circular arrangement matters, not individual seat assignments?
  b. Only the circular arrangements matter; the table at which a given circular arrangement appears is immaterial?

79. In a miniature version of the football half-time spelling of *OHIO*, 10 people are used to form each *O*, three for the *I* and each side of the *H*, and 1 for the crossbar in the *H*. In how many ways can this be done if
  a. Position is important in all letters?
  b. Position is not important in the *O*'s?
  c. Position is not important in the *O*'s, and the *H* and *I* can be inverted.

80. How many ways are there to arrange the numbers on a die in the shape of a
  a. Cube (6 sides)?
  b. Tetrahedra (4 sides)?
  c. Dodecahedra (12 sides)?

Section 11.6

Probability

Probability is the study of the laws of chance. Probability is used to test the effectiveness of new medications and to test claims of abilities in mental telepathy. Life insurance rates are based on the probability that individuals in certain classifications will live to a certain age. When the weather service tells us that the chance of rain is 10 percent, they are making a statement of probability. In this section we shall generally use the decimal or fractional equivalent of percent. Thus we would say that the chance or probability of rain is 0.1 or $\frac{1}{10}$. Questions of probability are also of interest in games of chance: games whose progress depends on chance events such as the outcome from tossing a coin, rolling dice or a single die, spinning a wheel, or dealing cards.

A fair coin is one that when tossed is equally likely to turn up heads as tails. (In the unlikely event that the coin is tossed for sharpshooting practice, we shall assume that the coin is missed and that it indeed falls to a surface and lands heads up or tails up with equal likelihood.) We say that the probability of turning up heads is $\frac{1}{2}$ and the probability of turning up tails is $\frac{1}{2}$. This means that if the coin is tossed a great many times, heads would turn up about half the time and tails would turn up about half the time. Heads will not necessarily turn up exactly half the time, though. We are concerned with probability or likelihood here, not precision or exactness.

Similarly, a fair die is one for which no number is more likely than another to turn up when the die is tossed or rolled. If the die is rolled a great many times, each number should turn up about $\frac{1}{6}$ of the time. We say that the probability of turning up a 4 on a single roll of the die is $\frac{1}{6}$.

Performing a task or succession of tasks whose results depend on chance will be called an **experiment**. The result of an experiment is called an **outcome**. The collection (or set) of all possible outcomes is called the **sample space** S. An **event** is a collection of one or more possible outcomes; it is a subset of S. Thus in the experiment of rolling a die the possible outcomes are 1, 2, 3, 4, 5, 6; we write $S = \{1, 2, 3, 4, 5, 6\}$. Turning up a 4 is an event; likewise, turning up a *4 or a 6* is an event. These events are denoted $E = \{4\}$ and $F = \{4, 6\}$, respectively. The sample space of the coin tossing experiment is $S = \{H, T\}$.

DEFINITION

Assuming that each possible outcome of an experiment is equally likely, the **probability** P of an event E is the number of ways in which E could occur $N(E)$ divided by the total number of possible outcomes T:

$$P(E) = \frac{N(E)}{T}$$

Needless to say, the counting techniques developed in the preceding section should be useful here. Note that the event could occur at most in every outcome; in this case $N(E) = T$ and $P(E) = 1$. This is called the certain event. At worst E could never occur in which case $N(E) = 0$ and $P(E) = 0$. This is called the impossible event. For example, rolling a 7 with one die is impossible; its probability is 0. On the other hand, we shall certainly turn up one of the numbers 1 to 6 with one die; the probability of rolling a 1, 2, 3, 4, 5, or 6 is 1. Thus

$$0 \leq P(E) \leq 1$$

EXAMPLE 1 Find the probability of rolling a 4 or a 6 with one die.

SOLUTION Think of rolling the die as a game that is won by rolling a 4 or a 6. The event in question is "winning the game." The number of possible outcomes is 6: $S =$

$\{1, 2, 3, 4, 5, 6\}$ and $T = 6$. There are two ways to win the game, by rolling a 4 or a 6: $E = \{4, 6\}$ and $N(E) = 2$. Thus

$$P(E) = \frac{N(E)}{T} = \frac{2}{6} = \frac{1}{3}$$

■

EXAMPLE 2 Find the probability of rolling a sum of eight with two dice.

SOLUTION If the dice are fair, it makes no difference whether they are thrown together or one after the other. Identify one as the first die and the other as the second. This is essential to counting the total number of possible outcomes. For instance, a sum of two can occur in only one way: as 1, 1 on the two dice. But a sum of three can occur in two ways: as 1, 2 or as 2, 1 on the first and second dice, respectively. Thus a sum of three is twice as likely to occur as a sum of two.

There are six ways in which the first die could turn up and six ways for the second die. Thus there are $6 \cdot 6 = 36$ possible outcomes from rolling first one die, then a second: $T = 36$. A sum of eight could be obtained from the combinations (2, 6), (3, 5), (4, 4), (5, 3), (6, 2). Thus for the event "sum = 8," $N(E) = 5$. Finally,

$$P(\text{sum} = 8) = \frac{N(E)}{T} = \frac{5}{36}$$

■

Independent Events

> **DEFINITION**
>
> Experiments are said to be **independent** if the outcome of one does not influence the outcome of any of the others. Events associated with independent experiments are also said to be independent.

Frequently, an event E can be expressed in terms of two or more independent events. For example, the event E of getting heads on each of two tossed coins can be expressed as

E_1:   "getting heads on the first coin"

and

E_2:   "getting heads on the second coin"

$E = (E_1 \text{ and } E_2)$. Clearly, $P(E_1) = \frac{1}{2}$ and $P(E_2) = \frac{1}{2}$. About half the time H will turn up on the first coin. Then for those times when H has turned up on the first coin, H can be expected only half the time on the second coin; $P(HH) = \frac{1}{2} \cdot \frac{1}{2} = \frac{1}{4}$. This is also verified by listing the possible outcomes: $S = \{HH, HT, TH, TT\}$.

> **PROBABILITY FOR INDEPENDENT EVENTS**
>
> For independent events $E_1, \cdots, E_n$
>
> $P(E_1 \text{ and } E_2 \text{ and } \cdots \text{ and } E_n) = P(E_1) \cdot P(E_2) \cdots \cdot P(E_n)$

EXAMPLE 3 Assume a card is drawn at random from a 52-card deck and returned to the deck. Then the deck is shuffled and again a card is drawn at random. Find the probability of drawing two hearts.

SOLUTION Getting a heart on the first and second draws form independent events E and F, since in each case we are drawing from the complete 52-card deck.

Since hearts comprise 13 of the 52 cards, $P(E) = P(F) = \frac{13}{52} = \frac{1}{4}$. Thus

$$P(\text{two hearts}) = P(E \text{ and } F) = P(E) \cdot P(F)$$
$$= \frac{1}{4} \cdot \frac{1}{4} = \frac{1}{16}$$

∎

P(E or F)

In Example 3, if we had asked for the probability of *drawing two hearts or two spades*, we could have expressed this event as E or F, where

$E:$ drawing two hearts
$F:$ drawing two spades

To investigate probabilities of this type, we illustrate the sample spaces (sets) using Venn diagrams. See Figure 11.8 for an illustration of "E and F" and of "E or F".

Figure 11.8

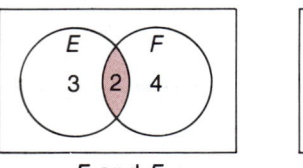

E and F $\qquad$ E or F

If the event "E and F" occurs, then certainly E occurs. Thus if we count the number of ways in which E occurs $N(E)$, we shall have included the number of ways in which "E and F" occurs: $N(E \text{ and } F)$. Similarly, $N(F)$ includes $N(E \text{ and } F)$. Thus $N(E) + N(F)$ includes $N(E \text{ and } F)$ twice; consequently,

$$N(E \text{ or } F) = N(E) + N(F) - N(E \text{ and } F) \qquad (1)$$

where the last term is subtracted in order to prevent those occurrences from being counted twice. This relationship is illustrated in Figure 11.8.

For example, in Figure 11.8, $N(E \text{ and } F) = 2$, $N(E) = 5 (= 3 + 2)$, and $N(F) = 6 (= 4 + 2)$. Then

$$N(E \text{ or } F) = 3 + 2 + 4 = (3 + 2) + (4 + 2) - 2$$
$$= N(E) + N(F) - N(E \text{ and } F)$$

Dividing (1) by the total number of possible outcomes T, we have

$$\boxed{P(E \text{ or } F) = P(E) + P(F) - P(E \text{ and } F)}$$

EXAMPLE 4 Three cards are drawn at random from a 52-card deck; the cards are not returned to the deck before drawing the next card. What is the probability of drawing *at least two aces* **or** *at least two clubs*?

SOLUTION We may assume that the cards are drawn simultaneously so that the order of the draw is unimportant.

T: Since there are 52 cards, the total number of distinct 3-card hands is

$$T = {}_{52}C_3$$

E: *Drawing at least two aces.*
We are interested in drawing either two or three aces. Since there are four aces, there are ${}_4C_3$ ways to select (or draw) exactly three of them. There are ${}_4C_2$ ways to draw exactly two aces and for each pair of aces there are 48 possible nonace third cards. Thus there are $48 \cdot {}_4C_2$ two-ace hands and as a result there are

$$N(E) = {}_4C_3 + 48 \cdot {}_4C_2$$

distinct hands having two or three aces:

$$P(E) = \frac{N(E)}{T} = \frac{{}_4C_3 + 48 \cdot {}_4C_2}{{}_{52}C_3}$$

F: *Drawing at least two clubs.*
Since there are 13 clubs, we obtain in a similar fashion

$$P(F) = \frac{{}_{13}C_3 + 39 \cdot {}_{13}C_2}{{}_{52}C_3}$$

E and F: *Drawing two aces **and** two clubs.*
Both of the events E and F occur when the cards drawn are A♣, another ace, and another club. There are 3 choices for the other ace and 12 choices for the other club. Thus

$$N(E \text{ and } F) = 3 \cdot 12 = 36$$

and

$$P(E \text{ and } F) = \frac{36}{{}_{52}C_3}$$

Finally

$$P(2 \text{ clubs or } 2 \text{ aces}) = P(E \text{ or } F) = P(E) + P(F) - P(E \text{ and } F)$$

$$= \frac{({}_4C_3 + 48 \cdot {}_4C_2) + ({}_{13}C_3 + 39 \cdot {}_{13}C_2) - 36}{{}_{52}C_3}$$

$$= \frac{\frac{4!}{1!3!} + 48 \cdot \frac{4!}{2!2!} + \frac{13!}{3!10!} + 39 \cdot \frac{13!}{2!11!} - 36}{\frac{52!}{3!49!}}$$

$$= \frac{4 + 48 \cdot 3 \cdot 2 + 13 \cdot 2 \cdot 11 + 39 \cdot 13 \cdot 6 - 36}{26 \cdot 17 \cdot 50}$$

$$= \frac{3584}{22100} = \frac{896}{5525}$$

$$\approx 0.162$$

Mutually Exclusive Events

DEFINITION

Events $E_1, E_2, \cdots, E_n$ are said to be **mutually exclusive** if no two of them can occur simultaneously; that is, if

$$P(E_i \text{ and } E_j) = 0 \quad \text{when } i \neq j$$

Mutually exclusive events are illustrated in Figure 11.9; E and F cannot occur simultaneously. For two events E and F which are mutually exclusive, we have $P(E \text{ and } F) = 0$ and hence

$$P(E \text{ or } F) = P(E) + P(F) - \overbrace{P(E \text{ and } F)}^{= 0}$$
$$= P(E) + P(F)$$

Figure 11.9

This property extends to more than two mutually exclusive events, as stated here.

PROBABILITY FOR MUTUALLY EXCLUSIVE EVENTS

For mutually exclusive events $E_1, E_2, \cdots, E_n$

a. $P(E_i \text{ and } E_j) = 0$ when $i \neq j$

b. $P(E_1 \text{ or } E_2 \text{ or } \cdots \text{ or } E_n) = P(E_1) + P(E_2) + \cdots + P(E_n)$

EXAMPLE 5 Assume a card is drawn at random from a 52-card deck and returned to the deck. Then the deck is shuffled and again a card is drawn at random. Find the probability of drawing two cards of the same suit.

SOLUTION Two cards of the same suit are obtained by drawing two spades or two clubs or two diamonds or two hearts. Call these events E_s, E_c, E_d, E_h, respectively. In Example 3, we found $P(E_h) = \frac{1}{16}$. Since there are 13 cards of *each* suit, we find

$$P(E_s) = P(E_c) = P(E_d) = P(E_h) = \frac{1}{16}$$

We cannot draw two hearts *and* two spades since only two cards can be drawn, not four. In fact, no two of these events can occur simultaneously. These are mutually exclusive events. Thus

$$P(\text{two cards of the same suit}) = P(E_s \text{ or } E_c \text{ or } E_d \text{ or } E_h)$$
$$= P(E_s) + P(E_c) + P(E_d) + P(E_h)$$
$$= \frac{1}{16} + \frac{1}{16} + \frac{1}{16} + \frac{1}{16} = \frac{1}{4} \quad \blacksquare$$

Complementary Events

In many situations the computation of probabilities can be very involved but the probability of the event not occurring (called the **complementary event**) may be easier to find. This situation will be illustrated in the following example. Before proceeding with the example, observe that if

T = total number of possible outcomes

and

$N(E)$ = number of ways in which the event E can occur

then the total number of ways in which the event would not occur ($\sim E$) is given by

$$N(\sim E) = T - N(E)$$

Dividing by T, we obtain

$$P(\sim E) = \frac{N(\sim E)}{T} = \frac{T - N(E)}{T} = 1 - \frac{N(E)}{T} = 1 - P(E)$$

$$\boxed{P(\sim E) = 1 - P(E)}$$

EXAMPLE 6

A couple plans to have four children. What is the probability that the family will include at least one boy and at least one girl?

SOLUTION

We shall assume that the chances of a male or a female child are equal. Thus for a given birth

$$P(\text{girl}) = P(\text{boy}) = \frac{1}{2}$$

There are several ways to approach this problem. We could list all possible four-child configurations (e.g., BGGB) and count $N(E)$ and T. Or we could calculate

P(exactly one girl)
P(exactly two girls and two boys)
P(exactly one boy)

and then add these since these events are mutually exclusive. This approach also involves counting or a careful analysis to get the correct probabilities. The simplest approach is to consider the complementary event: all boys or all girls. Since successive births are independent events,

$$P(GGGG) = \frac{1}{2} \cdot \frac{1}{2} \cdot \frac{1}{2} \cdot \frac{1}{2} = \frac{1}{16}$$

Similarly,

$$P(BBBB) = \frac{1}{16}$$

The events "all boys" and "all girls" are mutually exclusive. Thus

$$P(\text{all girls or all boys}) = \frac{1}{16} + \frac{1}{16} = \frac{1}{8}$$

In other words, about one of every eight four-child families are composed of a single gender. Thus

$$P(\text{mixed gender}) = 1 - P(\text{single gender})$$
$$= 1 - \frac{1}{8} = \frac{7}{8}$$

Note that we considered both independent and mutually exclusive events in the preceding example. It is important not to confuse the two concepts, since

1. Probabilities are multiplied for independent events [to find $P(E \text{ and } F)$].
2. Probabilities are added for mutually exclusive events [to find $P(E \text{ or } F)$].

Also note that two events can be neither independent nor mutually exclusive, as illustrated in the following example.

EXAMPLE 7 Draw a single card from a standard 52-card deck. *Do not replace the card.* Then draw another card. Find the probability that the second card is a spade.

SOLUTION Consider the events

E: "the first card is a spade"
F: "the second card is a spade"

Certainly, $P(F)$ depends on whether or not the first card was a spade; E and F are not independent. Since we could draw a spade each time, these events are likewise not mutually exclusive.

Since there are 13 spades in the 52-card deck, we have

$$P(E) = \frac{13}{52} = \frac{1}{4}$$

and

$$P(\sim E) = 1 - P(E) = \frac{3}{4}$$

If we draw a spade on the first card, then there are 12 spades left in the remaining 51 cards. If we do not draw a spade on the first card, then there are 13 spades left in the remaining 51 cards:

$\frac{1}{4}$ of the time we draw a spade on the first card and then $\frac{12}{51}$ of the time we get another spade on the second card

$\frac{3}{4}$ of the time we do not draw a spade on the first card and then $\frac{13}{51}$ of the time we get a spade on the second card

Thus

$P(\text{second card is a spade}) = P(F)$

$$= \frac{1}{4} \cdot \frac{12}{51} + \frac{3}{4} \cdot \frac{13}{51} = \frac{51}{4 \cdot 51}$$

$$= \frac{1}{4} \qquad \blacksquare$$

Odds

With sporting events, odds are usually given to indicate which team is favored to win and how strongly it is favored. For instance, to say "the odds are 3 to 2 (or $\frac{3}{2}$) that West will beat East" means that if West and East played five games, West could be expected to win three games and to lose two games. Thus for

West, $P(\text{win}) = \frac{3}{5}$ and $P(\text{lose}) = \frac{2}{5}$. Note that

$$\text{Odds for West to beat East} = \frac{3}{2} = \frac{\frac{3}{5}}{\frac{2}{5}} = \frac{P(\text{win})}{P(\text{lose})}$$

DEFINITION

The **odds** of an event E occurring are given by

$$\text{odds}(E) = \frac{P(E)}{P(\sim E)} = \frac{P(E)}{1 - P(E)}$$

where $P(\sim E)$ is the probability that the event E does not occur. Odds in favor of or against an event E can also be expressed as

$$\text{odds in favor} = \frac{\text{probability in favor}}{\text{probability against}}$$

or

$$\text{odds against} = \frac{\text{probability against}}{\text{probability in favor}}$$

Since

$$\frac{P(E)}{P(\sim E)} = \frac{N(E)/T}{N(\sim E)/T} = \frac{N(E)}{N(\sim E)}$$

we can write

$$\text{odds}(E) = \frac{N(E)}{N(\sim E)}$$

Historical Perspective

BLAISE PASCAL
(1623–1662)

The foundations for the theory of probability were developed by P. Fermat and Blaise Pascal in their correspondence regarding a gambling problem. Pascal (1623–1662) was plagued with ill health and pain most of his life. These problems caused him to become a religious neurotic. Several times his energies were abruptly converted from mathematics to religious contemplation, or vice-versa. Each time the change was based on some sort or "revelation" or "divine inspiration." Pascal is credited with the invention of the first computing machine when he was about 18 or 19 years old. He, however, did not discover Pascal's triangle, which is known to have been used by the Chinese as early as 1303 A.D. The triangle is named after Pascal because he discovered many of its properties and developed many of its applications. (Yes, it has properties and applications other than those relating to the binomial coefficients.) He is also credited with one of the earliest valid uses of mathematical induction in his published works regarding the triangle.

EXAMPLE 8 In a poker game with all cards dealt face down, you hold four clubs. What are the *odds against* completing a "flush" by drawing a club as your fifth card?

SOLUTION With all cards dealt face down (and no previous betting), the other players can be ignored. The problem is to deal you another card after you have already been dealt four clubs. This leaves 48 cards in the deck; only 9 of these are clubs, since you already hold 4 of the 13 clubs. Thus the remaining deck consists of

9 clubs

39 nonclubs

The odds against completing the flush are $\frac{39}{9} = \frac{13}{3}$, or 13 to 3. ■

Section 11.6 Exercises

1. In a shell game with a coin placed under 1 of 3 shells, what are your chances of guessing the correct shell?

2. Two dice are rolled. Find the probabilities of the following events.
   a. Exactly one 1
   b. Sum = 7
   c. Sum is odd
   d. Sum ≤ 9
   e. Sum is odd and ≤ 9
   f. Sum is odd or ≤ 9
   g. Sum ≥ 5
   h. 5 ≤ sum ≤ 9

3. In a set of dominoes containing 1 of each combination from double blank to double 6, find the probability of drawing each of the following.
   a. Double 3
   b. Any double
   c. At least one blank
   d. Exactly one blank
   e. A sum of 5
   f. A sum ≥ 5

4. Find the probability that the sum of 2 distinct numbers chosen at random from the digits 1 to 9 is
   a. Odd
   b. Even
   c. ≥ 9
   d. ≤ 5

5. With a 4-sided die (a tetrahedron), we count the side that is down. With 2 4-sided dice, find the probability of obtaining a sum
   a. = 4   b. ≥ 4

6. The dodecahedron (12-sided solid) can also be used as a gambling die. With 2 such dice, find the probability of obtaining a sum
   a. = 10   b. ≥ 10

7. What are the probabilities of obtaining a double from rolling 2 dice if the dice are
   a. Cubes?   b. Tetrahedra?   c. Dodecahedra?

8. In Exercise 7, find the probabilities of getting triples from 3 dice.

9. In a game called "Bye-Bye-Charlie," a red, a white, and a green marble are placed in a container. Each player draws one marble from the container. If the marble is red, it is "Bye-Bye-Charlie"; the player is out of the game. Otherwise the marble is replaced and the procedure is repeated. The game continues until only 1 player remains. For a given player, what is the probability of being in the game to play the fifth round?

10. Five white marbles and 5 red marbles are placed in an urn. A marble is drawn out and returned and then the procedure is repeated. What is the probability of drawing
    a. Two white marbles?   b. One of each color?

11. Answer Exercise 10 if the first marble is not returned to the urn.

12. A coin is tossed 5 times. Find the probabilities of the following events.
    a. Exactly 3 successive heads (and only 3 heads)
    b. At least 3 successive heads (and all heads in succession)
    c. At least 3 heads or 3 tails

13. Five people choose a digit in the range 1 to 9 at random. Find the probability that at least 2 people choose the same digit.

14. Find the probability that of 30 people, at least 2 have the same birthday. (*Hint:* Compute the complementary probability.)

15. In a telepathy experiment, 3 people are each to select one of 3 symbols that are "transmitted" by the subject. What is the probability that at least 2 could select the correct symbol simply by guessing?

16. In a family of 4 children, what is the probability of
    a. Exactly 2 boys and 2 girls?
    b. Exactly 3 children of the same sex?
    c. All children of the same sex?

17. In a family of 5 girls, what is the probability that the sixth child will be a boy?

18. A single card is drawn from a standard 52-card deck. What is the probability of drawing
    a. An ace?
    b. A diamond?
    c. The ace of diamonds?
    d. An ace or a diamond?
    e. An odd number?
    f. A face card?
    g. A black card?
    h. The ace of diamonds or a heart?
    i. The ace of diamonds or a face card?

19. Two cards are drawn (and not replaced) from a standard 52-card deck. Find the probability of
    a. At least 1 ace
    b. Two diamonds
    c. The ace of diamonds
    d. An ace or a diamond
    e. At least 1 odd number
    f. Two face cards
    g. At least 1 black card
    h. The ace of diamonds or a heart
    i. The ace of diamonds and a face card

20. In the game of Tripoley, the 10, J, Q, K and ace of hearts are each potential payoff cards. With four players each receiving 13 cards of a 52-card deck, what is the probability of a given player being dealt 1 or more of these payoff cards?

21. In Tripoley, the pair Q, K of hearts and any triple consisting of 8, 9, 10 in 1 suit are potential bonus payoffs. With 4 players each receiving 13 cards of a 52-card deck, what is the probability of a given player being dealt
    a. Q, K of hearts?
    b. 8, 9, 10 of clubs?
    c. Q, K of hearts and 8, 9, 10 of clubs?
    d. Q, K of hearts or 8, 9, 10 of clubs?

22. Answer Exercise 21 (a to c) for "at least 1 player" rather than "a given player."

23. What is the probability of exactly 1 ace in a
    a. 5-card poker hand?
    b. 13-card bridge hand?

In the game of pinochle, a deck of 48 cards is used that consists of two copies of each card in the range 9-Ace. Twelve cards are dealt to each of 4 players.

24. What is the probability that a given pinochle hand contains no aces and no face cards (only 9's and 10's)?

25. What is the probability that at least 1 of the 4 pinochle hands contains no aces and no face cards (only 9's and 10's)? (See Exercise 24.)

26. An important combination sought in pinochle is "double pinochle": J♦, J♦, Q♠, Q♠. What is the probability of double pinochle in a given hand?

27. Find the probability of double pinochle if pinochle partners are permitted to pool their hands. (*Hint:* See Exercise 26.)

28. If 10 people including Al and Bert are lined up at random, find the probability that
    a. Al precedes Bert

b. Bert precedes Al

c. Al and Bert are next to one another

d. Al and Bert are next to one another if the ends of the line are joined to form a circle.

29. What is the probability of guessing 10 correct answers on a 10-question true-false test?

30. Of 20,000 patients treated for a given disease with a certain drug, 16,000 recovered

   a. If you contract this disease, what is the probability that you could be cured with this medication?

   b. What are your odds for recovery using this medication?

31. Triominoes are triangles with a number ranging from 0 to 5 at each vertex. In a complete set ranging from triple 0 to triple 5, find the probabilities of drawing each of the following.

   a. A double

   b. A triple

   c. At least one 3

   d. A sum = 12

   For the purposes of this exercise, assume that the triomino can not be turned over; that is,

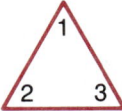 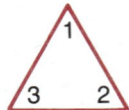

are distinct triominoes. (See Exercise 77, Section 11.5.)

32. If the probability of rain is 10%, what are the odds against rain?

33. Find the odds for rolling

   a. 5 with 1 die   b. Sum = 5 with 2 dice

34. If the odds for rain are 1 to 3 and the odds for snow are 1 to 5 and we assume that it will not both rain and snow, what are the odds for rain or snow?

35. If one card is drawn from a 52-card deck, what are the odds of drawing a face card?

36. Roulette wheels are numbered 00, 0, and 1–36 and pay $36 to the winner for each $1 bet. You cannot bet on 00 or 0 as these are reserved for the house.

   a. What are your odds for winning?

   b. What are the house's odd for keeping all bets?

37. If the odds for Speedy to win are 5 to 4, what is his probability of winning?

Section 11.7   Chapter Review

Terms and Concepts

Sequences or Progressions $a_1, a_2, \ldots$

Summation $\sum_{j=1}^{n} a_j = a_1 + \cdots + a_n$

Factorial $n! = 1 \cdot 2 \cdot \cdots \cdot n$

Permutations   Rearrangements in which order is important and repetitions are not permitted.

Combinations   Same as permutations, except that order is irrelevant.

Indistinguishable Permutations   Permutations that differ only in the arrangement of identical objects.

Probability

$$P(E) = \frac{\text{number of ways in which } E \text{ could occur}}{\text{total number of possible outcomes}}$$

Independent Events   No one of the events influences the outcome of any of the others.

Mutually Exclusive Events   No two of the events can occur simultaneously.

Rules and Formulas

Principle of Mathematical Induction   If 1. S_1 is true and
  2. The truth of S_k implies the truth of S_{k+1} for every k then S_n is true for every positive integer n.

Progressions

1. Arithmetic

$$a_n = a_{n-1} + d = a_1 + (n-1)d$$

$$a_1 + \cdots + a_n = \frac{n}{2}[a_1 + a_n]$$

2. Geometric

$$a_{n+1} = r \cdot a_n = ar^n$$

$$\sum_{j=0}^{n-1} ar^j = a\left(\frac{1-r^n}{1-r}\right), \quad r \neq 1$$

Geometric Series

$$\sum_{j=0}^{\infty} ar^j = \frac{a}{1-r} \quad \text{when } |r| < 1$$

Binomial Expansions

1. Binomial Theorem

$$(a+b)^n = \sum_{l=0}^{n} \frac{n!}{l!(n-l)!} a^{n-l} b^l = \sum_{l=0}^{n} {}_nC_l a^{n-l} b^l$$

2. Pascal's Triangle

$$(a+b)^0 = 1$$
$$(a+b)^1 = a + b$$
$$(a+b)^2 = a^2 + 2ab + b^2$$
$$(a+b)^3 = a^3 + 3a^2b + 3ab^2 + b^3$$
$$(a+b)^4$$
$$(a+b)^5$$
$$(a+b)^6$$
$$(a+b)^7$$

```
                1
              1   1
            1   2   1
          1   3   3   1
        1   4   6   4   1
      1   5  10  10   5   1
    1   6  15  20  15   6   1
  1   7  21  35  35  21   7   1
```

Permutations and Combinations

1. **Fundamental Counting Principle**

If N_k represents the number of possible outcomes for the kth event (after the first $k-1$ events have taken place), then the total number of possible outcomes from the sequence of n events is $N_1 \cdot N_2 \cdots N_n$

2. **Permutations**

The total number of permutations of n objects taken r at a time is

$${}_nP_r = n(n-1)\cdots(n-r+1) = \frac{n!}{(n-r)!}$$

3. **Combinations**

The total number of combinations of n objects taken r at a time is ${}_nC_r = \dfrac{n!}{(n-r)!r!}$

4. **Distinguishable Permutations**

If there are n_j identical objects of type j, the total number of distinguishable permutations on n objects is

$$\frac{n!}{n_1! n_2! \cdots n_k!}$$

Probability

1. $P(E_1 \text{ or } E_2) = P(E_1) + P(E_2) - P(E_1 \text{ and } E_2)$
2. $P(\sim E) = 1 - P(E)$
3. Independent Events $\quad P(E_1 \text{ and } \cdots \text{ and } E_n) = P(E_1) \cdot P(E_2) \cdots P(E_n)$
4. Mutually Exclusive Events $\quad P(E_1 \text{ or } \cdots \text{ or } E_n) = P(E_1) + P(E_2) + \cdots + P(E_n)$
5. Odds $(E) = \dfrac{P(E)}{P(\sim E)} = \dfrac{P(E)}{1 - P(E)}$

Section 11.8 Supplementary Exercises

Use mathematical induction to establish the following.

1. $2 + 5 + 10 + \cdots + (n^2 + 1) = \dfrac{n(2n^2 + 3n + 7)}{6}$

2. $1^2 + 3^2 + 5^2 + \cdots + (2n - 1)^2 = \dfrac{n(2n - 1)(2n + 1)}{3}$

3. $1^3 + 3^3 + 5^3 + \cdots + (2n - 1)^3 = n^2(2n^2 - 1)$

4. $3 + 9 + \cdots + 3^n = \dfrac{3(3^n - 1)}{2}$

5. $1 \cdot 3 + 2 \cdot 4 + \cdots + n(n + 2) = \dfrac{n(n + 1)(2n + 7)}{6}$

6. $2 \cdot 5 + 3 \cdot 6 + \cdots + (n + 1)(n + 4) = \dfrac{n(n + 4)(n + 5)}{3}$

7. $1 \cdot 4 + 2 \cdot 9 + \cdots + n(n + 1)^2 = \dfrac{n(n + 1)(n + 2)(3n + 5)}{12}$

8. $\dfrac{1}{1 \cdot 3} + \dfrac{1}{3 \cdot 5} + \cdots + \dfrac{1}{(2n - 1)(2n + 1)} = \dfrac{n}{2n + 1}$

9. $\dfrac{1}{2 \cdot 5} + \dfrac{1}{5 \cdot 8} + \cdots + \dfrac{1}{(3n - 1)(3n + 2)} = \dfrac{n}{2(3n + 2)}$

10. $8^n - 1$ is divisible by 7.

11. $4^{2n} - 1$ is divisible by 5.

12. $n(n^2 + 2)$ is divisible by 3.

Write out the first five terms and the twelfth term of the sequence whose nth term is given.

13. $n - 5$
14. $\dfrac{3n + 1}{2}$
15. $n^2 + 1$
16. $1 + (-2)^n$
17. $2n^2 - 1$

Write out the first five terms and the ninth term of the sequences whose nth term is given.

18. $\dfrac{2n}{3}$
19. $\dfrac{n}{n + 1}$
20. $\dfrac{n(n - 1)}{2}$
21. $\dfrac{3}{2n - 1}$
22. $\dfrac{n}{n^2 + n}$

Write out the first five terms of the sequences defined here.

23. $a_1 = 3,\ a_n = 2a_{n-1} + 1$
24. $a_1 = 2,\ a_n = 3a_{n-1} - 2$
25. $a_1 = -\sqrt{2},\ a_n = (a_{n-1})^2(-1)^n$
26. $a_1 = 0,\ a_n = (a_{n-1} + 2)^2$

Write out the first five terms and the tenth term of the arithmetic progression with the following data.

27. $a_1 = 4,\ d = 1$
28. $a_3 = 5,\ d = 4$
29. $a_4 = 16,\ d = 4$
30. $a_{20} = 100,\ d = -5$

Find the requested term of an arithmetic progression given the following information.

31. $a_2 = 7,\ a_6 = 23$, find a_7
32. $a_{10} = 32,\ a_{42} = 128$, find a_{167}
33. $a_8 = 0,\ a_{20} = -24$, find a_3
34. $a_{20} = 14,\ a_{38} = 86$, find a_7

Insert the requested number of arithmetic means between the following numbers.

35. 3 means between 3 and 19
36. 2 means between 67 and 76
37. 3 means between 42 and 58
38. 4 means between -6 and 14

Write out the first four terms and the eighth term of the geometric progression with the following data.

39. $a_1 = 1,\ r = 5$
40. $a_1 = 486,\ r = \dfrac{1}{3}$
41. $a_4 = 4,\ r = \sqrt{2}$
42. $a_6 = 250,\ r = 5$

Find the requested term of a geometric progression given the following information.

43. $a_3 = 18,\ a_6 = 486$, find a_8
44. $a_3 = -21,\ a_6 = 63\sqrt{3}$, find a_2
45. $a_6 = 512,\ a_{11} = 16$, find a_3
46. $a_5 = 75,\ a_9 = 1875$, find a_3

Insert the requested number of geometric means between the following numbers.

47. 2 terms between 12 and 96

48. 3 terms between 2 and 162
49. 5 terms between 67 and 536
50. 2 terms between 62 and 496

Write the following in Σ form.

51. $2 + 4 + 6 + 8$
52. $\dfrac{1}{2} + \dfrac{1}{4} + \dfrac{1}{8} + \dfrac{1}{16} + \dfrac{1}{32}$
53. $3 + 7 + 11 + 15 + 19 + 23$
54. $3 - 7 + 11 - 15 + 19 - 23$
55. $2!3 + 3!6 + 4!12 + 5!24 + 6!48 + 7!96$

Find the sum of the first n terms of the arithmetic progressions that satisfy the following.

56. $a_1 = 6, d = -3, n = 8$
57. $a_1 = 20, d = 8, n = 16$

Find the sum of the first n terms of the geometric progressions that satisfy the following.

58. $a = 3, r = -2, n = 6$
59. $a_1 = 4, r = 3, n = 5$

Find the sum of the given geometric series.

60. $3 - \dfrac{3}{2} + \dfrac{3}{4} - \dfrac{3}{8} + \cdots$
61. $5 + \dfrac{5}{4} + \dfrac{5}{16} + \dfrac{5}{64} + \cdots$
62. $\sum\limits_{j=1}^{\infty} 5\left(\dfrac{2}{3}\right)^j$
63. $\sum\limits_{j=1}^{\infty} 3\left(-\dfrac{2}{3}\right)^j$

Expand and evaluate the following.

64. $\sum\limits_{j=1}^{4} (2j - 1)$
65. $\sum\limits_{n=2}^{5} (n + 2)(n - 1)$
66. $\sum\limits_{l=3}^{6} (l + 2)l$
67. $\sum\limits_{m=1}^{3} (2m - 1)(2m + 1)$
68. $\sum\limits_{j=2}^{6} (-1)^j(j + 3)$
69. $\sum\limits_{k=3}^{10} (3k + 1)$
70. $\sum\limits_{j=1}^{16} (2 + j)$
71. $\sum\limits_{n=2}^{12} (n - 2)(n - 3)$
72. $\sum\limits_{n=2}^{10} 3 \cdot 2^n$
73. $\sum\limits_{n=0}^{10} (5 \cdot 2^n + 6n + 2)$

Evaluate the following factorial expressions.

74. $5!$
75. $\dfrac{11!}{9!}$
76. $\dfrac{5!}{4!3!}$
77. $\dfrac{16!5!}{13!}$

Use Pascal's triangle to expand the following.

78. $(a - b)^3$
79. $(a + 2)^9$
80. $(x + 3y)^7$
81. $(3x + 2y^2)^5$
82. $(\sqrt{a} - \sqrt{b})^6$
83. $\left(2\sqrt{x} - \dfrac{1}{x}\right)^6$
84. $\left(2a - \dfrac{1}{b^3}\right)^8$

Use the binomial theorem to find the indicated term of the following.

85. $(x - 2)^{10}$, fourth term
86. $(t - 2)^{20}$, sixth term
87. $(a + b)^{15}$, seventh term
88. $(2u + v)^{12}$, eleventh term
89. $(r^2 - s)^{10}$, sixth term

Use the binomial theorem to evaluate the following.

90. $(102)^4$
91. $(1.02)^3$
92. $(0.98)^6$
93. $(103)^4$

94. How many 9-digit Social Security numbers are possible if 0 is not permitted as a first digit?

95. How many pairs of shoes are required to stock a complete inventory of 6 styles in sizes 6 to 12 including half-sizes and widths AA, A, B, C, D, E, EE, EEE?

96. In selecting backpacking equipment, Cindy has narrowed her options down to 3 kinds of hiking boots, 4 sleeping bags, 3 tents, 3 backpacks, and 2 brands of cooking equipment. How many options does she have?

97. A rabbi, priest, minister, Muslim, agnostic, and atheist are scheduled to speak on a given evening. In how many ways can their talks be ordered?

98. For a race of 16 dogs, in how many ways can the first 3 places be won?

99. How many different "sheephead" hands of 6 cards can be dealt from a 24-card deck?

100. Answer Exercise 99 for the entire deal of 4 players.

101. A student is required to take core courses but must select 6 of 10 electives. In how many ways can this be done if only the list (and not the order) of the 6 courses is important.

102. An intrastate transit system services 7 cities. How many different kinds of tickets must it have printed including both one-way and round-trip tickets?

103. How many different ways are there to form a 4-person committee from a group of 5 men and 3 women if
    a. The committee members are titled so that order cannot be ignored.
    b. In addition to a, the second position must be filled by a female?
    c. In addition to a and b, the last position must be filled by a male?
    d. In addition to a, b, and c, the committee must be sexually balanced?
    e. In addition to a, b, c, and d, Lois must be a member of the committee?

104. How many different words can be formed from the letters in
    a. DAYTON   b. ZEBRA
    c. BANDANA d. MISSISSIPPI

105. Rework Exercise 104 under the restriction that the second and fourth letters must be vowels.

106. Two dice are rolled. Find the probabilities of the following events.
    a. Sum = 7 or 11
    b. Sum is even
    c. Sum is a multiple of 3
    d. Sum is even and a multiple of 3
    e. Sum is even or a multiple of 3

107. Two players each select a number from 1 to 50 at random. Find the probability that the sum is
    a. $=25$ b. ≤ 10 c. ≥ 50

108. Find the probabilities of obtaining triples when rolling 4 dice and the dice are
    a. Cubes   b. Tetrahedra   c. Dodecahedra

109. Answer Exercise 108 for quadruples.

110. Three white marbles and 5 red marbles are placed in an urn. A marble is drawn out and returned and then the procedure is repeated. What is the probability of drawing
    a. Two white marbles?
    b. Two red marbles?
    c. One of each color?

111. Answer Exercise 110 if the first marble is not returned.

112. A coin is tossed 4 times. Find the probability of
    a. Exactly 2 heads
    b. At least 2 heads
    c. The first toss is heads and at least 2 heads occur.

113. Five people choose a number 1 to 50 at random. Find the probability that at least 2 people choose the same number.

114. Of the first 40 U.S. presidents, what is the probability that at least 2 will share
    a. The same birthday?
    b. The same day of death?

115. If 5 math books, 4 history books, and 3 science books are shelved together in random order, what is the probability that the books are grouped according to subject matter? (See Exercise 59, Section 11.5.)

116. A coin is tossed 5 times. Find the probabilities of the following events.
    a. Exactly 3 heads
    b. At least 3 heads
    c. The first toss yields heads and at least 3 heads occur.
    d. Exactly 3 heads or 3 tails (See Exercise 12, Section 11.6.)

117. Two cards are drawn (and not replaced) from a standard 52-card deck. Find the probability of
    a. Two aces
    b. At least 1 diamond
    c. Two odd numbers
    d. At least 1 face card
    e. Two black cards
    f. The ace of diamonds and a heart
    g. The ace of diamonds or a face card

118. If a Tripoley player receives 6 cards, what is the probability of his being dealt 1 or more payoff cards? (See Exercise 20, Section 11.6.)

119. With 8 players in Tripoley receiving 6 cards each, what is the probability of a given player being dealt
    a. Q, K of hearts?
    b. 8, 9, 10 of diamonds?
    c. Q, K of hearts and 8, 9, 10 of diamonds?
    d. Q, K of hearts or 8, 9, 10 of diamonds?

120. Answer Exercise 119 (a to c) for "at least 1 player" rather than for "a given player."

121. What is the probability that 2 given bridge players are each dealt no aces and no face cards?

122. What is the probability that at least 1 of the 4 bridge hands contains no aces and no face cards?

123. If 2 pinochle players pool their hands, what is the probability that they corner all 12 cards in at least 1 suit? (See note preceding Exercise 24, Section 11.6.)

124. Of 1500 walnut seedlings planted on a tree farm, 1325 survived the first year.
    a. What is the probability of an individual plant surviving?
    b. What are its odds for survival?

125. Two red, 2 blue, and 1 white button are lined up at random.
    a. What is the probability that colors are grouped?
    b. Answer a for a circular arrangement.

126. Four men and 4 women are lined up at random.
    a. What is the probability that the line alternates sexes?
    b. Answer a if the people are arranged in a circle.

127. In a complete set of triominoes ranging from triple 0 to triple 5 in which circular order is ignored, (i.e., the triominoes pictured in Exercise 31, Section 11.6, are *not* distinct) find the probability of drawing
    a. A double
    b. A triple
    c. At least one 3
    d. A sum = 13

(See Exercise 77, Section 11.5 and Exercise 31, Section 11.6.)

Appendix A
Tables: Exponents, Logarithms, and Trigonometric Functions

For a detailed explanation regarding the use of tables, see Sections 6.4, 7.2, 7.3, 7.4, and 7.5.

Table I Natural Logarithms (ln x)

x	0	1	2	3	4	5	6	7	8	9
1.0	0.0000	0.0100	0.0198	0.0296	0.0392	0.0488	0.0583	0.0677	0.0770	0.0862
1.1	0.0953	0.1044	0.1133	0.1222	0.1310	0.1398	0.1484	0.1570	0.1655	0.1740
1.2	0.1823	0.1906	0.1989	0.2070	0.2151	0.2231	0.2311	0.2390	0.2469	0.2546
1.3	0.2624	0.2700	0.2776	0.2852	0.2927	0.3001	0.3075	0.3148	0.3221	0.3293
1.4	0.3365	0.3436	0.3507	0.3577	0.3646	0.3716	0.3784	0.3853	0.3920	0.3988
1.5	0.4055	0.4121	0.4187	0.4253	0.4318	0.4383	0.4447	0.4511	0.4574	0.4637
1.6	0.4700	0.4762	0.4824	0.4886	0.4947	0.5008	0.5068	0.5128	0.5188	0.5247
1.7	0.5306	0.5365	0.5423	0.5481	0.5539	0.5596	0.5653	0.5710	0.5766	0.5822
1.8	0.5878	0.5933	0.5988	0.6043	0.6098	0.6152	0.6206	0.6259	0.6313	0.6366
1.9	0.6419	0.6471	0.6523	0.6575	0.6627	0.6678	0.6729	0.6780	0.6831	0.6881
2.0	0.6931	0.6981	0.7031	0.7080	0.7130	0.7178	0.7227	0.7275	0.7324	0.7372
2.1	0.7419	0.7467	0.7514	0.7561	0.7608	0.7655	0.7701	0.7747	0.7793	0.7839
2.2	0.7885	0.7930	0.7975	0.8020	0.8065	0.8109	0.8154	0.8198	0.8242	0.8286
2.3	0.8329	0.8372	0.8416	0.8459	0.8502	0.8544	0.8587	0.8629	0.8671	0.8713
2.4	0.8755	0.8796	0.8838	0.8879	0.8920	0.8961	0.9002	0.9042	0.9083	0.9123
2.5	0.9163	0.9203	0.9243	0.9282	0.9322	0.9361	0.9400	0.9439	0.9478	0.9517
2.6	0.9555	0.9594	0.9632	0.9670	0.9708	0.9746	0.9783	0.9821	0.9858	0.9895
2.7	0.9933	0.9969	1.0006	1.0043	1.0080	1.0116	1.0152	1.0188	1.0255	1.0260
2.8	1.0296	1.0332	1.0367	1.0403	1.0438	1.0473	1.0508	1.0543	1.0578	1.0613
2.9	1.0647	1.0682	1.0716	1.0750	1.0784	1.0818	1.0852	1.0886	1.0919	1.0953
3.0	1.0986	1.1019	1.1053	1.1086	1.1119	1.1151	1.1184	1.1217	1.1249	1.1282
3.1	1.1314	1.1346	1.1378	1.1410	1.1442	1.1474	1.1506	1.1537	1.1569	1.1600
3.2	1.1632	1.1663	1.1694	1.1725	1.1756	1.1787	1.1817	1.1848	1.1878	1.1909
3.3	1.1939	1.1970	1.2000	1.2030	1.2060	1.2090	1.2119	1.2149	1.2179	1.2208
3.4	1.2238	1.2267	1.2296	1.2326	1.2355	1.2384	1.2413	1.2442	1.2470	1.2499
3.5	1.2528	1.2556	1.2585	1.2613	1.2641	1.2669	1.2698	1.2726	1.2754	1.2782
3.6	1.2809	1.2837	1.2865	1.2892	1.2920	1.2947	1.2975	1.3002	1.3029	1.3056
3.7	1.3083	1.3110	1.3137	1.3164	1.3191	1.3218	1.3244	1.3271	1.3297	1.3324
3.8	1.3350	1.3376	1.3403	1.3429	1.3455	1.3481	1.3507	1.3533	1.3558	1.3584
3.9	1.3610	1.3635	1.3661	1.3686	1.3712	1.3737	1.3762	1.3788	1.3813	1.3838
4.0	1.3863	1.3888	1.3913	1.3938	1.3962	1.3987	1.4012	1.4036	1.4061	1.4085
4.1	1.4110	1.4134	1.4159	1.4183	1.4207	1.4231	1.4255	1.4279	1.4303	1.4327
4.2	1.4351	1.4375	1.4398	1.4422	1.4446	1.4469	1.4493	1.4516	1.4540	1.4563
4.3	1.4586	1.4609	1.4633	1.4656	1.4679	1.4702	1.4725	1.4748	1.4770	1.4793
4.4	1.4816	1.4839	1.4861	1.4884	1.4907	1.4929	1.4952	1.4974	1.4996	1.5019
4.5	1.5041	1.5063	1.5085	1.5107	1.5129	1.5151	1.5173	1.5195	1.5217	1.5239
4.6	1.5261	1.5282	1.5304	1.5326	1.5347	1.5369	1.5390	1.5412	1.5433	1.5454
4.7	1.5476	1.5497	1.5518	1.5539	1.5560	1.5581	1.5602	1.5623	1.5644	1.5665
4.8	1.5686	1.5707	1.5728	1.5748	1.5769	1.5790	1.5810	1.5831	1.5851	1.5872
4.9	1.5892	1.5913	1.5933	1.5953	1.5974	1.5994	1.6014	1.6034	1.6054	1.6074
5.0	1.6094	1.6114	1.6134	1.6154	1.6174	1.6194	1.6214	1.6233	1.6253	1.6273
5.1	1.6292	1.6312	1.6332	1.6351	1.6371	1.6390	1.6409	1.6429	1.6448	1.6467
5.2	1.6487	1.6506	1.6525	1.6544	1.6563	1.6582	1.6601	1.6620	1.6639	1.6658
5.3	1.6677	1.6696	1.6715	1.6734	1.6752	1.6771	1.6790	1.6808	1.6827	1.6845
5.4	1.6864	1.6882	1.6901	1.6919	1.6938	1.6956	1.6974	1.6993	1.7011	1.7029

Table I Natural Logarithms (ln x) (continued)

x	0	1	2	3	4	5	6	7	8	9
5.5	1.7047	1.7066	1.7084	1.7102	1.7120	1.7138	1.7156	1.7174	1.7192	1.7210
5.6	1.7228	1.7246	1.7263	1.7281	1.7299	1.7317	1.7334	1.7352	1.7370	1.7387
5.7	1.7405	1.7422	1.7440	1.7457	1.7475	1.7492	1.7509	1.7527	1.7544	1.7561
5.8	1.7579	1.7596	1.7613	1.7630	1.7647	1.7664	1.7682	1.7699	1.7716	1.7733
5.9	1.7750	1.7766	1.7783	1.7800	1.7817	1.7834	1.7851	1.7867	1.7884	1.7901
6.0	1.7918	1.7934	1.7951	1.7967	1.7984	1.8001	1.8017	1.8034	1.8050	1.8066
6.1	1.8083	1.8099	1.8116	1.8132	1.8148	1.8165	1.8181	1.8197	1.8213	1.8229
6.2	1.8245	1.8262	1.8278	1.8294	1.8310	1.8326	1.8342	1.8358	1.8374	1.8390
6.3	1.8406	1.8421	1.8437	1.8453	1.8469	1.8485	1.8500	1.8516	1.8532	1.8547
6.4	1.8563	1.8579	1.8594	1.8610	1.8625	1.8641	1.8656	1.8672	1.8687	1.8703
6.5	1.8718	1.8733	1.8749	1.8764	1.8799	1.8795	1.8810	1.8825	1.8840	1.8856
6.6	1.8871	1.8886	1.8901	1.8916	1.8931	1.8946	1.8961	1.8976	1.8991	1.9006
6.7	1.9021	1.9036	1.9051	1.9066	1.9081	1.9095	1.9110	1.9125	1.9140	1.9155
6.8	1.9169	1.9184	1.9199	1.9213	1.9228	1.9242	1.9257	1.9272	1.9286	1.9301
6.9	1.9315	1.9330	1.9344	1.9359	1.9373	1.9387	1.9402	1.9416	1.9430	1.9445
7.0	1.9459	1.9473	1.9488	1.9502	1.9516	1.9530	1.9544	1.9559	1.9573	1.9587
7.1	1.9601	1.9615	1.9629	1.9643	1.9657	1.9671	1.9685	1.9699	1.9713	1.9727
7.2	1.9741	1.9755	1.9769	1.9782	1.9796	1.9810	1.9824	1.9838	1.9851	1.9865
7.3	1.9879	1.9892	1.9906	1.9920	1.9933	1.9947	1.9961	1.9974	1.9988	2.0001
7.4	2.0015	2.0028	2.0042	2.0055	2.0069	2.0082	2.0096	2.0109	2.0122	2.0136
7.5	2.0149	2.0162	2.0176	2.0189	2.0202	2.0215	2.0229	2.0242	2.0255	2.0268
7.6	2.0282	2.0295	2.0308	2.0321	2.0334	2.0347	2.0360	2.0373	2.0386	2.0399
7.7	2.0412	2.0425	2.0438	2.0451	2.0464	2.0477	2.0490	2.0503	2.0516	2.0528
7.8	2.0541	2.0554	2.0567	2.0580	2.0592	2.0605	2.0618	2.0631	2.0643	2.0656
7.9	2.0669	2.0681	2.0694	2.0707	2.0719	2.0732	2.0744	2.0757	2.0769	2.0782
8.0	2.0794	2.0807	2.0819	2.0832	2.0844	2.0857	2.0869	2.0882	2.0894	2.0906
8.1	2.0919	2.0931	2.0943	2.0956	2.0968	2.0980	2.0992	2.1005	2.1017	2.1029
8.2	2.1041	2.1054	2.1066	2.1078	2.1090	2.1102	2.1114	2.1126	2.1138	2.1150
8.3	2.1163	2.1175	2.1187	2.1199	2.1211	2.1223	2.1235	2.1247	2.1258	2.1270
8.4	2.1282	2.1294	2.1306	2.1318	2.1330	2.1342	2.1353	2.1365	2.1377	2.1389
8.5	2.1401	2.1412	2.1424	2.1436	2.1448	2.1459	2.1471	2.1483	2.1494	2.1506
8.6	2.1518	2.1529	2.1541	2.1552	2.1564	2.1576	2.1587	2.1599	2.1610	2.1622
8.7	2.1633	2.1645	2.1656	2.1668	2.1679	2.1691	2.1702	2.1713	2.1725	2.1736
8.8	2.1748	2.1759	2.1770	2.1782	2.1793	2.1804	2.1815	2.1827	2.1838	2.1849
8.9	2.1861	2.1872	2.1883	2.1894	2.1905	2.1917	2.1928	2.1939	2.1950	2.1961
9.0	2.1972	2.1983	2.1994	2.2006	2.2017	2.2028	2.2039	2.2050	2.2061	2.2072
9.1	2.2083	2.2094	2.2105	2.2116	2.2127	2.2138	2.2148	2.2159	2.2170	2.2181
9.2	2.2192	2.2203	2.2214	2.2225	2.2235	2.2246	2.2257	2.2268	2.2279	2.2289
9.3	2.2300	2.2311	2.2322	2.2332	2.2343	2.2354	2.2364	2.2375	2.2386	2.2396
9.4	2.2407	2.2418	2.2428	2.2439	2.2450	2.2460	2.2471	2.2481	2.2492	2.2502
9.5	2.2513	2.2523	2.2534	2.2544	2.2555	2.2565	2.2576	2.2586	2.2597	2.2607
9.6	2.2618	2.2628	2.2638	2.2649	2.2659	2.2670	2.2680	2.2690	2.2701	2.2711
9.7	2.2721	2.2732	2.2742	2.2752	2.2762	2.2773	2.2783	2.2793	2.2803	2.2814
9.8	2.2824	2.2834	2.2844	2.2854	2.2865	2.2875	2.2885	2.2895	2.2905	2.2915
9.9	2.2925	2.2935	2.2946	2.2956	2.2966	2.2976	2.2986	2.2996	2.3006	2.3016

For x not in the range listed in this table, use $\ln 10 \approx 2.3026$ and the laws of logarithms.

For example, $\ln(0.05) = \ln(5 \cdot 10^{-2}) = \ln 5 + \ln 10^{-2} = \ln 5 - 2 \ln 10$
$$\approx 1.6094 - 2(2.3026) \approx -2.9958$$

and $\ln 412 = \ln(4.12 \cdot 10^2) = \ln(4.12) + \ln 10^2 = \ln(4.12) + 2 \ln 10$
$$\approx 1.4159 + 2(2.3026) \approx 6.0211$$

Table 2 Common Logarithms ($\log_{10} x$)

x	0	1	2	3	4	5	6	7	8	9
1.0	.0000	.0043	.0086	.0128	.0170	.0212	.0253	.0294	.0334	.0374
1.1	.0414	.0453	.0492	.0531	.0569	.0607	.0645	.0682	.0719	.0755
1.2	.0792	.0828	.0864	.0899	.0934	.0969	.1004	.1038	.1072	.1106
1.3	.1139	.1173	.1206	.1239	.1271	.1303	.1335	.1367	.1399	.1430
1.4	.1461	.1492	.1523	.1553	.1584	.1614	.1644	.1673	.1703	.1732
1.5	.1761	.1790	.1818	.1847	.1875	.1903	.1931	.1959	.1987	.2014
1.6	.2041	.2068	.2095	.2122	.2148	.2175	.2201	.2227	.2253	.2279
1.7	.2304	.2330	.2355	.2380	.2405	.2430	.2455	.2480	.2504	.2529
1.8	.2553	.2577	.2601	.2625	.2648	.2672	.2695	.2718	.2742	.2765
1.9	.2788	.2810	.2833	.2856	.2878	.2900	.2923	.2945	.2967	.2989
2.0	.3010	.3032	.3054	.3075	.3096	.3118	.3139	.3160	.3181	.3201
2.1	.3222	.3243	.3263	.3284	.3304	.3324	.3345	.3365	.3385	.3404
2.2	.3424	.3444	.3464	.3483	.3502	.3522	.3541	.3560	.3579	.3598
2.3	.3617	.3636	.3655	.3674	.3692	.3711	.3729	.3747	.3766	.3784
2.4	.3802	.3820	.3838	.3856	.3874	.3892	.3909	.3927	.3945	.3962
2.5	.3979	.3997	.4014	.4031	.4048	.4065	.4082	.4099	.4116	.4133
2.6	.4150	.4166	.4183	.4200	.4216	.4232	.4249	.4265	.4281	.4298
2.7	.4314	.4330	.4346	.4362	.4378	.4393	.4409	.4425	.4440	.4456
2.8	.4472	.4487	.4502	.4518	.4533	.4548	.4564	.4579	.4594	.4609
2.9	.4624	.4639	.4654	.4669	.4683	.4698	.4713	.4728	.4742	.4757
3.0	.4771	.4786	.4800	.4814	.4829	.4843	.4857	.4871	.4886	.4900
3.1	.4914	.4928	.4942	.4955	.4969	.4983	.4997	.5011	.5024	.5038
3.2	.5051	.5065	.5079	.5092	.5105	.5119	.5132	.5145	.5159	.5172
3.3	.5185	.5198	.5211	.5224	.5237	.5250	.5263	.5276	.5289	.5302
3.4	.5315	.5328	.5340	.5353	.5366	.5378	.5391	.5403	.5416	.5428
3.5	.5441	.5453	.5465	.5478	.5490	.5502	.5514	.5527	.5539	.5551
3.6	.5563	.5575	.5587	.5599	.5611	.5623	.5635	.5647	.5658	.5670
3.7	.5682	.5694	.5705	.5717	.5729	.5740	.5752	.5763	.5775	.5786
3.8	.5798	.5809	.5821	.5832	.5843	.5855	.5866	.5877	.5888	.5899
3.9	.5911	.5922	.5933	.5944	.5955	.5966	.5977	.5988	.5999	.6010
4.0	.6021	.6031	.6042	.6053	.6064	.6075	.6085	.6096	.6107	.6117
4.1	.6128	.6138	.6149	.6160	.6170	.6180	.6191	.6201	.6212	.6222
4.2	.6232	.6243	.6253	.6263	.6274	.6284	.6294	.6304	.6314	.6325
4.3	.6335	.6345	.6355	.6365	.6375	.6385	.6395	.6405	.6415	.6425
4.4	.6435	.6444	.6454	.6464	.6474	.6484	.6493	.6503	.6513	.6522
4.5	.6532	.6542	.6551	.6561	.6571	.6580	.6590	.6599	.6609	.6618
4.6	.6628	.6637	.6646	.6656	.6665	.6675	.6684	.6693	.6702	.6712
4.7	.6721	.6730	.6739	.6749	.6758	.6767	.6776	.6785	.6794	.6803
4.8	.6812	.6821	.6830	.6839	.6848	.6857	.6866	.6875	.6884	.6893
4.9	.6902	.6911	.6920	.6928	.6937	.6946	.6955	.6964	.6972	.6981
5.0	.6990	.6998	.7007	.7016	.7024	.7033	.7042	.7050	.7059	.7067
5.1	.7076	.7084	.7093	.7101	.7110	.7118	.7126	.7135	.7143	.7152
5.2	.7160	.7168	.7177	.7185	.7193	.7202	.7210	.7218	.7226	.7235
5.3	.7243	.7251	.7259	.7267	.7275	.7284	.7292	.7300	.7308	.7316
5.4	.7324	.7332	.7340	.7348	.7356	.7364	.7372	.7380	.7388	.7396
5.5	.7404	.7412	.7419	.7427	.7435	.7443	.7451	.7459	.7466	.7474
5.6	.7482	.7490	.7497	.7505	.7513	.7520	.7528	.7536	.7543	.7551
5.7	.7559	.7566	.7574	.7582	.7589	.7597	.7604	.7612	.7619	.7627
5.8	.7634	.7642	.7649	.7657	.7664	.7672	.7679	.7686	.7694	.7701
5.9	.7709	.7716	.7723	.7731	.7738	.7745	.7752	.7760	.7767	.7774

Table 2 Common Logarithms ($\log_{10} x$) (continued)

x	0	1	2	3	4	5	6	7	8	9
6.0	.7782	.7789	.7796	.7803	.7810	.7818	.7825	.7832	.7839	.7846
6.1	.7853	.7860	.7868	.7875	.7882	.7889	.7896	.7903	.7910	.7917
6.2	.7924	.7931	.7938	.7945	.7952	.7959	.7966	.7973	.7980	.7987
6.3	.7993	.8000	.8007	.8014	.8021	.8028	.8035	.8041	.8048	.8055
6.4	.8062	.8069	.8075	.8082	.8089	.8096	.8102	.8109	.8116	.8122
6.5	.8129	.8136	.8142	.8149	.8156	.8162	.8169	.8176	.8182	.8189
6.6	.8195	.8202	.8209	.8215	.8222	.8228	.8235	.8241	.8248	.8254
6.7	.8261	.8267	.8274	.8280	.8287	.8293	.8299	.8306	.8312	.8319
6.8	.8325	.8331	.8338	.8344	.8351	.8357	.8363	.8370	.8376	.8382
6.9	.8388	.8395	.8401	.8407	.8414	.8420	.8426	.8432	.8439	.8445
7.0	.8451	.8457	.8463	.8470	.8476	.8482	.8488	.8494	.8500	.8506
7.1	.8513	.8519	.8525	.8531	.8537	.8543	.8549	.8555	.8561	.8567
7.2	.8573	.8579	.8585	.8591	.8597	.8603	.8609	.8615	.8621	.8627
7.3	.8633	.8639	.8645	.8651	.8657	.8663	.8669	.8675	.8681	.8686
7.4	.8692	.8698	.8704	.8710	.8716	.8722	.8727	.8733	.8739	.8745
7.5	.8751	.8756	.8762	.8768	.8774	.8779	.8785	.8791	.8797	.8802
7.6	.8808	.8814	.8820	.8825	.8831	.8837	.8842	.8848	.8854	.8859
7.7	.8865	.8871	.8876	.8882	.8887	.8893	.8899	.8904	.8910	.8915
7.8	.8921	.8927	.8932	.8938	.8943	.8949	.8954	.8960	.8965	.8971
7.9	.8976	.8982	.8987	.8993	.8998	.9004	.9009	.9015	.9020	.9025
8.0	.9031	.9036	.9042	.9047	.9053	.9058	.9063	.9069	.9074	.9079
8.1	.9085	.9090	.9096	.9101	.9106	.9112	.9117	.9122	.9128	.9133
8.2	.9138	.9143	.9149	.9154	.9159	.9165	.9170	.9175	.9180	.9186
8.3	.9191	.9196	.9201	.9206	.9212	.9217	.9222	.9227	.9232	.9238
8.4	.9243	.9248	.9253	.9258	.9263	.9269	.9274	.9279	.9284	.9289
8.5	.9294	.9299	.9304	.9309	.9315	.9320	.9325	.9330	.9335	.9340
8.6	.9345	.9350	.9355	.9360	.9365	.9370	.9375	.9380	.9385	.9390
8.7	.9395	.9400	.9405	.9410	.9415	.9420	.9425	.9430	.9435	.9440
8.8	.9445	.9450	.9455	.9460	.9465	.9469	.9474	.9479	.9484	.9489
8.9	.9494	.9499	.9504	.9509	.9513	.9518	.9523	.9528	.9533	.9538
9.0	.9542	.9547	.9552	.9557	.9562	.9566	.9571	.9576	.9581	.9586
9.1	.9590	.9595	.9600	.9605	.9609	.9614	.9619	.9624	.9628	.9633
9.2	.9638	.9643	.9647	.9652	.9657	.9661	.9666	.9671	.9675	.9680
9.3	.9685	.9689	.9694	.9699	.9703	.9708	.9713	.9717	.9722	.9727
9.4	.9731	.9736	.9741	.9745	.9750	.9754	.9759	.9763	.9768	.9773
9.5	.9777	.9782	.9786	.9791	.9795	.9800	.9805	.9809	.9814	.9818
9.6	.9823	.9827	.9832	.9836	.9841	.9845	.9850	.9854	.9859	.9863
9.7	.9868	.9872	.9877	.9881	.9886	.9890	.9894	.9899	.9903	.9908
9.8	.9912	.9917	.9921	.9926	.9930	.9934	.9939	.9943	.9948	.9952
9.9	.9956	.9961	.9965	.9969	.9974	.9978	.9983	.9987	.9991	.9996

Table 3 Powers of e

x	e^x	e^{-x}	e^x (.1)	e^{-x} (.1)	e^x (.2)	e^{-x} (.2)	e^x (.3)	e^{-x} (.3)	e^x (.4)	e^{-x} (.4)	e^x (.5)	e^{-x} (.5)	e^x (.6)	e^{-x} (.6)	e^x (.7)	e^{-x} (.7)	e^x (.8)	e^{-x} (.8)	e^x (.9)	e^{-x} (.9)
0.0	1.000	1.000	1.010	.990	1.020	.980	1.031	.970	1.041	.961	1.051	.951	1.062	.942	1.073	.932	1.083	.923	1.094	.914
0.1	1.105	.905	1.116	.896	1.127	.887	1.139	.878	1.150	.869	1.162	.861	1.174	.852	1.185	.844	1.197	.835	1.209	.827
0.2	1.221	.819	1.234	.811	1.246	.803	1.259	.795	1.271	.787	1.284	.779	1.297	.771	1.310	.763	1.323	.756	1.336	.748
0.3	1.350	.741	1.363	.733	1.377	.726	1.391	.719	1.405	.712	1.419	.705	1.433	.698	1.448	.691	1.462	.684	1.477	.677
0.4	1.492	.670	1.507	.664	1.522	.657	1.537	.651	1.553	.644	1.568	.638	1.584	.631	1.600	.625	1.616	.619	1.632	.613
0.5	1.649	.607	1.665	.600	1.682	.595	1.699	.589	1.716	.583	1.733	.577	1.751	.571	1.768	.566	1.786	.560	1.804	.554
0.6	1.822	.549	1.840	.543	1.859	.538	1.878	.533	1.896	.527	1.916	.522	1.935	.517	1.954	.512	1.974	.507	1.994	.502
0.7	2.014	.497	2.034	.492	2.054	.487	2.075	.482	2.096	.477	2.117	.472	2.138	.468	2.160	.463	2.182	.458	2.203	.454
0.8	2.226	.449	2.248	.445	2.270	.440	2.293	.436	2.316	.432	2.340	.427	2.363	.423	2.387	.419	2.411	.415	2.435	.411
0.9	2.460	.407	2.484	.403	2.509	.399	2.535	.395	2.560	.391	2.586	.387	2.612	.383	2.638	.379	2.664	.375	2.691	.372
1.0	2.718	.368	2.746	.364	2.773	.361	2.801	.357	2.829	.353	2.858	.350	2.886	.346	2.915	.343	2.945	.340	2.974	.336
1.1	3.004	.333	3.034	.330	3.065	.326	3.096	.323	3.127	.320	3.158	.317	3.190	.313	3.222	.310	3.254	.307	3.287	.304
1.2	3.320	.301	3.353	.298	3.387	.295	3.421	.292	3.456	.289	3.490	.287	3.525	.284	3.561	.281	3.597	.278	3.633	.275
1.3	3.669	.273	3.706	.270	3.743	.267	3.781	.264	3.819	.262	3.857	.259	3.896	.257	3.935	.254	3.975	.252	4.015	.249
1.4	4.055	.247	4.096	.244	4.137	.242	4.179	.239	4.221	.237	4.263	.235	4.306	.232	4.349	.230	4.393	.228	4.437	.225
1.5	4.482	.223	4.527	.221	4.572	.219	4.618	.217	4.665	.214	4.712	.212	4.759	.210	4.807	.208	4.855	.206	4.904	.204
1.6	4.953	.202	5.003	.200	5.053	.198	5.104	.196	5.155	.194	5.207	.192	5.259	.190	5.312	.188	5.366	.186	5.420	.185
1.7	5.474	.183	5.529	.181	5.585	.179	5.641	.177	5.697	.176	5.755	.174	5.812	.172	5.871	.170	5.930	.169	5.989	.167
1.8	6.050	.165	6.110	.164	6.172	.162	6.234	.160	6.297	.159	6.360	.157	6.424	.156	6.488	.154	6.553	.153	6.619	.151
1.9	6.686	.150	6.753	.148	6.821	.147	6.890	.145	6.959	.144	7.029	.142	7.099	.141	7.171	.139	7.243	.138	7.316	.137
2.0	7.389	.135	7.463	.134	7.538	.133	7.614	.131	7.691	.130	7.768	.129	7.846	.127	7.925	.126	8.004	.125	8.085	.124
2.1	8.166	.122	8.248	.121	8.331	.120	8.415	.119	8.499	.118	8.585	.116	8.671	.115	8.758	.114	8.846	.113	8.935	.112
2.2	9.025	.111	9.116	.110	9.207	.109	9.300	.108	9.393	.106	9.488	.105	9.583	.104	9.679	.103	9.777	.102	9.875	.101
2.3	9.974	.100	10.074	.099	10.176	.098	10.278	.097	10.381	.096	10.486	.095	10.591	.094	10.697	.093	10.805	.093	10.913	.092

For exponents not in this range use $e^{2.303} \approx 10$, $e^{4.605} \approx 100$, $e^{6.901} \approx 1000$, $e^{9.210} \approx 10000$ and the laws of exponents.

For example, $e^{3.52} \approx e^{2.303 + 1.117} = e^{2.303} e^{1.117} \approx 10 \cdot (3.058) = 30.58$

and $e^{-5.17} \approx e^{-4.605 - 0.565} = e^{-4.605} e^{-0.565} \approx (0.01)(0.568) = (0.00568)$

Table 4 The Circular Functions (θ in radians)

θ (radians)	sin θ	cos θ	tan θ	cot θ	sec θ	csc θ
.00	.0000	1.0000	.0000	. . . .	1.000	. . . .
.01	.0100	1.0000	.0100	99.997	1.000	100.00
.02	.0200	.9998	.0200	49.993	1.000	50.00
.03	.0300	.9996	.0300	33.323	1.000	33.34
.04	.0400	.9992	.0400	24.987	1.000	25.01
.05	.0500	.9988	.0500	19.983	1.001	20.01
.06	.0600	.9982	.0601	16.647	1.002	16.68
.07	.0699	.9976	.0701	14.262	1.002	14.30
.08	.0799	.9968	.0802	12.473	1.003	12.51
.09	.0899	.9960	.0902	11.081	1.004	11.13
.10	.0998	.9950	.1003	9.967	1.005	10.02
.11	.1098	.9940	.1104	9.054	1.006	9.109
.12	.1197	.9928	.1206	8.293	1.007	8.353
.13	.1296	.9916	.1307	7.649	1.009	7.714
.14	.1395	.9902	.1409	7.096	1.010	7.166
.15	.1494	.9888	.1511	6.617	1.011	6.692
.16	.1593	.9872	.1614	6.197	1.013	6.277
.17	.1692	.9856	.1717	5.826	1.015	5.911
.18	.1790	.9838	.1820	5.495	1.016	5.586
.19	.1889	.9820	.1923	5.200	1.018	5.295
.20	.1987	.9801	.2027	4.933	1.020	5.033
.21	.2085	.9780	.2131	4.692	1.022	4.797
.22	.2182	.9759	.2236	4.472	1.025	4.582
.23	.2280	.9737	.2341	4.271	1.027	4.386
.24	.2377	.9713	.2447	4.086	1.030	4.207
.25	.2474	.9689	.2553	3.916	1.032	4.042
.26	.2571	.9664	.2660	3.759	1.035	3.890
.27	.2667	.9638	.2768	3.613	1.038	3.749
.28	.2764	.9611	.2876	3.478	1.041	3.619
.29	.2860	.9582	.2984	3.351	1.044	3.497
.30	.2955	.9553	.3093	3.223	1.047	3.384
.31	.3051	.9523	.3203	3.122	1.050	3.278
.32	.3146	.9492	.3314	3.018	1.053	3.179
.33	.3240	.9460	.3425	2.920	1.057	3.086
.34	.3335	.9428	.3537	2.827	1.061	2.999
.35	.3429	.9394	.3650	2.740	1.065	2.916
.36	.3523	.9359	.3764	2.657	1.068	2.839
.37	.3616	.9323	.3879	2.578	1.073	2.765
.38	.3709	.9287	.3994	2.504	1.077	2.696
.39	.3802	.9249	.4111	2.433	1.081	2.630
.40	.3894	.9211	.4228	2.365	1.086	2.568
.41	.3986	.9171	.4346	2.301	1.090	2.509
.42	.4078	.9131	.4466	2.239	1.095	2.452
.43	.4169	.9090	.4586	2.180	1.100	2.399
.44	.4259	.9048	.4708	2.124	1.105	2.348
.45	.4350	.9004	.4831	2.070	1.111	2.229
.46	.4439	.8961	.4954	2.018	1.116	2.253
.47	.4529	.8916	.5080	1.969	1.122	2.208
.48	.4618	.8870	.5206	1.921	1.127	2.166
.49	.4706	.8823	.5334	1.875	1.133	2.125

Table 4 The Circular Functions (θ in radians) (continued)

θ (radians)	sin θ	cos θ	tan θ	cot θ	sec θ	csc θ
.50	.4794	.8776	.5463	1.830	1.139	2.086
.51	.4882	.8727	.5594	1.788	1.146	2.048
.52	.4969	.8678	.5726	1.747	1.152	2.013
.53	.5055	.8628	.5859	1.707	1.159	1.978
.54	.5141	.8577	.5994	1.668	1.166	1.945
.55	.5227	.8525	.6131	1.631	1.173	1.913
.56	.5312	.8473	.6269	1.595	1.180	1.883
.57	.5396	.8419	.6310	1.560	1.188	1.853
.58	.5480	.8365	.6552	1.526	1.196	1.825
.59	.5564	.8309	.6696	1.494	1.203	1.797
.60	.5646	.8253	.6841	1.462	1.212	1.771
.61	.5729	.8196	.6989	1.431	1.220	1.746
.62	.5810	.8239	.7139	1.401	1.229	1.721
.63	.5891	.8080	.7291	1.372	1.238	1.697
.64	.5972	.8021	.7445	1.343	1.247	1.674
.65	.6052	.7961	.7602	1.315	1.256	1.652
.66	.6131	.7900	.7761	1.288	1.266	1.631
.67	.6210	.7838	.7923	1.262	1.276	1.610
.68	.6288	.7776	.8087	1.237	1.286	1.590
.69	.6365	.7712	.8253	1.212	1.297	1.571
.70	.6442	.7648	.8423	1.187	1.307	1.552
.71	.6518	.7584	.8595	1.163	1.319	1.534
.72	.6594	.7518	.8771	1.140	1.330	1.517
.73	.6669	.7452	.8949	1.117	1.342	1.500
.74	.6743	.7358	.9131	1.095	1.354	1.483
.75	.6816	.7317	.8316	1.073	1.367	1.467
.76	.6889	.7248	.9505	1.052	1.380	1.452
.77	.6961	.7179	.9697	1.031	1.393	1.437
.78	.7033	.7109	.9893	1.011	1.407	1.422
.79	.7104	.7038	1.009	.9908	1.421	1.408
.80	.7174	.6967	1.030	.9712	1.435	1.394
.81	.7243	.6895	1.050	.9520	1.450	1.381
.82	.7311	.6822	1.072	.9331	1.466	1.368
.83	.7379	.6749	1.093	.9146	1.482	1.355
.84	.7446	.6675	1.116	.8964	1.498	1.343
.85	.7513	.6600	1.138	.8785	1.515	1.331
.86	.7578	.6524	1.162	.8609	1.533	1.320
.87	.7643	.6448	1.185	.8437	1.551	1.308
.88	.7707	.6372	1.210	.8267	1.569	1.297
.89	.7771	.6294	1.235	.8100	1.589	1.287
.90	.7833	.6216	1.260	.7936	1.609	1.277
.91	.7895	.6137	1.286	.7774	1.629	1.267
.92	.7956	.6058	1.313	.7615	1.651	1.257
.93	.8016	.5978	1.341	.7458	1.673	1.247
.94	.8076	.5898	1.369	.7303	1.696	1.238
.95	.8134	.5817	1.398	.7151	1.719	1.229
.96	.8192	.5735	1.428	.7001	1.744	1.221
.97	.8249	.5653	1.459	.6853	1.769	1.212
.98	.8305	.5570	1.491	.6707	1.795	1.204
.99	.8360	.5487	1.524	.6563	1.823	1.196

Table 4 The Circular Functions (θ in radians) (continued)

θ (radians)	sin θ	cos θ	tan θ	cot θ	sec θ	csc θ
1.00	.8415	.5403	1.557	.6421	1.851	1.118
1.01	.8468	.5319	1.592	.6281	1.880	1.181
1.02	.8521	.5234	1.628	.6142	1.911	1.174
1.03	.8573	.5148	1.665	.6005	1.942	1.166
1.04	.8624	.5062	1.704	.5870	1.975	1.160
1.05	.8674	.4976	1.743	.5736	2.010	1.153
1.06	.8724	.4889	1.784	.5604	2.046	1.146
1.07	.8772	.4801	1.827	.5473	2.083	1.140
1.08	.8820	.4713	1.871	.5344	2.122	1.134
1.09	.8866	.4625	1.917	.5216	2.162	1.128
1.10	.8912	.4536	1.965	.5090	2.205	1.122
1.11	.8957	.4447	2.014	.4964	2.249	1.116
1.12	.9001	.4357	2.066	.4840	2.295	1.111
1.13	.9044	.4267	2.120	.4718	2.344	1.106
1.14	.9086	.4176	2.176	.4596	2.395	1.101
1.15	.9128	.4085	2.234	.4475	2.448	1.096
1.16	.9168	.3993	2.296	.4356	2.504	1.091
1.17	.9208	.3902	2.360	.4237	2.563	1.086
1.18	.9246	.3809	2.427	.4120	2.625	1.082
1.19	.9284	.3717	2.498	.4003	2.691	1.077
1.20	.9320	.3624	2.572	.3888	2.760	1.073
1.21	.9356	.3530	2.650	.3373	2.833	1.069
1.22	.9391	.3436	2.733	.3659	2.910	1.065
1.23	.9425	.3342	2.820	.3546	2.992	1.061
1.24	.9458	.3248	2.912	.3434	3.079	1.057
1.25	.9490	.3153	3.010	.3323	3.171	1.054
1.26	.9521	.3058	3.113	.3212	3.270	1.050
1.27	.9551	.2963	3.224	.3102	3.375	1.047
1.28	.9580	.2867	3.341	.2993	3.488	1.044
1.29	.9608	.2771	3.467	.2884	3.609	1.041
1.30	.9636	.2675	3.602	.2776	3.738	1.038
1.31	.9662	.2579	3.747	.2669	3.878	1.035
1.32	.9687	.2482	3.903	2.562	4.029	1.032
1.33	.9711	.2385	4.072	.2456	4.193	1.030
1.34	.9735	.2288	4.256	.2350	4.372	1.027
1.35	.9757	.2190	4.455	.2245	4.566	1.025
1.36	.9779	.2092	4.673	.2140	4.779	1.023
1.37	.9799	.1994	4.913	.2035	5.014	1.021
1.38	.9819	.1896	5.177	.1931	5.273	1.018
1.39	.9837	.1798	5.471	,1828	5.561	1.017
1.40	.9854	.1700	5.798	.1725	5.883	1.015
1.41	.9871	.1601	6.165	.1622	6.246	1.013
1.42	.9887	.1502	6.581	.1519	6.657	1.011
1.43	.9901	.1403	7.055	.1417	7.126	1.010
1.44	.9915	.1304	7.602	.1315	7.667	1.009
1.45	.9927	.1205	8.238	.1214	8.299	1.007
1.46	.9939	.1106	8.989	.1113	9.044	1.006
1.47	.9949	.1006	9.887	.1011	9.938	1.005
1.48	.9959	.0907	0.938	.0910	11.029	1.004
1.49	.9967	.0807	12.350	.0810	12.390	1.003

Table 4   The Circular Functions (θ in radians) (continued)

θ (radians)	sin θ	cos θ	tan θ	cot θ	sec θ	csc θ
1.50	.9975	.0707	14.101	.0709	14.137	1.003
1.51	.9982	.0608	16.428	.0609	16.458	1.002
1.52	.9987	.0508	19.670	.0508	19.965	1.001
1.53	.9992	.0408	24.498	.0408	24.519	1.001
1.54	.9995	.0308	32.461	.0308	32.476	1.000
1.55	.9998	.0208	48.078	.0208	48.089	1.000
1.56	.9999	.0108	92.620	.0108	92.626	1.000
1.57	1.0000	.0008	1255.8	.0008	1255.8	1.000
1.58	1.0000	−.0092	−108.65	−.0092	−108.65	1.000
1.59	.9998	−.0192	−52.067	−.0192	−52.08	1.000
1.60	.9996	−.0292	−34.233	−.0292	−34.25	1.000

Table 5 The Trigonometric Functions (θ in degrees)

θ (degrees)	sin θ	cos θ	tan θ	cot θ	sec θ	csc θ	
0°00′	.0000	1.0000	.0000	—	1.000	—	90°00′
10	.0029	1.0000	.0029	343.8	1.000	343.8	50
20	.0058	1.0000	.0058	171.9	1.000	171.9	40
30	.0087	1.0000	.0087	114.6	1.000	114.6	30
40	.0116	.9999	.0116	85.94	1.000	85.95	20
50	.0145	.9999	.0145	68.75	1.000	68.76	10
1°00′	.0175	.9998	.0175	57.29	1.000	57.30	89°00′
10	.0204	.9998	.0204	49.10	1.000	49.11	50
20	.0233	.9997	.0233	42.96	1.000	42.98	40
30	.0262	.9997	.0262	38.19	1.000	38.20	30
40	.0291	.9996	.0291	34.37	1.000	34.38	20
50	.0320	.9995	.0320	31.24	1.001	31.26	10
2°00′	.0349	.9994	.0349	28.64	1.001	28.65	88°00′
10	.0378	.9993	.0378	26.43	1.001	26.45	50
20	.0407	.9992	.0407	24.54	1.001	24.56	40
30	.0436	.9990	.0437	22.90	1.001	22.93	30
40	.0465	.9989	.0466	21.47	1.001	21.49	20
50	.0494	.9988	.0495	20.21	1.001	20.23	10
3°00′	.0523	.9986	.0524	19.08	1.001	19.11	87°00′
10	.0552	.9985	.0553	18.07	1.002	18.10	50
20	.0581	.9983	.0582	17.17	1.002	17.20	40
30	.0610	.9981	.0612	16.35	1.002	16.38	30
40	.0640	.9980	.0641	15.60	1.002	15.64	20
50	.0669	.9978	.0670	14.92	1.002	14.96	10
4°00′	.0698	.9976	.0699	14.30	1.002	14.34	86°00′
10	.0727	.9974	.0729	13.73	1.003	13.76	50
20	.0756	.9971	.0758	13.20	1.003	13.23	40
30	.0785	.9969	.0787	12.71	1.003	12.75	30
40	.0814	.9967	.0816	12.25	1.003	12.29	20
50	.0843	.9964	.0846	11.83	1.004	11.87	10
5°00′	.0872	.9962	.0875	11.43	1.004	11.47	85°00′
10	.0901	.9959	.0904	11.06	1.004	11.10	50
20	.0929	.9957	.0934	10.71	1.004	10.76	40
30	.0958	.9954	.0963	10.39	1.005	10.43	30
40	.0987	.9951	.0992	10.08	1.005	10.13	20
50	.1016	.9948	.1022	9.788	1.005	9.839	10
6°00′	.1045	.9945	.1051	9.514	1.006	9.567	84°00′
10	.1074	.9942	.1080	9.255	1.006	9.309	50
20	.1103	.9939	.1110	9.010	1.006	9.065	40
30	.1132	.9936	.1139	8.777	1.006	8.834	30
40	.1161	.9932	.1169	8.556	1.007	8.614	20
50	.1190	.9929	.1198	8.345	1.007	8.405	10
7°00′	.1219	.9925	.1228	8.144	1.008	8.206	83°00′
10	.1248	.9922	.1257	7.953	1.008	8.016	50
20	.1276	.9918	.1287	7.770	1.008	7.834	40
30	.1305	.9914	.1317	7.596	1.009	7.661	30
40	.1334	.9911	.1346	7.429	1.009	7.496	20
50	.1363	.9907	.1376	7.269	1.009	7.337	10
8°00′	.1392	.9903	.1405	7.115	1.010	7.185	82°00′
10	.1421	.9899	.1435	6.968	1.010	7.040	50
20	.1449	.9894	.1465	6.827	1.011	6.900	40
30	.1478	.9890	.1495	6.691	1.011	6.765	30
40	.1507	.9886	.1524	6.561	1.012	6.636	20
50	.1536	.9881	.1554	6.435	1.012	6.512	10
	cos θ	sin θ	cot θ	tan θ	csc θ	sec θ	θ (degrees)

Table 5 The Trigonometric Functions (θ in degrees) (continued)

θ (degrees)	sin θ	cos θ	tan θ	cot θ	sec θ	csc θ	
9°00′	.1564	.9877	.1584	6.314	1.012	6.392	81°00′
10	.1593	.9872	.1614	6.197	1.013	6.277	50
20	.1622	.9868	.1644	6.084	1.013	6.166	40
30	.1650	.9863	.1673	5.976	1.014	6.059	30
40	.1679	.9858	.1703	5.871	1.014	5.955	20
50	.1708	.9853	.1733	5.769	1.015	5.855	10
10°00′	.1736	.9848	.1763	5.671	1.015	5.759	80°00′
10	.1765	.9843	.1793	5.576	1.016	5.665	50
20	.1794	.9838	.1823	5.485	1.016	5.575	40
30	.1822	.9833	.1853	5.396	1.017	5.487	30
40	.1851	.9827	.1883	5.309	1.018	5.403	20
50	.1880	.9822	.1914	5.226	1.018	5.320	10
11°00′	.1908	.9816	.1944	5.145	1.019	5.241	79°00′
10	.1937	.9811	.1974	5.066	1.019	5.164	50
20	.1965	.9805	.2004	4.989	1.020	5.089	40
30	.1994	.9799	.2035	4.915	1.020	5.016	30
40	.2022	.9793	.2065	4.843	1.021	4.945	20
50	.2051	.9787	.2095	4.773	1.022	4.876	10
12°00′	.2079	.9781	.2126	4.705	1.022	4.810	78°00′
10	.2108	.9775	.2156	4.638	1.023	4.745	50
20	.2136	.9769	.2186	4.574	1.024	4.682	40
30	.2164	.9763	.2217	4.511	1.024	4.620	30
40	.2193	.9757	.2247	4.449	1.025	4.560	20
50	.2221	.9750	.2278	4.390	1.026	4.502	10
13°00′	.2250	.9744	.2309	4.331	1.026	4.445	77°00′
10	.2278	.9737	.2339	4.275	1.027	4.390	50
20	.2306	.9730	.2370	4.219	1.028	4.336	40
30	.2334	.9724	.2401	4.165	1.028	4.284	30
40	.2363	.9717	.2432	4.113	1.029	4.232	20
50	.2391	.9710	.2462	4.061	1.030	4.182	10
14°00′	.2419	.9703	.2493	4.011	1.031	4.134	76°00′
10	.2447	.9696	.2524	3.962	1.031	4.086	50
20	.2476	.9689	.2555	3.914	1.032	4.039	40
30	.2504	.9681	.2586	3.867	1.033	3.994	30
40	.2532	.9674	.2617	3.821	1.034	3.950	20
50	.2560	.9667	.2648	3.776	1.034	3.906	10
15°00′	.2588	.9659	.2679	3.732	1.035	3.864	75°00′
10	.2616	.9652	.2711	3.689	1.036	3.822	50
20	.2644	.9644	.2742	3.647	1.037	3.782	40
30	.2672	.9636	.2773	3.606	1.038	3.742	30
40	.2700	.9628	.2805	3.566	1.039	3.703	20
50	.2728	.9621	.2836	3.526	1.039	3.665	10
16°00′	.2756	.9613	.2867	3.487	1.040	3.628	74°00′
10	.2784	.9605	.2899	3.450	1.041	3.592	50
20	.2812	.9596	.2931	3.412	1.042	3.556	40
30	.2840	.9588	.2962	3.376	1.043	3.521	30
40	.2868	.9580	.2994	3.340	1.044	3.487	20
50	.2896	.9572	.3026	3.305	1.045	3.453	10
17°00′	.2924	.9563	.3057	3.271	1.046	3.420	73°00′
10	.2952	.9555	.3089	3.237	1.047	3.388	50
20	.2979	.9546	.3121	3.204	1.048	3.356	40
30	.3007	.9537	.3153	3.172	1.049	3.326	30
40	.3035	.9528	.3185	3.140	1.049	3.295	20
50	.3062	.9520	.3217	3.108	1.050	3.265	10
	cos θ	sin θ	cot θ	tan θ	csc θ	sec θ	θ (degrees)

Table 5 The Trigonometric Functions (θ in degrees) (continued)

θ (degrees)	sin θ	cos θ	tan θ	cot θ	sec θ	csc θ	
18°00′	.3090	.9511	.3249	3.078	1.051	3.236	72°00′
10	.3118	.9502	.3281	3.047	1.052	3.207	50
20	.3145	.9492	.3314	3.018	1.053	3.179	40
30	.3173	.9483	.3346	2.989	1.054	3.152	30
40	.3201	.9474	.3378	2.960	1.056	3.124	20
50	.3228	.9465	.3411	2.932	1.057	3.098	10
19°00′	.3256	.9455	.3443	2.904	1.058	3.072	71°00′
10	.3283	.9446	.3476	2.877	1.059	3.046	50
20	.3311	.9436	.3508	2.850	1.060	3.021	40
30	.3338	.9426	.3541	2.824	1.061	2.996	30
40	.3365	.9417	.3574	2.798	1.062	2.971	20
50	.3393	.9407	.3607	2.773	1.063	2.947	10
20°00′	.3420	.9397	.3640	2.747	1.064	2.924	70°00′
10	.3448	.9387	.3673	2.723	1.065	2.901	50
20	.3475	.9377	.3706	2.699	1.066	2.878	40
30	.3502	.9367	.3739	2.675	1.068	2.855	30
40	.3529	.9356	.3772	2.651	1.069	2.833	20
50	.3557	.9346	.3805	2.628	1.070	2.812	10
21°00′	.3584	.9336	.3839	2.605	1.071	2.790	69°00′
10	.3611	.9325	.3872	2.583	1.072	2.769	50
20	.3638	.9315	.3906	2.560	1.074	2.749	40
30	.3665	.9304	.3939	2.539	1.075	2.729	30
40	.3692	.9293	.3973	2.517	1.076	2.709	20
50	.3719	.9283	.4006	2.496	1.077	2.689	10
22°00′	.3746	.9272	.4040	2.475	1.079	2.669	68°00′
10	.3773	.9261	.4074	2.455	1.080	2.650	50
20	.3800	.9250	.4108	2.434	1.081	2.632	40
30	.3827	.9239	.4142	2.414	1.082	2.613	30
40	.3854	.9228	.4176	2.394	1.084	2.595	20
50	.3881	.9216	.4210	2.375	1.085	2.557	10
23°00′	.3907	.9205	.4245	2.356	1.086	2.559	67°00′
10	.3934	.9194	.4279	2.337	1.088	2.542	50
20	.3961	.9182	.4314	2.318	1.089	2.525	40
30	.3987	.9171	.4348	2.300	1.090	2.508	30
40	.4014	.9159	.4383	2.282	1.092	2.491	20
50	.4041	.9147	.4417	2.264	1.093	2.475	10
24°00′	.4067	.9135	.4452	2.246	1.095	2.459	66°00′
10	.4094	.9124	.4487	2.229	1.096	2.443	50
20	.4120	.9112	.4522	2.211	1.097	2.427	40
30	.4147	.9100	.4557	2.194	1.099	2.411	30
40	.4173	.9088	.4592	2.177	1.100	2.396	20
50	.4200	.9075	.4628	2.161	1.102	2.381	10
25°00′	.4226	.9063	.4663	2.145	1.103	2.366	65°00′
10	.4253	.9051	.4699	2.128	1.105	2.352	50
20	.4279	.9038	.4734	2.112	1.106	2.337	40
30	.4305	.9026	.4770	2.097	1.108	2.323	30
40	.4331	.9013	.4806	2.081	1.109	2.309	20
50	.4358	.9001	.4841	2.066	1.111	2.295	10
26°00′	.4384	.8988	.4877	2.050	1.113	2.281	64°00′
10	.4410	.8975	.4913	2.035	1.114	2.268	50
20	.4436	.8962	.4950	2.020	1.116	2.254	40
30	.4462	.8949	.4986	2.006	1.117	2.241	30
40	.4488	.8936	.5022	1.991	1.119	2.228	20
50	.4514	.8923	.5059	1.977	1.121	2.215	10
	cos θ	sin θ	cot θ	tan θ	csc θ	sec θ	θ (degrees)

Table 5 The Trigonometric Functions (θ in degrees) (continued)

θ (degrees)	sin θ	cos θ	tan θ	cot θ	sec θ	csc θ	
27°00′	.4540	.8910	.5095	1.963	1.122	2.203	63°00′
10	.4566	.8897	.5132	1.949	1.124	2.190	50
20	.4592	.8884	.5169	1.935	1.126	2.178	40
30	.4617	.8870	.5206	1.921	1.127	2.166	30
40	.4643	.8857	.5243	1.907	1.129	2.154	20
50	.4669	.8843	.5280	1.894	1.131	2.142	10
28°00′	.4695	.8829	.5317	1.881	1.133	2.130	62°00′
10	.4720	.8816	.5354	1.868	1.134	2.118	50
20	.4746	.8802	.5392	1.855	1.136	2.107	40
30	.4772	.8788	.5430	1.842	1.138	2.096	30
40	.4797	.8774	.5467	1.829	1.140	2.085	20
50	.4823	.8760	.5505	1.816	1.142	2.074	10
29°00′	.4848	.8746	.5543	1.804	1.143	2.063	61°00′
10	.4874	.8732	.5581	1.792	1.145	2.052	50
20	.4899	.8718	.5619	1.780	1.147	2.041	40
30	.4924	.8704	.5658	1.767	1.149	2.031	30
40	.4950	.8689	.5696	1.756	1.151	2.020	20
50	.4975	.8675	.5735	1.744	1.153	2.010	10
30°00′	.5000	.8660	.5774	1.732	1.155	2.000	60°00′
10	.5025	.8646	.5812	1.720	1.157	1.990	50
20	.5050	.8631	.5851	1.709	1.159	1.980	40
30	.5075	.8616	.5890	1.698	1.161	1.970	30
40	.5100	.8601	.5930	1.686	1.163	1.961	20
50	.5125	.8587	.5969	1.675	1.165	1.951	10
31°00′	.5150	.8572	.6009	1.664	1.167	1.942	59°00′
10	.5175	.8557	.6048	1.653	1.169	1.932	50
20	.5200	.8542	.6088	1.643	1.171	1.923	40
30	.5225	.8526	.6128	1.632	1.173	1.914	30
40	.5250	.8511	.6168	1.621	1.175	1.905	20
50	.5275	.8496	.6208	1.611	1.177	1.896	10
32°00′	.5299	.8480	.6249	1.600	1.179	1.887	58°00′
10	.5324	.8465	.6289	1.590	1.181	1.878	50
20	.5348	.8450	.6330	1.580	1.184	1.870	40
30	.5375	.8434	.6371	1.570	1.186	1.861	30
40	.5398	.8418	.6412	1.560	1.188	1.853	20
50	.5422	.8403	.6453	1.550	1.190	1.844	10
33°00′	.5446	.8387	.6494	1.540	1.192	1.836	57°00′
10	.5471	.8371	.6536	1.530	1.195	1.828	50
20	.5495	.8355	.6577	1.520	1.197	1.820	40
30	.5519	.8339	.6619	1.511	1.199	1.812	30
40	.5544	.8323	.6661	1.501	1.202	1.804	20
50	.5568	.8307	.6703	1.492	1.204	1.796	10
34°00′	.5592	.8290	.6745	1.483	1.206	1.788	56°00′
10	.5616	.8274	.6787	1.473	1.209	1.781	50
20	.5640	.8258	.6830	1.464	1.211	1.773	40
30	.5664	.8241	.6873	1.455	1.213	1.766	30
40	.5688	.8225	.6916	1.446	1.216	1.758	20
50	.5712	.8208	.6959	1.437	1.218	1.751	10
	cos θ	sin θ	cot θ	tan θ	csc θ	sec θ	θ (degrees)

Table 5 The Trigonometric Functions (θ in degrees) (continued)

θ (degrees)	sin θ	cos θ	tan θ	cot θ	sec θ	csc θ	
35°00′	.5736	.8192	.7002	1.428	1.221	1.743	55°00′
10	.5760	.8175	.7046	1.419	1.223	1.736	50
20	.5783	.8158	.7089	1.411	1.226	1.729	40
30	.5807	.8141	.7133	1.402	1.228	1.722	30
40	.5831	.8124	.7177	1.393	1.231	1.715	20
50	.5854	.8107	.7221	1.385	1.233	1.708	10
36°00′	.5878	.8090	.7265	1.376	1.236	1.701	54°00′
10	.5901	.8073	.7310	1.368	1.239	1.695	50
20	.5925	.8056	.7355	1.360	1.241	1.688	40
30	.5948	.8039	.7400	1.351	1.244	1.681	30
40	.5972	.8021	.7445	1.343	1.247	1.675	20
50	.5995	.8004	.7490	1.335	1.249	1.668	10
37°00′	.6018	.7986	.7536	1.327	1.252	1.662	53°00′
10	.6041	.7969	.7581	1.319	1.255	1.655	50
20	.6065	.7951	.7627	1.311	1.258	1.649	40
30	.6088	.7934	.7673	1.303	1.260	1.643	30
40	.6111	.7916	.7720	1.295	1.263	1.636	20
50	.6134	.7898	.7766	1.288	1.266	1.630	10
38°00′	.6157	.7880	.7813	1.280	1.269	1.624	52°00′
10	.6180	.7862	.7860	1.272	1.272	1.618	50
20	.6202	.7844	.7907	1.265	1.275	1.612	40
30	.6225	.7826	.7954	1.257	1.278	1.606	30
40	.6248	.7808	.8002	1.250	1.281	1.601	20
50	.6271	.7790	.8050	1.242	1.284	1.595	10
39°00′	.6293	.7771	.8098	1.235	1.287	1.589	51°00′
10	.6313	.7753	.8146	1.228	1.290	1.583	50
20	.6338	.7735	.8195	1.220	1.293	1.578	40
30	.6361	.7716	.8243	1.213	1.296	1.572	30
40	.6383	.7698	.8292	1.206	1.299	1.567	20
50	.6406	.7679	.8342	1.199	1.302	1.561	10
40°00′	.6428	.7660	.8391	1.192	1.305	1.556	50°00′
10	.6450	.7642	.8441	1.185	1.309	1.550	50
20	.6472	.7623	.8491	1.178	1.312	1.545	40
30	.6494	.7604	.8541	1.171	1.315	1.540	30
40	.6517	.7585	.8591	1.164	1.318	1.535	20
50	.6539	.7566	.8642	1.157	1.322	1.529	10
41°00′	.6561	.7547	.8693	1.150	1.325	1.524	49°00′
10	.6583	.7528	.8744	1.144	1.328	1.519	50
20	.6604	.7509	.8796	1.137	1.332	1.514	40
30	.6626	.7490	.8847	1.130	1.335	1.509	30
40	.6648	.7470	.8899	1.124	1.339	1.504	20
50	.6670	.7451	.8952	1.117	1.342	1.499	10
42°00′	.6691	.7431	.9004	1.111	1.346	1.494	48°00′
10	.6713	.7412	.9057	1.104	1.349	1.490	50
20	.6734	.7392	.9110	1.098	1.353	1.485	40
30	.6756	.7373	.9163	1.091	1.356	1.480	30
40	.6777	.7353	.9217	1.085	1.360	1.476	20
50	.6799	.7333	.9271	1.079	1.364	1.471	10
	cos θ	sin θ	cot θ	tan θ	csc θ	sec θ	θ (degrees)

Table 5 The Trigonometric Functions (θ in degrees) (continued)

θ (degrees)	sin θ	cos θ	tan θ	cot θ	sec θ	csc θ	
43°00′	.6820	.7314	.9325	1.072	1.367	1.466	47°00′
10	.6841	.7294	.9380	1.066	1.371	1.462	50
20	.6862	.7274	.9435	1.060	1.375	1.457	40
30	.6884	.7254	.9490	1.054	1.379	1.453	30
40	.6905	.7234	.9545	1.048	1.382	1.448	20
50	.6926	.7214	.9601	1.042	1.386	1.444	10
44°00′	.6947	.7193	.9657	1.036	1.390	1.440	46°00′
10	.6967	.7173	.9713	1.030	1.394	1.435	50
20	.6988	.7153	.9770	1.024	1.398	1.431	40
30	.7009	.7133	.9827	1.018	1.402	1.427	30
40	.7030	.7112	.9884	1.012	1.406	1.423	20
50	.7050	.7092	.9942	1.006	1.410	1.418	10
45°00′	.7071	.7071	1.0000	1.000	1.414	1.414	45°00′
	cos θ	sin θ	cot θ	tan θ	csc θ	sec θ	θ (degrees)

Solutions

CHAPTER 1

Section 1.1
1. $2 \cdot 2 \cdot 2 \cdot 3$ 3. $2 \cdot 2 \cdot 3 \cdot 3 \cdot 11$ 5. $3 \cdot 3 \cdot 7 \cdot 13$
7. LCM = 225, GCD = 15   9. LCM = 645, GCD = 43
11. LCM = 210, GCD = 3   13. [number line: −1, 0 to 8, mark at 7.3]

15. [number line: mark at −3.2, 0]

17. [number line: 0, mark at 0.666...]   19. Multiplicative identity

21. Commutative property of addition
23. Multiplicative inverse   25. Distributive property
27. Associative property of multiplication   29. $\dfrac{5}{9}$

31. $\dfrac{3}{5}$ 33. $\dfrac{7}{8}$ 35. $-\dfrac{22}{15}$ 37. -6 39. 2
41. 48   43. 1   45. -52 47. -58 49. 22
51. 24   53. $\dfrac{5}{8}$ 55. $\dfrac{1}{6}$ 57. $\dfrac{1}{12}$ 59. $\dfrac{10}{9}$
61. $-\dfrac{3}{16}$ 63. 78   65. $\dfrac{1}{9}$ 67. $\dfrac{5}{2}$ 69. $\dfrac{14}{5}$ 71. $\dfrac{5}{2}$
73. 0.857   75. 2.091   77. 0.708   79. 0.500
81. 5.127   83. 672.346   85. 29.768
87. $(20-1)40 = 800 - 40 = 760$
89. $(25-1) \cdot 7 = 25 \cdot 7 - 7 = 175 - 7 = 168$
91. $8(20-1) = 160 - 8 = 152$ 93. $\{c, d, e, f, g\}$
95. $\{h, i, j, k\}$ 97. $\{a, c, d, e, f, g\}$ 99. $\{c, e, g, i, k\}$
101. $\varnothing$
103. No. Example: $2 - 3 \neq 3 - 2$
No. Example: $2 - (3 - 4) \neq (2 - 3) - 4$
105. No. Example: $4 \div (2+2) \neq (4 \div 2) + (4 \div 2)$
107. 160 bolts   109. 13.2 mi/gal
111. 5.1875; the first loaf   113. 2250
115. a. 60   b. -26 117. a. -15 b. 21
119. a. $0.03636\ldots$ b. $0.9090\ldots$
121. a. $0.141176\ldots$ b. $3.52941\ldots$ 123. a. 4   b. 40.30

Section 1.2
1. $2 < 7$ 3. $-2 > -6$ 5. $2 > -6$ 7. $-100 < 17$
9. $-2 > -7$ 11. $-4 > -12$ 13. $-4 < 12$
15. $-300 < 51$
17. (9) 1, (10) 0, (11) 2, (12) 6, (13) 2, (14) 3, (15) 3, (16) 12
19. $a \geq 4$ 21. $a \geq 0$ 23. $2 < x < 4$
25. $-5 < a < 2$ 27. $-4 < r < -2$ 29. $(-3, 5]$
31. $(10, \infty)$ 33. $(-2, 2)$ 35. $[-3, 5]$ 37. $[-2, 6]$
39. [number line: 0, 1, 2]   41. [number line: −2, 0]
43. [number line: 0, 4, 6]   45. [number line: 0, 3]
47. $(1, 2)$ 49. $(-4, 5)$ 51. $[3, 6]$ 53. $[4, 7)$
55. $[4, 5]$ 57. $[5, 8)$
59. a. $a \leq b$ if and only if $b - a \geq 0$ b. If $a \leq b$ and $b \leq c$, then $a \leq c$.   c. If $a \leq b$, then $(a+c) \leq (b+c)$.   d. If $a \leq b$, then $-a \geq -b$.   e. If $a \leq b$, and $x \geq 0$, then $ax \leq bx$.   f. If $a \leq b$ and $x = 0$, then $ax = bx$.   g. If $a \leq b$ and $x \leq 0$, then $ax \geq bx$.   h. If $0 < a \leq b$, then $0 < \dfrac{1}{b} \leq \dfrac{1}{a}$.

61. $-12 < -9 < 16 < 42$ 63. $-\dfrac{3}{5} < -\dfrac{1}{2} < \dfrac{1}{4} < \dfrac{3}{8}$
65. $-|-12| < -9 < |-15| < |-16|$
67. $4.8 < 4.87 < 4.8\overline{7} < 4.\overline{87} < 4.\overline{8}$ 69. 8   71. -4
73. -10 75. 17   77. $2 - x$ 79. $3a - 9$
81. $c^2 + 5$ 83. $x^2 - 25$ 85. 4   87. 9   89. 7
91. 27   93. $[12, 14]$
95. The 10-oz package @ 12.9¢/oz   97. The $3\tfrac{1}{2}$-lb box

Section 1.3

1. a^5
3. $(x-y)^4$
5. $\dfrac{(u+v)^3}{(w+z)^2}$
7. 3^5
9. 128
11. $\dfrac{1}{144}$
13. $5^9 = 1{,}953{,}125$
15. -2
17. 2
19. $\dfrac{1}{64}$
21. -32
23. 32
25. 64
27. -64
29. 64
31. 64
33. y^{17}
35. x^{30}
37. a^5
39. $\dfrac{y^2}{x}$
41. $\dfrac{y^3}{x^3}$
43. $12x^4 y^3 z^3$
45. $-\dfrac{14b^2}{a^4}$
47. $\dfrac{1}{x^{12}}$
49. $x^7 y^8 z^7$
51. $\dfrac{27}{x^2}$
53. $a^9 b^6$
55. y^{16}
57. $\dfrac{x^{12}}{y^8}$
59. $\dfrac{z}{x^7 y^4}$
61. $\dfrac{s^3 t^3}{8}$
63. s^4
65. $\dfrac{c^{18}}{a^{30} b^{24}}$
67. $\dfrac{x^{18} z^8}{y^{22}}$
69. $\dfrac{2x^2 + y}{x^2 y}$
71. $4x^2 + 2x$
73. $ab(2b + a)$
75. $27 - 8 \ne 1$
77. $9 + 9 \ne 36$
79. $\left(\dfrac{1}{25}\right) \ne \dfrac{1}{4} + \dfrac{1}{9}$
81. $10^{100},\ 10^{(10^{100})}$
83. 173,215
85. 0.0003512
87. 0.000,000,000,000,000,000,000,000,000,911
89. 30,000,000,000
91. $2.463(10^3)$
93. $(1.4317)(10^2)$
95. $1.32(10^{-3})$
97. $7.12(10^{-14})$
99. $3(10^3) = 3000$
101. $2(10^5) = 200{,}000$
103. $1.26263(10^{11})$ acres, $5.11(10^{10})$ hectares
105. -0.009040
107. 2106.126
109. $1.033(10^{-1})$
111. $6.59(10^{-6})$
113. $5.850(10^8)$
115. 89,280,000 mi
117. $4.40(10^{18})$ mi

Section 1.4

1. 7
3. $6\sqrt{2}$
5. 5
7. -2
9. 4
11. $\dfrac{\sqrt{2}}{2}$
13. 9
15. 3
17. 4
19. $\sqrt{5}$
21. 4
23. 3
25. 5
27. 7
29. 12
31. 8
33. 12
35. 15,000
37. $3r^2|s|$
39. $5x^2 y^4$
41. $4a^2 b^2$
43. $9|u|^3 v^4$
45. $\dfrac{|x|}{3|y||z|}$
47. $2uv^4$
49. $x^2 \sqrt{3}$
51. $25|xy|\sqrt{|y|}$
53. $5\sqrt{7}$
55. $7\sqrt{3}$
57. $\sqrt{11}$
59. $13\sqrt{3}$
61. $\sqrt[3]{3}$
63. $12\sqrt[3]{3}$
65. $9\sqrt[3]{6}$
67. $2\sqrt[3]{3} - 2\sqrt{2}$
69. -7
71. 45
73. $37 + 2\sqrt{3} + 15\sqrt{2}$
75. 1
77. 7
79. $y\sqrt{2} + 2y^2 \sqrt[4]{2}$
81. $\sqrt[3]{9} \ne 1 + 2$
83. $\sqrt{8} \ne 4$
85. $\dfrac{2\sqrt{3}}{3}$
87. $\dfrac{\sqrt{6}}{3}$
89. $\dfrac{\sqrt[3]{36}}{3}$
91. $\dfrac{\sqrt[4]{24}}{2}$
93. $\dfrac{3\sqrt{2} - 2\sqrt{3}}{6}$
95. $\dfrac{7\sqrt{3} + 3\sqrt{7}}{21}$
97. $\dfrac{5\sqrt{3} - 3\sqrt{5}}{15}$
99. $69.40 = 6.940(10^1)$
101. $733.57 = 7.3357(10^2)$
103. $11.87 = 1.187(10^1)$
105. $0.606 = 6.06(10^{-1})$
107. $\dfrac{7\sqrt{7}}{3} \approx 6.17$ mi/hr
109. $\dfrac{3\sqrt{10}}{5} \approx \$1.9$ million

Section 1.5

1. $\sqrt[5]{x^3}$
3. $\sqrt[4]{a^2 - b^2}$
5. $4\sqrt[3]{a^2 b^2}$
7. $8\sqrt[5]{u^3}$
9. $(x^2 + y^2)^{1/2}$
11. $(r^3 - s^3)^{4/3}$
13. $u^{7/5}$
15. $|v|^3$
17. 5
19. 2
21. -3
23. 8
25. -32
27. $-\dfrac{1}{3}$
29. $\dfrac{1}{9}$
31. 9
33. 9
35. 27
37. 3375
39. $-8x^3 y^2$
41. $u^2 v^4 w^{10}$
43. $-\dfrac{3}{x}$
45. $\dfrac{p^{1/4} r^4}{q^{1/2}}$
47. $a^{18} b^6 z^2$
49. $\dfrac{xz^{1/2}}{y^{1/4}}$
51. $x^4 y^6 z^3$
53. $\dfrac{1}{u^5 v^4 w^4}$
55. $3.59(10^2)$
57. $7.559(10^{-3})$
59. $3.39(10^2)$
61. $3.379(10^{-2})$
63. 33,636
65. a. \$10.82 trillion b. \$1.20 trillion
67. a. 4.76 mg b. 11.31 mg
69. 1047°C

Section 1.6

1. $6x^2 + 6x - 8$
3. $3a^4 + 4a^3 - 3a^2 + 9$
5. $14t^7 - 37t^6 + 12t^5 - 5t^3 + 4t^2 + 16$
7. $3v^5 - 5v^4 + 2v^3 + 2v - 5$
9. $x^2 + 6x$
11. $4v^3 - 3$
13. $6y^5 - 7y^4 - 4y^3 - 7y^2 + 5y$
15. $-2x^2 + xy + 3y^2$
17. $12r^2 + 36rs + 15s^2$
19. $z^3 + 4z^2 + 2z - 3$
21. $r^4 + r^3 + 4r^2 - 3r + 5$
23. $2s^5 + 11s^3 - 2s^2 + 12s - 8$
25. $t^3 - 7t + 6$
27. $p^4 - 8p^2 + 5p + 6$
29. Carry out the indicated multiplications.
31. $16a^2 - 16ab + 4b^2$
33. $4r^2 + 4rs + s^2$
35. $9x^2 - 12xy + 4y^2$
37. $4u^2 - 9v^2$
39. $36r^2 - 25s^2$
41. $81n^2 - 9k^2$
43. $125 + 75t + 15t^2 + t^3$
45. $8p^3 - 36p^2 q + 54pq^2 - 27q^3$
47. $64z^3 + 240z^2 y + 300zy^2 + 125y^3$
49. $12y^2 + 2y - 2$
51. $6y^2 + y - 12$
53. $6v^2 + 31v + 35$
55. $15s^2 + s - 2$
57. $x^2 - y^2 + 2yz - z^2$
59. $v^4 - 81$
61. $(70 + 5)(70 - 5) = 4900 - 25 = 4875$

Solutions A.19

63. $(15 + 3)(15 - 3) = 225 - 9 = 216$ 65. $2\sqrt{6} - 5$
67. $10 - 3\sqrt{10}$ 69. $\dfrac{3\sqrt{x} + 9}{x - 9}$ 71. $\dfrac{3\sqrt{x} - 3}{x - 3}$
73. $\dfrac{u - 2\sqrt{u}}{u - 4}$ 75. $\dfrac{a + b - 2\sqrt{ab}}{a - b}$ 77. $r + \sqrt{r^2 - 1}$

Section 1.7

1. $3(2z + 1)$ 3. $25(1 - 2x)$ 5. $3y(2y + 1 - 3a)$
7. $2rs^2(2r^2 - s^6 + 5r^3s^2 - 6r)$ 9. $(2u - v)(r^2 + s)$
11. $(r + 5)(r - 5)$ 13. $(6 + s)(6 - s)$
15. $(6x + 7y)(6x - 7y)$ 17. $(16 + w^2)(4 + w)(4 - w)$
19. $(u^4 + v^4)(u^2 + v^2)(u + v)(u - v)$ 21. $(2u - 3v)^2$
23. $(5r + 2s)^2$ 25. $(3a + 4b)^2$ 27. $(2r - 7s)^2$
29. $(5u + 6v)^2$ 31. $(y - 5)(y - 4)$ 33. $(t - 12)(t - 2)$
35. $(p - 5)(p + 4)$ 37. $(q - 6r)(q - 12r)$
39. $(2a - 3)(2a - 5)$ 41. $(w - 3)(w^2 + 3w + 9)$
43. $(2 - u)(4 + 2u + u^2)$ 45. $(s + 2)(s^2 - 2s + 4)$
47. $(2t + 3u)(4t^2 - 6tu + 9u^2)$
49. $(x + 2)(x - 2)(x + 1)(x - 1)$ 51. $(a + 1)^2(a - 1)^2$
53. $(1 - 2x)^2(1 + 2x)^2(1 + 4x^2)^2$
55. $(p - 2q)(p + 2q)(p^2 + 2pq + 4q^2)(p^2 - 2pq + 4q^2)$
57. $(x - 2)(x^2 + 2x + 4)(x + 1)(x^2 - x + 1)$
59. $(r + 2s)(2a + 3b)$ 61. $(y^2 + 1)(x + 2)(x - 2)$
63. $(x + 1)^2(x - 1)(x^2 - x - 1)$
65. $(y + 3)(y - 3)(x + 5)(x + 1)$
67. $(r - 2 + s)(r - 2 - s)$ 69. $(r^2 + 4 + r)(r^2 + 4 - r)$
71. $(u^4 + 2 + 2u^2)(u^4 + 2 - 2u^2)$
73. $(u^2 - 1 - 2u)(u^2 - 1 + 2u)$
75. $(z^2 + 3a^2 - 2az)(z^2 + 3a^2 + 2az)$
77. $(x^2 + y^2 + xy)(x^2 + y^2 - xy)$ 79. $(2a + 3b)^3$
81. $(3 - 2x)^3$ 83. Carry out the multiplications.
85. **a.** 144 ft   **b.** $-8(2t + 3)(t - 5) = 0$; at $t = 5$, the rocket hits the ground.

Section 1.8

1. $x^2 - 3x + 1$ 3. $s - 5$ 5. $a^2 - 3a + 9$ 7. $u + 7$
9. $z - 1$ 11. $\dfrac{m + 3}{m - 2}$ 13. $-\dfrac{1}{u}$ 15. $\dfrac{v - 5}{v - 4}$
17. $ab^2(a - b)$ 19. $(x - 2)^2$ 21. -1
23. $\dfrac{x + 5}{x - 3}$ 25. $\dfrac{1}{x - y}$ 27. $\dfrac{a^2(a + 2)}{(a - 2)^2}$ 29. $\dfrac{2t + 1}{t(t + 1)}$
31. $\dfrac{w + 1}{w(w - 1)}$ 33. $\dfrac{2r}{2 - r}$ 35. $\dfrac{-2}{x(x - 3)}$

37. $\dfrac{-1}{q + 3}$ 39. $\dfrac{5(v - u)(v + u)}{(v - 2u)(2v - u)}$ 41. $\dfrac{2}{(p - 2)(p + 1)}$
43. $\dfrac{1}{(a - 2)^2}$ 45. $\dfrac{8}{p + q}$ 47. $\dfrac{w(4w - 3)}{w^2 - 3w + 9}$
49. $\dfrac{9 - t^2}{3t}$ 51. $\dfrac{s + 3}{s}$ 53. $\dfrac{-(2x + y)}{2x}$ 55. $x + y$
57. $\dfrac{1}{rs}$ 59. $\dfrac{2a + 1}{2 - 3a}$ 61. $\dfrac{1 - 2w}{1 + 2w}$ 63. $\dfrac{m + 1}{m - 1}$
65. $\dfrac{x(x + y^2)}{y(y + x^2)}$ 67. $\dfrac{-(r^2 + rs + s^2)}{r + s}$ 69. $\dfrac{p}{p + 1}$
71. $\dfrac{-1}{q}$ 73. $6a^2 - 9a + 14$, $R = -6a - 14$
75. $2y - 3$, $R = 25y - 16$ 77. $x^3 - 13x + 12$
79. $\dfrac{7N + 11}{3}$
81. $7.41(10^{-11})$ on m_1, $8.34(10^{-11})$ on m_2, $2.41(10^{-11})$ on m_3
83. **a.** $\dfrac{xy}{y - x}$ **b.** 50.85 mm   **c.** 510 cm from the lens

Section 1.9

1. $3i$ 3. $6i$ 5. $11i$ 7. -10 9. -8 11. 2
13. 2 15. $3 - 2i$ 17. $5 - 4i$ 19. $4 + 7i$
21. $-2 + 2i$ 23. $-16 + 11i$ 25. $8 - 12i$
27. $-1 + 5i$ 29. $-1 - 13i$ 31. $25 + 0i$ 33. $3 + 4i$
35. $-11 - 2i$ 37. $0 + 0i$ 39. $1 + 0i$ 41. $1 + 2i$
43. $0 + 8i$ 45. $x = 5, y = 2$ 47. $x = 5, y = 0$
49. $x = -4, y = -7$ 51. $2 + 5i$ 53. $1 - \sqrt{2}i$
55. $-5 + 3i$ 57. $-2 - 6i$ 59. $-3 - 2i$
61. $-\dfrac{1}{5} + \dfrac{3}{5}i$ 63. $-\dfrac{3}{5} + \dfrac{6}{5}i$ 65. $-1 + 0i$
67. $0 - i$ 69. $0 + i$ 71. $\dfrac{7}{5} + \dfrac{11}{5}i$ 73. $\dfrac{3}{25} + \dfrac{4}{25}i$
75. $0 - i$ 77. $\dfrac{39}{10} + \dfrac{3}{10}i$ 79. $0 - 2i$ 81. $-\dfrac{7}{5} + \dfrac{6}{5}i$
83. $\overline{(a + bi)(c + di)} = \overline{(ac - bd) + (bc + ad)i}$
$= (ac - bd) - (bc + ad)i$;
$\overline{(a + bi)}\,\overline{(c + di)} = (a - bi)(c - di)$
$= (ac - bd) - (bc + ad)i$
85. $\overline{\overline{a + bi}} = \overline{a - bi} = a + bi$
87. $(a + bi) + \overline{(a + bi)} = (a + bi) + (a - bi) = 2a$
89. $\overline{(a + bi) - (c + di)} = \overline{(a - c) + (b - d)i}$
$= (a - c) - (b - d)i$;
$\overline{(a + bi)} - \overline{(c + di)} = (a - bi) - (c - di)$
$= (a - c) - (b - d)i$

Section 1.11

1. $2 \cdot 2 \cdot 19$ **3.** $2 \cdot 2 \cdot 2 \cdot 2 \cdot 2 \cdot 3 \cdot 5 \cdot 11$
5. GCF = 15, LCM = 19950   **7.** 1.25   **9.** $2.\overline{333}\ldots$

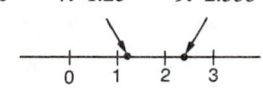

11. Commutative property of addition, multiplicative inverse

13. Multiplicative identity, commutative property of multiplication

15. $\dfrac{7}{22}$ **17.** $\dfrac{-5}{2}$ **19.** -15 **21.** -25 **23.** $\dfrac{3}{2}$

25. $\dfrac{23}{2}$ **27.** $\dfrac{119}{30}$ **29.** 3   **31.** $\dfrac{14}{27}$ **33.** 7

35. a. $7.\overline{63}\ldots$ **b.** $0.0\overline{530}\ldots$
37. a. 13.125 **b.** $0.\overline{0909}\ldots$ **39.** $\{2, 4, 5, 6, 7, 8, 9\}$
41. $\{0, 2, 4, 8, 10\}$ **43.** $\{6, 8, 9\}$
45. 0.889   **47.** 1.234   **49.** 15.207

51. $\dfrac{7}{11} < \dfrac{2}{3} < \dfrac{5}{7} < \dfrac{3}{4}$ **53.** $13.24\overline{5} > 13.2\overline{45} > 13.\overline{245}$

55. $-|-3| < -2 < |-2| < 4$ **57.** $-2 < y < 4$

59. ———(——————]———   **61.** ———+—(———
      -3 0   6   0 1

63. $[-3, \infty)$ **65.** $[-4, -3)$ **67.** $[2, 3]$
69. $(-2, 5)$ **71.** $(4, 8]$ **73.** $(4, 8)$ **75.** 1   **77.** 0
79. $-2v - 6$ **81.** $4 - r^2$ **83.** 5   **85.** 7
87. $(r + t)^3$ **89.** 64   **91.** $\dfrac{1}{2}$ **93.** $4|a|$ **95.** 10^{24}

97. 4   **99.** 13   **101.** $\dfrac{2(\sqrt{7} + \sqrt{2})}{5}$ **103.** $3 + 2\sqrt{2}$

105. 5   **107.** t^6 **109.** $\dfrac{1}{s^{12}}$ **111.** $8x^6$ **113.** $x^{48}y^{20}$

115. $\dfrac{1}{x^7}$ **117.** $12x^2yz$ **119.** $\dfrac{u}{v^8w^7}$

121. $\dfrac{-x^2y^2}{(x + y)(x - y)^3}$ **123.** $4|a^7|b^4$ **125.** $\dfrac{a^3}{y^{3/2}z^{5/2}}$

127. $\dfrac{x\sqrt{yz}}{y}$ **129.** $8 + 64 \neq 216$ **131.** $\dfrac{1}{4} - \dfrac{1}{9} \neq 1$

133. 6,500,000,000
135. 0.000,000,000,000,000,000,000,000,000,000,000,663
137. $2.372(10^{-1})$ **139.** $2.60014(10^1)$
141. $3.985(10^{-2}) = 0.03985$ **143.** -0.01112
145. $2.861(10^1) = 28.61$ **147.** $6r^2 + 11r - 35$

149. $\dfrac{2 - x}{x^2 + 2x + 4}$ **151.** $\dfrac{y + 9}{y + 5}$ **153.** $u - v$ **155.** $\dfrac{1}{uv}$

157. $\dfrac{y - x}{y + x}$ **159.** $4(4 + 3a)$ **161.** $(x - 5)(x + 4)$

163. $(3 - a)(3 + a)(9 + a^2)$
165. $(x^4 + 16)(x^2 + 4)(x + 2)(x - 2)$ **167.** $(2x + 3y)^2$
169. $(25 + z^2 - 5z)(25 + z^2 + 5z)$ **171.** $(2 - 3y)^3$
173. $(2x - y)(2x + y)(y - 1)(y + 1)$
175. $2r^2 - r + 4$, R = 1   **177.** $-2 + 3i$ **179.** $2 + 5i$

181. $0 + i$ **183.** $0 + 0i$ **185.** $-\dfrac{1}{10} - \dfrac{17}{10}i$

187. $1 + 0i$ **189.** $6 - 5i$ **191.** i **193.** 10,383 ft
195. $26.83   **197.** $2.54(10^{13})$ mi   **199.** 256 ft
201. 250

CHAPTER 2

Section 2.1

1. 5   **3.** -1 **5.** $\dfrac{18}{5}$ **7.** 6   **9.** 0   **11.** All t

13. 16   **15.** 26   **17.** 0   **19.** 4   **21.** 0   **23.** All s
25. 2   **27.** -2 **29.** 2   **31.** All w **33.** -4
35. 1   **37.** 2   **39.** -3 **41.** 16   **43.** -31
45. $\emptyset$ **47.** -6 **49.** 4   **51.** $\emptyset$ **53.** $\dfrac{-6}{7}$

55. $\dfrac{-7}{5}$ **57.** $\dfrac{9}{5}$ **59.** $\dfrac{12}{5}$ **61.** $\dfrac{-1}{5}$ **63.** $0 + 4i$

65. $0 - 2i$ **67.** $\dfrac{6 - 7i}{5}$

Section 2.2

1. a. $25°$ **b.** $-40°$ **c.** $C = \left(\dfrac{5}{9}\right)(F - 32)$

3. a. $700   **b.** $P = \dfrac{I}{rt}$, $r = \dfrac{I}{Pt}$, $t = \dfrac{I}{Pr}$ **5.** $y = \dfrac{rx}{(x - r)}$

7. a. $r = \dfrac{\sqrt{s\pi}}{2\pi}$ **b.** $r = \dfrac{\sqrt[3]{6\pi^2 V}}{2\pi}$

9. **a.** $l = \dfrac{A}{w} = \dfrac{P}{2} - w$ **b.** $w = \dfrac{A}{l} = \dfrac{P}{2} - l$

11. **a.** $600   **b.** $x = 1200 - 6T$ 13. $m = \dfrac{2K}{v^2}$

15. $y = \pm\sqrt{r^2 - x^2}$ 17. $x = \dfrac{m+y-2}{m}$

19. **a.** $P = \dfrac{V}{1+rt}$ **b.** $r = \dfrac{V-P}{Pt}$ 21. 89

23. 50 ft by 70 ft   25. $18   27. $360   29. $45

31. $12,000   33. 5 L   35. 18   37. $300,000

39. Bill 6 h, Jane 5 h; 300 mi

41. **a.** 6 P.M.   **b.** Eric 90 mi, Cindy 40 mi

43. **a.** 3 mi/h   **b.** 12 h   45. 1 h 12 min

47. 5 h 20 min

49. Through iv the computation yields $\dfrac{2x+4}{2} = x+2$; then v yields $x + 2 - x = 2$.

51. The corners are x, $x+2$, $x+14$, and $x+16$. The sum is $4x + 32 = 4(x+8)$. The mentalist divides by 4, then subtracts 8 to get the upper left corner entry. The rest is easy.

53. The numbers are x, $x+1$, $x+2$, $(x+7)$, $(x+7)+1$, $(x+7)+2$, $(x+14)$, $(x+14)+1$, and $(x+14)+2$. The sum is $9x + 3\cdot 7 + 3\cdot 14 + 3\cdot 1 + 3\cdot 2 = 9x + 72 = 9(x+8)$. From this formula the sum is obtained quickly for any given x.

55. Each eats $\tfrac{8}{3}$ sandwiches. The hungry man pays 80¢ for $\tfrac{8}{3}$ sandwiches, or 30¢ per sandwich. He buys $\tfrac{1}{3}$ and $\tfrac{7}{3}$ sandwiches, respectively, from his friends. Thus he pays them 10¢ and 70¢, respectively.

57. 84   59. 40   61. 6, 8   63. 3, 5, 7   65. 5 in²

Section 2.3

1. ± 2 3. $\pm 2i$ 5. 2, 3   7. -2 9. 2, -1

11. $\dfrac{-3}{2}, \dfrac{2}{3}$ 13. $\dfrac{-2}{3}$ 15. $\dfrac{1}{2}, -5$ 17. $\dfrac{1}{5}$

19. $\dfrac{-3}{2}, 5$ 21. $-7, 3$ 23. 4, -22 25. $-2 \pm \sqrt{2}$

27. $-1 \pm \dfrac{\sqrt{6}}{2}$ 29. 9, -8 31. 6, -2 33. $1 \pm \dfrac{\sqrt{10}}{2}$

35. $\dfrac{1}{2}$ 37. $\dfrac{3}{2}, -\dfrac{1}{2}$ 39. 1, -4 41. $\dfrac{2}{3}, -\dfrac{1}{3}$

43. $\dfrac{3}{2}$ 45. $-1, \dfrac{-2}{3}$ 47. $\dfrac{1 \pm i\sqrt{3}}{2}$ 49. $3 \pm 5i$

51. $2, \dfrac{-2}{3}$ 53. $2, \dfrac{-7}{5}$ 55. $1 \pm i\sqrt{3}$ 57. $-1 \pm i$

59. $-4 \pm 3i$ 61. $(3 \pm i\sqrt{7})$ 63. $4, \dfrac{2}{3}$ 65. $\dfrac{\sqrt{2}}{2} \pm i$

67. $-1 \pm 2i$ 69. $-2 \pm 3i$ 71. $1 \pm i$ 73. $-3 \pm i$

75. Two rational roots; $\sqrt{36}$ 77. A double root; $\sqrt{0}$

79. A double root; $\sqrt{0}$ 81. Two rational roots; $\sqrt{676}$

83. Two rational roots; $\sqrt{1}$ 85. $2i, \dfrac{-1}{3}$ 87. $\dfrac{-i \pm 5}{2}$

89. $-i, 4i$ 91. $t = \dfrac{-v \pm \sqrt{v^2 + 2gs}}{g}$

93. $r = s \pm \sqrt{s^2 + 3s}$ 95. $u = 0$ or $u = -4v$

97. $-3.42, -0.53$ 99. $-1.11, 5.90$

Section 2.4

1. $x = \pm\sqrt{5}, \pm 1$ 3. $\pm\sqrt{2}/2$ 5. $\pm\sqrt{7}, \pm 1$

7. $\sqrt[3]{5}, 1$ 9. $-\sqrt[3]{3}$ 11. $-\sqrt[3]{5}, \sqrt[3]{2}$ 13. $\pm\sqrt{2}, \pm\sqrt[4]{2}$

15. $\pm\sqrt[4]{2/3}, \pm 1$ 17. $\sqrt[5]{3}, -\sqrt[5]{2}$ 19. $\pm\sqrt[6]{3}$

21. $\pm\sqrt{-1 + \sqrt{7}}$ 23. $\pm\sqrt{3 + \sqrt{2}}, \pm\sqrt{3 - \sqrt{2}}$

25. $\pm\dfrac{\sqrt{-3 + \sqrt{29}}}{\sqrt{10}}$ 27. 8, 27   29. 32, -1 31. $\dfrac{3}{2}$

33. $\dfrac{1}{2}, 2$ 35. $\pm 2, \dfrac{\pm 1}{2}$ 37. $\pm 1, \pm 3$ 39. 9

41. 4   43. $\dfrac{1}{4}$ 45. $\dfrac{1}{4}$ 47. 256   49. 4   51. 14

53. 8   55. 7   57. $\varnothing$ 59. 5, 13   61. $-2, -3$

63. 3   65. ± 5 67. 3   69. -7 71. $-2, -4$

73. $0, \dfrac{2}{11}$ 75. 2, 13   77. $\dfrac{2}{3}, \dfrac{1}{2}$ 79. $\pm\sqrt{13}, \pm\sqrt{2}$

81. $\pm\sqrt{2}$ 83. $-1, -2\sqrt[3]{\dfrac{2}{3}}$ 85. ± 0.72

87. 5.52, 0.18   89. ± 64.04 91. $y = a - 2\sqrt{ax} + x$

93. $x = \pm[z^{2/3} - y^{2/3}]^{3/2}$ 95. $y = \dfrac{1}{(2x)}$

Section 2.5

1. 12, 14   3. 7, 8   5. 20 ft by 30 ft   7. 30 m by 45 m

9. 20 ft by 21 ft   11. 5 ft by 12 ft   13. 20 s after firing

15. **a.** 410 km   **b.** 418 km   17. 70 kg, or about 154 lb

19. 8 mi/h   21. 1 in   23. 15 ft by 30 ft   25. 5 ft

27. 16 cm by 13 cm   29. $\dfrac{1}{2}$ cm

31. $35 or $60, both are equilibrium prices

33. 8 scooters   35. 8, -3 37. 3, 6

39. 30 knots and 40 knots

41. Fred 60 mi/h, Mother 50 mi/h

43. a. 170 mi/h **b.** $2\frac{1}{2}$ h and $1\frac{3}{4}$ h   **45.** 30 h
47. Dan 42 days, Dale 56 days   **49.** 65 h and 104 h

Section 2.6

1. a. Tripled **b.** $y = \dfrac{x}{5}$ **c.** 6

3. a. Reduced to $\dfrac{1}{8}$ original **b.** $M = 4v^3$ **c.** 256

5. a. $s^* = \dfrac{s}{12}$ **b.** $s = 3t^2/\sqrt{uv^2}$ **c.** 1

7. a. $y^* = 48y$ **b.** $y = \dfrac{20w\sqrt{x}}{z^3}$ **c.** 900   **9.** 9 times

11. $\dfrac{2}{3}$ of original area   **13. a.** $d = \dfrac{9s^2}{484}$ **b.** 324 ft

15. a. $f = \dfrac{m}{40}$ **b.** 15 gal   **17. a.** $T = \dfrac{c}{16}$ **b.** $15

19. a. $C = 300\sqrt[3]{n}$ **b.** $15,000   **21. a.** $C = 2\sqrt{n}$ **b.** $80

23. a. $d = \sqrt{\dfrac{h}{3}}$ **b.** 50 mi   **25. a.** $I = \dfrac{2Pt}{25}$ **b.** $2400

27. $I^* = 4I$

29. a. Directly with respect to v and r^2 and inversely with respect to l **b.** $\dfrac{\sqrt{2}}{16}$

31. a. $L = \dfrac{10wd^2}{l}$ **b.** $w^* = 20$ in, $d^* = 24$ in   **c.** deeper

33. a. $m = k_1 S = k_1(k_2 h^2)$; $m = k_3 o = k_3(k_4 Vr) = k_3 k_4(k_5 h^3)r$; equating these two expressions for m and solving for r, we find $r = k_1 k_2 / k_3 k_4 k_5 h = K/h$.
b. $r\left(\dfrac{3}{2}\right) = \dfrac{2K}{3}$, $r(3) = \dfrac{K}{3} = \left(\dfrac{1}{2}\right)r\left(\dfrac{3}{2}\right)$, $r(6) = \dfrac{K}{6} = \left(\dfrac{1}{2}\right)r(3) = \left(\dfrac{1}{4}\right)r\left(\dfrac{3}{2}\right)$

Section 2.7

1. $(8, \infty)$ **3.** $(-\infty, 4)$ **5.** $(3, \infty)$ **7.** $\left(\dfrac{1}{2}, \infty\right)$
9. $(-\infty, -2]$ **11.** $[1, 2)$ **13.** $(-4, -1)$ **15.** $[3, 6]$
17. $\left(\dfrac{-2}{3}, 0\right]$ **19.** $\left(\dfrac{-1}{5}, 3\right)$ **21.** $\left(-\infty, \dfrac{1}{3}\right) \cup [3, \infty)$
23. $(-\infty, 4] \cup [9, \infty)$ **25.** $\left(-\infty, \dfrac{1}{2}\right] \cup \left[\dfrac{2}{3}, \infty\right)$
27. $(-2, 3)$ **29.** $\left(-\infty, \dfrac{-1}{2}\right] \cup [2, \infty)$

31. $(-\infty, -2) \cup (5, 6)$ **33.** $(-\infty, -2] \cup [2, 3]$
35. $(-\infty, -2] \cup \left[\dfrac{1}{2}, \infty\right)$ **37.** $(-3, 2) \cup (2, \infty)$
39. $[-1, 1]$ **41.** $(2, 4) \cup (4, 6)$ **43.** $(-\infty, -2) \cup (1, 3)$
45. $(-\infty, -2) \cup \left(\dfrac{1}{2}, \dfrac{3}{2}\right) \cup (3, \infty)$ **47.** $(-4, \infty)$
49. $(5, \infty)$ **51.** $(-\infty, 2) \cup [5, \infty)$
53. $(-\infty, -3) \cup (5, \infty)$ **55.** $(-1, 2)$
57. $(-1, 1] \cup (2, \infty)$ **59.** $(-5, -1) \cup (2, 3)$
61. $(-\infty, -3) \cup [-2, 2) \cup [4, \infty)$
63. $(-\infty, -4) \cup (-4, 1) \cup [2, 3]$
65. $(-\infty, -2) \cup (-2, -1) \cup (1, \infty)$
67. $(-6, 4) \cup (4, 5)$ **69.** $(-\infty, 3) \cup [6, \infty)$
71. $(-2, 1)$ **73.** $[-1, 0)$ **75.** $[1, 3)$
77. $[-2\sqrt{2}, -\sqrt{2}) \cup (\sqrt{2}, 2\sqrt{2}]$
79. $(-\infty, -24] \cup (-6, 3)$ **81.** $\left[\dfrac{19}{2}, \infty\right)$
83. $(-20, -2)$ **85.** $(-\infty, -3) \cup (1, 2) \cup (3, 7]$
87. Between 8 and 16 mg/h   **89.** Grade ≥ 80
91. $10 < h < 30$

Section 2.8

1. $|x - 4| \leq 7$ **3.** $|s - r| = 2$ **5.** $|w + 2| \geq 7$
7. $-12, 2$ **9.** $-7, 3$ **11.** $\varnothing$ **13.** $0, \pm 2$
15. $[4, \infty)$ **17.** $(-\infty, 3]$ **19.** 5   **21.** $\dfrac{11}{5}, 5$
23. $\pm\sqrt{2}, \pm\sqrt{5}$ **25.** $(-3, 4)$
27. $\left(-\infty, -\dfrac{2}{3}\right] \cup [8, \infty)$ **29.** $(-\infty, 0] \cup [8, \infty)$
31. $\varnothing$ **33.** $(-\infty, 0) \cup (5, \infty)$ **35.** $\left(-3, \dfrac{11}{3}\right)$
37. $(-\infty, -5] \cup [11, \infty)$ **39.** $(-7, -3)$ **41.** $[0, 3]$
43. $\left(-1, \dfrac{5}{3}\right)$ **45.** $(-2, 2) \cup (2, \infty)$
47. $(-3, -1) \cup (1, 3)$ **49.** $(-4, -1] \cup [3, 6)$
51. $[1, 3]$ **53.** $\varnothing$ **55. a.** $|T - 27| \leq 5$ **b.** $[22, 32]$
57. a. $(3.75)|T - 60| \leq 150$ **b.** $20 \leq T \leq 100$, $[20, 100]$

Section 2.10

1. -10 **3.** 16   **5.** 4   **7.** 6   **9.** 2
11. $x = 4, y = -2$ **13.** 3   **15.** $\dfrac{-1}{3}, -6$

17.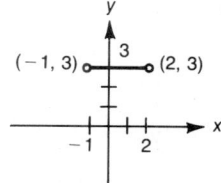
$-1 < x < 2, y = 3$

19.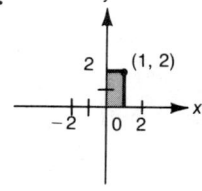
$0 \leq x \leq 1$
$0 \leq y \leq 2$

37.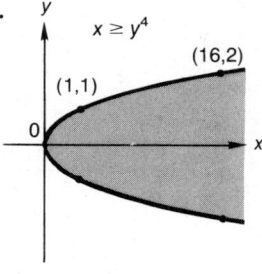
Symmetric about x-axis

39.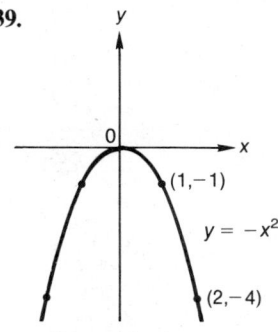
Symmetric about y-axis

21.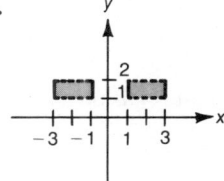
Symmetric about y-axis
$1 \leq |x| \leq 3$
$1 < y < 2$

23.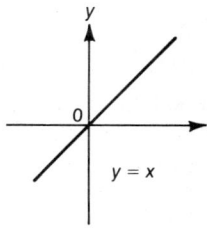
Symmetric about origin

41.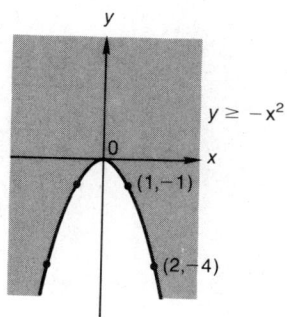
Symmetric about y-axis

25.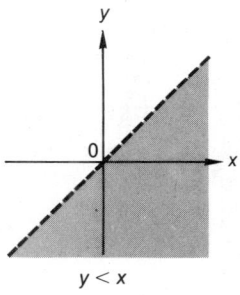
$y < x$

27.
$y \geq |x|$
Symmetric about y-axis

43.
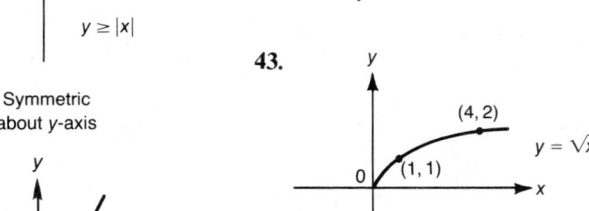

29.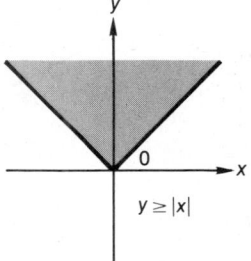
$y \geq |x|$
Symmetric about y-axis

31.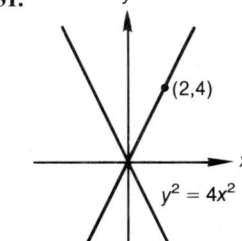
$y^2 = 4x^2$
Symmetric about x-axis, y-axis, and origin

45.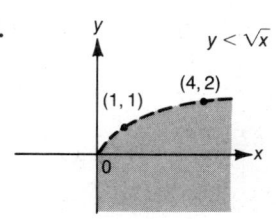

47.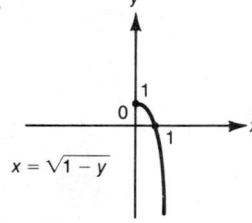

33.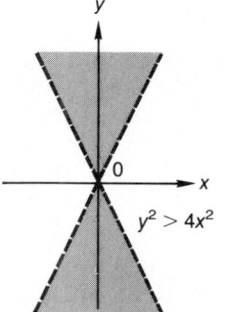
$y^2 > 4x^2$
Symmetric about x-axis, y-axis, and origin

35.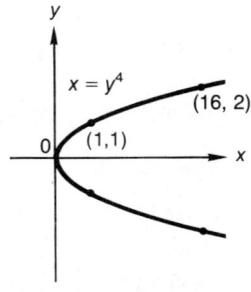
Symmetric about x-axis

49.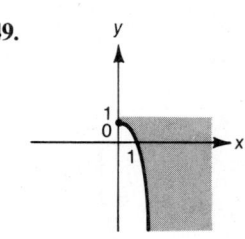
$x \geq \sqrt{1-y}$

51.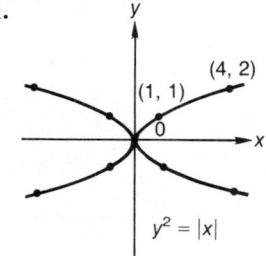
$y^2 = |x|$
Symmetric about x-axis, y-axis, and origin

53.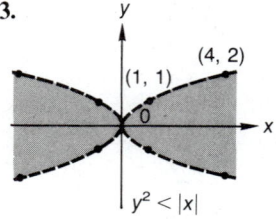
Symmetric about
x-axis, y-axis,
and origin

55.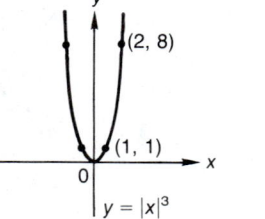
Symmetric about
y-axis

57.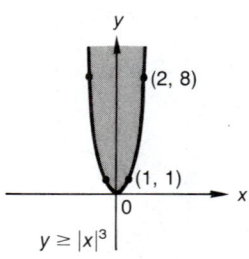
Symmetric about
y-axis

59.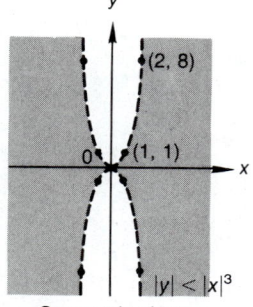
Symmetric about
x-axis, y-axis,
and origin

61.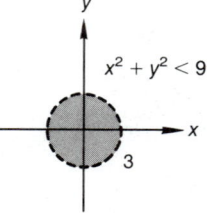
Symmetric about
x-axis, y-axis,
and origin

63.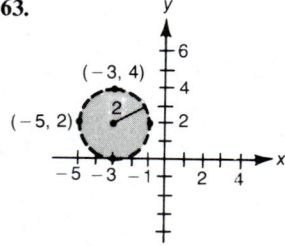

65.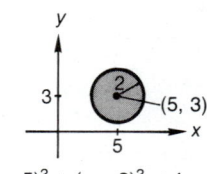

67.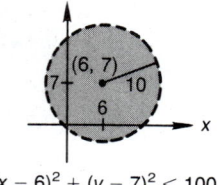

69.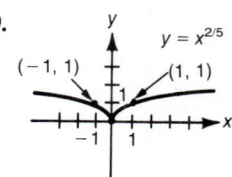

71.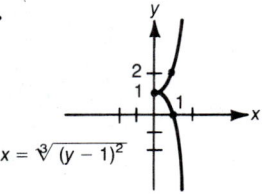

73.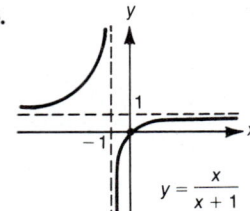

75.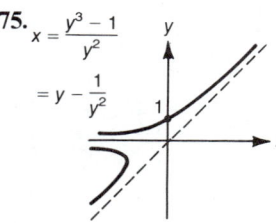

77. 71.5¢/dozen or 83.5¢/dozen; both are equilibrium prices.

79. In mid-1986, the increase in scrapping costs exceeded the savings in production costs.

81. Roadblocks should be placed on all roads 50 mi from Placerville.

83. $y = x$

Section 3.3

1.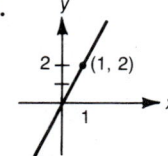

3.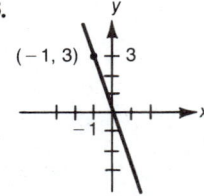

5.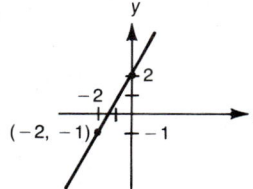

7.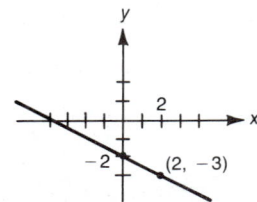

9. a. $m = 1$ **b.** $x - y + 5 = 0$
c. x-intercept $= -5$; y-intercept $= 5$

11. a. $m = -4$ **b.** $4x + y + 6 = 0$
c. x-intercept $= -\dfrac{3}{2}$; y-intercept $= -6$

13. a. No slope **b.** $x + 1 = 0$
c. x-intercept $= -1$; no y-intercept

15. a. $m = -2$ **b.** $2x + y - 6 = 0$
c. x-intercept $= 3$; y-intercept $= 6$

17. a. $m = 0$ **b.** $y - 1 = 0$
c. no x-intercept; y-intercept $= 1$

19. a. $m = \dfrac{1}{2}$ **b.** $x - 2y - 4 = 0$
c. x-intercept $= 4$; y-intercept $= -2$

21. a. No slope **b.** $x - 3 = 0$
c. x-intercept $= 3$; no y-intercept

23. a. $m = 7$ **b.** $7x - y - 14 = 0$
c. x-intercept $= 2$; y-intercept $= -14$

25. a. $m = 0$ b. $y + 1 = 0$
   c. no x-intercept; y-intercept $= -1$
27. $x - 3y + 12 = 0$; $(-12, 0)$, $(0, 4)$
29. $x + y - 2 = 0$; $(2, 0)$, $(0, 2)$
31. $y + 1 = 0$; no x-intercept, $(0, -1)$
33. $x + 3 = 0$; $(-3, 0)$, no y-intercept
35. $2x + y - 5 = 0$; $\left(\frac{5}{2}, 0\right)$, $(0, 5)$
37. $x + 2y + 12 = 0$; $(-12, 0)$, $(0, -6)$
39. $3x - y + 6 = 0$ 41. $2x - y - 12 = 0$
43. $10x + y + 4 = 0$ 45. $5x + y - 4 = 0$

47.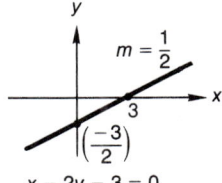
49.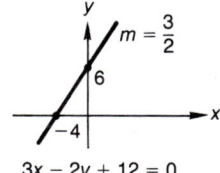

51.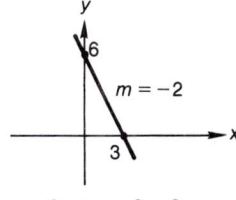
53.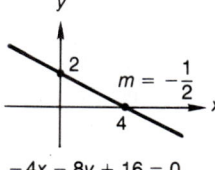

55.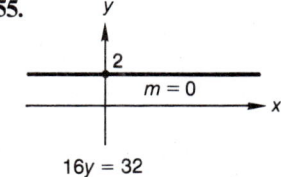
57.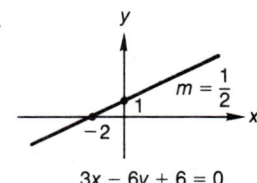

59. Perpendicular   61. Neither   63. Parallel
65. Perpendicular   67. Neither   69. Yes   71. No
73. Yes   75. $x + 1 = 0$; $y - 3 = 0$
77. $x + 3y + 7 = 0$; $3x - y + 1 = 0$
79. $2x + y + 3 = 0$; $x - 2y - 6 = 0$
81. $y - 2 = 0$; $x - 6 = 0$
83. $x + y - 29 = 0$; $x - y + 31 = 0$ 85. $x + y + 1 = 0$
87. $x - 4y + 10 = 0$ 89. $y + 1 = 0$
91. $2x - 4y + 9 = 0$ 93. $x + 1 = 0$
95. $2x + y - 8 = 0$ 97. $y + 2 = 0$
99. $x + 7y - 2 = 0$ 101. $x - 1 = 0$
103. $4x - 3y + 25 = 0$

105. $\frac{1}{a}x + \frac{1}{b}y - 1 = 0$ is the general form of a line; the line has intercepts $(a, 0)$ and $(0, b)$.

107. $5x - 3y - 15 = 0$, $m = \frac{5}{3}$

109. $2x + 3y + 6 = 0$, $m = -\frac{2}{3}$

111. $x + 2y - 2 = 0$, $m = -\frac{1}{2}$

113.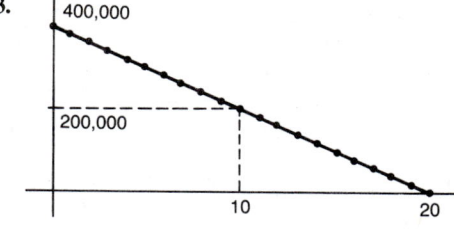

115.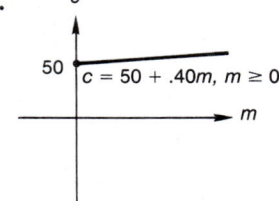
$c = 50 + .40m$, $m \geq 0$

117. a.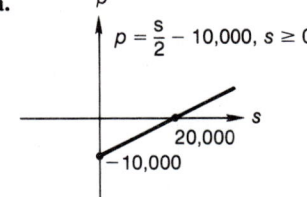
$p = \frac{s}{2} - 10,000$, $s \geq 0$

b. $20,000   c. $30,000

119. Either $0 = 0$ and every (x, y) satisfies the equation, or $c \neq 0$ and no (x, y) satisfies the equation.

Section 3.4

1.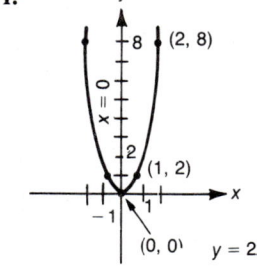
3.

A.28 Solutions

5.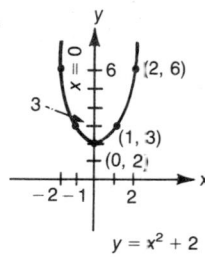

7.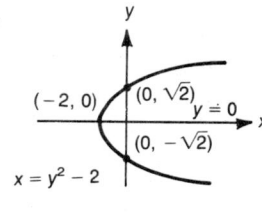

21.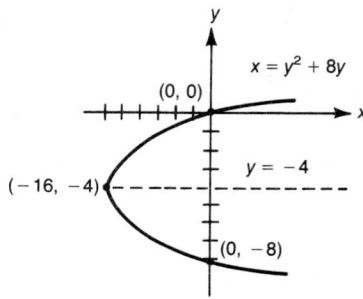

9.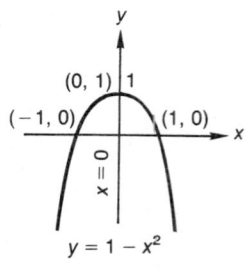

11.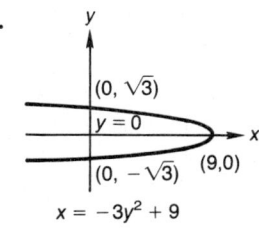

23.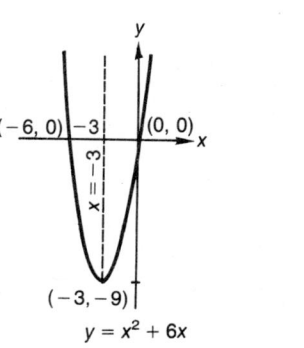

13.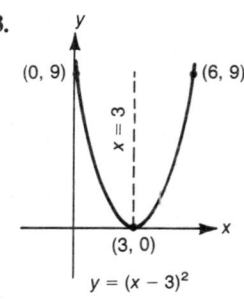

15.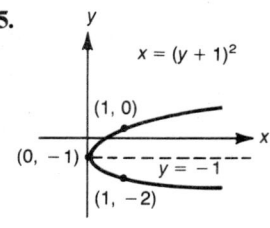

25.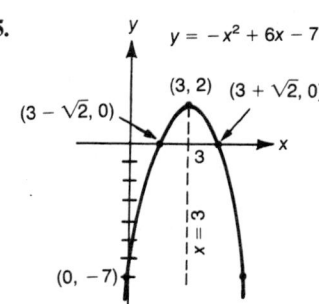

17.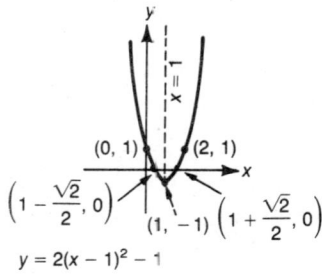

27.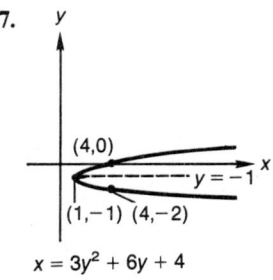

29.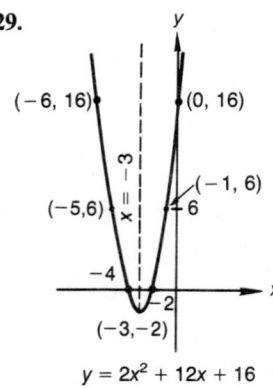

19.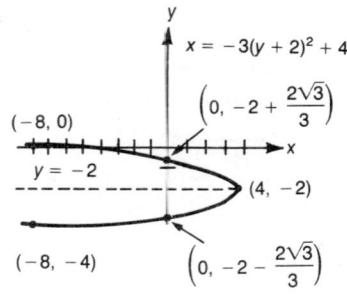

31.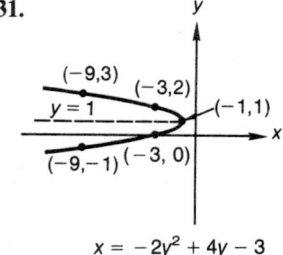

33.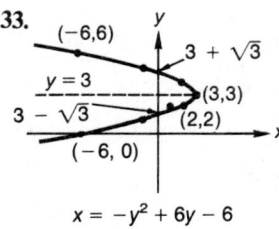

35.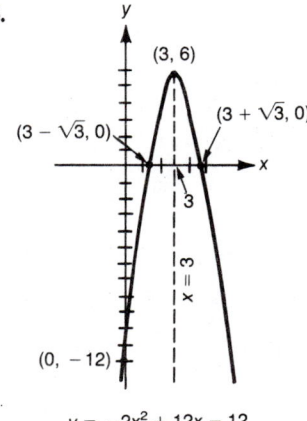

37.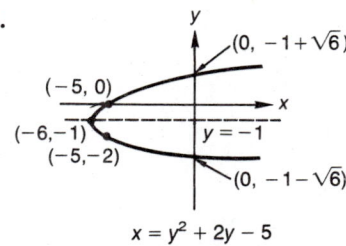

39.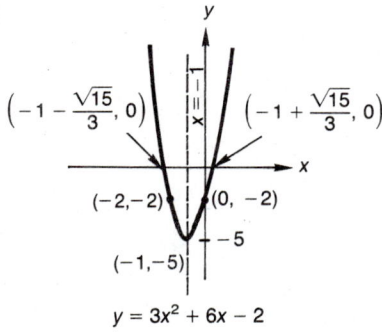

41. 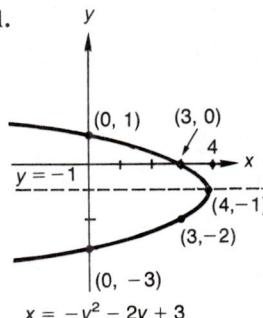   43.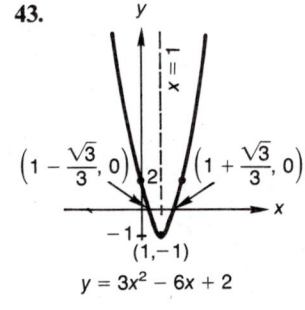

45. **a.** 6400 ft **b.** 20 s **c.** 40 s
47. **a.** 10,000 ft **b.** 20 s **c.** 45 s
49. **a.** 3600 ft **b.** 10 s **c.** 25 s   51. $\left[-1, \frac{1}{2}\right]$

53. $\left(\frac{1}{2}, \frac{5}{2}\right)$ 55. $\left[\frac{1}{3}, 2\right]$ 57. $(-\infty, \infty)$ 59. $(-1, 4)$
61. $[-3, 4]$ 63. $[-\sqrt[3]{12}, \sqrt[3]{2}]$ 65. $(9, \infty)$
67. Best selling price = $750   69. $10.50 per gallon
71. **a.** 4 in **b.** 32 in²   73. 25 ft by 25 ft
75. Between 30 and 40 widgets

Section 3.5

1.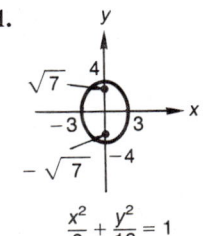

3.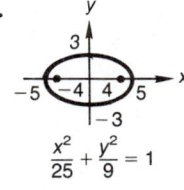

5.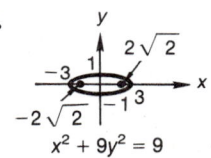

7.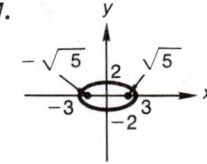

9.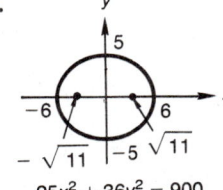

11.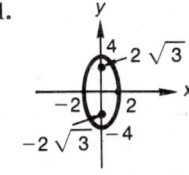

13.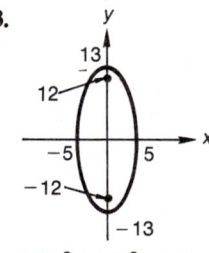

15. $\dfrac{x^2}{9} + \dfrac{y^2}{4} = 1$ 17. $\dfrac{x^2}{25} + \dfrac{y^2}{9} = 1$ 19. $\dfrac{x^2}{169} + \dfrac{y^2}{25} = 1$

21. $\dfrac{x^2}{625} + \dfrac{y^2}{576} = 1$ 23. $\dfrac{x^2}{25} + \dfrac{y^2}{169} = 1$

25. $\dfrac{x^2}{4} + \dfrac{y^2}{64} = 1$ 27. $\dfrac{x^2}{4} + \dfrac{y^2}{5} = 1$

A.30 Solutions

29.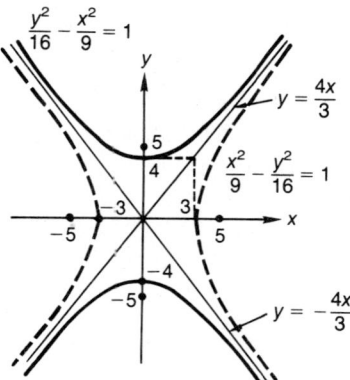

31.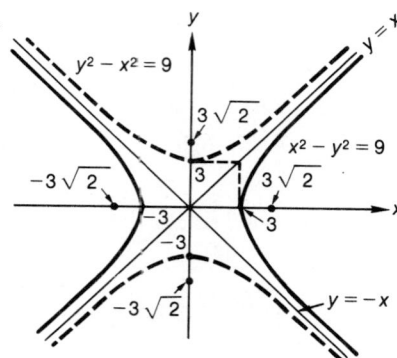

33.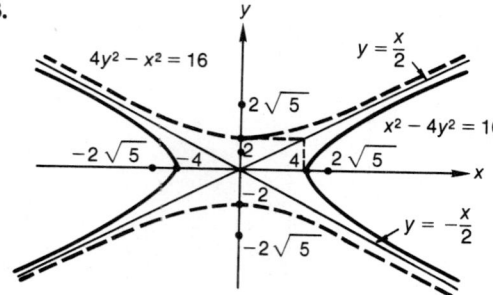

35.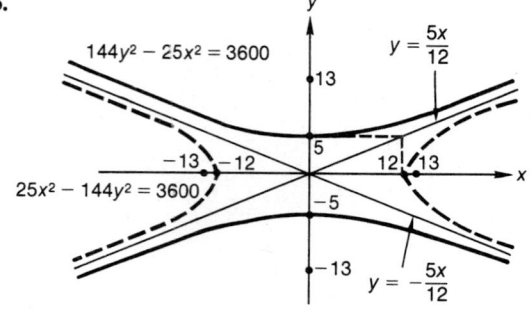

37.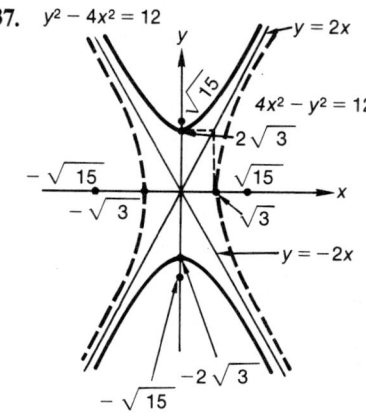

39.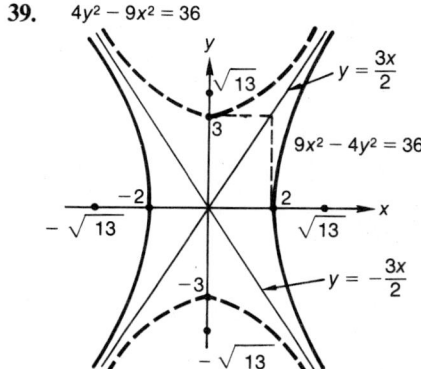

41.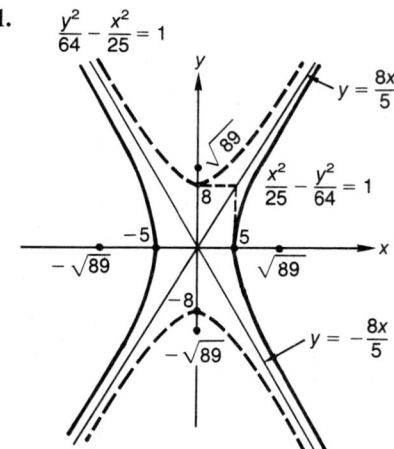

43. $\dfrac{x^2}{16} - \dfrac{y^2}{9} = 1$ **45.** $y^2 - 9x^2 = 1$ **47.** $\dfrac{x^2}{9} - \dfrac{y^2}{36} = 1$

49. $y^2 - 9x^2 = 16$ **51.** $\dfrac{x^2}{4} - \dfrac{y^2}{5} = 1$

53. $\dfrac{5x^2}{9} - \dfrac{5y^2}{36} = 1$ **55.** $\dfrac{y^2}{4} - \dfrac{x^2}{16} = 1$

57.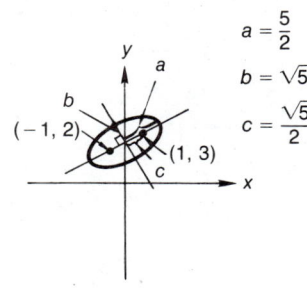

$21x^2 - 4xy + 24y^2 + 10x - 120y + 25 = 0$

27.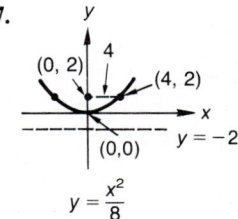

29.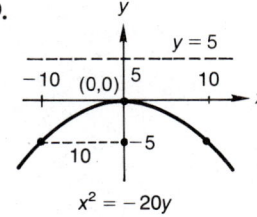

59.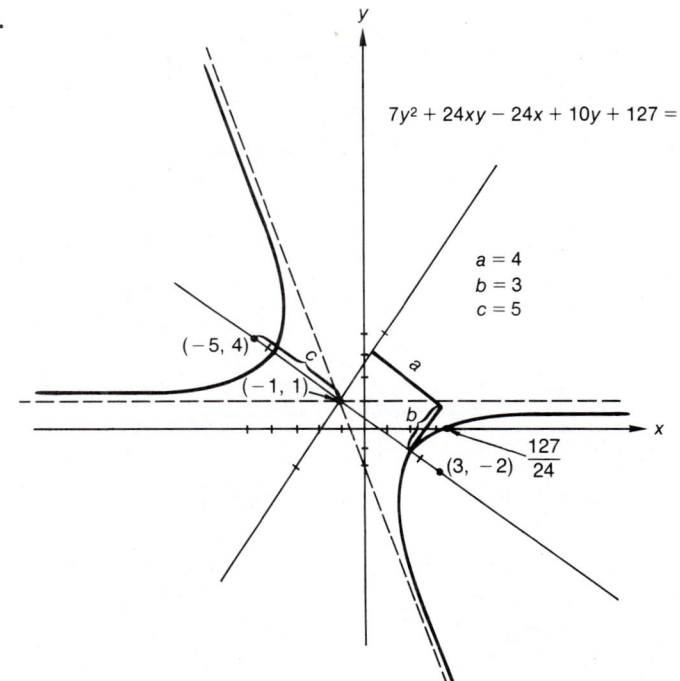

61.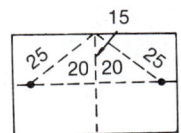

50 m of chain
Stakes located as shown.

31.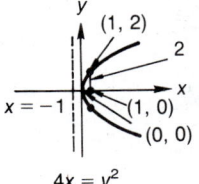

33.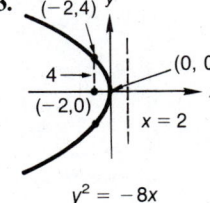

35.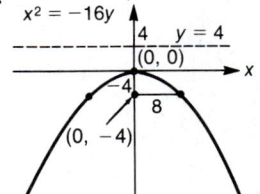

37.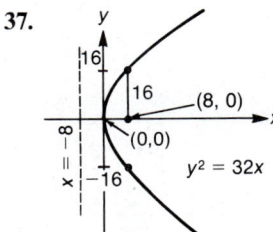

Section 3.6
1. $x - 2y + 3 = 0$ 3. $6x + 4y - 27 = 0$
5. $x - y + 4 = 0$ 7. $y + 1 = 0$ 9. $y^2 = 4x$
11. $x^2 = -20y$ 13. $y^2 = -8x$ 15. $x^2 = 28y$
17. $y^2 = 20x$ 19. $x^2 + 24y = 0$
21. $16x + y^2 + 2y = 31$ 23. $x^2 - 4x + 4y = 16$
25. $8x + y^2 + 14y = 7$

39. $8(x - 1) = -(y - 2)^2$ 41. $12y = -(x + 2)^2$

43. $4(x-1) = (y+1)^2$

45. $\dfrac{(x+2)^2}{25} + \dfrac{(y-1)^2}{16} = 1$

47. $-\dfrac{(x-3)^2}{4} + \dfrac{(y-1)^2}{25} = 1$

49. $\dfrac{(x-2)^2}{4} + (y-3)^2 = 1$

51. $\dfrac{(y+4)^2}{25} - (x+1)^2 = 1$;

conjugate: $(x+1)^2 - \dfrac{(y+4)^2}{25} = 1$

53. $\dfrac{(y-1)^2}{36} - \dfrac{(x-8)^2}{100} = 1$;

conjugate: $\dfrac{(x-8)^2}{100} - \dfrac{(y-1)^2}{36} = 1$

55. $\dfrac{(x+10)^2}{81} - \dfrac{(y-8)^2}{100} = 1$;

conjugate: $\dfrac{(y-8)^2}{100} - \dfrac{(x+10)^2}{81} = 1$;

57.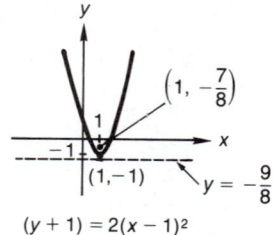

$(y+1) = 2(x-1)^2$

59.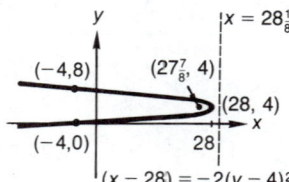

$\dfrac{(x+2)^2}{4} + 4(y-1)^2 = 1$

61.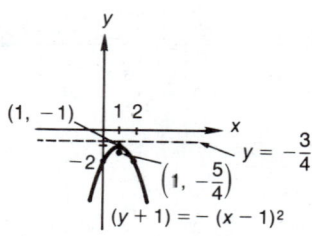

$(x-28) = -2(y-4)^2$

63.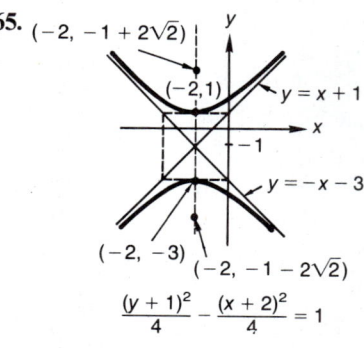

$(y+1) = -(x-1)^2$

65.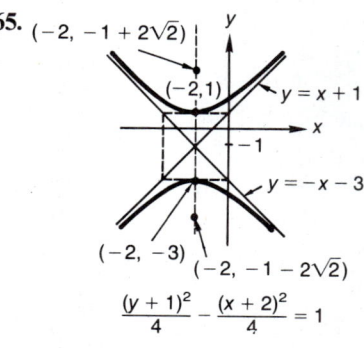

$\dfrac{(y+1)^2}{4} - \dfrac{(x+2)^2}{4} = 1$

67.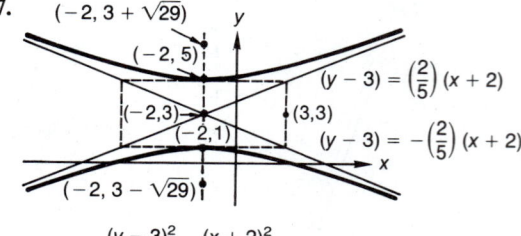

$\dfrac{(y-3)^2}{4} - \dfrac{(x+2)^2}{25} = 1$

69.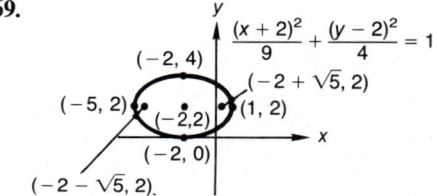

$\dfrac{(x+2)^2}{9} + \dfrac{(y-2)^2}{4} = 1$

71.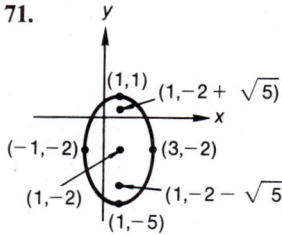

$\dfrac{(x-1)^2}{4} + \dfrac{(y+2)^2}{9} = 1$

73.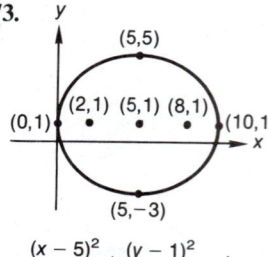

$\dfrac{(x-5)^2}{25} + \dfrac{(y-1)^2}{16} = 1$

Section 3.8

1. a. 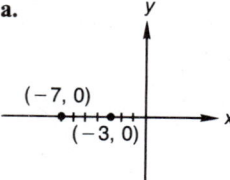 **b.** 4 **c.** $(-5, 0)$ **d.** $y = 0$
e. $x + 5 = 0$

3. a. 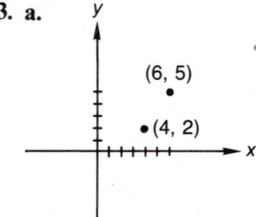 **b.** $\sqrt{13}$ **c.** $\left(5, \dfrac{7}{2}\right)$
d. $3x - 2y - 8 = 0$
e. $4x + 6y - 41 = 0$

5. a. 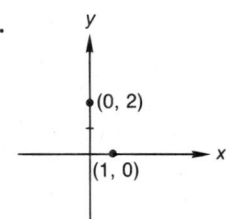 **b.** $\sqrt{5}$ **c.** $\left(\dfrac{1}{2}, 1\right)$
d. $2x + y - 2 = 0$
e. $2x - 4y + 3 = 0$

7. $x^2 + 6x + y^2 - 8y = 0$ **9.** $x^2 - 8x + y^2 - 2y + 12 = 0$
11. $x^2 + 2x + y^2 + 4y - 15 = 0$ **13.** Yes **15.** Yes
17. $6x + y = 3$ **19.** $y = 2$ **21.** $x + 2y = -14$
23. $4x - 3y + 12 = 0$ **25.** $3x + 4y = 25$ **27.** Parallel
29. Parallel **31.** Perpendicular **33.** $y = 5$; $x = -1$
35. $x = 0$; $y = 7$

37. 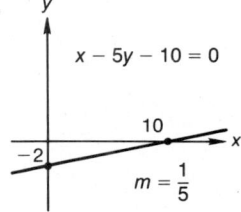 **39.**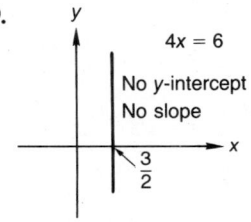

41. 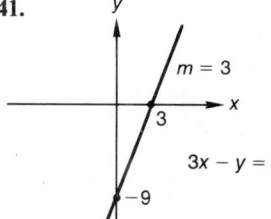 **43.**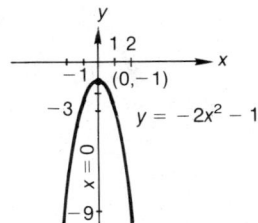

45.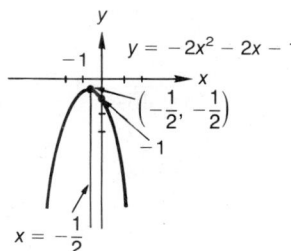

47.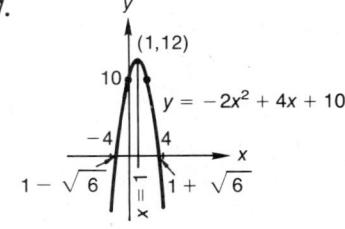

49.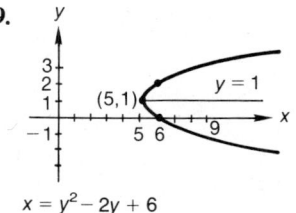

51. $16(x + 2) = (y + 1)^2$ **53.** $[-2, 3]$
55. $\left(-\infty, -\dfrac{5}{3}\right) \cup (4, \infty)$ **57.** $(-\sqrt{3}, \sqrt{3})$

59. 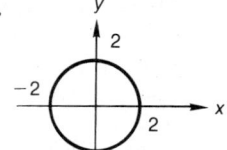 **61.**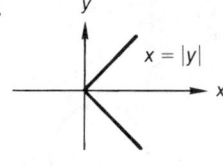

63. 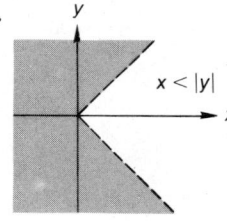 **65.**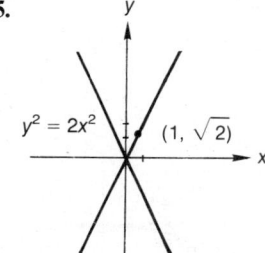

67.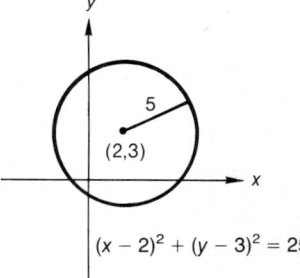

69.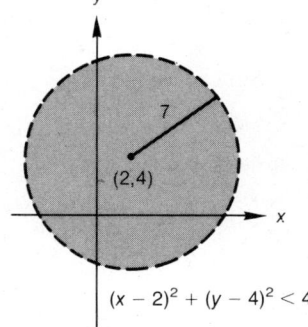

A.34 Solutions

71.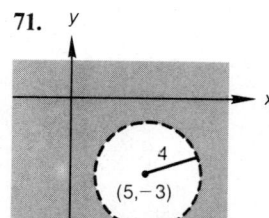
$(x-5)^2 + (y+3)^2 > 16$

73. $x - 1 = 0$

75. $2x - y - 7 = 0$

77. $12(y-1) = -(x-2)^2$

79.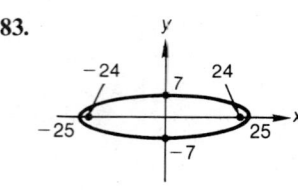
$\dfrac{x^2}{36} + \dfrac{y^2}{49} = 1$

81.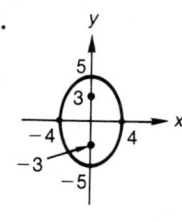
$25x^2 + 16y^2 = 400$

83.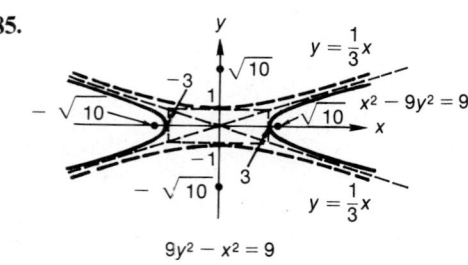
$49x^2 + 625y^2 = 30{,}625$

85.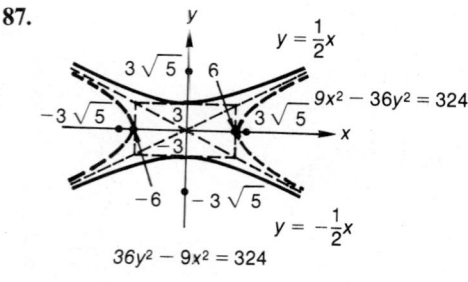
$9y^2 - x^2 = 9$

87.
$36y^2 - 9x^2 = 324$

89.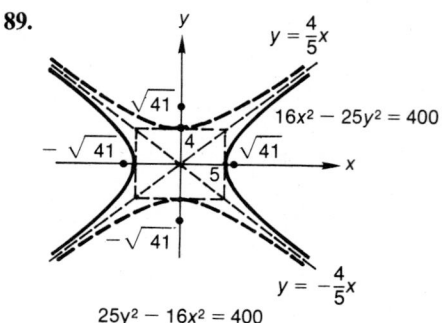
$16x^2 - 25y^2 = 400$
$25y^2 - 16x^2 = 400$

91.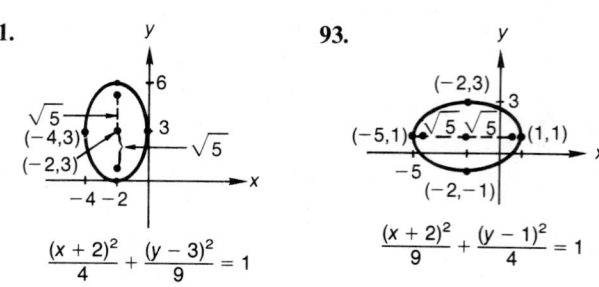
$\dfrac{(x+2)^2}{4} + \dfrac{(y-3)^2}{9} = 1$

93.
$\dfrac{(x+2)^2}{9} + \dfrac{(y-1)^2}{4} = 1$

95.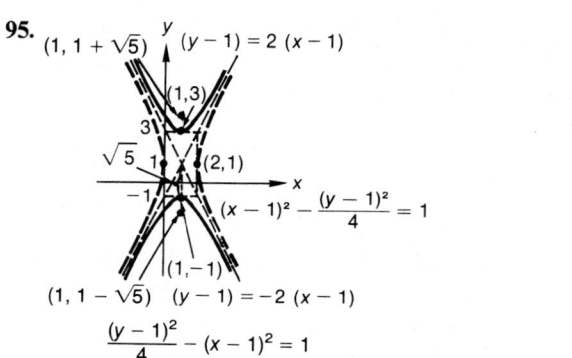
$(x-1)^2 - \dfrac{(y-1)^2}{4} = 1$
$\dfrac{(y-1)^2}{4} - (x-1)^2 = 1$

97.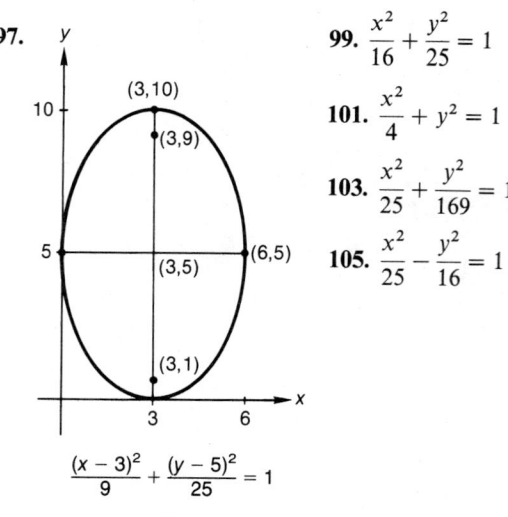
$\dfrac{(x-3)^2}{9} + \dfrac{(y-5)^2}{25} = 1$

99. $\dfrac{x^2}{16} + \dfrac{y^2}{25} = 1$

101. $\dfrac{x^2}{4} + y^2 = 1$

103. $\dfrac{x^2}{25} + \dfrac{y^2}{169} = 1$

105. $\dfrac{x^2}{25} - \dfrac{y^2}{16} = 1$

107. $\dfrac{y^2}{4} - \dfrac{5x^2}{16} = 1$

109. $\dfrac{x^2}{16} - \dfrac{3y^2}{100} = 1$

111. $y^2 - \dfrac{x^2}{9} = 1$

113. $\dfrac{(x+3)^2}{25} + \dfrac{(y-2)^2}{16} = 1$

115. $\dfrac{(x+1)^2}{25} + \dfrac{(y-8)^2}{169} = 1$ 117. $\dfrac{(y-5)^2}{81} - \dfrac{(x-3)^2}{9} = 1$ Conjugate: $\dfrac{(x-3)^2}{9} - \dfrac{(y-5)^2}{81} = 1$

119. $\dfrac{(y+6)^2}{9} - \dfrac{(x-4)^2}{49} = 1$ Conjugate: $\dfrac{(x-4)^2}{49} - \dfrac{(y+6)^2}{9} = 1$ 121. **a.** 144 ft **b.** $t = 1$ **c.** $t = 4$

CHAPTER 4

Section 4.1

1. $4, 9, -21$ 3. $-5, -5, 25$
5. $0, 21, 25$ 7. $1, 0, -3$
9. $3, 4, 1$ 11. Domain: $[1, \infty)$; range: $[0, \infty)$
13. Domain: $(-\infty, 2) \cup (2, \infty)$; range: $(0, \infty)$
15. Domain: $[-5, -2) \cup (-2, 5]$
17. Domain: $(-3, -2] \cup [2, 3)$
19. Domain: $[-6, -1) \cup (-1, 2) \cup (2, 6]$
21. Domain: $[0, 1] \cup [2, \infty)$
23. Domain: $(-\infty, -1] \cup (2, \infty)$

25. 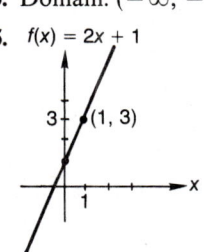   27.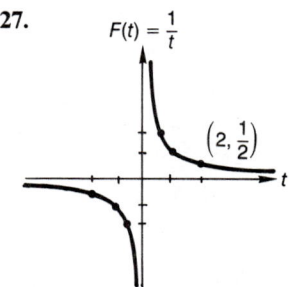

29. 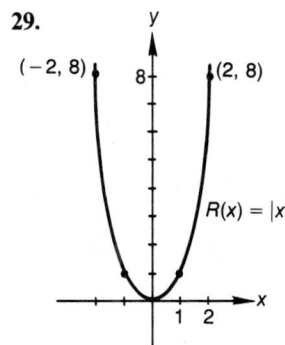   31.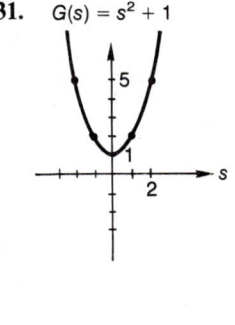

33. 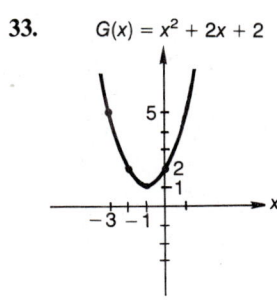   35.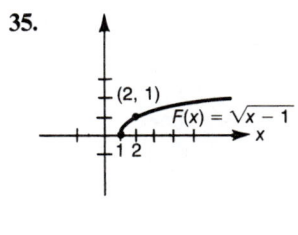

37.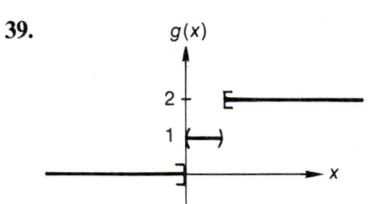

39.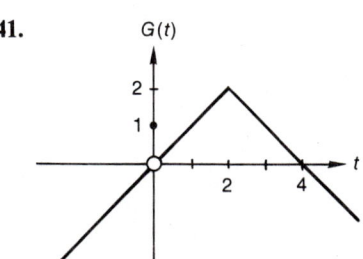

41.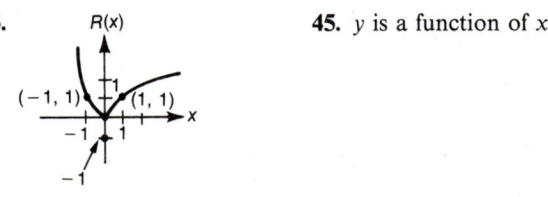

43. (graph of $R(x)$ with points $(-1, 1)$ and $(1, 1)$)   45. y is a function of x.

47. Neither is a function of the other.
49. x is a function of y.   51. Each is a function of the other.
53. x is a function of y.   55. Neither   57. $x = f(y)$
59. $y = f(x)$ 61. Both   63. $x = f(y)$
65. $y = f(x)$ 67. Both   69. Neither

71. **a.** $d = \begin{cases} 25t, & 0 \le t \le 3 \\ 75 + 55(t-3), & t > 3 \end{cases}$
   b. 9 h after the start of the journey **c.** 350 mi
   d. Domain: $[0, \infty)$; range: $[0, \infty)$

73. **a.** $C(n) = \begin{cases} 2n, & 0 \le n \le 9 \\ 1.75n, & 10 \le n \le 99 \\ 1.5n, & n \ge 100 \end{cases}$
   b. 9 cases **c.** 86 cases **d.** domain: integers $n \ge 0$;
   range: $2, 4, \ldots, 18, 17.5, 19.25, \ldots, 173.25, 150, 151.50, \ldots$

75. a. $E = 200m + 200 + 4000$, $I = 400m$, $P = I - E = 200m - 4200$ **b.** $m \geq 21$ **c.** domain: integers $m \geq 0$; range: $\{-4200, -4000, -3800, \ldots\}$

77. a. $r \geq 25f$ **b.** no

79. a. $h(x) = 120 - \dfrac{5x}{2}$ **b.** 20 mg **c.** $[0, 44)$ **d.** $(10, 120]$

Section 4.2

1. a.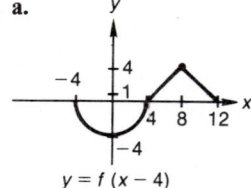
$y = f(x - 4)$

b.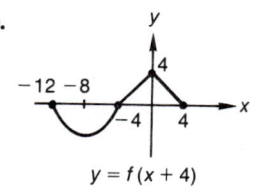
$y = f(x + 4)$

c.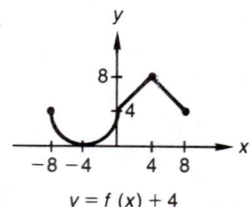
$y = f(x) + 4$

d.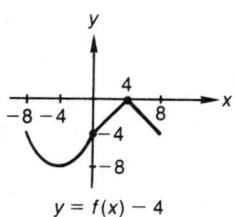
$y = f(x) - 4$

e.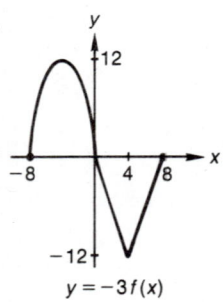
$y = -3f(x)$

f.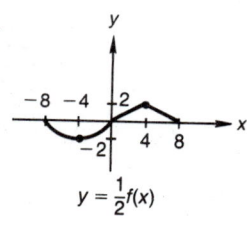
$y = \dfrac{1}{2} f(x)$

g.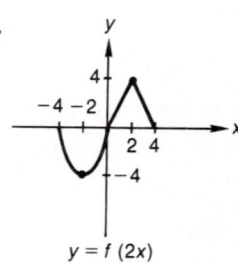
$y = f(2x)$

h.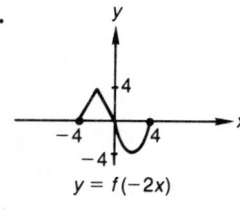
$y = f(-2x)$

i.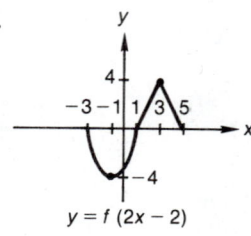
$y = f(2x - 2)$

j.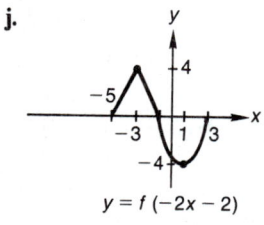
$y = f(-2x - 2)$

k.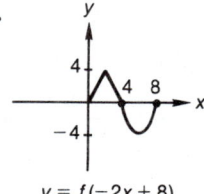
$y = f(-2x + 8)$

l.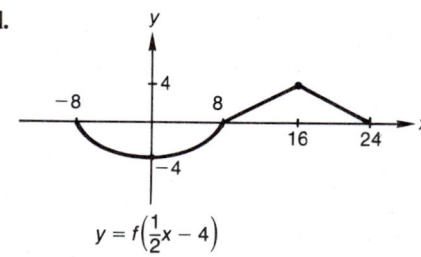
$y = f\left(\dfrac{1}{2}x - 4\right)$

m.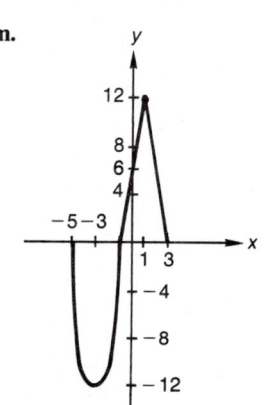
$y = 3f(2x + 2)$

n.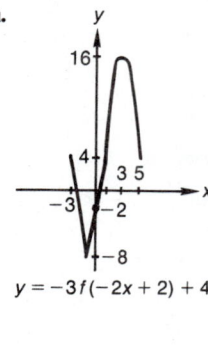
$y = -3f(-2x + 2) + 4$

o.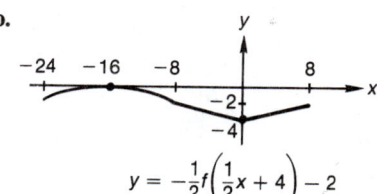
$y = -\dfrac{1}{2} f\left(\dfrac{1}{2}x + 4\right) - 2$

p.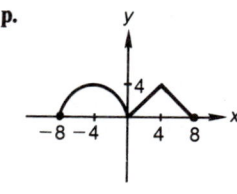
$y = |f(x)|$

q.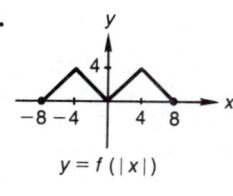
$y = f(|x|)$

3. a–c.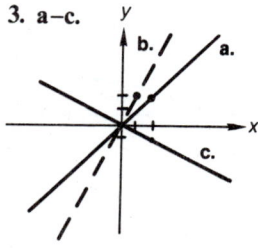

d–f.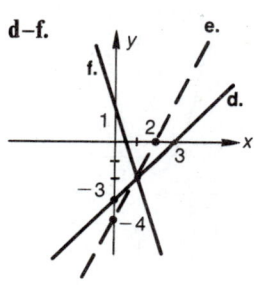

g–h.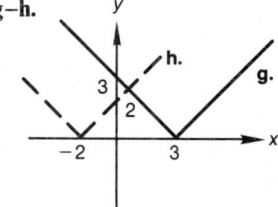

i–j.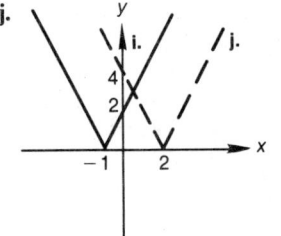

e.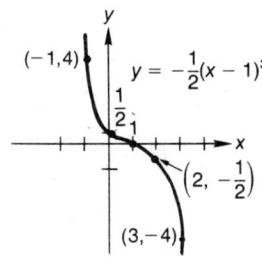

f.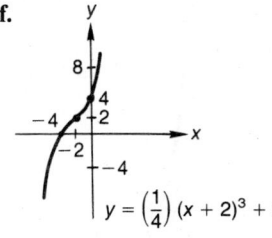

k.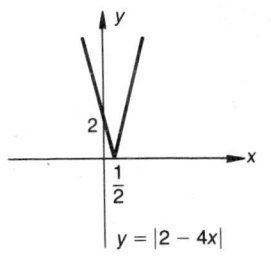
$y = |2 - 4x|$

l.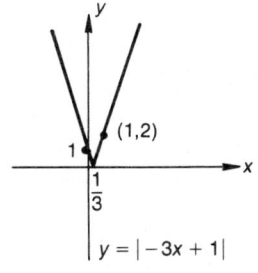
$y = |-3x + 1|$

g.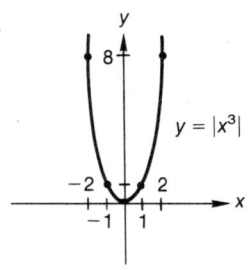
$y = |x^3|$

h.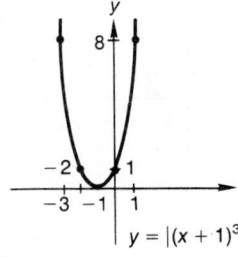
$y = |(x + 1)^3|$

m.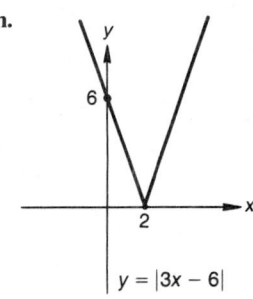
$y = |3x - 6|$

i.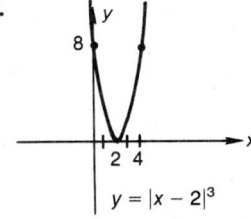
$y = |x - 2|^3$

5. a.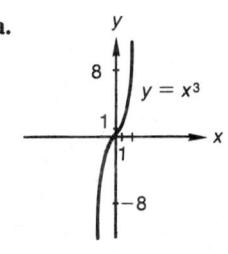
$y = x^3$

b.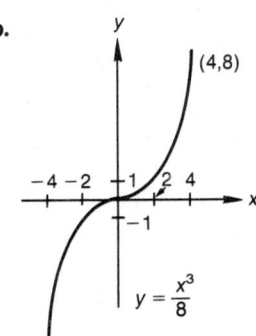
$y = \dfrac{x^3}{8}$

j.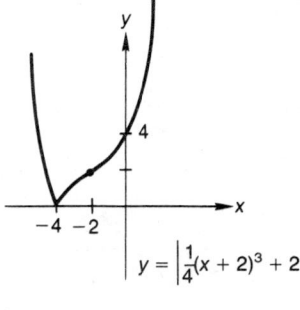
$y = \left|\dfrac{1}{4}(x + 2)^3 + 2\right|$

c.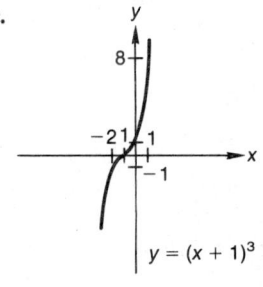
$y = (x + 1)^3$

d.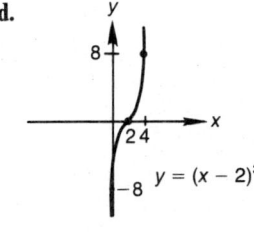
$y = (x - 2)^3$

k.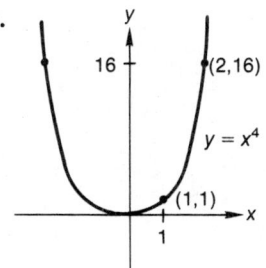
$y = x^4$

l.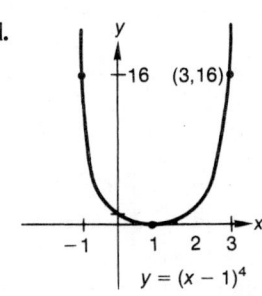
$y = (x - 1)^4$

m.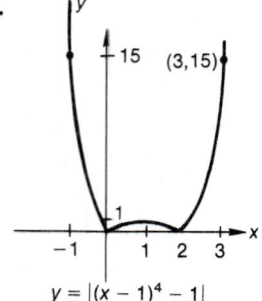

35.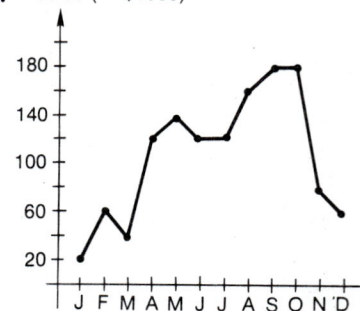

7. 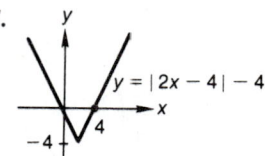   9.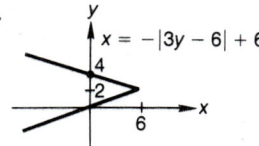

37.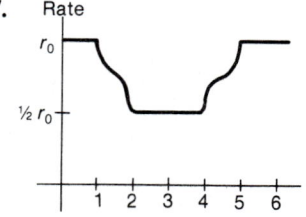

11.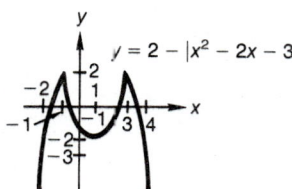

13.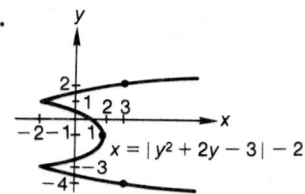

15.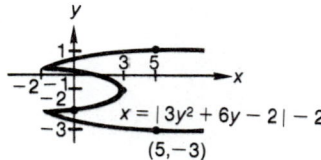

17. a. 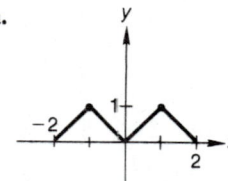   b.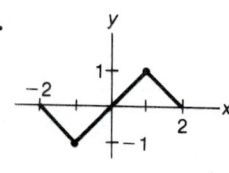

19. a. 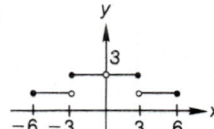   b.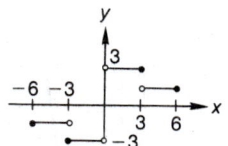

21. Even   23. Odd   25. Neither   27. Neither
29. Odd   31. Even

33. Both even: $f(-x)g(-x) = f(x)g(x)$;
both odd: $f(-x)g(-x) = [-f(x)] \cdot [-g(x)] = f(x)g(x)$

Section 4.3

1. $f(x) + g(x) = 2x^2 + x - 1$; $f(x) - g(x) = -x + 11$;
$f(x) \cdot g(x) = x^4 + x^3 - x^2 + 5x - 30$;
$\dfrac{f(x)}{g(x)} = (x^2 + 5)/(x^2 + x - 6)$, $x \neq -3, 2$;
$2f(x) = 2x^2 + 10$

3. $f(x) + g(x) = x^3 + 2x^2 - 2x - 6$;
$f(x) - g(x) = x^3 - 2x^2 - 4x + 6$;
$f(x) \cdot g(x) = 2x^5 + x^4 - 12x^3 - 3x^2 + 18x$;
$\dfrac{f(x)}{g(x)} = (x^3 - 3x)/(2x^2 + x - 6)$, $x \neq \dfrac{3}{2}, -2$;
$3f(x) = 3x^3 - 9x$

5. $f(x) + g(x) = x^3 - x$; $f(x) - g(x) = x^3 - 2x^2 - x + 2$;
$f(x) \cdot g(x) = x^5 - x^4 - 2x^3 + 2x^2 + x - 1$;
$\dfrac{f(x)}{g(x)} = x - 1$, $x \neq \pm 1$; $4f(x) = 4x^3 - 4x^2 - 4x + 4$

7. $f\left(\dfrac{1}{x}\right) = \left(\dfrac{1}{x}\right) + x$; $f(x^2) = x^2 + \left(\dfrac{1}{x^2}\right)$; $f(2x) = 2x + \left(\dfrac{1}{2x}\right)$

9. $g\left(\dfrac{1}{t}\right) = \left(\dfrac{3}{t^2}\right) - \left(\dfrac{1}{t}\right) + 2$; $g(-x) = 3x^2 + x + 2$;
$g(t + 2) = 3t^2 + 11t + 12$

11. a. $(f \circ g)(t) = 7 - 2t$; $(g \circ f)(t) = 2 - 2t$
b. $D_{f \circ g} = D_{g \circ f} = (-\infty, \infty)$
c. $(f \circ g)(-1) = 9$; $(g \circ f)(-1) = 4$

13. a. $(f \circ g)(t) = 6t^2$; $(g \circ f)(t) = 12t^2$
b. $D_{f \circ g} = D_{g \circ f} = (-\infty, \infty)$
c. $(f \circ g)(-2) = 24$; $(g \circ f)(-2) = 48$

15. a. $(f \circ g)(t) = 9t^2 - 9t + 2$; $(g \circ f)(t) = 3t^2 + 3t - 2$
b. $D_{f \circ g} = D_{g \circ f} = (-\infty, \infty)$ c. $(f \circ g)(1) = 2$; $(g \circ f)(1) = 4$

17. a. $(f \circ g)(t) = t^3 - 6t^2 + 13t - 9$; $(g \circ f)(t) = t^3 + t - 1$
b. $D_{f \circ g} = D_{g \circ f} = (-\infty, \infty)$ c. $(f \circ g)(2) = 1$; $(g \circ f)(2) = 9$

19. a. $(f \circ g)(t) = \dfrac{(2t-1)}{t^2}$; $(g \circ f)(t) = \dfrac{1}{2t - t^2}$
b. $D_{f \circ g} = (-\infty, 0) \cup (0, \infty)$;
$D_{g \circ f} = (-\infty, 0) \cup (0, 2) \cup (2, \infty)$
c. $(f \circ g)(4) = \dfrac{7}{16}$; $(g \circ f)(4) = -\dfrac{1}{8}$

21. a. $(f \circ g)(t) = t$; $(g \circ f)(t) = |t|$
b. $D_{f \circ g} = [0, \infty)$; $D_{g \circ f} = (-\infty, \infty)$
c. $(f \circ g)(-4)$ is not a real number; $(g \circ f)(-4) = 4$

23. a. $(f \circ g)(t) = \sqrt{\dfrac{1-t^2}{t^2}}$; $(g \circ f)(t) = \dfrac{1}{\sqrt{t^2-1}}$
b. $D_{f \circ g} = [-1, 0) \cup (0, 1]$; $D_{g \circ f} = (-\infty, -1) \cup (1, \infty)$
c. $(f \circ g)(2)$ is not a real number; $(g \circ f)(2) = \dfrac{1}{\sqrt{3}}$

25. a. $(f \circ g)(t) = -t^2$; $(g \circ f)(t) = t^2 + 2t$
b. $D_{f \circ g} = D_{g \circ f} = (-\infty, \infty)$
c. $(f \circ g)(-2) = -4$; $(g \circ f)(-2) = 0$

27. $f(t) = t - 1$; $g(u) = u^2$; $g \circ f(t) = (t-1)^2$

29. $f(n) = n + 1$, $g(x) = \dfrac{1}{x^2}$; $(g \circ f)(n) = \dfrac{1}{(n+1)^2}$

31. $f(u) = u^3 + 2$, $g(t) = 3t^2 + 2t + 4$;
$g \circ f(u) = 3(u^3 + 2)^2 + 2(u^3 + 2) + 4$

33. $G(t) = t^2 - t$; $G(x) = x^2 - x$; $G(0) = 0$;
$G(x + 1) = x^2 + x$

35. $P(t) = t^2$; $P(x) = x^2$; $P(2) = 4$; $P(x+2) = x^2 + 4x + 4$

37. $h(t) = t^2 + 2t - 3$; $h(x) = x^2 + 2x - 3$; $h(4) = 21$;
$h(x - 3) = x^2 - 4x$

39. a. $L(2^2) = L(4) = 8$, $[L(2)]^2 = 4$
b. $L(\sqrt{4}) = L(2) = 2$, $\sqrt{L(4)} = \sqrt{8} = 2\sqrt{2}$
c. $L(-1) = -7$, $-L(1) = 1$

41. a. $F(1 + 1) = F(2) = 5$, $F(1) + F(1) = 4$
b. $F(1 - 1) = F(0) = 1$, $F(1) - F(1) = 0$
c. $F(1 \cdot 1) = 2$, $F(1) \cdot F(1) = 4$

43. 0   **45.** -3 **47.** $4x + 2h$ **49.** $2x + 2 + h$

51. $-10x + 4 - 5h$ **53.** $4x^3 + 6x^2h + 4xh^2 + h^3$

55. $f(x) = (0.05)x$, $g(x) = (0.07)x$, $R(x) = f(x) + g(x) = (0.12)x$

57. $e(q) = 3q^2 + 2q$, $t(q) = 2q + 1$,
$d(q) = e(q) \cdot t(q) = 6q^3 + 7q^2 + 2q$

59. a. $C(s) = (0.5)s$; $B(C) = 1000 + (0.1)C$;
bonus $= B(C(s)) = 1000 + (0.1)(0.5)s$;
b. Income = bonus + commission = $1000 + (0.05)s + (0.5)s = 1000 + (0.55)s$

61. $V(r) = 20 \cdot 2000r = 40{,}000r$; $R(V) = V/200$;
$F(R) = (0.2)R = F(R(V(r))) = \dfrac{(0.2)V}{200} = \dfrac{(0.2) \cdot 40{,}000r}{200} = 40r$

63. No   **65.** Yes

Section 4.4

1. One-to-one   **3.** Not one-to-one   **5.** One-to-one
7. One-to-one   **9.** Not one-to-one   **11.** One-to-one
13. Not one-to-one   **15.** Not one-to-one
17. One-to-one   **19.** Not one-to-one

21. $f^{-1}(x) = \dfrac{(1-x)}{4}$ **23.** $g^{-1}(x) = \dfrac{\sqrt[5]{x}}{2}$

25. $P^{-1}(x) = \dfrac{\sqrt[4]{x}}{2}, x \geq 0$ **27.** No inverse

29. No inverse   **31.** No inverse

33. $s^{-1}(x) = 3 + \sqrt{x}, x \geq 0$

35. $F^{-1}(x) = F(x) = \dfrac{2x+1}{x-2}, x \neq 2$

37. $p^{-1}(t) = p(t) = \dfrac{3 - 2t}{2 - 3t}, t \neq \dfrac{2}{3}$

39–47. Insert the expressions for $u = f(x)$ and $v = g(x)$ into $g[u]$, $f[v]$, respectively.

49.
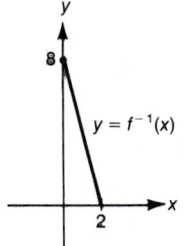
$D_f = R_{f^{-1}} = [0, 8]$
$D_{f^{-1}} = R_f = [0, 2]$

51.
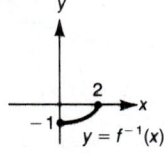
$D_f = R_{f^{-1}} = [-1, 0]$
$D_{f^{-1}} = R_f = [0, 2]$

53.
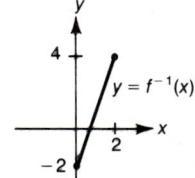
$D_f = R_{f^{-1}} = [-2, 4]$
$D_{f^{-1}} = R_f = [0, 2]$

55. (a and d)
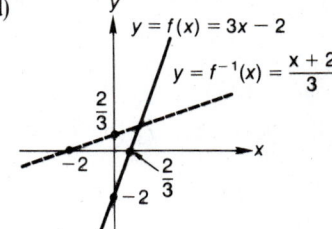

e. $D_f = R_f = D_{f^{-1}} = R_{f^{-1}} = R$

57. (a and d)
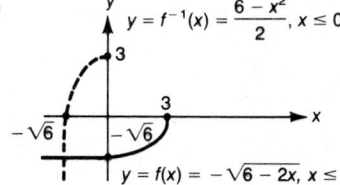

e. $D_f = R_{f^{-1}} = (-\infty, 3]$; $D_{f^{-1}} = R_f = (-\infty, 0]$

59. (a and d)

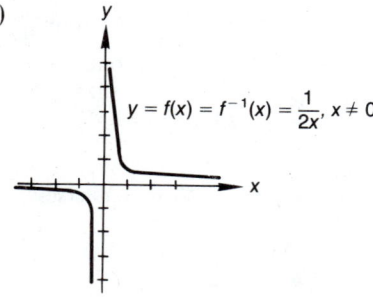

e. $D_f = R_f = D_{f^{-1}} = R_{f^{-1}} = (-\infty, 0) \cup (0, \infty)$

61. (a and d)

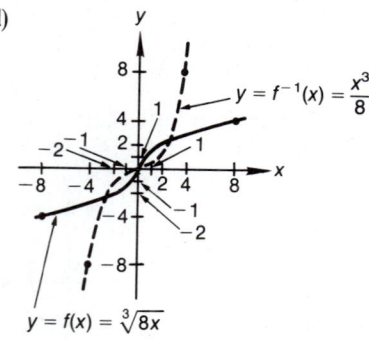

e. $D_f = R_f = D_{f^{-1}} = R_{f^{-1}} = R$

63. (a and d)

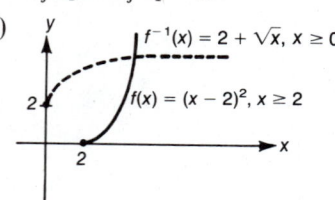

e. $D_f = R_{f^{-1}} = [2, \infty); D_{f^{-1}} = R_f = [0, \infty)$

65. a. $[-3, -2], [0, 4]$ **b.** $(-\infty, -3], [4, 7]$
c. $[-2, 0], [7, \infty)$ **d.** $[-3, 4], [7, \infty)$
e. $(-\infty, -3], [-2, 0], [4, \infty)$

67. a. $(-\infty, -4], [3, 8]$ **b.** $[-4, -2], [8, 10]$ **c.** $[-2, 3]$
d. $(-\infty, -4], [-2, 8]$ **e.** $[-4, 3], [8, 10)$

69. a. $(-\infty, -5)$ **b.** $(-5, 5), (5, \infty)$ **c.** nowhere
d. same as **a**   **e.** same as **b**

71. a. $[0, 5], [8, \infty)$ **b.** $(-15, -7], [5, 7]$ **c.** $[-7, 0]$
d. $[-7, 5], [8, \infty)$ **e.** $(-15, 0], [5, 8]$

73. Increasing on $(-\infty, \infty)$

75. Increasing on $(-\infty, 0]$, decreasing on $[0, \infty)$

77. Decreasing on $(-\infty, 2]$; increasing on $[2, \infty)$

79. Constant $(= 0)$ on $(-\infty, 0]$; increasing on $[0, \infty)$

81. $T = F(C) = \frac{9}{5}C + 32; C = \frac{5}{9}(T - 32) = F^{-1}(T)$

83. a. $I = (0.05)P = f(P)$ **b.** $P = \frac{I}{0.05} = 20I = f^{-1}(I)$

85. a. No   **b.** yes

Section 4.6

1. $x = f(y)$ **3.** Both   **5. a.** Both   **b.** $x = f(y)$
c. $y = f(x)$ **d.** neither   **7.** 36, 9, 1   **9.** $-1, \frac{5}{7}, 2$

11. $4x^2 + 4x + 3, x^4 + 2x^2 + 3, \dfrac{1 + 2x + 3x^2}{x^2}$

13. $g(x) = x^2 - 13x + 50; g(t) = t^2 - 13t + 50; g(1) = 38;$
$g(t - 1) = t^2 - 15t + 64$

15. a. 16 and 100   **b.** 10 and 4   **c.** $6x + 4$ and $6x + 8$
d. -2 and -10 **17.** $[-1, 1) \cup (1, \infty)$ **19.** $\emptyset$

21.

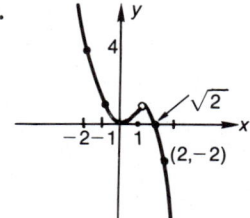

23.

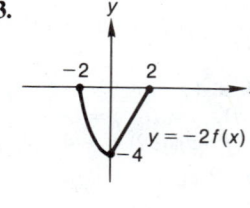

25.

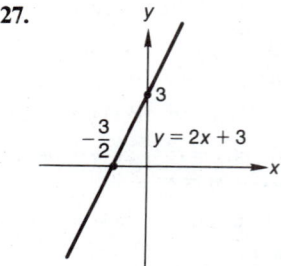

27.

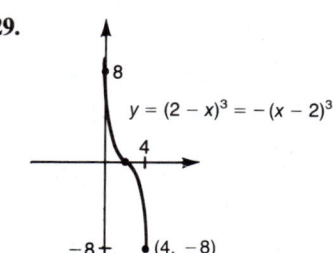

29.

31.

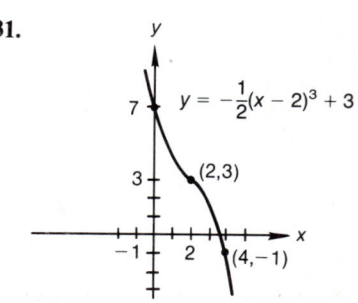

33.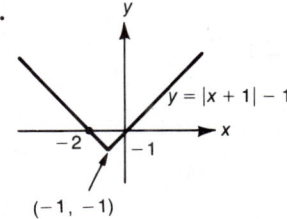

35.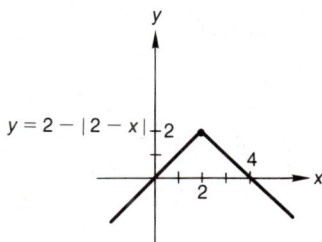

37.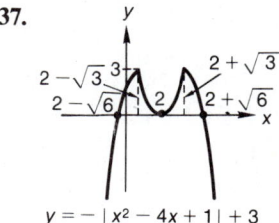

39. a. 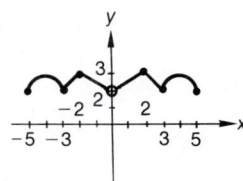   b.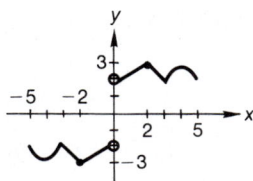

41. Odd   43. 0   45. $8x - 2 + 4h$ 47. $\dfrac{-(2x + h)}{x^2(x + h)^2}$

49. $f(x) + g(x) = x + 3$; $f(x) - g(x) = 5x - 3$;
$f(x) \cdot g(x) = 9x - 6x^2$; $\dfrac{f(x)}{g(x)} = \dfrac{3x}{3 - 2x}$, $x \neq \dfrac{3}{2}$;
$(f \circ g)(x) = 9 - 6x$; $(g \circ f)(x) = 3 - 6x$; $2f(x) = 6x$;
$(f \circ g)(2) = -3$; $(g \circ f)(2) = -9$

51. $f(x) + g(x) = x^2 + x$; $f(x) - g(x) = x^2 - 3x - 2$;
$f(x) \cdot g(x) = 2x^3 - x^2 - 3x - 1$; $\dfrac{f(x)}{g(x)} = \dfrac{x^2 - x - 1}{2x + 1}$,
$x \neq -\dfrac{1}{2}$: $f \circ g(x) = 4x^2 + 2x - 1$; $(g \circ f)(x) = 2x^2 - 2x - 1$;
$-3f(x) = -3x^2 + 3x + 3$; $(f \circ g)(-3) = 29$; $(g \circ f)(-3) = 23$

53. $f(t) = (t + 1)/(t - 1)$; $g(u) = u^2$;
$(g \circ f)(t) = \left(\dfrac{t + 1}{t - 1}\right)^2$

55. $f(u) = u^2 + 2u + 5$; $g(v) = v^{2/3}$;
$(g \circ f)(u) = (u^2 + 2u + 5)^{2/3}$

57. One-to-one   59. Not one-to-one

61. One-to-one   63. No   65. $p^{-1}(t) = \dfrac{t^7}{128}$

67. $g^{-1}(u) = (u - 1)/(u + 1)$, $u \neq -1$

69 and 71. Insert the expressions for $u = f(x)$ and $v = g(x)$ into $g[u]$, $f[v]$, respectively.

73.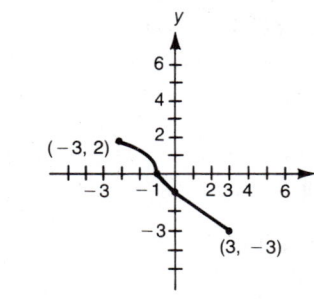

$D_f = R_{f^{-1}} = [-3, 2]$; $D_{f^{-1}} = R_f = [-3, 3]$

75. (a and d)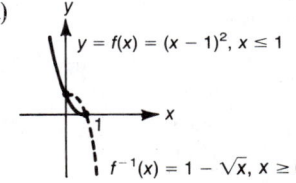

e. $D_f = R_{f^{-1}} = (-\infty, 1]$; $D_{f^{-1}} = R_f = [0, \infty)$

77. (a and d)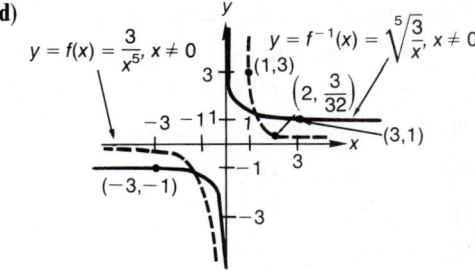

e. $D_f = R_{f^{-1}} = D_{f^{-1}} = R_f = (-\infty, 0) \cup (0, \infty)$

79. a. $(-5, -3], [0, 1]$ b. $[-3, -1), (-1, 0), [3, 4]$
c. $[1, 3]$ d. $(-5, -3], [0, 3]$ e. $[-3, -1), (-1, 0), [1, 4]$

81. Decreasing on $(-\infty, -1)$ and on $(-1, \infty)$

83. Decreasing on $(-\infty, -1]$ and on $[1, 3]$; increasing on $[-1, 1]$ and on $[3, \infty)$

CHAPTER 5

Section 5.1

1. $2x^3 + 4x^2 - 26x + 20$ 3. $-2x^4 + 68x^2 - 450$
5. $4x^3 - 16x^2 + 4x + 24$ 7. 24   9. 18

11. $(4x - 2)(x + 3) - 2$; $P(-3) = -2$
13. $(u + 3)(u + 2) + 1$; $P(-2) = 1$
15. $(t^2 - t + 2)(t - 1) - 2$; $P(1) = -2$

17. $(3y^2 + 5y + 17)(y - 3) + 41$; $P(3) = 41$
19. $(2t^3 - 5t^2 + 7t - 15)(t + 2) + 20$; $P(-2) = 20$
21. $(3w^3 + 2w - 5)(w + 2) + 7$; $P(-2) = 7$
23. $(v^3 + 7v^2 + 8v + 8)(v - 1) + 0$; $P(1) = 0$
25. $(x^5 + x)(x + 1) + 1$; $P(-1) = 1$ 27. Not a factor
29. Not a factor 31. $(b - 2)(b^3 - 4b^2 - 2b - 2)$
33. $(x - 2)(x^4 + 2x^3 + 4x^2 + 8x + 16)$
35. $(w + 2)(w^4 - 2w^3 - 6w^2 + 12w - 18)$
37. $(y + 2)(y^5 - 2y^4 + 2y^3 - 4y^2 + 8y - 6)$
39. $(p + 2)(p^5 - 2p^4 + 6p^3 - 10p^2 + 20p - 10)$

Section 5.2

1. $\{1, -2, 4\}$; $(v - 1)(v + 2)(v - 4)$
3. $\left\{-\dfrac{3}{2}, -\dfrac{1}{3}, \dfrac{1}{2}\right\}$; $(2s + 3)(3s + 1)(2s - 1)$
5. $\left\{-1, -\dfrac{1}{3}\right\}$; $(u + 1)(u + 1)(3u + 1)$
7. $\{2, -2\}$; $(t - 2)(t + 2)(t^2 + 2t + 2)$
9. Potential rational roots: $\pm 1, \pm 2$; none of these satisfies the equation $x^2 - 2 = 0$
11. Potential rational roots: $\pm 1, \pm 5$; none of these satisfies the equation $x^2 - 5 = 0$
13. 1 positive root, 0 or 2 negative roots
15. 0 or 2 positive roots, 1 or 3 negative roots
17. 0 or 2 positive roots, 1 negative root
19. 0, 2, or 4 positive roots, 1 negative root
21. $(-4, 2)$ 23. $(-1, 1)$ 25. $(-1, 2)$ 27. $(-3, 1]$
29. $P(-2) = -5 < 0$; $P(-1) = 1 > 0$
31. $P(0) = -2 < 0$; $P(1) = 2 > 0$
33. $P(-1) = -1 < 0$; $P(0) = 1 > 0$
35. $x[x(3x + 8) - 24] + 7$; 15
37. $x\{x[x(2x + 1) - 3] - 1\} + 4$; 1299
39. $x[x(x\{x[x(x - 1) + 1] - 1\} + 1) - 1] - 2$; 544
41. $u \approx -2.2$ 43. $y \approx -0.7$ 45. $y \approx -1.2$
47. $s \approx -0.9$ 49. $x = 3, x \approx 3.2, x \approx -1.2$ ($x = 1 \pm \sqrt{5}$)
51. $x = -\dfrac{1}{2}, \dfrac{5}{2}, -\dfrac{7}{2}$ 53. $x \approx -2.3$
55. $y = \dfrac{1}{2}, y \approx -3.1, y \approx -1.7$
57. $x = 1, x \approx \pm 1.7, x \approx \pm 1.4$ ($x = \pm\sqrt{3}, \pm\sqrt{2}$)
59. $s \approx 2.4, s \approx 2.7, s \approx -2.0$
61. About 26.8 ft. Use the method of successive approximations.
63. About 7.3 in by 7.3 in by 9.3 in

Section 5.3

1. $x^3 - x^2 + 4x - 4$ 3. $x^4 + 4$
5. $x^4 - 3x^3 + 3x^2 - 3x + 2$
7. $x^5 - 5x^4 + 10x^3 - 10x^2 + 9x - 5$
9. $x^6 - 6x^5 + 16x^4 - 24x^3 + 25x^2 - 18x + 10$
11. $(t - 3)(t^2 + 2t + 2)$; $\{3, -1 \pm i\}$
13. $(w^2 + 4)(w - 2)(w - 1)$; $\{1, 2, \pm 2i\}$
15. $(u^2 + 6u + 13)(3u^2 - 2u - 3)$; $\left\{-3 \pm 2i, \dfrac{1 \pm \sqrt{10}}{3}\right\}$
17. $(z^2 - 8z + 17)(z^2 - 2z + 2)$; $\{4 \pm i, 1 \pm i\}$
19. $(x^2 + 4)(x + 2)(x - 2)$; $\{\pm 2, \pm 2i\}$
21. $(y^2 + 3)(y + \sqrt{3})(y - \sqrt{3})$; $\{\pm\sqrt{3}, \pm\sqrt{3}i\}$
23. $(x - 1)(x^2 + x + 1)$; $\left\{1, \dfrac{-1 \pm \sqrt{3}i}{2}\right\}$
25. $(r + 2)(r^2 - 2r + 4)$; $\{-2, 1 \pm \sqrt{3}i\}$
27. $(x^2 + 2 + 2x)(x^2 + 2 - 2x)$; $\{-1 \pm i, 1 \pm i\}$
29. $(z^2 + 9 - 3\sqrt{2}z)(z^2 + 9 + 3\sqrt{2}z)$;
$\left\{\dfrac{3\sqrt{2} \pm 3\sqrt{2}i}{2}, \dfrac{-3\sqrt{2} \pm 3\sqrt{2}i}{2}\right\}$
31. $(t^2 + 25)(t + 3)$; $\{-3, \pm 5i\}$
33. $(w^2 + 1)(2w - 1)$; $\left\{\dfrac{1}{2}, \pm i\right\}$
35. $(x + 1)(4x^2 - 4x + 2)$; $\left\{-1, \dfrac{1 \pm i}{2}\right\}$

Section 5.4

1. [graph: $P(x) = x^4 - 1$]

3. [graph: $P(x) = x^4 + 2$, point $(1, 3)$]

5. [graph: $P(x) = 4 - x^4$, intercepts $\pm\sqrt{2}$]

7.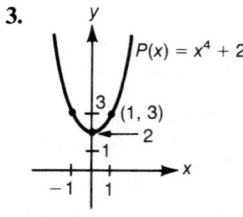
[graph: $P(x) = x^3 + 8$, point $(2, 16)$]

9.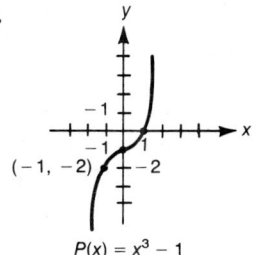
$P(x) = x^3 - 1$

11.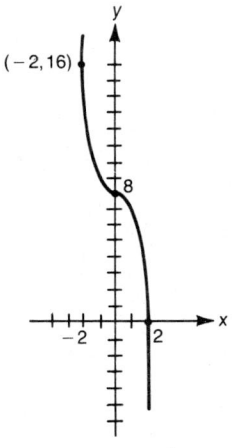
$P(x) = 8 - x^3$

25.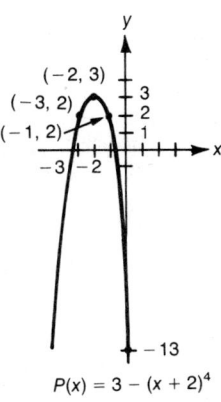
$P(x) = 3 - (x + 2)^4$

27.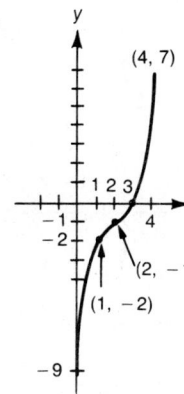
$P(x) = (x - 2)^3 - 1$

13.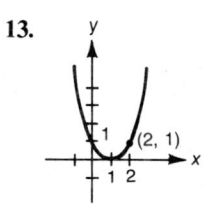
$P(x) = (x - 1)^4$

15.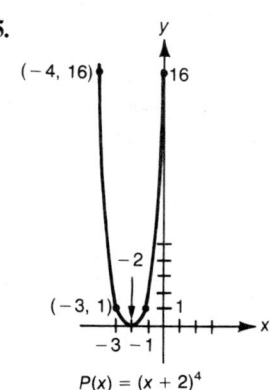
$P(x) = (x + 2)^4$

29.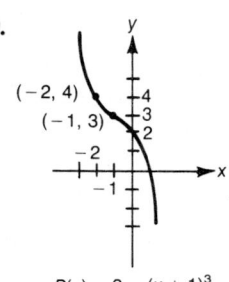
$P(x) = 3 - (x + 1)^3$

31.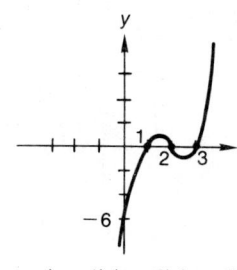
$y = (x - 1)(x - 2)(x - 3)$

33.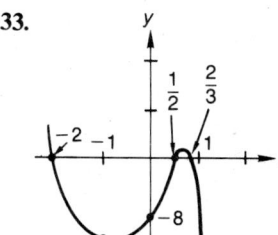
$y = (3x - 2)(1 - 2x)(2x + 4)$

35.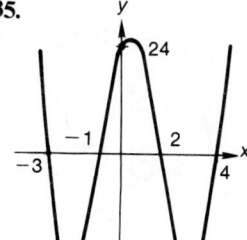
$y = (x + 1)(x - 2)(x + 3)(x - 4)$

17.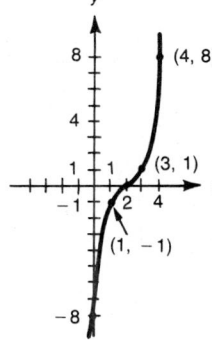
$P(x) = (x - 2)^3$

19.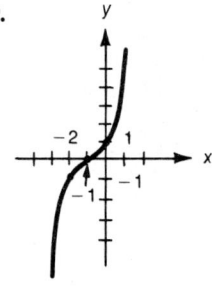
$P(x) = (x + 1)^3$

37.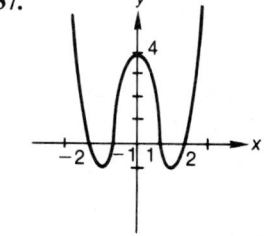
$y = (x^2 - 1)(x^2 - 4)$

39.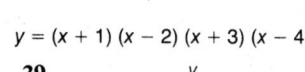
$y = (x^2 - 1)(x^2 - 4) - 4$

21.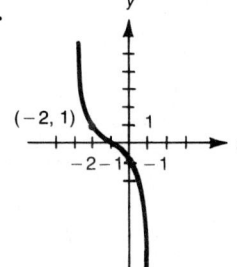
$P(x) = -(x + 1)^3$

23.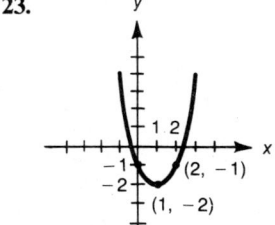
$P(x) = (x - 1)^4 - 2$

41.
$y = 4 - (x^2 - 1)(x^2 - 4)$

43.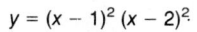
$y = (x - 1)^2 (x - 2)^2$

A.44 Solutions

45.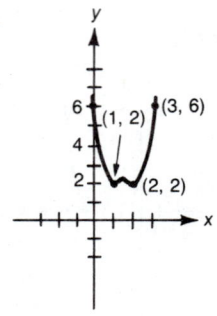
$y = (x-1)^2(x-2)^2 + 2$

47.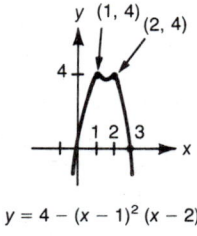
$y = 4 - (x-1)^2(x-2)^2$

Section 5.5

1.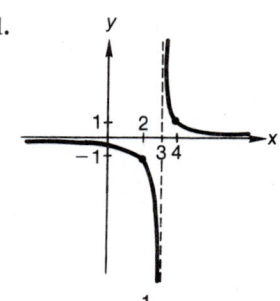
$y = \dfrac{1}{x-3}$

3.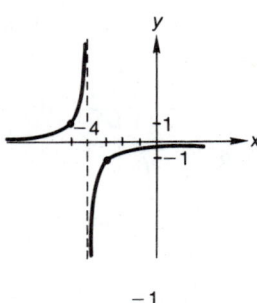
$y = \dfrac{-1}{x+4}$

49.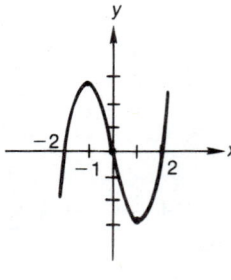
$y = x^3 - 4x$

51.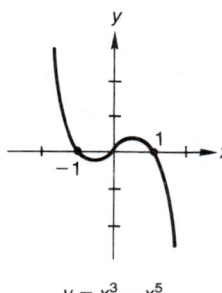
$y = x^3 - x^5$

5.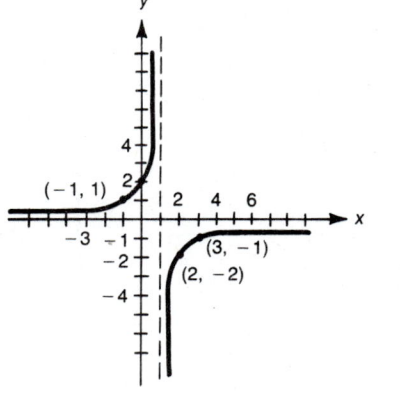
$y = \dfrac{2}{1-x}$

53.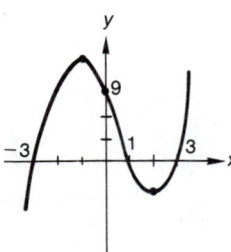
$y = x^3 - x^2 - 9x + 9$

55.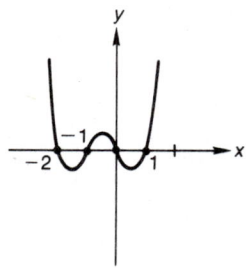
$y = 2x^4 + 4x^3 - 2x^2 - 4x$

7.
$y = \dfrac{2}{(x-1)^3}$

9.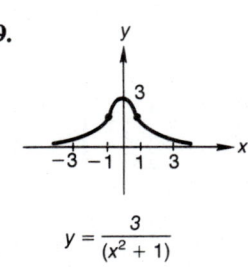
$y = \dfrac{3}{(x^2+1)}$

57.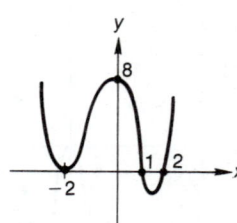
$y = x^4 + x^3 - 6x^2 - 4x + 8$

59.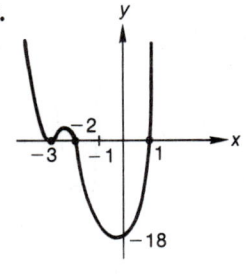
$y = x^4 + 7x^3 + 13x^2 - 3x - 18$

b. $[0, 15]$ **c.** 6.1

11.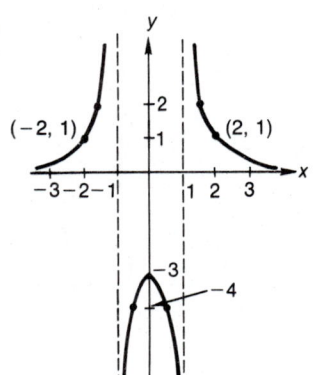
$y = \dfrac{3}{(x^2-1)}$

61. a.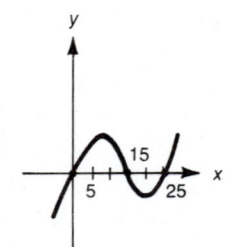
$P(x) = x(30 - 2x)(50 - 2x)$

13.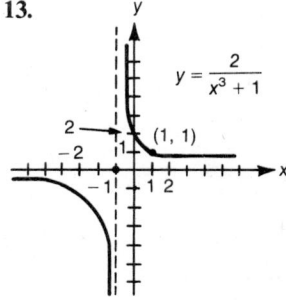

15.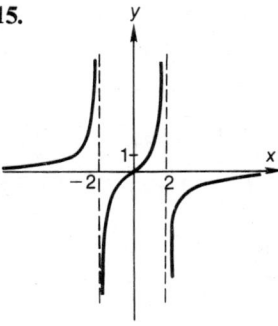

29.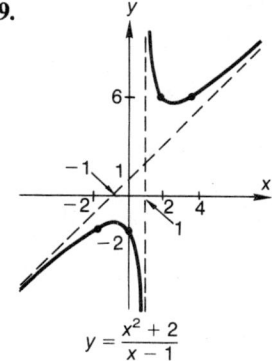

31.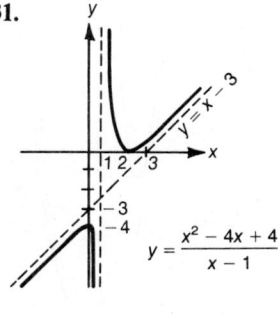

17.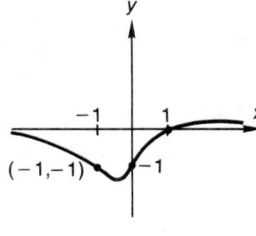

19.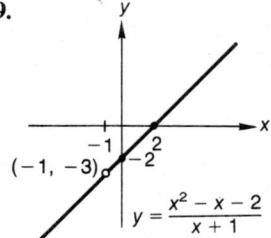

33.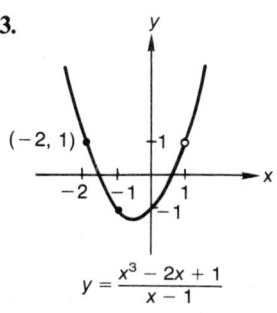

35.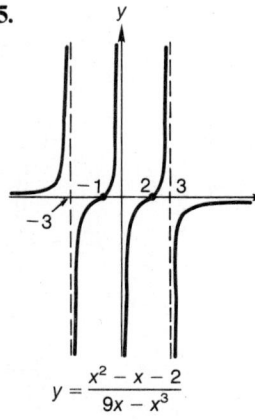

21.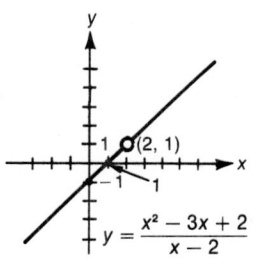

23.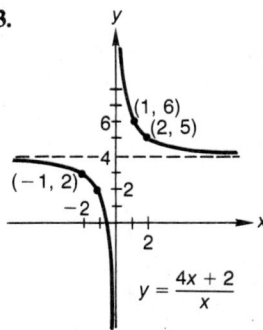

37.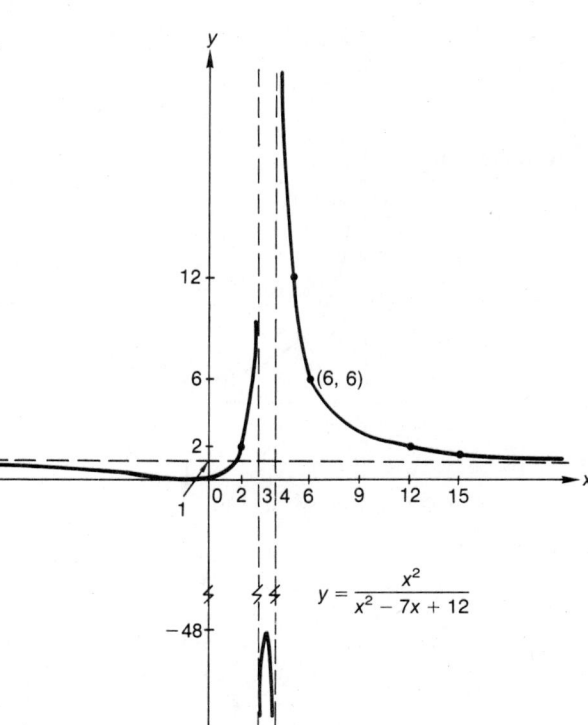

25.

27.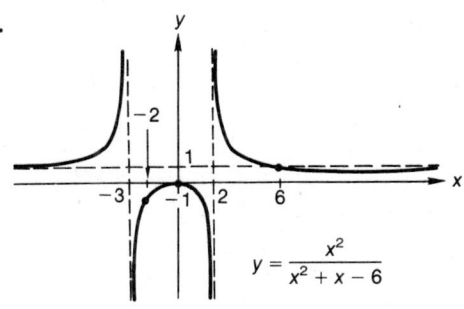

39.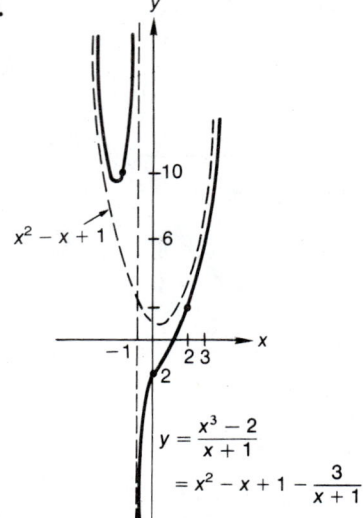

41.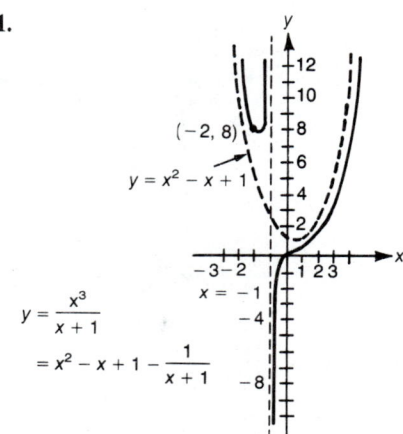

43. a. $(0, \infty)$ **b.**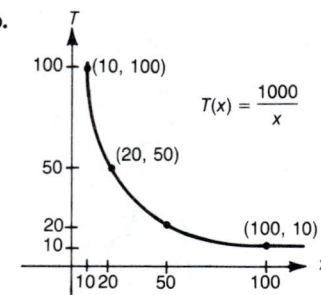

45. a. $C = \$(10{,}000 + 20x)$ **b.** $30
c. $R(x) = \dfrac{10{,}000 + 20x}{x}$ **d.**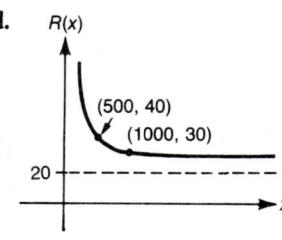

Section 5.7

1. $(v^3 - 4v^2 + 6v + 1)(v + 4) - 6$; $P(-4) = -6$
3. $(2s^4 - 2s^3 + s^2 - s + 4)(s + 1) - 6$; $P(-1) = -6$
5. Not a factor **7.** $(s + 1)(s^4 + 3s^3 - 2s^2 + 2s - 1)$
9. $\left\{-\dfrac{1}{3}, 2\right\}$; $(3x + 1)(x - 2)(x^2 + 2)$
11. No rational roots
13. 1 or 3 positive roots; no negative roots
15. 1, 3, or 5 positive roots; no negative roots
17. $(-2, 1)$ **19.** $(-1, 3)$
21. $P(0) = -10 < 0$; $P(1) = 4 > 0$
23. $P(-2) = 60 > 0$; $P(-1) = -20 < 0$ **25.** -0.8
27. 1.2 **29.** $\{0.8, -2.2, -0.6\}$ **31.** $\{-1.9, 0.3, 1.3\}$
33. $x^3 - 4x^2 - 2x + 20$ **35.** $x^4 - 2x^3 + x^2 + 2x - 2$
37. $\{2, \pm 2i\}$; $(z - 2)(z^2 + 4)$
39. $\{0, 1, \pm i\}$; $t(t - 1)(t^2 + 1)$
41. $\left\{-1 \pm i, \dfrac{-1 \pm \sqrt{5}}{2}\right\}$, $(w^2 + 2w + 2)(w^2 + w - 1)$
43. $(w - 4)(2w^2 - 4w - 3)$
45. $(v - 1)(v + 1)(v^2 + 2v + 2)$

47.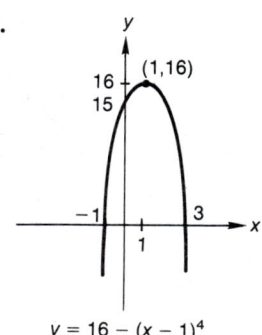

49.

51.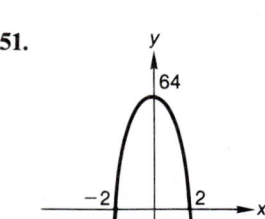

53.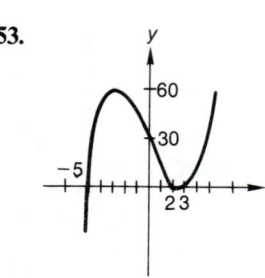

55.

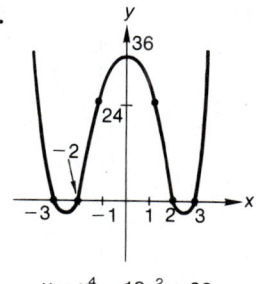

$y = x^4 - 13x^2 + 36$

59.

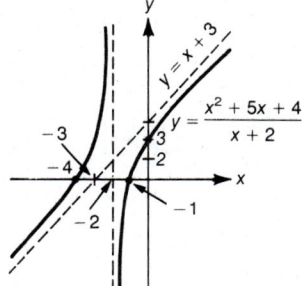

57.

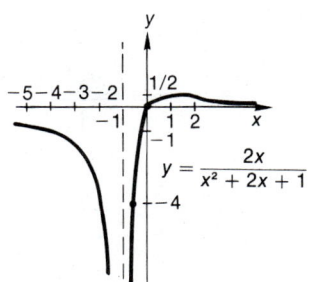

61.

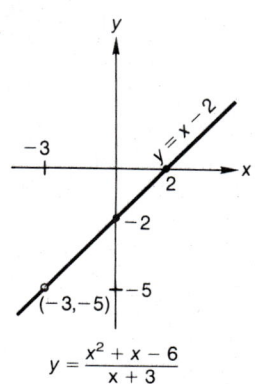

CHAPTER 6

Section 6.1

1. 3.322   **3.** 365.443   **5.** 16.552

7.

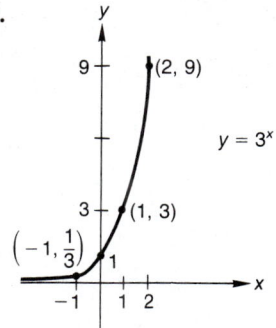

9 and 11.

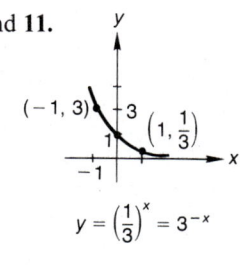

17.

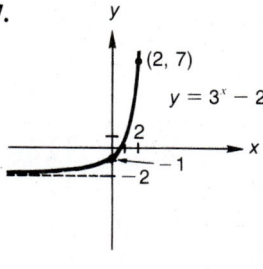

19.

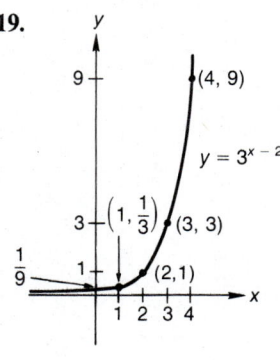

13.

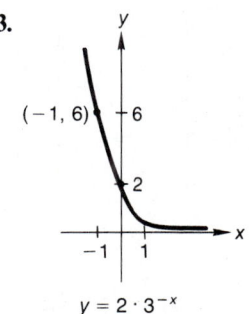

15.

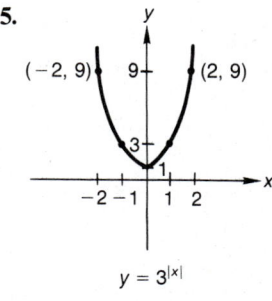

21.

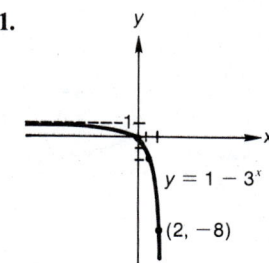

23.

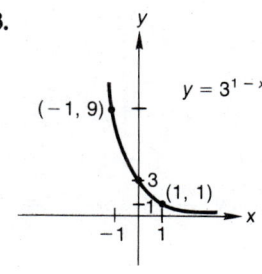

25.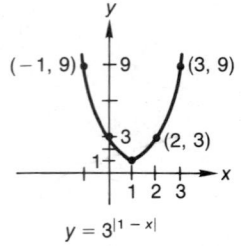
27.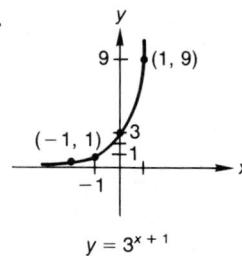
45.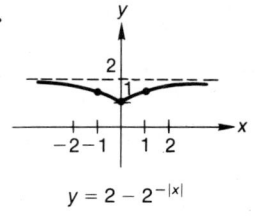
47.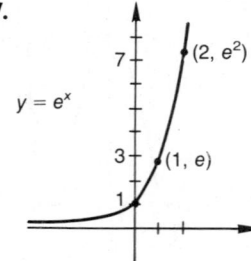

29.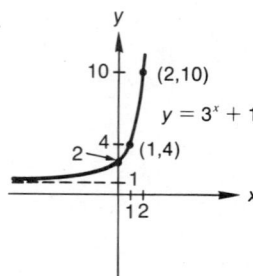
31.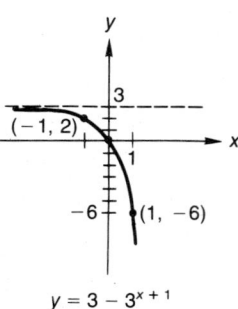
49.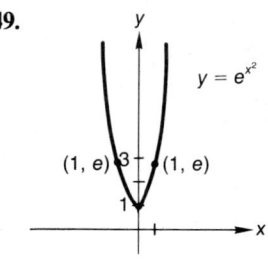
51.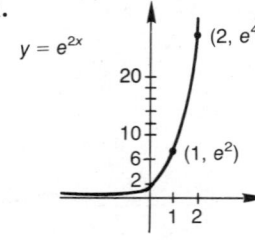

33.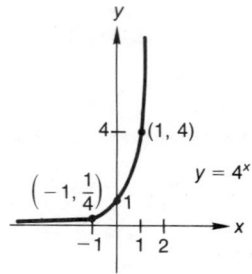
35.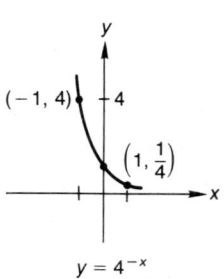
53.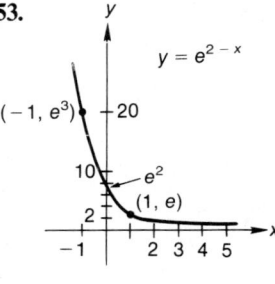
55.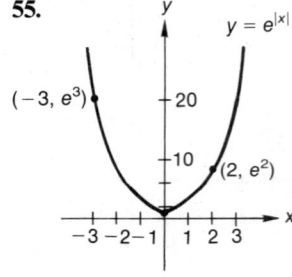

37.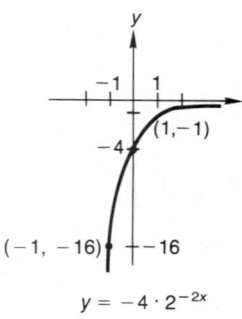
39.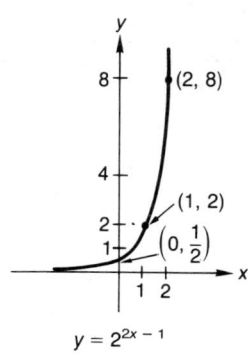
57.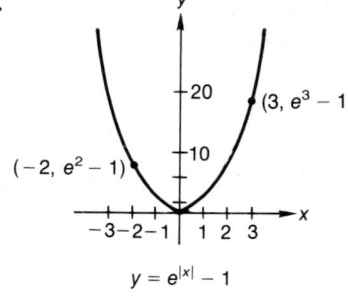

59. $x = \dfrac{1}{2}$ 61. $y = -3$ 63. $r = \dfrac{2}{3}$ 65. $t = -\dfrac{2}{3}$

67. $x = 4$ 69. $v = 2$ 71. $w = -2$ 73. $x = 0$

75. $z = 2$ 77. $x = 2$ 79. $x = -1$ 81. $r = 2$

83. $t = \pm 2$ 85. $v = \pm 4$ 87. $s = 4$ 89. $x = -\dfrac{5}{4}$

91. $u = 81$ 93. $v = 33$ 95. $x = 9, -3$

97. $s = -1, -2$

99. **a.** $\sim 55.78°F$, $\sim 53.36°F$
b. 1 ft: 80°F and 2.5°F; 3 ft: 61.25°F and 41.875°F

41.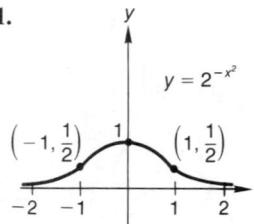
43.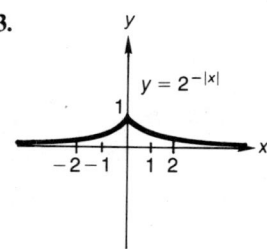

c.

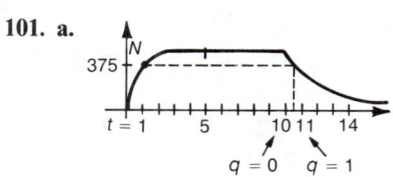

101. a.

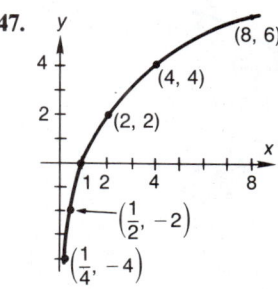

b. Almost immediately since $N(5) \approx 500$.
c. Less than 1/2 year; see dotted line on graph.

Section 6.2

1. $\log_{10} 100{,}000 = 5$ **3.** $\log_5\left(\dfrac{1}{125}\right) = -3$

5. $\log_{0.001} 1 = 0$ **7.** $\log_{32} 2 = \dfrac{1}{5}$ **9.** $3^2 = 9$

11. $4^{3/2} = 8$ **13.** $\left(\dfrac{1}{16}\right)^{1/2} = \dfrac{1}{4}$ **15.** $\left(\dfrac{1}{6}\right)^{-2} = 36$

17. $(16)^{-1/4} = \dfrac{1}{2}$ **19.** 8   **21.** 4   **23.** 2   **25.** $\dfrac{3}{2}$

27. -2 **29.** $\dfrac{3}{2}$ **31.** $\dfrac{8}{3}$ **33.** $-\dfrac{2}{3}$ **35.** 16   **37.** 7

39. 27   **41.** -8 **43.** $\dfrac{10}{3}$ **45.** 2

47.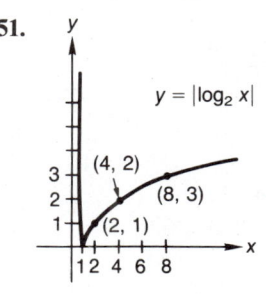

49.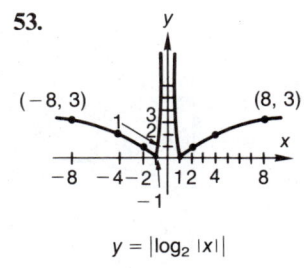

51. 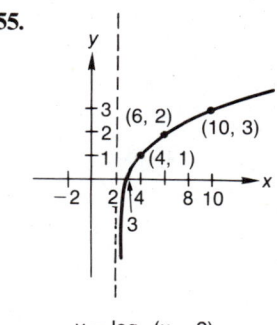

55.

$y = \log_2 (x - 2)$

57.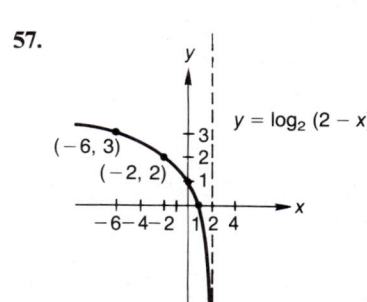

59.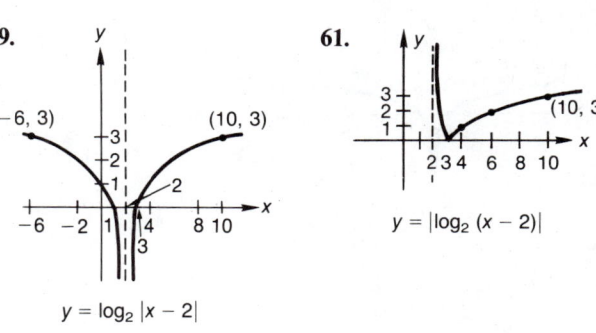

$y = \log_2 |x - 2|$

61.
$y = |\log_2 (x - 2)|$

63.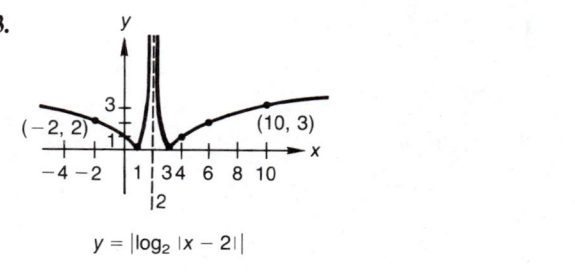

$y = |\log_2 |x - 2||$

65.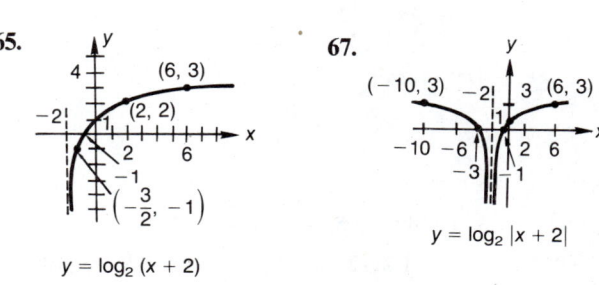

$y = \log_2 (x + 2)$

67.
$y = \log_2 |x + 2|$

69.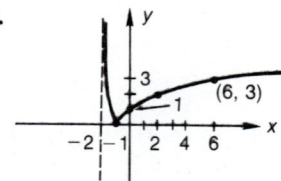
$y = |\log_2 (x + 2)|$

71.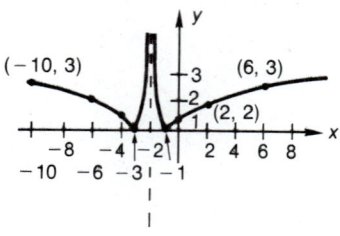
$y = |\log_2 |x + 2||$

73.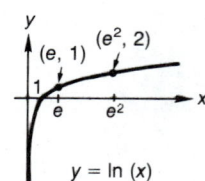
$y = \ln (x)$

75.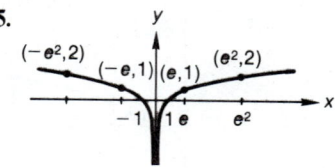
$y = \ln |x|$

77.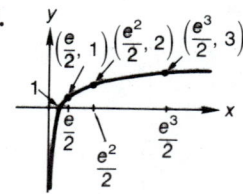
$y = \ln 2x$

79. a. 302 h, 2 h **b.**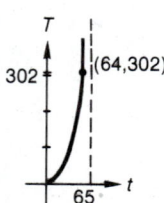

c. $t = 65 - 2^{(302-T)/50}$

Section 6.3

1. 1.602 **3.** 1.1549 **5.** −7.3313 **7.** 0.5187
9. −0.7571 **11.** 1.7712 **13.** 3.8928 **15.** 3.0391

17. 0.2519 **19.** 1.2851 **21.** 0
23. $-\log_b 8 = -3 \log_b 2$ **25.** $\log_2(4x^2 - 1)$ **27.** 3
29. $\log_b \dfrac{1}{(u + 1)^2} = -\log_b(u + 1)^2$ **31.** $x = \dfrac{1}{125}$
33. $v = \dfrac{-1}{32}$ **35.** $u = \pm 5$ **37.** $z = 81$ **39.** $t = \dfrac{1}{10}$
41. $k = 8$ **43.** $z = 8$ **45.** $w = \pm 3$
47. No solutions **49.** $r = -2$ **51.** $w = 5$
53. $y = -\sqrt{3}$ **55.** $s = 2$ **57.** $x = -1$ **59.** $y = 3$
61. $\dfrac{2 \log 3 - 3 \log 10}{\log 3 - \log 10} \approx 3.9125$
63. $\dfrac{5 \log 4 + \log 3}{2 \log 3 - \log 4} \approx 9.9023$
65. $\dfrac{2 \log 3 - \log 2}{\log 2 + \log 3} \approx 0.8394$ **67.** $\dfrac{\log 7 - \log 2}{\log 7 + \log 2} \approx 0.4747$
69. $\dfrac{2 \log 13 + \log 11}{\log 11 - \log 13} \approx -45.062$

71.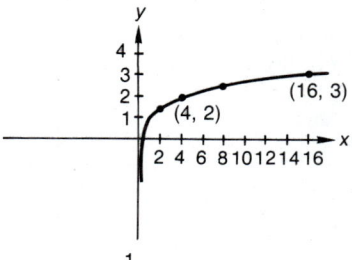
$y = \dfrac{1}{2} \log_2 4x$

73.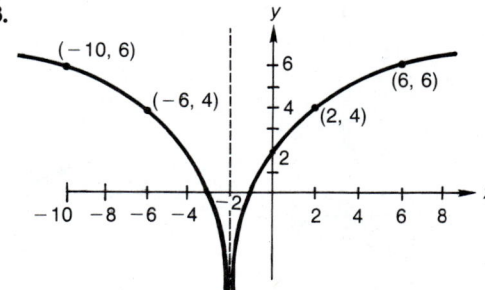
$y = \log_2 (x + 2)^2$

75.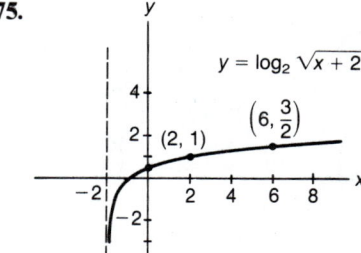
$y = \log_2 \sqrt{x} + 2$

77.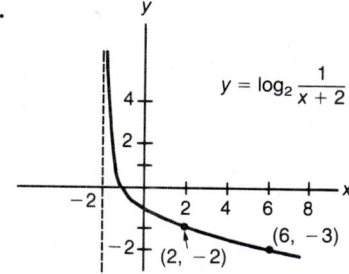

79.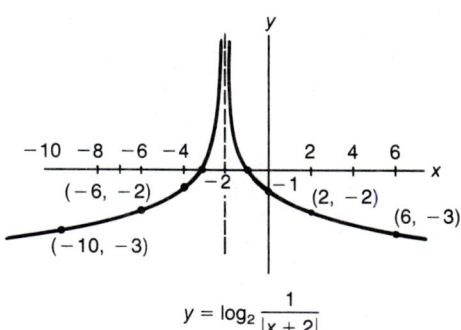

81.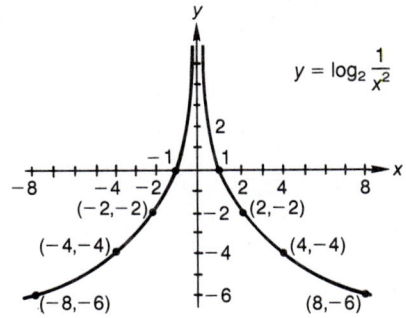

83.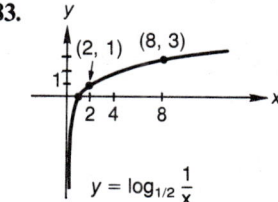

85. a. $5^5 \cdot \$10 = \$31,250$ **b.** 12 **c.** 14

87. 20 times as loud; 100 times as intense

89. a. $t = \dfrac{\log(T-70) - \log(28.6)}{\log(0.97)}$

b. $28\frac{1}{2}$ hours earlier, or 5:30 P.M. the previous day

Section 6.4

1. a. 2.0899 **b.** 4.0899 **c.** -2.9101

3. a. -0.2629 **b.** 5.8452 **c.** -35.6904

5. a. 1.2246 **b.** -1.2246 **c.** -0.5580

7. a. -6.3750 **b.** -5.6250 **c.** 1.0938

9. a. 14.5593 **b.** -0.3823 **c.** -0.9510

11. a. 5210 **b.** 52.1 **c.** 5,210,000

13. a. 0.00603 **b.** 0.000603 **c.** 0.00000603

15. a. 805 **b.** 8,050,000 **c.** 8050

17. a. 0.0911 **b.** 0.000911 **c.** 9110

19. a. 268 **b.** 26.8 **c.** 2.68 **21.** 0.4202 **23.** 2.8304

25. 4.4973 **27.** -3.1888 **29.** 1.671 **31.** 0.4539

33. 0.0001751 **35.** 0.00002078 **37.** 4,096,000

39. 0.009278

Section 6.5

1. a. 46,009 **b.** 12 yr 8 mo **c.** 34 yr 7 mo

3. $131,250 **5. a.** 225,000 **b.** 389,711

7. a. ~5.8 h **b.** ~6.6 h ago **c.** ~673 **d.** ~71

9. a. ~9.1 yr **b.** ~24.9 yr **c.** ~49.8 yr **d.** 300,338,967

11. a. 5.614% **b.** 8.16% **c.** 7.25% **d.** 7.25% **e.** 7.79%

13. a. $13,498.59 **b.** $8869.20

15. a. ≈ 14 yr **b.** $\approx 11\frac{3}{4}$ yr **c.** ≈ 10 yr

17. 6.93% **19. a.** $7871.71 **b.** $7830.69

21. a. $1487.11 **b.** $1481.64

23. a. ~$30,925,927,750, or about $30.9 billion
b. ~$32,524,707, or about $32.5 million

25. a. 2115 yr ago **b.** in 5315 yr **c.** 1957 units

27. a. 366.8 units **b.** 282.8 units **c.** 141.4 units
d. ~42.6 days

29. ~8066 yr old

31. a. In 3.2 min **b.** 2.6 min ago **c.** 43.5 coulombs
d. 70.7 coulombs

33. a. In 13.5 min **b.** 45 min from now **c.** 6.56 min ago
d. 561,167 ft³

35. a. ~1047°C **b.** ~356°C **c.** ~292 min, or 4 h 52 min

37. a. ~$5.2 trillion **b.** ~$10.8 trillion **c.** ~$1.2 trillion
d. ~$0.6 trillion **e.** ~9.46 yr

39. a. $T = 30(0.95)^{t/15}$
b. Never, theoretically, since $30(0.95)^{t/15} \neq 0$ for all $t \geq 0$

Section 6.7

1. $\log_5 125 = 3$ **3.** $\log_{16} 1 = 0$ **5.** $10^2 = 100$

7. $\left(\dfrac{1}{2}\right)^{-3} = 8$ **9.** 132 **11.** 25 **13.** -6 **15.** 2

17. $\dfrac{1}{3}$ **19.** $\dfrac{7}{2}$ **21.** 18 **23.** $\log_6 26{,}244$ **25.** 0
27. 1 **29.** 11.7305 **31.** -0.4911 **33.** -0.2519
35. 1.4650 **37.** 1.6992

39.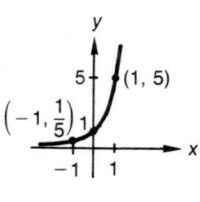

41.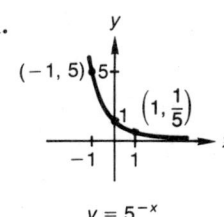

55.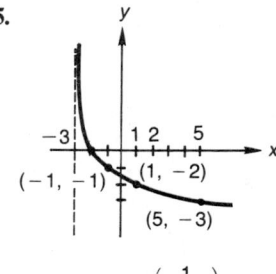

$y = \log_2\left(\dfrac{1}{x+3}\right)$

43.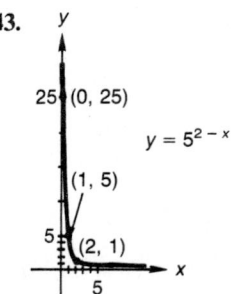

45.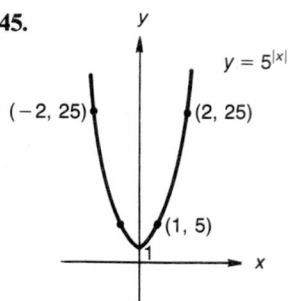

57.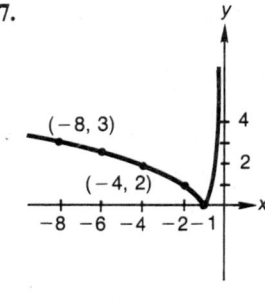

$y = |\log_2(-x)|$

47.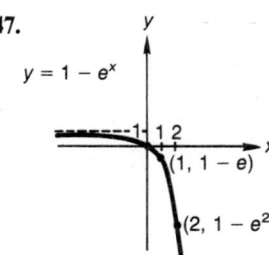

49.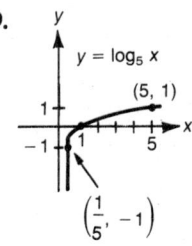

59. $s = 4$ **61.** $z = \dfrac{4}{3}$ **63.** $s = -5$ **65.** $w = 1$
67. $x = 1$ **69.** $u = 0, 1$ **71.** $r = 16, -2$
73. $w = \dfrac{1}{16}$ **75.** $r = 81$ **77.** $u = 4$ **79.** $t = 2$
81. $x = 15$ **83.** $x = 3$ **85.** $z = 8$ **87.** $u = 14$
89. $t = \dfrac{2 \log 2 - 3 \log 3}{\log 2 - \log 3} \approx 4.7095$
91. $y = \dfrac{\log 2 - \log 7}{\log 2 + \log 7} \approx -0.4747$
93. a. 24.6 coulombs **b.** 49.25 coulombs
   c. 20 min from now **d.** ~ 5.85 min ago
95. a. $22,550 **b.** $64,440
97. a. $1600.00 **b.** $1790.85 **c.** $1806.11 **d.** $1814.02
   e. $1822.03 **f.** $1822.12
99. a. $5906.25 **b.** $6093.37 **c.** $6121.42 **d.** $6135.93
   e. $6150.61 **f.** $6150.77
101. a. 1.4216 **b.** 3.4216 **c.** -1.5784
103. a. 2.4000 **b.** 0.9000 **c.** 7.6000
105. a. 47.7520 **b.** -22.2480 **c.** 27.7520
107. a. 2,160,000 **b.** 0.000216 **c.** 0.00216
109. a. 9,610,000 **b.** 9610 **c.** 0.00961
111. 0.7045 **113.** -1.7991 **115.** -4.3981
117. 89,440,000 **119.** 5695 **121.** 28130

53.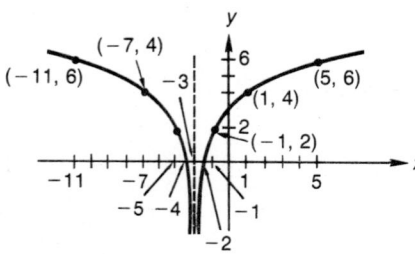

$y = \log_2(x+3)^2$

CHAPTER 7

Section 7.1

1.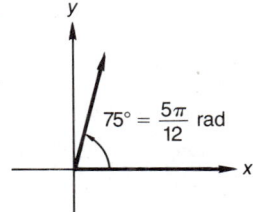
3.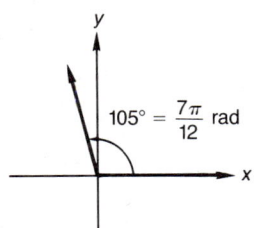
21.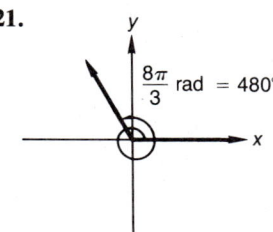
23.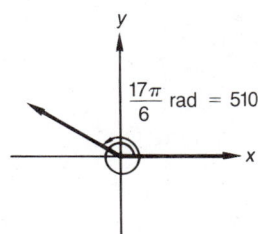

5.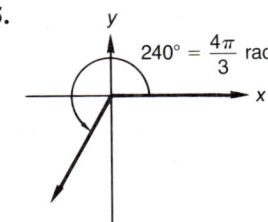
7.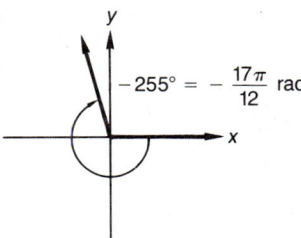
25.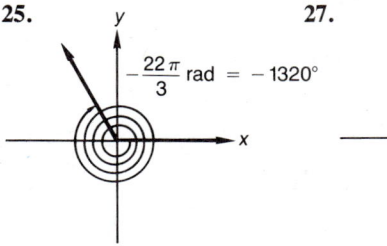
27.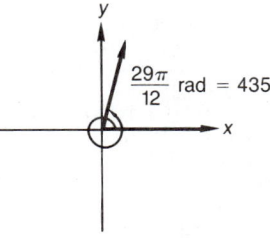

9.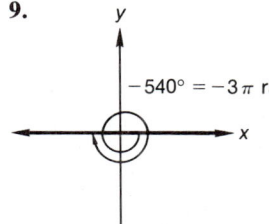
11.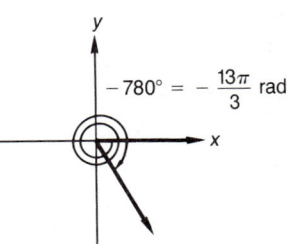
29.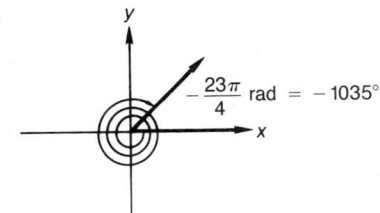

31. 114°35′30″   33. 131°46′49″   35. 303°40′3″
37. −187°55′49″   39. 113°26′44″   41. 0.406 rad
43. −3.678 rad   45. 5.237 rad   47. −1.369 rad
49. 4.376 rad   51. $s = \dfrac{5\pi}{6} \approx 2.62$ cm; $A = \dfrac{5\pi}{6}$ cm^2
53. $s = \dfrac{15\pi}{4}$ m; $A = \dfrac{75\pi}{8}$ m^2
55. $s = \dfrac{4\pi}{3}$ cm; $A = \dfrac{8\pi}{3}$ cm^2
57. $s = \dfrac{21\pi}{4}$ m; $A = \dfrac{189\pi}{8}$ m^2
59. $\theta = 2$ rad $\approx 114°35′30″$; $A = 25$ cm^2
61. $r = 3$; $A = 9$ in.^{2}; $\theta \approx 114°35′30″$
63. $\theta = 4$ rad $\approx 229°10′59″$; $s = 20$ km
65. $\theta = \dfrac{7\pi}{6}$ rad, $r = 3$ m, $s = 7\pi/2$ m
67. **a.** 4:30 P.M., **b.** 6 P.M.   69. At $\approx 7:38:11$
71. At $\approx 8:18:28$

13.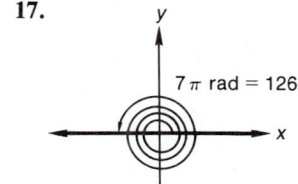
15.

17.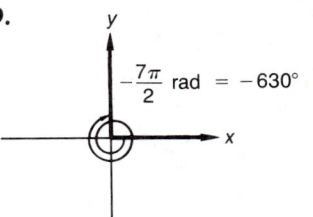
19.

73. a. 687.549 times around, **b.** 3.449 radians ≈ 197°38′24″

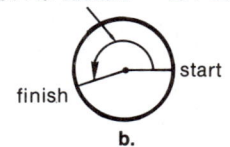

b.

75. 64°44′39″   **77.** (25.4 + 2.55π) m ≈ 33.41 m

Section 7.2

1. c = 25   **3.** b = 2√10   **5.** a = √33
7. c = 17   **9.** b = √21

	sin θ	cos θ	tan θ	cot θ	sec θ	csc θ
11.	$\frac{24}{25}$	$\frac{7}{25}$	$\frac{24}{7}$	$\frac{7}{24}$	$\frac{25}{7}$	$\frac{25}{24}$
13.	$\frac{2\sqrt{10}}{7}$	$\frac{3}{7}$	$\frac{2\sqrt{10}}{3}$	$\frac{3\sqrt{10}}{20}$	$\frac{7}{3}$	$\frac{7\sqrt{10}}{20}$
15.	$\frac{\sqrt{33}}{7}$	$\frac{4}{7}$	$\frac{\sqrt{33}}{4}$	$\frac{4\sqrt{33}}{33}$	$\frac{7}{4}$	$\frac{7\sqrt{33}}{33}$
17.	$\frac{15}{17}$	$\frac{8}{17}$	$\frac{15}{8}$	$\frac{8}{15}$	$\frac{17}{8}$	$\frac{17}{15}$
19.	$\frac{\sqrt{21}}{5}$	$\frac{2}{5}$	$\frac{\sqrt{21}}{2}$	$\frac{2\sqrt{21}}{21}$	$\frac{5}{2}$	$\frac{5\sqrt{21}}{21}$
21.	$\frac{7}{25}$	$\frac{24}{25}$	$\frac{7}{24}$	$\frac{24}{7}$	$\frac{25}{24}$	$\frac{25}{7}$
23.	$\frac{3}{7}$	$\frac{2\sqrt{10}}{7}$	$\frac{3\sqrt{10}}{20}$	$\frac{2\sqrt{10}}{3}$	$\frac{7\sqrt{10}}{20}$	$\frac{7}{3}$
25.	$\frac{4}{7}$	$\frac{\sqrt{33}}{7}$	$\frac{4\sqrt{33}}{33}$	$\frac{\sqrt{33}}{4}$	$\frac{7\sqrt{33}}{33}$	$\frac{7}{4}$
27.	$\frac{8}{17}$	$\frac{15}{17}$	$\frac{8}{15}$	$\frac{15}{8}$	$\frac{17}{15}$	$\frac{17}{8}$
29.	$\frac{2}{5}$	$\frac{\sqrt{21}}{5}$	$\frac{2\sqrt{21}}{21}$	$\frac{\sqrt{21}}{2}$	$\frac{5\sqrt{21}}{21}$	$\frac{5}{2}$
31.	$\frac{1}{2}$	$\frac{\sqrt{3}}{2}$	$\frac{\sqrt{3}}{3}$	$\sqrt{3}$	$\frac{2\sqrt{3}}{3}$	2
33.	$\frac{3\sqrt{13}}{13}$	$\frac{2\sqrt{13}}{13}$	$\frac{3}{2}$	$\frac{2}{3}$	$\frac{\sqrt{13}}{2}$	$\frac{\sqrt{13}}{3}$
35.	$\frac{\sqrt{21}}{5}$	$\frac{2}{5}$	$\frac{\sqrt{21}}{2}$	$\frac{2\sqrt{21}}{21}$	$\frac{5}{2}$	$\frac{5\sqrt{21}}{21}$
37.	$\frac{\sqrt{7}}{4}$	$\frac{3}{4}$	$\frac{\sqrt{7}}{3}$	$\frac{3\sqrt{7}}{7}$	$\frac{4}{3}$	$\frac{4\sqrt{7}}{7}$
39.	$\frac{3\sqrt{11}}{11}$	$\frac{\sqrt{22}}{11}$	$\frac{3\sqrt{2}}{2}$	$\frac{\sqrt{2}}{3}$	$\frac{\sqrt{22}}{2}$	$\frac{\sqrt{11}}{3}$
41.	$\frac{3}{7}$	$\frac{2\sqrt{10}}{7}$	$\frac{3\sqrt{10}}{20}$	$\frac{2\sqrt{10}}{3}$	$\frac{7\sqrt{10}}{20}$	$\frac{7}{3}$

43. $\frac{2\sqrt{3}}{3} - \sqrt{2}$ **45.** $\frac{\sqrt{6}}{3} - \frac{1}{2}$ **47.** 1

49. $4 - 2\sqrt{2} + \frac{2\sqrt{6}}{3} - \frac{2\sqrt{3}}{3}$

51. 0.6749   **53.** 3.747   **55.** 1.080   **57.** 1.311
59. 0.9874   **61.** 0.2436   **63.** 4.092   **65.** 0.9364
67. 0.7464

	sin θ	cos θ	tan θ	cot θ	sec θ	csc θ
69.	0.7808	0.6247	1.250	0.8000	1.601	1.281
71.	0.5539	0.8326	0.6652	1.503	1.201	1.805
73.	0.9480	0.3182	2.980	0.3356	3.143	1.055
75.	0.6322	0.7748	0.8160	1.226	1.291	1.582
77.	0.2986	0.9544	0.3128	3.197	1.048	3.349
79.	0.9939	0.1105	8.992	0.1112	9.048	1.006

Section 7.3

	α	β	a	b	h
1.	$\frac{\pi}{4}$	$\frac{\pi}{4}$	5	5	$5\sqrt{2}$
3.	$\frac{\pi}{6}$	$\frac{\pi}{3}$	6	$6\sqrt{3}$	12
5.	$\frac{\pi}{3}$	$\frac{\pi}{6}$	3	$\sqrt{3}$	$2\sqrt{3}$
7.	$\frac{\pi}{4}$	$\frac{\pi}{4}$	4	4	$4\sqrt{2}$
9.	$\frac{\pi}{6}$	$\frac{\pi}{3}$	3	$3\sqrt{3}$	6

11. 0.035 rad   **13.** 0.114 rad   **15.** 0.536 rad
17. 4°28′   **19.** 56°27′   **21.** 84°46′   **23.** 0.737 rad
25. 1.357 rad   **27.** 0.754 rad   **29.** 0.014 rad
31. 1.359 rad   **33.** 78°18′27″   **35.** 1°49′54″
37. 49°36′24″   **39.** 25°26′9″   **41.** 85°48′29″
43. α ≈ 54°3′, β ≈ 35°57′, b ≈ 3.7
45. β ≈ 75° 75° 0′;   a ≈ 3.1; b ≈ 11.6
47. α ≈ 52°24′; h ≈ 3.9; b ≈ 2.4
49. h ≈ 9.2; α ≈ 12°32′; β ≈ 77°28′
51. h ≈ 33.4, b ≈ 32.7; β ≈ 77°54′
53. α ≈ 69°58′; β ≈ 20°2′; α ≈ 13.7
55. h ≈ 6.8; b ≈ 2.6; β ≈ 22°30′
57. 22°37′12″ or 0.395 rad   **59.** 4°0′50″ or 0.070 rad
61. ≈41°49′
63. ≈332 ft 3 in.   **65.** ≈280 m   **67.** ≈423 ft 7 in.
69. ≈387.2 nautical miles   **71.** ≈8801 ft
73. ≈8174 ft
75. a. ≈1.918 (10^{13}) mi, ≈3.27 lt yr   **b.** ≈1.918 (10^{17}) mi
c. 0.0001 s

Section 7.4

	sin θ	cos θ	tan θ	cot θ	sec θ	csc θ
1.	0	−1	0	—	−1	—
3.	$\dfrac{1}{\sqrt{5}}$	$\dfrac{2}{\sqrt{5}}$	$\dfrac{1}{2}$	2	$\dfrac{\sqrt{5}}{2}$	$\sqrt{5}$
5.	$\dfrac{-5}{\sqrt{41}}$	$\dfrac{-4}{\sqrt{41}}$	$\dfrac{5}{4}$	$\dfrac{4}{5}$	$\dfrac{-\sqrt{41}}{4}$	$\dfrac{-\sqrt{41}}{5}$
7.	$\dfrac{2}{\sqrt{5}}$	$\dfrac{1}{\sqrt{5}}$	2	$\dfrac{1}{2}$	$\sqrt{5}$	$\dfrac{\sqrt{5}}{2}$
9.	$\dfrac{4}{\sqrt{41}}$	$\dfrac{-5}{\sqrt{41}}$	$\dfrac{-4}{5}$	$\dfrac{-5}{4}$	$\dfrac{-\sqrt{41}}{5}$	$\dfrac{\sqrt{41}}{4}$
11.	$\dfrac{-1}{\sqrt{10}}$	$\dfrac{-3}{\sqrt{10}}$	$\dfrac{1}{3}$	3	$\dfrac{-\sqrt{10}}{3}$	$-\sqrt{10}$

13.

Degrees (°)	Radians	sin θ	cos θ	tan θ	cot θ	sec θ	csc θ
120	$\dfrac{2\pi}{3}$	$\dfrac{\sqrt{3}}{2}$	$\dfrac{-1}{2}$	$-\sqrt{3}$	$\dfrac{-\sqrt{3}}{3}$	−2	$\dfrac{2\sqrt{3}}{3}$
135	$\dfrac{3\pi}{4}$	$\dfrac{\sqrt{2}}{2}$	$\dfrac{-\sqrt{2}}{2}$	−1	−1	$-\sqrt{2}$	$\sqrt{2}$
150	$\dfrac{5\pi}{6}$	$\dfrac{1}{2}$	$\dfrac{-\sqrt{3}}{2}$	$\dfrac{-\sqrt{3}}{3}$	$-\sqrt{3}$	$\dfrac{-2\sqrt{3}}{3}$	2
210	$\dfrac{7\pi}{6}$	$\dfrac{-1}{2}$	$\dfrac{-\sqrt{3}}{2}$	$\dfrac{\sqrt{3}}{3}$	$\sqrt{3}$	$\dfrac{-2\sqrt{3}}{3}$	−2
225	$\dfrac{5\pi}{4}$	$\dfrac{-\sqrt{2}}{2}$	$\dfrac{-\sqrt{2}}{2}$	1	1	$-\sqrt{2}$	$-\sqrt{2}$
240	$\dfrac{4\pi}{3}$	$\dfrac{-\sqrt{3}}{2}$	$\dfrac{-1}{2}$	$\sqrt{3}$	$\dfrac{\sqrt{3}}{3}$	−2	$\dfrac{-2\sqrt{3}}{3}$
300	$\dfrac{5\pi}{3}$	$\dfrac{-\sqrt{3}}{2}$	$\dfrac{1}{2}$	$-\sqrt{3}$	$\dfrac{-\sqrt{3}}{3}$	2	$\dfrac{-2\sqrt{3}}{3}$
315	$\dfrac{7\pi}{4}$	$\dfrac{-\sqrt{2}}{2}$	$\dfrac{\sqrt{2}}{2}$	−1	−1	$\sqrt{2}$	$-\sqrt{2}$
330	$\dfrac{11\pi}{6}$	$\dfrac{-1}{2}$	$\dfrac{\sqrt{3}}{2}$	$\dfrac{-\sqrt{3}}{3}$	$-\sqrt{3}$	$\dfrac{2\sqrt{3}}{3}$	−2

	sin θ	cos θ	tan θ	cot θ	sec θ	csc θ
15.	$\dfrac{-1}{2}$	$\dfrac{-\sqrt{3}}{2}$	$\dfrac{\sqrt{3}}{3}$	$\sqrt{3}$	$\dfrac{-2\sqrt{3}}{3}$	−2
17.	$\dfrac{\sqrt{3}}{2}$	$\dfrac{1}{2}$	$\sqrt{3}$	$\dfrac{\sqrt{3}}{3}$	2	$\dfrac{2\sqrt{3}}{3}$
19.	$\dfrac{-\sqrt{2}}{2}$	$\dfrac{\sqrt{2}}{2}$	−1	−1	$\sqrt{2}$	$-\sqrt{2}$
21.	$\dfrac{-\sqrt{3}}{2}$	$\dfrac{-1}{2}$	$\sqrt{3}$	$\dfrac{\sqrt{3}}{3}$	−2	$\dfrac{-2\sqrt{3}}{3}$
23.	$\dfrac{-\sqrt{2}}{2}$	$\dfrac{-\sqrt{2}}{2}$	1	1	$-\sqrt{2}$	$-\sqrt{2}$
25.	−1	0	undef.	0	undef.	−1
27.	0	1	0	undef.	1	undef.
29.	−1	0	undef.	0	undef.	−1
31.	$\dfrac{\sqrt{2}}{2}$	$\dfrac{-\sqrt{2}}{2}$	−1	−1	$-\sqrt{2}$	$\sqrt{2}$
33.	$\dfrac{-1}{2}$	$\dfrac{-\sqrt{3}}{2}$	$\dfrac{\sqrt{3}}{3}$	$\sqrt{3}$	$\dfrac{-2\sqrt{3}}{3}$	−2

35. $\dfrac{-\sqrt{3}}{4}$ 37. $-2 - \sqrt{3}$ 39. $\dfrac{12\sqrt{3} + 9}{13}$

41. $-1 - 2\sqrt{2}$ 43. $\dfrac{2 + \sqrt{3}}{4}$

	sin θ	cos θ	tan θ	cot θ	sec θ	csc θ
45.	$\dfrac{-4}{5}$	$\dfrac{3}{5}$	$\dfrac{-4}{3}$	$\dfrac{-3}{4}$	$\dfrac{5}{3}$	$\dfrac{-5}{4}$
47.	$\dfrac{-1}{\sqrt{10}}$	$\dfrac{3}{\sqrt{10}}$	$\dfrac{-1}{3}$	-3	$\dfrac{\sqrt{10}}{3}$	$-\sqrt{10}$
49.	$\dfrac{2\sqrt{2}}{3}$	$\dfrac{1}{3}$	$2\sqrt{2}$	$\dfrac{1}{2\sqrt{2}}$	3	$\dfrac{3}{2\sqrt{2}}$
51.	$\dfrac{-1}{3}$	$\dfrac{-2\sqrt{2}}{3}$	$\dfrac{1}{2\sqrt{2}}$	$2\sqrt{2}$	$\dfrac{-3}{2\sqrt{2}}$	-3
53.	$\dfrac{-1}{3}$	$\dfrac{-2\sqrt{2}}{3}$	$\dfrac{1}{2\sqrt{2}}$	$2\sqrt{2}$	$\dfrac{-3}{2\sqrt{2}}$	-3

55. $\dfrac{-3}{\sqrt{5}}$ 57. $\dfrac{3}{\sqrt{10}}$ 59. $\dfrac{3\sqrt{5}}{2}$

61. -0.7038 based on $16.5 = (3.142) \cdot 5 + 0.79$
63. 1.080 65. -0.1287 67. -4.092
69. -1.0000 71. 0.1956

	sin θ	cos θ	tan θ	cot θ	sec θ	csc θ	quadrant
73.	0.9826	0.1855	5.2964	0.1888	5.390	1.0177	I
75.	−0.9873	0.1586	−6.2266	−0.1606	6.3064	−1.0128	IV
77.	0.8174	−0.5760	−1.419	−0.7047	−1.736	1.223	II
79.	−0.9993	0.0368	−27.1207	−0.0369	27.139	−1.0007	IV
81.	−0.5349	−0.8449	0.6332	1.5794	−1.1836	−1.8694	III

Section 7.5

1.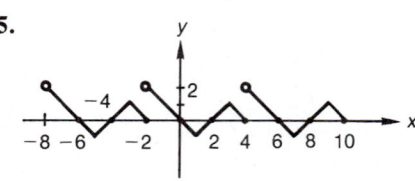

3.

5.

7. -10 9. 10 11. $-f(x)$ 13. 1 15. 1
17. 2 19. $f(x)$ 21. 2 23. 1 25. $f(x)$ 27. 0
29. -1 31. -2 33. $-f(x)$
35. 0.035 rad, 3.107 rad   37. 3.028 rad, 6.170 rad
39. 2.606 rad, 5.748 rad   41. 4°28′, 175°32′
43. 56°27′, 123°33′   45. 95°14′, 264°46′
47. 0.737 rad, 2.405 rad   49. 1.785 rad, 4.499 rad
51. 0.754 rad, 5.530 rad   53. 3.128 rad, 6.269 rad
55. 1.359 rad, 4.501 rad   57. 78°18′27″, 258°18′27″
59. 1°49′54″, 181°49′54″   61. 229°36′24″, 310°23′36″
63. 205°26′9″, 334°33′51″   65. 94°11′31″, 265°48′29″

67. $\csc \theta = \dfrac{r}{y} = \dfrac{1}{y/r} = \dfrac{1}{\sin \theta}$

69. $\cot \theta = \dfrac{x}{y} = \dfrac{x/r}{y/r} = \dfrac{\cos \theta}{\sin \theta}$

71.
	sin	cos
a.	0	1
b.	1	0
c.	$\dfrac{-1}{\sqrt{5}}$	$\dfrac{-2}{\sqrt{5}}$
d.	$\dfrac{-2}{\sqrt{13}}$	$\dfrac{3}{\sqrt{13}}$
e.	$\dfrac{5}{\sqrt{41}}$	$\dfrac{4}{\sqrt{41}}$
f.	$\dfrac{3}{\sqrt{13}}$	$\dfrac{-2}{\sqrt{13}}$

73.
	sin	cos
a.	1	0
b.	0	−1
c.	$\dfrac{-2}{\sqrt{5}}$	$\dfrac{1}{\sqrt{5}}$
d.	$\dfrac{3}{\sqrt{13}}$	$\dfrac{2}{\sqrt{13}}$
e.	$\dfrac{4}{\sqrt{41}}$	$\dfrac{-5}{\sqrt{41}}$
f.	$\dfrac{-2}{\sqrt{13}}$	$\dfrac{-3}{\sqrt{13}}$

	sin	cos	tan	cot	sec	csc
75.	$\dfrac{5}{\sqrt{29}}$	$\dfrac{2}{\sqrt{29}}$	$\dfrac{5}{2}$	$\dfrac{2}{5}$	$\dfrac{\sqrt{29}}{2}$	$\dfrac{\sqrt{29}}{5}$
77.	$\dfrac{1}{\sqrt{17}}$	$\dfrac{-4}{\sqrt{17}}$	$\dfrac{-1}{4}$	-4	$\dfrac{-\sqrt{17}}{4}$	$\sqrt{17}$
79.	$\dfrac{-2}{\sqrt{29}}$	$\dfrac{5}{\sqrt{29}}$	$\dfrac{-2}{5}$	$\dfrac{-5}{2}$	$\dfrac{\sqrt{29}}{5}$	$\dfrac{-\sqrt{29}}{2}$
81.	$\dfrac{4}{\sqrt{17}}$	$\dfrac{1}{\sqrt{17}}$	4	$\dfrac{1}{4}$	$\sqrt{17}$	$\dfrac{\sqrt{17}}{4}$

83. See Figure 7.37.

$\cos(\theta \pm \pi) = \dfrac{-x}{r} = -\cos\theta;$

$\sin(\theta \pm \pi) = \dfrac{-y}{r} = -\sin\theta;$

$\sec(\theta \pm \pi) = \dfrac{1}{\cos(\theta \pm \pi)} = \dfrac{1}{(-\cos\theta)} = -\sec\theta;$

$\csc(\theta \pm \pi) = \dfrac{1}{\sin(\theta \pm \pi)} = \dfrac{1}{(-\sin\theta)} = -\csc\theta$

85. See Figure 7.37.

$\tan\theta = \dfrac{y}{x}$ and $\tan(\theta \pm \pi) = \dfrac{-y}{-x} = \dfrac{y}{x} = \tan\theta;$

$\tan(\pi - \theta) = -\tan(\theta - \pi) = -\tan\theta;$

$\cot(\pi - \theta) = \dfrac{1}{\tan(\pi - \theta)} = \dfrac{1}{(-\tan\theta)} = -\cot\theta$

87. $\sin\left(\theta - \dfrac{\pi}{2}\right) = -\cos\theta;\ \cos\left(\theta - \dfrac{\pi}{2}\right) = \sin\theta;$

$\tan\left(\theta - \dfrac{\pi}{2}\right) = -\cot\theta;\ \cot\left(\theta - \dfrac{\pi}{2}\right) = -\tan\theta;$

$\sec\left(\theta - \dfrac{\pi}{2}\right) = \csc\theta;\ \csc\left(\theta - \dfrac{\pi}{2}\right) = -\sec\theta$

89. Solution outlined in hint.   **91.** π **93.** π

95. $f(0) = 0 = f(\pm 2) = f(\pm 4) = \cdots;$
$f(1) = f(-1) = -f(1);$ thus $f(\pm 1) = 0$ and then
$f(\pm 3) = 0 = f(\pm 5) = \cdots.$

Section 7.6

1.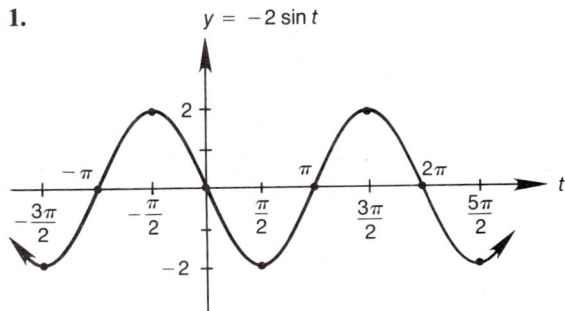
$y = -2\sin t$

3.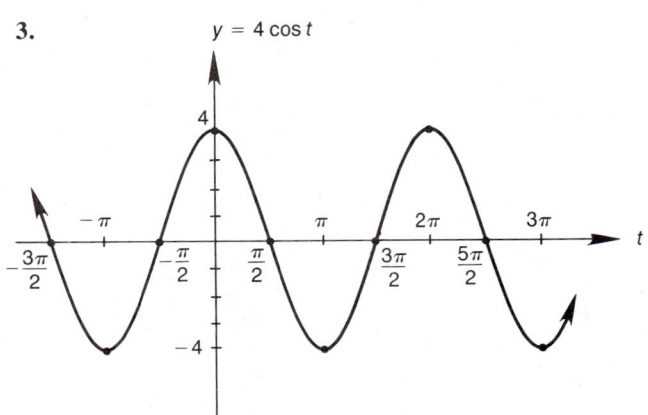
$y = 4\cos t$

5.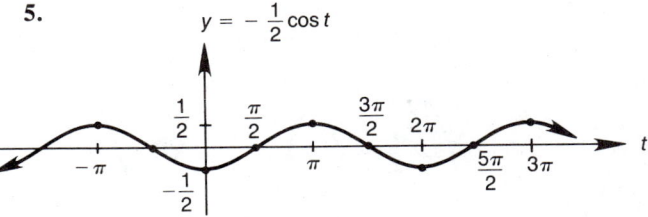
$y = -\dfrac{1}{2}\cos t$

7.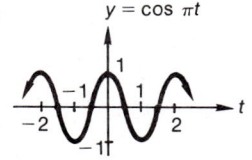
$y = \cos \pi t$

9.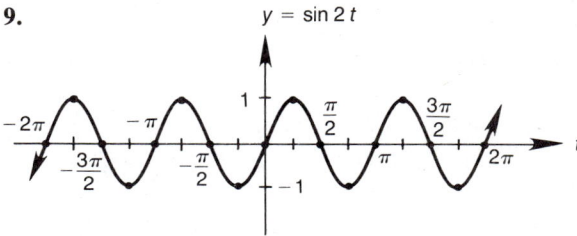
$y = \sin 2t$

11.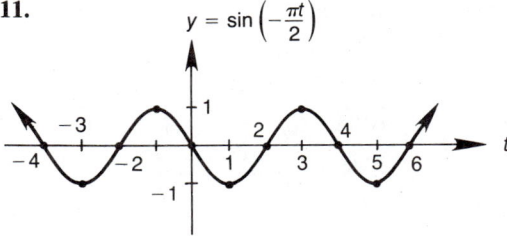
$y = \sin\left(-\dfrac{\pi t}{2}\right)$

13.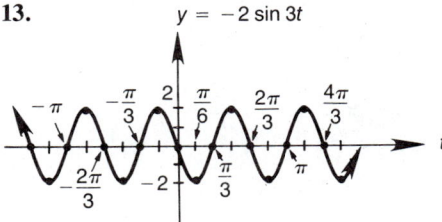
$y = -2\sin 3t$

15.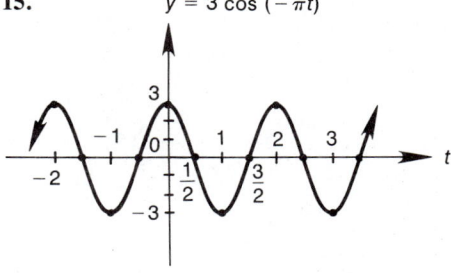
$y = 3\cos(-\pi t)$

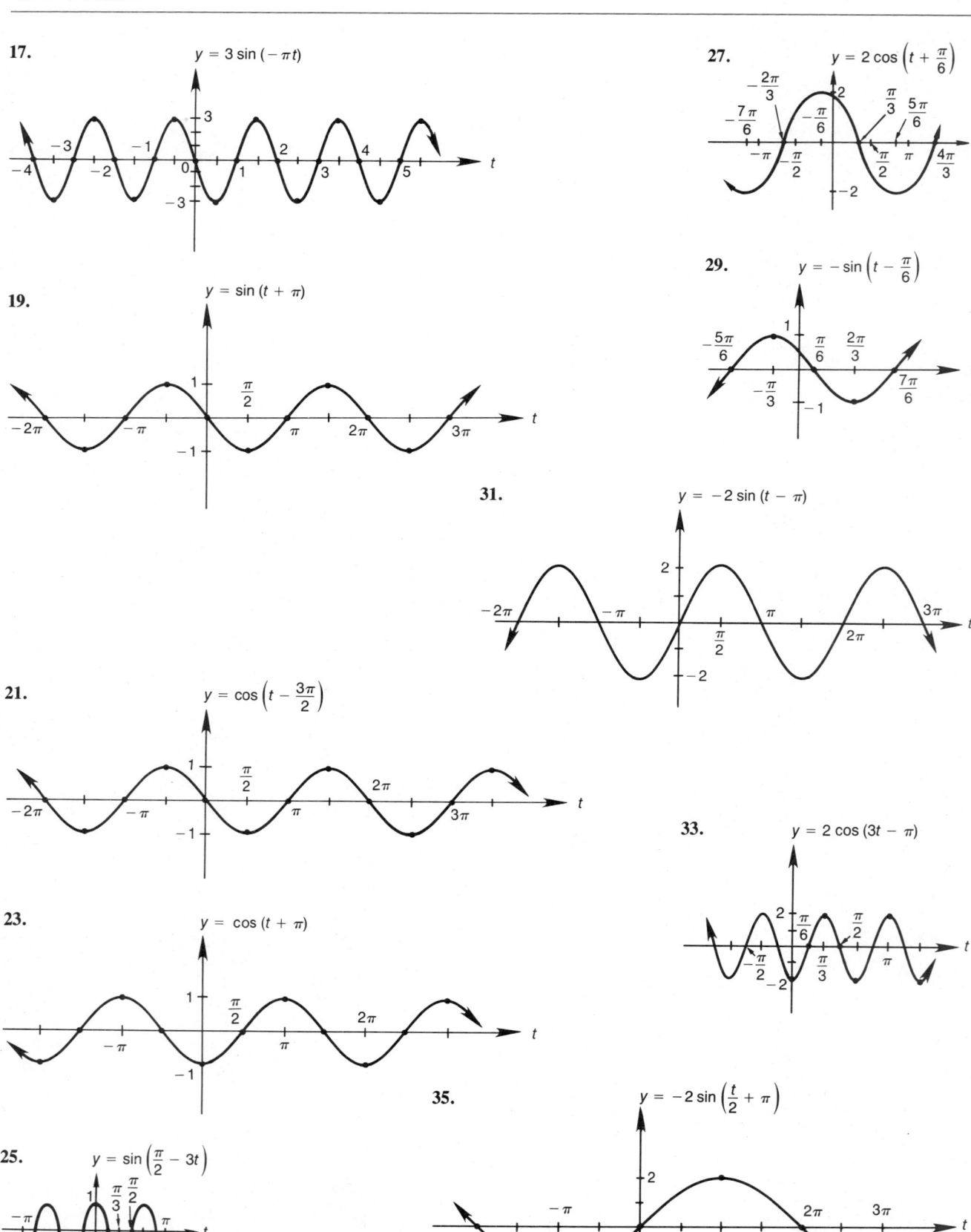

37.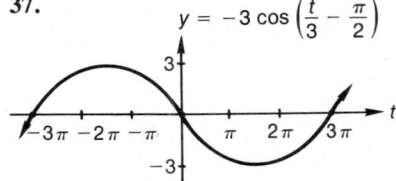

47.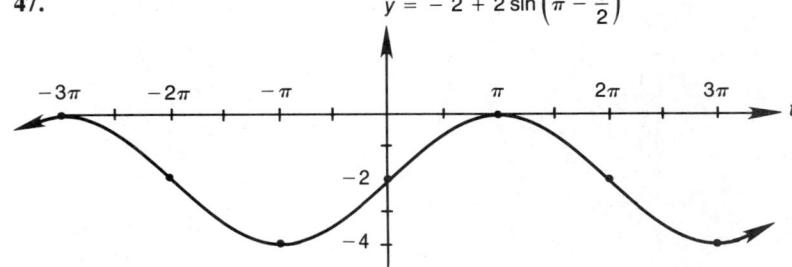

39.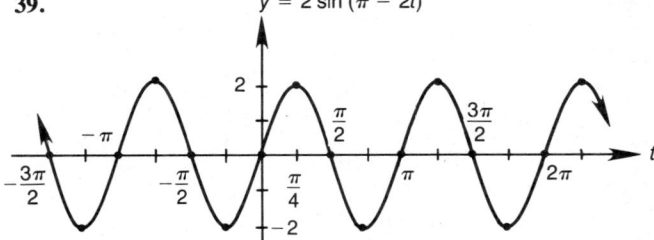

49.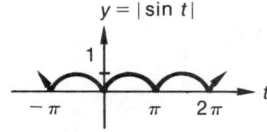

41.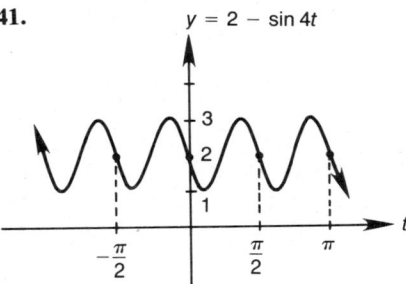

51.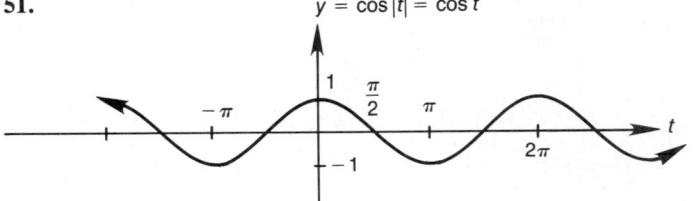

43.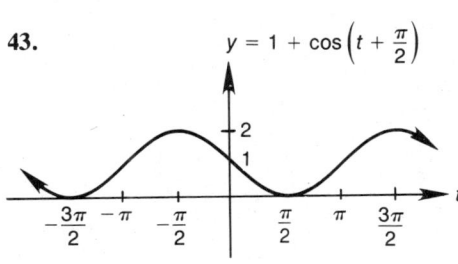

53.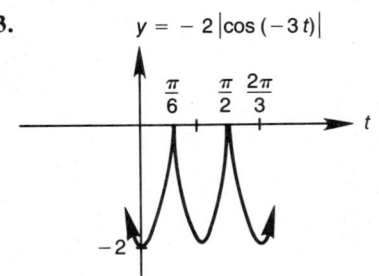

55.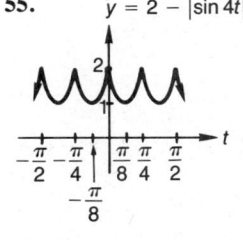

57.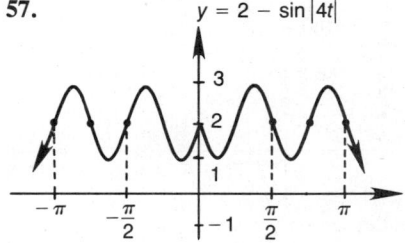

45.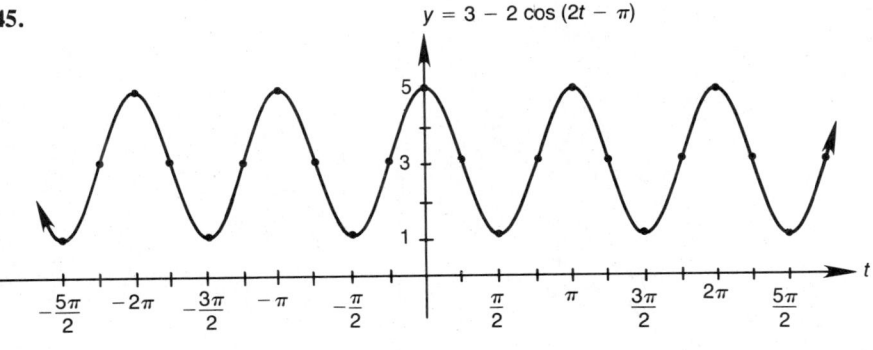

59.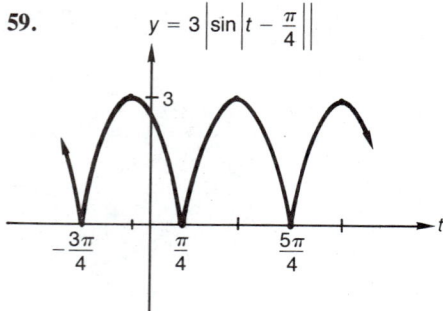

69.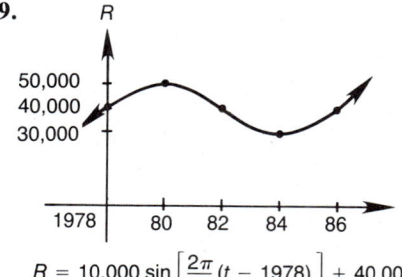

$R = 10{,}000 \sin\left[\dfrac{2\pi}{8}(t-1978)\right] + 40{,}000$

61. a.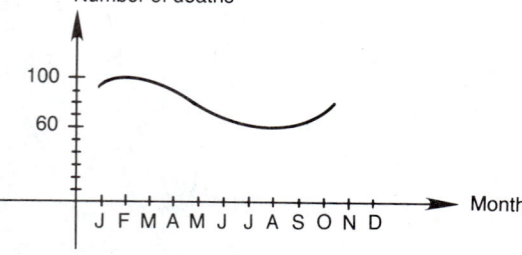

23-day cycle, 28-day cycle, 33-day cycle

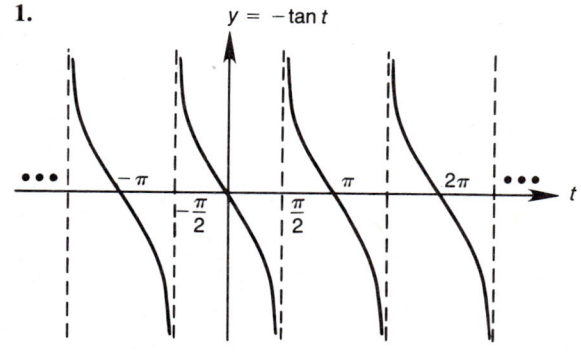

$F = 3000 \sin\left[\dfrac{2\pi}{8}(t-1979)\right] + 5000$

b. $11\tfrac{1}{2}$ days **c.** $16\tfrac{1}{2}$ days **d.** $6\tfrac{1}{2}$ days

e. at 33 days or 10 days after the end of the first bad period

63. a.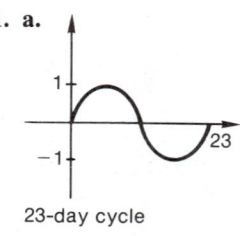

b. $y = 52 - 40 \cos\left[\left(\dfrac{2\pi}{365}\right)(t-20)\right]$

65. a.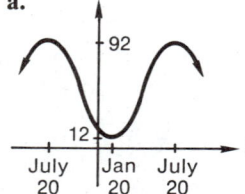

b. Let January = month 1. Deaths in month t
$= 80 - 20 \sin\left[\dfrac{\pi}{6}(t-5)\right]$.

67. Voltage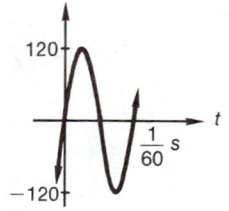

Section 7.7

1.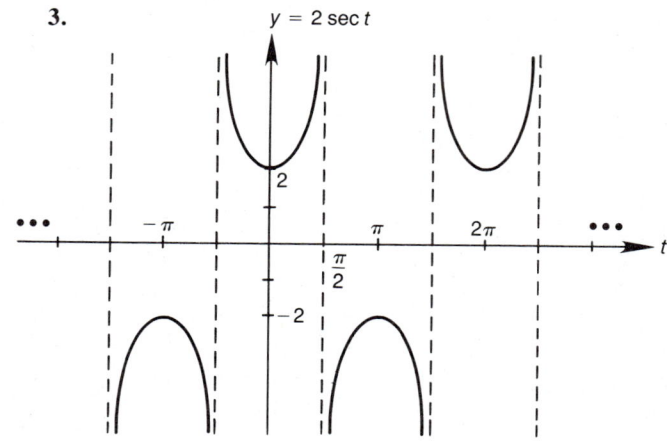

$y = -\tan t$

3. $y = 2 \sec t$

5. $y = \cot(-3t)$

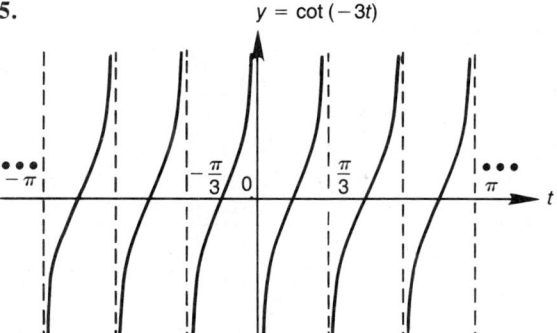

13. $y = 1 - \cot 2t$

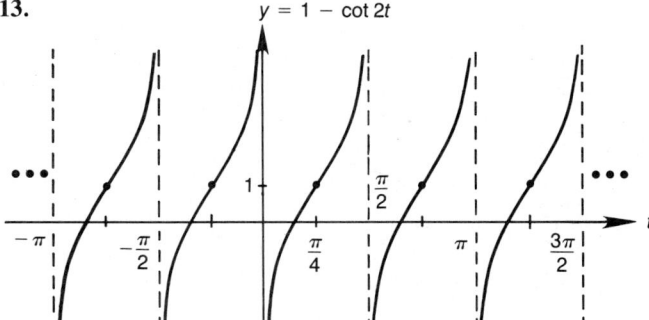

7. $y = \csc(-t)$

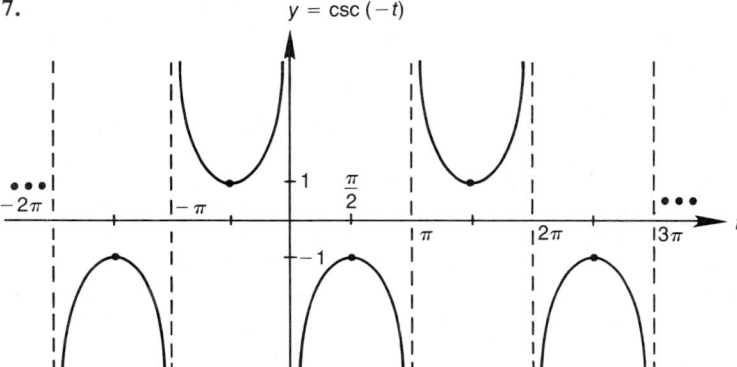

9. $y = \frac{1}{2}\tan 2t$

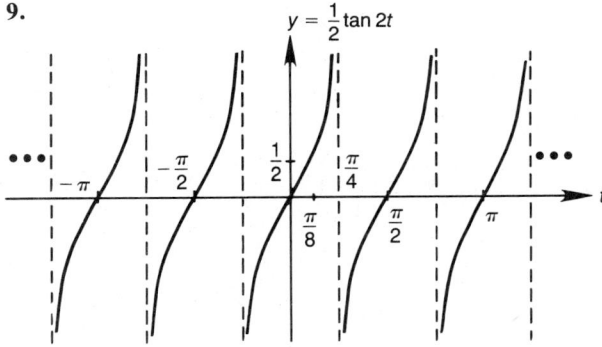

15. $y = 2\csc t - 2$

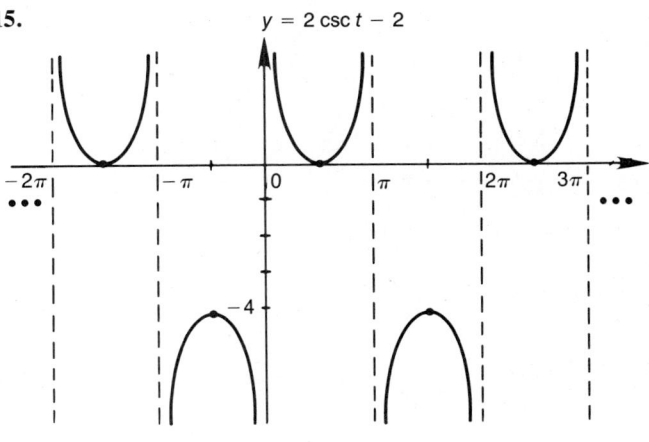

11. $y = -2\sec 2t$

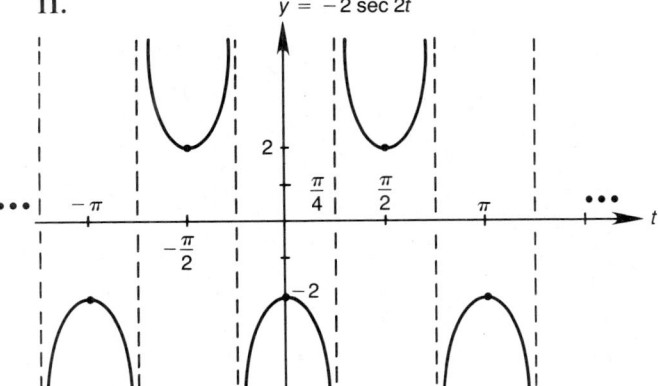

17. $y = \tan\left(t + \frac{\pi}{2}\right)$

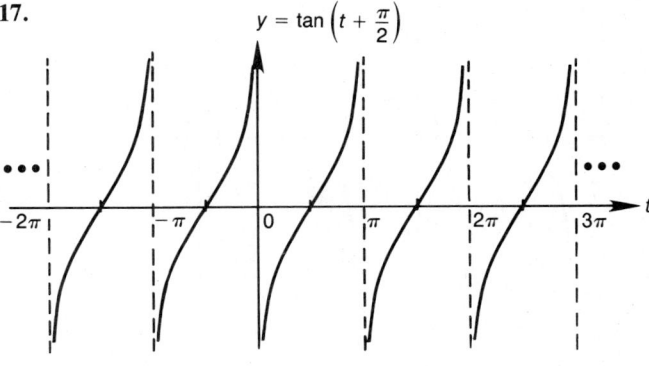

A.62 Solutions

19.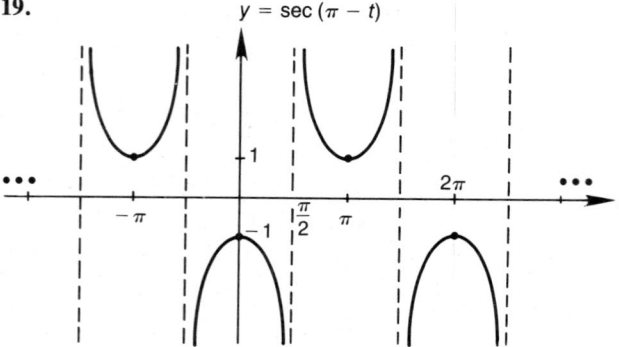
$y = \sec(\pi - t)$

21.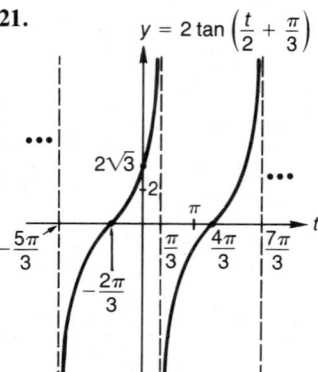
$y = 2\tan\left(\dfrac{t}{2} + \dfrac{\pi}{3}\right)$

23.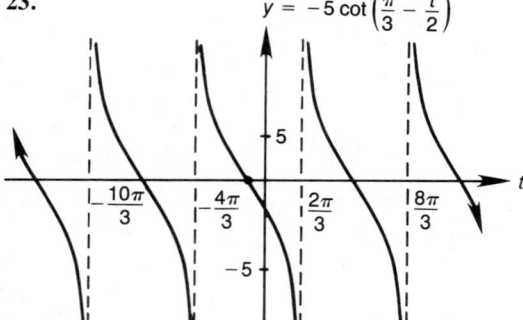
$y = -5\cot\left(\dfrac{\pi}{3} - \dfrac{t}{2}\right)$

25.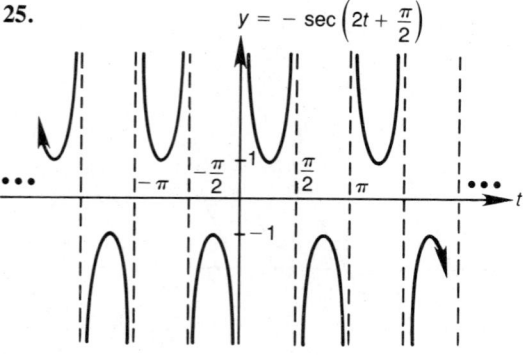
$y = -\sec\left(2t + \dfrac{\pi}{2}\right)$

27.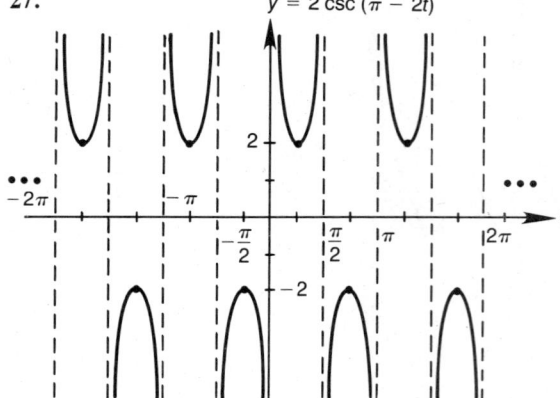
$y = 2\csc(\pi - 2t)$

29.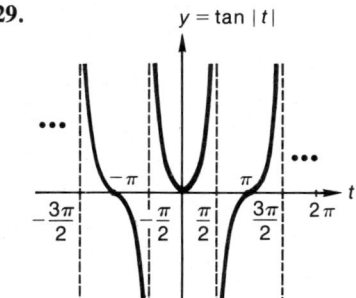
$y = \tan|t|$

31.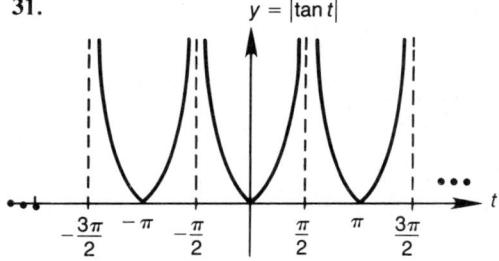
$y = |\tan t|$

33.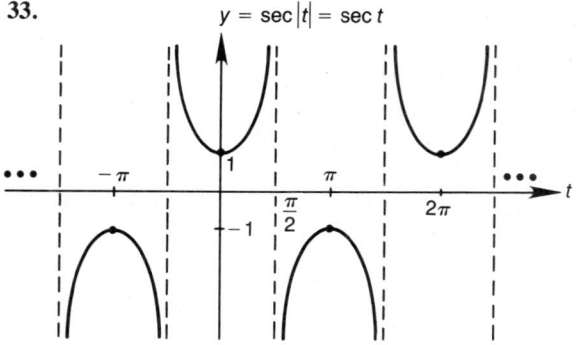
$y = \sec|t| = \sec t$

35. $y = |\sec t|$

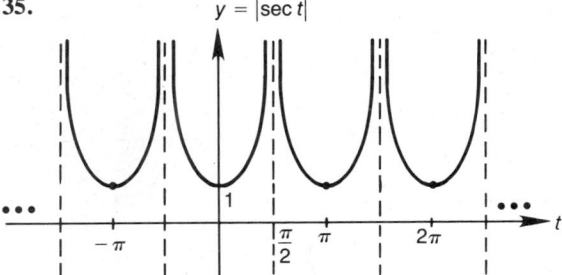

43. $y = t \sin t$

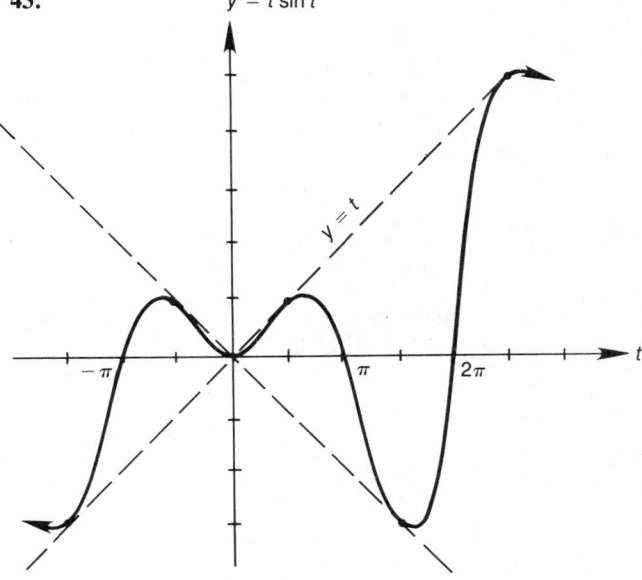

37. $y = \sin t - \cos t$

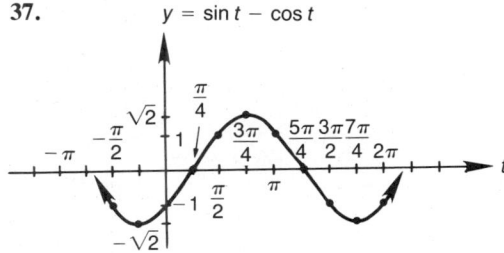

45. $y = e^{-t} \sin 2t$

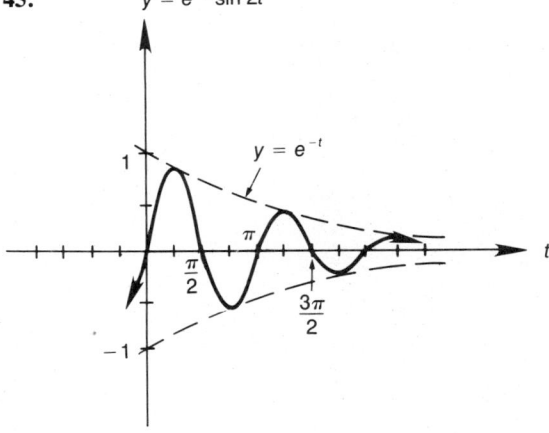

39. $y = \sin t + \sin 2t$

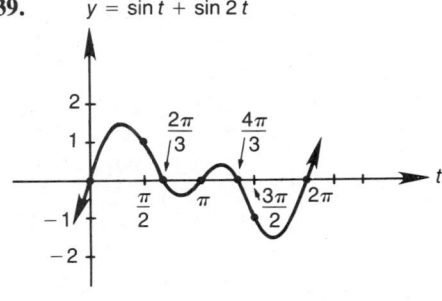

41. $y = 2 \sin t + \cos t$

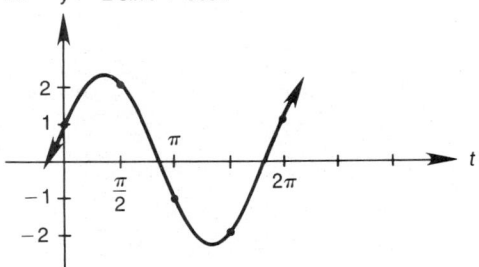

47. $y = \dfrac{t}{4} \sin(-2t)$

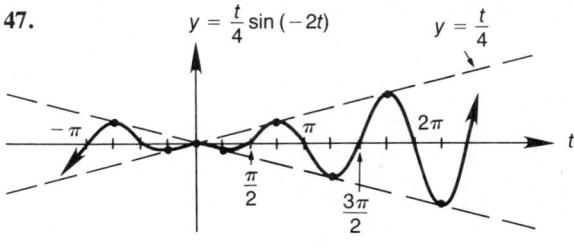

49.

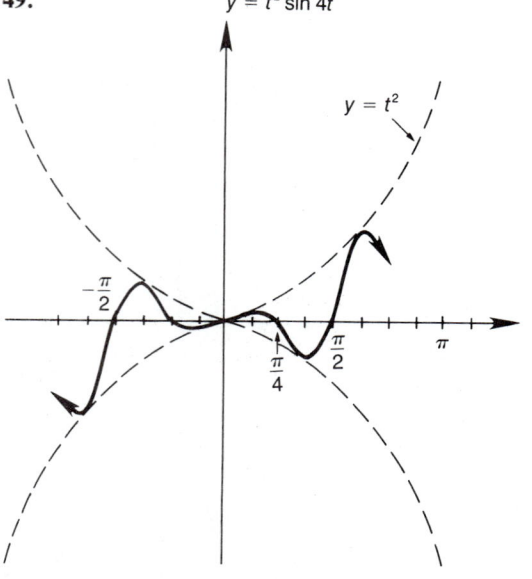

51. a.

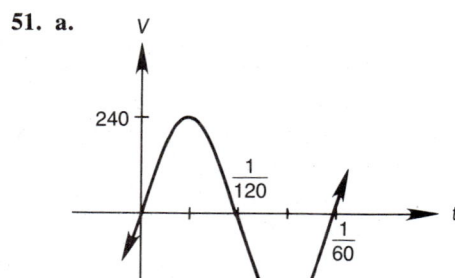

b. $v = 240 \sin(120\pi t)$, t in seconds

Section 7.9

1. $210° = \frac{7\pi}{6}$ rad

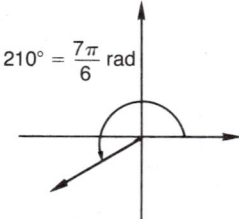

3. $-660° = -\frac{11\pi}{3}$ rad

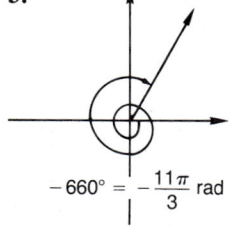

5. $915° = \frac{61\pi}{12}$ rad

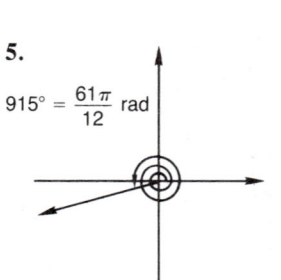

7. $\frac{23\pi}{12}$ rad = $345°$

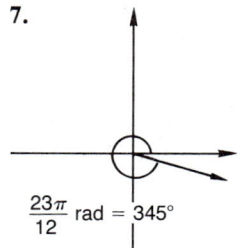

9. $\frac{29\pi}{2}$ rad = $2610°$

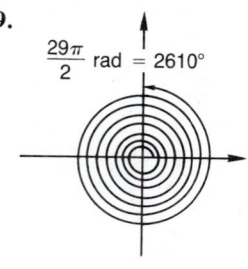

11. $135°47'28''$ **13.** $182°12'2''$ **15.** -0.627 rad

17. $s = \pi$ cm, $A = 5\pi$ cm²

19. $r = 10$ m; $\theta = 2$ rad $\approx 114°35'30''$

21. $\theta = \frac{5\pi}{4}$; $r = 8$ cm; $s = 10\pi$ cm

23. $\sin \theta = \frac{3}{\sqrt{10}}$; $\cos \theta = \frac{1}{\sqrt{10}}$; $\tan \theta = 3$

25. $\sin \theta = \frac{3}{\sqrt{13}}$; $\cos \theta = \frac{2}{\sqrt{13}}$; $\tan \theta = \frac{3}{2}$

27. $\beta \approx 65°23'$; $\alpha \approx 24°37'$; $b = 32.7$

29. $h \approx 7.0$; $\alpha \approx 36°42'$; $b \approx 5.6$

31. $h \approx 11.2$; $\beta \approx 63°26'$; $\alpha \approx 26°34'$

33. $\alpha \approx 6°6'$; $a \approx 0.5$; $b \approx 5.0$

	$\sin \theta$	$\cos \theta$	$\tan \theta$	$\cot \theta$	$\sec \theta$	$\csc \theta$
35.	$\frac{\sqrt{3}}{2}$	$\frac{1}{2}$	$\sqrt{3}$	$\frac{\sqrt{3}}{3}$	2	$\frac{2}{\sqrt{3}}$
37.	$\frac{\sqrt{3}}{2}$	$\frac{-1}{2}$	$-\sqrt{3}$	$\frac{-\sqrt{3}}{3}$	-2	$\frac{2}{\sqrt{3}}$
39.	0	-1	0	—	-1	—
41.	$\frac{-1}{\sqrt{2}}$	$\frac{-1}{\sqrt{2}}$	1	1	$-\sqrt{2}$	$-\sqrt{2}$
43.	$\frac{1}{2}$	$\frac{\sqrt{3}}{2}$	$\frac{\sqrt{3}}{3}$	$\sqrt{3}$	$\frac{2}{\sqrt{3}}$	2

45. $\cos \theta \approx 0.9559$ based on $6.582 - 2 \cdot (3.142) = 0.298$

47. $\tan \theta \approx 16.428$ based on $3.142 - 1.632 = 1.51$

49. $\cot \theta \approx -1.646$ based on $-3.688 + 3.142 = -0.546$

51. 0.5351 **53.** -2.017 **55.** -0.5882

	$\sin \theta$	$\cos \theta$	$\tan \theta$	$\cot \theta$	$\sec \theta$	$\csc \theta$	Quadrant
57.	-0.7476	-0.6642	1.1256	0.8884	-1.5056	-1.3377	III
59.	0.1535	0.9881	0.1553	6.4373	1.0120	6.5145	I
61.	-0.1333	-0.9911	0.1345	7.4355	-1.0090	-7.5025	III

63. 2.385 rad, 3.898 rad   **65.** 1.243 rad, 5.041 rad

67. 6.187 rad, 3.238 rad   **69.** $337°18'19''$, $157°18'19''$

71. $48°17'33''$, $228°17'33''$ **73.** $9°3'49''$, $170°56'11''$

	sin θ	cos θ	tan θ	cot θ	sec θ	csc θ
75.	$\frac{6}{\sqrt{37}}$	$\frac{-1}{\sqrt{37}}$	-6	$\frac{-1}{6}$	$-\sqrt{37}$	$\frac{\sqrt{37}}{6}$
77.	$\frac{4}{5}$	$\frac{3}{5}$	$\frac{4}{3}$	$\frac{3}{4}$	$\frac{5}{3}$	$\frac{5}{4}$
79.	$\frac{1}{\sqrt{2}}$	$\frac{-1}{\sqrt{2}}$	-1	-1	$-\sqrt{2}$	$\sqrt{2}$

		sin	cos
81.	a.	-1	0
	b.	$\frac{-6}{\sqrt{37}}$	$\frac{1}{\sqrt{37}}$
	c.	$\frac{1}{\sqrt{5}}$	$\frac{-2}{\sqrt{5}}$
83.	a.	0	1
	b.	$\frac{1}{\sqrt{37}}$	$\frac{6}{\sqrt{37}}$
	c.	$\frac{-2}{\sqrt{5}}$	$\frac{-1}{\sqrt{5}}$

85. $\sin\left(\theta + \frac{3\pi}{2}\right) = -\cos\theta$; $\cos\left(\theta + \frac{3\pi}{2}\right) = \sin\theta$;
$\tan\left(\theta + \frac{3\pi}{2}\right) = -\cot\theta$; $\cot\left(\theta + \frac{3\pi}{2}\right) = -\tan\theta$;
$\sec\left(\theta + \frac{3\pi}{2}\right) = \csc\theta$; $\csc\left(\theta + \frac{3\pi}{2}\right) = -\sec\theta$

		sin	cos	tan	cot	sec	csc
87.	a.	$\frac{3}{\sqrt{13}}$	$\frac{-2}{\sqrt{13}}$	$\frac{-3}{2}$	$\frac{-2}{3}$	$\frac{-\sqrt{13}}{2}$	$\frac{\sqrt{13}}{3}$
	b.	$\frac{-1}{\sqrt{5}}$	$\frac{-2}{\sqrt{5}}$	$\frac{1}{2}$	2	$\frac{-\sqrt{5}}{2}$	$-\sqrt{5}$
	c.	$\frac{2}{\sqrt{5}}$	$\frac{1}{\sqrt{5}}$	2	$\frac{1}{2}$	$\sqrt{5}$	$\frac{\sqrt{5}}{2}$
	d.	$\frac{-2}{\sqrt{13}}$	$\frac{3}{\sqrt{13}}$	$\frac{-2}{3}$	$\frac{-3}{2}$	$\frac{\sqrt{13}}{3}$	$\frac{-\sqrt{13}}{2}$
89.		$\frac{2}{3}$	$\frac{\sqrt{5}}{3}$	$\frac{2}{\sqrt{5}}$	$\frac{\sqrt{5}}{2}$	$\frac{3}{\sqrt{5}}$	$\frac{3}{2}$
91.		$\frac{-1}{10}$	$\frac{3\sqrt{11}}{10}$	$\frac{-1}{3\sqrt{11}}$	$-3\sqrt{11}$	$\frac{10}{3\sqrt{11}}$	-10
93.		$\frac{-2}{\sqrt{5}}$	$\frac{-1}{\sqrt{5}}$	2	$\frac{1}{2}$	$-\sqrt{5}$	$\frac{-\sqrt{5}}{2}$

95. $\frac{2\sqrt{6}}{5}$ 97. $\frac{\sqrt{7}}{3}$ 99. period $= \frac{2\pi}{3}$

101. period $= \pi$

103.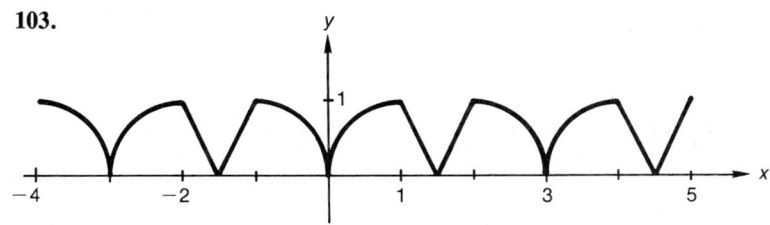

105. 5 107. -5 109.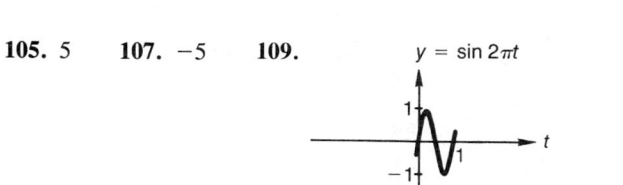
$y = \sin 2\pi t$

111.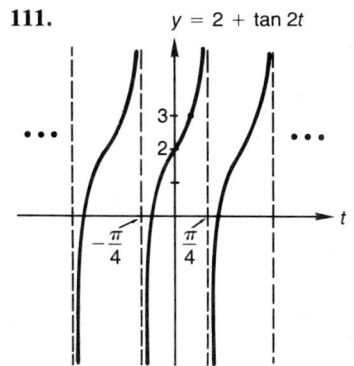
$y = 2 + \tan 2t$

113.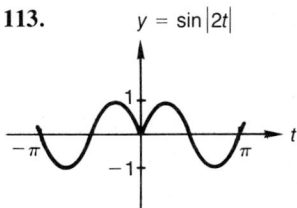
$y = \sin |2t|$

115.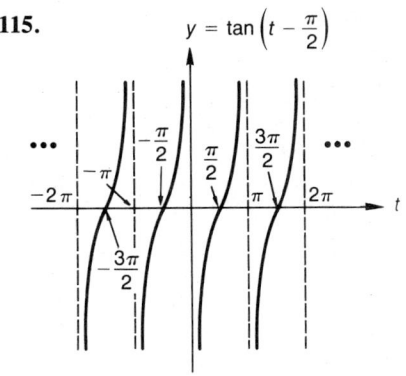
$y = \tan\left(t - \frac{\pi}{2}\right)$

117.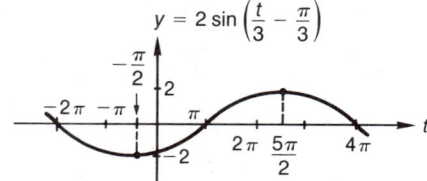

123.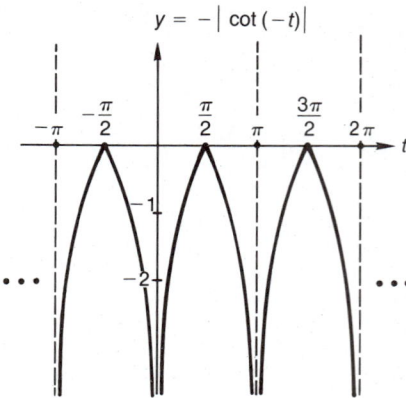

119.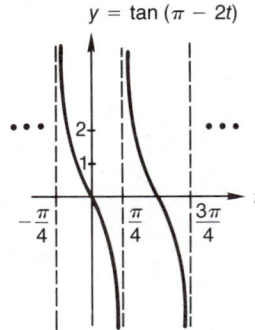

125.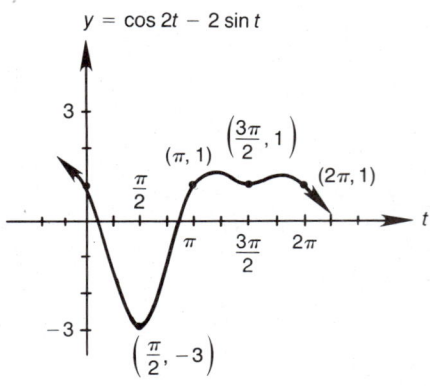

121.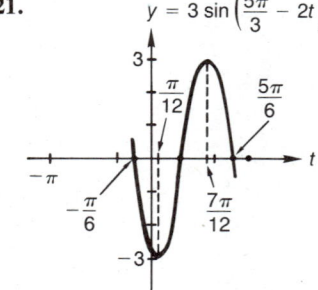

127.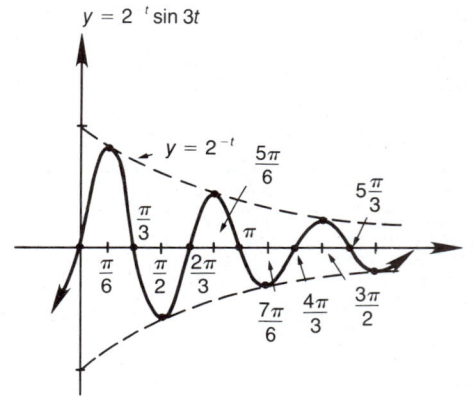

CHAPTER 8

Section 8.1

1. $\tan\left(\dfrac{\pi}{6}\right) + \cot\left(\dfrac{\pi}{6}\right) = \dfrac{4}{\sqrt{3}} \neq 2$; $\tan\left(\dfrac{\pi}{4}\right) + \cot\left(\dfrac{\pi}{4}\right) = 2$

3. $\tan^2\left(\dfrac{\pi}{4}\right) + \sin^2\left(\dfrac{\pi}{4}\right) = \dfrac{3}{2} \neq 0$; $\tan^2 0 + \sin^2 0 = 0$

5. $\sin^2\left(\dfrac{\pi}{4}\right) + \sec^2\left(\dfrac{\pi}{4}\right) = \dfrac{5}{2} \neq 1$; $\sin^2 0 + \sec^2 0 = 1$

7. $\cos\left(\dfrac{\pi}{2}\right) - \sin\left(\dfrac{\pi}{2}\right) = -1 \neq 1$; $\cos 0 - \sin 0 = 1$

9. $\cot^2\left(\dfrac{\pi}{4}\right) + \sin^2\left(\dfrac{\pi}{4}\right) = \dfrac{3}{2} \neq 1$; $\cot^2\left(\dfrac{\pi}{2}\right) + \sin^2\left(\dfrac{\pi}{2}\right) = 1$

11. $\dfrac{1}{\sin t} = \csc t$ 13. 1 15. $\dfrac{1 + \sin t}{\cos t}$

17. $\dfrac{1}{\cos^2 v} = \sec^2 v$ **19.** $\dfrac{1}{\sin^2 t} = \csc^2 t$

21–49. Only verifications were requested.

51. $-\sqrt{15}$ **53.** $-\sqrt{10}$ **55.** $-\sqrt{5}$ **57.** $2\sqrt{2}$
59. $\sqrt{5}$

Section 8.2

1–41. Only verifications were requested.

43. $A = 1, B = -1$ **45.** $A = 2$ **47.** $A = B = 1$
49. $A = -1, B = 2$ **51.** $A = C = 1, B = 2$

Section 8.3

1–21. Only verifications were requested.

23. $\csc 2\theta = \dfrac{\csc^2 \theta}{2 \cot \theta}$ **25.** $\tan 2\theta = \dfrac{2 \tan \theta}{1 - \tan^2 \theta}$

	sin	cos	tan
27.	$\dfrac{\sqrt{6} + \sqrt{2}}{4}$	$\dfrac{\sqrt{6} - \sqrt{2}}{4}$	$2 + \sqrt{3}$
29.	$\dfrac{\sqrt{2} - \sqrt{6}}{4}$	$\dfrac{-(\sqrt{2} + \sqrt{6})}{4}$	$2 - \sqrt{3}$
31.	$\dfrac{\sqrt{2} - \sqrt{6}}{4}$	$\dfrac{\sqrt{2} + \sqrt{6}}{4}$	$\sqrt{3} - 2$
33.	$\dfrac{\sqrt{6} + \sqrt{2}}{4}$	$\dfrac{\sqrt{2} - \sqrt{6}}{4}$	$-2 - \sqrt{3}$

35. $\dfrac{-1}{2}$ **37.** -1 **39.** $-\sqrt{3}$ **41.** $\dfrac{-\sqrt{3}}{2}$

		sin	cos	tan	Quadrant
43.	(for $\alpha + \beta$)	$\dfrac{-3 + 4\sqrt{3}}{10}$	$\dfrac{4 + 3\sqrt{3}}{10}$	$\dfrac{48 - 25\sqrt{3}}{11}$	I
	(for $\alpha - \beta$)	$\dfrac{-3 - 4\sqrt{3}}{10}$	$\dfrac{4 - 3\sqrt{3}}{10}$	$\dfrac{48 + 25\sqrt{3}}{11}$	III
45.	(for $\alpha + \beta$)	$\dfrac{-\sqrt{5} - 4\sqrt{2}}{9}$	$\dfrac{2(1 - \sqrt{10})}{9}$	$\dfrac{\sqrt{5} + \sqrt{2}}{2}$	III
	(for $\alpha - \beta$)	$\dfrac{-\sqrt{5} + 4\sqrt{2}}{9}$	$\dfrac{2(1 + \sqrt{10})}{9}$	$\dfrac{\sqrt{5} - \sqrt{2}}{2}$	I
47.	(for $\alpha + \beta$)	$\dfrac{3\sqrt{11} + \sqrt{15}}{40}$	$\dfrac{3\sqrt{165} - 1}{40}$	$\dfrac{25\sqrt{15} + 12\sqrt{11}}{371}$	I
	(for $\alpha - \beta$)	$\dfrac{3\sqrt{11} - \sqrt{15}}{40}$	$\dfrac{3\sqrt{165} + 1}{40}$	$\dfrac{25\sqrt{15} - 12\sqrt{11}}{371}$	I
49.	(for $\alpha + \beta$)	$\dfrac{4\sqrt{3} + 3\sqrt{6}}{15}$	$\dfrac{3\sqrt{3} - 4\sqrt{6}}{15}$	$\dfrac{-36 - 25\sqrt{2}}{23}$	II
	(for $\alpha - \beta$)	$\dfrac{4\sqrt{3} - 3\sqrt{6}}{15}$	$\dfrac{3\sqrt{3} + 4\sqrt{6}}{15}$	$\dfrac{-36 + 25\sqrt{2}}{23}$	IV

51–57. Only verifications were requested.

59. $y = \sin \theta - \cos \theta = -\sqrt{2} \cos(\theta + \pi/4)$

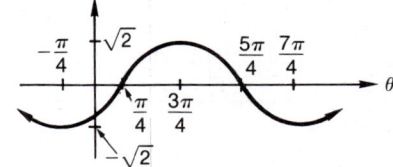

61. $y = \cos \theta - \sqrt{3} \sin \theta = 2 \cos(\theta + \pi/3)$

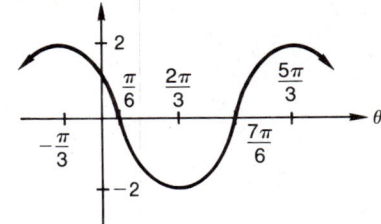

63. (for 58) $\sqrt{2} \sin\left(\theta + \dfrac{\pi}{4}\right)$ **65.** (for 60) $2 \sin\left(\theta + \dfrac{\pi}{6}\right)$

67. (for 62) $2 \sin\left(\theta - \dfrac{\pi}{3}\right)$ **69.** $\dfrac{\sqrt{6}}{2}$ **71.** $-\sqrt{2}$

Section 8.4

1. $\cos 3\theta = 4\cos^3\theta - 3\cos\theta$
3. $\cos 4\theta = 8\cos^4\theta - 8\cos^2\theta + 1$

	sin	cos	tan	Quadrant
5.	$\dfrac{-4\sqrt{2}}{9}$	$\dfrac{7}{9}$	$\dfrac{-4\sqrt{2}}{7}$	IV
7.	$\dfrac{-24}{25}$	$\dfrac{7}{25}$	$\dfrac{-24}{7}$	IV
9.	$\dfrac{24}{25}$	$\dfrac{-7}{25}$	$\dfrac{-24}{7}$	II
11.	$\dfrac{\sqrt{2-\sqrt{2}}}{2}$	$\dfrac{\sqrt{2+\sqrt{2}}}{2}$	$\sqrt{2}-1$	
13.	$\dfrac{-\sqrt{2-\sqrt{2}}}{2}$	$\dfrac{-\sqrt{2+\sqrt{2}}}{2}$	$\sqrt{2}-1$	
15.	$\dfrac{\sqrt{2+\sqrt{3}}}{2}$	$\dfrac{\sqrt{2-\sqrt{3}}}{2}$	$2+\sqrt{3}$	
17.	$\dfrac{-\sqrt{2+\sqrt{3}}}{2}$	$\dfrac{\sqrt{2-\sqrt{3}}}{2}$	$-(2+\sqrt{3})$	
19.	$\dfrac{\sqrt{2+\sqrt{3}}}{2}$	$\dfrac{-\sqrt{2-\sqrt{3}}}{2}$	$-(2+\sqrt{3})$	
21.	$\dfrac{\sqrt{6}}{6}$	$\dfrac{\sqrt{30}}{6}$	$\dfrac{\sqrt{5}}{5}$	
23.	$\dfrac{-\sqrt{6}}{4}$	$\dfrac{-\sqrt{10}}{4}$	$\dfrac{\sqrt{15}}{5}$	
25.	$\dfrac{-\sqrt{18+12\sqrt{2}}}{6}$	$\dfrac{\sqrt{18-12\sqrt{2}}}{6}$	$-3-2\sqrt{2}$	
27.	$\dfrac{-\sqrt{3}}{2}$	$\dfrac{1}{2}$	$-\sqrt{3}$	
29.	$\dfrac{\sqrt{6}}{3}$	$\dfrac{\sqrt{3}}{3}$	$\sqrt{2}$	

31–51. Only verfications were requested.

Section 8.5

1. Only verification was requested.
3. $\dfrac{1}{2}(\sin 11\alpha - \sin 3\alpha)$ 5. $\dfrac{1}{2}(\cos 14\gamma + \cos 6\gamma)$
7. $\dfrac{1}{2}(\sin 14v - \sin 6v)$ 9. $\dfrac{1}{2}(\cos 7\phi + \cos\phi)$
11. $\dfrac{1}{2}(\cos 3u - \cos 11u)$ 13. $\dfrac{\sqrt{3}+1}{4}$ 15. $\dfrac{2-\sqrt{2}}{4}$
17. $\dfrac{1+\sqrt{3}}{4}$ 19. $\dfrac{2-\sqrt{2}}{4}$ 21. $\dfrac{-(\sqrt{3}+2)}{4}$
23. $2\sin 2\beta\cos\beta$ 25. $2\cos\left(\dfrac{5\theta}{2}\right)\sin\left(\dfrac{3\theta}{2}\right)$
27. $2\cos 3\eta\cos\eta$ 29. $2\cos 3u\sin u$
31. $2\cos 2w\cos w$ 33. $-2\sin 2z\sin z$ 35. $\dfrac{\sqrt{6}}{2}$

37. $\dfrac{\sqrt{2}}{2}$ 39. $\dfrac{\sqrt{2}}{2}$ 41. $\dfrac{-\sqrt{2}}{2}$ 43. $\dfrac{-\sqrt{2}}{2}$
45. $\dfrac{\sqrt{6}}{2}$

47–55. Only verifications were requested.

57. $y = -2(\sin 3t)(\sin t) = \cos 4t - \cos 2t$

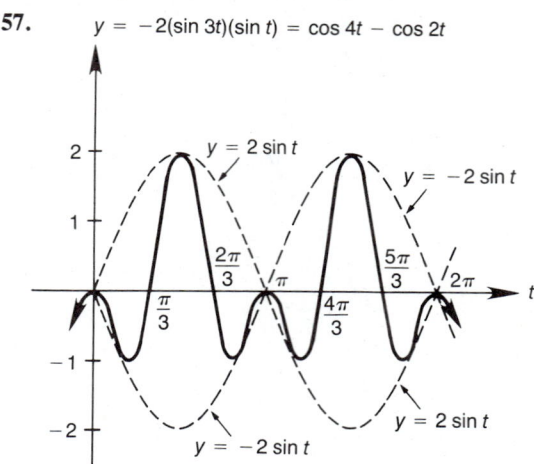

59. $y = 2(\cos 8t)(\sin 2t) = \sin 10t - \sin 6t$

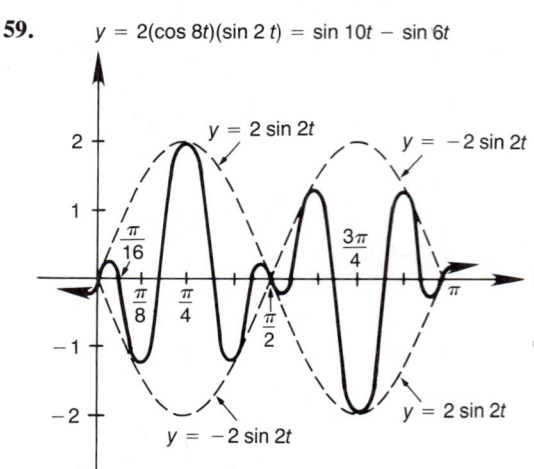

61. $y = 2(\cos 3t)(\cos t) = \cos 2t + \cos 4t$

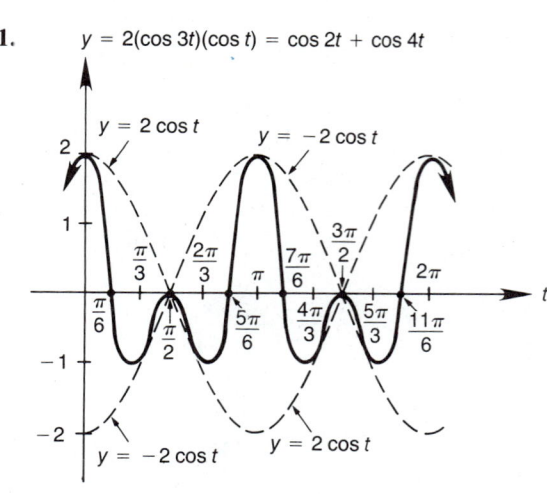

63. $y = 2(\sin 8t)(\cos 2t) = \sin 10t + \sin 6t$

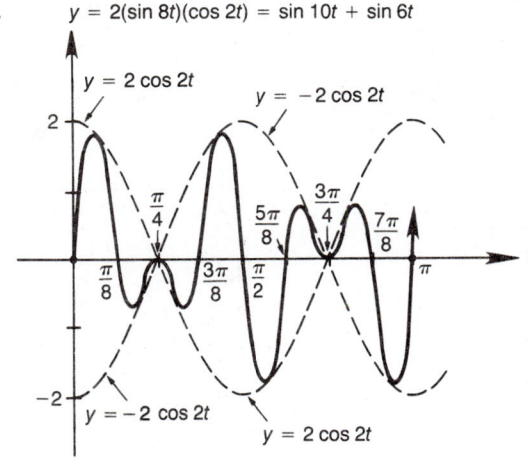

65. $y = 2(\cos 4t)(\sin t) = \sin 5t - \sin 3t$

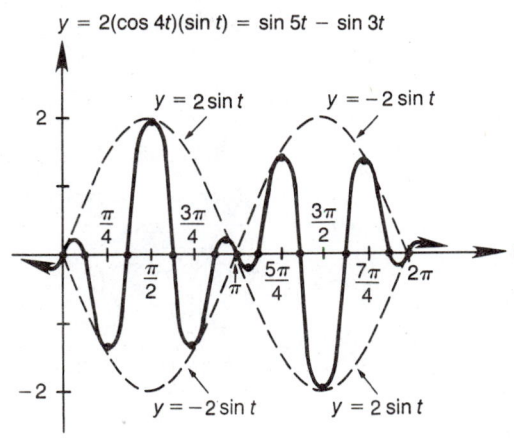

Section 8.6

1. $\left\{\dfrac{\pi}{6} + 2k\pi : k \in Z\right\} \cup \left\{\dfrac{5\pi}{6} + 2k\pi : k \in Z\right\}$
3. $\left\{\dfrac{\pi}{4} + k\pi : k \in Z\right\}$
5. $\left\{\dfrac{7\pi}{6} + 2k\pi : k \in Z\right\} \cup \left\{\dfrac{11\pi}{6} + 2k\pi : k \in Z\right\}$
7. $\left\{\dfrac{\pi}{6} + 2k\pi : k \in Z\right\} \cup \left\{\dfrac{5\pi}{6} + 2k\pi : k \in Z\right\}$
9. $\left\{\dfrac{\pi}{4} + \dfrac{k\pi}{2} : k \in Z\right\}$ 11. $\left\{\dfrac{\pi}{4} + \dfrac{k\pi}{2} : k \in Z\right\}$
13. $\left\{\dfrac{\pi}{4} + \dfrac{k\pi}{2} : k \in Z\right\}$ 15. $\left\{\dfrac{\pi}{12} + \dfrac{k\pi}{3} : k \in Z\right\}$
17. $\left\{\dfrac{5\pi}{12} + k\pi : k \in Z\right\} \cup \left\{\dfrac{7\pi}{12} + k\pi : k \in Z\right\}$
19. $\left\{\dfrac{10\pi}{3} + 4k\pi : k \in Z\right\} \cup \left\{\dfrac{8\pi}{3} + 4k\pi : k \in Z\right\}$
21. $\{4k\pi : k \in Z\}$
23. $\left\{\dfrac{\pi}{3} + 2k\pi : k \in Z\right\} \cup \left\{\dfrac{5\pi}{3} + 2k\pi : k \in Z\right\}$
25. $\left\{\dfrac{\pi}{4} + k\pi : k \in Z\right\} \cup \left\{\dfrac{\pi}{2} + 2k\pi : k \in Z\right\}$
27. $\left\{\dfrac{\pi}{3} + k\pi : k \in Z\right\} \cup \left\{\dfrac{\pi}{6} + 2k\pi : k \in Z\right\} \cup \left\{\dfrac{11\pi}{6} + 2k\pi k \in Z\right\}$
29. $\{k\pi : k \in Z\} \cup \left\{\dfrac{\pi}{4} + k\pi : k \in Z\right\}$
31. $\left\{\dfrac{\pi}{2} + k\pi : k \in Z\right\} \cup \left\{\dfrac{\pi}{6} + 2k\pi : k \in Z\right\} \cup \left\{\dfrac{5\pi}{6} + 2k\pi : k \in Z\right\}$
33. $\{k\pi : k \in Z\}$ 35. $\left\{\dfrac{k\pi}{2} : k \in Z\right\}$ 37. $\dfrac{\pi}{2}, \dfrac{7\pi}{6}, \dfrac{11\pi}{6}$
39. $\dfrac{\pi}{3}, \pi, \dfrac{5\pi}{3}$ 41. $\dfrac{\pi}{2}$ 43. $\dfrac{\pi}{6}, \dfrac{\pi}{2}, \dfrac{5\pi}{6}$ 45. $0, 2\pi$
47. $\dfrac{\pi}{6}, \dfrac{3\pi}{2}, \dfrac{5\pi}{6}$ 49. $\dfrac{\pi}{4}, \dfrac{5\pi}{4}$ 51. $\dfrac{\pi}{6}, \dfrac{5\pi}{6}$ 53. $\dfrac{2\pi}{3}, \dfrac{4\pi}{3}$
55. $\dfrac{\pi}{4}, \dfrac{3\pi}{4}, \dfrac{5\pi}{4}, \dfrac{7\pi}{4}, \dfrac{\pi}{6}, \dfrac{5\pi}{6}, \dfrac{7\pi}{6}, \dfrac{11\pi}{6}$
57. $\dfrac{\pi}{16}, \dfrac{5\pi}{16}, \dfrac{9\pi}{16}, \dfrac{13\pi}{16}, \dfrac{17\pi}{16}, \dfrac{21\pi}{16}, \dfrac{25\pi}{16}, \dfrac{29\pi}{16}$
59. $0, \dfrac{\pi}{2}, \pi, \dfrac{3\pi}{2}, 2\pi, \dfrac{\pi}{3}, \dfrac{2\pi}{3}, \dfrac{4\pi}{3}, \dfrac{5\pi}{3}$
61. $\dfrac{\pi}{3}, \dfrac{2\pi}{3}, \dfrac{4\pi}{3}, \dfrac{5\pi}{3}$ 63. $\dfrac{\pi}{2}, \dfrac{3\pi}{2}$
65. $\dfrac{\pi}{36}, \dfrac{5\pi}{36}, \dfrac{13\pi}{36}, \dfrac{17\pi}{36}, \dfrac{25\pi}{36}, \dfrac{29\pi}{36}, \dfrac{37\pi}{36}, \dfrac{41\pi}{36}, \dfrac{49\pi}{36}, \dfrac{53\pi}{36}, \dfrac{61\pi}{36}, \dfrac{65\pi}{36}$
67. $\dfrac{\pi}{3}, \dfrac{2\pi}{3}, \dfrac{4\pi}{3}, \dfrac{5\pi}{3}$ 69. $\left\{\dfrac{\pi}{2}, \dfrac{3\pi}{2}\right\}$
71. $\{2k\pi : k \in Z\} \cup \left\{\dfrac{\pi}{2} + 2k\pi : k \in Z\right\}$
73. $\left\{\dfrac{\pi}{4} + k\pi : k \in Z\right\} \cup \left\{\dfrac{\pi}{2} + k\pi : k \in Z\right\}$
75. $\{3.871, 5.553\}$ 77. $\{1.842, 4.441\}$
79. $\{0.785, 1.107, 3.927, 4.249\}$
81. $\{0.848, 2.294, 3.481, 5.943\}$
83. $\{0.322, 2.356, 3.463, 5.498\}$
85. During a single revolution of the slower wheel, the beep is heard 4 times: when $\theta = 0, \dfrac{2\pi}{3}, \pi, \dfrac{4\pi}{3}$. It beeps once (at $\theta = 0, 2\pi, \ldots$) followed by a triple burst ($\theta = \dfrac{2\pi}{3}, \pi, \dfrac{4\pi}{3}$, then $\dfrac{8\pi}{3}, 3\pi, \dfrac{10\pi}{3}, \ldots$).

Section 8.7

1. $\dfrac{3\pi}{4}$ **3.** $\dfrac{\pi}{4}$ **5.** $\dfrac{\pi}{6}$ **7.** 0.403 **9.** 0.752 **11.** 2.020

13. -0.215 **15.** 3.023 **17.** $\dfrac{-3}{4}$ **19.** $\dfrac{9}{\sqrt{77}}$ **21.** $\dfrac{4\sqrt{2}}{9}$

23. $\dfrac{56}{65}$ **25.** $\dfrac{-204}{253}$ **27.** $\dfrac{-13}{85}$ **29.** $\dfrac{56}{65}$ **31.** 1

33. $0 \le \text{Cos}^{-1} x \le \pi$, so $\sin(\text{Cos}^{-1} x) \ge 0$

35. See solution to Exercise 33. Then $\cot(\text{Cos}^{-1} x)$
$= \dfrac{\cos(\text{Cos}^{-1} x)}{\sin(\text{Cos}^{-1} x)} = \dfrac{x}{\sqrt{1 - \cos^2(\text{Cos}^{-1} x)}} = \dfrac{x}{\sqrt{1 - x^2}}$

37. $\csc(\text{Sin}^{-1} x) = \dfrac{1}{\sin(\text{Sin}^{-1} x)} = \dfrac{1}{x}$

39. $0 \le \text{Cos}^{-1} x \le \pi$; thus $\sin(\text{Cos}^{-1} x) \ge 0$ and
$\sin(\text{Cos}^{-1} x) = \sqrt{1 - \cos^2(\text{Cos}^{-1} x)} = \sqrt{1 - x^2}$;
$\tan(\text{Cos}^{-1} x) = \dfrac{\sin(\text{Cos}^{-1} x)}{\cos(\text{Cos}^{-1} x)} = \dfrac{\sqrt{1 - x^2}}{x}$

41. $\cot(\text{Sin}^{-1} x) = \dfrac{\cos(\text{Sin}^{-1} x)}{\sin(\text{Sin}^{-1} x)} = \dfrac{\sqrt{1 - x^2}}{x}$ (see Exercise 38)

43. $\dfrac{3\pi}{4}$ **45.** $\dfrac{\pi}{3}$ **47.** $\dfrac{\pi}{3}$

49.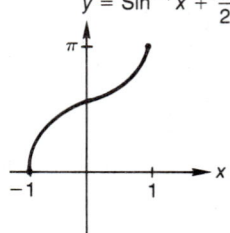

51.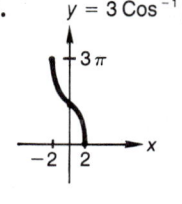

53.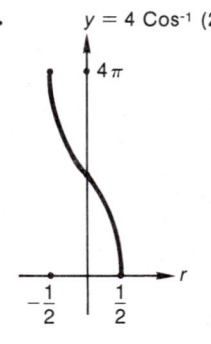

55.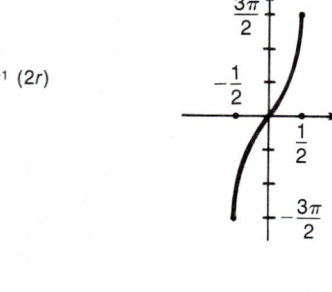

57.

59.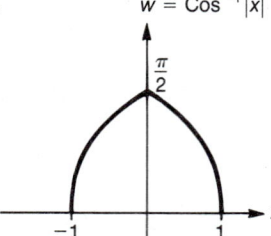

61.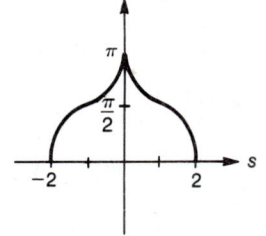

63

Section 8.8

1. $-\dfrac{\pi}{3}$ **3.** $\dfrac{\pi}{3}$ **5.** $\dfrac{\pi}{3}$ **7.** $-\dfrac{\pi}{6}$ **9.** 0.245 **11.** 1.633

13. 2.944 **15.** -0.340 **17.** $\dfrac{3\sqrt{10}}{10}$ **19.** $\dfrac{\sqrt{15}}{4}$

21. $\dfrac{\sqrt{899}}{30}$ **23.** $-2\sqrt{2}$ **25.** $\dfrac{\sqrt{15} - 2\sqrt{5}}{10}$ **27.** $-\dfrac{7}{9}$

29. $\dfrac{-(13\sqrt{5} + 27)}{8}$ **31.** $\dfrac{3\pi}{4}$ **33.** $0, 1, -1$ **35.** $\dfrac{56}{65}$

37. $\cot(\text{Tan}^{-1} x) = \dfrac{1}{\tan(\text{Tan}^{-1} x)} = \dfrac{1}{x}$

39. $\sin(\text{Csc}^{-1} x) = \dfrac{1}{\csc(\text{Csc}^{-1} x)} = \dfrac{1}{x}$

41. Let $\theta = \text{Tan}^{-1} x$.

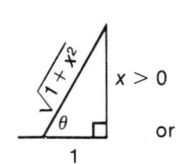

In either case, $\cos\theta = \dfrac{1}{\sqrt{1 + x^2}}$.

43. Let $\theta = \text{Cot}^{-1} x$.

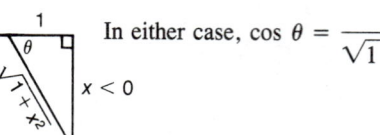

In either case, $\sec\theta = \dfrac{\sqrt{1 + x^2}}{x}$.

45.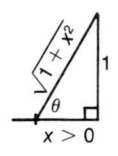

47. $s = \frac{\pi}{2} - \text{Tan}^{-1} r$

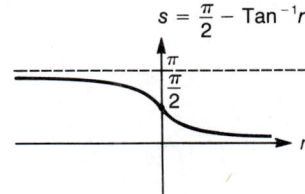

49. $y = -2\text{Tan}^{-1}(r - 1)$

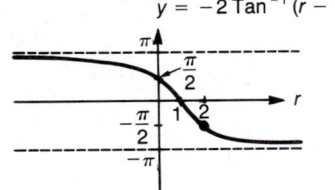

51. $u = 2\text{Sec}^{-1}|v|$

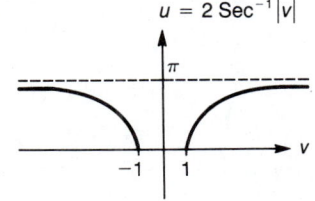

53.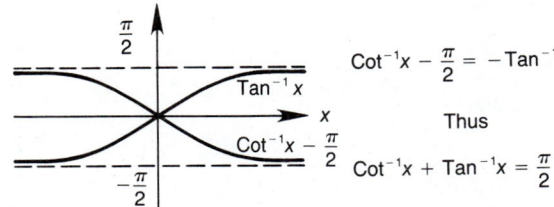

$\text{Cot}^{-1}x - \frac{\pi}{2} = -\text{Tan}^{-1}x$

Thus

$\text{Cot}^{-1}x + \text{Tan}^{-1}x = \frac{\pi}{2}$

Section 8.10

1. $\tan^2\left(\frac{\pi}{6}\right) - \cot^2\left(\frac{\pi}{6}\right) = -\frac{8}{3} \neq 0$;

$\tan^2\left(\frac{\pi}{4}\right) - \cot^2\left(\frac{\pi}{4}\right) = 0$

3–39. Only verfications were requested.   **41.** $A = B = 1$

43. $A = 1$

		sin	cos	tan	Quadrant
45.		$\frac{\sqrt{6} + \sqrt{2}}{4}$	$\frac{\sqrt{6} - \sqrt{2}}{4}$	$2 + \sqrt{3}$	
47.		$\frac{\sqrt{2} - \sqrt{6}}{4}$	$\frac{-(\sqrt{2} + \sqrt{6})}{4}$	$2 - \sqrt{3}$	
49.		$\frac{\sqrt{2} - \sqrt{6}}{4}$	$\frac{\sqrt{2} + \sqrt{6}}{4}$	$\sqrt{3} - 2$	
51.		$\frac{\sqrt{2} - \sqrt{2}}{2}$	$-\frac{\sqrt{2} + \sqrt{2}}{2}$	$1 - \sqrt{2}$	
53.		$-\frac{\sqrt{2} + \sqrt{2}}{2}$	$-\frac{\sqrt{2} - \sqrt{2}}{2}$	$1 + \sqrt{2}$	
55.		$-\frac{\sqrt{2} - \sqrt{3}}{2}$	$-\frac{\sqrt{2} + \sqrt{3}}{2}$	$2 - \sqrt{3}$	
57.		$\frac{-\sqrt{10}}{5}$	$\frac{\sqrt{15}}{5}$	$\frac{\sqrt{6}}{3}$	
59.		$\frac{\sqrt{18 + 6\sqrt{6}}}{6}$	$-\frac{\sqrt{18 - 6\sqrt{6}}}{6}$	$-\sqrt{3} - \sqrt{2}$	
61.		$\frac{\sqrt{10}}{10}$	$-\frac{3\sqrt{10}}{10}$	$-\frac{1}{3}$	
63.		$-\frac{24}{25}$	$\frac{7}{25}$	$-\frac{24}{7}$	IV
65.	(for $\alpha + \beta$)	$\frac{\sqrt{26} + 20\sqrt{13}}{78}$	$\frac{-4\sqrt{13} + 5\sqrt{26}}{78}$	$\frac{52\sqrt{2} + 45}{17}$	I
	(for $\alpha - \beta$)	$\frac{\sqrt{26} - 20\sqrt{13}}{78}$	$\frac{-4\sqrt{13} - 5\sqrt{26}}{78}$	$\frac{52\sqrt{2} - 45}{17}$	III
67.	(for $\alpha + \beta$)	$\frac{77}{85}$	$\frac{36}{85}$	$\frac{77}{36}$	I
	(for $\alpha - \beta$)	$-\frac{13}{85}$	$\frac{84}{85}$	$-\frac{13}{84}$	IV

69. $y = \cos\theta - \sin\theta = \sqrt{2}\cos\left(\dfrac{\pi}{4} + \theta\right)$

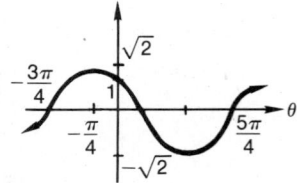

71. $y = \sqrt{3}\sin\theta + \cos\theta = 2\sin\left(\theta + \dfrac{\pi}{6}\right)$

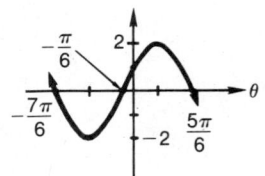

73. $\dfrac{-\sqrt{2}}{2}$ **75.** $\sqrt{2}$ **77.** $\dfrac{1}{2}(\cos 2\alpha - \cos 8\alpha)$

79. $\dfrac{1}{2}(\cos 8\gamma + \cos 2\gamma)$ **81.** $\dfrac{1}{2}(\sin 8\phi + \sin 2\phi)$

83. $-\dfrac{1}{4}$ **85.** $\dfrac{\sqrt{3}-1}{4}$ **87.** $2(\sin 3t)(\cos t)$

89. $-2(\sin 3v)(\sin v)$ **91.** $2\cos\left(\dfrac{5\gamma}{2}\right)\cos\left(\dfrac{\gamma}{2}\right)$

93. $\dfrac{\sqrt{6}}{2}$ **95.** $-\dfrac{\sqrt{2}}{2}$ **97.** $-\dfrac{\sqrt{2}}{2}$

99–101. Only verifications were requested.

103. $y = 2(\cos 2t)(\sin t) = \sin 3t - \sin t$

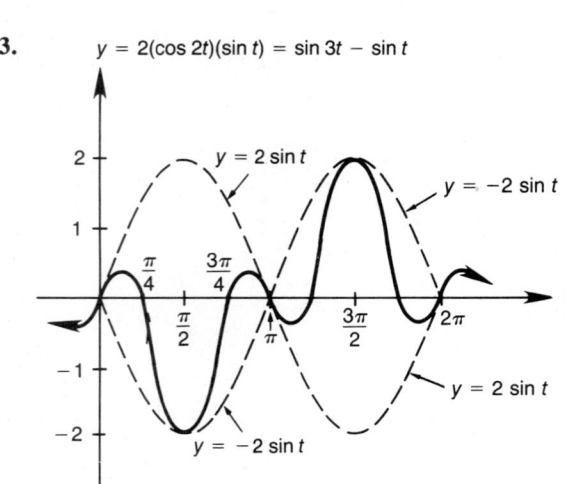

105. $y = -2(\sin 4t)(\sin t) = \cos 5t - \cos 3t$

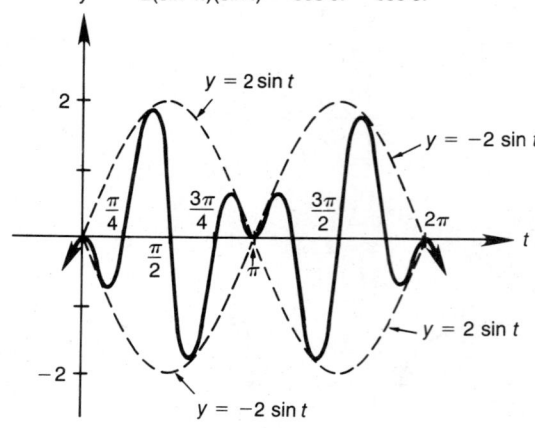

107. $\left\{\dfrac{\pi}{4}, \dfrac{5\pi}{4}\right\}$ **109.** $\left\{\dfrac{\pi}{12}, \dfrac{7\pi}{12}, \dfrac{13\pi}{12}, \dfrac{19\pi}{12}\right\}$

111. $\left\{\dfrac{\pi}{6}, \dfrac{\pi}{3}, \dfrac{2\pi}{3}, \dfrac{5\pi}{6}\right\}$ **113.** $\left\{\dfrac{\pi}{6}, \dfrac{5\pi}{6}\right\}$

115. $0,\ \pi,\ 2\pi,\ \mathrm{Sin}^{-1}\!\left(\dfrac{3}{4}\right),\ \left(\pi - \mathrm{Sin}^{-1}\!\left(\dfrac{3}{4}\right)\right)$

117. $\dfrac{2\pi}{3}, \dfrac{4\pi}{3}, 0, 2\pi$ **119.** $\dfrac{\pi}{6}, \dfrac{5\pi}{6}, \dfrac{3\pi}{2}$

121. $\mathrm{Sin}^{-1}\!\left(\dfrac{1}{\sqrt{10}}\right),\ \pi + \mathrm{Sin}^{-1}\!\left(\dfrac{1}{\sqrt{10}}\right),\ \dfrac{3\pi}{4}, \dfrac{7\pi}{4}$

123. $\dfrac{2\pi}{3},\ \pi,\ \dfrac{4\pi}{3}$,

125. $\dfrac{\pi}{2},\ \mathrm{Sin}^{-1}\!\left(\dfrac{-1}{4}\right) + 2\pi,\ \pi - \mathrm{Sin}^{-1}\!\left(\dfrac{-1}{4}\right)$

127. $\dfrac{\pi}{4}, \dfrac{5\pi}{4},\ \mathrm{Sin}^{-1}\!\left(\dfrac{1}{\sqrt{10}}\right),\ \pi + \mathrm{Sin}^{-1}\!\left(\dfrac{1}{\sqrt{10}}\right)$

129. $\dfrac{\pi}{2}, \dfrac{7\pi}{6}, \dfrac{11\pi}{6}$ **131.** $\dfrac{\pi}{2}, \dfrac{3\pi}{2}, \dfrac{\pi}{3}, \dfrac{2\pi}{3}, \dfrac{4\pi}{3}, \dfrac{5\pi}{3}$

133. $\dfrac{\pi}{2}, \dfrac{3\pi}{2}, \dfrac{\pi}{6}, \dfrac{5\pi}{6}$ **135.** $\dfrac{\pi}{2}, \dfrac{3\pi}{2}$

137. $\dfrac{\pi}{6}, \dfrac{5\pi}{6}, \dfrac{\pi}{3}, \dfrac{5\pi}{3}$ **139.** $\dfrac{3\pi}{4}$ **141.** $\dfrac{\pi}{3}$ **143.** $\dfrac{\pi}{6}$

145. $\dfrac{\pi}{4}$ **147.** 0.983   **149.** -0.390 **151.** 2.067

153. 1.271   **155.** $\dfrac{\sqrt{101}}{101}$ **157.** $\dfrac{2\sqrt{5}}{15}$ **159.** $\dfrac{\sqrt{17}}{17}$

161. $\dfrac{4}{5}$ **163.** $\dfrac{3 - \sqrt{105}}{16}$ **165.** $\dfrac{3\sqrt{10} - 4\sqrt{5}}{30}$

167. $\dfrac{2\pi}{3}$ **169.** $\dfrac{117}{125}$ **171.** 0

173.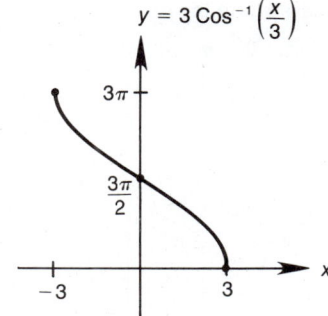

175.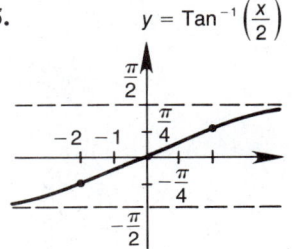

177.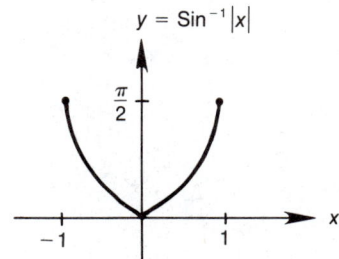

179. Let $\theta = \text{Sin}^{-1} x$.

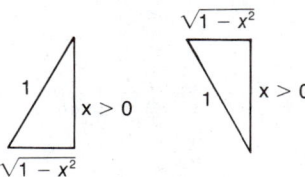

181.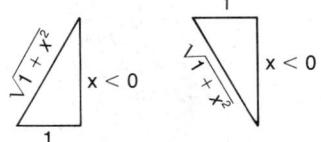

CHAPTER 9

Section 9.1

1. $C = 76°0'$, $a = 9.2$, $b = 14.3$
3. $b = 12.3$, $c = 14.7$, $C = 70°30'$
5. $B = 53°1'$, $A = 93°59'$, $a = 27.5$;
   or $B = 126°59'$, $A = 20°1'$, $a = 9.4$
7. $B = 27°37'$, $C = 40°23'$, $c = 29.3$
9. $A = 21°15'$, $B = 116°35'$, $b = 15.1$
11. No solutions 13. $C = 51°$, $a = 19.3$, $b = 28.2$
15. $A = 153°0'$, $c = 2.9$, $b = 2.3$
17. $C = 21°51'$, $A = 143°9'$, $a = 74.1$;
    or $C = 158°9'$, $A = 6°51'$, $a = 14.7$
19. $B = 24°43'$, $A = 12°17'$, $a = 16.3$
21. $B = 29°0'$, $b = 5.3$, $a = 10.2$
23. $C = 12°50'$, $a = 49.8$, $b = 53.2$
25. No solutions 27. 1683 ft
29. **a.** 285 ft **b.** 115 ft   31. 844 ft

33. **a.** At the second sighting, 7231 ft, or 1.37 mi, from the city. **b.** 633   35. 135 mi

Section 9.2

1. $a = 14.1$, $B = 77°1'$, $C = 40°59'$
3. $b = 68.9$, $C = 27°39'$, $A = 52°1'$
5. $C = 104°16'$, $B = 59°26'$, $A = 16°19'$
7. $b = 48.7$, $A = 70°28'$, $C = 50°52'$
9. $A = 80°26'$, $B = 44°18'$, $C = 55°16'$
11. $B = 88°57'$, $C = 66°25'$, $A = 24°38'$
13. $c = 189.2$, $A = 61°24'$, $B = 64°56'$
15. $A = 30°48'$, $B = 41°10'$, $C = 108°2'$ 17. 182.9 mi
19. **a.** 6.65 mi **b.** 4.7 mi wide   21. 6.38 acres
23. $7\frac{1}{2}$ h   25. 36,850 metric tons
27. $6\sqrt{3}$ in.²   29. 65.6 ft²

Section 9.3

1.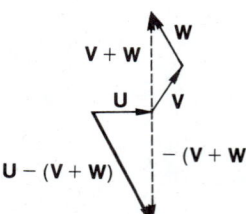

3. T − X − Z   **5.** R

7.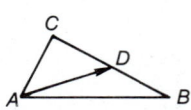

$\overrightarrow{AD} = \overrightarrow{AC} + \frac{1}{2}\overrightarrow{CB} = \overrightarrow{AC} + \frac{1}{2}(\overrightarrow{AB} - \overrightarrow{AC}) = \frac{1}{2}(\overrightarrow{AB} + \overrightarrow{AC})$

9.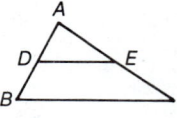

$\overrightarrow{DE} = \overrightarrow{DA} + \overrightarrow{AE} = \frac{1}{2}\overrightarrow{BA} + \frac{1}{2}\overrightarrow{AC} = \frac{1}{2}(\overrightarrow{BA} + \overrightarrow{AC}) = \frac{1}{2}\overrightarrow{BC}$

11.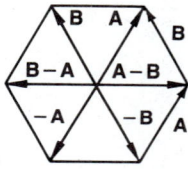

13. a. $\frac{1}{2}\overrightarrow{OB} + \frac{1}{2}\overrightarrow{OC} - \overrightarrow{OA}$ **b.** $\frac{1}{3}(\overrightarrow{OA} + \overrightarrow{OB} + \overrightarrow{OC})$

c. The results in (b) are the same. **d.** The medians of a triangle intersect in a point whose distance from each vertex is $\frac{2}{3}$ the length of the corresponding median.

15.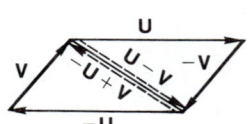

17. $i - 4j$ **19.** $i - 3j$ **21.** $i - j$

23. $\frac{13}{4}i + \frac{3}{4}j$

25. $\overrightarrow{AB} = -2i - 4j$, $\overrightarrow{AC} = i + 2j$, $\overrightarrow{AB} = -2\overrightarrow{AC}$; the points are collinear.

27. $\langle 3, -3 \rangle$ **29.** $\langle 3, -8 \rangle$ **31.** $\langle 7, 2 \rangle$ **33.** $\langle 2, 5 \rangle$
35. $-i - j$ **37.** $2i + 9j$ **39.** $\sqrt{2}$ **41.** 1
43. $\sqrt{65}$ **45.** $\frac{4}{5}i + \frac{3}{5}j$ **47.** $\frac{\sqrt{2}}{2}i + \frac{\sqrt{2}}{2}j$
49. $\langle \frac{5}{13}, \frac{12}{13} \rangle$ **51.** $\sqrt{37}$ **53.** 6

55. 245.58 knots, or 282.4 mi/h in the direction W49°31'S
57. 250 ft   **59. a.** 577.35 lb   **b.** 288.675 lb
61. a. 365.6 lb   **b.** 2977.64 lb

Section 9.4

1.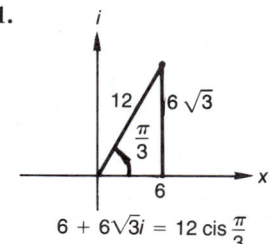

$6 + 6\sqrt{3}i = 12 \text{ cis } \frac{\pi}{3}$

3.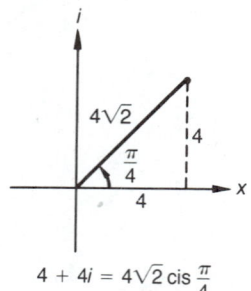

$4 + 4i = 4\sqrt{2} \text{ cis } \frac{\pi}{4}$

5.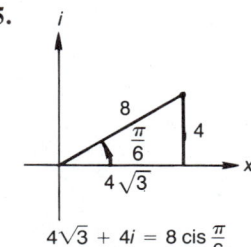

$4\sqrt{3} + 4i = 8 \text{ cis } \frac{\pi}{6}$

7.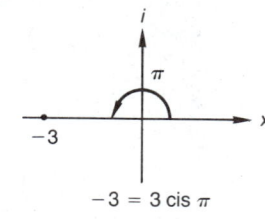

$-3 = 3 \text{ cis } \pi$

9.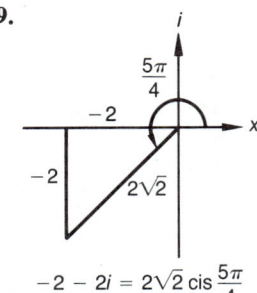

$-2 - 2i = 2\sqrt{2} \text{ cis } \frac{5\pi}{4}$

11.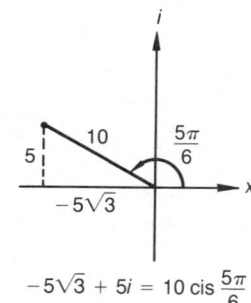

$-5\sqrt{3} + 5i = 10 \text{ cis } \frac{5\pi}{6}$

13.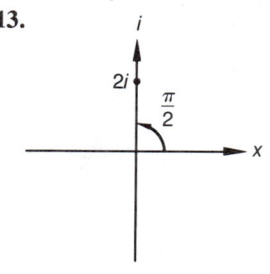

$2i = 2 \text{ cis } \frac{\pi}{2}$

15.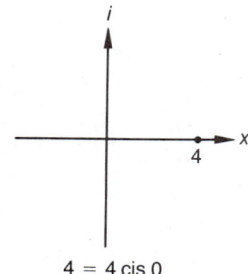

$4 = 4 \text{ cis } 0$

17.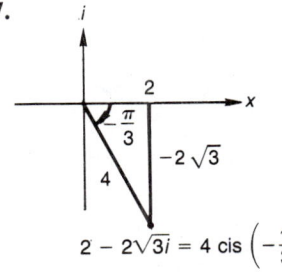

19.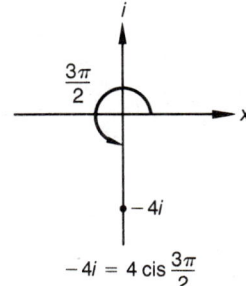

21.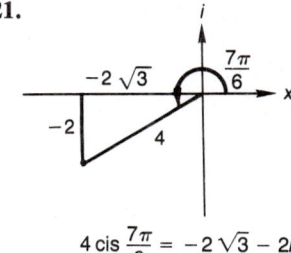

23.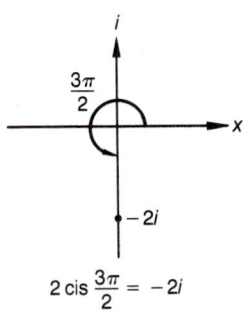

25.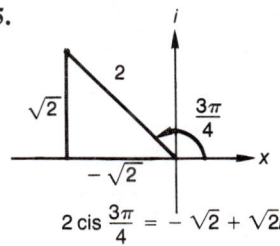

27.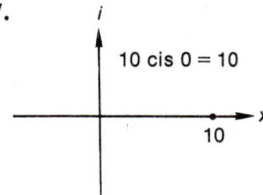

29.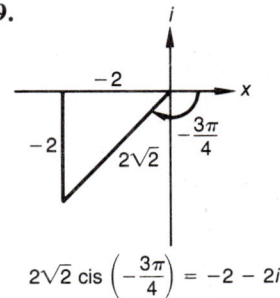

31.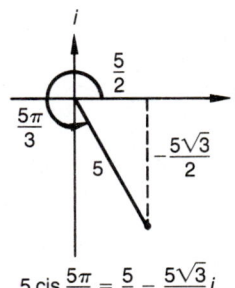

33.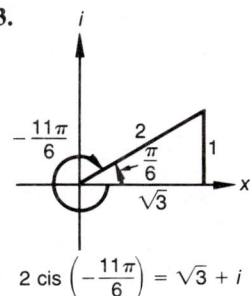

35.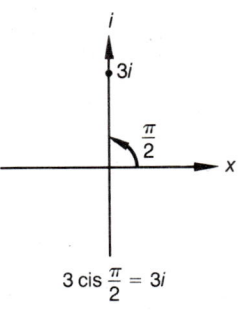

37.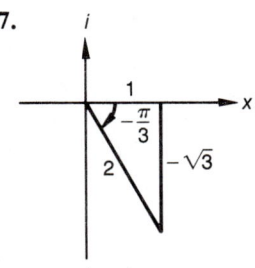

39. 16 cis 180° = −16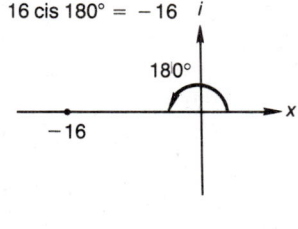

41. $|z_1| = \sqrt{2}$, $|z_2| = \sqrt{2}$, $|z_1 z_2| = 2$, $\left|\frac{z_1}{z_2}\right| = 1$, $|z_1 + z_2| = 2$

43. $|z_1| = \sqrt{2}$, $|z_2| = \sqrt{5}$, $|z_1 z_2| = 10$, $\left|\frac{z_1}{z_2}\right| = \frac{\sqrt{10}}{5}$, $|z_1 + z_2| = \sqrt{5}$

45. $|z_1| = \sqrt{5}$, $|z_2| = 5$, $|z_1 z_2| = 5\sqrt{5}$, $\left|\frac{z_1}{z_2}\right| = \frac{\sqrt{5}}{5}$, $|z_1 + z_2| = 5\sqrt{2}$

47. $|z_1| = 3\sqrt{2}$, $|z_2| = \sqrt{5}$, $|z_1 z_2| = 3\sqrt{10}$, $\left|\frac{z_1}{z_2}\right| = \frac{3\sqrt{10}}{5}$, $|z_1 + z_2| = \sqrt{41}$

49. $|z_1| = \sqrt{34}$, $|z_2| = 2\sqrt{2}$, $|z_1 z_2| = 4\sqrt{17}$, $\left|\frac{z_1}{z_2}\right| = \frac{\sqrt{17}}{2}$, $|z_1 + z_2| = \sqrt{74}$

51. $z_1 z_2 = 2$, $\frac{z_1}{z_2} = i$, $\frac{z_2}{z_1} = -i$

53. $z_1 z_2 = 2\sqrt{2}\operatorname{cis}\left(\frac{11\pi}{12}\right)$, $\frac{z_1}{z_2} = \frac{\sqrt{2}}{2}\operatorname{cis}\left(\frac{7\pi}{12}\right)$, $\frac{z_2}{z_1} = \sqrt{2}\operatorname{cis}\left(\frac{-7\pi}{12}\right)$

55. $z_1 z_2 = 32i$, $\frac{z_1}{z_2} = \frac{(\sqrt{3}+i)}{4}$, $\frac{z_2}{z_1} = \sqrt{3} - i$

57. $z_1 z_2 = 6\sqrt{2}\operatorname{cis}\left(\frac{-5\pi}{12}\right)$, $\frac{z_1}{z_2} = \left(\frac{3\sqrt{2}}{2}\right)\operatorname{cis}\left(\frac{-\pi}{12}\right)$, $\frac{z_2}{z_1} = \left(\frac{\sqrt{2}}{3}\right)\operatorname{cis}\frac{\pi}{12}$

59. $z_1 z_2 = 8\sqrt{2}\operatorname{cis}\left(\frac{-5\pi}{12}\right)$, $\frac{z_1}{z_2} = \sqrt{2}\operatorname{cis}\frac{\pi}{12}$, $\frac{z_2}{z_1} = \left(\frac{\sqrt{2}}{2}\right)\operatorname{cis}\left(\frac{-\pi}{12}\right)$

61. $\frac{1}{z} = \frac{\bar{z}}{z\bar{z}} = \frac{\bar{z}}{|z|^2}$

Section 9.5

1. 4096   **3.** $-4 + 4i$ **5.** $4 + 4i$ **7.** $32i$

9. $\frac{-(1+\sqrt{3}i)}{2}$ **11.** $-\left(\frac{243\sqrt{2}}{2}\right)(1+i)$ **13.** $-\frac{1}{64}$

A.76 Solutions

15. $\dfrac{1-i}{8}$ 17. $-16i$ 19. $\dfrac{-(\sqrt{3}+i)}{2}$

21. $64(\sqrt{3}+i)$ 23. $-i$ 25. $\sqrt{3}-i$

27. $2, -1 \pm \sqrt{3}i$ 29. $\pm 2, \pm 2i$ 31. ± 3

33. $\dfrac{\sqrt{2} \pm \sqrt{2}i}{2}, \dfrac{-(\sqrt{2} \pm \sqrt{2}i)}{2}$ 35. $-3, \dfrac{3(1 \pm \sqrt{3}i)}{2}$

37. $\dfrac{\pm(\sqrt{2}+\sqrt{2}i)}{2}$ 39. $i, \dfrac{(\pm\sqrt{3}-i)}{2}$

41. $\pm(\sqrt{2}-\sqrt{2}i)$ 43. $\pm(\sqrt{3}+i)$

45. $\pm(\sqrt{3}+i), \pm(1-\sqrt{3}i)$ 47. $\pm(1-\sqrt{3}i)$

49. $\sqrt{2}-\sqrt{2}i, 2\,\text{cis}\,\dfrac{5\pi}{12}, -2\,\text{cis}\,\dfrac{\pi}{12}$

51. Equate real and imaginary parts after expanding $\cos 3\theta + i\sin 3\theta = (\cos\theta + i\sin\theta)^3$; $\cos 3\theta = \cos^3\theta - 3\cos\theta\sin^2\theta$; $\sin 3\theta = 3\cos^2\theta\sin\theta - \sin^3\theta$.

53. $2, -2, 2i, -2i$ 55. $-1, \dfrac{1}{2}+\dfrac{\sqrt{3}}{2}i, \dfrac{1}{2}-\dfrac{\sqrt{3}}{2}i$

57. $3, 3\,\text{cis}\left(\dfrac{2\pi}{5}\right), 3\,\text{cis}\left(\dfrac{4\pi}{5}\right), 3\,\text{cis}\left(\dfrac{6\pi}{5}\right), 3\,\text{cis}\left(\dfrac{8\pi}{5}\right)$

59. $2\,\text{cis}\left(\dfrac{5\pi}{12}\right), 2\,\text{cis}\left(\dfrac{11\pi}{12}\right), 2\,\text{cis}\left(\dfrac{17\pi}{12}\right), 2\,\text{cis}\left(\dfrac{23\pi}{12}\right)$

Section 9.6

1.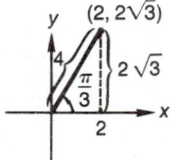

3.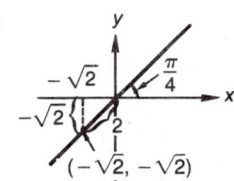

5.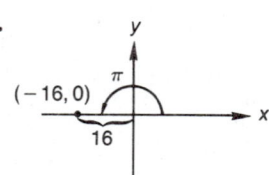

7.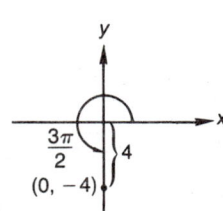

9.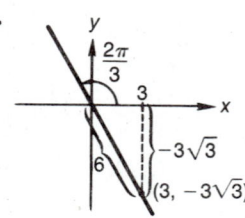

11.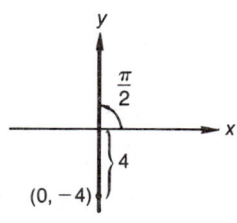

13.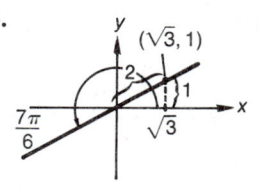

15.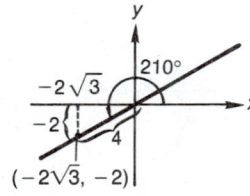

17.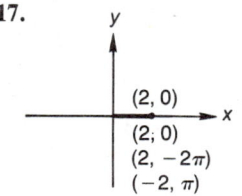

19.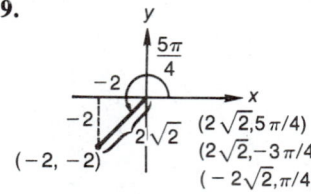

21.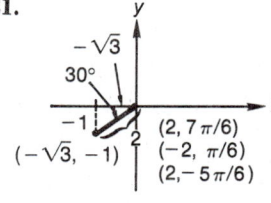

23.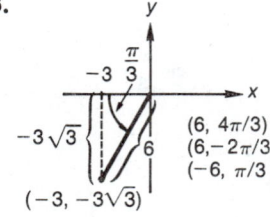

25.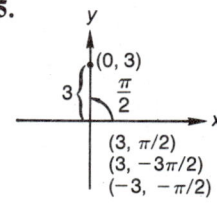

27.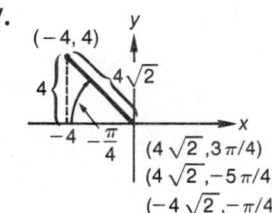

29.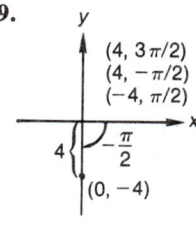

31.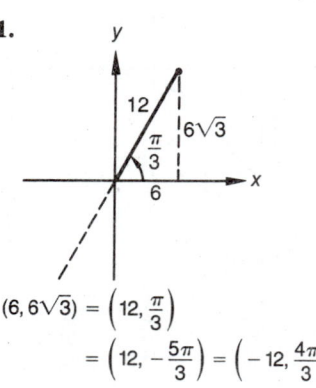

33.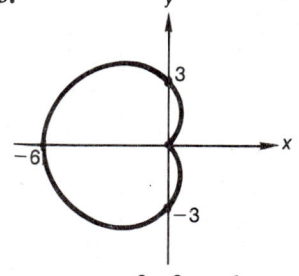

35.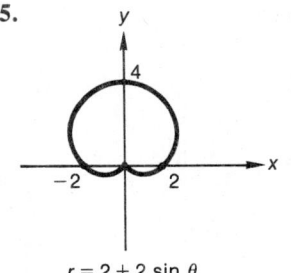
$r = 2 + 2 \sin \theta$

37.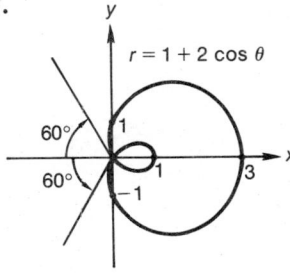
$r = 1 + 2 \cos \theta$

39.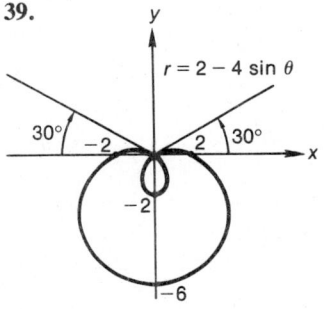
$r = 2 - 4 \sin \theta$

41.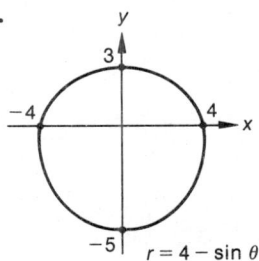
$r = 4 - \sin \theta$

43.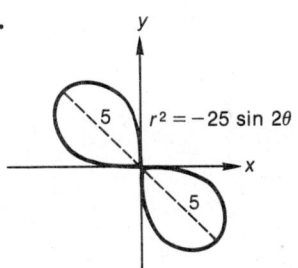
$r^2 = -25 \sin 2\theta$

45.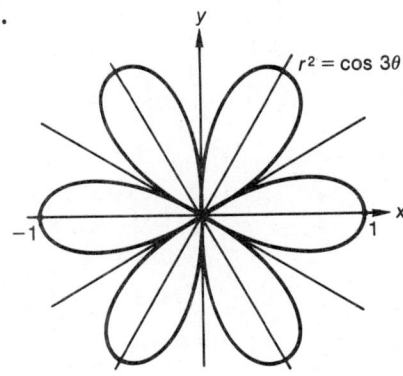
$r^2 = \cos 3\theta$

47.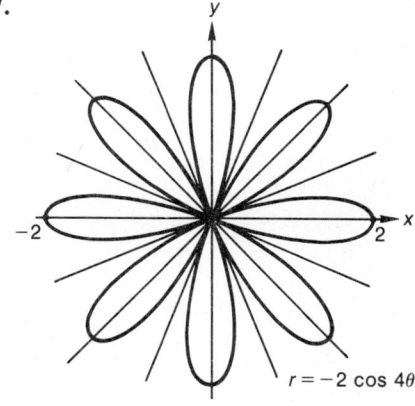
$r = -2 \cos 4\theta$

49.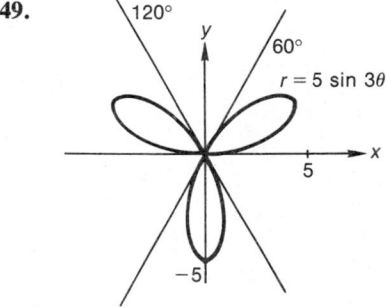
$r = 5 \sin 3\theta$

51.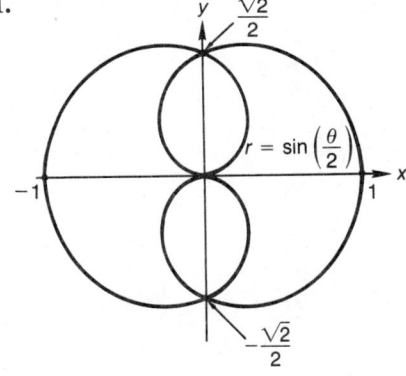
$r = \sin\left(\dfrac{\theta}{2}\right)$

53.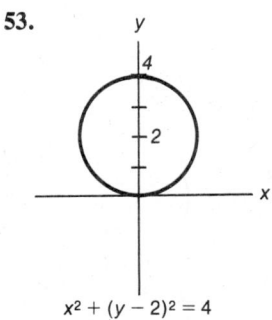
$x^2 + (y - 2)^2 = 4$

55.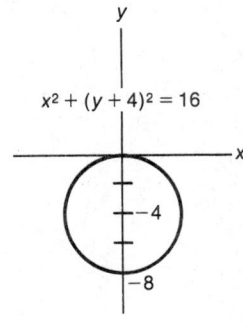
$x^2 + (y + 4)^2 = 16$

57.

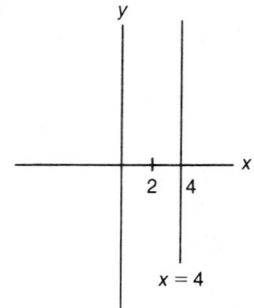

Section 9.8

1. $A = 75°0'$ $b = 13.2$, $c = 17.5$
3. $C = 81°0'$, $b = 5.9$, $c = 7.7$ 5. No solutions
7. $B = 41°0'$, $a = 7.4$, $c = 10.2$
9. $c = 8.0'$, $A = 31°5'$, $B = 12°25'$
11. $A = 27°40'$, $B = 40°32'$, $C = 111°48'$
13. No solutions   15. $A = 150°0'$, $a = 11.5$, $c = 7.9$
17. $B = 79°4'$, $C = 25°56'$, $c = 19.2$;
or $B = 100°56'$, $C = 4°4'$, $c = 3.1$
19. $a = 16.0'$, $C = 11°50'$, $c = 4.4$
21. $A = 113°25'$, $b = 75.5$, $C = 6°35'$ 23. 72.6 cm²
25. 2.6 mi²
27. a. $4i$ b. $2i + 4j$ c. $i + 6j$ d. $-3i - 10j$ 29. 13
31. $\sqrt{85}$ 33. $\langle \frac{3}{5}, -\frac{4}{5} \rangle$ 35. $i + j$

37.

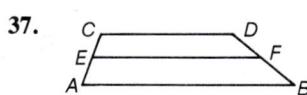

$\vec{EF} = \vec{EC} + \vec{CD} + \vec{DF} = \frac{1}{2}\vec{AC} + \vec{CD} + \frac{1}{2}\vec{DB} =$

$\frac{1}{2}(\vec{AC} + \vec{CD} + \vec{DB}) + \frac{1}{2}\vec{CD} =$

$\frac{1}{2}\vec{AB} + \frac{1}{2}\vec{CD} = \frac{1}{2}(\vec{AB} + \vec{CD})$.

Since $\vec{AB}$ and $\vec{CD}$ have the same direction, $\vec{EF}$ has this direction also.

39.

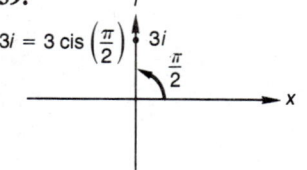

$3i = 3 \text{ cis}\left(\frac{\pi}{2}\right)$

41.

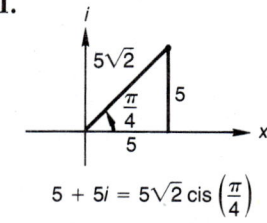

$5 + 5i = 5\sqrt{2} \text{ cis}\left(\frac{\pi}{4}\right)$

43.

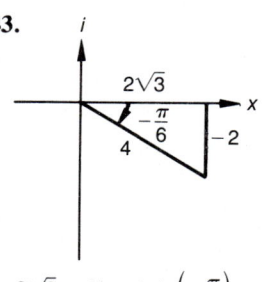

$2\sqrt{3} - 2i = 4 \text{ cis}\left(-\frac{\pi}{6}\right)$

45.

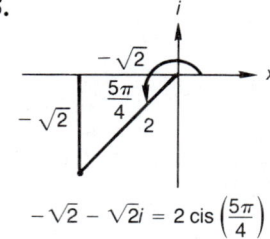

$-\sqrt{2} - \sqrt{2}i = 2 \text{ cis}\left(\frac{5\pi}{4}\right)$

47.

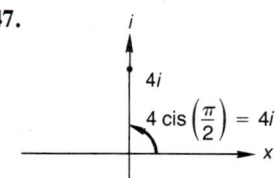

$4 \text{ cis}\left(\frac{\pi}{2}\right) = 4i$

49. $|z_1| = 1$, $|z_2| = \sqrt{2}$, $|z_1 z_2| = \sqrt{2}$, $\left|\frac{z_1}{z_2}\right| = \frac{\sqrt{2}}{2}$,
$|z_1 + z_2| = \sqrt{5}$
51. $|z_1| = 3\sqrt{10}$, $|z_2| = 3\sqrt{2}$, $|z_1 z_2| = 18\sqrt{5}$,
$\left|\frac{z_1}{z_2}\right| = \sqrt{5}$, $|z_1 + z_2| = 12$
53. $|z_1| = 2\sqrt{2}$, $|z_2| = \sqrt{2}$, $|z_1 z_2| = 4$, $\left|\frac{z_1}{z_2}\right| = 2$,
$|z_1 + z_2| = \sqrt{10}$
55. $z_1 z_2 = 24\sqrt{2} \text{ cis}\left(\frac{7\pi}{12}\right)$, $\frac{z_1}{z_2} = 3\sqrt{2} \text{ cis}\left(\frac{\pi}{12}\right)$
57. $z_1 z_2 = 2\sqrt{2} \text{ cis}\left(\frac{23\pi}{12}\right)$, $\frac{z_1}{z_2} = \frac{\sqrt{2}}{2} \text{ cis}\left(\frac{19\pi}{12}\right)$
59. $4\sqrt{2} \text{ cis}\left(\frac{15\pi}{4}\right) = 4 - 4i$
61. $16 \text{ cis}\left(\frac{16\pi}{3}\right) = -8 - 8\sqrt{3}i$
63. $\text{cis}\left(\frac{34\pi}{3}\right) = \left(-\frac{1}{2}\right) - \left(\frac{\sqrt{3}i}{2}\right)$ 65. $\frac{\sqrt{2}}{4} \text{ cis} \frac{19\pi}{12}$
67. $\frac{1}{2}$ 69. $\pm 4i$ 71. $\frac{\pm(-\sqrt{2} + \sqrt{2}i)}{2}$
73. ± 2, $\pm 2i$, $\sqrt{2} \pm \sqrt{2}i$, $-\sqrt{2} \pm \sqrt{2}i$
75. $\frac{\pm(\sqrt{6} + \sqrt{2}i)}{2}$ 77. $\pm(\sqrt{3} + i)$, $\pm(1 - \sqrt{3}i)$
79. $\left\{-3i, \frac{3\sqrt{3}}{2} + \frac{3}{2}i, \frac{-3\sqrt{3}}{2} + \frac{3}{2}i\right\}$

81.

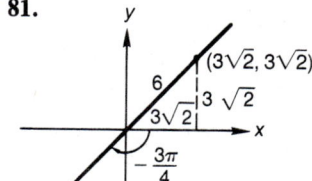

83.

85.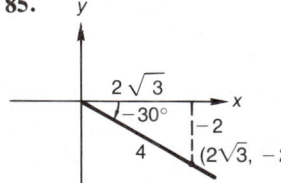

87.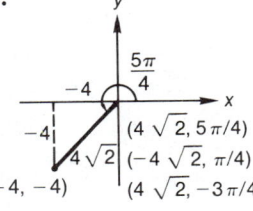

97.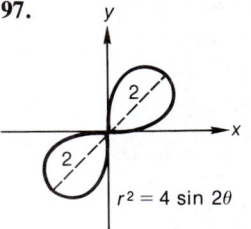

99.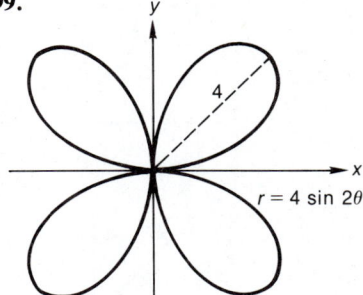

89.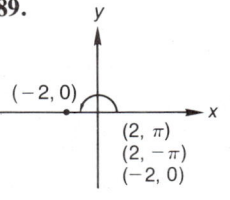

91.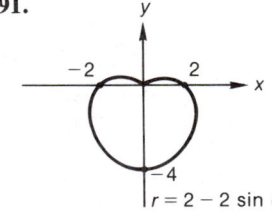

101.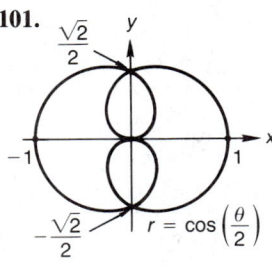

103.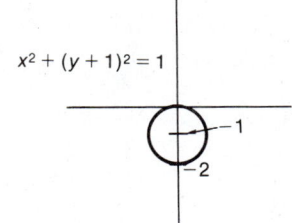

93.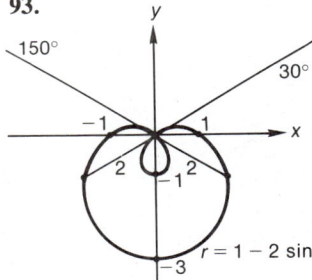

95.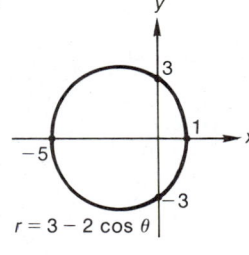

105.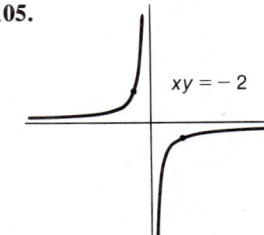

CHAPTER 10

Section 10.1

1. (a and c)

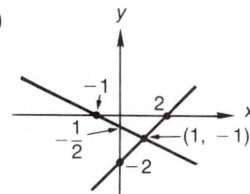

  b. Consistent **d.** $(x, y) = (1, -1)$

3. (a and c)

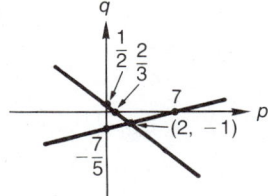

  b. Consistent **d.** $(p, q) = (2, -1)$

5. (a and c)

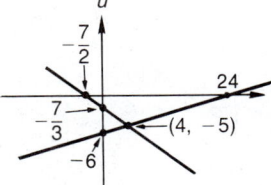

  b. Consistent **d.** $(t, u) = (4, -5)$

7. a.

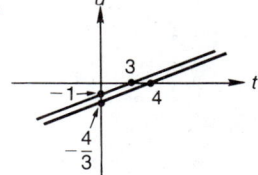

  b. Inconsistent

9. a. 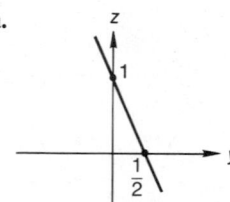 **b.** Dependent: $z = 1 - 2y$

11. (a and c)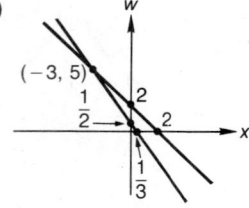

b. Consistent   **d.** $(x, w) = (-3, 5)$

13. (a and c)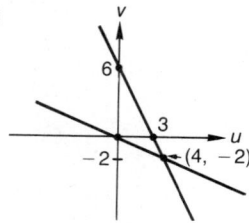

b. Consistent   **d.** $(u, v) = (4, -2)$

15. a.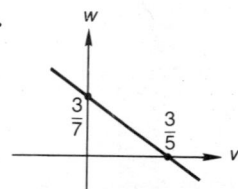

b. Dependent: $w = \dfrac{3 - 5v}{7}$

17. (a and c)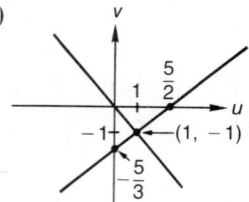

b. Consistent   **d.** $(u, v) = (1, -1)$

19. (a and c)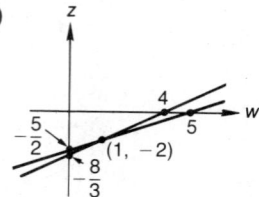

b. Consistent   **d.** $(w, z) = (1, -2)$

21. a. 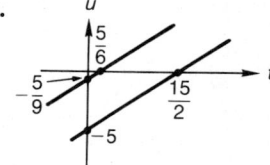   **b.** Inconsistent

23. (a and c)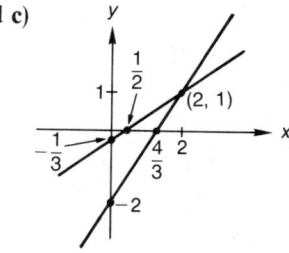

b. Consistent   **d.** $(x, y) = (2, 1)$

25. (a and c)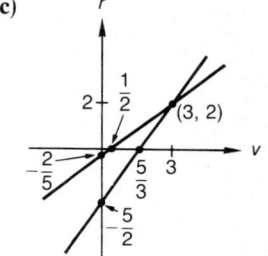

b. Consistent   **d.** $(v, r) = (3, 2)$

27. $(x, y) = (1, 2)$ **29.** $(x, y) = (2, 1)$ **31.** $(x, y) = (4, 11)$
33. $\{(5, 2), (2, -1)\}$ **35.** $\{(-7, -3), (-2, 2)\}$
37. $\{(26, 5), (5, -2)\}$ **39.** $\left\{(3, 2), \left(4, \dfrac{3}{2}\right)\right\}$
41. $\{(0, -5), (4, -3)\}$ **43.** $\{(2, 2)\}$
45. $\left\{(1, 1), \left(-\dfrac{7}{5}, -\dfrac{1}{5}\right)\right\}$
47. $\{(3, \sqrt{3}i), (3, -\sqrt{3}i), (-2, \sqrt{2}), (-2, -\sqrt{2})\}$
49. $\left\{\left(2, -\dfrac{1}{2}\right), \left(-\dfrac{1}{4}, 4\right)\right\}$ **51.** $\left\{(-2, -4), \left(-6, -\dfrac{4}{3}\right)\right\}$
53. $\{(0, 1), (1, 0)\}$ **55.** $\{(7, -4), (4, -1)\}$
57. $\left\{(2, -2), \left(-\dfrac{6}{5}, -\dfrac{2}{5}\right)\right\}$
59. $\{(2, 2\sqrt{2}), (2, -2\sqrt{2}), (-2, 2\sqrt{2}), (-2, -2\sqrt{2})\}$
61. $(1, 0)$ **63.** $\left(-\dfrac{1}{2}\ln 2, \dfrac{3}{\sqrt{2}}\right)$

65.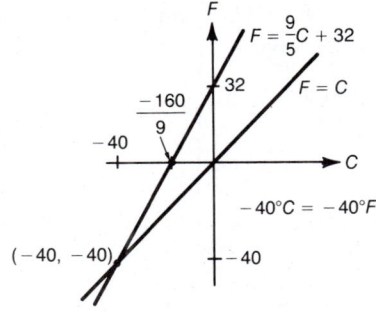

67. a. $\begin{cases} d = \dfrac{8}{5} + \dfrac{1}{5}t \\ d = 5t \end{cases}$ **b.** $t = \dfrac{1}{3}$ h $= 20$ min; $d = 1\dfrac{2}{3}$ mi

69. $V = 14$, $M = 36$ **71.** 52

73. 25 units of 18%; 20 units of 27%

75. (1, 7) and (2, 14)

77. 12 units of either chemical and 14 units of the other

Section 10.2

1. $(t, u) = (-1, 1)$ **3.** $(w, z) = (1, -1)$

5. $(x, y) = (2, -1)$ **7.** $(x, z) = (2, 3)$

9. Dependent; $z = 3 - \dfrac{3y}{2}$ **11.** $(t, v) = (5, 3)$

13. $(r, s) = (-4, -5)$ **15.** Inconsistent

17. $(x, y, z) = (-2, -1, -2)$ **19.** $(a, b, c) = (2, 0, -3)$

21. Inconsistent   **23.** $\{(2z - 3, z + 2, z): z \in R\}$

25. $(y, z, w) = (-1, -1, 4)$ **27.** $(x, y, z) = (-4, 3, 6)$

29. $(r, s, t) = (-2, -1, 2)$ **31.** $\{(2t + 1, 1 - t, t): t \in R\}$

33. Inconsistent   **35.** $(p, q, r) = (1, -1, -1)$

37. 30 in by 40 in   **39.** Fudge 90¢, mints 16¢

41. Laara 6, Mother 34

43. $5000 at 4%, $5000 at 5%, $10,000 at 8%

45. 273   **47.** 3 at 60¢, 6 at 16¢, 5 at 20¢

Section 10.3

1. $(x, y) = (1, -1)$ **3.** $(t, u) = \left(\dfrac{1}{4}, -\dfrac{1}{5}\right)$

5. $(u, v) = \left(\dfrac{1}{4}, -\dfrac{1}{2}\right)$ **7.** $\{(3, 2), (3, -2), (-3, 2), (-3, -2)\}$

9. $\{(3\sqrt{2}, 2), (3\sqrt{2}, -2), (-3\sqrt{2}, 2), (-3\sqrt{2}, -2)\}$

11. $\{(2, 3), (2, -3), (-2, 3), (-2, -3)\}$

13. $(x, y) = \left(\dfrac{1}{10}, 10\right)$ **15.** $(x, y) = (\ln 2, 0)$

17. $\{(2, -5), (-1, -2)\}$ **19.** $\left\{\left(\dfrac{4}{9}, -\dfrac{1}{3}\right), (1, -2)\right\}$

21. $\{(\sqrt{2}, -\sqrt{2}), (-\sqrt{2}, \sqrt{2})\}$ **23.** $(x, y) = (4, 2)$

25. $\{(2, 0), (3, \sqrt{\log_2 5 - 2}), (3, -\sqrt{\log_2 5 - 2})\}$

27. $\{(1, 2), (0, 1)\}$ **29.** $x = 2$ **31.** $x = 3$

33. $y = 2\sqrt{3}$, $x = \dfrac{\pi}{6} + 2k\pi$, $k \in z$

$y = -2\sqrt{3}$, $x = \pm\dfrac{5\pi}{6} + 2k\pi$, $k \in z$

35. $x = \dfrac{3}{2}$, $y = \pm\dfrac{\pi}{3} + 2k\pi$, $k \in z$

$x = -\dfrac{3}{2}$, $y = \pm\dfrac{2\pi}{3} + 2k\pi$, $k \in z$

37. $\{(2, 3), (1, 2)\}$ **39.** $\left\{(0, 3), (0, -3), (-1, 1), \left(\dfrac{4}{3}, 1\right)\right\}$

41. 8, 9   **43.** First pipe 36 h, second pipe 45 h

Section 10.4

1. $(z, t) = (3, -1)$ **3.** $(w, y) = \left(-\dfrac{3}{4}, \dfrac{1}{4}\right)$

5. $(x, y) = \left(\dfrac{3}{2}, \dfrac{1}{2}\right)$ **7.** $(r, s) = (7, 5)$ **9.** $(p, q) = (2, 1)$

11. $(a, b) = (4, -2)$ **13.** $(x, y, z) = (6, 9, 1)$

15. $(r, s, t) = (0, 0, 0)$ **17.** $\{(t + 1, 2 - t, t): t \in R\}$

19. $(p, q, r) = (3, -2, 1)$

21. $\{(-5 - 7y, y, 9 + 11y): y \in R\}$

23. $\{(19s - 16, s, 12 - 11s): s \in R\}$ **25.** $(u, v, w) = (0, 0, 0)$

27. $(x, y, z) = \left(\dfrac{1}{2}, 0, -\dfrac{3}{2}\right)$

29. $(r, s, t, u) = (-1, 2, 1, 1)$

31. $(r, s, t, u) = \left(\dfrac{8}{15}, -\dfrac{2}{3}, -\dfrac{11}{15}, 0\right)$

33. $(a, b, c, d) = (1, -2, 1, 1)$ **35.** $(r, s, t, u) = (1, 3, 2, 1)$

37. $\{(1, -1 - w, 2, w): w \in R\}$

39. $(u, v, w, x, y) = (-1, 1, -3, 2, 1)$

41. $(x - 1)^2 + (y - 2)^2 = 25$; center $(1, 2)$, $r = 5$

43. $(x - 24)^2 + (y - 7)^2 = 625$; center $(24, 7)$, $r = 25$

45. $(x - 7)^2 + (y + 4)^2 = 169$; center $(7, -4)$, $r = 13$

47. 320 acres @ $80, 240 acres @ $60, 400 arces @ $50

49. Pipe 1, 4 h; Pipe 2, 5 h; Pipe 3, 10 h

51. 105 nickels, 35 quarters, 84 half-dollars

53. 6 nickels, 4 dimes, 3 quarters   **55.** -7 and -8

57. 2, -7, -6 **59.** 154

Section 10.5

1. a. $M_{11} = \begin{vmatrix} 3 & 1 \\ -3 & 5 \end{vmatrix} = 18$; $A_{11} = 18$

b. $M_{23} = \begin{vmatrix} -1 & 2 \\ 4 & -3 \end{vmatrix} = -5$; $A_{23} = 5$

c. $M_{31} = \begin{vmatrix} 2 & -2 \\ 3 & 1 \end{vmatrix} = 8$; $A_{31} = 8$

3. a. $M_{12} = \begin{vmatrix} 1 & 5 & 1 \\ 6 & -6 & 2 \\ 7 & 3 & 7 \end{vmatrix} = -128$; $A_{12} = 128$

b. $M_{24} = \begin{vmatrix} 2 & 7 & -3 \\ 6 & 2 & -6 \\ 7 & 0 & 3 \end{vmatrix} = -366$; $A_{24} = -366$

c. $M_{43} = \begin{vmatrix} 2 & 7 & 6 \\ 1 & 0 & 1 \\ 6 & 2 & 2 \end{vmatrix} = 36$; $A_{43} = -36$

5. 5   **7.** -20 **9.** -13 **11.** -8 **13.** -32

15. Evaluate across the second row   **17.** Property 2

19. Property 3   **21.** Property 4   **23.** Property 5

25. Property 6   **27.** Property 3 applied twice

29. Property 5 applied twice

31. Property 6 applied twice   **33.** 415   **35.** -5

37. 60   **39.** -10 **41.** 63   **43.** 7   **45.** -8

47. 450   **49.** -6 **51.** -6 **53.** 832   **55.** 0

57. -18 **59.** $-x + y - 2 = 0$ **61.** $4x + y - 13 = 0$

63. $-x - y + 1 = 0$ **65.** 11   **67.** 7   **69.** 108

71. Expand one determinant across the first row and the other down the first column and match corresponding terms when finished.

Section 10.6

1. $(x, y) = (-5, -8)$ **3.** $(t, u) = \left(\dfrac{5}{2}, -\dfrac{5}{4}\right)$

5. $(u, v) = (11, 7)$ **7.** $(w, z) = (1, -1)$

9. $(t, y) = (1, -2)$ **11.** $(p, q) = \left(\dfrac{11}{26}, -\dfrac{3}{13}\right)$

13. $(r, s) = \left(-\dfrac{1}{11}, \dfrac{6}{11}\right)$ **15.** $(x, y) = \left(\dfrac{5}{3}, -\dfrac{1}{3}\right)$

17. $(r, u) = \left(\dfrac{3}{7}, \dfrac{5}{14}\right)$ **19.** $(x, y, z) = (3, -3, 2)$

21. $(x, y, z) = \left(-\dfrac{11}{9}, \dfrac{26}{9}, \dfrac{1}{3}\right)$ **23.** $(r, s, t) = (6, 9, 1)$

25. $(x, y, z) = \left(\dfrac{3}{4}, -\dfrac{1}{6}, -\dfrac{11}{12}\right)$ **27.** $(w, z, u) = \left(\dfrac{1}{2}, 0, -\dfrac{3}{2}\right)$

29. $(r, s, t, u) = (1, -2, -3, -1)$

31. $(a, b, c, d) = (-1, 1, 1, 1)$ **33.** 4 and 13

35. $5000 at 5%, $11,000 at 8%

37. Gideon is 5, Father is 26

Section 10.7

1. $\dfrac{3}{x-3} + \dfrac{5}{x+2}$ **3.** $\dfrac{-1/3}{x+1} + \dfrac{4/3}{x-2}$

5. $2 + \dfrac{1}{x-3} - \dfrac{3}{x+2}$ **7.** $\dfrac{3}{x-2} + \dfrac{2}{x+1}$

9. $\dfrac{1/2}{x} + \dfrac{1}{x+4} - \dfrac{1/2}{x-2}$ **11.** $\dfrac{1}{x} - \dfrac{2}{x^2} + \dfrac{3}{x-1}$

13. $\dfrac{1/2}{x} - \dfrac{1/2}{x+2} - \dfrac{1}{(x+2)^2}$ **15.** $\dfrac{4}{x-2} - \dfrac{1}{x+1} + \dfrac{3}{(x+1)^2}$

17. $2x + 1 - \dfrac{3}{x} + \dfrac{4}{x^2} + \dfrac{3}{x-3}$

19. $\dfrac{-4}{x-4} + \dfrac{2}{x-2} + \dfrac{1}{(x-2)^2}$ **21.** $\dfrac{1}{x} + \dfrac{1-x}{x^2-2}$

23. $\dfrac{1}{x+1} + \dfrac{2x}{x^2+2}$ **25.** $\dfrac{1}{x+2} + \dfrac{3x+1}{x^2+2}$

27. $\dfrac{9/2}{x+2} + \dfrac{1 - 9x/2}{x^2+2x+2}$ **29.** $\dfrac{2}{x+1} + \dfrac{3x-1}{x^2+2x+2}$

31. $\dfrac{1}{x} - \dfrac{x}{x^2+1} - \dfrac{x}{(x^2+1)^2}$

33. $\dfrac{1}{x} + \dfrac{2x}{x^2+3} - \dfrac{3}{(x^2+3)^2}$

35. $\dfrac{1}{x} + \dfrac{1-x}{x^2+2} - \dfrac{2}{(x^2+2)^2}$

37. $\dfrac{-1}{x-1} + \dfrac{x-1}{x^2+2} + \dfrac{x+1}{(x^2+2)^2}$

39. $\dfrac{1}{x-2} + \dfrac{3}{x^2+2} - \dfrac{x+10}{(x^2+2)^2}$

Section 10.8

1. $(x, y, u, v) = (3, 4, 1, 2)$

3. $(a, b, c, d, e, f) = (5, -6, 7, -1, 2, -3)$ **5.** $\begin{bmatrix} 1 & 3 \\ 3 & 2 \end{bmatrix}$

7. $\begin{bmatrix} 4 & 10 \\ 4 & 10 \end{bmatrix}$ **9.** $\begin{bmatrix} 8 & -1 \\ 12 & -4 \end{bmatrix}$ **11.** $\begin{bmatrix} 8 & 0 \\ 17 & 10 \end{bmatrix}$

13. $\begin{bmatrix} 1 & 1 \\ -19 & -51 \end{bmatrix}$ **15.** $\begin{bmatrix} 17 & 43 \\ 28 & 72 \end{bmatrix}$ **17.** $\begin{bmatrix} 3 & 1 \\ 4 & 1 \\ -1 & 2 \end{bmatrix}$

19. $\begin{bmatrix} 0 & 7 \\ 10 & 3 \end{bmatrix}$ **21.** $\begin{bmatrix} 5 & 4 & -1 \\ 10 & 2 & 2 \\ 4 & 8 & -4 \end{bmatrix}$ **23.** $\begin{bmatrix} -12 & 13 \\ 12 & 1 \end{bmatrix}$

25. $\begin{bmatrix} 7 & 2 & 1 \\ 29 & -14 & 19 \\ 5 & 46 & -29 \end{bmatrix}$

27. $\begin{bmatrix} -4 & -6 & -3 & 9 \\ 4 & 5 & 20 & -5 \end{bmatrix}$

29. $\begin{bmatrix} 5 & -1 & -1 & -6 \\ -6 & 1 & 0 & -10 \end{bmatrix}$

31. $\begin{bmatrix} -47 & -27 & 6 & 60 \\ -22 & -2 & 55 & 74 \end{bmatrix}$

33. $\begin{bmatrix} -6 & -1 & -3 & 14 \\ 1 & 4 & 20 & 0 \\ 9 & -3 & -1 & -13 \\ -18 & -8 & -1 & 18 \end{bmatrix}$ 35. $\begin{bmatrix} 3 & -5 \\ -1 & 2 \end{bmatrix}$

37. $\begin{bmatrix} 7 & 1 \\ 6 & 1 \end{bmatrix}$ 39. No inverse 41. $\begin{bmatrix} -1 & -2 & 3 \\ 2 & 3 & -5 \\ 3 & 8 & -10 \end{bmatrix}$

43. $\begin{bmatrix} 1 & -1 & 0 \\ -2 & 5 & 1 \\ -1 & 3 & 1 \end{bmatrix}$ 45. $(x, y) = (-9, 4)$

47. $(x, y) = (-5, -4)$ 49. $(x, y, z) = (4, -6, -15)$

51. $(x, y, z) = (2, -9, -5)$

53. **a.** $\begin{bmatrix} 2 & 1 & 2 \\ 4 & 2 & 1 \\ 1 & 3 & 1 \\ 1 & 2 & 3 \end{bmatrix} \begin{bmatrix} 10 \\ 5 \\ 20 \end{bmatrix} = \begin{bmatrix} 65 \\ 70 \\ 45 \\ 80 \end{bmatrix}$

b. stamper 65 hr, press 70 hr, welder 45 hr, painter 80 hr

55. **a.** $(0.6)(20 \quad 15 \quad 25 \quad 10)\begin{bmatrix} 2 & 1 & 2 \\ 4 & 2 & 1 \\ 1 & 3 & 1 \\ 1 & 2 & 3 \end{bmatrix}$

$+ (0.4)(10 \quad 12 \quad 20 \quad 10)\begin{bmatrix} 2 & 1 & 2 \\ 4 & 2 & 1 \\ 1 & 3 & 1 \\ 1 & 2 & 3 \end{bmatrix}$

$= (120.20, 132.60, 98.80)$

b. bicycle $120.20, moped $132.60, tricycle $98.80

Section 10.9

1.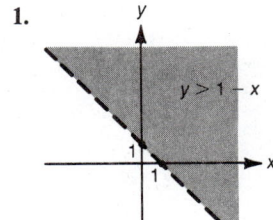

3.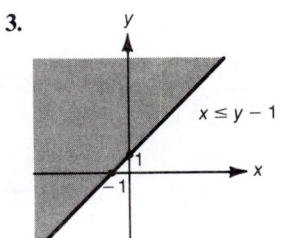

5.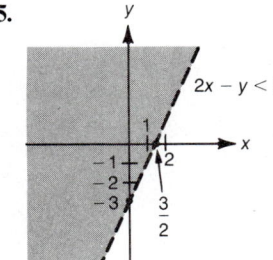

7.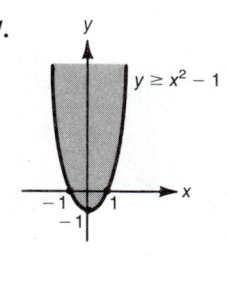

9.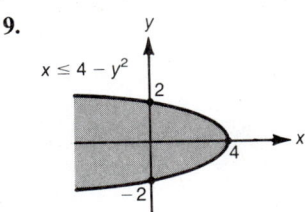

11.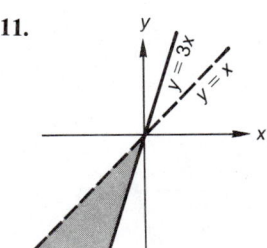

13.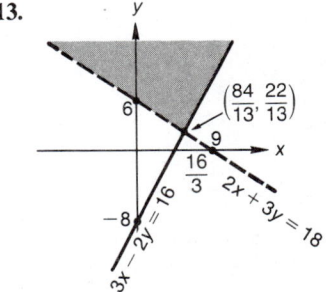

15.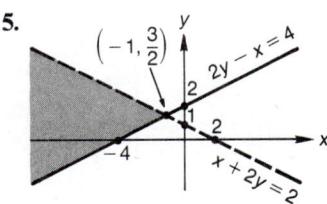

17.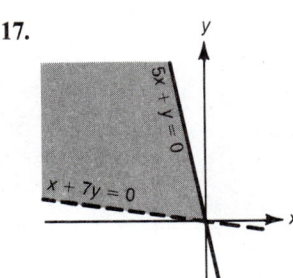

A.84 Solutions

19.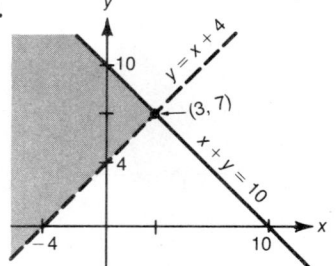

21.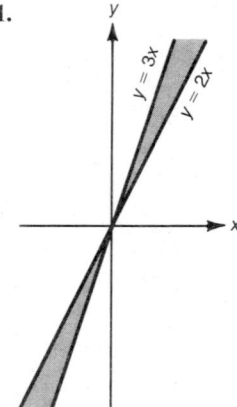

23.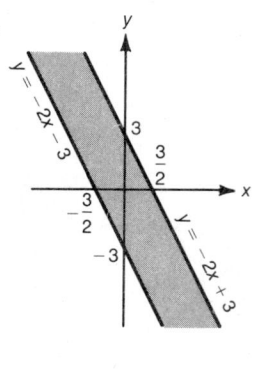

25.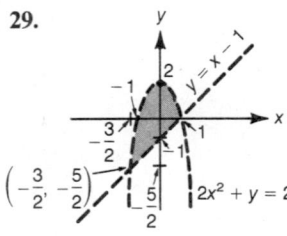

27.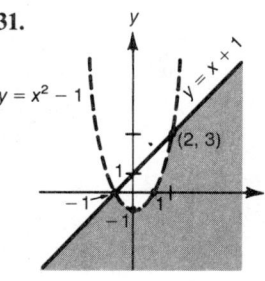

29.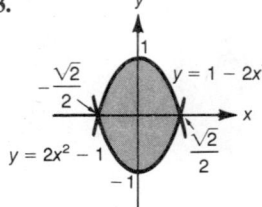

31.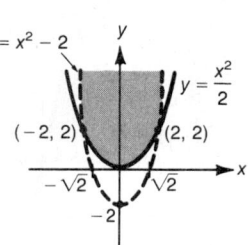

33.

35.

37.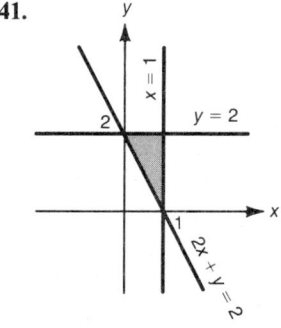

39.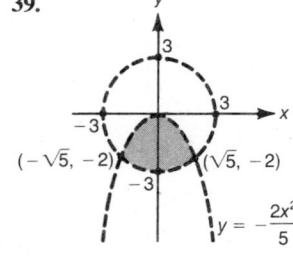

41.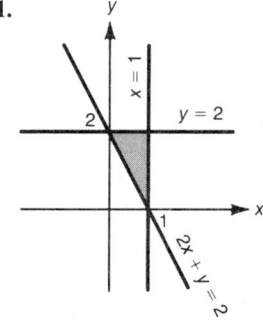

43.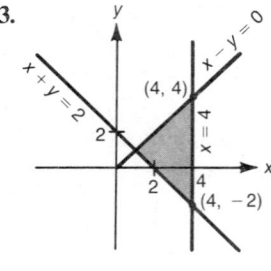

45.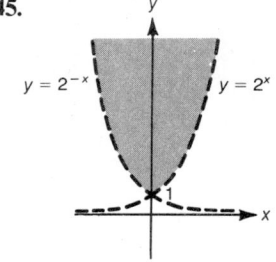

47.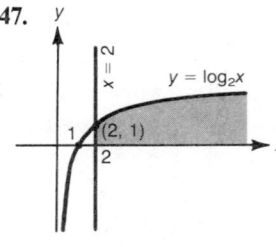

49.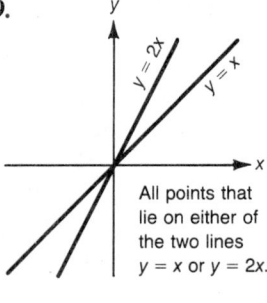
All points that lie on either of the two lines $y = x$ or $y = 2x$.

51.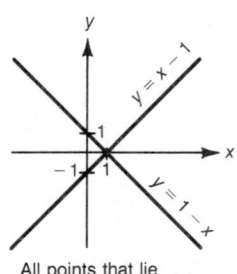
All points that lie on either of the lines $y = x - 1$ or $y = 1 - x$.

53.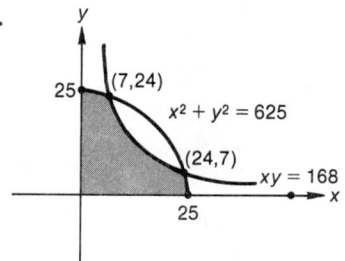

55.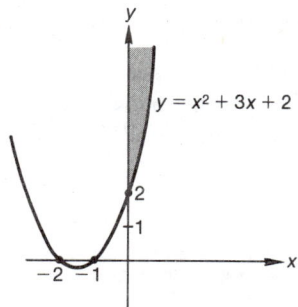

Section 10.10

1. Maximum $= 410$ at $(x, y) = (70, 20)$
3. Minimum $= 70$ at $(x, y) = (6, 10)$
5. Maximum $= 100$ at $(x, y) = (0, 50)$
7. Minimum $= 14$ at $(x, y) = (7, 0)$
9. Maximum $= 22$ at $(x, y) = (3, 2)$
11. Minimum $= 19$ at $(x, y) = (1, 3)$
13. Maximum $= 100$ at $(x, y) = (15, 25)$
15. Minimum $= 54$ at $(x, y) = (18, 0)$
17. Maximum $= 360$ at $(x, y) = (0, 40)$
19. Maximum $= 325$ at $(x, y) = (50, 25)$
21. **a.** 200 of each **b.** $900
23. **a.** 600 betas, no alphas **b.** $1500
25. **a.** Any combination of S semis and P pickups with $30 \leq S \leq 40$, $20 \geq P \geq 0$ and $2S + P = 80$ **b.** $12,000
27. **a.** 4 units F_1, 6 units F_2 **b.** $176
29. **a.** 2 units F_1, 9 units F_2, $148
31. **a.** Mine no. 1 40 days, mine no. 2 20 days **b.** $11,000
33. **a.** 80 homes, 160 apartment units **b.** $1,600,000
35. **a.** $1 million on print and $5 million on radio and TV **b.** $230 million
37. **a.** 2 units of Crunchy Bits, 6 units of Fluffy Flakes **b.** $2.00
39. **a.** 400 gal plain, 600 gal extra creamy **b.** $3180

Section 10.12

1. $(y, z) = (-1, 3)$ 3. Inconsistent
5. $(s, v) = (4, 8)$ 7. $(p, q) = (0, -1)$
9. $\left\{\left(x, \dfrac{1 - 3x}{2}\right) : x \in R\right\}$ 11. $\{(-5, -3), (3, 1)\}$
13. $(x, y) = (0, -2)$ 15. $(x, y) = (1, -2)$
17. $(t, u) = (2, 3)$ 19. $(z, y) = (-4, 6)$
21. $(a, b, c) = (4, 0, 1)$ 23. $(r, s, t) = (-1, -3, 2)$
25. $(u, v, w) = (1, -4, -2)$ 27. $(x, y, z) = (8, 10, -3)$

29. $\{(1, 2), (1, -2), (-1, 2), (-1, -2)\}$
31. $(x, y) = \left(1, \dfrac{-1}{2}\right)$ 33. $(x, y) = \left(\dfrac{1}{2}, 1\right)$
35. $\left\{\left(\sqrt{2}, \dfrac{1 + \log(\sqrt{2} + 1)}{2}\right), \left(-\sqrt{2}, \dfrac{1 - \log(\sqrt{2} + 1)}{2}\right), \left(0, \dfrac{1}{2}\right)\right\}$
37. Inconsistent 39. Dependent: $b = 0$, $a = c$
41. $(r, s, t) = (-3, 3, -4)$
43. $(x, y, z, w) = \left(\dfrac{4}{3}, 0, -\dfrac{1}{3}, \dfrac{11}{3}\right)$
45. $(p, q, r, s) = (1, -2, 0, 2)$ 47. $(u, v) = \left(\dfrac{19}{4}, -\dfrac{3}{4}\right)$
49. $(t, u) = (-1, 3)$ 51. $(x, y, z) = \left(\dfrac{2}{3}, -\dfrac{2}{3}, 3\right)$
53. $(t, u, v) = (0, -3, 2)$ 55. $(u, v, w, x) = \left(-1, 3, \dfrac{1}{2}, -2\right)$
57. 0 59. -74 61. -16 63. 23 65. -328
67. $(x + 2)^2 + (y - 2)^2 = 169$ center $(-2, 2)$, radius 13
69. $\dfrac{-1}{x + 2} + \dfrac{2}{x + 1}$ 71. $\dfrac{-2}{x} + \dfrac{2}{x + 1} + \dfrac{4}{(x + 1)^2}$
73. $\dfrac{2}{x + 1} + \dfrac{3x}{(x^2 + x + 1)^2}$ 75. $\begin{bmatrix} 6 & 11 \\ 17 & 2 \end{bmatrix}$
77. $\begin{bmatrix} 27 & 23 \\ -11 & 36 \end{bmatrix}$ 79. $\begin{bmatrix} 27 & 51 \\ 39 & 20 \end{bmatrix}$ 81. $\begin{bmatrix} 7 & 15 & 18 \\ 3 & -18 & 24 \\ 19 & 51 & 42 \end{bmatrix}$
83. $\begin{bmatrix} 80 & 15 & 83 \\ 125 & 30 & 47 \\ -8 & -12 & 124 \end{bmatrix}$ 85. $\begin{bmatrix} 3 & -1 \\ 5 & -2 \end{bmatrix}$ 87. $\begin{bmatrix} 6 & 3 & -2 \\ 1 & 1 & 0 \\ -1 & 0 & 1 \end{bmatrix}$
89. $(x, y) = (1, 1)$ 91. $(x, y, z) = (-2, -1, -1)$
93. 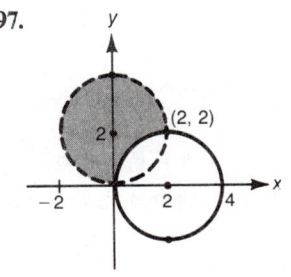 95.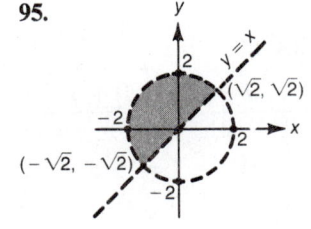
97. 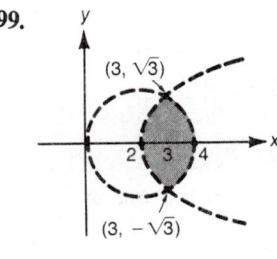 99.

101. 63 and 27 **103.** Boat = 10 mi/h, river = 5 mi/h

105. 674 **107.** Pork = $1.50, fish = $2.00, steak = $2.50

109. Maximum = 1300 @ $(x, y) = (80, 10)$

111. Any combination of x and y satisfying $20 \le x \le 80$, $30 \ge y \ge 0$, $x + 2y = 80$ for a minimum of 400

113. Minimum = 54 at $(x, y) = (4, 10)$

115. Maximum = 1000 at $(x, y) = (0, 50)$

117. Any combination of x and y satisfying $4 \le x \le 5$, $1/2 \ge y \ge 0$, $x + 2y = 5$ for a minimum of 10

CHAPTER 11

Section 11.1

1. Must eventually show that $(2n - 1)^2 + 8[(n + 1) - 1] = [2(n + 1) - 1]^2$

3. Must show that $\left[\dfrac{n(n + 1)}{2}\right]^2 + (n + 1)^3 = \{(n + 1)[(n + 1) + 1]/2\}^2$

5. Show that $2(2^n - 1) + 2^{n+1} = 2(2^{n+1} - 1)$

7. Show that $\left(1 - \dfrac{1}{2^n}\right) + \dfrac{1}{2^{n+1}} = 1 - \dfrac{1}{2^{n+1}}$

9. Show that $\dfrac{n[(n + 1)(n + 2)]}{3} + (n + 1)[(n + 1) + 1] = \dfrac{(n + 1)[(n + 1) + 1][(n + 1) + 2]}{3}$

11. Show that $\left[\dfrac{n(4n^2 + 6n - 1)}{3}\right] + [2(n + 1) - 1] \cdot [2(n + 1) + 1] = \dfrac{(n + 1)[4(n + 1)^2 + 6(n + 1) - 1]}{3}$

13. Show that $\left[\dfrac{n(n + 1)(n + 2)(n + 3)}{4}\right] + (n + 1)[(n + 1) + 1][(n + 1) + 2] = \dfrac{\{(n + 1)[(n + 1) + 1][(n + 1) + 2][(n + 1) + 3]\}}{4}$

15. Show that $\dfrac{n(n + 3)}{4(n + 1)(n + 2)} + \dfrac{1}{n(n + 1)(n + 2)} = \dfrac{(n + 1)[(n + 1) + 3]}{4[(n + 1) + 1][(n + 1) + 2]}$

17. $5^{n+1} - 1 = 5^n \cdot 5 - 1 = (5^n - 1) \cdot 5 + (5 - 1)$

19. $2^{2(n+1)+1} + 1 = 2^{2n+1} \cdot 4 + 1 = (2^{2n+1} + 1)4 - 4 + 1$

21. $(n + 1)^3 + 6(n + 1)^2 + 2(n + 1) = n^3 + 9n^2 + 17n + 9 = (n^3 + 6n^2 + 2n) + 3(n^2 + 5n + 3)$

23. $a^{n+1} - b^{n+1} = a^n a - b^n b = (a^n - b^n)a + (a - b)b^n$

25. $a^{2(n+1)-1} + b^{2(n+1)-1} = (a^{2n-1})a^2 + (b^{2n-1})b^2 = (a^{2n-1})(a^2 - b^2) + b^2(a^{2n-1} + b^{2n-1})$

27. $P_{n+1} = P_n(1 + r) = [P_0(1 + r)^n](1 + r) = P_0(1 + r)^{n+1}$

29. There is one message in T. There are three messages in

S
STS

Not counting the top E in the pyramid of three levels:

E
ESE
ESTSE

There are two ways to arrive at an S from an E; then the message can be completed from the S. Thus, without counting the top E, we have doubled the previous number of paths. There is exactly one way to complete a message that begins at the top E. Thus there are $2 \cdot 3 + 1 = 7$ paths in a pyramid of three levels.

Let P_n denote the number of paths in a pyramid of n levels. Then $P_n = 2^n - 1$ for $n = 1, 2, 3$. Assume $P_k = 2^k - 1$ and consider a pyramid of $k + 1$ levels.

Y
YXY
YX XY
YX XY
YX XY

If we do not count the top Y, then each X in the pyramid of level k can be reached from exactly 2 Y's; then the message can be completed from the X. From the top Y there is only one way to complete the message. Thus $P_{k+1} = 2P_k + 1 = 2(2^k - 1) + 1 = 2^{k+1} - 2 + 1 = 2^{k+1} - 1$. It follows that there are $P_n = 2^n - 1$ messages in a pyramid of n levels. In the given pyramid of 16 levels, there are $2^{16} - 1 = 65{,}535$ messages.

Section 11.2

1. 3, 6, 9, 12, 15; 30 **3.** 5, 7, 9, 11, 13; 23

5. 2, 8, 18, 32, 50; 200 **7.** 0, 2, 0, 2, 0; 2

9. $-2, 4, -8, 16, -32$; 1024 **11.** $0, \dfrac{1}{3}, \dfrac{1}{2}, \dfrac{3}{5}, \dfrac{7}{9}$

13. $\dfrac{2}{3}, \dfrac{3}{4}, \dfrac{4}{5}, \dfrac{5}{6}, \dfrac{9}{10}$ **15.** $\dfrac{1}{2}, \dfrac{3}{7}, \dfrac{3}{8}, \dfrac{1}{3}, \dfrac{3}{13}$

17. $-5, 7, 3, \dfrac{11}{5}, \dfrac{19}{13}$ **19.** $-1, \dfrac{1}{8}, -\dfrac{1}{27}, \dfrac{1}{64}, \dfrac{1}{512}$

21. $-2, 2, -2, 2, -2$ **23.** $1, 1, \dfrac{3}{2}, 3, \dfrac{15}{2}$

25. 2, 1, 0, 1, 0 **27.** 2, 5, 8, 11, 14; 44

29. $-5, -3, -1, 1, 3$; 23 **31.** 22, 20, 18, 16, 14; -6

33. 73 **35.** 21 **37.** -5 **39.** 9, 12, 15, 18, 21, 24, 27

41. 25, 30, 35, 40, 45, 50

43. 17.5, 22, 26.5, 31, 35.5, 40, 44.5   **45.** 2, 6, 18, 54; 4374

47. 3, 12, 48, 192; 49,152   **49.** $\frac{2}{3}, \frac{2\sqrt{3}}{3}, 2, 2\sqrt{3}; 18\sqrt{3}$

51. 1536   **53.** 243   **55.** -3

57. For $r = \sqrt{3}$: $21\sqrt{3}, 63, 63\sqrt{3}, 189, 189\sqrt{3}$;
for $r = -\sqrt{3}$: $-21\sqrt{3}, 63, -63\sqrt{3}, 189, -189\sqrt{3}$

59. $-54, 162$

61. For $r = \frac{1}{2}$: 1024, 512, 256, 128, 64, 32, 16, 8, 4;
for $r = -\frac{1}{2}$: $-1024, 512, -256, 128, -64, 32, -16, 8, -4$

63. Assuming that $a_n = a_1 + (n-1)d$, we have
$a_{n+1} = a_n + d = [a_1 + (n-1)d] + d = a_1 + [(n-1)d + d] = a_1 + nd$

65. a. $800, $750, $700, $650, $600, $550, $500, $450, $400, $350, $300, $250, $200, $150, $100, $50, $0   **b.** arithmetic

67. a. $a_n = h_0 - 16n^2$ **b.** $d_n = (2n - 1) \cdot 16$
c. $d_n = 16 + (n-1)32 = d_{n-1} + 32$ **d.** 304 ft, 624 ft

69. $6000, $4500, $3375, \ldots, $6000\left(\frac{3}{4}\right)^n, \ldots$ **b.** 450.51
c. geometric

Section 11.3

1. $\frac{25}{12}$ **3.** 44   **5.** 420   **7.** $\frac{3}{4}$ **9.** $\sum_{n=0}^{4}(5 + 2n)$

11. $\sum_{n=1}^{5}\left(\frac{1}{2n}\right)$ **13.** $\sum_{n=2}^{7} 10n$ **15.** $\sum_{n=2}^{99} n^n$ **17.** 175

19. 13,035   **21.** 649   **23.** 1430   **25.** 1358

27. 728   **29.** 1020   **31.** 1717   **33.** 820   **35.** 208

37. 160   **39.** 242   **41.** 215   **43.** 3   **45.** $\frac{2}{3}$

47. 6   **49.** $\frac{8}{5}$ **51.** 6   **53.** 465   **55.** $496

57. 13% of the original value

59. a. 3 echelons will more than suffice; two will not
b. 12 echelons will more than suffice; 11 will not

61. If $a_1 + \cdots + a_n = \frac{n}{2}(a_1 + a_n)$, then

$a_1 + \cdots + a_n + a_{n+1} = \frac{n}{2}(a_1 + a_n) + a_{n+1} = \frac{n}{2}(a_1 + a_n) +$

$(a_1 + nd) = \frac{n}{2}(a_1 + a_n + d) + (a_1 + \frac{n}{2}d) =$

$\frac{n}{2}(a_1 + a_{n+1}) + \frac{2a_1 + nd}{2} = \frac{n}{2}(a_1 + a_{n+1}) +$

$\frac{a_1 + a_1 + nd}{2} = \frac{n}{2}(a_1 + a_{n+1}) + \frac{1}{2}(a_1 + a_{n+1}) =$

$\frac{n+1}{2}(a_1 + a_{n+1})$

Section 11.4

1. 720   **3.** 40,320   **5.** 17,280   **7.** 241,920

9. 3024   **11.** 3   **13.** 5   **15.** 10   **17.** 25

19. $x^4 + 4x^3 + 6x^2 + 4x + 1$

21. $r^{10} - 10r^9s + 45r^8s^2 - 120r^7s^3 + 210r^6s^4 - 252r^5s^5 + 210r^4s^6 - 120r^3s^7 + 45r^2s^8 - 10rs^9 + s^{10}$

23. $64x^3 - 48x^2y + 12xy^2 - y^3$

25. $v^5 - 5v^4w^2 + 10v^3w^4 - 10v^2w^6 + 5vw^8 - w^{10}$

27. $16x^4 + 96x^3y^2 + 216x^2y^4 + 216xy^6 + 81y^8$

29. $a^{18} - 12a^{15}b^2 + 60a^{12}b^4 - 160a^9b^6 + 240a^6b^8 - 192a^3b^{10} + 64b^{12}$

31. $x^6 + 6x^4 + 15x^2 + 20 + \frac{15}{x^2} + \frac{6}{x^4} + \frac{1}{x^6}$

33. $t^8 - \frac{8t^7}{u^2} + \frac{28t^6}{u^4} - \frac{56t^5}{u^6} + \frac{70t^4}{u^8} - \frac{56t^3}{u^{10}} + \frac{28t^2}{u^{12}} - \frac{8t}{u^{14}} + \frac{1}{u^{16}}$

35. $32p^5 - \frac{160p^4}{q^2} + \frac{320p^3}{q^4} - \frac{320p^2}{q^6} + \frac{160p}{q^8} - \frac{32}{q^{10}}$

37. $x^5 - 5x^{7/2} + 10x^2 - 10x^{1/2} + \frac{5}{x} - \frac{1}{x^{5/2}}$

39. $(10 + 1)^5 = 161{,}051$ **41.** $(100 + 1)^4 = 104{,}060{,}401$

43. $(10 - 0.1)^3 = 970.299$

45. $(1 - 0.005)^3 = 0.985074875$ **47.** $280a^3b^4$

49. $9120v^3w^{17}$ **51.** $3003x^{10}y^{10}$ **53.** $-720a^2b^6$

55. $673{,}596x^6/y^6$ **57.** $35x^6y^8$ **59.** 924

61. $4060x^{12}$

63. $32{,}768x^{15} + 737{,}280x^{14}y + 7{,}741{,}440x^{13}y^2 + 50{,}319{,}360x^{12}y^3$

65. $1{,}934{,}445{,}240x^3y^{12} + 669{,}615{,}660x^2y^{13} + 143{,}489{,}070xy^{14} + 14{,}348{,}907y^{15}$

Section 11.5

1. 3   **3.** 1   **5.** 220   **7.** 4845   **9.** 7   **11.** 1225

13. 36   **15.** 24   **17.** 259,459,200   **19.** $1.326(10^{32})$

21. $2.179(10^{11})$ **23.** $3.467(10^{80})$

25. $9(10^6) = 9{,}000{,}000$ **27.** $2^{10} = 1024$

29. $3 \cdot 4 \cdot 2 \cdot 4 \cdot 3 = 288$ **31.** $3 \cdot 2 \cdot 2 \cdot 3 = 36$

33. $5 \cdot 3^5 \cdot 2 \cdot 3^5 \cdot 2^5 = 18{,}895{,}680$

35. $_{12}P_{12} = 12! = 479{,}001{,}600$ **37.** $_6P_6 = 6! = 720$

39. $_4P_4 = 4! = 24$

41. a. $_6P_6 = 6! = 720$
b. $(_3P_3)(_3P_3)(_2P_2) = (3!)(3!)(2!) = 72$
c. $(_2P_2)(_2P_2)(_2P_2)(_3P_3) = (2!)(2!)(2!)(3!) = 48$

43. $_{10}C_3 = 120$ **45.** $\frac{(_{12}C_6)(_6C_6)}{2} = 462$

47. $(_{24}C_5) \cdot 4 = 170{,}016$
49. $(_{24}C_5)(_{19}C_5)(_{14}C_5)(_9C_5) \cdot 4 \approx 4.987(10^{14})$
51. $(_{20}C_{13})(_{13}C_1) = 1{,}007{,}760$
53. a. $_{10}C_8 = 45$ b. $_5C_2 = 10$
    c. $(_5C_3)(_5C_4) + (_5C_4)(_5C_3) + (_5C_5)(_5C_2) = 110$
55. $_{10}C_2 - 10 = 35$
57. a. (6 triples) + (6 · 5 doubles) + $(_6C_3) = 56$
    b. (4 triples) + (4 · 3 doubles) + $(_4C_3) = 20$
    c. (12 triples) + (12 · 11 doubles) + $(_{12}C_3) = 364$
59. $(_3P_3)(_6P_6)(_4P_4)(_5P_5)(_4P_4) = 298{,}598{,}400$
61. $(_{52}C_2)(_{50}P_3) = 155{,}937{,}600$
63. a. $\dfrac{(_{16}C_4)(_{12}C_4)(_8C_4)(_4C_4)}{_4P_4} = 2{,}627{,}625$
    b. $\dfrac{(_{16}C_2)(_{14}C_2)(_{12}C_2)(_{10}C_2)(_8C_2)(_6C_2)(_4C_2)(_2C_2)}{(_8P_8)} = 2{,}027{,}025$
65. a. $\dfrac{(_{12}C_3)(_9C_3)(_6C_3)(_3C_3)}{(_4P_4)} = 15{,}400$
    b. $\dfrac{(_{12}C_6)(_6C_3)(_3C_3)}{(_2P_2)} = 9240$
67. a. $\dfrac{(_{10}C_2)(_8C_2) \cdot (_6C_3)(_3C_3)}{(_2P_2)} = 12{,}600$
    b. $(_{10}C_2)(_8C_2)(_6C_6) + 2(_{10}C_5)(_5C_2)(_3C_3) = 6300$
69. a. $(_{12}C_4)(_8C_4)(_4C_4) = 34{,}650$
    b. $(_3P_3)(_3P_3)(_3P_3)(_3P_3) = 1296$
71. $\dfrac{4!}{2!} = 12$
73. a. $\dfrac{9!}{1! \cdot 1! \cdot 2! \cdot 1! \cdot 2! \cdot 2!} = 45{,}360$ b. $\dfrac{7!}{3! \cdot 1! \cdot 2! \cdot 1!} = 420$
    c. $\dfrac{6!}{1! \cdot 3! \cdot 2!} = 60$ d. $\dfrac{10!}{1! \cdot 3! \cdot 1! \cdot 1! \cdot 2! \cdot 2!} = 151{,}200$
    e. $\dfrac{8!}{4! \cdot 1! \cdot 1! \cdot 1! \cdot 1!} = 1680$
    f. $\dfrac{10!}{1! \cdot 1! \cdot 1! \cdot 2! \cdot 1! \cdot 1! \cdot 1! \cdot 1! \cdot 1!} = 1{,}814{,}400$
75. a. $\dfrac{_6P_6}{6} = 120$ b. $\dfrac{\text{answer in } \mathbf{a}}{2} = 60$
77. (6 triples) + (6 · 5 doubles) + $\dfrac{(_6P_3)}{3} = 76$
79. a. $_{30}P_{30} = 30! \approx 2.653(10^{32})$
    b. $(_{30}C_{10})(_{20}C_{10})(_{10}P_{10}) \approx 2.014(10^{19})$
    c. $\dfrac{(_{30}C_{10})(_{20}C_{10})(_{10}P_{10})}{2 \cdot 2} \approx 5.036(10^{18})$

Section 11.6

1. $\dfrac{1}{3}$

3. $T = (7 \text{ doubles}) + (_7C_2) = 28$ a. $\dfrac{1}{28}$ b. $\dfrac{7}{28} = \dfrac{1}{4}$
   c. $\dfrac{7}{28} = \dfrac{1}{4}$ d. $\dfrac{6}{28} = \dfrac{3}{14}$ e. $\dfrac{3}{28}$ f. $P(\text{sum} = 4) = \dfrac{3}{28}$;
   $P(\text{sum} = 3) = P(\text{sum} = 2) = \dfrac{2}{28}$;
   $P(\text{sum} = 1) = P(\text{sum} = 0) = \dfrac{1}{28}$; $P(\text{sum} < 5) = \dfrac{9}{28}$;
   $P(\text{sum} \geq 5) = \dfrac{19}{28}$

5. $T = 4 \cdot 4 = 16$ a. $\dfrac{3}{16}$ b. $P(\text{sum} = 3) = \dfrac{2}{16}$;
   $P(\text{sum} = 2) = \dfrac{1}{16}$; $P(\text{sum} < 4) = \dfrac{3}{16}$; $P(\text{sum} \geq 4) = \dfrac{13}{16}$

7. a. $\dfrac{6}{36} = \dfrac{1}{6}$ b. $\dfrac{4}{16} = \dfrac{1}{4}$ c. $\dfrac{12}{12^2} = \dfrac{1}{12}$ 9. $\left(\dfrac{2}{3}\right)^4 = \dfrac{16}{81}$

11. a. $\dfrac{5}{10} \cdot \dfrac{4}{9} = \dfrac{2}{9}$ b. $\dfrac{5}{9}$ (the first marble doesn't matter)

13. $P(\text{all different}) = \dfrac{(9 \cdot 8 \cdot 7 \cdot 6 \cdot 5)}{9^5} \approx 0.26$;
    $P(\text{2 people choose same number}) \approx 1 - 0.26 = 0.74$

15. $P(0 \text{ correct}) = \left(\dfrac{2}{3}\right)^3 = \dfrac{8}{27}$;
    $P(1 \text{ correct}) = \left(\dfrac{2}{3}\right)\left(\dfrac{2}{3}\right)\left(\dfrac{1}{3}\right) \cdot 3 = \dfrac{4}{9}$;
    $P(\geq 2 \text{ correct}) = 1 - \dfrac{8}{27} - \dfrac{4}{9} = \dfrac{7}{27}$

17. $\dfrac{1}{2}$

19. a. $1 - \left(\dfrac{48}{52}\right)\left(\dfrac{47}{51}\right) = \dfrac{33}{221} \approx 0.149$ b. $\left(\dfrac{13}{52}\right)\left(\dfrac{12}{51}\right) = \dfrac{1}{17}$
    c. $\left(\dfrac{1}{52}\right) + \left(\dfrac{51}{52}\right)\left(\dfrac{1}{51}\right) = \dfrac{1}{26}$
    d. $1 - \left(\dfrac{36}{52}\right)\left(\dfrac{35}{51}\right) = \dfrac{116}{221} \approx 0.525$
    e. $1 - \left(\dfrac{36}{52}\right)\left(\dfrac{35}{51}\right) = \dfrac{116}{221} \approx 0.525$
    f. $\left(\dfrac{12}{52}\right)\left(\dfrac{11}{51}\right) = \dfrac{11}{221} \approx 0.050$
    g. $1 - \left(\dfrac{26}{52}\right)\left(\dfrac{25}{51}\right) = \dfrac{77}{102} \approx 0.755$
    h. $1 - \left(\dfrac{38}{52}\right)\left(\dfrac{37}{51}\right) = \dfrac{623}{1326} \approx 0.470$
    i. $\left(\dfrac{1}{52}\right)\left(\dfrac{12}{51}\right) + \left(\dfrac{12}{52}\right)\left(\dfrac{1}{51}\right) = \dfrac{2}{221} \approx 0.009$

21. a. $\dfrac{_{50}C_{11}}{_{52}C_{13}} = \dfrac{1}{17}$ b. $\dfrac{_{49}C_{10}}{_{52}C_{13}} = \dfrac{11}{850} \approx 0.013$
    c. $\dfrac{_{47}C_8}{_{52}C_{13}} = \dfrac{33}{66{,}640} \approx 0.0005$
    d. (answer for **a**) + (answer for **b**) − (answer for **c**) ≈ 0.071

23. a. $\dfrac{4 \cdot (_{48}C_4)}{_{52}C_5} = \dfrac{3243}{10{,}829} \approx 0.299$

b. $\dfrac{4 \cdot (_{48}C_{12})}{_{52}C_{13}} = \dfrac{9139}{20{,}825} \approx 0.439$

25. $\dfrac{4 \cdot (_{16}C_{12})}{_{48}C_{12}} \approx 0.000000104$ 27. $\dfrac{_{44}C_{20}}{_{48}C_{24}} \approx 0.0546$

29. $\left(\dfrac{1}{2}\right)^{10} = \dfrac{1}{1024}$

31. (See solution to Exercise 77, Section 8.5.)   a. $\dfrac{30}{76} = \dfrac{15}{38}$

b. $\dfrac{6}{76} = \dfrac{3}{38}$ c. $\dfrac{(1\text{ triple}) + (5\text{ doubles}) + 5 \cdot 5}{76} = \dfrac{31}{76}$

d. $\dfrac{4}{76} = \dfrac{1}{19}$

33. a. 1 to 5   b. (4 to 32) or (1 to 8)

35. (12 to 40) or (3 to 10)   37. $\dfrac{5}{9}$

Section 11.8

1. Must eventually show that $\dfrac{n(2n^2 + 3n + 7)}{6} + [(n+1)^2 + 1] = \dfrac{(n+1)[2(n+1)^2 + 3(n+1) + 7]}{6}$

3. Must show that $n^2(2n^2 - 1) + [2(n+1) - 1]^3 = (n+1)^2[2(n+1)^2 - 1]$

5. Show that $\dfrac{n(n+1)(2n+7)}{6} + (n+1)[(n+1)+2] = \dfrac{(n+1)[(n+1)+1][2(n+1)+7]}{6}$

7. Show that $\dfrac{n(n+1)(n+2)(3n+5)}{12} + (n+1)[(n+1)+1]^2$
$= \dfrac{(n+1)[(n+1)+1][(n+1)+2][3(n+1)+5]}{12}$

9. Show that $\dfrac{n}{2(3n+2)} + \dfrac{1}{(3[n+1]-1)(3[n+1]+2)} = \dfrac{[n+1]}{2(3[n+1]+2)}$

11. $4^{2(n+1)} - 1 = 4^{2n} \cdot 16 - 16 + 16 - 1 = 16(4^{2n} - 1) + (16 - 1)$

13. $-4, -3, -2, -1, 0; 7$ 15. $2, 5, 10, 17, 26; 145$

17. $1, 7, 17, 31, 49; 287$ 19. $\dfrac{1}{2}, \dfrac{2}{3}, \dfrac{3}{4}, \dfrac{4}{5}, \dfrac{5}{6}; \dfrac{9}{10}$

21. $3, 1, \dfrac{3}{5}, \dfrac{3}{7}, \dfrac{1}{3}; \dfrac{3}{17}$ 23. $3, 7, 15, 31, 63$

25. $-\sqrt{2}, 2, -4, 16, -256$ 27. $4, 5, 6, 7, 8; 13$

29. $4, 8, 12, 16, 20; 40$ 31. 27   33. 10   35. $7, 11, 15$

37. $46, 50, 54$ 39. $1, 5, 25, 125; 78{,}125$

41. $\sqrt{2}, 2, 2\sqrt{2}, 4; 16$ 43. 4374   45. 4096   47. 24, 48

49. For $r = \sqrt{2}$: $67\sqrt{2}, 134, 134\sqrt{2}, 268, 268\sqrt{2}$
For $r = -\sqrt{2}$: $-67\sqrt{2}, 134, -134\sqrt{2}, 268, -268\sqrt{2}$

51. $\sum_{1}^{4} 2n$ 53. $\sum_{0}^{5} (3 + 4n)$ 55. $\sum_{1}^{5} (n+1)! \cdot 3 \cdot 2^{n-1}$

57. 1280   59. 484   61. $\dfrac{20}{3}$ 63. $\dfrac{-6}{5}$ 65. 60

67. 53   69. 164   71. 330   73. 10,587

75. 110   77. 403,200

79. $a^9 + 18a^8 + 144a^7 + 672a^6 + 2016a^5 + 4032a^4 + 5376a^3 + 4608a^2 + 2304a + 512$

81. $243x^5 + 810x^4y^2 + 1080x^3y^4 + 720x^2y^6 + 240xy^8 + 32y^{10}$

83. $64x^3 - 192x^{3/2} + 240 - 160x^{-3/2} + 60x^{-3} - 12x^{-9/2} + x^{-6}$

85. $-960x^7$ 87. $5005a^9b^6$ 89. $-252r^{10}s^5$

91. 1.061208   93. 112,550,881   95. $6 \cdot 13 \cdot 8 = 624$

97. $_6P_6 = 6! = 720$ 99. $_{24}C_6 = 134{,}596$

101. $_{10}C_6 = 210$

103. a. $_8P_4 = 1680$ b. $3 \cdot (_7P_3) = 630$ c. $3 \cdot 5(_6P_2) = 450$
d. $(3 \cdot 5 \cdot 4 \cdot 2) \cdot 2 = 240$
e. $(1 \cdot 5 \cdot 4 \cdot 2 \cdot 2) + (3 \cdot 5 \cdot 4 \cdot 1 \cdot 2) = 200$

105. a. $4 \cdot 2 \cdot 3 \cdot 1 \cdot 2 \cdot 1 = 48$ b. $3 \cdot 2 \cdot 2 \cdot 1 \cdot 1 = 12$
c. $\dfrac{5!}{2!} = 60$ d. $\dfrac{9!}{2! \cdot 4! \cdot 2!} = 3780$

107. a. $\dfrac{24}{50^2} = \dfrac{12}{1225} \approx 0.010$

b. $\dfrac{9 + 8 + \cdots + 1}{50^2} = \dfrac{45}{50^2} = \dfrac{9}{500} \approx 0.018$

c. $\dfrac{(2 + 3 + \cdots + 50) + 50}{50^2} = \dfrac{331}{625} \approx 0.530$

109. a. $\dfrac{6}{6^4} = \dfrac{1}{6^3} = \dfrac{1}{216}$ b. $\dfrac{4}{4^4} = \dfrac{1}{4^3} = \dfrac{1}{64}$

c. $\dfrac{12}{12^4} = \dfrac{1}{12^3} = \dfrac{1}{1728}$

111. a. $\left(\dfrac{3}{8}\right)\left(\dfrac{2}{7}\right) = \dfrac{3}{28}$ b. $P(2\text{ red}) = \left(\dfrac{5}{8}\right)\left(\dfrac{4}{7}\right) = \dfrac{5}{14}$

c. $P(1\text{ of each}) = 1 - \dfrac{3}{28} - \dfrac{5}{14} = \dfrac{15}{28}$

113. $\dfrac{1 - 50 \cdot 49 \cdot 48 \cdot 47 \cdot 46}{50^5} = \dfrac{72{,}811}{390{,}625} \approx 0.186$

115. $\dfrac{(_5P_5)(_4P_4)(_3P_3)(_3P_3)}{_{12}P_{12}} = \dfrac{1}{4620}$

117. a. $\dfrac{_4C_2}{_{52}C_2} = \dfrac{1}{221} \approx 0.005$ b. $1 - \dfrac{_{39}C_2}{_{52}C_2} = \dfrac{15}{34} \approx 0.441$

c. $\dfrac{_{16}C_2}{_{52}C_2} = \dfrac{20}{221} \approx 0.090$ d. $1 - \dfrac{_{40}C_2}{_{52}C_2} = \dfrac{91}{221} \approx 0.412$

e. $\frac{_{26}C_2}{_{52}C_2} = \frac{25}{102} \approx 0.245$ **f.** $\frac{_{13}C_1}{_{52}C_2} = \frac{1}{102} \approx 0.010$

g. $1 - \frac{_{39}C_2}{_{52}C_2} = \frac{15}{34} \approx 0.441$

119. a. $\frac{_{50}C_4}{_{52}C_6} = \frac{5}{442} \approx 0.011$

b. $\frac{_{49}C_3}{_{52}C_6} = \frac{1}{1105} \approx 0.0009$

c. $\frac{_{47}C_1}{_{52}C_6} = \frac{1}{433,160} \approx 0.000002$

d. (answer for a) + (answer for b) − (answer for c) ≈ 0.012

121. $\frac{_{36}C_{26}}{_{52}C_{26}} = \frac{11}{37 \cdot 41 \cdot 43 \cdot 47 \cdot 7} \approx 0.0000005$

123. $\frac{4(_{36}C_{12})}{_{48}C_{24}} - \frac{_4C_2}{_{48}C_{24}} \approx 0.00016$

125. a. $\frac{(2 \cdot 2)(_3P_3)}{_5P_5} = \frac{1}{5}$ **b.** $\frac{(2 \cdot 2) \cdot 2}{5!/5} = \frac{1}{3}$

127. $T = (6 \text{ triples}) + (6 \cdot 5 \text{ doubles}) + (_6C_3) = 56$

a. $\frac{30}{56} = \frac{15}{28}$ **b.** $\frac{6}{56} = \frac{3}{28}$ **c.** $\frac{1 + 5 + 5 + _5C_2}{56} = \frac{21}{56}$ **d.** $\frac{2}{56} = \frac{1}{28}$

Index

Absolute value, 17–19
  of complex numbers, 524–526, 548
  in equations and inequalities, 138–143, 146
  rules for, 18, 141
Actuarial science, 539
Acute angle, 371, 374, 390, 394, 432, 500
Addition
  of complex numbers, 67
  formulas (trigonometry), 448–457, 494
  of matrices, 612–613, 639
  of ordinates, in graphing, 429–430
  of polynomials, 45
  of rational expressions, 57–58
  of real numbers, 1, 3, 6
  of vectors, 515, 517, 519, 521, 549
Additive identity, 3, 4
Additive inverse, 3, 4
Algebra, 1
Algebraic expression, 43, 49, 74
Alternating current, 361, 419
Ambiguous case, in solution of triangles, 502–503, 549
Amplitude, 411
Analytic geometry, 152
Ancient puzzles, 96
Angle, 361–371
  acute, 371, 374, 390, 394, 432, 500
  central, 363, 432
  complementary, 374, 403
  coterminal, 389
  degree measure of, 362–371, 432
  degree-radian conversion, 363–364, 433
  of depression, 385
  of elevation, 385
  important angles, 375–376, 392–393, 434
  initial side, 362
  measure, 362–371, 392, 432
  negative, 362
  obtuse, 371, 432, 500
  polar, 541
  positive, 362
  radian measure of, 362–371, 392, 432
  reference, 394, 432
  right, 154, 159, 181, 362, 371, 432, 500
  standard position, 389
  straight, 362, 371, 432
  sum, of a triangle, 371, 375, 500
  supplementary, 374
  terminal side, 362
  vertex, 362
Antilogarithms, 341
Approximately equal to, symbol for, 8
Approximation of roots, of a polynomial, 279, 306
Arc length, 363, 367–369, 392, 433, 449
Arccosecant function (Csc^{-1}), 491, 493
Arccosine function (Cos^{-1}), 482–484, 494
Arccotangent function (Cot^{-1}), 489–490, 494
Archimedes, 284, 366
Arcsecant function (Sec^{-1}), 491, 494
Arcsine function (Sin^{-1}), 480–482, 493
Arctangent function (Tan^{-1}), 489–490, 494
Area
  circle, 86, 123
  Heron's formula, 511–513, 549
  rectangle, 93, 123, 148
  sector of a circle, 367–369, 433
  sphere, 93, 123
  triangle, 22, 93, 123, 368, 511–513, 549
Argument, of a complex number, 526, 528–529
Arithmetic means, 654
Arithmetic operations, 1, 3
Arithmetic progression, 653–654, 662–664, 697
  common difference, 653
  sum of, 662, 697
Arrows, as vectors, 514
Associative property, 45, 67
Asymptotes, 195, 297, 305–306
  curved, 297, 306
  horizontal, 297, 305
  of a hyperbola, 195
  oblique, 297, 305
  of a rational function, 297, 305
  of trigonometric graphs, 421–427, 489, 491, 493, 494
  vertical, 297, 305
Augmented matrix, 578
Average, 94, 138

Axis (axes)
  coordinate, 151–153
  of ellipse, 191
  horizontal, 151
  i- or imaginary, 523
  major and minor, 191
  of parabola, 182–183, 204
  polar, 541
  real, 523
  semimajor and semiminor, 191
  translation of, 204–205
  vertical, 151
  x-, 151
  y-, 151

Babylonians, 2, 366, 463
Back substitution, 566
Base
  for exponents and exponential functions, 22, 311, 317
  for logarithms and logarithmic functions, 320
Binomial, 44
Binomial coefficients, 669, 697
Binomial perfect squares, factoring, 51, 74
Binomial theorem, 668–672, 697
Bisector, perpendicular, 180
Brahe, Tycho, 193
Briggs, Henry, 336

Calculator comments
  angles, 364
  evaluating polynomials, 280
  for exponents and exponential functions, 42, 317
  for finding logarithms, 331, 336
  for inverse functions, 383, 397, 481, 483, 491
  order of operations, 9
  for permutations and combinations, 682
  rounding and truncating, 8
  for scientific notation, 27
  for square roots, 36
  trigonometric functions, 378, 383, 397, 404–405, 484
Cancellation laws, 6–7, 56
Cantor, Georg, 10
Carbon 14 dating, 351
Cardioid, 544

Cartesian coordinate system, 151–153, 210, 519, 523
Caution boxes, 25, 39, 56, 66, 135, 337, 445, 460, 480, 486, 506
Cayley, Arthur, 619
Center
   of circle, 155, 201, 205, 210
   of ellipse, 191
   of hyperbola, 195
Central angle, 363, 432
Change-of-base formula, for logarithms, 328–329
Characteristic, of a logarithm, 342
Circle, 86, 123, 155–158, 201, 205, 210, 211, 363, 366, 370, 377, 433
   arc length, 363, 367–369, 392, 433
   central angle, 363, 432
   line tangent to, 180
   polar, 537
   sector, area of, 367, 433
   unit, 391–392, 441, 448
Circular functions, See Trigonometric functions.
Circular permutations, 682
Circumference, of a circle, 86, 123, 363, 366
   of the earth, 371
Cis θ, 527, 548
Closed interval, 16
Coefficient, 44, 79, 563
   binomial, 669, 697
   in an equation, 79
   leading, 44
Coefficient matrix, 578
Cofactor, 590, 638
Cofunctions, 374
   properties of, 374, 403–404, 434
Collinear points, 179
Column, of a matrix, 577
Combinations, 679, 682, 696, 697
Common denominator, 7
Common difference, 653
Common expansion formulas, 46, 47, 74
Common factor, 49
Common logarithms, 321, 336
Common ratio, 655
Commutative property, 3, 45, 67
Complementary angles, 374, 403
Complementary event, 690–691, 697
Complementary probability, 691, 697
Completing the square, 53, 74, 98–99, 145, 156–158, 182–183, 206–208
Complex fractions, 59
Complex numbers, 3, 65, 67, 73, 499, 548, 556
   absolute value of, 524–526, 548

addition of, 67
argument of, 526, 528–529
cis θ, 527, 548
conjugate, 69, 73, 524–525
DeMoivre's Theorem, 534–535, 549
division of, 69–70, 74, 530–540, 549
equality, 69
imaginary part, 67, 499
modulus, 524–526, 548
multiplication of, 67, 530–540, 549
nth roots of, 534, 536–540, 549
polar form, 526–540, 548
powers of, 534–535, 549
principal square root, 67, 537, 539
real part, 67, 499
triangle inequality, 526
Complex plane, 523–540, 548
Complex roots of polynomials, 102–103, 283–288
Components, of a vector, 519, 548
Composite number, 1, 72
Composition of functions, 240–244, 262
Compound interest, 346–350, 356
Compressing a graph, 229–230, 413–416, 425
Conditional equations, 79, 144, 439, 493
Cone, volume of, 93
Congruent triangles, 501
Conic sections, 200–201
Conjugate, of a complex number, 69, 73, 524–525
Conjugate hyperbolas, 196
Conjugate radicals, 48
Conjugate roots, 285
Consistent system of equations, 554, 563, 638
Constant, 43, 79, 144
   factorable, 662
   function, 255, 262
   polynomial, 44
Contradictions, 79
Coordinate(s)
   axes, 151–153
   Cartesian, 151–153, 210, 519, 541
   lines, 151
   polar, 541–548
   polar-rectangular conversion, 541–543, 548
   real and i-, 523–524
   rectangular, 519, 541
   system, 152, 523–524, 541
   x- and y-, 152
   vectors, 519–522, 548
Cosecant function, 361, 372, 390, 426–427, 433–435
Csc^{-1} function, 491, 493

Cosine function, 361, 372, 390, 409–410, 433–435
   sum and difference formulas for, 450, 494
Cos^{-1} function, 482–484, 494
Cosines, law of, 508–512, 549
Cotangent function, 361, 372, 390, 422–425, 433–435
Cot^{-1} function, 489–490, 494
Coterminal angles, 389
Counting Principle, Fundamental, 676
Cramer's Rule, 601–605, 680
Cube root, 3, 31
Curved asymptotes, 297, 306

Damping, 430–431, 469
Decibels, 335
Decimal fraction, 342
Decimal representations, 7, 8, 73
Decreasing function, 255, 262, 314
Degree
   as angle measure, 362–371, 432
   of an equation, 79
   of a polynomial, 44
   -radian conversion, 363–364, 433
DeMoivre, Abraham, 539
DeMoivre's theorem, 534–535, 549
Denominator, 2
   common, 7
   least common (LCD), 7, 57
   rationalizing, 35, 48
Dependent system of equations, 554, 563, 638
Dependent variable, 218, 262
Depreciation, 180
Depression, angle of, 385
Descartes, René, 66, 152, 278, 306
Descartes' rule of signs, 278, 306
Determinant, of a matrix, 589–601, 638–639
Difference, factoring
   of cubes, 53, 74
   of squares, 50, 74
Difference formulas for trigonometric functions, 448–457, 494
Difference quotient, 244, 262
Digits, significant, 27–28, 73
Diophantus, 96
Direct proportionality, 120, 144
Directed line segment, 514
Direct variation, 120, 144
Directrix, of a parabola, 202, 210
Discriminant, 101, 283
Distance formula, 154–155, 210, 449
   in the complex plane, 524, 526
   and equation of an ellipse, 190
   and equation of a hyperbola, 194

and equation of a parabola, 202
on the real number line, 18, 139
Distinguishable permutations, 680–681
Distributive property, 3, 45, 46, 49, 67
Dividend, 5
Divisible by, 650
Division, 1, 5–8
   of complex numbers, 69–70
      polar form, 530–540, 549
   of polynomials, long, 61–62, 74
   by radicals, 74
   by rational expressions, 57, 74
   synthetic, 272–273, 306
   by zero, 5
Divisor, 1, 5, 72
   greatest common (GCD), 1, 72
Domain, of a function, 218–219, 262
Domino principle, 646
Double angle formulas, 458, 494
Double root, 98, 284
Double subscripts, 577, 589
Dummy variable, 220, 660

e, 317, 355
Echelon form
   of a matrix, 579, 638
   of a system of equations, 566–567, 638
Effective interest rate, 348
Egyptians, 366
Einstein, Albert, 1, 93
Element
   of a set, 10
   of a matrix, 577
Elevation, angle of, 385
Elimination, method of solution, 562–571, 640
   matrix form of, 577–589, 640
Ellipse, 189–193, 206–207, 210–211
   as a locus, 189, 210
   summary box for, 191
Endpoints, 16, 361
Equal, identically, 311
Equality
   of complex numbers, 69
   of matrices, 611
   of vectors, 514
Equations, 79, 144
   absolute value, 138–143, 146
   of a circle, 155–158, 201, 211
   coefficients, 79
   completing the square, solution by, 98–99, 145
   conditional, 79, 144, 439, 493
   contradictions, 79
   degree of, 79
   double root, 98

of an ellipse, 190–192, 207, 211
equivalent, 82
exponential, 314–315
extraneous solutions of, 83, 107
first degree in one variable, 80–85, 144
fundamental rule for solving, 79
of a hyperbola, 194–196, 208, 212
identities, 79, 144, 439–448, 454, 459, 462, 469, 493–495
involving fractions, 82, 109, 145
of a parabola, 164, 181–183, 202–203, 211
of a plane, 563
involving radicals, 107–109
linear, 80, 175, 553, 562–563
literal, 85
logarithmic, 331–333
multiple roots, 278, 284
number of roots, 101–103, 283, 285, 306
quadratic, 96–112, 145, 181–188
roots or solutions of, 79, 144
solution by factoring, 97–98, 145
solvable by quadratic techniques, 104–112
system of, 553–605
translating word problems into, 87–92, 112–117
transposing in, 82
trigonometric, 439, 472–479, 484–487
variables in, 79
Equilateral triangle, 375
Equivalent equations, 82
Equivalent systems, of equations, 564
Euler, Leonhard, 318, 336
Even function, 233–234, 262, 402, 405–406, 434
Even-odd identities, 441, 494
Event, 676, 686
   complementary, 690–691, 697
   independent, 687, 696, 697
   mutually exclusive, 690, 696, 697
Expansion, of a determinant, 591–593
Expansion formulas, 46, 47, 74
Experiment, 686
Exponent(s), 22, 309
   base for, 22, 311, 317
   fractional, 38
   integral, 22
   negative, 25
   positive integer, 22
   rational, 38
   rules and formulas for, 23, 40, 310, 328, 355–356
   zero, 24

Exponent-logarithm relationships, examples of, 321
Exponential equations, 314–315
Exponential functions, 309–319, 345–356
   graphs of, 311–313, 356
   natural, 317
Exponential growth, 41, 43, 345, 356
Extraneous solutions (roots), 83, 107

Factor, 49, 72
   of an algebraic expression, 49, 74
   common, 49
   of a number, 1, 5, 72
Factor theorem, 271, 306
Factorable constant, 662
Factorial, 670, 696
Factoring
   of algebraic expressions, 49, 74
   binomial perfect squares, 51, 74
   by completing the square, 53, 74
   difference of cubes, 53, 74
   difference of squares, 50, 74
   FOIL, 51, 74
   formulas, 54, 74
   by grouping, 52, 74
   of numbers, 1
   of polynomials, 285
   and quadratic equations, 97–98, 145
   solving equations by, 97–98, 145
   sum of cubes, 53, 74
Feasible solution, linear programming, 631, 641
Fermat, P., 693
First degree equations in one variable, 80–85, 144
Focal properties
   ellipse, 193
   parabola, 204
Focus
   ellipse, 189, 210
   hyperbola, 193, 196, 210
   parabola, 202, 210
FOIL method, 47, 51, 74
Fractional exponents, 38
Fractional inequalities, 132–133, 146
   solution techniques, 132, 146
Fractions, 2, 6
   cancellation law, 6–7
   complex, 59
   decimal, 342
   denominator of, 2
   equations involving, 82, 109, 145
   least common denominator (LCD), 7, 57
   lowest terms, 7
   numerator of, 2

Fractions (*continued*)
   partial, 605–611, 639
   proper, 300–302
   rules for, 6
Function(s), 217–218
   algebraic combinations of, 239
   circular, 361, 372, 391–393, 432
   composition of, 240–244, 262
   constant, 255, 262
   decreasing, 255, 262, 314
   domain of, 218–219, 262
   even, 233–234, 262, 402, 405–406, 434
   exponential, 310–319, 345–355
   graph of, 221–236
   graph of inverse, 257–259, 263
   graphing techniques, 227–238
   horizontal line test, 221, 247, 249, 262–263
   increasing, 255, 262, 314
   inverse, 247–259, 262–263, 322
   inverse circular, 439, 479–494
   linear, 162, 168, 175, 221, 262, 269, 554–555, 563–564
   logarithmic, 310, 320–327, 345–355
   natural exponential, 317
   nondecreasing, 255
   nonincreasing, 255
   notation, 218
   odd, 233–234, 262, 402, 405–406, 435
   one-to-one, 247, 262
   operations on, 238–240, 262
   period of, 400
   periodic, 400–401
   piecewise definition of, 220
   polynomial, 269, 305
   quadratic, 221, 262, 269
   range of, 218, 262
   rational, 269, 297–305
   as a relation, 217, 221
   substituting into, 218
   trigonometric, 361, 372, 391–393, 432
   value of, 217–218
   vertical line test, 221, 262
Fundamental counting principle, 676
Fundamental rule for solving equations, 79
Fundamental theorem of algebra, 284, 306
Fundamental trigonometric identities, 440, 494

Gauss, C. F., 284
GCD. *See* Greatest Common Divisor.
General equation
   of a circle, 157
   of a line, 175, 178, 211
Geometric means, 656
Geometric progression, 655–656, 664–665, 697
   common ratio, 655
   sum of, 664, 697
Geometric proofs using vectors, 519–522
Geometric series, 665–666, 697
   sum of, 666, 697
Geometry, analytic, 152
Graph(s), 151–213 *See also* Graphing techniques.
   of absolute value of a function, 235, 263
   amplitude, 411
   of basic polynomials, 289
   of circular functions, 408–431, 435, 454–455, 469–471
   of conic sections, 191, 200–211
   of exponential functions, 311–313, 356
   of a function, 221–236
   horizontal line test, 221, 247, 249, 262–263
   of inequalities, 162–163, 212, 621–628, 640
   intercepts, 160, 210
   of inverse functions, 257–259, 263
   of inverse circular (trigonometric) functions, 481, 482, 489, 491, 493–494
   of linear functions, 162, 168, 175, 221, 554–555, 563–564
   of logarithmic functions, 323–326, 334, 357
   phase shift, 412, 435, 454–455
   in polar coordinates, 543–548
   of polynomial functions, 288–295, 306–307
   of quadratic functions, 164, 183–187, 221
   of rational functions, 297–304, 307
   of a relationship, 160, 210
   to solve a system of equations, 554–555, 562–563, 639
   to solve inequalities, 187, 212
   of trigonometric functions, 408–431, 435, 454–455, 469–471
   turnaround points, 293, 306
   vertical line test, 221, 262
Graphing techniques, 161, 164, 187, 212, 227–238, 257–259, 263–264
   addition of ordinates, 429–430
   completing the square, 156, 182–183, 206–208
   compression. *See* scaling.
   damping, 430–431, 469
   horizontal shift, 228
   plotting points, 161
   reflection, 228–229, 263, 408, 411, 413
   scaling, 229–230, 263, 410–416, 424–428, 486
   stretching. *See* scaling.
   summary of, 263–264, 294, 302, 306–307, 410, 411, 412, 415, 416, 423, 427, 428, 435
   symmetry, 164, 210, 212, 233, 263, 408–410, 422, 545, 549
   translation, 204–205, 227–228, 263, 412, 415, 416, 429
Greater than ($>$), 14
Greater than or equal ($\geq$), 15
Greatest common divisor (GCD), 1, 72
Greeks, 366
Grouping, factoring by, 52, 74

Half-angle formulas, 460–462, 494
Half-life, 350–351, 356
Half-open interval, 16
Harmonic motion, 417
Heron's area formula, 511–513, 549
Hierarchy of operations, 8–9, 23–24, 73
Hipparchus, 463
Hippasus, 2
Historical Perspectives, 2, 10, 66, 152, 193, 284, 318, 336, 366, 463, 539, 619, 693
Homogeneity, 662
Horizontal asymptotes, 297, 305, 489, 491, 493, 494
Horizontal line, equation of, 170
Horizontal line test, 221, 247, 249, 262–263
Horizontal shift, 228
Hyperbola(s), 189, 193–200, 207–208, 212
   asymptotes, 195
   conjugate, 196
   as a locus, 193, 210
   summary box for, 196
Hypotenuse, 2, 371

i, 66, 73
   powers of, 68
i, 519–521, 548
i-axis (imaginary axis), 523, 548
Identically equal, 311

Identities, 79, 144
  additive and multiplicative, 3, 4
  even-odd, 441, 494
  fundamental trigonometric, 440, 494
  Pythagorean, 442, 494
  Sum and difference formulas, 448–457, 494
  trigonometric, 439–448, 454, 459, 462, 469, 493–495
Identity matrix, 617, 638
Imaginary numbers, 66
Imaginary part of a complex number, 67, 499
Important angles, 375–376, 392–393, 434
Inconsistent system of equations, 554, 563, 638
Increasing function, 255, 262, 314
Independent events, 687, 696, 697
Independent variable, 218, 262
Index, of a radical, 31
Index of summation, 660
Indistinguishable permutations, 640, 696
Induction, mathematical, 645–652, 696
Inequalities
  involving absolute value, 138–143, 146
  algebraic, 14, 125–138
  fractional, 132–133, 146
  graphing, 162–163, 212, 621–628, 640
  higher order, 129–132, 145
  linear, 126–127, 145
  multiple, 128
  "or," 128–129
  quadratic, 187, 212
  rules for, 15, 126
  solution of, 126–127, 144–145, 621–628, 640
  solution by graphing, 187, 212
  systems of, 553, 621–629, 638, 640
  test points, 129, 291
  triangle, 19, 526
Inequality symbols ($>$, $\geq$, $<$, $\leq$), 14–15
Infinity (∞), 16
Initial point, of a vector, 514
Initial side, of an angle, 362
Integers, 2, 73
Intercepts, 160, 210
Interest, compound, 346–350, 356
Interest rate, effective, 348
Interpolation, linear, 340, 377, 378, 383, 397
Intersection
  of planes, 563–564
  of sets, 11

Intervals, 15–16, 73
Inverse
  additive, 3, 4
  of a matrix, 617–619, 638–639
  multiplicative, 3, 4
Inverse circular functions, 439, 479–494
  evaluation by calculator, 383, 384, 397, 481, 483, 491
  evaluation by tables, 382, 383, 481, 483
  graphs of, 481, 482, 489, 491, 493–494
Inverse functions, 247–259, 262–263, 322
  graphs of, 257–259, 263
Inverse proportion, 120, 144
Inverse trigonometric functions. See inverse circular functions.
Inverse variation, 120, 144
Irrational numbers, 2, 73
Irrational roots, of a polynomial, 279
Irreducible polynomial, 606
Isosceles triangle, 159, 375

j, 519–521, 548
Joint variation, 120, 144
Jointly proportional, 120, 144

Kaiser Rudolph II, 193
Kepler, Johannes, 193
Kronecker, Leopold, 10

Law of cosines, 508–512, 549
Law of sines, 499–506, 549
  ambiguous case, 502–503, 549
LCD. See Least common denominator.
LCM. See Least common multiple.
Leading coefficient, 44
Leading term, 566
Least common denominator (LCD), 7, 57, 74
Least common multiple (LCM), 1, 7, 57, 72
Legs, of a right triangle, 371
Lemniscate, 546
Length
  of arc, 363, 367–369, 392, 433
  of a vector, 514, 516, 519, 548
Less than ($<$), 14
Less than or equal ($\leq$), 15
Light year, 389
Like terms, 44
Line(s), 168, 211
  equations for, 170, 178, 211

  general equation of, 175, 178, 211
  as a graph, 162, 168, 175
  as a locus, 201, 210
  parallel, 176–178, 211
  perpendicular, 176–178, 181, 211, 375
  point-slope formula, 172, 178, 205, 211
  rise of, 168
  run of, 168
  segment, 3, 153, 514
    directed, 514
    midpoint formula, 519–521
  slope of, 168, 211
  slope-intercept formula, 173, 178, 211
  summary box for, 178
  two-intercept form, 180
  two-point formula, 181
Linear equation, 80, 175, 553, 562, 563
Linear function, 162, 168, 175, 221, 262, 269, 554–555, 563–564
Linear inequalities, 126–127, 145
Linear interpolation, 340, 377, 378, 383, 397
Linear programming, 629–638, 641
Literal equations, 85
Locus, 189, 193, 201
  circle as, 201, 210
  ellipse as, 189, 210
  hyperbola as, 193, 210
  lines as, 201, 210
  parabola as, 202, 210
Logarithmic equations, 331–333
Logarithmic functions, 310, 320–327, 345–355
  graphs of, 323–326, 334, 357
Logarithms, 310, 320, 336, 355
  antilogarithms, 341
  base for, 320
  change of base formula, 328–329
  characteristic, 342
  common, 321, 336
  mantissa, 342
  natural, 321
  properties and formulas, 328, 356
  use of tables, 339–345
Lowest terms, 7

$m \times n$ matrix, 577
Magnitude, of a vector, 514, 516, 519, 548
Main diagonal, of a matrix, 594
Major axis, 191
Mantissa, of a logarithm, 342
Mathematical induction, 645–652, 696

Matrix, 577
  addition of, 612–613, 639
  augmented, 578
  coefficient, 578
  cofactor of, 590, 638
  column of, 577
  determinant of, 589–601, 638
  echelon form, 578, 638
  element of, 577
  equality, 611
  identity, 617, 638
  inverse of, 617–619, 638–639
  $m \times n$, 577
  main diagonal of, 594
  method of solution, 577–589, 604, 640
  minor of, 590, 638
  multiplication, 613–616, 639
  multiplication by a scalar, 612–613, 639
  order of, 589
  reduced row echelon form, 581, 638
  row of, 577
  rules for, 639
  square, 577
  subtraction, 612–613, 639
  of a system of equations, 578
  transpose, 594
  zero, 612
Mean(s)
  arithmetic, 654
  geometric, 656
Member, of a set, 10
Midpoint formula, 153, 210, 519–521
Minor axis, 191
Minor, of a matrix or determinant, 590, 638
Minutes, in angle measure, 362
Modulus, of a complex number, 524–526, 548
Monomial, 44
Multiple, least common (LCM), 1, 7, 57, 72
Multiple angle formulas, 458, 494, 540
Multiple roots, 278, 284
Multiplication, 1, 3
  of complex numbers, 67
    polar form, 530–540, 549
  of matrices, 613–616, 639
  of polynomials, 46
  of rational expressions, 57, 74
  by a scalar, 516–517, 519, 521, 549, 612–613, 639
Multiplicative identity, 3, 4
Multiplicative inverse, 3, 4
Mutually exclusive events, 690, 696, 697

Napier, John, 336
Natural exponential function, 317
Natural logarithms, 321
Natural numbers, 1, 14, 72
Navigation, 515–516
Negative angle, 362
Negative exponent, 25
Negative integer, 2
Negative numbers, 2–5, 14
  rules and formulas, 4–5
  square roots of, 65–67, 73, 537, 539
Neugebauer, Otto, 366
Newton, Isaac, 284, 539
nth root
  of a complex number, 534, 536–540, 549
  principal, 537, 539
  of a real number, 31
Nondecreasing function, 255
Nonincreasing function, 255
Nonlinear systems, of equations, 558–560, 571–577
Numbers, 1
  basic properties of, 3–6
  complex, 3, 65, 73, 499, 548, 556
  composite, 1, 72
  difference of, 3
  divisor of, 1, 5, 72
  factor of, 1, 5, 72
  imaginary, 66
  integers, 2, 73
  irrational, 2, 73
  long division of, 7
  multiple of, 1
  natural, 1, 14, 72
  negative, 2–5, 14
  positive, 3, 14
  prime, 1, 72
  rational, 2, 7–8, 73
  real, 3, 73
  square of, 2
  zero, 2, 3, 14
Numerator, 2

Objective function, linear programming, 631, 641
Oblique asymptotes, 297, 305
Oblique triangle, 499, 549
  solution of, 499–512, 549
Obtuse angle, 371, 432, 500
Odd function, 233–234, 262, 402, 405–406, 435
Odds, 692
One-to-one function, 247, 262
Open interval, 16
Operations on functions, 238–240, 262
Operations, hierarchy of, 8–9, 23–24, 73

Optimal solution, linear programming, 631, 641
"Or" inequalities, 128–129
Order, of a determinant or matrix, 589
Order of operations, 8–9, 23–24, 73
Order relations, ($>$, $\geq$, $<$, $\leq$), 14, 15, 19, 141
  properties of, 15, 141
Ordered pair (a, b), 153
Origin, 3, 152
Oscillatory motion, 361, 417, 439
Outcome, 686

Pallas Athene, 96
Parabola, 164, 181–189, 202–206, 210–211
  as a locus, 202, 210
  summary box for, 203
Parallel lines, 176–178, 211
Parallelogram, 159
Parallelogram laws, addition and subtraction of vectors, 515, 517, 549
Parsec, 389
Period, of a function, 400
  of circular functions, 401, 415, 434, 435
Periodic function, 400–401
Partial fractions, 605–611, 639
Pascal, Blaise, 693
Pascal's triangle, 669, 697
Perimeter, of a rectangle, 89, 93, 123, 148
Permutations, 677–678, 682, 696–697
  circular, 682
  distinguishable, 680–681
  indistinguishable, 680, 696
Perpendicular lines, 176–178, 181, 211, 375
Perpendicular bisector, 180
pH, 338
Phase shift, 412, 435, 454–455
pi, π, 85, 362, 366
Piecewise defined function, 220
Plane(s), 563–564
  complex, 523–540, 548
  real or Cartesian, 152, 523
Plotting points, 161
Point-slope formula, 172, 178, 205, 211
Polar angle, 541
Polar axis, 541
Polar circle, 537
Polar coordinates, 541–548
  curve-sketching in, 543–548
  symmetry tests, 545, 549
Polar form, of a complex number, 526–540, 548

of multiplication and division, 530–540, 549
Polar ray, 541
Polar-rectangular conversion, 541–543, 548
Pole, 297, 541
Polynomials, 43, 44, 73
   addition of, 45
   approximation of roots, 279, 306
   behavior between roots, 290
   complex roots of, 102–103, 283–288
   conjugate roots, 285
   constant, 44
   degree of, 44
   factoring, 271, 285, 306
   functions, 269, 305
   graphing, 288–295, 306–307
   irrational roots, 279
   irreducible, 606
   LCM, 1, 7, 57, 72
   long division of, 61–62, 74
   multiplication of, 46
   number of roots, 101–103, 283, 285, 306
   number of turnaround points, 293, 306
   quotient of, 55–63, 109, 132, 145, 146, 305, 307
   range limitation on real roots, 279, 306
   rational roots, 275–278, 306
   real roots, 274–282
   root of, 102–103, 269, 278, 279, 283–288, 305–306
   subtraction of, 45
   zero, 44
   zero of, 269
Population growth, 41, 43, 345
Positive angle, 362
Positive integer exponents, 22
Positive number, 3, 14
Power, of a number, 22
Powers of i, 68
Present value, 358
Prime number, 1, 72
Principal axis, of a parabola, 182–183, 204
Principal nth root, 31, 537, 539
Principle of mathematical induction, 645–652, 696
Principal square root, of a complex number, 67
Probability, 539, 645, 685–686, 696
   complementary events, 690–691, 697
   independent events, 687, 696, 697
   mutually exclusive events, 690, 696, 697
   odds, 692, 697
   rules and formulas, 688, 697
   sample space, 686
Product
   of polynomials, 46
   of complex numbers, 67, 530–540, 549
   of matrices, 613–616, 639
   of natural numbers, 1
   of real numbers, 3
   rules, 126
Product-to-sum formulas, 465, 471, 494
Progressions, 653–664, 696–697
   arithmetic, 653–654, 662–664, 697
   common difference, 653
   common ratio, 655
   geometric, 655–656, 664–665, 697
   sum of arithmetic, 662, 697
   sum of geometric, 664, 697
Proper fraction, 300–302
Proportionality, 120–125, 144
Ptolemy, 463
Pythagoras, 2, 66
Pythagorean identities, 442, 494
Pythagorean theorem, 2, 113, 154, 372, 375, 381, 433, 509

Quadrant, 152
Quadratic equations, 96–112, 145, 164, 181–188, 221, 283
Quadratic formula, 100–103, 145
Quadratic functions, 164, 183–187, 221, 262, 269
Quadratic inequalities, 187, 212
Quotient, 1, 5–8, 269
   of complex numbers, 69–70, 74, 530–540, 549
   difference, 244, 262
   of polynomials, 55–63, 109, 132, 145, 146, 305, 307

Radian measure, of angles, 362–371, 392, 432
Radian-degree conversion, 363–364, 433
Radical(s), 31–33
   division by, 74
   equations involving, 107–109
   index of, 31
   rationalizing, 35, 48
   rules for, 32
Radicand, 31
Radioactive decay, 43, 78, 350
Radius, of a circle, 86, 155, 201, 205, 210, 363, 366, 367, 370, 377, 433
   of earth, 371
Range, of a function, 218, 262
Ratio, common, 655

Rational exponent, 38–40
Rational expressions, 55–63, 74, 109, 132, 145, 146
Rational functions, 269, 297–305
   asymptotes, 297, 305
   graphs of, 297–304, 307
   poles, 297
   roots or zeros, 297
   graphing techniques, summary of, 302, 307
Rational numbers, 2, 6–8, 73
   rules and formulas, 6
Rational root theorem, 275, 306
Rational roots, of polynomials, 275–278, 306
Rationalizing denominators, 35, 48
Ray, 361
   polar, 541
Real axis, 523
Real numbers, 1–8, 14, 73
   absolute value of, 17
   decimal representation, 7, 73
   nth root of, 31
   part of a complex number, 67, 499
   roots of polynomials, 102–103, 283–288
   square root of, 2, 3, 31
Real number line, 3
   distance on, 18, 139
Real plane, 152, 523
Real roots, of a polynomial, 274–282
Reciprocal relationships, 374, 400, 434
Rectangle, 159
   area, 93, 123, 148
   perimeter, 89, 93, 123, 148
Rectangular coordinates, 519, 541
Rectangular-polar conversion, 541–543, 548
Reduced row echelon form, 581, 638
Reference angle, 394, 432
Reflection, of a graph, 228–229, 263, 408, 411, 413
Relationship, 217
   as a function, 217, 221
   graph of, 160
Remainder, 62, 269
Remainder theorem, 269–270, 306
Repeating decimal, 7
Resultant, of a vector, 515
Rhind papyrus, 366, 463
Richter scale, 339
Right angle, 154, 159, 181, 362, 371, 432, 500
Right triangle(s), 2, 154, 159, 181, 361, 371–372, 500, 509
   solving, 381–389
   trigonometry, 371–392, 433

Rise, of a line, 168
Root(s)
   of complex numbers, 534, 536–540, 549
   conjugate, 285
   cube, 3, 31
   double, 98, 284
   of an equation, 79, 144
   extraneous, 83, 107
   irrational, 279
   multiple, multiplicity of, 278, 284
   nth, 31, 534, 536–540, 549
   number of, 101–103, 283, 285, 306, 537, 549
   of a polynomial, 102–103, 269, 279, 283–289, 305
   principal, 31, 537, 539
   rational, 275–278, 306
   of a rational function, 297
   of a real number, 31
   real, of a polynomial, 274–282
   square, 2, 3, 31, 65–67, 73, 537, 539
Rounding, 8, 28
Row, of a matrix, 577
Run, of a line, 168

Sample space, 686
Scalar multiple
   of a matrix, 612–613, 639
   of a vector, 516–517, 519, 521, 549
Scalar quantity, 514
Scaling a graph, 229–230, 263, 410–416, 424–428, 486
Scientific notation, 26–27, 73
Secant function, 361, 372, 390, 426–427, 433–435
Sec^{-1} function, 491, 494
Seconds, in angle measure, 362, 364, 366
Sector of a circle, area of, 367, 369, 433
Segment, of a line, 3, 153, 514
Semimajor axis, 191
Semiminor axis, 191
Semiperimeter, 512
Sequence, 652, 696
   term of, 652
Series, geometric, 665–666, 697
   sum of, 666, 697
Set, 10, 73
   difference, 11
   element of, 10
   intersection, 11
   member of, 10
   solution, 80
   subset, 10
   union, 11
   Venn diagrams, 11

Significant digits, 27–28, 73
Similar triangles, 168, 372, 501–502
Simplex method, 633
Simple harmonic motion, 417
Sine function, 361, 372, 390, 408–410, 433–435
   sum and difference formulas for, 450, 494
Sin^{-1} function, 480–482, 493
Sines, law of, 499–506, 549
   ambiguous case, 502–503, 549
Sketching a graph, 160, 543–548
Slope, 168, 211
Slope-intercept formula, 173, 178, 211
Solution(s)
   of equations, 79, 144
   extraneous, 83, 107
   of inequalities, 126–127, 144–145, 621–628, 640
   number of, 101–103, 283, 285, 306
   set, 80
   of systems of equations, 553
   of systems of inequalities, 553, 640
   of word problems, 88, 112, 145
Solving a triangle, 381–392, 499–512, 549
   ambiguous case, 502–503, 549
   law of cosines, 508–512, 549
   law of sines, 499–506, 549
   right triangles, 381–392
   techniques, summary of, 511, 549
Speed, 514
   of light, 389
Sphere
   surface area, 93, 123
   volume, 93, 121
Square matrix, 577
Square, of a number, 2
Square root, 2, 3, 31, 36, 65–67, 73
   of a negative number, 65–67, 73, 537, 539
Standard equation
   of circle, 155, 201, 211
   of ellipse, 190–192, 207, 211
   of hyperbola, 194–196, 208, 212
Standard position, of an angle, 389
Statistics, 539
Straight angle, 362, 371, 432
Stretching a graph, 229–230, 263, 413–414, 425, 427, 486
Subscripts, double, 577, 589
Subset, 10
Substituting into a function, 218
Substitution method of solution, 555–558, 639
Subtraction, 4
   of polynomials, 45
   of matrices, 612–613, 639

of rational expressions, 57–58, 74
of vectors, 517–518, 519, 521, 549
Sum, 1, 3, 6
   of angles of a triangle, 371, 375, 500
   of an arithmetic progression, 662, 697
   of complex numbers, 67
   of cubes, factoring, 53, ;74
   of a geometric progression, 664, 697
   of a geometric series, 666, 697
   of matrices, 612–613, 639
   of polynomials, 45
   of vectors, 515, 517, 519, 521, 549
Sum and difference formulas, for circular functions, 448–457, 494
Sum-to-product formulas, 467, 495
Summation notation, 659–668, 696
Supplementary angles, 374
Surveying, 361, 385–389
Symmetry
   in graphs, 164, 210, 212, 233, 263, 408–410, 422
   of trigonometric functions, 401–402, 434–435
   in polar coordinates, 545, 549
Symmetry line, parabola, 182–183, 204
Synthetic division, 272–273, 306
Systems of equations, 553–605
   consistent, 554, 563, 638
   Cramer's rule, 601–605, 640
   dependent, 554, 563, 638
   echelon form, 566–567, 638
   elimination method of solution, 562–571, 640
   equivalent, 564
   inconsistent, 554, 563, 638
   matrix method of solution, 577–589, 640
   matrix of, 578
   nonlinear, 558–560, 571–577
   solution of, 553
   solution by graphing, 554–555, 562–563, 639
   solution by matrix inverse, 618–619, 640
   substitution method of solution, 555–558, 639
System of inequalities, 553, 621–629, 638, 640

Tables, use of
   for logarithms, 339–345
   for trigonometric functions, 376–378, 382–383, 396–397, 404
   for inverse trigonometric functions, 382, 383, 481, 483
Tail, of a vector, 514
Tangent function, 361, 372, 390, 420–425, 433–435

sum and difference formulas for, 451, 494
Tan^{-1} function, 489–490, 494
Term(s)
   of an algebraic expression, 43
   leading, 566
   like, 44
   lowest, 7
   of a sequence, 652
Terminal point, of a vector, 514
Terminal side, of an angle, 362
Terminating decimal, 8
Test point, 129, 291
Three leaved rose, 545
Tip, of a vector, 514
Tower of Hanoi, 648
Translation of graphs, 204–205, 227–228, 263, 412, 415, 416, 429
Translation of axes, 204–205
Transposing
   in an equation, 82
   a matrix, 594
Triangle(s)
   angle sum of, 371, 375, 500
   area of, 22, 93, 123, 368, 511–513, 549
   congruent, 501
   equilateral, 375
   hypotenuse, 2, 371
   isosceles, 159, 375
   legs, of right, 371
   oblique, 499, 549
      solution of, 499–512, 549
   Pascal's, 669, 697
   Pythagorean theorem, 372, 375, 381, 433, 509
   right, 2, 154, 159, 181, 361, 371–372, 500–509
      solution of, 381–392
   similar, 168, 372, 501–502, 549
   solution techniques, summary of, 511, 549
Triangle inequalities, 19, 526
Trigonometric equations, 439, 472–479, 484–487
Trigonometric functions, 361, 372, 390–393, 432
Trigonometric identities, 439–448, 454, 459, 462, 469, 493–495
   of arbitrary angles, 390
   cofunctions, 374
   cofunction properties, 374, 403–404, 434
   definition of, 372, 390, 392
   double angle formulas, 458, 494
   elementary properties, 399–407
   evaluation
      by calculator, 376, 378–380, 383–387, 397, 404–405, 481, 483, 484, 491
      by geometry, 372–376, 382, 390–396
      by tables, 376–378, 382–383, 396–397, 404, 481, 483
   graphs of, 408–431, 435, 454–455, 469–471
   graphs of inverse, 481, 482, 489, 491, 493–494
   half-angle formulas, 460–462, 494
   identities, 439–448, 454, 459, 462, 469, 493–495
   important values of, 376, 392–393, 434
   inverse, 439, 479–494
   multiple angle formulas, 458, 494, 540
   periods of, 401, 415, 434, 435
   phase shift, 412, 435
   product-to-sum formulas, 465, 471, 494
   of real numbers, 392
   reciprocal relationships, 374, 400, 434
   relation to right triangles, 371–392, 433
   sum and difference formulas, 448–457, 494
   sum-to-product formulas, 467, 495
   summary of graphing techniques, 410, 411, 412, 415, 416, 423, 427, 428, 435
   symmetry properties, 401–402, 434–435
   values of, 376, 392–393, 410, 423, 427, 434
Turnaround points, 293, 306
Two-intercept formula, 180
Two-point formula, 181

Union, of sets, 11
Unit circle, 391–392, 441, 448
Unit vector, 514
Unknowns, 79

Value
   of circular functions, 376, 392–393, 410, 423, 427, 434
   of exponential functions, 311
   of an algebraic expression, 43
   of a function, 217–218
   of logarithmic functions, 325
   present, 358
Variables, 43, 79, 144
   dependent, 218, 262
   dummy, 220, 660
   in an equation, 79
   independent, 218, 262

Variation, 120, 144
Vectors, 499, 513–522, 548–549
   addition of, 515, 517, 519, 521, 549
   as arrows, 514
   components of, 519, 548
   coordinate, 519–522, 548
   direction of, 513–514
   equality, 514
   geometric proofs using, 519–522
   i and **j**, 519–521, 548
   initial point of, 514
   length or magnitude, 514, 516, 519, 548
   parallelogram laws, 515, 517, 549
   resultant, 515
   scalar multiples, 516, 517, 519, 521, 549
   subtraction, 517–519, 521, 549
   tail, 514
   terminal point, 514
   tip, 514
   unit, 514
   zero, 516
Velocity vector, 514–516
Venn diagrams, 11
Vertex
   of angle, 362
   of ellipse, 190
   of hyperbola, 195
   of parabola, 164, 183
Vertical asymptotes, 297, 305, 421–427
Vertical line, equation of, 170, 211
Vertical line test, 221, 262
Vibration, 417
Volume
   of a cone, 93
   of a sphere, 93, 121

Word problems, techniques, 87–92, 112–117, 145

x-axis, 151
x-coordinate, 152
x-intercepts, 160

y-axis, 151
y-coordinate, 152
y-intercepts, 160

Zero, 2, 3, 14
   division by, 5
   exponent, 24
   matrix, 612
   polynomial, 44
   of a polynomial, 269
   product principle, 97
   of a rational function, 297
   vector, 516

USEFUL FORMULAS

Triangles (2.2, 2.6, 7.2, 7.3)

1. $A + B + C = 180°$
2. Area $= \frac{1}{2}$ base $\cdot$ height
3. Isosceles triangle: $B = C$
4. Equilateral triangle $A = B = C = 60°$
5. Pythagorean theorem (1.1, 3.1, 7.2, 7.3, 9.2) If $A = 90°$, then $a^2 = b^2 + c^2$
6. Similar triangles
$$\frac{a}{a'} = \frac{b}{b'} = \frac{c}{c'}$$

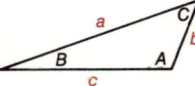

7. Simple triangles (7.2)

 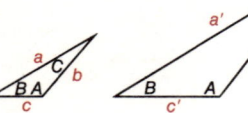

8. Law of sines (9.1)
$$\frac{\sin A}{a} = \frac{\sin B}{b} = \frac{\sin C}{c}$$
9. Law of cosines (9.2)
$a^2 = b^2 + c^2 - 2bc \cos A$
10. Heron's area formula (9.2)
Area $= \sqrt{s(s-a)(s-b)(s-c)}$
where $s = \frac{1}{2}(a + b + c)$

Trigonometric Functions (7.2, 7.4)

$\cos s = \cos \theta = \frac{x}{r}$

$\sin s = \sin \theta = \frac{y}{r}$

$\tan s = \tan \theta = \frac{y}{x} = \frac{\sin \theta}{\cos \theta}$

$\sec s = \sec \theta = \frac{r}{x} = \frac{1}{\cos \theta}$

$\csc s = \csc \theta = \frac{r}{y} = \frac{1}{\sin \theta}$

$\cot s = \cot \theta = \frac{x}{y} = \frac{\cos \theta}{\sin \theta} = \frac{1}{\tan \theta}$

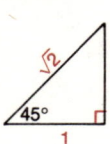 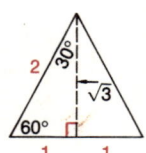

Basic identities (7.5, 8.1, 8.2)

1. $\cos^2 t + \sin^2 t = 1$
2. $1 + \tan^2 t = \sec^2 t$
3. $1 + \cot^2 t = \csc^2 t$
4. $\cos(-t) = \cos t$
5. $\sin(-t) = -\sin t$
6. $\tan(-t) = -\tan t$
7. $\cos(t \pm \pi) = -\cos t$
8. $\sin(t \pm \pi) = -\sin t$
9. $T(\theta + 2\pi) = T(\theta)$ for $T = \sin, \cos,$ sec, or csc
10. $T(\theta + \pi) = T(\theta)$ for $T = \tan$ or cot
11. $T(\theta) = T^*(\pi/2 - \theta)$ for cofunctions T, T^*

Sum & Difference Formulas (8.3)

1. $\cos(\alpha \pm \beta) = \cos \alpha \cos \beta \mp \sin \alpha \sin \beta$
2. $\sin(\alpha \pm \beta) = \sin \alpha \cos \beta \pm \cos \alpha \sin \beta$
3. $\tan(\alpha \pm \beta) = \dfrac{\tan \alpha \pm \tan \beta}{1 \mp \tan \alpha \tan \beta}$

Double-Angle Formulas (8.4)

1. $\sin 2\alpha = 2 \sin \alpha \cos \alpha$
2. $\cos 2\alpha = \cos^2 \alpha - \sin^2 \alpha$
   $= 1 - 2 \sin^2 \alpha$
   $= 2 \cos^2 \alpha - 1$
3. $\tan 2\alpha = \dfrac{2 \tan \alpha}{1 - \tan^2 \alpha}$

Half-Angle Formulas (8.4)

1. $\sin \dfrac{\alpha}{2} = \pm \sqrt{\dfrac{1 - \cos \alpha}{2}}$
2. $\cos \dfrac{\alpha}{2} = \pm \sqrt{\dfrac{1 + \cos \alpha}{2}}$
3. $\tan \dfrac{\alpha}{2} = \dfrac{1 - \cos \alpha}{\sin \alpha}$
   $= \dfrac{\sin \alpha}{1 + \cos \alpha}$

Product-Sum Formulas (8.5)

1. $\sin \alpha \cos \beta = \dfrac{1}{2}[\sin(\alpha + \beta) + \sin(\alpha - \beta)]$
2. $\cos \alpha \cos \beta = \dfrac{1}{2}[\cos(\alpha + \beta) + \cos(\alpha - \beta)]$
3. $\sin \alpha \sin \beta = \dfrac{1}{2}[\cos(\alpha - \beta) - \cos(\alpha + \beta)]$
4. $\sin \alpha + \sin \beta = 2 \sin \dfrac{\alpha + \beta}{2} \cos \dfrac{\alpha - \beta}{2}$
5. $\sin \alpha - \sin \beta = 2 \cos \dfrac{\alpha + \beta}{2} \sin \dfrac{\alpha - \beta}{2}$
6. $\cos \alpha + \cos \beta = 2 \cos \dfrac{\alpha + \beta}{2} \cos \dfrac{\alpha - \beta}{2}$
7. $\cos \alpha - \cos \beta = -2 \sin \dfrac{\alpha + \beta}{2} \sin \dfrac{\alpha - \beta}{2}$

Inverse Circular Functions (8.7, 8.8)

1. $y = \text{Sin}^{-1} x$ means $x = \sin y$ and $-\dfrac{\pi}{2} \le y \le \dfrac{\pi}{2}$
2. $y = \text{Cos}^{-1} x$ means $x = \cos y$ and $0 \le y \le \pi$
3. $y = \text{Tan}^{-1} x$ means $x = \tan y$ and $-\dfrac{\pi}{2} < y < \dfrac{\pi}{2}$
4. $y = \text{Csc}^{-1} x$ means $x = \csc y$ and $-\dfrac{\pi}{2} \le y \le \dfrac{\pi}{2}, \; y \ne 0$
5. $y = \text{Sec}^{-1} x$ means $x = \sec y$ and $0 < y < \pi, \; y \ne \dfrac{\pi}{2}$
6. $y = \text{Cot}^{-1} x$ means $x = \cot y$ and $0 < y < \pi$

17. $-1 \pm i$ 19. $6, -2$ 21. $\dfrac{2}{5}, -\dfrac{4}{3}$ 23. $\dfrac{1}{3}, 3$

25. $\dfrac{-1 \pm 3i}{2}$ 27. $\dfrac{(3 \pm 3i)}{2}$ 29. $2 \pm i$

31. $-3 \pm 2i$ 33. $-4 \pm 3i$ 35. $\pm\sqrt{2}, \pm i\sqrt{6}$

37. $\pm\sqrt{2}, \pm\sqrt[4]{2}, \pm i\sqrt{2}, \pm i\sqrt[4]{2}$

39. The real roots are ± 1.   41. $2, 3$ 43. $-1, 8$

45. $1, 16$ 47. $\dfrac{9}{2}, 2$ 49. 5 51. $\dfrac{1}{2}$ 53. 7

55. ± 2 57. $\pm\sqrt{7}$ 59. -1 61. $-2 - \dfrac{i}{2}$

63. $i \pm 2$ 65. $2i, -\dfrac{i}{3}$ 67. $\varnothing$ 69. $1, -2, 3$

71. $[-3, \infty)$ 73. $(-\infty, -5) \cup (1, \infty)$ 75. $(-2, 3)$

77. $(-\infty, -3] \cup [-1, 1] \cup [2, \infty)$

79. $(-3, -2] \cup [4, \infty)$ 81. $[-4, 0]$ 83. $(5, \infty)$

85. $(-\infty, -6] \cup [-2, \infty)$ 87. $\left(-\dfrac{5}{3}, 1\right)$

89. $(-\infty, -5] \cup [11, \infty)$ 91. $[-4, -2) \cup (2, 4]$

93. $(-6, -3) \cup (-1, 2)$ 95. $x = Ah/(1 - A)$

97. $v = \pm\sqrt{\dfrac{t + u}{r^2 s}}$ 99. a. Doubled   b. $r = \dfrac{10p^3 s}{\sqrt{q}}$ c. 40

101. $40¢$/mi   103. 300 m, 500 m   105. $\$16{,}000$

107. 15 109. a. 16 mi/h   b. $\dfrac{3}{4}$ h, $1\dfrac{1}{4}$ h   111. $\$4600$

113. 10 h, 8 h   115. 2 furlongs by 3 furlongs

117. The entries are $n, n + 1, n + 7$, and $(n + 7) + 1$. The sum of these is $4n + 16 = 4(n + 4)$. She divides by 4 and subtracts 4 to get the first entry. The rest is easy.

119. 17

CHAPTER 3

Section 3.1

1. $A = (2, 1); B = (-2, 3); C = (-1, -4); D = (6, -3); E = (4, 10); F = (-12, 2); G = (-12, -7); H = (3, -9); I = (12, 11); J = (-9, 10); K = (-8, -1); L = (10, -7); M = (11, 4); N = (-7, 6); O = (-5, -6)$

3–15

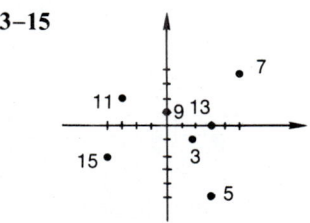

3. Quadrant IV   5. Quadrant IV   7. Quadrant I

9. On the line separating quadrants I and II

11. Quadrant II

13. On the line separating quadrants I and IV

15. Quadrant III

17. Burkettsville—F9, Cassella—L17, Celina—Q15, Chickasaw—O17, Coldwater—O11, Convoy—T12, Cranberry—J14, Ft. Recovery—N7, Maria Stein—L23, Montezuma—P18, Padua—P8, Philothea—N9, Sharpsburg—I7, St. Henry—M11, Van Wert—U14

19. $(2, 0)$ 21. $\left(\dfrac{1}{2}, 1\right)$ 23. $(0, 2)$

25. a. $\sqrt{82}$ b. 9.06 27. a. $\sqrt{122}$ b. 11.05

29. a. $\sqrt{2}$ b. 1.41 31. a. 3 b. 3 33. a. 8 b. 8

35. a. $3\sqrt{2}$ b. 4.24 37. $6\dfrac{1}{2}$ mi

39. $d[(-2, 4), (0, -1)] = d[(-2, 4), (3, 6)] = \sqrt{29}$ and the points are not collinear

41. $d[(-2, 1), (0, 4)] = d[(2, 2), (4, 5)] = \sqrt{13}$
$d[(-2, 1), (2, 2)] = d[(0, 4), (4, 5)] = \sqrt{17}$

43. $(0, 1)$ is the midpoint of both segments.

45. Place the vertices at $(0, 0), (0, b)$, and $(a, 0)$. The midpoint of the hypotenuse is then $\left(\dfrac{a}{2}, \dfrac{b}{2}\right)$; its distance from each vertex is $\dfrac{1}{2}\sqrt{a^2 + b^2}$.

47. $4x - 2y - 5 = 0$ 49. $\left(-\dfrac{6}{5}, -\dfrac{3}{5}\right), (-2, -1)$

51.

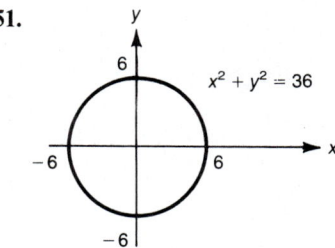

53.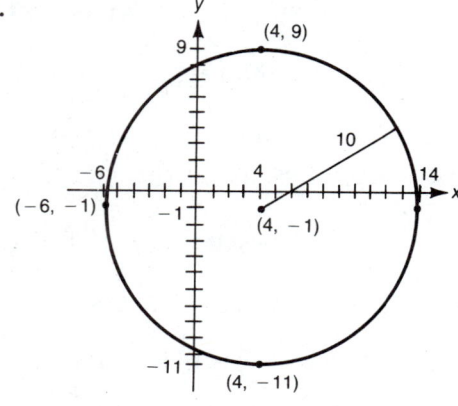
$(x - 4)^2 + (y + 1)^2 = 100$

55.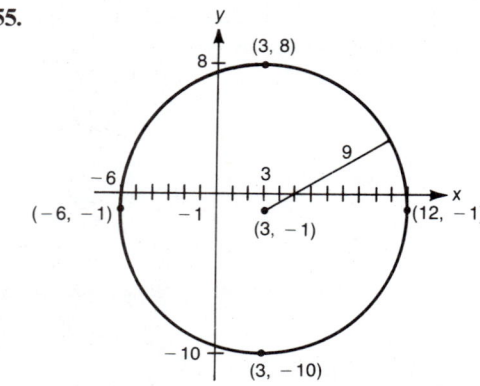
$(x - 3)^2 + (y + 1)^2 = 81$

57.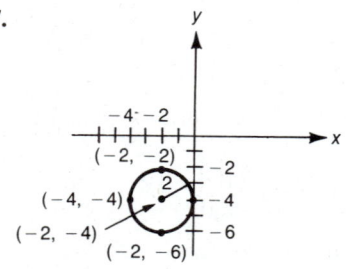
$(x + 2)^2 + (y + 4)^2 = 4$

59. $x^2 + y^2 - 8y - 33 = 0$
61. $x^2 + 2x + y^2 + 2y + 1 = 0$
63. $x^2 - 8x + y^2 - 4y + 11 = 0$
65. $x^2 + 4x + y^2 - 6y + 9 = 0$
67. $x^2 - 4x + y^2 + 8y - 205 = 0$
69. $x^2 + 4x + y^2 - 12y + 15 = 0$
71. $x^2 - 2x + y^2 - 10y - 143 = 0$
73. $x^2 + 4x + y^2 - 6y + 4 = 0$
75. $x^2 - 14x + y^2 - 2y - 575 = 0$
77. $x^2 - 2x + y^2 - 4y + 4 = 0$
79. Circle: center $(3, -1)$, $r = 5$
81. Circle: center $(-2, -4)$, $r = 7$
83. Circle: center $(5, -3)$, $r = 2$
85. Circle: center $(5, -5)$, $r = 5$
87. Single point $(3, -1)$
89. Circle: center $(-1, 2)$, $r = 3$
91. Circle: center $(2, -3)$, $r = 4$

Section 3.2

1.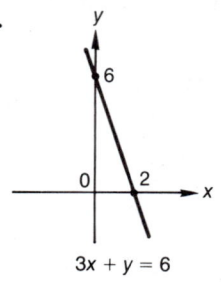
$3x + y = 6$

3.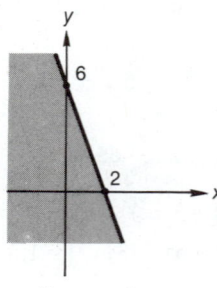
$3x + y \le 6$

5.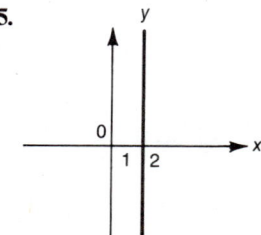
Symmetric about x-axis
$x = 2$

7.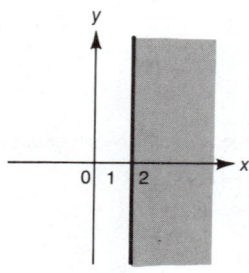
Symmetric about x-axis
$x \ge 2$

9.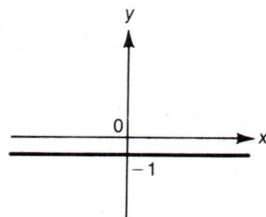
Symmetric about y-axis
$y = -1$

11.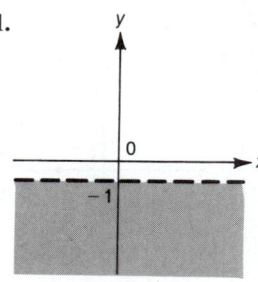
Symmetric about y-axis
$y < -1$

13.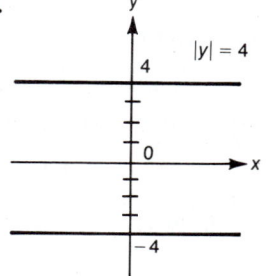
Symmetric about x-axis, y-axis, and origin

15.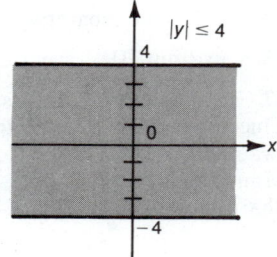
Symmetric about x-axis, y-axis, and origin